CELLULAR NEUROPHYSIOLOGY
A Source Book

CELLULAR NEUROPHYSIOLOGY
A Source Book

Edited by

IAN COOKE

MACK LIPKIN, JR.

Department of Medicine
University of North Carolina

HOLT, RINEHART AND WINSTON, INC.
New York Chicago San Francisco Atlanta Dallas Montreal Toronto London Sydney

Library of Congress Catalog Card Number: 74-138407
ISBN: 0-03-081221-6
Printed in the United States of America
45 008 9876543

CONTENTS

PREFACE

This collection was undertaken to make easily accessible to students and research workers a nucleus of original research reports in cellular neurophysiology. Most of the papers describe or quantify the short-term activity and behavior of individual neurons or the interactions between individual neurons and effectors. We have grouped papers under four topics: impulse conduction and the physical-chemical bases of the action potential, synaptic transmission, the physiology of sensory neurons, and neuronal integration. The ideas developed under the first two topics have general validity in the study of all nervous systems. We have chosen vision as an example for detailed development under the topic of sensory physiology, but here, too, the emphasis is on concepts with general applicability. Under integration we have restricted the material to studies of sequential single cell analysis of visual information. These topics represent a select subset from the domain of neurophysiology. It is our conviction, however, that they represent a fundamental core of observations which orients much of present research in cellular neurophysiology.

Within each topic enough papers have been included to introduce the major concepts, provide critical evidence supporting them, and reveal the techniques used to make the observations. Most papers contain factual observations of continuing importance to the general description of nerve functioning. We have attempted to select those examples of often-made observations that are experimentally simplest or clearest. Thus inevitably, invertebrate and cold-blooded vertebrate preparations are prominent. These "simpler" preparations frequently provide the basic models by which mammalian (including human) observations come to be interpreted. In addition, we have tried to choose papers which demonstrate excellent experimental design, careful reasoning, and lucid scientific reporting. We have sought to satisfy these criteria even at the cost of chronological priority. An enormous number of important papers are not mentioned. We feel, however, that each paper included deserves to be read as an essential part of the cumulative knowledge in the field.

We hope that this collection will prove a significant resource in the teaching of neurophysiology as well as a useful source book for research workers and teachers. In teaching basic neurophysiology to college, graduate, and medical students one senses an enormous quickening of interest when students study original papers. The student's sense of discovery becomes actively engaged as he analyzes the design of experiments and checks the reported results against the conclusions drawn. Students are astounded by the degree to which highly qualified conclusions become textbook and lecture dogma. Reading of original research reports makes students aware of the limitations imposed by the technique, the relations between experimental strategy, technique, and hypothesis, and the enormity of the gap between factual certainty and interpretation from experimental data.

The ability to read research papers critically is a skill that must be developed by any student of science. Yet undergraduate, medical, and even graduate students are insufficiently encouraged to read first-hand reports of experiments. This may be partially the result of the difficulty of placing journal articles in students' hands.

All these papers are reproduced photographically in their entirety. We have, in a few instances, reproduced the methods section of an earlier paper in a series where such methods are important for evaluating the work or seem intrinsically interesting. We rejected the possibility of including more papers or more topics by excerpting the articles because this lessens the value of the collection for training students to read critically and negates its usefulness as a source book for research workers.

Our introductions to the chapters are designed primarily to help orient those students who have had some preliminary exposure to neurophysiology, for example, through reading a text such as Katz, *Nerve, Muscle and Synapse* (1966). The editorial comments attempt to present in briefest outline the major concepts to which the papers contribute and to place the papers in some perspective. The introductions of the papers themselves are often valuable in this regard. We do provide mention of opposing or dissenting points of view. However, we intentionally limited the number of references and distinguished among these a selection for further reading. We hope this strategy will lift student confidence in our recommendations above reading threshold. Extensive categorized bibliographies of the literature up to 1962 are available in Bullock and Horridge (1965). In the comments, references to papers reproduced in this collection are distinguished by **bold-faced type**; and those particularly important for further reading by CAPITAL LETTERS. All references are collected in the final alphabetical bibliography where these typographic label distinctions are maintained.

The editors wish to note that this is a true collaborative production in which each editor has contributed equally although in different ways. There is no "senior" editor.

We wish to thank the authors whose work is reproduced in this book for their permission to reprint their papers and for their helpfulness in supplying precious personal reprints. Drs. P. Brown, T. N. Wiesel, W. A. Hagins and J. E. Dowling kindly provided original photographs to aid in the production of the book.

We thank the Countway Library of Medicine for making special arrangements for loaning journal volumes of articles for which reprints were unobtainable.

Dr. John L. Gliedman and Mrs. Claire Kramsch provided the translations from German.

We wish to thank those colleagues whose criticism of our selections and comments have substantially improved the book. We thank Drs. Michael Remler, John L. Gliedman, Margaret Anderson, and Michael D. Owen for their suggestions on the editorial text. Mrs. Martha W. Goldstone, Gill Kelly, and Miss Kathleen Horton aided in preparation of the manuscript.

It has been a pleasure to work with Mr. Gilbert Spears, Mr. Ian Baldwin, Mr. Joseph Campanella, and, especially, Miss Dorothy Garbose in the production of this book.

We thank the following journals for their generous permission to reproduce articles: *Journal of General Physiology; Journal of Neurophysiology; Journal of the Optical Society of America; National Academy of Sciences, Proceedings; Nature; Pflügers Archiv für die Gesammte Physiologie; Royal Society (London), Proceedings;* and *Science*. The *Journal of Physiology (London)* has also granted permission for the reproduction of articles.

May 1970

Ian M. Cooke
Mack Lipkin, Jr.

including some [illegible] Inc., Pasadena, [illegible] W. L. [illegible] and J. R. [illegible] kindly provided [illegible] and to the publication of this book.

We thank the [illegible] Library of [illegible] for [illegible] or [illegible] of articles [illegible] [illegible].

The [illegible] Mrs. [illegible] provided the translations from German.

We wish to thank [illegible] colleagues [illegible] of [illegible] and [illegible] have substantially improved the book. [illegible] Margaret [illegible] and [illegible] [illegible] of the [illegible] text [illegible] [illegible] in generalizing [illegible].

It is [illegible] with [illegible] Mrs. [illegible] and [illegible] Dorothy [illegible] this work.

[illegible] the following [illegible] for the [illegible] [illegible] [illegible] [illegible] [illegible] of the [illegible] National [illegible] [illegible] [illegible] [illegible] [illegible] [illegible] [illegible] [illegible] presented [illegible] of the [illegible].

[illegible] M. [illegible]

May 1950 [illegible]

CELLULAR NEUROPHYSIOLOGY

A Source Book

INTRODUCTION

Stephen W. Kuffler

There is no doubt that any student who is seriously interested in the nervous system should consult original sources. Only from the original literature can he get a genuine feeling for the reality of research, how it really is done by people who see and evaluate an interesting problem and attempt to solve it. The picture which emerges is likely to be different from that created by the rather abstract treatment in textbooks. From a series of well-chosen papers one gains more than perspective: one sees vividly how advances come about. These frequently occur when fresh and startling discoveries are made; but more often progress results when old concepts, built on much circumstantial plausible evidence, are confirmed by clear and decisive experiments or are shown to be false. In either case the issue itself loses most of its interest and one can proceed to the next problem.

The present collection offers many good illustrations. For example, in the mid-thirties, when many features of the "cable" properties of neurons had been studied, Hodgkin could demonstrate with great clarity and simplicity that the flow of electric current was indeed the mechanism by which nerve impulses propogate. From here onward, and as knowledge of the ionic basis of the nerve impulse expanded, one did not think anymore of current flow as something separate or extraneous but as part and parcel of neuronal activity.

Decisive experiments not only create new knowledge but they also significantly advance a field by creating new and higher standards of acceptable evidence. This in turn forces workers into clearer thinking and experimenting. An obvious example is provided by the comprehensive and compelling studies of Cole, Hodgkin, Huxley, and Katz on the ionic basis of the nerve impulse. These experiments brought a new degree of precision and depth of analysis to a difficult and long-standing problem. At the same time these studies provided a foundation for working out the mechanisms of postsynaptic excitation and inhibition. To see whether the principles underlying nerve impulse generation were applicable to excitation and inhibition was an obvious challenge. Katz, Eccles, Fatt, and colleagues were successful in demonstrating that the same scheme was relevant and worked out some of the fascinating variations which underlie the functionally significant differences between the nerve impulse and synaptic potentials and between synaptic excitation and inhibition. From there it was but a small step to understanding the mechanism by which synaptic excitatory and inhibitory processes interact on individual neurons, thereby determining their signaling. It is such interaction that provides the basis for integration at the cellular level.

At the majority of known synapses a chemical transmitter is secreted, forming the essential link by which excitation or inhibition is brought about between cells. Included in this selection are samples of the classical papers on transmitter physiology and chemistry by Loewi, Dale, and their colleagues who have opened a chapter that is far from complete. The more recent highlights in this story have been the discovery by Katz, Fatt, and del Castillo of the quantal nature of transmitter release, followed by experiments of Katz and Miledi demonstrating with great elegance the unexpected perfection to which the process of secretion by nerve terminals can be analyzed. It should be remembered, however, that electrical synapses revealed by Furshpan and Potter have now also been found in some vertebrates and they may yet become important in our thinking about the operation of the mammalian brain.

From the work of Hartline it was clear that features of a complex organization could be disclosed and analyzed by recording from single cells of the primitive *Limulus* eye. Yet many investigators did not expect that work on single neurons would contribute greatly to our concepts of how the connections in the mammalian brain are organized. After all, the numbers of cells in higher brains are discouragingly large. Yet the description of the somatosensory cortex, where cells of similar properties are arranged in vertical columns (Mountcastle), is one example; others are the studies on the mammalian visual system by Hubel and Wiesel who have demonstrated with surprising simplicity how information is processed, starting with the retina and progressing through various higher levels. Individual neurons in some visual areas are highly specialized and respond preferentially to some types of stimuli only; for example, a bar of light of a particular size falling on the retina at a certain angle will be preferentially "seen" by a cortical neuron which largely ignores diffuse illumination. There are different cells for different orientations, different shapes, and colors. The finding that a complex set of information can be brought together in one cell, after the component parts of such information have been conveyed and processed by thousands of other cells, forces us to postulate a very precise and specific network of connections or wiring.

One measure of the success of the studies of individual neurons is that they enable us to define more precisely areas of ignorance, and this should help the search for new experimental solutions. For example, the specificity of connections being known, renewed efforts are under way to uncover a corresponding specialization of the microchemistry of neurons. Another paramount challenge is the manner in which the connections are made in the first place and how they are modified during development and with use. These are just some of the questions, conditioned by recent developments, and it would be difficult to predict with confidence the direction of research that will be most rewarding.

Hopefully, the generation that is now flocking into the field of the nervous system from various disciplines will feel encouraged to fill the many gaps and open up new areas, and thereby provide the material for another edition of "selected readings" in not too many years.

I. The Nerve Impulse

The Nerve Impulse

The most conspicuous signs of activity in nervous tissue are electrical potential changes. The nerve impulse is a brief, all-or-nothing depolarization self-propagated at finite speed along tenuous nerve processes. It is the universal element of the pulse-code system by which the results of more subtle sensing and integrating activity of neurons are transmitted over long distances.

Although the existence of electrical changes associated with nervous activity was clearly established by the end of the nineteenth century (for a brief historical account, see Brazier, 1959), it remained to show whether the electrical activity played a direct role in the performance of nerve cells or was merely a by-product of other processes. If it plays a direct role, then it becomes important to measure the intrinsic electrical properties of nervous structures and to determine the mechanisms by which active responses are produced. In Part A of this section (I) are reproduced papers which show that the electrical currents of the nerve impulse are agents for its further propagation, and papers that describe measurements of the electrical properties of axons. In Part B are reproduced papers that have become the basis for current concepts of how active electrical responses are produced.

A. IMPULSE CONDUCTION AND ELECTRICAL PROPERTIES OF AXONS

Hodgkin (1937) was one of the first to provide convincing evidence that electrical currents are causal agents in nervous activity. The paper provides evidence for a local circuit theory. It shows that the nerve impulse involves electrical currents that normally excite the next length of axon and thus propagate the impulse. The findings in this paper are strictly applicable to nonmedullated ("nonmyelinated") axons, although these experiments were performed on nerve trunks composed of a population of medullated as well as nonmedullated axons. The all-or-nothing nature of the propagated impulse and the establishment of electric current as a normal form of stimulation imply that some self-reinforcing, regenerative response of the axon membrane is brought about by electrical stimulation.

Hodgkin (1938) clearly describes and distinguishes active and passive responses of an axon by using isolated, nonmedullated axons. The electrical change under a stimulating anode is passive and always proportional to the applied stimulus. In contrast, above a certain (threshold) intensity electrical changes at the cathode cease to mirror those at the anode. At intensities approaching threshold for all-or-none responses there are sustained, nonpropagated responses. These have important implications for complex integration by neurons. Katz (1937) had earlier arrived at similar conclusions from less direct observations.

The demonstration that active responses of neuronal membranes are affected by electrical potentials makes the intrinsic electrical properties of the neuron and of the surrounding structures and medium of importance because they influence the magnitude, time course, and spatial distribution of electrical potentials. Hodgkin (1939) clearly demonstrated the influence of the surrounding medium on conduction velocity as predicted by the local circuit theory. Hodgkin and Rushton (1946) considered the axon to be electrically analogous to a marine cable. Thus the mathematical theory developed for the cable could be applied to the description of axon properties. (An enormous monograph concerning cable analysis has been published by Lorente de No, 1947.) In **Fatt and Katz (1951)**, reproduced here, the measurement of cable properties utilizing intracellular recording techniques is described. Unfortunately, cable properties can rarely be measured where the most interesting integrative events occur. For example, the shape of a neuron and the spatial arrangement of synapses on the soma or dendrites will have an influence on the integrative performance of the neuron simply because of passive electrical properties. The mathematical modeling work of Rall represents an exciting approach to these problems (e.g., RALL, 1967). In some neurons the nerve impulse arises at a distinct "trigger zone," which is the area in which the varied inputs are finally summed. The location of the trigger zone is often predictable from considerations of topography and cable properties (e.g., the axon hillock of a vertebrate motor neuron).

Study of the effect of fiber diameter on the conduction speed of impulses shows that predictions from the cable equations are valid only among axons of similar function in the same species. The properties of the sheath formed by satellite cells around axons alter cable properties. A thicker sheath results in a faster conduction velocity despite smaller total diameter. It is important to realize that analyses of cable properties treat the axon membrane and associated sheath as a single entity.

Application of cable theory to medullated axons of vertebrates provides a striking example of the contribution of satellite cells to nerve function. The concentrically wrapped membranes of the Schwann cell form the myelin sheath which insulates the axon between the evenly spaced nodes or gaps between Schwann cells. The local circuit from the point of active response at a node is thus channeled to the next node where it produces excitation. Thus the impulse skips from node to node (saltatory conduction), and the intervening membrane remains passive. This process both increases the speed and decreases the energy requirement for propagation of an impulse. Thus a 15μ frog myelinated fiber achieves a conduction velocity (about 20 meters per second) as great as a 500μ squid giant axon (both at room temperature).

Tasaki was one of the first to utilize dissected single myelinated axons and has contributed heavily since the 1930s to evidence on the mechanism of conduction in medullated nerve (for a review, see Tasaki, 1959). Two papers (Tasaki and Takeuchi,. 1941, 1942) were published in wartime Germany and were not available in England when **Huxley and Stämpfli (1949)** were performing their similar experiments (see HODGKIN, 1964, Chapter IV). These three papers provided crucial proof of the hypothesis of saltatory conduction.

[*Reprinted from the Journal of Physiology,*
1937, Vol. **90**, No. 2, p. 183.]

PRINTED IN GREAT BRITAIN

612.816.2

EVIDENCE FOR ELECTRICAL TRANSMISSION IN NERVE. PART I

BY A. L. HODGKIN

From the Department of Physiology, Cambridge

(*Received* 18 *March* 1937)

THIS paper is concerned with the way in which activity is transmitted in medullated nerve. The most widely accepted theory is that transmission depends upon excitation by the action current. The theory is a plausible one, since the nervous impulse and electrical change travel at the same speed and have many similar properties. Moreover, the amplitude and duration of the action current are not very different from those of an effective electrical stimulus. The theory may be formulated more precisely by saying that each section of nerve is excited by the local electric circuits produced by the activity of adjacent parts. This process is possible, since the direction of the electric change is such that the local circuits set up by an active region would excite a resting one. Many properties of nerve and muscle can be explained on this basis, but it is difficult to accept the theory until more is known about the nature of the local circuits on which transmission is supposed to depend. The fundamental question is to decide whether the local circuits set up by an active region of a nerve fibre are able to excite an adjacent part. The present work was undertaken with the object of answering this question.

The method which has been employed consisted in blocking conduction in one section of nerve and observing any changes which were transmitted through the block. The starting point for the research was provided by an observation which has since been recorded by Blair & Erlanger [1936]. It was found that an impulse which arrived at a blocked region could increase excitability in the nerve beyond the block. Another way of describing the phenomenon is to say that a blocked impulse can sum with a subthreshold electric shock. My own observations were originally made in the following way. A nerve was blocked by the action of local cold. When the block was complete a maximal shock S_1 applied to A produced no response at C (Fig. 1). A subthreshold shock S_2 was then applied to B; since this was below threshold, it also produced no response. However, if the two shocks were combined so that S_2 followed S_1 by 1 or 2 msec., impulses were set up and could be detected at C. Clearly the volley of blocked impulses increased the excitability beyond the block and enabled S_2 to excite. Subsequent experiments showed that the increase in excitability lasted for a few milliseconds, and that it might involve a decrease of 90 p.c. in the electrical threshold.

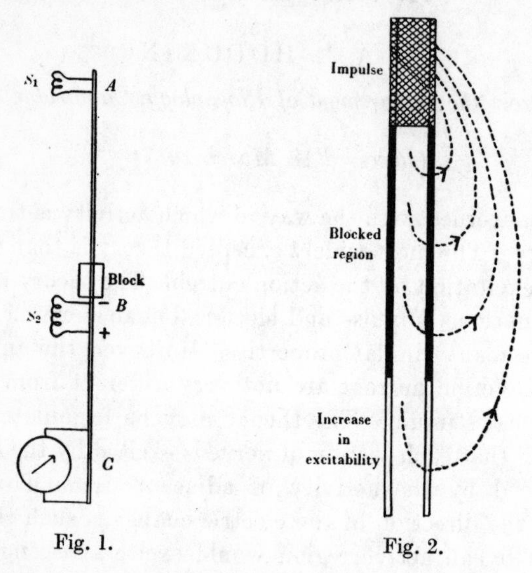

Fig. 1. Fig. 2.

This kind of summation can be observed in various ways; thus Blair & Erlanger [1936] found that it occurred in anodally polarized nerves, and I have observed it at pressure as well as at cold blocks. The important point about the observation is that it provides a method of testing theories of nervous transmission. According to an electrical theory the explanation of summation would be that eddy currents spread through the block and increase excitability in the nerve beyond. Fig. 2 shows how the conventional local circuit diagram can be extended to explain the interaction between blocked impulse and subthreshold shock. If the flow of current beyond the block was just below threshold, an impulse would not traverse the block, but it would produce a large increase in excitability. The electrical view of nervous transmission would be considerably strengthened if it could be shown that the increase in excitability was produced in this way. If the local circuits set up by a blocked impulse caused a 90 p.c. decrease in threshold, there is a strong probability that those associated with a normal impulse would produce a decrease of at least 100 p.c. and could therefore transmit activity electrically. The general aim of this research is to discover how far propagation depends upon excitation by electric currents; its immediate concern is to decide whether the increase in excitability near a block is produced by local circuits.

Apparatus and method

I. *Nerves employed*

Sciatic nerves of Hungarian frogs (*Rana esculenta*) or of English *Rana temporaria* were employed. These were dissected from the spine to the knee, and sometimes the dissection of peroneal and tibial branches was continued to the ankle.

II. *Methods of blocking nerve conduction*

(*a*) *Cold block.* Boyd & Ets [1934] have shown that two kinds of low-temperature block may be distinguished. If supercooling is avoided, conduction is abolished at a temperature of about $-1°$ C. and returns at once when the nerve is rewarmed. If supercooling is allowed, the nerve may be cooled to $-6°$ C. before it freezes, and in this case recovery only occurs after long delay. In the present work the first, reversible, type of block was employed. One end of a silver rod was cooled in ice and salt, the other projected into the moist chamber and made contact with the nerve. The temperature was controlled by adjusting the length of rod in contact with the freezing mixture. The end of the rod was tapered, so that different lengths of nerve (from 3 to 5 mm.) could be cooled. Boyd & Ets, who used an arrangement of this kind, showed that certain precautions must be observed if supercooling is to be avoided. They found that supercooling occurred when the rod was sealed with vaseline into an ebonite chamber, and that reversible blocks could be obtained only if the rod fitted loosely into the chamber wall. Their explanation of this result is that condensed water first freezes near the cold end of the rod, and that the film of ice which subsequently creeps up the rod prevents supercooling of the nerve. A vaseline seal interrupts the spread of the ice film and supercooling occurs beyond it. In my experiments the rod was embedded in a wax chamber, and in order to avoid supercooling a slit was cut in the chamber wall, in such a way as to expose one side of the rod to the air throughout its entire length.

(*b*) *Pressure block.* The nerve was compressed between two blocks of ebonite which were covered with thin rubber pads. Pressure was applied by resting weights on a platform connected to one of the blocks. Two or three millimetres of nerve were compressed and a weight of about 30 g. used. With this arrangement conduction was abolished after about 20 min. This type of block was only partially reversible, so that the time available for experiments was often rather short.

Electrode system. Bright silver was used for stimulating and recording electrodes. The use of metallic electrodes was necessitated by the rather unusual requirements of the research. In order to obtain the curves described on p. 197 it was necessary to measure the distribution of potential over a few millimetres of nerve. Satisfactory results could only be obtained if the whole operation was made within a few minutes, and if the conditions at the block remained constant. For these reasons it was impracticable to use a single movable electrode, and it was necessary to lead off from the nerve with several electrodes. It is difficult to arrange that four or five calomel electrodes make contact within a centimetre of nerve. With the ordinary systems fluid tends to collect round the electrodes, and this is obviously undesirable when the leads are only a millimetre apart. The accumulation of fluid might be avoided if cotton strands soaked in saline or fine capillary tubes were used as leads. These would necessarily have a high resistance and would therefore increase the danger of recording artefacts [see Bishop *et al.* 1926]. For these reasons metallic electrodes were employed, and potentials near the block were recorded by resting the nerve on a grid of fine silver strips, each 0·2 mm. thick and 1–2 mm. apart. The whole system of electrodes and block was sealed into a paraffin wax chamber. This was washed and dried after each experiment, and as an additional precaution against electrical leaks the surface was rewaxed after the chamber had been used for a few weeks.

There are two important objections to the use of metallic electrodes. In the first place, polarization may distort the wave form of the potential recorded by the amplifier. Now the input resistance of the amplifier was 0·5 megohm, and the potentials recorded were for the most part about 1 mV. in amplitude and 1–10 msec. in duration. Thus the total charge crossing the electrodes was of the order of 10^{-11} coulomb, and it is unlikely that this could have produced much polarization. The possibility that the electrodes might introduce some wave-form distortion was tested in the following way. A potential difference, 5 mV. in amplitude and 2·5 msec. in duration, was applied to the amplifier through the nerve and electrodes. The quantity of current crossing the electrodes was, therefore, about the same as that drawn from the nerve in activity. If polarization affected the form of the action potential, it should also have modified the form of the rectangular pulse. The record obtained by calibrating the amplifier through the nerve was compared with one obtained by calibrating through 15,000 ohms. Since these were found to be identical, it was concluded that polarization did not introduce any serious wave form distortion.

The second objection to the use of metallic electrodes is that polarization may affect the form of the stimulating current. This leads to serious errors in some experiments, but the objection does not apply to the present research where the exact form of the stimulus was of no importance.

Stimulating apparatus. Condenser discharges were sometimes used for stimulating, but for the most part break shocks from induction coils were employed. The coils contained iron cores which were removed when brief shocks were required. The strength of the shocks was graded by the insertion of precision resistances in the primary circuits. When iron-cored coils are used, the current in the secondary coil may not be proportional to that in the primary. This possibility was eliminated by a calibration which showed that the primary and secondary currents were directly proportional over the range used experimentally.

The keys controlling the stimuli and the oscillograph time sweep consisted of magneto contact breakers with platinum points. Each was mounted on an adjustable arm and was broken by a cam rotating at a velocity of about 5 m./sec. The time interval between successive stimuli was controlled by moving the arms on which the keys were mounted. Examination with the oscillograph showed that the timing arrangement was satisfactory, and that the break of the contacts was free from chatter. The nerve was stimulated repetitively at rates of 5–10/sec.

Recording system. A resistance capacity coupled amplifier of conventional design was employed; 2 or 4 μF. coupling condensers and 0·5 megohm grid leaks were used. In the earliest experiments the amplifier was connected to a Matthews oscillograph. The arrangement of oscillograph and output stage was similar to the one used by Matthews [1928], and needs no further description. During the greater part of the work a Cossor cathode-ray oscillograph was employed. A linear traverse was obtained by allowing a condenser connected to one pair of deflectors to discharge through a saturated diode. The discharge was started by a contact on the interruptor used for stimulating; thus potentials produced by the nerve could be viewed as a standing wave. The time axis was calibrated with an oscillatory discharge which was obtained by allowing a condenser to discharge through an inductance. The linearity of the recording system was tested by applying rectangular pulses to the input and photographing the response of the oscillograph. Careful measurement showed that the system was linear over the range employed experimentally. The temporal characteristics of the amplifier were sufficient for the requirements of this research. When a rectangular pulse was applied to the input the deflexion was 95 p.c. complete in 100 μsec.; it subsided to 95 p.c. of its maximum in about 10 msec.

The oscillograph was mounted on a turn-table and could either be viewed directly or swung into the field of a camera. Photographs of single transits were obtained by running film through the camera in a direction opposite to that of the oscillograph traverse. Since the nerve was stimulated repetititively the film had to move on one frame (3 cm.) during each revolution of the interruptor. The time scale of the records obtained was determined by the sum of the velocities of traverse and film. The speed of the traverse was much greater than that of the film, and the records were therefore only slightly influenced by variations in film speed.

Results

(a) *Preliminary experiments*

This paper is primarily concerned with the way in which a blocked impulse sums with a subthreshold electric shock. The experiments described in this section do not throw much light upon the mechanism of summation, but they must precede any detailed analysis of the process. In the first place it is necessary to show that summation represents an interaction between blocked impulse and subthreshold shock. All that can be observed is that two shocks, which by themselves are ineffective, may sum to produce an impulse. It is conceivable that this might depend upon a direct electrical interaction between the shocks. The possibility of summation being produced in this way was excluded by the following experiments.

(1) The effect disappeared when the nerve was crushed between *A* (Fig. 1) and the block. If the increase in excitability depended upon the nervous impulse, it would have been abolished by crushing, but if it was caused by an electrical leak there is no reason why it should have been affected.

(2) Summation might be caused by a spread of electrotonus from the stimulating electrodes. The first control does not exclude this possibility, since electrotonus is abolished by crushing. If this explanation were

correct, summation would depend upon the direction of S_1; thus there would be an increase of excitability with a descending current and a decrease with an ascending one. This observation was contradicted by experiment, for summation was obtained with ascending as well as with descending currents.

(3) The view that summation depends upon the nervous impulse is strengthened by the relation between the strength of S_1 and the magnitude of the increase in excitability. It was found that when S_1 was subthreshold there was no increase in excitability: and that increasing S_1 beyond its maximal value led to no further change in the amount of summation. Clearly summation depends upon the nervous impulses set up by S_1 and not upon the shock itself.

(4) The most convincing control depends upon the time relations of the effect. When the two shocks were applied simultaneously, no summation occurred, and the increase in excitability only developed when S_2 followed S_1 by an interval which corresponded to the conduction time. This is illustrated by the results given in Fig. 3. The abscissa indicates the interval between the moments of application of S_1 and S_2. The ordinate represents the increase in excitability produced by an impulse. This was measured by observing the variation in the response to a threshold stimulus. S_2 was adjusted until it just produced an action potential at *C*. It was then combined with S_1. If the nervous impulses set up by S_1 produced any increase in excitability beyond the block, S_2 would excite more fibres and so S_1+S_2 would give a larger action potential than S_2 alone. The increase in action potential provides a convenient and rapid index of the increase in excitability, and it has been used in this experiment. Fig. 3 shows that there is a brief interval between the moment of application of S_1 and the beginning of the increase in excitability. The obvious explanation of this latency is that it corresponds to the time taken by impulses to travel between *A* and the block. If this is correct, the latency should be increased by moving the *A* electrodes away from the block. This deduction was verified; thus curve I was made with an electrode-block distance of 14 mm., and curve II with a distance of 45 mm. In the first case the latency was about 0·4 msec., and in the second it increased to about 1·25 msec. Records of action potentials proximal to the block showed that these latencies corresponded to the actual conduction times. The times at which impulses reached the block are marked by the arrows in Fig. 3; they show the crest and foot of each action potential. This experiment shows that summation starts at about the moment when impulses reach the block,

and that the latency is determined by the conduction time. It is therefore impossible to believe that the increase in excitability depends upon any kind of direct interaction between the two shocks.

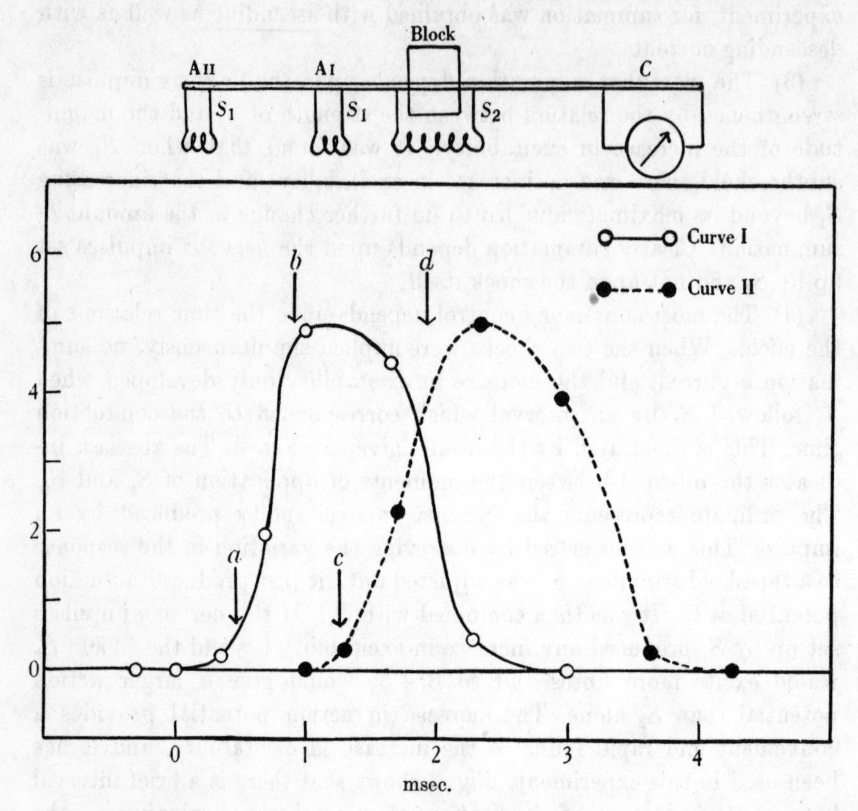

Fig. 3. 4 mm. pressure block. Abscissa: interval by which S_2 follows S_1. Ordinate: increase in action potential at C produced by combining S_1 with S_2. Curve I: S_1 applied 14 mm. proximal to block. Curve II: S_1 applied 45 mm. proximal to block. The ordinate scale for I is 1 unit = 0·18 mV and for II, 1 unit = 0·37 mV. The difference is due to the fact that the effect diminished between making curve II and curve I. Arrows: show time of arrival of impulses at a point 3 mm. proximal to block. a = beginning, b = crest of action potential starting at A I. c, d = beginning and crest of potential starting at A II. The beginning is defined as the moment when the potential rises to one-twentieth of its maximum.

Conditions necessary for obtaining summation.

It was important to use blocks whose intensity could be delicately graded. If the intensity of the block was increased beyond the point at which conduction failed, the increase in excitability was reduced and might be abolished. No increase could be detected if the nerve was crushed at the block, or if conduction was abolished by tying a ligature round the nerve. It will be shown later that the increase in excitability declines fairly rapidly along the nerve distal to the block. Consequently we should not expect to obtain summation with blocks of more than a certain width. Summation was observed with 5 mm. cold blocks, but no experiments have been tried with blocks wider than this.

The position of the anode.

The electrodes at the block may be arranged in various ways. In Fig. 1 the anode was connected with a point distal to the block; an alternative, and in some ways a more convenient arrangement, was with the anode proximal to the block. There did not appear to be any great difference between the magnitudes of the change in threshold obtained with the two electrode arrangements. The change usually seemed to be larger when the anode was proximal to the block, but the difference was not significant in the two experiments where the percentage change in threshold was carefully measured.

The magnitude of the increase in excitability.

The largest changes in threshold were observed when the block had abolished conduction in the majority of fibres, but when a few impulses were still transmitted through it. The measurement of summation in partially blocked nerve was complicated by the fact that S_1 alone produced a response at C. The threshold for S_2 in presence of a blocked impulse was determined by observing whether the addition of S_2 to S_1 led to a response which was larger than that produced by S_1 alone. A 90 p.c. decrease in threshold implies that, in presence of an impulse, S_2 only had to be $\frac{1}{10}$ threshold intensity to excite. Actually this means that the response produced by S_1+S_2 was significantly larger than that due to S_1 alone as long as the intensity of S_2 was more than a tenth of its normal threshold.

The decrease in threshold usually amounted to between 50 and 70 p.c. and greater changes were sometimes observed. The following figures show the largest values obtained.

Pressure block:	9. xii. 1936	...	84 p.c.
	10. xii. 1936	...	95 p.c.
Cold block:	12. xi. 1936	...	78 p.c.
	14. xii. 1936	...	76 p.c.

(*b*) *The electrical changes near a blocked region*

The experiments described in the preceding section show that summation depends upon an interaction between blocked impulse and subthreshold shock, but they do not throw any light upon the way in which

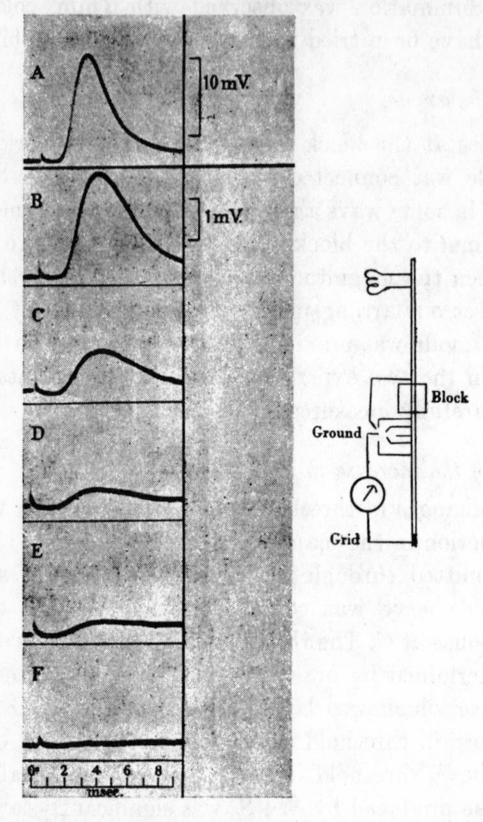

Fig. 4. Potentials proximal and distal to block. 3 mm. cold block. A, action potential 2 mm. proximal to block. B–F, extrinsic potentials distal to block; all at same amplification which was about five times greater than in A. B, 1·4 mm. distal to block; C, 2·5 mm. distal to block; D, 4·1 mm. distal to block; E, 5·5 mm. distal to block; F, 8·3 mm. distal to block. The time scale applies to all the records.

it is produced. The most plausible theory is that the increase in excitability is caused by local electric circuits spreading through the blocked region. Now current spread of this kind would be associated with a potential change which could be observed experimentally. This suggests that an investigation of the electrical changes near the block might throw light upon the mechanism of summation. If the experimental results showed that the increase in excitability was invariably accompanied by an electrical disturbance, the view that it depended upon local circuits would be strengthened. On the other hand, if the increase could be dissociated from any potential change, we could no longer maintain that it was electrical in origin. The electrical changes near the block were investigated and are described in this section. The experiments were made in the following way. A nerve was blocked with cold or pressure, and potentials were led off from various points beyond the block. Under these conditions a response could be recorded from just beyond the

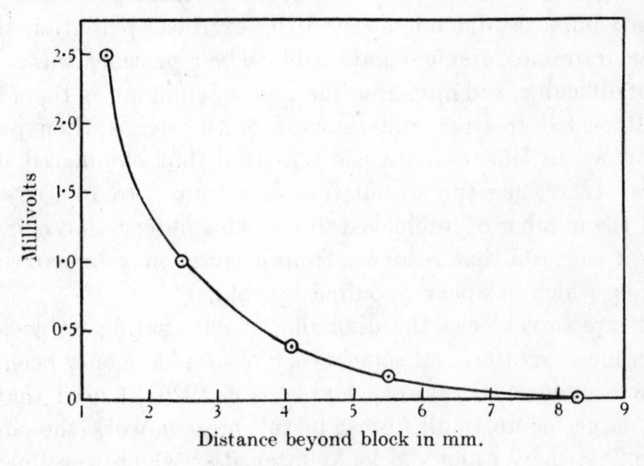

Fig. 5. Potential spread beyond block. Measurements were made from the records illustrated in Fig. 4.

block, although no impulses were transmitted through it. This was distinguished from an ordinary action potential by the fact that it was not propagated, but declined exponentially along the nerve. The records of Fig. 4 illustrate the phenomenon. The response beyond the block has roughly the same form as an action potential from the proximal side of the block, but it is smaller in amplitude. The prolonged duration of the wave was due to the low temperature of the nerve near the block; for the potential was actively produced by cold nerve, although it may have been led off at some distance beyond the block. The form of the potential gradient is shown in Fig. 5, where the amplitude of the potential wave is plotted against the distance from the block. The electrical change illustrated by these records has been observed in more than eighty nerves. It is not caused by the activity of nerve under the lead and it seems to be produced by a spread of current along the nerve fibres themselves. In order to distinguish the change from a propagated action potential, it will be called the extrinsic potential. This term has been employed because Erlanger *et al.* [1926] used extrinsic phase to describe the part of the action wave which is due to current flowing in front of the active region.

A small action potential, due to unblocked impulses was usually superimposed upon the extrinsic potential. This is illustrated by the records in Fig. 6, where a response due to blocked and to unblocked impulses may be distinguished. The distinction has been emphasized by marking the probable time course of the extrinsic potential. Although unblocked impulses did not obscure the extrinsic potential, they often made accurate measurement impossible. Their presence was a continual source of difficulty, and numerous methods of eliminating them have been tried. Repeated freezing and thawing of the nerve often produced a condition which left the extrinsic potential, but eliminated unblocked impulses. Increasing the stimulation rate from 5 to 10/sec. sometimes reduced the number of unblocked fibres. This observation is interesting, because it suggests that recovery from activity may be extremely slow in a region which has been impaired by cold.

With pressure blocks the difficulty of eliminating unblocked fibres has been much greater, and satisfactory results have only been obtained on a few occasions. Gasser & Erlanger [1929] showed that pressure affected large before small fibres; in the present work the small fibres were eliminated by using shocks at intensities which were maximal for the α group, but below the threshold of smaller fibres.

The extrinsic potential may be observed in a different way. It has been known for a long time that cooling a section of nerve increases the interval for muscular summation, and that this effect is due to failure of the second impulse in the cold nerve. The cold region behaves as though it had a prolonged absolute refractory period, and acts as a block to the second but not to the first of two impulses. This property of cold nerve provides the basis for a method of abolishing transmission; for the intensity of the block can be adjusted so that one nearly maximal action potential is transmitted but a second is almost completely extinguished. Under these conditions the second impulse produces an extrinsic potential which is similar to that produced by a single blocked impulse. Fig. 7 illustrates this. The records show the descending phase of the first action potential and, following this, a smaller wave which dies out along the nerve. This experiment is interesting for two reasons. In

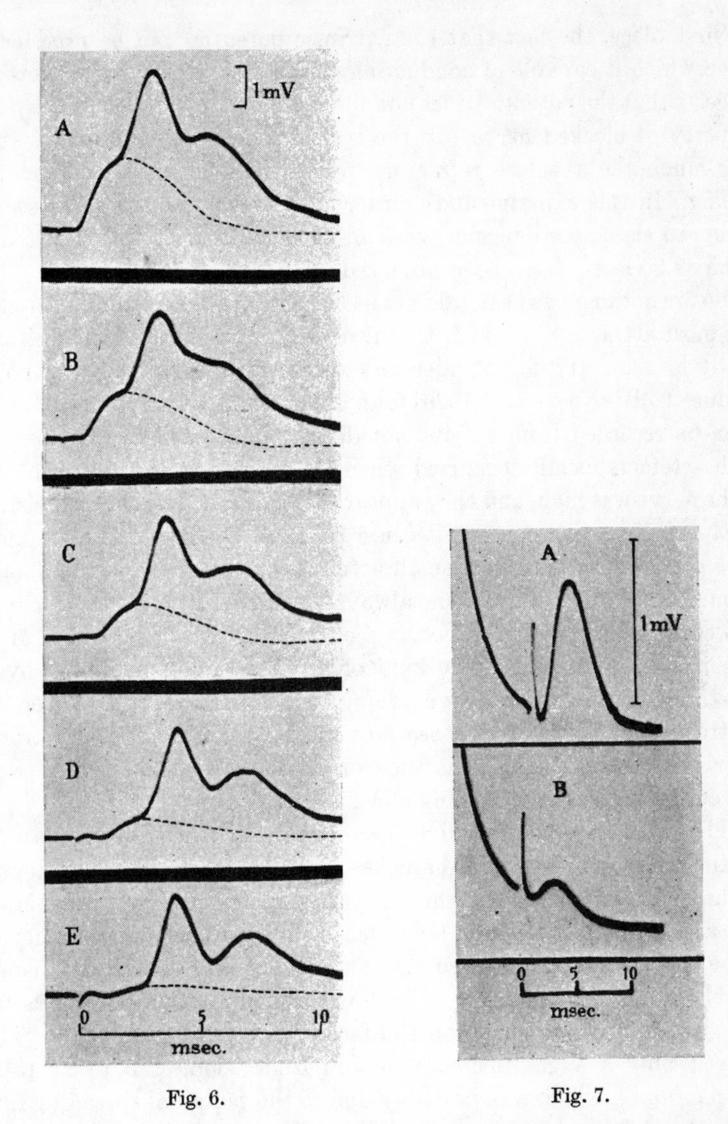

Fig. 6.

Fig. 7.

Fig. 6. Extrinsic potential and propagated action potentials. 3·5 mm. cold block. A, 1·2 mm. distal to block; B, 2·4 mm. distal to block; C, 3·7 mm. distal to block; D, 5·4 mm. distal to block; E, 7·3 mm. distal to block. Amplification and time scale same throughout.

Fig. 7. 3 mm. cold block. Temperature adjusted so that the first but not the second of two impulses was transmitted. Descending phase of first impulse shown; followed by extrinsic potential. The nicks in the upper record are due to interference. A, 1·4 mm. distal to block; B, 4·2 mm. distal to block.

the first place, the fact that the extrinsic potential can be produced in nerve which is capable of conducting a nearly maximal action potential suggests that the potential does not depend upon any abnormal or specific property of blocked nerve. In the second place, the experiment shows how much the absolute refractory period may be prolonged by local cooling. In this experiment the interval between the two impulses was about 15 msec., and blocks with an absolute refractory period of as much as 50 msec. have been observed.

So far nothing has been said about the origin of the extrinsic potential. The most attractive theory is that it depends upon local electric circuits, but it is necessary to consider any other ways in which it might be produced. Bishop *et al.* [1926] found that small action potentials could often be recorded from regions not directly connected to the amplifier. Such artefacts usually occurred when the resistance of the ground lead to the nerve was high, and they appeared to be caused by a high resistance or capacitative connexion between a point on the nerve and ground. I have occasionally observed small artefacts of this kind, but unlike the extrinsic potential, they were always associated with a poor contact between the nerve and recording electrodes. The possibility of the extrinsic potential being any kind of artefact is best excluded by the fact that it was abolished by crushing between the block and recording electrodes. It is difficult to see how crushing could affect an artefact depending upon a high resistance or capacitative leak, but it would prevent local circuits spreading along a nerve fibre.

The extrinsic potential might be produced in the following way. The sciatic nerve gives off small branches, and if these were less susceptible to the action of cold or pressure, impulses might traverse the block and stop at the cut ends of fibres. In order to eliminate such a possibility, the nerve was always arranged so that an unbranching stretch lay distal to the block. At the beginning of each experiment, action potentials from this stretch were measured on the face of the oscillograph tube. These showed only a slight decrease in amplitude along the nerve (about 2–5 p.c./mm.), which was probably due to the temporal spreading introduced by differences in conduction rate. The possibility that the decrement might depend upon impulses stopping at cut branches was excluded in a different way. The nerve was stimulated peripherally, and potentials were recorded from the central end. In this case a potential gradient which depended upon cut branches would be impossible, since any fibres stimulated must have run the whole length of the nerve. The experimental result showed that peripheral stimulation produced a potential gradient which was very like that obtained in the ordinary way.

Another possibility is that the potential gradient might be produced by impulses which emerged from the region where cold or pressure was applied and failed in the distal stretch of nerve. This argument would be tenable if the decrement occupied only a millimetre, but a consideration of dimensions makes it unlikely. It is difficult to see how an impulse could pass a region where the cold was most severe, run 6 or 7 mm. in more normal nerve and then fail. A process of this kind is even less likely to occur at a compression block, where the effect of the pressure is sharply localized. Moreover, this hypothesis provides no satisfactory explanation of records like those shown in Fig. 6. Here there is a component which is caused by propagated impulses, but it shows no sign of decrement. Consequently it is difficult to suppose that the extrinsic potential itself could be made up of propagated impulses.

The nature of the extrinsic potential may now be considered more closely. The shape of the decrement suggests that it might be fitted by a formula of the type

$$p = p_0 \, e^{-x/L}, \qquad \ldots\ldots(1)$$

where p is the potential at a distance x from a point ($x=0$), where $p=p_0$. L is a constant with the dimensions of a length; the potential falls to $1/e$ of its value in a distance L.

In order to illustrate the exponential nature of the decrement several experiments are plotted in semi-logarithmic co-ordinates in Fig. 8. If the decrement obeys an exponential formula the results should fall on a straight line. This is approximately true. There are deviations, but they are mostly of a random kind and may be due to experimental errors. The principal sources of error were of two kinds. First, the determination of potentials at distances greater than 4 mm. was often made rather uncertain by the presence of unblocked impulses. The second and more serious type of error was introduced in the measurement of the abscissa. The position of the electrodes was fixed and could be accurately measured, but a slightly asymmetrical collection of fluid might alter their effective point of contact with the nerve and errors of at least half a millimetre may have arisen in this way.

The principal interest of these results lies not in the exponential nature of the curves, but in the value of L, the space constant involved. Fig. 8 shows that there is a considerable amount of variation in the slope of the lines and hence in the value of L. Table I indicates the value of L obtained in the most accurate experiments. In each case the results

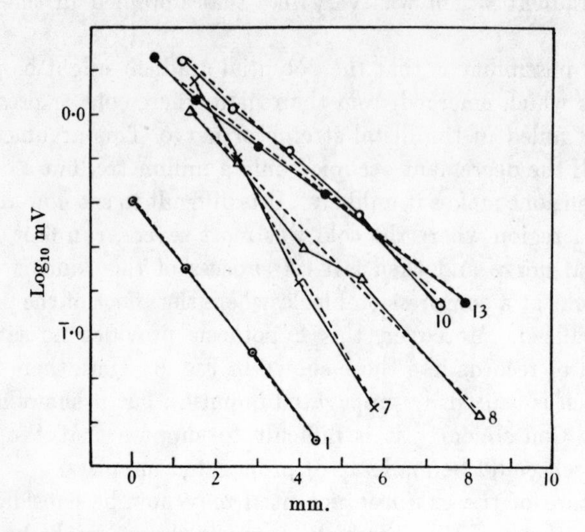

Fig. 8. Abscissa: distance in mm. from block. Ordinate: extrinsic potential, $\log_{10}$ mV. The numbers on the curves refer to the experiments in Table I below, where details of the type of block used are given.

TABLE I. The extent of spread of the extrinsic potential

Several sets of determinations have often been made during the course of one experiment; since these were fairly consistent, one set of values has been chosen to represent the whole experiment. With the exception of No. 2, the potentials were measured from photographic records. In three cases (6, 10 and 11) the potential was measured at a constant interval after the stimulus artefact; in the remainder the maximum of the wave was measured. In the experiments marked with an asterisk, the nerves were subjected to prolonged action of the block.

Experimental details, type of block	Range of measurement: Distance from block mm.	Range of measurement: Amplitude of potential mV.	L (mm.)
(1) 10. iii. 36: 4 mm. pressure	0–4·4	0·42–0·037	1·9
(2) 20. iii. 36: 2 mm. pressure	0–4·6	1·4–0·07	1·8
(3) 25. v. 36: 3 mm. cold	1·4–6·0	0·87–0·058	1·7
(4) 29. v. 36: 3 mm. cold	1·4–6·0	0·51–0·057	2·1
(5) 8. vi. 36: 3 mm. cold	1·4–6·0	0·88–0·074	2·0
(6)* 11. vi. 36: 3 mm. cold	1·4–6·0	1·35–0·078	1·6
(7)* 16. vi. 36: 3 mm. cold	1·5–5·8	1·41–0·046	1·25
(8) 10. vii. 36: 3 mm. cold	1·4–8·3	1·07–0·043	2·15
(9)* 11. vii. 36: 3 mm. cold	1·4–5·5	2·5–0·11	1·55
(10) 25. x. 36: 4 mm. cold	1·2–7·3	1·74–0·14	2·4
(11) 6. xi. 36: 4 mm. cold	1·2–5·4	0·86–0·11	2·0
(12) 18. xi. 36: 4 mm. cold	1·2–5·4	2·1–0·34	2·2
(13) 4. xii. 36: 3 mm. pressure	0·5–7·9	1·8–0·16	2·9

Average value for $L = 2{\cdot}0$ mm.

were plotted in semilogarithmic co-ordinates, and L obtained from the slope of a straight line drawn through the points. When Briggsian logarithms are used L is of course given by the distance in which the potential falls by 0·434 ($\log_{10}e$).

The electrotonic nature of the extrinsic potential.

The properties of the extrinsic potential suggest that it may be similar to an electrotonic potential. This is probable on theoretical

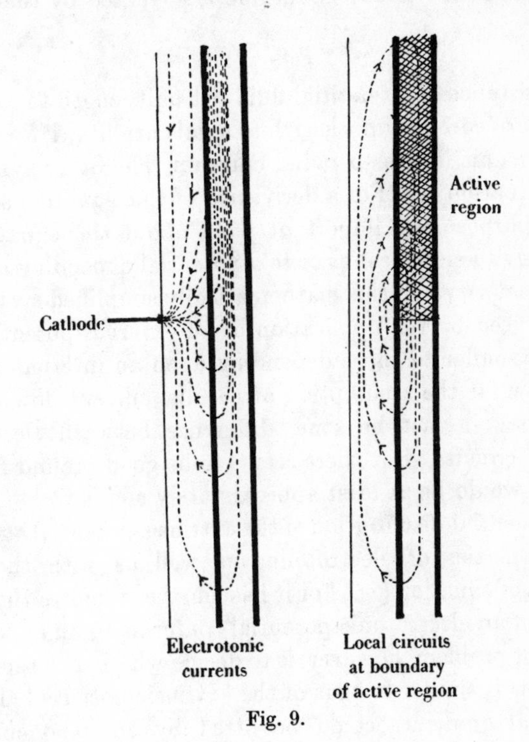

Fig. 9.

grounds, for the type of current flow described by the conventional local circuit diagram is essentially similar to the current flow which is supposed to be responsible for electrotonus. Both involve local circuits, in which current in the core is associated with an equal and opposite component in the interstitial fluid. A high-resistance sheath has the effect of spreading out electrotonic currents, and this should operate equally for the currents produced in activity. Fig. 9 illustrates this argument.

The expectation that the extrinsic potential would be of an electrotonic nature is strengthened by considering the equations which determine the distribution of current in cable-like systems. In order to treat electrotonus mathematically it is usual to consider the current flow in core and interstitial fluid as strictly parallel. An alternative statement of the same assumption is that in any transverse section the potential is uniform throughout the core and throughout the interstitial fluid. When this assumption is made, it can be shown that the distribution of electrotonic potential in the extrapolar region is governed by the equation

$$p = p_0 e^{-x} \Big/ \sqrt{\frac{R}{r+\sigma}}, \qquad \ldots\ldots(2)$$

where r = resistance of interstitial fluid per unit length to axial currents; σ = resistance of core per unit length to axial currents; $1/R$ = conductivity of sheath per unit length to radial currents. The other symbols are the same as in equation (1). For a derivation of this equation see Appendix.

For our purpose, the important point about this equation is that it applies equally to electrotonus or to a potential depending upon currents produced in activity. In this mathematically simplified system, the local circuits produced by the application of an external potential difference are precisely similar to those depending upon an internal change. It is uncertain how far the assumption of parallel current flow is legitimate, and there may in fact be some difference between electrotonic and activity local circuits. But there seems to be good ground for supposing that the two would be at least approximately similar.

The exponential distribution of the extrinsic potential agrees with the observed properties of electrotonus as well as with the theoretical requirements of equation (2). For it has long been known that the spatial distribution of an electrotonic potential conforms to an exponential law. The important problem, however, is to decide whether the space constant of electrotonus is similar to that of the extrinsic potential. The fact that both potential gradients could be fitted by an exponential formula would be of no significance if the logarithmic slope of the gradients were completely different. At first it appears that the decline of the extrinsic potential is too rapid for an electrotonic potential. The average value for L was 2·0 mm., and extremes of 1·2 and 2·9 mm. were obtained. Now the general opinion is that the exponential space constant of electrotonus is about 3 mm. Curves published by Hecht [1931 *a*] and by Bishop *et al.* [1926] indicate that the potential falls to $1/e$ of its value in 2·6 and in 2·8 mm. Schultz [1924] obtained values ranging from 2·1 to 5·6 mm., and Bogue & Rosenberg [1934] give values of 3·6 mm. at

20° C. and 2·4 mm. at 2·5° C.[1] With the exception of the last, all these measurements were made with constant currents lasting many seconds, and the results may have been complicated by slow after-variations. These seem to be small in the case of catelectrotonus [e.g. Hecht, 1931*b*], and a space constant of about 3 mm. probably applies to potentials lasting only a few milliseconds. This indicates that the decline of electrotonus is more gradual than that of the extrinsic potential. However, the difference is not very great, and it might conceivably be due to abnormal conditions near the block.

The question of the electrotonic nature of the extrinsic potential can only be settled by comparing the distribution of the two potentials under similar experimental conditions. The work of Bogue & Rosenberg [1934] and of Harris *et al.* [1936] indicates that the distribution of an electrotonic potential depends upon its wave form. Consequently it is desirable to compare the extrinsic potential with an electrotonic potential of similar form. A current pulse of suitable shape was obtained from a circuit involving two condensers, and was applied to the nerve through electrodes x and y (Fig. 10). In this way electrotonus could be measured at the same time and in the same stretch of nerve as the extrinsic potential. The leading-in electrodes consisted of cotton wicks soaked in gelatine Ringer and joined to chlorinated silver wires. When pressure blocks were used the current was applied through a moist cotton thread which lay in a groove cut in one of the pieces of ebonite. A gap of at least 2 mm. was left between the cathode y and the recording electrodes. In experiments with brief pulses of current, it is easy to record potentials which have nothing to do with electrotonus, but are similar to some of the artefacts responsible for stimulus escape. These were reduced by careful screening and by passing the double condenser discharge through a transformer. The circuit employed is shown in Fig. 10, and with this

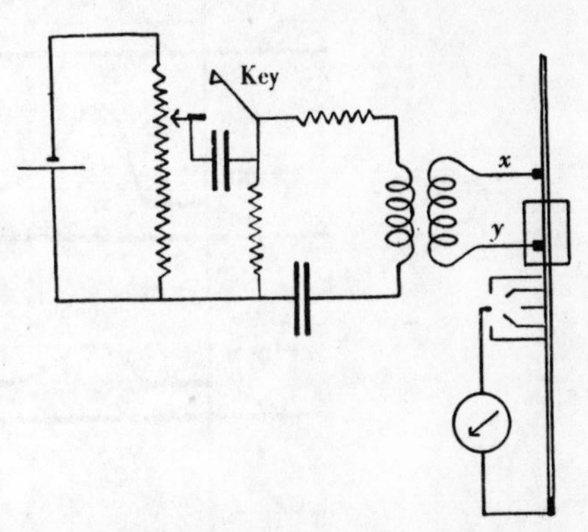

Fig. 10. Circuit diagram for applied electrotonic potential measurements. On break of the key this gives a current pulse of the form illustrated by Fig. 11.

[1] Schultz and Bogue & Rosenberg calculate a constant α which is defined by the equation $p = p_0 e^{-\alpha x}$ (other symbols are the same as in equation (1)). α is the reciprocal of the space constant which is used here.

arrangement artefacts did not appear as long as the electrodes made good contact with the nerve. The electrotonic nature of the potentials recorded was established by the fact that when increased in intensity they reached threshold and produced action potentials, and by the fact

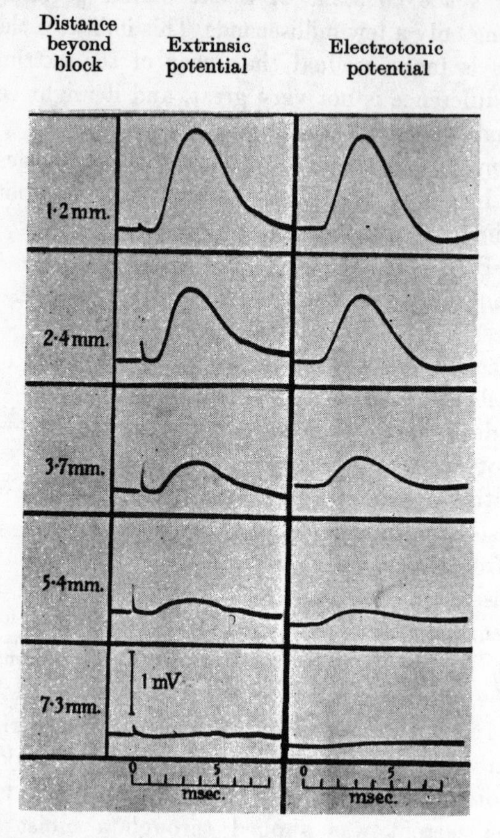

Fig. 11. Extrinsic potential and applied electrotonic potential at corresponding distances beyond block. 3·5 mm. cold block. Amplification and time scale same throughout.

that they were abolished by crushing between the block and recording leads. The actual course of an experiment was as follows. The intensity of the block was increased until the extrinsic potential was freed from propagated impulses. The applied potential, led in through x and y, was adjusted so that electrotonus had the same amplitude as the extrinsic potential. The potentials from different points along the nerve were then recorded photographically. At the end of an experiment the nerve was usually crushed between block and electrodes. The records reproduced in Fig. 11 were obtained in an experiment of this kind; they show that the distribution of the extrinsic potential was almost identical with that of the electrotonic potential.

The records in Fig. 11 call for comment in two respects. At 7·3 mm. from the block, the extrinsic potential has practically vanished, but there is still a small and irregular disturbance of the base line. This component was due to unblocked impulses (cf. Fig. 6), and it can be traced through all the records.

The diphasic shape of the electrotonic potential was due to the fact that the double condenser discharge was passed through a transformer. The records show that the positive phase falls off much more rapidly along the nerve than the negative phase which precedes it. This may be explained by the fact that electrotonus rises and falls more slowly the greater the distance from the polarizing electrodes. At a distance from the block the smaller positive phase is swamped by catelectrotonic potential which lasts longer than the negative phase of the applied potential. This point will be further discussed on p. 205.

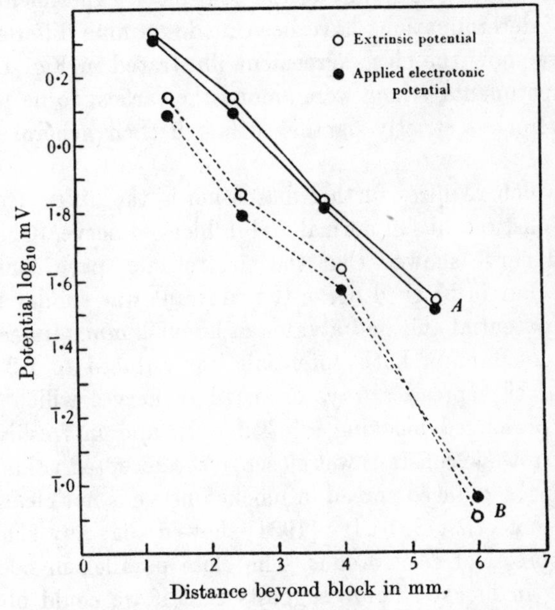

Fig. 12. Distribution of extrinsic and applied electrotonic potentials. *A*, 3·5 mm. cold block, 18 November 1936; *B*, 3 mm. cold block, 11 June 1936.

The results of two experiments of this kind are plotted in Fig. 12. Not only is the general slope of the two curves similar, but the same irregularities are reproduced in both cases. This is easy to understand if we suppose that the experimental error lies mainly in the measurement of the abscissa. If, for example, accumulation of fluid altered the effective point of contact of an electrode with the nerve, the error introduced would have affected both determinations equally. Or the irregularities may have depended upon local variations in the resistance of the interstitial fluid, and this again would have produced deviations of the same kind in both determinations. The agreement shown in Fig. 12 could only be obtained if the two determinations were made in rapid succession. Any experimental manipulation, such as that involved in blotting off moisture which condensed on the nerve near the block, destroyed the precise agreement of the curves, although they still had the same general slope. The parallelism in spatial distribution seems to hold for pressure as well as for cold blocks. Owing to the difficulty of eliminating unblocked fibres, only one experiment was performed with a compression block, but in this case the agreement between the two potentials was as good as in any of the cold block experiments. In all, eleven sets of determinations have been made on nine different nerves. Eight of these show the close agreement illustrated in Fig. 12. In the remaining experiments, which were among the earliest to be performed, the curves were less strictly parallel although their general slope was similar.

A point which requires further discussion is the difference between the spread of electrotonus in normal and in blocked nerve. Observations on unblocked nerve showed that the electrotonic space constant was about 3 mm., but in blocked nerve the constant was similar to that of the extrinsic potential and had a value of about 2 mm. Indeed, on one occasion the constant of both potentials was reduced to 1·2 mm. The low values for the spread always occurred in nerves which had been subjected to prolonged blocking (cf. Table I), and in freshly blocked nerve the electrotonic constant was closer to the accepted value of 3 mm. The reason for the reduced spread in blocked nerve is not clear, but it is interesting to note that Schultz [1924] showed that any slight injury reduced the spread of electrotonus. The close parallelism between the two potentials in blocked nerve suggests that if we could observe the extrinsic potential in normal nerve it would still be similar to electrotonus and would have a space constant of about 3 mm.

The time relation of the potential at different distances from the block.

If a rectangular pulse of current is applied to nerve, electrotonus does not reach its maximum instantaneously, but lags behind the applied potential by an amount which depends upon the distance from the cathode [Bogue & Rosenberg, 1934]. In view of the similarity between electrotonic and extrinsic potentials we should expect the form of both to change as the distance from the block was increased. The potential curves at a distance should rise and fall more slowly and should have a later maximum than those recorded close to the block. A small effect of this kind can be observed in the records of Figs. 4 and 11, but before discussing it we must consider some of the difficulties introduced by such a comparison. In the first place the change in form with distance is not likely to be large; for at a cold block the rate of change of the potential is small in comparison with the electrotonic response to a rectangular pulse. At a pressure block the potential changes are more rapid and a greater modification in form is to be expected. Unfortunately it was impossible to eliminate the unblocked impulses completely, and since these were propagated they affected the form of the potential to an extent which depended upon the electrode block distance. This can be allowed for, but the correction is unsatisfactory since the unblocked fibre component is rather variable. In nerves blocked with cold it was occasionally possible to obtain a response which was quite free from unblocked impulses. An example of this is provided by the records in Fig. 4. Here the potential wave has approximately the same form at different points, but small differences can be brought out by a careful examination. The waves at 1·4 and 2·5 mm. were traced and replotted so that they had the same amplitude. This showed that the maximum at 2·5 mm. occurred $\frac{1}{2}$ msec. later than at 1·4 mm. However, the principal difference was that the potential wave at the longer distance had a slower falling phase. The same kind of modification may be seen in the records of the electrotonic potential (Fig. 11). Close to the block the second positive wave is quite distinct, but at the longer distance the falling phase of the negative wave is prolonged and neutralizes the positive phase. Another difference was that both potentials started to rise slightly later at the longer distances than close to the block. This effect was small, but it was usually quite definite. In view of the difficulty of obtaining accurate records of the extrinsic potential, I do not wish to lay much stress on an exact comparison of its wave form at different points. However, it is clear that the wave form is approximately constant, and that the differences which do occur are consistent with the electrotonic nature of the potential.

Discussion

The investigation of the electrical changes near a block was undertaken as a step in the analysis of the increase in excitability produced by an impulse. The results show that an action potential causes a spread of electrotonic potential in the nerve beyond a block. Since the potential was observed with the same type of block and under the same conditions as the increase in excitability, the experiments suggest that the spread of potential may be responsible for the increase in excitability. Although the results are consistent with an electrical theory, they do not afford any kind of proof that the excitability and potential change are causally related. For the electrotonic potential might be no more than an incidental accompaniment of activity, and its association with the increase in excitability a matter of chance. The importance of the results is that they provide a basis for further experimental analysis. If the increase in excitability was produced electrically, it should have the same temporal and spatial properties as the electrotonic potential. Hence both electrical and excitability change should last for the same time, and both should spread for the same distance beyond the block. The temporal and spatial characteristics of the increase in excitability have been investigated and compared with those of the electrotonic potential. An account of these experiments will be given in a subsequent paper, and further discussion of the origin of the increase in excitability must be reserved until they have been described.

Although the main interest of the electrical changes at a block lies in their possible relation to the increase in excitability, they are, in themselves, worth some attention. In the first place, it is necessary to consider how far the results are consistent with previous work. Davis *et al.* [1926] examined the way in which an action potential was extinguished at a narcotized region. They showed that if any potential gradient was produced by an impulse it did not extend for more than 7 mm. into a narcotized region. This can be reconciled with my results, since the potential change at 7 mm. would be small and might not have been detected by their string galvanometer. Erlanger & Blair's [1934] work on axon segmentation provides an entirely different method of observing how potential spreads along the inactive parts of a nerve fibre. They found that, under the influence of anodal polarization, the action potential in a single fibre loses its normal continuous shape and acquires a characteristic notched configuration. The notches involve a sharp inflexion and are spaced at intervals of about 0·5 msec. The only satisfactory way of explaining these action potentials is to assume, first, that several internodes of each axon contribute to the potential which is recorded at any point, and secondly, that in polarized nerve activity is transmitted across each node of Ranvier with a significant delay. Since the response is determined by the activity of several internodes, the delay at each node leads to a notch, and each inflexion is caused by the activity of a new segment. A detailed account of the evidence for this hypothesis would be out of place; for our purpose the important point is the conclusion that the response is determined by the activity of several segments. It follows from this that an active segment produces a potential gradient in the inactive part of an axon. Erlanger & Blair estimated the extent of the gradient from the contributions of segments at different distances from the recording lead. They found that the potential fell to approximately half its value in one internode, which, assuming an exponential decline and 1·25 mm. for the segment distance, gave a value of 1·8 mm. for the space constant of the gradient. This is close to my average value for the potential spread near a block. Erlanger & Blair do not discuss how the potential spread occurs, although they compare it with electrotonus. It seems very likely that it depends upon the same electrotonic mechanism as the extrinsic potential which can be observed near a cold or compression block.

At first the properties of the extrinsic potential seem to be in conflict with what is known about the rising phase of the action potential. Since electrotonic currents spread in front of the action potential at a block, we should expect the rising phase to have an electrotonic component. The initial rise should be exponential, and its rate should be determined by the conduction velocity and the electrotonic space constant. The initial rise of the axon action wave seems to be exponential, but it is considerably steeper than would be expected from the assumption of full electrotonic spread. Erlanger & Blair's [1934] record of an axon wave shows that the exponential time constant of the initial rise is about 40 μsec. Assuming a conduction velocity of 25 m./sec. this implies that in the front of the wave the potential falls to $1/e$ of its value in 1 mm. The reduced spread of potential may be explained in two ways. In the first place, my estimate of 3 mm. for the potential spread was for a large nerve trunk, and it is possible that the value for the electrotonic constant might be much smaller in the thin bundles studied by Erlanger & Blair. In the second place, it is important to consider the temporal as well as the spatial properties of electrotonus. Bogue & Rosenberg [1934] have shown that electrotonus rises more slowly the greater the distance from the polarizing electrode, and this implies that electrotonus takes an appreciable time to spread to its full extent. Consequently we should not expect that the potential could spread far in front of an active region which was advancing with a velocity comparable to that of the electrotonus itself. The difference between the electrotonic spread at a block and in front of the action wave is due to the difference in the velocity of the active region. When the impulse is propagated, there is no time for full electrotonic spread, but when it is checked at a block electrotonus can spread for some distance beyond the active region.

Summary

1. When an impulse arrives at a cold or compression block, it can decrease the electrical threshold beyond the block by as much as 80 to 90 p.c.

2. An action potential on one side of a block produces a transient potential gradient in the stretch of nerve beyond the block. The gradient is exponential in shape and the potential falls to $1/e$ of its value in about 2 mm.

3. The spatial distribution of the potential is almost identical with that of an electrotonic potential of similar form.

4. It is concluded that an impulse produces a spread of electrotonic potential in the nerve beyond a block.

5. It is suggested that the local circuits which produce the electrotonic potential are the cause of the increase in excitability.

APPENDIX

A mathematical basis for the spread of electrotonus was first provided by Hermann [1905]. He derived equations for a core conductor model, where one of the conductors—core or interstitial fluid—was supposed to have zero resistance. The object of this appendix is to give a more general treatment, which takes into account the resistance of both core and interstitial fluid. The method is essentially the same as that of Rushton [1934], and indeed the problem is only an example of the general case which he considered.

Assumptions

A nerve fibre is assumed to have a cable-like structure with a conducting core and a resistant sheath. Any capacity across the sheath may be ignored, since a case of steady current distribution is considered. The system is assumed to obey Ohm's Law. It is assumed that in any transverse section the potential is uniform throughout the core and throughout the interstitial fluid. This cannot be strictly true, but calculations based on this assumption will be valid as long as the potential difference across core and interstitial fluid is small in comparison with that across the sheath.

Let x = distance along nerve, p = potential of interstitial fluid, v = potential of core, i = current in interstitial fluid, r = resistance of interstitial fluid per unit length to axial currents, σ = resistance of core per unit length to axial currents, $1/R$ = conductivity of sheath per unit length to radial currents. Since the nerve does not form part of an external circuit, the current in the interstitial fluid must be equal and opposite to that in the core

Hence
$$i = -\frac{1}{r}\frac{dp}{dx} = \frac{1}{\sigma}\frac{dv}{dx}. \quad \text{......(1)}$$

Now the current leaving the interstitial fluid at any point must be proportional to the potential difference across the sheath at that point:

$$-\frac{di}{dx} = \frac{p-v}{R}. \quad \text{......(2)}$$

Differentiating

$$-\frac{d^2i}{dx^2} = \frac{1}{R}\left(\frac{dp}{dx} - \frac{dv}{dx}\right). \quad \text{......(3)}$$

On combining with (1) this becomes

$$\frac{d^2i}{dx^2} - i\Big/\frac{R}{r+\sigma} = 0. \quad \text{......(4)}$$

Now the general solution of this equation is

$$i = Ae^{x\big/\sqrt{\frac{R}{r+\sigma}}} + Be^{-x\big/\sqrt{\frac{R}{r+\sigma}}}. \quad \text{......(5)}$$

The source of potential will be considered to be in the direction of $-\infty$; at $x = +\infty$ the current flow will therefore be zero. Hence A must be zero and

$$i = Be^{-x\big/\sqrt{\frac{R}{r+\sigma}}}. \quad \text{......(6)}$$

Combining (6) with (1) and integrating between x and ∞

$$p - p_\infty = Ce^{-x\big/\sqrt{\frac{R}{r+\sigma}}}, \quad \text{......(7)}$$

where C is another constant.

Potentials are measured with reference to a remote point on the nerve, which is considered to have zero potential; thus p_∞ is zero. C must be equal to p_0, the value of p at $x=0$. Hence

$$p = p_0 e^{-x\big/\sqrt{\frac{R}{r+\sigma}}}. \quad \text{......(8)}$$

This result has been reached without referring to the nature of the source of current, and thus it applies equally to the steady distribution of an electrotonic potential or to that of an extrinsic potential produced by an action current.

The application of this equation is not affected by the fact that resting nerve probably has a potential difference in series with the sheath. The resting potential can be allowed for by adding an extra term to equation (2); but this term does not vary with distance, so that it disappears when (2) is differentiated to give (3).

REFERENCES

Bishop, G. H., Erlanger, J. & Gasser, H. S. (1926). *Amer. J. Physiol.* **78**, 592.
Blair, E. A. & Erlanger, J. (1936). *Ibid.* **117**, 355.
Bogue, J. Y. & Rosenberg, H. (1934). *J. Physiol.* **82**, 353.
Boyd, T. E. & Ets, H. N. (1934). *Amer. J. Physiol.* **107**, 76.
Davis, H., Forbes, A., Brunswick, D. & Hopkins, A. McH. (1926). *Ibid.* **76**, 448.
Erlanger, J. & Blair, E. A. (1934). *Ibid.* **110**, 287.
Erlanger, J., Bishop, G. H. & Gasser, H. S. (1926). *Ibid.* **78**, 537.
Gasser, H. S. & Erlanger, J. (1929). *Ibid.* **88**, 581.
Harris, D. T., Rosenberg, H. & Sager, O. (1936). *J. Physiol.* **86**, 4*P*.
Hecht, K. (1931*a*). *Z. Biol.* **91**, 252.
Hecht, K. (1931*b*). *Ibid.* **91**, 231.
Hermann, L. (1905). *Pflügers Arch.* **109**, 95.
Matthews, B. H. C. (1928). *J. Physiol.* **65**, 225.
Rushton, W. A. H. (1934). *Ibid.* **82**, 332.
Schultz, E. (1924). *Cremer's Beiträge Physiol.* **2**, 107. Quoted from: Cremer, M. (1929). *Handbuch der normalen u. pathologischen Physiologie*, **9**, 278. Berlin: Springer.

[*Reprinted from the Journal of Physiology,*
1937, Vol. **90**, No. 2, p. 211.]
PRINTED IN GREAT BRITAIN

612.816.3

EVIDENCE FOR ELECTRICAL TRANSMISSION IN NERVE. PART II

By A. L. HODGKIN

From the Department of Physiology, Cambridge

(*Received* 18 *March* 1937)

In a previous paper it was shown that when a nervous impulse was extinguished by local cold or pressure, there resulted two changes in the region just beyond the block. In the first place there was a spread of electrotonus, and in the second, a localized increase in excitability. It is possible that these two facts are related, and the intention of the experiments described here is to decide whether spread of electrotonic current is, in fact, the cause of the increase in excitability.

The first piece of evidence is that the association of the electrotonic potential with the increase in excitability is extremely constant. This is to be expected if the two processes are causally related, but if the association were a matter of chance we should expect to find certain conditions which would allow transmission of the excitability change but not of the electrotonic potential. There is, of course, no way of proving that it is impossible to dissociate the two processes, but experiments have shown that this is extremely difficult. For electrical and excitability change have been observed in more than seventy nerves, and in no case was it possible to obtain the increase in excitability without a concomitant potential change.

Evidence of a more definite kind comes from a study of the time relations of the two processes. The temporal configuration of the increase in excitability has been determined and can be compared with that of the potential. This allows much more certain conclusions than have hitherto been possible. For we should have to abandon an electrical theory if the duration of the increase in excitability was found to be quite different from that of the potential. On the other hand the theory would be considerably strengthened, if the results showed that the time course of the excitability change could be accurately predicted from the potential time curve.

Comparison of excitability and potential time curve

The method of measuring the excitability time curve was essentially similar to that described in the previous paper [Hodgkin, 1937, p. 189]. The stimuli consisted of brief shocks from coreless induction coils. Examination with the oscillograph showed that the duration of these was about 40 μsec. The interval between the two shocks was adjusted by varying the setting of the contacts on the interruptor. Possible errors in timing may be introduced by fluctuations in the speed of the interruptor and by changes in the zero setting of the contacts. These were controlled by a direct calibration, which was carried out as soon as possible after each experiment. Two methods of calibration were employed. In one case the shock interval for different settings of the contacts was measured from photographic records of the shocks and of the time base. The other method consisted of a direct comparison of the shock interval with the time base. The circuits were arranged so that the key for producing the first stimulus could also be used for starting an oscillatory discharge of known frequency. When a calibration was required, the key was allowed to start the discharge, which appeared as a standing wave on the oscillograph. The second shock could be superimposed upon this, and the shock interval obtained by counting the number of oscillations between the beginning of the discharge and the appearance of the shock.

The principal difficulty in the determination of the excitability time curve was introduced by variation in the condition of the nerve at the block. In order to avoid progressive changes in threshold, the measurements of excitability must be as rapid as possible and there is no time to determine the percentage change in threshold for each point on the curve. Consequently the variation in response to a threshold stimulus was used as a measure of the increase in excitability. The principle of the method has already been described [Hodgkin, 1937, p. 189]; the main objection to it is that it gives only a rough indication of the behaviour of the individual nerve fibres. We should not, therefore, expect to obtain a very exact correspondence between potential and excitability time curves. The excitability curve was usually made from two sets of values, one with increasing and the other with decreasing shock intervals. The results were discarded if there was any serious disagreement between the two determinations, but repeatable results could usually be obtained if the values were determined in quick succession. As soon as the excitability measurements were complete the potential beyond the block was recorded photographically. Here again there were considerable difficulties, for it was essential to obtain a record of the electrotonic potential undistorted by action potentials in unblocked fibres. It was occasionally possible to determine the excitability and potential time curves in nerves which were quite free from unblocked impulses. When this was impossible, the distortion introduced by action potentials was reduced by recording with electrodes only 1 or 2 mm. apart. In this way the propagated potentials were reduced to about a tenth of their former amplitude, but on account of its decremental nature, the electrotonic potential was much less affected, and since it had approximately the same form at all points, diphasic recording gave a fairly accurate picture of its time relations.

The results of two experiments with cold blocks are shown in Fig. 1 *a* and *b*. Potential and increase in excitability are plotted on the same time scale, the continuous line represents potential and the dotted line the increase in excitability. If the two processes had similar time relations it would be possible to superimpose the curves by correct scaling of their ordinates. This has been done and it is evident that the curves have approximately the same shape. Altogether ten experiments with cold blocks have been completed, and on the whole there has been good agreement between potential and excitability. In six cases the coincidence was as close as that shown in Fig. 1. In the remaining experiments there was more divergence, but with one exception this was not large enough to require special comment. In this case both potential and excitability were unusually prolonged. The duration of the increase in excitability was 10 msec. but the potential lasted for about 20 msec. Unfortunately, it is uncertain whether this divergence was genuine, or whether it depended upon modification of the electrotonic potential by unblocked fibres. The records showed that there were certainly some unblocked fibres present, and these may have produced more serious distortion than was realized at the time of the experiment. Another possible explanation of this case of divergence is discussed later.

At a pressure block the potential curves were much more rapid, and it is satisfactory to find that the time scale of the increase in excitability was correspondingly reduced. Fig. 1 *c* shows the result of a typical experiment. The prolonged tail in the excitability curve was an unexpected phenomenon, but it corresponds exactly with the unusually large negative after-potential shown by the electrotonic wave. The after-potential appears to depend upon the influence of the compression, since it was obtained under conditions when the nerve proximal to the block

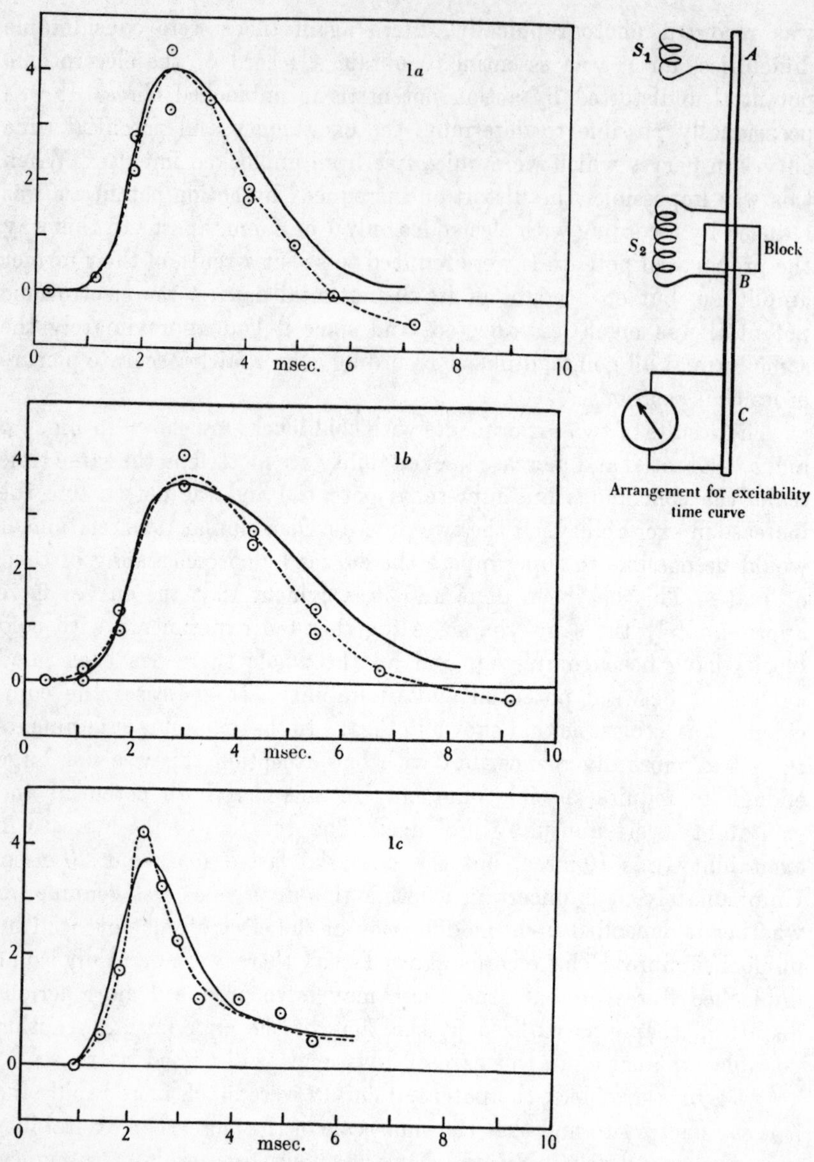

Fig. 1. Potential and excitability time curve. Abscissæ: time in milliseconds; zero time application of S_1. Ordinates: continuous curve, electrotonic potential produced by impulse. Dotted curve, circles, increase in excitability measured as increase in action potential at C. Scale, linear but arbitrary. *a*, 3·5 mm. cold block 12 November 1936. Potential recorded, ground lead 1·2 mm. distal to block, grid 30 mm. distal. Potential scale; 1 unit = 0·4 mV. *b*, 3·5 mm. cold block. 4 November 1936. Leads as in *a*. Scale 1 unit = 0·15 mV. *c*, 3 mm. pressure block. 20 November 1936. Ground 0 mm. distal, grid 2·9 mm. distal. Scale 1 unit = 0·14 mV.

gave a perfectly normal response. The tail in the increase in excitability was shown by four out of the five pressure-block experiments, and it has been observed in many nerves which were not examined quantitatively. It was difficult to form an exact estimate of its duration, but in one case it lasted for at least 20 msec. This observation is interesting in connexion with the work of Blair & Erlanger [1936]. They have put forward evidence to show that a blocked impulse can increase excitability in an adjacent section for as much as 100 msec. It is possible that this is also due to a negative after-potential.

If the action of the current produced by an impulse was similar to that of an applied current, we should expect the nerve to show some sign of accommodation. The excitability should decline more rapidly from an earlier maximum than the potential, and the period of increased excitability ought to be followed by a phase of depression lasting many milliseconds. These effects would not have been large, since the potentials were too rapid to produce much accommodation. Moreover, accommodation is likely to be slow in the vicinity of a cold block, for Katz [1936] has shown that cooling decreases the rate of accommodation. A divergence which may be due to accommodation can be seen in the curves of Fig. 1 *a* and *b*. The phase of depression following the increase in excitability was certainly an experimental fact, but it is too small to afford a convincing demonstration of accommodation. It is possible that the one case of divergence between potential and excitability curves may have been partly due to accommodation. For on this occasion the potential had an unusually long duration and there would have been more opportunity for accommodation to occur.

The important point about these experiments is the general similarity between potential and excitability. The measurements were not very accurate, and there were often small differences between the two processes. These may have been due to experimental error, but some difference is to be expected on theoretical grounds, since a current must flow for a certain time in order to produce a change in excitability. The general similarity in time relations can hardly be fortuitous, since the individual peculiarities of each potential wave were reflected in the excitability curves. Thus it is difficult to believe that the coincidence between curves of such an unusual shape as those in Fig. 1 *c* could have been due to chance. The only alternative is to suppose that the two processes are causally related. The experiments therefore afford strong evidence for the view that the current spread which is responsible for the electrotonic potential is the cause of the increase in excitability.

A possible criticism of this conclusion is that owing to variation in conduction rate, impulses in different fibres do not reach the block at the same time. The observed potential and excitability curves must therefore be longer and flatter than the curves in the axons themselves. If the temporal spread introduced by differences in conduction rate were large compared to the duration of the axon response at the block, the composite time curves might be more or less independent of the shape of the axon curves. Hence it would be possible to have agreement between potential and excitability, without the two processes having the same time relations in the axons themselves. At a cold block it is out of the question for the temporal spread to be large compared to the duration of the axon response. The conduction time to the block was often less than 1 msec., whereas the response at the block lasted 5–8 msec. Hence the electrotonic response beyond the block could not have differed by more than about a millisecond from the average axon response. This objection has more force when applied to pressure block experiments, but it cannot explain the correspondence between the negative after-potential and the tail in the excitability curve. This type of criticism suggests that it would be unwise to lay much stress on any exact agreement between potential and excitability, but it cannot seriously detract from the evidential value of the general similarity in temporal properties.

The spatial relations of the increase in excitability and potential

In the previous paper it was shown that the current which spread through the block was electrotonic in nature, and that it occupied several millimetres of nerve. If the increase in excitability was produced by the flow of current, it should have the same kind of spatial distribution as the potential. This suggests that a good way of testing the electrical theory would be to compare the spatial distribution of the change in threshold with that of the potential. The measurement of the increase in excitability at different points on the nerve raises certain difficulties. In the first place it is desirable to use a more quantitative measure of excitability than the variation in response to a threshold stimulus; for the increase in response produced by a given amount of electrotonic potential depends upon the statistical distribution of thresholds in different fibres, which would not be the same at different points on the nerve. It was therefore necessary to determine the change in threshold directly. In order to facilitate this measurement, the following procedure was adopted. The stimulating circuit was arranged so that by movement of a single switch the stimuli applied to the nerve were changed from:

(1) S_2 alone.

to

(2) S_1, followed at 1–3 msec. by S_2, which was now reduced by an amount y. y was determined by the magnitude of a variable resistance.

S_2 was increased till an action potential of suitable size (usually about 1/10th maximal) was produced. It was then only necessary to increase y

to the point where changing the position of the switch produced no alteration in the height of the action potential. At this point the increase in excitability produced by the impulse was balanced by the reduction in the strength of S_2. In this way it was possible to measure the decrease in threshold very rapidly, and the errors introduced by progressive changes in the nerve were reduced.

The second difficulty was connected with the inequality in the resting excitability of different points on the nerve. Excitability was not initially uniform, because in addition to the ordinary local variations, the thresholds were raised at the electrodes nearest the block. This introduces the question of the way in which the increase in excitability is to be expressed. For many purposes it is sufficient to express the change in threshold as a percentage of the resting threshold. But when comparing the influence of the impulse at different points on the nerve, it is obviously correct to consider the absolute change in threshold current. For the absolute change is independent of the resting excitability, whereas the percentage change must be less when the resting threshold is high. The measurements of the change in threshold at different points on the nerve is complicated by the fact that the alteration of interpolar length introduces differences in the resistance of the secondary circuit. Hence the change in threshold current cannot be regarded as proportional to the primary current, which was the ordinary measure of excitability employed. In some experiments this difficulty was overcome by placing a high resistance in series with the nerve. In others the current in the secondary was measured directly by placing the amplifier leads across a resistance in series with the nerve. Actually it was found convenient to measure, not the peak current, but the total quantity of current produced by the shock. This was obtained by converting the amplifier into a ballistic instrument by means of a condenser across the input leads.

Fig. 2 illustrates the result of an experiment with a pressure block. Here the distribution of the potential is compared with that of the change and of the percentage change in threshold. The values for the potential were determined from photographic records which were made as soon as the excitability measurements were complete. The ordinates have been scaled so as to superimpose the curves as far as their shape allows. The exact coincidence of the change in threshold and potential curves means that the two quantities have the same spatial distribution. The fact that the ratio

$$\frac{\text{Percentage change in threshold}}{\text{Potential}}$$

decreases near the block is presumably due to the thresholds being higher in this region. Six experiments of this kind were completed, and the results showed that the change in threshold always had approximately the same distribution as the potential. The experiments therefore afford further support for the view that a spread of electrotonic current is responsible for the increase in excitability.

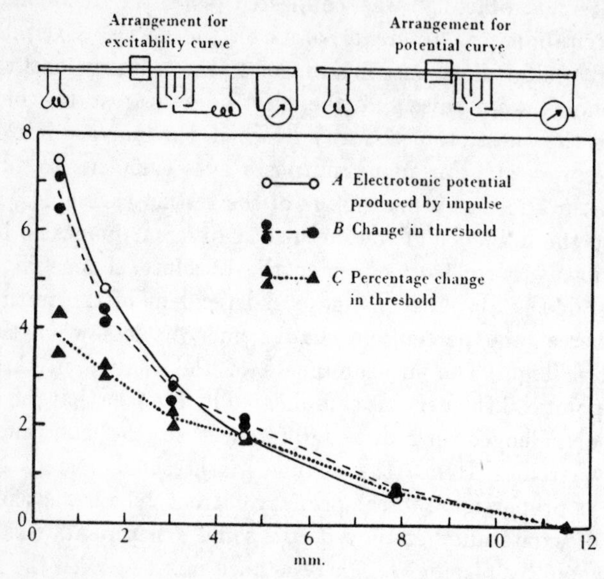

Fig. 2. Spatial distribution of potential and increase in excitability. 3 mm. pressure block. Abscissa: distance beyond block in millimetres. Ordinates: A, 1 unit = 0·24 mV.; B, change in threshold measured as change in threshold quantity of electricity: 1 unit = $1{\cdot}8 \times 10^{-11}$ coulomb; C, 1 unit = 5 p.c.

Evidence for the electrical origin of the increase in excitability was obtained in a slightly different way. It has been shown that the distribution of the extrinsic potential is similar to that of an applied electrotonic potential, and it follows from this that the change in threshold produced by the extrinsic potential should have the same distribution as the change produced by an electrotonic potential. This deduction was tested in the following way. First of all the increase in excitability produced by the nervous impulse was measured at different distances from the block. A current pulse with a wave form similar to that of the extrinsic potential was then applied to the blocked section of nerve. The circuit for producing this has been described in the previous paper. The applied current produced a spread of electrotonic potential.

which in turn caused an increase in excitability. The distribution of this was then measured in exactly the same way as the change produced by the impulse. The method had one great advantage, since it was only necessary to observe changes in excitability and there was no need to determine values for the extrinsic potential. Hence the experiment could be performed with partially blocked nerve, where the increase in excitability was larger and easier to measure. Actually the most accurate

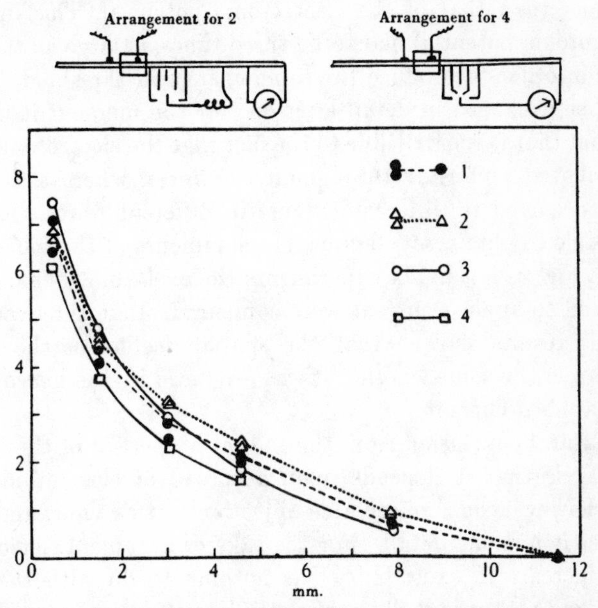

Fig. 3. Experimental details as in Fig. 2. Electrode arrangements for curves 1 and 3 are the same as in Fig. 2: arrangements for 2 and 4 shown. Abscissa: distance beyond block in millimetres. Ordinates: (1) Change in threshold produced by impulse. 1 unit $=1{\cdot}8\times10^{-11}$ coulomb. Data same as those in Fig. 2. (2) Change in threshold produced by applied current. 1 unit $=1{\cdot}8\times10^{-11}$ coulomb. (3) Extrinsic potential. 1 unit $=0{\cdot}24$ mV. (4) Applied electrotonic potential. 1 unit $=0{\cdot}8$ mV.

results were obtained in a nerve in which it was possible to make successive measurements of:

(1) Change in threshold produced by an impulse.

(2) Change produced by an applied current.

(3) Electrotonic potential produced by an impulse (extrinsic potential).

(4) Electrotonic potential produced by an applied current.

The results of this experiment are plotted in Fig. 3. It is clear that all four processes have similar distributions, and this affords strong evidence for the view that the change in threshold is electrotonic in origin. There is, however, one important difference between the action of an impulse and of an applied current. The similarity of the potential curves indicates that the values for the two potentials were in a fixed proportion to one another, but it does not mean that their absolute magnitudes were identical. This would only be true if the ordinate scales were the same, whereas the scale of the extrinsic potential was actually about three times larger than that of the electrotonic potential. This implies that the electrotonic potential had to be three times as large as the extrinsic potential in order to produce the same change in threshold. The difference will be discussed in detail later on; for the moment it is sufficient to point out that it is partly due to the fact that the electrotonic potential was distributed uniformly throughout the fibres, whereas the extrinsic potential occurred to different extents in different fibres. It was only once possible to obtain satisfactory measurements of the four quantities plotted in Fig. 3, but in six experiments the excitability changes due to impulse and to applied current were compared. Like the experiment of Fig. 3 the results showed that the spatial decline of the change in threshold was the same whether it was produced by the nervous impulse or by an applied current.

The natural conclusion from the spatial properties of the increase in excitability is that it depends upon a spread of electrotonic current. There is, however, one conceivable objection to this view, and it is well to consider it in some detail. For the sake of argument, let us suppose that the increase in excitability has nothing to do with the extrinsic potential, and that it is really confined to an exceedingly small length of nerve just beyond the point where the impulse fails. The fact that a change in threshold is observed when a stimulus is applied several millimetres distal to this can be explained by supposing that catelectrotonus spreads from the stimulating electrode and sums with the impulse. If the electrode is moved farther away from the block, there is less catelectrotonus and so a smaller change in threshold is observed. On this view, the apparent decline of excitability along the nerve is a property not of a process set up by the impulse but of the extent of spread of the stimulus. There are several reasons why this hypothesis in its most extreme form is untenable, but it is best excluded by the following experiment. The S_2 electrodes were arranged so that the anode was interposed between the cathode and the block (Fig. 4A). There was now no possibility of catelectrotonic potential reaching the block itself, and any change in threshold must depend upon an actual change at or within a millimetre of the cathode. Under these conditions it was found that the impulse still produced a large change in threshold. The results of an experiment with a pressure block were as follows:

TABLE I. 3 mm. pressure block. 9 December 1936

Electrode arrangement A (Fig. 4) Anode 0·5 mm. distal to block Cathode 3·0 mm. distal	11·6 p.c. decrease in threshold
Arrangement B (Fig. 4) Anode 18 mm. distal Cathode 3·0 mm. distal	15·5–14·6 p.c.

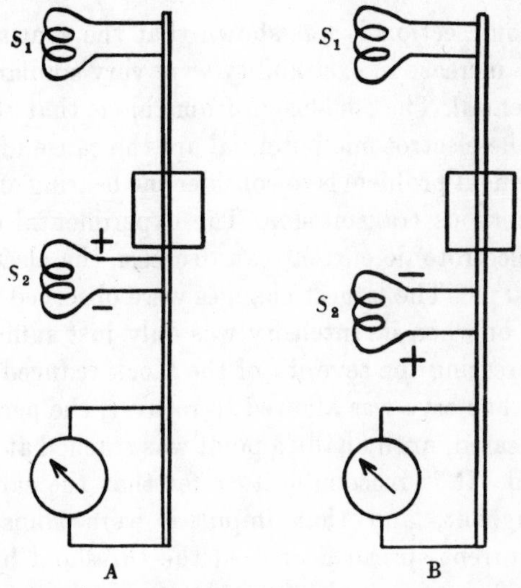

Fig. 4. See text.

This experiment shows that the two percentage changes were not identical, and it was found that the change was always lower with the arrangement of Fig. 4 A. This was unexpected, since the position of the cathode was fixed and variations in the local excitability ought not to have affected the results.

There are various possible explanations of this difference, but it is unnecessary to discuss them, since exactly the same result was obtained when an applied current was substituted for the impulse. The measurements in Table II were made at the same time as those in Table I. It is clear that the difference between the two electrode arrangements depends upon something which is common both to an impulse and to an applied current. It is therefore unnecessary to analyse the phenomenon further, for the object of this paper is to discover how far the influence of the

TABLE II. Details as above, except that increase in excitability was produced by applying pulse of current to blocked nerve, instead of by blocked impulses

Arrangement A	13·4 p.c.
Arrangement B	14·3–16·1 p.c.

nervous impulse can be explained in terms of the known properties of an electrical stimulus; it is not concerned with the mechanism of electrical stimulation.

THE RELATIVE MAGNITUDES OF INCREASE IN EXCITABILITY AND POTENTIAL

In the previous section it was shown that the spatial and temporal properties of the increase in excitability were very similar to those of the electrotonic potential. The conclusion from this is that the local circuits which produce the electrotonic potential are the cause of the increase in excitability. The next problem is to consider the bearing of this conclusion on theories of nervous transmission. The experimental evidence shows that spread of electrotonic current can decrease the electrical threshold by as much as 90 p.c. The largest changes were observed when the block was incomplete, or when its intensity was only just sufficient to abolish conduction. Increasing the severity of the block reduced the increase in excitability. If the nerve was allowed to recover, the percentage change in threshold increased, until finally a point was reached at which impulses were transmitted. It is reasonable to infer that the same process was operating throughout, and that impulses were transmitted at the moment when current spread decreased the threshold by 100 p.c. This affords strong evidence for an electrical theory of transmission. For local circuits should be more than sufficient for normal propagation, if they can transmit activity through a region impaired by cold or pressure. The argument may be given a more definite form by considering the relative magnitudes of potential and excitability. If it was found that current spread, which was associated with a potential difference of 1 mV., caused a 90 p.c. change in threshold, we should be able to conclude that an action potential of 20 mV., could be propagated electrically with a wide margin of safety.

The measurement of the relation between potential and increase in excitability raises various problems. In the first place, the results obtained with a block do not necessarily apply to normal nerve. For the thresholds are raised in the vicinity of the block, and so the ratio

$$\frac{\text{Percentage change in threshold}}{\text{Potential}}$$

would be smaller than in unblocked nerve. There is a more serious difficulty than this. If the relation between potential and change in threshold is to have any meaning, the measurements must refer to the same nerve fibres, and if the increase in excitability is measured in the ordinary way this will not be the case. Suppose, for example, that there is a large spread of electrotonus in a few fibres. If the stimulating current is adjusted to produce a threshold action potential, a large increase in excitability would be observed, although the average increase in excitability might be very small. The potential is made up of contributions

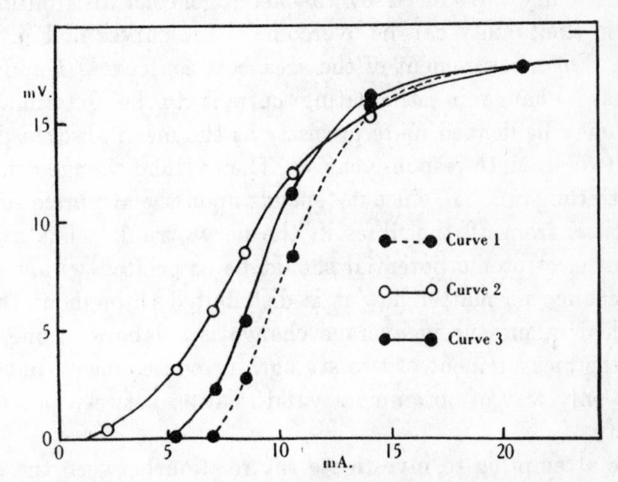

Fig. 5. 3·5 mm. cold block. S_2 applied 1·2 mm. distal to block. Abscissa: strength of S_2 shock. Primary current in milliamperes. Ordinates: Action potential in millivolts, produced by: (1) S_2 alone; (2) S_2 in presence of impulses set up by S_1; (3) S_2 in presence of electrotonic potential produced by applying pulse of current to blocked section of nerve (for arrangement see Fig. 3).

from all the fibres in the nerve trunk, and it would be illogical to compare it with an excitability change which was limited to a few fibres. The following experiment illustrates this difficulty, and at the same time it suggests a possible solution.

The relation between the strength of a stimulus (S_2) and the action potential it produces was determined at a point beyond the block. This gave curve 1, fig. 5. Curve 2 was obtained by repeating the operation in presence of a blocked impulse. Since the excitability was increased, the strength response curve was shifted to the left. Now it has been shown that the potential is proportional to change rather than to percentage change in threshold. Hence the two curves ought to have the same shape, if the potential were distributed uniformly throughout the fibres. Actually curve 2 is flatter than curve 1, and the change in stimulating current was greater for a threshold response than for a half-maximal one. The experiment suggests that the increase in excitability occurs to different extents in different fibres, and this conclusion is supported by the shape of curve 3. Here the change in excitability was produced by an applied current. This seems to produce a more uniform distribution of electrotonic potential in the fibres, for the curve obtained has the same shape as the initial strength-response curve.

The difficulty introduced by the heterogeneous distribution of the increase in excitability can be overcome if the curves in Fig. 5 can be obtained. For measurement of the area between curves 1 and 2 allows the average change in stimulating current to be determined. This quantity may be defined more precisely as the mean abscissa difference between two strength-response curves. The average change can be compared with the potential, since it depends upon the algebraic sum of the contributions from all the fibres in the nerve trunk. Thus a constant amount of electrotonic potential should be associated with a constant average change no matter how it is distributed throughout the fibres. The method of measuring average change is a laborious one, since it requires the measurement of two strength-response curves, but it seems to be the only way of obtaining a valid relation between potential and excitability.

Before attempting to investigate the relation between the extrinsic potential and the increase in excitability, it is well to consider another problem which is raised by the curves in Fig. 5. This is the question of the relative efficiency of extrinsic potential and applied electrotonic potentials in producing an increase in excitability. The experiment of Fig. 3 showed that the electrotonic potential had to be three times as large as the extrinsic potential to produce the same change in threshold. This must have been partly due to the difference in the distribution of the two potentials. For a potential which is uniformly distributed throughout the fibres would produce a smaller change in threshold than one with a random distribution. The important problem is not whether the potentials produce the same threshold change, but whether the average changes in excitability are the same. In the experiment of Fig. 5 the value for the applied electrotonic potential was 2·2 mV. and for the extrinsic potential 1·7 mV. The average changes in stimulating current were 1·1 and 1·9 units. The extrinsic potential was therefore about twice as efficient as the electrotonic potential in producing an increase in excitability. Several experiments were performed with the object of discovering the relative efficiencies of the two potentials.

It was necessary to determine three curves like those in Fig. 5. These were not measured separately, but the values for each strength of current were observed simultaneously. The average change in excitability was always measured from curves made with both ascending and descending current strengths. The observations in Fig. 5 were for descending strengths. The ascending curves had essentially the same shape, but they have not been included because they did not coincide exactly with the others and would have confused the graph. The current pulse for producing the electrotonic potential was applied to the blocked nerve in the same way as before (p. 219). The shape of the applied electrotonic potential was similar to that of the extrinsic potential. Records of it were given in a former paper [Hodgkin, 1937, p. 202]. The potentials were measured at the moment when the testing shock (S_2) was applied; this was usually at about the maximum of extrinsic and applied potential waves. The values for the two potentials were observed at the beginning and end of the excitability measurements. All the experiments were made with 3·5 mm. cold blocks.

The results of these experiments are summarized in Table III. The figures in the right hand column indicate the relative effects on excitability of extrinsic and applied electrotonic potentials. They were obtained from four measurements, namely,

(*a*) average change in stimulating current produced by extrinsic potential,

(*b*) extrinsic potential,

(*c*) average change in stimulating current produced by applied electrotonic potential,

(*d*) applied electrotonic potential.

The relative effect of the two potentials is given by $a/b \div c/d$, and this is the quantity tabulated in the right-hand column.

TABLE III. Effect of extrinsic and applied electrotonic potentials on excitability

Experimental details. Date and distance from block	Excitability change per mV. extrinsic potential / Excitability change per mV. applied electrotonic potential
16. xi. 36: 1·2 mm. distal	1·6
18. xi. 36: 1·2 mm. distal	2·3
18. xi. 36: 1·2 mm. distal	1·6
14. xii. 36: 2·4 mm. distal	1·4
14. xii. 36: 1·2 mm. distal	1·4
15. xii. 36: 2·4 mm. distal	1·5
15. xii. 36: 2·4 mm. distal	1·5
18. xii. 36: 1·2 mm. distal	2·3

Table III shows that the extrinsic potential always produces relatively larger changes in excitability than an applied electrotonic potential. A detailed discussion of the reason for this difference is beyond the scope of this paper, but two possibilities may be mentioned. In the first place it is important to notice that the wave form of the two potentials was not identical. Part of the difference in action may be explained on this basis, but it is difficult to attribute all the difference to this. For both potentials lasted several milliseconds, and their stimulating efficacy should have depended more upon amplitude than upon wave form. In the second place, it is possible that an electrotonic potential may not be entirely due to spread of current along nerve fibres, but may be partly determined by connective tissue sheaths and other structures in the nerve trunk. If this is correct we should expect to find a difference between the action of an electrotonic potential produced by an impulse and one produced by an applied current. For a potential of internal origin ought to depend much more upon current spread along the axons than one of external origin. Hence the extrinsic potential should cause larger changes in excitability than an applied electrotonic potential.

It will by this time have become evident that there is no constant numerical relation between the extrinsic potential and the increase in excitability. This is due primarily to variability in the resting excitability of nerve beyond the block. In a freshly blocked nerve the resting threshold does not differ markedly from normal, but it may be increased many times in a nerve which has been subjected to the prolonged action of the block. In the first case, a potential of 1 mV. causes a large percentage change in excitability, but in the second the change may be too small to be measured. The interesting question is the size of the ratio

$$\frac{\text{Percentage change in excitability}}{\text{Extrinsic potential}}$$

in normal nerve.

If we could estimate this ratio, we should be able to form some idea of the fraction of the action potential required for excitation. The importance of this quantity is that it determines the size of the margin of safety in the electrical transmission of impulses. If only a small fraction of the action potential was required for excitation, there would clearly be a wide margin of safety; for action potential or excitability would have to be considerably reduced before conduction failed. On the other hand, if a large fraction of the action potential were required to excite, transmission would fail after quite a small reduction had occurred. The final experiments in this research were an attempt to estimate the fraction of the action potential required for the excitation of normal nerve. The method employed consisted in observing potential and excitability in nerves which had been subjected to a minimal amount of blocking.

In order to illustrate the principle of the method we must again refer to the experiment in Fig. 5. Here the average change in excitability produced by the impulse could be determined from the area between curves 1 and 2; it was 1·9 units. This was produced by an extrinsic potential of 1·7 mV. Hence 1·7 mV. potential was equivalent to 1·9 units of current. Curve I indicates that the resting threshold was 7·1 units. If we assume that the change in threshold is directly proportional to the potential it follows that a potential of $1{\cdot}7 \times 7{\cdot}1/1{\cdot}9 = 6{\cdot}4$ mV. would be equivalent to a threshold stimulus. Now this result cannot be applied to normal nerve, since these measurements were made after the block had been in action for some time. Table V summarizes the result of a series of similar experiments with freshly blocked nerve. Cold blocks were used throughout and the experiments were discontinued unless the necessary measurements could be made within about 20 min. after the cooling had been started. The curves for determining the average change in stimulating current were made in exactly the same way as those in Fig. 5. With the exception of the values marked with an asterisk, the figures in Table V are for the mean of two observations. The order in which the measurements were made was as follows:

(1) Extrinsic potential.

(2) Curves for determining average change in stimulus, made with increasing current strengths.

(3) Curves made with descending current strengths.

(4) Extrinsic potential.

Table V. Relation between extrinsic potential and excitability change in freshly blocked nerve

Experimental details. Date and distance from block	Extrinsic potential mV.	Change in stimulus / Threshold stimulus	Potential required to excite (calculated) mV.
16. xi. 36: 1·2 mm. distal	0·7	0·24*	2·9
18. xi. 36: 1·2 mm. distal	1·0	0·37*	2·7
14. xii. 36: 2·4 mm. distal	0·56	0·24	2·3
14. xii. 36: 1·2 mm. distal	0·74	0·24	3·1
15. xii. 36: 2·4 mm. distal	0·65	0·21	3·1
16. xii. 36: 2·4 mm. distal	0·57	0·22	2·6

The unstable condition of freshly blocked nerve prohibited accurate measurement, but the determinations usually agreed to within about 20 p.c. The average change in stimulating current produced by the impulse has been expressed as a fraction of the initial threshold (col. 3). The calculated values for the potential required to excite are given in col. 4; they were obtained by dividing col. 2 by col. 3.

Table V suggests that a potential of between 2 and 3 mV. may be sufficient for excitation. In these nerves the amplitude of the action potential after conduction to the block was about 20 mV. Hence it appears that only about a tenth of the action potential may be required for excitation. It is still uncertain if this is the smallest possible fraction required for excitation. Near a cold block the duration of the extrinsic potential was 4 or 5 msec., so that it could not go far above the rheobase without exciting; nevertheless it is possible that an even smaller potential would have sufficed if it had a longer duration and a more abrupt onset.

If the argument of the preceding paragraphs is correct, it follows that the extrinsic potential beyond the block ought not to exceed a value of 2 or 3 mV. without propagating impulses. This deduction has been verified. It is possible to obtain extrinsic potentials of 3 or 4 mV. which are practically free from unblocked impulses, but these only occur in nerves where the threshold has been raised by prolonged blocking. In fresh nerves impulses are always propagated through the block as soon as the potential beyond it rises above 1 or 2 mV. The potential was recorded at a lead which was 1·2 mm. from the distal side of the cooling rod, so that the potential at the edge of the block may have been larger. However there is evidence to show that the margin of inexcitable nerve was often very close to this lead. I have observed that in fresh nerves block does not occur until ice has formed over the whole of the cooled section (3·5 mm.) and that it often has to spread for $\frac{1}{2}$ mm. beyond. It is reasonable to suppose that the whole of the frozen region is inexcitable, and in this case the inexcitable nerve must extend to within about $\frac{1}{2}$ mm. of the recording lead. Now the potential only falls by about 20 p.c., in this distance, so that the potential recorded would not have been very different from that at the edge of the block. If this is correct it follows that the potential at the edge of the block cannot exceed 1 or 2 mV. without propagating impulses. The method is obviously a very crude one, but the results are consistent with the view that a small fraction—approximately one-tenth—of the action potential is sufficient to excite. Like the previous method the experiments are open to the objection that the boundary between excitable and inexcitable nerve is almost certainly not a sharp one. Hence even under the most favourable conditions, the potential at the edge of the block would have less influence than in normal nerve.

It is clear that the estimate of one-tenth for the fraction required to excite is at best only a first approximation. There are so many possible sources of error in experiments with mixed nerve trunks that any quantitative conclusions must be regarded with great diffidence. The main justification for including these final experiments is that they give a definite, although qualitative, indication of a wide margin of safety in the electrical transmission of impulses.

At first sight this conclusion seems to be in conflict with the properties of electrotonus. Records published by Schmitz & Schaefer [1933] show that an electrotonic potential of 12 mV. is required for excitation at the cathode of a rheobasic stimulus. Now experiments described in this paper suggest

(*a*) That the change in threshold produced by the extrinsic potential is 1·4–2·3 times as large as that produced by an applied electrotonic potential.

(*b*) That an extrinsic potential of 2–3 mV. is required to excite.

If these conclusions are correct we should expect that an applied electrotonic potential of about 4 mV. would excite. The explanation may be that Schmitz & Schaefer were recording the electrotonic potential at the cathode of the applied current, whereas in my experiments potential and excitability were observed at a distance of 2 or 3 mm. from the cathode. It is possible that the ratio

$$\frac{\text{Change in threshold}}{\text{Potential}}$$

may be less in the immediate vicinity of the cathode than at a distance of 2 or 3 mm. This is to be expected if electrotonus is partly due to polarization at the connective tissue sheath, for the effect of this would be greatest at the point where current was led into the nerve.

Discussion

The main conclusion from this research is that it is possible for nervous impulses to be transmitted by the electrotonic currents produced in activity. The experiments do not prove that this is the only process involved in propagation, but the assumption of an additional mechanism would be unnecessary if all the properties of the nervous impulse could be explained in terms of an electrotonic theory. A general discussion of the theory is beyond the scope of this paper, but it is well to consider whether there are any serious objections to the view that impulses are transmitted by electrotonic currents.

The accepted explanation of electrotonic currents is that they depend, at least in part, upon the cable-like structure of nerve fibres. Owing to the high resistance of the myelin sheath current cannot enter an axon at one point, and so local circuits spread for some distance along it. Hence the essential point in an electrotonic theory is that transmission depends upon local circuits, with current flowing in one direction in the core and in the other direction in the interstitial fluid. This is, of course, exactly the theory that has been put forward by Lillie [1923], Cremer [1929] and many others. Erlanger & Blair [1934] think unfavourably of a mechanism involving eddy currents flowing around nerve fibres. They consider that conduction is much more likely to depend upon a process operating within the axon and they have put forward evidence to show "that the eddy currents that are supposed to develop in nerve are without physiological significance". The evidence depends upon their failure to demonstrate any increase in excitability in resting fibres lying parallel to active ones [Blair & Erlanger, 1932]. I do not think that there is necessarily any conflict between this result and my own. My evidence suggests that an active section produces electrotonic currents in a resting part of the same axon; theirs indicates that an active region cannot produce appreciable electrotonic currents in neighbouring fibres. The electrical structure of nerve is so complicated that it is difficult to form an exact picture of the current distribution in the two cases, but it is not unreasonable to suppose that electrotonic current could spread along one fibre without affecting others. This would be true if the resistance of the interstitial fluid were low compared to that of the cores of the axons. For in this case the current in the external circuit of an active fibre would flow mainly through the interstitial fluid, and would hardly affect adjacent axons. At the moment it is impossible to decide whether this is what actually does happen, and it is therefore best to leave the question until more is known about the details of current distribution in nerve.

The most serious problem for an electrical theory is whether a process of restimulation by the action current allows a sufficient margin of safety. It is certain that conduction can occur after excitability and action potential have been considerably reduced. Gasser [1931] showed that the threshold was increased fivefold at the beginning of the relative refractory period, and Gasser & Erlanger's [1925] work implies that the axon action potentials were reduced by at least 50 p.c. in the relative refractory period. Thus in the relative refractory period an action potential of half-normal size can be propagated in nerve where the threshold is five times normal. Taken at its face value this suggests that in normal nerve at least one-tenth of the action potential would suffice for excitation. This computation assumes that the whole of the action potential is available for stimulation. Theoretical considerations suggest that this is unlikely, so that the critical fraction would probably have to be less than one-tenth.

Other evidence pointing towards a large safety factor comes from the question of anode block. If the action potential was only just sufficient to excite, it should be possible to block impulses with a very small anodal shock. On the other hand, if the amplitude of the action potential was much larger than the potential required for excitation, impulses could only be blocked by a current many times greater than threshold. There does not seem to be any quantitative information about the strength of current necessary to block impulses, but it is clear that block often does not occur until the intensity of the current is 10–15 times threshold.

This estimate is based on a few preliminary experiments which were made in the following way. A sciatic gastrocnemius preparation was employed and a condenser discharge with a very long time constant (>1 msec.) used for stimulating and blocking. The threshold was first found with a descending current and electrodes about 20 mm. apart. The current was then reversed and increased until significant anode block occurred. In this way the ratio of blocking to exciting voltages could be determined. The estimate of ten to fifteen must be regarded as a minimum. At the beginning of an experiment the ratio often seemed to be higher, but it was very variable and difficult to measure. Repeatable measurements could only be obtained after a few trials, when the ratio had settled down to a steady level of between seven and fifteen.

In these experiments impulses reached the anode about 1 msec. after the make of the current, so that there would not have been time for any gradual or cumulative anelectrotonic effects to develop. It is not unreasonable to suppose that this type of block depends upon a simple neutralization of the anelectrotonic potential produced by the applied current and the catelectrotonic potential produced by the impulse. If this is correct, the block/threshold ratio indicates the amount by which an action potential exceeds a threshold stimulus.

The phenomena of anode block and of conduction in the relative refractory period suggest that there must be a large safety factor in the electrical transmission of impulses. The tentative conclusion from my experiments was that only a small fraction of the action potential is required to excite, and this implies that electrotonic currents can transmit activity with a considerable margin of safety. At present there is not enough information to decide whether the electrotonic theory allows a sufficient margin of safety, and future work may demonstrate the necessity for another mechanism besides the electrotonic one. If such a mechanism exists, it does not seem to play any part in propagation through cold and compression blocks, since all my observations have been entirely consistent with the view that impulses are transmitted by electrotonic currents.

SUMMARY

1. In a previous paper it was found:

(*a*) That a nervous impulse produces a large increase in excitability beyond a block.

(*b*) That a nervous impulse produces an "extrinsic potential" beyond a block. This was shown to depend upon spread of electrotonic current.

2. In this paper evidence is brought forward to show that the electrotonic current which produces the extrinsic potential is the cause of the increase in excitability.

The evidence is as follows:

(*a*) The time relations of the increase in excitability are similar to those of the extrinsic potential. Not only is the general shape of the two time curves similar, but the individual peculiarities of the potential wave are reflected in the excitability curve. This is illustrated by the difference between the curves obtained with cold and compression blocks.

(*b*) The change in threshold beyond the block falls off with distance at the same rate as the extrinsic potential.

(*c*) The spatial decline of the change in threshold beyond the block is the same whether it is produced by an impulse or by an applied electrotonic potential.

3. The average change in excitability produced by the extrinsic potential was 1·4–2·3 times larger than that produced by an applied electrotonic potential of the same form.

4. It was estimated that in freshly blocked nerve an extrinsic potential of 2–3 mV. would be required to excite.

5. The general conclusion is that it is possible for nervous impulses to be transmitted by electrotonic currents.

REFERENCES

Blair, E. A. & Erlanger, J. (1932). *Amer. J. Physiol.* **101**, 559.

Blair, E. A. & Erlanger, J. (1936). *Ibid.* **117**, 355.

Cremer, M. (1929). *Handbuch der normalen u. pathologischen Physiologie*, **9**, 244. Berlin: Springer.

Erlanger, J. & Blair, E. A. (1934). *Amer. J. Physiol.* **110**, 287.

Gasser, H. S. (1931). *Ibid.* **97**, 254.

Gasser, H. S. & Erlanger, J. (1925). *Ibid.* **73**, 613.

Hodgkin, A. L. (1937). *J. Physiol.* **90**, 183.

Katz, B. (1936). *Ibid.* **88**, 239.

Lillie, R. S. (1923). *Protoplasmic Action and Nervous Action.* Chicago University Press.

Schmitz, W. & Schaefer, H. (1933). *Pflügers Arch.* **233**, 229.

2

Reprinted from 'Proceedings of the Royal Society of London'
Series B No. 842 *vol.* 126 *pp.* 87–121 *September* 1938 612.816.1

The subthreshold potentials in a crustacean nerve fibre

By A. L. Hodgkin
From the Physiological Laboratory, Cambridge, and the Laboratories of the Rockefeller Institute for Medical Research, New York

(*Communicated by E. D. Adrian, F.R.S.—Received* 26 *May* 1938)

[Plates 4–6]

In a recent paper Katz (1937) has described a number of observations which are inconsistent with classical theories of excitation, but which may be explained by assuming that a subthreshold shock can elicit a small and localized action potential in the cathodic part of a nerve fibre. At first the assumption of a subthreshold response seems to be in conflict with the all-or-nothing law, but there is in reality no contradiction, since the all-or-nothing principle refers to the propagated disturbance, and does not exclude the possibility of graded reactions in the stimulated region. Indeed, the existence of some kind of subliminal action potential is to be expected from the local circuit theory of nervous transmission; for the current generated by the activity of a very short length of nerve would be too weak to excite surrounding regions, so that a shock which activated less than a certain length would fail to produce a self-propagating impulse and would leave behind only a small and localized response (cf. Rosenberg 1937, Rushton 1937). The theory of a local response is plausible *a priori*, and it would explain a number of observations in widely different fields; but it is open to one serious objection, since many workers have searched and failed to find any electrical sign of subthreshold activity. It is true that Katz (1937) showed that alternating shocks produced a residual negativity, which could be detected with a sensitive galvanometer, but it is uncertain if this was due to the summation of transient local action potentials, rather than to a persistent depolarization of the stimulated region. At present, the only direct evidence for a subliminal response is provided by Auger and Fessard's (1935) studies on the excitable cells of the plant *Chara*. They observed the potentials in the stimulated region and were able to record a small, wave-like response, which grew up to a propagated action potential when it reached a certain size. The present paper contains an account of experiments which show that a similar result can be obtained in crustacean nerve. Non-medullated nerve was used, because it was hoped that the electrical properties of the excitable membranes would be more conspicuous in the absence of a thick myelin sheath. At first the whole nerve trunk or bundles of nerve fibres were employed, but these experiments failed to give any clear evidence of a subthreshold response. However, during the course of the work a chance experiment showed that it was not difficult to isolate and to record action potentials from a single fibre. In an isolated fibre it is comparatively easy to record from the stimulated region, and the evolution of the spike can be watched without using a balancing circuit. This made the search for a local response much easier, and experiments with single fibres soon gave definite evidence of something like a subliminal action potential.

Methods

Material

Single fibres were usually obtained from the limb nerves of the shore crab, *Carcinus maenas*. Most of the axons in the nerve trunk would have been too small and fragile to handle, but there are a few large fibres which can be isolated without much difficulty. These consist of a transparent axis cylinder surrounded by concentric layers of connective tissue (cf. Young 1936). The diameter of the axis cylinder is about 25μ and the thickness of the connective tissue about 5μ. A photomicrograph of a living axon is reproduced in fig. 1, Plate 4.

In a few experiments single fibres were isolated from blue crabs, *Callinectes sapidus*, or lobsters, *Homarus vulgaris*. Axons from these animals behave in the same way as those from *Carcinus*, but they are harder to dissect and are usually covered with several layers of diffuse connective tissue which stick to the electrodes and make it difficult to obtain a clean recording lead.

Isolation of fibres

The nerve trunk was dissected from the meropodite of a walking leg and transferred to a Petri dish of sea water which was mounted on an illuminated dissecting stage. Crustacean nerve is only loosely held together by connective tissue, and when immersed in sea water it floats apart into a number of thin strands, which can be separated with fine forceps or needles. Most of the strands consist of small fibres and must be rejected. The large fibres were identified with a microscope (magnification × 150), which was mounted on an adjustable arm, so that it could be swung over any part of the Petri dish. When a large fibre had been found, the strand containing it was split into two by pulling with dissecting needles. The two parts of the strand were again examined with the microscope, and the half containing the large fibre was set aside for further subdivision. This process was repeated until a single fibre was obtained. Sometimes the isolation of an axon could be completed in a few minutes, but often repeated subdivision was necessary. The fibres were handled only by their ends, so that injury to the main body of the axon was avoided. It was necessary to use a microscope for identifying the large fibres, but it was found best to view the nerve with the naked eye during the actual dissection of the fibres.

Electrical recording

The essential features of the electrode system are shown in fig. 2. The electrodes consisted of fine platinum hooks (A) made from 100 or 200μ diameter wire. They were sealed into glass tubes (B), which in turn were mounted in adjustable bakelite bosses (C). The nerve fibre was held in position by two pairs of screw-controlled forceps (D). The electrode holders and forceps were mounted on a glass bar which could be moved horizontally about the pivot (E) or raised vertically in a Palmer adjustable stand.

When the dissection was complete, the electrodes were swung over the Petri dish and lowered into the sea water. One end of the fibre was placed between the tips of one pair of forceps and gripped by tightening up the screw (F). The other end was seized with a needle and the fibre drawn across the tips of the electrodes. When almost taut, it was lowered into position and then gripped in the second pair of forceps. The final result of this manœuvre was to stretch the fibre horizontally above the tips of the hook electrodes. It was important to keep the fibre from touching the electrodes, since it might stick to them so firmly that it could not be pulled free without injury.

In order to record from the fibre, it must be kept in an insulating medium. At first it was raised out of the sea water into air. Action potentials could be recorded in this way, but the method was unsatisfactory, since the fibre dried quickly and could not be kept out of sea water for more than a minute at a time. After a few trials it was found possible to use the fibre

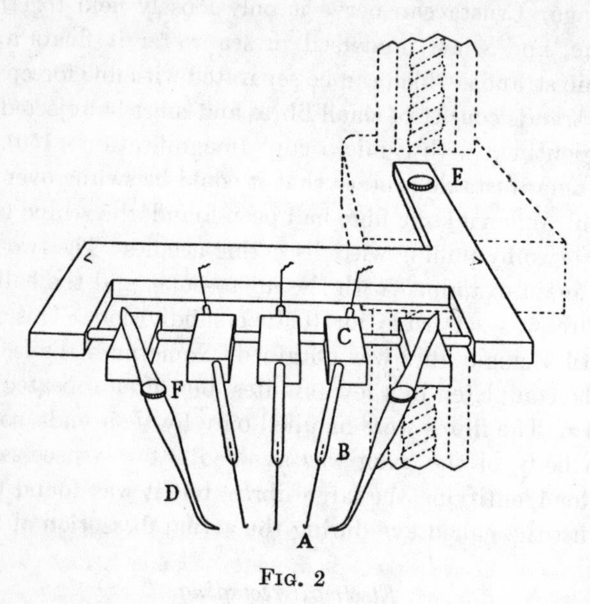

FIG. 2

in aerated paraffin oil. Paraffin does not seem to have any harmful effects, for fibres immersed in it remain excitable and transmit impulses for several hours. A layer of oil was poured on to the sea water, and the electrodes and nerve fibre were then slowly raised into it by turning the screw on the Palmer stand. As the axon was brought through the interface, the surface tension caused it to stick firmly to the platinum hooks, which therefore made good electrical contact despite the surrounding paraffin.

The principal difficulty in recording electrical changes from a single fibre is that its high resistance slows the response of the first stage of the amplifier. The input resistance was kept as low as possible by working with short interelectrode distances, but it could not be reduced below about 0·5 megohm, and with an ordinary amplifier a resistance of this magnitude introduces serious distortion at comparatively low frequencies. The difficulty was finally overcome by using a specially fast amplifier which was designed by Dr Toennies and which will be described by him elsewhere. Tests made by calibrating through a high resistance showed that this was sufficiently rapid to follow any ordinary nerve potentials. Thus when a rectangular pulse was applied through 2 megohms, the oscillograph deflexion was 90 % complete in 60 μsec. (fig. 3, Plate 4). Calibrations made with an oscillator under similar conditions showed that the response was reduced by 7 % at 5000 cycles.

Grid leaks of 15 megohms were used in the first stage of the amplifier, so that no appreciable quantity of current was drawn from the nerve.*

In the Cambridge experiments the fibre was stimulated with coreless induction coils, but in New York these were abandoned in favour of the thyratron discharge circuit used in Dr Gasser's laboratory. This is essentially similar to the arrangement described by Schmitt and Schmitt (1932), except that the intensity of the shock is controlled by a potentiometer in the primary, instead of in the secondary circuit of the transformer. Measurement showed that the shock produced in this way was 95 % complete in 60 μsec.

During the early stages of the work various artifacts and sources of error were encountered. Thus under certain conditions the nerve fibre might be stimulated by pick-up from the sweep circuit of the oscillograph, or occasionally a strong shock might stimulate through the ground lead of the amplifier. Another difficulty was that the action potential might record through the capacity to ground of an idle electrode or through one of the forceps used for holding the fibre. These difficulties were overcome by keeping the capacity to ground of all points connected to the nerve as low as possible. Another important precaution is to maintain a very high degree of insulation between the two pairs of forceps at each end of the fibre. Thus the forceps were originally attached to bakelite bosses on a metal bar, but subsequent experiments showed that a large part of the shock artifact could be traced to a leak across the bakelite insulation. The metal bar was therefore replaced by one made of glass, and in this way the artifact was greatly reduced.

* Records of electrical changes in a single fibre were published in a preliminary communication (Hodgkin 1937). These were made with an amplifier which was too slow to follow rapid changes produced by a high-resistance preparation, and the results, therefore, differ in a number of ways from those given in the present paper. These differences are not entirely due to the relative speed of the two recording systems, but are partly to be attributed to the fact that the earlier amplifier drew an appreciative amount of current from the fibre.

It is important to reverse the direction of the shock by changing the polarity of the primary circuit of the transformer or induction coil. If the stimulating leads are reversed, the balance of the secondary circuit with respect to ground is altered, so that the intensity of corresponding anodic and cathodic shocks may not be exactly equal. The stimulating leads were connected through large condensers (1 μF) in order to avoid polarization of the nerve by potential differences developed at the electrodes.

Experimental results

The propagated action potential in a single fibre

The propagated action potential is of constant size and form and travels without decrement at a uniform velocity of 3–5 m./sec.* This result agrees with that of Bogue and Rosenberg (1936), who showed that the fastest fibres in the limb nerve of *Maia* conduct at 5 m./sec. The duration of the spike usually lies between 0·8 and 1·0 msec., so that its wave-length is only about 4 mm. In the *A* fibres of a frog or mammal the wave-length is about 40 mm. This difference is interesting, because it may be connected with the fact that the spread of the polarization potential (electrotonus) is much less extensive in crab fibres than it is in medullated nerve. The form of the spike is essentially similar to that in a medullated axon, as may be seen by comparing the record in fig. 4 with the one published by Gasser and

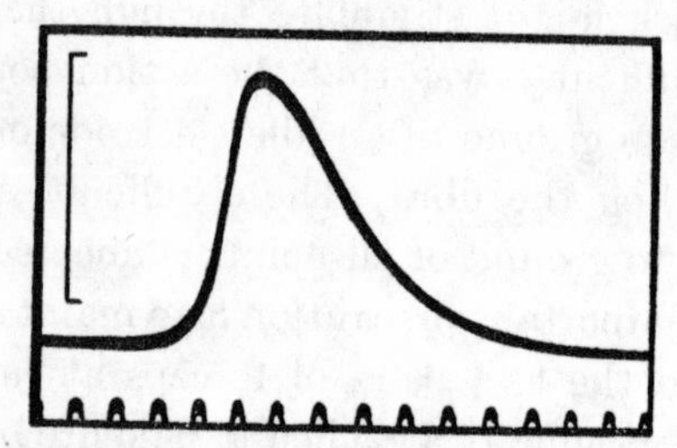

Fig. 4. Propagated action potential. Fibre made monophasic by crushing. Time 0·1 msec. Amplitude of action potential, 55 mV.

Grundfest (1936). The absolutely refractory period lasts for about 1 msec. and the fibre can transmit a series of impulses at a frequency of over 500/sec. Fig. 5 *b* shows a train of spikes following one another at 580/sec. A comparison of these action potentials with one evoked by a single shock (fig. 5 *a*) shows that the potentials were reduced by about 25 % during the tetanus.

* All the experiments described in this paper were made at a temperature of 20–25° C.

The most surprising fact about an isolated fibre is the size of the action potential. Spikes of 40 mV were regularly observed and potentials of 50–60 mV were recorded on several occasions. These large values suggest that the action potentials which are ordinarily recorded from a nerve trunk must be greatly reduced by the short-circuiting effect of the inactive tissue and interstitial fluid. In a single fibre preparation much of the inactive tissue is removed, so that a higher proportion of the potential generated by the fibre is recorded. But despite its large size, there is reason to believe that the recorded spike still falls far below the potential change at the nerve

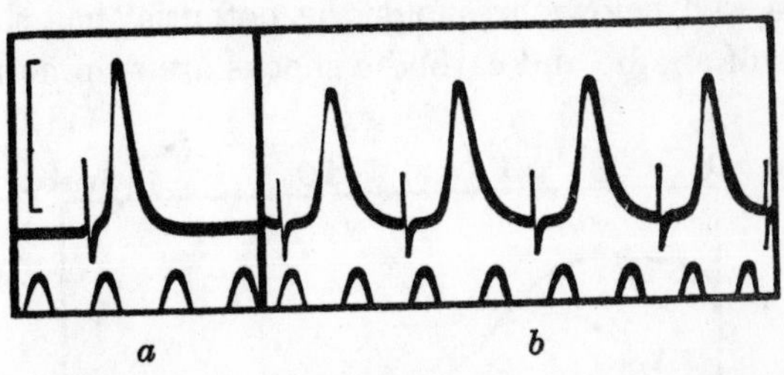

FIG. 5. *a*, monophasic spike. *b*, same experiment; train of spikes recorded during a tetanus. Time 1 msec.

membrane. The recorded potential would only be equal to the internal potential change if the resistance outside the nerve membrane were high compared to that of the axis cylinder, and this condition was almost certainly not realized in the present experiments. For the axis cylinder was surrounded by layers of diffuse connective tissue which must have had a considerable short-circuiting effect, and this would have been increased still further by the film of sea water which adhered to the fibre after it had been raised into paraffin.

The electrical changes in the stimulated region

Fig. 6, Plate 5, shows the different kinds of electrical change which can be observed in the stimulated region. The stimulus was a brief thyratron shock, and the potentials were recorded by connecting the distal stimulating lead to the ground lead of the amplifier.

When the shock is well above threshold, it produces a large diphasic action potential which arises with no appreciable latency. As the stimulus is weakened, the spike starts later and finally disappears, revealing a smaller subthreshold potential. The properties of this potential are quite different from those of the electrotonic potential in a nerve trunk, and they suggest that it must be a combination of a passive polarization process and a second effect which behaves like a subliminal response of the nerve fibre.

These two processes may be dissociated by varying the strength of the stimulus. When the shock is weak or anodic, it affects only the polarization process, and the potential has a simple time course which is readily explained by the capacitative prcperties of the nerve. During the period in which the shock is applied, the nerve membrane is charged very rapidly so that the potential rises abruptly, reaching a maximum in 50 μsec.; after the end of the shock it subsides relatively slowly, because the charge takes an appreciable time to leak away from the membrane capacity. Thus the potential has the rapid rise and gradual fall shown in G and H. As long as the stimulus is well below threshold, the potential has this simple shape and the effects of anodic and cathodic shocks are symmetrical. However,

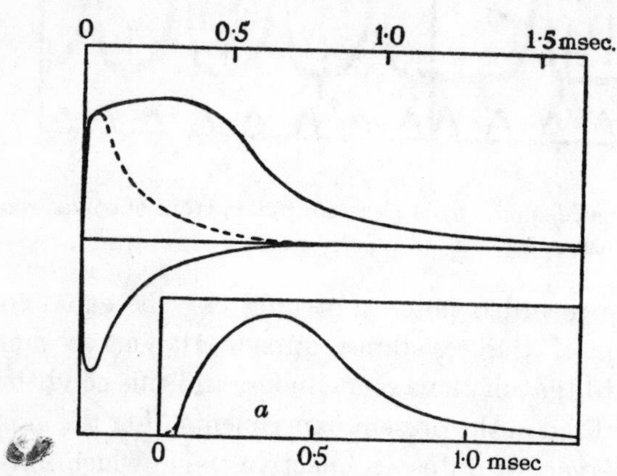

FIG. 7. Cathodic and anodic potentials, traced from records in fig. 6 E and F. Cathodic potential analysed into polarization potential and local response.

FIG. 7 *a* (inset). Local response obtained by subtracting polarization from total cathodic effect.

as soon as the shock is brought close to threshold, the cathodic potential changes in a striking and characteristic way. In the first place its duration is greatly increased, and in the second it has a rounded form which is quite unlike that of the corresponding anodic polarization. The simplest way of explaining these changes is to assume that the nerve produces a small and localized wave of negativity which adds to the polarization potential. According to this view, we may analyse the rounded curve in E into a cathodic polarization potential which is symmetrical with the anodic one, and an additional wave of negativity which will be called a local response. Fig. 7 illustrates an analysis of this kind and shows that this particular local response had a rising phase of 270 μsec. and a total duration of about 1500 μsec.

The development of the local response is illustrated by the typical experiment in fig. 8*a*. It is clear that shocks of less than 0·5 threshold affect only the polarization process, because:

(1) After 100 μsec. the potential has a simple, approximately exponential decline.

(2) The potential is directly proportional to the strength of the stimulus.

(3) The effects of anodic and cathodic shocks are symmetrical.

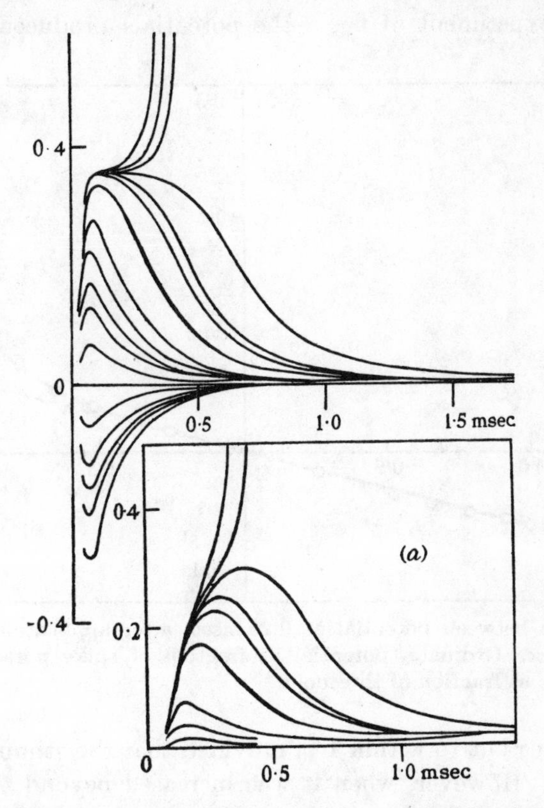

Fig. 8. Electrical changes at stimulating electrode produced by shocks with relative strengths, successively from above, 1·00 (upper 6 curves), 0·96, 0·85, 0·71, 0·57, 0·43, 0·21, −0·21, −0·43, −0·57, −0·71, −1·00. The ordinate scale gives the potential as a fraction of the propagated spike, which was about 40 mV in amplitude. The 0·96 curve is thicker than the others, because the local response had begun to fluctuate very slightly at this strength. The width of the line indicates the extent of fluctuation.

Fig. 8*a* (inset). Responses produced by shocks with strengths, successively from above, 1·00 (upper 5 curves), 0·96, 0·85, 0·71, 0·57; obtained from curves in fig. 8 by subtracting anodic changes from corresponding cathodic curves. Two of the anodic curves necessary for this analysis were recorded, but are not shown in fig. 8. Ordinate, as above.

When the stimulus is increased beyond 0·5 threshold it produces a local response, the size and time relations of which are shown in fig. 8*a*. It is clear that these curves are similar to those which Katz (1937) obtained from excitability measurements. In particular, both types of experiment show that the subliminal response increases in duration as well as in amplitude, and that its rising phase becomes longer as the stimulus increases.

Another important property of the local response is that it starts to vary in amplitude and duration when it has almost reached propagating size. Thus in the experiment of fig. 8 the potentials produced by successive

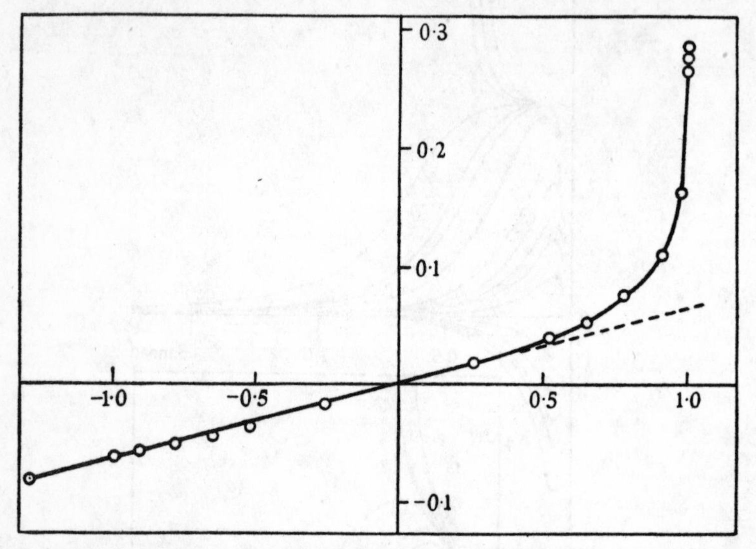

FIG. 9. Relation between potential at 0·29 msec. after application of shock, and strength of shock. Ordinate, potential as fraction of spike potential. Abscissa, strength of shock as fraction of threshold.

shocks were constant to within 1 % provided that the stimulus was below 0·96 threshold. However, when it was increased beyond this point, the responses started to fluctuate over a wide range and ultimately a few succeeded in growing up into propagated spikes. The transition from local to propagated responses seems to occur at a constant level; for if the responses succeed in reaching a critical size, they turn into propagated spikes, whereas when they fail to attain it they die out as localized monophasic waves (cf. especially fig. 10). Although it is impossible to give a precise explanation of the fluctuations in the subliminal response, it is not difficult to understand their general nature. For it has been known since the work of Blair and Erlanger (1933) that the threshold varies continually

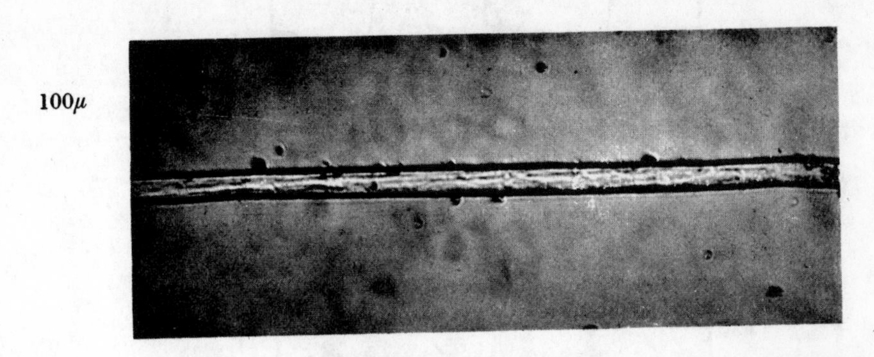

Fig. 1. Living axon from *Carcinus* in sea water. The light intensity was adjusted so as to produce the maximum contrast between the opaque connective tissue and the transparent axis cylinder.

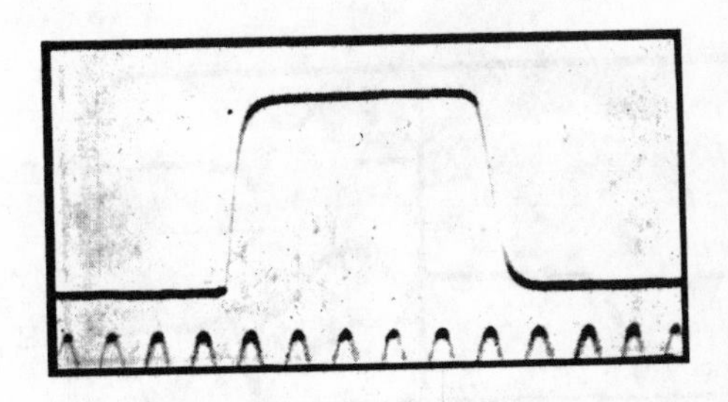

Fig. 3. 10 mV applied to amplifier through 2 megohms. Time 0·1 msec.

(*Facing p.* 96)

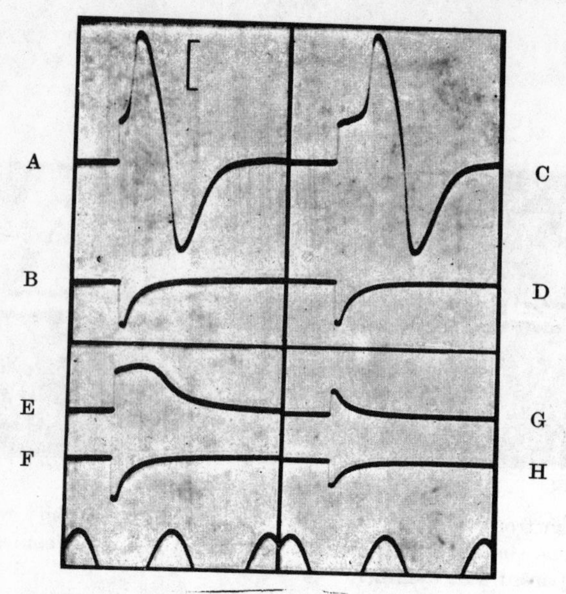

FIG. 6. Electrical changes at stimulating electrode, produced by shocks with relative strengths: A, 1·05; B, −1·05; C, 1·00; D, −1·00; E, 1·00; F, −1·00; G, 0·61; H, −0·61. Time 1 msec. The difference between C and E was due to a slight fluctuation in excitability, since the shock strength was not changed between making these two records. Scale, 15 mV.

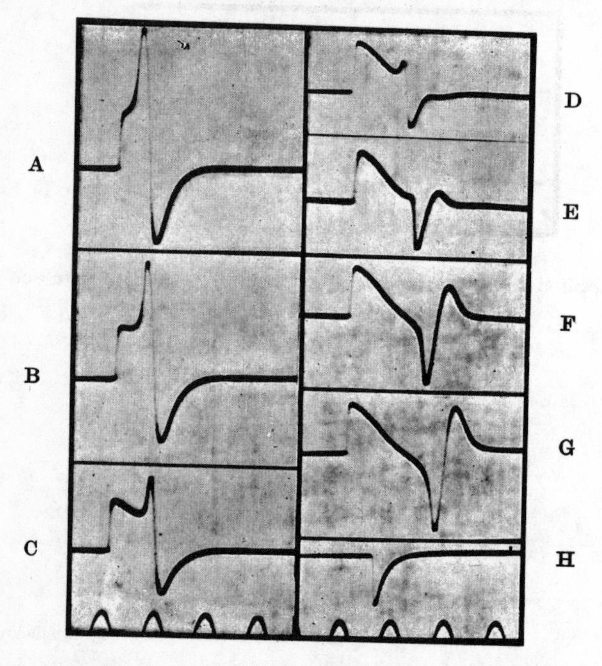

FIG. 19. Spikes recorded diphasically with ground lead connected to cathode and grid lead 1·0 mm. distal. A, shock = 1·08 threshold; B, 1·03; C–G, 1·00; H, anodic polarization produced by shock with strength −1·00. Time, I msec

in a nerve fibre, and it is clear that a slight fluctuation in excitability might cause large changes in the local response, since this increases very rapidly near threshold. Blair and Erlanger (1936) found that the threshold varied spontaneously over a range of 10%. In an isolated crustacean axon the threshold does not usually fluctuate over more than $\frac{1}{2}$%, but some play can always be detected. Now this slight shift in excitability would not produce appreciable variations when the stimulus is small, but it should cause large fluctuations near threshold, since a slight change in stimulus then produces a relatively enormous change in the local response.

The rapid increase in the local response as the shock approaches threshold is illustrated in fig. 9. It is clear that anodic or weak cathodic shocks produce only polarization, since the potential is at first directly proportional to the strength of the stimulus. The response seems to start at about 0·5 threshold, but it is rather hard to tell exactly where it begins since the initial curvature in fig. 9 is very gradual. At 0·7 threshold the response is quite definite, and after this it increases at accelerating rate until it finally reaches propagating size.

Local response and latency

One of the most important properties of the local response is that its time relations determine the latency of the nerve fibre. The reason for this is that the spike initially follows the same time course as the local response, so that the latency of a response which just succeeds in propagating is approximately equal to the time of rise of one which just fails. This point is illustrated by fig. 8, but it is more clearly shown when the latency is long, as in fig. 10. Here it is evident that both local and propagated responses start out in the same way and may not separate until nearly 500 μsec. after the shock. The final division of propagated and local responses occurs at a critical potential, for all the responses which reach this potential turn into propagated spikes, whereas those which fail to attain it die out as localized monophasic waves.

The association between the latency and the form of the local response is very constant, and has been checked in a large number of experiments. Usually the spike does not start later than the crest of the response, but occasionally it may rise when the wave has fallen a little below its maximum. An example is shown in fig. 11. Here the latency is increased to over 1 msec. and the local response has a correspondingly long duration. In this case the response does not follow a smooth curve, but is prolonged by the development of a hump on the falling phase. This secondary wave seems to be associated with the long latency, since the delayed spikes start near its crest. At first it is rather difficult to see why the spike should start after the peak of the local response, and this problem becomes more acute when we remember that the response is superimposed upon the decaying cathodic polarization, so that the spike actually starts when the potential has fallen far below its maximum (fig. 11*a*). One possible explanation is that both the secondary humps and the spikes which rise from their crest are due to the development of activity at a short distance from the cathode.

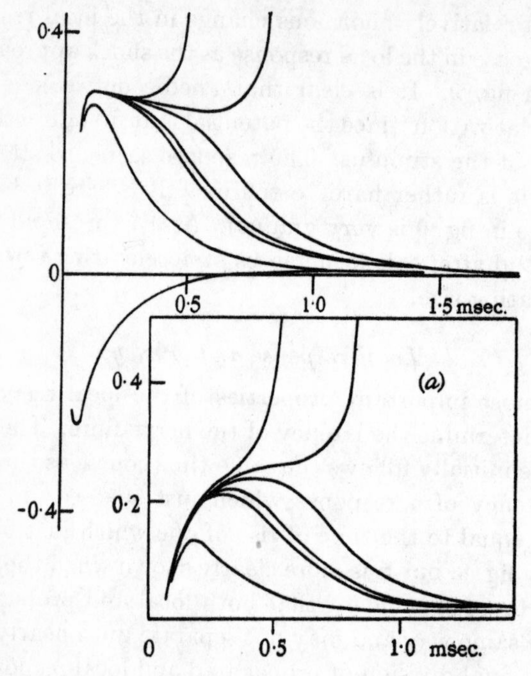

FIG. 10. Electrical changes produced by shocks with relative strengths 1·00 (upper 5 curves), 0·97, −0·97, −1·00 (these two anodic curves coincide).

FIG. 10*a* (inset). Responses evoked by cathodic shocks in fig. 10. The ordinate scale in both figures gives the potential as a fraction of the spike.

This would account for the fact that the spike begins when the potential at the cathode is falling, since it does not actually start from there, but at a distant point where the potential is still rising, or just reaching its crest. The evidence for this hypothesis will be presented later, but it is well to point out now that the short latency type of response is probably the more normal one. Thus the long latencies are usually found in fibres which have been badly dissected or which have been stimulated for some time. Figs. 8 and 10 illustrate the effect of stimulation on the latency, since both were made on the same axon, but at different times, fig. 8 being made at the beginning of the experiment, and fig. 10 after 20 min. of stimulation.

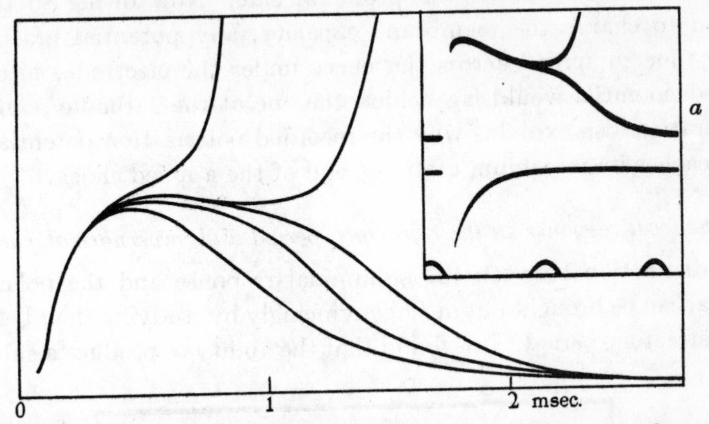

FIG. 11. Local and propagated responses in a nerve with a very long latency, obtained in the usual way by subtracting polarization from total effect produced by cathodic shock. All curves made with same strength of stimulus.

FIG. 11*a* (inset). Three of the original records used for making analysed curves in fig. 11. Upper two curves: effects produced by threshold cathodic shocks. Lower curve: anodic polarization produced by reversing shock. Time 1 msec. The hump on the local response seems to be larger in the original records than in the analysed curves. This is an illusion depending on the fact that the inflexion on the local response is exaggerated by being superimposed on the concave polarization potential.

Possible latency of local response

The experiments described in the previous section do not give a complete picture of nerve latency, but they suggest that it is connected with the time required for the local response to grow into a self-propagating impulse, rather than with delay in the activation process itself. Thus it is clear from fig. 10*a* that only a very small part of the spike latency could possibly be due to a delay in the initiation of the local response; most of it is due to the time taken by the response to reach its maximum.

The shape of the local response at first suggests that it may have a very small but measurable latency, since it starts with a delay or inflexion lasting for 50–80 μsec. It is possible that some of this inflexion is due to a latency or inertia in the activation process, but a large part of it must arise in other ways. In the first place, we should expect the local response to have a gradual rather than an abrupt beginning, because the shock does not charge the nerve instantaneously but over a period of about 60 μsec. Secondly, even if the response had an abrupt start, we should not be able to measure it, because the lag in the amplifier would delay the rise of the potential during the first 50 μsec. Thirdly, it is important to remember that potentials are recorded from the distal side of the cathode, whereas the stimulus first acts on the proximal side. Now owing to the time required to charge the membrane capacity, any potential must take a certain time to spread across the nerve under the electrode, so that the recorded potential would lag behind the one at the cathodic point. This consideration also explains why the recorded polarization potential sometimes reaches its maximum after the end of the applied shock.

The local response in the refractory period and supernormal phase

The distinction between the subliminal response and the polarization potential can be brought out most convincingly by studying their behaviour in the refractory period. It is found that the ability to produce a subliminal

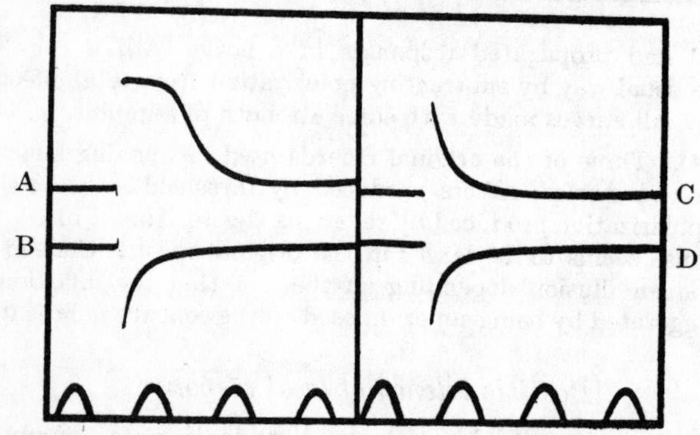

FIG. 12. A. Effect produced by just subthreshold shock in a resting nerve. B. Anodic polarization produced by reversing shock in A. C. Same shock as in A applied to refractory nerve. D. Same shock as in B applied to refractory nerve. Time, 1 msec. The nerve was made refractory by a shock which preceded the testing stimulus by 2·5 msec. At the moment when the second shock was applied the threshold was 119% normal. Potentials recorded at stimulating electrode.

response is greatly reduced during the refractory period, so that a shock which normally evokes a large local response may produce only polarization if applied to refractory nerve. This is illustrated by fig. 12. Record A shows the electrical change produced by a just subthreshold shock, and B the corresponding anodic polarization. C and D were made a few seconds later and show the effect of applying the same shocks in the relatively refractory period. It is clear that the local response is practically abolished by making the nerve refractory, for the cathodic potential no longer has a rounded form and is only slightly larger than the corresponding anodic potential. On the other hand, the polarizable properties of the nerve do not seem to be much changed, since the anodic potentials in B and D are almost identical. Actually the potential in D is 2 or 3 % less than that in B, and it has usually been found that the polarizability is slightly reduced for several milliseconds after the spike.

In this experiment the fibre was left for several hours in sea water before being used. This was done in order to prolong the relatively refractory period, so that records C and D should not be complicated by the conducted phase of the conditioning spike.

The excitability following an action potential usually goes through a pronounced supernormal phase, so that it is easy to observe the effect of supernormality on the local response. As we should expect, the ability to respond is greater during the supernormal phase, and a large local response can be produced by shocks which are too weak to elicit any response in

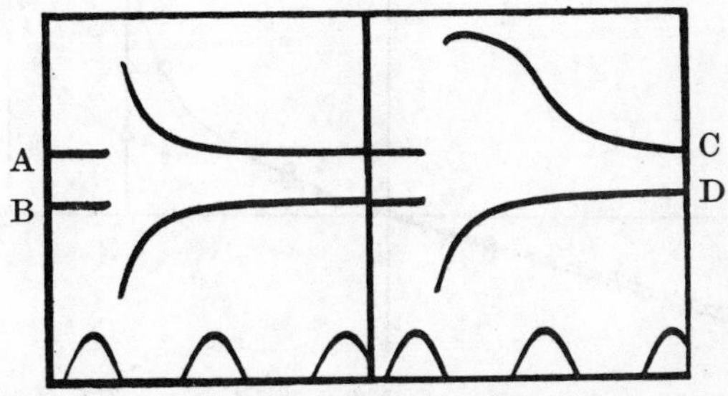

FIG. 13. Potentials produced by: A. Shock with strength 0·76 resting threshold applied in resting nerve. B. Anodic polarization produced by reversing shock in A. C. Same shock as in A applied in supernormal phase. D. Same shock as in B applied in supernormal phase. Time, 1 msec. Nerve made supernormal by shock applied 6·7 msec. before second stimulus; the threshold was 76% normal at moment when second shock was applied. Potentials recorded at stimulating electrode.

normal nerve. Fig. 13 illustrates this point. In this experiment there was practically no difference between the anodic potentials in normal and in supernormal nerve, but in others the polarizability was sometimes slightly reduced during supernormality.

The evolution of the local response in the refractory period is essentially similar to that in normal nerve, although of course it does not develop until the shock is increased beyond the normal threshold. This is illustrated by fig. 14, which shows that the potential increases rapidly before the refractory threshold, in the same way as it does before the normal one. This experiment brings out another interesting point, since it shows that the potential at which the local response turns into a propagated action potential is increased during the refractory period. This seems reasonable, since it must be harder for the action potential to excite refractory nerve, and therefore the local action potential would have to be larger in order to propagate away from the cathodic region.

The position in the supernormal phase is the exact opposite of that in the refractory period, since it is then relatively easy to set up a local response and this propagates at a potential which is lower than that in a normal nerve. This is illustrated by fig. 15.

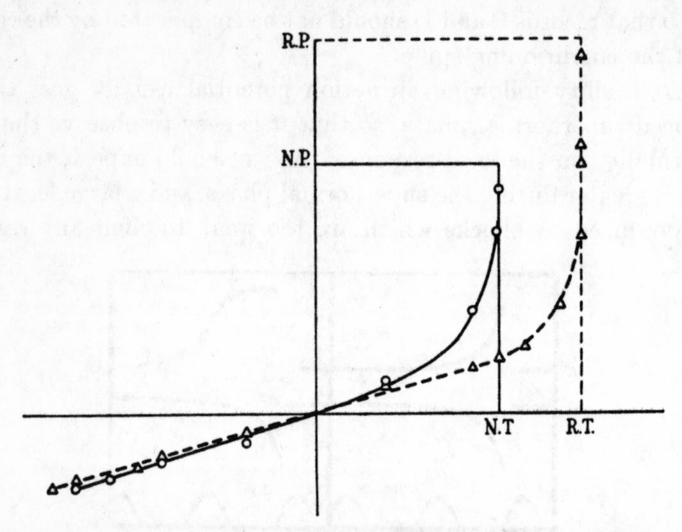

FIG. 14. Broken curve, triangles. Subthreshold potentials in refractory nerve. Continuous curves, circles. Subthreshold potentials in resting nerve. Abscissa: strength of shock. Ordinate: potential, measured at 0·5 msec. after the shock. Vertical lines *N.T.* and *R.T.* give the normal and refractory thresholds respectively. The horizontal lines *N.P.* and *R.P.* show, approximately, the potentials at which the local response starts to propagate in normal and refractory nerve.

The relation between the potential at which the local response propagates and the excitability is illustrated by an experiment in which threshold responses were measured successively in supernormal, resting, and refractory nerve (fig. 16). In the first two cases the form of the response was obtained in the ordinary way, by subtracting the polarization from the total effect produced by the cathodic shock. The procedure for measuring the refractory response was slightly more complicated, because the potentials were superimposed upon the curved base line left by the conditioning spike. It was therefore necessary to combine three records in the following way:

$$(A-B)+(C-B),$$

where A = first spike + cathodic polarization + response; B = first spike alone; C = first spike + anodic polarization. The curves in fig. 16 show

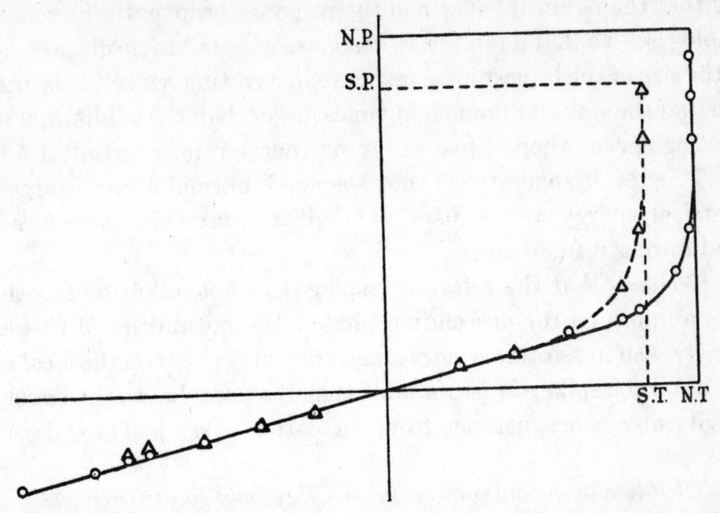

FIG. 15. Broken curve, triangles. Subthreshold potentials in supernormal period. Continuous curve, circles. Potentials in resting nerve. Abscissa: strength of shock. Ordinate: potential measured at 0·47 msec. after shock. *N.T.* and *S.T.* are normal and supernormal thresholds. *N.P.* and *S.P.* indicate potentials at which response propagates in normal and in supernormal nerve.

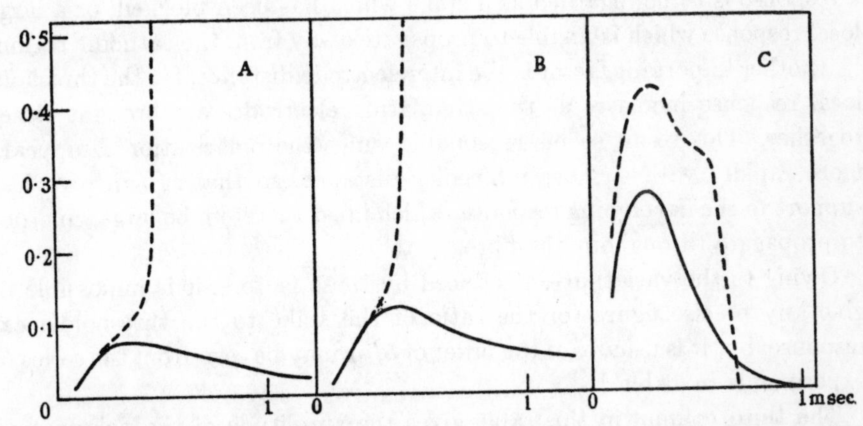

FIG. 16. Local and propagated responses, recorded at cathode, under different experimental conditions. Response measured by subtracting anodic polarization from total effect produced by cathodic shock. Dotted lines: spikes which just succeed in propagating. Continuous curves: largest local responses observed. A. Supernormal nerve, 1·7 msec. after beginning of conditioning spike. Strength of shock = 82 % resting threshold. B. Resting nerve. C. Relatively refractory nerve, 0·9 msec. after beginning of conditioning spike. Strength of shock = 218 % resting threshold. The ordinate gives the potential as a fraction of the spike in a resting nerve. The spike amplitude was about 40 mV.

clearly that the potential at which the response propagates increases as the threshold rises, so that a refractory response may fail to propagate, although twice the size of the largest local response in a resting nerve. It is interesting to note that the spike is about eight times larger than the subliminal response in a resting nerve, whereas in a refractory nerve it may be only 1·5 times as great. This result suggests (1) that the spike normally has a large reserve of electrical energy at its disposal, and (2) that this reserve is greatly reduced during refractoriness.

Fig. 16 shows that the refractory spike does not follow a smooth curve, but has a hump on the descending phase. Discontinuities of this kind are frequently seen in refractory nerve, and they may occur on the local response as well as on the spike. It is probable that they are due to the development of activity at a short distance from the cathode (see p. 114 *et seq.*).

Relative amplitudes of spike and threshold local response

The difference between the propagated and the local action potentials can be reduced in other ways besides making the nerve refractory. Thus it decreases if the fibres are placed in K-rich solutions or left to soak in sea water, or in fact if they are maltreated in any way. Under these conditions conduction ultimately fails, and it is then often difficult to decide whether a response is to be regarded as a spike which has been blocked, or a large local response which is unable to propagate away from the cathodic region.

Another important factor is the interelectrode distance, for the threshold local response increases as the stimulating electrodes are brought closer together. This result seems reasonable, since the polarization disappears more rapidly with a short interpolar distance, so that it can give less support to the developing response, which must therefore be larger in order to propagate throughout the fibre.

Owing to the variations introduced by these factors, it is impossible to give any precise figure for the ratio of the spike to the threshold local response, but it is usually of the order of 5, as may be seen from the series of experiments in Table I.

The third column in this table gives the amplitude of the polarization produced by a threshold shock. This cannot be measured directly, since the peak of the cathodic polarization is complicated by the beginning of the local response. It may, however, be estimated quite accurately, by reversing the direction of the shock, and measuring the amplitude of the pure anodic polarization. This procedure is justifiable, since the cathodic and anodic polarizations produced by weak shocks are always mirror images of one another.

Another quantity which is of some interest is the potential at which the local response first becomes appreciable. This can be measured directly by recording the electrical changes produced by increasing strengths of shock. Thus in the experiment of fig. 8 the local response started when the shocks were increased to 0·57 threshold, and at this strength they produced a cathodic polarization which was equal to 0·17 of the spike height. Direct measurements of this kind are given in some of the experiments in column 4; in the others a complete series of subthreshold curves was not made, so that the potential could not be measured directly. However, a rough estimate could be formed from the figures in column 3; for the local response always begins somewhere between 0·5 and 0·8 threshold, so that it must start at a polarization which lies between 50 and 80% of the threshold one.

TABLE I

Data obtained from experiments made during December 1937 and January 1938. All experiments in which the necessary measurements were made have been included. 7 *a* and *b* were obtained from the same axon at different times; otherwise measurements made on separate fibres. Column 3 obtained by reversing threshold shock and measuring anodic polarization. Column 4, ratio of polarization at which local response starts, to spike potential; unbracketed figures measured directly; bracketed figures, limits calculated as 0·5–0·8 threshold polarization.

	1	2	3	4
	Spike amplitude mV	Local response / Spike	Threshold polarization / Spike	
1	39	0·21	0·21	(0·10–0·16)
2	36	0·17	0·21	0·11
3	52	0·15	0·28	0·14
4	34	0·20	0·23	0·14
5	52	0·19	0·20	(0·10–0·16)
6	44	0·18	0·20	(0·10–0·16)
7*a*	—	0·30	0·29	0·17
7*b*	—	0·24	0·25	0·17
8	43	0·29	0·31	(0·15–0·24)
9	40	0·12	0·16	(0·08–0·13)
10	56	0·18	0·20	(0·10–0·16)

The results in column 4 are interesting in connexion with the rising phase of the spike. If transmission is electrical, the initial part of the spike should be due to current spreading ahead of the active region, and the activity under the lead should not start until the nerve is depolarized to a critical extent. Column 4 indicates that, with an external stimulus, the activity does not start until the nerve is depolarized to about one-seventh spike height. If this result is applied to a propagated impulse, it follows that the active generation of potential should start when the action potential rises to one-seventh of its maximum, and that below this the rise in potential should be solely due to the physical spread of polarization ahead of the active region.

The estimate of one-seventh is likely to be too large, because controls with dead fibres show that about 20 % of the polarization which is recorded arises in the stimulating electrode and not in the nerve fibre itself. However, it seemed best to give an estimate which was an average of experimental values, rather than to add an uncertain correction for electrode polarization.

Possible sources of artifact—Electrode polarization

The fact that the local response can be abolished by making the nerve refractory proves that it is a product of the nerve, but this does not exclude the possibility of the polarization potential being complicated by artifacts. It is, therefore, important to consider the various types of artifact which could be observed. In the first place, there might be a capacitative disturbance due to coupling between the leads. This effect was nearly always present and was rather serious in the earlier experiments, but it could be distinguished from the polarization potential, since it consisted of a quick spike lasting for less than the shock duration. As the work progressed, this was gradually reduced and it might be completely absent when the interelectrode resistances were low. Thus the records in fig. 6, Plate 5 were made with a particularly large fibre, and as a result they are free from any initial artifact. In the other experiments illustrated there was an initial disturbance in the original records, but this was too faint for reproduction, so that there is a gap of 30 or 40 μsec. in the oscillograph line.

The second possibility which must be considered is that the potentials might be complicated by polarization in the common stimulating and recording electrodes. In order to determine the importance of electrode polarization, various control experiments were made on dead nerves. It was found that if the fibres were left for 7 or 8 hr. in sea water, they became completely inexcitable and failed to give any trace of a local or propagated response. Under these conditions the shocks produced a perfectly symmetrical polarization potential which might not be much smaller than that in a normal nerve. This potential gradually declined as the fibre was left in sea water, and after prolonged soaking or after treatment with dilute alcohol it was reduced to 15–25 % of its original value, so that at least 75 % of the polarization observed must have been in the fibre and not in the electrodes. The remaining 25 % was probably due to electrode polarization, but it is difficult to be sure that there is not some residual polarization in a dead fibre. These experiments show that most of the polarization is in the nerve, but they do not prove that the accumulation of charge occurs at the membrane involved in excitation. Information on this point was obtained by comparing the polarizability during the resting and active states. It was found that a shock applied at the crest of the spike produced less than half the normal polarization. This proves that a large part of the polarizability of the fibre must be due to the excitable membrane, since it would not otherwise be affected during activity. The experiment also has a more general interest, as it affords new evidence for the theory that the action potential arises by the breakdown of a polarized membrane.

The spatial distribution of the subliminal response

The principal difficulty in measuring the distribution of the subliminal potentials in a single fibre arose from the fact that they occupy an extremely short length of nerve. It was therefore impossible to use a series of fixed electrodes, and it was necessary to record from a single electrode which could be moved along the nerve fibre. A satisfactory movable electrode was made by attaching one of the electrode holders (B) (fig. 2) to a strip of phosphor bronze, which in turn was connected to the bakelite boss (C). The motion of the electrode could then be smoothly controlled by means of a screw which was arranged to press against the phosphor bronze strip. This arrangement worked fairly well, provided that the fibre could be dissected free from all loose connective tissue, but if this was present the electrode stuck to the fibre and could not be moved without injuring it. The distance between the stimulating and recording electrodes was measured by observing them with a microscope into which a micrometer scale had been inserted. This method was an accurate one, but the measurement of distances less than a millimetre is uncertain since the electrodes were $100\,\mu$ in diameter, and were, moreover, effectively broadened by the film of sea water which collected at their point of contact with the nerve. The distance between the electrodes was measured between their distal edges, since the potentials probably record from these points.

At first the distribution of potentials was measured with an ordinary grounded amplifier, but this method proved to be unsatisfactory, on account of the inevitable capacity between the stimulating circuit and the ground lead. This capacity led to two undesirable results. In the first place it had the effect of making the strength of the shock dependent upon the position of the ground lead, so that the stimulus was altered by a small but unknown amount as the ground lead was moved along the nerve. This effect depends on the fact that the stimulating current does not flow entirely through the interpolar stretch of nerve, but may also flow in the extrapolar stretch as far as the ground lead and then back to the other side of the stimulating circuit through a capacitative connexion.

The second source of error arose from the fact that the capacity current flowing through the ground lead might polarize the nerve appreciably, so that the potential due to the spread of current along the fibre might be complicated by the additional current flowing through capacitative connexions outside it. Both these difficulties were overcome by keeping the distal stimulating electrode at ground potential and recording with a balanced amplifier. This method of recording potentials sometimes introduced other types of artifact, but on the whole it proved much more satisfactory than the first one. The amplifier employed was similar to that described by Toennies (1938) and need not be discussed here.

In comparing the size of the local response at different points on the nerve a difficulty arose from the fact that the threshold responses varied from shock to shock, so that it would have been meaningless to compare single responses from different points. In order to avoid this difficulty a number of responses were photographed at each point and the largest ones selected for comparison. This procedure probably yields valid results, because the size of the local response is sharply limited by the condition that it turns into a propagated spike whenever it reaches a critical potential.

The potentials at the cathode and at two other points on the nerve are shown in fig. 17. The excitability altered slightly during the course of the experiment, so that it was necessary to change the strength of the shocks in order to keep the local responses at threshold size. This must have altered the polarization to a slight extent, but the changes introduced in this way would have been too small to affect the results appreciably. The anodic curves in fig. 17 show that the spread of the polarization is much more restricted than in a medullated nerve trunk. In medullated nerve the polarization falls to $1/e$ of its value in 2–3 mm., whereas here it declines to approximately $1/e$ in only 0·5 mm. The potential gradient may not always be so abrupt as this, but it is invariably much steeper than in medullated nerve. This result supports the idea that the myelin sheath has a high electrical resistance and is responsible for the extensive spread of polarization in vertebrate nerve. It is also interesting because it suggests that the difference between the action wave-lengths in medullated and in non-medullated nerve may be connected with the difference in the spatial spread of the polarization.

Fig. 17 shows that the form of the local response changes as it travels along the nerve, for the potential at 0·5 mm. has a later maximum and a more gradual rising phase than when recorded at the cathode. Comparable but less striking changes occur in the form of the anodic polarization, since the maximum occurs 70 μsec. later at 0·5 mm. and the falling phase is slower at this point than at the stimulating electrode. These changes in form are an obvious consequence of the cable-like structure of the nerve fibre, since the potential at a distant point must always lag behind one recorded closer to the cathode, on account of the time required to charge the intermediate membrane capacities.

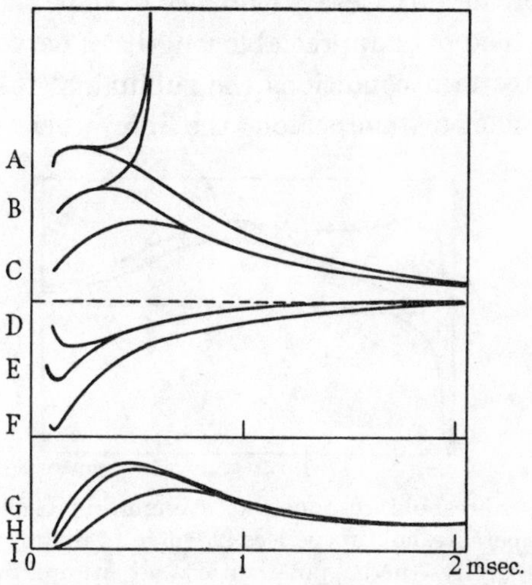

FIG. 17. A. Propagated spike and subthreshold potential recorded at cathode. Strength of shock 1·00. B. Ditto, recorded at 0·2 mm. from cathode. Strength of shock 1·04. C. Subthreshold potential at 0·5 mm. Strength of shock 0·95. D, E, F. Potentials corresponding to C, B, A, but with direction of shock reversed. G, H, I. Subthreshold responses at cathode, 0·2 and 0·5 mm., obtained from previous curves by subtraction.

The spread of the cathodic polarization seems to be identical with that of the anodic one, since it has been found that the potential curves are always symmetrical until the stimulus is increased beyond 50 % of threshold.

It is interesting to notice that the subliminal response spreads further along the nerve than the physical polarization, for the anodic potential in fig. 17 falls to 1/3 in 0·5 mm., whereas the response declines to only 2/3 in that distance. This result can be explained most simply on the basis that the response lasts longer and so has time to spread further than the anodic polarization. There is another possibility, but at present it can only be stated in vague terms. Thus it may be that the spread of the local response is due to an increase in length of the active region, as well as to the passive current spread which operates in the case of the anodic polarization. The extension of the active region must in the first instance depend upon a spread of polarization, but it would clearly enable the response to spread much further than a polarization process alone. If this view is correct, it follows that the spread of the local response must be fundamentally unlike that of the polarization, and is more closely allied to the propagation of the spike.

It is difficult to distinguish experimentally between the active and passive types of spread, but there is evidence to show that the active region may sometimes extend over appreciable lengths of nerve. Thus it has been found that under certain conditions the subliminal response may spread for relatively enormous distances along the fibre, whereas the spread of the

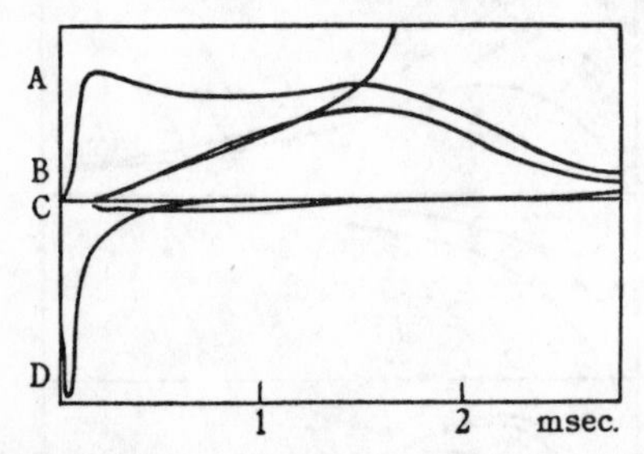

Fig. 18. A. Just subthreshold response at stimulating electrode. B. Just subthreshold and just superthreshold responses at 1·5 mm. distal. Strength of shock = 98% of that in A. C, D. Anodic polarizations at stimulating electrode and at 1·5 mm. produced by reversing shocks in B and A.

anodic polarization remains perfectly constant. An example of this is shown in fig. 18. Here the response at 1·5 mm. is nearly as large as it was at the cathode, whereas the anodic polarization is so small that it can hardly be measured. The prolonged duration and peculiar shape of the response in A are characteristic of fibres in which the subliminal wave spreads for a long way. This experiment is satisfactory in so far as it illustrates the difference between the polarization and the subliminal response, but it immediately raises the question as to why the response should sometimes spread to such unusual extents. One possible explanation is that there may occasionally be a gradient of excitability along the fibre. In this case, the normal tendency to decrement would be counterbalanced by the fact that the response was continually working itself into more excitable nerve and so could more easily induce a response in the next section.

Spread of local response and long latency

The widely spreading subliminal responses described in the previous section are interesting because they only occur in fibres which have a long latency. Fig. 18 B illustrates this association; for the maximum delay of the spike was 1·3 msec., and of this only 0·3 msec. could have been due to conduction time because the recording lead was 1·5 mm. from the cathode. At first the connexion between an increase in latency and an increase in spatial spread seems obscure, but the two curves in fig. 18 B suggest a plausible explanation for the association. Thus it is possible that the subliminal response might travel along the nerve and finally work up into a spike in the more excitable nerve at a distance from the cathode. This hypothesis would account for the fact that the spike arises near the crest of the subliminal response in B. More direct evidence may be obtained by recording with both leads close to the cathode, since it is then possible to determine the place from which the spike starts. The records in fig. 19, Plate 5, show the results of an experiment in which the ground lead was connected to the cathode and the grid was placed 1·0 mm. away. Now if the spike arises at, say, 0·7 mm. from the cathode, it must reach the grid lead before it has time to propagate back to the cathode, so that instead of obtaining a normal diphasic impulse, we shall record one in which the two phases are reversed. This effect can be seen in F and G, but it occurs only when the latency is exceptionally long; as this decreases, the response starts progressively closer to the cathode. Theoretically we should expect to record only a minute change when the response starts midway between the leads; but this result is never obtained in practice, since the spike is always slightly larger at one lead or the other, so that a potential like D is recorded. The wide shift in the point of origin of the spike is at first a little surprising, since we might expect that the spike would always start from the distant points, if these were more excitable than the rest of the nerve. The reason for the shift is that the crest of the subliminal response travels very much slower than the spike, so that if the shock is strong enough to excite at the cathode, the spike starts from there with a short latency and travels along the nerve before the subliminal response has had time to spread for any distance.

Records like those in fig. 19 could be observed in most experiments in which the latency was longer than 500 μsec., but exceptions have occasionally been found. These can be explained quite simply by supposing that the spike is still starting at a distance from the cathode, but in the interpolar, instead of in the extrapolar region. If this happened, it would clearly be impossible to observe any "reversed" action potentials, since the activity must always reach the ground electrode before it arrives at the grid. If this explanation is correct, it should become possible to record a reverse action potential effect on interchanging the grid and anode leads (fig. 20). This deduction has been verified in the few cases in which it has been possible to test it. In most cases, however, there has been no need to make the test, since the delayed spikes usually start in the extrapolar stretch of nerve.

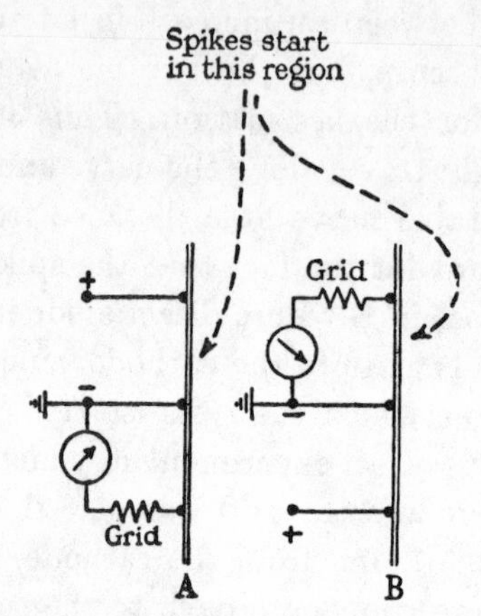

FIG. 20. A. Spike starting in interpolar region, no reversal of phases possible. B. Grid and anode leads interchanged, now possible to record reverse spikes.

The experimental analysis of the long latencies may be summarized by saying that the following changes occur in association:

(1) The local response spreads for a long way.

(2) It is unusually prolonged and often has a hump on the descending phase.

(3) Delayed spikes start at a distance from the cathode.

(4) The latency is long.

It is easy to account for all these changes, if we assume that they occur only when the excitability beneath the cathode is slightly lower than in a neighbouring part of the nerve. In this case, the local response would spread for a long way, because it would be working into more excitable regions; it would last for a long time, because the negative wave at one point would be maintained by the development of activity in a more distal part; the spike would tend to start at a distant point, because the excitability there would be greater than underneath the cathode; and finally there would be an increased latency, because the local response at a distance takes a long time to work up to propagating size.

This explanation receives some additional support from the nature of the conditions that promote long latencies. Thus they tend to occur after prolonged stimulation or in fibres which have been stretched too tightly across the electrodes. Another way of obtaining long latencies is to place the leads near the killed end of the fibre. Now we should expect that all these conditions would tend to give an irregular distribution of excitability. In the first case the cathode might become depressed as a result of some locally injurious effect of the stimulus, and in the second, as a result of direct mechanical contact with the electrode. In the third case we should expect a continuous gradient of excitability rather than a local depression at the cathode, but it is clear that this should be equally effective in promoting a long latency.

The fact that spikes which start with a long latency arise at a distance from the cathode, raises the question of whether there may not always be a similar but smaller shift in the point of origin of the spike. It is true that in a normal nerve there should be no gradient of excitability to favour points at a distance from the cathode, but we should nevertheless expect some effect of this kind. For the local response must be due to the activity of a certain length of nerve, and the spike should arise at the edge of this region rather than from the centre, which may indeed be completely active already. In order to obtain information on this point, a few experiments were made with the recording leads very close together, but the results soon showed that it was impossible to measure the exact point of origin of the spike, unless the latency was exceptionally long. Owing to surface tension effects it was difficult to bring the grid lead closer to the cathode than 0·3 mm., and at this distance "reversed" action potentials were not obtained until the latency exceeded 500–800 μsec. These spikes must rise at about 0·15 mm. from the cathode, so that action potentials starting with a normal latency of 200 μsec. would almost certainly rise within less than 0·1 mm. of the cathode—a distance too small to introduce any detectable interaction or reversal of the two phases of a diphasic spike. The question of the exact point of origin of the short latency spikes must, therefore, be left open for the moment.

Further properties of refractory local responses

In an earlier section, it was shown that it was more difficult to elicit a subliminal response in the relatively refractory period, and that this had to be larger before it could propagate. Another important property of the refractory local response is that it may spread for relatively large distances. This effect was first discovered in an experiment in which the refractory local response was occasionally conducted as far as the grid electrode, thus producing a diphasic subthreshold wave. Similar results obtained in a later experiment are shown in fig. 21. These records illustrate certain other properties of the refractory response which must now be discussed. In the first place it is clear that the variation in the size and form of the local response is much greater in the refractory period than in resting nerve. This is to be expected, for the refractory responses must be influenced by random variations in the rate of recovery as well as by the play of excitability. The second point is that the local response may either be perfectly smooth, as in C and D, or it may have a secondary wave rising from its crest, as in E–I. These extra waves find a parallel in the humps which are occasionally recorded in resting nerve (cf. fig. 11), and it now seems that they may be due to the development of activity at a short distance from the cathode. In a normal resting nerve the response decrements rapidly, but in the refractory period this tendency is counteracted by the fact that the response is continually working itself into more excitable nerve. If recovery were perfectly uniform throughout the fibre, we should not anticipate such definite humps as those in fig. 21, but we should expect to find them if the local response suddenly encountered a patch of nerve where the recovery was much more advanced than elsewhere. In this case the activity would immediately start to spread faster and the potential would rise steeply in the whole of the cathodic region. If the rise in potential were large enough, the response would be able to sustain itself by restimulation and would ultimately propagate throughout the fibre (I). On the other hand, if it were not sufficiently strong to overcome the resistance offered by the more refractory nerve on the other side of the excitable patch, the wave would die out and would produce a monophasic hump like that in F. The experimental evidence for supposing that the secondary humps are due to activity spreading away from the cathode is as follows. The positive wave in G can only be explained by supposing that activity has spread over to the grid electrode. The negative humps occurring in E and F are plainly related to the positive waves in G and H. Thus both have the same general shape and the production of a positive wave depends upon the occurrence of a large negative hump. It therefore seems reasonable to suppose that both the negative and the positive humps are due to spread of activity and that the only difference between them is that the activity occurs closer to the cathode in the first case, and to the grid lead in the second.

This experiment shows very clearly that there is no fundamental difference between a wave of activity which just succeeds in propagating and one which just fails. Thus the response in I at first follows a course which is similar to that in H, and it is only when activity has spread to the grid electrode that the greater initial size of I finally enables it to grow up into a propagated spike. Indeed, it is impossible to be absolutely certain that H failed to propagate, since it is conceivable that it might have developed into a spike at some point beyond the grid electrode.

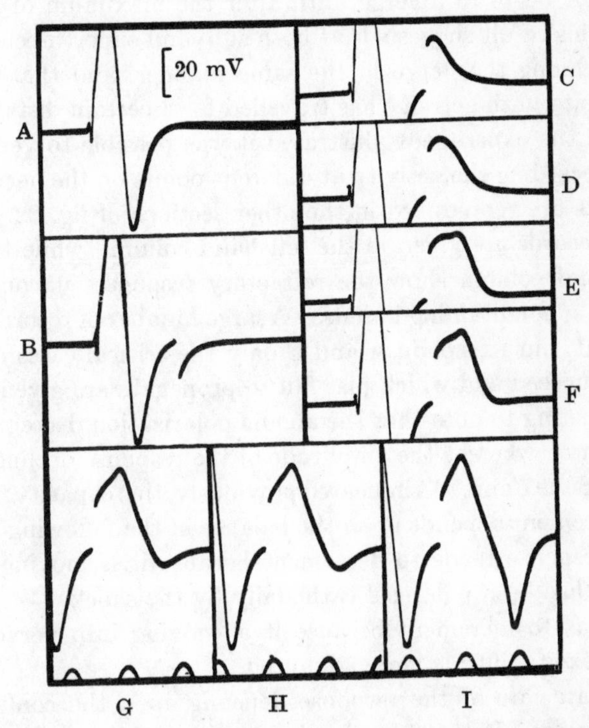

FIG. 21. A. Conditioning spike alone. B. Conditioning spike + anodic polarization produced by shock with strength − 1·5 resting threshold. C–I. Conditioning spike + effect produced by shock with strength 1·5 resting threshold. Varying amounts of response. Time, 1 msec. The conditioning spike was 50 mV in amplitude. The grid lead was 1·5 mm. distal to the ground lead. The conditioning spike has been retouched.

The similarity between subthreshold and superthreshold activity becomes most striking when we study the action potentials set up early in the relatively refractory period. In a resting nerve the propagated action potential is about five times as large as the subthreshold one, but in the refractory period the normal safety margin may be reduced to a point where a superthreshold wave is imperceptibly larger than a subthreshold one. An example of this is shown in fig. 22 *a*. These curves were made with a shock of constant strength, but owing to slight changes in the rate of recovery, the size of the response varied from shock to shock. The superthreshold waves are sharply distinguished by the presence of a conducted phase, but apart from this it is impossible to see any fundamental difference between propagated and local responses. Indeed, in the limiting cases of the waves which just succeed and which just fail to propagate, the two curves do not begin to diverge until after the maximum of the cathodic response. This result suggests that both sub- and superthreshold waves at first spread along the nerve in the same manner, and that they do not finally separate until activity has travelled for a certain distance from the cathode. In the experiment illustrated it was possible to verify this conclusion by recording the activity at different points on the nerve, and some of the results are reproduced in the other sections of fig. 22. Tracings of the actual records are given in the left-hand column, while the curves in the right-hand column show the refractory responses uncomplicated by polarization or conditioning impulse. A large number of records were made at each point, but except in A and C only the critical cases of the waves which just succeed and which just fail to propagate are given.

It is interesting to note that the anodic polarization decrements rapidly along the nerve, whereas the amplitude of the response declines very little during the first 0·7 mm. As indicated previously, this capacity for spreading without decrement depends upon the balance of the following two factors:

(1) The response tends to decrement because it is moving away from nerve which has been polarized cathodally by the shock.

(2) It tends to increment because it is moving into nerve where the recovery of excitability is more advanced.

The ultimate fate of the response depends upon the conflict between these two factors. If the first one dominates, the activity dies out as a monophasic wave, whereas if the second gains the upper hand the activity flares up into a propagated disturbance.

It was not always possible to obtain complete convergence of the subthreshold and superthreshold responses during the refractory period. Often the superthreshold wave had a hump which was not shown by the subthreshold one, so that the situation at the cathode resembled that at 0·4 mm. in fig. 22 *c*. This type of response seemed to occur more frequently when the recovery of excitability was very rapid, and we should expect to find it under these conditions, since the sub- and superthreshold waves would then separate much sooner and much closer to the cathode than they do in fig. 22. When this happens there can be no possibility of any convergence, since the superthreshold wave at the cathode must always be

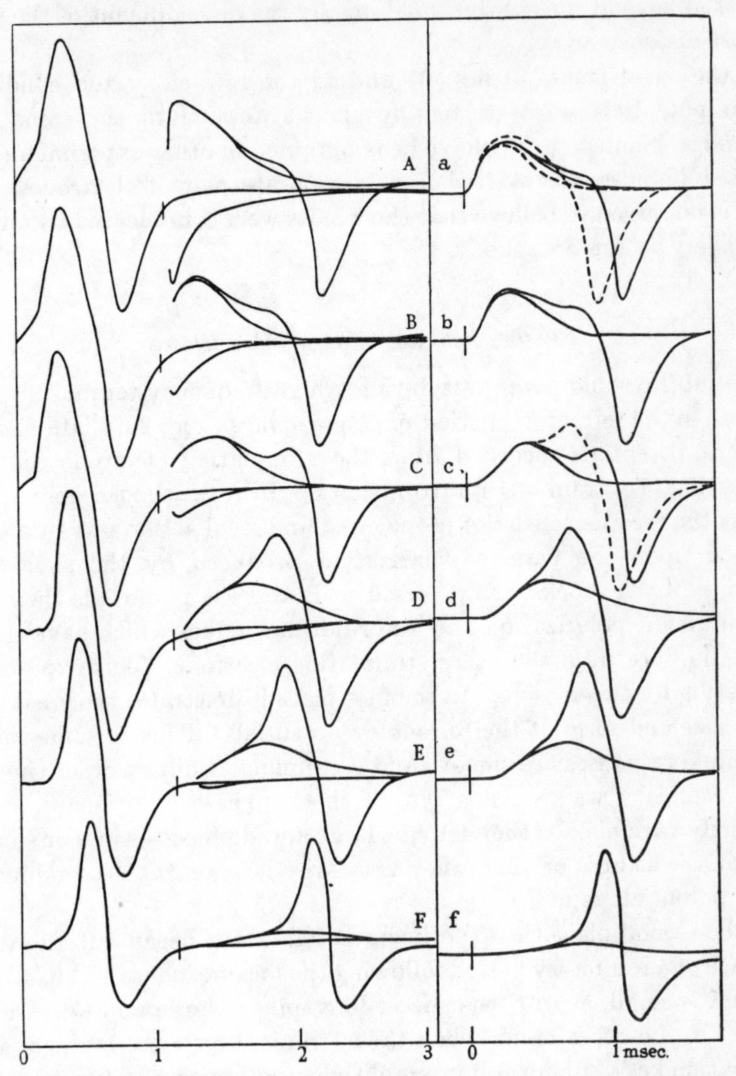

FIG. 22. Refractory responses and polarization recorded at different distance from cathode. A–E. Tracings of records showing: upper line, conditioning spike + cathodic polarization + response; middle line, conditioning spike alone; lower line, conditioning spike + anodic polarization. *a–e*. Corresponding responses obtained by subtraction. A, *a*, at cathode; B, *b*, 0·3 mm. distal; C, *c*, 0·4 mm. distal; D, *d*, 0·7 mm. distal; E, *e*, 1·0 mm. distal; F, *f*, 1·4 mm. distal. The strength of the second shock was 3·8 × threshold. Except in *a* and *c* only the critical cases of the waves which just succeed and which just fail to propagate are given. The additional responses in A and C are marked by dotted lines.

increased beyond the subthreshold one by the development of the spike at a short distance away.

In the experiments of figs. 21 and 22 the refractory and conditioning action potentials were started by shocks applied to the same pair of electrodes. Similar results have been obtained in other experiments, where the first impulse was started from a separate pair of electrodes, so that there is no reason to believe that the results were complicated by any after-effects left by the first shock.

Subthreshold activity during a tetanus

The subthreshold potentials during a high frequency tetanus are similar to those in the refractory period of single spikes. Fig. 23, Plate 6, shows a series of potentials recorded from the stimulating electrode during the application of a train of thyratron shocks. In A the shocks were cathodic, so that the records consist of propagated and local action potentials superimposed upon the passive polarization produced by the shocks. The direction of the shocks was reversed in B and the potentials then consist only of anodic polarization and all-or-nothing spikes which have travelled along the fibre from the other stimulating electrode. This experiment is interesting for two reasons. In the first place it illustrates the great variety in the size and form of the refractory potentials. Thus in some cases the subliminal responses are smooth and continuous, while at other times they have secondary waves rising out of them. These additional waves are extremely variable, for they may be large and diphasic as in *a* or small and monophasic as in *b*, or again they may arise late, on the descending phase of the potential, as in *d*.

In the second place the experiment is important because it affords clear evidence of a refractory period following the local response. Thus when the response is small, as in *c*, the succeeding spike is large and arises with no delay. On the other hand, when the subliminal response is large, as in *b*, the next spike is subnormal in amplitude and starts after an appreciable latency. Finally, when the subliminal wave is large enough to spread to the grid electrode (*a*), the succeeding shock fails to excite and produces only a small local response.

The results obtained with high frequency stimulation recall an observation made by Gasser and Grundfest (1936). They studied the effect of tetanizing mammalian nerve and showed that it could follow 2000 shocks per sec. in the cathodic region, whereas more distant parts failed to respond to these high frequencies. It now seems possible that their cathodic

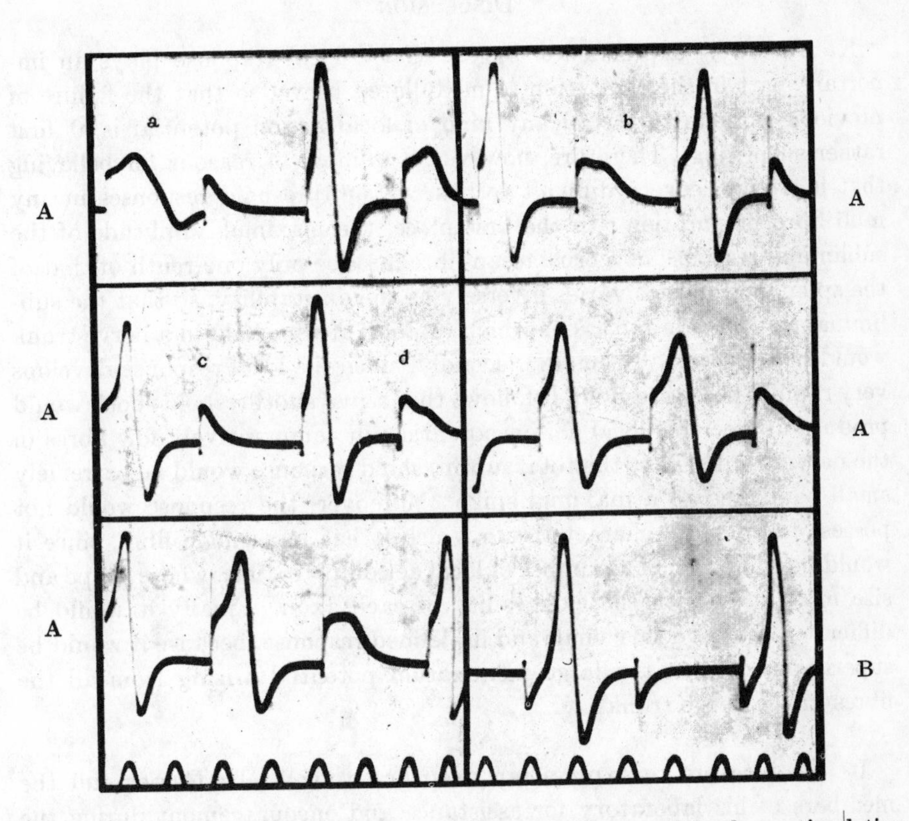

FIG. 23. Potentials during a tetanus, recorded with ground lead on stimulating electrode, grid lead approximately 1·5 mm. distal. The records were photographed at intervals of 2 or 3 sec. A, cathodic shocks; B, anodic shocks. Time, 1 msec.

response must have been due to local action potentials of the same type as those observed in the present research.

Discussion

Katz's (1937) experiments suggest that a local response plays an important part in the excitation of medullated nerve, so that the failure of previous workers to record any kind of local action potential is at first rather surprising. There are, however, a number of reasons for believing that it would be very difficult to observe subthreshold responses in any multifibre preparation. In the first place, the maximal amplitude of the subliminal response in a crustacean fibre may be only one-tenth of that of the spike, and its size varies inversely as the excitability, so that the subliminal responses in the axons which respond at threshold in a nerve trunk would be exceptionally small. Secondly, since the local response develops very rapidly near threshold, it follows that a just subthreshold shock would produce appreciable local action potentials in comparatively few fibres in the nerve trunk. Thus the total subthreshold response would be extremely small compared to a maximal spike. Moreover, the response would not possess the definite characteristics which it has in a single fibre, since it would be composed of a number of local action potentials, whose shape and size must vary with the excitability of each axon. Finally it would be difficult to observe this small and ill-defined response, because it would be superimposed upon the large polarization potential arising from all the fibres in the nerve trunk.

It is a pleasure to express my deep gratitude to Dr Gasser and the members of his laboratory for assistance and encouragement during the course of this work. I also wish to thank Dr Pantin for demonstrating the dissection of the limb nerve in *Carcinus*.

Summary

Experiments with isolated crustacean nerve fibres show that subthreshold shocks produce two quite distinct effects. When the stimulus is weak and cathodic, or anodic and of any strength, it produces a "polarization" potential which behaves as though it were due to the passive accumulation of charge at the nerve membrane. On the other hand, when the stimulus approaches threshold, it produces an additional wave of negativity which behaves like a subliminal response of the nerve fibre. These two processes have entirely different properties. Thus:

(1) The size of the polarization potential is directly proportional to the strength of the shock, whereas the subliminal response increases at an accelerating rate as the shock approaches threshold.

(2) The time relations of the processes are different. The polarization potential rises during the period in which the shock is applied and then declines in an approximately exponential way. On the other hand, the subliminal response starts with an initial inflexion and continues to grow for as much as 500 μsec. after the end of the shock.

(3) The time relations of the polarization potential are independent of the strength of the shock; whereas the duration of the local response increases and its maximum occurs later as the shock approaches threshold.

(4) When the local response has almost reached propagating size, it becomes unstable and varies from shock to shock. On the other hand, the polarization produced by successive shocks is always quite constant.

(5) The local response can be greatly reduced or abolished by making the nerve refractory, whereas the polarizability is only slightly reduced by refractoriness.

(6) The local response can be increased by making the nerve supernormal, whereas the polarization potential is slightly reduced or unaffected during the supernormal phase.

(7) The spatial spread of the two processes is different. The polarization potential falls to one-third of its value in approximately 0·5 mm., whereas the local response usually declines to about two-thirds in that distance. Under certain conditions the local response may spread very much further than the polarization potential.

The latency of the nerve fibre is determined by the time relations of the local response. The latency is short when the local wave has a rapid rise and fall, whereas it is long when the wave has slower time relations. It is suggested that the latency of the fibre is due to the time required for the local response to reach propagating size. Under certain conditions the latency may be enormously increased. When this happens it is found: (1) that the local response is prolonged, often by the development of a secondary hump on the falling phase; (2) that the response spreads for an unusually long distance; (3) that threshold spikes start at a distance from the cathode. It is, therefore, suggested that the long latencies occur when the excitability is depressed beneath the cathode, so that the local response runs along the fibre and works up to propagating size at a short distance from the cathode.

The potential at which the local response starts to propagate is increased during the relatively refractory period and decreased during the supernormal phase. In a normal nerve the response starts to propagate when its amplitude is equal to about one-fifth of the spike. During refractoriness this large safety margin may be reduced to the point at which a propagated action potential is only slightly larger than one which just fails to propagate. Under these conditions both sub- and superthreshold waves of activity at first spread along the nerve in the same way and do not finally separate until they have travelled a short distance from the cathode.

References

Auger, D. and Fessard, A. 1935 *C.R. Soc. Biol., Paris*, **118**, 1059.
Blair, E. A. and Erlanger, J. 1933 *Amer. J. Physiol.* **106**, 524.
— — 1936 *Amer. J. Physiol.* **114**, 309.
Bogue, J. Y. and Rosenberg, H. 1936 *J. Physiol.* **87**, 158.
Gasser, H. S. and Grundfest, H. 1936 *Amer. J. Physiol.* **117**, 113.
Hodgkin, A. L. 1937 *J. Physiol.* **91**, 5 *P*.
Katz, B. 1937 *Proc. Roy. Soc.* B, **124**, 244.
Rosenberg, H. 1937 *Proc. Roy. Soc.* B, **124**, 308.
Rushton, W. A. H. 1937 *Proc. Roy. Soc.* B, **124**, 210.
Schmitt, Otto H. A. and Schmitt, Francis O. 1932 *Science*, **76**, 328.
Toennies, J. F. 1938 *Rev. Sci. Instrum.* **9**, 95.
Young, J. Z. 1936 *Proc. Roy. Soc.* B, **121**, 318.

PRINTED IN GREAT BRITAIN BY W. LEWIS, M.A., AT THE CAMBRIDGE UNIVERSITY PRESS

3

Reprinted without change of pagination from the
Proceedings of the Royal Society, B, *volume* 133, 1946

The electrical constants of a crustacean nerve fibre

By A. L. Hodgkin and W. A. H. Rushton
The Physiological Laboratory, Cambridge

(*Communicated by E. D. Adrian, F.R.S.—Received* 3 *October* 1945)

Theoretical equations are derived for the response of a nerve fibre to the sudden application of a weak current. The equations describe the behaviour of the nerve fibre in terms of the membrane resistance and capacity, the axoplasm resistance and the resistance of the external fluid. Expressions are given which allow these four constants to be calculated from experimental observations.

Axons from *Carcinus maenas* were used in preliminary experiments. Quantitative determinations were made on a new single-fibre preparation—the 75μ diameter axon from the walking leg of the lobster (*Homarus vulgaris*). Currents with a strength of one-third to one-half threshold were used in the quantitative determinations.

The behaviour of lobster axons agreed with theoretical predictions in the following respects: (*a*) the steady extrapolar potential declined exponentially with distance; (*b*) the voltage gradient midway between two distant electrodes was uniform; (*c*) the rise and fall of the extrapolar potential at different distances conformed to the correct theoretical curves.

The extrapolar potential disappeared when the axon was treated with a solution of chloroform, indicating that the surface membrane was destroyed by this treatment, and that the potential recorded was in fact derived from the membrane.

The ratio of the internal to external resistance per unit length was found to be about 0·7.

The absolute magnitude of the action potential at the surface membrane was estimated at about 110 mV.

The specific resistance of the axoplasm had an average value of 60Ω cm., which was roughly three times that of the surrounding sea water.

The calculated resistance of one square centimetre of membrane was found to vary from 600 to 7000Ω in thirteen experiments.

The membrane capacity was of the order of $1{\cdot}3\mu$F cm.$^{-2}$.

No trace of inductive behaviour could be observed in the majority of the experiments. But three axons with low membrane resistances showed effects which could be attributed either to inductance or to a small local response. The absence of inductive behaviour in axons with high membrane resistance does not prove the absence of an inductive element. Currents with a strength several times greater than threshold often produced oscillating potentials at the cathode.

A local response was always observed when the strength of current approached threshold. The response had a striking inflected form if the current strength was near threshold and its duration less than the utilization time.

Indirect evidence indicates that the membrane resistance falls to a low value during activity.

Experiments with non-medullated nerve fibres have shown that a sub-threshold electric current produces two quite distinct effects (Hodgkin 1938; Pumphrey, Schmitt & Young 1940). Currents with a strength less than half-threshold produce a voltage which behaves as though it were due to the passive accumulation of charge at the nerve membrane. This voltage varies linearly with the applied current and is sometimes called an electrotonic potential. Currents with a strength greater than half-threshold evoke an additional wave of negativity which is non-linear and which behaves as though it were a subliminal response in the cathodic part of the nerve fibre. The present paper is concerned with an analysis of the first of these effects and contains only qualitative observations of the second. There are several reasons for believing in the importance of such an analysis. In the first place a physical understanding of the passive behaviour of nerve is essential to any theory of excitation. Thus a strength-duration curve cannot be explained until the time course of the voltage across the excitable membrane is known. Nor can the mechanism of excitation by the action potential be fully understood until there exists a thorough knowledge of the effect of applied currents on a single nerve fibre. A physical analysis is also interesting because it provides an insight into the structure of the surface membrane. Physical chemists are now able to prepare very thin films of lipoid material between two aqueous phases (Dean 1939), and it is clearly of the utmost importance to compare the electrical resistance and capacity of such films with those of the surface membrane in the living cell. The membrane resistance is also interesting from a more general point of view. Many biological processes depend upon the movement of ions through cell membranes, and the rate at which ions are transferred across a membrane should be related to its electrical resistance. Our results may, therefore, be of use to those who study the ionic movements that occur in the processes of growth, secretion and respiration. But perhaps the most important reason for making an analysis of the passive properties of a nerve fibre is that such an analysis must precede an understanding of the more complicated electrical changes which make up the nervous impulse itself.

Certain assumptions about the electrical structure of a nerve fibre must be made before any analysis can be started. The basic assumption of our work is that the structure of a non-medullated nerve fibre is similar to that of other cells which are known to have an interior of conducting protoplasm and a thin surface membrane with a high leakage resistance and a large capacity per unit area (Höber 1910; Fricke & Morse 1925; Cole 1937, etc.). If this general type of structure is granted, it follows that the passive behaviour of a new fibre must be governed by the equations of cable theory (Cremer 1899; Hermann 1905; Rushton 1934; Bogue & Rosenberg 1934; Cole & Curtis 1938, and others). The quantitative behaviour of the fibre should be determined by four electrical constants, viz.

(1) The electrical resistance of the fluid outside the nerve fibre.

(2) The electrical resistance of the axoplasm.

(3) The electrical capacity of the surface membrane.

(4) The electrical resistance of the surface membrane.

One other parameter, the membrane inductance, may have to be added (Cole 1941), but will not be considered in the initial stages of this paper. There already exists a considerable amount of information about the magnitude of three of these quantities. The external resistance can be calculated from the volume and conductivity of the fluid bathing the nerve fibre; the cell interior appears to have a resistivity two or three times as great as that of the external fluid and the surface membrane to have a capacity of about 1 μF/sq.cm. Very little is known about the membrane resistance and measurements have so far been confined to a few plant cells (Blinks 1937) and one animal cell, the giant nerve fibre of the squid (Cole & Hodgkin 1939; Cole 1940).* The determination of the membrane resistance was therefore the first aim of our experiments and measurement of the remaining constants was originally regarded as of secondary importance. But it so happens that one constant cannot be determined without making measurements of at least two others. An attempt was therefore made to determine all four quantities simultaneously on a single fibre. Four sets of measurements had to be made since there were four unknowns to be evaluated, and after several trials we chose the following methods:

(1) The extent of spread of potential in the extrapolar region.

(2) The rate of rise of potential in the extrapolar region.

(3) The ratio of the applied current to the voltage recorded between cathode or anode and a distant extrapolar point.

(4) The voltage gradient in the region midway between two distant electrodes.

Axons with a diameter of 30 μ from *Carcinus maenas* (Hodgkin 1938) were used initially. This work served to develop the experimental technique, but the extent of spread of potential was thought too small for accurate measurement. A search was therefore made for a fibre with a larger diameter and a suitable preparation was eventually found in the walking legs of the lobster (*Homarus vulgaris*). The meropodite of the walking legs contains a few fibres which have a diameter of 75 μ and are robust enough to permit isolation without damage. This preparation was used in the majority of experiments. Electrical measurements were made by applying rectangular pulses of current and recording the potential response photographically. About fifteen sets of film were obtained in May and June of 1939, and a preliminary analysis was started during the following months. The work was then abandoned and the records and notes stored for six years. A final analysis was made in 1945 and forms the basis of this paper. A certain amount of biophysical work has proceeded during the interval, but no one seems to have repeated these particular experiments.

Nomenclature

The passive spread of potential which occurs in nerve fibres is sometimes described by the term polarization potential and sometimes electrotonus or electrotonic potential. Both words are unfortunate. Electrotonus implies that the

* An estimate of the plasma membrane resistance in the frog's egg has been made by Cole & Guttman (1942).

axon is in a state of enhanced physiological activity: polarization potential that the change of voltage is due to an alteration of ionic concentration in the vicinity of the membrane. Weak currents do not necessarily evoke an active or tonic response; nor is it at all certain that there is a significant polarization in the sense of Nernst (1908) or Warburg (1899). For it seems likely that the change of voltage with current is due to a frictional resistance opposing the motion of ions and not to changes in ionic concentration. We have therefore avoided the use of both terms so far as is possible and have used instead words such as membrane potential or extrapolar potential according to the context.

Wherever possible we have used the same symbols as Cole and his colleagues. λ has been employed as a space constant and should not be confused with the λ of Hill's theory of excitation (Hill 1936). The dual use of symbols is unfortunate but cannot be avoided, since both American and British writers have used λ as a space constant (e.g. Cole & Curtis 1938; Rushton 1934).

Theoretical section

Assumptions

(1) The axon has a uniform cable-like structure with a conducting core, an external conducting path and a surface membrane with resistance and capacity.

(2) The axon is sufficiently thin and the membrane resistance sufficiently high for the flow of current in core and interstitial fluid to be strictly parallel. An alternative statement of this assumption is that at any given distance along the nerve the potential is constant throughout the core or throughout the external fluid.

(3) The axoplasm and external fluid behave as pure ohmic resistances.

(4) The membrane resistance is constant when the current density through the membrane is small.

(5) The membrane capacity behaves like a pure dielectric with no loss.

These are general assumptions which allow the differential equations for current or potential to be written. Each assumption is really an approximation, but we shall show later that no very serious errors are likely to result from their use. Certain experimental conditions must also be defined in order to allow the differential equations to be solved. These may be stated in the following way:

(1) The extrapolar and interpolar lengths are sufficiently long to be taken as infinite.

(2) The electrodes are sufficiently fine to be considered of zero breadth.

(3) A current of rectangular wave form is passed through the nerve.

Symbols and definitions

Variables

x is distance along axon in cm.

t is time in seconds.

i_1 is the current in amperes flowing through the external fluid (figure 1).

i_2 is the current in amperes flowing through the axis cylinder.

I is the total current in amperes flowing through the fibre and external fluid ($I = i_1 + i_2$).

i_m is the current penetrating the surface membrane at any point in ampere cm.$^{-1}$.

V_1 is the potential in volts of the external fluid with respect to a distant point $\left(V_1 = -\int_{-\infty}^{x} r_1 i_1 dx\right)$.

V_2 is the potential in volts of the axis cylinder with respect to a distant point $\left(V_2 = -\int_{-\infty}^{x} r_2 i_2 dx\right)$.

V_m is the change in potential difference across the surface membrane which results from the flow of current ($V_m = V_1 - V_2$).

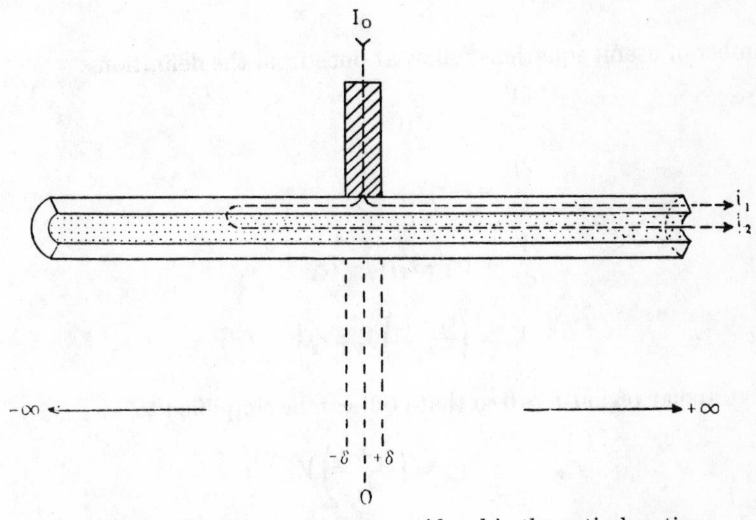

FIGURE 1. Geometry of system considered in theoretical section.

Basic constants

a is the radius of the axis cylinder in cm.

R_2 is the specific resistivity of the axoplasm in Ω cm.

R_4 is the resistance $\times$ unit area of the surface membrane in Ω cm.^{2}.

C_M is the capacity per unit area of the surface membrane in F cm.$^{-2}$.

Practical constants

r_1 is the resistance per unit length of the external fluid in Ω cm.$^{-1}$.

r_2 is the resistance per unit length of the axis cylinder in Ω cm.$^{-1}$ ($r_2 = R_2/\pi a^2$).

r_4 is the resistance $\times$ unit length of the surface membrane in the axon in Ω cm. ($r_4 = R_4/2\pi a$).

c is the capacity per unit length of the surface membrane in the axon in F cm.$^{-1}$ ($c = C_M \times 2\pi a$).

$\lambda = \sqrt{(r_4/r_1+r_2)}$, and is the characteristic length in cm.

$m = r_1 r_2/r_1+r_2$, and is the parallel resistance of the axis cylinder and external fluid in Ω cm.$^{-1}$.

$y = m\lambda r_1/2r_2 = r_1^2\lambda/2(r_1+r_2)$.

$\tau_m = r_4 c = R_4 C_M$, and is the characteristic time of the surface membrane in seconds.

Miscellaneous

δ is equal to half the electrode width (figure 1). This quantity will eventually be made vanishingly small, but is introduced in order to deal with discontinuities. X, T, U, ξ and I_0 are best defined as they are introduced.

Theory

A number of useful equations follow at once from the definitions:

$$\frac{\partial V_1}{\partial x} = -r_1 i_1, \tag{1·0}$$

$$\frac{\partial V_2}{\partial x} = -r_2 i_2, \tag{1·1}$$

$$\frac{\partial V_m}{\partial x} = (r_1+r_2)\, i_2 - I r_1, \tag{1·2}$$

$$V_m = \left(\frac{r_1+r_2}{r_1}\right) V_1 + r_2 \int_{-\infty}^{x} I\, dx. \tag{1·3}$$

In the extrapolar region $I = 0$ so that (1·3) can be simplified to

$$V_m = \left(\frac{r_1+r_2}{r_1}\right) V_1. \tag{1·4}$$

The total current through the membrane can be obtained in two ways:

$$i_m = \frac{\partial i_2}{\partial x}, \tag{1·5}$$

$$i_m = \frac{V_m}{r_4} + c\frac{\partial V_m}{\partial t}. \tag{1·6}$$

Hence

$$\frac{V_m}{r_4} + c\frac{\partial V_m}{\partial t} = \frac{\partial i_2}{\partial x}, \tag{1·7}$$

and on substituting from (1·2)

$$\frac{V_m}{r_4} + c\frac{\partial V_m}{\partial t} = \frac{1}{r_1+r_2}\frac{\partial^2 V_m}{\partial x^2} + \frac{r_1}{r_1+r_2}\frac{\partial I}{\partial x}, \tag{2·0}$$

or

$$-\lambda^2\frac{\partial^2 V_m}{\partial x^2} + \tau_m\frac{\partial V_m}{\partial t} + V_m = r_1\lambda^2\frac{\partial I}{\partial x}. \tag{2·1}$$

Now $\partial I/\partial x$ vanishes except at the electrode, since $I = 0$ for $-\infty < x < -\delta$ and $I = I_0$ for $\delta < x < \infty$. Hence the following equation (2·2) applies to the regions $-\infty < x < -\delta$, $\delta < x < \infty$:

$$-\lambda^2 \frac{\partial^2 V_m}{\partial x^2} + \tau_m \frac{\partial V_m}{\partial t} + V_m = 0. \tag{2·2}$$

This equation must now be solved for the particular case where I is a constant current I_0 starting abruptly at $t = 0$.

The boundary conditions are

$$V_m = 0 \text{ everywhere } -\infty < t < 0, \quad V_m = 0 \text{ always when } x = \pm\infty.$$

There are also two continuity conditions. First, V_m is always a continuous function of x, since a discontinuity in V_m would mean that an infinite current must flow through the nerve. Secondly, i_2 is also a continuous function of x since the current density through the membrane cannot be infinite when $t \neq 0$. Introduce new variables $X = x/\lambda$, $T = t/\tau_m$, $U = V_m e^T$. Equation (2·2) can now be written

$$-\frac{\partial^2 V_m}{\partial X^2} + \frac{\partial V_m}{\partial T} + V_m = 0, \tag{2·3}$$

or

$$-\frac{\partial^2 U}{\partial X^2} + \frac{\partial U}{\partial T} = 0. \tag{2·4}$$

The operator q^2 may be substituted directly for $\partial/\partial T$ since $U = 0$ when $T = 0$. Hence

$$\frac{\partial^2 U}{\partial X^2} = q^2 U. \tag{3·0}$$

The solutions of (3·0) are

$$U = Ae^{qX} + Be^{-qX}, \quad \text{when} \quad -\infty < X < -\delta/\lambda,$$
$$U = A_1 e^{qX} + B_1 e^{-qX}, \quad \text{when} \quad \delta/\lambda < X < \infty.$$

But the second boundary condition indicates that $U \neq \infty$ when $X = \pm\infty$, so $B = 0 = A_1$.

From the first continuity condition it follows that $A = B_1$, since

$$U_{X=\delta/\lambda} = U_{X=-\delta/\lambda},$$

when δ/λ is made vanishingly small. Hence

$$U = Ae^{qX} \quad \text{for} \quad -\infty < X < -\delta/\lambda, \tag{3·1}$$
$$U = Ae^{-qX} \quad \text{for} \quad \delta/\lambda < X < \infty. \tag{3·2}$$

The value of A can be found by applying the continuity of i_2 to equation (1·2). For

$$\left(\frac{\partial V_m}{\partial x}\right)_{x=\delta} - \left(\frac{\partial V_m}{\partial x}\right)_{x=-\delta} = (r_1 + r_2)\{(i_2)_{x=\delta} - (i_2)_{x=-\delta}\} - r_1(I_{x=\delta} - I_{x=-\delta}),$$

whence

$$\left(\frac{\partial U}{\partial X}\right)_{X=\delta/\lambda} - \left(\frac{\partial U}{\partial X}\right)_{X=-\delta/\lambda} = -r_1 I_0 \lambda e^T,$$

which in the operational form becomes $-r_1\lambda I_0\dfrac{q^2}{q^2-1}$. But from (3·1) and (3·2)

$$\left(\frac{\partial U}{\partial X}\right)_{X=\delta/\lambda}-\left(\frac{\partial U}{\partial X}\right)_{X=-\delta/\lambda}=-2qA.$$

So
$$U=\frac{r_1\lambda I_0}{4}\left\{\frac{1}{q-1}+\frac{1}{q+1}\right\}e^{qX}\quad\text{for}\quad -\infty<X<-\delta/\lambda, \tag{3·3}$$

and
$$U=\frac{r_1\lambda I_0}{4}\left\{\frac{1}{q-1}+\frac{1}{q+1}\right\}e^{-qX}\quad\text{for}\quad \delta/\lambda<X<\infty. \tag{3·4}$$

The interpretations of the operational expressions in (3·3) and (3·4) are known (Jeffreys 1931). When they are substituted, the following equations for V_m are obtained:

$$V_m=\frac{r_1\lambda I_0}{4}\{e^X[1+\operatorname{erf}(X/2\sqrt{T}+\sqrt{T})]-e^{-X}[1+\operatorname{erf}(X/2\sqrt{T}-\sqrt{T})]\},$$
$$\text{when}\quad -\infty<X<0, \tag{4·0}$$

and
$$V_m=\frac{r_1\lambda I_0}{4}\{e^{-X}[1-\operatorname{erf}(X/2\sqrt{T}-\sqrt{T})]-e^{X}[1-\operatorname{erf}(X/2\sqrt{T}+\sqrt{T})]\},$$
$$\text{when}\quad 0<X<\infty, \tag{4·1}$$

where
$$\operatorname{erf}Z=\frac{2}{\sqrt{\pi}}\int_0^Z e^{-\omega^2}d\omega.$$

These expressions satisfy equation (2·3) and the boundary conditions. Campbell & Foster (1931, p. 162) give an expression which is equivalent to (4·1) for the response of a non-inductive cable to the sudden application of current.

The solutions for the case when the applied current is maintained for a long time and then broken suddenly at $t=0$ can be written down at once from the superposition theorem. They are

$$V_m=\frac{r_1\lambda I_0}{4}\{e^X[1-\operatorname{erf}(X/2\sqrt{T}+\sqrt{T})]+e^{-X}[1+\operatorname{erf}(X/2\sqrt{T}-\sqrt{T})]\},$$
$$\text{when}\quad -\infty<X<0, \tag{4·2}$$

and
$$V_m=\frac{r_1\lambda I_0}{4}\{e^{-X}[1+\operatorname{erf}(X/2\sqrt{T}-\sqrt{T})]+e^{X}[1-\operatorname{erf}(X/2\sqrt{T}+\sqrt{T})]\},$$
$$\text{when}\quad 0<X<\infty. \tag{4·3}$$

Equations (4·0), (4·1) and (4·2), (4·3) are symmetrical pairs differing only in the sign of X. Thus it is necessary to compute only one set of curves in order to describe the distribution of potential for the make or break of a constant current. And the curves for the break of current can be obtained from those for the make by a direct application of the superposition theorem. Equation (4·1) is the most convenient to compute, since it deals with positive values of X. An evaluation of the essential part of (4·1) is given in table 1 and the results are plotted graphically in figure 2.

Table 1. Table of the function $\{e^{-X}[1-\mathrm{erf}\,(X/2\sqrt{T}-\sqrt{T})]-e^{X}[1-\mathrm{erf}\,(X/2\sqrt{T}+\sqrt{T})]\}$ for different values of X and T

	$T=0{\cdot}01$	0·04	0·16	0·36	0·64	1·0	1·44	1·96	2·56	3·24	4·00	6·25	∞
$X=0$	0·2249	0·4454	0·8567	1·208	1·484	1·685	1·821	1·904	1·953	1·987	1·991	1·999	2·000
0·1	0·0795	0·2743	0·673	1·020	1·294	1·496	1·631	1·714	1·763	1·788	1·801	1·800	1·810
0·2	0·0200	0·1561	0·5201	0·855	1·126	1·325	1·459	1·542	1·590	1·616	1·628	1·636	1·637
0·3	0·0035	0·0816	0·3926	0·712	0·976	1·172	1·305	1·387	1·435	1·460	1·473	1·481	1·482
0·4	0·0004	0·0390	0·2921	0·588	0·842	1·034	1·165	1·247	1·294	1·320	1·332	1·340	1·341
0·5	—	0·0170	0·2125	0·483	0·725	0·911	1·040	1·120	1·167	1·192	1·204	1·212	1·213
0·6	—	0·0067	0·1513	0·392	0·621	0·801	0·926	1·006	1·052	1·077	1·089	1·097	1·098
0·7	—	0·0024	0·1055	0·316	0·530	0·702	0·824	0·903	0·948	0·972	0·984	0·992	0·993
0·8	—	0·00076	0·0718	0·252	0·450	0·614	0·732	0·809	0·853	0·878	0·890	0·898	0·899
0·9	—	—	0·0474	0·2000	0·381	0·537	0·651	0·725	0·769	0·793	0·804	0·812	0·813
1·0	—	—	0·0311	0·1554	0·319	0·467	0·577	0·649	0·691	0·715	0·727	0·735	0·736
1·2	—	—	0·0122	0·0937	0·224	0·351	0·451	0·517	0·560	0·583	0·594	0·601	0·602
1·5	—	—	0·0026	0·0401	0·125	0·225	0·309	0·369	0·407	0·428	0·438	0·445	0·446
2·0	—	—	0·0001	0·0086	0·0435	0·1008	0·159	0·205	0·236	0·254	0·263	0·270	0·271
2·5	—	—	—	0·0012	0·0129	0·0415	0·0773	0·1113	0·1347	0·1494	0·1574	0·1636	0·1642
3·0	—	—	—	—	0·0033	0·0174	0·0364	0·0581	0·0754	0·0870	0·0937	0·0988	0·0996

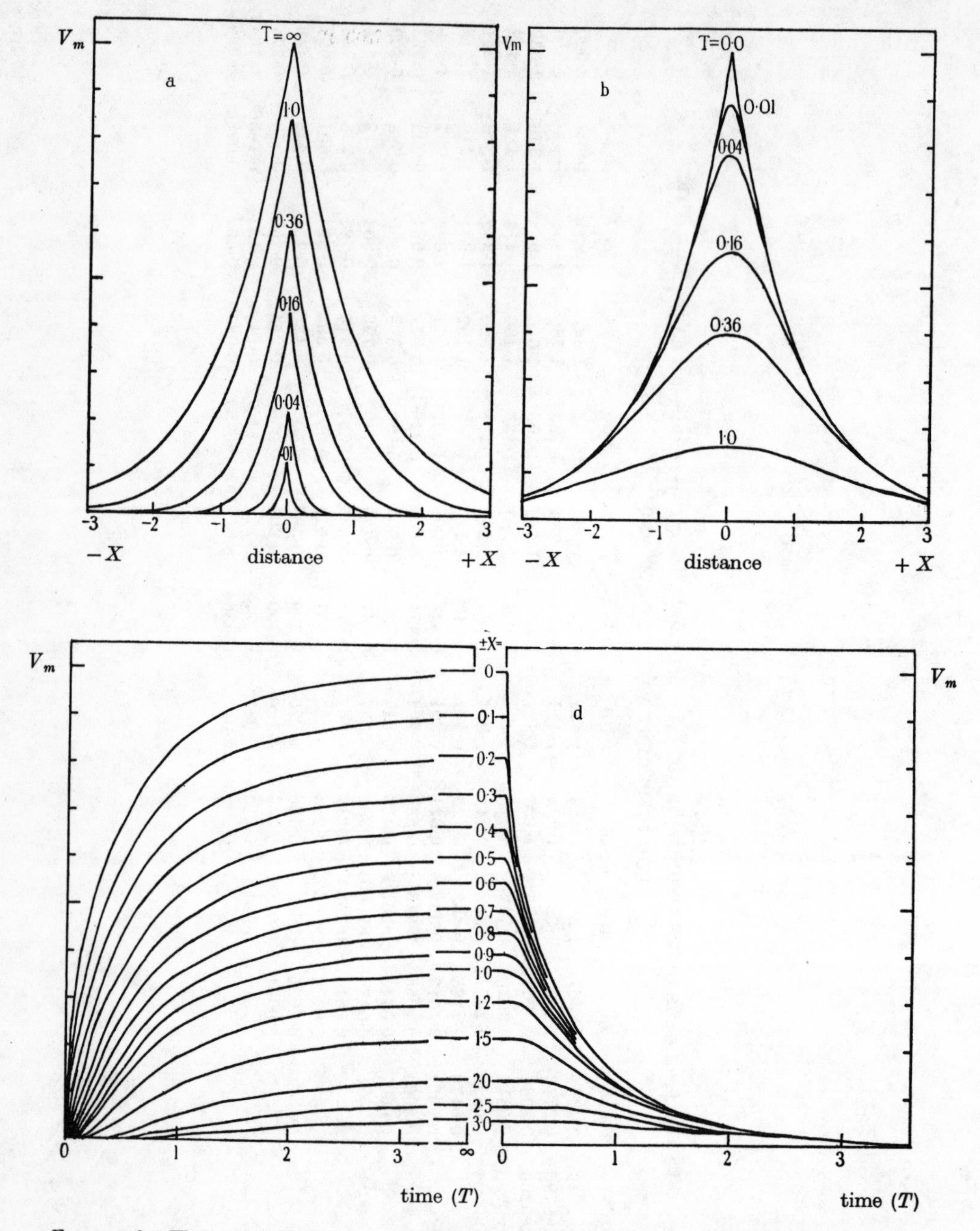

FIGURE 2. Theoretical behaviour of potential difference across nerve membrane (V_m). *a*, *b*, spatial distribution of potential at different times; *c*, *d*, time course of potential at different distances from electrode; *a*, *c*, current made at $T = 0$; *b*, *d*, current maintained for a long time and then broken at $T = 0$.

A great simplification of the mathematical theory can be achieved by considering the total charge in the region of the electrode instead of the membrane potential. Define total charge by a new variable

$$\xi = c\int_{-\infty}^{\beta} V_m\, dx, \tag{5·0}$$

where β is sufficiently large to allow the integration to include all the charge in the electrode region. β can be considered as infinite provided that the integration does not include the region of the second electrode. Integration of equation (2·1) from $-\infty$ to β gives

$$\tau_m \frac{\partial \xi}{\partial t} + \xi = r_1 c \lambda^2 I_0, \tag{5·1}$$

since $$\int_{-\infty}^{\beta} \partial V_m/\partial t\, dx = \frac{\partial}{\partial t}\left[\int_{-\infty}^{\beta} V_m\, dx\right]$$

and $\frac{\partial V_m}{\partial x} = 0$ when $x = -\infty$ and $x = +\beta$. The solutions of (5·1) are

$$\xi = r_1 c \lambda^2 I_0 (1 - e^{-t/\tau_m}) \tag{5·2}$$

for a constant current made at $t = 0$ and

$$\xi = r_1 c \lambda^2 I_0 e^{-t/\tau_m} \tag{5·3}$$

for a constant current broken at $t = 0$. Unfortunately, these simple equations are of little practical use, since ξ can only be obtained indirectly from the experimental results.

The extrapolar potential

Equations (4·0) and (4·2) can be applied directly to the experimental results since $V_m = \frac{r_1 + r_2}{r_1} V_1$, $X = x/\lambda$ and $T = t/\tau_m$.

The most convenient expressions for the make of current are

$$V_1 = (V_1)_{\substack{t=\infty\\x=0}} \tfrac{1}{2}\{e^X[1 + \mathrm{erf}(X/2\sqrt{T} + \sqrt{T})] - e^{-X}[1 + \mathrm{erf}(X/2\sqrt{T} - \sqrt{T})]\}, \tag{6·0}$$

where $$(V_1)_{\substack{t=\infty\\x=0}} = \frac{r_1^2 \lambda I_0}{2(r_1 + r_2)} = y I_0, \tag{6·1}$$

y being thus defined in the table of practical constants,

$$(V_1)_{t=\infty} = (V_1)_{\substack{t=\infty\\x=0}} e^X, \tag{6·2}$$

and $$(V_1)_{x=0} = (V_1)_{\substack{t=\infty\\x=0}} \mathrm{erf}(\sqrt{T}). \tag{6·3}$$

For the break of current the relevant expressions are

$$V_1 = (V_1)_{\substack{t=0\\x=0}} \tfrac{1}{2}\{e^X[1 - \mathrm{erf}(X/2\sqrt{T} + \sqrt{T})] + e^{-X}[1 + \mathrm{erf}(X/2\sqrt{T} - \sqrt{T})]\}, \tag{6·4}$$

and $$(V_1)_{x=0} = (V_1)_{\substack{t=0\\x=0}} [1 - \mathrm{erf}(\sqrt{T})]. \tag{6·5}$$

The mid-interpolar gradient

The expressions given in the preceding paragraph allow λ, τ_m and y to be determined from experimental observations of the extrapolar potential. The constant m can be obtained from a measurement of the voltage gradient in the interpolar region at a large distance from either electrode. For equation (4·1) shows that $\left(\frac{\partial V_m}{\partial x}\right)_{x=\beta} = 0$ when $\beta \gg \lambda$. Hence differentiation of (1·3) gives

$$-\left(\frac{\partial V_1}{\partial x}\right)_{x=\beta} \Big/ I_0 = \frac{r_1 r_2}{r_1 + r_2} = m. \tag{7·0}$$

Determination of basic constants

Convenient expressions for determining the basic constants are

$$R_2 = \pi a^2 m(1 + m\lambda/2y), \tag{8·0}$$

$$R_4 = 2\pi a\lambda^2 m(2 + m\lambda/2y + 2y/m\lambda), \tag{8·1}$$

$$C_M = \tau_m / R_4. \tag{8·2}$$

An expression for the resistivity of the external fluid is not given because there was no easy way of determining the volume of fluid surrounding the nerve fibre. Nor would this quantity have been of great interest. But it is desirable to have an index of the amount of short-circuiting introduced by the external fluid and the ratio r_2/r_1 has been used for this purpose. It can be computed by the relation

$$r_2/r_1 = m\lambda/2y. \tag{8·3}$$

Validity of assumptions

We are now in a better position to assess the errors which are introduced by the approximations made in the theory. The assumption of parallel current flow is not likely to involve any serious error provided that the current spreads over a length which is several times greater than the diameter of the axon. This condition was satisfied experimentally, since the average value for the space constant λ was twenty times greater than the axon diameter. The assumption of zero breadth for the electrode can be justified in the same way, since λ/δ was also of the order of twenty. Both approximations are doubtful at short time intervals. Thus figure 2 shows that the effective space constant is only $\lambda/5$ when t/τ_m is 0·04. But it can be said that the cable equations apply with reasonable accuracy provided $t/\tau_m > 0·04$.

In practice anode and cathode were separated by about 8 mm. of nerve; theory assumes them to be an infinite distance apart. But interference between the two electrode regions must have been negligible, since 8 mm. was equivalent to 5λ in an average experiment and e^{-5} is 0·007. A similar argument applies to the recording electrodes which were also 8 mm. apart.

The assumption that the internal and external resistances obey Ohm's law is fully justified by earlier work (see, for instance, Cole & Hodgkin 1939) and finds further confirmation in the measurements of mid-interpolar gradient which will be described presently. The constancy of the membrane resistance might be questioned in view of Cole & Curtis's (1941) demonstration of the rectifying properties of the surface membrane. But any rectifier behaves as a linear element if it is examined with a sufficiently weak current. And we shall show later that the measuring currents used were probably small enough to keep the membrane in a linear part of its characteristic.

Some error must have been introduced by assuming that the membrane capacity behaved like a pure dielectric. The magnitude of the error cannot be estimated in any simple way, but it is not likely to have been very large. For a.c. measurements give a value of 76° for the phase angle of the dielectric of the membrane in the squid axon (Curtis & Cole 1938). This suggests that the membrane capacity would be reduced by 30 % when the frequency was increased tenfold. Most of the records dealt with here could be reproduced fairly accurately by a Fourier synthesis containing a tenfold range of frequencies and so would not have been greatly affected by imperfections in the membrane capacity.

Method

Material

Single-nerve fibres with a diameter of 60–80μ were obtained from the walking legs of the common lobster (*Homarus vulgaris*). Live lobsters were bought from a fishmonger and kept in an aquarium filled with circulating sea water. The animals were in poor condition when first obtained, but they recovered after a few hours in the aquarium and were able to live there for several weeks. Axons were obtained from the first two pairs of walking legs which are chelate and appear to be better supplied with large fibres than the last two which have no terminal claw. The nerve was dissected from the meropodite and teased apart in a Petri dish of sea water. *Homarus* nerve contains much connective tissue, and separation of a single fibre proved to be a more laborious process than in a *Carcinus* preparation. More time had to be spent in cutting away connective tissue, and no attempt could be made to pull fibres apart until they had been freed from the strands of connective tissue which bound them together. All loose material was removed from the isolated axon whose length varied from 25 to 40 mm. Fibres with branches were never employed.

The method of isolating *Carcinus* axons was similar to that employed in earlier work (Hodgkin 1938) and need not be described again.

Apparatus

A general plan of the equipment used is shown in figure 3. The axon was kept in paraffin oil and was gripped at each end by the tips of insulated forceps (AA'). It was held in a horizontal position and could be observed from above by means of a binocular microscope. The axon rested on the wick electrodes B, D, and made contact with the tip of electrode C. Electrodes B and D made contact over a length of about 250μ, and electrode C over approximately 100μ. These three electrodes consisted of small glass tubes containing sea water and silver wires which had been coated electrolytically with chloride. One end of the glass tube was sealed with wax; the other was drawn out into a coarse capillary and plugged with agar sea water. Connexion to the nerve was made through fine agar wicks which projected for about 5 mm. beyond the tip of the glass capillary. The wicks were built by allowing agar sea water to solidify around a fine silk thread. Silver chloride electrodes were sufficiently non-polarizable, since a 5MΩ resistance in series with the electrodes ensured that the current was entirely unaffected by residual electrode polarization. Electrode E did not need to be non-polarizable, since it was used only for recording transient pulses with an amplifier of high input impedance. This electrode consisted of a fine glass tube into which a platinum wire was sealed; one end of the tube was drawn out into a fine glass capillary, ground square and the whole filled with sea water. The diameter of the tip was about 50μ and the region of contact with the nerve fibre of the same order of magnitude.

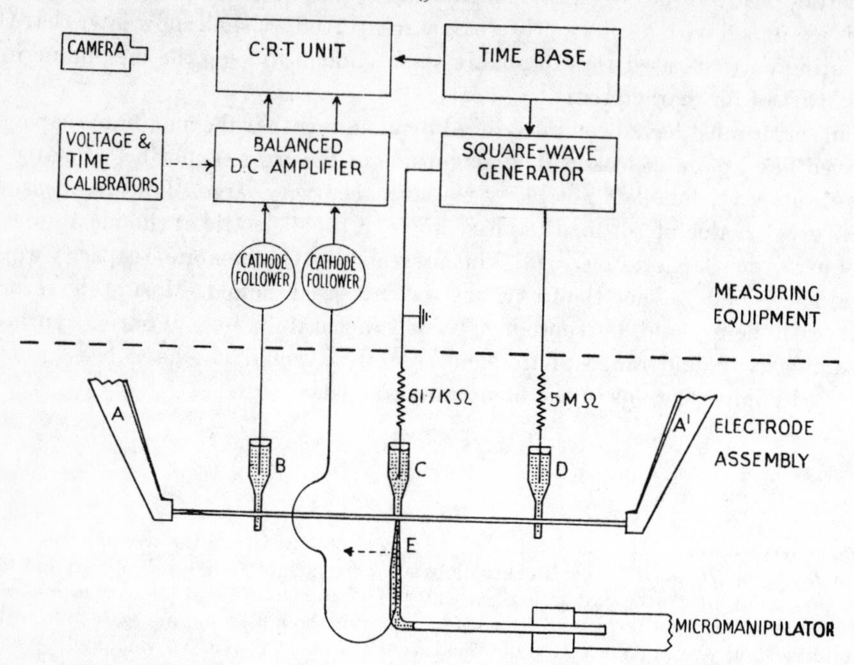

FIGURE 3. General plan of equipment. For letter references, see text.

Electrode E was held in a micromanipulator carriage and could be moved along the nerve fibre by turning one of the vernier controls on the manipulator. The electrode slid smoothly along the fibre provided that all loose connective tissue had been removed and that the direction of movement was parallel to the axis of the fibre. The position of the electrode was determined by a scale on the manipulator which was calibrated to read in fractions of a millimetre. This method of measurement was checked periodically by observing the motion of the electrode under the binocular microscope. The movement of the electrode was found to be the same as that given by the scale on the screw adjustment. Back-lash could be taken as zero, since it was less than 10 micra.

All electrical measurements were made by applying a rectangular pulse of current to the axon and recording the resulting potential changes with an amplifier and oscillograph. The rectangular pulse was generated by means of an arrangement of thyratrons (R.C.A. 885) and a multivibrator of the type described by Schmitt (1938). The wave form of the rectangular pulses was tested by connecting the pulse output to the plates of a cathode-ray tube. This showed that the deflexion was 90 % complete in less than 10μsec. The pulse could be synchronized with the sweep circuit and its duration varied between 10 and $10^6\mu$sec. A low-resistance attenuator was used for varying the magnitude of the pulse applied to the nerve. One terminal of the pulse generator was connected to earth; the other became positive for the duration of the pulse. The positive-going terminal was connected to electrode D through $5\text{M}\Omega$ and the other terminal (earth) to electrode C through a monitoring resistance of $61{,}700\Omega$. The 5-megohm resistance ensured that a constant current was passed through the nerve, while the monitoring resistance was used to measure the current through the nerve fibre. The pulses of current were repeated at a rate of about one a second.

30-2

Electrical changes were recorded with a balanced d.c. amplifier designed by Dr Rawdon Smith of the Psychological Laboratory, Cambridge. This consisted of three pairs of pentodes with separate anode loads and common cathode resistances. The line voltages were arranged so that the anode of one stage could be connected to the grid of the next. In this way the undesirable resistance chains usually associated with d.c. amplifiers were avoided. Occasional checks showed that the differential action of the amplifier was better than one part in five hundred (i.e. when both inputs were raised 1 V above earth the oscillograph deflexion was equivalent to less than 2 mV difference between inputs). Initial checks with a signal generator indicated that the response of the amplifier was substantially flat between 0 and 50 kcyc./sec. In order to increase the input impedance the recording leads were connected to the grids of two cathode followers which were placed at a distance of 15 cm. from the preparation. Calibrations of the input stage and the whole amplifier were made by applying the rectangular pulse to the grids of the input stage through a resistance of the same magnitude as that involved in recording from the nerve. This test showed that the deflexion produced was 90 % complete in about 30μsec., and the system was therefore sufficiently rapid for the investigation of phenomena lasting several milliseconds. The d.c. input impedance was greater than $10^{10}\,\Omega$ and the grid current less than 10^{-10} amp.

The time base was calibrated by applying the output from a 500 cyc./sec. oscillator to the amplifier. Voltage calibrations were made by photographing the series of oscillograph lines produced by varying the position of a decade resistance attenuator. In this way a calibration grid was obtained and could be compared with the experimental results. In general the experimental records fell in a region which was linear to within 2 % and so could be analysed without correction. Corrections had occasionally to be made but did not materially affect the results, since the amount of instrumental distortion rarely exceeded 5 %. All photographic records were taken on film and were traced on to graph paper after they had been enlarged about ten times.

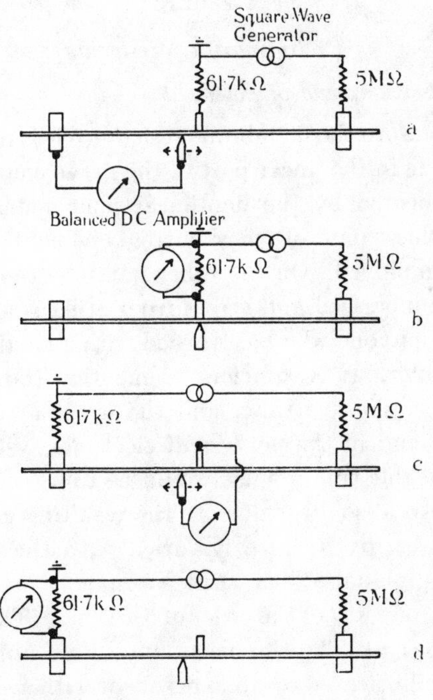

FIGURE 4. Arrangement of leads and electrodes employed in a quantitative experiment. *a*, system used for determining λ and τ_m; *b*, system used in conjunction with *a* for determining *y*; *c*, *d*, system used for determining *m*.

A typical experiment

The sequence of events in a quantitative experiment must now be described. The isolated axon was mounted on the electrode system and raised into a layer of aerated mineral oil which floated on the surface of the sea water. The recording electrode was brought into contact with the axon and a preliminary test made to ensure that the action potential was propagated normally throughout the whole fibre. The strength of current was reduced until it was below half-threshold and the potential response observed visually on the C.R.T. The duration of the rectangular wave was adjusted until it was sufficient to allow the membrane voltage to reach its equilibrium value. A test was made to ensure that the recording electrode slid smoothly along the axon. A series of photographic records of the extrapolar potential was then obtained with the arrangement of electrodes shown in figure 4*a*; in general these were similar to those in figure 6. One set of records was made with the movable electrode receding from the cathode and another with it approaching. There was sometimes a difference of 5 or 10 % between the two sets of records, but as a rule they agreed closely with one another. The current through the axon was determined with the arrangement of electrodes shown in figure 4*b* and a typical record is given in figure 6. This observation also provided a routine check of the squareness of the current wave form through the nerve fibre. The next operation was to determine the voltage gradient in the mid-interpolar stretch using the arrangement of figure 4*c*. The recording electrode was moved along the axon and a series of records similar to those in figure 9*a* obtained. The current through the axon was again determined; the arrangement of leads being that of figure 4*d* and a typical record that of figure 9*b*. At the end of each experiment the fibre diameter was measured in the following way. The axon was lowered into sea water and transferred to a hollow-ground slide; it was then examined with a microscope using a $\frac{1}{8}$ in. objective and an eyepiece micrometer. Some variation in diameter was always encountered, but this rarely exceeded 5 %.

Results

Preliminary experiments

Local response and passive spread of potential

In attempting to measure the membrane resistance it is important to ensure that measurements are made in the linear part of the nerve characteristic, and that the results are not complicated by the non-linear phenomena of local response and rectification. From this point of view currents which are much weaker than threshold should be employed. On the other hand, as the current is reduced the amplification must be increased and errors from other sources increase. This fact will be appreciated by anyone who has worked with a single-fibre preparation and a high-gain d.c. amplifier. It is sufficient to mention the difficulties which arise from stray interference, shock artifact and the irregular drifts in voltage which occur in the amplifier and in the nerve and electrode system. Preliminary tests indicated that a reasonable compromise would be to use currents with a strength of 0·4–0·5 threshold. An absolute value for the resulting current density through the membrane cannot be given, since it varied with the excitability and membrane resistance of individual axons. But a rough estimate is that the current density under the electrode was of the order of $5\,\mu A\,cm.^{-2}$. The total current through the axon was roughly $0{\cdot}1\,\mu A$. The absence of any significant response in the region below half-threshold is illustrated by an experiment with a *Carcinus* fibre (figure 5). Here the behaviour of the axon is shown for different strengths of applied current. Anodic or weak cathodic currents appear to affect only the passive charging process;

for all the curves have the same shape and their amplitude is roughly proportional to the applied current. And the shape of the curves is of a type which is to be expected from a process involving passive charge and discharge of the membrane capacity. The picture changed completely when the applied current approached threshold. At 0·9 threshold the cathodic potential showed a fast creep, and at 1·0 the curves turned upwards as if to give rise to a propagated impulse. But a true

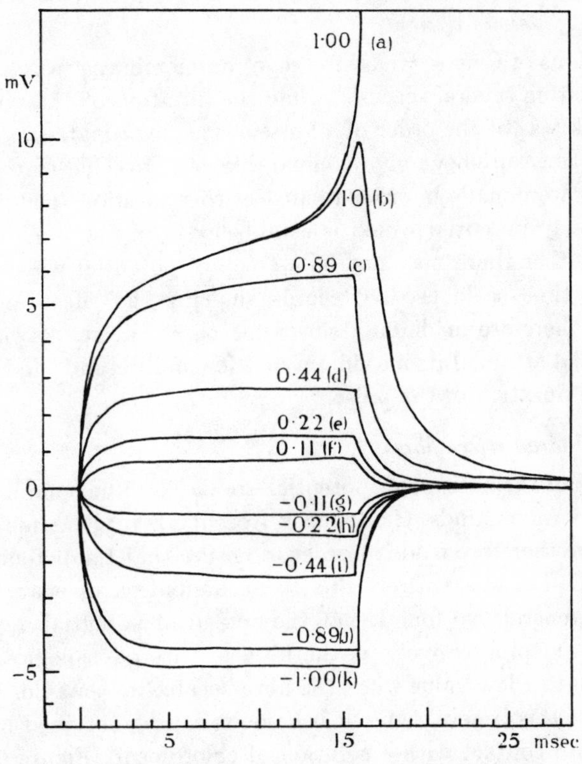

FIGURE 5. Response of *Carcinus* axon to rectangular waves of current of different intensity; recorded at a polarizing electrode of width about 200μ. The numbers on each record give the strength of current relative to threshold. Depolarization of the nerve is shown as positive.

action potential did not result in every case. Owing to the spontaneous play in excitability, a stimulus does not invariably evoke an impulse until its strength is slightly greater than threshold. In fact, a threshold shock is normally defined as one which produces impulses on 50 % of occasions. Record *b* shows what happens when a threshold shock failed to evoke an impulse. The potential turned upwards as if to give rise to a spike, but it failed to reach a critical level and died out as a localized wave. Inflected local responses of this kind only occurred when the current was nearly threshold but their onset was completely gradual. Thus all transitions between records *b* and *c* could be obtained by careful adjustment of the strength of current.

The striking form of local response illustrated by this experiment was observed on a large number of occasions and will be described in greater detail later. For the moment our chief concern is that it did not occur when the current was less than half-threshold.

Measurement of the curves in figure 5 indicated that there were small deviations from linearity in the region below half-threshold. At present there is no evidence to show whether these deviations were reproducible, and they may well have been instrumental in origin.

The effect of long pulses of current

The observations of Cole & Hodgkin (1939) on membrane resistance were made with currents lasting several seconds, while the duration of the currents used in the present work was of the order of 20 msec. It is legitimate to ask whether the two methods of measurement give comparable results. One or two experiments with long pulses were made in order to answer this question. The point at issue is whether the steady potential which is established in a few milliseconds is really constant, or whether there may not be a creep of potential which is too slow to register on the time scale used. Records showing the effect of pulses lasting 300 msec. were therefore made on a slow time base. The result was unequivocal, since the potential attained its maximum in a few milliseconds and then remained constant for the duration of the pulse.

Experiments with dead nerve fibres

Measurements of the extrapolar potential are liable to be complicated by errors and artifacts of various kinds (cf. Bogue & Rosenberg 1934). A number of control experiments were therefore made in order to ensure that the potential recorded in the extrapolar region was entirely due to accumulation of charge at the nerve membrane. In general, we found that the spread of potential in the extrapolar region was reduced progressively as the fibre lost its physiological activity, and that it finally fell to a low value when the fibre became inexcitable. A very striking demonstration of this general type of behaviour can be obtained by allowing the axon to come into contact with a solution of chloroform. Figure 6 illustrates an experiment of this kind. Records *b–g* show the spread of potential in the extrapolar region of a normal axon, and demonstrate the passive accumulation of charge at the surface membrane. The fibre was then dipped into sea water which had been shaken with chloroform. It was left in this solution for 1 or 2 min. and raised into oil. The result was extremely striking; for the potential change at the cathode was reduced to one-twentieth of its former value and was abolished at all other points. Records *a* and *A* are an index of the current through the axon, which was unchanged by the chloroform treatment. This experiment illustrates the delicate nature of the surface membrane and provides a convincing demonstration of the virtual absence of artifacts. The small potential which is recorded at the cathode in *B* may be attributed either to a residual membrane resistance or to the finite thickness of the nerve fibre. Close examination of the original records revealed a rapid spike which occurred at the beginning and end of the square wave, but was too faint for reproduction. This persisted after chloroform treatment and must be regarded as an artifact caused by capacitative coupling between the polarizing and recording leads. The spike was ignored in analysing the records, but served a useful purpose in defining precisely the beginning and end of the applied current.

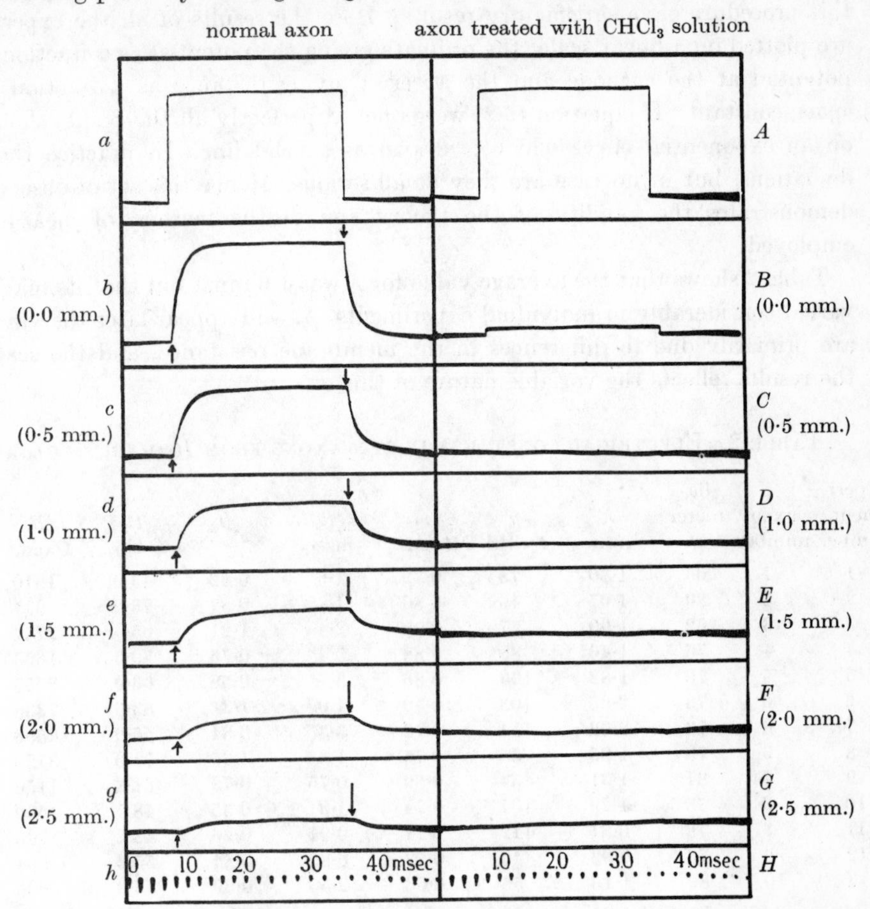

FIGURE 6. Effect of chloroform solution on spread of membrane potential in *Homarus* axon. *a*, *A*, current through normal and chloroform treated axon, measured as voltage across 61,700Ω resistance in series with axon; *b–g*, potential recorded in extrapolar region of normal axon; the distance from the cathode is shown by the figures in brackets; *B–G*, potentials recorded in the same way after application of chloroform solution; *h*, *H*, 500 cyc./sec. time calibration. The vertical arrows indicate the beginning and end of the square wave of current and were marked from a capacitative artifact which appeared on the original records. Records *a* to *c* have been retouched. The amplification was the same in all records and the amplitude of the wave in *b* was approximately 4·5 mV. Records taken from experiment 13.

The measurement of λ

Equation (6·2) shows that there should be an exponential relation between the steady potential in the extrapolar region and the distance from the cathode. Hence a straight line with slope $(\log_{10} e)/\lambda$ should result when the $\log_{10}$ of the potential is plotted against distance. This method was used in all the experiments and is illustrated by figure 7. In drawing a straight line through the experimental points, more weight was placed on observations near the cathode, since the percentage error increased as the recorded voltage decreased. Figure 8 proves that this procedure gave satisfactory results. Here the results of all the experiments are plotted on a linear scale: the ordinate giving the potential as a fraction of the potential at the cathode and the abscissa giving distance as a fraction of the space constant. If equation (6·2) were obeyed perfectly all the points should fall on an exponential curve which is drawn as a solid line. In practice there are deviations, but in no case are they at all serious. Hence this set of observations demonstrates the validity of the theory and of the method of measurement employed.

Table 2 shows that the average value for λ was 1·6 mm., but that its magnitude varied considerably in individual experiments. As will appear later the variations are primarily due to differences in the membrane resistance, and the scatter in the results reflects the variable nature of this quantity.

TABLE 2. ELECTRICAL CONSTANTS IN TEN AXONS FROM *HOMARUS VULGARIS*

experiment number	axon number	diameter μ	λ mm.	y $\Omega \times 10^3$	m MΩ cm.$^{-1}$	τ_m msec.	r_2/r_1	R_2 Ω cm.	R_4 Ω cm.^{2}	C_M μF cm.$^{-2}$
1	1	65	1·80	78	0·72	1·6	0·82	43·6	1910	0·83
2	2	80	1·07	49	0·80	1·8	0·87	75·2	927	1·94
3	3	62	1·90	77	0·98	2·4	1·21	65·4	2784	0·87
4	4	76	1·40	80	0·88	5·4	0·76	70·6	1655	3·24
5	5	76	1·82	134	0·86	3·7	0·59	63·0	2955	1·25
[6	6	73	2·95	103	0·90	4·0	1·3	83·6	7330	0·55
{7	6	73	2·62	114	0·73	3·3	0·84	55·9	4590	0·71
8}]	6	73	1·95	55	0·76	1·3	1·35	74·6	2720	0·46
9	7	87	1·31	54	0·59	0·76	0·72	61·2	1150	0·66
{10	8	78	1·29	137	0·74	1·9	0·35	48·1	1590	1·23
11}	8	78	0·81	112	0·71	0·91	0·26	43·2	706	1·29
12	9	73	0·92	40	0·72	0·89	0·84	55·6	564	1·58
13	10	80	1·15	55	0·66	2·5	0·69	56·6	905	2·73
average value		75	1·61	81	0·83	2·3	0·81	60·5	2290	1·33

Square brackets indicate that successive measurements were made on the same nerve fibre; curved brackets that they were made on the same stretch of the same fibre. Temperature: 15–20° C. Strength of current: 0·4–0·5 threshold. The values given for τ_m are the mean of four measurements.

The measurement of y

The constant y has the dimensions of a resistance and is given by the ratio of the steady voltage at the cathode to the applied current (see equation (6·1)). The method of measurement is clarified by referring to figure 6. Here b gives the voltage at the cathode and a the voltage across 61,700 Ω. Hence $y = 61{,}700 \times b/a\ \Omega$, where b/a is the ratio of the observed voltages. In this case y was 55,400 Ω, which was rather smaller than that usually obtained (see table 2).

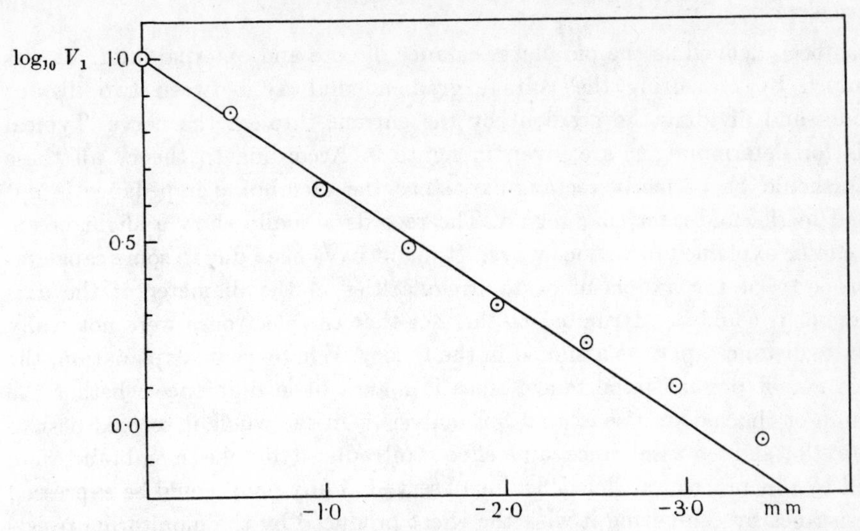

FIGURE 7. Equilibrium distribution of extrapolar potential. Ordinate: $\log_{10}$ potential. Abscissa: distance of recording electrode from cathode in mm. The distance is shown as negative in order to conform to the convention used in the theoretical section.

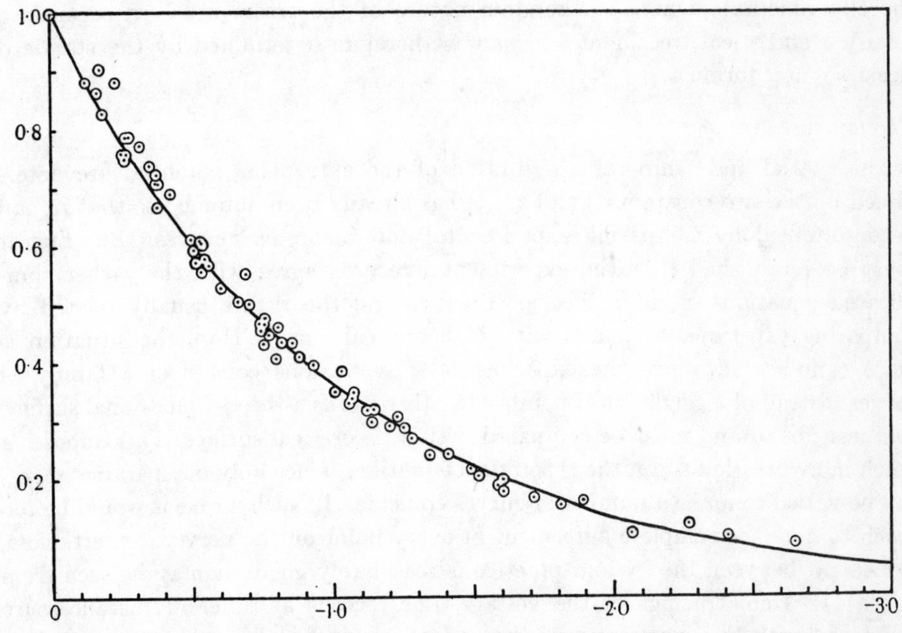

FIGURE 8. Equilibrium distribution of extrapolar potential in thirteen experiments. Ordinate: potential as a fraction of the potential at the cathode. Abscissa: distance as a fraction of the measured space constant λ. The solid line is drawn according to equation (6·2).

The measurement of m

m has been defined as the parallel resistance of core and external fluid. It was determined by measuring the voltage gradient midway between two distant electrodes and dividing the gradient by the current through the nerve. Typical records for determining m are given in figure 9. According to theory all these records should be perfectly rectangular, since the membrane impedance is not involved in the mid-interpolar region. The records actually show a slight creep, which can be explained in various ways. It might have been due to some capacitative property of the axoplasm or to irregularities in the diameter of the axis cylinder; or it could be attributed to the fact that the electrodes were not really an infinite distance apart as assumed in the theory. Whatever its explanation, the effect is not of present importance, since it makes little difference whether the maximum or the sudden rise is used for analysis. On the whole it seemed best to measure the sudden rise, since any effects introduced by the membrane were avoided by this procedure. The deflexion observed at any point could be expressed as a resistance by comparing it with the effect produced by the monitoring resistance. It was therefore possible to plot resistance against electrode separation as has been done in figure 10, which illustrates three typical experiments. The observed points fall very close to straight lines as they should according to theory. A direct measurement of m is given by the slope of the best straight line through the experimental points. The random nature of the errors involved seemed to justify a statistical treatment and m was therefore determined by the standard 'least square' formula.

The measurement of τ_m

The spatial and temporal distribution of the extrapolar potential are determined by the two constants λ and τ_m. λ has already been obtained so that τ_m can be determined by comparing experimental and theoretical curves. But first it must be established that the experimental records agree with the rather complicated equations of cable theory. Practice and theory are usually related by comparing experimental points with a theoretical curve. Here the situation is more complicated, since the experimental observations consist of a family of curves instead of a single set of points. In other words a three-dimensional surface has been found and must be compared with a theoretical surface. This imposes a much more drastic test on the theoretical equations, since only one parameter, τ_m, can be varied to make a number of curves coincide. In such a case it would be too much to hope for complete agreement at every point on the nerve. Nevertheless, agreement between theory and practice is reasonably good, as may be seen from figure 11. Here tracings of the voltage-time records at different distances are compared with the corresponding theoretical curves for those distances. Only a finite number of theoretical curves was computed and it was therefore impossible to use a theoretical curve which corresponded exactly with the experimental one. Thus C is the experimental curve for $x/\lambda = 0{\cdot}38$ and d the theoretical curve for

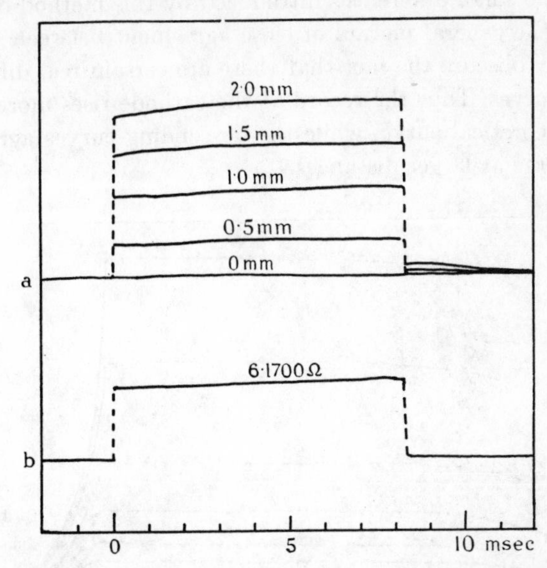

FIGURE 9. *a*, voltage gradient in mid-interpolar region. Records obtained with arrangement of figure 4*c* and with measuring electrodes separated by distances of 0–2·0 mm. *b*, voltage across 61,700 Ω using the same strength of current as that in *a*. Electrode arrangement as in figure 4*d*.

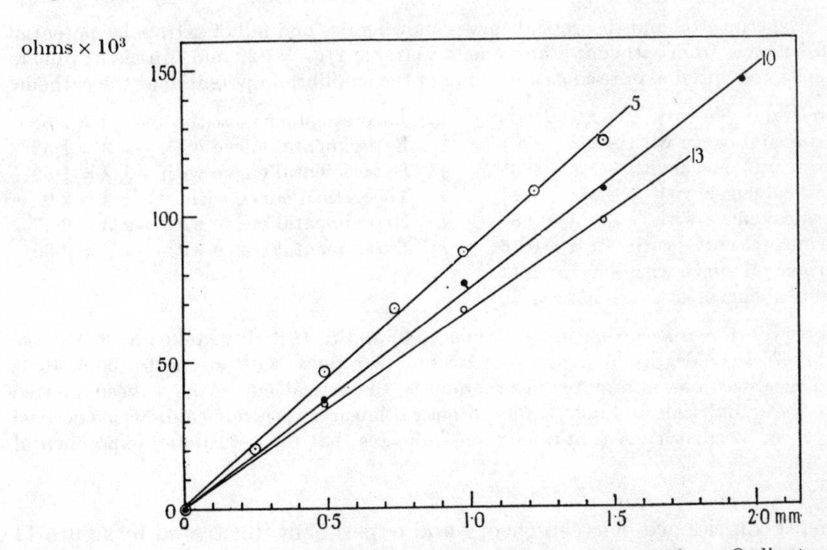

FIGURE 10. Resistance length relation in the mid-interpolar region. Ordinate: resistance, measured from records of the type shown in figure 9. Abscissa: distance between recording leads. The numbers on the straight lines refer to the experiments in table 2. The current was led into the nerve through electrodes about 16 mm. apart.

$x/\lambda = 0{\cdot}4$. But the small differences introduced by this method of plotting do not materially alter the general picture of close agreement between theory and practice. Nor do they obscure the fact that there are certain real differences between the two sets of curves. Thus the record at the cathode rises more slowly than the corresponding theoretical curve, while the descending curves agree closely at the cathode but diverge at larger distances.

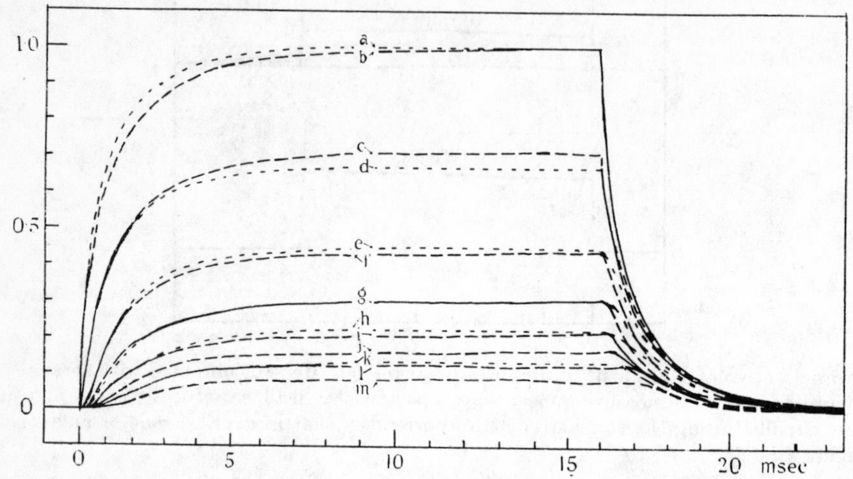

FIGURE 11. Experimental and theoretical curves showing rise and fall of extrapolar potential at different distances from cathode. Experiment 10 (table 2); $\lambda = 1{\cdot}29$ mm. Abscissa: time in msec. Ordinate: potential expressed as a fraction of the equilibrium potential at the cathode.

a. Theoretical curve with $-x/\lambda = 0{\cdot}0$
b. Experimental curve with $-x/\lambda = 0{\cdot}0$
c. Experimental curve with $-x/\lambda = 0{\cdot}38$
d. Theoretical curve with $-x/\lambda = 0{\cdot}4$
e. Theoretical curve with $-x/\lambda = 0{\cdot}8$
f. Experimental curve with $-x/\lambda = 0{\cdot}76$
g. Experimental curve with $-x/\lambda = 1{\cdot}14$
Theoretical curve with $-x/\lambda = 1{\cdot}2$
h. Theoretical curve with $-x/\lambda = 1{\cdot}5$
i. Experimental curve with $-x/\lambda = 1{\cdot}52$
j. Experimental curve with $-x/\lambda = 1{\cdot}89$
k. Theoretical curve with $-x/\lambda = 2{\cdot}0$
l. Experimental curve with $-x/\lambda = 2{\cdot}27$
m. Experimental curve with $-x/\lambda = 2{\cdot}65$

Theoretical curves drawn according to equations (6·0) and (6·4) with τ_m taken as 2·10 msec. Arrangement of electrodes as in figure 4*a*. Rectangular pulse with strength about 40 % threshold. The abscissa is not quite linear and the theoretical curves have been plotted according to the actual scale and not to a hypothetical linear scale; time calibrations derived from 500 cyc./sec. oscillator. A continuous line indicates that theoretical and experimental curves coincide.

The general coincidence between theory and experiment illustrated by figure 11 was only obtained because the theoretical curves were plotted with the correct time constant which in this case happened to be 2·10 msec. This value was obtained by a laborious process of trial and error which was too cumbersome for use in every experiment. It was therefore necessary to find a swifter method of computation. One possibility is to make use of the equations for total charge. This method was of little general use, but will be described briefly because it is of considerable theoretical interest. Equations (5·2) and (5·3) show that the total charge obeys simple exponential laws. It follows immediately that the total extrapolar charge, which is proportional to $\int_{-\infty}^{0} V_1 dx$, must also obey exponential charging laws. This quantity can be obtained by graphical integration of the potential in the extrapolar region and may then be plotted against time. The result of such an analysis is given in figure 12. Here the theoretical curve for the rise of a charge is drawn with a time constant of 2·02 msec. and for the fall with a time constant of 1·65 msec. The charging process obviously agrees closely with theory, but there is a definite deviation in the process of discharge. Further, the time constant for the charging process agrees with that found previously (2·10 msec.), whereas the discharge constant is appreciably smaller. The reason for these discrepancies is not clear, but they may arise from an apparently trivial circumstance. During the charging process the potential is relatively large and occupies a small area, whereas the converse situation holds during the period of discharge (see figures 2 *a*, *b*). This means that graphical integration is much less susceptible to cumulative errors in the former case than it is in the latter. The discharge curve may therefore be a less reliable index of the behaviour of the nerve than the corresponding charging curve. Whatever the explanation, this method will not be pursued further, since it proved too laborious for use in more than one experiment.

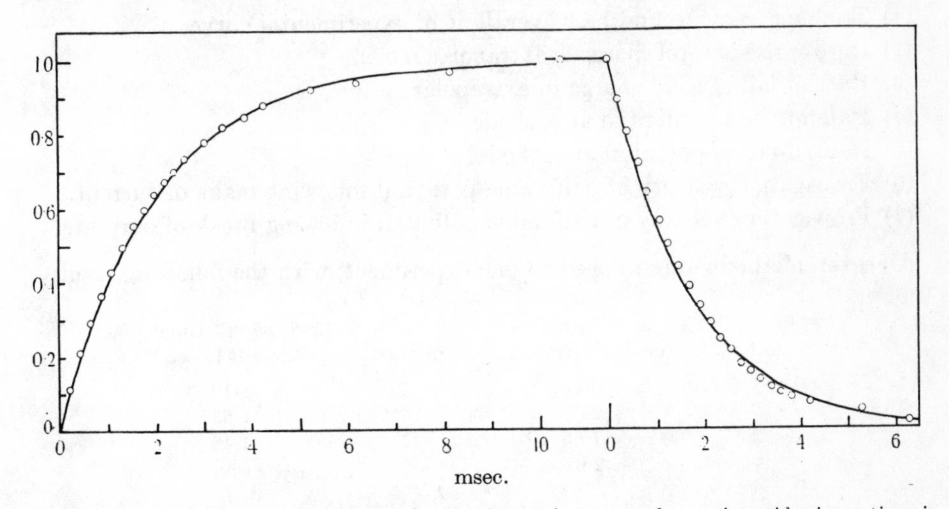

FIGURE 12. Time course of total membrane charge in extrapolar region. Abscissa: time in msec. Ordinate: $\int_{-\infty}^{0} V_1 dx$ in arbitrary units. The circles are experimental points computed by graphical integration of photographic records from experiment 10. The solid line is a plot of equations (5·2) and (5·3) with $\tau_m = 2{\cdot}02$ and 1·65 msec. for the rise and fall, respectively.

A simple method of measuring the membrane time constant is to ignore all observations except those at the cathode and find the time constant by comparing a single theoretical curve with the correct equation ((6·3) or (6·5)). This can be done for both charging and discharging processes and has the advantage of simplicity. But it suffers from two serious disadvantages. In the first place it ignores a great deal of valuable information, and in the second it is liable to magnify the errors which arise from the finite width of the electrode. The general effect of electrode width is to lengthen the apparent time constant; for the effective cathode occurs on the interpolar side of the electrode and the effective recording point on the extrapolar side. With a cathode of width $\lambda/10$ the apparent time constant should be 10 % larger than the true time constant. But no worth-while correction can be made, since the exact current distribution at the electrode is unknown. Time constants measured by this method should therefore be regarded as only approximately correct.

Another method of measuring the time constant depends upon a remarkable property of equations (6·0) and (6·4). If the time to reach half-maximum is plotted against distance, a curve is obtained which is very nearly a straight line with a slope of $2\lambda/\tau_m$. An alternative statement of this result is that the half-value potential propagates at a constant velocity of $2\lambda/\tau_m$.* Since λ is known, τ_m can be obtained by measuring the velocity of propagation from the experimental records.

Seven methods of measuring τ_m from the experimental data have now been described:

(1) Trial and error to find best overall fit of experimental curves.
(2) Rate of rise of total charge in extrapolar region.
(3) Rate of fall of total charge in extrapolar region.
(4) Rate of rise of potential at cathode.
(5) Rate of fall of potential at cathode.
(6) Propagation velocity of half-value potential following make of current.
(7) Propagation velocity of half-value potential following break of current.

All seven methods were applied to one experiment, with the following results:

method	apparent time constant (msec.)	method	apparent time constant (msec.)
1	2·10	5	2·10
2	2·02	6	1·63
3	1·65	7	1·40
4	2·67	average	1·93

The last four methods were applied as a routine procedure to all the experiments, with results which are shown in table 3. The agreement between different methods is often poor and the variations seem to be entirely random in nature. But there is little doubt as to the order of magnitude of the time constant, and it is this that is of interest at the moment.

* A footnote in Bogue & Rosenberg's (1934) paper suggests that this relation was known to Cremer.

TABLE 3. VALUES OF MEMBRANE TIME CONSTANT (τ_m) OBTAINED BY FOUR DIFFERENT METHODS

experiment number	τ_m in msec. determined by method 4	5	6	7	τ_m in msec. average
1	1·97	1·71	1·28	1·40	1·6
2	1·28	1·49	2·48	1·96	1·8
3	3·09	2·34	2·49	1·80	2·4
4	3·85	6·22	6·00	5·40	5·4
5	4·51	3·91	3·33	3·06	3·7
6	2·90	2·67	5·64	4·86	4·0
7	3·46	3·11	3·28	3·18	3·3
8	0·81	0·84	1·81	1·59	1·3
9	0·73	0·91	0·78	0·63	0·76
10	2·67	2·10	1·63	1·40	1·9
11	1·24	1·22	0·65	0·54	0·91
12	1·17	1·36	0·56	0·49	0·89
13	2·43	3·50	1·82	2·12	2·5
average	2·31	2·40	2·44	2·19	2·34

The relative magnitude of internal and external resistances

The ratio of the internal to external resistance per unit length (r_2/r_1) is important, because it allows us to estimate the absolute magnitude of potential changes at the nerve membrane. Equation (1·4) was derived without reference to the properties of the surface membrane, and it may therefore be applied to any region of nerve which does not form part of an external circuit. In general

$$\text{(potential change at membrane)} = \text{(potential change recorded externally)} \times (1 + r_2/r_1).$$

r_2/r_1 was obtained from the experimental results by equation (8·3) and calculated values are given in table 2. Action potentials were measured in five of these experiments, and the absolute magnitude of the electrical change at the surface membrane could therefore be estimated. The average value for the membrane action potential was found to be 110 mV and the extremes 135 and 87 mV. This result is in good agreement with the direct measurements which have been made with a micro-electrode in squid axons (Curtis & Cole 1942; Hodgkin & Huxley 1939).

The measurement of r_2/r_1 was subject to a small systematic error. In the theory it was assumed that the electrode was infinitesimal in width, whereas it actually had an effective width of 100–150μ. The measured value for r_2/r_1 would therefore exceed the true value by an amount which we estimate roughly at 10 %.

The axoplasm resistivity (R_2)

The resistivity of the axoplasm can be computed by equation (8·0):

$$R_2 = \pi a^2 m(1 + m\lambda/2y).$$

It would be unwise to expect great accuracy or consistency in the calculated value of R_2, since four separate measurements enter into its determination, and the final result is subject to the errors which arise from the assumption of infinitesimal electrode width. A rough estimate of the total error in R_2 is that it amounts to ± 30 %. Table 2 shows that the average value of R_2 was $60{\cdot}5\Omega$ cm. and the limits $43{\cdot}2$ and $83{\cdot}6\Omega$ cm. The average value of the axoplasm resistivity was, therefore, about three times as great as that of the surrounding sea water. This result is similar to those obtained for other cells. Measurements with transverse electrodes gave an average value of four times sea water for the resistivity of squid axoplasm (Cole & Curtis 1938), and observations with axial electrodes an average of 1·4 times sea water for the same material (Cole & Hodgkin 1939). Red and white blood corpuscles have a resistivity of twice plasma, frog's sartorius muscle one of about three times Ringer and various echinoderm eggs a resistivity of four to eleven times sea water (for references see Cole & Cole (1936) and Bozler & Cole (1935)).

The membrane resistance

The resistance × unit area of the surface membrane is determined by equation (8·1):

$$R_4 = 2\pi a\lambda^2 m(2 + m\lambda/2y + 2y/m\lambda).$$

Table 2 shows that the ratio r_2/r_1 which is equal to the factor $m\lambda/2y$ usually lies between $\frac{2}{3}$ and $\frac{3}{2}$. This means that a large error in $m\lambda/2y$ will have only a small effect on R_4. Suppose, for example, that the true value of $m\lambda/2y$ is 1·0 and that it is measured as 1·5. In the first case the factor in brackets in (8·1) would be 4·0 and in the second 4·17; hence the error in R_4 would only be 4 %. A similar line of argument shows that the measured value of R_4 will only be very slightly affected by the assumption of infinitesimal electrode width. The accuracy of the R_4 determination is, therefore, primarily controlled by the measurement of λ^2, a and m. The errors in λ^2 are likely to be of the order of ± 30 %, and almost certainly swamp the errors in a and m. A conservative estimate of the accuracy of the measurements in table 2 is that the values given for R_4 are correct to within 50 %. The observed variation was much greater than this, and successive measurements on one axon showed that the membrane resistance declined progressively during the course of an experiment. Thus axons 6 and 8 had initial resistances of 7330 and 1590Ω cm.^{2}, while their final resistances were 2720 and 706Ω cm.^{2}. The variable properties of the surface membrane mean that an average or standard value cannot be given for its resistance. All that can be said is that axons with resistances varying from 600 to 7000Ω cm.^{2} are capable of conducting nervous impulses in a normal manner. It is equally impossible to estimate the value of the membrane resistance in the living animal. The natural membrane resistance is not likely to be less than that found *in vitro*, but it may be much higher since Blinks's (1930) work on *Valonia* indicates that the surface resistance falls when cells are handled.

The values for R_4 given in table 2 are considerably larger than those recorded in the squid axon. Cole & Hodgkin (1939) reported values ranging from 400 to 1100Ω cm.^{2} on the basis of resistance-length measurements with direct current, while Cole & Baker (1941*a*) obtained an upper limit of 200Ω cm.^{2} from measurements with a.c. and transverse electrodes. On the other hand, Cole & Curtis (1941) give an average value of only 23Ω cm.^{2} from measurements with an internal electrode and d.c. pulses. Finally, Cole & Baker (1941*b*) calculated a value of 350Ω cm.$^{-2}$ from the result of a.c. measurements with longitudinal electrodes and the assumption of a membrane capacity of 1·1μF cm.^{2}. Cole (1941) appears to regard 300Ω cm.^{2} as a more or less average value. The low value of 23Ω cm.^{2} was attributed by Cole & Curtis (1941) to the poor physiological condition of impaled axons, but as they point out it may also have been due to the fact that two constants required in the analysis were assumed and not measured. In any case, there seems to be no doubt that the membrane resistance of 75μ lobster axons is several times larger than it is in 500μ squid axons. This difference may have some functional significance, since the rate of attaining ionic equilibrium tends to increase with surface-volume ratio, other things being equal. The membrane resistance would therefore need to decrease as the diameter increased if the cell economy demands a constant rate of approach to equilibrium.

The values of membrane resistance encountered in our work suggest that the permeability to ions must be rather low. Some idea of this may be gained by supposing that potassium ions alone can diffuse through the membrane and that permeability is studied by replacing the potassium in the external solution with a radioactive isotope. In this case it is fairly easy to show that approximately 30 min. would elapse before an 80μ fibre with a membrane resistance of 7000Ω cm.^{2} reached a state in which one-tenth of its internal potassium was replaced by the radioactive isotope. It would be interesting to see whether the rate of penetration of potassium is of this general order of magnitude.

Our values for the membrane resistance may be compared with those obtained by Dean, Curtis & Cole (1940) on artificial films containing lipoid and protein molecules. These films were of the right electrical thickness, since their capacity was about 1μF cm.$^{-2}$, but their electrical resistance was only 50–100Ω cm.^{2}. It is too early to try to correlate this difference with chemical structure, but there is some hope that future work will show what sort of structure is needed to produce a membrane of high resistance.

The magnitude of the membrane capacity

The membrane capacity was determined by the relation

$$C_M = \tau_m / R_4.$$

Both τ_m and R_4 are subject to large errors, so that little confidence can be placed on the exact numerical values obtained for C_M. In fact, it is possible that the variation encountered in table 2 was entirely due to experimental error. But there can be little doubt that the membrane capacity was of the order of 0·5–2·0μF cm.$^{-2}$. A value of this kind has been obtained in a wide variety of living cells; well-known examples are red blood cells 0·95μF cm.$^{-2}$, yeast 0·60μF cm.$^{-2}$, echinoderm eggs

0·87–3·1 μF cm.$^{-2}$, frog's sartorius muscle *c.* 1 μF cm.$^{-2}$, squid nerve 1·1 μF cm.$^{-2}$, and *Nitella* 0·94 μF cm.$^{-2}$ (for references and qualifications see Cole 1940).

All these results depend on the use of a.c., transverse electrodes and a theory based on Maxwell's application of Laplace's equation to a suspension of spheres. Our observations were made with pulses of d.c., longitudinal electrodes and a theory based on Kelvin's equations for the submarine cable. So it is pleasing to find even a broad agreement between the two sets of results.

The implications of the membrane capacity of 1 μF cm.$^{-2}$ are too well known to be repeated. All that need be said is that the result suggests the presence either of a very thin membrane, or of one with a large dielectric constant. If the dielectric constant were 3, the membrane thickness would be 27 A; and if the thickness were 1 μ the dielectric constant would be 1100.

Possible membrane inductance

Cole & Baker (1941*b*) have presented experimental evidence which suggests that an inductive element is present in the surface membrane of the squid axon. No sign of inductive behaviour could be observed in the majority of our experiments. But the two sets of observations do not conflict in spite of the apparent contradiction. Cole & Baker's axons had a membrane resistance of about 300Ω cm.^{2}, ours an average of 2300Ω cm.^{2}. The effect of an inductive element would have been profoundly influenced by the value of the membrane resistance, since Cole & Baker's work indicates that the two elements are in series. To take a specific example: assume that the membrane has the equivalent circuit suggested by Cole & Baker, that the capacity is 1 μF, the inductance 0·2 H and the resistance 300Ω cm.^{2}. When a rectangular current is applied to this circuit, the voltage response is oscillatory and the first overshoot is 75 % greater than the final steady value. The response is entirely different if the resistance is increased to 2500Ω cm.^{2}. In this case the wave form is no longer oscillatory, it does not overshoot the steady value, and it differs from a simple exponential solution by less than 0·2 %. The absence of inductive or oscillatory behaviour therefore agrees with Cole & Baker's hypothesis, although it clearly cannot be used in evidence one way or the other. But some of the axons studied had low membrane resistances and should have shown signs of inductive behaviour, if Cole & Baker's picture is correct. This, in fact, is what happened. Figure 13*c* gives the response of an axon with a resistance of 700Ω cm.^{2} and shows that there is an overshoot of 5 %. There is no equation with which to compare this record, but a theory for total charge can be developed by the method used in deriving (5·1). The resulting expressions allow the membrane inductance (L) to be calculated from the overshoot and predict that the response will only be oscillatory when $L > R_4^2 C_m/4$. Experiments 9, 11 and 12 (table 2) showed a small overshoot and gave an average value of 0·3 H for the membrane inductance. No overshoot was observed in the remaining experiments, and this is to be expected since the factor $R_4^2 C_M/4$ always exceeded 0·4. Our results are therefore consistent with the existence of an inductive element of about 0·2 H cm.^{2}.

But there is an entirely different way of explaining the experimental facts and this must now be considered. In figure 13 *a* and *b* the current had been increased until it was of just threshold strength, which means that it was strong enough to produce propagated spikes on 50 % of occasions. The propagated response is shown by *a* and the critical local response by *b*. It is arguable that the local response is of the same general nature as the spike, and that the discontinuity in nerve arises because the response to a superthreshold shock is large enough to involve the whole fibre by local circuit action, whereas the subthreshold response cannot spread beyond the cathodic region. It is also arguable that the small overshoot produced by the weak current is of the same general nature as the larger overshoot produced by the threshold current. And the similarity of the two lower curves in figure 13 suggests rather strongly that a common process is involved. According to this train of reasoning the overshoot seen in figure 13*c* is to be regarded as a vestige of the normal action potential. In this case it cannot be considered as an inductive effect. For the process underlying the action potential must involve energy liberation by the nerve, whereas a pure inductive overshoot would not. The two theories are therefore quite distinct, although no attempt can be made to decide between them until there are precise concepts to replace the general notions of inductance and energy liberation.

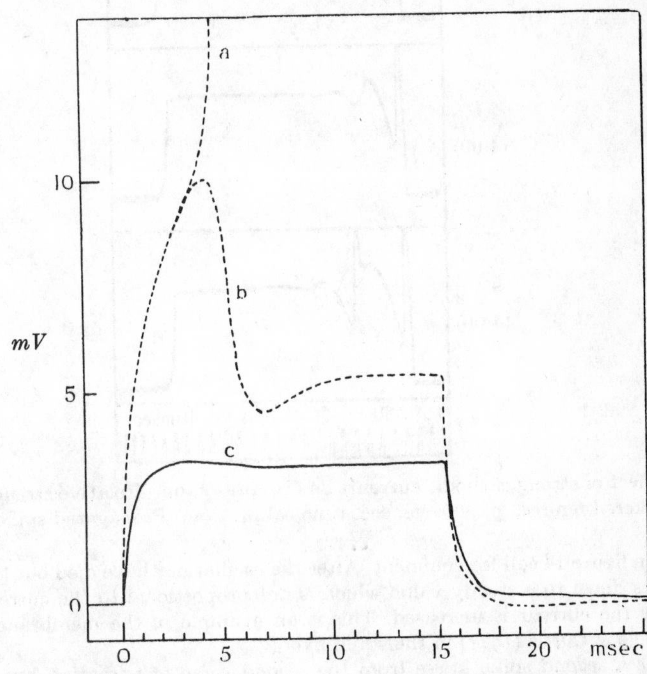

FIGURE 13. Potential recorded at cathode in axon with membrane resistance of 700Ω cm.² (experiment 11). *a*, propagated response produced by current of strength 1·00; *b*, local response produced by current of strength 1·00; *c*, potential produced by current of strength 0·49. The absolute values given on the ordinate are approximate, but the scale is linear.

The idea of a membrane inductance is certainly useful, whatever its ultimate truth or falsehood. One application was found in the attempt to explain the difference between the action potential and the resting potential (Curtis & Cole 1942; Hodgkin & Huxley 1945). Another is illustrated by figure 14, which shows the effect of strong cathodic currents on a *Carcinus* axon. The records indicate that the wave form of the cathodic potential becomes increasingly oscillatory as

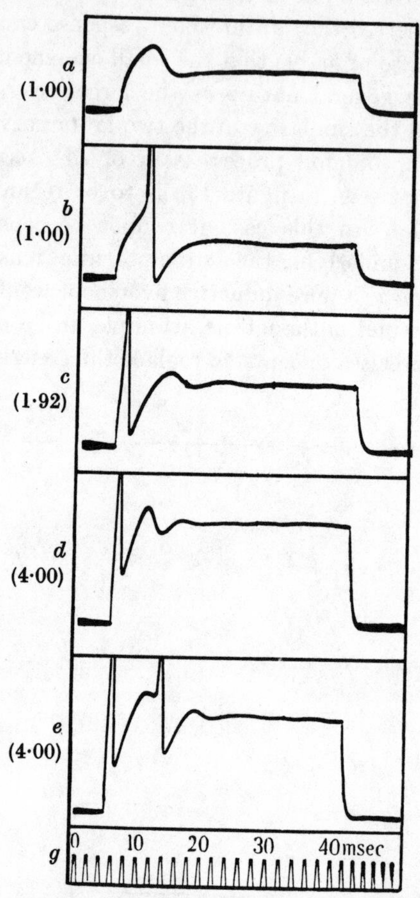

FIGURE 14. Effect of strong cathodic currents on *Carcinus* axon. Relative strength of current shown by bracketed figures. *g*, 500 cyc./sec. time calibration. Propagated spikes retouched.

Two points in figure 14 call for comment. After the oscillations have died out the membrane potential settles down to a steady value which is not proportional to the current but varies more slowly as the current is increased. This is an example of the membrane rectification described by Cole & Curtis (1941) in the squid axon.

In figure 14*e* a second spike arises from the second wave of potential, but a little after its crest. Hodgkin (1938) showed that a spike always started at a distance from the cathode when it arose later than the crest of the local response. Examination of the original records suggested that the same thing was occurring here, although no positive evidence for this conclusion was obtained.

the strength of current is increased. Similar results have been reported by Arvanitaki (1939) in *Sepia* and are to be expected from Cole & Baker's hypothesis. For the membrane resistance decreases progressively as the nerve is depolarized (Cole & Curtis 1941), and the response therefore becomes increasingly oscillatory as the current is raised. The frequency of the oscillations in figure 14 is consistent with an inductance of 0·3 H cm.^{2} and a capacity of 2μF cm.$^{-2}$.

Observations on the local response

The most striking features of the local response were the inflexion and miniature spike which occurred when the duration of the rectangular wave was less than the utilization time. One example has already been described (figure 5), and a more general picture is given by figure 15 which shows the effect produced by threshold pulses of different duration. A large number of photographs were taken and with two exceptions only the responses which just succeed or just fail to propagate have been reproduced. The effect of a current longer than the utilization time is given by c and C. In this case the record shows first the passive charging of the nerve membrane and then a slow creep, which must be regarded as a local response, since it is absent from the anodic wave form (a). If the local activity succeeded in reaching a critical level it turned upwards and gave rise to a propagated action potential. When the critical level was not reached the response died out as a monophasic wave of low amplitude. The form of the propagating responses was not very different when the duration of the rectangular wave was less than the utilization time (D to H), but the responses which failed to propagate showed the characteristic inflexion and miniature spike (d to h). This type of response persisted as the duration of the current was reduced, but at very short times it changed to that characteristic of excitation by short shocks (cf. Hodgkin 1938). An example is given in h and H, but the details of the record cannot be appreciated on the slow time scale used. This set of records suggests that the condition for excitation by currents of different duration is that a critical potential must be reached; they also illustrate the reversibility of the process responsible for the action potential. One is accustomed to think that nothing can stop an impulse once the potential has begun to turn upwards into a spike. Our records indicate that the potential wave may fail to propagate, although it has shown the inflexion normally associated with a spike.

The records which have just been described were obtained from a *Carcinus* axon and may be regarded as typical of this preparation. *Homarus* fibres behaved in a similar manner, but the utilization time was considerably shorter and the rheobasic local response had a more conspicuous humped form. We obtained the impression that the long utilization time and flat local response were associated with a high membrane resistance, and that axons with a low resistance gave the short utilization time and humped local response characterized by figure 13.

In comparing our results with those obtained in whole nerve trunks it should be remembered that the amplitude of the subthreshold potentials was small compared to the spike. Thus the propagated potential was ten times larger than the subthreshold potential shown in figure 15. The effects we have described would therefore be difficult to observe in preparations giving spikes only 100 μV in amplitude.

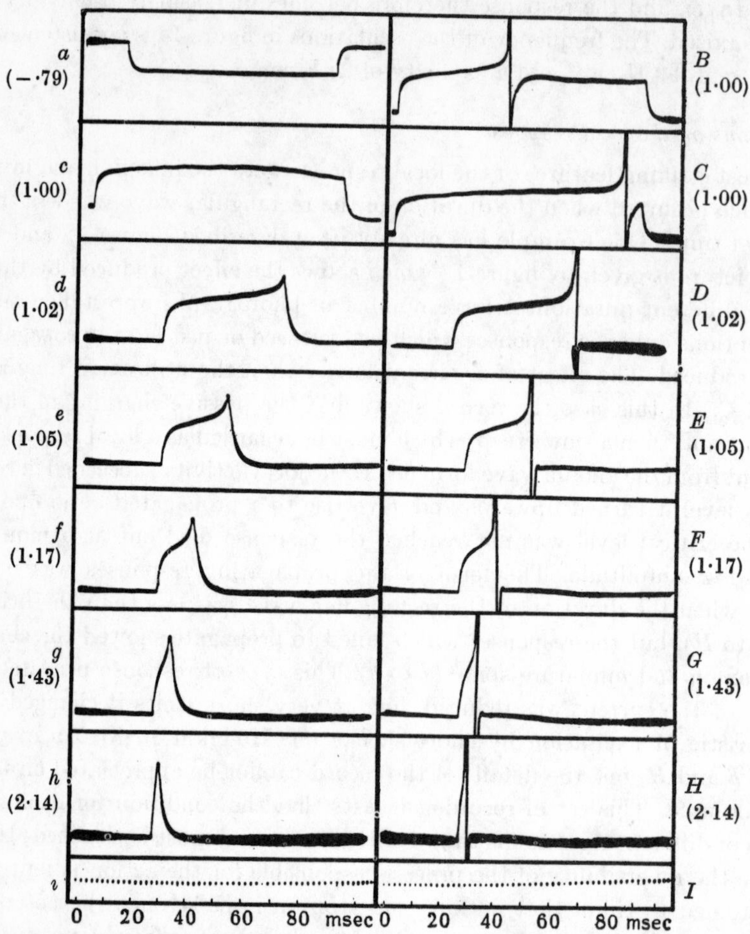

FIGURE 15. Effect of rectangular currents on *Carcinus* axon recorded at polarizing electrode. *a*, polarization produced by anodic current; *c–h*, local responses produced by threshold currents; *B–H*, propagated spikes (retouched) produced by threshold currents; *i*, *I*, 500 cyc./sec. time calibration. The strength of the current relative to the rheobase is indicated by the bracketed figures. The strength and duration of the current was identical in the pairs *c*, *C*, *d*, *D*, ..., *h*, *H*.

The change of membrane resistance during activity

A transient decrease of membrane resistance during activity has been proved by the well-known experiments of Cole & Curtis (1939). One result of this phenomenon is shown in figure 15*B*. Here the diphasic action potential arose before the end of the rectangular wave and lasted for about 2 msec. The membrane capacity should be discharged during the spike and must charge again when the resistance returns to its normal value. The spike should therefore be followed by a charging process similar to that which occurred at the beginning of the rectangular wave. This effect is clearly shown by record *B* but is absent when the action potential arises at the end of the applied current (*D* to *H*). In this case the charge disappeared rapidly during the spike and did not reform, because the applied current was removed.

Discussion

The implications of the resistance and capacity measurements have already been discussed. It remains to consider the bearing of our results on studies of electric excitation. The local responses observed in our experiments agree in a remarkable way with the instability described by one of us. Rushton (1932) studied the excitation process in medullated nerve by superimposing a short shock on a rectangular wave. A plot of excitability against time showed that the excitation process followed an inflected time course very similar to that observed in our records of local response. This is another example of the general similarity between the results of excitability studies on medullated nerves and the electrical records obtained in non-medullated nerve fibres. The phenomenon of latency and the excitability effects described by Rushton and by Katz (1937) all find an explanation in the electrical behaviour of isolated crustacean axons. The obvious conclusion is that similar electric effects exist in medullated axons, but that they are too small to be detected in studies of whole nerve trunks. This conclusion is not generally accepted and is likely to remain in dispute until satisfactory records can be obtained from an isolated medullated axon.

Hill (1936) and others have shown that many phenomena can be explained by supposing that the process of excitation is equivalent to the charging of a leaky condenser. This theory is useful in co-ordinating a wide range of observations, but extra assumptions have to be introduced to deal with the phenomena of accommodation, latency and the decay of excitability following a brief stimulus. Our results indicate that the processes underlying excitation are of great mathematical complexity. When the current is weak its spatial and temporal distribution is determined by the cumbersome equations of cable theory; when it is strong an immense complication is introduced by the non-linear effect of the local response. Hill's equations must therefore be regarded as largely empirical in nature. But there can be no doubt that certain facts seem to agree better with Hill's theory than with the cable equations. To take a specific example. It is universally agreed that the criterion for excitation by short shocks is that a fixed quantity of electricity must flow through the nerve. This follows at once from Hill's theory, but not from the equations of cable theory. For the condition which allows a short pulse to produce a constant potential at the cathode in a cable-like system is that a pulse of constant energy must flow through the electrodes. This difficulty and others of a similar kind can be resolved in the following way. The condition for excitation seems to be that the cathodic response must reach a potential at which it can propagate through the nerve by local circuit action. It is easy to suppose that the criterion for propagation is related not to the membrane potential at the cathode but to the total membrane charge in the region of the electrode. In this case a constant quantity relation would be obtained and the behaviour of nerve would approximate to that of a leaky condenser in many respects. According to this view, Hill's 'local potential' is to be identified with the total charge in the electrode region and Hill's constant k with the membrane time constant. The true situation is obviously much more complicated, but this hypothesis provides a simple and convenient way of looking at the excitation process.

We wish to express our indebtedness to the Rockefeller Foundation for defraying the expenses associated with this work and to Professor Gray for allowing us to use the aquarium in the Zoological Laboratory.

References

Arvanitaki, A. 1939 *Arch. int. Physiol.* **49**, 209.
Blinks, L. R. 1930 *J. Gen. Physiol.* **13**, 361.
Blinks, L. R. 1937 *Trans. Faraday Soc.* **33**, 991.
Bogue, J. G. & Rosenberg, H. 1934 *J. Physiol.* **82**, 353.
Bozler, E. & Cole, K. S. 1935 *J. Cell. Comp. Physiol.* **6**, 229.
Campbell, G. A. & Foster, R. M. 1931 *Fourier Integrals for practical applications.* Bell Telephone system Technical publications monograph, B. 584, 162.
Cole, K. S. 1937 *Trans. Faraday Soc.* **33**, 966.
Cole, K. S. 1940 *Cold. Spr. Harb. Symp. Quant. Biol.* **8**, 110.
Cole, K. S. 1941 *J. Gen. Physiol.* **25**, 29.
Cole, K. S. & Baker, R. F. 1941*a* *J. Gen. Physiol.* **24**, 535.
Cole, K. S. & Baker, R. F. 1941*b* *J. Gen. Physiol.* **24**, 771.
Cole, K. S. & Cole, R. H. 1936 *J. Gen. Physiol.* **19**, 609.
Cole, K. S. & Curtis, H. J. 1938 *J. Gen. Physiol.* **22**, 37.
Cole, K. S. & Curtis, H. J. 1939 *J. Gen. Physiol.* **22**, 649.
Cole, K. S. & Curtis, H. J. 1941 *J. Gen. Physiol.* **24**, 551.
Cole, K. S. & Guttman, R. M. 1942 *J. Gen. Physiol.* **25**, 765.
Cole, K. S. & Hodgkin, A. L. 1939 *J. Gen. Physiol.* **22**, 671.
Cremer, M. 1899 *Z. Biol.* **37**, 550.
Curtis, H. J. & Cole, K. S. 1938 *J. Gen. Physiol.* **21**, 757.
Curtis, H. J. & Cole, K. S. 1942 *J. Cell. Comp. Physiol.* **19**, 135.
Dean, R. B. 1939 *Nature*, **144**, 32.
Dean, R. B., Curtis, H. J. & Cole, K. S. 1940 *Science*, **91**, 50.
Fricke, H. & Morse, S. 1925 *J. Gen. Physiol.* **9**, 153.
Hermann, L. 1905 *Pflüg. Arch. ges. Physiol.* **109**, 95.
Hill, A. V. 1936 *Proc. Roy. Soc.* B, **119**, 305.
Höber, R. 1910 *Pflüg. Arch. ges. Physiol.* **133**, 237.
Hodgkin, A. L. 1938 *Proc. Roy. Soc.* B, **126**, 87.
Hodgkin, A. L. & Huxley, A. F. 1939 *Nature*, **144**, 710.
Hodgkin, A. L. & Huxley, A. F. 1945 *J. Physiol.* **104**, 176.
Jeffreys, H. 1931 *Operational methods in mathematical physics*, 2nd ed. Camb. Univ. Press.
Katz, B. 1937 *Proc. Roy. Soc.* B, **124**, 244.
Nernst, W. 1908 *Pflüg. Arch. ges. Physiol.* **122**, 275.
Pumphrey, R. J., Schmitt, O. H. & Young, J. Z. 1940 *J. Physiol.* **98**, 47.
Rushton, W. A. H. 1932 *J. Physiol.* **75**, 16*P*.
Rushton, W. A. H. 1934 *J. Physiol.* **82**, 332.
Schmitt, O. H. 1938 *J. Sci. Instrum.* **15**, 24.
Warburg, E. 1899 *Ann. Phys., Lpz.*, **67**, 493.

PRINTED IN GREAT BRITAIN AT THE UNIVERSITY PRESS, CAMBRIDGE
(BROOKE CRUTCHLEY, UNIVERSITY PRINTER)

J. Physiol. (1949) 108, 315–339

612.816.3

EVIDENCE FOR SALTATORY CONDUCTION IN PERIPHERAL MYELINATED NERVE FIBRES

BY A. F. HUXLEY AND R. STÄMPFLI

From the Physiological Laboratory, University of Cambridge, and the Physiological Institute, Berne

(*Received* 12 *June* 1948)

Lillie (1925) suggested that, in myelinated nerve fibres, excitation and the processes which maintain the propagated action potential take place only at the nodes of Ranvier. On this view, the myelin is an insulator, and its function is to increase the conduction velocity by making the local circuits act at a considerable distance ahead of the active region. Much evidence in favour of this theory has accumulated since that date. Thus, many agents which cause stimulation or affect conduction have a stronger action at the nodes than in the internodal regions. This has been shown for electrical stimulation (Kubo, Ono & Toyoda, 1934; Tasaki, 1940), for blocking by electrical polarization (Erlanger & Blair, 1934; Takeuchi & Tasaki, 1942), and for blocking by various ions, ion-free solutions and narcotics (Erlanger & Blair, 1934, 1938; Tasaki, Amikura & Mizushima, 1936). Tasaki & Takeuchi (1941) obtained action currents from a short length of an isolated fibre between two narcotized regions if, and only if, the unnarcotized stretch contained a node of Ranvier. Pfaffmann (1940) obtained larger action potentials from nodes than from internodal regions.

These results all support the theory of saltatory conduction, but there are two respects in which the evidence they provide is not compelling. In the first place, they are consistent also with the hypothesis that only the axis cylinder is concerned with conduction, and that it is shielded by the myelin against external agents at all points except the nodes. In the second place, all except Pfaffmann's results refer only to the initiation or blocking of an impulse, and not directly to normal conduction. If, however, the results are taken together with the evidence that the impulse is propagated by local circuits (e.g. Hodgkin, 1937*a*, *b*, 1939; Tasaki, 1939) they provide strong evidence in favour of saltatory conduction. On the other hand, there are certain difficulties which have prevented the theory from being universally adopted. Thus, Sanders & Whitteridge (1946) conclude that conduction velocity does not depend on the spacing of the nodes, while a simplified theory of saltatory conduction (Offner, Weinberg & Young, 1940) predicts that the velocity will increase with node spacing. Another difficulty is that, according to many authors (e.g. Maximow & Bloom, 1942; Grundfest, 1947), the myelin sheaths of fibres of the central nervous system are uninterrupted except at points where the fibres branch. If this is true, the saltatory theory cannot apply to central fibres. Bielschowsky (1928), however, insists that many authors have described interruptions in the sheaths of central fibres, and regards their existence as certain. But whichever view may be correct, this point cannot be decisive in a question which concerns peripheral nerve fibres, since it is still possible, though unlikely, that the mechanism of conduction is different in the myelinated fibres of the central and peripheral nervous systems. The other objection is likewise an indirect inference, and cannot stand against direct evidence.

Fig. 1. Diagram illustrating principle of method.

On balance, the evidence seemed to favour the theory of saltatory conduction, but was not sufficiently direct for a certain conclusion to be reached. The object of the main experiment described in this paper (already briefly reported elsewhere, Huxley & Stämpfli, 1948) was to test the theory further by observing the distribution of current around a single fibre during the passage of an impulse. The principle of the method was suggested by Mr A. L. Hodgkin of Cambridge, who pointed out that, if current can enter or leave the axis cylinder only at the nodes of Ranvier, the current along the axis cylinder must be the same at all points in any one internode at any one moment. If a single fibre is used and the recording system passes no appreciable current, the longitudinal current outside the fibre must be equal and opposite to that in the axis cylinder. This external current can be detected by raising the external resistance over a short length of the fibre, and amplifying the potential difference which is developed across this resistance (Fig. 1). If this recording stretch can be made short compared with the length of an internode, the longitudinal current can be observed at different positions in each internode by moving the recording stretch along the fibre. Records from different positions in one internode should then be identical, while records from different internodes should be similar in form but displaced in time.

In addition, a simple experiment giving further evidence that the impulse is transmitted by local circuits is described.

METHOD

Preparation of single nerve fibres. Single myelinated fibres were isolated from the sciatic nerves of large specimens of *Rana temporaria* and *R. esculenta* by the method described by Kato (1934) and modified by one of us (Stämpfli, 1946). A few further modifications were introduced. Thus, dark-ground illumination was employed, making the fibres clearly visible whatever their direction. The oblique illumination from above that was previously used only showed up fibres that ran nearly perpendicularly to the direction of illumination. Also, the motor branch from which the fibre was to be isolated was not separated from the sensory branch with which it runs. This eliminated a difficult step in the preparation, and considerably reduced the time required.

Fibres were usually isolated for about 15 mm. After the dissection the nerve trunk was moved so that the fibre lay straight on the slide, which was placed on an ordinary microscope. The positions of the nodes of Ranvier were determined by means of the mechanical stage, and the external diameter of the fibre was measured with an eyepiece micrometer, using a 4 mm. objective and 20 × eyepiece.

The distance between adjacent nodes was fairly regular in each fibre (within ±20% except for one or two instances), but the mean distance varied from about 1·5 to 3 mm. in different fibres. It appeared not to depend on fibre diameter, which usually lay between 12 and 15 μ., while one fibre had a diameter of 18 μ.

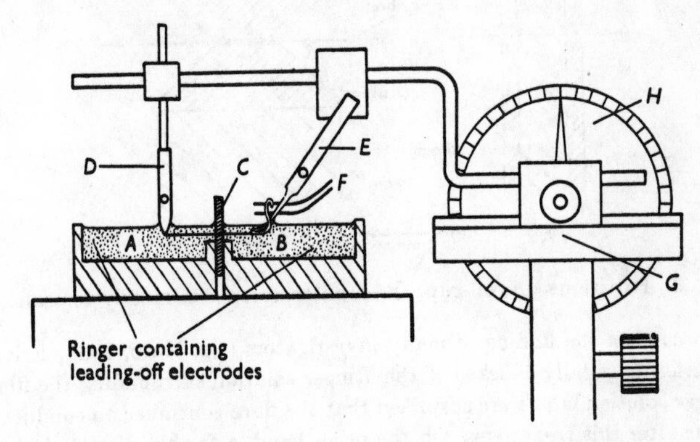

Fig. 2. General arrangement of apparatus. *A* and *B*, troughs cut in 'Perspex' blocks. *C*, partition. *D* and *E*, forceps. *F*, stimulating electrodes. *G*, micromanipulator. *H*, dial.

Apparatus. The general arrangement of the apparatus is shown in Fig. 2. The troughs *A* and *B* were cut in 'Perspex' blocks, and the channel between them was closed by a partition *C*. The fibre lay in the Ringer solution in the two troughs, passing through a hole in the partition. In order to draw the nerve fibre through the hole, a fibre of nylon or silk was pushed through, knotted round the distal end of the nerve fibre, and pulled back. The nylon or silk fibre was gripped in the forceps *D*, and the cut end of a branch of the nerve trunk was gripped in the forceps *E*. The free (central) end of the nerve trunk was lifted out of the Ringer solution and placed in contact with the stimulating electrodes *F*. These were made of silver wire, and were attached to the forceps *E*.

Both pairs of forceps were mounted on a bar carried on the horizontal movement of a micromanipulator *G*. Thus, by operating the rack and pinion, the fibre could be moved forward or back through the hole. Displacements of the fibre were measured on a dial *H* attached to the pinion shaft. The scale was divided to 0·1 mm., and intermediate values could be estimated to 0·01 mm. The forceps *E* could be moved along the bar by another rack and pinion (not shown) in order to get the fibre just stretched out. Excessive tension damaged the fibre immediately.

Trough *A* was fixed to a base-plate, while trough *B* could be moved by means of a screw. The partition was sealed in place by smearing vaseline on the opposed ends of the blocks, placing the partition in between with its hole in line with the open ends of the troughs, and bringing the blocks together with the screw.

The partition was designed so that the resistance to current passing between the troughs outside the fibre should be fairly high (0·5–10 megohms), but that the high-resistance part of the path should extend for only 0·4–0·8 mm. along the fibre. The two ways in which this was done are shown in Fig. 3. The earlier type of partition ('oil-gap') is represented by diagram A. It consisted of two coverslips, cemented together along three of their sides with spacing pieces. A hole of diameter between 80 and 400 μ. was drilled through both, and the space between them filled with liquid paraffin. The nerve appeared to move through the holes without being damaged by touching the sides. This type of partition proved to have the following disadvantages:

(a) The resistance of the film of Ringer solution outside the fibre was greatly affected by small pieces of connective tissue, etc., adhering to the fibre.

(b) The resistance of the film was so high (of the order of 10 megohms) that stray capacities distorted the record. The distortion was prevented by inserting an external shunt, but this must have affected the distribution of current crossing the sheath in the region surrounded by oil.

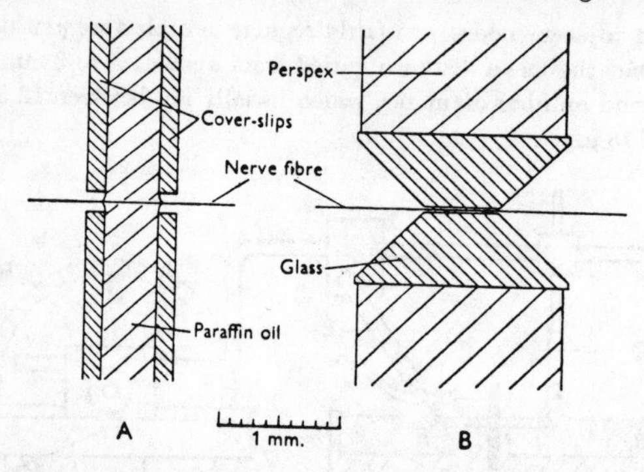

Fig. 3. Partitions. A, oil-gap. B, capillary. Approximately to scale.

(c) We tried to confirm the finding of many investigators (e.g. Kato, 1936; Erlanger & Blair, 1938) that conduction is rapidly blocked if the Ringer solution surrounding the fibre is replaced by an isotonic sugar solution, and were surprised that the fibre continued to conduct normally for about half an hour after this treatment. On the other hand, a freshly dissected fibre, which had not been in contact with paraffin oil, was blocked within 1 sec. We concluded that the oil had in some way hindered the diffusion of ions away from the film of fluid surrounding the fibre. If this was so, it was likely that the electrical properties of the fibre would also have been affected.

For these reasons the later experiments were made with partitions of the 'capillary' type shown in diagram B (Fig. 3). A piece of glass capillary tubing was drawn out so that its internal diameter was about 40 μ. Its external diameter was measured, and a hole of the same diameter was drilled through a piece of 'Perspex' sheet 1·7 mm. thick. The capillary was cemented into this hole and cut off flush with each side of the 'Perspex'. The two ends of the capillary were then opened out with a conical drill until only the central 0·5 mm. or so had the original diameter. This type of partition had a resistance of about 0·5 megohm, giving a rather low signal/noise ratio. Capillaries of smaller diameter were tried, but the fibres appeared to be damaged in passing through them.

Electrical recording system. We used the amplifier and cathode-ray oscillograph described by Hodgkin & Huxley (1945). This was built as a direct-coupled balanced amplifier. Since we were concerned only with rapid changes of potential, one of the stages was coupled by resistance and capacity with a time constant of about 0·5 sec. It was also found unnecessary to use it as a balanced instrument, and one of the inputs was connected to earth throughout these experiments.

The input stage was a cathode follower, placed with the grid cap of the valve within 5 cm. of the preparation. Fig. 4 A shows the circuit finally employed when the oil-gap partition was in use.

The effects of stray capacities were reduced by the following means. The capacity of the control grid to earth through the screen grid and anode was reduced by connecting the screen grid to the cathode through an h.t. battery of the appropriate voltage. The potential of the screen grid was thus made to follow the changes in potential of the control grid, so that practically no current passed through the capacity between them. The effect of the capacity of the live electrode to the stand and other earthed objects was similarly reduced by connecting the stand, micromanipulator, etc., to the cathode instead of to earth. Finally, the preparation was shunted by a resistance of 2·7 megohms, which brought the time constant down to about 20 μsec. Further reduction of the shunt resistance did not appear to affect the shape of the action-current record.

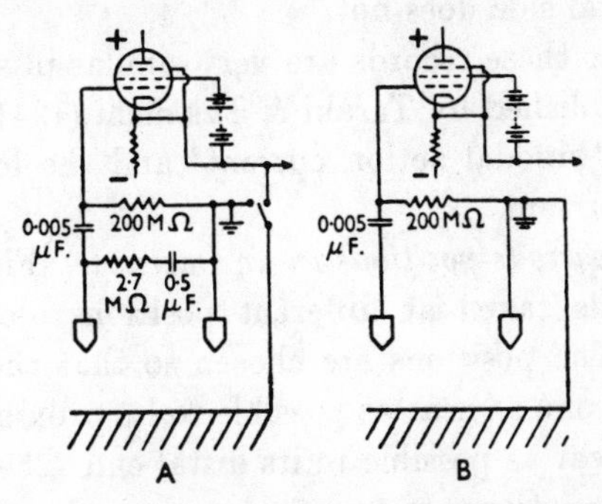

Fig. 4. Input circuits used (A) with oil-gap and (B) with capillary type of partition.

Condensers were inserted in the positions shown to prevent steady currents from passing through the preparation as a result of either the grid current of the valve or the residual potential difference between the electrodes. The grid leak was 200 megohms.

Fig. 4B shows the input circuit used with the capillary partition. The resistance of the preparation was only about 0·5 megohm in this case, so that the stray capacities did not produce serious effects.

The electrodes were pieces of silver sheet, coated electrolytically with silver chloride. The earth electrode was fixed in trough *B*, while the leading-off electrode was fixed in trough *A* (Fig. 2).

The nerve trunk was stimulated by short thyratron discharges at a frequency of about 40 shocks/min. throughout the experiment. The strength of the shocks was adjusted to about twice the threshold for the isolated fibre.

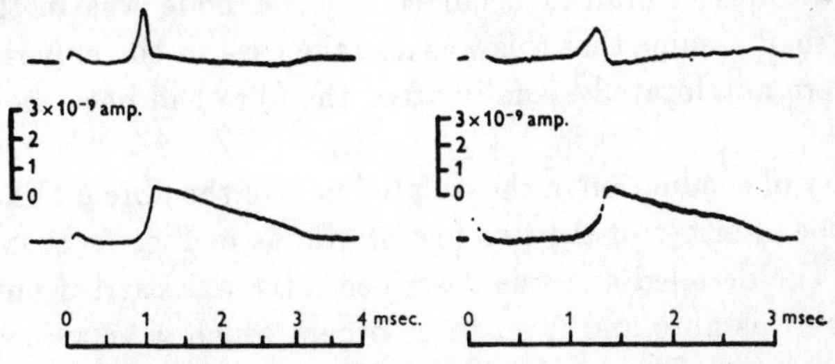

Fig. 5. Records of longitudinal current taken (left) with oil-gap and (right) with capillary type of partition. In each case, upper record taken just proximal to, and lower record just distal to, last working node.

RESULTS

Shape of action-current records. Typical records of the action current at about the middle of an internode are shown in Fig. 5. The left-hand pair of records was obtained with the 'oil-gap', the right-hand with the 'capillary' partition.

The upper record in each case is taken from a point on the fibre which is far enough from the damaged distal end to give a normal action current. The lower records are from points beyond which there is no activity, and the impulse is conducted decrementally for a few millimetres by electrotonic spread. On the theory of saltatory conduction, it would be said that, in the upper records, the node at each side of the recording stretch becomes active, while in the lower records the node of the proximal side of the recording stretch becomes active while that of the distal side does not.

It will be seen that these records are very similar in shape, amplitude and duration to those published by Tasaki & Takeuchi (1941). The upper records correspond to their 'binodal action current' and the lower records to their 'mononodal action current'.

Action currents at various positions in an internode. Fig. 6 shows a series of action-current records taken at different positions along a fibre with the capillary partition. The positions are chosen so that there are three records from each internode, one as near as possible to its proximal end, one near the middle, and one as near as possible to its distal end. There is never a node of Ranvier within the recording stretch.

It will be seen that the three records from any one internode are practically synchronous, while records from different internodes are displaced in time. This is also shown in Fig. 7, where the times of certain features of the first phase of the record, measured from the shock artefact, are plotted against distance. These conduction times increase discontinuously at certain definite positions on the fibre. This was an invariable finding, and the spacing between the discontinuities always agreed with the measured spacing between the nodes. In a number of cases the nodes were located with a microscope while the fibre was in the apparatus and records were being taken. It was then found that the discontinuities occurred when a node was in the recording stretch. We shall assume that this was also the case in the experiments where the nodes were not located visually after the fibre had been mounted in the apparatus.

The velocity of conduction in the isolated part of the fibre is the reciprocal of the mean slope of either of the two lowest graphs in Fig. 7. It is found to be 23·2 m./sec. The detailed analysis described later was carried out on records obtained from this fibre and from three others which gave velocities of 22·2, 24·3 and 23·1 m./sec. These values are normal for frog fibres of 12–15 μ. diameter, at temperatures of 18–20° C. (Erlanger & Gasser, 1937), indicating that the fibres cannot have been seriously damaged by the dissection and other experimental procedures. Most other fibres gave somewhat lower velocities (not less than 12 m./sec.). It is possible that these had been damaged, but in all qualitative respects the results they gave were similar to those described here.

So far, this is what would be expected if current entered or left the axis cylinder only at the nodes. But if that were actually the case, the records would be identical in shape and amplitude, as well as in time, at different positions in one internode. The records in Fig. 6 show that this is not the case. The amplitude

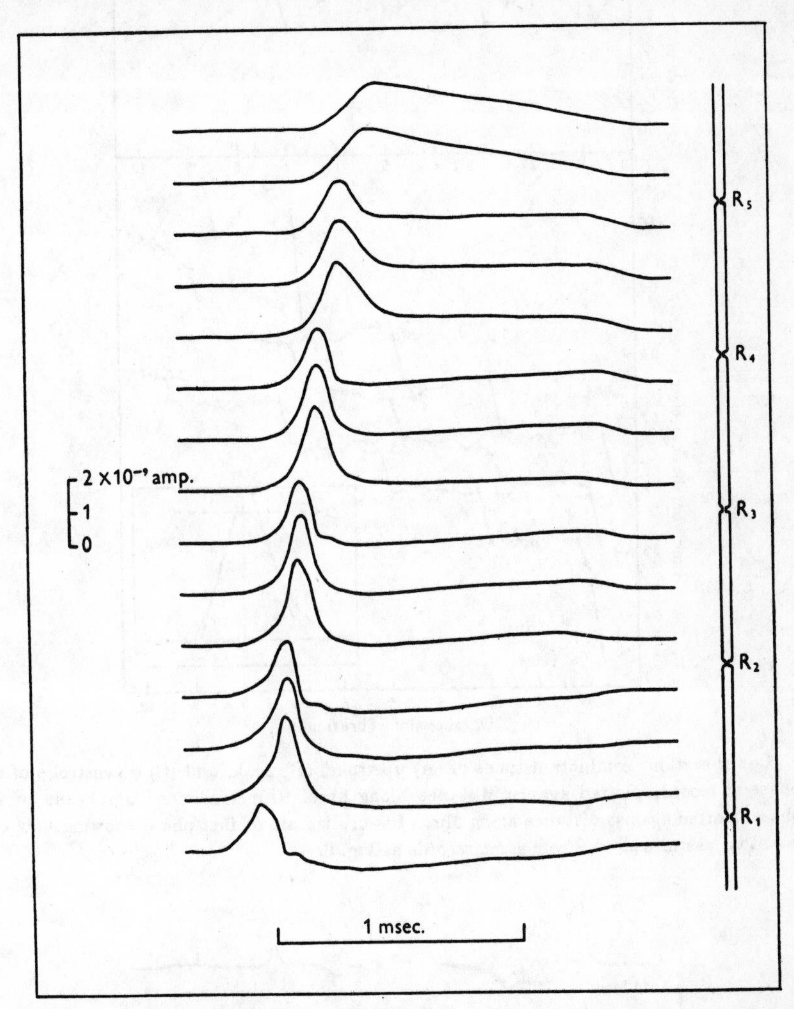

Fig. 6. Tracings of records obtained at a series of positions along one fibre, with capillary partition. Stimulus artefact has been subtracted. Diagram of fibre on right-hand side shows position where each record was taken.

of the first phase decreases from the proximal towards the distal end of each internode. This is better seen in the upper graph of Fig. 7. Also, the shape of the record is different at the different positions. The peak of the first phase is sharpest at the proximal end of each internode, while the angle where the record becomes flat at the end of the first phase is sharpest at the distal end of

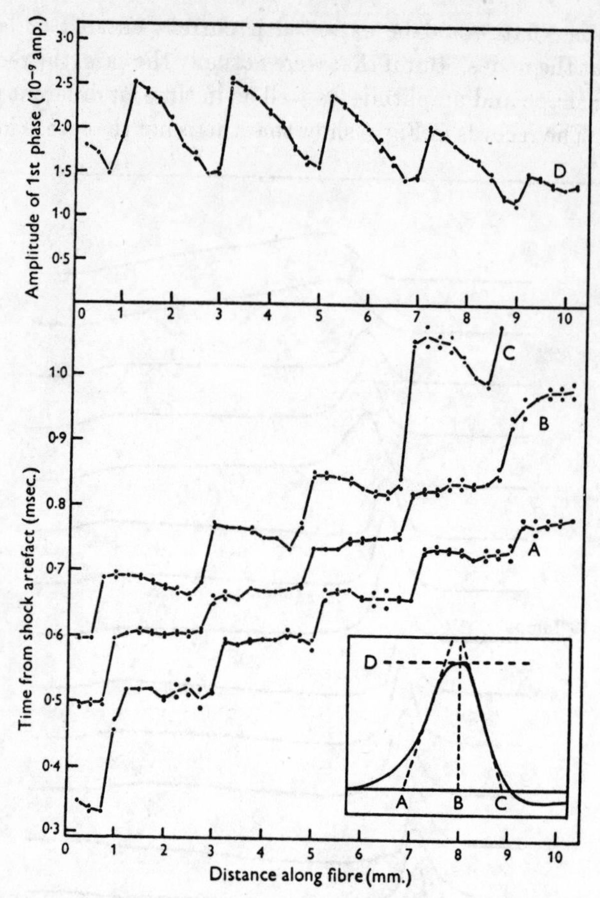

Fig. 7. Lower section: conduction times of (A) upstroke, (B) peak, and (C) downstroke of first phase of record, plotted against distance along fibre. Upper section: amplitude of first phase, plotted against distance along fibre. Inset: diagram of first phase showing how each quantity was measured. From same records as Fig. 6.

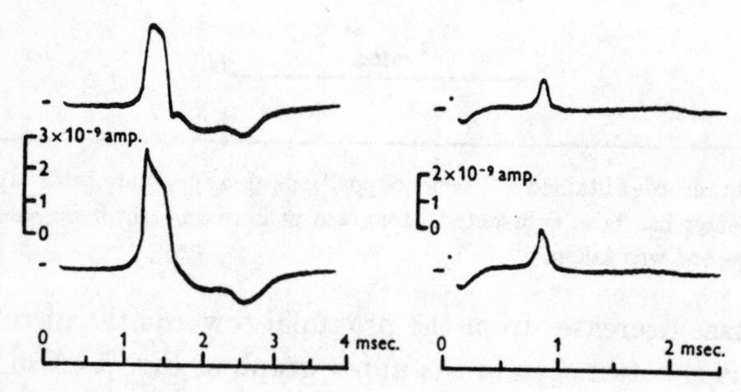

Fig. 8. Longitudinal current at proximal end (lower record) and distal end (upper record) of one internode. Left-hand pair of records obtained with oil-gap, right-hand with capillary type of partition.

the internode. Two pairs of records showing this difference clearly are shown in Fig. 8. These differences between the records obtained at different positions within one internode mean that some current does pass through the myelin sheath. Without further analysis these results are therefore not unequivocal evidence in favour of saltatory conduction. It might be, for instance, that conduction is extremely rapid in each internode, and that a delay occurs at each node because the membrane capacity and conductance are higher there than in the internodes.

The analysis required to clear up this point is carried out in the section entitled 'Determination of membrane current', and shows that the current through the myelin can be explained as a passive current driven through a resistance and a capacity in parallel by the change in potential in the axis cylinder. On this basis the fact that the graphs in Fig. 7 are not horizontal straight lines in each internode can be interpreted as follows. As regards the graph of amplitudes the potential in the axis cylinder is rising and causing current to flow outwards through the electrostatic capacity of the myelin during the first phase of the current record. The logitudinal current is directed forwards in the axis cylinder, so that outward current through the myelin makes the amplitude of the longitudinal current decrease from the proximal to the distal end of each internode.

The graphs of times, with the surprising feature that the descending phase occurs earlier at the distal than at the proximal end of the internode, are best understood by considering the spread of longitudinal current due to the potential change at one node. The rapid rise of potential in the axis cylinder at the node causes, in the axis cylinder of the internode on the distal side, an increase in forward current whose peak is roughly indicated by the peak of the first phase of the record. In the more proximal internode, however, it causes a decrease in the forward current whose peak is given approximately by the end of the first phase, or by the time which is plotted as graph C. Thus, graph C in one internode and graph B in the next more distal internode represent different aspects of the same disturbance spreading symmetrically from the node separating them. This spread takes place with a finite velocity (not necessarily constant), so that graph B becomes later, and graph C earlier, towards the distal end of each internode.

Results when recording stretch contains a node. Consider first the results that would be expected when the recording stretch contains a node, on the hypothesis that current enters and leaves the axis cylinder principally at the nodes. The potential difference across the recording stretch is built up partly by the current in the more proximal of the two internodes separated by this node, and partly by that in the more distal one. As a simplified case we shall first assume that the longitudinal current is the same at all points in one internode at any one moment. The situation is illustrated by Fig. 9.

Let s = length of recording stretch,
y = distance of node from proximal end of recording stretch,
i_a = longitudinal current in axis cylinder of more proximal internode,
i_b = longitudinal current in axis cylinder of more distal internode,
r_1 = resistance per unit length of fluid surrounding fibre,
v = potential on distal side − potential on proximal side.

Then v = potential drop between C and B + potential drop between B and A

$$= r_1(s-y)\,i_b + r_1 y i_a$$
$$= r_1 s\,(i_b + (i_a - i_b)\,y/s).$$

v is therefore a linear function of y, and is equal to $r_1 s i_b$ when $y=0$, and to $r_1 s i_a$ when $y=s$. If the recording stretch is shunted by a resistance, it is easy to show that the same result holds, except that the coefficient $r_1 s$ is replaced by the parallel resistance of $r_1 s$ and the shunt.

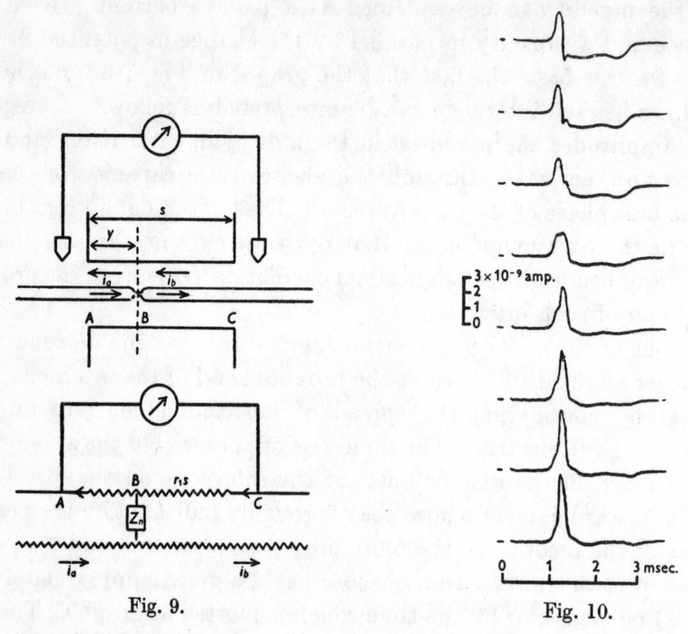

Fig. 9. Fig. 10.

Fig. 9. Diagram of situation when recording stretch contains a node. Z_n = impedance of membrane at node. For meaning of other symbols, see text. Proximal end of fibre is to the left.

Fig. 10. Records taken as a node passed through the recording stretch. Fibre moved 0·1 mm. between successive records. Oil-gap type of partition.

This relation cannot be expected to hold exactly in a real case for three reasons. The first is that some current does cross the myelin, so that i_a and i_b are not constants, but depend on distance from the node. The second is that unless r_1 is small compared with the resistance per unit length of the axis cylinder, the values of i_a and i_b at given positions in the fibre will change as the fibre is drawn through the recording stretch. The third is that r_1 may not be the same at all points on the fibre.

The first of these factors is probably unimportant. If i_a and i_b can be sufficiently represented as linear functions of the distance x along the fibre, then the expression for v contains a term in $y(s-y)$ proportional to $(di_a/dx - di_b/dx)$. The values of these differential coefficients can be obtained from records taken in the internodes, and calculation shows that the resulting deviation from linearity of the relation between v and y is negligible.

The second factor probably has an appreciable effect on records taken with the oil-gap, but not on those taken with the capillary, in which the value of r_1 is of the order of $\frac{1}{30}$ of the resistance per unit length of the axis cylinder.

The third factor probably also caused appreciable errors with the oil-gap but not with the capillary. With the oil-gap, the resistance of the external fluid film was affected by local variation in fibre diameter, and, probably more important, by connective tissue fibrils, etc., adhering to the fibre. This factor would be expected to cause irregularities in the relation between observed potential and distance also in the internodes; this is sometimes detectable in records taken with the oil-gap, but not with the capillary.

These sources of error are therefore probably not serious, but may cause deviations from the quantitative predictions of the simple theory when the oil-gap partition is used. We should thus expect that, as the node goes from one side of the recording stretch to the other, the measured potential will change steadily from its value in the proximal to that in the distal internode, but that the change may not be exactly linear, especially in records taken with the oil-gap partition.

Fig. 10 shows a series of records taken at various positions as a node of Ranvier passed through the recording stretch. In order to see whether the transition between the two forms of action potential takes place as predicted,

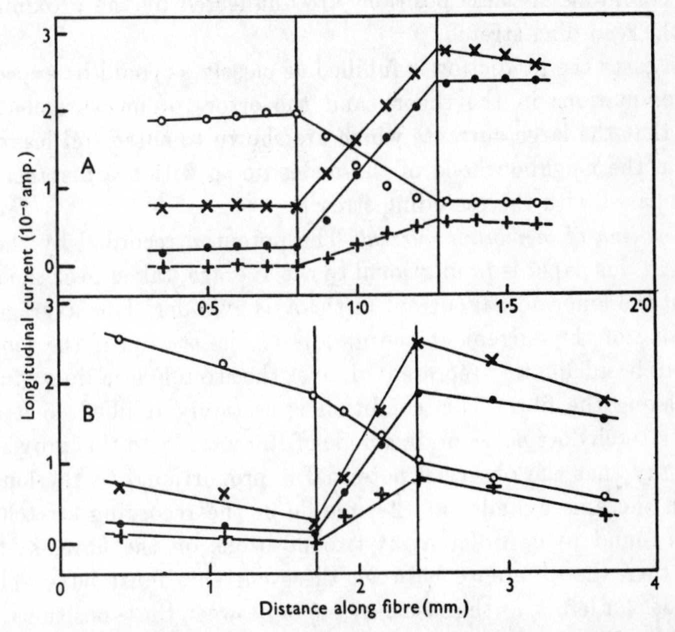

Fig. 11. Longitudinal current plotted against distance along fibre, as a node passes through recording stretch. Each graph corresponds to a constant time after the stimulus. A, oil-gap; B, capillary type of partition. ○—○, near peak of first phase in proximal internode; ×—×, near peak of first phase in distal internode; ●—●, during downstroke of first phase in distal internode; +—+, near end of first phase in distal internode. In A, scale of ordinates is only approximate.

the observed potential is plotted in Fig. 11 against distance along the fibre. Each graph in Fig. 11 corresponds to a particular time after the stimulus. The ordinates are the deflexions at that time in the records taken at varying positions as the fibre was drawn through the recording stretch. The records from which Fig. 11 A was made were taken with the oil-gap, at intervals of 0·1 mm. The exact length of the recording stretch could not be determined, as the menisci of the oil could not be seen. The positions at which the node enters and leaves the recording stretch are, however, clearly seen on the graphs, and on the simple theory the points in between these should lie on the straight lines joining the ordinates at those positions. Fig. 11 B was constructed from records taken with the capillary partition at intervals of 0·25 mm. The capillary was 0·60 mm. long, but end-effects would be expected to increase its apparent length to about 0·68 mm. The vertical lines in the figure are drawn at this distance apart, and on the simple theory, the points between them should lie on the straight lines which have been drawn, joining the ordinates on the vertical lines. By good fortune, records were obtained when the node was only just outside each end of the recording stretch. It is evident fron the graphs that the potentials at these positions are unaffected by the proximity of the node to the recording stretch.

In both cases the prediction is fulfilled as closely as could be expected from the approximations in the theory and the errors of measurement. This is evidence that the large currents which are shown to enter and leave the axis cylinder in the neighbourhood of the nodes do so within a distance which is short compared with the recording stretch.

Determination of membrane current. The potential recorded by the method described in this paper is proportional to the average, taken over the recording stretch, of the longitudinal current in the axis cylinder. This average is equal to the value of the current at the middle of the stretch if the longitudinal current can be adequately represented, over the stretch, as a linear function of distance along the fibre. This condition is certainly fulfilled so long as the recording stretch does not contain a node of Ranvier. With this proviso, we can therefore say that the observed potential is proportional to the longitudinal current in the axis cylinder at the middle of the recording stretch. If this current is found to be different at two positions on the fibre, at the same moment, then the difference between these currents must have entered the axis cylinder (or left it, as the case may be) in between these positions. We shall refer to this difference as the 'membrane current'.

Thus we can find the current entering or leaving the axis cylinder by taking the difference between the potentials recorded at two nearby positions on the fibre. With our apparatus it was not possible to lead off potentials simultaneously from two stretches of the nerve fibre. We therefore took a record at one position, moved the fibre, and took another record. We then took the difference between the potentials in these two records at the same time after the stimulus. This procedure may have introduced some errors, since the action currents at any one position on the fibre may not have been identical when the two records were taken. In particular, with the oil-gap method, the position of the recording stretch on the fibre probably affected the current distribution. This objection probably does not arise with the capillary method, since the resistance of the fluid

outside the fibre in the recording stretch was small compared with that of the axis cylinder. The results obtained with the two types of partition are very similar, but we shall rely chiefly on the capillary method both because of this objection to the oil-gap method, and because of the other objections mentioned under the heading 'Apparatus'.

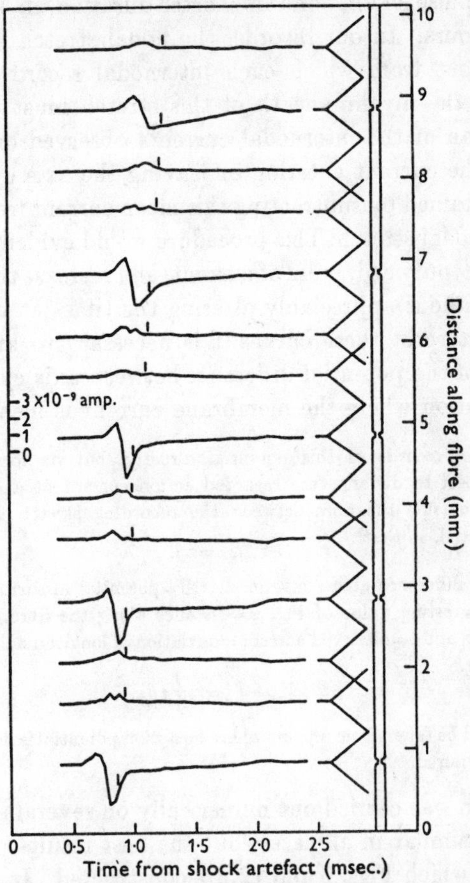

Fig. 12. Membrane currents. Each curve shows the difference between the longitudinal currents at two points 0·75 mm. apart on the fibre. The positions of those two points relative to the nodes is indicated on the diagrammatic fibre on the right. The vertical mark above each graph shows the time when the change in membrane potential reached its peak at that position on the fibre. Outward current is plotted upwards.

Fig. 12 shows a series of results obtained with the capillary partition. It is at once obvious that each graph which refers to a stretch of the fibre containing a node of Ranvier has a large inward component, while the other graphs do not. These curves correspond to the records obtained with tripolar recording by Tasaki & Takeuchi (1941 for nodal, 1942 for internodal stretch), as regards both the principle of the method and the form of the results. With the apparatus used by those authors, however, it was not possible to obtain series of records from different points on the same fibre. Also their recording stretches were longer than ours, so that a record containing the membrane current at a node contained also the current through the myelin sheath of a greater length of the fibre. This makes the first, outward, current pulse larger, and the second, inward, current pulse smaller and of shorter duration, in Tasaki & Takeuchi's records than in ours. In our records, the nodal stretch includes a length of fibre equal to that from which each internodal record was obtained. The current through the myelin sheath of this stretch must therefore be nearly equal to the mean of the internodal currents observed on either side of the nodal stretch. The current entering or leaving the axis cylinder at the node itself could be obtained by subtracting this mean current from the total current observed in the nodal stretch. This procedure would evidently decrease further the size of the first, outward, pulse of current, and increase the size of the second, inward, pulse, without appreciably altering the times at which they occur.

In order to interpret these curves it is necessary to know approximately the time course of the potential difference between axis cylinder and external fluid at each position where the membrane current is measured.

We did not obtain records of this potential directly, but its form can be obtained by integrating with respect to distance the recorded action current at a constant time after the stimulus. The fluid outside the fibre between the recording stretch and the distal end was practically equipotential, so that

$$\partial V_m/\partial x = r_2 i,$$

where V_m = potential difference across myelin sheath = potential in surrounding fluid − potential in axis cylinder, V_0 = resting value of V_m, x = distance along the fibre, r_2 = resistance per unit length of axis cylinder, and i = observed current = deviation of longitudinal current in axis cylinder from its resting value.

Hence

$$V_m - V_0 = r_2 \int i\,dx.$$

The integration should be taken from a point where no action potential is detectable to the position where the result is required.

This integration was carried out numerically on several sets of records. The conclusions were similar in all cases, but the best results were obtained with the records from which Figs. 6 and 12 are constructed. In this case there was some decremental spread of action current beyond the position where the most distal record was taken, and an allowance for this had to be estimated and added to the result of integration. The main conclusion which is drawn, however, is quite independent of these estimates.

The time at which the potential change across the membrane reached its maximum was read from the curve obtained for each position, and is marked on the corresponding membrane current curve in Fig. 12.

Each of the curves in Fig. 12 taken from a length of the fibre which does not include a node is closely similar to the current which would flow if the calculated membrane potential change were applied to a resistance and capacity in parallel. In the rising phase, the currents through both elements are outward, and add to give the definite outward pulse which is seen in the records. During the falling phase, however, the current through the capacity is inward, while that through the resistance is still outward. The net current in this phase is therefore small and may be either inward or outward, according to the rate at which the potential falls and the size of the product of the values of the resistance and capacity.

All these 'internodal' records can be explained in this way. In each record from a length of fibre which contains a node, however, there is a large inward pulse of current which begins and indeed reaches its maximum while the internal potential is still rising. This cannot be explained as a passive current being driven through any ordinary circuit element by the change in membrane potential, and must be regarded as current produced 'actively' by the nerve.

Determination of resistance and capacity of myelin. The time course both of the current through the myelin sheath and of the change in the potential difference across it were determined by the methods described in the previous section. The results suggested that the myelin behaves like a membrane with finite conductance and capacitance in parallel. If it is assumed that this interpretation is correct, and, further, that the conductance and capacitance are linear and constant at each point on a fibre, it should be possible to calculate the values of these circuit elements at each position where the current and potential difference curves have been found. This was done at a series of points on each of four fibres. The method used was as follows:

Let i_m = current entering axis cylinder per unit area of myelin,
G_m = conductance of myelin per unit area,
C_m = capacitance of myelin per unit area.

V_m, V_0, and r_2 have the same significance as in the preceding section.

Then
$$i_m = G_m (V_m - V_0) + C_m \frac{\partial V_m}{\partial t}.$$

Integrating this with respect to time,
$$\int_{t_1}^{t_2} i_m \, dt = G_m \int_{t_1}^{t_2} (V_m - V_0) \, dt + C_m \left[(V_m)_{t_2} - (V_m)_{t_1} \right].$$

To find the conductance, the integration was taken from the beginning of the action potential ('foot') to the time when the potential difference had returned to its initial value ('end'). Then
$$(V_m)_{t_1} = (V_m)_{t_2} \quad \text{and} \quad G_m = \int_{\text{foot}}^{\text{end}} i_m \, dt \Big/ \int_{\text{foot}}^{\text{end}} (V_m - V_0) \, dt.$$

To find the capacitance, the integration was taken from the beginning of the action potential to the time when the potential change was a maximum ('peak'). Then
$$(V_m)_{t_2} - (V_m)_{t_1} = \text{height of action potential},$$

and
$$C_m = \frac{\int_{\text{foot}}^{\text{peak}} i_m \, dt - G_m \int_{\text{foot}}^{\text{peak}} (V_m - V_0) \, dt}{\text{height of action potential}}.$$

The value used for G_m in this equation was that determined at the same position on the fibre.

Membrane currents were obtained by differencing the observed longitudinal current, as described in the previous section. Each value was divided by the area of the myelin from which the current was obtained in order to give i_m, the membrane current per unit area.

The values of $(V_m - V_0)$ and of the height of the action potential were determined by the integration method described in the previous section. The results of this method are proportional to the value taken for r_2, the longitudinal resistance per unit length of the axis cylinder. This was not measured, and the value used was calculated from the dimensions of the axis cylinder, using the specific resistance of Ringer solution. The values of the potential change thus obtained are too small by the factor α by which the specific resistance of the axoplasm exceeds that of Ringer solution. The values of conductance and capacitance must therefore be divided by this unknown factor α. The specific resistance of Ringer solution was taken as 94 ohm-cm. at 18° C., 92 ohm-cm at 19° C. and 90 ohm-cm. at 20° C. These values were based on data kindly provided by Dr B. Katz, of University College, London.

The results of this analysis are given in Table 1, together with other particulars of the data on which it was carried out.

TABLE 1. Results of complete analysis of membrane currents of four fibres. α = (specific resistance of axoplasm)/(specific resistance of Ringer solution, taken as 90 ohm-cm. at 20° C.)

Date of experiment ...	5. xii. 47	17. xii. 47	4. i. 48	7. i. 48
Type of partition used ...	Oil-gap	Oil-gap	Capillary	Capillary
Species of frog ...	*Esculenta*	*Temporaria*	*Esculenta*	*Temporaria*
External diameter of fibre (μ.)	14·5	15·0	14·5	12·0
Thickness of myelin assumed (μ.)	2·0	2·0	2·0	1·5
Temperature (° C.)	18	19	20	20
Conduction velocity (m./sec.)	22·2	24·3	23·1	23·2
Highest action potential (mV.)	106α	93α	79α	109α
Mean capacity of myelin sheath (μF./cm.^{2})	0·0023/α	0·0030/α	0·0035/α	0·0035/α
Dielectric constant of myelin	5·1/α	6·8/α	7·9/α	6·0/α
Resistance of myelin sheath (megohms-cm.^{2}) (reciprocal of mean conductance)	0·082α	0·109α	0·167α	0·158α
Specific resistance of myelin (ohm-cm. × 10^8)	4·2α	5·5α	8·4α	10·5α
No. of positions on fibre where resistance and capacity were measured	5	8	5	6
Standard error of mean capacity (% of mean)	34	10	9	3
Standard error of mean conductance (% of mean)	18	12	12	30

Experiment to test local circuit theory. Two microscope slides were supported end to end on insulating blocks screwed to an earthed base-plate. A single-fibre preparation was chosen in which stimulation of the nerve trunk, proximal to the isolated stretch, caused a visible contraction in the muscle, which was left attached to the nerve. The preparation was laid in a pool of Ringer solution on the slides, so that the muscle lay on one slide and the nerve trunk on the other, with the isolated fibre crossing the junction. The slides were drawn apart, making a gap of 1–2 mm. which was bridged by the isolated fibre. Care was taken that the part of the fibre in the air-gap should not contain a node of Ranvier. The pools of Ringer solution on the two slides could also be connected by laying a thread, moistened with Ringer solution, across the gap. The nerve trunk was stimulated by means of galvanic forceps. It was found that the muscle twitched when the nerve was stimulated if, but only if, the thread connecting the fluids on the two sides of the gap was in place.

The only effect of putting the thread in place was to make an electrical connexion between the two pools of Ringer solution. Conduction was thus impossible if the longitudinal resistance outside the fibre was above a certain value. This demonstrates that the transmission of the nervous impulse depends on currents flowing outside the myelin sheath, the circuit being presumably completed by the axis cylinder. In non-medullated nerve fibres, whose properties are the same at all points along their length, a change in external resistance affects only the velocity of conduction, and transmission should still be possible however high the external resistance. In a discontinuous system this is not the case. The explanation of this difference of behaviour is as follows.

In a continuous system, the velocity is proportional to $1/(r_1+r_2)^{\frac{1}{2}}$, where r_1 = resistance per unit length of external fluid, and r_2 = resistance per unit length of the axis cylinder. This relation has been derived for certain particular sets of assumptions by Rushton (1937) and by Offner *et al.* (1940), but can also be shown to be a direct consequence of uniform propagation by local circuits, as mentioned by Hodgkin (1947). The time course of the potential change is unaltered, so that its length scale is also proportional to $1/(r_1+r_2)^{\frac{1}{2}}$. The potential gradient is therefore increased in proportion to $(r_1+r_2)^{\frac{1}{2}}$, and the longitudinal current, which is equal to (potential gradient)/(r_1+r_2), is reduced in proportion to $1/(r_1+r_2)^{\frac{1}{2}}$. The length of nerve in front of the active region which has to be depolarized in order to excite it is, however, also reduced in the same ratio, so that the current is sufficient. But in a discontinuous system the membrane (at a node) which has to be depolarized has a fixed capacity and conductance. Its distance from the active node is also fixed, so that the longitudinal current is proportional to $1/(r_1+r_2)$. If r_1 is increased sufficiently, this current will therefore become insufficient to excite.

The experimental result that an increase of external resistance can cause a block might therefore be taken as evidence not only that the impulse is transmitted by local circuits, but also that the system is discontinuous. This point will not be pressed, however, for two reasons. In the first place, it is possible that the part of the fibre which was in air might have been damaged, so that the fibre was made discontinuous by the experimental conditions. In the second place, although a continuous system could in principle conduct in a region where the external resistance was indefinitely high but uniform, it might block at a point where the external resistance drops suddenly from a high to a low value.

DISCUSSION

Evidence for saltatory conduction. The results described in this paper include three more or less independent pieces of evidence in favour of the view that 'activity' takes place only at the nodes of Ranvier.

(*a*) The amplitude and time of occurrence of the main action current wave change discontinuously at each node of Ranvier.

(*b*) The membrane current through the myelin can be explained as a passive current through a resistance and capacity in parallel, but the membrane current at a node of Ranvier has a component which cannot be so explained.

(*c*) Conduction can be blocked by raising the external resistance.

It was pointed out in the 'Results' section that alternative explanations could be produced for observations (*a*) and (*c*). We do not believe that these alternative explanations are correct, but the fact that they exist reduces the value of these pieces of evidence. We shall therefore rely chiefly on the analysis of the distribution of membrane current during the passage of an impulse.

In the region covered by the nerve impulse, the interior of a nerve fibre is more positively charged relative to the exterior than when it is in its normal state. The sheath of the fibre has a finite conductance, so that this charge tends to leak away. For non-decremental conduction, this leakage must be replaced, and the process by which this takes place is called 'activity'. It occurs predominantly in the front of the region carrying the charge, so that the impulse moves forward. If this inward transfer of positive charge could be detected as such, it would be the most direct criterion of activity. But we can only measure the total current entering or leaving the axis cylinder (membrane current), and this contains the currents through the conductance and capacitance of the sheath as well as the current due to 'activity'. If, therefore, we are to locate the regions in a nerve fibre where 'activity' takes place by observing the membrane current, we must first deduce relations in terms of membrane current which are characteristic of 'activity'.

We could evidently say that a region of a nerve fibre does not take part in 'activity' if all the currents in it during an impulse can be explained by the observed potential change acting on known passive conductances and capacities. The resistance and capacity of the myelin sheath have not yet been measured, so that the most that could be said now is that there is no evidence of activity if the currents during an impulse can be explained by the observed potential change acting on plausible values of resistance (not necessarily linear or constant) and capacity. By 'plausible' is meant, for instance, that the capacity should not be so low as to imply an improbable value for the dielectric constant of the myelin.

It would, however, be unsatisfactory to locate 'activity' solely by the failure to satisfy this condition. We shall therefore try to find a characteristic way in which the membrane current of an 'active' region differs from the current to be expected in the absence of activity.

Consider the membrane current to be expected during non-decremental conduction of an impulse in a core conductor whose sheath has resistance and capacity. These passive elements are assumed to be in parallel with each other and with the mechanism which produces the 'active' current; this schematic circuit is not fundamentally different from that which represents 'activity' as a change in the e.m.f. in series with the resistance, but is somewhat more general.

Let x = distance along fibre,

r_1 = longitudinal resistance per unit length of fluid around fibre,

r_2 = longitudinal resistance per unit length of axis cylinder,

i_1 = longitudinal current outside fibre, positive in direction of increasing x,

i_2 = longitudinal current in axis cylinder, positive in direction of increasing x,

V_1 = potential in fluid surrounding fibre,
V_2 = potential in axis cylinder,
G_m = conductance of sheath per unit length,
C_m = capacitance of sheath per unit length,
i_m = current entering axis cylinder per unit length,
V_m = potential difference across sheath $= V_1 - V_2$,
V_0 = resting value of V_m,
u = velocity of conduction, assumed positive so that impulse travels in direction of increasing x.

We shall take first the case of a fibre whose properties are uniform along its length. We assume that during the action potential, $(V_m - V_0)$ is negative and has only one minimum. Then

$$r_1 i_m = -r_1 \partial i_1/\partial x = \partial^2 V_1/\partial x^2 \quad \text{and} \quad r_2 i_m = r_2 \partial i_2/\partial x = -\partial^2 V_2/\partial x^2,$$

so that
$$\partial^2 V_m/\partial x^2 = \partial^2 V_1/\partial x^2 - \partial^2 V_2/\partial x^2 = (r_1 + r_2)\, i_m.$$

During steady conduction, $V_m = f(t - x/u)$, so that

$$\partial^2 V_m/\partial x^2 = 1/u^2 . \partial^2 V_m/\partial t^2 \quad \text{and} \quad i_m = 1/(r_1 + r_2)\, u^2 . \partial^2 V_m/\partial t^2.$$

The membrane current is therefore outward when the graph of V_m against t (the action potential plotted with its peak downwards) has a downward curvature, and inward when it has an upward curvature. The curvature begins downward, but must become upward before the peak of the potential change in order for $\partial V_m/\partial t$ to be zero at the peak. Hence, there must be inward membrane current before the peak of the potential change. During the whole of this phase the currents through the membrane conductance and capacitance are both outward, since both $(V_m - V_0)$ and $\partial V_m/\partial t$ are negative. Hence the inward membrane current must be produced by some 'active' process. This active process must, of course, produce a greater inward current than this, since it is only the excess over the currents through the conductance and capacitance which can be observed as 'membrane current'.

Now consider the more difficult case of a nerve fibre in which activity takes place only at certain small areas of the sheath, which we shall refer to as 'nodes'. We shall assume as before that $(V_m - V_0)$ is negative during the impulse and has only one minimum, and that the time course of the potential change is the same at each position where activity occurs, though occurring later the greater the value of x. The distribution of potential along the fibre will then be roughly as shown in Fig. 13. In each internodal stretch,

$$\partial^2 V_m/\partial x^2 = (r_1 + r_2)\, i_m = (r_1 + r_2)\, \{G_m(V_m - V_0) + C_m \partial V_m/\partial t\},$$

which must be negative throughout the rising phase, so that the curvature of the graph is downward between each two successive nodes.

When the value of V_m at a particular node is at its minimum, and for a finite time before and after, the point on Fig. 13 representing it must lie below the straight line joining the points representing the values of V_m at the neighbouring nodes. Hence $\partial V_m/\partial x$ must be algebraically greater just beyond the node than just before it, and the same is true of i_2, since $i_2 = 1/(r_1+r_2) . \partial V_m/\partial x$. The membrane current at the node itself must therefore be inward, by Kirchhoff's first law. Hence, at the points where activity occurs, the membrane current becomes inward before the peak of the potential change at that point, while at all other points this is not the case.

This argument depends on the assumption that the amount of activity is the same at all points where activity occurs. But if, at points between the nodes, a slight degree of activity existed which was insufficient to make the membrane current inward before the peak of the potential change, then the argument and its conclusion would be unchanged. We can therefore conclude that points

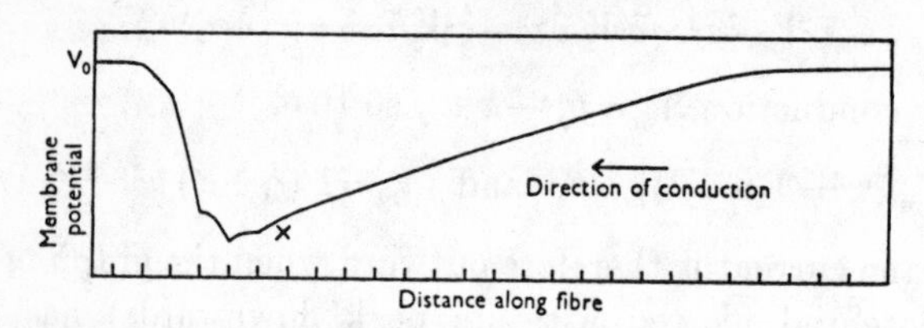

Fig. 13. Diagram showing form of distribution of membrane potential along a nerve fibre with nodes. Positions of nodes marked on horizontal axis. The part of the curve to the left of the point X was obtained by the integration method from actual records of longitudinal current.

must exist where the membrane current becomes inward before the peak of the potential change, and that these are the points where at least the main part of the activity is located. It was shown in the 'Results' section that the membrane current at each node of Ranvier behaved exactly in this way, while that through the myelin sheath did not. We take this as proof that the main part of the 'activity' in these nerve fibres occurs at the nodes of Ranvier.

The question remains whether or not a slight amount of activity occurs in the internodal regions. The 'active current', if any, would be sufficient only to reduce the losses, and not to contribute directly to the current which passes forward in the axis cylinder to depolarize the sheath ahead of the active region. The only criterion we can apply in this case is whether or not the membrane current through the myelin can be explained without postulating activity. As stated in the 'Results' section, the membrane current agrees qualitatively with the current which would flow if the membrane potential change were applied to a resistance and capacity in parallel. The agreement is in fact closer than was indicated at that point. The two peaks frequently seen in the membrane current curves during the rising phase of the potential change correspond to two detectable maxima in the rate of rise of potential, as determined by our integration method. This agrees with the finding of Tasaki & Takeuchi (1942) that each of these peaks is due to activity at the node at one end of the internode under observation, and not in the internode itself. Also, the small pulse of inward current at the very end of the impulse corresponds to a rapid fall in the calculated potential curves.

Inward current due to activity would make the conductance of the myelin appear lower than its true value. If, as would be expected, the activity occurred chiefly in the rising phase of the potential change, it would also reduce the apparent value of the capacity. It would therefore be evidence of 'activity' in the myelin if the apparent capacity or conductance of the myelin were improbably low for a passive membrane of its thickness. Values of these quantities deduced from our records, and given in Table 1, are uncertain in that they include an unknown factor α, the ratio of the specific resistance of the axoplasm to that of the surrounding medium. The value of this ratio has been determined for axons of certain invertebrates; it is about 1·4 in the giant axon of *Loligo* (Cole & Hodgkin, 1939), 3·0 in *Homarus* axons (Hodgkin & Rushton, 1946) and 4·5 in *Carcinus* axons (Hodgkin, 1947). A value between 1·4 and 1·8 seems likely for the fibres used in our experiments; this would imply that the action potential is about 130–170 mV. in amplitude (Table 1). But even if α is as high as 2·2, and the action potential is about 210 mV. the specific resistance of the myelin is only about 2×10^9 ohm-cm., which does not seem improbably high, and the apparent dielectric constant is about 3. The value found by our method for the capacity, and hence that for the dielectric constant, would be considerably affected if the resistance of the myelin sheath changed during the impulse or were non-linear. If either of these effects existed and were in the same direction as was observed in *Loligo* by Cole & Curtis (1939) and Cole & Baker (1941), it would make the capacity appear smaller than its true value. Thus, our results make it unlikely that the dielectric constant of the myelin was less than 3, which does not seem improbably low. We conclude that the currents through the myelin are adequately explained without postulating an 'active' process there as well as at the nodes.

All our analysis so far has been made on the assumptions that the longitudinal current in the axis cylinder is equal and opposite in direction to that in the fluid outside the myelin, and that the axis cylinder is a uniform ohmic conductor. These are standard assumptions of the membrane theory, but it will be well to see if there is any reason to doubt them, and, if so, whether our analysis is affected.

As regards the first of these assumptions, it is conceivable that some current flows either between a surface membrane of the axis cylinder and the myelin or between the lipoid layers of the myelin itself. If these were merely passive currents, in parallel with either that in the axis cylinder or with that in the external fluid, they would be negligible in magnitude because of the extremely high resistance of such pathways. On the other hand, such currents might be produced locally by a conduction process taking place entirely within the myelin sheath and axis cylinder. It was in order to test this possibility that we carried out the experiment described in the last part of the 'Results' section. This showed that conduction could be prevented by raising the external resistance. This result agrees with those of Hodgkin (1937*a*, *b*, 1939) and Tasaki (1939) in indicating that conduction depends on external currents. We conclude that there is no conduction process entirely within the myelin sheath and axis cylinder, and that it is justifiable to assume that the current in the axis cylinder is equal and opposite to that flowing in the external fluid.

As regards the second assumption, the axis cylinder is not uniform since it is markedly constricted at the nodes of Ranvier. The constrictions extend for so short a distance along the fibre, however, that it is justifiable to neglect their effect on resistance. Our analysis might, however, be seriously affected if the axis cylinder is interrupted by transverse membranes across which potential differences can be developed, as suggested by von Muralt (1945). It is difficult to say how this theory would alter our analysis, since it has not yet been given quantitative expression. We are inclined to think that the theory would predict action currents similar qualitatively to those required by the usual theory of a membrane concentric with the axis of the fibre, so that our results do not provide evidence either way concerning the theory.

Node spacing and conduction velocity. Sanders & Whitteridge (1946) found that a large change in node spacing could occur without an appreciable change in conduction velocity. This is generally regarded as evidence against the theory of saltatory conduction. This argument depends on the assumption that a saltatory system will always conduct faster the greater the distance between nodes, as in the simplified case treated by Offner *et al.* (1940). The following examples show that this is by no means necessarily so.

Consider first the case where the conductance and capacitance of the myelin and the conductance of the membrane at a node are negligible. The active node has then to charge the membrane capacity of adjacent nodes through the longitudinal resistances of the axoplasm and the fluid surrounding the fibre.

Let C_n = capacity of membrane at a node,
l = distance between adjacent nodes.
V_m, V_0, r_1, r_2 have the same significance as in the preceding section.

Suppose that as soon as V_m at a node reaches a fixed value V_c it drops instantaneously to a lower value V_a and stays at that value for a time long enough for subsequent changes to have no appreciable effect on the conduction velocity. If, then, the equations and boundary conditions defining the time course of $(V_m - V_0)/(V_a - V_0)$ are expressed in terms of a time variable $T = t/(r_1 + r_2)\, lC_n$, they do not contain r_1, r_2, l or C_n. The interval between the moments at which successive nodes reach V_c is therefore constant in units of T, whatever values those parameters may have. It will depend only on $(V_a-V_0)/(V_c-V_0)$, so that we can say

$$\Delta T = f\left\{\frac{V_a-V_0}{V_c-V_0}\right\}.$$

Hence

$$\Delta t = (r_1+r_2)\, l C_n f\left\{\frac{V_a-V_0}{V_c-V_0}\right\},$$

and the velocity is

$$l/\Delta t = \frac{1}{(r_1+r_2)\, C_n f\left\{\dfrac{V_a-V_0}{V_c-V_0}\right\}},$$

which is independent of node spacing.

The simplifications on which this result depends are certainly too drastic both for very short and for very long internodes. In the former case, our assumptions would require the currents to become very large. Real conditions will therefore be represented better if we imagine that the source of potential which charges each node from V_c to V_a has an internal resistance r. The dimensional argument will now apply only so long as $r/(r_1+r_2)\, l$ is constant, and then, for given values of V_c and V_a,

$$\Delta T = \phi\left\{\frac{r}{(r_1+r_2)\, l}\right\}.$$

Clearly an increase in r will delay the charging of the next node, so that $\phi\left\{\frac{r}{(r_1+r_2)\, l}\right\}$ increases as $r/(r_1+r_2)\, l$ increases. The velocity is

$$\frac{l}{\Delta t} = \frac{l}{(r_1+r_2)\, l\, C_n \phi\left\{\dfrac{r}{(r_1+r_2)\, l}\right\}} = \frac{1}{(r_1+r_2)\, C_n \phi\left\{\dfrac{r}{(r_1+r_2)\, l}\right\}},$$

which increases with l, approaching for large l the constant value it would have if $r=0$.

When the internodes are long, the distributed capacity of the myelin must become important, and in the limit the capacity of the nodes will be negligible in comparison with it. In that case,

$$1/(r_1+r_2) \,.\, \partial^2 V_m/\partial x^2 = C_m \partial V_m/\partial t,$$

where C_m = capacity of myelin sheath per unit length of the fibre. If we now define

$$X = x/l \quad \text{and} \quad T = t/(r_1+r_2)\, l^2 C_m,$$

this equation becomes

$$\partial^2 V_m/\partial X^2 = \partial V_m/\partial T,$$

and the boundary conditions do not involve r_1, r_2, l or C_m. Hence the time per internodal distance is constant in units of T, and proportional to $(r_1+r_2)\, l^2 C_m$

in units of t. The conduction velocity is therefore proportional to $1/(r_1+r_2)\,lC_m$, which actually decreases as node spacing increases. At still greater internodal spacing, the conductance, both of the myelin and of the nodes, may become important, reducing velocity still further and eventually causing a block.

On this set of assumptions, therefore, one would expect that the conduction velocity will have a maximum at a particular node spacing, and that it might be a very flat maximum. Natural selection will probably have made the normal spacing fall near this optimum, so that considerable deviations from the normal spacing would cause only small changes in velocity.

Sanders & Whitteridge compared the velocity in normal fibres with that in fibres with considerably reduced internodal lengths. We should expect the difference in velocity to be least if the normal spacing were somewhat above, and the reduced spacing below, the value which would give maximum velocity. There is, in fact, some evidence to suggest that, at least in frog and toad fibres, the normal spacing is above the value for maximum velocity. Thus, Tasaki & Takeuchi (1941) recorded 'mononodal' action currents from a node both when only one and when both of the neighbouring nodes were narcotized. The records were practically identical, suggesting that the internal resistance of the active node, the factor which makes conduction velocity increase with node spacing, is negligible. Further, our membrane current records (Fig. 12) show that the capacity of the myelin in an internode is greater than the capacity of the nodal membrane, so that the conditions approach those in which it was shown above that velocity is inversely proportional to node spacing.

SUMMARY

1. The longitudinal current which flows in the external fluid when an impulse passes was recorded at various positions along a fibre isolated from the sciatic nerve of *Rana esculenta* or *R. temporaria*.

2. The conduction time of the longitudinal action current was found to be practically constant in each internode, and to increase stepwise as each node of Ranvier was passed.

3. The amplitude of the first phase of the longitudinal current was found to fall steadily from the proximal to the distal end of each internode, and to increase suddenly as each node was passed.

4. The current crossing the myelin sheath during the impulse could be explained as a passive current due to the potential change acting on a resistance and capacity in parallel.

5. At each node, positive current began to enter the axis cylinder before the potential change had reached its maximum. This relation is impossible in a system of resistances and capacities and is shown to be a necessary characteristic of the points which maintain decrementless conduction in a cable-like structure. It is concluded that the process which gives rise to the action

potential takes place at the nodes of Ranvier, confirming the theory of saltatory conduction.

6. Conduction was blocked reversibly by increasing the external resistance between two adjacent nodes. It is concluded that the action potential at each node excites the next node by current flowing forward in the axis cylinder and back in the fluid outside the myelin sheath.

7. The finding of Sanders & Whitteridge that a large decrease in node spacing can occur without a drop in conduction velocity is shown not to conflict with the theory of saltatory conduction.

This work was made possible by a grant from the Rockefeller Foundation, to whom we wish to express our gratitude. We are also deeply indebted to Mr A. L. Hodgkin for much valuable discussion and criticism, and for the loan of apparatus.

REFERENCES

Bielschowsky, M. (1928). *Handbuch der Mikroskopischen Anatomie des Menschen*, **4**, 98 et seq. Berlin: Springer.
Cole, K. S. & Baker, R. F. (1941). *J. gen. Physiol.* **24**, 535.
Cole, K. S. & Curtis, H. J. (1939). *J. gen. Physiol.* **22**, 649.
Cole, K. S. & Hodgkin, A. L. (1939). *J. gen. Physiol.* **22**, 671.
Erlanger, J. & Blair, E. A. (1934). *Amer. J. Physiol.* **110**, 287.
Erlanger, J. & Blair, E. A. (1938). *Amer. J. Physiol.* **124**, 341.
Erlanger, J. & Gasser, H. S. (1937). *Electrical Signs of Nervous Activity*, p. 29. Philadelphia: University of Pennsylvania Press.
Grundfest, H. (1947). *Ann. Rev. Physiol.* **9**, 488.
Hodgkin, A. L. (1937*a*). *J. Physiol.* **90**, 183.
Hodgkin, A. L. (1937*b*). *J. Physiol.* **90**, 211.
Hodgkin, A. L. (1939). *J. Physiol.* **94**, 560.
Hodgkin, A. L. (1947). *J. Physiol.* **106**, 305.
Hodgkin, A. L. & Huxley, A. F. (1945). *J. Physiol.* **104**, 176.
Hodgkin, A. L. & Rushton, W. A. H. (1946). *Proc. Roy. Soc.* B, **133**, 444.
Huxley, A. F. & Stämpfli, R. (1948). *Helv. physiol. pharmacol. Acta*, **6**, C 22.
Kato, G. (1934). *Microphysiology of Nerve.* Tokyo: Maruzen.
Kato, G. (1936). *Cold Spr. Harb. Sym. quant. Biol.* **4**, 202.
Kubo, M., Ono, S. & Toyoda, H. (1934), *Jap. J. med. Sci. Biophys.* **3**, 213.
Lillie, R. S. (1925). *J. gen. Physiol.* **7**, 473.
Maximow, A. A. & Bloom, W. (1942). *Textbook of Histology*, 4th ed. p. 196. Philadelphia: Saunders.
von Muralt, A. (1945). *Die Signalübermittlung im Nerven*, pp. 229 et seq. Basle: Birkhäuser.
Offner, F., Weinberg, A. & Young, G. (1940). *Bull. math. Biophys.* **2**, 89.
Pfaffmann, C. (1940). *J. cell. comp. Physiol.* **16**, 407.
Rushton, W. A. H. (1937). *Proc. Roy. Soc.* B, **124**, 210.
Sanders, F. K. & Whitteridge, D. (1946). *J. Physiol.* **105**, 152.
Stämpfli, R. (1946). *Helv. physiol. pharmacol. Acta*, **4**, 411.
Takeuchi, T. & Tasaki, I. (1942). *Pflüg. Arch. ges. Physiol.* **246**, 32.
Tasaki, I. (1939). *Amer. J. Physiol.* **127**, 211.
Tasaki, I. (1940). *Pflüg. Arch. ges. Physiol.* **244**, 125.
Tasaki, I., Amikura, H. & Mizushima, S. (1936). *Jap. J. med. Sci. Biophys.* **4**, 53.
Tasaki, I. & Takeuchi, T. (1941). *Pflüg. Arch. ges. Physiol.* **244**, 696.
Tasaki, I. & Takeuchi, T. (1942). *Pflüg. Arch. ges. Physiol.* **245**, 764.

B. THE MEMBRANE THEORY: THE PHYSICAL AND CHEMICAL BASES OF THE ACTION POTENTIAL

Electrical potentials of nerve cells result from metabolically maintained differences in the concentration of ions between the inside and outside of the cell, together with specific differences in the ability of these ions to pass through the membrane. That complex potential changes in nerve cells result from passive diffusion of ions along pre-established concentration gradients, following changes of membrane permeability to specific ions, has become a unifying hypothesis for interpreting many biological potentials. While this membrane theory, or ionic hypothesis, has been and continues to be challenged (e.g., Tasaki, 1968), judged by its fruitfulness in stimulating research and its ability to unify experimental data, it remains one of the most important general hypotheses in cellular neurophysiology (see Cole, 1968).

Bernstein (1902) is credited with first proposing a membrane theory. He suggested that the membrane, at rest, is a barrier to Na^+ which allows the establishment of a K^+ concentration potential (resting potential). He further postulated a breakdown of permeability barriers during the action potential. Rigorous tests of Bernstein's proposals became possible after the rediscovery of giant axons in squid. They also required considerable growth in sophistication of electronic techniques for measurement. COLE AND CURTIS (1939) demonstrated a large decrease in the electrical impedance of the squid axon membrane during the nerve impulse, as predicted by Bernstein's hypothesis. Hodgkin and Huxley (1939) and Curtis and Cole (1942) took advantage of these giant axons to measure directly the electrical potential across the membrane. It then became clear that another prediction of Bernstein's hypotheses did not hold: the maximum amplitude of the action potential was not equal to the resting potential, but was considerably larger. The significance of this overshoot was clarified by **Hodgkin and Katz (1949)** who proposed the sodium hypothesis. They suggested that a large increase in conductance to Na^+ and consequent inward movement of Na^+ down its pre-existing electrochemical gradient led to the depolarization and overshoot of the action potential.

Underlying such modern theories of membrane potentials is an equation, derived from physical-chemical principles, known as the Nernst equation. This equation defines, in terms of the concentrations of a specific ion on either side of a membrane, the potential across the membrane at which there will be no net movement of that ion from one side to the other. In other words, the equation defines the potential required to oppose exactly the movement of a particular ion down its concentration gradient. If the membrane is exclusively permeable to a specific ion, then the membrane potential will be given by the equilibrium potential for that ion. For any ion the equation is

$$V = \frac{RT}{z\mathfrak{F}} \ln \frac{(X)_{out}}{(X)_{in}}$$

where V is the equilibrium potential for the ion under consideration, defined in the sense of internal potential minus external potential, $(X)_{in}$ and $(X)_{out}$ are the concentrations (or, more accurately, the activities) inside and outside of, for example, an axon, z is the valence of the ion, R the gas constant, T the absolute temperature of the system, and $\mathfrak{F}$ the Farady constant.

Using the K^+ and Na^+ concentrations observed in squid axoplasm and blood, the Nernst equation gives values for the equilibrium potentials of −75 mV for K^+ and 55 mV for Na^+. **Hodgkin and Katz (1949)** proposed that at rest the membrane is predominantly permeable to K^+, and hence the resting potential is close to the V_K, but that during the action potential the membrane becomes predominantly permeable to Na^+, and hence the potential shifts toward the Na^+ equilibrium potential. Termination of the increased Na^+ permeability is in some cases accompanied by a marked increase in K^+ permeability which hastens the return of the potential toward the K^+ equilibrium potential.

Evidence to support the sodium hypothesis in detail required further technological advance. Keynes and Lewis (1951) used radiotracer techniques to demonstrate that the amounts of Na^+ and K^+ exchanged during activity of squid axons is sufficient to produce the membrane action potential change.

A critical observation to be explained by the sodium hypothesis is **Hodgkin's** observation **(1937)** that membrane potential is a controlling factor in nerve activity. Growing sophistication in electronics permitted the development of the voltage clamping technique. This method involves the electrical isolation of a portion of the axon (or muscle cell) in such a way that no currents flow along the inside or outside parallel to the membrane, but rather all current flows across the isolated portion of membrane. A feedback amplifier permits the experimenter to impose a sudden change in voltage across the membrane; it compares this instantaneously with the existing membrane voltage and supplies current across the membrane appropriate to cancel the voltage difference. The magnitude and time course of the current supplied is measured directly. The variables of voltage change with distance along the axon, and of voltage change with time, are thus eliminated. From such data, the membrane permeability, with respect to time and membrane potential, can be calculated. One remarkable and important conclusion which emerges is that membrane potential is the independent variable; it controls the sequence of events underlying the action potential. As an introduction to this work the reviews by HODGKIN (1958, 1964) are recommended. Note that the conventions for presenting voltage clamp current data are reversed from those used in the series of five papers reproduced here in these reviews and in all subsequent work of this kind.

The paper by **Hodgkin, Huxley, and Katz (1952)** and those by **Hodgkin and Huxley (1952a, b, c, d)** describe the voltage clamping technique and its application to a quantitative analysis of membrane events in squid nerve. These studies show that Na^+ permeability is increased when the membrane is rapidly depolarized; the extent of the increase is greater for larger depolarizations. Thus, in an unclamped axon, depolarization leads to a self-reinforcing change: the increased Na^+ permeability leads to inward Na^+ movement along the Na^+ concentration gradient and to further depolarization. This leads to further permeability increase. Threshold is reached when depolarization results in increased Na^+ conductance and consequently inward Na^+ movement. When this exceeds the outward K^+ movement (resulting from the departure of membrane potential from K^+ equilibrium), the inward Na^+ movement reinforces the depolarization and the regenerative events underlying the action potential are set off.

The voltage clamping experiments revealed a process referred to as sodium inactivation. After depolarization the increased Na^+ permeability, due to depolarization, becomes reduced in time, even if the depolarization is maintained. This inactivation process combines with a delayed increase in K^+ permeability, in response to depolarization, to repolarize the membrane. These membrane potential and time dependent processes provide a basis for understanding accommodation to maintained stimuli, refractory periods, and limitations in the repetitive responsiveness of some axons.

Hodgkin and Huxley (1952d) used the quantitative data relating membrane conductance, membrane potential, and time to reconstruct mathematically the membrane potential changes during an action potential. Their formulation also allowed the prediction of a number of membrane phenomena in a variety of tissues and species thus supporting the generality of the hypothesis (see Noble, 1966, and, e.g., Binstock and Goldman, 1969). Although a number of variations are found, they adhere, in general, to the theme. Perhaps the most significant variation is shown by crustacean muscle (e.g., Fatt and Ginsborg. 1958; Hagiwara and Naka, 1964) and certain molluscan neurons (e.g., Geduldig and Junge, 1968) in which the rapid depolarization involves inward Ca^{++} movement. The influence of Ca^{++} on the properties of the squid axon membrane is described by Frankenhaeuser and Hodgkin (1957). The role of Ca^{++}, although obviously of fundamental importance, is not yet clearly defined.

In the voltage clamp experiments, Hodgkin and Huxley were able to provide rather direct evidence that Na^+ carries the initial inward current of the action potential. Direct evidence that K^+ carries the later outward current is provided by **Hodgkin and Huxley (1953)**. They measure the radiotracer efflux of K^+ simultaneously with current and show that it accounts exactly for the charge transfer.

Early proponents of a membrane theory could not explain how steady ionic concentration gradients were maintained during prolonged activity in the face of the postulated ionic exchanges. **Hodgkin and Keynes** (1955) demonstrated that a metabolically driven pump extrudes Na^+ and accumulates K^+ across the squid axon membrane. Recent work indicates that the pump is not always electrically neutral (a point which Hodgkin and Keynes discuss very skillfully). Thus it can contribute to the integrative performance of neurons (e.g., Ritchie and Straub, 1957; Nakajima and Takahashi, 1966; Carpenter and Alving, 1968).

A new period in the quantitative investigation of membrane biophysics is in progress now that giant axons and barnacle giant muscle fibers can be perfused internally with artificial solutions (**Baker, Hodgkin and Shaw, 1962**; Hagiwara, Hayashi, and Takahashi, 1969). These methods have provided further confirmation of the ionic hypothesis (for a contrary view see, e.g., Tasaki, Singer, and Takenaka, 1965). The use of unnatural internal and external bathing solutions in the study of membrane permeability changes has led to observations that are still under discussion (e.g., Chandler, Hodgkin, and Meves, 1965).

Hodgkin and Huxley made the assumption that the Na^+ and K^+ conductances behave independently. This assumption has now been justified by evidence such as that reported by **Moore, Blaustein, Anderson and Narahashi (1967)**. They show that tetrodotoxin (TTX, a fish neurotoxin) specifically blocks the early, membrane potential dependent conductance increase to Na^+. The great specificity of TTX makes it a powerful analytical tool in separating from impulse generation, neuron responses involving processes independent of membrane potential (e.g., sensory generatory potentials, Loewenstein, Terzuolo, and Washizu, 1963; postsynaptic responses, e.g., Katz and Miledi, 1967a). Further evidence of Na^+ and K^+ conductance changes comes from examination of the effects of tetraethylammonium (TEA). This ion specifically blocks the delayed conductance increase to K^+ (e.g., Hille, 1967). The further information that TTX acts on the exterior of squid axons (Narahashi, Anderson, and Moore, 1967) and TEA is effective on the interior of the squid membrane (Armstrong and Binstock, 1965) suggests that separate and specific but unknown membrane components mediate these ionic conductances.

There are as yet no explanations of the mechanisms by which changes in membrane potential lead to the complex conductance changes that occur during an action potential. New techniques for studying the changes occurring in the membrane at macromolecular dimensions are being actively sought (e.g., Mueller and Rudin, 1968).

J. Physiol. (1949) 108, 37–77

612.813

THE EFFECT OF SODIUM IONS ON THE ELECTRICAL ACTIVITY OF THE GIANT AXON OF THE SQUID

By A. L. HODGKIN and B. KATZ

From the Laboratory of the Marine Biological Association, Plymouth, and the Physiological Laboratory, University of Cambridge

(*Received* 15 *January* 1948)

Experiments with internal electrodes suggest that the active reaction of nerve is not a simple depolarization of the kind postulated by Bernstein (1912) and Lillie (1923). In the giant axon of the squid, the resting membrane potential appears to be about 50 mV. whereas the action potential is of the order of 100 mV. (Curtis & Cole, 1942; Hodgkin and Huxley, 1939, 1945). This result implies that the surface membrane undergoes a transient reversal of potential difference during the passage of the nervous impulse. The magnitude of the reversal cannot be measured precisely, because of uncertainties concerning the liquid junction potential between the axoplasm and the internal recording electrode. But there is now little doubt that the membrane potential of certain types of nerve fibre does undergo an apparent reversal which cannot be reconciled with the classical form of the membrane theory. Several attempts have been made to provide a theoretical basis for this result (Curtis & Cole, 1942; Hodgkin & Huxley, 1945; Höber, 1946; Grundfest, 1947), but the explanations so far advanced are speculative and suffer from the disadvantage that they are not easily subject to experimental test. A simpler type of hypothesis has recently been worked out, in collaboration with Mr Huxley, and forms the theoretical background of this paper. The hypothesis is based upon a comparison of the ionic composition of the axoplasm of a squid nerve with that of the sea water in which experimental preparations are normally immersed. The potassium concentration of fresh squid axoplasm appears to be some twenty to forty times greater than that in sea water, whereas the sodium and chloride ions may be present in concentrations which are less than one-tenth of those in sea water (Steinbach, 1941; Steinbach & Spiegelman, 1943). The resting membrane potential is supposed to arise in a manner which is essentially similar to that postulated in Bernstein's form of the membrane theory. The resting membrane is assumed to be permeable to potassium and possibly to chloride ions, but is only very sparingly permeable to sodium. There should, therefore, be a potential difference of the correct sign and magnitude across the surface membrane of a resting nerve fibre.

According to the membrane theory excitation leads to a loss of the normal selectively permeable character of the membrane, with the result that the resting potential falls towards zero during activity. This aspect of the theory is at variance with modern observations and must be rejected. However, a large reversal of membrane potential can be obtained if it is assumed that the active membrane does not lose its selective permeability, but reverses the resting conditions by becoming highly and specifically permeable to sodium. The reversed potential difference which could be obtained by a mechanism of this kind might be as great as 60 mV. in a nerve with an internal sodium concentration equal to one-tenth of that outside. The essential point in the hypothesis is that the permeability to sodium must rise to a value which is much higher than that to potassium and chloride. Unless this occurs the potential difference which should arise from the sodium concentration difference would be abolished by the contributions of potassium and chloride ions to the membrane potential. The hypothesis therefore presupposes the existence of a special mechanism which allows sodium ions to traverse the active membrane at a much higher rate than either potassium or chloride ions.

A simple consequence of the hypothesis is that the magnitude of the action potential should be greatly influenced by the concentration of sodium in the external fluid. Thus the active membrane should no longer be capable of giving a reversed e.m.f. if the external sodium concentration were made equal to the internal concentration. On the other hand, an increase of membrane reversal would occur if the external sodium concentration could be raised without damaging the axon by osmotic effects. Experiments of this kind are difficult to make when external electrodes are used for recording; for the sodium content of the external medium cannot be varied without changing the electrical resistance of the extra-cellular fluid, and this would in itself cause a large alteration in the magnitude of the recorded action potential. We have therefore studied the influence of sodium concentration on the form and size of the action potential recorded with an internal electrode in the giant axon of the squid.

APPARATUS

The general plan of the equipment was essentially similar to that used by Hodgkin & Huxley (1945) and need not be described in detail. A diagram of the recording cell is shown (Fig. 1) in order to facilitate description of the experimental procedure. The walls of the cell were made of glass or Perspex. The Perspex was at first coated with a thin film of paraffin wax, but no adverse effects were observed when this precaution was omitted. The cell was illuminated from the side and this had the advantage that a double image of the axon could be obtained in the microscope by using a single mirror instead of the more complicated arrangement employed by Hodgkin & Huxley (1945). The mirror was removed from the cell as soon as the microelectrode had been inserted to the correct distance.

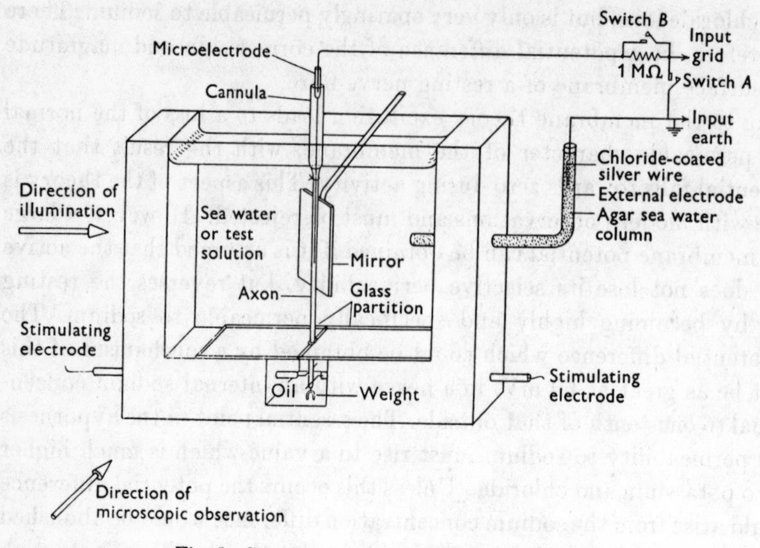

Fig. 1. Simplified diagram of recording cell.

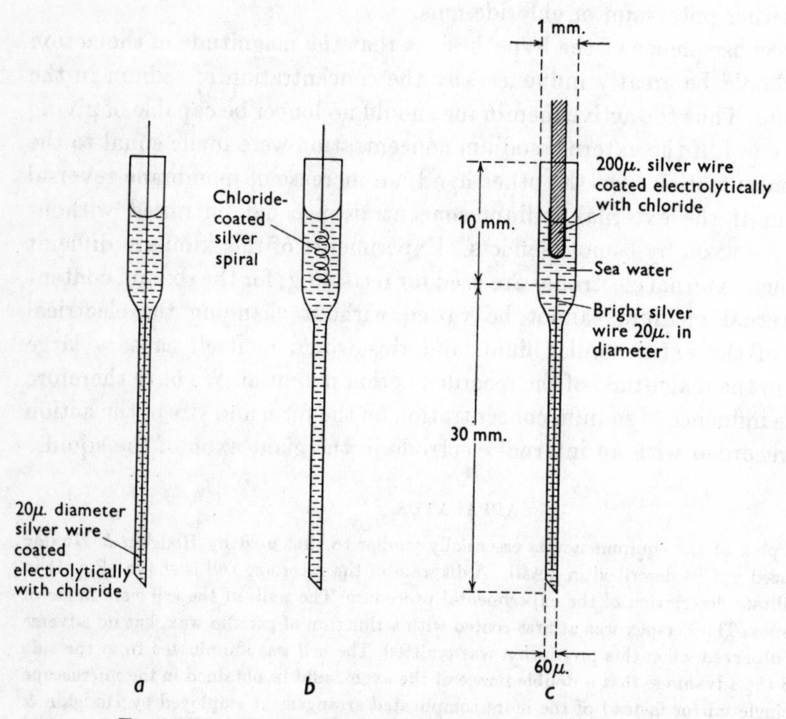

Fig. 2. Diagrams of microelectrodes: *a* and *b*, types formerly employed; *c*, type used in present work.

Microelectrode. The microelectrodes employed in earlier work (Hodgkin & Huxley, 1945) are shown in Fig. 2*a* and *b*. Type *a* had a reasonably low resistance, but it was unsuitable for prolonged experiments because the electrode potential was unsteady. The electrode therefore had to be withdrawn from the fibre at frequent intervals in order to allow its potential to be measured against that of the external electrode. Type *b* gave a steadier potential, but its resistance was so high that action potentials were reduced in magnitude by the stray capacity of the input circuit. The electrode used in the present work is shown in Fig. 2*c* and was designed to combine the advantages of both the first and second types. A relatively thick silver wire made contact with the sea water in the shank of the tube, and was electrolytically coated with chloride. In addition, a bright silver wire was thrust down the capillary to within 1·5 mm. of the tip. When a steady potential was measured this electrode acted in the same way as the second type of electrode. The bright silver wire did not influence the steady potential, because it was effectively shunted by the column of sea water in the capillary. Under these conditions the electrode resistance was determined by the whole length of the column of sea water which amounted to several megohms. However, this resistance did not affect the measurements, because the input resistance of the amplifier was greater than 10^9 Ω. When a transient potential difference such as a spike was measured, the column of sea water was momentarily short-circuited by the polarizable silver wire which acted like a condenser connected between the tip and base of the microelectrode. The electrode therefore behaved as though it had a relatively low resistance and thus avoided the errors which would otherwise have been introduced by stray capacity. It can be shown theoretically that an electrode of this kind should not introduce any distortion of the action potential, provided that the polarization capacity of the bright silver wire is large compared to the stray capacity of the input circuit. But these calculations need not be presented, since the performance of the electrode could be tested directly in a manner which will now be described.

The first test consisted in the sudden application of a potential difference to the tip of the microelectrode. A rectangular wave of current was passed through a 5 kΩ. resistance connected between the external electrode of the recording cell and earth. The terminal of the external electrode was connected directly to the input stage, and the resulting deflexion of the cathode ray tube recorded photographically. This test showed that the amplifier and input stage operated with an exponential lag of 4 μsec. The tip of the microelectrode was then lowered into the sea water, the input lead transferred from the external electrode to the microelectrode and a second photograph obtained. The two records were found to differ by less than 5%, thus demonstrating that the total lag of the recording system was of the order of 4 μsec. This lag may be neglected with safety, since the rising phase of the action potential occupied about 200 μsec.

A test which was essentially similar to this was also made using rectangular steps of current lasting many seconds, and it was again found that the microelectrode introduced no perceptible change in the size or shape of the potential recorded.

Amplifier and recording system. The characteristics of the d.c. amplifier and input stage were essentially similar to those described by Hodgkin & Huxley (1945), as were the methods of calibration employed. Errors which might have arisen from non-linearity or cathode-ray tube curvature were eliminated by comparing the records directly with a calibration grid which was obtained by photographing the series of oscillograph lines resulting from successive application of 10 mV. steps of potential to the input.

Records of the rate of change of membrane voltage during the action potential were obtained by electrical differentiation. This was achieved by introducing a single stage of condenser coupling into the d.c. amplifier. The time constant of the condenser and resistance used for differentiating was approximately 13 μsec. Under these conditions the output of the amplifier should be proportional to the rate of change of the input. What is measured is not exactly equal to the instantaneous rate of change at any given moment, but is more nearly equivalent to the average rate of change over a period which is reasonably short compared to the action potential. The rate circuit and amplifier were tested with rectangular inputs or sine waves of known frequency and amplitude. The rate amplifier was calibrated by comparing the absolute magnitude of the action potential in millivolts with the values obtained by graphical integration of the first phase of the rate record. The quotient of the two quantities gives the scale factor for the rate amplifier.

Rates were occasionally measured by graphical differentiation of the action potential. This procedure was laborious, and subject to considerable uncertainty unless the action potential was recorded on an expanded time-scale. The difference between electrical and graphical measurements usually amounted to about 10%, and there may be an error of 10% or even more in the absolute magnitude of the rates quoted in this paper. For the most part we shall be concerned not with absolute rates but with the relative magnitude of the rates in solutions of varying sodium concentration, and the error here is likely to be less than 5%.

Artificial solutions. Test solutions were usually made by mixing isotonic solutions in different proportions. The values used in preparing isotonic solutions are given in Table 1 and were based on cryoscopic data in International Critical Tables. Concentrations were chosen to give a freezing-point depression of 1·88° C. which appears to be the correct value for sea water of salinity 3·45% (Glazebrook, 1923). No cryoscopic data could be obtained for choline chloride and the figure of 0·6 M must be regarded as a guess.

Sea water was used as a normal medium, since no Ringer's solution applicable to the squid has yet been developed. An artificial sea-water solution was also employed on certain occasions, and was made according to the formula in Table 1. No appreciable difference could be detected between the action of this solution and that of sea water.

Sodium-rich solutions were made by adding solid sodium chloride to sea water.

Tests with indicators showed that all solutions employed had approximately the same pH as that of sea water.

Dextrose solutions were made up at frequent intervals and were stored in a refrigerator when not in use.

TABLE 1. Composition of stock solutions

I. Isotonic solutions

Solution	Solute	Molality	g./1000 g. H_2O
a	NaCl	0·56	32·7
b	KCl	0·56	41·8
c	$CaCl_2$	0·38	42·1
d	$MgCl_2$	0·36	34·3
e	$NaHCO_3$	0·56	47·1
f	Dextrose	0·98	177·0
g	Sucrose	0·93	319·0
h	Choline chloride	0·60	83·6

II. Artificial sea water

804 parts *a*: 18 parts *b*: 28 parts *c*: 146 parts *d*: 4·6 parts *e*.

Liquid junction potentials in the external circuit. The action of a test solution was examined by sucking out the sea water from the recording cell and running in a new solution. The test solution did not alter the potential of the external electrode, since the silver chloride surface was separated from the recording cell by a long column of agar sea water. However, the test solution set up a small liquid junction potential at the edge of the agar sea-water column and this had to be measured before the effect of a test solution on the resting potential could be evaluated.

Junction potentials were measured by dipping a silver chloride electrode (in some cases the microelectrode itself) into a beaker filled with sea water which was connected to the recording cell by means of a saturated KCl bridge. The system employed was, therefore,

Ag.AgCl	Sea water	Saturated KCl	Test solution in recording cell	Agar sea-water column in external electrode	AgCl.Ag

In measuring junction potentials an attempt was made to reproduce, as far as possible, the experimental conditions used in examining a living nerve. The data obtained are, therefore, not strictly comparable to those given by standard physicochemical methods, but should provide the right corrections for the present research. The saturated KCl bridge method is known to be unsatisfactory in certain respects, but it probably gives results of an accuracy sufficient for the present purpose.

The results obtained are shown in Table 2, and give the corrections which have to be subtracted from any apparent change in resting potential produced by the solution in question. No value is given for isotonic dextrose, since this solution gave an unsteady potential which increased with time to a large value.

Liquid junction potentials for solutions of intermediate strength (e.g. 0·7 sea water, 0·3 dextrose) were obtained by interpolation.

TABLE 2. Liquid junction potentials measured by saturated KCl bridge method

Test solution	E sea water − E test solution (mV.)
1 part sea water: 1 part isotonic dextrose	+2·6
1 part sea water: 4 parts isotonic dextrose	+6·0
1000 c.c. sea water + 15 g. NaCl	−0·7
1 part sea water: 1 part 0·6 M-choline chloride	−0·7

Experimental procedure. Giant axons, with a diameter of 500–700 μ., were obtained from the hindmost stellar nerve of *Loligo forbesi*. The methods of mounting the axon and of inserting the microelectrode require no description, since they were essentially similar to those employed by Hodgkin & Huxley (1945). Before introducing the microelectrode, a value was obtained for the small potential difference between the microelectrode and the external recording electrode. The potential difference was obtained by dipping the microelectrode into the sea water in the recording cell (which was normally connected to earth by the external electrode) and comparing the position of the base-line with the value obtained by 'earthing' the input lead. The potential difference between the two electrodes usually amounted to several millivolts and this value had to be subtracted from the apparent resting potential. Errors which might have arisen from amplifier drift were avoided by repeated checks of the amplifier zero, but this procedure did not obviate errors caused by changes in the microelectrode potential. The microelectrode could not be withdrawn during the course of an experiment, and we therefore had to rely on the stability of its potential. In the most complete experiments the electrode did not drift by more than 4 mV. in about 4 hr., but changes equivalent to 2 mV./hr. were sometimes encountered. We attempted to allow for changes in microelectrode potential by a sliding correction, but measurements of the resting potential cannot be presented with the same confidence as can those relating to the amplitude of the spike. The method of obtaining the amplifier zero requires comment since this was not such a simple operation as might at first be supposed. In the interests of stability it was desirable, first, that the input circuit should never be open-circuited; and secondly, that the nerve membrane should never be short-circuited. The following procedure was therefore adopted. First, a photographic record of the action potential and resting base-line was obtained with switch A open and switch B closed (Fig. 1). Switch B was opened then switch A closed and a second record obtained. This operation gave the amplifier zero but did not short-circuit the membrane, since this was protected first by the resistance of the microelectrode and secondly by the 1 MΩ. resistance. The switching procedure was reversed when the amplifier zero had been obtained. In a few experiments, switch B was left open throughout. This increased the recording lag from 4 μsec. to 11 μsec., but it did not cause any measurable change in the form of the action potential or its derivative.

The giant fibre was normally stimulated at 40 per min. throughout the entire period of experimental test.

RESULTS

Electrical properties of axons immersed in sea water

The magnitudes of the action potential, resting potential and positive phase were measured as a matter of routine at the beginning of each experiment, and are shown in Table 3. Approximate values for the maximum rates of rise and fall of the spike are also included. A few axons gave spikes less than 80 mV., but were not used for quantitative measurements because they deteriorated rapidly. The values for spike height are in good agreement with those obtained by Hodgkin & Huxley (1945), but are considerably smaller than those reported

TABLE 3. Electrical properties of axons in sea water

	Temperature (° C.)	Resting potential (mV.)	Action potential (mV.)	Membrane reversal (mV.)	Positive phase (mV.)	Maximum rate rise (V.sec.$^{-1}$)	Maximum rate fall (V.sec.$^{-1}$)
	22	46	85	39	14	—	—
	20	52	93	39	11	—	—
	21	52	86	34	10	490	290
	24	51	83	32	13	580	380
	22	50	86	36	15	650	400
	22	49	93	44	15	770	460
	20	40	87	47	15	560	330
	20	51	98	47	15	630	390
	—	48	87	39	15	520	340
	21	46	89	43	16	600	360
	20	53	99	46	14	1000*	530*
	20	46	85	39	14	620	480
	19	42	85	43	15	490	330
	20	45	86	41	16	590	360
	21	45	82	37	15	680*	350*
Average	21	48	88	40	14	630	380

* Indicates that these values were obtained by graphical differentiation. The values for resting potential are those observed with a microelectrode containing sea water. No correction has been made for the junction potential between axoplasm and sea water.

by Curtis & Cole (1942). The average value for the resting potential (48 mV.) is slightly smaller than that given by Curtis & Cole (51 mV.), but a difference of this kind is to be expected since Curtis & Cole used KCl in the microelectrode, whereas we employed sea water. The average action potential was about 20 mV. smaller than that given by Curtis & Cole. But a more serious discrepancy arises from the fact that we have never observed action potentials greater than 100 mV. at 18–23° C., whereas Curtis & Cole describe a spike as large as 168 mV. in a fibre which gave a resting potential of 58 mV. The matter is not one that can be lightly dismissed, because the existence of a fibre capable of giving an overshoot of 110 mV. has far-reaching implications. We are no longer inclined to think that our relatively small action potentials can be attributed to the poor condition of the experimental animals, since a number of the squids employed were extremely lively and in perfect condition. Nor does it seem likely that axons were damaged in the process of isolation, since microelectrodes were sometimes inserted into axons which were still surrounded by a greater part of the nerve trunk and had been subjected to a minimum amount of dissection. Hodgkin & Huxley's (1945) experiments indicate that the process of inserting a microelectrode did not in itself reduce the action potential, so that the possibility of damage at this stage may also be reasonably dismissed. Curtis & Cole's experiments may have been made at a different temperature, but this does not account for the discrepancy, since the action potential increases by 5–10% when the nerve is cooled from 20 to 0° C. and decreases as the temperature is raised above 20° C. (unpublished results). Apart from the possibility of instrumental error, the only explanation which can be offered is that there is a real difference between the properties of *L. peali* used at Woods Hole and *L. forbesi* used at Plymouth.

Sodium-free solutions

Many years ago Overton (1902) demonstrated that frog muscles became inexcitable when they were immersed in isotonic solutions containing less than 10% of the normal sodium-chloride concentration. He also showed that chloride ions were not an essential constituent of Ringer's solution, since excitability was maintained in solutions of sodium nitrate, bromide, sulphate, phosphate, bicarbonate, benzoate, etc. On the other hand, lithium was found to be the only kation which provided a reasonably effective substitute for sodium. Overton was unable to repeat his experiment with a frog's sciatic nerve, which maintained its excitability for long periods of time in salt-free solutions. But it now seems likely that this result was due to retention of salt in the interstitial spaces of the nerve trunk. Thus Kato (1936) found that application of isotonic dextrose to single medullated fibres of the frog caused a rapid but reversible loss of excitability, and a similar result was obtained by Erlanger & Blair (1938) on the sensory rootlets of the bull-frog. Kato's result has also been confirmed by Huxley & Stämpfli (unpublished experiments), who applied both isotonic sucrose and isotonic dextrose to single medullated fibres of the frog and found that conduction is blocked reversibly within a few seconds when the saline content falls below about 0·011 M. Katz (1947) has shown that isotonic sucrose mixtures abolish the action potential of *Carcinus* axons if the sodium-chloride concentration is less than 10–15% of that normally present in sea water. Further experiments on the effect of sodium-free solutions on *Carcinus* axons were made by one of us and will be summarized, because they provide a useful addition to the work with squid axons. In the first place, the action of isotonic dextrose on a single *Carcinus* axon is exceedingly rapid. The action potential is blocked in a few seconds and is restored in a similar space of time by restoration of saline. The speed at which these solutions act is not surprising, since solute molecules have to diffuse across a distance of only a few micra of loose connective tissue in order to reach the surface membrane of the axis cylinder. Further evidence can be obtained for the conclusion of Overton

(1902) and Lorente de Nó (1944, 1947) that it is the sodium and not the chloride ion which is essential for propagation. Thus axons are blocked by a mixture of 50 % isotonic choline chloride and 50 % dextrose, although this solution contains three or four times as much chloride as that present in a solution containing the minimum amount of sodium in the form of sodium chloride. The blocking effect of the first solution is not due to any harmful property of choline, since propagation occurs satisfactorily through a mixture of 50 % choline chloride

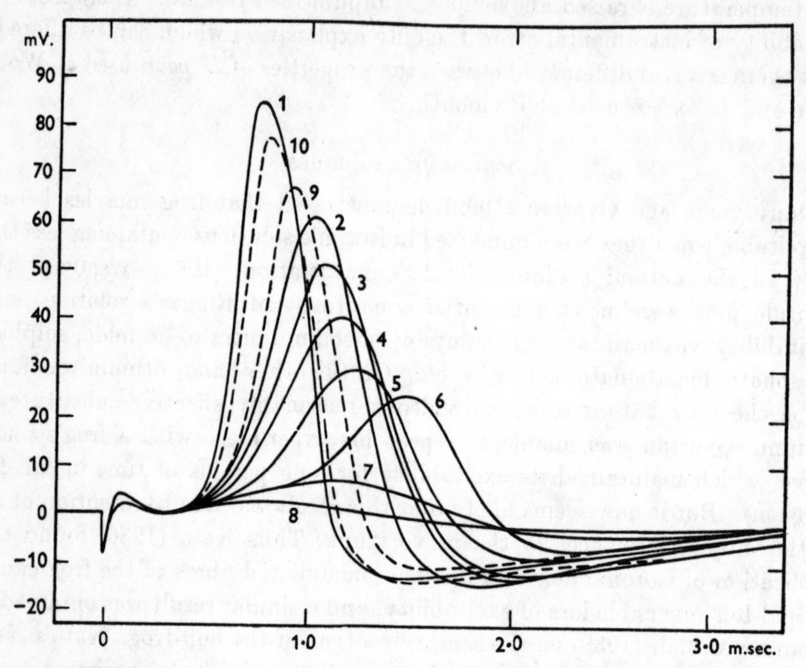

Fig. 3. Action of isotonic dextrose. Record 1: action potential in sea water just before application of dextrose. 2–8: records taken at following times after arbitrary zero, defined as moment of application of dextrose: 2, 30 sec.; 3, 46 sec.; 4, 62 sec.; 5, 86 sec.; 6, 102 sec.; 7, 107 sec.; 8, 118 sec. Record 9 taken 30 sec. after reapplication of sea water; 10, record at 90 and 500 sec. after reapplication of sea water (only one curve is drawn since the responses at these times were almost identical).

and 50 % sea water. Another point is that propagation is not affected by replacing the chloride in sea water with sulphate. All these experiments support the view that removal of sodium is the primary cause of block in salt-free solutions. A subsidiary factor may be the removal of calcium, since *Carcinus* axons do not survive for any length of time in a medium from which all traces of calcium have been removed.

The records in Fig. 3 show what happens to the membrane action potential when Overton's experiment is repeated on the giant axon of the squid. Curve 1 shows the electrical response of an axon immersed in sea water. Isotonic dextrose was substituted for sea water as soon as this record had been obtained. The operation of changing solutions took 30–60 sec. and the zero time to which subsequent records are referred was defined at a somewhat arbitrary point during this process. Records 2–8 show how the action potential changed as the preparation came into diffusion equilibrium with the new medium. The spike amplitude dropped progressively and eventually fell to a very small value (records 7 and 8). This residual deflexion was almost certainly due to electrotonic spread from the part of the axon surrounded by oil which was not affected by the test solution. Removal of salt had a very striking effect on the rate of rise of the action potential which decreased to about 1/12 of its former amplitude after 107 sec. On the other hand, the rate of fall and the positive phase changed much less rapidly. The resting potential appeared to increase with time, but this effect may be attributed to the external liquid junction potential which could not be evaluated in this experiment; all records have therefore been traced from the same base-line. At 6 min. after zero, sea water was restored, with the result that the action potential recovered rapidly to a value which was close to that observed initially. The effect of isotonic dextrose thus appears to be almost completely reversible.

The action potential was also found to be abolished reversibly by a mixture of 50% isotonic dextrose and 50% choline chloride. Only one satisfactory experiment was performed, but this gave a result which was essentially similar to that in Fig. 3. On the other hand, the action potential was maintained at a value of about 70 mV. in a solution containing 50% isotonic dextrose and 50% sea water, or in one containing 50% choline chloride and 50% sea water. Axons from the squid therefore behave like those of *Carcinus*, in that a certain amount of external sodium is necessary for production of the action potential.

Fig. 3 shows that the action of isotonic dextrose was considerably slower in the giant axon of the squid than it was in axons from *Carcinus*. The difference is not surprising since *Carcinus* axons are surrounded by only 3 μ. of connective tissue, whereas the squid axons were rarely dissected cleanly and in this experiment the axon was left with a layer of tissue about 110 μ. in thickness. Such a thickness of external tissue is of the right order of magnitude to account for the delay in terms of diffusion. A detailed analysis of the process of equilibration has not been attempted, but a rough calculation suggests that the delay may reasonably be attributed to diffusion of sodium chloride from the adventitious tissue surrounding the axon into the large volume of isotonic dextrose in the recording cell. After the records in Fig. 3 had been obtained, the axon was immersed in a solution containing 20% sea water and 80% isotonic dextrose. In this solution the action potential fell rapidly to a value corresponding to that in record 5, and underwent only a small reduction during the subsequent period of 14 min. It therefore appears that the action of salt-free solution was 80% complete in about 90 sec. This figure can be used to calculate an apparent diffusion constant if it is assumed that the fluid outside the preparation was completely stirred and that the diffusion process operated in the same manner as for a single substance diffusing into a slab of tissue 110 μ. in thickness. The value found for this experiment was about 0·1 cm.²/day and values of this order of magnitude have been obtained in other cases. The diffusion constant for sodium chloride in water is 1·0 cm.²/day and for dextrose 0·5 cm.²/day (Landolt-Börnstein, 1931). The lag in the action of salt-free solutions can be explained if diffusion through the connective tissue and interstitial spaces in the remains of the nerve trunk are assumed to be about 1/7 of those in water. This is a reasonable assumption, since Stella (1928) concluded that diffusion of phosphate through the extra-cellular part of the frog's sartorius muscle was very slow compared to that in free solution. Another factor which may have retarded diffusion in the later stages is that the external solution was not stirred mechanically after the initial process of applying the test solution had been completed.

The effect of solutions of reduced sodium content on the resting potential and action potential

The general action of solutions containing a low sodium concentration is illustrated by Fig. 4. Record *a* 1 shows the action potential of an axon immersed in sea water. The base-line has been displaced from the zero of the calibration scale by an amount which corresponds to the resting potential. The zero was determined by short-circuiting the amplifier input and subtracting the small difference of potential which existed between the two recording electrodes. The zero therefore occurs at the potential which would have been observed if the microelectrode had been withdrawn and placed in the sea water outside the axon. Record *a* 2 shows the resting potential and action potential recorded after 16 min. in a solution containing 33% sea water and 67% isotonic dextrose. The method of obtaining the resting potential was similar to that formerly employed, except that an additional correction for the liquid junction potential has been introduced. The resting potential would have appeared to be 4·2 mV. larger if no such allowance had been made. The zero on the record now corresponds to the potential which would have been observed if the microelectrode had been withdrawn and connected to the solution in the recording cell by means of a saturated KCl bridge. Record *a* 3 was obtained 14 min. after replacing sea water in the recording cell. The action potential was 5 mV. less than that at the beginning of the experiment, but the difference was small compared to the decrease shown by *a* 2. The spike also arose with a greater delay, although the form and rate of rise were close to those observed originally. An effect of this kind is inevitable because the test solution diffused from the upper part of the recording cell into the region of nerve surrounded by oil. The total conduction time was increased by this process, and the effect was only very slowly removed by application of sea water to the upper part of the

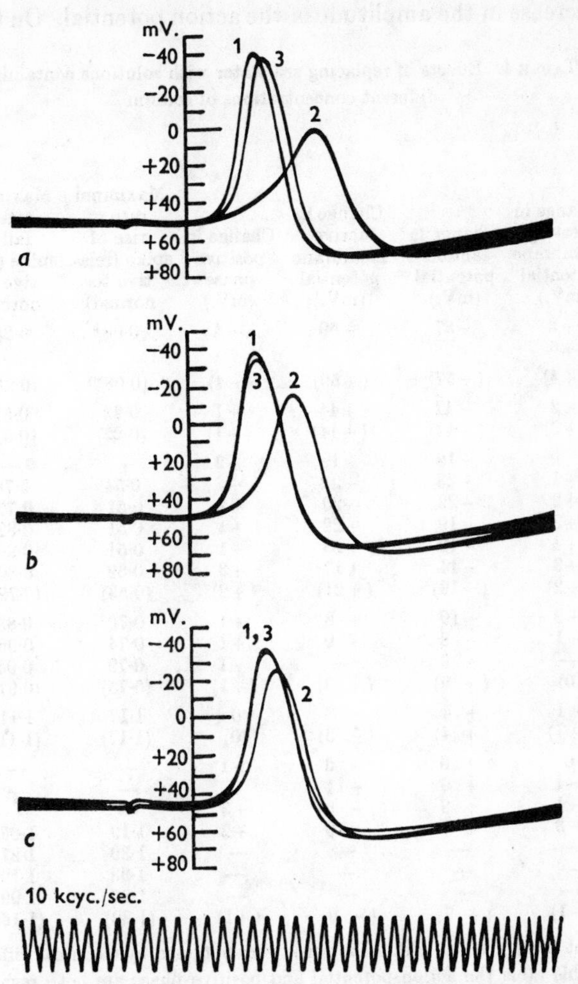

Fig. 4. Action of sodium-deficient solutions on the resting and action potential. *a*1, response in sea water; *a* 2, after 16 min. in 33 % sea water, 67 % isotonic dextrose; *a* 3, 13 min. after reapplication of sea water. *b* 1, response in sea water, *b* 2, after 15 min. in 50 % sea water, 50 % isotonic dextrose; *b* 3, 6 min. after reapplication of sea water. *c* 1, response in sea water; *c* 2, after 16 min. in 71 % sea water, 29 % isotonic dextrose; *c* 3, 7 min. after reapplication of sea water. The scale gives the potential difference across the nerve membrane (outside – inside) with no allowance for the junction potential between the axoplasm and the sea water in the microelectrode.

recording cell. Records *b*1, *b*2, and *b*3 or *c*1, *c*2, and *c*3 were made in a comparable manner, except that the test solutions consisted of 50 or 71 % sea water. The effects produced by these solutions were smaller, but of the same general type as those illustrated by *a*1, *a*2 and *a*3.

This experiment illustrates a number of important points. In the first place it shows that dilution of sea water with isotonic dextrose caused a large and reversible decrease in the amplitude of the action potential. On the other hand

TABLE 4. Effects of replacing sea water with solutions containing different concentrations of sodium

Concentration of sodium in test solution (expressed as fraction of concentration in sea water)	Change in resting membrane potential (mV.)	Change in action potential (mV.)	Change in active membrane potential (mV.)	Change in positive phase (mV.)	Maximum rate of rise of spike (relative to normal).	Maximum rate of fall of spike (relative to normal)	Action potential amplitude (relative to normal)
0·2	+2	−57	+59	−4	0·08*	0·25*	0·24
	+6						
	(+4)	(−57)	(+59)	(−4)	(0·08*)	(0·25*)	(0·24)
0·33	+2	−42	+44	+1	0·22	0·51	0·54
	(+2)	(−42)	(+44)	(+1)	(0·22)	(0·51)	(0·54)
0·50	0	−19	+19	+2	—	—	0·79
	−1	−23	+22	+2	0·54	0·78	0·73
	+2	−22	+20	+2	0·51	0·77	0·76
	+3	−19	+22	+1	0·51	0·82	0·76
	+4	−19	+23	+1	0·51	0·80	0·75
	+3	−14	+17	+3	0·59	0·80	0·84
	(+2)	(−19)	(+21)	(+2)	(0·53)	(0·79)	(0·77)
0·715	−2	−10	+ 8	+1	0·70	0·83	0·87
	+1	− 8	+ 9	+1	0·74	0·96	0·90
	—	− 9	—	+1	0·79	0·95	0·89
	(0)	(− 9)	(+ 9)	(+1)	(0·75)	(0·91)	(0·89)
1·26	+1	+ 4	− 3	0	1·17	1·11	1·05
	(+1)	(+ 4)	(− 3)	(0)	(1·17)	(1·11)	(1·05)
1·56	0	+ 6	− 6	+1	—	—	1·07
	−1	+ 9	−11	0	—	—	1·12
	−1	+ 8	− 9	+3	—	—	1·10
	−3	+ 7	−10	+2	1·19	1·09	1·09
	—	—	—	—	1·39	1·27	—
	—	—	—	—	1·33	1·19	—
	—	—	—	—	1·27	1·09	—
	(−1)	(+ 7)	(− 9)	(+1)	(1·30)	(1·16)	(1·09)

All rates except those marked with an asterisk were obtained by electrical differentiation. For the purpose of this table the action potential and positive phase are both regarded as positive quantities. A positive change in membrane potential means that the outside of the nerve becomes more positive with respect to the inside. Sodium-deficient solutions were made by diluting sea water with isotonic dextrose, sodium-rich solutions by adding solid sodium chloride to sea water. Average values are enclosed in parentheses. Changes were measured with reference to a normal value in sea water which was obtained in each case from the mean of determinations made before and after application of a test solution.

the resting potential was altered to such a small extent that no difference can be seen in Fig. 4. There was usually a small increase in resting potential, as may be seen from the figures in Table 4, but the change was always small compared to the change in spike amplitude. The constancy of the resting potential means that removal of external sodium reduces the action potential by decreasing the overshoot or membrane reversal. In fact, in 33% sea water, the overshoot had disappeared and the action potential was then slightly smaller than the resting potential. Another interesting point is that the rate of rise of the action potential was markedly affected by sodium-deficient solutions, whereas the rate of fall changed only in proportion to the amplitude. It can also be seen that the positive phase was only slightly affected by removal of sodium.

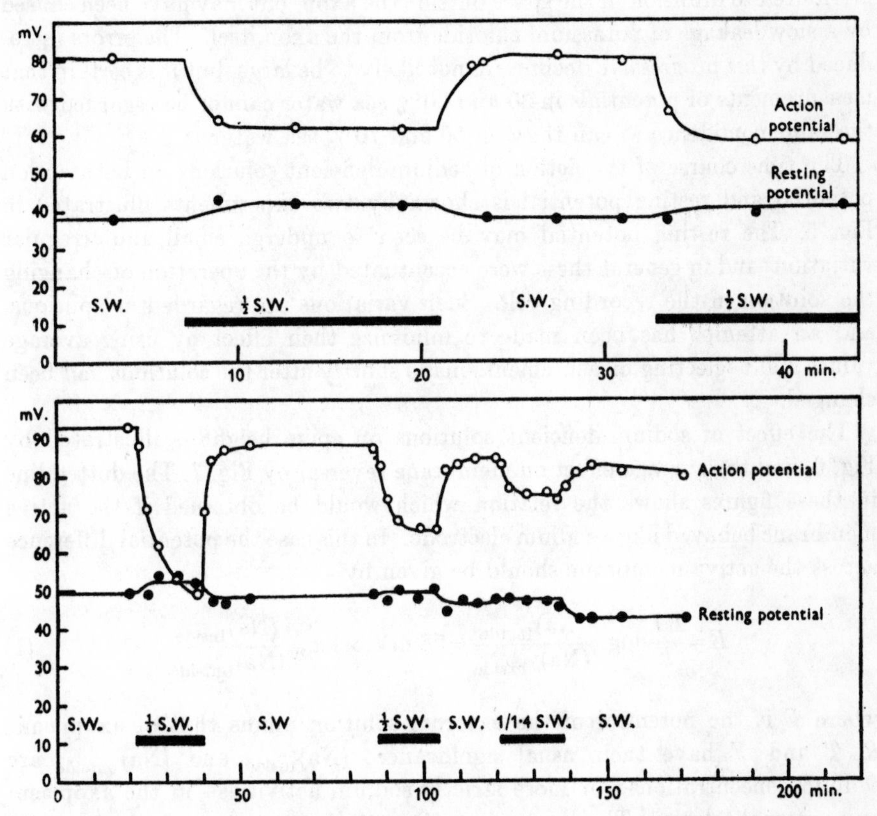

Fig. 5. Time course of action of sodium-deficient solutions made by diluting sea water with isotonic dextrose. Resting potentials are apparent potentials corrected for liquid junction potentials in the external circuit, but not corrected for the junction potential between axoplasm and sea water at the tip of the microelectrode. Both action potential and resting potential are treated as positive quantities.

The quantitative results obtained with sodium-deficient solutions are shown in Table 4. The principal difficulty in making these measurements was connected with the fact that diffusion times prevented the sodium-deficient solutions from acting instantaneously, and it was essential that quantitative measurements should not be made until equilibrium had been obtained. Photographic records were usually made at intervals of 2, 5, 10 and 15 min. after application of the new solution. This procedure gave satisfactory results when solutions containing more than 50% sea water were employed. The results at 10 and 15 min. rarely differed by more than 2 mV., and equilibrium was sometimes attained after 5 min. On the other hand, measurements in solutions containing less than 50% sodium were unsatisfactory, because there was always a progressive decline in the action potential and in the resting potential. This could not be attributed to diffusion in the space outside the axon, but may have been caused by a slow leakage of potassium chloride from the axon itself. The errors introduced by this progressive decline are not likely to be large, but it is certain that measurements of potentials in 30 and 20% sea water cannot be regarded with the same confidence as can those in 50 and 70% sea water.

The time course of the action of sodium-deficient solutions on both action potential and resting potential is shown by two experiments illustrated in Fig. 5. The resting potential may be seen to undergo small and irregular variations and in general these were accentuated by the operation of changing the solution in the recording cell. Such variations are regarded as spurious, and an attempt has been made to minimize their effect by using average values and neglecting measurements made shortly after the solutions had been changed.

The effect of sodium-deficient solutions on spike height is illustrated by Fig. 6, and the average effect on membrane reversal by Fig. 7. The dotted line in these figures shows the relation which would be obtained if the active membrane behaved like a sodium electrode. In this case the potential difference across the active membrane should be given by

$$E=\frac{RT}{F}\log_e\frac{(\mathrm{Na})_{\mathrm{inside}}}{(\mathrm{Na})_{\mathrm{outside}}}=58\ \mathrm{mV.}\times\log_{10}\frac{(\mathrm{Na})_{\mathrm{inside}}}{(\mathrm{Na})_{\mathrm{outside}}}, \qquad (1)$$

where E is the potential of the external solution minus that of axoplasm; R, T and F have their usual significance; $(\mathrm{Na})_{\mathrm{inside}}$ and $(\mathrm{Na})_{\mathrm{outside}}$ are sodium concentrations—or more strictly sodium activities—in the axoplasm and external solution. The change in active membrane potential which results from an alteration of external sodium should be given by equation 2, since it may be assumed that the internal concentration of sodium does not change, or changes only very slowly when the external sodium is altered.

$$\Delta E=E_{\mathrm{test}}-E_{\mathrm{sea\ water}}=\frac{RT}{F}\log_e\frac{(\mathrm{Na})_{\mathrm{sea\ water}}}{(\mathrm{Na})_{\mathrm{test}}}. \qquad (2)$$

The absolute magnitude of the action potential is equal to the difference between the membrane potentials of resting and active nerve. Since the resting potential is only slightly altered by dilution of sea water, equation 2

4—2

should also apply to the change in spike height. (The sign of the change must be reversed if the spike is regarded as a positive quantity.) The data in Table 4, Figs. 6 and 7 show that equation 2 is obeyed reasonably by solutions containing 50 and 70% of the normal sodium concentration. The rough agreement must not be pressed, because the behaviour of the active membrane is likely to be much more complicated than that of a sodium electrode. Another reason for caution is that there is no certain information about the activity coefficient of

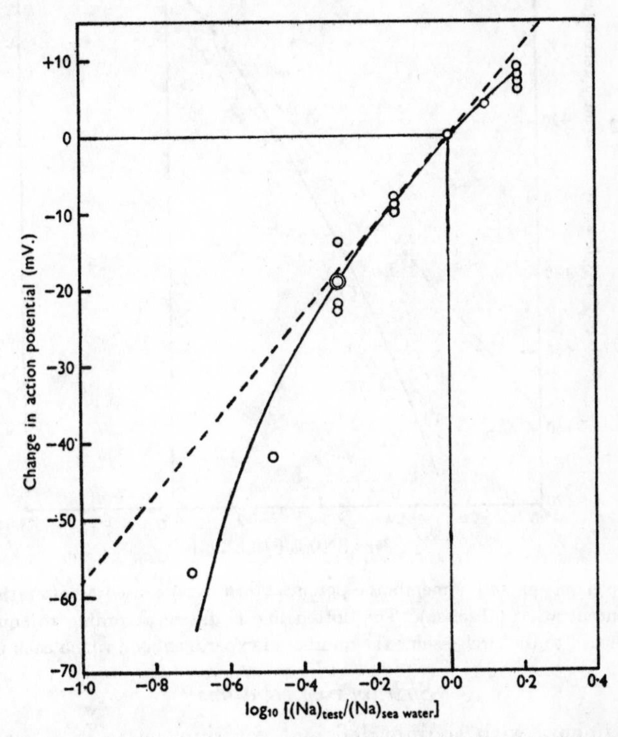

Fig. 6. Change in amplitude of action potential (ordinate) caused by alteration of external sodium concentration (abscissa). The dotted line is drawn through the origin with a slope of 58 mV.

the sodium ion in dextrose mixtures. Preliminary measurements indicate that an allowance for changes in activity coefficients would make the slope of the theoretical line about 10% less than that in Figs. 6 and 7. But the corrections for activity coefficients were so uncertain that we have preferred to use concentrations in equation 2. The results in Figs. 6 and 7 indicate that the depression of action potential height was disproportionately large in solutions containing 20 and 33% sea water. An effect of this kind can be explained if it is assumed that the permeability to sodium increases with the depolarization of the membrane. The action potentials in solutions of 20 or 33% sodium were much smaller than normal, so that it is plausible to suppose that the mechanism responsible for transporting sodium might not be operating at full efficiency in these solutions.

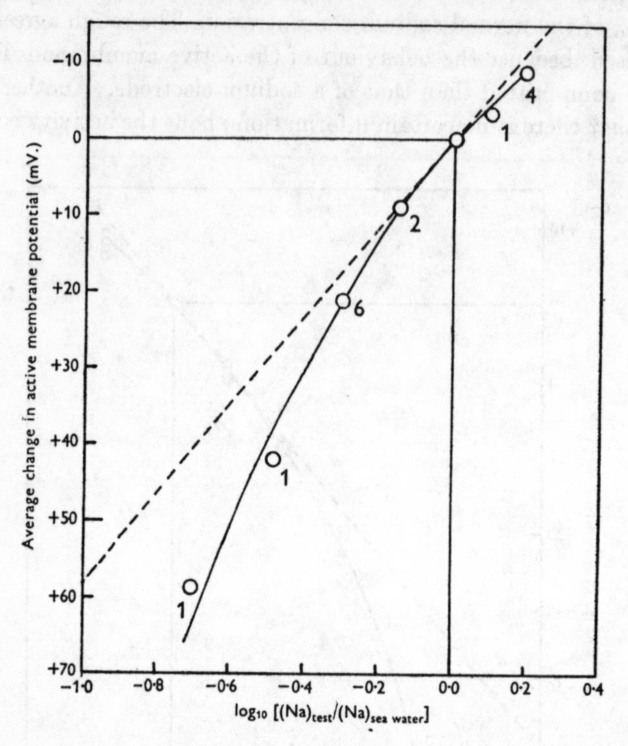

Fig. 7. Average change in active membrane potential (ordinate) caused by alteration of external sodium concentration (abscissa). The dotted line is drawn according to equation 2. The figures attached to the circles show the number of experiments on which each point is based.

Sodium-rich solutions

The experiments with sodium-deficient solutions were in good agreement with the simple hypothesis which they were designed to test. But it might be argued that the results observed were due to the abnormal nature of the external media rather than to any specific effect of the sodium ion. The regular and reversible nature of the changes speaks against this view, but there is a more compelling reason for rejecting it. The concentration of sodium chloride was increased from 455 to 711 mM. by dissolving 15 g. of solid NaCl in 1 l. of sea water. This solution was strongly hypertonic and damaged the axon by osmotic effects in 5–15 min. But before the osmotic effects became apparent the axon gave an increased action potential with characteristics which were the converse of those in sodium-deficient solutions. The effect of sodium-rich solutions is best illustrated by the behaviour of an axon from which almost all the external tissue had been removed by dissection. In this axon the thickness of the external tissue was about 25 μ. so that diffusion times were relatively short. Fig. 8*a* shows the action potential observed when this axon was immersed in sea water. A sodium-rich solution was applied 2 min. after curve *a* had been obtained, and curve *b* recorded 50 sec. later. The height of the action potential, the overshoot and the rate of rise all show small but quite definite increases. Measurements indicate that the action potential increased from 86 to 95 mV., while the active membrane potential changed from -42 to -53 mV. These values were maintained for 4 min. and, on replacing sea water, returned to

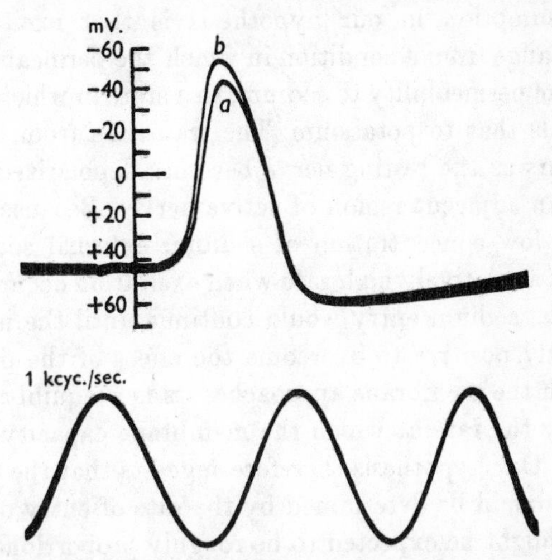

Fig. 8. Action of sodium-rich solution on the resting and action potential. *a*, response in sea water; *b*, response 50 sec. after application of sea water containing additional quantity of NaCl. (The sodium concentration of this solution was 1·56 times that of sea water.) The scale gives the potential difference across the nerve membrane (outside − inside) with no allowance for the junction potential between the axoplasm and the sea water in the micro-electrode.

84 and -41 mV. The changes are not large, but the increase in overshoot is close to that predicted by equation 2. The sodium concentration of this solution was 1·56 times that in sea water so that the theoretical change in overshoot is

$$\Delta E = E_{\text{test}} - E_{\text{sea water}} = 58 \text{ mV.} \times \log_{10} 1/1{\cdot}56 = -11 \text{ mV.}$$

A control with a solution containing 0·5 mol. dextrose dissolved in 1 l. of sea water gave no immediate increase in spike height but only a very small and gradual decrease which must be regarded as an osmotic effect.

Other experiments with sodium-rich solutions gave results which were essentially similar to those in Fig. 8, although the changes observed were somewhat smaller, as may be seen in Table 4. It was also found that the period of increased spike height was rarely maintained for more than a few minutes, and was followed by a period of progressive deterioration which was only partially reversible. Control experiments with solutions containing extra dextrose showed the phase of progressive deterioration, but never gave the initial increase in spike height, or rate of rise. There is therefore some reason for believing that the changes produced by excess of sodium would have been rather larger if the action of extra sodium could have been dissociated from the osmotic effect of the solutions.

The rate of rise of the action potential

The basic assumption in our hypothesis is that excitation causes the membrane to change from a condition in which the permeability to potassium is greater than the permeability to sodium, to a state in which the permeability to sodium exceeds that to potassium. The transition from the resting to the active state occurs as the resting nerve becomes depolarized by local circuits spreading from an adjacent region of active nerve. Because the inside of the axon contains a low concentration of sodium, external sodium ions should enter the axon at a relatively high rate when excitation occurs. In the absence of other processes, sodium entry would continue until the inside of the axon became sufficiently positive to overcome the effect of the diffusion gradient. The rate at which the membrane approaches its new equilibrium value should be determined by the rate at which the membrane capacity is discharged by entry of sodium. Our hypothesis therefore suggests that the rate of rise of the action potential should be determined by the rate of entry of sodium, and on a simple view it might be expected to be roughly proportional to the external concentration of sodium.

A quantitative basis for part of this argument can be provided in the following way. The membrane current during the action potential is proportional to the second derivative of potential with respect to time, and is therefore zero when the first derivative is at a maximum or a minimum. The current passing through the membrane consists of capacity current ($C\, \partial V/\partial t$), which involves a change of ion density at its outer and inner surface, and an ionic current due to transport of ions across the membrane. These two components must be equal and opposite when the total membrane current is zero. The following relation therefore applies at the moments when the rate of change of membrane voltage is at a maximum or minimum

$$-C\, \partial V/\partial t = I, \tag{3}$$

where I is the net inward current per sq.cm. due to transfer of ions from outside to inside, C is the membrane capacity per sq.cm., V is the potential difference across the nerve membrane (outside potential − inside potential).

The simple nature of equation (3) indicates that the most valuable type of rate measurement is a determination of the maximum rate. This can be obtained by graphical analysis of the action potential, but is best recorded directly by electrical differentiation. Typical records obtained by this procedure are given in Figs. 9 and 10. These show two distinct phases and not three, as might have been expected from the diphasic character of the squid action potential. The reason for this is that the rate at which the positive phase disappears is small compared to the rates at which the initial part of the spike

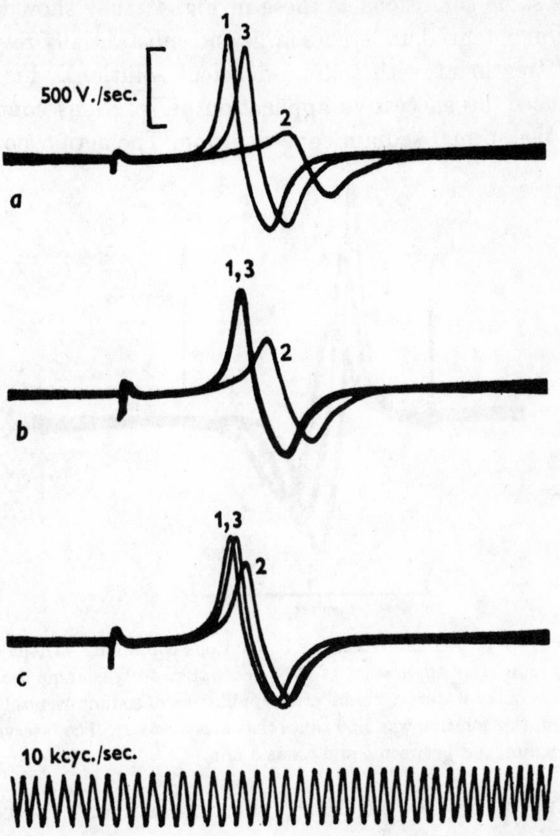

Fig. 9. Action of sodium-deficient solutions on rate of change of membrane voltage. *a* 2, *b* 2, *c* 2, response in 33, 50 and 71 % sea water, remainder in pure sea water. The times at which records were obtained are approximately the same as those in Fig. 4. The calibration scale in *a* applies to all records.

rises and falls. The peak of the rate record is proportional to the positive ionic current entering the axon and the trough to the positive ionic current leaving the axon. The absolute value of these currents can be estimated roughly, since the membrane capacity of the squid axon has been determined as 1·1 μF.cm.$^{-2}$ (Cole & Curtis, 1938) or 1·8 μF.cm.$^{-2}$ (Cole & Curtis, 1939), and may be taken as 1·5 μF.cm.$^{-2}$. The average values for the maximum rates of rise and fall of the spike were 630 and 380 V.sec.$^{-1}$, so that the ionic current entering the axon during the rising phase was of the order of 0·9 mA.cm.$^{-2}$, whereas the ionic current leaving during the falling phase was of the order of 0·6 mA.cm.$^{-2}$. Corresponding figures in terms of the rate of entry or exit of a monovalent kation are 10^{-8} and $0{\cdot}6 \times 10^{-8}$ mol.cm.$^{-2}$ sec.$^{-1}$.

Fig. 9 shows how the first derivative of the action potential is affected by sodium-deficient solutions. These records were obtained from the same axon and under the same conditions as those in Fig. 4; they show that the rate of rise of the action potential undergoes large and substantially reversible changes as a result of treatment with sodium-deficient solutions. Fig. 10 shows the changes produced by successive application of solutions containing 50, 100 and 156% of the normal sodium concentration. The action potential reached

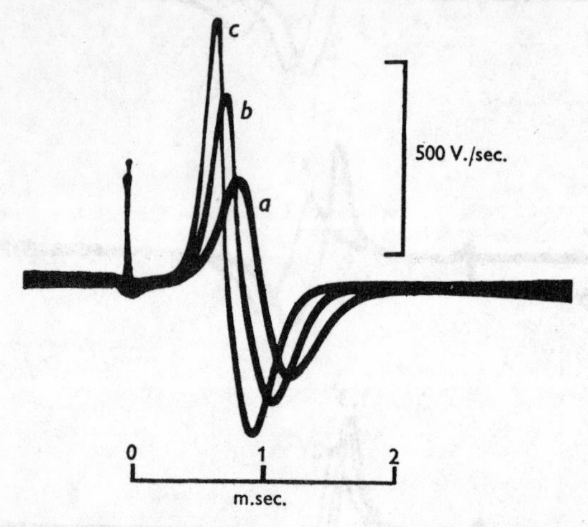

Fig. 10. Rate of change of membrane voltage in solutions containing various concentrations of sodium: *a*, 12 min. after application of 50% sea water, 50% isotonic dextrose; *b*, 16 min. after application of sea water; *c*, 2 min. after application of sodium-rich solution (the sodium concentration in this solution was 1·56 times that in sea water). The interval between record *a* and *b* was 18 min. and between *b* and *c* was 3 min.

a constant value in the 50 and 100% solutions so that the change in rate shown by record *c* was certainly a genuine increase and not merely a recovery from the previous immersion in the 50% solution. Data from other axons are collected in Table 4 and plotted graphically in Fig. 11. It will be seen that the rate of rise is proportional to sodium concentration over the range 50–100% sea water, but that the rate falls off rapidly below 50%. This effect is almost certainly related to the disproportionately large changes in action potential observed in solutions containing 20 and 33% sea water.

The rates of rise showed substantial increases in solutions containing extra sodium, but the effects were no longer proportional to the sodium concentration. Thus the largest increase encountered in a solution containing 1·56 times the normal sodium was 1·39, and the average value was only 1·3. This result may be attributed to the deleterious action of the hypertonic solutions, but it is also possible that there may be a genuine lack of proportionality in solutions containing an excess of sodium. However, there is good evidence to show that simple proportionality does hold over a limited region and it is certain that the rate of rise is altered reversibly over a wide range by changes in the external sodium concentration.

The rate of fall of the action potential is also influenced by sodium, but to a lesser extent (Table 4). Thus the average change in rate of rise in a 50% solution was 0·53, whereas the average change in rate of fall was 0·8. The rates of fall appear to change in proportion to the height of the action potential, as

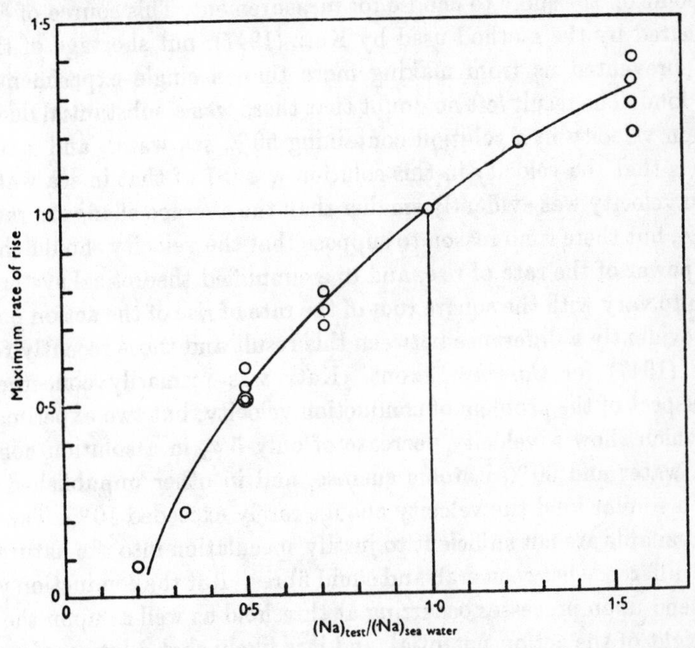

Fig. 11. Ordinate: maximum rate of rise of spike in test solution/maximum rate of rise in sea water. Abscissa: sodium concentration of test solution/sodium concentration in sea water.

may be seen by comparing the average ratios in the last two columns of Table 4. This result suggests that changes in external sodium affect the rate of fall indirectly by altering the amplitude of the spike. A change in the rate of fall is a natural consequence of a change in spike height; for it is to be expected that the rate at which the potential reverts to its resting value should depend upon the extent to which activity has displaced the membrane from its resting level.

Changes in conduction velocity

The velocity of transmission should be reduced by sodium-deficient solutions, since these solutions lower the rate of rise of the action potential. This deduction could not be tested in the experiments with internal electrodes, since a large part of the conduction time arose in the lower part of the nerve which was immersed in oil. A single experiment with external electrodes was made in order to find out if there was an appreciable change of velocity in a large volume of 50% sea water. The axon was arranged in such a way that the conduction stretch between stimulating and recording electrodes could be dipped into sea water or into a test solution (cf. Hodgkin, 1939). With an arrangement of this kind the absolute changes in conduction time could be measured with considerable accuracy, but there was some uncertainty in determining the velocity because it was difficult to know which was the correct point on the spike to choose for measurement. This source of error can be eliminated by the method used by Katz (1947), but shortage of time and material prevented us from making more than a single experiment of the simplest kind. The result left no doubt that there was a substantial decrease of conduction velocity in a solution containing 50% sea water, and a tentative estimate is that the velocity in this solution was 0·7 of that in sea water. The change in velocity was evidently smaller than the average change in rate given in Table 4, but there is no reason to suppose that the velocity should change as the first power of the rate of rise, and in a simplified theoretical system it can be shown to vary with the square root of the rate of rise of the action potential. There is evidently a difference between this result and those recently reported by Katz (1947) for *Carcinus* axons. Katz was primarily concerned with another aspect of the problem of conduction velocity, but two experiments are quoted which show a velocity decrease of only 5% in a solution containing 50% sea water and 50% isotonic sucrose, and in other unpublished experiments of a similar kind the velocity change rarely exceeded 10%. The data at present available are not sufficient to justify speculation into the nature of this apparent difference between crab and squid fibres. But the conduction velocity must depend upon processes occurring at threshold as well as upon the rate of rise or height of the action potential, and it is likely that dilution of sea water would give different overall effects in different types of axon. In this connexion it should be remembered that dilution of sea water with sugar solutions alters the concentration of other ions besides that of sodium, and it is conceivable that the effect of sodium removal may sometimes be balanced by an increase in excitability resulting from the simultaneous reduction of calcium concentration. Apart from the numerical discrepancy, the results of Katz are in general agreement with those reported here. Thus the velocity of conduction in *Carcinus* axons was found to undergo a substantial decrease in solutions containing less than 30% sea water and block occurred when the sea-water content was less than 10%.

Specificity of sodium action

The reduction in spike height which results from mixing sea water with isotonic dextrose has been attributed to dilution of sodium, but it is conceivable that the observed effects might have been partly due to dilution of other ions such as Ca, K or Cl. This possibility was examined by comparing the effects of two solutions. The first solution was made by mixing artificial sea water with isotonic dextrose, while the second was made in such a way that all components except the sodium and chloride ions were maintained at about their normal level. The composition of the two solutions and the results obtained with them are given in Table 5. It will be seen that solution 2 gave a smaller action potential than solution 1, but that this drop was almost entirely due to a 5 mV. diminution in resting potential which probably arose from the increased potassium concentration of the second solution. Since both solutions contained the same concentration of sodium, equation (2) predicts that the active membrane potential should remain constant and the figures in Table 5 show that this prediction is verified. The rate of rise in the second solution was 20% less than that in the first, and this effect may be attributed either to the higher calcium and magnesium content of solution 2 or to the lowered resting potential resulting from the increase in potassium concentration.

TABLE 5.

Operation	Change in resting membrane potential (mV.)	Change in action potential (mV.)	Change in active membrane potential (mV.)	Change in positive phase (mV.)	Maximum rate of rise (relative to previous condition)	Maximum rate of fall (relative to previous condition)	Action potential amplitude (relative to previous condition)
From artificial sea water to solution 1	+4	−19	+23	+1	0·51	0·80	0·75]
From solution 1 to solution 2	−5	−5	0	0	0·81	0·92	0·90]
From artificial sea water to solution 1	+3	−14	+17	+3	0·59	0·80	0·84 }
From solution 1 to solution 3	+1	+2	−1	0	1·17	1·09	1·03 }

Square brackets indicate that measurements were made on the same axon but were separated by a considerable time interval; curved brackets that they were obtained on the same axon at approximately the same time. All figures are average values determined in the usual way. The compositions of the solutions are given below:

Solution	Description	Na	K	Ca	Mg	Cl	HCO_3
		Concentration as fraction of concentration in artificial sea water					
1	Artificial sea water diluted 1:1 with isotonic dextrose	0·5	0·5	0·5	0·5	0·5	0·5
2	Solution 1 + K, Ca, Mg, HCO_3	0·5	1·0	1·0	1·0	0·5	1·0
3	Artificial sea water diluted 1:1 with 0·6 M-choline chloride	0·5	0·5	0·5	0·5	1·0	0·5

The previous experiment indicates that the changes in active membrane potential and spike height were primarily due to alterations in the concentrations of either the chloride or the sodium ion. The effect of these two ions may be separated by diluting sea water with isotonic choline chloride instead of isotonic dextrose. The results obtained in a single experiment of this type are illustrated by the effects given in Table 5 for solutions 1 and 3. It will be seen that the general action of these solutions was similar and, in particular, that the active membrane potentials differed by less than 1 mV., although the chloride concentrations of the two solutions were widely different. The only anomalous point is that the rate of rise was found to be appreciably greater in the solution containing choline chloride than in the one containing dextrose. Part of this difference may be attributed to a small change in resting potential, but it seems unreasonable to ascribe all the increase to this cause.

A single experiment with an artificial sea-water solution containing lithium instead of sodium indicated that the action of these two ions was almost identical. This result is supported by unpublished experiments with *Carcinus* axons which show that propagation occurs satisfactorily for at least 1 hr. in a solution containing lithium but no sodium. Gallego & Lorente de Nó (1947) report that medullated nerve becomes depolarized and inexcitable after immersion in lithium solutions for several hours. We must therefore suppose either that the reactions of vertebrate nerve to lithium differ from those of invertebrate nerve, or that our experiments were not maintained for sufficient time to reveal the effects described by Gallego & Lorente de Nó.

Preliminary experiments with isotonic sucrose mixtures show that the action of this sugar was similar to that of dextrose.

No perceptible changes occurred when the oxygen tension of the sea water was increased fivefold.

Effect of varying potassium concentration

The action potential may be regarded as being made up of a component due to the resting potential, which is only slightly altered by dilution of sea water with isotonic dextrose, and an overshoot which is directly influenced by the external sodium concentration. It is known that variations in the external potassium concentration alter the resting potential, and on a simple view it is to be expected that these variations would change the amplitude of the action potential but not the reversed potential difference of the active membrane. This hypothesis cannot be pressed, because increasing the potassium concentration causes nerve fibres to become inexcitable long before they are completely depolarized (Curtis & Cole, 1942). There is also the experimental difficulty that the changes in resting potential are small over the range in which excitability is retained. In practice, this meant that the values of resting potential and spike height had to be measured to a degree of precision which was near the experimental limit.

The effect of changing from a potassium-free solution to one containing the normal potassium content is shown in Fig. 12. It will be seen that the action potential was slightly greater in the potassium-free solution than it was in sea water, and that this effect was largely due to a change in the resting potential. The numerical results obtained in this and other experiments are given in Table 6 and are more reliable than values obtained by comparison of single records of the type shown in Fig. 12. They show that the action potential in a potassium-free solution was 4·5 mV. greater than normal and that the resting

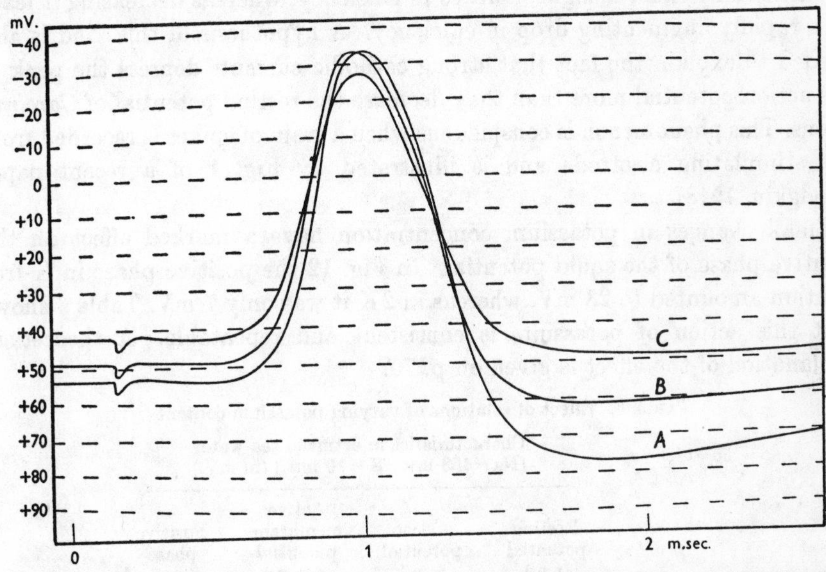

Fig. 12. Effect of varying K concentration on action potential and resting potential. Curve *A*, K-free artificial sea water (K = 0, Na = 463 mM.); *B*, artificial sea water (K = 10 mM., Na = 453 mM.); *C*, K-rich artificial sea water (K = 20 mM., Na = 443 mM.). The dotted lines were traced from a projection of the calibration grid and give the potential difference across the nerve membrane (outside − inside) without correction for the junction potential between the axoplasm and the sea water in the microelectrode.

potential was 3·1 mV. larger. The active membrane was therefore 1·4 mV. more negative in a potassium-free solution than in sea water. The sodium concentration of the first solution was 2·2% greater than that of the second, so that 0·5 mV. of the difference must be attributed to this cause. The remaining difference is not greater than the experimental error, but is probably real because the converse effect is seen with potassium-rich solutions. Thus, in a solution containing twice the usual amount of potassium, the decrease in resting potential accounts for less than half the decrease in action potential, and further increases in potassium concentration cause the spike to drop rapidly to zero, although there is still a substantial resting potential. These facts indicate that it is an over-simplification to suppose that the active membrane always reaches the equilibrium potential of a sodium electrode. Instead, we must assume that the sodium permeability does not rise to a value high enough to swamp the contribution of potassium and chloride to the active membrane potential. Under these circumstances anything which interferes with the sodium transport mechanism must result in a diminution of spike height. If this view is adopted the changes in Table 6 can be explained by supposing that the efficiency of the sodium transport mechanism depends upon the membrane potential of the resting nerve. Raising the resting membrane potential may cause a slight increase in efficiency, whereas decreasing it leads to a rapidly augmenting drop in efficiency. A hypothesis of this kind is also needed to explain the fact that strong cathodic currents depress the peak of the action potential more than they decrease the resting potential of *Carcinus* axons. This phenomenon is conspicuous when a train of spikes is recorded from the stimulating electrode and is illustrated by Fig. 1 of a recent paper (Hodgkin, 1948).

Small changes in potassium concentration have a marked effect on the positive phase of the squid potential. In Fig. 12 the positive phase in K-free solution amounted to 23 mV. whereas in 2 K it was only 7 mV. Table 6 shows that this action of potassium is consistent and repeatable. A theoretical explanation of the effect is given on p. 70.

TABLE 6. Effect of solutions of varying potassium content

	Characteristics in artificial sea water [Na = 453 mM., K = 10 mM.] (in mV.)			
	Resting potential (r.p.)	Action potential (a.p.)	Active membrane potential (a.m.p.)	Positive phase (p.p.)
	+46	+86	−40	+12
	+41	+83	−42	+15
	+48	+82	−34	+14
	+48	+83	−35	+12
Average	+46	+84	−38	+13

Change in potential on substituting test solution for artificial sea water (in mV.)

	K-free [Na = 463 mM., K = 0.]				$1\frac{1}{2}$K [Na = 448 mM., K = 15 mM.]				2K [Na = 443 mM., K = 20 mM.]			
	Δr.p.	Δa.p.	Δa.m.p.	Δp.p.	Δr.p.	Δa.p.	Δa.m.p.	Δp.p.	Δr.p.	Δa.p.	Δa.m.p.	Δp.p.
	+2·5	+4·2	−1·7	+10·0	−1·7	−2·7	+1·0	−3·2	−3·6	−6·5	+2·9	−6·2
	+1·7	+4·5	−2·8	+10·9					−2·1	−7·3	+5·2	−5·7
	+4·2	+4·6	−0·4	+9·6	−3·0	−4·5	+1·5	−4·2				
	+3·9	+4·7	−0·8	+10·3					−5·5	−12·1	+6·6	−5·9
Average	+3·1	+4·5	−1·4	+10·2	−2·3	−3·6	+1·2	−3·7	−3·7	−8·6	+4·9	−5·9

DISCUSSION

The experiments described in this paper are clearly consistent with the view that the active membrane becomes selectively permeable to sodium, and thereby allows a reversed membrane e.m.f. to be established. The evidence is indirect, and the sodium hypothesis cannot be pressed until more is known about the ionic exchanges associated with nervous transmission. But the hypothesis does provide a good working basis for future experiments, and it gives a satisfactory explanation of several observations which cannot be reconciled with the classical membrane theory. On the other hand, the hypothesis encounters a number of difficulties of which only a few can be mentioned here. One of the most serious objections arises from the fact that Curtis & Cole (1942) describe an experiment in which the active membrane reversed by 110 mV., whereas Steinbach & Spiegelman's (1943) figures indicate that the sodium concentration of fresh axoplasm is about one-tenth of that in sea water. The maximum overshoot allowed by a tenfold ratio is 58 mV. and the ratio would have to be nearly 100 in order to produce an overshoot of 110 mV. The discrepancy is all the more serious because it is exceedingly unlikely that the membrane potential could reach the theoretical maximum for a sodium electrode. The difficulty does not arise in our experiments, since the reversed membrane e.m.f. has always been well below the limit allowed by Steinbach & Spiegelman's figures. The only alternatives which remain if Curtis & Cole's figure of 110 mV. is accepted are: first, that the sodium hypothesis is incorrect or incomplete; and secondly, that the sodium-ion activity in certain axons may be less than one-hundredth of that in the external fluid. Another possible criticism is that many agents affect the amplitude of the action potential without causing much change in the magnitude of the resting potential. Examples are afforded by cocaine or amyl alcohol, which block conduction at concentrations that cause a slight increase in resting potential (Bishop, 1932). Observations of this kind can be explained by assuming that the mechanism for transporting sodium is of a highly specialized nature, and is readily put out of action by agents which have little effect on the resting potential. Another possibility is that certain substances may act on the secretory mechanism which normally keeps the internal sodium at a low level.

For many years physiologists have known that the action potential of medullated nerve is ultimately abolished by anoxia or by agents which interfere with oxidative processes (Gerard, 1932; Schmitt, 1930; Schmitt & Schmitt, 1931; Lorente de Nó, 1947). But agents of this type also reduce the resting potential and, in such cases, the action potential of medullated nerve can be restored by repolarizing the nerve with an anodic current (Lorente de Nó, 1947). There is therefore little reason to believe that the processes directly concerned with the generation of the action potential are of an oxidative nature. The converse view is expressed by Arvanitaki & Chalazonitis (1947) as a result of an interesting investigation into the effect of metabolic inhibitors on *Sepia* nerve. But the axons used in these experiments were surrounded by a relatively small amount of external fluid and stimulation frequencies of the order of 100 per sec. were employed. Under these conditions secretory activity may be of great importance for the maintenance of ionic concentration differences, and hence for the maintenance of normal excitability. There is, in any case, no direct conflict between the views of Arvanitaki & Chalazonitis and our own, since it is conceivable that oxidative metabolism may be essential for the proper operation of the mechanism responsible for transport of sodium.

The last objection to be mentioned is of a different kind. It has been assumed that the resting membrane is permeable to potassium and to chloride, but impermeable or only sparingly permeable to sodium. This is a plausible assumption since sodium is a more heavily hydrated ion than potassium or chloride. On the other hand, it is much more difficult to accept the assumption that the active membrane can become selectively permeable to sodium. We therefore suggest that sodium does not cross the membrane in ionic form, but enters into combination with a lipoid soluble carrier in the membrane which is only free to move when the membrane is depolarized. Potassium ions cannot cross the membrane by this route, because their affinity for the carrier is assumed to be small. An assumption of this kind is speculative but not unreasonable, since there is already some indication that a specific, enzyme-like process is concerned with the transport of sodium through cell membranes (Davson & Reiner, 1942; Krogh, 1946; Ussing, 1947). In this connexion it is interesting to read that the permeability of the erythrocyte of the cat to sodium may be five to ten times greater than the permeability to potassium (Davson & Reiner, 1942), and that sodium permeability is reduced to zero by concentrations of amyl alcohol which cause an increase in potassium permeability (Davson, 1940).

In formulating our hypothesis we have been careful to avoid making any quantitative assumptions about the relative permeabilities of the membrane to sodium and potassium. The resting membrane has been considered as more permeable to potassium than sodium, and this condition was regarded as reversed during activity. It is natural to inquire whether any limit can be set to the degree of selective permeability actually present in the resting and active membranes. Some light can be thrown on this problem if the observed potentials are compared with those predicted by a simple equation. In order to interpret the results in terms of selective permeability we need to know the potential difference which would arise across a membrane separating different concentrations of potassium, chloride and sodium. Thermodynamic equations cannot be applied because the system is not in equilibrium, while the theories of Planck (1890*a*, *b*) and Henderson (1907, 1908) make assumptions which are almost certainly not valid for a thin membrane of high resistance. A simple equation has been derived by Goldman (1943). He assumes that the voltage gradient through the membrane may be regarded as constant and that ions move under the influence of diffusion and the electric field. Goldman also makes the tacit assumption that the concentrations of ions at the edges of the membrane are directly proportional to those in the aqueous solutions. In the Appendix we show that these assumptions give the following expression for the membrane potential:

$$E=\frac{RT}{F}\log_e\left[\frac{P_K\,(K)_i+P_{Na}\,(Na)_i+P_{Cl}\,(Cl)_o}{P_K\,(K)_o+P_{Na}\,(Na)_o+P_{Cl}\,(Cl)_i}\right], \tag{4}$$

where $(K)_i$, $(Na)_i$ and $(Cl)_i$ are activities inside the axon; $(K)_o$, $(Na)_o$ and $(Cl)_o$ are activities outside the axon; P_K, P_{Na} and P_{Cl} are permeability constants for the individual ions. The relative magnitudes of the permeability constants depend upon the relative mobilities and solubilities of the ions in the membrane. Thus

$$P_K=\frac{RT}{Fa}u_K\,b_K;\quad P_{Na}=\frac{RT}{Fa}u_{Na}\,b_{Na};\quad P_{Cl}=\frac{RT}{Fa}u_{Cl}\,b_{Cl},$$

where a is the thickness of the membrane; u_K, u_{Na} and u_{Cl} are mobilities of the ions in the membrane; b_K, b_{Na} and b_{Cl} are the partition coefficients between the membrane and the aqueous solution. E is the potential difference across the membrane (outside – inside) in the absence of any net ionic current.

There are many reasons for supposing that this equation is no more than a rough approximation, and it clearly cannot give exact results if ions enter into chemical combination with carrier molecules in the membrane or if appreciable quantities of current are transported by ions other than K, Na or Cl. On the other hand, the equation has two important advantages. In the first place it is extremely simple, and in the second it reduces to the thermodynamically correct forms when any one permeability constant is made large compared to the others.

In order to apply this equation we must first adopt values for the internal concentrations of K, Cl and Na, and for this purpose the data of Steinbach (1941) and Steinbach & Spiegelman (1943) will be employed. These writers give values for freshly isolated axons and for those treated with sea water for 2–4 hr. The physiological condition of the axons used in the present work is thought to be intermediate between these two conditions and we therefore propose that the following values should be used:

$(K)_i=345$ mM. (mean of average values in table 4, in Steinbach & Spiegelman, 1943);

$(Na)_i=72$ mM. (mean of average values in table 4, in Steinbach & Spiegelman, 1943);

$(Cl)_i=61$ mM. (mean of tables 1 and 2 in Steinbach, 1941).

The experiments of Steinbach (1941) and Steinbach & Spiegelman (1943) suggest that the squid axon is permeable to chloride, sodium and potassium, but they give little information about the relative permeabilities to these three ions. It is extremely unlikely that the permeability ratios can be determined from electrical measurements with any degree of certainty, since the values adopted for the internal concentration are subject to considerable error, and equation 4 cannot be regarded as more than a rough approximation. Our object is to show that a large number of observations can be fitted into a coherent picture by the use of appropriate permeability ratios for resting, active and refractory nerve. The experimental data against which equation (4) must be tested are summarized in Table 7, which shows the average change in membrane potential produced by substituting a test solution for sea water or

TABLE 7.

		Composition of test solution			Change in membrane potential on substituting test solution for sea water or artificial sea water		Permeability coefficients used in calculation		
State of nerve	Solution	K mM.	Na mM.	Cl mM.	Observed mV.	Calculated mV.	P_K	P_{Na}	P_{Cl}
Resting	A	0	465	587	+ 3	+ 5			
	B	15	450	587	− 2	− 2			
	C	20	445	587	− 4	− 4			
	D	7	324	384	0	+ 1			
	E	5	227	270	+ 2	+ 2	1	0·04	0·45
	F	3	152	180	+ 2	+ 2			
	G	2	91	108	+ 4	+ 3			
	H	10	573	658	+ 1	0			
	I	10	711	796	− 2	0			
Active (peak of spike)	A	0	465	587	− 1	− 1			
	B	15	450	587	+ 1	0			
	C	20	445	587	+ 5	+ 1			
	D	7	324	384	+ 9	+ 8			
	E	5	227	270	+21	+16	1	20	0·45
	F	3	152	180	+44	+25			
	G	2	91	108	+59	+38			
	H	10	573	658	− 3	− 5			
	I	10	711	796	− 9	−10			
Refractory (maximum of positive phase)	A	0	465	587	+13	+12			
	B	15	450	587	− 6	− 5			
	C	20	445	587	−10	− 9			
	D	7	324	384	+ 1	+ 1			
	E	5	227	270	+ 4	+ 2	1·8	0	0·45
	F	3	152	180	+ 4	+ 3			
	G	2	91	108	0	+ 3			
	H	10	573	658	+ 1	+ 1			
	I	10	711	796	0	+ 3			
Membrane potential at rest in sea water					+48 + J	+59			
Membrane potential at height of activity in sea water					−40 + J	−38			
Membrane potential at maximum of positive phase					+62 + J	+74	As above		
Action potential in sea water					88	97			
Positive phase in sea water					14	15			

Solutions A, B and C were tested against an artificial sea-water solution containing 10 mM-K 455 mM-Na, 587 mM-Cl. Solutions D–I were tested against sea water containing approximately 10mM-K, 455 mM-Na, 540 mM-Cl. Calculated potentials were obtained from equation 4 using values of $(K)_i = 345$ mM., $(Na)_i = 72$ mM., $(Cl)_i = 61$ mM. J is the liquid junction potential between the axoplasm and the sea water in the microelectrode.

5—2

artificial sea water. Solution A is potassium-free artificial sea water, solutions B and C are potassium-rich artificial sea water; D, E, F, G are sea-water solutions diluted with isotonic dextrose while H and I were made by adding solid sodium chloride to sea water. It will be seen that there is reasonable agreement between all the results obtained with resting nerve and those predicted by the theory for $P_K:P_{Na}:P_{Cl}=1:0{\cdot}04:0{\cdot}45$. These coefficients were

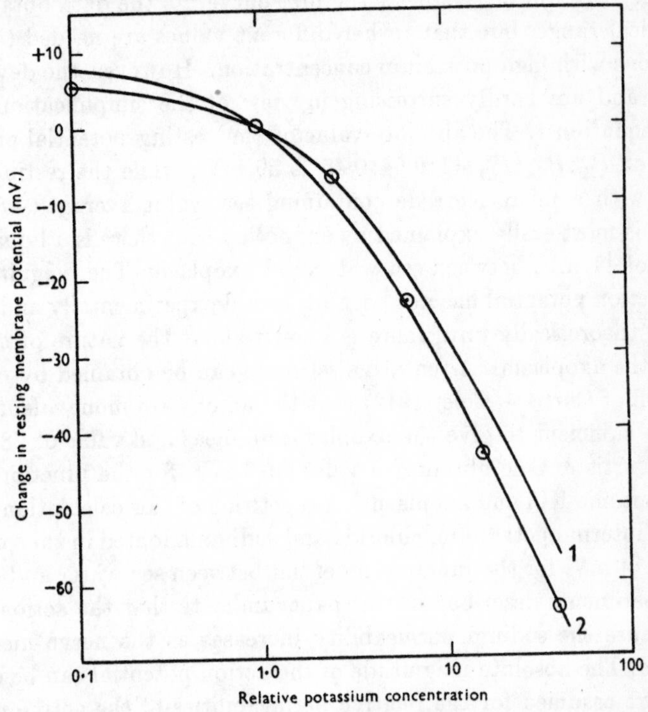

Fig. 13. Data from Curtis & Cole (1942, fig. 2). Ordinate: change in resting membrane potential produced by substitution of test solution for standard sea-water solution containing 13 mM-K. Abscissa: potassium concentration of test solution in multiples of potassium concentration in standard solution (13 mM.); logarithmic scale. Circles were obtained from Curtis & Cole's experimental points. Smooth curves were drawn according to equation 4 with

$P_K:P_{Na}:P_{Cl}=1:0{\cdot}04:0{\cdot}45$ (curve 1) and $P_K:P_{Na}:P_{Cl}=1:0{\cdot}025:0{\cdot}3$ (curve 2).

Internal concentrations assumed to remain constant; $(Cl)_i$ taken as 61 mM. and $(Na)_o$ obtained from the proportion of isotonic sodium chloride present in the test solution.

found by trial and error, and serious deviations occur if values differing by more than 50 % are employed. Thus, if the chloride permeability is made zero, an increase of 17 mV. in the resting potential is predicted for a solution of 50 % sea water, while if it is made equal to the potassium permeability a decrease of 5 mV. is predicted. The average change observed experimentally was +2 mV. and this is the value predicted by the coefficients which have been adopted. The variations in resting potential were not large, because the external potassium concentration was kept well within the physiological range. However, the relation between potassium concentration and resting potential has been determined by Curtis & Cole (1942) and their data are supported by unpublished results obtained with Mr Huxley in 1939. Curtis & Cole's data are shown by the hollow circles in Fig. 13, while curves 1 and 2 were plotted according to the theory for $P_K : P_{Na} : P_{Cl} = 1 : 0{\cdot}04 : 0{\cdot}45$ and $1 : 0{\cdot}025 : 0{\cdot}3$ respectively. It will be seen that the first curve fits the data obtained in the physiological range, but that rather different values are needed to cover the observations with high potassium concentration. However, the deviations are not large and are hardly surprising in view of the simplifications made in deriving equation 4. The absolute value of the resting potential predicted on the basis of $P_K : P_{Na} : P_{Cl} = 1 : 0{\cdot}04 : 0{\cdot}45$ is 59 mV., while the resting potential observed with a microelectrode containing sea water averaged 48 mV. The difference is most easily explained by supposing that there is a liquid junction potential of 11 mV. between sea water and axoplasm. The magnitude of the liquid junction potential has not been measured experimentally and cannot be computed theoretically until more is known about the nature of the organic anions in the axoplasm. A tentative estimate can be obtained by making the assumption of Curtis & Cole (1942) that the anions are monovalent and have a mobility sufficient to give the axoplasm its measured value of 28 Ω.cm. In this way Curtis & Cole obtained a value of 6 mV. for the junction potential between isotonic KCl and axoplasm. A repetition of this calculation, using the figures for internal potassium, chloride and sodium adopted in this paper, gave a value of 14 mV. for the junction potential between sea water and axoplasm.

The experiments described in this paper indicate that the action potential arises because the sodium permeability increases as the nerve membrane is depolarized. The absolute magnitude of the action potential can be calculated if values are assumed for the relative permeabilities of the active membrane to sodium, potassium and chloride ions. If the permeabilities are assumed to change from a resting condition in which $P_K : P_{Na} : P_{Cl} = 1 : 0{\cdot}04 : 0{\cdot}45$ to an active condition in which $P_K : P_{Na} : P_{Cl} = 1 : 20 : 0{\cdot}45$ an action potential of 97 mV. is obtained. This is 9 mV. larger than the average value observed experimentally, but it must be remembered that equation 4 only applies if there is no external current through the membrane. The difference of 9 mV. appears to be a safe allowance for the potential difference arising from current flow, since the membrane current density at the height of activity is about 0·2 mA.cm.$^{-2}$ (Hodgkin & Huxley, 1945), and the active membrane resistance is about 25 Ω.cm.^{2} (Cole & Curtis, 1939). The new values of the permeability coefficients may be used to predict the changes in potential which would arise from the applications of solutions A–I and these are compared with the average experimental results in Table 7. It will be seen that there is reasonable agreement over most of the range, but that deviations occur in the case of solution C (twice normal potassium), E (2/3 isotonic dextrose, 1/3 sea water) and F (4/5 isotonic dextrose, 1/5 sea water). The nature of these deviations has already been discussed and requires little further comment. In order to account for them we must assume that, for one reason or another, the active membrane does not acquire its full sodium permeability. Thus a change of 58 mV. would have been predicted for solution G if we had assumed that in this solution the permeability coefficients were 1:2·5:0·45 instead of 1:20:0·45 as in the normal active membrane. Although the sodium permeability has been assumed to be twenty times the potassium permeability, the active membrane potential is still 8 mV. below the theoretical maximum for a sodium electrode. This indicates that equation 1 is useful only in so far as it gives an upper limit to the reversed membrane potential. On the other hand, equation 2 remains a reasonably good approximation since it predicts changes which are within 10% of the calculated values in Table 7.

The third block of figures in Table 7 give the changes in membrane potential recorded at the height of the positive phase. In this condition the nerve is in a refractory state, so that there is no reason to assume that the permeability ratios are intermediate between those in the resting and active states. If the sodium permeability remained at its active level of 20, the nerve could show no recovery from the crest of the action potential. The sodium permeability must therefore be reduced by exhaustion or inactivation of the special mechanism which comes into play when the nerve is first depolarized. We now assume that the sodium permeability is reduced to zero and that it does not recover its normal value until the end of the relative refractory period. If this assumption is made, we find that the nerve would show a positive phase of 10 mV. This is not far from that recorded experimentally, but there are still considerable deviations between theory and experiment which can be resolved by making $P_K:P_{Na}:P_{Cl}=1{\cdot}8:0:0{\cdot}45$. These values have been adopted in Table 7 and give good agreement both with respect to the absolute magnitude of the positive phase and to its variation in solutions of different potassium content. The agreement may be fortuitous and can hardly be used as evidence for a differential action on potassium and chloride permeability. On the other hand, the assumption that the sodium permeability is reduced to a subnormal value during the recovery process appears to be in keeping with the general nature of the refractory period, and provides a plausible explanation of the characteristic diphasic appearance of the squid action potential. The positive phase is not seen in other single fibre preparations, but it must be remembered that the assumptions which have been made only lead to a positive phase when there is an appreciable leakage of sodium in the resting condition. A fibre with low-sodium leakage and with potassium and chloride ions distributed according to a Donnan ratio would have a membrane potential close to the theoretical maximum for a potassium electrode, which would be relatively insensitive to a decrease in the amount of sodium leakage. In such fibres the action potential would return to the resting level without showing any appreciable positive phase.

The preceding arguments suggest that an isolated squid axon is not in a steady state, but is gaining sodium and leaking potassium at a rate determined by the permeability of the membrane and the concentration differences across it. An exchange of this kind has been observed by Steinbach & Spiegelman (1943), and it is interesting to compare their result with that predicted by the constant field theory on which equation 4 is based. Steinbach & Spiegelman's figures show an average increase of 50 mM-Na and an average decrease of 72 mM-K during a period of 3 hr. These figures may be expressed in terms of the flow of ions through 1 sq.cm. of membrane, since the average axon diameter cannot have been far from 500 μ. Adopting this value for the diameter we find that the entry of sodium through 1 sq.cm. was 6×10^{-11} mol.sec.$^{-1}$ while the exit of potassium was 8×10^{-11} mol.sec.$^{-1}$. In order to calculate a theoretical flow from the constant field theory we need to know the concentration differences across the membrane, the permeability ratios and the absolute value of the membrane conductance. The relation between these quantities is given by equation 7·0, 7·1 or 7·2 of the Appendix and numerical values can be obtained by adopting the concentrations and permeability coefficients used previously, with a value of 1000 Ω.cm.^{2} for the membrane resistance (Cole & Hodgkin, 1939). The following theoretical rates are obtained: entry Na, $8{\cdot}4 \times 10^{-11}$ mol.cm.$^{-2}$sec.$^{-1}$; exit K, $10{\cdot}6 \times 10^{-11}$ mol.cm.$^{-2}$sec.$^{-1}$; exit Cl, $2{\cdot}2 \times 10^{-11}$ mol.cm.$^{-2}$sec.$^{-1}$. Steinbach & Spiegelman (1943) give no figures for the flow of chloride, but Steinbach (1941) states that the chloride concentration of squid axoplasm shows a rise from an initial value of 36 mM. to one of about 75 mM. at which level the concentration remains constant for long periods of time. If a value of 36 mM. had been adopted for the chloride concentration a substantial entry of chloride would have been predicted, and this may explain the initial rise in chloride concentration observed by Steinbach. The difference between the theoretical rates for sodium and potassium and those observed by Steinbach & Spiegelman is not larger than would be expected from the nature of the calculations used in making the comparison. However, a difference of this kind is to be expected, since it is likely that entry of sodium would be partly compensated by the active extrusion process normally responsible for maintaining a low internal sodium concentration in the living animal.

The experiments described in this paper suggest that sodium ions enter the axon during the rising phase of the spike, and that the rate of rise is determined by the speed at which the charge on the membrane capacity is altered by entry of sodium. It is natural to inquire how large the sodium permeability would have to be in order to give a rate of rise comparable to that observed experimentally. The problem may be formulated in a different way. The maximum rate of rise of the spike is of the order of 600 V./sec. and, for a membrane capacity of 1·5 μF.cm.$^{-2}$, this corresponds to an ionic current density of 0·9 mA.cm.$^{-2}$. The maximum rate occurs approximately at zero membrane potential, and we may suppose that at this moment the permeability coefficients have already assumed their fully active values of $P_K:P_{Na}:P_{Cl}=1:20:0{\cdot}45$. We are also given the fact that the resistance of the resting membrane is roughly 1000 Ω.cm.^{2}, and in this condition we assume as before that $P_K:P_{Na}:P_{Cl}=1:0{\cdot}04:0{\cdot}45$. This information allows the total ionic current to be calculated by the methods described in the Appendix. We find

(1) an inward sodium current of 1·3 mA.cm.$^{-2}$;
(2) an outward potassium current of 0·06 mA.cm.$^{-2}$;
(3) an outward chloride current of 0·04 mA.cm.$^{-2}$;
(4) a net inward current of 1·2 mA.cm.$^{-2}$.

The total inward current is of the same order as that obtained experimentally so that there is no difficulty in accounting for the rate of rise of the action potential in terms of our hypothesis.

The preceding calculation suggests that the inward sodium current greatly exceeds the outward potassium current during the rising phase of the action potential, and we should expect that this situation would be reversed during the falling phase of the spike. A minimum value for the quantity of sodium entering the axon can be obtained by assuming that the period of sodium entry does not overlap to any appreciable extent with the period of potassium exit. In this case the total quantity of sodium entering the axon would be given by the product of the membrane action potential and the membrane capacity divided by the Faraday. Thus $1{\cdot}5\times10^{-12}$ mol. must be transferred through a membrane of capacity 1·5 μF. in order to change its potential difference from +50 mV. to −45 mV. More sodium would enter if there was a simultaneous exchange of potassium and sodium, but the quantity entering could not be less than $1{\cdot}5\times10^{-12}$ mol. unless some other mechanism assists in the active process. A crucial test of the sodium hypothesis would therefore be to measure the quantity of sodium entering the axon in one impulse. This experiment has never been performed in a satisfactory way, although the work of Fenn & Cobb (1936), Tipton (1938) and v. Euler, v. Euler & Hevesy (1946) provides some indication of sodium entry during activity. The total charge moving out through the membrane during the falling phase must be approximately equal to the charge transferred during the rising phase. The outward charge would be carried primarily by potassium ions if the permeability of the active membrane is greater to potassium than to chloride. Under these conditions a minimum potassium leakage of $1{\cdot}5\times10^{-12}$ mol. is to be expected. This is not far from the value obtained by Hodgkin & Huxley (1947), who gave an average value of

$1{\cdot}7 \times 10^{-12}$ moles in *Carcinus* axons with an average membrane capacity of $1{\cdot}3$ μF.cm.$^{-2}$. The average action potential in *Carcinus* axons has been estimated at about 120 mV. (Hodgkin, 1947) so that the theoretical minimum for the potassium leakage would be

$$\frac{120 \text{ mV.} \times 1{\cdot}3 \ \mu\text{F.cm.}^{-2}}{96{,}500 \text{ coulomb mol.}^{-1}} = 1{\cdot}6 \times 10^{-12} \text{ mol.cm.}^{-2}.$$

The close agreement is unlikely to be more than a coincidence, but the similarity in order of magnitude may be significant, since Keynes (1948) has recently obtained comparable results by the use of radioactive tracers.

SUMMARY

The reversal of membrane potential during the action potential can be explained if it is assumed that the permeability conditions of the membrane in the active state are the reverse of those in the resting state. The resting membrane is taken to be more permeable to potassium than sodium, and the active membrane more permeable to sodium than to potassium. (It is suggested that the reversal of permeability is brought about by a large increase in sodium permeability and that the potassium permeability remains unaltered or undergoes a relatively small change.) A reversed membrane potential can arise in a system of this kind if the concentration of sodium in the external solution is greater than that in the axoplasm.

This hypothesis is supported by the following observations made with a microelectrode in squid giant axons:

1. The action potential is abolished by sodium-free solutions, but returns to its former value when sea water is replaced.

2. Dilution of sea water with isotonic dextrose produces a slight increase in resting potential, but a large and reversible decrease in the height of the action potential. The reversed potential difference of the active membrane depends upon the sodium concentration in the external fluid and is reduced to zero by solutions containing less than about 30% of the normal sodium concentration.

3. The height of the action potential is increased by a hypertonic solution containing additional sodium chloride, but is not increased by addition of dextrose to sea water. The resting potential is unaffected or slightly reduced by sodium-rich solutions.

4. The changes in active membrane potential which result from increases or decreases of external sodium are of the same order of magnitude as those for a sodium electrode.

5. The rate of rise of the action potential can be increased to 140% of its normal value and reduced to 10% by altering the concentration of sodium in

the external solution. To a first approximation, the rate of rise is directly proportional to the external concentration of sodium.

6. The conduction velocity undergoes a substantial decrease in solutions of low-sodium content.

7. The changes produced by dilution of sea water with isotonic dextrose appear to be caused by reduction of the sodium concentration and not by alterations in the concentrations of other ions.

Removal of external potassium causes a small increase in action potential which is almost entirely due to an increase in the resting potential, the reversed potential difference of the active membrane remaining substantially constant. Increasing the external potassium causes a depression of both action potential and resting potential, but the former is affected to a much greater extent than the latter. The positive phase of the squid action potential is markedly increased by potassium-free solutions and decreased by potassium-rich solutions.

The effects of a large number of solutions on the membrane potential in the resting, active and refractory state are shown to be consistent with a quantitative formulation of the sodium hypothesis.

We wish to express our gratitude to the Rockefeller Foundation for Medical Research for financial aid; to the director and staff of the Laboratory of the Marine Biological Association, Plymouth, for their assistance at all stages of the experimental work; and to Mr A. F. Huxley for much helpful and stimulating discussion.

APPENDIX

This section contains a brief description of the way in which constant field equations may be derived and applied to practical problems. The treatment is essentially similar to that of Goldman (1943) but is summarized here for the convenience of the reader.

The basic assumptions are (1) that ions in the membrane move under the influence of diffusion and the electric field in a manner which is essentially similar to that in free solution; (2) that the electric field may be regarded as constant throughout the membrane; (3) that the concentrations of ions at the edges of the membrane are directly proportional to those in the aqueous solutions bounding the membrane; and (4) that the membrane is homogeneous.

Assumption (1) leads to the following equations for the current carried by ions:

$$-I_K = RTu_K \frac{dC_K}{dx} + C_K u_K F \frac{d\psi}{dx}, \qquad (1\cdot1)$$

$$-I_{Na} = RTu_{Na} \frac{dC_{Na}}{dx} + C_{Na} u_{Na} F \frac{d\psi}{dx}, \qquad (1\cdot2)$$

$$-I_{Cl} = -RTu_{Cl} \frac{dC_{Cl}}{dx} + C_{Cl} u_{Cl} F \frac{d\psi}{dx}. \qquad (1\cdot3)$$

Here I_K, I_{Na} and I_{Cl} are the contributions of potassium, sodium and chloride to the total inward current density through the membrane. C_K, C_{Na} and C_{Cl} are the concentrations of ions in the membrane and u_K, u_{Na} and u_{Cl} are their mobilities; x is the distance through the membrane from the outer boundary defined as $x=0$. The inner boundary is defined as $x=a$. ψ is the potential at a point x; R, T and F have their usual significance. In the steady state I_K, I_{Na} and I_{Cl} must be constant throughout the membrane; $d\psi/dx$ is also regarded as constant and equal to $-V/a$, where V is the potential of the outside solution minus that of the inside solution. Equations (1·1), (1·2) and (1·3) may therefore be integrated directly. Thus (1·1) gives

$$\left[\frac{aI_K e^{-VFx/RTa}}{VFu_K}\right]_0^a = \left[C_K e^{-VFx/RTa}\right]_0^a. \tag{2·1}$$

Hence

$$I_K = \frac{u_K F V}{a} \frac{(C_K)_o - (C_K)_a e^{-VF/RT}}{1 - e^{-VF/RT}}. \tag{2·2}$$

Now the concentration $(C_K)_o$ at the outer edge of the membrane is regarded as directly proportional to the concentration $(K)_o$ of potassium in the external fluid. Hence

$$(C_K)_o = \beta_K (K)_o \quad \text{and} \quad (C_K)_a = \beta_K (K)_i,$$

where β_K is the partition coefficient between the membrane and the aqueous solution; $(K)_i$ is the concentration in the axoplasm.

Equation (2·2) then becomes

$$I_K = P_K \frac{F^2 V}{RT} \frac{(K)_o - (K)_i e^{-VF/RT}}{1 - e^{-VF/RT}}, \tag{2·3}$$

where P_K is a permeability coefficient defined as $u_K \beta_K RT/aF$.

In a similar way we obtain

$$I_{Na} = P_{Na} \frac{F^2 V}{RT} \frac{(Na)_o - (Na)_i e^{-VF/RT}}{1 - e^{-VF/RT}}, \tag{2·4}$$

and

$$I_{Cl} = P_{Cl} \frac{F^2 V}{RT} \frac{(Cl)_i - (Cl)_o e^{-VF/RT}}{1 - e^{-VF/RT}}. \tag{2·5}$$

The total ionic current density through the membrane is therefore given by

$$I = \frac{F^2 V P_K}{RT} \frac{w - ye^{-VF/RT}}{1 - e^{-VF/RT}}, \tag{3·0}$$

where

$$w = (K)_o + \frac{P_{Na}}{P_K}(Na)_o + \frac{P_{Cl}}{P_K}(Cl)_i,$$

$$y = (K)_i + \frac{P_{Na}}{P_K}(Na)_i + \frac{P_{Cl}}{P_K}(Cl)_o.$$

The potential difference across the membrane in the absence of ionic current will be designated E. $V = E$ when $I = 0$. Hence

$$E = \frac{RT}{F} \log_e \frac{y}{w}, \tag{4·0}$$

which is equivalent to equation (4) used in the text. The membrane conductance G is defined as $(dI/dV)_{I\to 0}$ and is given by

$$G = \frac{F^2 P_K}{RT} \left\{ V \frac{d}{dV} \left[\frac{w - ye^{-VF/RT}}{1 - e^{-VF/RT}}\right] + \left[\frac{w - ye^{-VF/RT}}{1 - e^{-VF/RT}}\right] \right\}. \tag{5·0}$$

The second term in this expression is zero when $I = 0$ and $V = E$. After differentiation V may be equated to E. Hence

$$G = \frac{F^3}{(RT)^2} E P_K \left\{\frac{wy}{y - w}\right\}. \tag{6·0}$$

This expression allows us to compute the numerical values of the permeability coefficient P_K provided that the ratios P_{Na}/P_K and P_{Cl}/P_K are known. For the case considered in the text P_K is found to be $1{\cdot}8 \times 10^{-6}$ cm.sec.$^{-1}$. The individual ionic currents may be determined by using this value in applying equations (2·3), (2·4), and (2·5) to any particular set of experimental conditions.

When $I=0$ and $V=E$ a more convenient method is to use the following relations which may be obtained from (2·3), (2·4), (2·5), (4·0) and (6·0):

$$I_K=\frac{RT}{F}G\left\{\frac{(K)_o}{w}-\frac{(K)_i}{y}\right\}, \tag{7·0}$$

$$I_{Na}=\frac{RT}{F}G\frac{P_{Na}}{P_K}\left\{\frac{(Na)_o}{w}-\frac{(Na)_i}{y}\right\}, \tag{7·1}$$

$$I_{Cl}=\frac{RT}{F}G\frac{P_{Cl}}{P_K}\left\{\frac{(Cl)_i}{w}-\frac{(Cl)_o}{y}\right\}. \tag{7·2}$$

These equations were used in the calculation given on p. 71.

The constant field equations may be applied to the rising phase of the spike if it is assumed that the rate of change of potential is low enough to allow the ionic currents to attain their steady state value. At the moment when the rate of rise of the spike is at a maximum the total membrane current is zero, but there is a large ionic current which is equal and opposite to the capacity current through the membrane dielectric. In this case we cannot use (7·0), (7·1) and (7·2), but must return to (2·3), (2·4) and (2·5). Since the maximum rate of rise occurs at approximately the time when $V=0$ these equations may be simplified to

$$I_K=P_K F\,[(K)_o-(K)_i], \tag{8·0}$$

$$I_{Na}=P_{Na} F\,[(Na)_o-(Na)_i], \tag{8·1}$$

$$I_{Cl}=P_{Cl} F\,[(Cl)_i-(Cl)_o]. \tag{8·2}$$

In making the calculation on p. 72 we assumed that when $V=0$, P_K and P_{Cl} had the same values as in the resting nerve, but that P_{Na} was $20P_K$ instead of $0{\cdot}04P_K$.

REFERENCES

Arvanitaki, A. & Chalazonitis, N. (1947). *Arch. int. Physiol.* **54**, 406.

Bernstein, J. (1912). *Electrobiologie.* Braunschweig: Vieweg.

Bishop, G. H. (1932). *J. cell. comp. Physiol.* **1**, 177.

Cole, K. S. & Curtis, H. J. (1938). *J. gen. Physiol.* **21**, 757.

Cole, K. S. & Curtis, H. J. (1939). *J. gen. Physiol.* **22**, 649.

Cole, K. S. & Hodgkin, A. L. (1939). *J. gen. Physiol.* **22**, 671.

Curtis, H. J. & Cole, K. S. (1942). *J. cell. comp. Physiol.* **19**, 135.

Davson, H. (1940). *J. cell. comp. Physiol.* **15**, 317.

Davson, H. & Reiner, M. (1942). *J. cell. comp. Physiol.* **20**, 325.

Erlanger, J. & Blair, E. A. (1938). *Amer. J. Physiol.* **124**, 341.

Fenn, W. O. & Cobb, D. M. (1936). *Amer. J. Physiol.* **115**, 345.

Gallego, A. & Lorente de Nó (1947). *J. cell. comp. Physiol.* **29**, 189.

Gerard, R. W. (1932). *Physiol. Rev.* **12**, 469.

Glazebrook, Sir R. (1923). *A Dictionary of Applied Physics*, vol. 3. London: Macmillan.

Goldman, D. E. (1943). *J. gen. Physiol.* **27**, 37.

Grundfest, H. (1947). *Ann. Rev. Physiol.* **9**, 477.

Henderson, P. (1907). *Z. phys. Chem.* **59**, 118.

Henderson, P. (1908). *Z. phys. Chem.* **63**, 325.

Höber, R. (1946). *Ann. N.Y. Acad. Sci.* **47**, 381.

Hodgkin, A. L. (1939). *J. Physiol.* **94**, 560.

Hodgkin, A. L. (1947). *J. Physiol.* **106**, 305.

Hodgkin, A. L. (1948). *J. Physiol.* **107**, 165.

Hodgkin, A. L. & Huxley, A. F. (1939). *Nature, Lond.*, **144**, 710.

Hodgkin, A. L. & Huxley, A. F. (1945). *J. Physiol.* **104**, 176.

Hodgkin, A. L. & Huxley, A. F. (1947). *J. Physiol.* **106**, 341.

Kato, G. (1936). *Cold. Spr. Harb. Symp. quant. Biol.* **4**, 202.
Katz, B. (1947). *J. Physiol.* **106**, 411.
Keynes, R. D. (1948). *J. Physiol.* **107**, 35 *P*.
Krogh, A. (1946). *Proc. Roy. Soc.* B, **133**, 123.
Landolt-Börnstein (1931). *Physicalisch-chemische Tabellen*, 5th ed. Ergänzungsband IIa, 189. Berlin: Springer.
Lillie, R. S. (1923). *Protoplasmic Action and Nervous Action.* Chicago: University Press.
Lorente de Nó, R. (1944). *J. cell. comp. Physiol.* **24**, 85.
Lorente de Nó, R. (1947). *A Study of Nerve Physiology*, vols. 1 and 2 in *Studies from the Rockefeller Institute for Medical Research*, vols. **131** and **132**. New York.
Overton, E. (1902). *Pflüg. Arch. ges. Physiol.* **92**, 346.
Planck, M. (1890 *a*). *Ann. Phys., Lpz.*, **39**, 161.
Planck, M. (1890 *b*). *Ann. Phys., Lpz.*, **40**, 561.
Schmitt, F. O. (1930). *Amer. J. Physiol.* **95**, 650.
Schmitt, F. O. & Schmitt, O. H. A. (1931). *Amer. J. Physiol.* **97**, 302.
Steinbach, H. B. (1941). *J. cell. comp. Physiol.* **17**, 57.
Steinbach, H. B. & Spiegelman, S. (1943). *J. cell. comp. Physiol.* **22**, 187.
Stella, G. (1928). *J. Physiol.* **66**, 19.
Tipton, S. R. (1938). *Amer. J. Physiol.* **124**, 322.
Ussing, H. H. (1947). *Nature, Lond.*, **160**, 262.
v. Euler, H. V., v. Euler, U. S. & Hevesy, G. (1946). *Acta physiol. Scand.* **12**, 261.

6

J. Physiol. (1952) 116, 424–448

MEASUREMENT OF CURRENT-VOLTAGE RELATIONS IN THE MEMBRANE OF THE GIANT AXON OF *LOLIGO*

By A. L. HODGKIN, A. F. HUXLEY and B. KATZ

From the Laboratory of the Marine Biological Association, Plymouth, and the Physiological Laboratory, University of Cambridge

(*Received* 24 *October* 1951)

The importance of ionic movements in excitable tissues has been emphasized by a number of recent experiments. On the one hand, there is the finding that the nervous impulse is associated with an inflow of sodium and an outflow of potassium (e.g. Rothenberg, 1950; Keynes & Lewis, 1951). On the other, there are experiments which show that the rate of rise and amplitude of the action potential are determined by the concentration of sodium in the external medium (e.g. Hodgkin & Katz, 1949*a*; Huxley & Stämpfli, 1951). Both groups of experiments are consistent with the theory that nervous conduction depends on a specific increase in permeability which allows sodium ions to move from the more concentrated solution outside a nerve fibre to the more dilute solution inside it. This movement of charge makes the inside of the fibre positive and provides a satisfactory explanation for the rising phase of the spike. Repolarization during the falling phase probably depends on an outflow of potassium ions and may be accelerated by a process which increases the potassium permeability after the action potential has reached its crest (Hodgkin, Huxley & Katz, 1949).

Outline of experiments

The general aim of this series of papers is to determine the laws which govern movements of ions during electrical activity. The experimental method was based on that of Cole (1949) and Marmont (1949), and consisted in measuring the flow of current through a definite area of the membrane of a giant axon from *Loligo*, when the membrane potential was kept uniform over this area and was changed in a stepwise manner by a feed-back amplifier. Two internal electrodes consisting of fine silver wires were thrust down the axis of the fibre for a distance of about 30 mm. One of these electrodes recorded the membrane potential, and the feed-back amplifier regulated the current entering the other electrode in such a way as to change the membrane potential suddenly and hold it at the new level. Under these conditions it was found that the membrane current consisted of a nearly instantaneous surge of capacity current, associated with the sudden change of potential, and an ionic current during the period of maintained potential. The ionic current could be resolved into a transient component associated with movement of sodium ions, and a prolonged phase of 'potassium current'. Both currents varied with the permeability of the membrane to sodium or potassium and with the electrical and osmotic driving force. They could be distinguished by studying the effect of changing the concentration of sodium in the external medium.

The first paper of this series deals with the experimental method and with the behaviour of the membrane in a normal ionic environment. The second (Hodgkin & Huxley, 1952*a*) is concerned with the effect of changes in sodium concentration and with a resolution of the ionic current into sodium and potassium currents. Permeability to these ions may conveniently be expressed in units of ionic conductance. The third paper (Hodgkin & Huxley, 1952*b*) describes the effect of sudden changes in potential on the time course of the ionic conductances, while the fourth (Hodgkin & Huxley, 1952*c*) deals with the inactivation process which reduces sodium permeability during the falling phase of the spike. The fifth paper (Hodgkin & Huxley, 1952*d*) concludes the series and shows that the form and velocity of the action potential may be calculated from the results described previously.

A report of preliminary experiments of the type described here was given at the symposium on electrophysiology in Paris (Hodgkin *et al.* 1949).

Nomenclature

In this series of papers we shall regard the resting potential as a positive quantity and the action potential as a negative variation. V is used to denote displacements of the membrane potential from its resting value. Thus

$$V=E-E_r,$$

where E is the absolute value of the membrane potential and E_r is the absolute value of the resting potential, with signs taken in the sense outside potential minus inside potential. With this choice of signs it is logical to take $+I$ for inward current density through the membrane. These definitions make membrane current positive under an external anode and agree with the accepted use of the terms negative and positive after-potential. They conflict with the common practice of showing action potentials as an upward deflexion and are inconvenient in experiments in which an internal electrode measures potentials with respect to an external earth. Lower-case symbols (v_n) are employed when it is necessary to give potentials with respect to earth, but no confusion should arise since this usage is confined to the sections dealing with the experimental method.

Theory

Although the results described in this paper do not depend on any particular assumption about the electrical properties of the surface membrane, it may be helpful to begin by stating the theoretical assumption which determined the design and analysis of the experiments. This is that the membrane current may be divided into a capacity current which involves a change in ion density at the outer and inner surfaces of the membrane, and an ionic current which depends on the movement of charged particles through the membrane. Equation 1 applies to such a system, provided that the behaviour of the membrane capacity is reasonably close to that of a perfect condenser:

$$I = C_M \frac{\partial V}{\partial t} + I_i, \tag{1}$$

where I is the total current density through the membrane, I_i is the ionic current density, C_M is the membrane capacity per unit area, and t is time. In most of our experiments $\partial V/\partial t = 0$, so that the ionic current can be obtained directly from the experimental records. This is the most obvious reason for using electronic feed-back to keep the membrane potential constant. Other advantages will appear as the experimental results are described.

EXPERIMENTAL METHOD

The essential features of the electrode system are illustrated by Fig. 1. Two long silver wires, each 20 μ. in diameter, were thrust down the axis of a giant axon for a distance of 20–30 mm. The greater part of these wires was insulated but the terminal portions were exposed in the manner shown in Fig. 1. The axon was surrounded by a 'guard ring' system which contained the external electrodes. Current was applied between the current wire (a) and an earth (e), while the potential difference across the membrane could be recorded from the voltage wire (b) and an external electrode (c). The advantage of using two wires inside the nerve is that polarization of the current wire does not affect the potential recorded by the voltage wire. The current wire was exposed for a length which corresponded to the total height of the guard-system, while the voltage wire was exposed only for the height of the central channel. The guard system ensured that the current crossing the membrane between the partitions A_2 and A_3 flowed down the channel C. This component of the current was determined by recording the potential difference between the external electrodes c and d.

Internal electrode assembly

In practice it would be difficult to introduce two silver wires into an axon without using some form of support. Another requirement is that the electrode must be compact, since previous experience showed that axons do not survive well unless the width of an internal electrode is less than 150 μ. (Hodgkin & Huxley, 1945). After numerous trials the design shown in Fig. 4 was adopted. The first operation in making such an electrode was to push a length of the voltage wire through a 70 μ. glass capillary and twist it round the capillary in a spiral which started at the tip and proceeded toward the shank of the capillary. The spiral was wound by rotating the shank of the capillary in a small chuck attached to a long screw. During this process the free end of the wire was pulled taut by a weight while the capillary was supported, against the pull of the wire, by a fine glass hook. A second hook controlled the angle at which the wire left the capillary. When sufficient wire had been wound it was attached to the capillary by application of shellac solution, cut close to the capillary and insulated with shellac in the appropriate regions (Fig. 4). The next operation was to wind on the current wire, starting from the shank and proceeding to the tip. Correct spacing of current and voltage wires was maintained by making small adjustments in the position of the second glass hook. When the current wire had been wound to the tip it was attached to the capillary, cut short and insulated as before. The whole operation was carried out under a binocular microscope. Shellac was applied as an alcoholic solution and was dried and hardened

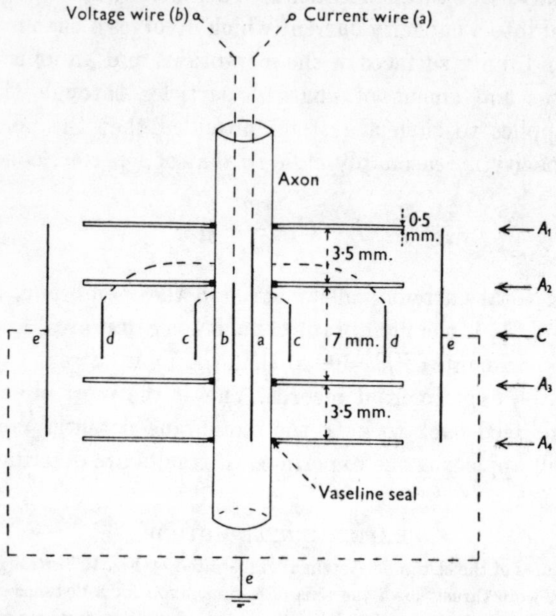

Fig. 1. Diagram illustrating arrangement of internal and external electrodes. A_1, A_2, A_3 and A_4 are Perspex partitions. *a*, *b*, *c*, *d* and *e* are electrodes. Insulated wires are shown by dotted lines. For sections through *A* and *C*, see Figs. 2 and 3.

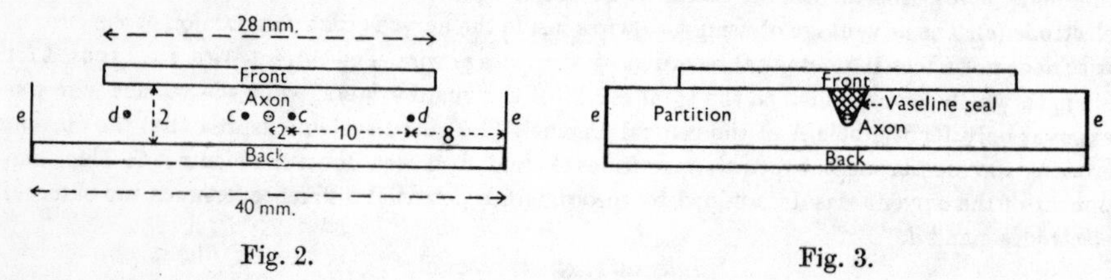

Fig. 2. Fig. 3.

Fig. 2. Central channel of guard system. Section through *C*, Fig. 1. *c* and *d* are silver wires, *e* is a silver sheet. All dimensions are in mm.

Fig. 3. Partition of guard system. Section through A_1, A_2, A_3 or A_4, Fig. 1.

by baking for several hours under a lamp. Insulation between wires and across the shellac was tested so as to ensure that the film of shellac was complete and that the wires did not touch at any point. The exposed portion of the wires was then coated electrolytically with chloride. The electrode was first made an anode in order to deposit chloride and was then made a cathode in order to reduce some of the chloride and obtain a large surface of silver. This process was repeated a number of times ending with an application of current in the direction to deposit chloride. In this way an electrode of low polarization resistance was obtained.

In order to test the performance of the electrode it was immersed in salt solution and the current wire polarized by application of an electric current. In theory this should have caused no change in the potential difference between the voltage wire and the solution in which it was immersed. In practice we observed a very small change in potential which will be called 'mutual polarization'. Leakage between wires was a possible cause of this effect, but other explanations cannot be excluded.

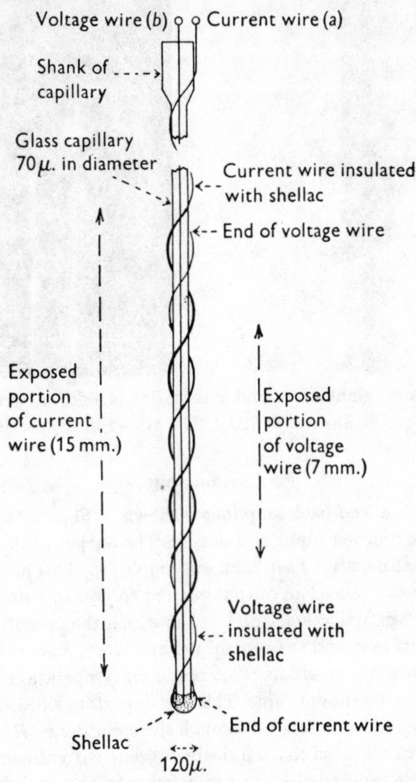

Fig. 4. Diagram of internal electrode (not to scale). The pitch of each spiral was 0·5 mm. The exposed portions of the wires are shown by heavy lines.

The general appearance of the electrode inside a giant axon is illustrated by Fig. 5. These photographs were obtained at an early stage of the investigation, and the axon was cleaned less carefully than in later experiments. The internal electrode differed from those finally employed in that both wires were wound from the shank of the electrode and that the pitch was somewhat greater.

Guard system

The general form of the guard system is shown in Figs. 1–3. It consisted of a flat box made out of Perspex which was divided into three compartments by two partitions A_2 and A_3 and closed with walls A_1 and A_4. The front of the box was removable and was made from a thin sheet of Perspex which could be sealed into position with vaseline. V-shaped notches were made in the two end walls and in the partitions. The partitions were greased and the notches filled with an oil-vaseline mixture in order to prevent leakage between compartments (Figs. 1 and 3). The guard ring assembly was mounted on a micromanipulator so that it could be manoeuvred into position after the electrode had been inserted. The outer electrode (*e*) was made from silver sheet while the inner electrodes (*c*) and (*d*) were made from 0·5 mm. silver wire. Exposed portions of the electrodes were coated electrolytically with chloride and the wires connecting the electrodes with external terminals were insulated with shellac.

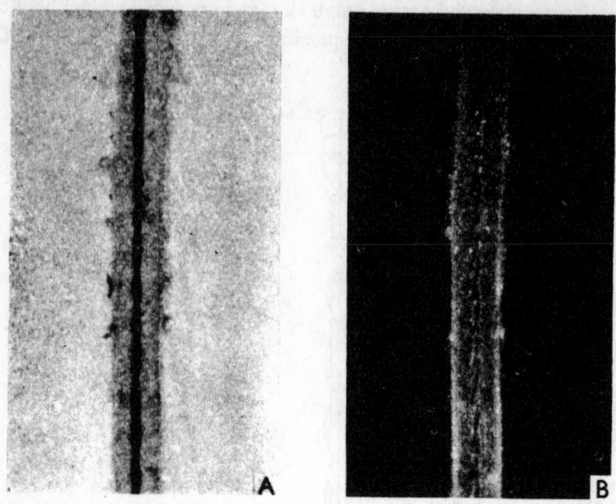

Fig. 5. Photomicrographs of giant axon and internal electrode. A, transmitted light; B, dark ground. The axon diameter was about 600 μ. The glass rod supporting the wires is not clearly seen.

Feed-back amplifier

A simplified diagram of the feed-back amplifier is shown in Fig. 6. It consisted of a differential d.c. amplifier with cathode follower input and output. The output of the amplifier was coupled to the input in such a way that negative feed-back was employed. This meant that any spontaneous change in membrane potential caused an output current to flow in a direction which restored the membrane potential to its original value. The level at which the potential was held constant was determined by the bias voltage v_3 and the control voltage v_4. v_3 was set so that no current passed through the nerve in the resting condition. This preliminary operation was carried out with the protective resistance R_f at its maximum value. This was important since an incorrect setting would otherwise have caused a large current to flow through the membrane. R_f was gradually reduced to zero; at the same time v_3 was adjusted to keep the membrane current zero. In order to change the membrane potential a rectangular pulse $\pm v_4$ was fed into the second stage of the amplifier. A large current then flowed into the membrane and changed its potential abruptly to a new level determined by

$$v_1 - v_2 = \beta v_4, \tag{2}$$

where v_1 and v_2 are the two input voltages and v_4 is the control voltage; β is a constant determined by resistance values and valve characteristics. Its value was of the order of 0·001. Any tendency to depart from Equation 2 was neutralized by a large output current which promptly restored the equilibrium condition defined by this relation.

In the majority of the experiments the slider of the potentiometer *P* was set to zero. Under these conditions the potential difference between the internal and external recording electrodes $(v_b - v_c)$ was directly proportional to $(v_1 - v_2)$. If α is the voltage gain of the cathode followers (about 0·9), then

$$v_b - v_c = \frac{2}{\alpha}(v_1 - v_2) = \frac{2\beta v_4}{\alpha}. \tag{3}$$

The performance of the feed-back amplifier was tested in each experiment by recording the time course of the potential difference between the internal and external electrodes. This showed that the recorded potential followed the control voltage with a time lag of about 1 μsec. and an accuracy of 1–2%. It is therefore unnecessary to discuss the numerous approximations which have to be made in order to derive Equation 2.

The voltage gain of the feed-back amplifier and cathode followers was about 400 in the steady state. At high frequencies the gain was about 1200, since the condenser C_1 increased the gain under transient conditions. The mutual conductance of the feed-back system $\left[\frac{\partial i}{\partial (v_b - v_c)}\right]$ was about 1 mho in the steady state and 3 mhos at high frequencies. The maximum current that the amplifier could deliver was about 5 mA.

The method described would be entirely satisfactory if there were no resistance, apart from that of the membrane, between internal and external electrodes. In practice there was a small series

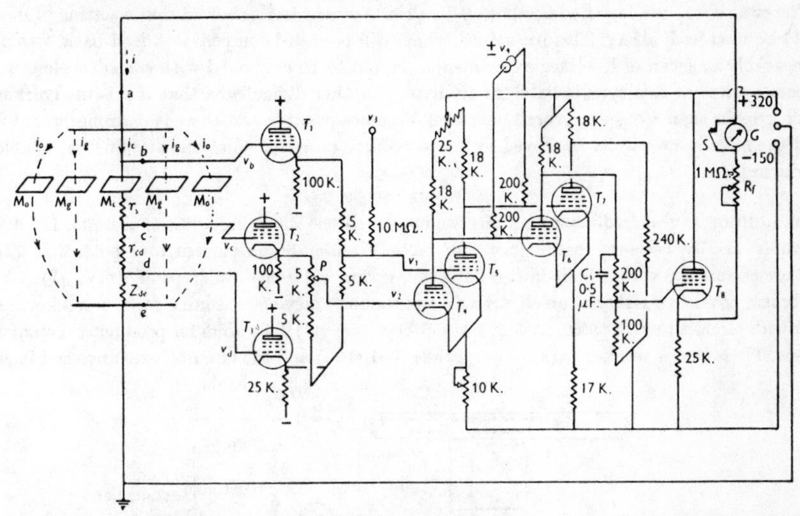

Fig. 6. Schematic diagram of feed-back amplifier. Screen resistances, grid stoppers and other minor circuit elements have been omitted. T_1, T_2, T_3 and T_8 are cathode followers; T_4, T_5, T_6 and T_7 are d.c. amplifiers. All valves were 6AK5 except T_1 and T_2 which were 1223. G is a microammeter used in setting-up. S is a switch for short-circuiting G. M_c is the membrane in the central section of the guard system. M_g, membrane in guard channels. M_o, membrane outside guard system. i_c, i_g and i_o are currents through these elements. r_{cd}, fluid-resistance used to measure current (74 Ω. at 20° C.). r_s, resistance in series with membrane (about 52 Ω. at 20° C.). z_{de}, impedance of large earthed electrode and sea water between d and e. Potentials are given with respect to earth.

resistance, represented by r_s in Fig. 6 and discussed further on p. 444. This meant that the true membrane potential was in error by the quantity $r_s i_c$. Thus

$$v_i - v_o = v_b - v_c - r_s i_c = 2\beta v_4/\alpha - r_s i_c, \tag{4}$$

where $v_i - v_o$ is the potential difference between the inner and outer surfaces of the membrane, r_s is the resistance in series with the membrane and i_c is the current flowing through the central area of membrane.

In principle the error introduced by r_s can be abolished by setting the potentiometer P to an appropriate value. All three cathode followers (T_1, T_2, T_3) had the same gain so that v_1 and v_2 were determined by the following equations:

$$v_1 = \tfrac{1}{2}\alpha\,(v_b + v_d), \tag{5}$$

$$v_2 = \tfrac{1}{2}\alpha\,[v_c + v_d + p\,(v_c - v_d)], \tag{6}$$

and

$$v_1 - v_2 = \tfrac{1}{2}\alpha\,[v_b - v_c - p\,(v_c - v_d)], \tag{7}$$

where p is proportional to the setting of the potentiometer P and varied between extremes of 0 and 1 and v_d is the potential of electrode d.

From Ohm's law

$$v_c - v_d = r_{cd} i_c, \tag{8}$$

where r_{cd} is the resistance of the central channel between electrodes c and d.

From Equations 4, 7 and 8

$$v_1 - v_2 = \tfrac{1}{2}\alpha\,[v_i - v_o + i_c\,(r_s - p r_{cd})]. \tag{9}$$

If $p = r_s/r_{cd}$

$$v_i - v_o = \frac{2}{\alpha}(v_1 - v_2) = \frac{2\beta v_4}{\alpha}. \tag{10}$$

The ratio r_s/r_{cd} was found to be about 0·7 and subsequent trials showed that a setting of $p=0{\cdot}6$ could be used with safety. This procedure, which will be called compensated feed-back, was used successfully in seven of the later experiments. It had to be employed with considerable caution since a system of this type is liable to oscillate. Another difficulty is that if p is inadvertently made greater than r_s/r_{cd} the overall feed-back becomes positive and there is a strong probability that the membrane will be destroyed by the very large currents which the amplifier is capable of producing.

Auxiliary equipment

In addition to the feed-back amplifier we employed the following additional units: (1) A d.c. amplifier and cathode-ray oscillograph for recording membrane current and potential. (2) A voltage calibrator, with a built-in standard cell, giving ±110 mV. in steps of 1 mV. (3) A time calibrator consisting either of an electrically maintained 1 kcyc./sec. tuning fork, or a 4 kcyc./sec. fork with circuits to give pulses at 4, 2, 1 or 0·5 kcyc./sec. (4) Two units for producing rectangular pulses. These pulses were of variable amplitude (0–100 V.) and the circuits were arranged in such

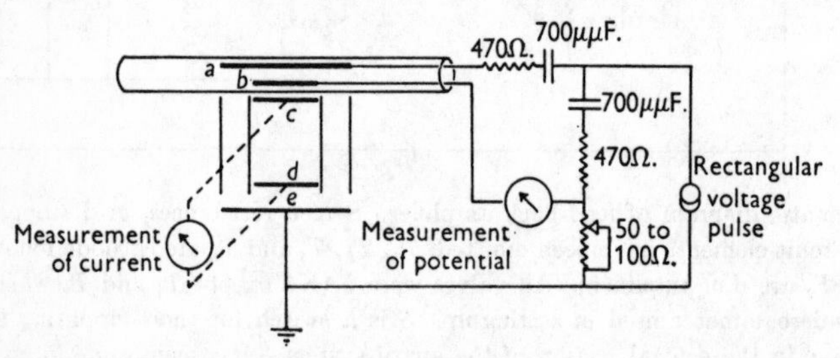

Fig. 7. Diagram of arrangement for recording response of membrane to short shock.

a way that the outputs of each generator were symmetrical with respect to earth. A single pulse generator was used in the early experiments, and its output was applied to the feed-back amplifier in the manner shown in Fig. 6. When required, the output of a second pulse generator was applied in parallel through a second pair of resistances. (5) An electrically operated refrigerator unit for cooling the preparation. All these items were of conventional design and need no detailed description.

Stimulation with brief currents

In the early stages of the work it was important to prove that the membrane was capable of giving an action potential of normal size. For this purpose we disconnected the feed-back amplifier and employed the arrangement shown in Fig. 7. A rectangular voltage step v_4 was applied to one internal electrode through a 700 μμF. condenser. The total area of membrane exposed to current flow from the electrode was about 0·3 cm.^{2} (1·5 cm. × π × 0·06 cm.). It therefore had a capacity of about 0·3 μF. When v_4 was suddenly changed by 10 V. the membrane potential was displaced by about 23 mV. (10 V. × 700/300,000). With this arrangement the membrane current consisted of very brief currents at the beginning and end of the voltage step. The size of the current could be varied by altering the size of the step, while the membrane current in the central channel of the guard-system could be measured by recording the potential difference between electrodes c and d. The potential difference between the voltage wire (b) and the external electrode (c) was equal to the sum of the membrane potential and the potential difference across the ohmic resistance in series with the membrane. The second component was eliminated by the bridge circuit illustrated in Fig. 7.

Experimental procedure

Giant axons with a diameter of 400–800 μ. were obtained from the hindmost stellar nerve of *Loligo forbesi* and freed from all adherent tissue. Careful cleaning was important since the guard system did not operate satisfactorily if the axon was left with small nerve fibres attached to it. A further advantage in using cleaned axons was that the time required for equilibration in a test solution was greatly reduced by removing adherent tissue.

The axon was cannulated and mounted in the same type of cell as that described by Hodgkin & Huxley (1945) and Hodgkin & Katz (1949 *a*). A conventional type of internal electrode, consisting of a long glass capillary, was thrust down the axon for a distance of 25–30 mm. This was then removed and the double wire electrode inserted in its place. Action potentials and resting potentials were recorded from the first electrode and the axon was rejected if these were not reasonably uniform over a distance of 20 mm. Another reason for starting with a conventional type of electrode was that the double wire electrode, in spite of the rigidity of its glass support, could not be inserted without buckling unless the axon had first been drilled with the glass capillary.

When the wire electrode was in position the guard system was brought into place by means of a micromanipulator. This operation was observed through a binocular microscope and care was taken to ensure that the central channel coincided exactly with the exposed portion of the internal voltage wire. The front of the guard-ring box was gently pressed into position and finally sealed by firm pressure with a pair of forceps. Before applying the front, spots of a vaseline-oil mixture were placed in such a position that they completed the seal round the axon when the front was pressed home (Figs. 1 and 3).

After the axon was sealed into position cold sea water (3–11° C.) was run into the cell and this temperature was maintained by means of a cooling coil which dipped into the cell. Air was bubbled through the cell in order to stir the contents and obtain a uniform temperature.

Before proceeding to study the behaviour of the axon under conditions of constant voltage its response to a short shock was observed. The experiment was discontinued if the action potential recorded in this way was less than about 85 mV. If the axon passed this test it was connected to the feed-back amplifier in the manner described previously.

Solutions were changed by running all the fluid from the cell and removing it from the guard-ring assembly with the aid of a curved capillary attached to a suction pump. A new solution was then run into the cell and was drawn into the guard rings by applying suction at appropriate places.

Calibration

The amplifier was calibrated at the end of each experiment, and all photographic records were analysed by projecting them on to a calibration grid. The readings obtained in this way were converted into current by dividing the potential difference between the two external electrodes c and d by the resistance between these electrodes (r_{cd}). This resistance was determined by blocking up the outer compartments of the guard-ring assembly and filling the central channel with sea water or with one of the standard test fluids. A silver wire was coated with silver chloride and inserted into the position normally occupied by the axon (Fig. 1). A known current was applied between the central wire and the outer electrode (e). The resistance between the two external electrodes c and d could then be obtained by measuring the change in potential difference resulting from a given application of current. It was found that the resistance between these electrodes was 74 Ω. when the central chamber was filled with sea water at 20° C. This value was close to that calculated from the dimensions of the system.

Membrane currents were converted to current densities by dividing them by the area of membrane exposed to current flow in the central compartment. The area was calculated from the measured axon diameter and the distance between the partitions A_2 and A_3 (Fig. 1).

RESULTS

Stimulation with brief currents

Before investigating the effect of a constant voltage it was important to establish that the membranes studied were capable of giving normal action potentials. This was done by applying a brief shock to one internal electrode and recording changes in membrane potential with the other. Details of the method are given on p. 431; typical results are shown in Fig. 8 (23° C.) and Fig. 9 (6° C.).

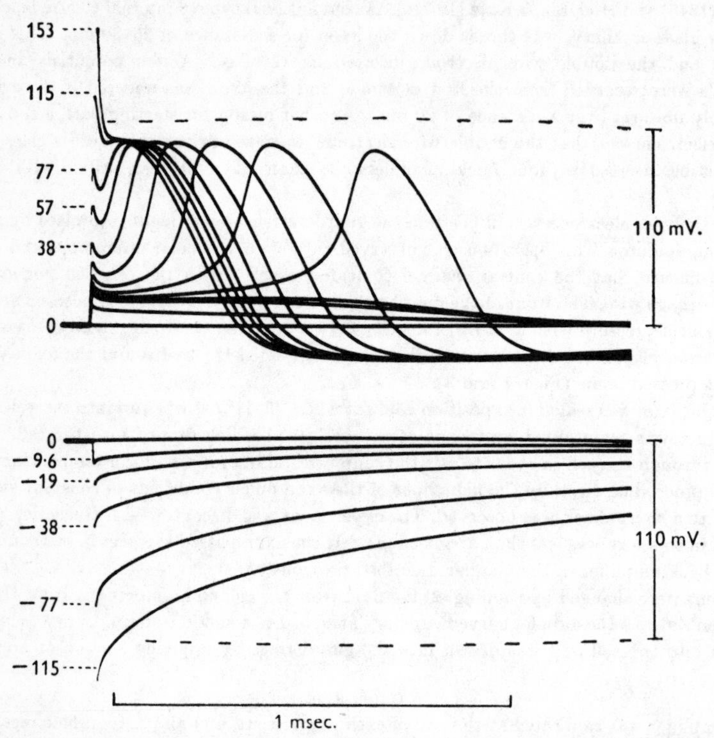

Fig. 8. Time course of membrane potential following a short shock at 23° C. Depolarizations shown upwards. Axon 18. The numbers attached to the curves give the strength of shock in mμcoulomb/cm.². Shock strengths for unlabelled curves are 29, 23, 19·2, 17·3, 16·7, 15·3, 9·6.

The shock used to displace the membrane potential was calibrated by recording the membrane current in the central channel (Fig. 7). This test showed that the current pulse consisted of a brief surge which was 95% complete in about 8 μsec. and reached a peak amplitude of about 50 mA./cm.² at the highest strengths. The total quantity of current passing through the central channel was evaluated by integrating the current record and was used to define the strength of the shock. The numbers attached to the records in Figs. 8 and 9 give the charge applied per unit area in mμcoulomb/cm.². It will be seen that the initial displacement of membrane potential was proportional to the charge applied and that it corresponded to a membrane capacity of about 0·9 μF./cm.². Values obtained by this method are given in Table 1.

Although the initial charging process was linear, the subsequent behaviour of the membrane potential varied with the strength of shock in a characteristic

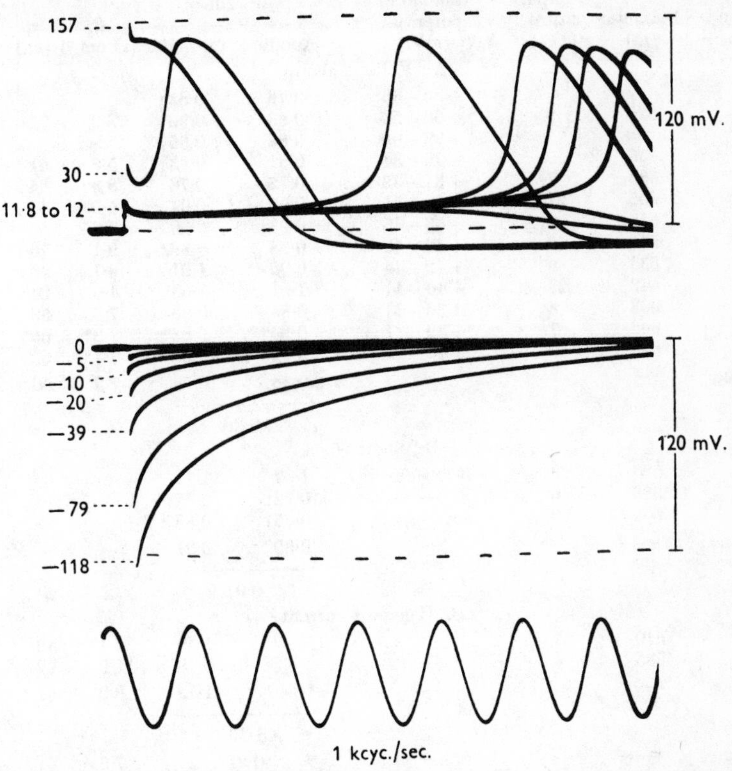

Fig. 9. Time course of membrane potential following a short shock at 6° C. Depolarization shown upwards. Axon 17. The numbers attached to the curves give the strength of shock in mμcoulomb/cm.². The initial displacement in the case of the uppermost curve cannot be seen; its value was about 200 mV.

manner. All the anodal records had the same general shape, but depolarizations of more than a few millivolts gave non-linear responses. If the depolarization was more than 15 mV. (Fig. 8) or 12 mV. (Fig. 9) the response became regenerative and produced an action potential of about 100 mV. If it was less than 12 or 15 mV. it was followed by a subthreshold response similar to that seen in most excitable tissues. If the potential was displaced to the threshold level it might remain in a state of unstable equilibrium for considerable periods of time. This is illustrated by Fig. 9 which shows the effect of a small variation of shock strength in the region of threshold.

Records such as those in Fig. 9 may be used to estimate the relation between membrane potential and ionic current. The total membrane current density (I) is negligible at times greater than 200 μsec. after application of the short

TABLE 1. Membrane capacities

Axon no.	Diameter (μ.)	Temperature (° C.)	Change in potential (mV.)	Membrane capacity (μF./cm.^{2}) Anodic	Cathodic	R_s (Ω.cm.^{2})	r_s (Ω.)	r_s/r_{cd}
			A. Voltage clamp					
13	520	9	+36 − 36	0·76	0·83	8·2	72	0·77
			+56 − 56	0·83	0·90			
			+98 − 98	0·83	0·96			
14	430	9	+36 − 34	0·81	0·83	5·8	61	0·65
17	588	7	+31 − 32	0·72	0·76	8·3	64	0·65
18	605	21	+30 − 31	0·92	0·91	5·5	41	0·57
19	515	8	+43 − 45	0·93	0·90	7·8	69	0·73
20	545	6	+42 − 43	0·88	0·86	9·1	76	0·77
21	533	9	+42 − 44	0·98	1·01	9·1	78	0·84
22	542	23	+40 − 41	1·01	1·03	4·0	34	0·50
25	603	8	+39 − 41	0·88	0·86	7·0	53	0·57
25*	603	7	+39 − 41	0·84*	0·82*	8·8*	66*	0·55*
26	675	20	+40 − 42	0·97	0·93	7·7	52	0·70
Average	—	—	—	0·88	0·90	7·3	60	0·68
				0·89				
			B. Short shock					
13	520	9	+58 − 50	1·07	1·11	—	—	—
17†	588	6	—	0·79†	0·74†	—	—	—
18†	605	23	—	0·85†	0·88†	—	—	—
Average	—	—	—	0·90	0·91	—	—	—
				0·91				
			C. Constant current					
29	540	21	—	—	1·49	6·4	42	0·57
41	585	4	—	—	0·78	11·9	92	0·88
Average	—	—	—	—	1·13	9·2	67	0·73
				1·13				
Average by all methods	—	—	—	0·91		7·6	61	0·68

* In this experiment choline was substituted for sodium in the external solution. The values obtained are excluded from the averages.

† In these experiments the shock strength was not measured directly but was obtained from the calibration for axon 13. The values for C_M are means obtained from a wide range of shock strengths.

shock. This means that the ionic current density (I_i) must be equal to the product of the membrane capacity per unit area (C_M) and the rate of depolarization. Thus if $I=0$, Equation 1 becomes

$$I_i = -C_M \frac{\partial V}{\partial t}. \tag{11}$$

Fig. 10 illustrates the relation between membrane potential and ionic current at a fixed time (290 μsec.) after application of the stimulus. It shows that ionic current and membrane potential are related by a continuous curve which crosses the zero current axis at $V=0$, $V=-12$ mV. and $V=-110$ mV. Ionic current is inward over the regions -110 mV. $< V < -12$ mV. and $V>0$, and is outward for $V<-110$ mV. and -12 mV. $< V <0$. $\partial I_i/\partial V$ is negative for -76 mV. $< V < -6$ mV. and is positive elsewhere.

A curve of this type can be used to describe most of the initial effects seen in Figs. 8 and 9. When the membrane potential is increased by anodal shocks the ionic current associated with the change in potential is in the inward direction. This means that the original membrane potential must be restored by an inward transfer of positive charge through the membrane. If the

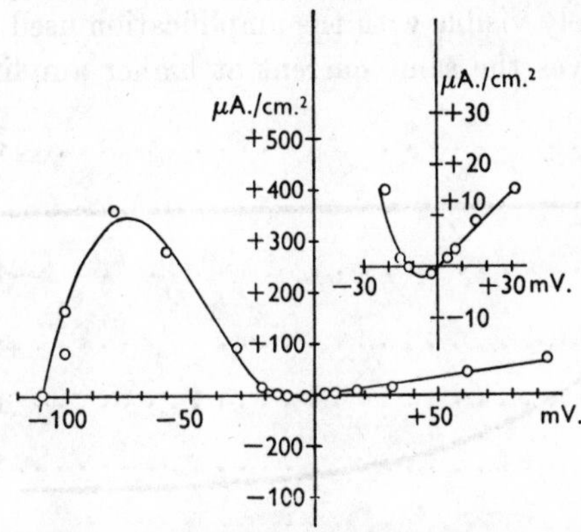

Fig. 10. Relation between ionic current density (I_i) and displacement of membrane potentia (V). Abscissa: displacement of membrane potential from its resting value (anodal displacement shown positive). Ordinate: ionic current density obtained from $-C_M\frac{dV}{dt}$ (inward current shown positive). Inset: curve in region of origin drawn with tenfold increase in vertical scale. Axon 17; $C_M=0{\cdot}74\,\mu$F./cm.²; temperature 6·3° C. Measurements made 0·29 msec. after application of shock.

membrane potential is depolarized by less than 12 mV., ionic current is outward and again restores the resting condition by repolarizing the membrane capacity. At $V=-12$ mV., I_i is zero so that the membrane potential can remain in a state of unstable equilibrium. Between $V=-110$ mV. and $V=-12$ mV., I_i is inward so that the membrane continues to depolarize until it reaches $V=-110$ mV. If the initial depolarization is greater than 110 mV. I_i is outward which means that it will repolarize the membrane towards $V=-110$ mV. These effects are clearly seen in Figs. 8 and 9.

Membrane current under conditions of controlled potential

General description

The behaviour of the membrane under a 'voltage clamp' is illustrated by the pair of records in Fig. 11. These show the membrane current which flowed as a result of a sudden displacement of the potential from its resting value to a new level at which it was held constant by electronic feed-back. In the upper record the membrane potential was increased by 65 mV.; in the lower record it was decreased by the same amount. The amplification was the same in both cases.

The first event in both records is a slight gap, caused by the surge of 'capacity current' which flowed when the membrane potential was altered suddenly. The surge was too rapid to be visible on these records, but was examined in other experiments in which low gain and high time base speed were employed (see p. 442). The ionic current during the period of constant potential was small when the membrane potential was displaced in the anodal direction, and is barely visible with the amplification used in Fig. 11. The top record in Fig. 12 gives the same current at higher amplification and shows

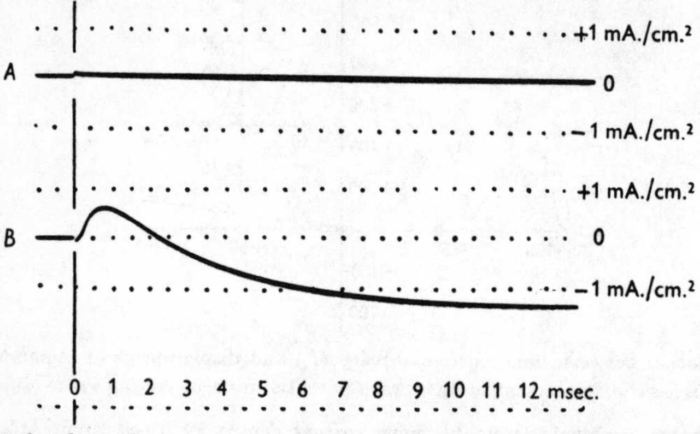

Fig. 11. Records of membrane current under a voltage clamp. At zero time the membrane potential was increased by 65 mV. (record *A*) or decreased by 65 mV. (record *B*); this level was then maintained constant throughout the record. Inward current is shown as an upward deflexion. Axon 41; diameter 585 μ. Temperature 3·8° C. Compensated feed-back.

that an increase of 65 mV. in the membrane potential was associated with an inward ionic current of about 30 μA./cm.^{2} which did not vary markedly with time. The sequence of events was entirely different when the membrane potential was reduced by 65 mV. (Fig. 11*B*). In this case the current changed sign during the course of the record and reached maximum amplitudes of +600 and −1300 μA./cm.^{2}. The initial phase of ionic current was inward and was therefore in the opposite direction to that expected in a stable system. If it had not been drawn off by the feed-back amplifier it would have continued to depolarize the membrane at a rate given by Equation 11. In this experiment C_M was 0·8 μF./cm.^{2} and I_i had a maximum value of 600 μA./cm.^{2}. The rate of depolarization in the absence of feed-back would therefore have been 750 V./sec., which is of the same general order as the maximum rate of rise of the spike (Hodgkin & Katz, 1949*a*, *b*). The phase of inward current was not maintained but changed fairly rapidly into a prolonged period of outward current. In the absence of feed-back this current would have repolarized the membrane at a rate substantially greater than that observed during the falling phase of the spike. The outward current appeared to be maintained for an indefinite period if the membrane was not depolarized by more than 30 mV. With greater depolarization it declined slowly as a result of a polarization effect discussed on p. 445.

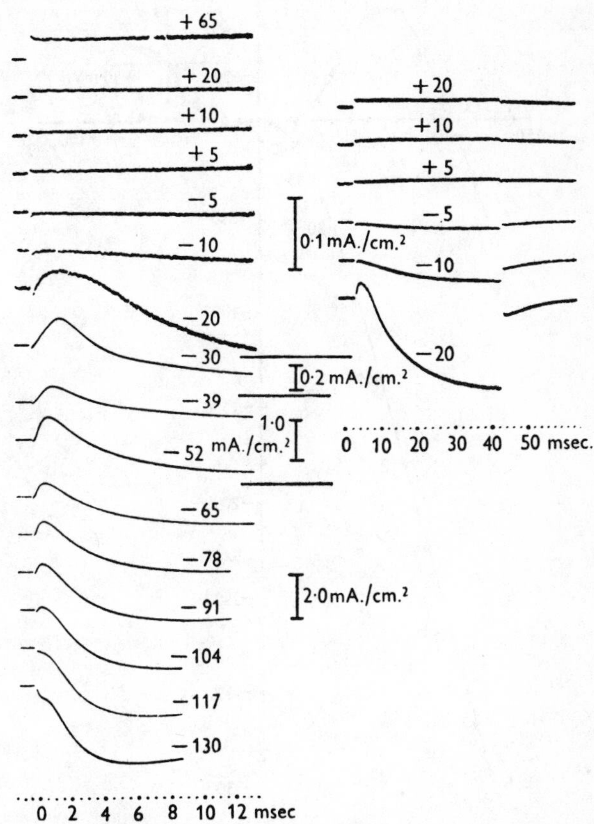

Fig. 12. Records of membrane current under a voltage clamp. The displacement of membrane potential (V) is given in millivolts by the number attached to each record. Inward current is shown as an upward deflexion. Six records at a lower time base speed are given in the right-hand column. Experimental details as in Fig. 11.

The features illustrated in Fig. 11 *B* were found over a wide range of voltages as may be seen from the complete family of curves in Fig. 12. An initial phase of inward current was conspicuous with depolarizations of 20–100 mV., while the delayed rise in outward current was present in all cathodal records. A convenient way of examining these curves is to plot ionic current density against membrane potential. This has been done in Fig. 13, in which the abscissa gives the displacement of membrane potential and the ordinate gives the ionic current density at a short time (curve *A*) and in the 'steady state' (curve *B*). It will be seen that there is a continuous relation over the whole range, but that small changes in membrane potential are associated with large changes in current. At short times the relation between ionic current density

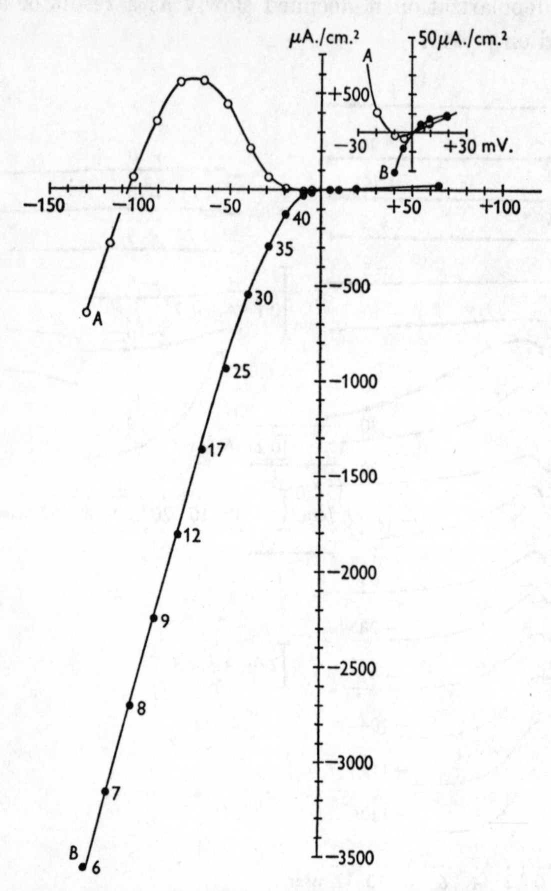

Fig. 13. Relation between membrane current density and membrane potential. Abscissa: displacement of membrane potential from its resting value in mV. Ordinate: membrane current density at 0·63 msec. after beginning of voltage step (curve *A*) and in 'steady state' (curve *B*). The numbers attached to curve *B* indicate the times in msec. at which the measurements were made. Inset: curves in region of origin drawn with a tenfold increase in the vertical scale. Inward current density is taken as positive and the membrane potential is given in the sense external potential minus the internal potential. Measurements were made from the records reproduced in Fig. 12 (3·8° C.).

and membrane potential is qualitatively similar to that obtained indirectly in Fig. 10. Ionic current is inward over the region -106 mV. $< V < -12$ mV. and for $V > 0$; it is outward for $V < -106$ mV. and for -12 mV. $< V < 0$. $\partial I_i/\partial V$ is negative for -70 mV. $< V < -7$ mV. and is positive elsewhere. More quantitative comparisons are invalidated by the fact that the ionic current is a function of time as well as of membrane potential. At long times depolarization is invariably associated with an outward current and $\partial I_i/\partial V$ is always positive.

The electrical resistance of the membrane varied markedly with membrane potential. In Fig. 13, $\partial V/\partial I_i$ is about 2500 Ω.cm.^{2} for $V > 30$ mV. For $V = -110$ mV. it is 35 Ω.cm.^{2} (curve *A*) or 30 Ω.cm.^{2} (curve *B*). At $V = 0$, $\partial V/\partial I_i$ is 2300 Ω.cm.^{2} at short times and 650 Ω.cm.^{2} in the steady state. These results are comparable with those obtained by other methods (Cole & Curtis, 1939; Cole & Hodgkin, 1939).

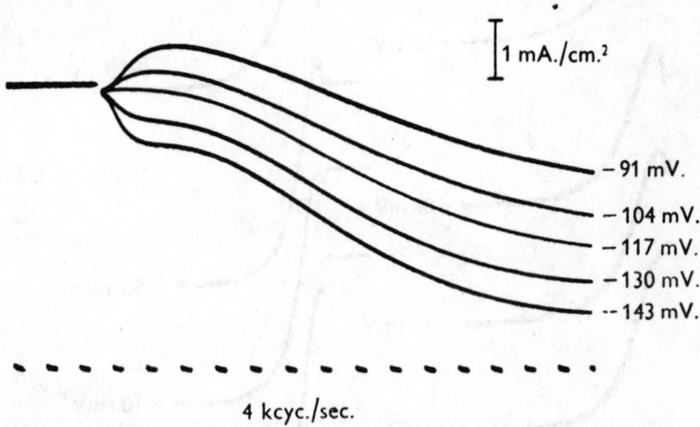

Fig. 14. Time course of membrane current during large depolarizations. Abscissa: time. Ordinate: inward current density. The numbers attached to the records give the displacement of membrane potential from its resting value. Axon 41; temperature 3·5° C. Compensated feed-back.

Fig. 14 illustrates the initial phase of ionic current at large depolarizations in greater detail. These records were obtained from the same axon as Fig. 12 but at an earlier stage of the experiment. They show that the initial 'hump' of ionic current changed sign at a potential of -117 mV. At -130 mV. the initial hump consists of outward current while it is plainly inward at -104 mV. The curve at -117 mV. satisfies the condition that $\partial I_i/\partial t = 0$ at short times and has no sign of the initial hump seen in the other records. It will be shown later that this potential probably corresponds to the equilibrium potential for sodium and that it varies with the concentration of sodium in the external medium (Hodgkin & Huxley, 1952*a*).

The effect of temperature

The influence of temperature on the ionic currents under a voltage clamp is illustrated by the records in Fig. 15. These were obtained from a pair of axons from the same squid. The first axon isolated was examined at 6° C. and gave the series of records shown in the left-hand column. About 5 hr. later the second axon, which had been kept at 5° C. in order to retard deterioration, was examined in a similar manner at 22° C. Its physiological condition is likely to have been less normal than that of the first but the difference is not thought to be large since the two axons gave propagated action potentials of amplitude 107 and 103 mV. respectively, both measured at 22° C. The resting potentials were 55 mV. in both cases. The results obtained with the second axon are given in the right-hand column of Fig. 15. It will be seen that the general form and

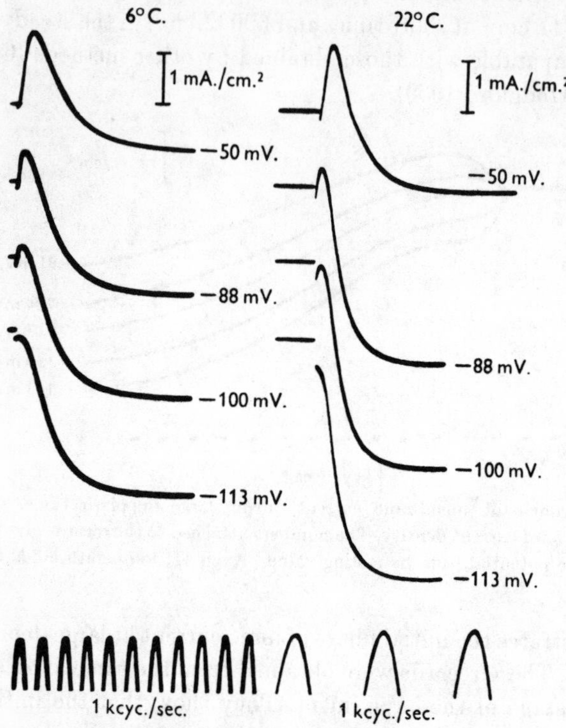

Fig. 15. Membrane currents at different temperatures. Axons 17 (6° C.) and 18 (22° C.), from the same squid. Inward current is shown as an upward deflexion. The numbers attached to each curve give the displacement of membrane potential. Uncompensated feed-back was employed.

amplitude of the two sets of records are similar but that the rate at which the ionic current changes with time was increased about sixfold by the rise in temperature of 16° C. It was found that the two families could be roughly superposed by assuming a Q_{10} of 3 for the rate at which ionic current changes with time. Values between 2·7 and 3·5 were found by analysing a number of experimental records obtained under similar conditions, but with different axons at different temperatures. In the absence of more precise information we shall use a temperature coefficient of 3 when it is necessary to compare rates measured at different temperatures.

The absolute magnitude of the current attained at any voltage probably varies with temperature, but much less than the time scale. In the experiment of Fig. 15 a rise of 16° C. increased the outward current about 1·5-fold, while the inward current at −50 mV. was approximately the same in the two records. Since the initial phase of inward current declined relatively rapidly as the axon deteriorated it is possible that a temperature coefficient of about 1·3 per 10° C. applies to both components of the current. Temperature coefficients of the order of 1·0–1·5 were also obtained by examining a number of results obtained with other axons.

The capacity current

The surge of current associated with the sudden change in membrane potential was examined by taking records at high time-base speed and low amplification. Tracings of a typical result are shown in Fig. 16. It will be seen

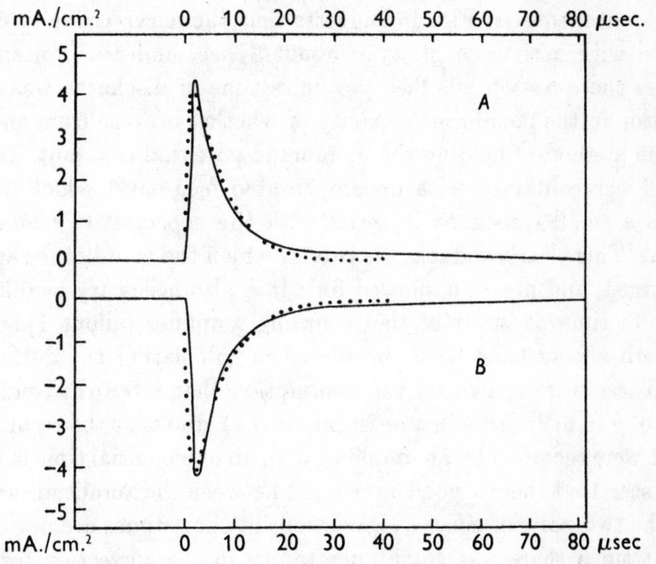

Fig. 16. Current through capacitative element of membrane during a voltage clamp. Abscissa: time in μsec. Ordinate: membrane current density (mA./cm.²) with inward current taken as positive. At $t=0$ the potential difference between external and internal electrodes was displaced +40 mV. in curve *A* or −40 mV. in curve *B*. The continuous curves were traced from experimental records. The dotted curves were calculated according to the equation

$$I^* = 6{\cdot}8\,[\exp(-0{\cdot}159t) - \exp(-t)],$$

where I^* is the current in mA./cm.² and t is time in μsec. This follows from the assumptions given in the text. Axon 25; temperature 8° C.

that the current records for anodal and cathodal changes are almost symmetrical and that the charging process is virtually complete in 50 μsec. In the anodal record the observed current declined from a peak of 4·5 mA./cm.² to a steady level of about 0·04 mA./cm.². The steady current is barely visible in the tracing but could be seen more clearly at higher amplification and lower time-base speed; records taken under these conditions were similar to those in Fig. 12.

The membrane capacity was obtained from the change in potential and the area under the curves. A small correction was made for ionic current but the resting membrane conductance was sufficiently low for uncertainties here to be unimportant. In the experiment illustrated by Fig. 16 *A* the charge entering the membrane capacity in 60 μsec. was 35 mμcoulomb/cm.^{2}, while the change in potential was 40 mV. Hence the membrane capacity per unit area was about 0·9 μF./cm.^{2}. Table 1 (p. 435) gives the results of other experiments of this kind. It also shows that replacement of all the sodium in sea water by choline had little effect on the membrane capacity and that there was no large change of capacity with temperature.

If a perfect condenser is short-circuited through zero resistance it loses its charge instantaneously. Fig. 16 suggests that the nerve capacity charged or discharged with a time constant of about 6 μsec. under a 'voltage clamp'. This raises the question whether the finite time of discharge was due to an imperfection in the membrane capacity or whether it arose from an imperfection in the method of holding the membrane potential constant. The records in Fig. 16 were obtained with uncompensated feed-back, which means that there was a small resistance in series with the capacitative element of the membrane. This clearly reduces the rate at which the membrane capacity can be discharged, and must be allowed for. It is also necessary to take account of the finite time constant of the recording amplifier (about 1 μsec. at this gain). Both effects have been considered in calculating the dotted lines in Fig. 16. These were drawn on the assumption that a 0·9 μF. condenser was charged to ± 40 mV. through a resistance of 7 Ω. and that the resulting pulses of current were recorded by an amplifier with an exponential time lag of 1 μsec. It will be seen that there is good agreement between the amplitude and general form of the two pairs of curves. Deviations at short times are not considered important since there was some uncertainty in the correction for amplifier delay.

At relatively long times ($> 25\,\mu$sec.) the current record shows a 'tail' which is not explained by the presence of a series resistance. This effect was present in all records and was larger at higher temperatures. It can be explained by supposing that the membrane capacity was not perfect but behaved in the manner described by Curtis & Cole (1938). The records in Fig. 16 are roughly consistent with a constant phase angle of 80°, while those at higher temperatures require somewhat lower values. These statements must be regarded as tentative since our experiments were not designed to measure the phase angle and do not give good data for quantitative analysis. For the time being the principal point to be emphasized is that the surge associated with a sudden change in potential is adequately described by assuming that the membrane has a capacity of about 1 μF./cm.^{2} and a series resistance of about 7 Ω. cm.^{2}.

The surge of capacity current was larger in amplitude and occupied a shorter time when compensated feed-back was employed. These experiments were not suitable for analysis, since the charging current was oscillatory and could not be adequately recorded by the camera employed. All that could be seen in records of ionic current is a gap, as in Fig. 14.

Our values for the membrane capacity are in reasonable agreement with those obtained previously. Using transverse electrodes 5·6 mm. in length, Curtis & Cole (1938) obtained the following values in twenty-two experiments: average membrane capacity at 1 kcyc./sec., 1·1 μF./cm.^{2}, range 0·66 μF./cm.^{2} to 1·60 μF./cm.^{2}; average phase angle, 76°, range 64–85°.

The values for membrane capacity in the upper part of Table 1 were obtained by integrating the initial surge of current over a total time of about 50 μsec. If the phase angle is assumed to be 76° at all frequencies the average value of 0·89 μF./cm.^{2} obtained by this method is equivalent to one of 1·03 μF./cm.^{2} at 1 kcyc./sec. This is clearly in good agreement with the figures given by Curtis & Cole (1938), but is substantially less than the value of 1·8 μF./cm.^{2} mentioned in a later paper (Cole & Curtis, 1939). However, as Cole & Curtis point out, the second measurement is likely to be too large since the electrode length was only 0·57 mm. and no allowance was made for end-effects.

The series resistance

Between the internal and external electrodes there is a membrane with a resting resistance of about 1000 Ω. cm.^{2}. This resistance is shunted by a condenser with a capacity of about 1 μF./cm.^{2}. In series with the condenser, and presumably in series with the membrane as a whole, there is a small resistance which, in the experiment illustrated by Fig. 16, had an approximate value of 7 Ω. cm.^{2}. This 'series resistance' can be estimated without fitting the complete theoretical curve shown in that figure. A satisfactory approximation is to divide the time constant determining the decline of the capacitative curve by the measured value of the membrane capacity. This procedure was followed in calculating the values for the series resistance given in the upper part of Table 1. The symbol r_s gives the actual resistance in series with the central area of membrane, while R_s is the same quantity multiplied by the area of membrane exposed to current flow in the central channel of the guard system. The last column gives the ratio of r_s to the resistance (r_{cd}) between the current measuring electrodes, *c* and *d*. This ratio is of interest since it determined the potentiometer setting required to give fully compensated feed-back (pp. 430–1).

Another method of measuring the series resistance was to apply a rectangular pulse of current to the nerve and to record the potential difference (v_{bc}) between the internal electrode *b* and the external electrode *c* as a function of time. The current in the central channel of the guard system was also obtained by recording the potential difference (v_{cd}) between the external electrodes *c* and *d*. The two records were rounded to the same extent by amplifier delay so that the series resistance and the membrane capacity could be determined by fitting the record obtained from the internal electrode by the following equation

$$v_{bc}=\frac{r_s}{r_{cd}}\,v_{cd}+\frac{1}{r_{cd}c}\cdot\int_0^t v_{cd}\,\mathrm{d}t,$$

where *c* is the capacity of the area of membrane exposed to current flow in the central channel.

This analysis was made with two axons and gave satisfactory agreement between observed and calculated values of v_{bc}, with values of r_s/r_{cd} and C_M which were similar to those obtained by the voltage clamp method (see Table 1).

The observed value of the series resistance ($r_s = 61\,\Omega$.) cannot be explained solely by convergence of current between the electrodes used to measure membrane potential. Only about 30% was due to convergence of current between electrode *c* and the surface of the nerve, while convergence between the membrane and the internal electrode should not account for more than 25%, unless the specific resistance of axoplasm was much greater than that found by Cole & Hodgkin (1939). The axons used in the present work were surrounded by a dense layer of connective tissue, 5–20 μ. in thickness, which adheres tightly and cannot easily be removed by dissection. According to Bear, Schmitt & Young (1937) the inner layer of this sheath has special optical properties and may be lipoid in nature. It seems reasonable to suppose that one or other of these external sheaths may have sufficient resistance to account for 45% of the series resistance. There was, in fact, some evidence that the greater part of the series resistance was external to the main barrier to ionic movement. Substitution of choline sea water for normal sea water increased r_{cd} by 23%, but it reduced r_s/r_{cd} by only 3·5% (Table 1, axon 25). This suggests that about 80% of r_s varied directly with the specific resistance of the external medium. Since the composition of axoplasm probably does not change when choline is substituted for sodium (Keynes & Lewis, 1951) it seems likely that most of the series resistance is located outside the main barrier to ionic movement. Further experiments are needed to establish this point and also to determine whether the resistive layer has any measurable capacity.

Accuracy of method

The effect of the series resistance. The error introduced by the series resistance (r_s) was discussed on p. 430. Its magnitude was assessed by comparing records obtained with uncompensated feed-back ($p=0$) with those obtained with compensated feed-back ($p=0·6$). The effect of compensation was most conspicuous with a depolarization of about 30 mV. Fig. 17 shows typical curves in this region. *A*, *B* and *C* were obtained with uncompensated feed-back; α, β and γ with compensated feed-back. *A* gives the potential difference between external and internal electrodes. *B* is the potential difference between the external electrodes used to measure current and is equal to the product of the membrane current and the resistance r_{cd}. The true membrane potential differs from *A* by the voltage drop across r_s which is equal to (r_s/r_{cd}) *B*. *C* shows the membrane potential calculated on the assumption that r_s/r_{cd} had its average value of 0·68. α, β and γ were obtained in exactly the same manner as *A*, *B* and *C*, except that the potentiometer setting (p) was increased from 0 to 0·6. It will be seen that this altered the form of the upper record in a manner which compensates for the effect of current flow. The error in *C* is about 20%, while γ deviates by only about 2·5%. Hence any error present in β is likely to be increased eightfold in *B*. Since the two records are not grossly different, β may be taken as a reasonably faithful record of membrane current under a voltage clamp.

Experiments of this type indicated that use of uncompensated feed-back introduced errors but that it did not alter the general form of the current record. Since the method of compensated feed-back was liable to damage axons it was not employed in experiments in which the preparation had to be kept in good condition for long periods of time.

Polarization effects. The outward current associated with a large and prolonged reduction of membrane potential was not maintained, but declined slowly as a result of a 'polarization effect'. The beginning of this decline can be seen in the lower records in Fig. 12. It occurred under conditions which had little physiological significance, for an axon does not normally remain with a membrane potential of -100 mV. for more than 1 msec. Nor does the total current through the membrane approach that in Fig. 12.

In order to explain the effect it may be supposed either: (1) that 'mutual polarization' of the electrode (p. 428) is substantially greater inside the axon than in sea water; (2) that currents may cause appreciable changes in ionic concentration near the membrane; (3) that some structure in series with the membrane may undergo a slow polarization. We were unable to distinguish between these suggestions, but it was clear that the 'polarization effect' had little to do with the active changes, since it was also present in moribund axons and was little affected by temperature.

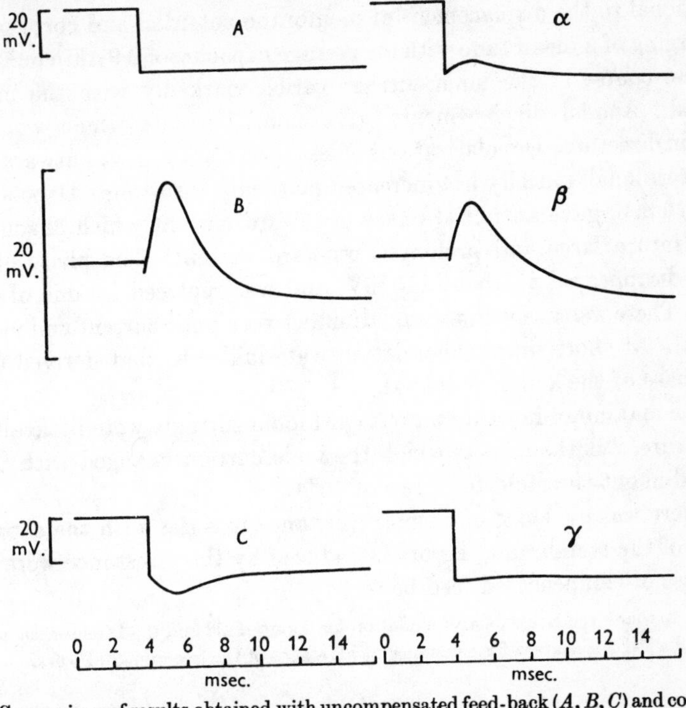

Fig. 17. Comparison of results obtained with uncompensated feed-back (*A*, *B*, *C*) and compensated feed-back (α, β, γ). *A*, α: potential difference between external and internal electrodes ($v_c - v_b$). *B*, β: potential difference between current measuring electrodes ($v_d - v_c$). *C*, γ: membrane potential calculated as $C = A - 0{\cdot}68B$, or $\gamma = \alpha - 0{\cdot}68\beta$. Records *B* and β may be converted into membrane current density by dividing by $11{\cdot}9\,\Omega.\text{cm.}^2$. Temperature 4° C. Axon 34.

SUMMARY

1. An experimental method for controlling membrane potential in the giant axon of *Loligo* is described. This depended on the use of an internal electrode consisting of two silver wires, a guard system for measuring membrane current and a 'feed-back' amplifier for clamping the membrane potential at any desired level.

2. Axons impaled with the double electrode gave 'all-or-nothing' action potentials of about 100 mV. when stimulated with a brief shock. The action potential had a well-defined threshold at a critical depolarization of about 15 mV. Depolarizations less than 10–15 mV. gave graded responses similar to those seen in other excitable tissues.

3. The feed-back amplifier was arranged to make the membrane potential undergo a sudden displacement to a new level at which it was held constant for

10–50 msec. Under these conditions the membrane current consisted of a brief surge of capacity current, associated with the sudden change in potential, and an ionic current during the period of maintained potential. The brief surge was proportional to the displacement of membrane potential and corresponded to the charging of a membrane with an average capacity of $0{\cdot}9\,\mu$F./cm.^{2}. The sign and time course of the ionic current varied markedly with the membrane potential. Anodal displacements gave small currents which were always inward in direction. Depolarizations of less than 15 mV. gave outward currents which were small initially but increased markedly with time. Depolarizations of 15–110 mV. gave an initial phase of inward current which changed fairly rapidly into a large and prolonged outward current. The phase of inward current disappeared at about 110 mV. and was replaced by one of outward current. There was a continuous relation between ionic current and membrane potential. At short times this relation was similar to that derived from the rising phase of the action potential.

4. The maximum inward and outward ionic currents were little altered by temperature, but the rate at which the ionic current changed with time was increased about threefold for a rise of 10° C.

5. There was evidence of a small resistance in series with the capacitative element of the membrane. Errors introduced by this resistance were reduced by the use of compensated feed-back.

We wish to thank the Rockefeller Foundation for financial aid and the Director and staff of the Marine Biological Association for assistance at all stages of the experimental work.

REFERENCES

Bear, R. S., Schmitt, F. O. & Young, J. Z. (1937). The sheath components of the giant nerve fibres of the squid. *Proc. Roy. Soc.* B, **123**, 496–529.

Cole, K. S. (1949). Dynamic electrical characteristics of the squid axon membrane. *Arch. Sci. physiol.* **3**, 253–258.

Cole, K. S. & Curtis, H. J. (1939). Electric impedance of the squid giant axon during activity. *J. gen. Physiol.* **22**, 649–670.

Cole, K. S. & Hodgkin, A. L. (1939). Membrane and protoplasm resistance in the squid giant axon. *J. gen. Physiol.* **22**, 671–687.

Curtis, H. J. & Cole, K. S. (1938). Transverse electric impedance of the squid giant axon. *J. gen. Physiol.* **21**, 757–765.

Hodgkin, A. L. & Huxley, A. F. (1945). Resting and action potentials in single nerve fibres. *J. Physiol.* **104**, 176–195.

Hodgkin, A. L. & Huxley, A. F. (1952*a*). Currents carried by sodium and potassium ions through the membrane of the giant axon of *Loligo*. *J. Physiol.* **116**, 449–472.

Hodgkin, A. L. & Huxley, A. F. (1952*b*). The components of membrane conductance in the giant axon of *Loligo*. *J. Physiol.* **116**, 473–496.

Hodgkin, A. L. & Huxley, A. F. (1952*c*). The dual effect of membrane potential on sodium conductance in the giant axon of *Loligo*. *J. Physiol.* **116**, 497–506.

Hodgkin, A. L. & Huxley, A. F. (1952*d*). A quantitative description of membrane current and its application to conduction and excitation in nerve. *J. Physiol.* (in the press).

Hodgkin, A. L., Huxley, A. F. & Katz, B. (1949). Ionic currents underlying activity in the giant axon of the squid. *Arch. Sci. physiol.* **3**, 129–150.

Hodgkin, A. L. & Katz, B. (1949*a*). The effect of sodium ions on the electrical activity of the giant axon of the squid. *J. Physiol.* **108**, 37–77.

HODGKIN, A. L. & KATZ, B. (1949b). The effect of temperature on the electrical activity of the giant axon of the squid. *J. Physiol.* **109**, 240–249.

HUXLEY, A. F. & STÄMPFLI, R. (1951). Effect of potassium and sodium on resting and action potentials of single myelinated nerve fibres. *J. Physiol.* **112**, 496–508.

KEYNES, R. D. & LEWIS, P. R. (1951). The sodium and potassium content of cephalod nerve fibres. *J. Physiol.* **114**, 151–182.

MARMONT, G. (1949). Studies on the axon membrane. *J. cell. comp. Physiol.* **34**, 351–382.

ROTHENBERG, M. A. (1950). Studies on permeability in relation to nerve function. II. Ionic movements across axonal membranes. *Biochim. biophys. acta*, **4**, 96–114.

7

J. Physiol. (1952) 116, 449–472

CURRENTS CARRIED BY SODIUM AND POTASSIUM IONS THROUGH THE MEMBRANE OF THE GIANT AXON OF *LOLIGO*

BY A. L. HODGKIN AND A. F. HUXLEY

From the Laboratory of the Marine Biological Association, Plymouth, and the Physiological Laboratory, University of Cambridge

(*Received* 24 *October* 1951)

In the preceding paper (Hodgkin, Huxley & Katz, 1952) we gave a general description of the time course of the current which flows through the membrane of the squid giant axon when the potential difference across the membrane is suddenly changed from its resting value, and held at the new level by a feed-back circuit ('voltage clamp' procedure). This article is chiefly concerned with the identity of the ions which carry the various phases of the membrane current.

One of the most striking features of the records of membrane current obtained under these conditions was that when the membrane potential was lowered from its resting value by an amount between about 10 and 100 mV. the initial current (after completion of the quick pulse through the membrane capacity) was in the inward direction, that is to say, the reverse of the direction of the current which the same voltage change would have caused to flow in an ohmic resistance. The inward current was of the right order of magnitude, and occurred over the right range of membrane potentials, to be the current responsible for charging the membrane capacity during the rising phase of an action potential. This suggested that the phase of inward current in the voltage clamp records might be carried by sodium ions, since there is much evidence (reviewed by Hodgkin, 1951) that the rising phase of the action potential is caused by the entry of these ions, moving under the influence of concentration and potential differences. To investigate this possibility, we carried out voltage clamp runs with the axon surrounded by solutions with reduced sodium concentration. Choline was used as an inert cation since replacement of sodium with this ion makes the squid axon completely inexcitable, but does not reduce the resting potential (Hodgkin & Katz, 1949; Hodgkin, Huxley & Katz, 1949).

METHOD

The apparatus and experimental procedure are fully described in the preceding paper (Hodgkin *et al.* 1952). 'Uncompensated feed-back' was employed.

Sea water was used as a normal solution. Sodium-deficient solutions were made by mixing sea water in varying proportions with isotonic 'choline sea water' of the following composition:

Ion	g. ions/kg. H_2O	Ion	g. ions/kg. H_2O
Choline$^+$	484	Mg^{++}	54
K^+	10	Cl^-	621
Ca^{++}	11	HCO_3^-	3

The mixtures are referred to by their sodium content, expressed as a percentage of that in sea water (30 % Na sea water, etc.).

RESULTS

Voltage clamps in sodium-free solution

Fig. 1 shows the main differences between voltage clamp records taken with the axon surrounded by sea water and by a sodium-free solution. Each record gives the current which crossed the membrane when it was depolarized by 65 mV. After the top record was made, the sea water surrounding the axon was replaced by choline sea water, and the middle record was taken. The fluid was again changed to sea water, and the bottom record taken. The amplifier gain was the same in all three records, but a given deflexion represents a smaller current in the choline solution, since the current was detected by the potential drop along a channel filled with the fluid which surrounded the axon, and the specific resistance of the choline sea water was about 23 % higher than that of ordinary sea water.

The most important features shown in Fig. 1 are the following: (1) When the external sodium concentration was reduced to zero, the inward current disappeared and was replaced by an early hump in the outward current. (2) The late outward current was only slightly altered, the steady level being 15–20 % less in the sodium-free solution. (3) The changes were reversed when sea water was replaced. The currents are slightly smaller in the bottom record than in the top one, but the change is not attributable to an action of the choline since a similar drop occurred when an axon was kept in sea water for an equal length of time.

A series of similar records with different strengths of depolarization is shown in Fig. 2. The features described in connexion with Fig. 1 are seen at all strengths between -28 and -84 mV. At the weakest depolarization (-14 mV.) the early phase of outward current in the sodium-free record is too small to be detected. At the highest strengths the early current is outward even in sea water, and is then increased in the sodium-free solution.

These results are in qualitative agreement with the hypothesis that the inward current is carried by sodium ions which, as an early result of the decrease in membrane potential, are permitted to cross the membrane in both directions under a driving force which is the resultant of the effects of the concentration difference and the electrical potential difference across the membrane. When the axon is in sea water, the concentration of sodium outside the membrane $[\mathrm{Na}]_o$ is 5–10 times greater than that inside, $[\mathrm{Na}]_i$. This tends to make the inward flux exceed the outward. The electrical potential

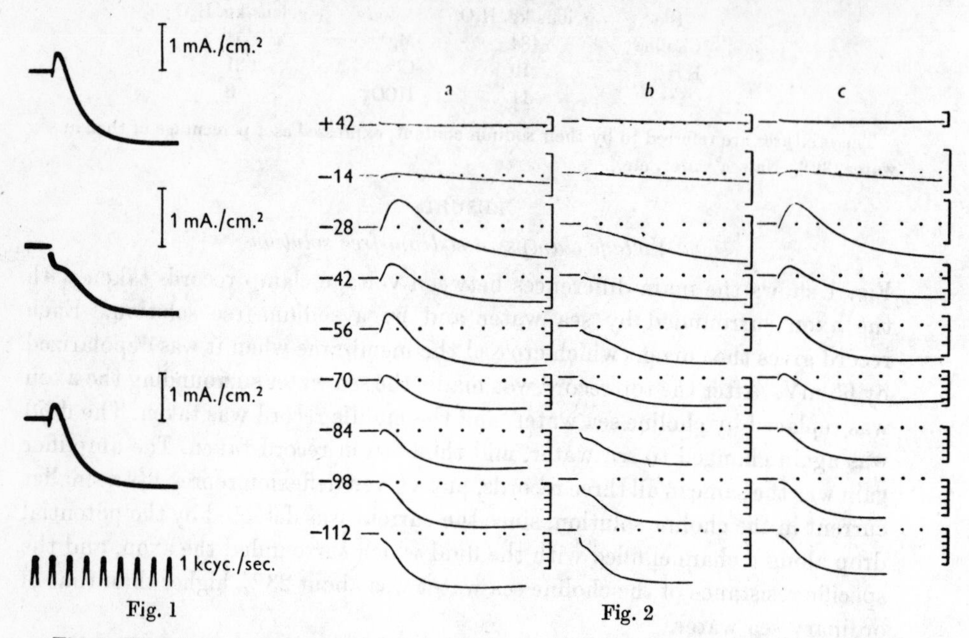

Fig. 1

Fig. 2

Fig. 1. Records of membrane current during 'voltage clamps' in which membrane potential was lowered by 65 mV. Top record: axon in sea water. Centre record: axon in choline sea water. Bottom record: after replacing sea water. Axon no. 15; temperature 11° C. Inward current is shown upwards in this and all other figures.

Fig. 2. Records of membrane current during 'voltage clamps'. *a*, axon in sea water; *b*, axon in choline sea water; *c*, after replacing sea water. Displacement of membrane potential indicated in mV. Axon no. 21; temperature 8·5° C. Vertical scale: 1 division is 0·5 mA./cm.². Horizontal scale: interval between dots is 1 msec.

difference E also helps the inward and hinders the outward flux so long as it is positive, i.e. in the same direction as the resting potential. The net current carried by the positive charge of the sodium ions is therefore inward unless the depolarization is strong enough to bring E to a sufficiently large negative value to overcome the effect of the concentration difference. The critical value of E at which the fluxes are equal, and the net sodium current is therefore zero, will be called the 'sodium potential', E_{Na}. Its value should be given by the Nernst equation

$$E_{\mathrm{Na}} = \frac{RT}{F} \log_e \frac{[\mathrm{Na}]_i}{[\mathrm{Na}]_o}. \qquad (1)$$

With values of E more negative than this, the net sodium flux is outward, causing the early phase of the outward current seen in the lowest record of the first and third columns of Fig. 2, where the axon was in sea water and was depolarized by 112 mV. A family of voltage clamp records which shows particularly well this transition from an initial rise to an initial fall as the strength of depolarization is increased is reproduced as Fig. 14 of the preceding paper.

When the axon is placed in a sodium-free medium, such as the 'choline sea water', there can be no inward flux of sodium, and the sodium current must always be outward. This will account for the early hump on the outward current which is seen at all but the lowest strength of depolarization in the centre column of Fig. 2.

Voltage clamps with reduced sodium concentration

The results of reducing the sodium concentration to 30 and 10% of the value in sea water are shown in Figs. 3 and 4 respectively. These figures do not show actual records of current through the membrane. The curves are graphs of ionic current against time, obtained by subtracting the current through the capacity from the recorded total current. The initial surge in an anodal record was assumed to consist only of capacity current, and the capacity current at other strengths was estimated by scaling this in proportion to the amplitude of the applied voltage change.

As would be expected, the results are intermediate between those shown in Fig. 2 for an axon in sea water and in choline sea water. Inward current is present, but only over a range of membrane potentials which decreases with the sodium concentration, and within that range, the strength of the current is reduced. A definite sodium potential still exists beyond which the early hump of ionic current is outward, but the strength of depolarization required to reach it decreases with the sodium concentration. Thus, in the first column of Fig. 3, with the axon in 30% sodium sea water, the sodium potential is almost exactly reached by a depolarization of 79 mV. In the second column, with sea water surrounding the axon, the sodium potential is just exceeded by a depolarization of 108 mV. In column 3, after re-introducing 30% sodium sea water, the sodium potential is slightly exceeded by a depolarization of 79 mV. Similarly, in Fig. 4, the sodium potentials are almost exactly reached by depolarizations of 105, 49 and 98 mV. in the three columns, the axon being in sea water, 10% sodium sea water and sea water respectively. The sequence of changes in the form of the curves as the sodium potential is passed is remarkably similar in all cases.

The external sodium concentration and the 'sodium potential'

Estimation of the 'sodium potential' in solutions with different sodium concentrations is of particular importance because it leads to a quantitative

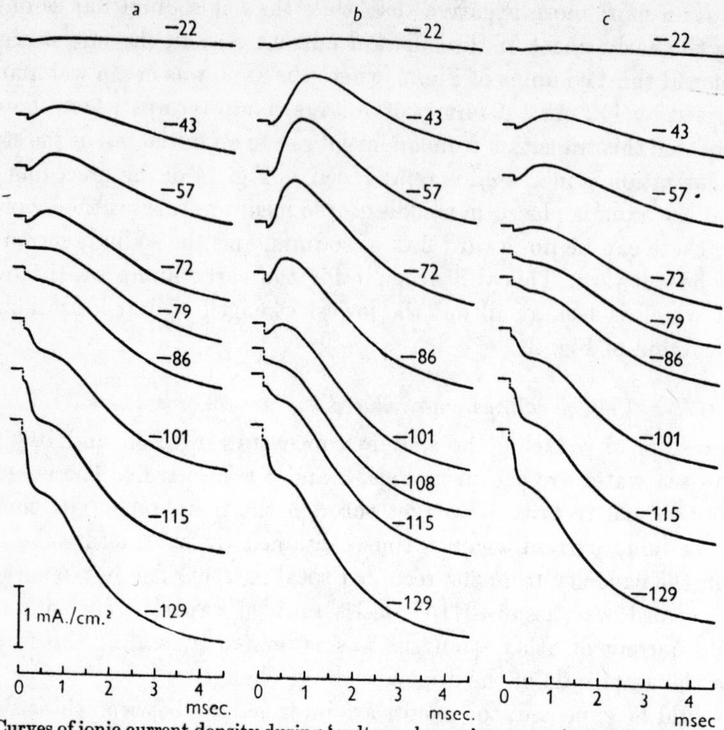

Fig. 3. Curves of ionic current density during 'voltage clamps'. *a*, axon in 30% sodium sea water; *b*, axon in sea water; *c*, after replacing 30% sodium sea water. Displacement of membrane potential indicated in millivolts. Axon no. 20; temperature 6·3° C.

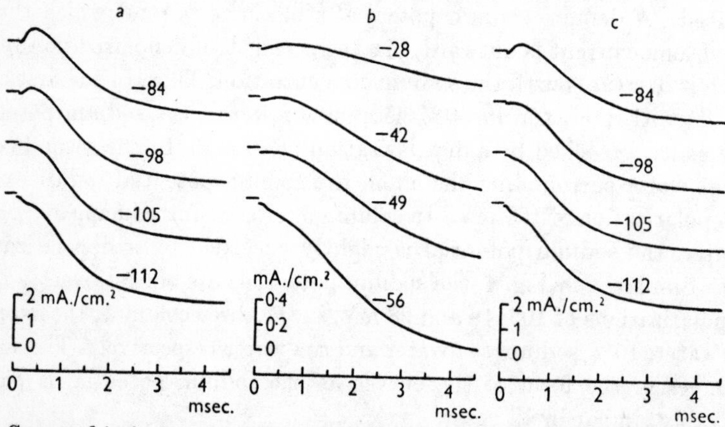

Fig. 4. Curves of ionic current density during voltage clamps in neighbourhood of sodium potential. *a*, axon in sea water; *b*, axon in 10% sodium sea water; *c*, after replacing sea water. Note that ordinate scale is larger in *b* than in *a* and *c*. Displacement of membrane potential in millivolts indicated for each curve. Axon no. 21; temperature 8·5° C.

test of our hypothesis. Equation (1) gives the sodium potential in sea water (E_{Na}), and the corresponding quantity (E'_{Na}) when the external sodium concentration is reduced to $[Na]'_o$ is given by

$$E'_{Na} = \frac{RT}{F} \log_e \frac{[Na]_i}{[Na]'_o}.$$

$$\text{Hence} \quad E'_{Na} - E_{Na} = \frac{RT}{F} \left\{ \log_e \frac{[Na]_i}{[Na]'_o} - \log_e \frac{[Na]_i}{[Na]_o} \right\} = \frac{RT}{F} \log_e \frac{[Na]_o}{[Na]'_o}. \tag{2}$$

The displacements of membrane potential, V, corresponding to these values are $V_{Na} = E_{Na} - E_r$ and $V'_{Na} = E'_{Na} - E'_r$, where E_r and E'_r are the values of the resting potential in sea water and in the test solution respectively. Hence

$$(V'_{Na} - V_{Na}) + (E'_r - E_r) = \frac{RT}{F} \log_e \frac{[Na]_o}{[Na]'_o}. \tag{3}$$

Each term in this equation can be determined experimentally, and data were obtained in four experiments on two axons. The results are given in Table 1, where the observed shift in sodium potential is compared with that predicted from the change in sodium concentration by Equation (3). It will be

TABLE 1. Comparison of observed and theoretical change in sodium potential when the fluid surrounding an axon is changed from sea water to a low sodium solution. Observed change: $E'_{Na} - E_{Na} = (V'_{Na} - V_{Na}) + (E'_r - E_r)$; theoretical change $= \frac{RT}{F} \log_e \frac{[Na]_o}{[Na]'_o}$.

Axon no.	Temp. (° C.)	$[Na]'_o/[Na]_o$	V_{Na} (mV.)	V'_{Na} (mV.)	$(E'_r - E_r)$ (mV.)	Sodium potential shift: Observed (mV.)	Sodium potential shift: Theoretical (mV.)
20	6·3	0·3	−105	−78	+3	+30	+28·9
20	6·3	0·1	−96	−45	+4	+55	+55·3
21	8·5	0·1	−100	−48	+4	+56	+55·6
21	8·5	0·1	−95	−45	+4	+54	+55·6

seen that there is good agreement, providing strong evidence that the early rise or fall in the recorded ionic current is carried by sodium ions, moving under the influence of their concentration difference and of the electric potential difference across the membrane.

Details of the estimation of the quantities which enter into Equation (3) are given in the following paragraphs.

Determination of V_{Na}. At the sodium potential there is neither inward sodium current, shown by an initial rise in the ionic current, nor outward sodium current, shown by an early hump in the outward current. It was found that these two criteria did in fact define the sodium potential very sharply, i.e. a hump appeared as soon as the ionic current showed an initial fall. It was therefore permissible to take as V_{Na} the strength of depolarization which gave an ionic current curve which started horizontally. This criterion was much more convenient to apply than the absence of a hump, since records were taken at fairly wide intervals of V (usually 7 mV.) and an interpolation procedure was necessary in order to estimate V_{Na} to the nearest 0·5 mV.

Change in resting potential. Experiments with ordinary capillary internal electrodes showed that the resting potential increased on the average by 4 mV. when the sea water surrounding the axon was replaced by choline sea water (a correction of 1·5 mV. for junction potentials in the external solutions is included in this figure). With intermediate sodium concentrations, the change in resting potential was assumed to be proportional to the change in sodium concentration. For instance, the resting potential in 30 % sodium sea water was taken as 2·8 mV. higher than that in sea water.

Slow change in condition of axon. When an axon is kept in sea water, its sodium content rises (Steinbach & Spiegelman, 1943; Keynes & Lewis, 1951) and its resting potential falls. Both of these effects bring E_r and E_{Na} closer together, diminishing the absolute magnitude of V_{Na}. In comparing V_{Na} in two solutions, it was therefore necessary to determine V_{Na} first in one solution, then in the other and finally in the first solution again. The second value of V_{Na} was then compared with the mean of the first and third.

The internal sodium concentration and the sodium potential

In freshly mounted fibres the average difference between the sodium potential and the resting potential was found to be -109 mV. (ten axons with a range of -95 to -119 mV. at an average temperature of 8° C.). The average resting potential in these fibres was 56 mV. when measured with a microelectrode containing sea water. By the time the sodium potential was measured the resting potential had probably declined by a few millivolts and may be taken as 50 mV. Allowing 10–15 mV. for the junction potential between sea water and axoplasm (Curtis & Cole, 1942; Hodgkin & Katz, 1949) this gives an absolute resting potential of 60–65 mV. The absolute value of the sodium potential would then be -45 to -50 mV. The sodium concentration in sea water is about 460 m.mol./kg. H_2O (Webb, 1939, 1940) so that the internal concentration of sodium would have to be 60–70 m.mol./kg. H_2O in order to satisfy Equation 1. This seems to be a very reasonable estimate since the sodium concentration in freshly dissected axons is about 50 m.mol./kg. H_2O while that in axons kept for 2 or 3 hr. is about 100 m.mol./kg. H_2O (Steinbach & Spiegelman, 1943; Keynes & Lewis, 1951; Manery, 1939, for fraction of water in axoplasm).

Outward currents at long times

So far, this paper has been concerned with the earliest phases of the membrane current that flows during a voltage clamp. The only current which has the opposite sign from the applied voltage pulse is the inward current which occurs over a certain range of depolarizations when the surrounding medium contains sodium ions. This inward current is always transient, passing over into outward current after a time which depends on the strength of depolarization and on the temperature. The current at long times resembles that in an ohmic resistance in having the same sign as the applied voltage change, but differs in that the outward current due to depolarization rises with a delay to a density which may be 50–100 times greater than that associated with a similar increase in membrane potential. Figs. 1–3 show that this late current is not greatly affected by the concentration of sodium in the fluid surrounding the axon.

An outward current which arises with a delay after a fall in the membrane potential is clearly what is required in order to explain the falling phase of the action potential. The outward currents reached in a voltage clamp may considerably exceed the maximum which occurs in an action potential; this may well be because the duration of an action potential is not sufficient to allow the outward current to reach its maximum value. These facts suggest that the outward current associated with prolonged depolarization is the same current which causes the falling phase of the action potential. The evidence (reviewed by Hodgkin, 1951) that the latter is caused by potassium ions leaving the axon is therefore a suggestion that the former is also carried by potassium ions. Direct evidence that such long-continued and outwardly directed membrane currents are carried by potassium ions has now been obtained in *Sepia* axons by a tracer technique (unpublished experiments). We shall therefore assume that this delayed outward current is carried by potassium ions, and we shall refer to it as 'potassium current', I_K. Since it is outward, it is not appreciably affected by the external potassium concentration, and evidence for or against potassium being the carrier cannot easily be obtained by means of experiments analogous to those which have just been described with altered external sodium concentration.

I_K *in sea water and choline.* As has been mentioned, the later part of the current record during a constant depolarization is much the same whether the axon is surrounded by sea water or by one of the solutions with reduced sodium concentration. There are, however, certain differences. For a given strength of depolarization, the maximum outward current is smaller by some 10 or 20% in the low-sodium solution, and at the higher strengths where the outward current is not fully maintained, the maximum occurs earlier in the low-sodium solution. Part of the difference in amplitude is explained by the difference of resting potential. Since the resting potential is greater in the low-sodium medium, a higher strength of depolarization is needed to reach a given membrane potential during the voltage clamp. This difference can be allowed for by interpolation between the actual strengths employed in one of the solutions. In most cases, this procedure did not entirely remove the difference between the amplitudes. There are, however, two other effects which are likely to contribute. In the first place, the effect of not using 'compensated feed-back' is probably greater in the low-sodium solution (see preceding paper, p. 445). This further reduces the amplitude of the voltage change which actually occurs across the membrane. In the second place, the fact that the current reached its maximum earlier suggests that 'polarization' (preceding paper, p. 445) had a greater effect in the low-sodium solution. We do not know enough about either of these effects to estimate the amount by which they may have reduced the potassium current. It does seem at least possible that they account for the whole of the discrepancy, and we therefore assume provisionally that substituting choline sea water for sea water has no direct effect on the potassium current.

Separation of ionic current into I_{Na} and I_K

The results so far described suggest that the ionic current during a depolarization consists of two more or less independent components in parallel, an early transient phase of current carried by sodium ions, and a delayed long-lasting phase of current carried by potassium ions. In each case, the direction of the current is determined by the gradient of the electro-chemical potential of the ion concerned. It will clearly be of great interest if it is possible to estimate separately the time courses of these two components. There is enough information for doing this in data such as are presented in Fig. 2, if we make certain assumptions about the effect of changing the solution around the axon. If we compare the currents when the axon is in the low-sodium solution with those in sea water, the membrane potential during the voltage clamp being the same in both cases, then our assumptions are:

(1) The time course of the potassium current is the same in both cases.

(2) The time course of the sodium current is similar in the two cases, the amplitude and sometimes the direction being changed, but not the time scale or the form of the time course.

(3) $\frac{dI_K}{dt}=0$ initially for a period about one-third of that taken by I_{Na} to reach its maximum.

The first two of these assumptions are the simplest that can be made, and do not conflict with any of the results we have described, while the third is strongly suggested by the form of records near the sodium potential, as pointed out on p. 454. These points are sufficient reason for trying this set of assumptions first, but their justification can only come from the consistency of the results to which they lead. The differences between the effects of lack of compensation, and of the polarization phenomenon, in the two solutions, referred to at the end of the last section, will of course lead to certain errors in the analysis in the later stages of the ionic current.

The procedure by which we carried out this analysis was as follows:

(1) Three series of voltage clamp records at a range of strengths were taken, the first with the axon in one of the solutions chosen for the comparison, the second with the axon in the other solution, and the third with the first solution again. Such a set of records is reproduced in Fig. 2.

(2) Each record was projected on to a grid in which the lines corresponded to equal intervals of time and current, and the current was measured at a series of time intervals after the beginning of the voltage change.

(3) The time course of the initial pulse of current through the membrane capacity was determined from anodal records as described on p. 452 above, and subtracted from the measured total currents. Different corrections were needed in the two solutions, because the capacity current had a slower time course in the low sodium solutions, perhaps as a result of their lower conductivity. This procedure yielded a family of curves of ionic current against time such as is shown in Fig. 3.

(4) Each pair of curves in the first and third series at the same strength was averaged, in order to allow for the slow deterioration in the condition of the axon that took place during the experiment.

(5) The difference in resting potential was allowed for by interpolating between consecutive curves in either the second series or the series of averaged curves.

(6) We have now obtained curves of ionic current against time in the two solutions, with strengths of depolarization which reach the same membrane potential during the voltage clamp. The ionic current will be called I_i in sea water and I'_i in the low-sodium solution. The components carried by sodium and potassium in the two cases will be called I_{Na}, I'_{Na}, I_K and I'_K respectively. The next step was to plot I'_i against I_i, and to measure the initial slope k of the resulting graph (corresponding to the beginning of the voltage clamp).

Since we assume that initially $dI_K/dt=0$, and that I_{Na} and I'_{Na} have similar time courses, $k=I'_{Na}/I_{Na}$. Further, since we assume that $I_K=I'_K$,

$$I_i-I'_i=I_{Na}-I'_{Na}=I_{Na}(1-k).$$

Hence

$$I_{Na}=(I_i-I'_i)/(1-k), \tag{4}$$

$$I'_{Na}=k(I_i-I'_i)/(1-k), \tag{5}$$

and

$$I_K=I'_K=I_i-I_{Na}=(I'_i-kI_i)/(1-k). \tag{6}$$

These equations give the values of the component currents at any time in terms of the known quantities I_i and I'_i at that time. Curves of I_{Na} and I_K against time could therefore be constructed by means of these equations.

This procedure is illustrated in Fig. 5, which shows two pairs of ionic current curves together with the deduced curves of I_{Na}, I'_{Na} and I_K against time. The complete family of I_K curves from this experiment is shown in Fig. 6*b*, while Fig. 6*a* shows the family derived by the same procedure from another experiment. A satisfactory feature of these curves, which is to some extent a check on the validity of the assumptions, is that the general shape is the same at all strengths. If the time courses of I_{Na} and I'_{Na} had not been of similar form, Equation (6) would not have removed sodium current correctly. It would then have been unlikely that the curve of potassium current at a potential away from the sodium potential would have been similar to that at the sodium potential, where the sodium current is zero and Equation (6) reduces to $I_K=I_i$ because $k=\infty$.

On the other hand, it is clearly inconsistent that the I_{Na} and I'_{Na} curves in the lower part of Fig. 5 reverse their direction at 2 msec. after the beginning of the pulse. This is a direct consequence of the fact, discussed on p. 456 above, that the late outward current is somewhat greater in sea water than in the low-sodium solutions, even when allowance is made for the resting potential shift.

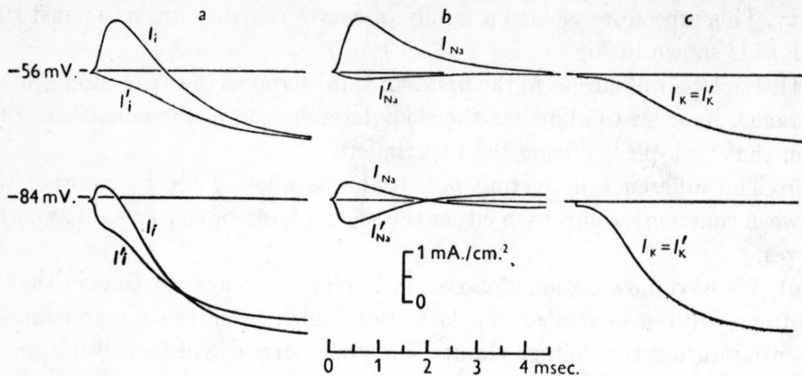

Fig. 5. Curves illustrating separation of ionic current into I_{Na} and I_K. Upper part of figure. *a*, ionic currents: I_i, axon in sea water, membrane potential lowered by 56 mV.; I'_i, axon in 10% sodium sea water, membrane potential lowered by 60 mV. (average of curves taken before and after I_i). *b*, sodium currents: I_{Na}, sodium current in sea water; I'_{Na}, sodium current in 10% sodium sea water. *c*, potassium current, same in both solutions. Lower part of figure. Same, but membrane potential lowered by 84 mV. in sea water and 88 mV. in 10% sodium sea water. Current and time scales same for all curves. Axon no. 21; temperature 8·5° C.

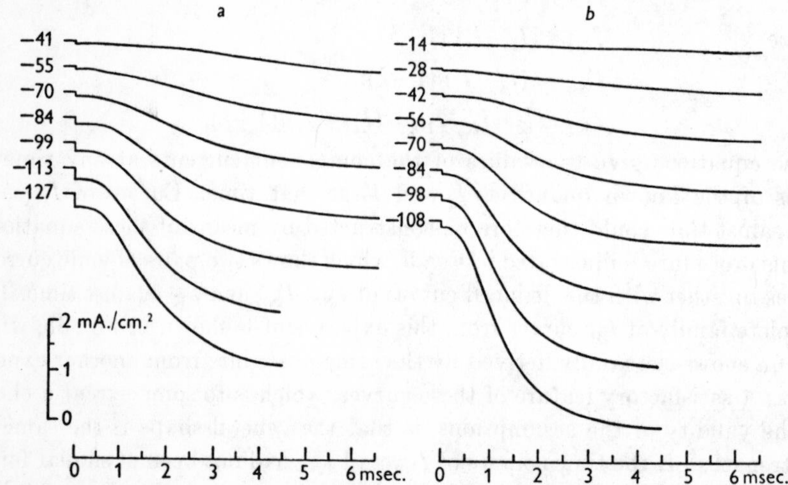

Fig. 6. Curves of potassium current against time for various strengths of depolarization. Displacement of membrane potential when axon is in sea water is indicated for each curve, in millivolts. *a*, derived from voltage clamps with axon in 30% sodium sea water, sea water and 30% sodium sea water. Axon no. 20; temperature 6·3° C. *b*, derived from voltage clamps with axon in 10% sodium sea water, sea water and 10% sodium sea water. Axon no. 21; temperature 8·5° C.

It was pointed out there that the difference may well be due to lack of compensation and to 'polarization'. Until these effects can be eliminated, estimates of sodium current at the longer times will be quite unreliable, and the corresponding estimates of potassium current will be somewhat reduced by these errors.

All the sodium current curves agree in showing that I_{Na} rises to a peak and then falls. With weak depolarizations (less than 40 mV.) the steady state value is definitely in the same direction as the peak, but at higher strengths the measured I_{Na} tends to a value which may have either direction. Since the sources of error mentioned in the last paragraph can cause an apparent reversal of I_{Na} during the pulse, it is possible that if these errors were larger than we suppose the whole of the apparent drop of I_{Na} from its peak value might also be spurious. At the time that an account of preliminary work with this technique was published (Hodgkin *et al.* 1949) we were unable to decide this point, and assumed provisionally that I_{Na} did not fall after reaching its maximum value. We are now convinced that this fall is genuine: (1) because of improvements in technique; (2) because of further experiments of other kinds which are described in the next two papers of this series (Hodgkin & Huxley, 1952*a*, *b*); and (3) because we occasionally observed records of the kind shown in Fig. 7. This is a record of membrane current during a voltage clamp in which an axon in choline sea water was depolarized by 84 mV. It will be seen that the early hump of outward current (due to sodium ions) was so marked that the total current reached a maximum at about 0·2 msec. and then fell before finally rising to the plateau attributable to movement of potassium ions. Unless we make the quite unwarrantable assumption that I_K itself has this double-humped form, this curve can only be explained by supposing that I_{Na} (outward in this case) falls after passing through a maximum value. The fact that such a clear maximum was not regularly observed was no doubt due to I_{Na} usually being smaller in relation to I_K than in this case.

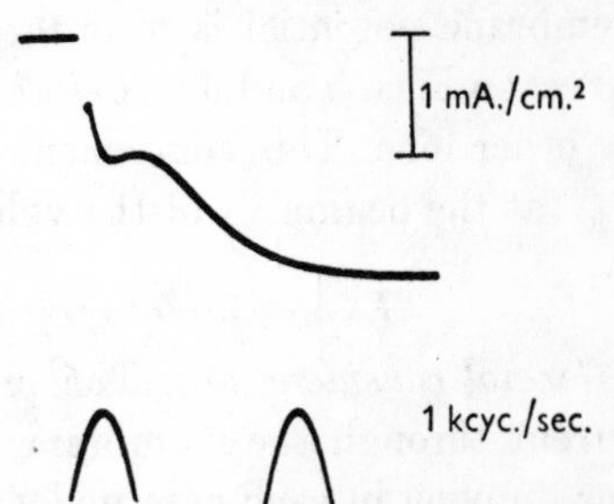

Fig. 7. Record of membrane current during a voltage clamp with axon in choline sea water, showing early maximum of outward current. Displacement of membrane potential during clamp = −84 mV. Axon no. 24; temperature 20° C.

We do not present a family of I_{Na} curves here, because the sequence of the curves is interrupted at the sodium potential. For this reason, the information is better given in the curves of 'sodium conductance' which are derived later in this paper (pp. 461–2 and Fig. 8). The variation of peak sodium current with strength of depolarization is shown in Fig. 13 for axons both in sea water and in low-sodium solutions.

Current carried by other ions. It seems to be possible to account for the variation of current with time during the voltage clamp by variations in the currents carried across the membrane by two ions, namely sodium and potassium. If, however, the membrane allowed constant fluxes of one or more other ion species, the current carried by these would form part of the 'I_K' which is deduced by our procedure, since this current would be independent both of time and of sodium concentration, and I_K is defined by its satisfying these criteria during the earliest part of the pulse. Reasons will be given in the next paper (Hodgkin & Huxley, 1952*a*) for supposing that the current carried by other ions is appreciable, though not of great importance except when the membrane potential is near to or above its resting value. Each of the I_K curves in Figs. 5 and 6 therefore includes a small constant component carried by other ions. This component probably accounts for much of the step in 'I_K' at the beginning of the voltage pulse.

Expression of ionic currents in terms of conductances

General considerations. The preceding sections have shown that the ionic current through the membrane is chiefly carried by sodium and potassium ions, moving in each case under a driving force which is the resultant of the concentration difference of the ion on the two sides of the membrane, and of the electrical potential difference across the membrane. This driving force alone determines the direction of the current carried by each ionic species, but the magnitude of the current depends also on the freedom with which the membrane allows the ions to pass. This last factor is a true measure of the 'permeability' of the membrane to the ion species in question. As pointed out by Teorell (1949*a*), a definition of permeability which takes no account of electrical forces is meaningless in connexion with the movements of ions, though it may well be appropriate for uncharged solutes.

The driving force for a particular ion species is clearly zero at the equilibrium potential for that ion. The driving force may therefore be measured as the difference between the membrane potential and the equilibrium potential. Using the same symbols as in Equations (1)–(3), the driving force for sodium ions will be $(E-E_{Na})$, which is also equal to $(V-V_{Na})$. The permeability of the membrane to sodium ions may therefore be measured by $I_{Na}/(E-E_{Na})$. This quotient, which we denote by g_{Na}, has the dimensions of a conductance (current divided by potential difference), and will therefore be referred to as the sodium conductance of the membrane. Similarly, the permeability of the membrane to potassium ions is measured by the potassium conductance g_K, which is defined as $I_K/(E-E_K)$. Conductances defined in this way may be called chord conductances and must be distinguished from slope conductances (G) defined as $\partial I/\partial E$.

These definitions are valid whatever the relation between I_{Na} and $(E-E_{Na})$, or between I_K and $(E - E_K)$, but the usefulness of the definitions, and the degree to which they measure real properties of the membrane, will clearly be much increased if each of these relations is a direct proportionality, so that g_{Na} and g_K are independent of the strength of the driving force under which they are measured. It will be shown in the next paper (Hodgkin & Huxley, 1952*a*) that this is the case, for both sodium and potassium currents, in an axon surrounded by sea water, when the measurement is made so rapidly that the condition of the membrane has no time to change.

Application to measured sodium and potassium currents. The determination of sodium current, potassium current and sodium potential have been described in earlier sections of the present paper. The method by which the potassium potential, E_K, was found is described in the next paper (Hodgkin & Huxley, 1952*a*), and the values used here are taken from that paper. We have

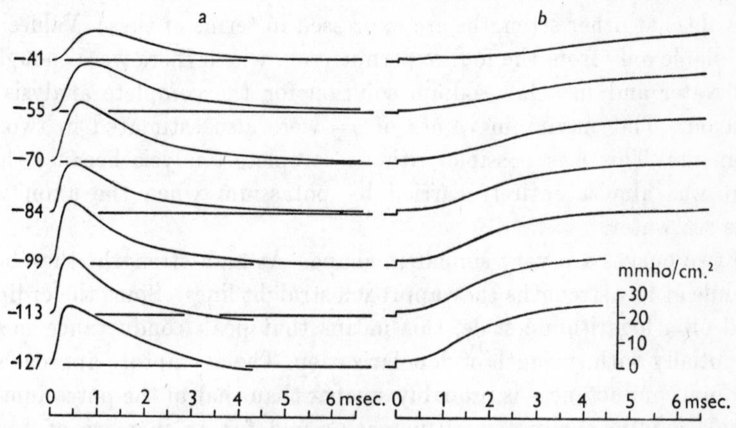

Fig. 8. Curves of sodium conductance (*a*) and potassium conductance (*b*). Displacement of membrane potential (millivolts) when axon was in sea water is indicated on each curve. Curves of I_i and I_K in same experiment are shown in Figs. 3 and 6*a* respectively. Axon no. 20; temperature 6·3° C.

therefore sufficient data to estimate g_{Na} and g_K as functions of time during a voltage clamp. Families of g_{Na} and g_K curves, for various strengths of depolarization, are shown in Fig. 8. The sodium conductances are calculated from the sodium currents in sea water, divided by the difference between membrane potential and sodium potential in sea water. If the same procedure had been applied to the corresponding quantities in the low-sodium solution, a similar family would have been obtained, but the relative amplitudes of the members of the family would have been slightly different. The values obtained from the sea water figures are the more interesting, both because they refer to a more normal condition, and because it is only in this case that the instantaneous relation between sodium current and voltage is linear

(Hodgkin & Huxley, 1952*a*). The corresponding distinction does not arise with g_K, since both I_K and E_K are the same in both solutions.

The shapes of individual curves in Fig. 8 are of course similar to those of curves of I_{Na} or I_K, such as are shown in Figs. 5 and 6, since the driving force for each ion is constant during any one voltage clamp. The change of amplitude of the curves with strength of depolarization is, however, less marked than with the current curves. For potassium, this can be seen by comparing Figs. 6*a* and 8*b*, which refer to the same experiment. For sodium, it is clear from Fig. 8*a* that the conductance curves undergo no marked change at the sodium potential, while the current curves reverse their direction at this point.

Membrane potential and magnitude of conductance. The effect of strength of depolarization on the magnitude of the conductances is shown in Figs. 9 and 10. For each experiment, the maximum values of g_{Na} and g_K reached in a voltage clamp of strength about 100 mV. are taken as unity, and the maximum values at other strengths are expressed in terms of these. Values of g_{Na} are available only from the four experiments in which there were enough data in sea water and in a low-sodium solution for the complete analysis to be carried out. The maximum values of g_K were also estimated in two other experiments. This was possible without complete analysis because the late current was almost entirely carried by potassium when the axon was in choline sea water.

The two curves are very similar in shape. At high strengths they become flat, while at low strengths they approach straight lines. Since the ordinate is plotted on a logarithmic scale, this means that peak conductance increases exponentially with strength of depolarization. The asymptote approached by the sodium conductances is probably steeper than that of the potassium data; the peak sodium conductances increase e-fold for an increase of 4 mV. in strength of depolarization; for potassium the corresponding figure is 5 mV.

The values of conductance at a depolarization of 100 mV., which are represented as unity in Figs. 9 and 10, are given in Table 2. In all these cases where enough measurements were made to construct a curve, the axon had been used for other observations before we took the records on which the analysis is based. In several cases, it was possible to estimate one or two values of sodium and potassium peak conductance at the beginning of the same experiment, and these were considerably higher than the corresponding values in Table 2, which must therefore be depressed by deterioration of the fibres.

More representative values of the peak g_K and g_{Na} at high strengths were estimated at the beginning of experiments on several fibres. Potassium current at long times can be estimated without difficulty, since I_{Na} is then negligible, especially at these depolarizations which are near the sodium potential. Nine fibres at 3–11° C. gave peak values of g_K at -100 mV. ranging from 22 to 41 m.mho/cm.², with a mean of 28; five fibres at 19–23° C. gave a range of

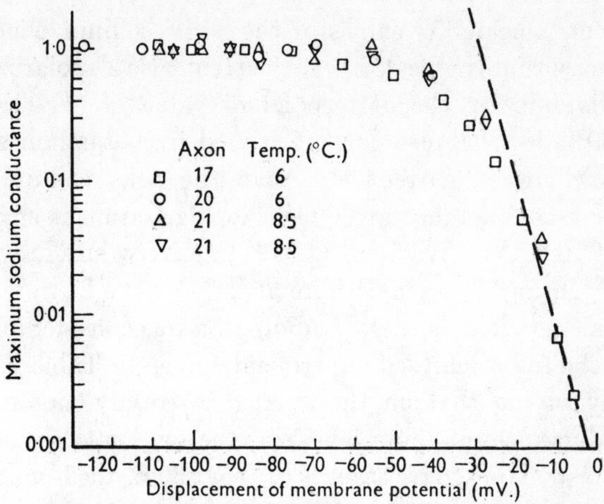

Fig. 9. Maximum sodium conductance reached during a voltage clamp. Ordinate: peak conductance relative to value reached with depolarization of 100 mV., logarithmic scale. Abscissa: displacement of membrane potential from resting value (depolarization negative).

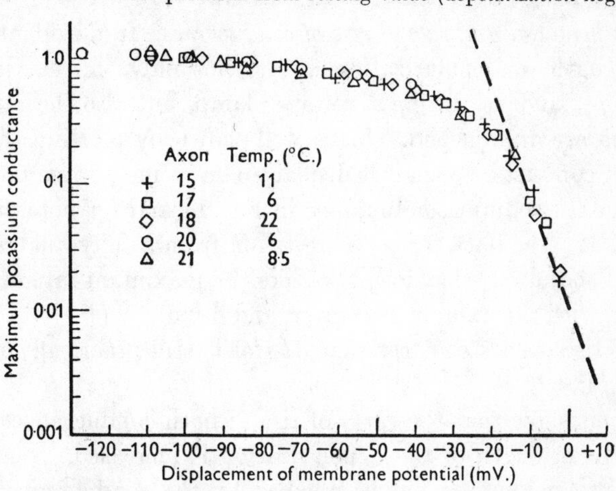

Fig. 10. Maximum potassium conductance reached during a voltage clamp. Ordinate: maximum conductance relative to value reached with depolarization of 100 mV., logarithmic scale. Abscissa: displacement of membrane potential from resting value (depolarization negative).

TABLE 2. Peak values of sodium and potassium conductance at a depolarization of 100 mV. Same experiments as Figs. 9 and 10. In each case, the value given in this table is represented as unity in Fig. 9 or Fig. 10.

Axon no.	Temp. (° C.)	Peak conductances at − 100 mV. Sodium (m.mho/cm.²)	Potassium (m.mho/cm.²)
15	11	—	21
17	6	18	20
18	21	—	28
20	6	22	23
21	8·5	23	31
21	8·5	17	—
	Mean	20	25

33–37 m.mho/cm.^{2}, mean 35. Values of the peak sodium conductance were obtained by measuring the peak inward current at a depolarization of about 60 mV. and dividing by the corresponding value of $(V - V_{Na})$. They are probably 10–20% low because current carried by potassium and other ions makes the peak inward current less than the peak sodium current, and because the peak conductance at 60 mV. depolarization is slightly less than that reached at 100 mV. Five fibres at 3–9° C. gave values ranging from 22 to 48 m.mho/cm.^{2}, mean 30; a single fibre at 22° C. gave 24 m.mho/cm.^{2}.

These results show that both g_K and g_{Na} can rise considerably higher than the values for the fully analysed experiments given in Table 2. They may be summarized by saying that on the average a freshly mounted fibre gives maximum conductances of about 30–35 m.mho/cm.^{2} both for sodium and for potassium, corresponding to resistances of about 30 Ω. for 1 cm.^{2} of membrane. This value may be compared with the resting resistance of about 1000 Ω. cm.^{2} (Cole & Hodgkin, 1939), and the resistance at the peak of an action potential, which is about 25 Ω. cm.^{2} (Cole & Curtis, 1939).

Membrane potential and rate of rise of conductance. It is evident from Fig. 8 that the strength of depolarization affects not only the maximum values attained by g_{Na} and g_K during a voltage clamp, but also the rates at which these maxima are approached. This is well shown by plotting the maximum rate of rise of conductance against displacement of membrane potential. This has been done for sodium conductance in Fig. 11 and for potassium conductance in Fig. 12. The data for g_K were taken from a fully analysed run, but in the case of sodium it is sufficient to take the maximum rate of rise of total ionic current, with the axon in sea water, and divide by $(V - V_{Na})$. The maximum rate of rise occurs so early that dI_K/dt is still practically zero, so that $dI_i/dt = dI_{Na}/dt$.

These graphs show that the rates of rise of both conductances continue to increase as the strength of the depolarization is increased, even beyond the point where the maximum values reached by the conductances themselves have become practically constant.

DISCUSSION

Only two aspects of the results described in this paper will be discussed at this stage. The first is the relationship between sodium current and external sodium concentration; the second is the application of the results to the interpretation of the action potential. Further discussion will be reserved for the final paper of this series (Hodgkin & Huxley, 1952*c*).

Sodium current and external sodium concentration

General considerations and theory. We have shown in the earlier parts of this paper that there is good reason for believing that the component of membrane current that we refer to as I_{Na} is carried by sodium ions which move down their own electrochemical gradient, the speed of their movement, and therefore the magnitude of the current, being also determined by changes in the freedom

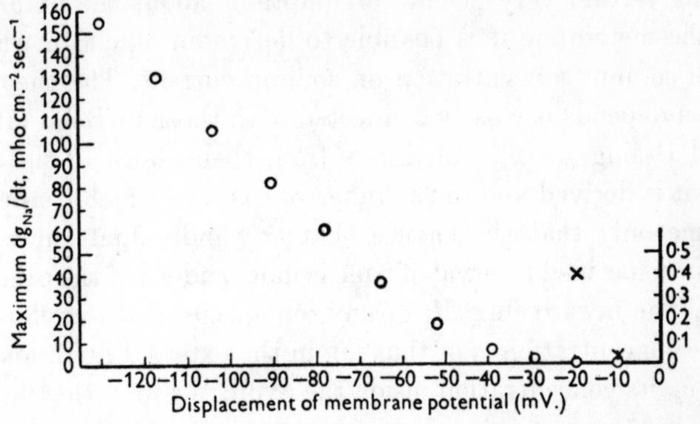

Fig. 11. Maximum rates of rise of sodium conductance during voltage clamps plotted against displacement of membrane potential. Circles are to be read with the scale on the left-hand side. The two lowest points are also re-plotted as crosses on 100 times the vertical scale, and are to be read with the scale on the right-hand side. The peak sodium conductance reached at high strengths of depolarization was 16 m.mho/cm.^{2}. Axon no. 41; temperature 3·5° C. Compensated feed-back.

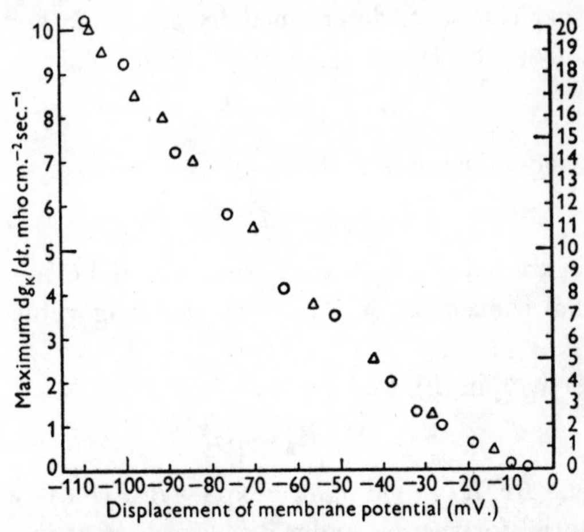

Fig. 12. Maximum rates of rise of potassium conductance during voltage clamps, plotted against displacement of membrane potential. Circles, left-hand scale: axon no. 17; temperature 6° C. Triangles, right-hand scale: axon no. 21; temperature 8·5° C. At −100 mV. the maximum potassium conductance was 20 m.mho/cm.^{2} for axon no. 17 and 31 m.mho/cm.^{2} for axon no. 21.

with which they are permitted to cross the membrane under this driving force. If this is in fact the case, we should expect that sodium ions would cross the membrane in both directions, the observed I_{Na} being the difference between the opposing currents carried by these two fluxes. At the sodium potential the fluxes would be equal, making I_{Na} zero; as the membrane potential is increased from this value, the ratio of inward to outward flux would increase, making I_{Na} positive, and vice versa.

By making certain very general assumptions about the manner in which ions cross the membrane it is possible to derive an equation which predicts the effect of sodium concentration on sodium current. The theory on which this equation depends is closely connected with those of Behn (1897), Teorell (1949*b*) and Ussing (1949), but differs from them, both in the assumptions from which it is derived and in the range of cases to which it applies.

We assume only that the chance that any individual ion will cross the membrane in a specified interval of time is independent of the other ions which are present. The inward flux M_1 of any ion species will therefore be proportional to the concentration c_1 of that ion in the external fluid, and will not be affected by c_2, its concentration inside the axon. We may therefore write

$$M_1 = k_1 c_1, \tag{7}$$

where k_1 is a constant which depends on the condition of the membrane and on the potential difference across it. Similarly, the outward flux M_2 is given by

$$M_2 = k_2 c_2, \tag{8}$$

where k_2 is another constant, determined by the same factors as k_1 but in general different from it. Hence

$$M_1/M_2 = k_1 c_1 / k_2 c_2. \tag{9}$$

The condition for equilibrium is that $M_1 = M_2$, so that

$$k_2/k_1 = c_1^*/c_2,$$

where c_1^* is the external concentration that would be in equilibrium with the (fixed) internal concentration, under the existing value of E, the membrane potential.

Substituting for k_1/k_2 in (9), we have

$$M_1/M_2 = c_1/c_1^*. \tag{10}$$

Now $c_1^*/c_2 = \exp(-EF/RT)$ and $c_1/c_2 = \exp(-E^*F/RT)$, where E^* is the equilibrium potential for the ion under discussion, so that

$$c_1/c_1^* = \exp(E - E^*)\, F/RT$$

and

$$M_1/M_2 = \exp(E - E^*)\, F/RT. \tag{11}$$

We now have in Equations (7), (8) and (11) three simple relations between M_1, M_2, c_1 and E. The effect of membrane potential on either of the fluxes alone is not specified by these equations, but is immaterial for our purpose.

If we wish to compare the sodium currents when the axon is immersed first in sea water, with sodium concentration $[Na]_o$, and then in a low-sodium solution with sodium concentration $[Na]'_o$, the membrane potential having the same value E in both cases, we have:

$$\frac{I'_{Na}}{I_{Na}} = \frac{M'_{Na_1} - M'_{Na_2}}{M_{Na_1} - M_{Na_2}}.$$

From (7), $M'_{Na_1}/M_{Na_1} = [Na]'_o/[Na]_o$ and from (8) $M'_{Na_2} = M_{Na_2}$. Using these relations and Equation (11)

$$\frac{I'_{Na}}{I_{Na}} = \frac{([Na]'_o/[Na]_o) \exp (E - E_{Na}) F/RT - 1}{\exp (E - E_{Na}) F/RT - 1}. \quad (12)$$

Strictly, activities should have been used instead of concentrations throughout. In the final Equation (12), however, concentrations appear only in the ratio of the sodium concentrations in sea water and the sodium-deficient solution. The total ionic strength was the same in these two solutions, so that the ratio of activities should be very close to the ratio of concentrations. The activity coefficient in axoplasm may well be different, but this does not affect Equation (12).

Equation (11) is equivalent to the relation deduced by Ussing (1949) and is a special case of the more general equation derived by Behn (1897) and Teorell (1949*b*). All these authors start from the assumption that each ion moves under the influence of an electric field, a concentration gradient and a frictional resistance proportional to the velocity of the ion in the membrane. This derivation is more general than ours in the respect that it is still applicable if, for instance, a change in c_1 alters the form of the electric field in the membrane and therefore alters M_2; in this case, Equations (8) and therefore (12) are not obeyed. On the other hand, it is more restricted than our derivation in that it specifies the nature of the resistance to movement of the ions.

Agreement with experimental results. Equation (12) is tested against experimental results in Fig. 13. Section (*a*) shows data from the experiment illustrated in Figs. 3 and 6*a*. The values of the sodium current in sea water (I_{Na}) and in 30% Na sea water (I'_{Na}) were derived by the procedure described in the 'Results' section. The crosses are the peak values of I_{Na}, plotted against V, the displacement of membrane potential during the voltage clamp. A smooth curve has been fitted to them by eye. V_{Na} was taken as the position at which the axis of V was cut by this curve. Since

$$V = E - E_r, \quad (E - E_{Na}) = (V - V_{Na}),$$

and, for each point on the smoothed curve of I_{Na} against V, a corresponding value of I'_{Na} was calculated by means of Equation (12). These values are plotted as curve B. The experimentally determined peak values of I'_{Na} are shown as circles. These are seen to form a curve of shape similar to B, but of greater amplitude. They are well fitted by curve C, which was obtained from B by multiplying all ordinates by the factor 1·20.

Fig. 13*b*, *c* were obtained in the same way from experiments in which the low-sodium solutions were 10% Na sea water and choline sea water respectively. In each case, the peak values of I'_{Na} are well fitted by the values predicted by means of Equation (12), after multiplying by constant factors of 1·333 and 1·60 in (*b*) and (*c*) respectively.

These constant factors appear at first sight to indicate a disagreement with the theory, but they are explained quantitatively by an effect which is described in the fourth paper of this series (Hodgkin & Huxley, 1952*b*). The

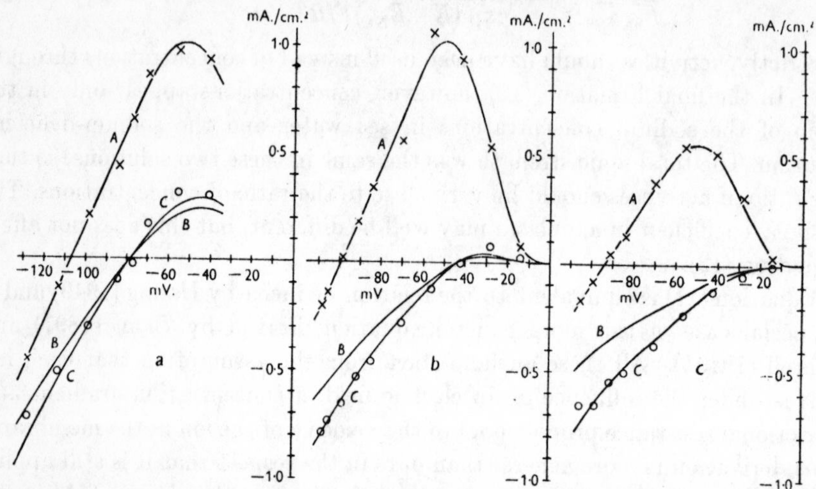

Fig. 13. Test of 'independence principle'. Three experiments. Crosses: peak sodium current density during voltage clamp; axon in sea water. Curve *A* fitted by eye. Circles: peak sodium current density during voltage clamp; axon in low sodium sea water. Curve *B*: peak sodium current density in low sodium sea water, predicted from curve *A* by Equation (12). Curve *C*: as curve *B*, but all ordinates multiplied by a factor *f*. Abscissa: membrane potential measured from resting potential in sea water. (*a*) Axon no. 20; temperature 6° C. Data from voltage clamps in (1) 30% sodium sea water, (2) sea water, (3) 30% sodium sea water. Circles are average values from runs (1) and (3). Value of V_{Na} inserted in Equation (12): − 106·8 mV. Factor $f = 1·20$. (*b*) Axon no. 21; temperature 8·5° C. Data from voltage clamps in (1) 10% sodium sea water, (2) sea water, (3) 10% sodium sea water. Circles are average values from runs (1) and (3). $V_{Na} = -98·8$ mV., $f = 1·333$. (*c*) Axon no. 21; temperature 8·5° C. Data from voltage clamps in (1) sea water, (2) choline sea water, (3) sea water, taken later than (*b*). Crosses are average values from runs (1) and (3). $V_{Na} = -93·8$ mV., $f = 1·60$.

resting potential was higher in the low-sodium solutions than in sea water, and it is shown in that paper that increasing the membrane potential by current flow allows a subsequent depolarization to produce greater sodium currents than it would otherwise have done. The factor by which the sodium currents are thus increased is greater the lower the sodium concentration, and the poorer the condition of the fibre. The first of these effects explains why the factor is greater in (*b*) than in (*a*), while the second explains why it is greater in (*c*) than in (*b*). The experiments in Fig. 13*b*, *c* were performed on the same fibre, and the deterioration between the experiments is shown by the fact that the I_{Na} values are only about half as great in (*c*) as in (*b*).

We can therefore say that, within experimental error, the sodium currents in sea water and in low-sodium solutions are connected by Equation (12), suggesting that the 'independence principle' from which this equation was derived is applicable to the manner in which the ions cross the membrane. This does not tell us much about the physical mechanism involved, since the 'independence' relations would be obeyed by several quite different systems. Examples are the 'constant field' system discussed by Goldman (1943), where the electric field through the membrane is assumed to be uniform and unaffected by the concentrations of ions present; and any system involving combination with carrier molecules in the membrane, so long as only a small proportion of the carrier is combined with the ion at any moment.

Origin of the action potential

The main conclusions that were drawn from the analysis presented in the 'Results' section of this paper may be summarized as follows. When the membrane potential is suddenly reduced (depolarization), the initial pulse of current through the capacity of the membrane is followed by large currents carried by ions (chiefly sodium and potassium), moving down their own electrochemical gradients. The current carried by sodium ions rises rapidly to a peak and then decays to a low value; that carried by potassium ions rises much more slowly along an S-shaped curve, reaching a plateau which is maintained with little change until the membrane potential is restored to its resting value.

These two components of the membrane current are enough to account qualitatively for the propagation of an action potential, the sequence of events at each point on the nerve fibre being as follows: (1) Current from a neighbouring active region depolarizes the membrane by spread along the cable structure of the fibre ('local circuits'). (2) As a result of this depolarization, sodium current is allowed to flow. Since the external sodium concentration is several times greater than the internal, this current is directed inwards and depolarizes the membrane still further, until the membrane potential reverses its sign and approaches the value at which sodium ions are in equilibrium. (3) As a delayed result of the depolarization, the potassium current increases and the ability of the membrane to pass sodium current decreases. Since the internal potassium concentration is greater than the external, the potassium current is directed outwards. When it exceeds the sodium current, it repolarizes the membrane, raising the membrane potential to the neighbourhood of the resting potential, at which potassium ions inside and outside the fibre are near to equilibrium.

The further changes which restore the membrane to a condition in which it can propagate another impulse have also been studied by the 'voltage clamp' technique and are described in subsequent papers (Hodgkin & Huxley, 1952*a*, *b*). In the final paper of the series (Hodgkin & Huxley, 1952*c*), we show that an action potential can be predicted quantitatively from the voltage clamp results, by carrying through numerically the procedure which has just been outlined.

SUMMARY

1. The effect of sodium ions on the current through the membrane of the giant axon of *Loligo* was investigated by the 'voltage-clamp' method.

2. The initial phase of inward current, normally associated with depolarizations of 10–100 mV., was reversed in sign by replacing the sodium in the external medium with choline.

3. Provided that sodium ions were present in the external medium it was possible to find a critical potential above which the initial phase of ionic current was inward and below which it was outward. This potential was normally reached by a depolarization of 110 mV., and varied with external sodium concentration in the same way as the potential of a sodium electrode.

4. These results support the view that depolarization leads to a rapid increase in permeability which allows sodium ions to move in either direction through the membrane. These movements carry the initial phase of ionic current, which may be inward or outward, according to the difference between the sodium concentration and the electrical potential of the inside and outside of the fibre.

5. The delayed outward current associated with prolonged depolarization was little affected by replacing sodium ions with choline ions. Reasons are given for supposing that this component of the current is largely carried by potassium ions.

6. By making certain simple assumptions it is possible to resolve the total ionic current into sodium and potassium currents. The time course of the sodium or potassium permeability when the axon is held in the depolarized condition is found by using conductance as a measure of permeability.

7. It is shown that the sodium conductance rises rapidly to a maximum and then declines along an approximately exponential curve. The potassium conductance rises more slowly along an S-shaped curve and is maintained at a high level for long periods of time. The maximum sodium and potassium conductances were normally of the order of 30 m.mho/cm.^{2} at a depolarization of 100 mV.

8. The relation between sodium concentration and sodium current agrees with a theoretical equation based on the assumption that ions cross the membrane independently of one another.

REFERENCES

Behn, U. (1897). Ueber wechselseitige Diffusion von Elektrolyten in verdünnten wässerigen Lösungen, insbesondere über Diffusion gegen das Concentrationsgefälle. *Ann. Phys., Lpz.*, N.F. **62**, 54–67.

Cole, K. S. & Curtis, H. J. (1939). Electric impedance of the squid giant axon during activity. *J. gen. Physiol.* **22**, 649–670.

Cole, K. S. & Hodgkin, A. L. (1939). Membrane and protoplasm resistance in the squid giant axon. *J. gen. Physiol.* **22**, 671–687.

Curtis, H. J. & Cole, K. S. (1942). Membrane resting and action potentials from the squid giant axon. *J. cell. comp. Physiol.* **19**, 135–144.

Goldman, D. E. (1943). Potential, impedance, and rectification in membranes. *J. gen. Physiol.* **27**, 37–60.

Hodgkin, A. L. (1951). The ionic basis of electrical activity in nerve and muscle. *Biol. Rev.* **26**, 339–409.

Hodgkin, A. L. & Huxley, A. F. (1952*a*). The components of membrane conductance in the giant axon of *Loligo*. *J. Physiol.* **116**, 473–496.

Hodgkin, A. L. & Huxley, A. F. (1952*b*). The dual effect of membrane potential on sodium conductance in the giant axon of *Loligo*. *J. Physiol.* **116**, 497–506.

Hodgkin, A. L. & Huxley, A. F. (1952*c*). A quantitative description of membrane current and its application to conduction and excitation in nerve. *J. Physiol.* (in the press).

Hodgkin, A. L., Huxley, A. F. & Katz, B. (1949). Ionic currents underlying activity in the giant axon of the squid. *Arch. Sci. physiol.* **3**, 129–150.

Hodgkin, A. L., Huxley, A. F. & Katz, B. (1952). Measurement of current-voltage relations in the membrane of the giant axon of *Loligo*. *J. Physiol.* **116**, 424–448.

Hodgkin, A. L. & Katz, B. (1949). The effect of sodium ions on the electrical activity of the giant axon of the squid. *J. Physiol.* **108**, 37–77.

Keynes, R. D. & Lewis, P. R. (1951). The sodium and potassium content of cephalopod nerve fibres. *J. Physiol.* **114**, 151–182.

Manery, J. F. (1939). Electrolytes in squid blood and muscle. *J. cell. comp. Physiol.* **14**, 365–369.

Steinbach, H. B. & Spiegelman, S. (1943). The sodium and potassium balance in squid nerve axoplasm. *J. cell. comp. Physiol.* **22**, 187–196.

Teorell, T. (1949*a*). *Annu. Rev. Physiol.* **11**, 545–564.

Teorell, T. (1949*b*). Membrane electrophoresis in relation to bio-electrical polarization effects. *Arch. Sci. physiol.* **3**, 205–218.

Ussing, H. H. (1949). The distinction by means of tracers between active transport and diffusion. *Acta physiol. scand.* **19**, 43–56.

Webb, D. A. (1939). The sodium and potassium content of sea water. *J. exp. Biol.* **16**, 178–183.

Webb, D. A. (1940). Ionic regulation in *Carcinus maenas*. *Proc. Roy. Soc.* B, **129**, 107–135.

J. Physiol. (1952) 116, 473–496

THE COMPONENTS OF MEMBRANE CONDUCTANCE IN THE GIANT AXON OF *LOLIGO*

By A. L. HODGKIN and A. F. HUXLEY

From the Laboratory of the Marine Biological Association, Plymouth, and the Physiological Laboratory, University of Cambridge

(*Received* 24 *October* 1951)

The flow of current associated with depolarizations of the giant axon of *Loligo* has been described in two previous papers (Hodgkin, Huxley & Katz, 1952; Hodgkin & Huxley, 1952). These experiments were concerned with the effect of sudden displacements of the membrane potential from its resting level ($V=0$) to a new level ($V=V_1$). This paper describes the converse situation in which the membrane potential is suddenly restored from $V=V_1$ to $V=0$. It also deals with certain aspects of the more general case in which V is changed suddenly from V_1 to a new value V_2. The experiments may be conveniently divided into those in which the period of depolarization is brief compared to the time scale of the nerve and those in which it is relatively long. The first group is largely concerned with movements of sodium ions and the second with movements of potassium ions.

METHODS

The apparatus and method were similar to those described by Hodgkin *et al.* (1952). The only new technique employed was that on some occasions two pulses, beginning at the same moment but lasting for different times, were applied to the feed-back amplifier in order to give a wave form of the type shown in Fig. 6. The amplitude of the shorter pulse was proportional to $V_1 - V_2$, while the amplitude of the longer pulse was proportional to V_2. The resulting changes in membrane potential consisted of a step of amplitude V_1, during the period when the two pulses overlap, followed by a second step of amplitude V_2.

RESULTS

Experiments with relatively brief depolarizations

Discontinuities in the sodium current

The effect of restoring the membrane potential after a brief period of depolarization is illustrated by Fig. 1. Record A gives the current associated with a maintained depolarization of 41 mV. As in previous experiments, this consisted of a wave of inward current followed by a maintained phase of outward current. Only the beginning of the second phase can be seen at the relatively high time base speed employed. At 0·85 msec. the ionic current reached a value of 1·4 mA./cm.². Record B shows the effect of cutting short the period of depolarization at this time. The sudden change in potential was associated with a rapid surge of capacity current which is barely visible on the time scale employed. This was followed by a 'tail' of ionic current which

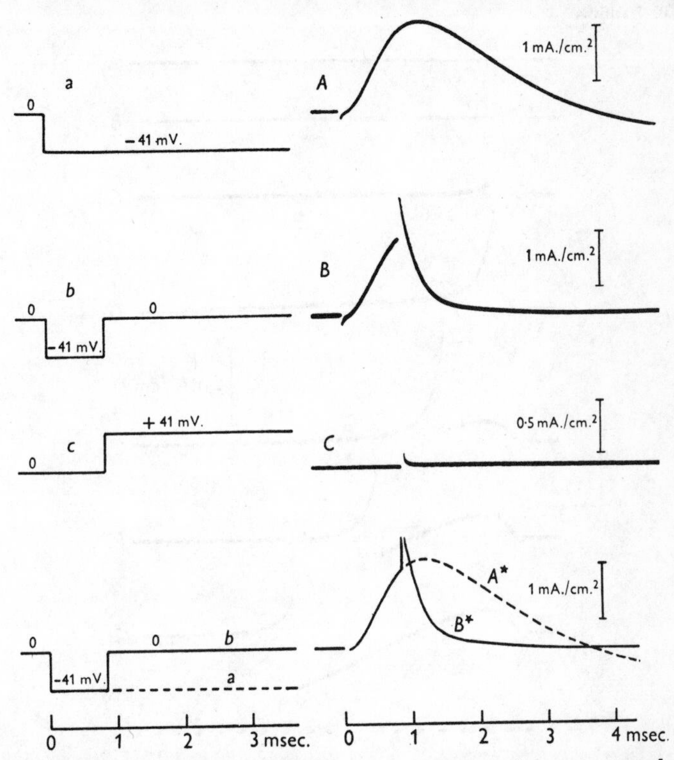

Fig. 1. Left-hand column: *a*, *b*, *c*, time course of potential difference between external and internal electrode. Right-hand column: *A*, *B*, *C*, records of membrane current associated with changes in membrane potential shown in left-hand column. (The amplification in *C* was 90% greater than that in *A* and *B*.) *A**, *B**, time course of ionic currents obtained by subtracting capacity current in *C* from *A* and *B*. Axon 25; temperature 5° C.; uncompensated feed-back. Inward current is shown upward in this and all other figures except Fig. 13.

started at about 2·2 mA./cm.² and declined to zero with a time constant of 0·27 msec. The residual effects of the capacitative surge were small and could be eliminated by subtracting the record obtained with a corresponding anodal displacement (C). Curves corrected by this method are shown in A^* and B^*.

The first point which emerges from this experiment is that the total period of inward current is greatly reduced by cutting short the period of depolarization. This suggests that the process underlying the increase in sodium permeability is reversible, and that repolarization causes the sodium current to fall more rapidly than it would with a maintained depolarization. Further experiments dealing with this phenomenon are described on p. 482. At present our principal concern is with the discontinuity in ionic current associated with a sudden change of membrane potential. Fig. 2*D* illustrates the discontinuity in a more striking manner. In this experiment the nerve was depolarized nearly to the sodium potential, so that the ionic current was relatively small during the pulse.

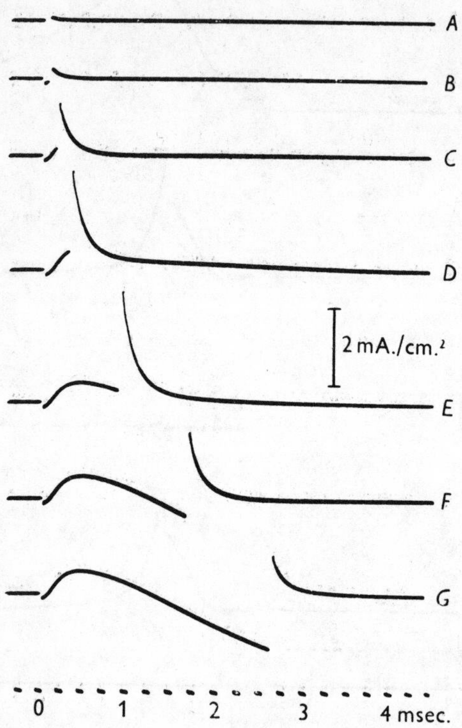

Fig. 2. Records of membrane current associated with depolarization of 97·5 mV. lasting, 0·05, 0·08, 0·19, 0·32, 0·91, 1·6 and 2·6 msec. The time and current calibration apply to all records. Axon 41; temperature 3·5° C.; compensated feed-back.

The other records in Fig. 2 illustrate the effect of altering the duration of the pulse. The surge of ionic current was small when the pulse was very short; it reached a maximum at a duration of 0·5 msec. and then declined with a time constant of about 1·4 msec. For durations less than 0·3 msec. the surge of ionic current was roughly proportional to the inward current at the end of the pulse. Since previous experiments suggest that this inward current is carried by sodium ions (Hodgkin & Huxley, 1952), it seems likely that the tail of inward current after the pulse also consists of sodium current. Fig. 3 illustrates an experiment to test this point. In *A*, the membrane was initially depolarized to the sodium potential. The ionic current was very small during the pulse but the usual tail followed the restoration of the resting potential. The sequence of events was entirely different when choline was substituted for the sodium in the external fluid (Fig. 3*B*). In this case there was a phase of outward current during the pulse but no tail of ionic current when the membrane potential was restored. The absence of ionic current after the pulse is proved by the fact that the capacitative surges obtained with anodal and cathodal displacements were almost perfectly symmetrical (records *B* and *C*). These effects are explained quite simply by supposing that sodium permeability rises when the membrane is depolarized and falls exponentially after it has been repolarized.

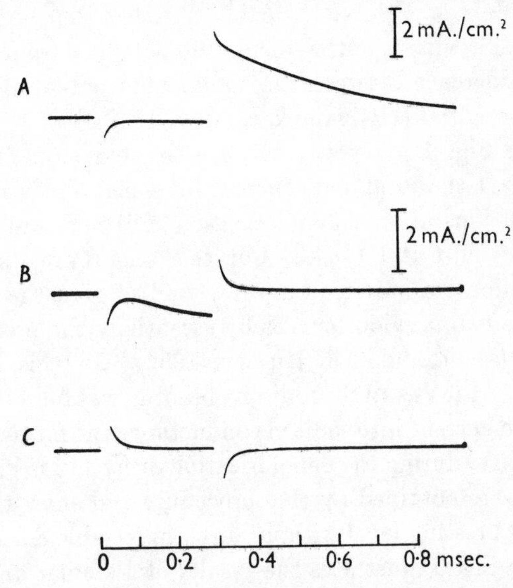

Fig. 3. *A*, membrane current associated with depolarization of 110 mV. lasting 0·28 msec.; nerve in sea water. *B*, same, but with nerve in choline sea water. *C*, membrane currents associated with an increase of 110 mV. in membrane potential; nerve in choline sea water. Axon 25; temperature 5° C.; uncompensated feed-back.

In record *A* the increase in permeability did not lead to any current during the pulse, since inward and outward movements of sodium are equal at the sodium potential. After the pulse the tendency of external sodium ions to enter the fibre is very much greater than that of internal sodium ions to leave. This means that there must be a large inward current after the pulse unless the sodium permeability reverts instantaneously to a low value. Record *B* is different because there were no external sodium ions to carry the current in an inward direction. The increase in sodium permeability therefore gave a substantial outward current during the period of depolarization but no inward current after the pulse. One might expect to see a 'tail' of outward current in *B* corresponding to the tail of inward current in *A*. However, the tendency of the internal sodium ions to leave the fibre against the resting potential difference would be so small that the resulting outward current would be indistinguishable from the capacitative surge. According to the 'independence principle' (Hodgkin & Huxley, 1952, equation 12), the outward current in B should be only 1/97 of the inward current in A.

Continuity of sodium conductance

Discontinuities such as those in Figs. 1, 2 and 3 A disappear if the results are expressed in terms of the sodium conductance (g_{Na}). This quantity was defined previously by the following equation:

$$g_{Na} = I_{Na}/(V - V_{Na}), \quad (1)$$

where V is the displacement of the membrane potential from its resting value and V_{Na} is the difference between the equilibrium potential for sodium ions and the resting potential (Hodgkin & Huxley, 1952).

The records in Fig. 4 allow g_{Na} to be estimated as a function of time. Curves α and A give the total ionic current for a nerve in sea water. Curve α was obtained with a maintained depolarization of 51 mV. and A with the same depolarization cut short at 1·1 msec. Curves β and B are a similar pair with the nerve in choline sea water. Curves γ and C give the sodium current obtained from the two previous curves by essentially the method used in the preceding paper (see Hodgkin & Huxley, 1952). In this experiment the depolarization was 51 mV. and the sodium potential was found to be -112 mV. To convert sodium current into sodium conductance the former must therefore be divided by 61 mV. during the depolarization or by 112 mV. after the pulse. Curves δ and D were obtained by this procedure and show that the conductance reverts to its resting level without any appreciable discontinuity at the end of the pulse. Fig. 5 illustrates the results of a similar analysis using the records shown in Fig. 2. In this experiment no tests were made in choline sea water, but the early part of the curve of sodium current was obtained by assuming that sodium current was zero initially and that the contribution of other ions remained at the level observed at the beginning of the pulse. Records made at the sodium potential (-117 mV.) indicated that the error introduced by this approximation should not exceed 5% for pulses shorter than 0·5 msec.

The instantaneous relation between ionic current and membrane potential

The results described in the preceding section suggest that the membrane obeys Ohm's law if the ionic current is measured immediately after a sudden change in membrane potential. In order to establish this point we carried out the more complicated experiment illustrated by Fig. 6. Two rectangular pulses were fed into the feed-back amplifier in order to produce a double step of membrane potential of the type shown inset in Fig. 6. The first step had a duration of 1·53 msec. and an amplitude of -29 mV. The second step was relatively long and its amplitude was varied between -60 mV. and $+30$ mV.

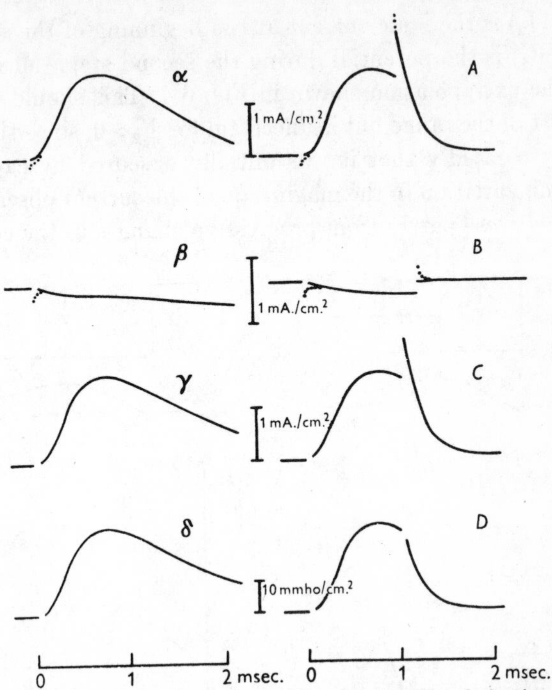

Fig. 4. α, ionic current in sea water associated with maintained depolarization of 51 mV. applied at $t=0$. (The dotted line shows the form of the original record before correcting for capacity current.) β, same in choline sea water. γ, sodium current estimated as $(\alpha-\beta)\times 0{\cdot}92$. δ, sodium conductance estimated as γ/61 mV. *A*, *B*, same as α and β respectively, but with depolarization lasting about 1·1 msec. *C*, sodium current estimated as $(A-B)\times 0{\cdot}92$ during pulse or $(A-B)\times 0{\cdot}99$ after pulse. *D*, sodium conductance estimated as *C*/61 mV. during pulse or *C*/112 mV. after pulse. The factors 0·92 and 0·99 allow for the outward sodium current in choline sea water and were obtained from the 'independence principle'. Axon 17; temperature 6° C.; V_{Na} in sea water $= -112$ mV.; uncompensated feed-back.

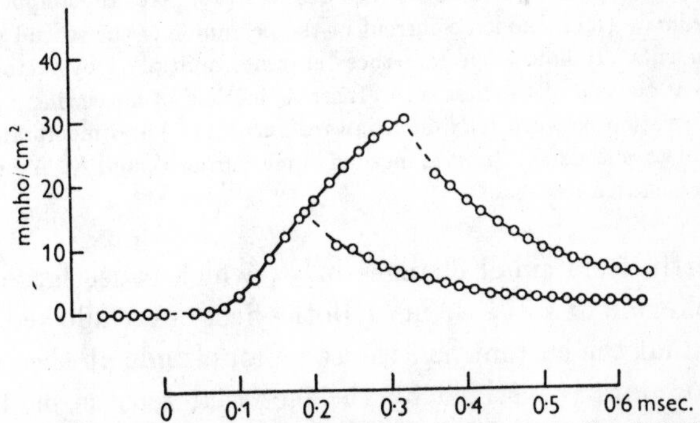

Fig. 5. Time course of sodium conductance estimated from records *C* and *D* (Fig. 2) by method described in text. At zero time the membrane potential was reduced by 97·5 mV. and was restored to its resting value at 0·19 msec. (lower curve) or 0·32 msec. (upper curve). The broken part of the curve has been interpolated in the region occupied by the capacitative surge. Axon 41; temperature 3·5° C.; compensated feed-back; $V_{Na} = -117$ mV.

The ordinate (I_2) is the ionic current at the beginning of the second step and the abscissa (V_2) is the potential during the second step. Measurement of I_2 depends on the extrapolation shown in Fig. 6*A*. This should introduce little error over most of the range but is uncertain for $V_2 > 0$, since the ionic current then declined so rapidly that it was initially obscured by capacity current. There was some variation in the magnitude of the current observed during the first pulse. This arose partly from progressive changes in the condition of the

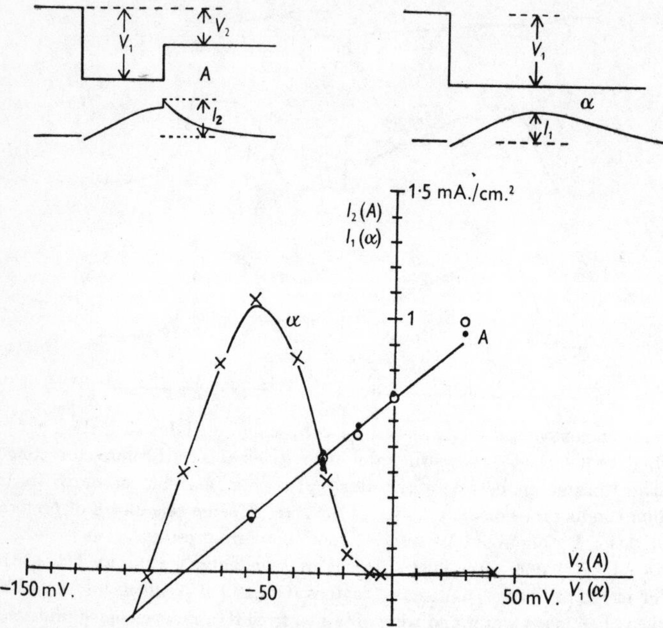

Fig. 6. Line *A*, instantaneous current-voltage relation. The first step had an amplitude of −29 mV. and a duration of 1·53 msec. The abscissa (V_2) gives the amplitude of the second step. The ordinate (I_2) is the ionic current at the beginning of the second step. The dots are observed currents. Hollow circles are these currents multiplied by factors which equalize the currents at the end of the first step. Inset *A*, method of measuring V_2 and I_2. Curve α and crosses, relation between maximum inward current (I_1) and membrane potential using single pulse of amplitude V_1. Inset α, method of measuring V_1 and I_1. Axon 31; temperature 4° C.; uncompensated feed-back.

nerve and partly from small changes in V_1 which cause large variations in current in the region of $V = -29$ mV. Both effects were allowed for by scaling all records so that the current had the same amplitude at the end of the first step. This procedure is justified by the fact that records made with $V_2 = 0$ show that the amplitude of the current immediately after the step was directly proportional to the current immediately before it.

The results are plotted in curve *A* and show that the relation between I_2 and V_2 is approximately linear. This is in striking contrast to the extremely non-linear relation obtained when the current is measured at longer intervals. An example of the second type is provided by curve α which shows how the maximum inward current varied with membrane potential in the same axon. In this case only a single pulse of variable amplitude was employed and current was measured at times of 0·5–2·0 msec. Under these conditions the sodium conductance had time to reach the value appropriate to each depolarization and the current-voltage relation is therefore far from linear.

The line A and the curve α intersect at -29 mV. since the two methods of measurement are identical if $V_2 = V_1$. A second intersection occurs at -106 mV. which is close to the sodium potential in this fibre.

A similar pair of curves obtained with a larger initial depolarization is shown by A and α in Fig. 7. In this case the nerve was depolarized to the sodium potential so that one would expect the line A to be tangential to the curve α. This is approximately true, although any exact comparison is invalidated by the fact that the two curves could not be obtained at exactly the same time.

The instantaneous current-voltage relation in sodium-free solution

The measurements described in the preceding section indicate that the instantaneous behaviour of the membrane is linear when the nerve is in sea water. The conclusion cannot be expected to apply for all sodium concentrations. The method of defining a chord conductance breaks down altogether if there is no sodium in the external medium. In this case $V_{Na} = \infty$ and g_{Na} must be zero if the sodium current is to be finite. This condition could not be realized in practice but the theoretical possibility of its existence indicates that the concept of sodium conductance must be used with caution.

The lower part of Fig. 7 illustrates an attempt to determine the instantaneous current-voltage relation in a sodium-free solution. The upper curves (A and α) were measured in sea water and have already been described. The crosses in the lower part of the figure give the instantaneous currents in choline sea water, determined in the same way as the circles which give the corresponding relation in sea water. The effect of the change in resting potential has been allowed for by shifting the origin to the right by 4 mV. (see Hodgkin & Huxley, 1952). The series of records from which these measurements were made was started shortly after replacing normal sea water by choline sea water and was continued, in the order shown, with an interval of about 40 sec. between records. On analysis it was found that the earliest records (e.g. 1) showed a small inward current, whereas records taken later (e.g. 11 or 15) gave no such effect. It is evident that the series was started before all the sodium had diffused away from the nerve and that only the later records (e.g. 6–15) can be regarded as representative of a nerve in a sodium-free solution. Nevertheless, it is clear that the instantaneous current-voltage relation shows a marked curvature and is quite different from the linear relation in sea water.

The results are, in fact, reasonably close to those predicted by the 'independence principle'. This is illustrated by a comparison of the crosses in Fig. 7 with the theoretical curves *B* and *C* which were calculated from *A* on the assumption that the independence principle holds and that the sodium

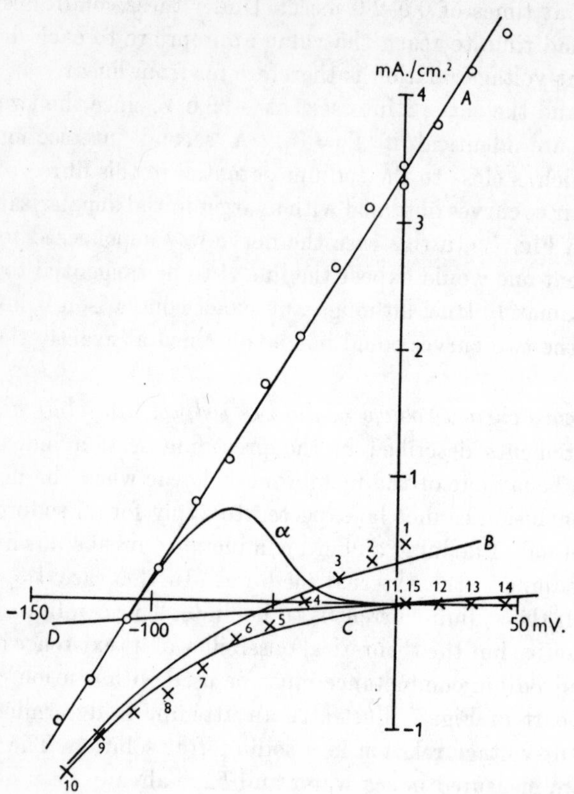

Fig. 7. Current-voltage relations in sea water and choline sea water. Ordinate: current density. Abscissa: displacement of membrane potential from resting potential in sea water. Line *A* and curve α were obtained in sea water in the same way as in Fig. 6, except that the current for α was not measured at the maximum but at a fixed time (0·28 msec.) after application of a single step. The initial depolarization for *A* was 110 mV. and the duration of the first step was 0·28 msec. The crosses give the instantaneous currents in choline sea water, determined in the same way as the circles in *A*. The numbers show the order in which the measurements were made. *B* and *C*, instantaneous current in 10% sodium sea water and in choline sea water respectively, derived from *A* by means of the 'independence principle' using the equations

$$\frac{(I_{Na})_B}{(I_{Na})_A}=\frac{0{\cdot}1\exp(V-V_{Na})/24-1}{\exp(V-V_{Na})/24-1}$$

and

$$\frac{(I_{Na})_C}{(I_{Na})_A}=\frac{-1}{\exp(V-V_{Na})/24-1},$$

V_{Na} = sodium potential in sea water = −110 mV. Sodium currents measured from the line *D* which passes through the origin and the point for the small current observed at the sodium potential in sea water. Axon 25; temperature 5° C.; uncompensated feed-back.

concentrations in the external solution were 10% (*B*) and 0% (*C*) of that in *A*. It will be seen that there is general agreement between calculated and observed results, although the change in external sodium and the possibility of progressive changes invalidates any exact comparison. In the preceding paper it was shown that the observed sodium currents in a choline solution were usually larger than those calculated from the independence principle. This deviation is not seen here, probably because the measurements in choline were made later than those in sea water and no attempt was made to correct for deterioration, which is likely to have reduced the currents by 30% between the sea water and choline runs.

The experiment described in the preceding paragraph indicates that the linear relation between current and voltage observed in sea water is not a general property of the membrane since it fails in sodium-free solutions. This does not greatly detract from the usefulness of the result, for the primary concern of this paper is to determine the laws governing ionic movements under conditions which allow a normal action potential to be propagated.

The reversible nature of the change in sodium conductance

The results described in the first part of this paper show that the sodium conductance reverts rapidly to a low value when the membrane potential is restored to its resting value. Figs. 2 and 5 suggest that this is true at all stages of the response and that the rate at which the conductance declines is roughly proportional to the value of the conductance. A rate constant (b_{Na}) can be defined by fitting a curve of the form exp $(-b_{Na}t)$ to the experimental results. Values obtained by this method are given in Table 1.

In order to investigate the effect of repolarizing the membrane to different levels on the rate of decline of sodium conductance we carried out the experiment illustrated by Fig. 8. The curves in the left-hand column are tracings of the membrane current while those on the right give the sodium conductance, calculated on the assumption that the contribution of ions other than sodium is negligible (records made in a solution containing 10% of the normal sodium concentration show that the error introduced by this approximation should not exceed 5% of the maximum current). The initial depolarization was 29 mV. and the sodium conductance reached its maximum value in 1·53 msec. When the membrane potential was restored to its resting level the conductance fell towards zero with a rate constant of about 4·3 msec.$^{-1}$ (curve γ). If V_2 was made +28 mV. the rate constant increased to about 10 msec.$^{-1}$ and a further increase to 15 msec.$^{-1}$ occurred with $V_2 = +57$ mV. On the other hand, if V_2 was reduced to −14 mV. the conductance returned with a rate constant of only 1·6 msec.$^{-1}$. When $V_2 = -57$ mV. the conductance no longer fell but increased towards an 'equilibrium' value which was greater than that attained at −29 mV. (The curve cannot be followed beyond about 2 msec.

TABLE 1. Apparent values of rate constant determining decline of sodium conductance following repolarization to resting potential

Axon	Membrane potential during pulse (mV.)	Duration of pulse (msec.)	Temperature (° C.)	Average rate constant (msec.$^{-1}$)	Rate constant at 6° C. (msec.$^{-1}$)
15	−32	0·4–1·1	11	9·4	5·4
15	−91	0·1–0·5	11	9·0	5·2
17	−32	0·7–1·6	6	5·9	5·8
17	−51	0·2–1·0	6	6·7	6·6
24	−42	0·2	20	18·5	4·1
24	−84	0·1	20	17·2	3·9
25	−41	1·0	4	3·8	4·8
25	−110	0·3	4	3·3	4·0
31	−100	0·3	4	3·0	3·8
31	−29	1·5	4	4·2	5·3
32	−116	0·2	5	6·3*	6·9*
32	−67	0·7	5	6·3*	6·8*
41	−98	0·1–4	3	7·1*	9·6*
41	−117	0·1–3	3	7·7*	10·5*

Results marked with an asterisk were obtained with compensated feed-back. The last column is calculated on the assumption that the temperature coefficient (Q_{10}) of the rate constant is 3.

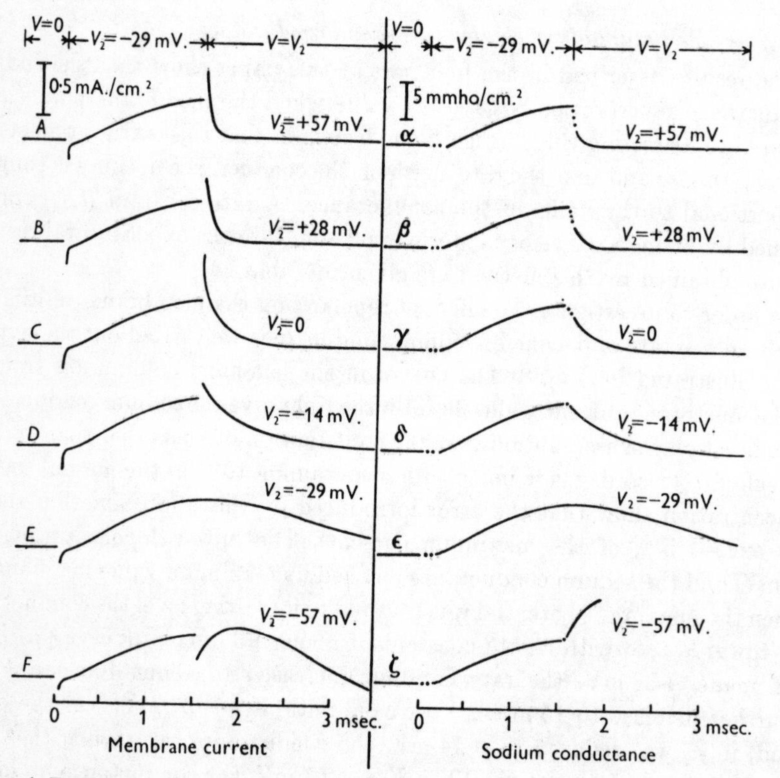

Fig. 8. *A–F*, time course of membrane current associated with change in membrane potential shown at top of figure. α–ζ; time course of sodium conductance obtained by dividing *A–F* by $V + 100$ mV. Axon 31; temperature 4° C.; uncompensated feed-back.

because the contribution of potassium ions soon becomes important at large depolarizations.) The whole family of curves suggests that the conductance reached at any depolarization depends on the balance of two processes occurring at rates which vary in opposite directions with membrane potential.

The observation that the rate constant increases with membrane potential does not depend on the details of the method used to estimate sodium conductance, for the tracings of current in the left-hand column of Fig. 8 show exactly the same phenomenon. Similar results were obtained in all the experiments of this type, and are plotted against V_2 in Fig. 9. It will be seen that there is good agreement between different experiments, and that a tenfold increase of rate constant occurs between $V_2 = -20$ and $+50$ mV.

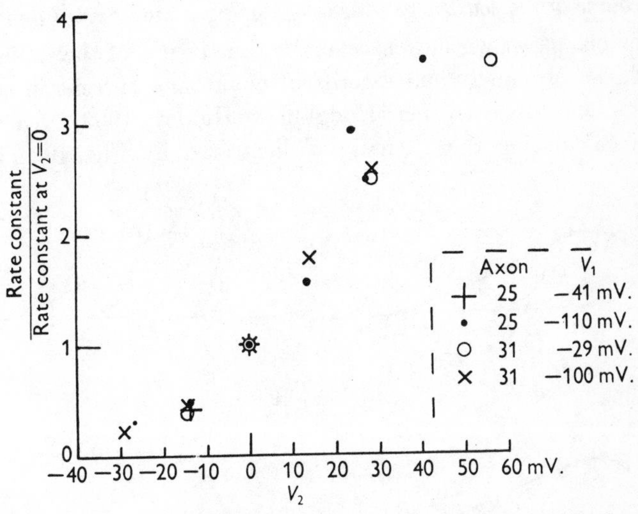

Fig. 9. Relation between rate constant determining decline of sodium conductance and potential to which membrane is repolarized. Abscissa: membrane potential during second step (V_2). Ordinate: relative value of rate constant.

Errors due to the series resistance

Most of the experiments in this paper were obtained with uncompensated feed-back and must therefore have been somewhat affected by the small resistance in series with the membrane (Hodgkin *et al.* 1952). The linearity of the relation between current and voltage illustrated by Figs. 6 and 7 can clearly stand, for the effect of a series resistance would simply be to change the slope of the straight line and not to introduce any curvature. From our estimates of the value of the series resistance it can be shown that the true slopes in these two figures should be 7 and 30% greater than those shown. A more serious error is introduced in the measurement of the rate constant. In Fig. 8 the total current at the beginning of the record was about 0·5 mA./cm.². This means that the true membrane potential was not zero but about −4 mV. At this potential the rate of return of membrane conductance would be slowed by about 8%. In some of the experiments used in compiling Table 1 this error may be as great as 50%. However, it should be small in axons 32 and 41 which were examined with compensated feed-back. We are also uncertain about the extent to which the rate of return of sodium conductance can be regarded as exponential. Axon 41, which was in excellent condition and was examined with compensated feed-back, showed clear departures from exponential behaviour in that the initial fall of conductance was too rapid (see Figs. 2 and 5). In all other experiments the curves of current against time were reasonably close to exponentials, but in many cases this may have been due to an error introduced by the series resistance.

Errors due to polarization effects

If the membrane was maintained in the depolarized condition for long periods of time the outward current declined as a result of a 'polarization effect' (Hodgkin *et al.* 1952). At the end of such a pulse a phase of inward current was observed and was found to be roughly proportional to the amount of 'polarization'. This was quite distinct from the inward current described in the preceding sections, since it only appeared with long pulses and was unaffected by removing external sodium. With the possible exception of Fig. 2*G*, the results described in the preceding sections are unlikely to have been affected by polarization since the duration of the pulse was always kept short.

The time course of the sodium conductance during a maintained depolarization

In a previous paper we showed that the time course of the sodium conductance could be obtained from records of membrane current in solutions of different sodium concentration (Hodgkin & Huxley, 1952). An alternative method of calculating these curves is illustrated by Fig. 10. The method

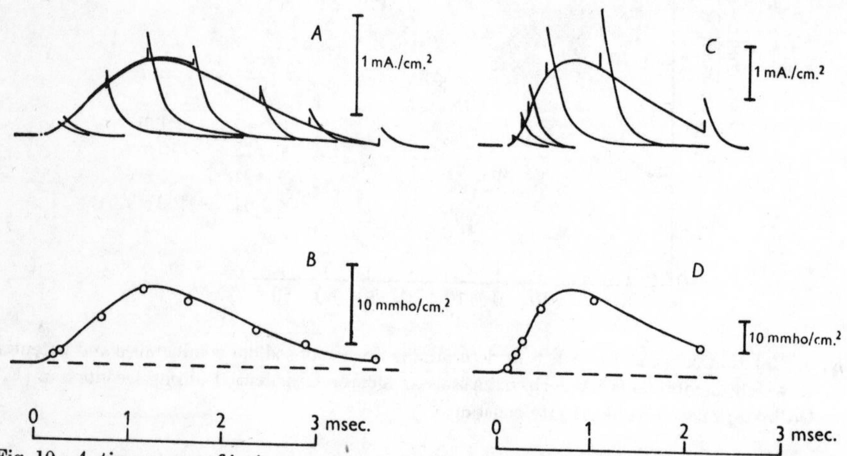

Fig. 10. *A*, time course of ionic currents associated with depolarizations of 32 mV. lasting from $t=0$ to times indicated by vertical strokes on tracings. Nerve in sea water. *B*, time course of sodium conductance. The circles were obtained by dividing the peak currents in *A* by 112 mV. and the continuous curve from the difference between the continuous curve in *A* and a similar curve in choline sea water (see text and legend to Fig. 4). *C*, same as *A* but employing depolarization of 51 mV. *D*, sodium conductance obtained from *C* by similar methods to those used for *B*. (The smooth curve is the same as that shown in Fig. 4.) Axon 17; temperature 6° C.; V_{Na} in sea water = -112 mV.; uncompensated feed-back.

depends on the fact that the inward current immediately after a pulse is proportional to the sodium conductance at the end of the preceding depolarization. The variation of inward current with pulse duration is illustrated by the tracings in Fig. 10*A*. The time course of the sodium conductance can be measured by determining the maximum ionic current associated with repolarization and dividing this quantity by the difference between the resting potential and the sodium potential (about −112 mV. in this experiment). A series of points obtained by this method is shown in Fig. 10*B*. These may be compared with the smooth curve, which represents sodium conductance obtained by the method described in the preceding paper (subtraction of ionic current in choline from that in sea water). Good agreement was obtained, and also when other depolarizations were employed, for example *C* and *D* at −51 mV. The only occasions on which the two methods did not agree were those in which the sodium conductance was measured at long times, with a large depolarization. In these experiments the subtraction method sometimes gave an apparent negative conductance which we regarded as an error due to slight differences between the potassium currents in the two records. This conclusion was confirmed by the fact that the alternative method never showed a 'negative conductance' but only a residual positive conductance.

Experiments with relatively long depolarizations

The instantaneous relation between potassium current and membrane potential

In a previous paper (Hodgkin & Huxley, 1952) we gave reasons for thinking that potassium ions were largely responsible for carrying the maintained outward current associated with prolonged depolarization of the membrane. In order to investigate the instantaneous relation between potassium current and membrane potential it is necessary to employ depolarizations lasting for much longer periods than those used to study sodium current. Polarization effects made such experiments difficult at 5° C., but errors from this cause could be greatly reduced by working at 20° C. In this case the polarization effect occurred at the same rate but the potassium conductance rose in about one-fifth of the time required at 5° C.

A typical experiment with a nerve in choline sea water is illustrated by Fig. 11. Its general object is to measure the ionic currents associated with repolarization of the membrane when the potassium conductance is much greater than the sodium conductance. The amplitude of the first step was −84 mV. and its duration 0·63 msec., which is equivalent to about 4 msec. at 5° C. Under these conditions 90–95 % of the outward current should be potassium current and only 5–10 % should be sodium current (see Fig. 10 for an indication of the rate of fall of sodium conductance from its initial maximum). After the pulse, sodium current should be negligible since the nerve was in choline sea water (see Fig. 3).

The simplest record in Fig. 11 is *E*, in which the membrane potential was restored to its resting value at the end of the first step. The sequence of events was as follows. At $t=0$ the membrane was depolarized by 84 mV. and was held at this level until $t=0{\cdot}63$ msec. The current was outward during the whole period and consisted of a hump of sodium current followed by a rise of potassium current which reached 1·83 mA./cm.^{2} at $t=0{\cdot}63$ msec. At this moment the membrane potential was restored to its resting value. The sudden increase in potential was associated with a brief capacity current in an inward direction. This was followed by an outward current which declined exponentially to zero. A record at higher amplification (*e*) shows this 'tail' of outward current more clearly. The dots give the ionic current extrapolated to

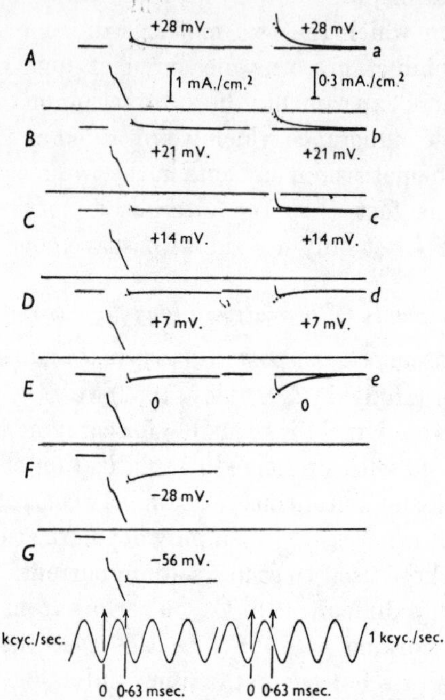

Fig. 11. Membrane currents associated with depolarization of 84 mV. followed by repolarization to value shown on each record. The duration of the first step was 0·63 msec. The second step lasted longer than these records. *A* to *G*, records at low amplification showing current during both steps. *a–e*, records at higher amplification showing only the current during the second step. The dots give the ionic current after correcting for capacity current. Axon 26 in choline sea water; temperature 20° C.; uncompensated feed-back.

$t=0{\cdot}63$ msec. after correcting for the residual effect of the capacitative surge. The 'tail' of outward current can be explained by supposing that the equilibrium potential for potassium is about 12 mV. greater than the resting potential in choline sea water, and that the instantaneous value of the potassium conductance (g_K) is independent of V. At $t=0{\cdot}63$ msec. the current is 1·83 mA./cm.^{2} when $V=-84$ mV., or 0·22 mA./cm.^{2} when $V=0$. Taking the potassium potential (V_K) as $+12$ mV. and neglecting the contribution of chloride and other ions it follows that the potassium conductance (g_K) was approximately the same in the two cases. Thus

$$g_K = I_K/(V - V_K), \tag{2}$$

so that

$$g_K = \frac{-1{\cdot}83 \text{ mA./cm.}^2}{-96 \text{ mV.}} = 19 \text{ m.mho/cm.}^2 \quad \text{when } V = -84 \text{ mV.},$$

and
$$g_K = \frac{-0{\cdot}22 \text{ mA./cm.}^2}{-12 \text{ mV.}} = 18 \text{ m.mho/cm.}^2 \quad \text{when } V = 0.$$

As soon as the membrane potential is restored to its resting value the potassium conductance reverts exponentially to its resting level and therefore gives the tail of current seen in *E* and *e*. If this explanation is right, the exponential tail of current should disappear at $V = V_K$ and should be reversed in sign when $V > V_K$. Records *a* to *d* show that the current is inward for $V > 21$ mV. and is outward for $V < 7$ mV. There is practically no inward current at $V = 14$ mV. and 13 ± 1 mV. would seem to be a reasonable estimate of V_K.

This method of measuring the potassium potential depends on the assumption that potassium ions are the only charged particles responsible for the component of the current which varies with time after the end of the pulse. It is not affected by the fact that chloride and other ions may carry appreciable quantities of current, provided that the resistance to the motion of these ions is constant at any given value of membrane potential. The magnitude of the 'leak' due to chloride and other ions may be estimated from the current needed to maintain the membrane at the potassium potential. In the experiment illustrated by Fig. 11 this current was about 0·008 mA./cm.^{2} which is small compared with the maximum potassium current at $V = 0$ or $V = +28$ mV.

Fig. 12 was prepared from the records in Fig. 11 by essentially the same method as that used in studying the instantaneous relation between sodium current and membrane potential. Curve α gives the relation between current and voltage 0·63 msec. after the application of a single step of amplitude V_1. In curve *A*, V_1 was fixed at -84 mV. and the potential was changed suddenly to a new level V_2. The abscissa is V_2, while the ordinate is the ionic current immediately after the sudden change. The experimental points in *A* are seen to fall very close to a straight line which crosses curve α at the potassium potential ($+13$ mV.). In this experiment no measurements were made with $V_2 < V_1$ but records obtained with other fibres showed that the instantaneous current-voltage relation was linear for $V_2 < V_1$ as well as for $V_2 > V_1$.

In the experiment considered in the previous paragraphs the initial rise of potassium current was obscured by sodium current and the plateau was not reached because the pulse was kept short in order to reduce possible errors from the 'polarization effect'. A clearer picture of the sequence of events is provided by Fig. 13. In this experiment the amplitude of the pulse was -25 mV. and its duration nearly 5 msec. The polarization effect was not appreciable since the current density was relatively small. Sodium current was also small since the nerve was in choline sea water, and the depolarization

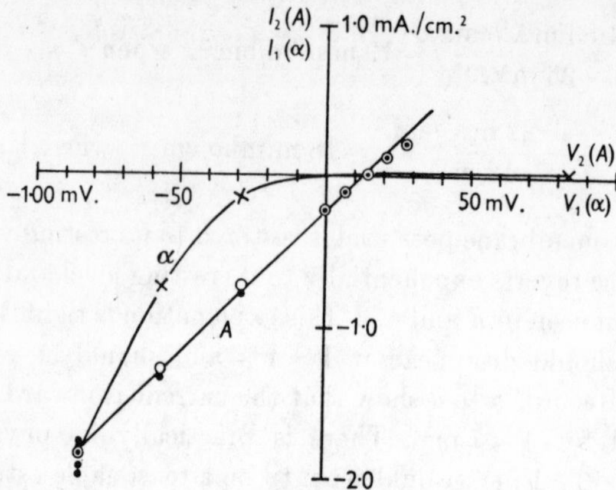

Fig. 12. Current-voltage relations during period of high potassium permeability. Line A, instantaneous current-voltage relation determined by changing membrane potential in two steps. The first step had a constant amplitude of − 84 mV. and a constant duration of 0·63 msec. The abscissa (V_2) gives the amplitude of the second step in millivolts. The ordinate (I_2) is the ionic current density at the beginning of the second step. The dots are observed currents. Hollow circles are these currents multiplied by factors which equalize the currents at the end of the first step. Curve α and crosses, relation between current and membrane potential at 0·63 msec. after beginning of single step of amplitude V_1. Experimental details are as in Fig. 11.

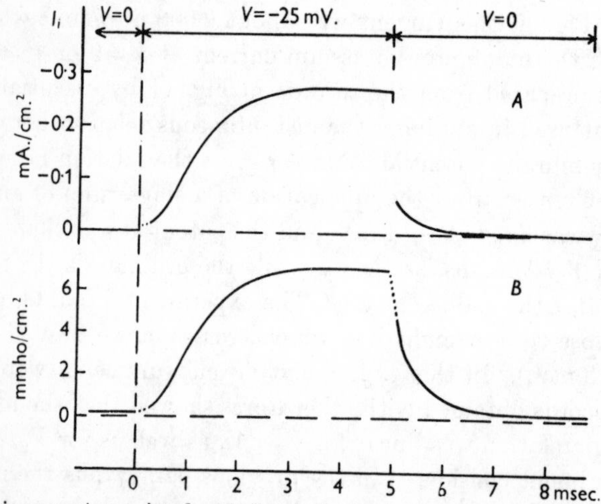

Fig. 13. A, ionic current associated with depolarization of 25 mV. lasting 4·9 msec. Axon 18 in choline sea water at a temperature of 21° C. The curve is a direct replot of the original current record except in the regions 0–0·3 msec. and 4·9–5·2 msec., where corrections for capacity current were made by the usual method. Outward current shown upward. B, potassium conductance estimated from A by the equation $g_K = I_K/(V - V_K)$, where V_K is 12 mV. and I_K is taken as the ionic current (I_i) minus a leakage current of 0·5 m.mho/cm.² × ($V + 4$ mV.).

was less than that at which a hump of outward sodium current first became appreciable. On the other hand, it was desirable to make a small correction for the leakage current due to ions other than sodium and potassium. The method of estimating this current at different voltages is indicated on p. 494. The experiment shows that whereas the potassium conductance rises with a marked delay it falls along an exponential type of curve which has no inflexion corresponding to that on the rising phase. This difference was present in all records except, possibly, those with very small depolarizations. It was also present in the curves calculated for the rise and fall of sodium conductance (e.g. Figs. 4, 5 and 8).

The rate constant determining the decline of potassium conductance

The experiments described in the preceding sections indicate that the potassium conductance returns to a low level when the membrane is repolarized to its resting value. The restoration of the condition of low conductance leads to a 'tail' of potassium current which can be fitted with reasonable accuracy by a curve of the form $\exp(-b_K t)$. Table 2 gives the values of the rate constant (b_K) determined by this method. It shows that b_K varies markedly with temperature and also to some extent with the amplitude of the step used to depolarize the axon. The second effect was particularly noticeable in axon 1 which was in poor condition and had a high potassium conductance in the resting state.

The effect of repolarizing the membrane to different levels is shown in Fig. 14. For $V_2 > -20$ mV. the rate constant increased with membrane potential but the relation is less steep than in the corresponding curve for sodium conductance (Fig. 9). Thus, changing V_2 from 0 to +40 mV. increases b_{Na} about 3·2-fold and b_K about 1·6-fold. Another important difference between the two processes is that b_{Na} is about 30 times greater than b_K at the resting potential.

The potassium potential

Table 3 summarizes a number of measurements of the potential at which potassium current reverses its direction. At 22° C. the apparent potassium potential is about 19 mV. higher than the resting potential if the axon is in sea water and about 13 mV. higher if it is in choline sea water. Corresponding figures at 6–11° C. are 13 mV. in sea water and 8 mV. in choline sea water. Since the resting potential is about 4 mV. higher in choline sea water (Hodgkin & Huxley, 1952) it seems likely that the absolute value of the potassium potential is unaffected by substituting choline for sodium ions. At 20° C. the absolute value of the potassium potential would be 80–85 mV. if the resting potential is taken as 60–65 mV. (Hodgkin & Huxley, 1952). This is nearly equal to the potential of 91 mV. estimated from chemical analyses

(Hodgkin, 1951). A similar conclusion applies to the results at 6–11° C. In squid fibres, cooling from 20 to 8° C. either has no effect or increases the resting potential by 1 or 2 mV. (Hodgkin & Katz, 1949). The observed

TABLE 2. Rate constant determining decline of potassium conductance following repolarization to resting potential

Group	Axon	Depolarization (mV.)	Temperature (° C.)	Rate constant (msec.$^{-1}$)	Rate constant at 6° C. (msec.$^{-1}$)	Average rate constant at 6° C. (msec.$^{-1}$)
A	1	6	23	1·2	0·14	0·17
	1	13	23	1·3	0·15	
	1	21	23	1·3	0·16	
	15	13	11	0·36	0·19	
	15	20	11	0·35	0·19	
	17	10	6	0·20	0·20	
	18	6	22	1·5	0·20	
	18	13	22	1·6	0·22	
	20	21	6	0·17	0·17	
	21	14	7	0·19	0·16	
	38*	10	5	0·12	0·13	
	39*	20	19	1·0	0·20	
	39*	10	19	0·83	0·16	
	39*	10	3	0·10	0·15	
	41*	20	4	0·10	0·12	
	41*	10	4	0·11	0·14	
B	1	36	23	1·5	0·18	0·22
	1	54	23	1·8	0·22	
	18	50	22	2·0	0·27	
	18	63	22	1·7	0·23	
	18	112	22	1·8	0·24	
	27	28	21	1·4	0·22	
	28	28	21	1·3	0·20	
C	15	13	11	0·49	0·26	0·23
	17	10	6	0·21	0·21	
	18	6	22	1·6	0·21	
	18	13	22	1·7	0·23	
	18	19	22	1·9	0·25	
D	18	25	22	2·0	0·27	0·28
	18	50	22	2·1	0·28	
	18	63	22	2·1	0·28	
	23	84	21	1·8	0·28	
	24	84	20	2·0	0·35	
	26	84	20	1·8	0·30	
	27	28	21	1·3	0·19	

Groups A and B in sea water; C and D in choline sea water. Groups A and C: depolarization less than 25 mV.; B and D: depolarization greater than 25 mV. An asterisk denotes the use of compensated feed-back. Rate constants at 6° C. are calculated for a Q_{10} of 3·5, which was found suitable for groups A and C.

potassium potential should therefore be taken as 75–80 mV. while the theoretical potassium potential would be reduced from 91 to 87 mV.

The effect of changing the external concentration of potassium on the potassium potential

Experimental determinations of V_K such as that illustrated by Fig. 11 were made in choline solutions containing different concentrations of potassium.

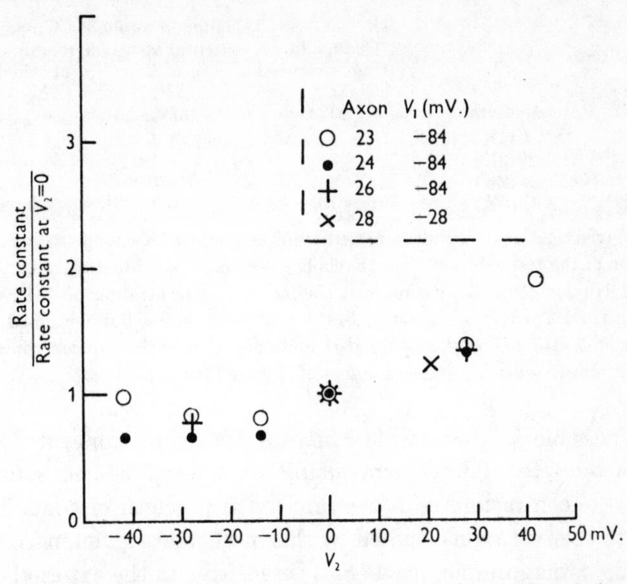

Fig. 14. Effect of membrane potential on the rate constant determining decline of potassium conductance. Abscissa (V_2): membrane potential during second step. Ordinate: relative value of rate constant.

TABLE 3. Apparent values of potassium potential

Axon	Medium	Temperature (°C.)	Potassium potential minus resting potential (mV.)	Average (mV.)
1	S	23	18	19
28	S	21	19	
15	S	11	14	11
20	S	6	10	
21	S	8	9	
18	C	22	12	13
23	C	21	14	
24	C	20	7	
26	C	20	13	
27	C	21	15	
28	C	21	16	
15	C	11	8	8
17	C	6	7	
20	0·7C:0·3S	6	8	
21	0·9C:0·1S	8	10	

S denotes sea water; C choline sea water. In axons 23–28 the potassium potential was found by the method illustrated by Fig. 12. In other cases it was taken as the potential at which the steady state current-voltage curve intersects the line joining the potassium currents before and after repolarization to the resting potential (e.g. line A and curve α intersect at 13 mV. in Fig. 12).

TABLE 4. Effect of potassium concentration on the apparent value of the potassium potential

Axon	Medium	Change in resting potential (ΔE_r) (mV.)	Change in applied potential at which I_K is zero (ΔV_K) (mV.)	Change in absolute membrane potential at which I_K is zero ($\Delta E_K = \Delta E_r + \Delta V_K$) (mV.)
27	A (½K)	+2	+6	+8
27	B (2K)	−3	−6	−9
28	B (2K)	−3	−7	−10
28	C (5K)	−13	−9	−22

Changes are given relative to the mean potentials observed in a 1K choline sea water before and after application of the test solution. The 1K choline sea water was identical with that described by Hodgkin & Huxley (1952) and contained choline at a concentration of 484 and potassium at 10 g.ions/kg. H_2O. The test solutions A, B, C and D were similar but contained potassium at concentrations of 5, 10, 20, 50 g.ions/kg. H_2O and correspondingly reduced concentrations of choline. Potentials are given as 'outside potential' minus 'inside potential'.

It was not possible to use a wide range of potassium concentrations, since squid axons tend to undergo irreversible changes if left in solutions containing high concentrations of potassium for any length of time. The results obtained with the two axons studied by this method are given in Table 4. They show that the potassium potential (E_K) is sensitive to the external concentration of potassium but that it changes by only about half the amount calculated for a concentration cell. Thus solutions *A*, *B*, *C* should give changes of +17, −17 and −41 mV. if E_K obeyed the ordinary equation for a concentration cell.

One possible explanation of this result is that potassium ions are not the only charged particles responsible for the delayed rise in conductance associated with depolarization (delayed rectification). Thus the discrepancy would be explained if choline or sodium, which are present in relatively high concentrations in the external solution, take part in the process with an affinity only 5% of that of potassium. This explanation might be consistent with the evidence which suggests that potassium ions are responsible for carrying most of the outward current through a depolarized membrane. For the concentration of potassium inside a fibre is about 10 times greater than that of sodium and the internal concentration of choline is almost certainly negligible. The participation of chloride ions in the process responsible for delayed rectification can probably be eliminated since one experiment showed that replacing all the choline chloride and two-thirds of the magnesium chloride in the choline sea water by dextrose gave an apparent *increase* of 3 mV. in the 'potassium potential'. The magnitude of the change was less certain in this experiment, since the solution employed gave a junction potential of 5–7 mV. which had to be allowed for in estimating the shift in resting potential. But it was clear that any change in E_K was small compared with the *reduction* of 45 mV. expected on the hypothesis that delayed rectification is entirely due to chloride ions.

Another way of accounting for the relatively small changes seen in Table 4 is to suppose that the potassium concentration in the immediate vicinity of the surface membrane is not the same as that in the external solution. Isolated cephalopod axons leak potassium ions at a fairly high rate and these must diffuse through layers of connective tissue and other structures between the excitable membrane and the external solution. This leakage is likely to increase in potassium-deficient solutions and to decrease in potassium-rich solutions. Hence the changes in effective potassium concentration might be less than those in Table 4. A related possibility is that the potassium concentration may be raised locally by the large outward currents used in these experiments. This hypothesis is of interest since it might account for the slow polarization effect which is not otherwise explained except in terms of a complicated polarization process at the internal electrode. In a former paper (Hodgkin *et al.* 1952) we obtained evidence of an external layer with a resistance of about 3 Ω.cm.². The transient change in potassium concentration due to current cannot be calculated without knowing the thickness of this layer. The steady change due to leakage might be large enough to explain the deviations in Table 4, if the leakage of potassium had been several times greater than that found by Steinbach & Spiegelman (1943).

It may be asked why effects similar to those discussed in the preceding paragraph do not upset the relation between the external sodium concentration and V_{Na}. The answer, probably, is that similar effects are present but that they are small because the sodium concentration in sea water is 45 times greater than that of potassium. Changes in concentration due to current would also be smaller in the case of sodium because the sodium currents are of relatively short duration.

The contribution of ions other than sodium and potassium

The experimental results described in this series of papers point to the existence of special mechanisms which allow first sodium and then potassium to cross the membrane at a high rate when it is depolarized. In addition it is likely that charge can be carried through the membrane by other means. Steinbach's (1941) experiments suggest that chloride ions can cross the membrane and there is probably a small leakage of sodium, potassium and choline through cut branches

TABLE 5. Tentative values of leakage conductance and 'equilibrium' potential for leakage current. Five nerves in choline sea water at 6–22° C.

	Average	Range
Leakage conductance (g_l) (m.mho/cm.²)	0·26	0·13 to 0·50
Equilibrium potential for leakage current (V_l) (mV.)	−11	−4 to −22
Resting potassium conductance $(g_K)_r$ (m.mho/cm.²)	0·23	0·12 to 0·39
Equilibrium potential for potassium (V_K)	+10	+7 to +13

or through parts of the membrane other than those concerned with the selective system. All these minor currents may be thought of as contributing towards a leakage current (I_l) which has a conductance (g_l) and an apparent equilibrium potential (V_l) at which I_l is zero. In this leakage current we should probably also include ions transferred by metabolism against concentration gradients. So many processes may contribute towards a leakage current that measurement of its properties is unlikely to give useful information about the nature of the charged particles on which it depends. Nevertheless, a knowledge of the approximate magnitude of g_l and V_l is important since it is needed for any calculation of threshold or electrical stability. Various methods of measurement were tried but only the simplest will be considered since the orders of magnitude of g_l and V_l were unaffected by the precise method employed. In the experiment of Fig. 11 the steady current needed to maintain the membrane at the potassium potential (+13 mV.) was 8 μA./cm.². According to our definitions this inward current must have been almost entirely leakage current, for the nerve was in choline sea water and $I_K = 0$ when $V = V_K$. Hence

$$(13\text{ mV.} - V_l)\, g_l = 8\ \mu\text{A./cm.}^2.$$

In order to estimate g_l we make use of the fact that the inward current associated with $V = +84$ mV. was not appreciably affected by a fourfold change of potassium concentration (from 5 to 20 mM.). We therefore assume that the potassium conductance was reduced to a negligible value at this membrane potential and that the inward current of 24 μA./cm.² was entirely leakage current. Hence

$$(84\text{ mV.} - V_l)\, g_l = 24\ \mu\text{A./cm.}^2.$$

From these two equations we find a value of −22 mV. for V_l and one of 0·23 m.mho/cm.² for g_l. An estimate of the resting value of g_K may also be obtained by this method. At the resting potential in choline sea water

$$V_l g_l + V_K (g_K)_r = 0.$$

Hence

$$(g_K)_r = 0{\cdot}39\text{ m.mho/cm.}^2.$$

Tentative values obtained by this type of method are given in Table 5.

DISCUSSION

At this stage all that will be attempted by way of a discussion is a brief comparison of the processes underlying the changes in sodium and potassium conductance. The main points of resemblance are: (1) both sodium and potassium conductances rise along an inflected curve when the membrane is depolarized and fall without any appreciable inflexion when the membrane is repolarized; (2) the rate of rise of conductance increases continuously as the membrane potential is reduced whereas the rate of fall associated with repolarization increases continuously as the membrane potential is raised; (3) the rates at which the conductances rise or fall have high temperature coefficients whereas the absolute values attained depend only slightly on temperature; (4) the instantaneous relation between sodium or potassium current and membrane potential normally consists of a straight line with zero current at the sodium or potassium potential.

The main differences are: (1) the rise and fall of sodium conductance occurs 10–30 times faster than the corresponding rates for potassium; (2) the variation of peak conductance with membrane potential is greater for sodium than for potassium; (3) if the axon is held in the depolarized condition the potassium conductance is maintained but the sodium conductance declines to a low level after reaching its peak.

SUMMARY

1. Repolarization of the giant axon of *Loligo* during the period of high sodium permeability is associated with a large inward current which declines rapidly along an approximately exponential curve.

2. The 'tail' of inward current disappears if sodium ions are removed from the external medium.

3. These results are explained quantitatively by supposing that the sodium conductance is a continuous function of time which rises when the membrane is depolarized and falls when it is repolarized.

4. For nerves in sea water the instantaneous relation between sodium current and membrane potential is a straight line passing through zero current about 110 mV. below the resting potential.

5. The rate at which sodium conductance is reduced when the fibre is repolarized increases markedly with membrane potential.

6. The time course of the sodium conductance during a voltage clamp can be calculated from the variation of the 'tail' of inward current with the duration of depolarization. The curves obtained by this method agree with those described in previous paper.

7. Repolarization of the membrane during the period of high potassium permeability is associated with a 'tail' of current which is outward at the

resting potential and inward above a critical potential about 10–20 mV. above the resting potential.

8. The instantaneous relation between potassium current and membrane potential is a straight line passing through zero at 10–20 mV. above the resting potential.

9. These results suggest that the potassium conductance is a continuous function of time which rises when the nerve is depolarized and falls when it is repolarized.

10. The rate at which the potassium conductance is reduced on repolarization increases with membrane potential.

11. The critical potential at which the 'potassium current' appears to reverse in sign varies with external potassium concentration but less steeply than the theoretical potential of a potassium electrode.

REFERENCES

Hodgkin, A. L. (1951). The ionic basis of electrical activity in nerve and muscle. *Biol. Rev.* **26**, 339–409.

Hodgkin, A. L. & Huxley, A. F. (1952). Currents carried by sodium and potassium ions through the membrane of the giant axon of *Loligo*. *J. Physiol.* **116**, 449–272.

Hodgkin, A. L., Huxley, A. F. & Katz, B. (1952). Measurement of current-voltage relations in the membrane of the giant axon of *Loligo*. *J. Physiol.* **116**, 424–448.

Hodgkin, A. L. & Katz, B. (1949). The effect of temperature on the electrical activity of the giant axon of the squid. *J. Physiol.* **109**, 240–249.

Steinbach, H. B. (1941). Chloride in the giant axons of the squid. *J. cell. comp. Physiol.* **17**, 57–64.

Steinbach, H. B. & Spiegelman, S. (1943). The sodium and potassium balance in squid nerve axoplasm. *J. cell. comp. Physiol.* **22**, 187–196.

9

J. Physiol. (1952) 116, 497–506

THE DUAL EFFECT OF MEMBRANE POTENTIAL ON SODIUM CONDUCTANCE IN THE GIANT AXON OF *LOLIGO*

BY A. L. HODGKIN AND A. F. HUXLEY

From the Laboratory of the Marine Biological Association, Plymouth, and the Physiological Laboratory, University of Cambridge

(*Received* 24 *October* 1951)

This paper contains a further account of the electrical properties of the giant axon of *Loligo*. It deals with the 'inactivation' process which gradually reduces sodium permeability after it has undergone the initial rise associated with depolarization. Experiments described previously (Hodgkin & Huxley, 1952*a*, *b*) show that the sodium conductance always declines from its initial maximum, but they leave a number of important points unresolved. Thus they give no information about the rate at which repolarization restores the ability of the membrane to respond with its characteristic increase of sodium conductance. Nor do they provide much quantitative evidence about the influence of membrane potential on the process responsible for inactivation. These are the main problems with which this paper is concerned. The experimental method needs no special description, since it was essentially the same as that used previously (Hodgkin, Huxley & Katz, 1952; Hodgkin & Huxley, 1952*b*).

RESULTS

The influence of a small change in membrane potential on the ability of the membrane to undergo its increase in sodium permeability is illustrated by Fig. 1. In this experiment the membrane potential was changed in two steps. The amplitude of the first step was -8 mV. and its duration varied between 0 and 50 msec. This step will be called the conditioning voltage (V_1). It was followed by a second step called the test voltage (V_2) which was kept at a constant amplitude of -44 mV.

Record *A* gives the current observed with the test voltage alone. *B*–*F* show the effect of preceding this by a conditioning pulse of varying duration. Although the depolarization of 8 mV. was not associated with any appreciable inward current it greatly altered the subsequent response of the nerve. Thus, if the conditioning voltage lasted longer than 20 msec., it reduced the inward current during the test pulse by about 40%. At intermediate durations the inward current decreased along a smooth exponential curve with a time constant of about 7 msec. The outward current, on the other hand, evidently behaved in a different manner; for it may be seen to approach a final level which was independent of the duration of the conditioning step. This is consistent with the observation that depolarization is associated with a maintained increase in potassium conductance (Hodgkin & Huxley, 1952*a*).

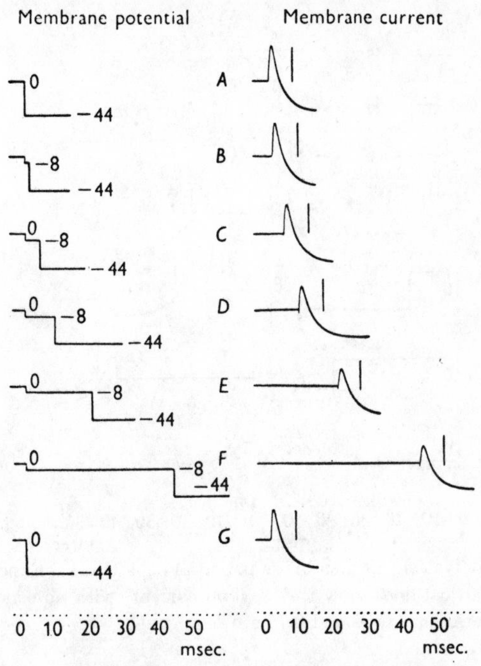

Fig. 1. Development of 'inactivation' during constant depolarization of 8 mV. Left-hand column: time course of membrane potential (the numbers show the displacement of the membrane potential from its resting value in mV.). Right-hand column: time course of membrane current density. Inward current is shown as an upward deflexion. (The vertical lines show the 'sodium current' expected in the absence of a conditioning step; they vary between 1·03 mA./cm.^{2} in *A* and 0·87 mA./cm.^{2} in *G*). Axon 38; compensated feed-back; temperature 5° C.

Fig. 2 illustrates the converse process of raising the membrane potential before applying the test pulse. In this case the conditioning voltage improved the state of the nerve for the inward current increased by about 70% if the first step lasted longer than 15 msec. This finding is not altogether surprising, for the resting potential of isolated squid axons is less than that of other excitable cells (Hodgkin, 1951) and is probably lower than that in the living animal.

A convenient way of expressing these results is to plot the amplitude of the sodium current during the test pulse against the duration of the conditioning pulse. For this purpose we used the simple method of measurement illustrated by Fig. 3 (inset). This procedure avoids the error introduced by variations of potassium conductance during the first step and should give reasonable results for $V > -15$ mV. With larger depolarizations both the method of measurement and the interpretation of the results become somewhat doubtful, since there may be appreciable sodium current during the conditioning period. Two

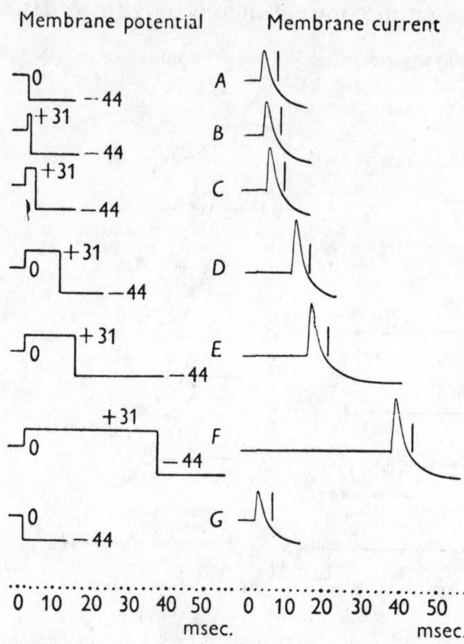

Fig. 2. Removal of 'inactivation' at membrane potential of +31 mV. Experimental details are as in Fig. 1. The vertical lines show the 'sodium current' with no conditioning step; they vary between 0·82 mA./cm.² in *A* and 0·75 mA./cm.² in *G*.

of the curves in Fig. 3 were obtained from the families of records illustrated in Figs. 1 and 2. The other two were determined from similar families obtained on the same axon. All four curves show that inactivation developed or was removed in an approximately exponential manner with a time constant which varied with membrane potential and had a maximum near $V=0$. They also indicate that inactivation tended to a definite steady level at any particular membrane potential. Values of the exponential time-constant (τ_h) of the inactivation process are given in Table 1.

The influence of membrane potential on the steady level of inactivation is illustrated by the records in Fig. 4. In this experiment the conditioning step lasted long enough to allow inactivation to attain its final level at all voltages. Its amplitude was varied between +46 and −29 mV., while that of the test step was again kept constant at −44 mV. The effect of a small progressive change was allowed for in calculating the vertical lines on each record. These give the inward current expected in the absence of a conditioning step and

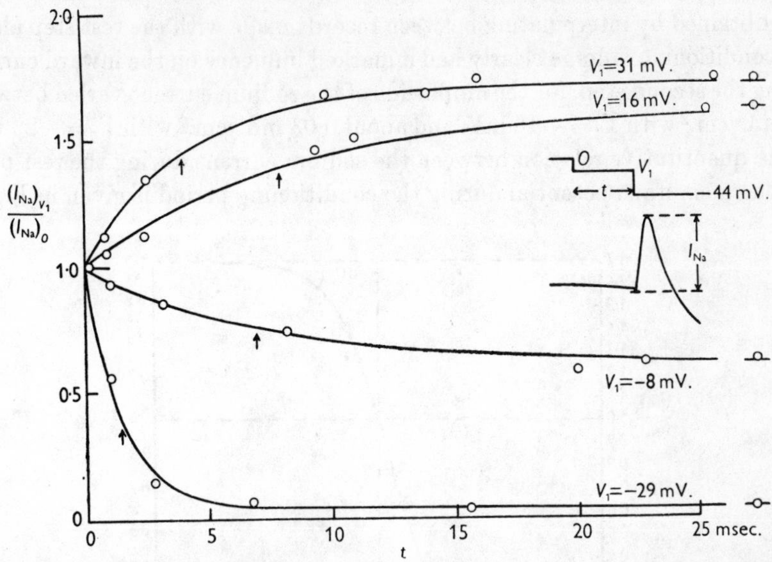

Fig. 3. Time course of inactivation at four different membrane potentials. Abscissa: duration of conditioning step. Ordinate: circles, sodium current (measured as inset) relative to normal sodium current; smooth curve, $y = y_\infty - (y_\infty - 1)\exp(-t/\tau_h)$, where y_∞ is the ordinate at $t = \infty$ and τ_h is the time constant (shown by arrows). Experimental details as in Figs. 1 and 2.

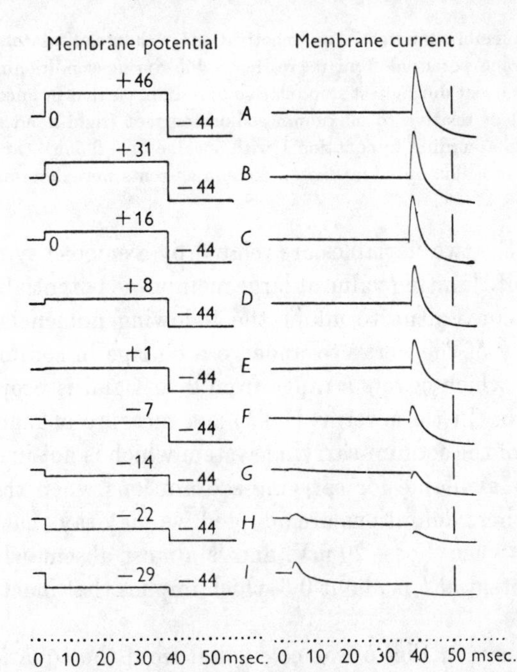

Fig. 4. Influence of membrane potential on 'inactivation' in the steady state. Experimental details are as in Fig. 1. The vertical lines show the sodium current with no conditioning step; they vary between 0·74 mA./cm.² in A and 0·70 mA./cm.² in I.

were obtained by interpolating between records made with the test step alone. The conditioning voltage clearly had a marked influence on the inward current during the second step, for the amplitude of the sodium current varied between 1·3 mA./cm.² with $V_1 = +46$ mV. and about 0·03 mA./cm.² with $V_1 = -29$ mV.

The quantitative relation between the sodium current during the test pulse and the membrane potential during the conditioning period is given in Fig. 5.

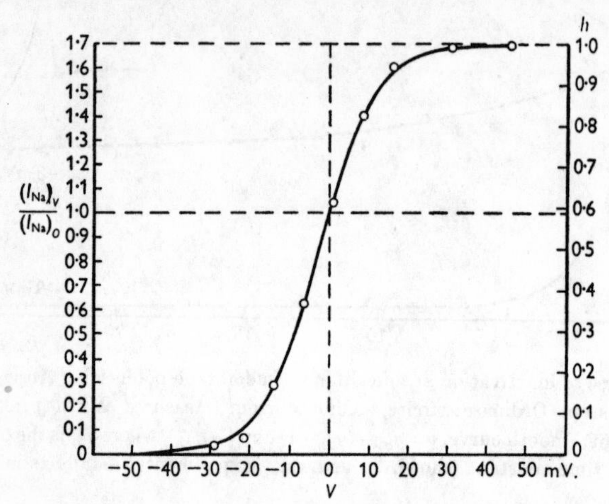

Fig. 5. Influence of membrane potential on 'inactivation' in the steady state. Abscissa: displacement of membrane potential from its resting value during conditioning step. Ordinate: circles, sodium current during test step relative to sodium current in unconditioned test step (left-hand scale) or relative to maximum sodium current (right-hand scale). The smooth curve was drawn according to equation 1 with a value of $-2·5$ mV. for V_h. This graph is based on the records illustrated in Fig. 4. Sodium currents were determined in the manner shown in Fig. 3

This shows that the two variables are related by a smooth symmetrical curve which has a definite limiting value at large membrane potentials. In discussing this curve it is convenient to adopt the following nomenclature. We shall denote the ability of the nerve to undergo a change in sodium permeability by a variable, h, which covers a range from 0 to 1 and is proportional to the ordinate in Fig. 5. In these terms $(1-h)$ is a measure of inactivation, while h is the fraction of the sodium-carrying system which is not inactivated and is therefore rapidly available for carrying sodium ions when the membrane is depolarized. If these definitions are adopted we may say that inactivation is almost complete when $V < -20$ mV. and is almost absent when $V > 30$ mV. At the resting potential h is about 0·6 which implies that inactivation is 40% complete.

The smooth curve in Fig. 5 was calculated from the equation

$$(h)_{\text{steady state}} = \frac{1}{1+\exp(V_h - V)/7}, \tag{1}$$

where V is expressed in millivolts and V_h is the value of V at which $h=\frac{1}{2}$ in the steady state. The same equation gave a satisfactory fit in all experiments but there was some variation in the value of V_h. Five experiments with three fresh fibres gave resting values of h between 0·55 and 0·62. In these cases V_h varied between $-1·5$ and $-3·5$ mV. On the other hand, two experiments with a fibre which had been used for some time gave a resting h of only about 0·25; V_h was then $+7·5$ mV. Since the resting potential was found to decline by 10–15 mV. during the course of a long experiment it is reasonable to suppose that the change in V_h arose solely from this cause and that the relation between h and the absolute membrane potential was independent of the condition of the fibre.

In a former paper we examined the relation between the concentration of sodium ions in the external medium and the sodium current through the membrane (Hodgkin & Huxley, 1952*a*). The results were reasonably close to those predicted by the 'independence principle' except that the currents were 20–60% too large in the sodium-deficient solutions. This effect was attributed to the small increase in resting potential associated with the substitution of choline ions for sodium ions. This explanation now seems very reasonable. The resting potential probably increased by about 4 mV. in choline sea water and this would raise h from 0·6 to 0·73 in a fresh fibre and from 0·25 to 0·37 in a fibre which had been used for some time.

The quantitative results obtained in this series of experiments are summarized in Table 1. Most of the experiments were made at 3–7° C. but a temperature of 19° C. was used on one occasion. The results suggest that temperature has little effect on the equilibrium relation between h and V, but greatly alters the rate at which this equilibrium is attained. The Q_{10} of the rate constants cannot be stated with certainty but is clearly of the order of 3.

Two-pulse experiment

This section deals with a single experiment which gave an independent measurement of the time constant of inactivation.

Two pulses of amplitude -44 mV. and duration 1·8 msec. were applied to the membrane. Fig. 6*A* is a record obtained with the second pulse alone. The ionic current was inward and reached a maximum of about 0·25 mA./cm.^{2}. As in all other records, the inward current was not maintained but declined as a result of inactivation. Restoration of the normal membrane potential was associated with a tail of inward current due to the rapid fall of sodium conductance (see Hodgkin & Huxley, 1952*b*). When two pulses were applied in quick succession the effect of the first was similar to that in *A*, but the inward current during the second was reduced to about one half (record *B*). A gradual recovery to the normal level is shown in records *C–G*.

The curve in Fig. 7 was obtained by estimating sodium current in the manner

TABLE 1. Experiments with conditioning voltage

Axon	Temperature (° C.)	Variable	Displacement of membrane potential (mV.)									
			−29	−22	−14	−8	−7	0	9	16	31	46
38	5	h* (steady state)	0·02	0·04	0·17	—	0·37	0·59	0·82	0·94	0·99	1·00
39	19		0·02	0·04	0·09	—	0·28	0·55	0·83	0·94	0·98	0·99
39†	3		0·01	0·03	0·04	—	0·11	0·26	0·50	0·69	0·93	0·99
38	5	h‡ (steady state)	0·02	—	—	0·43	—	0·58	—	0·92	0·99	—
39	19		0·03	—	—	0·40	—	0·61	—	0·94	—	—
39†	3		—	—	—	—	—	0·22	—	0·75	0·93	—
37	3		—	0·04	—	0·34	—	0·55	0·81	0·96	—	—
38	5	τ_h‡ (msec.)	1·5	—	—	7	—	[8–10]	—	8	4	—
39	19		0·35	—	—	1·5	—	[1·7–2·1]	—	1·8	—	—
39†	3		—	—	—	—	—	—	—	13	7	—
37	3		—	3	—	6	—	[8–10]	9	7	—	—
38	6	τ_h§ (msec.)	1·3	—	—	6	—	[7–9]	—	7	3·6	—
39	6		1·5	—	—	6	—	[7–9]	—	8	—	—
39†	6		—	—	—	—	—	—	—	9	5	—
37	6		—	2·2	—	4	—	[6–7]	7	5	—	—

Two-pulse experiment

Axon 31 at 4·5° C.	$\tau_h = 1{\cdot}8$ msec. at $V = -44$ mV.	$\tau_h = 12$ msec. at $V = 0$
Axon 31 at 6° C.	τ_h§ $= 1{\cdot}5$ msec. at $V = -44$ mV.	τ_h§ $= 10$ msec. at $V = 0$

* Measurements made by methods illustrated in Figs. 4 and 5.
† The axon had been used for some time and was in poor condition when these measurements were made.
‡ Methods illustrated in Figs. 1–3.
§ Calculated from above assuming Q_{10} of 3.
[ ] Interpolated.
h is the fraction of the sodium system which is rapidly available, $(1-h)$ is the fraction inactivated.
τ_h determines the rate at which h approaches its steady state.

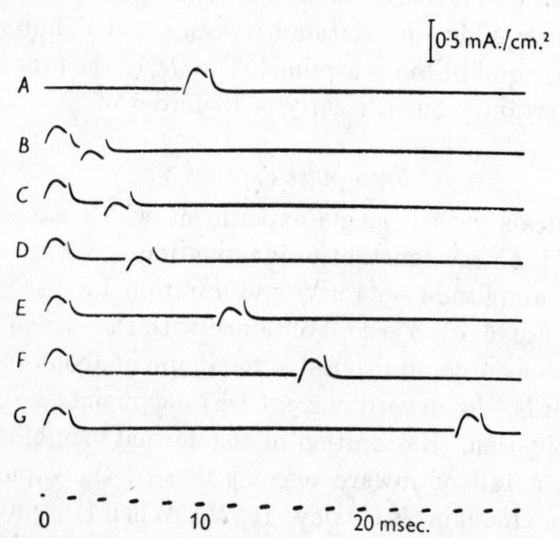

Fig. 6. Membrane currents associated with two square waves applied in succession. The amplitude of each square wave was −44 mV. and the duration 1·8 msec. Record *A* shows the second square wave alone, *B–G* both square waves at various intervals. Axon 31; uncompensated feed-back; temperature 4·5° C.

shown in Fig. 3 and plotting this against the interval between the two pulses. It will be seen that recovery from inactivation took place in an approximately exponential manner with a time constant of about 12 msec. A similar curve and a similar time constant were obtained by plotting

$$\left(\frac{\mathrm{d}I_{\mathrm{Na}}}{\mathrm{d}t}\right)_{\mathrm{max.}} \quad \text{instead of} \quad (I_{\mathrm{Na}})_{\mathrm{max.}}.$$

This time constant is clearly of the same order as that given by the method using weak conditioning voltages (see Table 1). An estimate of the inactivation time constant at -44 mV. may be obtained by extrapolating the curve in Fig. 7 to zero time. This indicates that the available fraction of the sodium-carrying system was reduced to 0·37 at the end of a pulse of amplitude -44 mV. and duration 1·8 msec. Hence the inactivation time constant at -44 mV. is about 1·8 msec., which is of the same order as the values obtained with large depolarizations by the first method (Table 1). It is also in satisfactory agreement with the time constant obtained by fitting a curve to the variation of sodium conductance during a maintained depolarization of 40–50 mV. (Hodgkin & Huxley, 1952*c*).

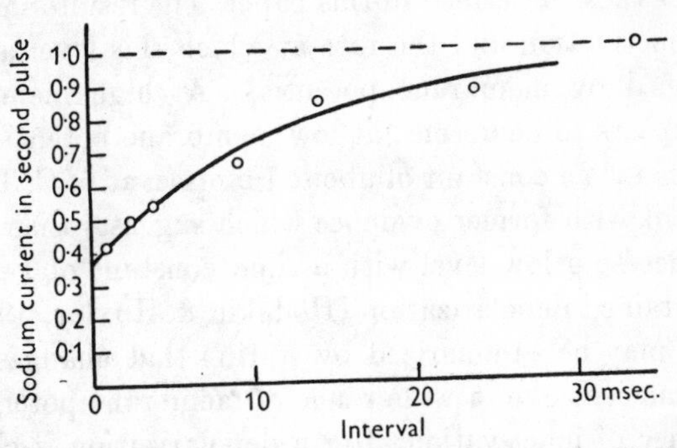

Fig. 7. Recovery from inactivation. Abscissa: interval between end of first pulse and beginning of second pulse. Ordinate: sodium current in second pulse measured as shown in Fig. 3 and expressed as a fraction of the sodium current in an unconditioned pulse. The circles are experimental points derived from the records in Fig. 6. The smooth curve is drawn according to the expression $1-0{\cdot}63 \exp(-t/\tau_h)$, where $\tau_h = 12$ msec.

The two-pulse experiment is interesting because it emphasizes the difference between the rapid fall of sodium conductance associated with repolarization and the slower decline during a maintained depolarization. Both events lead to a decrease in sodium current, but the underlying mechanisms are clearly different. In the first case it must be supposed that repolarization converts active membrane into resting membrane; in the second that prolonged depolarization turns it into a refractory or inactivated condition from which it recovers at a relatively slow rate when the fibre is repolarized. It cannot be argued that repolarization reduces sodium conductance by making the active membrane refractory. If this were so, one would expect that the inward current during the second pulse would be reduced to zero at short intervals, instead of to 37% as in Fig. 7. The reduction to 37% is clearly associated with the incomplete decline of sodium conductance during the first pulse and not with the rapid and complete decline due to repolarization at the end of the first pulse.

DISCUSSION

The experimental evidence in this paper and in those which precede it (Hodgkin & Huxley, 1952*a*, *b*) suggests that the membrane potential has two distinct influences on the system which allows sodium ions to flow through the membrane. The early effects of changes in membrane potential are a rapid increase in sodium conductance when the fibre is depolarized and a rapid decrease when it is repolarized. The late effects are a slow onset of a refractory or inactive condition during a maintained depolarization and a slow recovery following repolarization. A membrane in the refractory or inactive condition resembles one in the resting state in having a low sodium conductance. It differs in that it cannot undergo an increase in sodium conductance if the fibre is depolarized. The difference allows inactivation to be measured by methods such as those described in this paper. The results show that both the final level of inactivation and the rate at which this level is approached are greatly influenced by membrane potential. At high membrane potentials inactivation appears to be absent, at low membrane potentials it approaches completion with a time constant of about 1·5 msec. at 6° C. This conclusion is clearly consistent with former evidence which suggests that the sodium conductance declines to a low level with a time constant of 1–2 msec. during a large and maintained depolarization (Hodgkin & Huxley, 1952*a*). Both sets of experiments may be summarized by stating that changes in sodium conductance are transient over a wide range of membrane potentials.

The persistence of inactivation after a depolarization is clearly connected with the existence of a refractory state and with accommodation. It is not the only factor concerned, since the persistence of the raised potassium conductance will also help to hold the membrane potential at a positive value and will therefore tend to make the fibre inexcitable. The relative importance of the two processes can only be judged by numerical analysis of the type described in the final paper of this series (Hodgkin & Huxley, 1952*c*).

SUMMARY

1. Small changes in the membrane potential of the giant axon of *Loligo* are associated with large alterations in the ability of the surface membrane to undergo its characteristic increase in sodium conductance.

2. A steady depolarization of 10 mV. reduces the sodium current associated with a sudden depolarization of 45 mV. by about 60%. A steady rise of 10 mV.

increases the sodium current associated with subsequent depolarization by about 50%.

3. These effects are described by stating that depolarization gradually inactivates the system which enables sodium ions to cross the membrane.

4. In the steady state, inactivation appears to be almost complete if the membrane potential is reduced by 30 mV. and is almost absent if it is increased by 30 mV. Between these limits the amount of inactivation is determined by a smooth symmetrical curve and is about 40% complete in a resting fibre at the beginning of an experiment.

5. At 6° C. the time constant of the inactivation process is about 10 msec. with $V=0$, about 1·5 msec. with $V=-30$ mV. and about 5 msec. at $V=+30$ mV.

REFERENCES

HODGKIN, A. L. (1951). The ionic basis of electrical activity in nerve and muscle. *Biol. Rev.* **26**, 339–409.

HODGKIN, A. L. & HUXLEY, A. F. (1952*a*). Currents carried by sodium and potassium ions through the membrane of the giant axon of *Loligo*. *J. Physiol.* **116**, 449–472.

HODGKIN, A. L. & HUXLEY, A. F. (1952*b*). The components of membrane conductance in the giant axon of *Loligo*. *J. Physiol.* **116**, 473–496.

HODGKIN, A. L. & HUXLEY, A. F. (1952*c*). A quantitative description of membrane current and its application to conduction and excitation in nerve. *J. Physiol.* (in the press).

HODGKIN, A. L., HUXLEY, A. F. & KATZ, B. (1952). Measurement of current-voltage relations in the membrane of the giant axon of *Loligo*. *J. Physiol.* **116**, 424–448.

10

J. Physiol. (1952) 117, 500–544

A QUANTITATIVE DESCRIPTION OF MEMBRANE CURRENT AND ITS APPLICATION TO CONDUCTION AND EXCITATION IN NERVE

By A. L. HODGKIN and A. F. HUXLEY

From the Physiological Laboratory, University of Cambridge

(*Received* 10 *March* 1952)

This article concludes a series of papers concerned with the flow of electric current through the surface membrane of a giant nerve fibre (Hodgkin, Huxley & Katz, 1952; Hodgkin & Huxley, 1952 *a–c*). Its general object is to discuss the results of the preceding papers (Part I), to put them into mathematical form (Part II) and to show that they will account for conduction and excitation in quantitative terms (Part III).

PART I. DISCUSSION OF EXPERIMENTAL RESULTS

The results described in the preceding papers suggest that the electrical behaviour of the membrane may be represented by the network shown in Fig. 1. Current can be carried through the membrane either by charging the membrane capacity or by movement of ions through the resistances in parallel with the capacity. The ionic current is divided into components carried by sodium and potassium ions (I_{Na} and I_K), and a small 'leakage current' (I_l) made up by chloride and other ions. Each component of the ionic current is determined by a driving force which may conveniently be measured as an electrical potential difference and a permeability coefficient which has the dimensions of a conductance. Thus the sodium current (I_{Na}) is equal to the sodium conductance (g_{Na}) multiplied by the difference between the membrane potential (E) and the equilibrium potential for the sodium ion (E_{Na}). Similar equations apply to I_K and I_l and are collected on p. 505.

Our experiments suggest that g_{Na} and g_K are functions of time and membrane potential, but that E_{Na}, E_K, E_l, C_M and $\bar{g}_l$ may be taken as constant. The influence of membrane potential on permeability can be summarized by stating: first, that depolarization causes a transient increase in sodium conductance and a slower but maintained increase in potassium conductance; secondly, that these changes are graded and that they can be reversed by repolarizing the membrane. In order to decide whether these effects are sufficient to account for complicated phenomena such as the action potential and refractory period, it is necessary to obtain expressions relating the sodium and potassium conductances to time and membrane potential. Before attempting this we shall consider briefly what types of physical system are likely to be consistent with the observed changes in permeability.

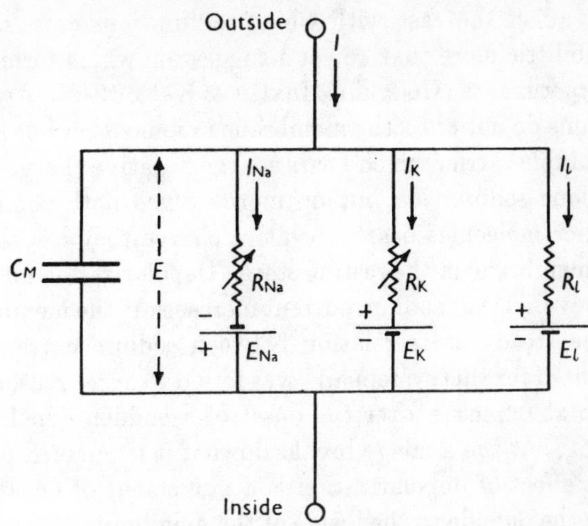

Fig. 1. Electrical circuit representing membrane. $R_{Na} = 1/g_{Na}$; $R_K = 1/g_K$; $R_l = 1/\bar{g}_l$. R_{Na} and R_K vary with time and membrane potential; the other components are constant.

The nature of the permeability changes

At present the thickness and composition of the excitable membrane are unknown. Our experiments are therefore unlikely to give any certain information about the nature of the molecular events underlying changes in permeability. The object of this section is to show that certain types of theory are excluded by our experiments and that others are consistent with them.

The first point which emerges is that the changes in permeability appear to depend on membrane potential and not on membrane current. At a fixed depolarization the sodium current follows a time course whose form is independent of the current through the membrane. If the sodium concentration is such that $E_{Na} < E$, the sodium current is inward; if it is reduced until $E_{Na} > E$ the current changes in sign but still appears to follow the same time course. Further support for the view that membrane potential is the variable controlling permeability is provided by the observation that restoration of the normal membrane potential causes the sodium or potassium conductance to decline to a low value at any stage of the response.

The dependence of g_{Na} and g_K on membrane potential suggests that the permeability changes arise from the effect of the electric field on the distribution or orientation of molecules with a charge or dipole moment. By this we do not mean to exclude chemical reactions, for the rate at which these occur might depend on the position of a charged substrate or catalyst. All that is intended is that small changes in membrane potential would be most unlikely to cause large alterations in the state of a membrane which was composed entirely of electrically neutral molecules.

The next question to consider is how changes in the distribution of a charged particle might affect the ease with which sodium ions cross the membrane. Here we can do little more than reject a suggestion which formed the original basis of our experiments (Hodgkin, Huxley & Katz, 1949). According to this view, sodium ions do not cross the membrane in ionic form but in combination with a lipoid soluble carrier which bears a large negative charge and which can combine with one sodium ion but no more. Since both combined and uncombined carrier molecules bear a negative charge they are attracted to the outside of the membrane in the resting state. Depolarization allows the carrier molecules to move, so that sodium current increases as the membrane potential is reduced. The steady state relation between sodium current and voltage could be calculated for this system and was found to agree reasonably with the observed curve at 0·2 msec after the onset of a sudden depolarization. This was encouraging, but the analogy breaks down if it is pursued further. In the model the first effect of depolarization is a movement of negatively charged molecules from the outside to the inside of the membrane. This gives an initial outward current, and an inward current does not occur until combined carriers lose sodium to the internal solution and return to the outside of the membrane. In our original treatment the initial outward current was reduced to vanishingly small proportions by assuming a low density of carriers and a high rate of movement and combination. Since we now know that the sodium current takes an appreciable time to reach its maximum, it is necessary to suppose that there are more carriers and that they react or move more slowly. This means that any inward current should be preceded by a large outward current. Our experiments show no sign of a component large enough to be consistent with the model. This invalidates the detailed mechanism assumed for the permeability change but it does not exclude the more general possibility that sodium ions cross the membrane in combination with a lipoid soluble carrier.

A different form of hypothesis is to suppose that sodium movement depends on the distribution of charged particles which do not act as carriers in the usual sense, but which allow sodium to pass through the membrane when they occupy particular sites in the membrane. On this view the rate of movement of the activating particles determines the rate at which the sodium conductance approaches its maximum but has little effect on the magnitude of the conductance. It is therefore reasonable to find that temperature has a large effect on the rate of rise of sodium conductance but a relatively small effect on its maximum value. In terms of this hypothesis one might explain the transient nature of the rise in sodium conductance by supposing that the activating particles undergo a chemical change after moving from the position which they occupy when the membrane potential is high. An alternative is to attribute the decline of sodium conductance to the relatively slow movement of another particle which blocks the flow of sodium ions when it reaches a certain position in the membrane.

Much of what has been said about the changes in sodium permeability applies equally to the mechanism underlying the change in potassium permeability. In this case one might suppose that there is a completely separate system which differs from the sodium system in the following respects: (1) the activating molecules have an affinity for potassium but not for sodium; (2) they move more slowly; (3) they are not blocked or inactivated. An alternative hypothesis is that only one system is present but that its selectivity changes soon after the membrane is depolarized. A situation of this kind would arise if inactivation of the particles selective for sodium converted them into particles selective for potassium. However, this hypothesis cannot be applied in a simple form since the potassium conductance rises too slowly for a direct conversion from a state of sodium permeability to one of potassium permeability.

One of the most striking properties of the membrane is the extreme steepness of the relation between ionic conductance and membrane potential. Thus g_{Na} may be increased e-fold by a reduction of only 4 mV, while the corresponding figure for g_K is 5–6 mV (Hodgkin & Huxley, 1952*a*, figs. 9, 10). In order to illustrate the possible meaning of this result we shall suppose that a charged molecule which has some special affinity for sodium may rest either on the inside or the outside of the membrane but is present in negligible concentrations elsewhere. We shall also suppose that the sodium conductance is proportional to the number of such molecules on the inside of the membrane but is independent of the number on the outside. From Boltzmann's principle the proportion P_i of the molecules on the inside of the membrane is related to the proportion on the outside, P_o, by

$$\frac{P_i}{P_o} = \exp[(w + zeE)/kT],$$

where E is the potential difference between the outside and the inside of the membrane, w is the work required to move the molecule from the inside to the outside of the membrane when $E=0$, e is the absolute value of the electronic charge, z is the valency of the molecule (i.e. the number of positive electronic charges on it), k is Boltzmann's constant and T is the absolute temperature. Since we have assumed that $P_i + P_o = 1$ the expression for P_i is

$$P_i = 1 \Bigg/ \left[1 + \exp - \left(\frac{w + zeE}{kT}\right)\right].$$

For negative values of z and with E sufficiently large and positive this gives

$$P_i = \text{constant} \times \exp[zeE/kT].$$

In order to explain our results z must be about -6 since $\frac{kT}{e}\left(=\frac{RT}{F}\right)$ is 25 mV at room temperature and $g_{Na} \propto \exp - E/4$ for E large. This suggests that the particle whose distribution changes must bear six negative electronic charges, or, if a similar theory is developed in terms of the orientation of a long molecule with a dipole moment, it must have at least three negative charges on one end and three positive charges on the other. A different but related approach is to suppose that sodium movement depends on the presence of six singly charged molecules at a particular site near the inside of the membrane. The proportion of the time that each of the charged molecules spends at the inside is determined by $\exp - E/25$ so that the proportion of sites at which all six are at the inside is $\exp - E/4{\cdot}17$. This suggestion may be given plausibility but not mathematical simplicity by imagining that a number of charges form a bridge or chain which allows sodium ions to flow through the membrane when it is depolarized. Details of the mechanism will probably not be settled for some time, but it seems difficult to escape the conclusion that the changes in ionic permeability depend on the movement of some component of the membrane which behaves as though it had a large charge or dipole moment. If such components exist it is necessary to suppose that their density is relatively low and that a number of sodium ions cross the membrane at a single active patch. Unless this were true one would expect the increase in sodium permeability to be accompanied by an outward current comparable in magnitude to the current carried by sodium ions. For movement of any charged particle in the membrane should contribute to the total current and the effect would be particularly marked with a molecule, or aggregate, bearing a large charge. As was mentioned earlier, there is no evidence from our experiments of any current associated with the change in sodium permeability, apart from the contribution of the sodium ion itself. We cannot set a definite upper limit to this hypothetical current, but it could hardly have been more than a few per cent of the maximum sodium current without producing a conspicuous effect at the sodium potential.

PART II. MATHEMATICAL DESCRIPTION OF MEMBRANE CURRENT DURING A VOLTAGE CLAMP

Total membrane current

The first step in our analysis is to divide the total membrane current into a capacity current and an ionic current. Thus

$$I = C_M \frac{dV}{dt} + I_i, \tag{1}$$

where

- I is the total membrane current density (inward current positive);
- I_i is the ionic current density (inward current positive);
- V is the displacement of the membrane potential from its resting value (depolarization negative);
- C_M is the membrane capacity per unit area (assumed constant);
- t is time.

The justification for this equation is that it is the simplest which can be used and that it gives values for the membrane capacity which are independent of the magnitude or sign of V and are little affected by the time course of V (see, for example, table 1 of Hodgkin *et al.* 1952). Evidence that the capacity current and ionic current are in parallel (as suggested by eqn. (1)) is provided by the similarity between ionic currents measured with $\frac{dV}{dt}=0$ and those calculated from $-C_M \frac{dV}{dt}$ with $I=0$ (Hodgkin *et al.* 1952).

The only major reservation which must be made about eqn. (1) is that it takes no account of dielectric loss in the membrane. There is no simple way of estimating the error introduced by this approximation, but it is not thought to be large since the time course of the capacitative surge was reasonably close to that calculated for a perfect condenser (Hodgkin *et al.* 1952).

The ionic current

A further subdivision of the membrane current can be made by splitting the ionic current into components carried by sodium ions (I_{Na}), potassium ions (I_K) and other ions (I_l):

$$I_i = I_{Na} + I_K + I_l. \tag{2}$$

The individual ionic currents

In the third paper of this series (Hodgkin & Huxley, 1952*b*), we showed that the ionic permeability of the membrane could be satisfactorily expressed in terms of ionic conductances (g_{Na}, g_K and $\bar{g}_l$). The individual ionic currents are obtained from these by the relations

$$I_{Na} = g_{Na}(E - E_{Na}),$$
$$I_K = g_K(E - E_K),$$
$$I_l = \bar{g}_l(E - E_l),$$

where E_{Na} and E_K are the equilibrium potentials for the sodium and potassium ions. E_l is the potential at which the 'leakage current' due to chloride and other ions is zero. For practical application it is convenient to write these equations in the form

$$I_{Na} = g_{Na}(V - V_{Na}), \tag{3}$$
$$I_K = g_K(V - V_K), \tag{4}$$
$$I_l = \bar{g}_l(V - V_l), \tag{5}$$

where

$$V = E - E_r,$$
$$V_{Na} = E_{Na} - E_r,$$
$$V_K = E_K - E_r,$$
$$V_l = E_l - E_r,$$

and E_r is the absolute value of the resting potential. V, V_{Na}, V_K and V_l can then be measured directly as displacements from the resting potential.

The ionic conductances

The discussion in Part I shows that there is little hope of calculating the time course of the sodium and potassium conductances from first principles. Our object here is to find equations which describe the conductances with reasonable accuracy and are sufficiently simple for theoretical calculation of the action potential and refractory period. For the sake of illustration we shall try to provide a physical basis for the equations, but must emphasize that the interpretation given is unlikely to provide a correct picture of the membrane.

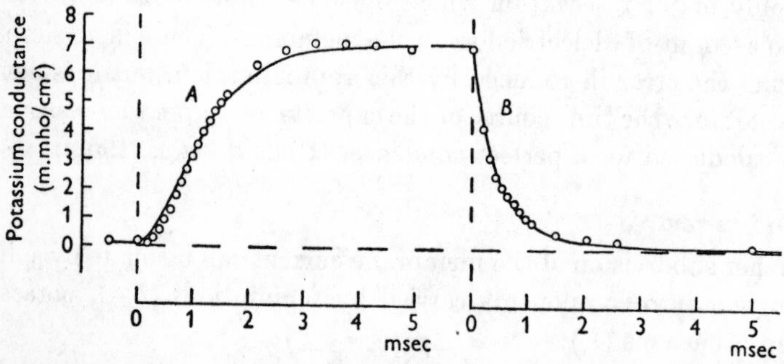

Fig. 2. *A*, rise of potassium conductance associated with depolarization of 25 mV; *B*, fall of potassium conductance associated with repolarization to the resting potential. Circles: experimental points replotted from Hodgkin & Huxley (1952*b*, Fig. 13). The last point of *A* is the same as the first point in *B*. Axon 18, 21° C in choline sea water. The smooth curve is drawn according to eqn. (11) with the following parameters:

	Curve *A* ($V = -25$ mV)	Curve *B* ($V = 0$)
g_{K0}	0·09 m.mho/cm²	7·06 m.mho/cm²
$g_{K\infty}$	7·06 m.mho/cm²	0·09 m.mho/cm²
τ_n	0·75 msec	1·1 msec

At the outset there is the difficulty that both sodium and potassium conductances increase with a delay when the axon is depolarized but fall with no appreciable inflexion when it is repolarized. This is illustrated by the circles in Fig. 2, which shows the change in potassium conductance associated with a depolarization of 25 mV lasting 4·9 msec. If g_K is used as a variable the end of the record can be fitted by a first-order equation but a third- or fourth-order equation is needed to describe the beginning. A useful simplification is achieved by supposing that g_K is proportional to the fourth power of a variable which obeys a first-order equation. In this case the rise of potassium conductance from zero to a finite value is described by $(1-\exp(-t))^4$, while the fall is given by $\exp(-4t)$. The rise in conductance therefore shows a marked inflexion, while the fall is a simple exponential. A similar assumption using a cube instead of a fourth power describes the initial rise of sodium conductance, but a term representing inactivation must be included to cover the behaviour at long times.

The potassium conductance

The formal assumptions used to describe the potassium conductance are:

$$g_K = \bar{g}_K n^4, \tag{6}$$

$$\frac{dn}{dt} = \alpha_n (1-n) - \beta_n n, \tag{7}$$

where $\bar{g}_K$ is a constant with the dimensions of conductance/cm^2, α_n and β_n are rate constants which vary with voltage but not with time and have dimensions of $[\text{time}]^{-1}$, n is a dimensionless variable which can vary between 0 and 1.

These equations may be given a physical basis if we assume that potassium ions can only cross the membrane when four similar particles occupy a certain region of the membrane. n represents the proportion of the particles in a certain position (for example at the inside of the membrane) and $1-n$ represents the proportion that are somewhere else (for example at the outside of the membrane). α_n determines the rate of transfer from outside to inside, while β_n determines the transfer in the opposite direction. If the particle has a negative charge α_n should increase and β_n should decrease when the membrane is depolarized.

Application of these equations will be discussed in terms of the family of curves in Fig. 3. Here the circles are experimental observations of the rise of potassium conductance associated with depolarization, while the smooth curves are theoretical solutions of eqns. (6) and (7).

In the resting state, defined by $V=0$, n has a resting value given by

$$n_0 = \frac{\alpha_{n0}}{\alpha_{n0}+\beta_{n0}}.$$

If V is changed suddenly α_n and β_n instantly take up values appropriate to the new voltage. The solution of (7) which satisfies the boundary condition that $n=n_0$ when $t=0$ is

$$n = n_\infty - (n_\infty - n_0)\exp(-t/\tau_n), \tag{8}$$

where

$$n_\infty = \alpha_n/(\alpha_n+\beta_n), \tag{9}$$

and

$$\tau_n = 1/(\alpha_n+\beta_n). \tag{10}$$

From eqn. (6) this may be transformed into a form suitable for comparison with the experimental results, i.e.

$$g_K = \{(g_{K\infty})^{\frac{1}{4}} - [(g_{K\infty})^{\frac{1}{4}} - (g_{K0})^{\frac{1}{4}}] \exp(-t/\tau_n)\}^4, \tag{11}$$

where $g_{K\infty}$ is the value which the conductance finally attains and g_{K0} is the conductance at $t=0$. The smooth curves in Fig. 3 were calculated from

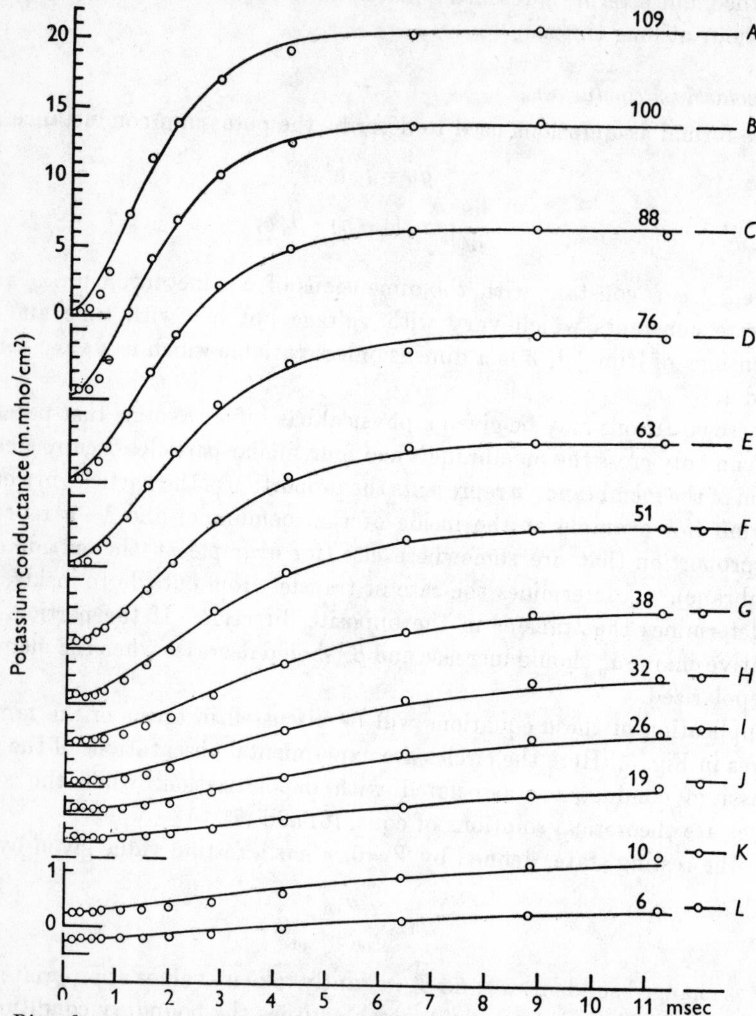

Fig. 3. Rise of potassium conductance associated with different depolarizations. The circles are experimental points obtained on axon 17, temperature 6–7° C, using observations in sea water and choline sea water (see Hodgkin & Huxley, 1952*a*). The smooth curves were drawn from eqn. (11) with $g_{K0} = 0{\cdot}24$ m.mho/cm² and other parameters as shown in Table 1. The time scale applies to all records. The ordinate scale is the same in the upper ten curves (*A* to *J*) and is increased fourfold in the lower two curves (*K* and *L*). The number on each curve gives the depolarization in mV.

eqn. (11) with a value of τ_n chosen to give the best fit. It will be seen that there is reasonable agreement between theoretical and experimental curves, except that the latter show more initial delay. Better agreement might have been obtained with a fifth or sixth power, but the improvement was not considered to be worth the additional complication.

The rate constants α_n *and* β_n. At large depolarizations $g_{K\infty}$ seems to approach an asymptote about 20–50% greater than the conductance at −100 mV.

TABLE 1. Analysis of curves in Fig. 3

Curve	V (mV) (1)	$g_{K\infty}$ (m.mho/cm²) (2)	n_∞ (3)	τ_n (msec) (4)	α_n (msec⁻¹) (5)	β_n (msec⁻¹) (6)
—	(−∞)	(24·31)	(1·000)		—	—
A	−109	20·70	0·961	1·05	0·915	0·037
B	−100	20·00	0·953	1·10	0·866	0·043
C	−88	18·60	0·935	1·25	0·748	0·052
D	−76	17·00	0·915	1·50	0·610	0·057
E	−63	15·30	0·891	1·70	0·524	0·064
F	−51	13·27	0·859	2·05	0·419	0·069
G	−38	10·29	0·806	2·60	0·310	0·075
H	−32	8·62	0·772	3·20	0·241	0·071
I	−26	6·84	0·728	3·80	0·192	0·072
J	−19	5·00	0·674	4·50	0·150	0·072
K	−10	1·47	0·496	5·25	0·095	0·096
L	−6	0·98	0·448	5·25	0·085	0·105
—	(0)	(0·24)	(0·315)	—	—	—

Col. 1 shows depolarization in mV; col. 2, final potassium conductance; col. 3, $n_\infty = (g_{K\infty}/\bar{g}_K)^{\frac{1}{4}}$; col. 4, time constant used to compute curve; col. 5, $\alpha_n = n_\infty/\tau_n$; col. 6, $\beta_n = (1-n_\infty)/\tau_n$. The figure of 24·31 was chosen for $\bar{g}_K$ because it made the asymptotic value of n_∞ 5% greater than the value at −100 mV.

For the purpose of calculation we assume that $n=1$ at the asymptote which is taken as about 20% greater than the value of $g_{K\infty}$ at $V=-100$ mV. These assumptions are somewhat arbitrary, but should introduce little error since we are not concerned with the behaviour of g_K at depolarizations greater than about 110 mV. In the experiment illustrated by Fig. 3, $g_{K\infty}=20$ m.mho/cm² at $V=-100$ mV. $\bar{g}_K$ was therefore chosen to be near 24 m.mho/cm². This value was used to calculate n_∞ at various voltages by means of eqn. (6). α_n and β_n could then be obtained from the following relations which are derived from eqns. (9) and (10):

$$\alpha_n = n_\infty/\tau_n,$$

$$\beta_n = (1-n_\infty)/\tau_n.$$

The results of analysing the curves in Fig. 3 by this method are shown in Table 1.

An estimate of the resting values of α_n and β_n could be obtained from the decline in potassium conductance associated with repolarization. The procedure was essentially the same but the results were approximate because the resting value of the potassium conductance was not known with any accuracy when the membrane potential was high. Fig. 2 illustrates an experiment in which the membrane potential was restored to its resting value after a depolarization of 25 mV. It will be seen that both the rise and fall of the potassium conductance agree reasonably with theoretical curves calculated from eqn. (11) after an appropriate choice of parameters. The rate constants derived from these parameters were (in msec^{-1}): $\alpha_n = 0{\cdot}21$, $\beta_n = 0{\cdot}70$ when $V = 0$ and $\alpha_n = 0{\cdot}90$, $\beta_n = 0{\cdot}43$ when $V = -25$ mV.

In order to find functions connecting α_n and β_n with membrane potential we collected all our measurements and plotted them against V, as in Fig. 4. Differences in temperature were allowed for by adopting a temperature coefficient of 3 (Hodgkin *et al.* 1952) and scaling to 6° C. The effect of replacing sodium by choline on the resting potential was taken into account by displacing the origin for values in choline sea water by +4 mV. The continuous curves, which are clearly a good fit to the experimental data, were calculated from the following expressions:

$$\alpha_n = 0{\cdot}01\,(V+10) \Big/ \left[\exp\frac{V+10}{10} - 1\right], \tag{12}$$

$$\beta_n = 0{\cdot}125 \exp(V/80), \tag{13}$$

where α_n and β_n are given in reciprocal msec and V is the displacement of the membrane potential from its resting value in mV.

These expressions should also give a satisfactory formula for the steady potassium conductance ($g_{K\infty}$) at any membrane potential (V), for this relation is implicit in the measurement of α_n and β_n. This is illustrated by Fig. 5, in which the abscissa is the membrane potential and the ordinate is $(g_{K\infty}/\bar{g}_K)^{\frac{1}{4}}$. The smooth curve was calculated from eqn. (9) with α_n and β_n substituted from eqns. (12) and (13).

Fig. 4 shows that β_n is small compared to α_n over most of the range; we therefore do not attach much weight to the curve relating β_n to V and have used the simplest expression which gave a reasonable fit. The function for α_n was chosen for two reasons. First, it is one of the simplest which fits the experimental results and, secondly, it bears a close resemblance to the equation derived by Goldman (1943) for the movements of a charged particle in a constant field. Our equations can therefore be given a qualitative physical basis if it is supposed that the variation of α and β with membrane potential arises from the effect of the electric field on the movement of a negatively charged particle which rests on the outside of the membrane when V is large and positive, and on the inside when it is large and negative. The analogy cannot be pressed since α and β are not symmetrical about $E = 0$, as they should be if Goldman's theory held in a simple form. Better agreement might

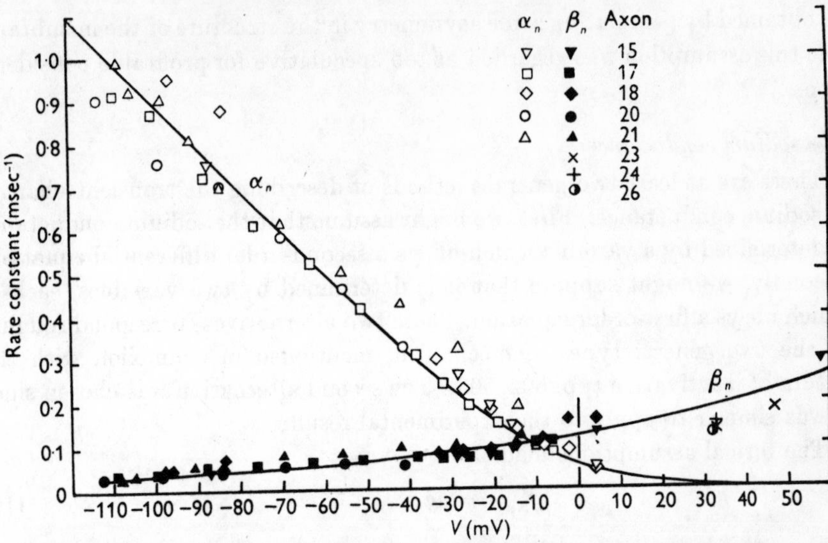

Fig. 4. Abscissa: membrane potential minus resting potential in sea water. Ordinate: rate constants determining rise (α_n) or fall (β_n) of potassium conductance at 6° C. The resting potential was assumed to be 4 mV higher in choline sea water than in ordinary sea water. Temperature differences were allowed for by assuming a Q_{10} of 3. All values for $V < 0$ were obtained by the method illustrated by Fig. 3 and Table 1; those for $V > 0$ were obtained from the decline of potassium conductance associated with an increase of membrane potential or from repolarization to the resting potential in choline sea water (e.g. Fig. 2). Axons 17–21 at 6–11° C, the remainder at about 20° C. The smooth curves were drawn from eqns. (12) and (13).

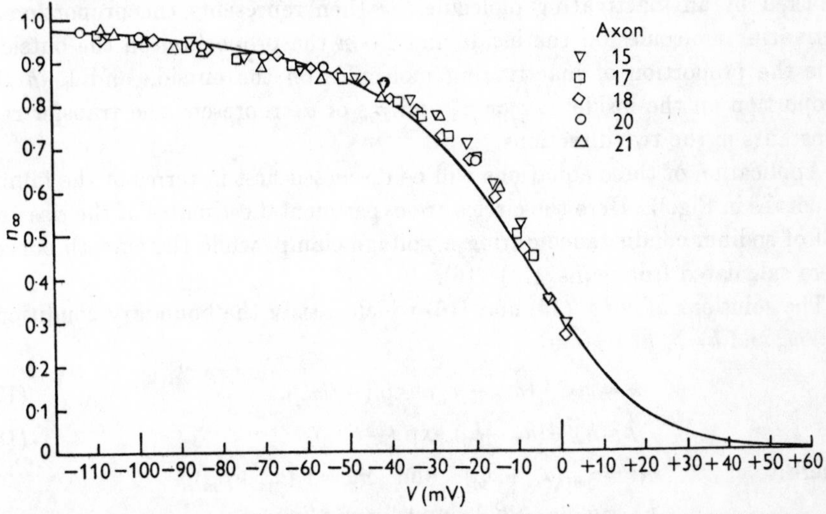

Fig. 5. Abscissa: membrane potential minus resting potential in sea water. Ordinate: experimental measurements of n_∞ calculated from the steady potassium conductance by the relation $n_\infty = \sqrt[4]{(g_{K\infty}/\bar{g}_K)}$, where $\bar{g}_K$ is the 'maximum' potassium conductance. The smooth curve is drawn according to eqn. (9).

be obtained by postulating some asymmetry in the structure of the membrane, but this assumption was regarded as too speculative for profitable consideration.

The sodium conductance

There are at least two general methods of describing the transient changes in sodium conductance. First, we might assume that the sodium conductance is determined by a variable which obeys a second-order differential equation. Secondly, we might suppose that it is determined by two variables, each of which obeys a first-order equation. These two alternatives correspond roughly to the two general types of mechanism mentioned in connexion with the nature of inactivation (pp. 502–503). The second alternative was chosen since it was simpler to apply to the experimental results.

The formal assumptions made are:

$$g_{Na}=m^3h\bar{g}_{Na}, \tag{14}$$

$$\frac{dm}{dt}=\alpha_m(1-m)-\beta_m m, \tag{15}$$

$$\frac{dh}{dt}=\alpha_h(1-h)-\beta_h h, \tag{16}$$

where $\bar{g}_{Na}$ is a constant and the α's and β's are functions of V but not of t.

These equations may be given a physical basis if sodium conductance is assumed to be proportional to the number of sites on the inside of the membrane which are occupied simultaneously by three activating molecules but are not blocked by an inactivating molecule. m then represents the proportion of activating molecules on the inside and $1-m$ the proportion on the outside; h is the proportion of inactivating molecules on the outside and $1-h$ the proportion on the inside. α_m or β_h and β_m or α_h represent the transfer rate constants in the two directions.

Application of these equations will be discussed first in terms of the family of curves in Fig. 6. Here the circles are experimental estimates of the rise and fall of sodium conductance during a voltage clamp, while the smooth curves were calculated from eqns. (14)–(16).

The solutions of eqns. (15) and (16) which satisfy the boundary conditions $m=m_0$ and $h=h_0$ at $t=0$ are

$$m=m_\infty-(m_\infty-m_0)\exp(-t/\tau_m), \tag{17}$$

$$h=h_\infty-(h_\infty-h_0)\exp(-t/\tau_h), \tag{18}$$

where

$$m_\infty=\alpha_m/(\alpha_m+\beta_m) \quad \text{and} \quad \tau_m=1/(\alpha_m+\beta_m),$$

$$h_\infty=\alpha_h/(\alpha_h+\beta_h) \quad \text{and} \quad \tau_h=1/(\alpha_h+\beta_h).$$

In the resting state the sodium conductance is very small compared with the value attained during a large depolarization. We therefore neglect m_0 if the depolarization is greater than 30 mV. Further, inactivation is very nearly complete if $V < -30$ mV so that h_∞ may also be neglected. The expression for the sodium conductance then becomes

$$g_{Na} = g'_{Na} \left[1 - \exp\left(-t/\tau_m\right)\right]^3 \exp\left(-t/\tau_h\right), \tag{19}$$

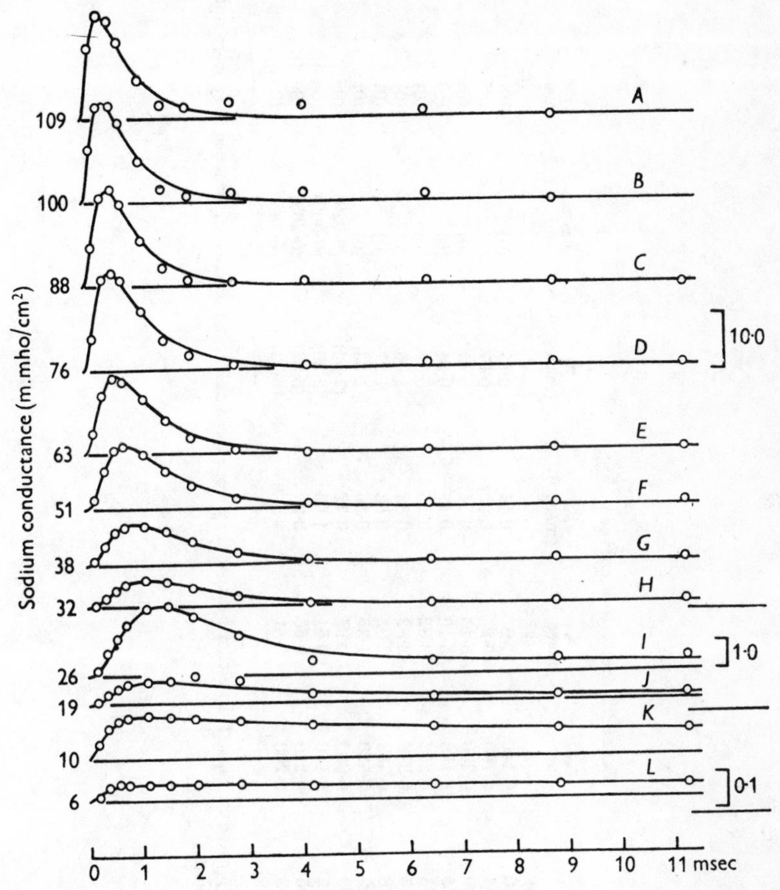

Fig. 6. Changes of sodium conductance associated with different depolarizations. The circles are experimental estimates of sodium conductance obtained on axon 17, temperature 6–7° C (cf. Fig. 3). The smooth curves are theoretical curves with parameters shown in Table 2; *A* to *H* drawn from eqn. 19, *I* to *L* from 14, 17, 18 with $\bar{g}_{Na} = 70{\cdot}7$ m.mho/cm². The ordinate scales on the right are given in m.mho/cm². The numbers on the left show the depolarization in mV. The time scale applies to all curves.

where $g'_{Na} = \bar{g}_{Na} m_\infty^3 h_0$ and is the value which the sodium conductance would attain if h remained at its resting level (h_0). Eqn. (19) was fitted to an experimental curve by plotting the latter on double log paper and comparing it with a similar plot of a family of theoretical curves drawn with different ratios of τ_m to τ_h. Curves *A* to *H* in Fig. 6 were obtained by this method and gave the

TABLE 2. Analysis of curves in Fig. 6

Curve	V (mV)	g'_{Na} (m.mho/cm²)	m_∞	τ_m (msec)	α_m (msec⁻¹)	β_m (msec⁻¹)	τ_h (msec)	h_∞	α_h (msec⁻¹)	β_h (msec⁻¹)
—	(−∞)	(42·9)	(1·00)	—	—	—	—	—	—	—
A	−109	40·3	0·980	0·140	7·0	(0·14)	0·67	(0)	(0)	1·50
B	−100	42·6	0·997	0·160	6·2	(0·02)	0·67	(0)	(0)	1·50
C	− 88	46·8	1·029	0·200	5·15	(−0·14)	0·67	(0)	(0)	1·50
D	− 76	39·5	0·975	0·189	5·15	0·13	0·84	(0)	(0)	1·19
E	− 63	38·2	0·963	0·252	3·82	0·15	0·84	(0)	(0)	1·19
F	− 51	30·7	0·895	0·318	2·82	0·33	1·06	(0)	(0)	0·94
G	− 38	20·0	0·778	0·382	2·03	0·58	1·27	(0)	(0)	0·79
H	− 32	15·3	0·709	0·520	1·36	0·56	1·33	(0)	(0)	0·75
I	− 26	7·90	0·569	0·600	0·95	0·72	(1·50)	(0·029)	(0·02)	(0·65)
J	− 19	1·44	0·323	0·400	0·81	1·69	(2·30)	(0·069)	(0·03)	(0·40)
K	− 10	0·13	0·145	0·220	0·66	3·9	(5·52)	(0·263)	(0·05)	(0·13)
L	− 6	0·046	0·103	0·200	0·51	4·5	(6·73)	(0·388)	(0·06)	(0·09)
—	(0)	(0·0033)	(0·042)	—	—	—	—	(0·608)	—	—

Values enclosed in brackets were not plotted in Figs. 7–10 either because they were too small to be reliable or because they were not independent measurements obtained in this experiment.

values of g'_{Na}, τ_m and τ_h shown in Table 2. Curves I to L were obtained from eqns. (17) and (18) assuming that h_∞ and τ_h had values calculated from experiments described in a previous paper (Hodgkin & Huxley, 1952 *c*).

The rate constants α_m and β_m. Having fitted theoretical curves to the experimental points, α_m and β_m were found by a procedure similar to that used with α_n and β_n, i.e.

$$\alpha_m = m_\infty/\tau_m, \quad \beta_m = (1 - m_\infty)/\tau_m,$$

the value of m_∞ being obtained from $\sqrt[3]{g'_{Na}}$ on the basis that m_∞ approaches unity at large depolarizations.

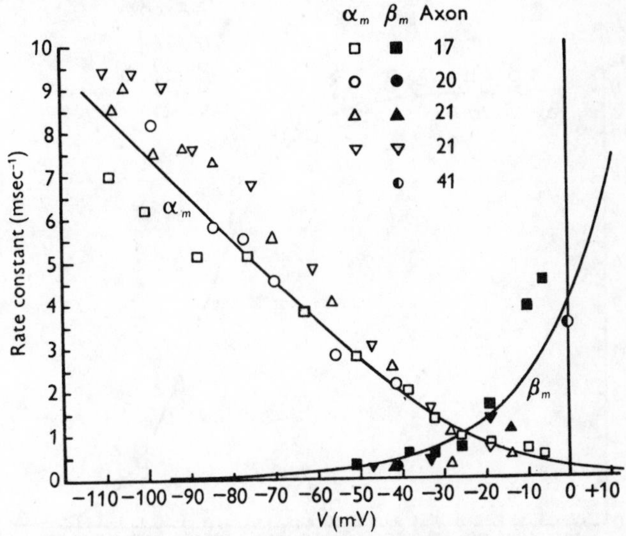

Fig. 7. Abscissa: membrane potential minus resting potential in sea water. Ordinate: rate constants (α_m and β_m) determining initial changes in sodium conductance at 6° C. All values for $V < 0$ were obtained by the method illustrated by Fig. 6 and Table 2; the value at $V = 0$ was obtained from the decline in sodium conductance associated with repolarization to the resting potential. The temperature varied between 3 and 11° C and was allowed for by assuming a Q_{10} of 3. The smooth curves were drawn from eqns. (20) and (21).

Values of α_m and β_m were collected from different experiments, reduced to a temperature of 6° C by adopting a Q_{10} of 3 and plotted in the manner shown in Fig. 7. The point for $V = 0$ was obtained from what we regard as the most reliable estimate of the rate constant determining the decline of sodium conductance when the membrane is repolarized (Hodgkin & Huxley, 1952 *b*, table 1, axon 41). The smooth curves in Fig. 7 were drawn according to the equations:

$$\alpha_m = 0{\cdot}1\,(V + 25) \Big/ \left(\exp\frac{V + 25}{10} - 1\right), \tag{20}$$

$$\beta_m = 4\exp(V/18), \tag{21}$$

where α_m and β_m are expressed in $msec^{-1}$ and V is in mV.

Fig. 8 illustrates the relation between m_∞ and V. The symbols are experimental estimates and the smooth curve was calculated from the equation

$$m_\infty = \alpha_m/(\alpha_m + \beta_m), \tag{22}$$

where α_m and β_m have the values given by eqns. (20) and (21).

The rate constants α_h and β_h. The rate constants for the inactivation process were calculated from the expressions

$$\alpha_h = h_\infty/\tau_h,$$
$$\beta_h = (1 - h_\infty)/\tau_h.$$

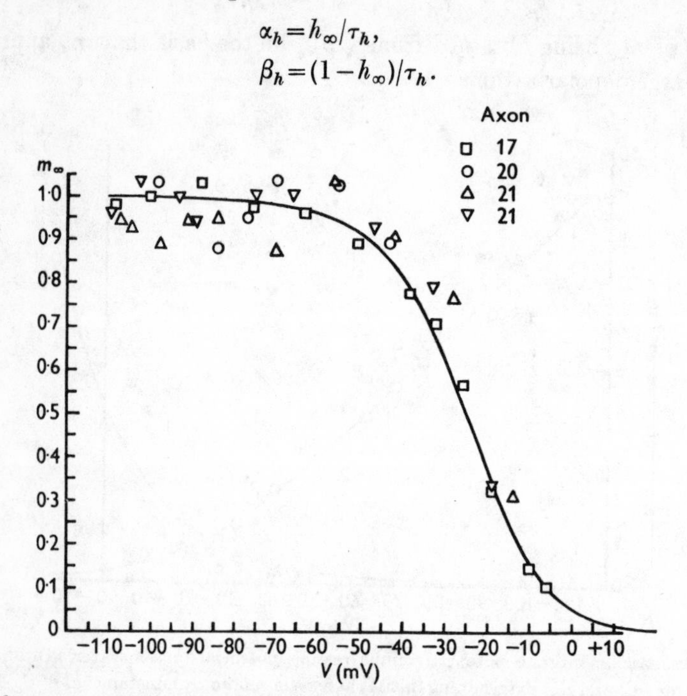

Fig. 8. Abscissa: membrane potential minus resting potential in sea water. Ordinate: m_∞ obtained by fitting curves to observed changes in sodium conductance at different depolarizations (e.g. Fig. 6 and Table 2). The smooth curve is drawn according to eqn. (22). The experimental points are proportional to the cube root of the sodium conductance which would have been obtained if there were no inactivation.

Values obtained by these equations are plotted against membrane potential in Fig. 9. The points for $V < -30$ mV were derived from the analysis described in this paper (e.g. Table 2), while those for $V > -30$ mV were obtained from the results given in a previous paper (Hodgkin & Huxley, 1952 *c*). A temperature coefficient of 3 was assumed and differences in resting potential were allowed for by taking the origin at a potential corresponding to $h_\infty = 0{\cdot}6$.

The smooth curves in this figure were calculated from the expressions

$$\alpha_h = 0{\cdot}07 \exp(V/20), \tag{23}$$

and

$$\beta_h = 1\bigg/\left(\exp\frac{V+30}{10} + 1\right). \tag{24}$$

The steady state relation between h_∞ and V is shown in Fig. 10. The smooth curve is calculated from the relation

$$h_\infty = \alpha_h/(\alpha_h + \beta_h), \quad (25)$$

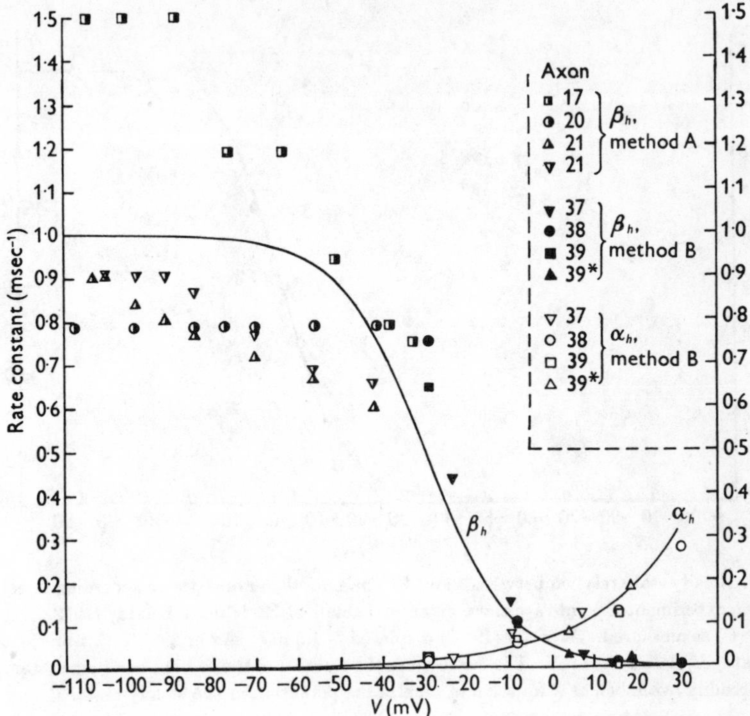

Fig. 9. Rate constants of inactivation (α_h and β_h) as functions of membrane potential (V). The smooth curves were calculated from eqns. (23) and (24). The experimental values of α_h and β_h were obtained from data such as those in Table 2 of this paper (method A) or from the values of τ_h and h_∞ given in Table 1 of Hodgkin & Huxley (1952*c*) (method B). Temperature differences were allowed for by scaling with a Q_{10} of 3. Axon 39 was at 19° C; all others at 3–9° C. The values for axons 37 and 39* were displaced by −1·5 and −12 mV in order to give $h_\infty = 0{\cdot}6$ at $V = 0$.

with α_h and β_h given by eqns. (23) and (24). If $V > -30$ mV this expression approximates to the simple expression used in a previous paper (Hodgkin & Huxley, 1952 *c*), i.e.

$$h_\infty = 1\bigg/\left(1 + \exp\frac{V_h - V}{7}\right),$$

where V_h is about −2 and is the potential at which $h_\infty = 0{\cdot}5$. This equation is the same as that giving the effect of a potential difference on the proportion of negatively charged particles on the outside of a membrane to the total number of such particles on both sides of the membrane (see p. 503). It is therefore consistent with the suggestion that inactivation might be due to the movement of a negatively charged particle which blocks the flow of sodium ions when it reaches the inside of the membrane. This is encouraging, but it must be mentioned that a physical theory of this kind does not lead to satisfactory functions for α_h and β_h without further *ad hoc* assumptions.

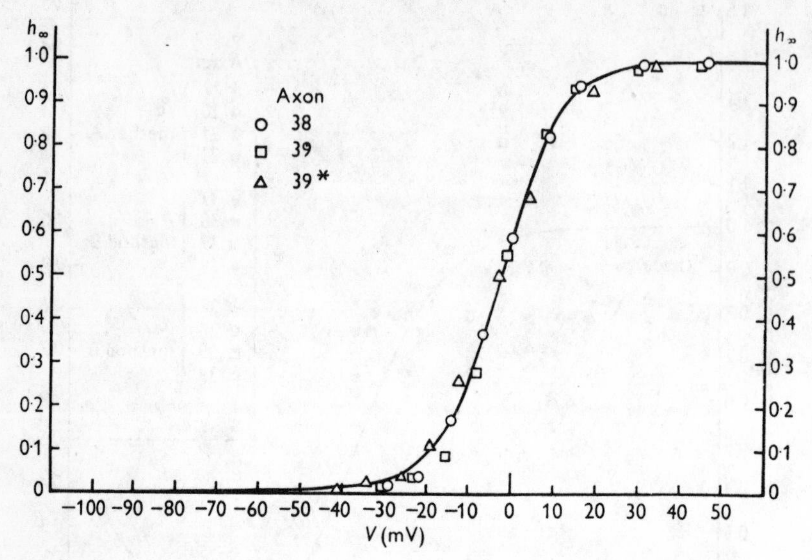

Fig. 10. Steady state relation between h and V. The smooth curve is drawn according to eqn. (25). The experimental points are those given in Table 1 of Hodgkin & Huxley (1952*c*). Axon 38 (5° C) as measured. Axon 39 (19° C) displaced −1·5 mV. Axon 39* (3° C, fibre in derelict state) displaced −12 mV. The curve gives the fraction of the sodium-carrying system which is readily available, as a function of membrane potential, in the steady state.

PART III. RECONSTRUCTION OF NERVE BEHAVIOUR

The remainder of this paper will be devoted to calculations of the electrical behaviour of a model nerve whose properties are defined by the equations which were fitted in Part II to the voltage clamp records described in the earlier papers of this series.

Summary of equations and parameters

We may first collect the equations which give the total membrane current I as a function of time and voltage. These are:

$$I = C_M \frac{dV}{dt} + \bar{g}_K n^4 (V - V_K) + \bar{g}_{Na} m^3 h (V - V_{Na}) + \bar{g}_l (V - V_l), \quad (26)$$

where
$$dn/dt = \alpha_n(1-n) - \beta_n n, \quad (7)$$
$$dm/dt = \alpha_m(1-m) - \beta_m m, \quad (15)$$
$$dh/dt = \alpha_h(1-h) - \beta_h h, \quad (16)$$

and

$$\alpha_n = 0{\cdot}01\,(V+10)\Big/\left(\exp\frac{V+10}{10}-1\right), \tag{12}$$

$$\beta_n = 0{\cdot}125\,\exp\,(V/80), \tag{13}$$

$$\alpha_m = 0{\cdot}1\,(V+25)\Big/\left(\exp\frac{V+25}{10}-1\right), \tag{20}$$

$$\beta_m = 4\,\exp\,(V/18), \tag{21}$$

$$\alpha_h = 0{\cdot}07\,\exp\,(V/20), \tag{23}$$

$$\beta_h = 1\Big/\left(\exp\frac{V+30}{10}+1\right). \tag{24}$$

Equation (26) is derived simply from eqns. (1)–(6) and (14) in Part II. The four terms on the right-hand side give respectively the capacity current, the current carried by K ions, the current carried by Na ions and the leak current, for 1 cm^2 of membrane. These four components are in parallel and add up to give the total current density through the membrane I. The conductances to K and Na are given by the constants $\bar{g}_K$ and $\bar{g}_{Na}$, together with the dimensionless quantities n, m and h, whose variation with time after a change of membrane potential is determined by the three subsidiary equations (7), (15) and (16). The α's and β's in these equations depend only on the instantaneous value of the membrane potential, and are given by the remaining six equations.

Potentials are given in mV, current density in $\mu A/cm^2$, conductances in m.mho/cm^2, capacity in $\mu F/cm^2$, and time in msec. The expressions for the α's and β's are appropriate to a temperature of 6·3° C; for other temperatures they must be scaled with a Q_{10} of 3.

The constants in eqn. (26) are taken as independent of temperature. The values chosen are given in Table 3, column 2, and may be compared with the experimental values in columns 3 and 4.

Membrane currents during a voltage clamp

Before applying eqn. (26) to the action potential it is well to check that it predicts correctly the total current during a voltage clamp. At constant voltage $dV/dt=0$ and the coefficients α and β are constant. The solution is then obtained directly in terms of the expressions already given for n, m and h (eqns. (8), (17) and (18)). The total ionic current was computed from these for a number of different voltages and is compared with a series of experimental curves in Fig. 11. The only important difference is that the theoretical current has too little delay at the sodium potential; this reflects the inability of our equations to account fully for the delay in the rise of g_K (p. 509).

'Membrane' and propagated action potentials

By a 'membrane' action potential is meant one in which the membrane potential is uniform, at each instant, over the whole of the length of fibre

TABLE 3

Constant (1)	Value chosen (2)	Experimental values: Mean (3)	Experimental values: Range (4)	Reference (5)
C_M (μF/cm²)	1·0	0·91	0·8 to 1·5	Table 1, Hodgkin *et al.* (1952)
V_{Na} (mV)	−115	−109	−95 to −119	p. 455, Hodgkin & Huxley (1952*a*)
V_K (mV)	+12	+11	+9 to +14	Table 3, values for low temperature in sea water, Hodgkin & Huxley (1952*b*)
V_l (mV)	−10·613*	−11	−4 to −22	Table 5, Hodgkin & Huxley (1952*b*)
$\bar{g}_{Na}$ (m.mho/cm²)	120	80	65 to 90	Fully analysed results, Table 2† } Hodgkin & Huxley (1952*a*)
		160	120 to 260	Fresh fibres, p. 465† } Hodgkin & Huxley (1952*a*)
$\bar{g}_K$ (m.mho/cm²)	36	34	26 to 49	p. 463, Hodgkin & Huxley (1952*a*)
$\bar{g}_l$ (m.mho/cm²)	0·3	0·26	0·13 to 0·50	Table 5, Hodgkin & Huxley (1952*b*)

* Exact value chosen to make the total ionic current zero at the resting potential ($V=0$).

† The experimental values for $\bar{g}_{Na}$ were obtained by multiplying the peak sodium conductances by factors derived from the values chosen for α_m, β_m, α_h, and β_h.

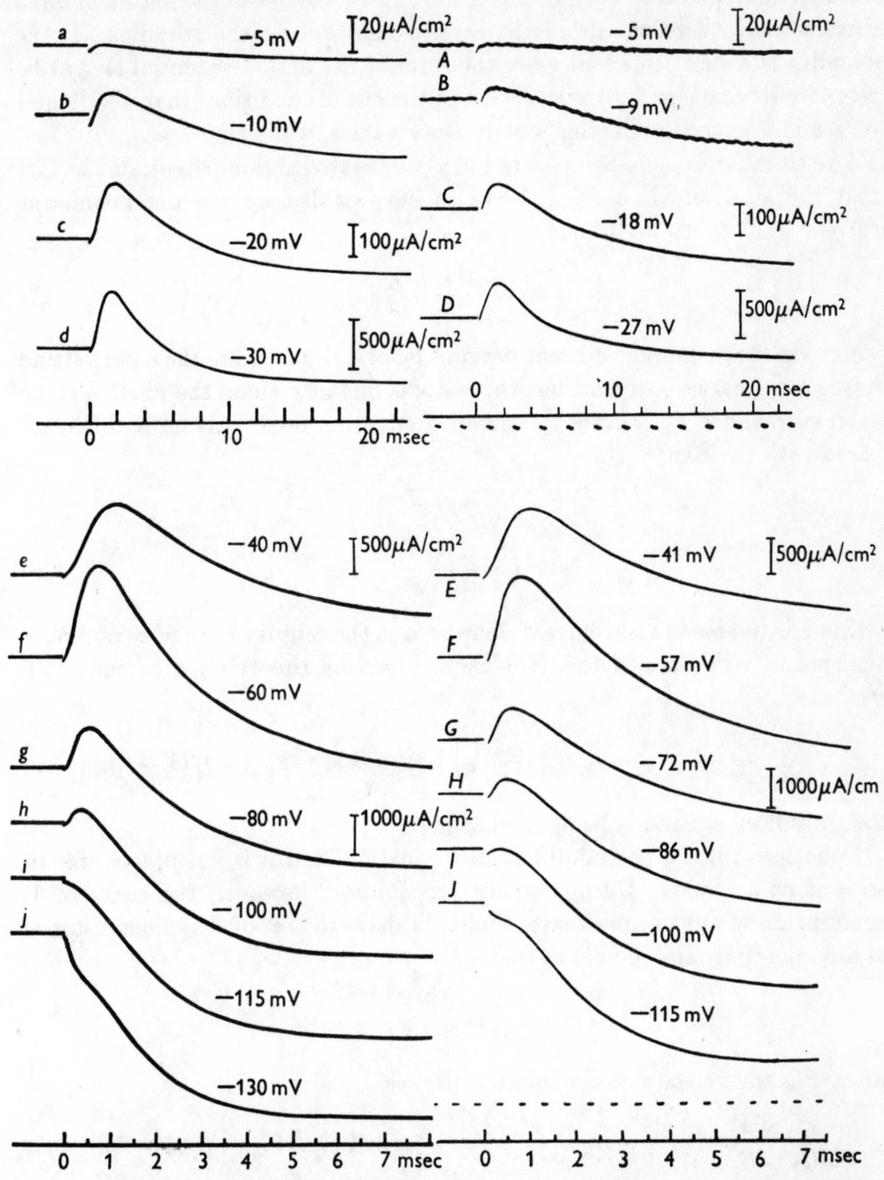

Fig. 11. Left-hand column: time course of membrane current during voltage clamp, calculated for temperature of 4° C from eqn. (26) and subsidiaries and plotted on the same scale as the experimental curves in the right-hand column. Right-hand column: observed time course of membrane currents during voltage clamp. Axon 31 at 4° C; compensated feedback. The time scale changes between *d*, *D* and *e*, *E*. The current scale changes after *b*, *B*; *c*, *C*; *d*, *D* and *f*, *F*.

considered. There is no current along the axis cylinder and the net membrane current must therefore always be zero, except during the stimulus. If the stimulus is a short shock at $t=0$, the form of the action potential should be given by solving eqn. (26) with $I=0$ and the initial conditions that $V=V_0$ and m, n and h have their resting steady state values, when $t=0$.

The situation is more complicated in a propagated action potential. The fact that the local circuit currents have to be provided by the net membrane current leads to the well-known relation

$$i=\frac{1}{r_1+r_2}\frac{\partial^2 V}{\partial x^2}, \tag{27}$$

where i is the membrane current per unit length, r_1 and r_2 are the external and internal resistances per unit length, and x is distance along the fibre. For an axon surrounded by a large volume of conducting fluid, r_1 is negligible compared with r_2. Hence

$$i=\frac{1}{r_2}\frac{\partial^2 V}{\partial x^2},$$

or

$$I=\frac{a}{2R_2}\frac{\partial^2 V}{\partial x^2}, \tag{28}$$

where I is the membrane current density, a is the radius of the fibre and R_2 is the specific resistance of the axoplasm. Inserting this relation in eqn. (26), we have

$$\frac{a}{2R_2}\frac{\partial^2 V}{\partial x^2}=C_M\frac{\partial V}{\partial t}+\bar{g}_K n^4(V-V_K)+\bar{g}_{Na}m^3h(V-V_{Na})+\bar{g}_l(V-V_l), \tag{29}$$

the subsidiary equations being unchanged.

Equation (29) is a partial differential equation, and it is not practicable to solve it as it stands. During steady propagation, however, the curve of V against time at any one position is similar in shape to that of V against distance at any one time, and it follows that

$$\frac{\partial^2 V}{\partial x^2}=\frac{1}{\theta^2}\frac{\partial^2 V}{\partial t^2},$$

where θ is the velocity of conduction. Hence

$$\frac{a}{2R_2\theta^2}\frac{d^2 V}{dt^2}=C_M\frac{dV}{dt}+\bar{g}_K n^4(V-V_K)+\bar{g}_{Na}m^3h(V-V_{Na})+\bar{g}_l(V-V_l). \tag{30}$$

This is an ordinary differential equation and can be solved numerically, but the procedure is still complicated by the fact that θ is not known in advance. It is necessary to guess a value of θ, insert it in eqn. (30) and carry out the numerical solution starting from the resting state at the foot of the action potential. It is then found that V goes off towards either $+\infty$ or $-\infty$, according as the guessed θ was too small or too large. A new value of θ is then chosen and the procedure repeated, and so on. The correct value brings V back to zero (the resting condition) when the action potential is over.

The solutions which go towards $\pm\infty$ correspond to action potentials travelling slower than normal under a travelling anode or faster than normal under a travelling cathode. We suspect that a system which tends to $-\infty$ for all values of θ after an initial negative displacement of V is one which is incapable of propagating an action potential.

NUMERICAL METHODS

Membrane action potentials

Integration procedure. The equations to be solved are the four simultaneous first-order equations (26), (7), (15), and (16) (p. 518). After slight rearrangement (which will be omitted in this description) these were integrated by the method of Hartree (1932–3). Denoting the beginning and end of a step by t_0 and t_1 ($=t_0+\delta t$) the procedure for each step was as follows:

(1) Estimate V_1 from V_0 and its backward differences.

(2) Estimate n_1 from n_0 and its backward differences.

(3) Calculate $(\mathrm{d}n/\mathrm{d}t)_1$ from eqn. 7 using the estimated n_1 and the values of α_n and β_n appropriate to the estimated V_1.

(4) Calculate n_1 from the equation

$$n_1 - n_0 = \frac{\delta t}{2}\left\{\left(\frac{\mathrm{d}n}{\mathrm{d}t}\right)_0 + \left(\frac{\mathrm{d}n}{\mathrm{d}t}\right)_1 - \frac{1}{12}\left[\Delta^2\left(\frac{\mathrm{d}n}{\mathrm{d}t}\right)_0 + \Delta^2\left(\frac{\mathrm{d}n}{\mathrm{d}t}\right)_1\right]\right\};$$

$\Delta^2(\mathrm{d}n/\mathrm{d}t)$ is the second difference of $\mathrm{d}n/\mathrm{d}t$; its value at t_1 has to be estimated.

(5) If this value of n_1 differs from that estimated in (2), repeat (3) and (4) using the new n_1. If necessary, repeat again until successive values of n_1 are the same.

(6) Find m_1 and h_1 by procedures analogous to steps (2)–(5).

(7) Calculate $\bar{g}_K n_1^4$ and $\bar{g}_{Na} m_1^3 h_1$.

(8) Calculate $(\mathrm{d}V/\mathrm{d}t)_1$ from eqn. 26 using the values found in (7) and the originally estimated V_1.

(9) Calculate a corrected V_1 by procedures analogous to steps (4) and (5). This result never differed enough from the original estimated value to necessitate repeating the whole procedure from step (3) onwards.

The step value had to be very small initially (since there are no differences at $t=0$) and it also had to be changed repeatedly during a run, because the differences became unmanageable if it was too large. It varied between about 0·01 msec at the beginning of a run or 0·02 msec during the rising phase of the action potential, and 1 msec during the small oscillations which follow the spike.

Accuracy. The last digit retained in V corresponded to microvolts. Sufficient digits were kept in the other variables for the resulting errors in the change of V at each step to be only occasionally as large as 1 μV. It is difficult to estimate the degree to which the errors at successive steps accumulate, but we are confident that the overall errors are not large enough to be detected in the illustrations of this paper.

Temperature differences. In calculating the action potential it was convenient to use tables giving the α's and β's at intervals of 1 mV. The tabulated values were appropriate to a fibre at 6·3° C. To obtain the action potential at some other temperature T' °C the direct method would be to multiply all α's and β's by a factor $\phi = 3^{(T'-6\cdot3)/10}$, this being correct for a Q_{10} of 3. Inspection of eqn. 26 shows that the same result is achieved by calculating the action potential at 6·3° C with a membrane capacity of ϕC_M μF/cm^2, the unit of time being $1/\phi$ msec. This method was adopted since it saved recalculating the tables.

Propagated action potential

Equations. The main equation for a propagated action potential is eqn. (30). Introducing a quantity $K = 2R_2\theta^2 C_M/a$, this becomes

$$\frac{d^2V}{dt^2} = K\left\{\frac{dV}{dt} + \frac{1}{C_M}\left[\bar{g}_K n^4 (V - V_K) + \bar{g}_{Na} m^3 h (V - V_{Na}) + \bar{g}_l (V - V_l)\right]\right\}. \tag{31}$$

The subsidiary equations (7), (15) and (16), and the α's and β's, are the same as for the membrane equation.

Integration procedure. Steps (1)–(7) were the same as for the membrane action potential. After that the procedure was as follows:

(8) Estimate $(dV/dt)_1$ from $(dV/dt)_0$ and its backward differences.

(9) Calculate $(d^2V/dt^2)_1$ from eqn. (31), using the values found in (7) and the estimated values of V_1 and $(dV/dt)_1$.

(10) Calculate a corrected $(dV/dt)_1$ by procedures analogous to steps (4) and (5).

(11) Calculate a corrected V_1 by a procedure analogous to step (4), using the corrected $(dV/dt)_1$.

(12) If necessary, repeat (9)–(11) using the new V_1 and $(dV/dt)_1$, until successive values of V_1 agree.

Starting conditions. In practice it is necessary to start with V deviating from zero by a finite amount (0·1 mV was used). The first few values of V, and hence the differences, were obtained as follows. Neglecting the changes in g_K and g_{Na}, eqn. (31) is

$$\frac{d^2V}{dt^2} = K\left\{\frac{dV}{dt} + \frac{g_0}{C_M} V\right\},$$

where g_0 is the resting conductance of the membrane. The solution of this equation is $V = V_0 e^{\mu t}$, where μ is a solution of

$$\mu^2 - K\mu - Kg_0/C_M = 0. \tag{32}$$

When K has been chosen, μ can thus be found and hence V_1, V_2, etc. ($V_0 e^{\mu t_1}$, $V_0 e^{\mu t_2}$, etc.).

After several runs had been calculated, so that K was known within fairly narrow limits, time was saved by starting new runs not from near $V = 0$ but from a set of values interpolated between corresponding points on a run which had gone towards $+\infty$ and another which had gone towards $-\infty$.

Choice of K. The value of K chosen for the first run makes no difference to the final result, but the nearer it is to the correct value the fewer runs will need to be evaluated. The starting value was found by inserting in eqn. (32) a value of μ found by measuring the foot of an observed action potential.

Calculation of falling phase. The procedure outlined above is satisfactory for the rising phase and peak of the action potential but becomes excessively tedious in the falling phase and the oscillations which follow the spike. A different method, which for other reasons is not applicable in the earlier phases, was therefore employed. The solution was continued as a membrane action potential, and the value of d^2V/dt^2 calculated at each step from the differences of dV/dt. From these it was possible to derive an estimate of the values (denoted by z) that d^2V/dt^2 would have taken in a propagated action potential. The membrane solution was then re-calculated using the following equation instead of eqn. (31):

$$\frac{dV}{dt} = -\frac{1}{C_M}\{\bar{g}_K n^4 (V - V_K) + \bar{g}_{Na} m^3 h (V - V_{Na}) + \bar{g}_l (V - V_l)\} + \frac{z}{K}. \tag{33}$$

This was repeated until the z's assumed for a particular run agreed with the d^2V/dt^2's derived from the same run. When this is the case, eqn. (33) is identical with eqn. (31), the main equation for the propagated action potential.

RESULTS

Membrane action potentials

Form of action potential at 6° C. Three calculated membrane action potentials, with different strengths of stimulus, are shown in the upper part of Fig. 12. Only one, in which the initial displacement of membrane potential was 15 mV, is complete; in the other two the calculation was not carried beyond the middle of the falling phase because of the labour involved and because the solution

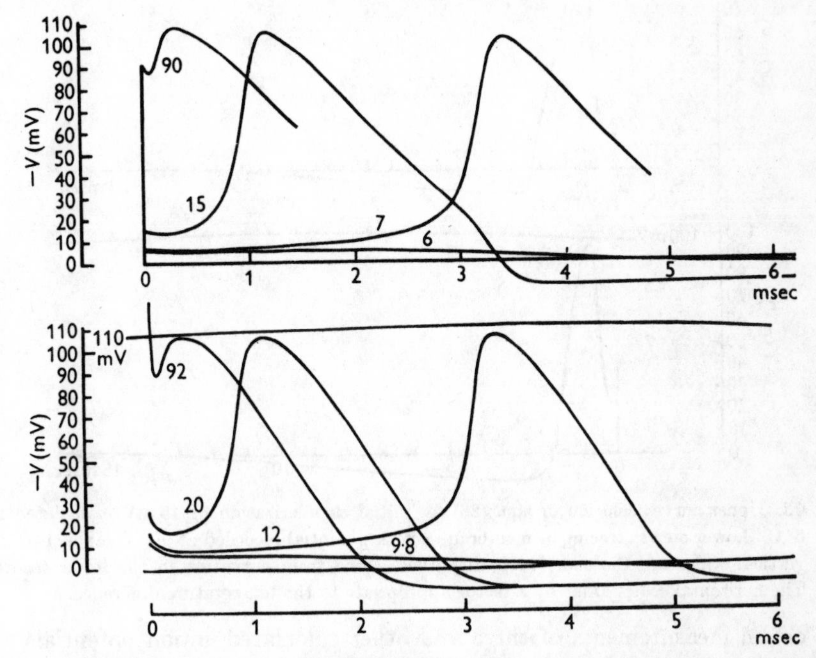

Fig. 12. Upper family: solutions of eqn. (26) for initial depolarizations of 90, 15, 7 and 6 mV (calculated for 6° C). Lower family: tracings of membrane action potentials recorded at 6° C from axon 17. The numbers attached to the curves give the shock strength in mμcoulomb/cm^2. The vertical and horizontal scales are the same in both families (apart from the slight curvature indicated by the 110 mV calibration line). In this and all subsequent figures depolarizations (or negative displacements of V) are plotted upwards.

had become almost identical with the 15 mV action potential, apart from the displacement in time. One solution for a stimulus just below threshold is also shown.

The lower half of Fig. 12 shows a corresponding series of experimental membrane action potentials. It will be seen that the general agreement is good, as regards amplitude, form and time-scale. The calculated action potentials do, however, differ from the experimental in the following respects: (1) The drop during the first 0·1 msec is smaller. (2) The peaks are sharper.

(3) There is a small hump in the lower part of the falling phase. (4) The ending of the falling phase is too sharp. The extent to which these differences are the result of known shortcomings in our formulation will be discussed on pp. 542–3.

The positive phase of the calculated action potential has approximately the correct form and duration, as may be seen from Fig. 13 in which a pair of curves are plotted on a slower time scale.

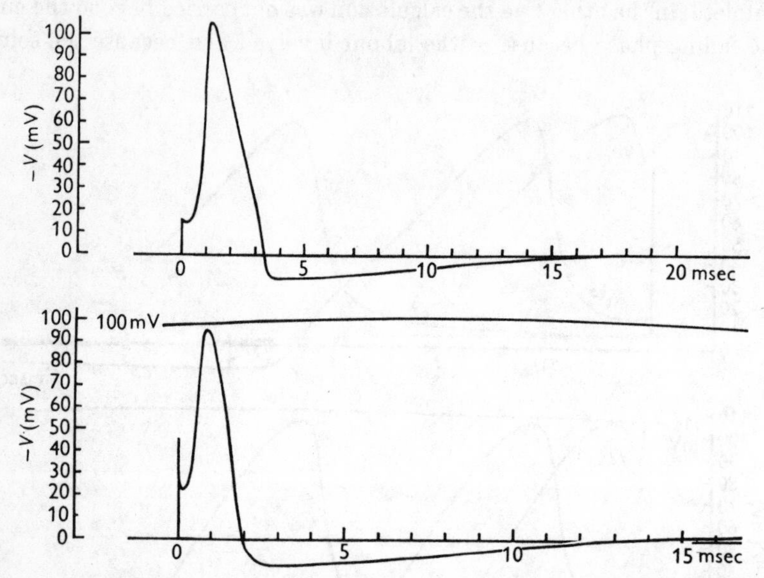

Fig. 13. Upper curve: solution of eqn. (26) for initial depolarization of 15 mV, calculated for 6° C. Lower curve: tracing of membrane action potential recorded at 9·1° C (axon 14). The vertical scales are the same in both curves (apart from curvature in the lower record). The horizontal scales differ by a factor appropriate to the temperature difference.

Certain measurements of these and other calculated action potentials are collected in Table 4.

Form of action potential at 18·5° C. Fig. 14 shows a comparison between a calculated membrane action potential at 18·5° C and an experimental one at 20·5° C. The same differences can be seen as at the low temperature, but, except for the initial drop, they are less marked. In both the calculated and the experimental case, the rise of temperature has greatly reduced the duration of the spike, the difference being more marked in the falling than in the rising phase (Table 4), as was shown in propagated action potentials by Hodgkin & Katz (1949).

The durations of both falling phase and positive phase are reduced at the higher temperature by factors which are not far short of that (3·84) by which the rate constants of the permeability changes are raised ($Q_{10} = 3{\cdot}0$). This is the justification for the differences in time scale between the upper and lower parts in Figs. 13 and 14.

TABLE 4. Characteristics of calculated action potentials

Type of action potential	Temperature (°C)	Stimulus	Spike height (mV)	Amplitude of positive phase (mV)	Peak conductance (m.mho/cm²)	Duration of rising phase, 20 mV to peak (msec)	Duration of falling phase, peak to $V=0$ (msec)	Duration of positive phase (msec)	Interval from peak of potential to peak of conductance (msec)	Max. rate of rise (V/sec)
Propagated	18·5	—	90·5	9·7	32·6	0·252	0·67	5·20	−0·016	431
Membrane	18·5	15 mV depolarization	96·8	10·5	30·7	0·275	0·61	5·09	+0·012	564
Membrane	6·3	100 mV depolarization	108·8	—	45·5	—	—	—	+0·16	—
Membrane	6·3	90 mV depolarization	108·5	—	44·8	—	—	—	+0·15	—
Membrane	6·3	15 mV depolarization	105·4	11·2	37·0	0·59	2·21	14·15	+0·15	311
Membrane	6·3	7 mV depolarization	102·1	—	33·4	0·62	—	—	+0·16	277
Membrane	6·3	Anode break	112·1	11·2	53·4	0·50	2·54	14·4	+0·14	414

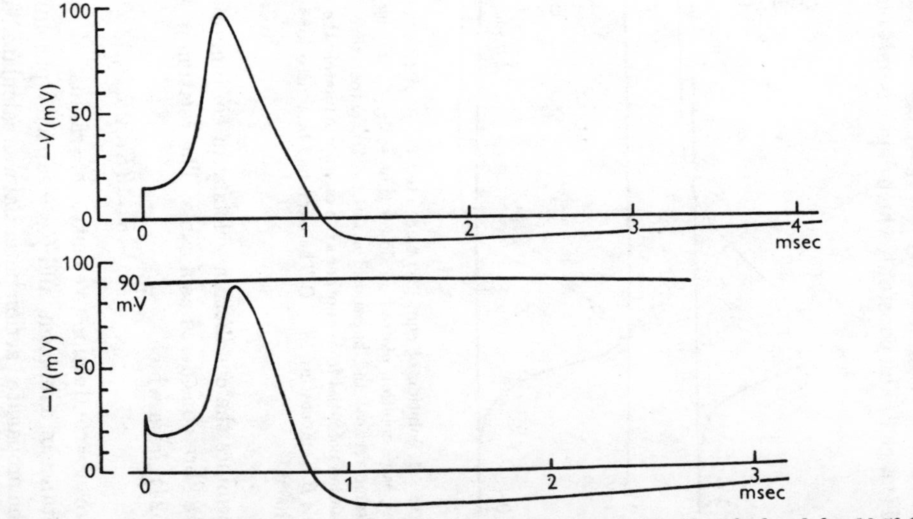

Fig. 14. Upper curve: solution of eqn. (26) for initial depolarization of 15 mV, calculated for 18·5° C. Lower curve: tracing of membrane action potential recorded at 20·5° C (axon 11). Vertical scales are similar. Horizontal scales differ by a factor appropriate to the temperature difference.

Propagated action potential

Form of propagated action potential. Fig. 15 compares the calculated propagated action potential, at 18·5° C, with experimental records on both fast and slow time bases. As in the case of the membrane action potential, the only differences are in certain details of the form of the spike.

Velocity of conduction. The value of the constant K that was found to be needed in the equation for the propagated action potential (eqn. 31) was 10·47 $msec^{-1}$. This constant, which depends only on properties of the membrane,

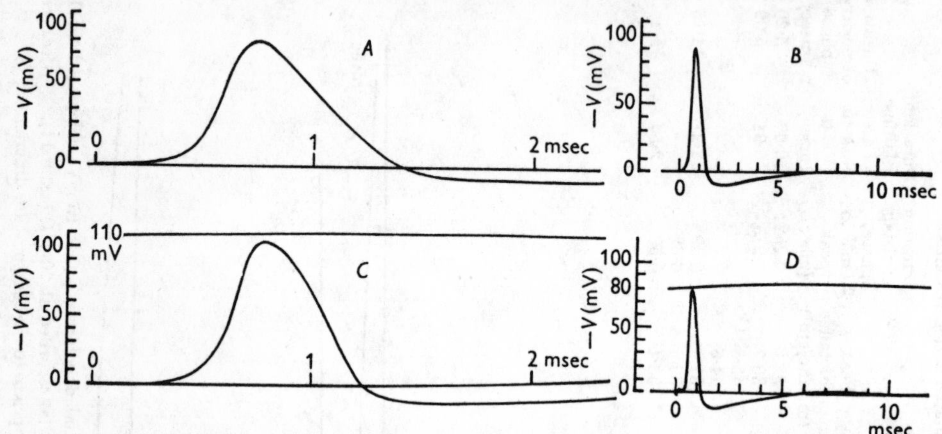

Fig. 15. *A*, solution of eqn. (31) calculated for K of 10·47 $msec^{-1}$ and temperature of 18·5° C. *B*, same solution plotted on slower time scale. *C*, tracing of propagated action potential on same vertical and horizontal scales as *A*. Temperature 18·5° C. *D*, tracing of propagated action potential from another axon on approximately the same vertical and horizontal scales as *B*. Temperature 19·2° C. This axon had been used for several hours; its spike was initially 100 mV.

determines the conduction velocity in conjunction with the constants of the nerve fibre considered as a cable. The relation is given by the definition of K (p. 524), from which

$$\theta = \sqrt{(Ka/2R_2C_M)}, \qquad (34)$$

where θ = conduction velocity, a = radius of axis cylinder, R_2 = specific resistance of axoplasm, and C_M = capacity per unit area of membrane.

The propagated action potential was calculated for the temperature at which the record C of Fig. 15 was obtained, and with the value of C_M (1·0 $\mu F/cm^2$) that was measured on the fibre from which that record was made. Since θ, a and R_2 were also measured on that fibre, a direct comparison between calculated and observed velocities is possible. The values of a and R_2 were 238 μ and 35·4 Ω. cm respectively. Hence the calculated conduction velocity is

$$(10470 \times 0{\cdot}0238/2 \times 35{\cdot}4 \times 10^{-6})^{\frac{1}{2}} \text{ cm/sec} = 18{\cdot}8 \text{ m/sec}.$$

The velocity found experimentally in this fibre was 21·2 m/sec.

Impedance changes

Time course of conductance change. Cole & Curtis (1939) showed that the impedance of the membrane fell during a spike, and that the fall was due to a great increase in the conductance which is in parallel with the membrane capacity. An effect of this kind is to be expected on our formulation, since the entry of Na^+ which causes the rising phase, and the loss of K^+ which causes the falling phase, are consequent on increases in the conductance of the membrane to currents carried by these ions. These component conductances are evaluated during the calculation, and the total conductance is obtained by adding them and the constant 'leak conductance', $\bar{g}_l$.

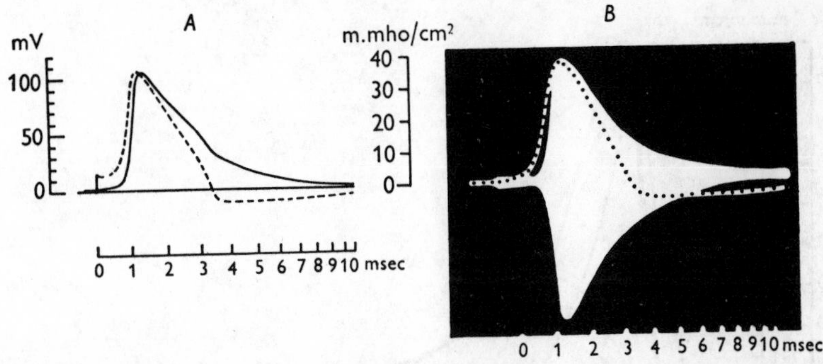

Fig. 16. *A*, solution of eqn. (26) for initial depolarization of 15 mV at a temperature of 6° C. The broken curve shows the membrane action potential in mV; the continuous curve shows the total membrane conductance ($g_{Na} + g_K + \bar{g}_l$) as a function of time. *B*, records of propagated action potential (dotted curve) and conductance change reproduced from Cole & Curtis (1939). The time scales are the same in *A* and *B*.

Fig. 16*A* shows the membrane potential and conductance in a calculated membrane action potential. For comparison, Fig. 16*B* shows superposed records of potential and impedance bridge output (proportional to conductance change), taken from Cole & Curtis's paper. The time scale is the same in *B* as in *A*, and the curves have been drawn with the same peak height. It will be seen that the main features of Cole & Curtis's record are reproduced in the calculated curve. Thus (1) the main rise in conductance begins later than the rise of potential; (2) the conductance does not fall to its resting value until late in the positive phase; and (3) the peak of the conductance change occurs at nearly the same time as the peak of potential. The exact time relation between the peaks depends on the conditions, as can be seen from Table 4.

We chose a membrane action potential for the comparison in Fig. 16 because the spike duration shows that the experimental records were obtained at about 6° C, and our propagated action potential was calculated for 18·5° C. The conductance during the latter is plotted together with the potential in Fig. 17. The same features are seen as in the membrane action potential, the delay between the rise of potential and the rise of conductance being even more marked.

Absolute value of peak conductance. The agreement between the height of the conductance peak in Fig. 16*A* and the half-amplitude of the bridge output in Fig. 16*B* is due simply to the choice of scale. Nevertheless, our calculated action potentials agree well with Cole & Curtis's results in this respect. These authors found that the average membrane resistance at the peak of the impedance change was 25 Ω.cm², corresponding to a conductance of 40 m.mho/cm². The peak conductances in our calculated action potentials ranged from 31 to 53 m.mho/cm² according to the conditions, as shown in Table 4.

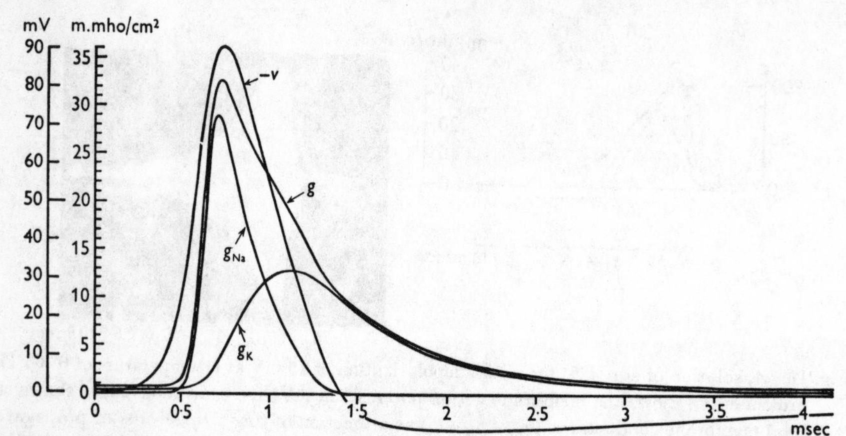

Fig. 17. Numerical solution of eqn. (31) showing components of membrane conductance (g) during propagated action potential ($-V$). Details of the analysis are as in Fig. 15.

Components of conductance change. The manner in which the conductances to Na^+ and K^+ contribute to the change in total conductance is shown in Fig. 17 for the calculated propagated action potential. The rapid rise is due almost entirely to sodium conductance, but after the peak the potassium conductance takes a progressively larger share until, by the beginning of the positive phase, the sodium conductance has become negligible. The tail of raised conductance that falls away gradually during the positive phase is due solely to potassium conductance, the small constant leak conductance being of course present throughout.

Ionic movements

Time course of ionic currents. The time course of the components of membrane current carried by sodium and potassium ions during the calculated propagated spike is shown in Fig. 18 *C*. The total ionic current contains also a small contribution from 'leak current' which is not plotted separately.

Two courses are open to current which is carried into the axis cylinder by ions crossing the membrane: it may leave the axis cylinder again by altering the charge on the membrane capacity, or it may turn either way along the axis cylinder making a net contribution, I, to the local circuit current. The magnitudes of these two terms during steady propagation are $-C_M \mathrm{d}V/\mathrm{d}t$ and $(C_M/K)\, \mathrm{d}^2V/\mathrm{d}t^2$ respectively, and the manner in which the ionic current is divided between them at the different stages of the spike is shown in Fig. 18*B*. It will be seen that the ionic current is very small until the potential is well beyond the threshold level, which is shown by Fig. 12*A* to be about 6 mV.

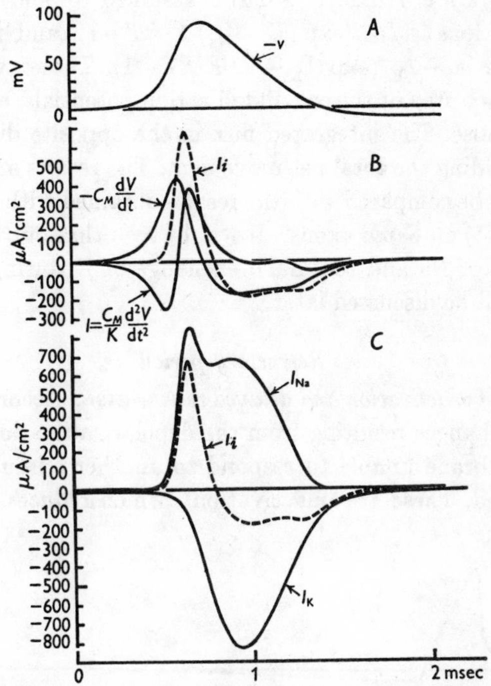

Fig. 18. Numerical solution of eqn. (31) showing components of membrane current during propagated action potential. *A*, membrane potential $(-V)$. *B*, ionic current (I_i), capacity current $\left(-C_M \frac{\mathrm{d}V}{\mathrm{d}t}\right)$ and total membrane current $\left(I = \frac{C_M}{K}\frac{\mathrm{d}^2V}{\mathrm{d}t^2}\right)$. *C*, ionic current (I_i), sodium current (I_{Na}) and potassium current (I_K). The time scale applies to all the curves. Details of the analysis are as in Fig. 15.

During this period the current for charging the membrane capacity comes almost entirely from local circuits. The fact that the ionic current does not become appreciable as soon as the threshold depolarization is passed is due partly to the smallness of the currents reached in any circumstances near the threshold, and partly to the delay with which sodium conductance rises when the potential is lowered.

Total movements of ions. The total entry of sodium and loss of potassium can be obtained by integrating the corresponding ionic currents over the whole impulse. This has been done for the four complete action potentials that we calculated, and the results are given in Table 5. It will be seen that the results at 18·5° C are in good agreement with the values found experimentally by Keynes (1951) and Keynes & Lewis (1951), which were obtained at comparable temperatures.

Ionic fluxes. The flux in either direction of an ion can be obtained from the net current and the equilibrium potential for that ion, if the independence principle (Hodgkin & Huxley, 1952*a*) is assumed to hold. Thus the outward flux of sodium ions is $I_{Na}/(\exp(V-V_{Na})F/RT-1)$, and the inward flux of potassium ions is $-I_K/(\exp(V_K-V)F/RT-1)$. These two quantities were evaluated at each step of the calculated action potentials, and integrated over the whole impulse. The integrated flux in the opposite direction is given in each case by adding the total net movement. The results are given in Table 5, where they can be compared with the results obtained with radioactive tracers by Keynes (1951) on *Sepia* axons. It will be seen that our theory predicts too little exchange of Na and too much exchange of K during an impulse. This discrepancy will be discussed later.

Refractory period

Time course of inactivation and delayed rectification. According to our theory, there are two changes resulting from the depolarization during a spike which make the membrane unable to respond to another stimulus until a certain time has elapsed. These are 'inactivation', which reduces the level to which

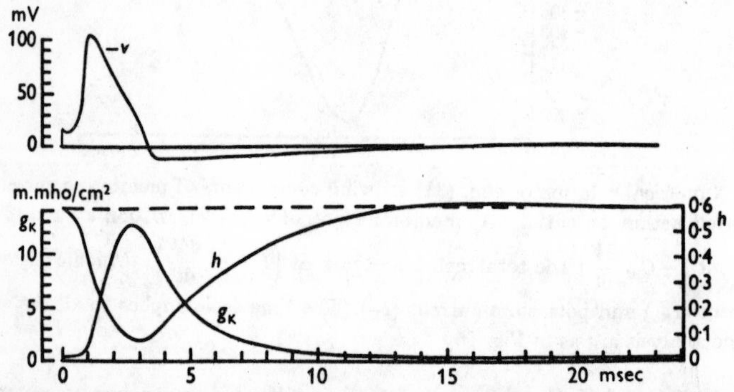

Fig. 19. Numerical solution of eqn. (26) for initial depolarization of 15 mV and temperature of 6° C. Upper curve: membrane potential, as in Fig. 13. Lower curves show time course of g_K and h during action potential and refractory period.

the sodium conductance can be raised by a depolarization, and the delayed rise in potassium conductance, which tends to hold the membrane potential near to the equilibrium value for potassium ions. These two effects are shown in Fig. 19 for the calculated membrane action potential at 6° C. Both curves reach their normal levels again near the end of the positive phase, and finally

TABLE 5. Ionic movements during an impulse. All values are expressed in $\mu\mu$mole/cm² and represent the excess over the corresponding movement in the resting state. In the theoretical cases the integration is taken as far as the 3rd intersection with the base line after the spike; it is begun in case (1) when $V = 0{\cdot}1$ mV; (2) and (3) at the stimulus; (4) when $V = 0$ before the spike. Experimental data from Keynes (1951) for row 6 and from Keynes & Lewis (1951) for rows 5 and 7.

	Type of action potential	Temp. (° C)	Stimulus (mV)	Sodium Influx	Sodium Outflux	Sodium Net entry	Potassium Influx	Potassium Outflux	Potassium Net loss
Theoretical (*Loligo*):									
1	Propagated	18·5	—	5·42	1·09	4·33	1·72	5·98	4·26
2	Membrane	18·5	15	5·01	1·02	3·99	1·71	5·78	4·07
3	Membrane	6·3	15	19·30	4·84	14·46	6·17	20·49	14·32
4	Membrane	6·3	Anode break	26·61	9·45	17·16	6·64	23·41	16·77
Experimental:									
5	Propagated (*Loligo*)	22	—	—	—	3·5	—	—	3·0
6	Propagated (*Sepia*)	14	—	10·3	6·6	3·7	0·39	4·7	4·3
7	Propagated (*Sepia*)	22	—	—	—	3·8	—	—	3·6

settle down after a heavily damped oscillation of small amplitude which is not seen in the figure.

Responses to stimuli during positive phase. We calculated the responses of the membrane when it was suddenly depolarized by 90 mV at various times during the positive phase of the membrane action potential at 6° C. These are shown by the upper curves in Fig. 20. After the earliest stimulus the

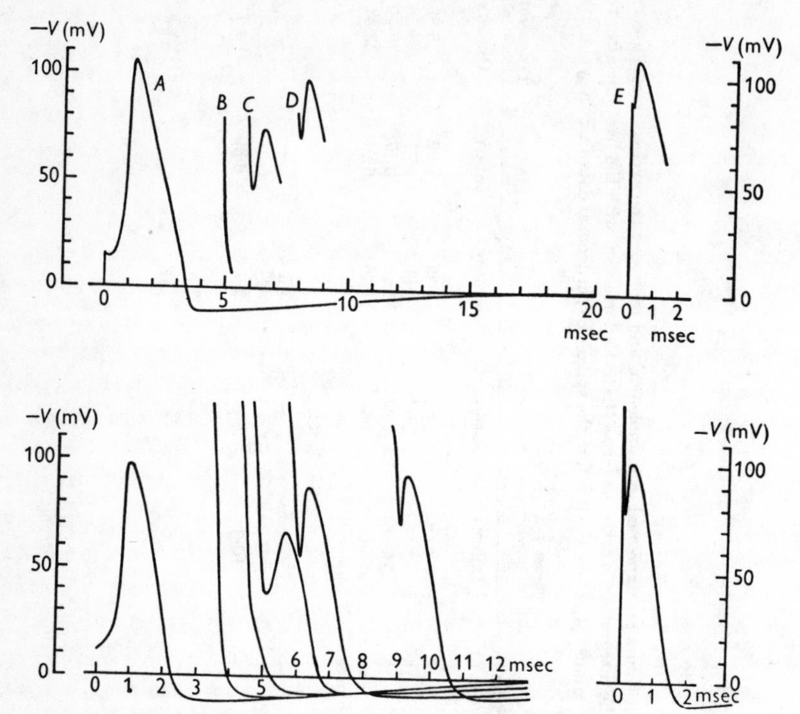

Fig. 20. Theoretical basis of refractory period. Upper curves: numerical solutions of eqn. (26) for temperature of 6° C. Curve *A* gives the response to 15 mμcoulomb/cm² applied instantaneously at $t=0$. Curve *E* gives the response to 90 mμcoulomb/cm² again applied in the resting state. Curves *B* to *D* show effect of applying 90 mμcoulomb/cm² at various times after curve *A*. Lower curves: a similar experiment with an actual nerve, temperature 9° C. The voltage scales are the same throughout. The time scales differ by a factor appropriate to the temperature difference.

membrane potential falls again with hardly a sign of activity, and the membrane can be said to be in the 'absolute refractory period'. The later stimuli produce action potentials of increasing amplitude, but still smaller than the control; these are in the 'relative refractory period'. Corresponding experimental curves are shown in the lower part of Fig. 20. The agreement is good, as regards both the duration of the absolute refractory period and the changes in shape of the spike as recovery progresses.

Excitation

Our calculations of excitation processes were all made for the case where the membrane potential is uniform over the whole area considered, and not for the case of local stimulation of a whole nerve. There were two reasons for this: first, that such data from the squid giant fibre as we had for comparison were obtained by uniform stimulation of the membrane with the long electrode; and, secondly, that calculations for the whole nerve case would have been extremely laborious since the main equation is then a partial differential equation.

Threshold. The curves in Figs. 12 and 21 show that the theoretical 'membrane' has a definite threshold when stimulated by a sudden displacement of membrane potential. Since the initial fall after the stimulus is much less marked in these than in the experimental curves, it is relevant to compare the lowest point reached in a just threshold curve, rather than the magnitude of the original displacement. In the calculated series this is about 6 mV and in the experimental about 8 mV. This agreement is satisfactory, especially as the value for the calculated series must depend critically on such things as the leak conductance, whose value was not very well determined experimentally.

The agreement might have been somewhat less good if the comparison had been made at a higher temperature. The calculated value would have been much the same, but the experimental value in the series at 23° C shown in Fig. 8 of Hodgkin *et al.* (1952) is about 15 mV. However, this fibre had been stored for 5 hr before use and was therefore not in exactly the same state as those on which our measurements were based.

Subthreshold responses. When the displacement of membrane potential was less than the threshold for setting up a spike, characteristic subthreshold responses were seen. One such response is shown in Fig. 12, while several are plotted on a larger scale in Fig. 21*B*. Fig. 21*A* shows for comparison the corresponding calculated responses of our model. The only appreciable differences, in the size of the initial fall and in the threshold level, have been mentioned already in other connexions.

During the positive phase which follows each calculated subthreshold response, the potassium conductance is raised and there is a higher degree of 'inactivation' than in the resting state. The threshold must therefore be raised in the same way as it is during the relative refractory period following a spike. This agrees with the experimental findings of Pumphrey, Schmitt & Young (1940).

Anode break excitation. Our axons with the long electrode in place often gave anode break responses at the end of a period during which current was made to flow inward through the membrane. The corresponding response of out theoretical model was calculated for the case in which a current sufficient to bring the membrane potential to 30 mV above the resting potential was suddenly stopped after passing for a time long compared with all the time-constants of the membrane. To do this, eqn. (26) was solved with $I=0$ and the initial conditions that $V=+30$ mV, and m, n and h have their steady state values for $V=+30$ mV, when $t=0$. The calculation was made for a temperature

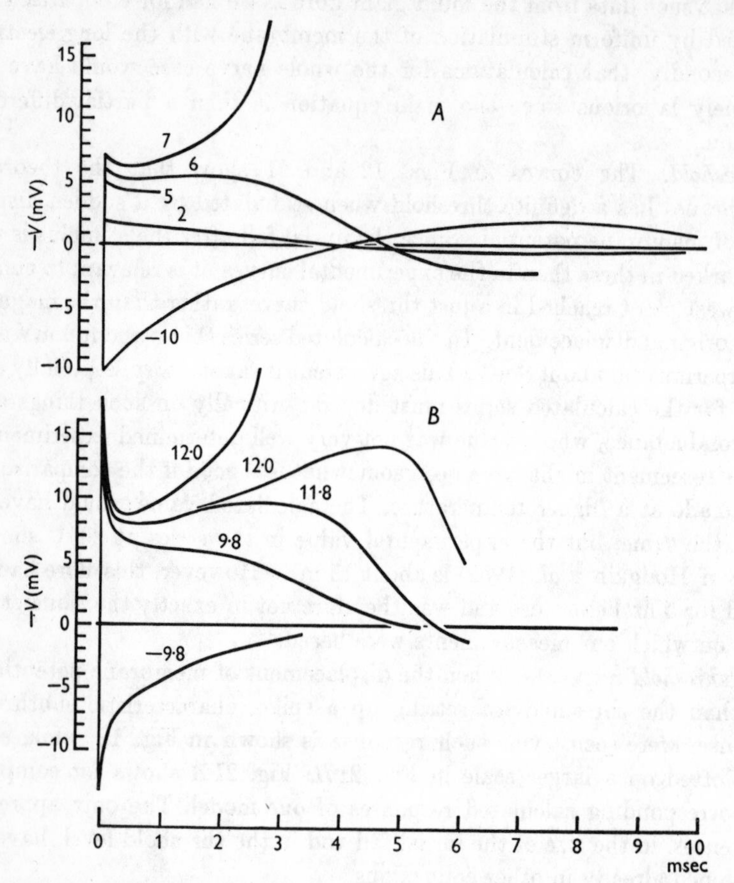

Fig. 21. *A*, numerical solutions of eqn. (26) for 6° C. The numbers attached to the curves give the initial depolarization in mV (also the quantity of charge applied in mμcoulomb/cm²). *B*, response of nerve membrane at 6° C to short shocks; the numbers show the charge applied in mμcoulomb/cm². The curves have been replotted from records taken at low amplification and a relatively high time-base speed.

of 6·3° C. A spike resulted, and the time course of membrane potential is plotted in Fig. 22*A*. A tracing of an experimental anode break response is shown in Fig. 22*B*; the temperature is 18·5° C, no record near 6° being available. It will be seen that there is good general agreement. (The oscillations after the positive phase in Fig. 22*B* are exceptionally large; the response of this axon to a small constant current was also unusually oscillatory as shown in Fig. 23.)

The basis of the anode break excitation is that anodal polarization decreases the potassium conductance and removes inactivation. These effects persist for an appreciable time so that the membrane potential reaches its resting value with a reduced outward potassium current and an increased inward sodium current. The total ionic current is therefore inward at $V=0$ and the membrane undergoes a depolarization which rapidly becomes regenerative.

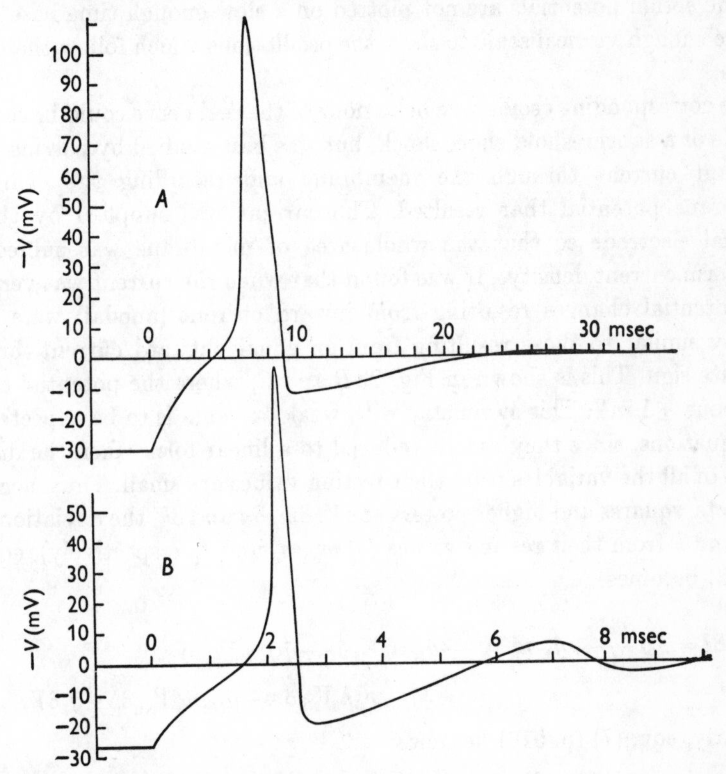

Fig. 22. Theoretical basis of anode break excitation. *A*, numerical solution of eqn. (26) for boundary condition $-V=-30$ mV for $t<0$; temperature 6° C. *B*, anode break excitation following sudden cessation of external current which had raised the membrane potential by 26·5 mV; giant axon with long electrode at 18·5° C. Time scales differ by a factor appropriate to the temperature difference.

Accommodation. No measurements of accommodation were made nor did we make any corresponding calculations for our model. It is clear, however, that the model will show 'accommodation' in appropriate cases. This may be shown in two ways. First, during the passage of a constant cathodal current through the membrane, the potassium conductance and the degree of inactivation will rise, both factors raising the threshold. Secondly, the steady state ionic current at all strengths of depolarization is outward (Fig. 11), so that an applied cathodal current which rises sufficiently slowly will never evoke a regenerative response from the membrane, and excitation will not occur.

Oscillations

In all the calculated action potentials and subthreshold responses the membrane potential finally returns to its resting value with a heavily damped oscillation. This is well seen after subthreshold stimuli in Figs. 21 *A* and 24, but the action potentials are not plotted on a slow enough time base or with a large enough vertical scale to show the oscillations which follow the positive phase.

The corresponding oscillatory behaviour of the real nerve could be seen after a spike or a subthreshold short shock, but was best studied by passing a small constant current through the membrane and recording the changes of membrane potential that resulted. The current was supplied by the long internal electrode so that the whole area of membrane was subjected to a uniform current density. It was found that when the current was very weak the potential changes resulting from inward current (anodal) were almost exactly similar to those resulting from an equal outward current, but with opposite sign. This is shown in Fig. 23 *B* and *C*, where the potential changes are about ± 1 mV. This symmetry with weak currents is to be expected from our equations, since they can be reduced to a linear form when the displacements of all the variables from their resting values are small. Thus, neglecting products, squares and higher powers of δV, δm, δn and δh, the deviations of V, m, n and h from their resting values (0, m_0, n_0 and h_0 respectively), eqn. (26) (p. 518) becomes

$$\delta I = C_M \frac{\mathrm{d}\delta V}{\mathrm{d}t} + \bar{g}_K n_0^4 \delta V - 4\bar{g}_K n_0^3 V_K \delta n + \bar{g}_{Na} m_0^3 h_0 \delta V - 3\bar{g}_{Na} m_0^2 h_0 V_{Na} \delta m - \bar{g}_{Na} m_0^3 V_{Na} \delta h + \bar{g}_l \delta V. \quad (35)$$

Similarly, eqn. (7) (p. 518) becomes

$$\frac{\mathrm{d}\delta n}{\mathrm{d}t} = \frac{\partial \alpha_n}{\partial V} \delta V - (\alpha_n + \beta_n)\,\delta n - n_0 \frac{\partial(\alpha_n + \beta_n)}{\partial V} \delta V,$$

or

$$(p + \alpha_n + \beta_n)\,\delta n = \left\{\frac{\partial \alpha_n}{\partial V} - n_0 \frac{\partial(\alpha_n + \beta_n)}{\partial V}\right\} \delta V, \quad (36)$$

where p represents d/dt, the operation of differentiating with respect to time.

The quantity δn can be eliminated between eqns. (35) and (36). This process is repeated for δm and δh, yielding a fourth-order linear differential equation with constant coefficients for δV. This can be solved by standard methods for any particular time course of the applied current density δI.

Fig. 23 *A* shows the response of the membrane to a constant current pulse calculated in this way. The constants in the equations are chosen to be appropriate to a temperature of 18·5° C so as to make the result comparable with the tracings of experimental records shown in *B* and *C*. It will be seen that the calculated curve agrees well with the records in *B*, while those in *C*, obtained from another axon, are much less heavily damped and show a higher

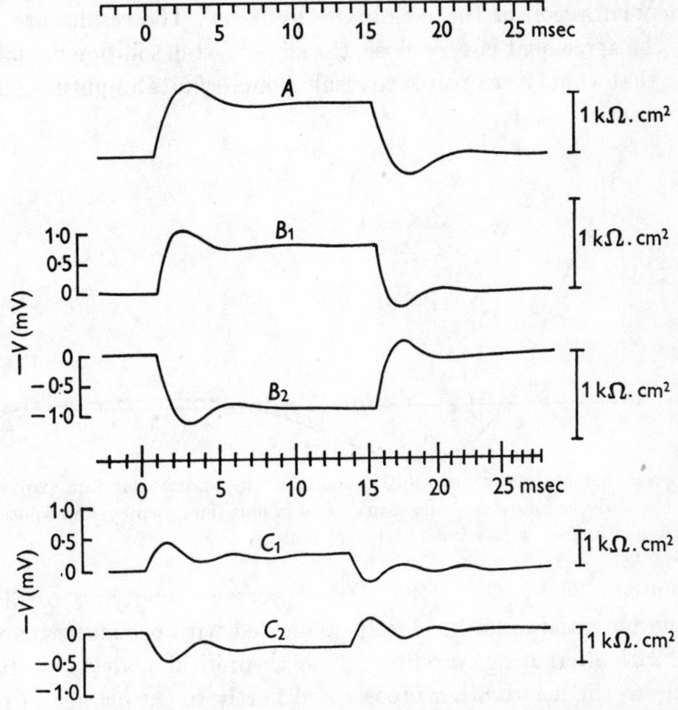

Fig. 23. *A*, solution of eqn. (35) for small constant current pulse; temperature 18·5° C; linear approximation. The curve shows $\delta V/\delta I$ (plotted upwards) as a function of time. *B*, changes in membrane potential associated with application of weak constant currents of duration 15 msec and strength $\pm 1{\cdot}49\,\mu A/cm^2$. B_1, cathodic current; B_2, anodic current. Depolarization is shown upward. Temperature 19° C. *C*, similar records from another fibre enlarged to have same time scale. Current strengths are $\pm 0{\cdot}55\,\mu A/cm^2$. Temperature 18° C. The response is unusually oscillatory.

frequency of oscillation. A fair degree of variability is to be expected in these respects since both frequency and damping depend on the values of the components of the resting conductance. Of these, g_{Na} and g_K depend critically on the resting potential, while $\bar{g}_l$ is very variable from one fibre to another.

Both theory and experiment indicate a greater degree of oscillatory behaviour than is usually seen in a cephalopod nerve in a medium of normal ionic composition. We believe that this is largely a direct result of using the long internal electrode. If current is applied to a whole nerve through a point electrode, neighbouring points on the membrane will have different membrane potentials and the resulting currents in the axis cylinder will increase the damping.

The linear solution for the behaviour of the theoretical membrane at small displacements provided a convenient check on our step-by-step numerical procedure. The response of the membrane at 6·3° C to a small short shock was calculated by this means and compared with the step-by-step solution for an initial depolarization of the membrane by 2 mV. The results are plotted in Fig. 24. The agreement is very close, the step-by-step solution deviating in the direction that would be expected to result from its finite amplitude (cf. Fig. 21).

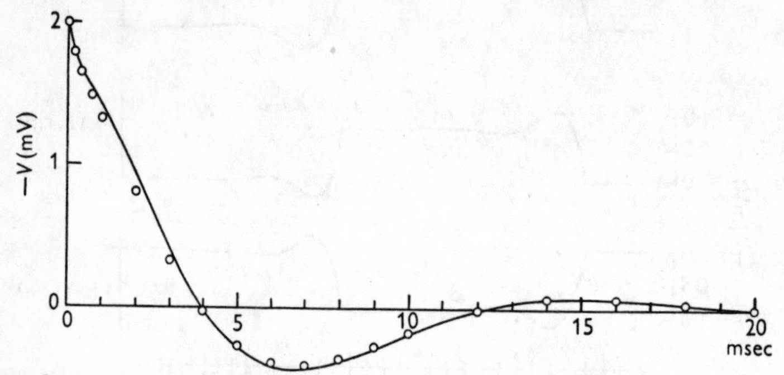

Fig. 24. Comparison of step-by-step solution and linear approximation. Eqn. (26), temperature 6° C; initial displacement of $-V=2$ mV. Continuous line: step-by-step solution. Circles: linear approximation with same initial displacement.

As pointed out by Cole (1941), the process underlying oscillations in membrane potential must be closely connected with the inductive reactance observed with alternating currents. In our theoretical model the inductance is due partly to the inactivation process and partly to the change in potassium conductance, the latter being somewhat more important. For small displacements of the resting potential the variations in potassium current in 1 cm² of membrane are identical with those in a circuit containing a resistance of 820 Ω in series with an inductance which is shunted by a resistance of 1900 Ω. The value of the inductance is 0·39 H at 25° C, which is of the same order as the 0·2 H found by Cole & Baker (1941). The calculated inductance increases 3-fold for a 10° C fall in temperature and decreases rapidly as the membrane potential is increased; it disappears at the potassium potential and is replaced by a capacity for $E > E_K$.

DISCUSSION

The results presented here show that the equations derived in Part II of this paper predict with fair accuracy many of the electrical properties of the squid giant axon: the form, duration and amplitude of spike, both 'membrane' and propagated; the conduction velocity; the impedance changes during the spike; the refractory period; ionic exchanges; subthreshold responses; and oscillations. In addition, they account at least qualitatively for many of the phenomena of excitation, including anode break excitation and accommodation. This is a satisfactory degree of agreement, since the equations and constants were derived entirely from 'voltage clamp' records, without any adjustment to make them fit the phenomena to which they were subsequently applied. Indeed any such adjustment would be extremely difficult, because in most cases it is impossible to tell in advance what effect a given change in one of the equations will have on the final solution.

The agreement must not be taken as evidence that our equations are anything more than an empirical description of the time-course of the changes in permeability to sodium and potassium. An equally satisfactory description of the voltage clamp data could no doubt have been achieved with equations of very different form, which would probably have been equally successful in predicting the electrical behaviour of the membrane. It was pointed out in Part II of this paper that certain features of our equations were capable of a physical interpretation, but the success of the equations is no evidence in favour of the mechanism of permeability change that we tentatively had in mind when formulating them.

The point that we do consider to be established is that fairly simple permeability changes in response to alterations in membrane potential, of the kind deduced from the voltage clamp results, are a sufficient explanation of the wide range of phenomena that have been fitted by solutions of the equations.

Range of applicability of the equations

The range of phenomena to which our equations are relevant is limited in two respects: in the first place, they cover only the short-term responses of the membrane, and in the second, they apply in their present form only to the isolated squid giant axon.

Slow changes. A nerve fibre whose membrane was described by our equations would run down gradually, since even in the resting state potassium leaves and sodium enters the axis cylinder, and both processes are accelerated by activity. This is no defect in describing the isolated squid giant axon, which does in fact run down in this way, but some additional process must take place in a nerve in the living animal to maintain the ionic gradients which are the immediate source of the energy used in impulse conduction.

After-potentials. Our equations give no account of after-potentials, apart from the positive phase and subsequent oscillations.

Conditions of isolated giant axon. There are many reasons for supposing that the resting potential of the squid giant axon is considerably lower after isolation than when it is intact in the living animal. Further evidence for this view is provided by the observation (Hodgkin & Huxley, 1952*c*) that the maximum inward current that the membrane can pass on depolarization is increased by previously raising the resting potential by 10–20 mV by means of anodally directed current. Our equations could easily be modified to increase the resting potential (e.g. by reducing the leak conductance and adding a small outward current representing metabolic extrusion of sodium ions). We have not made any calculations for such a case, but certain qualitative results are evident from inspection of other solutions. If, for instance, the resting potential were raised (by 12 mV) to the potassium potential, the positive phase and subsequent oscillations after the spike would disappear, the rate of rise of the spike would be increased, the exchange of internal and external sodium in a spike would be increased, the membrane would not be oscillatory unless depolarized, and accommodation and the tendency to give anode break responses would be greatly reduced. Several of these phenomena have been observed when the resting potential of frog nerve is raised (Lorente de Nó, 1947), but no corresponding information exists about the squid giant axon.

Applicability to other tissues. The similarity of the effects of changing the concentrations of sodium and potassium on the resting and action potentials of many excitable tissues (Hodgkin, 1951) suggests that the basic mechanism of conduction may be the same as implied by our equations, but the great differences in the shape of action potentials show that even if equations of the same form as ours are applicable in other cases, some at least of the parameters must have very different values.

Differences between calculated and observed behaviour

In the Results section, a number of points were noted on which the calculated behaviour of our model did not agree with the experimental results. We shall now discuss the extent to which these discrepancies can be attributed to known shortcomings in our equations. Two such shortcomings were pointed out in Part II of this paper, and were accepted for the sake of keeping the equations simple. One was that the membrane capacity was assumed to behave as a 'perfect' condenser (phase angle 90°; p. 505), and the other was that the equations governing the potassium conductance do not give as much delay in the conductance rise on depolarization (e.g. to the sodium potential) as was observed in voltage clamps (p. 509).

The assumption of a perfect capacity probably accounts for the fact that the initial fall in potential after application of a short shock is much less marked in the calculated than in the experimental curves (Figs. 12 and 21). Some of the initial drop in the experimental curves may also be due to end-effects, the guard system being designed for the voltage clamp procedure but not for stimulation by short shocks.

The inadequacy of the delay in the rise of potassium conductance has several effects. In the first place the falling phase of the spike develops too early, reducing the spike amplitude slightly and making the peak too pointed in shape (p. 525). In the membrane action potentials these effects become more marked the smaller the stimulus, since the potassium conductance begins to rise during the latent period. This causes the spike amplitude to decrease more in the calculated than in the experimental curves (Fig. 12).

The low calculated value for the exchange of internal and external sodium ions is probably due to this cause. Most of the sodium exchange occurs near the peak of the spike, when the potential is close to the sodium potential. The early rise of potassium conductance prevents the potential from getting as close to the sodium potential, and from staying there for as long a time, as it should.

A check on these points is provided by the 'anode break' action potential. Until the break of the applied current, the quantity n has the steady state value appropriate to $V = +30$ mV, i.e. it is much smaller than in the usual resting condition. This greatly increases the delay in the rise of potassium conductance when the membrane is depolarized. It was found that the spike height was greater (Table 4), the peak was more rounded, and the exchange of internal and external sodium was greater (Table 5), than in an action potential which followed a cathodal short shock.

The other important respect in which the model results disagreed with the experimental was that the calculated exchange of internal and external potassium ions per impulse was too large. This exchange took place largely during the positive phase, when the potential is close to the potassium potential and the potassium conductance is still fairly high. We have no satisfactory explanation for this discrepancy, but it is probably connected with the fact that the value of the potassium potential was less strongly affected by changes in external potassium concentration than is required by the Nernst equation.

SUMMARY

1. The voltage clamp data obtained previously are used to find equations which describe the changes in sodium and potassium conductance associated with an alteration of membrane potential. The parameters in these equations were determined by fitting solutions to the experimental curves relating sodium or potassium conductance to time at various membrane potentials.

2. The equations, given on pp. 518–19, were used to predict the quantitative behaviour of a model nerve under a variety of conditions which corresponded to those in actual experiments. Good agreement was obtained in the following cases:

(a) The form, amplitude and threshold of an action potential under zero membrane current at two temperatures.

(b) The form, amplitude and velocity of a propagated action potential.

(*c*) The form and amplitude of the impedance changes associated with an action potential.

(*d*) The total inward movement of sodium ions and the total outward movement of potassium ions associated with an impulse.

(*e*) The threshold and response during the refractory period.

(*f*) The existence and form of subthreshold responses.

(*g*) The existence and form of an anode break response.

(*h*) The properties of the subthreshold oscillations seen in cephalopod axons.

3. The theory also predicts that a direct current will not excite if it rises sufficiently slowly.

4. Of the minor defects the only one for which there is no fairly simple explanation is that the calculated exchange of potassium ions is higher than that found in *Sepia* axons.

5. It is concluded that the responses of an isolated giant axon of *Loligo* to electrical stimuli are due to reversible alterations in sodium and potassium permeability arising from changes in membrane potential.

REFERENCES

COLE, K. S. (1941). Rectification and inductance in the squid giant axon. *J. gen. Physiol.* **25**, 29–51.

COLE, K. S. & BAKER, R. F. (1941). Longitudinal impedance of the squid giant axon. *J. gen. Physiol.* **24**, 771–788.

COLE, K. S. & CURTIS, H. J. (1939). Electric impedance of the squid giant axon during activity. *J. gen. Physiol.* **22**, 649–670.

GOLDMAN, D. E. (1943). Potential, impedance, and rectification in membranes. *J. gen. Physiol.* **27**, 37–60.

HARTREE, D. R. (1932–3). A practical method for the numerical solution of differential equations. *Mem. Manchr lit. phil. Soc.* **77**, 91–107.

HODGKIN, A. L. (1951). The ionic basis of electrical activity in nerve and muscle. *Biol. Rev.* **26**, 339–409.

HODGKIN, A. L. & HUXLEY, A. F. (1952*a*). Currents carried by sodium and potassium ions through the membrane of the giant axon of *Loligo*. *J. Physiol.* **116**, 449–472.

HODGKIN, A. L. & HUXLEY, A. F. (1952*b*). The components of membrane conductance in the giant axon of *Loligo*. *J. Physiol.* **116**, 473–496.

HODGKIN, A. L. & HUXLEY, A. F. (1952*c*). The dual effect of membrane potential on sodium conductance in the giant axon of *Loligo*. *J. Physiol.* **116**, 497–506.

HODGKIN, A. L., HUXLEY, A. F. & KATZ, B. (1949). Ionic currents underlying activity in the giant axon of the squid. *Arch. Sci. physiol.* **3**, 129–150.

HODGKIN, A. L., HUXLEY, A. F. & KATZ, B. (1952). Measurement of current-voltage relations in the membrane of the giant axon of *Loligo*. *J. Physiol.* **116**, 424–448.

HODGKIN, A. L. & KATZ, B. (1949). The effect of temperature on the electrical activity of the giant axon of the squid. *J. Physiol.* **109**, 240–249.

KEYNES, R. D. (1951). The ionic movements during nervous activity. *J. Physiol.* **114**, 119–150.

KEYNES, R. D. & LEWIS, P. R. (1951). The sodium and potassium content of cephalopod nerve fibres. *J. Physiol.* **114**, 151–182.

LORENTE DE NÓ, R. (1947). A study of nerve physiology. *Stud. Rockefeller Inst. med. Res.* **131**, **132**.

PUMPHREY, R. J., SCHMITT, O. H. & YOUNG, J. Z. (1940). Correlation of local excitability with local physiological response in the giant axon of the squid (*Loligo*). *J. Physiol.* **98**, 47–72.

J. Physiol. (1953) 121, 403-414

MOVEMENT OF RADIOACTIVE POTASSIUM AND MEMBRANE CURRENT IN A GIANT AXON

By A. L. HODGKIN and A. F. HUXLEY
From the Physiological Laboratory, University of Cambridge

(*Received* 23 *February* 1953)

One remarkable property of nerve fibres is that they are capable of passing large outwardly-directed currents for considerable periods of time when depolarized by 10–50 mV (Cole & Curtis, 1941; Hodgkin & Huxley, 1952). These currents may be nearly one hundred times greater than those associated with a corresponding increase in membrane potential. They are interesting physiologically because they are of the right sign and magnitude to explain the rapid recharging of the membrane capacity during the falling phase of the action potential. Since there is evidence that potassium ions move outwards during activity it has been assumed that the prolonged outward current associated with depolarization is carried by potassium ions (Hodgkin & Huxley, 1952). The experiments described here were designed to test this point and the affirmative answer which they provide has already been mentioned in earlier papers.

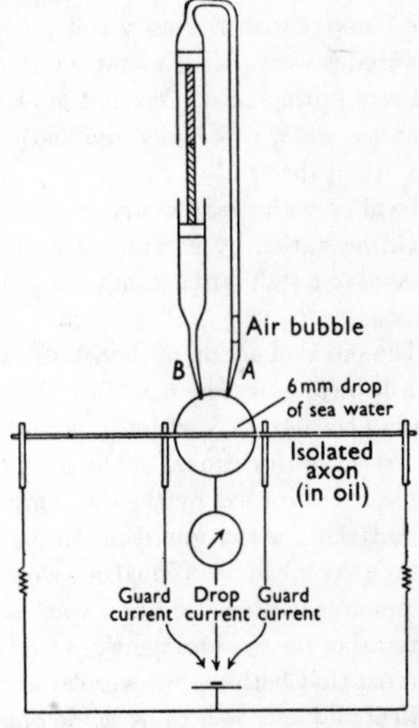

Fig. 1. Diagram illustrating method.

The principle of the method is illustrated by Fig. 1. A single nerve fibre from *Sepia* was isolated and soaked in a solution containing radioactive potassium for a few hours. It was then mounted in oil with the central portion in a drop of sea water about 6 mm in diameter. The contents of this drop were changed periodically by operating a pair of syringes coupled in 'push-pull'. After each change the fluid collected in one syringe was ejected on to a nickel dish, dried and analysed for ^{42}K with a Geiger counter. The outward flux of potassium at rest and in the presence of current was calculated from the quantities of ^{42}K leaving the nerve in unit time and from measurements of specific activity made at the end of the experiment. The difference between the two fluxes gave the extra leakage of internal potassium associated with any particular current. The electrical quantity with which this must be compared is not the total current through the nerve and external fluid but the component which leaves the axis cylinder in the region occupied by the drop. In the present experiments sufficient current was supplied to the guard electrodes to make their ends equipotential with the drop. Under these conditions no current could flow along the outside of the nerve between the drop and the guards so that the current crossing the membrane in the region of the drop was necessarily equal to that recorded by the galvanometer.

METHOD

Material

Giant axons 170–260 μ in diameter and 40–70 mm in length were isolated from *Sepia officinalis* by the usual methods (Keynes, 1951).

Electrode system and electrical connexions

The method of supplying current to the nerve fibre is illustrated in Fig. 1, and, in greater detail but still diagrammatically, in Fig. 2. The nerve was held at either end by forceps and was mounted about 0·5 mm above the bottom of a Perspex chamber which was filled with oil. A drop of sea water 6 mm in diameter and about 2 mm in height was located and stabilized by a circular groove in the bottom of the Perspex chamber. Electrical contact with the drop was made by two holes (80 μ in diameter) which were plugged with agar sea water and connected through tubes drilled in the Perspex with the sea-water pools P_1 and P_2, which contained Ag-AgCl electrodes. The guard electrodes were placed about 0·6 mm on either side of the drop and consisted of agar wicks about 0·4 mm in thickness. One end of each wick was brought into contact with a small hole filled with agar sea water which communicated through a wider tube with an open pool of sea water (P_3 or P_4). The other end of each wick was connected to the common guard electrode through a glass tube filled with agar sea water and a rubber tube filled with sea water. The resistance of the latter could be varied by compressing the rubber tube with a screw. In practice one tube was compressed to a standard extent giving the fixed resistance R_{g1} while the other was used as the variable resistance R_{g2}.

The mode of action of the whole system is best described by considering the procedure followed in a typical experiment. After the fibre had been mounted, a standard current was applied by closing the switch S with the potentiometer R_1 set to a suitable value. The potentiometer R_2 was increased until removal of the fluid bridge F_1 gave a deflexion of less than 50 μV in the potential difference recorded by the d.c. amplifier. The potential of the drop was then equal to the mean potential of the two guard electrodes. If the nerve had been perfectly uniform and if the resistance of the guards had been equal one would expect that the two guard electrodes would be equipotential. In practice neither condition was easy to realize, and it was therefore necessary to vary R_{g2} until removal of the fluid bridge F_2 as well as F_1 caused no detectable shift in potential. This adjustment ensured that both guards were at the same potential as the drop. The setting of R_{g2} was not at all critical and only had to be made once in each experiment. The position of R_2, on the other hand, had to be altered fairly frequently since the fraction of the total current which flowed into the drop varied with the membrane conductance.

If the circuit for supplying current was disconnected it was often possible to detect a change in p.d. when the fluid bridge F_1 was removed. This effect was due to small differences in resting potential and its sign was usually such as to make the potential of the guards lower than that of the drop. Under these conditions current must have been flowing from drop to guards along the fluid outside the nerve. This means that the drop region was not in a condition of zero membrane current, as it should have been if the analysis is to be exact. The difficulty was overcome by taking resting measurements with the lead to the drop disconnected and supplying the guards with sufficient current to bring them to the same potential as the drop.

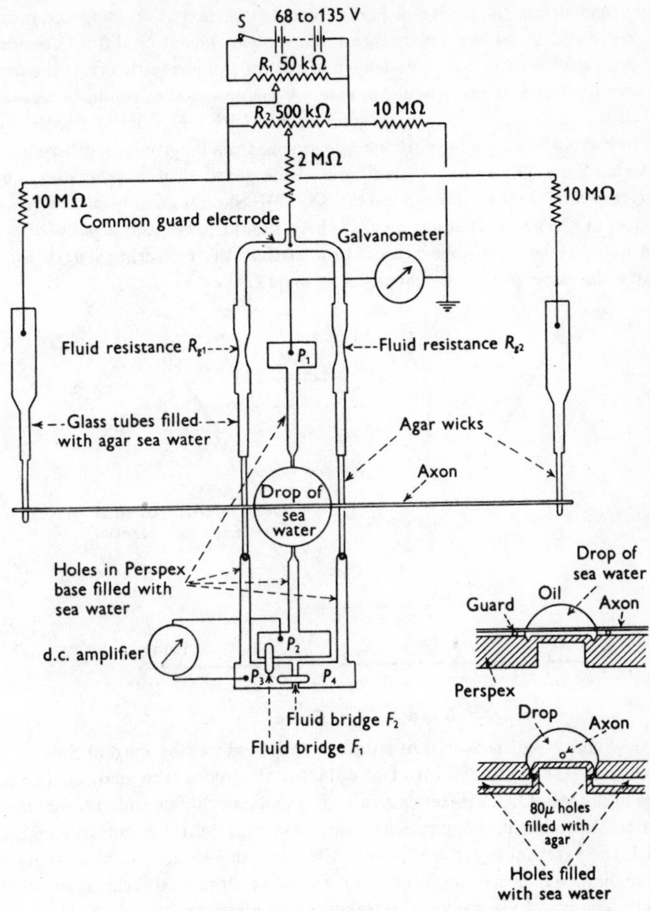

Fig. 2. Diagram of nerve chamber and electrical connexions. Inset: vertical sections through drop, in plane of nerve (above) and at right angles (below). The Perspex was cut away below the drop in order to allow β rays to reach a Geiger counter.

The reason for using fluid bridges (F_1 and F_2) and fluid resistors (R_{g1} and R_{g2}) instead of the metal equivalent of these elements is that it makes the operation of the whole apparatus independent of the potential differences across the Ag-AgCl electrodes used for supplying current or recording potential. It was therefore unnecessary to take any special precautions to obtain uniform electrodes or to prevent them polarizing when currents were applied.

Mode of action of guard system

In considering the action of the guard electrodes it is important to know whether their operation will be upset by the use of a finite distance (0·6 mm) between guard and drop instead of an infinitesimal one as assumed in the simplified discussion on p. 404. This is best done by treating the nerve as a linear cable and calculating the distribution of potential and membrane current by standard methods. The results of such an analysis are given in Fig. 3. A fairly low value for the membrane resistance has been chosen in order to represent the condition of a depolarized axon. It will be seen that the external voltage gradient is zero about half-way between the edge of the drop and the guards. (The exact distance is 0·303 mm from the edge of the drop.) These points of zero voltage gradient define the length of nerve over which current crossing the membrane goes to the central electrode. Outside these limits any current in the external fluid necessarily goes to the guards. This means that the collecting length of the central electrode is not the drop itself but the drop plus about 0·3 mm of axon on either side. A correction here would be uncertain since it would depend on the variation of membrane conductance with membrane potential. Fortunately this source of error was offset by a very similar one arising from longitudinal diffusion of potassium ions in the external fluid. The cross-sectional area of the guard electrodes was large compared to that of the external fluid so that the concentration of ^{42}K should have been nearly as low at the guarded points as in the drop. This means that there should have been a point of zero diffusion gradient about half-way between guards and drop so that the collecting length for ^{42}K should have been nearly the same as that for membrane current.

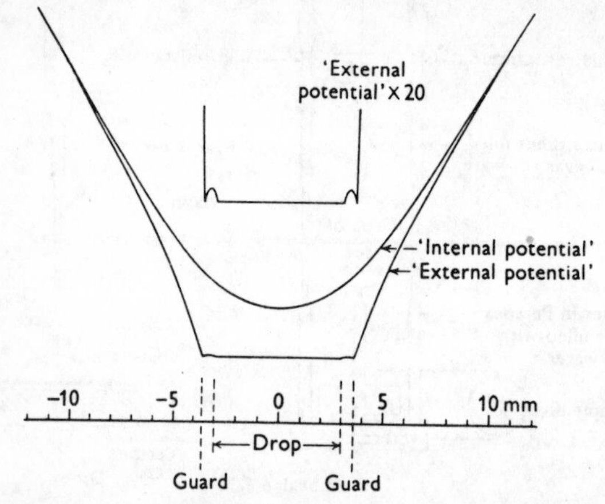

Fig. 3. Theoretical changes of potential in a linear cable produced by current flowing into a large central electrode (the drop) and into two adjacent electrodes (the guards). The curves show the change in potential from its resting value in the axis cylinder and external fluid and were calculated on the following assumptions: width of drop, 6 mm; standard equations of cable theory with the external resistance zero in the drop and twice the internal resistance elsewhere; axon diameter 200 μ; specific resistance of axis cylinder, 50 Ω cm; membrane resistance 1000 Ω cm^2. The guard electrodes were taken as infinitesimal and the cable as infinite. The ratio of guard to drop current is determined by the condition that guard and drop are equipotential.

Modified method for large cathodal currents

The guard system worked well for currents in which the component into the drop was less than 2 μA but was unsatisfactory for higher currents. With increasing cathodal currents a progressively larger fraction of the total current went to the guards and eventually damaged the nerve. The change in the relative proportions of drop current to guard current is a necessary consequence of the rectifying properties of the membrane which make the membrane resistance fall when the axon is depolarized. In theory it should be possible to improve matters by reducing the gap between guards and drop. But this was not practicable because surface tension effects tended to constrict the fibre if the gap was too short. In any case short gaps were unstable mechanically and therefore unsatisfactory.

In order to avoid these difficulties the guard electrodes were removed and current outside the drop was reduced to a relatively low value by wiping the fibre periodically with a solution of high resistivity. The composition of this solution was: 750 mM-dextrose, 10 mM-K, 11 mM-Ca, 37 mM-Mg, 103 mM-Cl and 3 mM-HCO_3; its specific resistance was 6·85 times that of sea water. The drop itself contained sea water, as in the previous method. In order to treat the parts of the fibre immediately adjacent to the sea-water drop, it was necessary to draw the small drop of sugar solution along the nerve until it finally coalesced with the sea-water drop. The sea water in the drop was changed after each wipe so as to avoid contamination from any ^{42}K collected by the drop of sugar solution.

This method was intended for use with large cathodal currents which were mainly concentrated at the edges of the drop. If the effective space constant is small compared with the width of the drop it can be shown that the fraction of the total current which crosses the membrane in the drop is $r_1/\{r_1+r_2+\sqrt{[r_2(r_1+r_2)]}\}$, where r_1 and r_2 are the resistances per unit length of the external fluid and axis cylinder. This statement can be proved by a method similar to that used by Cole & Curtis (1941) without assuming that the membrane conductance is constant. According to Weidmann (1951) $r_1/r_2=1\cdot9$ for an oil-immersed axon which has previously been in sea water. In the present case the wiping solution was 6·85 times more resistant than sea water so that r_1/r_2 is taken as 13 and the drop current is then 0·73 times the total current. This ratio was adopted in calculating the currents enclosed in parentheses in Table 2.

Drop-changer

This was built from two matched syringes of bore 0·7 cm and capacity 1·5 ml. The plungers were operated by a rack and pinion in order to give a smooth motion. When collecting from the drop, both nozzles were placed below its surface and the plungers were moved downwards until the air bubble had travelled from the tip of tube *A* (Fig. 1) to a position about half-way up. This operation transferred the contents of the drop to tube *A* and replaced it with an equal amount of fresh sea water from *B*. The drop-changer was then raised and swung into a position suitable for ejection and refilling. Nozzle *A* was placed above a nickel dish while *B* was dipped into fresh sea water. The plungers were then moved upwards so that all the fluid below the air bubble was ejected on to the nickel dish and *B* was refilled with fresh sea water. The volume of fluid used to wash out the drop was 0·6 ml. which was about 10 times the volume of the drop. The apparatus was tested in the absence of a nerve by starting with a solution of known radioactivity in the drop and measuring the amount left after operating the drop-changer. This showed that 90–95 % of the radioactivity was removed in a single change. The collecting efficiency sometimes appeared to be less good when the nerve was in position and two changes were normally made after a period of current flow. Errors due to inefficient collection would have had little effect on the estimates of resting leakage because these were usually based on a number of measurements made with the nerve in an approximately steady state. Under these conditions a 'carry-over' from one drop to the next would not influence the result.

Radioactive tracer methods

Samples of K_2CO_3 or $KHCO_3$ were irradiated at A.E.R.E., Harwell, turning some of the potassium into ^{42}K, and were subsequently converted into ^{42}KCl by the methods mentioned by Hodgkin & Keynes (1953). Artificial sea water containing ^{42}K was made up with a K concentration of 20 mM and the concentration of other ions approximately as stated by Keynes (1951, table 1). Axons were left in the radioactive sea water for 1–4 hr and were then transferred to ordinary sea water (10 mM-K) for about 10 min in order to wash off extracellular ^{42}K. The next operations were to mount the fibre in the measuring chamber and to determine the amount of labelled potassium in the central part of the axon with a screened Geiger counter placed below the measuring cell. This determination influenced the choice of a suitable time for collecting ^{42}K in the drop of sea water but was not used in the final calculation.

The ^{42}K content of each drop was determined by drying the drop on a nickel dish and counting the β particles emitted with an end-window Geiger counter of conventional design. The results were standardized by comparing them with the counting rates produced by a weighed quantity of a ^{42}K solution of known concentration.

Since potassium ions were collected from only 6 mm of nerve the counting rates were low, and it was sometimes necessary to count each sample for 30–60 min. The labour and loss of efficiency resulting from prolonged manual counting led us to design a simple form of automatic counter which handled twelve samples without attention. A description of this device would be out of place since it was not finished until the experiments were nearly complete.

At the end of each experiment about 15 mm of the central part of the axon was cut out and dried on a quartz thread. The total quantity of ^{42}K was obtained by counting the fragment of nerve in the same way as the dried drop of sea water.

Potassium analyses

After the ^{42}K content of the dried nerve had been measured it was stored in a quartz tube and subsequently analysed for total potassium. These determinations were carried out with the help of Dr Keynes by the method of activation analysis (Keynes & Lewis, 1951).The method gave both sodium and potassium but about half the sodium was extracellular since we did not soak the fibres in a choline solution. Potassium concentrations were calculated from the total potassium and the axon diameter using a correction for a layer of extracellular fluid $20\,\mu$ in thickness. This figure, rather than $13\,\mu$ (Keynes & Lewis, 1951), was chosen because it gave a reasonable value for the internal sodium concentration. However, the total correction for extracellular potassium was only 2% so that the difference is unimportant.

EXPERIMENTS AND RESULTS

A typical experiment and method of calculation

Although the sequence of operations altered slightly during the course of the work, it is easiest to describe the procedure by considering a single experiment in detail. The essential results are given in Table 1; qualitatively this shows that the outward flux of potassium is reduced by an anodal current and increased by a cathodal one.

The method of calculation is illustrated by considering the extra leakage due to a current of $1\,\mu$A lasting 9 min (sample 8). Samples 7 and 10 gave a mean resting leakage of 1·065 counts/min per min which is in good agreement with values obtained in the rest of the experiment. Most of the extra leakage due to current occurred in sample 8, but about 10% was not removed in a single change and appeared in sample 9. The extra leakage due to current was therefore taken as the total quantity in samples 8 and 9 minus the amount due to resting leakage, i.e. $52{\cdot}0 + 9{\cdot}8 - 14{\cdot}8 \times 1{\cdot}065 = 46{\cdot}0$ counts/min. At the end of the experiment the central 15 mm of axon was found to give 783 counts/min (all counting rates have been corrected for decay). Two months later activation analysis showed that this piece of axon contained 115,000 p.mole of potassium (1 p.mole $\equiv$ 1 $\mu\mu$mole $\equiv 10^{-12}$ mole). The last two measurements give the counting rate of potassium in the nerve, and this ratio is assumed to apply to potassium ions which crossed the membrane during the experiment as well as to those left in the axon (see p. 411 for discussion of possible errors). On this basis the extra leakage of potassium associated with outward transport of charge is $46 \times 115{,}000/783 = 6760$ p.mole. The total charge crossing the membrane is $1 \cdot 07 \times 9 \times 60 = 578$ μcoulomb or 6000 p.mole of monovalent cation. The average current density in the drop can be obtained by dividing figures such as these by the duration of current and the area of membrane in the drop. Resting fluxes are obtained in a similar manner. Since the counter was standardized with a weighed sample of the solution used to make the nerve radioactive it was also possible to calculate the fraction of labelled potassium present. The complete results of this experiment, together with others, are summarized in Table 2.

TABLE 1. Results of a typical experiment

Axon 3: Diameter 177 μ

Time (min)	Time interval (min)	Condition of nerve	Sample	Counting rate of sample (counts/min)	Resting leakage (counts/min per min)
5	Excitability test: threshold 27				
22	9·6	Resting	1	10·2	1·06
32	10·4	Resting	2	11·5	1·10
42	10·0	0·107 μA (*a*) for 9·25 min	3	6·9	—
52	10·0	Resting	4	11·6	1·16
62	10·0	0·545 μA (*c*) for 9·25 min	5	26·9	—
69	3·5	Resting	6	6·0	—
76	10·0	Resting	7	11·8	1·18
86	10·0	1·07 μA (*c*) for 9·0 min	8	52·0	—
93	4·8	Resting	9	9·8	—
101	10·0	Resting	10	9·5	0·95
111	10·0	0·545 μA (*c*) for 9·25 min	11	33·2	—
118	3·8	Resting	12	4·0	—
125	10·2	Resting	13	6·5	0·64
135	Excitability test: threshold 26				
141	15 mm cut out and dried				
—	Counting rate in excised 15 mm determined later as 783 counts/min				
—	Total potassium in excised 15 mm determined later as 115,000 p.mole				

Notes. (*a*) indicates an anodal current; (*c*) a cathodal current. All counting rates have been corrected for background and decay. The times in the first column refer to the mid-time of each operation.

Results

Six complete experiments were carried out by the guard method and three by the sugar method. The results are given in Table 2 and are seen to be similar to those described in the previous paragraph. An instructive way of examining the data is to plot the increment in the outward flux of potassium against the current density as in Fig. 4. The straight line in this graph was drawn through the origin with a slope given by the Faraday, and the fit to the cathodal points is evidence that steady cathodal currents are carried mainly by potassium ions moving outward through the membrane. A similar conclusion may be drawn from the fact that the ratios in column (8) of Table 2 are near unity.

TABLE 2. Collected results

(1) Axon	(2) Resting leakage (p.mole cm^{-2} sec^{-1})	(3) Current (μA)	(4) Mean current density (μA/cm^2)	(5) Duration of current (sec)	(6) Chemical equivalent of total current (p.equiv)	(7) Extra K leakage (p.mole)	(8) Ratio (7)/(6)	(9) Mean increase in outward K flux (p.mole cm^{-2} sec^{-1})	(10) S.E. of column 7 (as %)
1	58	0·545	16·8	570	3220	3820	1·19	207	4
		−0·107	−3·3	550	−610	−428	(0·70)	−24	17
2	45	0·545	11·6	530	2990	4125	1·38	165	16
		−0·107	−2·3	545	−604	−746	(1·24)	−29	56
		1·07	22·7	540	6000	5470	0·91	215	14
3	83	−0·107	−3·2	565	−626	−639	(1·02)	−34	34
		0·545	16·3	565	3190	2600	0·81	138	15
		1·07	32·0	540	6000	6760	1·13	375	8
		0·545	16·3	565	3190	3850	1·21	204	12
4	54	1·07	23·9	550	6080	6540	1·08	267	7
		2·58	58·0	225	6020	6330	1·05	630	7
		1·07	23·9	560	6200	5340	0·86	214	11
5	40	−0·107	−2·18	565	−626	−469	(0·75)	−17	23
		2·08	42·4	260	5600	6730	1·20	528	7
6	40	(0·78)	(17·9)	560	(4510)	3780	0·84	155	9
		(4·45)	(102)	120	(5540)	5330	0·96	1020	6
7	52	(0·78)	(17·5)	573	(4620)	5410	1·17	212	5
		(3·72)	(83·6)	120	(4630)	4590	0·99	860	5
8	85	(8·31)	(230)	60	(5160)	4300	0·83	1980	17
9	35	−0·26	−6·6	1170	−3150	−714	(0·23)	−16	59
Mean	55						1·04		

Additional measurements

Axon number	1	2	3	4	5	6	7	8	9	Mean
Diameter (μ)	172	250	177	237	260	232	236	192	208	218
Potassium concentration (m.mole/l.)	304	244	306	254	306	251	276	306	288	282
Fraction of potassium exchanged ...	0·168	0·079	0·065	0·214	0·051	0·15	0·065	0·041	0·076	

Notes. All axons except 4 were excitable at the end of the experiment. Axons 6, 7, 8 were studied by the 'sugar' method: the bracketed values in cols. (3) and (4) were calculated by the method described in the text. A minus sign in cols. (3) and (4) indicates an anodal current. Col. (6) is col. (3) × col. (5)/96500. Current densities and fluxes are average values obtained by dividing drop current by the area of membrane in the drop. Anodal ratios have been omitted in calculating the mean of col. (8). Temperature 16–20° C. Col. (10) gives the standard error due to the counting rate of the samples.

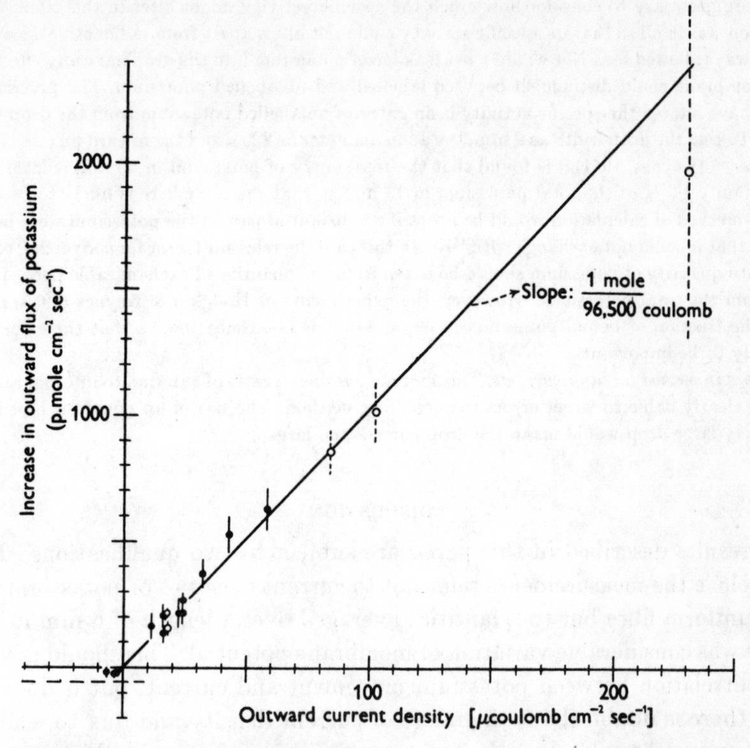

Fig. 4. Abscissa: mean outward membrane current density in drop (=total membrane current in drop divided by area of membrane in drop). Ordinate: mean increment in potassium outflux associated with flow of current (=outflow ÷ by area of membrane in drop). The vertical lines show ± twice the S.E. estimated from the observed counting rates. Full circles and continuous lines were obtained by the guard method; hollow circles and dotted lines by the modified method for large currents. The horizontal line at -55 p.mole cm^{-2} sec^{-1} is drawn at a level corresponding to complete suppression of the average resting outflux.

Sources of error

Counting errors. These were estimated in each experiment and the results are given in Table 2, column (10). The large standard error of the extreme point in Fig. 4 arose because only freshly isolated axons will remain excitable after very large currents. This meant that the duration of treatment in ^{42}K was shorter than we should otherwise have employed.

Errors in collecting potassium. A trace of ^{42}K must have been lost by diffusion into the small holes used to make electrical connexion with the drop: the error here was calculated as about 0·3%. The effect of diffusion from the parts in oil was probably offset by a similar error in measuring current (see p. 406).

Errors in potassium analysis. Keynes & Lewis (1951) estimated the standard error of their method as ±2%. Our measurements may have been somewhat less accurate since we used smaller quantities of axoplasm.

Errors in measuring specific activity. In working out the results it was assumed that the specific activity of the potassium which left the axon during the experiment could be obtained from the ratio of ^{42}K to total potassium at the end of the experiment. The time between the application of current and the end of the experiment varied between 30 and 110 min and averaged 50 min. It is therefore necessary to consider how much the specific activity might alter in this time. Most of the axon was in oil so that its specific activity could not alter, apart from radioactive decay which was always allowed for. Nor would a net leakage of potassium into the drop have any effect unless the membrane could distinguish between labelled and unlabelled potassium. The process which could have altered the specific activity is an entry of unlabelled potassium from the drop into the axon. Taking the drop width as 6 mm, the axon diameter as 220 μ and the inward flux as 17 p.mole cm^{-2} sec^{-1} (Keynes, 1951) it is found that the total entry of potassium in 50 min is 2000 p.mole. This is only 1·3% of the total potassium in 15 mm so that the error here is likely to be small.

The method of calculation would be upset if a substantial part of the potassium were bound so firmly that it could not exchange with ^{42}K. In this case the relevant factor for converting counting rate into quantity of potassium should be taken from the quantity of exchangeable potassium and not from the total potassium. However, the experiments of Hodgkin & Keynes (1953) indicate that the fraction of bound potassium in *Sepia* axons is less than 10% so that the error here is unlikely to be important.

Errors in method for large currents. This method assumes a ratio of external to internal resistance and is clearly liable to larger errors than the first method. The use of an equation based on an infinitely large drop would make the drop current too large.

DISCUSSION

The results described in this paper are subject to two qualifications. In the first place the measurements refer not to current densities or potassium fluxes in a uniform fibre but to quantities averaged over a length of 6 mm in which there was considerable variation of membrane potential. This should not upset the correlation between potassium movement and current, but it does mean that there is doubt about the range of current density and flux to which the measurements apply. In order to form some idea of the way in which membrane current varied over the drop we assumed that the steady state relation between membrane potential and current density was similar to that in *Loligo* (Hodgkin, Huxley & Katz, 1952, fig. 13). On this basis it can be shown, by a method similar to that of Cole & Curtis (1941), that an average current density of 100 $\mu A/cm^2$ over 6 mm would correspond to a current density of about 1000 $\mu A/cm^2$ at the edge and a maximum depolarization of roughly 50 mV. This implies that the outward flux of potassium at the edge would be about 10,000 p.mole cm^{-2} sec^{-1}, which is 200 times greater than the resting outflux. These estimates are obviously rather uncertain but they indicate that the experiments described here apply to the range of current density and membrane potential which we used with *Loligo*.

The second reservation which must be made is that the approximate equivalence of outward potassium flux and membrane current does not necessarily mean that potassium is the only ion concerned in carrying current. In order to establish this point in a rigorous manner it would be necessary not only to make more accurate measurements but also to study potassium influx at the same time as the outflux. If there were no change in permeability one would expect that depolarization would increase the outward potassium flux and decrease the influx. In this case the contribution of potassium ions to the current would be larger than that found by our method. On the other hand, since the permeability to potassium almost certainly rises when the fibre is depolarized there may be an increase in influx, so that our method may overestimate the contribution of potassium ions. Although we cannot eliminate this possibility it is not thought to be particularly important. With the larger cathodal currents the membrane potential would be near zero and the ratio of inward to outward flux ought not to exceed the concentration ratio across the membrane. In this case the influx would be less than 4% of the outflux so that it may reasonably be neglected. There is more doubt about the weaker cathodal currents, but it seems unlikely that the influx could have exceeded one-fifth of the outflux. On general grounds one would expect the transport number of potassium to approach unity over the range of membrane potentials in which the membrane behaves like a potassium electrode but to fall off markedly as the potential approaches its resting value.

Little need be said about the subsidiary results in Table 2. The resting fluxes are similar to Keynes's (1951) average value but somewhat greater than the figures which he gives for fresh fibres. This is not surprising since we found it necessary to soak axons in ^{42}K sea water for long periods in order to obtain reasonable counting rates. The average figure for the internal potassium concentration also agrees reasonably with those given by Keynes & Lewis (1951) for fibres which had been isolated for several hours. Any comment on the contribution of potassium movement to anodal currents would be premature since our method gave no information about variations in potassium influx, which are likely to be of considerable importance under these conditions.

SUMMARY

1. A method for comparing membrane current and potassium outflux was applied to isolated axons from *Sepia officinalis*.

2. The outward flux of potassium was decreased under an anode and increased under a cathode.

3. Over a wide range of cathodal currents the quantity of potassium leaving 6 mm of axon was equivalent to the total electric charge passing through the same area of membrane in the same time.

4. It is concluded that the steady outward current associated with depolarization is mainly carried by potassium ions.

We are greatly indebted to Dr R. D. Keynes for help with the tracer methods and potassium analyses. The expenses of the work were met by grants from the Rockefeller and Nuffield Foundations.

REFERENCES

COLE, K. S. & CURTIS, H. J. (1941). Membrane potential of the squid giant axon during current flow. *J. gen. Physiol.* **24**, 551–563.

HODGKIN, A. L. & HUXLEY, A. F. (1952). Currents carried by sodium and potassium ions through the membrane of the giant axon of *Loligo*. *J. Physiol.* **116**, 449–472.

HODGKIN, A. L., HUXLEY, A. F. & KATZ, B. (1952). Measurement of current–voltage relations in the membrane of the giant axon of *Loligo*. *J. Physiol.* **116**, 424–448.

HODGKIN, A. L. & KEYNES, R. D. (1953). The mobility and diffusion coefficient of potassium in giant axons from *Sepia*. *J. Physiol.* **119**, 513–528.

KEYNES, R. D. (1951). The ionic movements during nervous activity. *J. Physiol.* **114**, 119–150.

KEYNES, R. D. & LEWIS, P. R. (1951). The sodium and potassium content of cephalopod nerve fibres. *J. Physiol.* **114**, 151–182.

WEIDMANN, S. (1951). Electrical characteristics of *Sepia* axons. *J. Physiol.* **114**, 372–381.

12

J. Physiol. (1955) 128, 28–60

ACTIVE TRANSPORT OF CATIONS IN GIANT AXONS FROM *SEPIA* AND *LOLIGO*

BY A. L. HODGKIN AND R. D. KEYNES

From the Physiological Laboratory, University of Cambridge, and the Laboratory of the Marine Biological Association, Plymouth

(*Received* 30 *August* 1954)

Like many other living cells, nerve and muscle fibres use metabolic energy to move sodium and potassium ions against concentration gradients. In excitable tissues this process is of particular interest because it is essential for building up the concentration differences on which the conduction of impulses depends. When a nerve fibre is stimulated it undergoes rapid changes in permeability which allow first sodium and then potassium to move down concentration gradients. The effect of a train of impulses is therefore to raise the sodium and to lower the potassium concentration inside the cell. In giant nerve fibres from cephalopods the concentration changes are small unless the fibres are made to conduct many thousand impulses. Nevertheless, if the fibres are to be of permanent use to the animal, they must be equipped with a mechanism for reversing the ionic interchange which occurs during the action potential. In contrast to the conduction mechanism, where the ions move down pre-existing concentration gradients, the recovery process requires a supply of metabolic energy, since the ions move from weak to strong solutions. This forced movement uphill is commonly called active transport. It can conveniently be studied with radioactive tracers, and the present paper describes experiments in which ^{24}Na and ^{42}K were used to observe the extrusion of sodium and the absorption of potassium by giant axons dissected from squids and cuttlefish. It is followed (Hodgkin & Keynes, 1955) by an experimental analysis of the passive potassium movements which remain when the secretory uptake of potassium has been inhibited by dinitrophenol.

Preliminary accounts of the experiments described here were given at a meeting of the Physiological Society and at the symposium on active transport held by the Society for Experimental Biology (Hodgkin & Keynes, 1953*a*, *b*, 1954).

METHODS

Materials

Most of the experiments were done with unbranched axons, 160–300 μ in diameter and 40–70 mm in length, from *Sepia officinalis*. On a few occasions 60–100 mm lengths of 450–650 μ axons from *Loligo forbesi* were used; these have small branches which must be cut at a point well away from the main axon. All fibres were carefully cleaned from adherent tissue, and we tried to have at least 15 mm of intact axon on each side of the stretch used for measurement. Most fibres were stimulated at 50/sec for 5–10 min at the beginning of the experiment. At first this was done in order to load the fibres with ^{24}Na, but eventually the same procedure was adopted in all experiments, in order to ensure that the recovery process was working under approximately standard conditions.

Measurement of influxes

The apparatus used to measure the influxes of sodium and potassium was that employed by Keynes (1951*a*, fig. 1). The fibre was mounted on a movable assembly so that it could be placed above a Geiger counter or swung to one side and dipped into a dish containing ^{24}Na or ^{42}K sea water. For temperature experiments the dish containing the tracer solution was immersed in a water-bath kept near 0° C or at room temperature. The procedures for washing off extracellular ions and for calibrating the equipment were those described by Keynes (1951*a*). The method was quite straightforward when the potassium influx was measured, but when it was used to determine the resting sodium entry, there was often difficulty in deciding how much of the observed gain of radioactivity was due to intracellular ^{24}Na. The values given in this paper were obtained by taking counts for about an hour after dipping the fibre in ^{24}Na sea water. After correcting for the decay of the isotope, these results were plotted on a logarithmic scale, and the influx was obtained by linear extrapolation from the observations made 20–60 min after removal from the radioactive solution. This was done because the shape of the curve for loss of radioactivity, and the effect on it of inhibitors, suggested that the continued presence of some extracellular sodium might affect the answer during the first 20 min of washing. This time is much longer than would be expected from the rate at which fibres become inexcitable in a sodium-free medium, but it seems reasonable to suppose that a certain quantity of ^{24}Na sticks rather tenaciously to the connective tissue. The uncertainty caused by extracellular sodium was much less when the ^{24}Na was introduced by stimulation, since this greatly increased the amount entering the axis cylinder without altering the component attributed to sodium in the extracellular space.

The observed gain of radioactivity was corrected in every case to allow for the slight loss of radioactivity occurring during the period of exposure to the radioactive ions (see the procedure described by Keynes, 1951*a*, 1954), using an appropriate value for the rate constant of ionic exchange. Since this period was invariably short compared with the exchange time constant, the correction was usually less than 5%.

Loading with ^{24}Na *or* ^{42}K

Before effluxes could be investigated it was necessary to introduce the radioactive ions into the axis cylinder. With sodium the standard method was to stimulate at 50/sec for 5–10 min in about 3 ml. of an artificial sea water containing 486 mM of Na labelled as ^{24}Na. (The mixture of ^{23}Na and ^{24}Na in this solution will be called labelled sodium and designated Na*.) The axon was held at either end in forceps and looped in a V under one glass hook, or in a flat U under two hooks. The extreme ends of the axon were lifted out of the solution in order to stimulate and record electrically. The stimulus was applied between one pair of forceps and an earthed platinum wire in the tracer solution, while the action potential was recorded between the other pair of forceps and the common earth. The ends of the axon were kept moist by dipping them into the tracer solution once every 2 min, or, if it was desirable to keep radioactivity away from the ends, by dipping the whole fibre into inactive sea water.

For experiments with potassium, the fibres were soaked either for 2–4 hr in about 25 ml. of an artificial sea water containing 20·7 mM-K labelled as ^{42}K, or for 1–2 hr in one containing 52 mM-K.

The temperature was usually kept near 0° C when 52 mM-K was employed; this was not done with 20·7 mM-K because the influx from this solution depends largely on the metabolic activity of the fibre, and has a high temperature coefficient (see p. 44).

Determination of effluxes of radioactive ions

A. *Measurement of total radioactivity of nerve*

The first efflux measurements were made by the method described by Keynes (1951*a*). After the axon had been loaded with radioactive ions it was set up in flowing sea water above a Geiger counter. The efflux of labelled ions could then be obtained from the rate at which the radioactivity declined with time. This method is convenient if one wishes to compare influxes and effluxes in a single experiment, but owing to counting errors it is less accurate than the alternative method of collecting the radioactive ions which emerge from the fibre and measuring their radioactivity

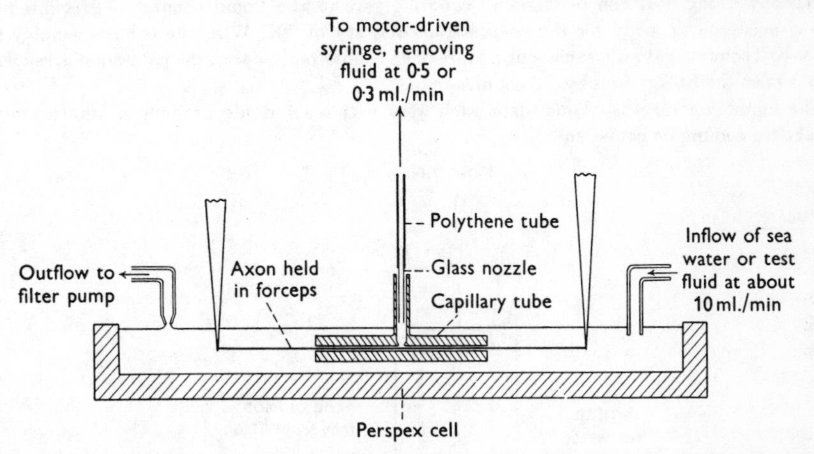

Fig. 1. Apparatus used for measuring effluxes by the 'flow method'. The overall length of the capillary was 17, 29 or 32 mm. For squid axons its internal diameter was 1 mm; for *Sepia* axons it was 600 μ.

directly. For example, if two consecutive 10 min counts of the fibre are made, and if radioactivity is lost at 10 counts/min^2 out of a total of 10,000 counts/min, the estimate of the rate of loss of radioactivity will have a standard error of $\frac{\sqrt{(100{,}000+99{,}000)}}{10\times 10}=4\cdot46$ counts/min^2. On the other hand, if the effluent radioactivity is collected for 10 min and then counted for 10 min, the standard error will be $\frac{\sqrt{1000}}{10\times 10}=0\cdot32$ counts/min^2. This calculation shows that measurements of the total radioactivity in the nerve will not give accurate figures for the efflux unless the observations are made over an appreciable fraction of the time constant for the loss of radioactivity. When sodium effluxes are investigated, this condition is fairly easy to satisfy, and the method was used extensively in the early stages of the work.

B. *Flow method*

Fig. 1 illustrates the method in the form in which it was finally employed; an earlier version is shown in Hodgkin & Keynes (1954, fig. 1). After the axon had been loaded with radioactive ions it was drawn into a short length of capillary tubing (17, 29 or 32 mm of 600 or 1000 μ bore were employed). Fluid was withdrawn from the side arm at 0·3 or 0·5 ml./min by means of a motor-driven syringe. Radioactivity was thus collected from the length of fibre in the tube, but not from the ends, which were washed in a stream of inactive sea water flowing at about 10 ml./min. When about 6 ml. of fluid had been removed, the collecting tube was taken out of the side arm, and the contents of the syringe were ejected into a test-tube. The polythene connecting tube was removed from the syringe, and the small amount of radioactive solution inside it was blown into the test-tube containing the rest of the sample. The connecting tube and the syringe were then thoroughly washed with sea water. Throughout these operations, which took about 2 min, a stand-by motor-driven syringe was sucking fluid from the side arm in order to prevent radioactive ions from accumulating in the capillary tube or side arm.

The radioactivity of the effluent was measured in a liquid counter of conventional design (20th Century Electronics, Type M6M). This was washed out between samples by pouring in saline with a trace of detergent, and removing the washing fluid through a fine polythene tube attached to a suction pump. Three washes were sufficient to remove all traces of radioactivity from the counter. The collecting volume of 6 ml. was chosen because it was close to the volume for maximum sensitivity.

When working near the background counting rate of the liquid counter (9·0 counts/min), it was necessary to allow for the natural radioactivity of ^{40}K. With the counter employed, a 1·0 M-KCl solution gave a mean counting rate of 48 counts/min, so that the potassium in sea water would raise the background by 0·5 count/min.

The liquid counter was standardized each week with a solution containing a known amount of labelled sodium or potassium.

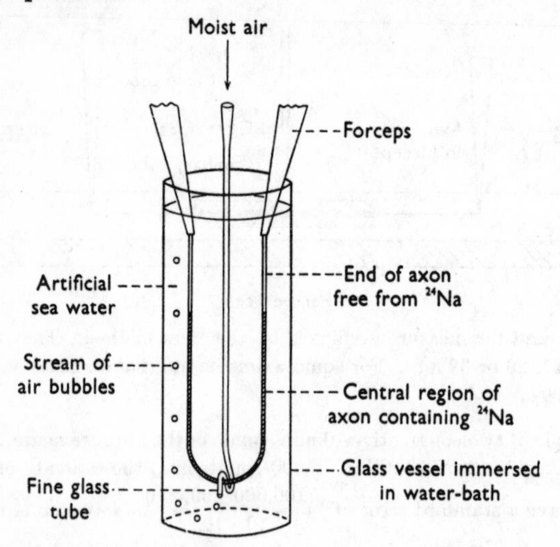

Fig. 2. Apparatus used for measuring effluxes by the 'static method'.

C. '*Static*' *method for investigating sodium effluxes*

The flow method had the disadvantage that it used large volumes of solution and that, in the form employed by us, it was difficult to reach temperatures below 4° C. Lower temperatures could no doubt have been achieved by improving the thermal insulation, but it proved simpler to use the method shown diagrammatically in Fig. 2. The axon was looped in a V under a glass tube, with its ends held in forceps. It was dipped for periods of about 10 min into small vessels containing 6 ml. of sea water at any desired temperature. After remaining for 10 min in one vessel, the fibre was transferred to another, and the contents of the first one were pipetted into the liquid counter. The fluid was stirred by a fine jet of air, and leakage of ^{24}Na from the ends was avoided by loading radioactivity only into the central part of the axon. During the course of the experiment ^{24}Na must have diffused along the axis cylinder, but this did not matter provided the radioactive ions remained in a uniform length of intact nerve. This method was extremely simple, but it had the disadvantage that the length of axon from which ^{24}Na was collected was not well defined. It was therefore not employed in experiments in which it was important to know the absolute magnitude of the sodium efflux. Nor was it used in experiments with ^{42}K, since there was a danger that diffusion towards the ends might have given errors as a result of the very large potassium leakage occurring near the cut end of a nerve.

Determination of the efflux in absolute units

The methods described in the previous sections gave the efflux in terms of the rate at which labelled Na or K left the fibre. For many purposes this was all that was required, but it was also desirable to be able to convert this figure into the total number of ions leaving unit area of membrane in unit time. This required a knowledge of the proportion of labelled ions to the total quantity of sodium or potassium in the axoplasm. For *Sepia* axons the procedure at the end of the experiment was to soak the fibre for 10 min in potassium-free choline sea water, measure the axon diameter, cut out the central portion, and suspend it on a quartz thread. The quantity of ^{24}Na or ^{42}K in this sample was determined with a Geiger counter, and the total content of sodium or potassium was measured by activation analysis (Keynes & Lewis, 1951). With *Loligo* axons the procedure was similar, except that the axoplasm was finally extruded on to a quartz thread and weighed, before being counted and analysed.

TABLE 1. Composition of solutions

	Normal artificial sea water (mg-ions/l.)	Choline sea water (mg-ions/l.)	Dextrose sea water (mg-ions/l.)
Na^+	486	—	—
K^+	10·4	10·4	10·4
Ca^{++}	10·7	10·7	10·7
Mg^{++}	55·2	55·2	55·2
Cl^-	567	567	84
SO_4^{--}	29·2	29·2	29·2
HPO_4^{--} } $H_2PO_4^-$ }	1·5	—	—
$Choline^+$	—	483	—
Dextrose	—	—	855

Solutions

The compositions of some of the solutions used in the experiments are shown in Table 1. For the sake of brevity the normal artificial sea water which was our standard medium is referred to elsewhere in this paper as 'sea water'. The pH of all solutions was checked with indicators, and was always close to 7. In the potassium-free versions of the solutions, 10·4 mM-KCl was omitted and the other constituents were left unchanged. In potassium-rich sea water the amount of NaCl was reduced by the amount of extra KCl added. Salt-free isotonic dextrose was a 1·0 molal solution (180 g dextrose/kg H_2O); on a few occasions 10·7 mM-Ca was added to this.

As the solutions contained little or no buffer, the inhibitors were carefully neutralized before being added to them. Thus dinitrophenol was first ground up in dilute NaOH to make a neutral 10 mM solution, and an appropriate volume of this was added to the sea water.

^{24}Na and ^{42}K samples were treated as described by Keynes (1954), and were used to make up radioactive solutions identical in composition with the inactive ones.

Electrical methods

Action potentials and resting potentials were determined by the following methods: (*a*) external electrodes with axons immersed in oil or air, (*b*) for *Sepia* axons, transverse impalement with 0·5 μ electrodes filled with 3 M-KCl (Ling & Gerard, 1949; Nastuk & Hodgkin, 1950), (*c*) for *Loligo* axons, longitudinal impalement with 50–100 μ capillaries filled with isotonic KCl or sea water (Hodgkin & Huxley, 1945; Hodgkin & Katz, 1949*a*).

At the beginning and end of all tracer experiments the excitability was checked by lifting the ends of the axon into air, stimulating between one pair of forceps and an earthed electrode in the bath, and recording from the second pair of forceps at the other end of the fibre.

RESULTS

Sodium and potassium fluxes in sea water

During the course of the work we accumulated a number of observations from which the absolute values of the normal fluxes of sodium and potassium could be calculated. Most of the measurements were made on *Sepia* fibres which had been stimulated for 5–10 min at 50/sec, and these results are given in Table 2 A. All the axons were in sea water at the time that the measurements were made, but many had been subjected to the action of reversible inhibitors such as dinitrophenol or cyanide. Experiments in which there was any sign of loss of excitability in or near the region studied have been excluded from the averages. This restriction and general improvements in technique probably account for both the sodium influx and the potassium efflux being about half as great as was found by Keynes (1951*a*).

TABLE 2. Ionic fluxes for *Sepia* axons in artificial sea water

Quantity	Mean value	S.D. of each observation	S.E. of mean	No. of observations
A. Sodium and potassium fluxes in axons which have carried 10,000–40,000 impulses, measured 0·5–4 hr after stimulation				
Na influx	32 pmole/cm^2 sec	16	4	14
Na efflux	39 pmole/cm^2 sec	11	3	21
K influx	21 pmole/cm^2 sec	4	1	14
K efflux	28 pmole/cm^2 sec	15	5	12
Internal [Na] at end of experiment	77 m-mole/l. axoplasm	37	8	22
Internal [K] at end of experiment	267 m-mole/l. axoplasm	52	11	24
Axon diameter	197 μ	31	5	27
Temperature	18° C	1·6	0·3	26
B. Potassium fluxes in lightly stimulated axons				
K influx	13 pmole/cm^2 sec	1·3	0·6	5
K efflux	29 pmole/cm^2 sec	10	5	5
Internal [K] at end of experiment	300 m-mole/l. axoplasm	19	10	5
Axon diameter	207 μ	25	13	5
Temperature	17° C	2·0	1	5

The results in Table 2B are for axons which had not been stimulated more than was necessary to check their excitability. They show that the potassium influx was significantly less in lightly stimulated fibres than in those which had carried a large number of impulses. Individual experiments like those in Table 7A suggest that the potassium influx was greatest immediately after stimulation, and returned towards its resting level with a time constant of 2–3 hr. We suspect that the sodium efflux behaves in a similar manner (see Table 3), but have not made enough measurements with lightly stimulated fibres to substantiate this conclusion.

Although axons which have been isolated for many hours often have high sodium and low potassium contents, our present results do not support the idea that isolated *Sepia* axons which are kept at room temperature always gain sodium steadily as inferred previously (Keynes, 1951*a*; Keynes & Lewis, 1951). In many cases the total internal sodium concentration, determined by analysis at the end of the experiment, was less than the internal concentration of labelled sodium immediately after stimulation (e.g. Table 3). This indicates that isolated fibres are able to recover from stimulation by pumping out the sodium which enters during activity. The average sodium concentration in six fibres which had not been treated with inhibitors was 43 mM, 7 hr after killing the animal, and it seems likely that this level can be maintained for considerable periods of time.

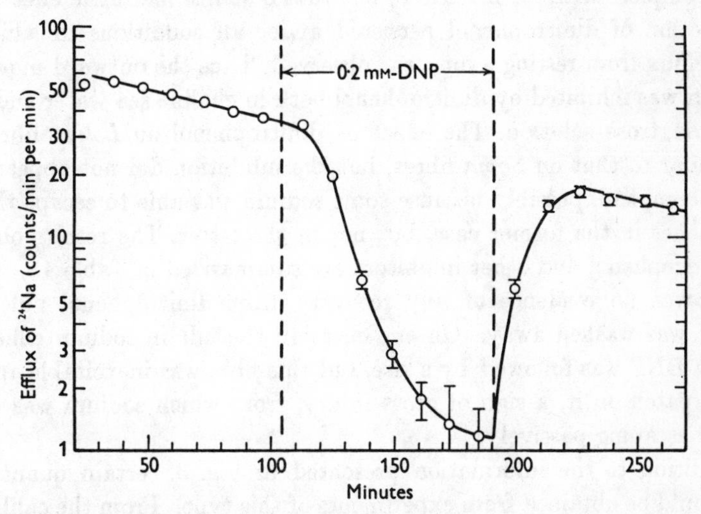

Fig. 3. Sodium efflux from a *Sepia* axon during treatment with dinitrophenol. At the beginning and end of the experiment the axon was in artificial sea water. Abscissa: time after end of stimulation in ^{24}Na sea water. Ordinate: rate at which ^{24}Na leaves axon. Vertical lines are $\pm 2 \times$ S.E. Temperature 18° C.

The effect of metabolic inhibitors on the sodium efflux

A. *Dinitrophenol*

The action of 2:4-dinitrophenol (DNP) on a *Sepia* axon is shown in Fig. 3. This fibre was first stimulated at 156/sec for 4 min in sea water made up with ^{24}Na. It was then set up in the apparatus shown in Fig. 1, and samples of the fluid drawn past the axon were collected and counted at intervals of about 12 min. In normal sea water the efflux of labelled sodium declined exponentially as the ^{24}Na which had entered during stimulation was progressively eliminated. This exponential decline appears as a straight line in Fig. 3 because the ordinate is plotted on a logarithmic scale. When enough counts of the normal efflux of ^{24}Na had been taken, the sea water flowing past the axon was replaced by a similar solution containing 0·2 mM-dinitrophenol. After an initial delay of about 10 min the efflux began to decline more rapidly, and fell in an hour to about one-twentieth of the value which it had in the absence of dinitrophenol. This effect was largely reversed by washing the inhibitor away—as may be seen from the recovery which took place at the end of the experiment. The inhibitory action of dinitrophenol was observed in every experiment in which it was tested by this method. Similar results were obtained by examining the total radioactivity of the nerve instead of that in the external fluid. Such an experiment is illustrated in Fig. 9, where it may be seen that the ^{24}Na content of the axon remained almost constant after the inhibitor had taken effect. Other experiments of this type indicated that inhibition was usually complete with 0·1 mM-DNP, but that 0·02 mM had little effect.

The action of dinitrophenol persisted under all conditions in which the sodium efflux from resting axons was observed. Thus the outward movement of sodium was inhibited by dinitrophenol both in choline sea water and in an isotonic dextrose solution. The effect of dinitrophenol on *Loligo* fibres was very similar to that on *Sepia* fibres, but the inhibition did not appear to be quite as complete, possibly because some sodium was able to escape through cut branches in the former case, but not in the latter. The results obtained with dinitrophenol and other inhibitors are summarized in Table 4.

There was no evidence of any recovery from dinitrophenol unless the inhibitor was washed away. On one occasion the fall in sodium efflux produced by DNP was followed by a rise, but this fibre was inexcitable and had a brown patch on it, a sign of gross injury, from which sodium was almost certainly escaping passively.

In addition to the information presented in Fig. 3, certain quantitative results could be obtained from experiments of this type. From the calibration of the Geiger counter it was easy to calculate the quantity of labelled sodium which corresponded to any given counting rate. In the experiment of Fig. 3, 1 count/min was equivalent to 6 pmole of the labelled sodium [Na*] in the solution used to make the fibre radioactive. The surface area of the membrane from which ^{24}Na was collected could be worked out from the axon diameter and the total length of the collecting tube. These two figures led to a value of 7 pmole/cm^2 sec for the efflux of Na* at the end of the experiment, and to one of 34 pmole/cm^2 sec at a time 30 min after the end of stimulation. The Na* efflux immediately after stimulation was found by extrapolation to be about 40 pmole/cm^2 sec. At the end of the period in dinitrophenol it was 0·7 pmole/cm^2 sec. In order to convert these figures into absolute effluxes it was necessary to know the ratio of [Na*] to total [Na] in the axoplasm at any given moment. This ratio was obtained at the end of the experiment by washing the fibre with potassium-free choline sea water, cutting out the central 22 mm, and drying it on a quartz thread. Labelled sodium was then determined with a Geiger counter, and total sodium was determined by activation analysis

(Keynes & Lewis, 1951). These measurements showed that the ratio of Na* to total Na was 0·30 at the end of the experiment, the total concentration of sodium in the axoplasm being 50 mM. Values for the total sodium at other times could be estimated by solving numerically the equations

$$\frac{d[\mathrm{Na}]_i}{dt} = \frac{a}{2}(M^{\mathrm{Na}}_{\mathrm{in}} - M^{\mathrm{Na}}_{\mathrm{out}}), \tag{1}$$

and

$$M^{\mathrm{Na}}_{\mathrm{out}} = \frac{[\mathrm{Na}]_i}{[\mathrm{Na}^*]_i} M^{\mathrm{Na}^*}_{\mathrm{out}} \tag{2}$$

where $[\mathrm{Na}^*]_i$ and $[\mathrm{Na}]_i$ are the concentrations of labelled and total sodium respectively, $M^{\mathrm{Na}^*}_{\mathrm{out}}$ and $M^{\mathrm{Na}}_{\mathrm{out}}$ are the corresponding effluxes, a is the fibre radius, and $M^{\mathrm{Na}}_{\mathrm{n}}$ is the influx, assumed to be constant. The value chosen for

Table 3. Analysis of experiment illustrated in Fig. 3. Axon diameter was 184 μ. Between $t = -4$ min and $t = 0$ the axon was stimulated at 156/sec in Na* sea water. From $t = 0$ onwards the axon was in inactive sea water. 0·2 mM-DNP was added at $t = 104$ min and removed at 191 min.

	(1) t: time after end of stimulation (min)	(2) [Na*] (mM)	(3) [Na] (mM)	(4) $M^{\mathrm{Na}^*}_{\mathrm{out}}$: efflux of Na* (pmole/cm² sec)	(5) $M^{\mathrm{Na}}_{\mathrm{out}}$: efflux of Na (pmole/cm² sec)
	−4	0	(40)	—	—
	0	72	82	—	—
	20	62	74	37	44
	40	53	67	32	40
	60	45	61	28	38
	80	38	56	24	35
	100	33	51	21	33
↑	120	27	47	14·4	24
0·2 mM-DNP	140	25	47	3·2	5·9
	160	25	50	1·1	2·2
↓	180	25	54	0·74	1·6
	200	24	57	2·9	6·8
	220	23	58	9·0	23
	240	20	56	9·2	25
	260	18	53	8·15	24
	280	16	51	7·2	23
	293	15	50	—	—

Influx for $t > 0$ = 15·3 pmole/cm² sec. Sodium efflux per impulse = 10 pmole/cm².
Sodium influx per impulse = 15 pmole/cm². Net entry of sodium per impulse = 5 pmole/cm².

Method of calculation and assumptions:

Column (4) is obtained from the experimental curve in Fig. 3.

Column (2) is obtained by adding the terminal [Na*] to the value obtained by integrating column (4) over the appropriate time interval.

In column (3) the first figure (which has little effect on the rest of the table) is taken from Keynes & Lewis (1951) and from other analyses of fresh fibres. The second figure is calculated from [Na*] = 72 mM at $t = 0$ and [Na] = 40 mM at $t = -4$, on the assumption that the ratio of influx to efflux during the period of stimulation was 1·5 as found by Keynes (1951a). This also gave the figures shown at the foot of the table. Subsequent values in column (3) are calculated by equations (1) and (2), using an influx of 15·3 pmole/cm² sec.

Column (5) = column (4) × column (3)/column (2).

M_{in}^{Na} was adjusted by trial and error until the numerical solution made the final value of $[Na]_i$ equal to that observed experimentally. The initial value of $[Na]_i$ could be obtained without serious error from the initial value of $[Na^*]_i$ by the method outlined at the foot of Table 3.

The results of this somewhat indirect analysis are given in Table 3. They show that the total sodium rose from about 40 to 82 mM after 4 min stimulation at 156/sec. The total sodium in the axoplasm was then slowly reduced by being pumped out faster than it entered. The rate of sodium extrusion declined as the internal sodium was reduced, and had a value of 33 pmole/cm^2 sec at the time when DNP was applied. It then fell fairly rapidly to about 1·6 pmole/cm^2 sec, with the result that the internal sodium concentration rose by about 7 mM during the period in DNP. The efflux recovered to about 25 pmole/cm^2 sec when the DNP was removed, and the sodium concentration then fell slowly to its final value of 50 mM.

B. *Cyanide*

The action of cyanide at concentrations of 1–10 mM was very similar to that of dinitrophenol. A typical experiment is illustrated in Fig. 4, while other results are summarized in Table 4. A single experiment performed some time ago by the less sensitive method of taking counts from the nerve, failed to give any effect in an hour (Keynes, 1951 *b*), possibly because the temperature was only 13° C and the inhibitor was not applied for long enough.

C. *Azide*

Sodium azide acted in a similar way to cyanide and dinitrophenol, but its effect took longer to develop, perhaps because it penetrated the fibre less quickly. Only two experiments were done with azide, but both gave very similar results. Fig. 5 is taken from the first experiment.

The effect of metabolic inhibitors on the sodium influx

The action of DNP on the resting influx of sodium is illustrated by the observations summarized in Table 5. These experiments occupied about 6 hr, and it is probable that the relatively high values in column *c* were caused by deterioration of the fibre rather than by the removal of DNP. If this supposition was correct, the conclusion would be that DNP had little effect on the sodium influx. Even if no allowance is made for a change for the worse in the condition of the fibre, it is clear that the reduction in influx caused by DNP is trivial compared with the effect on the efflux of sodium.

The effect of metabolic inhibitors on the potassium efflux

These experiments are summarized in Table 6. They showed that concentrations of DNP or cyanide which blocked the extrusion of sodium, had relatively little effect on the resting outflow of potassium. The experiments suggest that

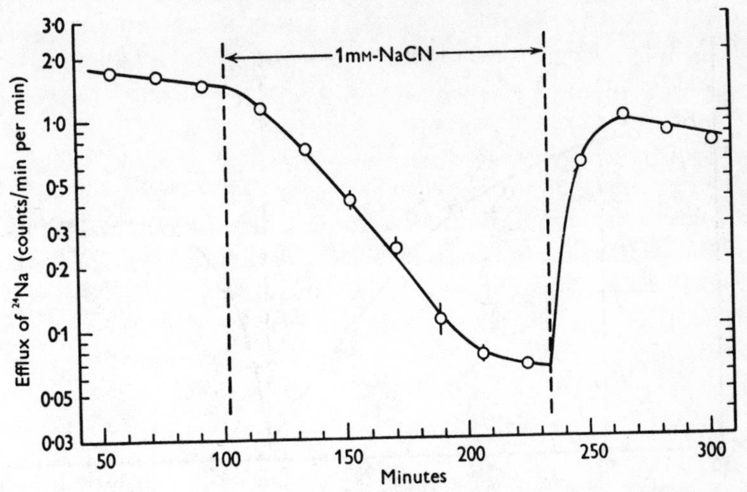

Fig. 4. Sodium efflux from a *Sepia* axon during treatment with cyanide. Temperature about 16° C. Other details as in Fig. 3.

TABLE 4. Effect of metabolic inhibitors on sodium efflux

Expt.	Axon diam. (μ)	Inhibitor	Concentration (mM)	Ratio of efflux in poisoned axon to normal efflux	Time to reduce efflux to half (min)	Time to restore efflux to half (min)	Normal efflux (pmole/cm² sec)
			A. *Sepia* axons				
1	214	DNP	0·1	<0·15	23	20	—
2	196	DNP	0·2	<0·1	35	28	33
3	184	DNP	0·2	0·06	22	20	28
4	170	DNP	0·2	0·05	24	15	—
5	165	DNP in choline sea water	0·2	0–0·15	—	—	—
6	216	DNP in isotonic dextrose	0·2	0·1–0·3	—	—	—
7	196	Cyanide	1	0·06	36	12	42
8	176	Cyanide	10	0·08	33	40	49
9	214	Cyanide	10	<0·1	40	35	43
10	181	Azide	3	<0·08	63	23	38
11	176	Azide	3	<0·17	50	14	—
			B. *Loligo* axons				
12	565	DNP	0·2	0·16	14	22	—
13	490	DNP	0·2	0·11	10	20	50
14	475	Cyanide	2	<0·10	14	—	—
15	534	Cyanide	2	0·10	15	12	—

The < sign means that the inhibitor was not applied for long enough to produce a complete effect.

In Expt. 5 the DNP was made up in choline sea water (see Table 1) containing 1·5 mM-potassium phosphate buffer; the effect is given relative to the efflux in sea water.

In Expt. 6 the DNP was made up in isotonic dextrose containing no salts other than 0·2 mM-Na to neutralize it; the effect is given relative to the efflux into the same solution without DNP.

Expts. 5 and 6 were made by taking counts from the nerve (Method A, p. 30), Expts. 2–4 and 7–13 by the flow method (B, p. 30), and Expts. 1, 14 and 15 by a method similar to C (p. 31).

^{24}Na was introduced by stimulation except in Expt. 15, where it was injected (Hodgkin & Keynes, to be published).

Temperature 15–20° C.

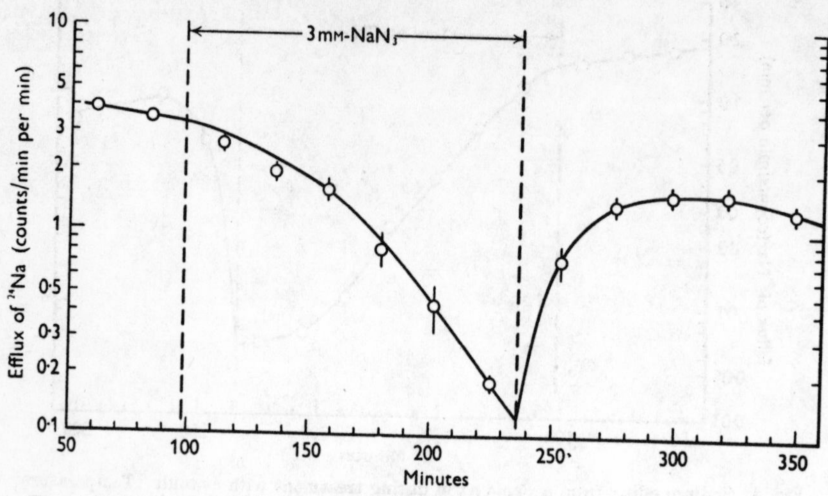

Fig. 5. Sodium efflux from a *Sepia* axon during treatment with azide. Temperature 16° C. Other details as in Fig. 3.

TABLE 5. The effect of dinitrophenol on sodium influx in *Sepia* axons

Expt.	Axon diam. (μ)	Influxes in pmole/cm² sec (a) Normal	(b) DNP	(c) Normal	$\sqrt{(a \times c)}$ Mean normal
1	205	20·6	17·8	20·1	20·4
2	212	7·7	13·5	11·3	9·3
3	167	15·5	13·2	44·4	26·2
4	184	24·1	17·2	73·0	42·0
5	148	16·6	15·0	21·7	19·0
Mean		17	15	34	23

The concentration of DNP was 0·2 mM except in Expt. 1, where it was 0·1 mM.

Expts. 1 and 2 were done with lightly stimulated axons, the remainder with axons stimulated for 5–10 min at 50/sec.

Temperature 18–20° C.

TABLE 6. Effect of inhibitors on potassium efflux (in pmole/cm² sec) in *Sepia* axons

Normal	0·2 mM-DNP	Normal	Ratio (DNP/normal)
—	—	—	1·4
30	67	52	1·6
18	31	22	1·6
38	34	25	1·1
	2 mM-CN		(CN/normal)
32	33	29	1·1

The measurements in the second column were made about 1 hr after applying the inhibitor. Temperature 15–18° C.

DNP may increase the efflux of potassium slightly, but it is difficult to be sure of this result, since the efflux was highly variable, and the action of DNP had to be examined on a somewhat unsteady base-line.

In the next paper the average efflux of potassium in fibres poisoned with DNP is given as 27 pmole/cm^2 sec (Hodgkin & Keynes, 1955, table 3), which is almost the same as the value of 28 pmole/cm^2 sec for normal fibres (Table 2 of this paper). The figure for the poisoned fibres is based on averages of 25, 39 and 18 pmole/cm^2 sec obtained by different methods. Since the method which gave 18 pmole/cm^2 sec was not used for normal fibres, we regard the agreement between the overall means as a coincidence, and attach more weight to the experiments in Table 6 where the normal and poisoned fluxes were obtained for the same preparation. However, as has been explained above, the only certain conclusion is that DNP does not have any large effect on the potassium efflux.

The effect of inhibitors on the potassium influx

The most striking results were obtained in those experiments in which DNP or cyanide was applied to fibres which were recovering from 5–10 min stimulation at 50/sec. The figures listed in Table 7A suggest that, in the absence of an inhibitor, the potassium influx declined as the effect of stimulation wore off—the slow decrease being parallel to the decline in sodium efflux seen in Table 3. 0·2 mM-Dinitrophenol or 2 mM-cyanide reversibly reduced the potassium influx from about 22 to 3 pmole/cm^2 sec. This reduction cannot be attributed to a decrease in permeability to potassium, since the potassium efflux remained unchanged or was even increased by DNP. Nor can it be caused by a change in resting potential, since this was not appreciably altered by DNP (see p. 47). The explanation appears to be that most of the potassium influx from sea water in fibres which are recovering from stimulation is an active absorption, and that not more than 3 pmole/cm^2 sec should be attributed to an inward movement through the passive permeability channel. This finding was not what we originally expected, for it had previously seemed simpler to suppose that the potassium movements in nerve and muscle were largely passive, and that potassium ions were drawn into the cell against the concentration gradient by the electrical potential difference arising from the extrusion of sodium (Hodgkin, 1951; Keynes, 1951*a*). Our new results bring cephalopod axons into line with mammalian erythrocytes, in which there is clear evidence for an active uptake of potassium (see Ponder 1950), and they also agree with the conclusion of Shanes (1951*b*) that potassium uptake in nerve is active. The evidence which led us to think that the potassium influx was passive is examined in detail in the next paper (Hodgkin & Keynes, 1955).

Expts. 6 and 7 in Table 7B were made with fibres which had not been stimulated more than was needed to test their excitability. The action of DNP was still present, but was less marked than with the other fibres, the normal influx being lower and the poisoned influx higher than in fibres recovering from stimulation. The former difference has already been noticed in connexion

TABLE 7. The effect of inhibitors on the potassium influx in *Sepia* axons

Experiment	Axon diam. (μ)	Temperature (° C)	Time after stimulation (min)	Potassium influx (pmole/ cm^2 sec)	Conditions
			A. Main experiments		
1	211	20	56	27·3	Normal
			125	26·0	Normal
			203	3·4	DNP (43 min pre-treatment)
			293	22·9	Normal (after 60 min recovery)
2	202	20	27	30·4	Normal
			109	21·8	Normal
			214	2·0	DNP (60 min pre-treatment)
			329	15·2	Normal (after 70 min recovery)
3	178	18	68	3·0	DNP (68 min pre-treatment)
			189	22·5	Normal (after 70 min recovery)
			276	21·2	Normal
4	234	16	123	3·3	DNP (60 min pre-treatment)
			411	2·1	DNP (380 min pre-treatment)
5	183	17	101	3·0	Cyanide (70 min pre-treatment)
			224	16·2	Normal (after 62 min recovery)
			B. Subsidiary experiments		
6	190	18	—	14·1	Normal
			—	13·3	Normal
			—	7·1	DNP (52 min pre-treatment)
			—	14·6	Normal (after 54 min recovery)
7	232	18	—	13·5	Normal
			—	14·9	Normal
			—	6·3	DNP (44 min pre-treatment)
			—	18·4	Normal (after 43 min recovery)
8	199	18	79	3·1	DNP (63 min pre-treatment)
			183	5·0	DNP in dextrose sea water
			279	3·0	DNP
			361	16·8	Normal (after 72 min recovery)
			432	16·8	Normal
9	150	19	96	2·6	DNP (66 min pre-treatment)
			201	4·5	DNP in dextrose sea water
			312	3·4	DNP
			408	13·3	Normal (after 62 min recovery)
10	199	16	70	4·3	DNP (54 min pre-treatment)
			136	123	DNP with 52 mM-K
11	234	16	91	3·2	DNP (60 min pre-treatment)
			171	150	DNP with 52 mM-K

All axons except 6 and 7 were stimulated for 5–10 min at 50/sec at the beginning of the experiment.

The concentration of DNP was 0·2 mM throughout; in Expt. 5 the concentration of cyanide was 2 mM.

Unless otherwise stated the external potassium concentration was 10·4 mM.

with Table 2, while the latter may indicate a reduced susceptibility to DNP or a genuine variation in the passive component of the influx between lightly and heavily stimulated axons.

The object of Expts. 8 and 9 was to see whether the potassium influx in a poisoned fibre could be made to recover by removing the external sodium, and substituting dextrose for NaCl in the DNP solution. It will be seen that there was a small increase in the influx in the sodium-free solution—possibly due to a change in resting potential—but nothing comparable to the increase on finally removing the DNP.

The last two experiments are discussed on p. 45. They show that increasing the external potassium concentration from 10·4 to 52 mM for a DNP-poisoned fibre caused a 30- or 40-fold increase in the inward movement of potassium.

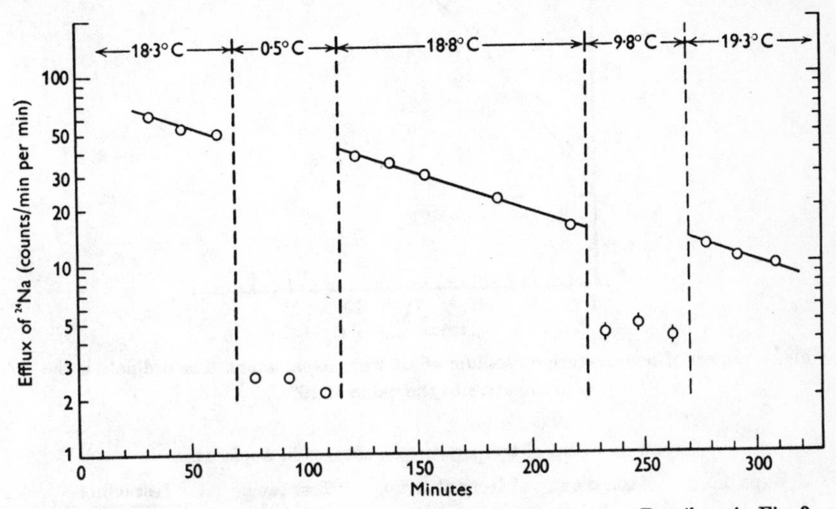

Fig. 6. Effect of temperature on sodium efflux from a *Sepia* axon. Details as in Fig. 3.

Temperature and sodium efflux

The effect of cooling a *Sepia* axon on the activity of the sodium pump was examined by the method described on p. 31. A typical result is shown in Fig. 6. After three observations of the efflux had been made at 18° C, the axon was immersed in sea water at a temperature of 0·5° C. The efflux fell at once to about one-eighteenth of its previous value, and remained low as long as the fibre was kept cold. Prompt recovery occurred when the temperature was restored to its initial value. A similar but less marked reduction took place when an intermediate temperature of 9·8° C was employed. In no case was there any sign of a lag in the development of the effect such as that found with metabolic inhibitors.

Table 8A and Fig. 7 summarize the results of all the experiments of this type. In Fig. 7 the abscissa is temperature and the ordinate gives the value of the efflux relative to that at 19° C. Since a logarithmic scale has been used, the points would fall on a straight line if the temperature coefficient were constant. Although it is hard to be sure about the exact shape of the curve, there is little doubt that the points are not fitted by a straight line, and that the temperature coefficient is greatest near 0° C. There is a marked resemblance between these results and the relation reported by Hodgkin & Katz (1949*b*) for the effect of temperature on the rate of fall of the action potential. This is

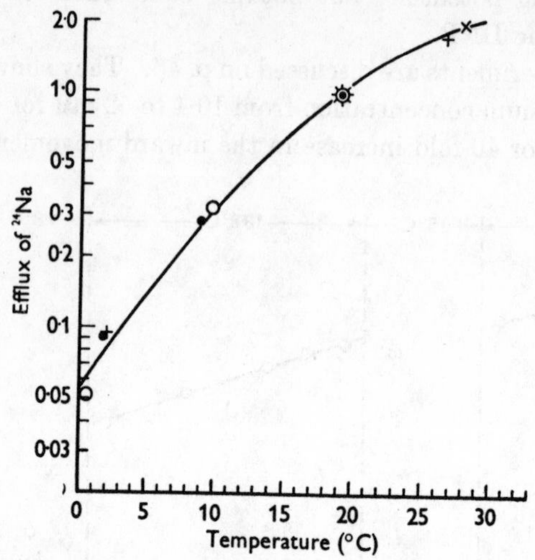

Fig. 7. Effect of temperature on sodium efflux from *Sepia* axons. The ordinate is the efflux relative to the value at 19° C.

TABLE 8. Effect of temperature on effluxes in *Sepia* axons

Expt. no.	Axon diam. (μ)	Normal temp. (° C)	Test temp. (° C)	Test efflux / Normal efflux
		A. Sodium efflux		
1	190	18·6	1·7	0·098
		19·2	8·9	0·28
2	162	19·0	27·8	2·0
3	176	18·6	0·5	0·056
		19·1	9·8	0·32
4	199	18·8	26·5	1·79
		19·1	2·0	0·095
		B. Potassium efflux		
5	222	17·0	3·0	0·92
6	195	17·7	4·0	0·84

In Expt. 6 the average potassium efflux was 34 pmole/cm² sec at 17·7° C and 29 pmole/cm² sec at 4·0° C. Absolute values were not determined in the other experiments.

probably a coincidence, but a causal connexion cannot be entirely excluded, since both the pumping out of sodium and the falling phase of the action potential might depend on chemical reactions in which sodium-selective molecules are converted into potassium-selective molecules.

The effect of cooling on K *influx*, K *efflux and* Na *influx*

Tables 8B, 9A and 9B summarize the experiments in which the effect of temperature on the other three fluxes was examined. They show that the uptake of potassium from a solution containing 10·4 mM-K was reduced from 20 to about 2 pmole/cm² sec when the fibre was cooled from 20 to 1·3° C ($Q_{10} \doteqdot 3{\cdot}3$). In contrast, the movements down concentration gradients had low temperature coefficients, the average Q_{10}'s being about 1·1 for potassium efflux

TABLE 9. Effect of temperature on influxes in *Sepia* axons

Expt. no.	Axon diam. (μ)	Temperature (° C)	Influx (pmole/cm² sec)	Q_{10}
		A. Sodium influx from sea water (with 10·4 mM-K)		
1	239	3·0	16·4	
		16·2	25·6	1·3
		3·0	18·4	
2	266	3·0	10·8	
		16·2	15·2	1·2
		3·0	12·5	
3	231	2·1	13·0	
		15·8	25·3	1·6
		1·5	12·3	
		B. Potassium influx from sea water (with 10·4 mM-K)		
4	150	1·4	2·9	
		19·7	20·5	2·9
5	249	19·6	21·6	
		1·2	1·6	3·7
		20·4	16·1	
		C. Potassium influx from potassium-rich sea water (with 52 mM-K)		
6	206	18·7	221	
		1·8	215	1·0
		20·8	193	
7	220	2·6	167	
		20·9	163	0·9
		1·2	232	

and 1·4 for sodium influx. This is not altogether surprising, since Hodgkin, Huxley & Katz (1952) found that although temperature had a large effect on the rate at which the sodium and potassium conductances changed with time, it did not have much effect on the maximum sodium or potassium conductance associated with a given depolarization.

These observations are consistent with those recently reported by Shanes (1954) for the nerves of *Libinia emarginata* and *Loligo pealii*. Here cooling increased the total potassium loss associated with each impulse in about the proportion calculated by Hodgkin & Huxley (1952), but in *Libinia* reduced the rate of reabsorption of potassium during recovery. A similar effect of temperature on the potassium fluxes has been found by Calkins, Taylor & Hastings (1954) in rat diaphragm muscle, where cooling to 2° C markedly decreased the potassium influx without much change in the efflux.

Distinction between active and passive uptake of potassium

The results described in the previous sections showed that the influx of potassium from solutions containing 10·4 mM-K had a high temperature coefficient and was greatly reduced by agents which inhibited the sodium pump. In these respects it differed from the potassium efflux, which had a low Q_{10} and was little affected by DNP or cyanide. It may be concluded that the potassium influx from sea water is mainly an active absorption, and that the passive component is relatively small. This is only true if the external potassium concentration is low. When axons were depolarized by raising the external potassium to 52 mM, the membrane became much more permeable

TABLE 10. Potassium fluxes in *Sepia* axons with an external potassium concentration of 52 mM

Quantity	Value (pmole/cm² sec)	S.E. of mean (pmole/cm² sec)	No. of observations
K influx, normal	160	11	9
K efflux, normal	181	11	9
K influx in 0·2 mM-DNP	128	9	4
K efflux in 0·2 mM-DNP	134	22	4

Most of the values for unpoisoned axons were obtained in the course of experiments on the extent of potassium exchange in *Sepia* fibres (see Hodgkin & Keynes, 1953*c*). The remaining values in the table are from experiments quoted in other contexts in this paper and the next (Hodgkin & Keynes, 1955). Temperature 16–21° C.

to potassium, and permitted large and roughly equal fluxes to pass in both directions. Under these conditions the influx appeared to be largely passive, since it had a low temperature coefficient and was little changed by DNP. This is demonstrated by the experiments in Table 9C, and by the average results in Table 10. The effects of potassium concentration and membrane potential on the passive movements of potassium are considered further in the following paper (Hodgkin & Keynes, 1955).

The electrical properties of poisoned axons

Inhibition of the sodium pump by dinitrophenol or cyanide is not accompanied by loss of excitability, nor are there any marked changes in the electrical properties of the fibres. Action potentials could still be obtained after many hours of exposure to 0·2 mM-DNP or 1–10 mM-cyanide, and on one occasion a *Sepia* axon 180 μ in diameter conducted impulses for 70 min at 50/sec in sea water containing 0·1 mM-DNP. The ability of giant axons to conduct impulses at high repetition rates after their recovery mechanism has been blocked with DNP is not really surprising. At 50 impulses/sec the rate of gain of sodium is much higher than the maximum rate at which the fibre can eject sodium, so that it makes little difference whether or not the pump is working. In most of our experiments the axons were surrounded by large volumes of sea water, so that potassium could not accumulate outside. If the fibre is kept in a small volume, potassium accumulation may cause inexcitability, as described by Shanes (1951*a*) in connexion with the effect of anoxia on the squid axon. A good example of this was provided by the following experiment. A cleaned *Sepia* axon was set up in oil and stimulated at 6·25/sec for 40 min. Under these conditions the spike declined slowly, but there was no block or intermittent failure of conduction. The axon was then treated with 0·2 mM-DNP for 1 hr, and the experiment was repeated. On this occasion the spike declined more rapidly, and block occurred after only 17 min. This was probably caused by potassium accumulation in the external fluid, since the spike was promptly restored by a dip into a large volume of sea water still containing 0·2 mM-DNP. On repeating the stimulation in oil, the spike again declined, and was blocked after 16 min. As before, the block was immediately relieved by dipping the fibre into a large volume of DNP sea water. The fibre was then allowed to recover in sea water (with no DNP) for 40 min, and was again tested in oil at 6·25/sec. It now conducted impulses at the full rate for 84 min, and continued to conduct with only occasional failures for a further period of 36 min, at the end of which time the experiment had to be brought to an end. These observations suggest that metabolism helps to keep the potassium concentration at a low level in the external fluid, and that potassium accumulation may cause block if the recovery processes are inhibited with dinitrophenol. The time to block agrees well with estimates of the potassium leakage. Most, but not all, fibres become inexcitable in a solution containing 50 mM-K, so that it is reasonable to take this as the external concentration at which block occurred. Taking the thickness of the layer of external fluid as $12\,\mu$ (Weidmann, 1951; Keynes & Lewis, 1951), a blocking time of 17 min would correspond to a mean potassium leakage of 47 pmole/cm^2 sec. This is consistent with a net leakage per impulse of 3·6 pmole/cm^2 (cf. Keynes & Lewis, 1951), and a resting leakage in DNP of 24 pmole/cm^2 sec.

The lack of any marked effect from 0·2 mM-DNP on the electrical behaviour of cephalopod axons is illustrated by Fig. 8, which was obtained with a long internal electrode inserted into a $500\,\mu$ axon from *Loligo*. The only noticeable effects were (1) a slight increase in the rate of decline of the spike such as might occur if sodium were slowly accumulating inside the axon, and (2) an initial rise of about 2 mV in the resting potential, followed by a slow decline. The initial increase was not seen in all the experiments, and the overall change in potential was too small to be measured satisfactorily.

There would be definite advantages in repeating this type of experiment on *Sepia* axons, because these have a membrane resistance which is 5 or 10 times higher than that of *Loligo* axons (cf. Weidmann, 1951; Cole & Hodgkin, 1939). One would therefore expect any effect of DNP on the resting potential to show up better with *Sepia* than with *Loligo*. Unfortunately, *Sepia* axons are too small for the long-electrode technique to be practicable, and it is necessary to use the less reliable method of impaling transversely with 0·5 μ electrodes. The upshot of some measurements of this kind was that the resting potential in four DNP-treated fibres was found to average 62 mV, as against 65 mV in three different untreated fibres. This difference in potential was not statistically significant.

While 0·2 mM-DNP had no obvious effect on excitability, it is possible that higher concentrations of the poison may impair the action-potential mechanism. Our only evidence on this point is that on one occasion a 1 mM solution blocked a *Sepia* axon reversibly in 10 min. This concentration did not block a *Loligo* axon in 30 min, although it reduced the action potential.

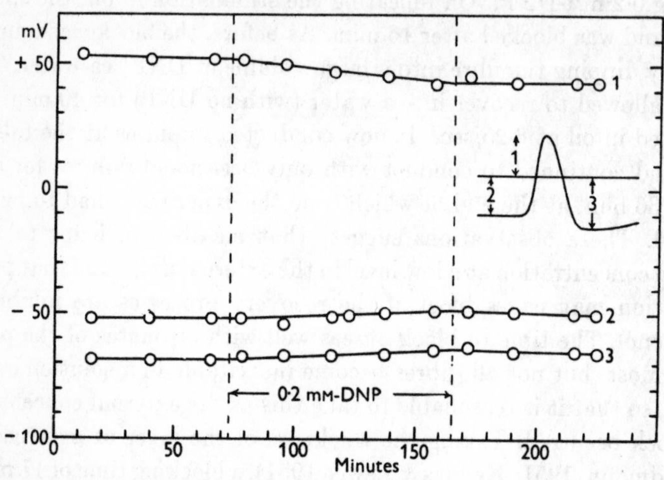

Fig. 8. Effect of 0·2 mM-DNP on the membrane potentials in a squid axon. The potential of a long 100 μ internal electrode was determined at rest (2), at the crest of the spike (1), and at the maximum of the positive phase (3). At the beginning and end of the experiment the axon was in normal artificial sea water. Temperature 18° C. The volume of external fluid was about 100 ml.

Sodium movements during electrical activity in DNP-*treated axons*

The observations described in the previous section are all consistent with the view that metabolic energy is needed for recovery, but not for the conduction of impulses, which depends on the movements of ions down concentration gradients. In order to test this point further, we used ^{24}Na to find out whether the rapid movements of sodium which accompany electrical activity are appreciably affected by inhibiting the sodium pump with dinitrophenol. Fig. 9 shows the effect of 0·2 mM-DNP on the inward movement of sodium during the nerve impulse in a giant axon from *Loligo*. The fibre was in flowing sea water above a Geiger counter except for short periods when it was dipped into sea water made up with ^{24}Na. The first dip gave the resting entry of sodium and the second the entry during stimulation at 50/sec. Stimulation resulted in a large uptake of labelled sodium, the mean entry of ^{24}Na being about 10 times greater than at rest. After each of these dips the radioactivity of the fibre gradually declined, showing that labelled sodium was being extruded from the axon. This was checked when 0·2 mM-DNP was applied, and the counting rate then became practically constant. After the dinitrophenol had taken effect, the fibre was stimulated in ^{24}Na sea water containing 0·2 mM-DNP. The inhibitor

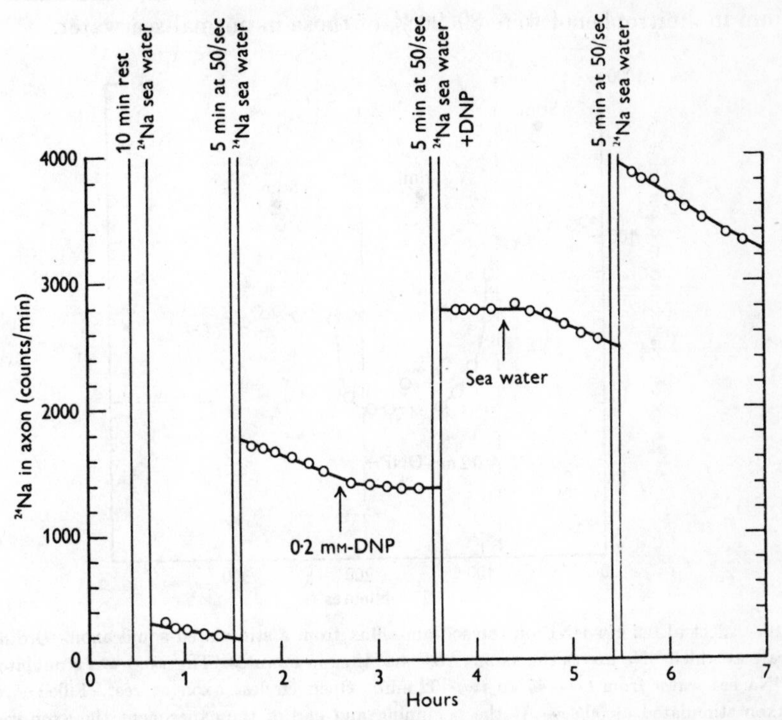

Fig. 9. Effect of 0·2 mM-DNP on sodium entry during stimulation of a squid axon. For description of experimental procedure see text. Resting sodium influx during first immersion in ^{24}Na sea water was 50 pmole/cm^2 sec; other numerical results are given in Table 11. Temperature 17° C.

evidently had little effect on the entry of sodium, because the increase in radioactivity was about the same as that observed in the absence of dinitrophenol. A final set of measurements, made when the fibre had recovered from being poisoned, established that it was in a steady condition throughout the experiment.

Fig. 10 shows the effect of DNP on the outward movement of sodium during electrical activity. At the beginning of the experiment the fibre was loaded with ^{24}Na by stimulation at 100/sec for 15 min in radioactive sea water. It was then set up in the flow apparatus (Fig. 1), and the ^{24}Na leaving the axon was collected and counted in the usual way. It will be seen that 0·2 mM-DNP caused a large reduction in the resting extrusion of sodium, but had little effect on the extra efflux associated with the passage of impulses.

Table 11 gives the results of working out these two experiments. It shows that a concentration of dinitrophenol which inhibited the sodium pump had no marked action on the rapid movements of sodium during the spike. This conclusion was supported by two further experiments which were not worked out in detail, but which indicated that the stimulated inflow and outflow of sodium in dinitrophenol were 80–90% of those in normal sea water.

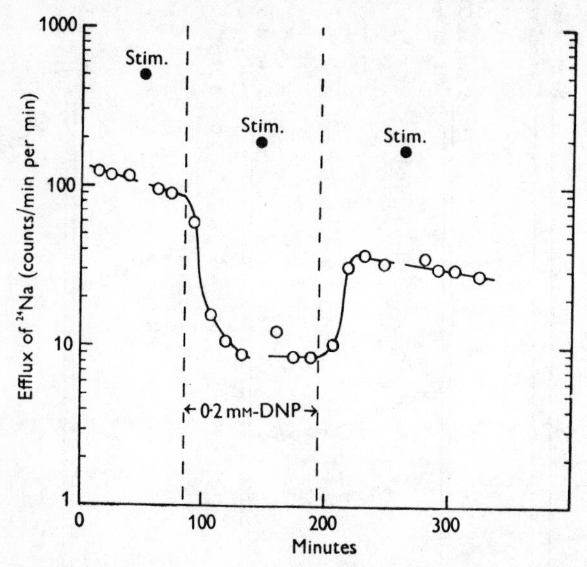

Fig. 10. Effect of 0·2 mM-DNP on the sodium efflux from a stimulated squid axon. Ordinate: rate at which ^{24}Na leaves the axon. Abscissa: time in minutes. The axon was stimulated in ^{24}Na sea water from $t = -47$ to $t = -32$ min. Open circles: axon at rest. Filled circles: axon stimulated at 50/sec. At the beginning and end of the experiment the axon was in normal artificial sea water. Temperature 18° C.

TABLE 11. Effect of 0·2 mM-DNP on sodium movements during electrical activity of axons from *Loligo forbesi*

	Inflow (pmole/cm² sec)	Outflow (pmole/cm² sec)	Net entry (pmole/cm² sec)
Normal	12·0	5·9	6·1
0·2 mM-DNP	10·9	4·6	6·3
Normal	12·4	5·4	7·0

The first column was worked out by the method used by Keynes (1951*a*) from the experiment of Fig. 9, and the second from Fig. 10 by the method used for Table 3.

The effect of external potassium concentration on the sodium efflux

Steinbach (1940, 1951, 1952) and later Desmedt (1953) showed that when frog muscles are soaked in potassium-free Ringer's solution they gain sodium and lose potassium, and that they are able to extrude some of their internal sodium if they are subsequently transferred to potassium-rich Ringer's solution. Working with tracers, Keynes (1954) found that the sodium efflux from frog muscles was reduced in a potassium-free medium and increased in potassium-rich Ringer. A similar but smaller effect has been noticed in human erythrocytes by Harris & Maizels (1951) and Glynn (1954), and in horse erythrocytes by Shaw (1954). Active transport of sodium through frog skin is also greatly reduced by removing potassium from the fluid bathing the skin (Ussing, 1954).

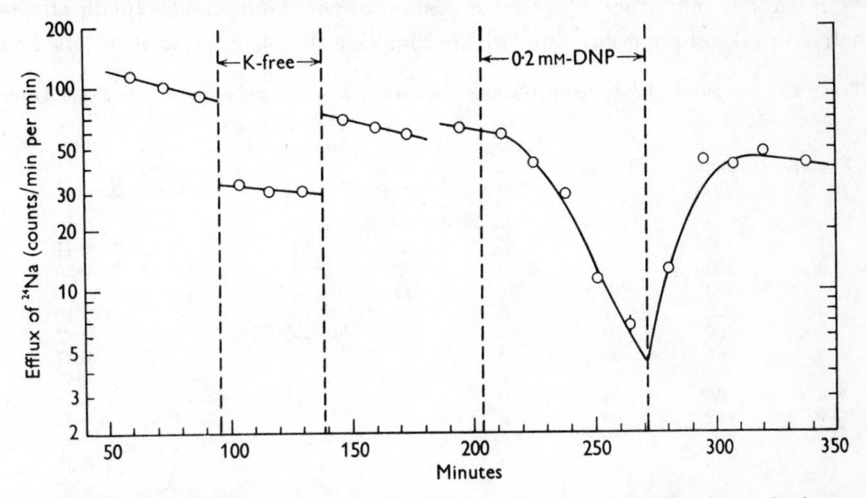

Fig. 11. Effects of potassium-free sea water and 0·2 mM-DNP on sodium efflux from a *Sepia* axon. When not in a test solution the axon was in artificial sea water containing 10·4 mM-K. A break has been left in the curve at 180 min because the excitability was tested at this point, and the length of the fibre in the tube may have been altered slightly. Temperature 17° C.

As shown in Fig. 11 and Table 12A, *Sepia* axons behave in the same way. In potassium-free sea water the sodium efflux is, on the average, reduced to about 0·3 of its normal value, the effect being immediate in contrast to the delayed action of inhibitors. The difference in time course between the actions of potassium-free and DNP solutions is well illustrated by applying the two agents consecutively as in the experiment of Fig. 11. It will be seen that the lag in the development of the potassium-free effect did not exceed 2 or 3 min, whereas the DNP effect was still increasing after 70 min and 50% inhibition required 35 min.

Raising the external potassium above 10·4 mM caused an increase in sodium efflux, but not a large one, since the efflux was only about 40% above normal in a solution containing 52 mM-K.

Since the internal potassium changes only very slowly when external potassium is removed, it cannot be argued that the potassium-free effect on sodium efflux is due to a decrease in internal potassium concentration. In absolute units the reduction of sodium efflux on removal of potassium amounted to about 20 pmole/cm² sec, as may be seen from the fully analysed experiments in Table 12. The reduction was thus roughly equal to the potassium influx in fibres recovering from stimulation (see Table 2). Since most of the uptake of potassium under these conditions was removed by agents which inhibited the sodium pump, it is not unreasonable to conclude that metabolism drives a cycle (Fig. 13) in which one Na^+ ion is extruded for each K^+ ion absorbed, and that this cycle can be interrupted by removal of external potassium as well as by metabolic inhibitors. A coupled system of this kind could explain all the active uptake of potassium, but without making the assumption mentioned on p. 56 it does not account for all the sodium extruded, since the sodium efflux is greater than the potassium influx, and does not drop to zero when potassium is removed from the external medium.

TABLE 12. The effect of external potassium concentration on the sodium fluxes in *Sepia* axons

Experiment	Axon diam. (μ)	Temp. (° C)	[K] in test fluid (mM)	Flux values (pmole/cm² sec) Normal	Test	Normal	Ratio Test flux / Normal flux
			A. Sodium efflux				
1	196	17	0	28	11	26	0·41
2	193	18	0	31	8	33	0·25
	193	18	52	33	46	31	1·44
3	220	17	0	—	—	—	0·19
4	199	19	0	—	—	—	0·22
5	201	20	0	—	—	—	0·57
	201	20	1	—	—	—	0·67
6	208	20	0	—	—	—	0·21
7	182	20	0	—	—	—	0·33
	182	20	52	—	—	—	1·38
8	252	20	0	—	—	—	0·21
						Mean, K-free	0·30
						Mean, 52 mM-K	1·41
			B. Sodium influx				
9	148	20	0	35	40	57	0·9
10	184	21	0	33	36	27	1·2

Expts. 1–4 were done by the flow method (B, p. 30).
Expts. 5–8 were done by determining the rate constant for loss of total radioactivity (A, p. 30).
In calculating the ratio in the last column, the normal value was always taken as the geometric mean of two figures determined before and after the test value.

We have done only two experiments on the effect of removing external potassium on the sodium influx. These are summarized in Table 12B, and showed no marked change.

Membrane potential and sodium efflux

One possible explanation of the effect of applying a potassium-free solution is that it might arise from an increase in resting potential. In potassium-free sea water the resting potential of *Sepia* axons has been found to increase by 5–10 mV (Hodgkin & Keynes, 1955). It is conceivable, though perhaps not very likely, that an increase of 10 mV in the potential gradient against which sodium has to be extruded might slow up the transport system enough to give a threefold reduction in efflux. This possibility was investigated by measuring the rate at which sodium was extruded from a 6 mm length of *Sepia* axon whose membrane potential could be raised by anodal polarization. These experiments have already been described in sufficient detail in a previous article (Hodgkin & Keynes, 1954). They showed that increasing the membrane potential by 10–30 mV did not alter the efflux of sodium by more than 10%. (In ten experiments the average efflux during polarization was 0·99 of normal, the S.E. of the mean being 0·04, for an increase in membrane potential averaging 18 mV.) This indicates that the effect of removing external potassium cannot be due to a change in membrane potential.

All the experiments with polarizing currents were done in sea water containing 10·4 mM-K, and it is possible that an effect of membrane potential on sodium efflux might be observed if the same experiments were repeated with potassium-free solutions.

External sodium concentration and sodium efflux

The effect of replacing the sodium in sea water by choline was to increase the efflux of sodium by 20–50% (Expts. 5–8, Table 13). The change was relatively larger if potassium was first removed from the external medium (Expts. 1–4), but the absolute difference was probably about the same. Thus substitution of choline for external sodium increased the sodium efflux by about one-third of its value in normal sea water both when [K] = 10·4 mM and when [K] = 0 mM. The action appeared to be immediate, like that of changing the external potassium concentration, and was not delayed as with metabolic inhibitors. This is illustrated in Fig. 12.

The experiments in Table 13 seem to rule out the possibility that the outward movement of labelled sodium into a potassium-free medium is an exchange of internal for external sodium of the type discussed by Ussing (1949). For if it were an 'exchange diffusion', removal of sodium as well as potassium from sea water should reduce the efflux to zero, instead of increasing it as in Table 13. The outward movement of sodium into potassium-free choline sea water does not appear to be an exchange with choline, since similar or slightly larger effluxes were observed in potassium-free dextrose sea water. Nor can it be regarded as a passive leak, since the sodium efflux into choline sea water or salt-free sugar solutions was inhibited by dinitrophenol (Table 4).

The ability of *Sepia* axons to extrude sodium into isotonic solutions of dextrose means that under these conditions Na^+ is either exchanging with H^+ or moving out with an anion. The former possibility was excluded by measuring the pH of a small volume of isotonic dextrose solution into which sodium had been extruded. This showed no detectable change in pH, although to judge from similar experiments with tracers the concentration of sodium in the solution should have reached about 0·5 mM. It follows that sodium ions extruded into a salt-free solution must be accompanied by anions from the nerve. Owing to the very small quantities involved, chemical analysis of these anions is difficult, and we have no evidence as to their identity.

TABLE 13. The effect of external sodium concentration on sodium efflux in *Sepia* axons

Experiment	Axon diam. (μ)	Test solution	Concentrations in mM K	Na	Ca	Mg	Efflux in test solution / Efflux in sea water
1	199	K-free sea water	0	486	11	55	0·22
		K-free choline sea water	0	0	11	55	0·42
2	208	K-free sea water	0	486	11	55	0·2
		K-free choline sea water	0	0	11	55	0·6
3	220	K-free sea water	0	486	11	55	0·19
		K-free dextrose sea water	0	0	11	55	0·82
4	252	K-free sea water	0	486	11	55	0·2
		K-free dextrose sea water	0	0	11	55	0·8
5	216	Choline sea water	10	0	11	55	1·35
6	194	Choline sea water	10	0	11	55	1·23
7	159	Choline sea water	10	0	11	55	1·4
8	181	Choline sea water	10	0	11	55	1·5
9	202	Isotonic dextrose	0	0	0	0	0·9
10	216	Isotonic dextrose plus calcium	0	0	11	0	0·9

Expts. 1, 3, 5 and 6 were done by the flow method (B, p. 30), the remainder by the less accurate method of determining the rate constant for loss of total radioactivity (A, p. 30).

The efflux in sea water was taken as the geometric mean of determinations made in normal sea water at the beginning and end of each experiment.

Temperature 16–20° C.

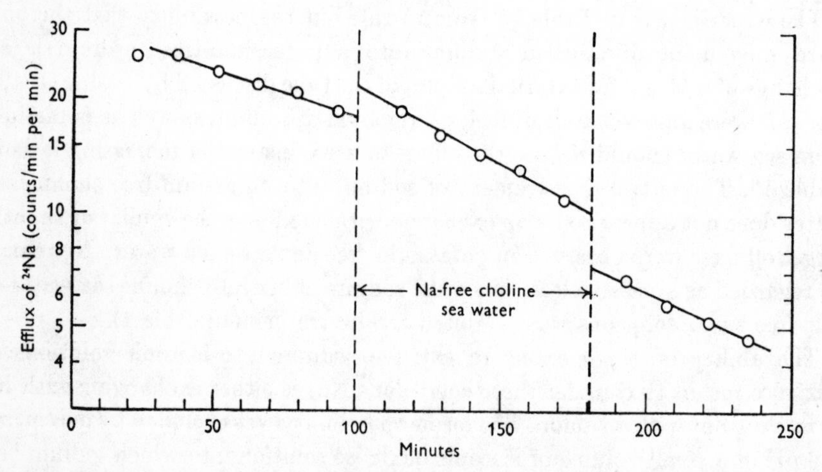

Fig. 12. Effect of a sodium-free solution (choline sea water) on the sodium efflux from a *Sepia* axon. When not in choline the axon was in sea water containing 486 mM-Na (see Table 1). Temperature 15° C.

Axons which had been kept for long periods in pure dextrose solutions always became irreversibly inexcitable, whereas those kept in dextrose plus 10·7 mM-Ca recovered their excitability when sodium ions were restored to the external medium. Since there was no marked difference in the sodium efflux under the two conditions, it appears that the action potential mechanism can be irreversibly damaged without greatly affecting the sodium pump.

DISCUSSION

The distinction between active and passive transport through the nerve membrane

The main conclusion from our experiments can be summarized with the help of the diagram shown in Fig. 13. In this scheme, the sodium and potassium concentration differences are built up during recovery by a cyclical process which absorbs potassium and extrudes sodium against the electrochemical potential gradient. (The broken arrow allows for the fact that the amount of sodium extruded exceeds the amount of potassium absorbed.) In parallel with this system are the channels which allow sodium and potassium ions to move passively through the membrane under the influence of electrochemical

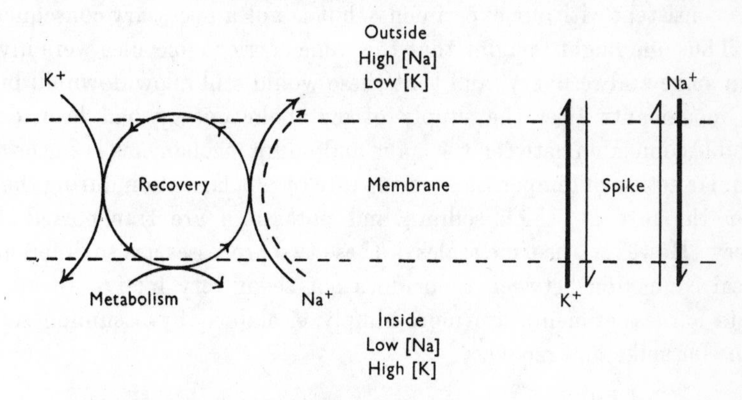

Fig. 13. Diagram illustrating spike and recovery mechanisms.

potential differences. Normally these channels only allow a trickle of ions to leak through the membrane, but they open up when the membrane is depolarized, and large ionic movements then take place. These generate the action potential, and the fibre pays for it by an accumulation of sodium and a loss of potassium.

These two mechanisms for active and passive transport of ions through the membrane are entirely different in their properties, and it is possible to differentiate between them in a number of ways. Thus the secretory movements have high temperature coefficients, are inhibited by dinitrophenol, and are unaltered by increases in membrane potential. Passive movements, on the other hand, have low temperature coefficients, and are little affected by dinitrophenol, but vary greatly with membrane potential. Another distinction is that the maximum rate at which the secretory mechanism can move ions is about 50 pmole/cm^2 sec, whereas the spike mechanism allows peak movements of the order of 10,000 pmole/cm^2 sec (Hodgkin *et al.* 1952).

Dissociation of spike and recovery mechanisms

Among the mechanisms proposed for the action potential, one tentative suggestion discussed by Grundfest (1950) is that the net inward movement of sodium during the spike might arise from a transient inhibition of the process which normally extrudes sodium. In addition to the difficulties mentioned by Katz (1952), this hypothesis now has to meet the objection that axons in which the pump has been completely inhibited with dinitrophenol or cyanide are nevertheless capable of conducting normal action potentials. These observations seem to exclude the pump from direct participation in the spike, but raise no difficulty if one assumes that the latter arises from increases in permeability which allow sodium and potassium ions to move down concentration gradients.

It is tempting to go further, and assume that the spike and recovery mechanisms are separate systems as indicated in Fig. 13. A scheme of this kind is consistent with our experiments, but is not a necessary consequence of them. Thus one might imagine that the same carrier molecules were involved both in spike and recovery, and that these would still allow downhill but not uphill movements after the supply of metabolic energy had been cut off. A possible connexion between the spike and pump mechanisms is suggested by the similar action of temperature on the rate of repolarization during the spike and on the rate at which sodium and potassium are transported during recovery. However, the time scales of these two processes are so different that a causal connexion between them does not seem very likely. In all other respects our experiments are more simply explained by assuming separate systems for spike and recovery.

The linkage between sodium extrusion and potassium absorption

The idea of a coupled system which ejects sodium from the axon on one limb of a cycle, and absorbs potassium on the other, receives strong support from our experiments. Thus if the potassium influx is abolished by removing external potassium, the sodium efflux drops by an amount roughly equal to the original potassium influx. There is also evidence that most of the uptake of potassium by fibres recovering from stimulation is not a passive movement, but is an active one requiring a supply of energy. Thus cooling to 1° C, or agents like DNP and cyanide, which greatly reduce the sodium efflux, also remove most of the potassium influx. The fact that the potassium influx and probably the sodium efflux increase after a period of stimulation, and then slowly return towards their resting values, may have a simple explanation in terms of a coupled pump. For it seems possible that both effects, and perhaps an increase in metabolism as well, may arise from the increase in internal sodium concentration associated with stimulation.

A coupled system which absorbed one K^+ for each Na^+ ejected would be electrically neutral in the sense that it would transfer no charge across the membrane. With such a mechanism one would expect changes in the activity of the pump to have no immediate effect on membrane potential, while alterations in membrane potential ought not to have much effect on the activity of the pump. Both predictions agree with our observations, but this is not really good evidence that the pump is neutral. If a chemical reaction involved in the sodium pump gave 10,000 cal/mole (as in the hydrolysis of ATP), it would be capable of driving ions against an electrochemical potential difference of 430 mV. In this case the change of 20 mV produced by applied currents might reduce the efflux of sodium by an insignificant amount. A reservation must also be made about the lack of any change in membrane potential when the sodium pump is blocked. The argument is clearest for the squid axon, which has a resting membrane resistance of only about 1000 ohm cm^2 (Cole & Hodgkin, 1939; Hodgkin *et al.* 1952). The sodium efflux from the squid axon is about 50 pmole/cm^2 sec, and it may be assumed that, as in *Sepia* axons, two-thirds of this is balanced by an active uptake of potassium. If the remaining third consists of a stream of sodium ejected as ions, the pump would generate a current of $1{\cdot}8\,\mu A$, and its direct contribution to the membrane potential would be only 1·8 mV. The lack of any marked change in resting potential on applying DNP to the squid axon is therefore not conclusive evidence that the pump is neutral.

Although we have presented evidence for a loose or partial coupling between the secretory movements of sodium and potassium, it seems that there cannot be a rigid linkage between sodium efflux and potassium influx. Removing the potassium from the external medium reduces the sodium efflux by a factor of about 3, but much greater reductions are observed at low temperatures or under the influence of metabolic inhibitors. The residual sodium efflux into potassium-free solutions is not an exchange for external sodium, since it is increased by substituting choline or dextrose for sodium. When sodium ions are extruded into a salt-free solution they must be accompanied by anions, but there is no indication as to the nature of these substances, and we do not know whether they leave the nerve passively or are actively extruded with sodium.

A possibility which has not been excluded is that the net leakage of potassium occurring in potassium-free solutions may prevent the concentration of potassium immediately outside the membrane from falling below 1 or 2 mM. A local accumulation of this kind might allow even a tightly coupled pump to continue at a reduced rate in potassium-free sea water, but it would still be necessary to assume that the sodium and potassium ions leaving the axon in dextrose solutions were accompanied by anions.

Comparison with other tissues

The action of metabolic inhibitors on cephalopod axons is consistent with results on other tissues. Poisons like dinitrophenol and cyanide block the active transport of ions in tissues such as gastric mucosa (Davies, 1951), kidney slices (Mudge, 1951), frog skin (Fuhrman, 1952), and chicken erythrocytes (Maizels, 1954), as well as in plants (Robertson, Wilkins & Weeks, 1951; Scott & Hayward, 1954). Two preparations which do not seem to be affected by these agents are mammalian erythrocytes (Maizels, 1951) and amphibian muscle (Keynes & Maisel, 1954). In the former case the lack of effect is not surprising, since it may be that cyanide and DNP have little blocking action on the glycolytic system which is apparently the principal metabolic pathway in mammalian blood. This explanation is less satisfactory for frog muscle, since muscles doubly poisoned with cyanide and iodoacetate—which should block both respiration and glycolysis—show no striking reduction in sodium efflux after several hours treatment at room temperature (Keynes & Maisel, 1954), and no potassium leakage at 0° C (Ling, 1952). Whether the sodium pump can be driven by the store of phosphocreatine, or whether there is some other energy reserve in muscle, is not yet clear, but the evidence at present available does not seem to us sufficiently compelling to justify Ling's (1952) conclusion that there is no active transport of cations in muscle, and that the ionic state is maintained by selective binding at fixed sites in the myoplasm. Apart from the many other difficulties raised by this hypothesis, there is now strong evidence that in cephalopod nerve (whose electrical and ionic properties are not unlike those of frog muscle) sodium ions do not leave the cell passively but are driven out by metabolism.

The influence of temperature on the active transport of sodium and potassium in *Sepia* nerve is in qualitative agreement with the observations of Steinbach (1954) or Harris (1950) on frog muscle, of Solomon (1952) on red cells, and of Shanes (1954) on crab nerve. However, it is difficult to make any exact comparison, since the temperature coefficient of the sodium efflux in *Sepia* fibres is not constant, but is very high near 0° C and declines progressively as the temperature is raised.

The inability of *Sepia* axons to pump sodium effectively near 0° C fits in with the geographical distribution of these animals. *S. officinalis* is scarce to the north of Britain, more common to the south, and exceedingly abundant in the Mediterranean (Forbes & Hanley, 1853). At Plymouth it appears in large numbers in September or October, and becomes infrequent after January. Many factors besides temperature must govern the distribution and migration of *Sepia*, but the animals are certainly not well adapted to low temperatures, as we found to our cost during a cold spell when a consignment arrived at Cambridge with all eight specimens dead and the sea water in the containers at a temperature of 4° C. (The mortality for similar journeys at temperatures of 10–15° C was normally about 10%.) At present there is no evidence for a connexion between the habitat of a species and the temperature dependence of its secretory mechanisms, but since many animals live happily at 0° C, there must presumably be a different relation between temperature and sodium efflux in cold-water forms.

SUMMARY

1. The rate at which sodium is extruded by giant axons from *Sepia* or *Loligo* is markedly reduced by dinitrophenol, cyanide and azide, or by cooling to 1° C. Sodium influx is not much altered under these conditions.

2. Dinitrophenol, cyanide and low temperatures greatly reduce the uptake of potassium from sea water (containing 10·4 mM-K), but have relatively little effect on the efflux of potassium, or on the influx from potassium-rich solutions (containing 52 mM-K).

3. Removal of potassium ions from the external fluid reduces the sodium efflux to about one-third, the effect being immediate in contrast to the delayed action of inhibitors.

4. The decrease in potassium influx produced by dinitrophenol is about equal to the decrease in sodium efflux produced by removing external potassium.

5. Increasing the membrane potential by anodal currents does not alter the sodium efflux into sea water containing 10·4 mM-K.

6. Sodium can be extruded into solutions from which all the sodium and potassium ions have been removed. Removing external sodium at constant external potassium increases the efflux of sodium.

7. The resting potential and action potential are little changed by dinitrophenol or cyanide.

8. The rapid sodium movements associated with the passage of impulses are little altered by concentrations of dinitrophenol which prevent sodium extrusion during recovery.

9. It is concluded that in addition to a permeability system which allows ions to move down electrochemical gradients during electrical activity, there is a secretory mechanism driven by metabolism which ejects sodium and absorbs potassium against the electrochemical gradients. Conduction of impulses, but not recovery, can take place if the secretory mechanism is put out of action with inhibitors. Sodium efflux and potassium influx are coupled, but do not seem to be linked rigidly.

We wish to express our thanks to the Director and staff of the laboratory of the Marine Biological Association at Plymouth for much assistance, to Mr. A. F. Huxley for helpful discussion, and to the Rockefeller and Nuffield Foundations for financial support.

REFERENCES

Calkins, E., Taylor, I. M. & Hastings, A. B. (1954). Potassium exchange in the isolated rat diaphragm; effect of anoxia and cold. *Amer. J. Physiol.* **177**, 211–218.

Cole, K. S. & Hodgkin, A. L. (1939). Membrane and protoplasm resistance in the squid giant axon. *J. gen. Physiol.* **22**, 671–687.

Davies, R. E. (1951). The mechanism of hydrochloric acid production by the stomach. *Biol. Rev.* **26**, 87–120.

Desmedt, J. E. (1953). Electrical activity and intracellular sodium concentration in frog muscle. *J. Physiol.* **121**, 191–205.

Forbes, E. & Hanley, S. (1853). *A History of British Mollusca*, **4**, 240. London: van Voorst.

Fuhrman, F. A. (1952). Inhibition of active sodium transport in the isolated frog skin. *Amer. J. Physiol.* **171**, 266–278.

Glynn, I. M. (1954). Linked sodium and potassium movements in human red cells. *J. Physiol.* **126**, 35*P*.

Grundfest, H. (1950). General neurophysiology. *Progr. Neurol. Psychiat.* **5**, 16–42.

Harris, E. J. (1950). The transfer of sodium and potassium between muscle and the surrounding medium. Part II. The sodium flux. *Trans. Faraday Soc.* **46**, 872–882.

Harris, E. J. & Maizels, M. (1951). The permeability of human erythrocytes to sodium. *J. Physiol.* **113**, 506–524.

Hodgkin, A. L. (1951). The ionic basis of electrical activity in nerve and muscle. *Biol. Rev.* **26**, 339–409.

Hodgkin, A. L. & Huxley, A. F. (1945). Resting and action potentials in single nerve fibres. *J. Physiol.* **104**, 176–195.

Hodgkin, A. L. & Huxley, A. F. (1952). A quantitative description of membrane current and its application to conduction and excitation in nerve. *J. Physiol.* **117**, 500–544.

Hodgkin, A. L., Huxley, A. F. & Katz, B. (1952). Measurement of current-voltage relations in the membrane of the giant axon of *Loligo*. *J. Physiol.* **116**, 424–448.

Hodgkin, A. L. & Katz, B. (1949*a*). The effect of sodium ions on the electrical activity of the giant axon of the squid. *J. Physiol.* **108**, 37–77.

Hodgkin, A. L. & Katz. B. (1949*b*). The effect of temperature on the electrical activity of the giant axon of the squid. *J. Physiol.* **109**, 240–249.

Hodgkin, A. L. & Keynes, R. D. (1953*a*). Metabolic inhibitors and sodium movements in giant axons. *J. Physiol.* **120**, 45–46*P*.

Hodgkin, A. L. & Keynes, R. D. (1953*b*). Sodium extrusion and potassium absorption in *Sepia* axons. *J. Physiol.* **120**, 46–47*P*.

Hodgkin, A. L. & Keynes, R. D. (1953*c*). The mobility and diffusion coefficient of potassium in giant axons from *Sepia*. *J. Physiol.* **119**, 513–528.

Hodgkin, A. L. & Keynes, R. D. (1954). Movements of cations during recovery in nerve. *Symp. Soc. exp. Biol.* **8**, 423–437.

Hodgkin, A. L. & Keynes, R. D. (1955). The potassium permeability of a giant nerve fibre. *J. Physiol.* **128**, 61–88.

Katz, B. (1952). The properties of the nerve membrane and its relation to the conduction of impulses. *Symp. Soc. exp. Biol.* **6**, 16–38.

Keynes, R. D. (1951*a*). The ionic movements during nervous activity. *J. Physiol.* **114**, 119–150.

Keynes, R. D. (1951*b*). *The Role of Electrolytes in Excitable Tissues.* Rio de Janeiro: Instituto de Biofisica, Universidade do Brasil.

Keynes, R. D. (1954). The ionic fluxes in frog muscle. *Proc. Roy. Soc.* B, **142**, 359–382.

Keynes, R. D. & Lewis, P. R. (1951). The sodium and potassium content of cephalopod nerve fibres. *J. Physiol.* **114**, 151–182.

Keynes, R. D. & Maisel, G. W. (1954). The energy requirement for sodium extrusion from a frog muscle. *Proc. Roy. Soc.* B, **142**, 383–392.

Ling, G. (1952). The role of phosphate in the maintenance of the resting potential and selective ionic accumulation in frog muscle cells. In *Phosphorus Metabolism*, vol. **2**, ed. McElroy, W. D. & Glass, B. Baltimore: Johns Hopkins Press.

Ling, G. & Gerard, R. W. (1949). The normal membrane potential of frog sartorius fibres. *J. cell. comp. Physiol.* **34**, 383–396.

Maizels, M. (1951). Factors in the active transport of cations. *J. Physiol.* **112**, 59–83.

Maizels, M. (1954). Active cation transport in erythrocytes. *Symp. Soc. exp. Biol.* **8**, 202–227.

Mudge, G. H. (1951). Electrolyte and water metabolism of rabbit kidney slices; effect of metabolic inhibitors. *Amer. J. Physiol.* **167**, 206—223.

Nastuk, W. L. & Hodgkin, A. L. (1950). The electrical activity of single muscle fibres. *J. cell. comp. Physiol.* **35**, 39–73.

Ponder, E. (1950). Accumulation of potassium by human red cells. *J. gen. Physiol.* **33**, 745–757.

Robertson, R. N., Wilkins, M. J. & Weeks, D. C. (1951). Studies in the metabolism of plant cells. IX. The effects of 2,4-dinitrophenol on salt accumulation and salt respiration. *Aust. J. sci. Res.* **4**, 248–264.

Scott, G. T. & Hayward, H. R. (1954). Evidence for the presence of separate mechanisms regulating potassium and sodium distribution in *Ulva lactuca*. *J. gen. Physiol.* **37**, 601–620.

Shanes, A. M. (1951*a*). Potassium movement in relation to nerve activity. *J. gen. Physiol.* **34**, 795–807.

Shanes, A. M. (1951*b*). Factors in nerve functioning. *Fed. Proc.* **10**, 611–621.

Shanes, A. M. (1954). Effect of temperature on potassium liberation during nerve activity. *Fed. Proc.* **13**, 134.

Shaw, T. I. (1954). Sodium and potassium movements in red cells. Ph.D. Thesis, Cambridge University.

Solomon, A. K. (1952). The permeability of the human erythrocyte to sodium and potassium. *J. gen. Physiol.* **36**, 57–110.

Steinbach, H. B. (1940). Sodium and potassium in frog muscle. *J. biol. Chem.* **133**, 695–701.

Steinbach, H. B. (1951). Sodium extrusion from isolated frog muscle. *Amer. J. Physiol.* **167**, 284–287.

Steinbach, H. B. (1952). On the sodium and potassium balance of isolated frog muscles. *Proc. nat. Acad. Sci., Wash.*, **38**, 451–455.

Steinbach, H. B. (1954). The regulation of sodium and potassium in muscle fibres. *Symp. Soc. exp. Biol.* **8**, 438–452.

Ussing, H. H. (1949). Transport of ions across cellular membranes. *Physiol. Rev.* **29**, 127–155.

Ussing, H. H. (1954). Ion transport across biological membranes. In *Ion Transport Across Membranes*, ed. Clarke, H. T. & Nachmansohn, D. New York: Academic Press.

Weidmann, S. (1951). Electrical characteristics of *Sepia* axons. *J. Physiol.* **114**, 372–381.

13

J. Physiol. (1962), **164**, 355–374
With 9 *text-figures*
Printed in Great Britain

THE EFFECTS OF CHANGES IN INTERNAL IONIC CONCENTRATIONS ON THE ELECTRICAL PROPERTIES OF PERFUSED GIANT AXONS

By P. F. BAKER, A. L. HODGKIN and T. I. SHAW
From the Laboratory of the Marine Biological Association, Plymouth, and the Physiological and Biochemical Laboratories, University of Cambridge

(*Received* 6 *June* 1962)

This paper is concerned with the effects of altering the composition of the internal fluid on the electrical properties of giant axons from which the bulk of the axoplasm had been removed by extrusion (Baker, Hodgkin & Shaw, 1962).* To begin with, we shall consider a series of experiments which show that the difference in potassium concentration between the internal and external fluid provides the main electromotive force for generating the resting potential. In order to reduce the error introduced by junction potentials and to avoid making uncertain estimates of the activity coefficient of K^+ in K_2SO_4 it was simplest to work with an internal solution consisting of mixtures of KCl and NaCl. The high concentration of chloride did not have any markedly deleterious effect, for fibres filled with isotonic KCl were usually excitable and had a resting potential only about 5 mV less than in those filled with isotonic potassium sulphate.

RESULTS

The effect of replacing internal K by Na on the resting potential

Figure 1 shows the time course of the change in membrane potential caused by a temporary replacement of KCl with NaCl. The resting potential was reduced from about -50 mV to zero and returned to within a few millivolts of its original value when KCl was restored. The asymmetry of the curve, i.e. the slow depolarization and rapid repolarization, is explained by the non-linear relation between $[K]_i$ and membrane potential. From curve *a* in Fig. 2 the potassium concentration at which the resting potential is reduced to half its normal value may be taken as 60 mM. Since this is only 10 % of the initial concentration, depolarization should be much slower than repolarization. The time for complete depolarization probably represents the time required for a flow of about 50 μl./min to remove 99 % of the original fluid from a total volume of roughly 50 μl. (30 μl. dead

[*See pp. 396–403 for "Methods" from Baker, Hodgkin and Shaw, 1962 (Eds.).]

space, 20 μl. axon). Shorter time delays were obtained in experiments in which the dead space was reduced by introducing solutions through the upper cannula. In one such test the depolarization was largely complete in 20 sec and the repolarization in 2 sec.

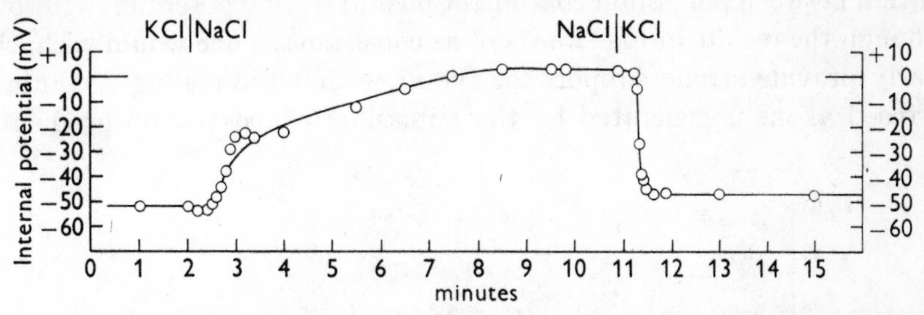

Fig. 1. Effect on the resting potential of replacing isotonic KCl with isotonic NaCl inside axon. The results have not been corrected for the junction potential between KCl and NaCl; correction would shift the final points in NaCl upwards by 4 mV.

The composition of the isotonic KCl is given against solution *E* in Table 1 of Baker *et al.* (1962); isotonic NaCl was the same with Na replacing K on a mole-for-mole basis. This experiment was done with a doubly cannulated axon with the apparatus shown in Text-fig. 5 of Baker *et al.* (1962). The total volume of isotonic NaCl passed through the axon was about 0·5 ml. Axon 81; 16° C; external solution, sea water.

The curves in Fig. 2 give the relation between the internal potassium concentration and membrane potential at three different external potassium concentrations. Curve *a*, which is the relation for normal sea water, will be considered first. With NaCl inside the fibre the potential was close to zero but it increased rapidly with $[K]_i$, reaching 40–50 mV at 150 mM-K. A further rise from 150–600 mM-K increased the resting potential by only 5–10 mV as against the 35 mV expected for a potassium electrode. This saturation effect probably occurs because the potassium permeability falls as the resting potential rises (see p. 369). The corresponding curve with sulphate as the anion is given in Fig. 4; it is similar to that observed with chloride but displaced about 10 mV in the negative direction.

With isotonic KCl outside and isotonic NaCl inside, the normal resting potential was reversed, the inside being about 50 mV positive to the outside (Fig. 2, curve *c*). Details of one experiment are as follows. To begin with, the external solution was sea water and the internal solution was isotonic K_2SO_4. On replacing the latter with NaCl the internal potential changed from −60 to −2 mV. The sea water outside the fibre was then replaced with isotonic KCl containing 50 mM-$CaCl_2$. The internal potential rose in about 30 sec to a maximum of +64 mV and then slowly declined, the value after 1 min being +54 mV. It seems likely that under the highly abnormal conditions of a reversed membrane potential the membrane loses its selective permeability to potassium rather rapidly and that this limits the extent to which the inside of the fibre can be made positive. A similar but more gradual loss of selectivity seems to occur whenever the fibre is depolarized and makes it difficult to determine the exact form of the relation between potassium concentration and resting potential. However, although the results in Fig. 2 are not as consistent as one would wish, they clearly provide strong support for the view that the resting potential of perfused axons is generated by the potassium concentration gradient.

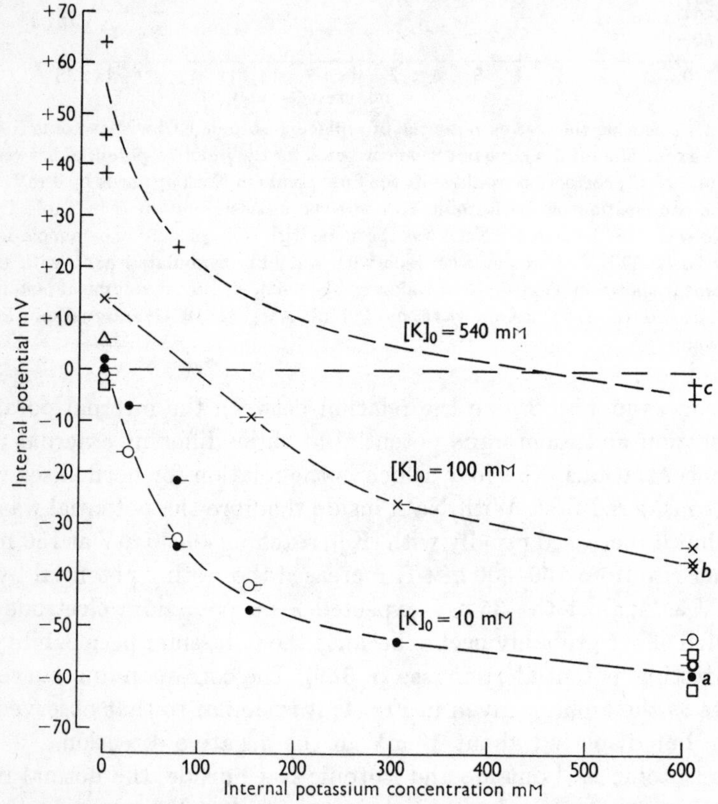

Fig. 2. Effect of varying the internal potassium concentration by replacing isotonic KCl with isotonic NaCl in perfused axons surrounded by solutions containing 10, (*a*); 100, (*b*); or 540 mM-K, (*c*). *a*; ○, △, □, natural sea water with 10 mM-K, 470 mM Na and 550 mM-Cl; each symbol represents a different axon. —○, after increasing internal potassium concentration, ○—, after decreasing internal potassium concentration. ●, artificial sea water with 10 mM-K, 526 mM-Na, 633 mM-Cl, 50 mM-Ca. *b*; ×; artificial sea water with 100 mM-K, 436 mM-Na, 633 mM-Cl, 50 mM-Ca; one axon in poor condition. *c*; +; high-K artificial sea water with 538 mM-K, 635 mM-Cl, 50 mM-Ca; 6 different axons. Corrections for junction potential of 1–4 mV have been applied when necessary. The chloride concentration of the internal solutions was 560 mM.

Similar solutions on both sides of the membrane

If the membrane potential of a perfused fibre depends only on ionic concentration differences it should be zero when identical solutions are on both sides of the membrane. Curves *a*, *b* and *c* in Fig. 2 indicate that the potential was close to zero in the absence of appreciable gradients of K, Cl or Na. However, in these experiments the external and internal solutions were not identical because the former contained Ca whereas the latter did not.

TABLE 1. Membrane potentials with identical solutions on both sides of membrane

Axon	Solution X	Potential difference with X on both sides of membrane (mV)	Potential difference with K_2SO_4 inside and sea water outside, corrected for junction potential: Before test (mV)	After test (mV)
73	K_2SO_4	−1	−59	−21
74	K_2SO_4	−1	−59	−26
195	K_2SO_4	0	−62	—
193	KCl	−0·5	−64	−43
187	Na_2SO_4	0	−51	−35*
190	NaCl	+1	−64	−47†
189	Sea water	0	−54	−30

* External solution Na_2SO_4. † External solution NaCl. Except in the first two rows the P.D. with X on both sides was determined directly as the change in potential associated with transference of the micro-electrode from the inside of the membrane to the outside. In the first two cases the electrode was not withdrawn until the end of the experiment and the usual corrections for junction potentials were applied. For the composition of the solutions see Table 1 of Baker *et al.* (1962). Temperature 17–20° C.

Table 1 summarizes experiments in which identical solutions were placed on both sides of the membrane and shows that the membrane potential under these conditions was always within 1 mV of zero. Recovery after such tests was necessarily poor because they required either that the external solution be calcium-free or that the internal solution should contain calcium. Both conditions tend to destroy the selective properties of the membrane, so it is not easy to obtain fully reversible effects in experiments with identical solutions inside and outside the membrane.

Effect of changing the internal anion on the resting potential

Substitution of isotonic KCl for isotonic K_2SO_4 inside an axon immersed in sea water reduced the resting potential by a few millivolts, the average change after correcting for the difference in junction potential between the two solutions being 5 mV (Table 2). This suggests that chloride ions do not make a substantial contribution to the resting potential of perfused axons.

The effect of replacing isotonic potassium sulphate by isotonic potassium methylsulphate was studied on two axons at the end of the experi-

TABLE 2. Effect of replacing isotonic K_2SO_4 by isotonic KCl inside axon on resting potential and action potential (mV)

Axon	Internal solution	Resting potential: Uncorrected	Resting potential: Corrected	Action potential	V_A	V_U	Change in resting potential (corrected)
73	K_2SO_4	−55	−62	98	+36	−83	
	KCl	−60	−60	96	+36	−81	+2
	K_2SO_4	−55	−62	96	+34	−82	
74	K_2SO_4	−51	−58	91	+33	−78	
	KCl	−54	−54	90	+36	−74	+4
75	K_2SO_4	−55	−62	106	+44	−82	
	KCl	−59	−59	106	+47	−80	+3
81	K_2SO_4	−57	−64	100	+36	−80	
	KCl	−58	−58	96	+38	−77	+6

V_A is the internal potential at the peak of the spike, V_U at the bottom of the underswing, both corrected for junction potential. The external solution was sea water.

In these experiments resting potentials were measured with reference to isotonic KCl connected to the sea water through a 3 M-KCl bridge. A correction of 7 mV has been made for the junction potential between isotonic K_2SO_4 and isotonic KCl. Temperature 16–18° C. For details of solutions see Table 1, Baker *et al.* (1962).

TABLE 3. Effect of replacing isotonic potassium sulphate by isotonic potassium methylsulphate on resting potential and action potential

Axon	Internal solution	Resting potential (corrected; mV)	Action potential (mV)	V_A (mV)	V_U (mV)	Max. rate of rise (V/sec)	Max. rate of fall (V/sec)
212	K_2SO_4	−59	93	+34	−78	500	240
	KCH_3SO_4	−68	115	+47	−77	670	320
	K_2SO_4	−66	98	+32	−81	545	280
214	K_2SO_4	−61	85	+24	−76	400	225
	KCH_3SO_4	−65	114	+49	−73·5	625	300

V_A is the internal potential at the peak of the spike, V_U at the bottom of the underswing. For details of solutions see Table 1, Baker *et al.* (1962). Temperature 16° C; external solution, sea water.

mental period. These results which are given in Table 3 suggest that the resting potential was increased by the solution containing methylsulphate. This is surprising because the activity of potassium in isotonic KCH_3SO_4 is 0·38 as against 0·42 in K_2SO_4 and if the membrane were impermeable to both sulphate and methylsulphate the resting potential should be slightly higher with potassium sulphate inside the axon. The simplest explanation is that sulphate leaks out of holes in the membrane, or through cut branches and that the magnitude of the electrical leak is reduced when sulphate is replaced by an ion of lower mobility such as methylsulphate. A leakage of sulphate ions would also explain why dilution of K_2SO_4 with isotonic glucose sometimes increased the resting potential (Table 5).

Effect on the action potential of replacing K by Na

These experiments were carried out by replacing isotonic K_2SO_4 with mixtures of isotonic K_2SO_4 and Na_2SO_4. Figure 3 and Table 4, which give the results obtained in the most complete experiment, show that increasing the internal sodium reduced the overshoot in a reversible manner. The effect took place without much change in resting potential and so was unlikely to be caused by increasing inactivation of the sodium-carrying system. Nor could it be attributed to a decrease in internal potassium concentration, since replacing half the potassium sulphate with isotonic glucose tended to increase rather than to decrease the overshoot. The simplest explanation is that raising the internal sodium concentration decreased the concentration gradient on which the overshoot depends.

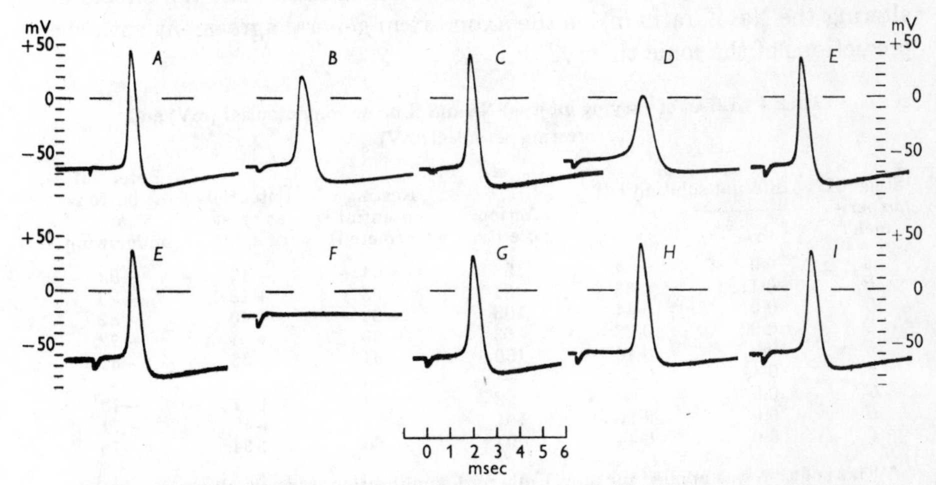

Fig. 3. Effect of varying internal sodium concentration. The external solution was sea water throughout; the internal solution was: *A*, *C*, *E*, *G*, *I*, isotonic K_2SO_4; *B*, $\frac{1}{4}$K replaced by Na; *D*, $\frac{1}{2}$K replaced by Na; *F*, isotonic Na_2SO_4; *H*, $\frac{1}{2}$ isotonic K_2SO_4, $\frac{1}{2}$ isotonic glucose. For the activity of Na and K in these solutions see Table 4. Except in *F*, records were taken about 5 min after the change in solutions; *F* was taken at 1 min. Time and voltage scales apply to all records; the position of the voltage scale has been corrected for the junction potential. Axon 64, 20·8° C.

The upper curve in Fig. 4 shows the membrane potential at the crest of the spike plotted against the sodium activity in the external solution; the potassium activity is also shown on the abscissa. The continuous line is drawn according to the equation

$$V = \frac{RT}{F} \ln \frac{[K]_o + b[Na]_o}{[K]_i + b[Na]_i}, \quad (1)$$

where [ ] are activities, R, T and F have their usual significance and b is the permeability ratio, P_{Na}/P_K, which has been given the value of 7 for the active membrane. For comparison, the experimental results and theoretical curves for the resting potential and for the potential at the peak of the underswing have been included. For resting nerve, b has been given the value of 0·08 and for refractory nerve, i.e. at the peak of the underswing, b = 0·03. Values obtained with an axon which gave a larger action potential (120 mV) were: b (resting) = 0·06, b (active) = 14, b (refractory) = 0·03.

Obvious objections to the use of eqn. (1) are that the effect of inward current at the crest of the spike has been neglected and that no allowance has been made for the variation of permeability coefficients with membrane potential. Another factor which has been ignored is the contribution of anions to leakage through the membrane. However, in spite of its limitations the simple theoretical treatment does show that the effects of altering the Na:K ratio inside the axon are in general agreement with the predictions of the ionic theory.

TABLE 4. Effect of varying internal Na and K on action potential (mV) and resting potential (mV)

Stage of experiment	Internal solution (M) a_{Na}	a_K	Action potential	Resting potential (corrected)	Potential at crest of spike	Potential at bottom of underswing
A	0·0	0·44	106	−64	+42	−83
B	0·11	0·33	81	−63	+18	−79
C	0·0	0·44	105	−65	+40	−82
D	0·22	0·22	63	−59	+ 4	−72
E	0·0	0·44	100	−64	+36	−81
F	0·44	0·0	—	−23*	—	—
G	0·0	0·44	93	−62	+31	−78
H	0·0	0·26	101	−59	+42	−71
I	0·0	0·44	94	−60	+34	−75

* This solution was applied for only 1 min and equilibration was probably not complete; in other cases measurements were made on records taken at 5 min after change. Activities in external solution (sea water) were a_{Na} = 0·32 and a_K = 0·0068 molar. All activities are based on Table 2 of Baker *et al.* (1962). Experiment of Fig. 3: measurements are from the mean of several records.

In addition to the rapid and reversible effects which have just been described, sodium-rich solutions inside the axon may also have a slower and irreversible effect. Fibres seemed to deteriorate more rapidly when the internal sodium concentration exceeded 100 mM and they often did not recover if an isotonic sodium solution was perfused for 5–10 min. It is not clear whether the second effect was due to the high concentration of sodium or to the prolonged depolarization of the membrane.

The effects of diluting isotonic K_2SO_4 with isotonic glucose on the action potential and resting potential

From eqn. (1) it is evident that when the term $b[\mathrm{Na}]_i$ is small compared with $[\mathrm{K}]_i$ the overshoot of the action potential should be limited by the internal potassium concentration. With $[\mathrm{Na}]_i = 0$ and $[\mathrm{Na}]_o \gg [\mathrm{K}]_o$ the equation reduces to

$$V = \frac{RT}{F} \ln \frac{b[\mathrm{Na}]_o}{[\mathrm{K}]_i}. \tag{2}$$

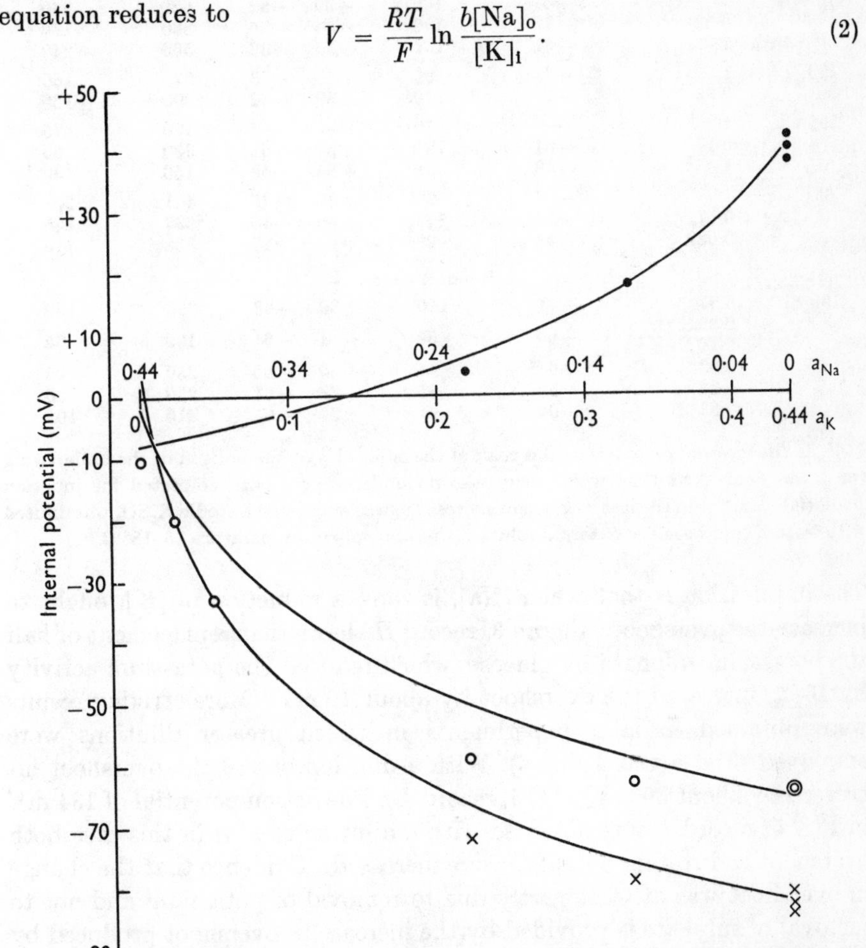

Fig. 4. Effect of replacing K by Na in internal solution on active membrane potential, resting potential and underswing. Abscissa; activity (moles/l.) of potassium and sodium in internal solution; the scale for potassium reads from left to right and for sodium from right to left. Ordinate; internal potential, ●, at crest of spike, ○ in resting state, × at bottom of underswing. The smooth curves were drawn according to equation (1) with $P_{Na}/P_K = 7$, upper curve; $P_{Na}/P_K = 0{\cdot}08$, middle curve; $P_{Na}/P_K = 0{\cdot}03$, lower curve. Data from axon 64 Table 4 except for the point at $a_K = 0$ which is the mean of 4 experiments. The external solution was sea water in which $a_{Na} = 0{\cdot}32$ molar and $a_K = 0{\cdot}0068$ molar.

TABLE 5. Effect of replacing isotonic K_2SO_4 with isotonic glucose

Axon	Fraction isotonic K_2SO_4	Resting potential (corrected; mV)	Action potential (mV)	V_A (mV)	V_U (mV)	Max. rate of rise (V/sec)	Max. rate of fall (V/sec)
198*a*	1	−66	103	+37	−88	590	270
	0·33	−79	134	+55	−76	910	190
	1	−67	106	+39	−82	620	240
	0·33	−76	130	+54	−75	880	170
	1	−67	102	+35	−83	560	240
199	1	−75	95	+20	−73	270	160
	0·33	−71	110	+39	−62	390	120
198*b*	1	—	70	—	—	100	75
	0·17	−64	124	+60	−61	330	65
	1	−58	92	+34	−69	150	90
200	1	−57	93	+36	−75	441	208
	0·17	−50	116	+66	−46	637	135
	0·083	−33	93	+60	−27	368	110
			Related tests				
199	0·33	−71	110	+39	−62	390	120
	0·23K, 0·10Na	−60	59	−1	−61	100	65
	0·33	−64	94	+30	−62	250	105
	0·33	−59	81	+22	−61	250	90
	0·33(S)	−64	91	+27	−64	315	100

V_A is the internal potential at the peak of the spike, V_U at the bottom of the underswing (or at an equivalent time when there was no underswing); both corrected for junction potential. Except in the last row where sucrose (S) was employed isotonic K_2SO_4 was diluted with isotonic glucose. The external solution was sea water; temperature 16–18° C.

The implication is that when $[Na]_i$ is zero, a reduction of $[K]_i$ ought to increase the overshoot. Figure 3, record *H* shows that replacement of half the potassium sulphate by glucose, which reduced the potassium activity by 40 %, increased the overshoot by about 10 mV. More striking results were obtained in later experiments in which greater dilutions were employed (Fig. 5 and Table 5). With a dilution of 1:6 the overshoot increased by about 30 mV (Fig. 5, record *E*). The action potential of 134 mV in Fig. 5, record *B* was obtained with a dilution of 1:3; in this case both overshoot and resting potential were increased. Evidence that the change in overshoot was at least partly due to removal of potassium and not to removal of sulphate is provided by the increase in overshoot produced by diluting isotonic potassium sulphate with isotonic caesium sulphate (Fig. 6, record *G*). However, since replacing potassium sulphate with potassium methylsulphate increased the overshoot by 15 mV (Table 3), we do not feel confident that dilution of potassium ions is the only factor responsible for the large overshoots produced by mixing isotonic glucose with isotonic potassium sulphate. Other factors which might be important are the change in ionic strength or some deleterious effect of the sulphate ion.

As would be expected the resting potential fell if the concentration of K_2SO_4 was reduced sufficiently, for example, from -57 to -33 mV when 11/12 of the K_2SO_4 was replaced by glucose. With a dilution of 1:3 the resting potential sometimes increased. This may be because a leakage of sulphate ions keeps the resting potential below the potassium equilibrium potential. Table 6 shows that this explanation is possible, since the resting potential was always less than the potassium equilibrium potential. It also shows that the potential at the bottom of the underswing, V_U, decreased progressively as the internal potassium concentration was reduced.

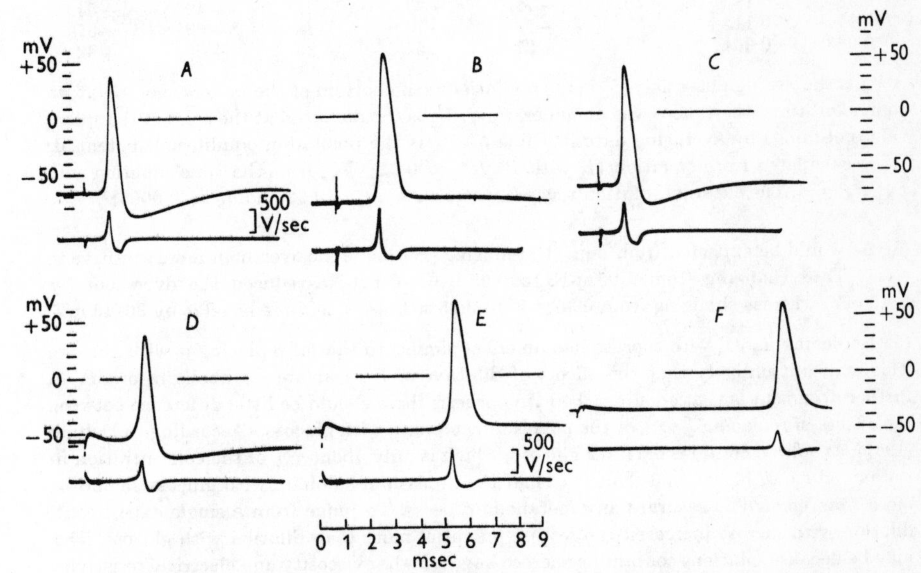

Fig. 5. Effect of diluting isotonic K_2SO_4 with isotonic glucose on action potential and resting potential. The upper trace is the internal potential and the lower trace is the rate of change of potential. *A, C, D*; isotonic K_2SO_4. *B*; 0·33 K_2SO_4, 0·67 glucose. *E*; 0·17 K_2SO_4, 0·83 glucose. *F*; 0·083 K_2SO_4, 0·917 glucose. *A, B, C*, axon 198, 16° C; *D, E, F*, axon 200, 18° C, both at same amplification and time scale; the position of the voltage scale has been corrected for junction potential. Records were taken 3–5 min after changing internal solutions; the external solution was sea water throughout. For the activity of K in the solutions see Table 6.

There are some other points about the dilution experiments which can only be mentioned briefly.

(1) The conduction velocity was reduced by the high electrical resistance of the solutions containing little K_2SO_4.

(2) Replacement of K_2SO_4 with glucose increased the rate of rise of the spike and decreased the rate of fall. The reduction in rate of fall is to be expected, since the outward potassium current should decrease when the internal potassium concentration is reduced. The increased rate of rise is more surprising, particularly in those instances in which the increase

in rate of rise accompanied a decrease in resting potential. From this it would seem that the internal potassium concentration may affect the level of membrane potential at which the sodium-carrying system is inactivated.

(3) Fibres sometimes fired repetitively for a minute or two when a solution such as $\frac{1}{3}K_2SO_4 + \frac{2}{3}$ glucose was applied.

TABLE 6. Comparison of membrane potentials (mV) with 'equilibrium potentials' (mV) for different dilutions of isotonic K_2SO_4

Fraction isotonic K_2SO_4	$[a_K]_i$ (M)	V_R	V_U	V_K	V'_K	V_A	V'_{Na}
1	0·42	−66	−80	−104	−82	+33	+33
0·33	0·18	−75	−71	−83	−61	+49	+54
0·17	0·11	−57	−53	−70	−48	+63	+67
0·083	0·061	−33	−27	−55	−33	+60	+82

V_R is the resting potential; V_U is the potential at the bottom of the underswing, or at an equivalent time when there was no underswing; V_A is the potential at the crest of the spike; values obtained by averaging data in Table 5. V_K is the potassium equilibrium potential; V'_K is calculated from equation (1) with $P_{Na}/P_K = 0{\cdot}03$; V'_{Na} from the same equation with $P_{Na}/P_K = 5$. The external solution was sea water with $a_{Na} = 0{\cdot}32$ and $a_K = 0{\cdot}0068$ M.

(4) As would be expected from eqn. (1), reducing $[K]_i$ made the overshoot more sensitive to $[Na]_i$. Thus changing from 1·05 M-K to 0·95 M-K+0·1 M-Na reduced the overshoot by 5–10 mV, whereas changing from 0·35 M-K to 0·25 M-K+0·1 M-Na reduced it by 30–40 mV.

(5) Replacing K_2SO_4 with sucrose had an effect similar to that of replacing it with glucose. This makes it unlikely that the effects of dilution with sugar are metabolic in origin. A further argument is that on a metabolic hypothesis there should be little difference between the effects of replacing $\frac{1}{2}$ or $\frac{5}{6}$ of the potassium sulphate with glucose. According to Deffner (1961), axoplasm contains 0·24 mM glucose, which is only about $\frac{1}{400}$ of the concentration in isotonic glucose; hence replacement of half the potassium sulphate with glucose should be more than enough to saturate any metabolic process. To judge from a single experiment, dilution with sucrose increased the resting potential more than dilution with glucose. This may be because solutions containing sucrose have a higher viscosity and electrical resistivity than those containing glucose and are therefore more effective in reducing electrical leaks through the membrane.

Tests with lithium, rubidium and caesium as the internal cation

A single test showed that substituting isotonic Li_2SO_4 for isotonic K_2SO_4 first reduced the overshoot and then depolarized the fibre to about the same extent as Na_2SO_4. This is consistent with the generally accepted view that lithium resembles sodium in having a low permeability coefficient in the resting membrane and a high one in the active membrane.

Table 7 gives the effect of the five alkali metals on the resting potential. Their electrogenic action is clearly in the order $K > Rb > Cs > Na \doteqdot Li$.

Figure 6 shows the change in action potential produced by replacing K_2SO_4 with Rb_2SO_4 and Cs_2SO_4. The caesium solution first prolonged the action potential and then, as the resting potential fell, abolished it. Before the caesium had replaced all the potassium the overshoot of the action potential increased by about 13 mV, although the membrane was already somewhat depolarized. As has been mentioned previously (p. 363), the probable explanation is that caesium ions do not pass through the potassium channel, so that the sodium selectivity of the active membrane is increased

TABLE 7. Effect of alkali metals on the resting potential in sea water

Axon	Internal solution	Membrane potential (mV)	Axon	Internal solution	Membrane potential (mV)
189	K_2SO_4	−70	192	K_2SO_4	−61
	Rb_2SO_4	−55		Na_2SO_4	−14
	K_2SO_4	−64		Cs_2SO_4	−15
	K_2SO_4	−62		K_2SO_4	−55
	Cs_2SO_4	−22		Na_2SO_4	−11
	K_2SO_4	−62		Rb_2SO_4	−48
				K_2SO_4	−54
194	K_2SO_4	−73			
	Li_2SO_4	−11			
	K_2SO_4	−55			

All solutions were similar to isotonic K_2SO_4 (solution *D* in Table 1, Baker *et al.* 1962) with the test element replacing K on a mole-for-mole basis.

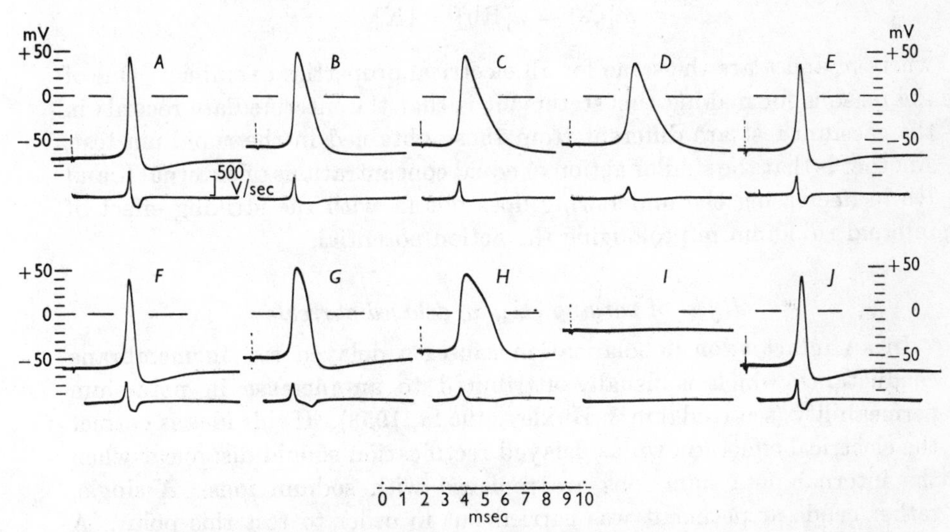

Fig. 6. Effect of replacing isotonic potassium sulphate by isotonic rubidium sulphate and isotonic caesium sulphate. The upper trace is the internal potential and the lower trace is the rate of change of potential. *A*, K_2SO_4 inside axon a few seconds before applying Rb_2SO_4. *B, C, D*, 14, 25 and 60 sec after applying Rb_2SO_4. *E*, 7 min after restoring K_2SO_4. *F*, in K_2SO_4 (18 min after *E*, 20 sec before *G*). *G, H, I*, 4, 7 and 50 sec after applying Cs_2SO_4. *J*, 3·3 min after restoring K_2SO_4. Axon 189; temperature 21° C; corrected for junction potential; external solution, sea water.

when caesium rather than potassium is inside the fibre. It is not clear whether the caesium effects are caused by the presence of this particular ion or whether the change in the shape of the action potential is a consequence of removal of potassium ions. Rubidium sulphate also prolonged the action potential, but here the axon remained excitable for some time and the shape of the spike was different from those recorded in the intermediate stages of the caesium test.

In one experiment the effect on the underswing of replacing potassium ions in the external solution with Na, Cs or Rb was examined. The result was that 10 mM-Cs had about the same effect in reducing the underswing as 3 ± 1 mM-K and that 10 mM-Rb was indistinguishable from 10 mM-K.

The conclusion from these preliminary experiments with alkali metal ions is that lithium is very like sodium, that caesium has a slight potassium-like action and that rubidium resembles potassium in some respects but is not a complete substitute. Neither caesium nor rubidium seem to pass through the sodium channel in the active membrane to any appreciable extent. Although Rb and Cs have some potassium-like effects it seems unlikely that their influence can be expressed by any simple numerical relation such as

$$m[\mathrm{Cs}] = n[\mathrm{Rb}] = [\mathrm{K}],$$

where m and n are the same for all electrical properties examined. One of the reasons for making this statement is that the intermediate records in the caesium test are different from those obtained in the rubidium test. Another is that the similar action of equal concentrations of external K and Rb in decreasing the underswing does not fit with the striking effect of internal rubidium in prolonging the action potential.

Effect of varying $[K]_i$ on delayed currents

In an intact axon depolarization causes a delayed rise in membrane conductance which is usually attributed to an increase in potassium permeability (see Hodgkin & Huxley, 1952*a*, 1953). If this idea is correct the electrical effect known as delayed rectification should disappear when the internal potassium ions are replaced with sodium ions. A single, rather crude, experiment was carried out in order to test this point. A short length of axon (about 3 cm) was employed and current was applied through the lower cannula; the membrane potential was recorded with an internal electrode at a distance of 1–2 cm from the lower cannula. At the time the experiment was done the axon gave no action potential unless hyperpolarized before a cathodal pulse, but it had a resting potential of about 50 mV and showed a marked rectification when filled with isotonic K_2SO_4. Thus in the left-hand part of Fig. 7 an inward current hyperpolarized the fibre by 50 mV, whereas an outward current of equal magnitude depolarized it by only 4 mV. On replacing potassium sulphate with sodium sulphate the resting potential fell by 35 mV and the rectification practically disappeared, the change produced by the inward current being 35 mV and by the outward current, 31 mV. Replacing potassium sulphate restored the membrane to its original condition. The reason why the hyperpolarization with sodium sulphate inside was smaller than that with potassium sulphate inside may be that isotonic sodium sulphate has a higher electrical resistivity than isotonic potassium sulphate; this would reduce the size of the electrotonic potential recorded at some distance from the current electrode. The experiment should clearly be repeated, preferably with a voltage clamp technique, but it does nevertheless afford additional evidence that the delayed currents are carried by potassium ions.

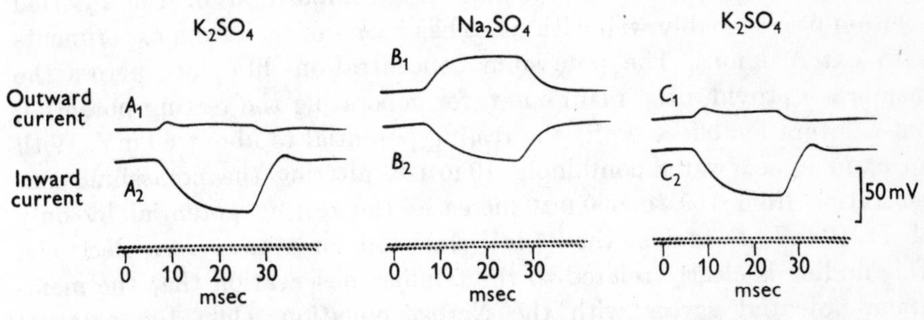

Fig. 7. Changes in membrane potential produced by equal and opposite currents on axon filled either with isotonic K_2SO_4 (*A* and *C*) or isotonic Na_2SO_4 (*B*). Depolarization is shown upwards. This axon was inexcitable unless previously hyperpolarized. The external solution was sea water. Current pulses were applied through a 100 kΩ resistance connected to the lower cannula by a platinum wire (see Text-fig. 5 of Baker *et al.* 1962). Current pulses were approximately rectangular. Correction for junction potentials would shift records B_1 and B_2 3 mV upward.

DISCUSSION

The experiments described here establish that axons which have been extruded and inflated with isotonic solutions of potassium salts conduct impulses of essentially the normal type. The action potential and resting potential of perfused fibres were of the usual magnitude and such differences as occurred were of the kind expected from the composition of the artificial axoplasm. Examples of differences which could have been predicted are the variation of conduction velocity with the electrical conductivity of the internal solution, the absence of an underswing in axons perfused with solutions of reduced potassium content and the effect of raising internal sodium concentration in reducing the overshoot. We have as yet no quantitative information about membrane currents or conductances, and investigations along these lines may show up deficiencies in the present method. It should be kept in mind that there is a large safety margin for propagation and that the size of the spike is not a sensitive index of the condition of a nerve fibre. Some axons which gave spikes of more than 100 mV were clearly not free from injury, for there was often a step on the rising phase of the kind expected when an impulse passes through a region of high threshold. For these reasons it seems highly desirable that other methods of perfusing fibres should be explored; our reaction to the present technique is to be surprised that it should work as well as it does, rather than to be disappointed that it does not work invariably or perfectly.

Effect of changing internal ions

The electrical effects of changing the composition of the internal solution fit reasonably well with what has been deduced from experiments with external ions. The potassium concentration difference across the membrane provides the main e.m.f. for generating the resting potential, but a saturation effect limits the resting potential to about 60 mV. With an axon in sea water containing 10 mM-K altering the potassium concentration from 150 to 600 mM increases the resting potential by only about 10 mV, as against the 35 mV expected for a potassium electrode. This finding is clearly related to the familiar observation that the membrane potential agrees with the Nernst equation when the external potassium concentration is high and the membrane is depolarized, but deviates if $[K]_o$ is less than 50 mM and the resting potential is more than about 40 mV (Curtis & Cole, 1942). It also agrees with the conclusion drawn by Grundfest, Kao & Altamirano (1954) from their studies of the effects of injecting concentrated solutions of potassium salts into giant axons. As Stämpfli (1959) has suggested, the probable explanation is that the potassium permeability falls as the membrane potential increases and that this limits the resting potential to a value of 50–70 mV. To allow for the variation of potassium permeability in a quantitative manner, curves such as those in Fig. 8 were calculated from the equations of Hodgkin & Huxley (1952*b*) with the additional assumptions that ions move independently and that the leakage current is carried by chloride ions. Curve *a*, which is for an axon in sea water, is clearly a good fit to the experimental points and we regard this as establishing that the change in potassium permeability can account satisfactorily for the observed relation between internal potassium concentration and membrane potential. The agreement was less good when the external potassium concentration was 100 or 540 mM, but the deviation here is probably explained by the gradual loss of selectivity that occurs when axons are kept depolarized for any length of time.

The additional assumptions which had to be made in calculating the curves in Fig. 8 are clearly somewhat suspect. As far as the sodium currents are concerned the assumption of independent movement is supported by experiments on the squid axon (Hodgkin & Huxley, 1952*a*); for potassium currents the assumption is less justifiable, although Frankenhaeuser (1962) has shown that potassium currents at the node of Ranvier change with external potassium concentration in the manner expected from the independence principle. It is true that tracer measurements show an interaction between potassium influx and efflux in *Sepia* axons (Hodgkin & Keynes, 1955), but this type of interaction might appear only in isotopic experiments and does not necessarily imply that Frankenhaeuser's evidence is inapplicable

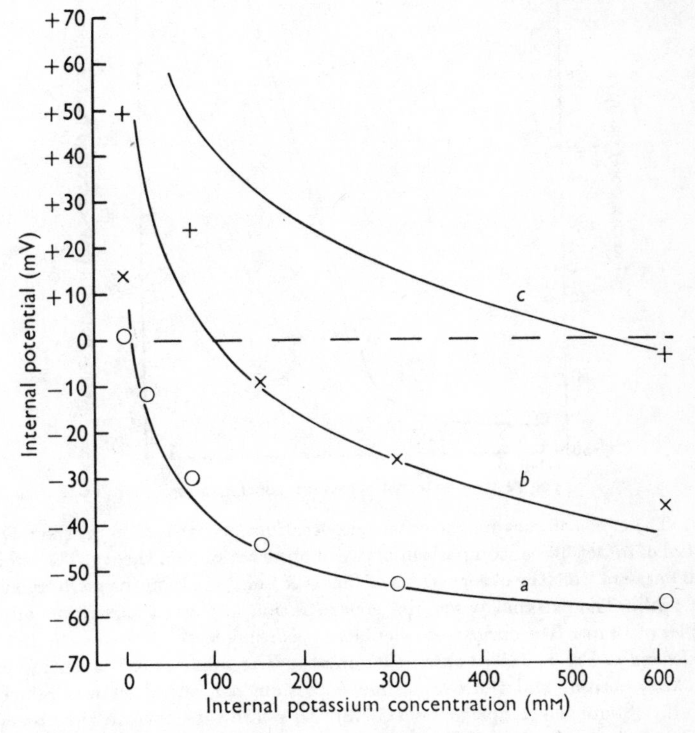

Fig. 8. Comparison between the observed and calculated resting potentials of perfused fibres. The data are the means of values shown in Fig. 2 and the external potassium concentrations are indicated thus: ○, 10 mM; ×, 100 mM; +, 540 mM.

The theoretical curves were calculated on the assumption that the ionic currents vary with membrane potential according to the equations of Hodgkin & Huxley (1952*b*) and that for a given resting potential the currents vary with concentration as for ions moving independently (Hodgkin & Huxley, 1952*a*). The 'leakage' current has been attributed to chloride and the internal ionic concentrations of intact axons have been selected to be consistent with the respective equilibrium potentials. The resting potential of intact axons has been taken as 62 mV. Curves *a*, *b* and *c* are for external potassium concentrations of 10, 100 and 540 mM, respectively.

to cephalopod axons. The identification of the leakage current with a chloride current is at best a poor approximation, since the influx of chloride is only 20 % of that calculated by attributing all the leakage to independently moving chloride ions (Caldwell & Keynes, 1960). However, the assumption about the leakage path is probably not critical, because a further calculation showed that a fivefold increase of leakage conductance decreased the resting potential by only 9 mV.

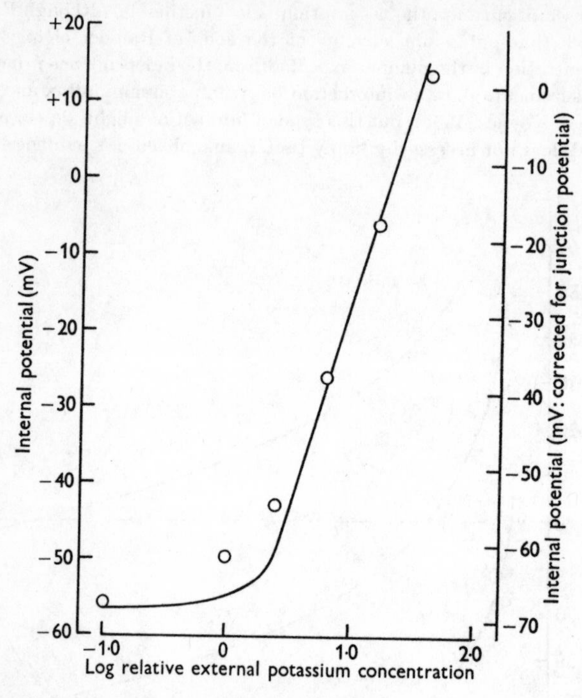

Fig. 9. The effect of varying the external potassium concentration on the resting potential of intact fibres; comparison between observation and theory. The experimental data are from the observations of Curtis & Cole (1942) on the giant axons of *Loligo pealii*. The external potassium concentration is plotted logarithmically in multiples of 13 mM. The curve was calculated according to the procedure indicated in the legend to Fig. 8, except that no assumption was made as to the ion carrying the leakage current and the axoplasmic potassium concentration was taken as 369 mM/kg (Steinbach & Spiegelman, 1943). To relate the curve to the observed potentials (left-hand ordinate) a 12 mV junction potential was allowed for.

A similar calculation was made for the effect of varying the external potassium concentration on the resting potential of intact axons. In this case no assumption need be made as to the nature of the leakage current. The theoretical curve is shown in Fig. 9, together with the experimental values obtained by Curtis & Cole (1942). In contrast to the constant-field equation which predicts less curvature than the experimental results (Goldman, 1943) the theoretical curve has a corner that is too sharp. This is probably not a serious defect, since axons vary considerably with respect to the exact shape of the potassium-concentration curve in the physiological region (see Hodgkin & Keynes, 1955, Table 2).

The relatively small difference between the resting potential of axons filled with isotonic KCl and K_2SO_4 indicates that other ions besides Cl^- must contribute to the leakage current and that Cl^- has much less effect on the resting potential of squid nerve than that of frog muscle (Hodgkin & Horowicz, 1959). A movement of Na^+ and K^+ through parts of the membrane other than those occupied by the selective system is one explanation of that part of the leakage current which is not due to chloride. On this basis the factor which keeps the resting potential of an axon filled with isotonic K_2SO_4 at -60 to -70 mV and prevents it approaching the equilibrium potential for chloride ($V_{Cl} = -\infty$) or potassium ($V_K = -104$ mV) is the inward leak of sodium through the membrane. An alternative explanation, which receives support from the finding that replacing isotonic potassium sulphate with isotonic potassium methylsulphate increased the resting potential, is that the leakage current is partly an outward movement of sulphate, possibly through cut branches or holes in the membrane.

The rapid and reversible changes in overshoot produced by variations in the concentration of sodium inside the membrane are an obvious corollary to the results obtained when the external concentration of sodium is varied, and support the idea that the membrane becomes selectively permeable to sodium during the rising phase of the action potential. Evidence that the delayed rise in permeability associated with depolarization involves the potassium but not the sodium ion was provided by the experiment which showed that 'delayed rectification' disappeared when potassium ions were replaced with sodium ions.

Nature of permeability changes

At present two views are held as to the nature of the permeability changes that occur during the action potential. In one type of hypothesis it is assumed that depolarization causes increases in sodium and potassium permeability because charged particles or carriers move to new positions in the membrane under the influences of the electric field; when the membrane repolarizes, the charged particles return to their original positions and the only alteration in the system is that some Na, K and possibly other ions have moved down their concentration gradients. In the second type of hypothesis it is assumed that the permeability changes are brought about by chemical reactions set in motion by the alteration in membrane potential. It is usually assumed, although this is not strictly necessary, that one of the reactions, for example, the hydrolysis of acetylcholine in Nachmansohn's (1959) theory, involves the degradation of free energy and must be reversed at a later stage. Our results do not distinguish between the two ideas but they impose certain restrictions on the chemical theory. We find that axons, from which at least 95 % of the axoplasm has been removed and which have been thoroughly perfused with isotonic potassium sulphate, can conduct about 3×10^5 impulses. This is clearly compatible with the first type of hypothesis, since the ionic concentration differences provide all the energy required and the working parts of the system are built into the membrane. It does not exclude a chemical type of hypothesis, but in that case one must suppose either that the reactants and catalysts are rather tightly bound to the membrane or that the reactions occur behind permeability barriers which allow ions to pass but prevent the reactants escaping into the axoplasm or external fluid.

SUMMARY

1. Replacing isotonic K_2SO_4 with isotonic KCl inside a perfused axon reduced the resting potential by about 5 mV.

2. Replacing isotonic KCl by isotonic NaCl inside the axon reduced the resting potential of about 55 mV to near zero.

3. As K replaced Na in the internal solution the resting potential first increased rapidly and then reached a saturating value of 50–60 mV.

4. The inside of the fibre became 40–60 mV positive if isotonic KCl was outside the fibre and isotonic NaCl inside it.

5. The membrane potential was within 1 mV of zero if identical solutions were placed on both sides of the membrane.

6. Replacement of K by Na reduced the overshoot of the action potential and eventually blocked the fibre; the block was reversible if the internal Na was not kept at a high level for more than about 10 min.

7. The changes in action potential produced by replacing internal K with Li, Rb or Cs, and the effects of partially replacing K_2SO_4 with glucose, are described.

8. The general conclusion is that concentration differences of K and Na across the membrane normally provide the immediate source of energy for the impulse and that the action potential arises from changes in permeability of the kind deduced from previous experiments with intact axons.

We are indebted to the following for assistance: Dr J. S. Alexandrowicz, histology; Dr V. J. Howarth, vapour-pressure measurements; Dr R. D. Keynes, activity measurements; Mr R. H. Cook, design and construction of apparatus; Mr L. Hummerstone, preparation of salts by ion exchange. Our thanks are also due to the Nuffield Foundation, the Medical Research Council, the Royal Society and the Wellcome Trust for financial support.

REFERENCES

Baker, P. F., Hodgkin, A. L. & Shaw, T. I. (1962). Replacement of the axoplasm of giant nerve fibres with artificial solutions. *J. Physiol.* **164**, 330–354.

Caldwell, P. C. & Keynes, R. D. (1960). The permeability of the squid giant axon to radioactive potassium and chloride ions. *J. Physiol.* **154**, 177–189.

Curtis, H. J. & Cole, K. S. (1942). Membrane resting and action potentials from the squid giant axon. *J. cell. comp. Physiol.* **19**, 135–144.

Deffner, G. G. J. (1961). The dialyzable free organic constituents of squid blood; a comparison with nerve axoplasm. *Biochim. biophys. acta*, **47**, 378–388.

Frankenhaeuser, B. (1962). Delayed currents in myelinated nerve fibres of *Xenopus laevis* investigated with voltage clamp technique. *J. Physiol.* **160**, 40–45.

Goldman, D. E. (1943). Potential, impedance and rectification in membranes. *J. gen. Physiol.* **27**, 37–60.

Grundfest, H., Kao, C. Y. & Altamirano, M. (1954). Biolectric effects of ions microinjected into the giant axon of *Loligo*. *J. gen. Physiol.* **38**, 245–282.

Hodgkin, A. L. & Horowicz, P. (1959). The influence of potassium and chloride ions on the membrane potential of single muscle fibres. *J. Physiol.* **148**, 127–160.

Hodgkin, A. L. & Huxley, A. F. (1952*a*). Currents carried by sodium and potassium ions through the membrane of the giant axon of *Loligo*. *J. Physiol.* **116**, 449–472.

Hodgkin, A. L. & Huxley, A. F. (1952*b*). A quantitative description of membrane current and its application to conduction and excitation in nerve. *J. Physiol.* **117**, 500–544.

Hodgkin, A. L. & Huxley, A. F. (1953). Movement of radioactive potassium and membrane current in a giant axon. *J. Physiol.* **121**, 403–414.

Hodgkin, A. L. & Keynes, R. D. (1955). The potassium permeability of a giant nerve fibre. *J. Physiol.* **128**, 61–88.

Nachmansohn, D. (1959). *Chemical and Molecular Basis of Nerve Activity.* New York: Academic Press.

Stämpfli, R. (1959). Is the resting potential of Ranvier nodes a potassium potential? *Ann. N.Y. Acad. Sci.* **81**, 265–284.

Steinbach, H. B. & Spiegelman, S. (1943). The sodium and potassium balance in squid nerve axoplasm. *J. cell. comp. Physiol.* **22**, 187–196.

J. Physiol. (1962) **164,** 330–337

METHODS FROM

REPLACEMENT OF THE AXOPLASM OF GIANT NERVE FIBRES WITH ARTIFICIAL SOLUTIONS

BY P. F. BAKER, A. L. HODGKIN AND T. I. SHAW

From the Laboratory of the Marine Biological Association, Plymouth, and the Physiological and Biochemical Laboratories, University of Cambridge

(*Received* 6 *June* 1962)

Material

Giant axons from large specimens of *Loligo forbesi* were used throughout the investigation. The mantle lengths of the squid were usually 30–50 cm, and the diameters of the axon 700–900 μ. Living squid were used occasionally but as a rule we employed refrigerated mantles. In this technique live squid are decapitated as soon as the trawl is brought up and the central 30 cm of mantle is dissected and placed in a large Thermos flask filled with ice-cold sea water. The dissection of the nerves in the laboratory is started about 2 hr later.

Giant axons were left in the nerve trunk and were not cleaned.

Experimental procedures

Extrusion and perfusion. A cannula filled with perfusion fluid was tied into the distal end of a giant axon of length 6–8 cm. The axon was placed on a rubber pad and axoplasm was extruded by passing a rubber-covered roller over it in a series of sweeps. The first sweep started at about 1·5 cm from the cut end, the second at 3 cm and so on (Text-fig. 1). At the end of this operation all the axoplasm had been removed except for a column 5–10 mm in length near the cannula. The axon was then suspended vertically in a large beaker of sea water. The cannula was connected to a mechanically driven 'Agla' syringe and perfusion fluid was forced into the axon at about 6 μl./min. This moved the plug of axoplasm along the axon as shown in Text-fig. 2 and after about 4 min a stream of perfusion fluid (which was denser than sea water) could be seen flowing out of the cut end of the axon. Occasionally, when the axon narrowed in the middle, the plug of axoplasm stuck and in these instances the experiment usually had to be abandoned.

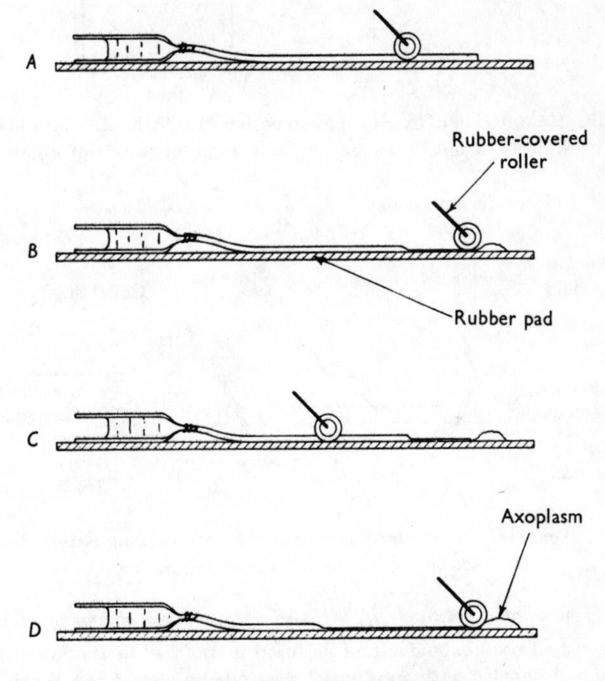

Text-fig. 1. Extrusion of axoplasm.

After establishing a flow through the axon, its excitability was tested by using the apparatus shown in Text-fig. 3. Out of a total of 207 attempts 147 axons were excitable after extrusion and inflation. By excitable we mean that there was an all-or-nothing action potential over at least 2 cm of axon. This is not a critical test since it is well known that the action potential can propagate through damaged regions. However, about 60 % of the excitable axons continued to give action potentials for 1–5 hr and in a number of instances the failure of the axon could reasonably be attributed to use of a faulty solution or excessive perfusion pressure. Our general impression is that with squid of mantle length greater than 30 cm the method gave a reliable preparation in about 50 % of trials.

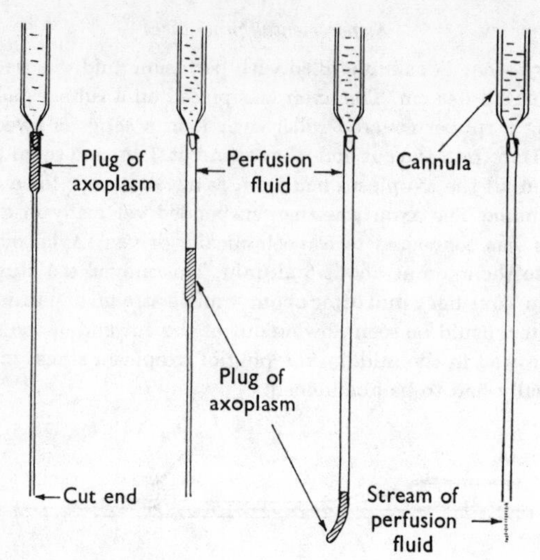

Text-fig. 2. Re-inflation of extruded axon with perfusion fluid. The fluid was driven into the cannula by a motor driven 'Agla' syringe at a rate of about 6 μl./min.

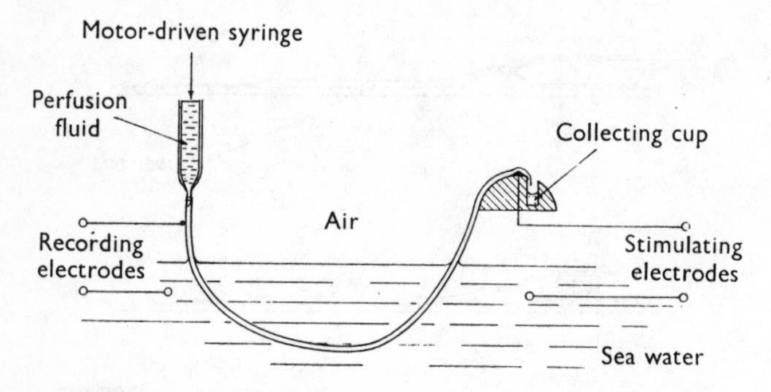

Text-fig. 3. Standard arrangement for recording external action potentials from perfused axons.

Measurement of membrane potentials in perfused fibres without change of solution

After an axon had been extruded and perfused it was tied at the lower end, filled with perfusion fluid and impaled with an internal electrode by almost exactly the same method as that used with an intact axon. The apparatus was similar to that described by Hodgkin & Katz (1949*a*) and the only variation in technique was in the method of lowering the micro-electrode. The axoplasm of an intact axon is solid and a sideways movement of the bottom of the fibre bends both internal electrode and axon but does not move one relative to the other. An electrode is steered down an intact axon by bending the fibre below the electrode in such a way that, when moved down, the electrode remains near the centre of the fibre. As would be expected, perfused fibres behaved like sheaths filled with fluid and there was no evidence of any internal structure. Steering, therefore, involved relatively small movements of the lower part of the axon and unless the fibre was both straight and fully inflated the electrode often touched the surface. One might suppose that this would be disastrous and axons were occasionally damaged by the internal electrode. However, in many instances we recorded action potentials of 100–110 mV for several hours after lowering an electrode into a shrunk and slightly curved axon, with virtually no steering. Since the micro-electrode appeared to touch the surface at several points it would seem that the membrane is protected, perhaps by a thin film of axoplasm. Another indication of the robustness of the inner surface is provided by the observation that although passing air bubbles through axons reduced the action potential it did not make them inexcitable (see p. 352).

Measurement of membrane potentials with change of solution

Method A. A relatively simple way of changing solutions with the internal electrode in position is shown diagrammatically in Text-fig. 4. In this system the fibre was open at the bottom and was not kept fully inflated. To introduce a new solution the standard solution was first removed from the cannula, which was then refilled with the test solution. Application of a negative pressure of a few centimetres H_2O to the cell caused fluid to enter the fibre and, after about 30 sec, to flow out of the cut end.

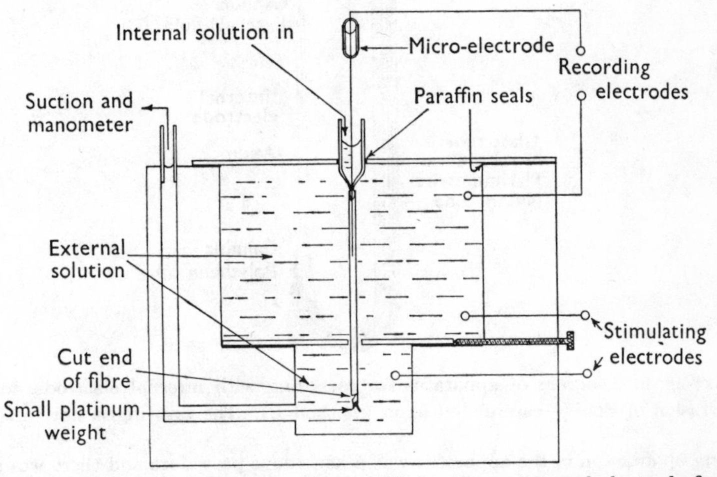

Text-fig. 4. Diagram of apparatus for recording with an internal electrode from perfused axon (method *A*). Application of negative pressure to the cell caused fluid to flow through the axon. Negative pressure was produced by pulling gently on the piston of a large, greased, air-filled syringe connected to the cell and to a manometer by polythene tubing.

The method had the advantage that the dead space was small and that setting up the fibre was a relatively simple operation. Its disadvantages were that the volume of the fibre was not constant and that the internal electrode might touch the membrane.

Method B. Text-figure 5 is a diagram of the fibre in position. Fluid was introduced at the left by a syringe and was withdrawn at the right by a nylon tube attached to a suction device. In order to change the internal solution, fluid was removed from the left-hand reservoir and immediately replaced with the new solution. Fluid then drained through the axon until the system came to rest with fluid in both cannula and reservoir at the same level; this height was 1–3 cm and was sufficient to keep the axon inflated. Removing fluid from the left-hand reservoir did not cause a back-flow because the axon collapsed on to the lower cannula and stopped the flow.

The procedure for setting up the axon was as follows: The axon was cannulated, extruded and perfused in the usual way. After tying off the open end and inflating with solution, the axon was cut close to the tie and a second cannula introduced and tied in. The doubly cannulated axon was mounted by pushing the polythene tube, which was permanently attached to the small cannula, over the inverted end of the U-tube. The polythene tube was then tied in position and the whole assembly lowered into a large cell filled with sea water. When introducing the internal electrode an attempt was made to keep it in the centre of the axon by adjusting the position, in the horizontal plane, of the lower cannula relative to the fixed upper cannula.

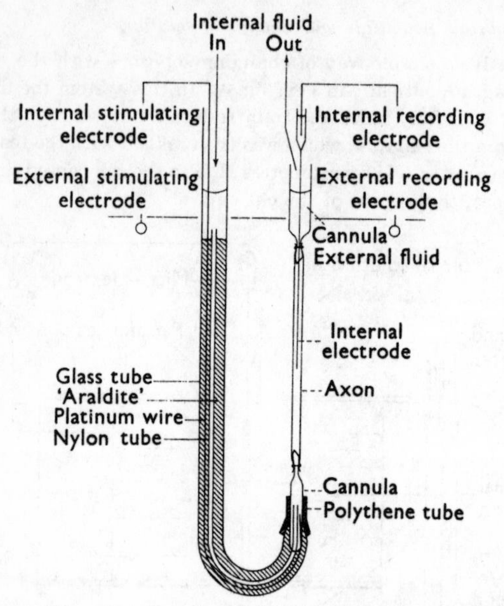

Text-fig. 5. Diagram of apparatus for recording with internal electrode during perfusion of doubly cannulated axon (method *B*). The drawing is not to scale.

Setting up an axon in the apparatus was a hazardous procedure and there was always a danger that one or other cannula would be blocked by pieces of axoplasm. It was important to make sure that there was a free flow through the axon before introducing the second cannula. Formation of an air bubble at the final stage of mounting the axon was avoided by allowing fluid to drain both from the lower cannula and from the end of the U-tube at the moment when the union between them was made.

Solutions

Table 1 gives the compositions of some typical solutions. The tonicity of the internal solutions was adjusted to be the same as that of sea water by the vapour-pressure method of Hill (1930), modified as described by Krogh (1939). Analar or 'specpure' salts were used if available; potassium methylsulphate was supplied by Hopkin & Williams; potassium isethionate (twice recrystallized) was normally prepared from sodium isethionate (British Drug Houses) by ion exchange with Amberlite IR 120.

Activity measurements

A solution of K_2SO_4 which is isotonic with Plymouth sea water contains 1·05 equiv K/l. and has an ionic strength which is so high that the activity coefficient of K^+ (γ_K) cannot be computed reliably from published values of the mean activity coefficient. Approximate measurements of the activity coefficients of K and Na in the main solutions employed were therefore made with glass electrodes selective to K or Na (Table 2). The electrodes were supplied by Electronic Industries Ltd. and potentials were read with a Vibron electrometer, the general procedure being very similar to that used when determining pH with a glass electrode; junction potentials were 'abolished' with a saturated KCl bridge. Since the solutions contained either K or Na but not both ions the results did not depend on the

TABLE 1. Composition of solutions. Concentrations in mg ions/l. solution

	External solutions						
	K^+	Na^+	Ca^{2+}	Mg^{2+}	Cl^-	SO_4^{2-}	HCO_3^-
A Sea water (s.w.)	10	470	10	54	550	29	—
B Artificial sea water (A.S.W.)	10	526	50	—	633	—	2·5
C High-K A.S.W.	538	—	50	—	635	—	2·5

	Internal solutions		Phosphate*
	K^+	Main anion	as $H_2PO_4^-$
D K_2SO_4	1050	500	30
E KCl	610	560	30
F K isethionate†	*ca.* 610	*ca.* 560	30
G K methylsulphate	720	670	30
H Isotonic glucose	0·98 mole/l. H_2O		

* pH adjusted to about 7·7 adding by KOH to KH_2PO_4 in final solution.

† Early isethionate solutions contained 5 mM-Mg and some HCO_3, Cl and Na.

Internal solutions containing sodium were made by replacing K by Na on a mole-for-mole basis.

TABLE 2. Activity coefficients of Na and K in various solutions

Row	1 Solution X = K or Na G = glucose	2 Concen-tration K or Na (equiv/l.)	3 K-electrode ΔV (mV)	4 Na-electrode ΔV (mV)	5 From ΔV γ_K	6 From ΔV γ_{Na}	7 From $\gamma\pm$ γ_K	8 From $\gamma\pm$ γ_{Na}
A	0·5 M-XCl	0·50	0.	0	(0·64)	(0·68)	0·64	0·68
B	0·1 M-XCl	0·10	−37	−38	0·74	0·75	0·76	0·78
C	1·0 M-XCl	1·00	+16·5	+19	0·62	0·70	0·60	0·66
D	Sea water	0·47 (Na) 0·01 (K)	—	− 1·6	(0·68)	0·68	—	—
E	NaCl	0·61	—	+ 4·5	—	0·66	—	—
F	X_2SO_4	1·05	+ 6·9	+ 7·0	0·40	0·43	0·48	0·51
G	$\frac{1}{2}X_2SO_4\frac{1}{2}G$	0·525	− 5	− 5·6	0·50	0·52	0·55	0·57
H	$\frac{1}{3}X_2SO_4\frac{2}{3}G$	0·350	−14	−15	0·53	0·53	0·58	0·61
I	$\frac{1}{6}X_2SO_4\frac{5}{6}G$	0·175	−27	—	0·63	—	0·66	0·67
J	$\frac{1}{12}X_2SO_4\frac{11}{12}G$	0·088	−42	—	0·69	—	0·72	0·73
K	X methylsulphate	0·72	+ 3·7	+ 8·5	0·52	0·66	—	—

The electrode potential was read to about 0·5 mV with 0·5 M-KCl or NaCl as the reference solution; the temperature was about 20° C. Solutions *E*–*K* were isotonic with sea water and contained phosphate buffer (see Table 1). Values enclosed in brackets were assumed. In columns 7 and 8, γ_+ was calculated from values of $\gamma_\pm$ in Taylor (1931) by the formula $\gamma_+ = \gamma_\pm^{z_+/z_-}$ where z_+ is the valency of the cation and z_- of the anion. This formula cannot be expected to apply to concentrated solutions of an asymmetrical electrolyte such as K_2SO_4. The difference between γ_K and γ_{Na} in row *K* has not been checked and might be due to water in the salts used. In calculating a_K and a_{Na} in mixtures of isotonic K_2SO_4 and isotonic Na_2SO_4 the approximation $\gamma_K = \gamma_{Na} = 0·42$ will be used.

ability of the electrodes to discriminate between Na and K. We are indebted to Dr R. D. Keynes for assistance with these measurements, some of which were made at the A.R.C. Institute of Physiology, Babraham.

Recording electrodes

The external electrode was a silver wire, coated electrolytically with chloride, which made contact either with sea water or, in some cases, with a 0·6 M-KCl solution connected to the external fluid by a 3 M-KCl-agar bridge.

The internal electrode consisted of a 100 μ glass capillary filled with 0·6 M-KCl and containing a 20 μ silver wire, to reduce the high-frequency impedance (see Hodgkin & Katz, 1949*a*, Fig. 2*c*). A silver–silver-chloride electrode made contact with the KCl solution in the shank of the electrode but did not touch the 20 μ silver wire. Tests showed that the presence of the 20 μ silver wire in the capillary did not make the electrode sensitive to the concentration of chloride in the fluid in which the electrode was immersed. However, it was necessary to allow for junction potentials at the tip of the electrode.

Table 3. Corrections for junction potential between 0·6 M-KCl and test solution

Test solution		$E_{0\cdot6\,\mathrm{M\text{-}KCl}} - E_{\mathrm{test\ solution}}$ (mV)
0·6 M-KCl		0
0·6 M-NaCl		−4
Sea water		−4
Isotonic solution:	K_2SO_4	+7
	Rb_2SO_4	+7
	Cs_2SO_4	+7
	Na_2SO_4	+4
	Li_2SO_4	+4
Isotonic K_2SO_4 (parts)	isotonic glucose (parts)	
1	1	+5
1	2	+4·5
1	5	+4
1	11	+2
Isotonic K isethionate		+4
Isotonic K methylsulphate		+3

3 M-KCl bridges and an internal electrode filled with 0·6 M-KCl were used in making these measurements.

Junction potentials

In a few experiments the resting potential was determined as the difference between the potential of the micro-electrode (1) in solution *X* inside the axon and (2) outside the axon in contact with a small volume of solution *X* which was connected to the external solution by a 3 M-KCl bridge. On the assumption that 3 M-KCl abolished the junction potential this should give the resting potential directly. An equivalent procedure was to measure the potential difference between the internal and external solutions and to apply a correction for the junction potential between the 0·6 M-KCl in the electrode and the internal or external solution. The necessary corrections, which were determined with 3 M-KCl bridges, are given in Table 3. As an example of the method of applying the correction, suppose that the potential of the 0·6 M-KCl in the micro-electrode changes from 0 to −60 mV on transferring it from sea water in the recording cell to the inside of an axon containing isotonic potassium sulphate. From Table 3 it follows that the corrected potential of the sea water is +4 mV and that the K_2SO_4 solution inside the axon is at −67 mV. The corrected potential difference is therefore taken as −71 mV.

Electron microscopy

At least 2 hr elapsed between the death of the squid and the fixation of the giant fibre. During this period, the mantle or nerve was kept in refrigerated sea water. Nerves kept in this way showed little change in structure even after 24 hr.

All the axons studied were electrically excitable before fixation. For a general comparison of intact, extruded and perfused nerves, 1½ cm lengths of fibre were removed from the peripheral end of the nerve trunk, loosely tied to glass supports and fixed by immersion for 1–2 hr in ice-cold 1 % osmium tetroxide solution, buffered with acetate-veronal at pH 8·0 and made isotonic by addition of a suitable volume of three-times-isotonic artificial sea water (Palade, 1952; Villegas & Villegas, 1960*b*). In the case of perfused nerves it was possible to replace the artificial axoplasm with an ice-cold perfusion fluid containing osmium tetroxide buffered with acetate-veronal at pH 7·8 and made isotonic by addition of KCl. This method proved very satisfactory and a series of nerves was fixed in this way, in some instances osmium also being included in the external solution. Ten axons, all of which had been perfused initially with isotonic K_2SO_4 (solution *D*, Table 1), were examined after OsO_4 fixation.

A few perfused nerves were also fixed by the following four methods: 1, Vapour fixation by suspending the axon over buffered (pH 8·0) isotonic 2 % osmium tetroxide at 4° C for 4 hr. 2, Vapour fixation by suspending the axon over citrate-buffered (pH 6·8), isotonic formalin solution for 4–6 hr at 4° C, usually followed by a short treatment in 1 % osmium tetroxide solution. 3, Unbuffered, isotonic 3 %, potassium permanganate at room temperature (Zebrun & Mollenhauer, 1960), 4, Buffered, isotonic 1 % potassium permanganate at 4° C (Luft, 1956).

Vapour fixation was used in an attempt to minimize leaching out of any residual protoplasm from perfused nerve fibres. A check on the efficacy of fixation, which is largely independent of the experimental procedures applied to the giant fibre, is obtained by studying the small nerve fibres which are always present in the preparation. Both osmium tetroxide solution and vapour and formaldehyde vapour gave essentially similar results. Potassium permanganate gave very poor fixation. Even in the best instances it caused appreciable swelling of the Schwann cell layer of the small fibres and, in these same preparations, markedly damaged the Schwann cell layer of the giant axon.

After osmium fixation, nerves were dehydrated in successive changes of ethanol and, in some instances, tissue blocks were stained by substituting a 1 % solution of uranyl nitrate in absolute ethanol for the final alcohol treatment in the dehydration procedure. Material was embedded in araldite (Glauert & Glauert, 1958) and transverse sections were cut on an A. F. Huxley microtome. Those sections showing silver interference colours were selected and either picked up directly on to a bare grid or, if further staining was expected, were mounted on a 'formvar' film. Contrast was frequently increased, in sections which had not previously been stained, by floating the grid section-side downwards for periods up to 30 min on a warm 1 % solution of uranyl nitrate in 75 % ethanol. After staining, the grids were thoroughly washed by agitation in clean 75 % ethanol. The sections were examined in a Siemens Elmiskop 1 electron microscope at 60 kV.

14

J. Gen. Physiol. (1967), 50, 1401–1411

Basis of Tetrodotoxin's Selectivity in Blockage of Squid Axons

JOHN W. MOORE, MORDECAI P. BLAUSTEIN,
NELS C. ANDERSON, and TOSHIO NARAHASHI

From the Department of Physiology, Duke University Medical Center, Durham, North Carolina, and the Naval Medical Research Institute, Bethesda, Maryland. Dr. Blaustein's present address is Physiological Laboratories, University of Cambridge, England

ABSTRACT The blockage of nerve activity by tetrodotoxin is unusually potent and specific. Our experiments were designed to distinguish whether its specificity of action was based on the identification of ions, the direction of cation flow, or differences in the early transient and late steady conductance pathways. Alkali cations were substituted for sodium in the sea water, bathing an "artificial node" in a voltage-clamped squid axon. When tetrodotoxin was added to the artificial sea waters at a concentration of 100 to 150 nM, it was found to always block the flow of cations through the early transient channel, both inward and outward, but it never blocked the flow of ions using the late steady pathway. We conclude that the selectivity of tetrodotoxin is based on some difference in these two channels.

It is now well established that tetrodotoxin (hereafter to be called TTX), the neurotoxin extracted from puffer fish,[1] blocks nerve impulse transmission by selectively interfering with the early transient conductance increase for sodium (Narahashi, Moore, and Scott, 1964; Nakamura, Nakajima, and Grundfest, 1965; Moore, 1965; Takata, Moore, Kao, and Fuhrman, 1966). TTX affects neither the late steady conductance changes, nor the kinetics of either the early or late conductance changes (Takata et al., 1966).

The present study was initiated in order to reevaluate the effect of TTX on the early transient current during large membrane potential steps because of some differences in results between lobster and squid axons when the axoplasm was stepped to potentials more positive than the sodium equilibrium potential E_{Na}. In unpublished experiments by one of us (MPB), it appeared that the outflow of sodium current which normally occurs in this potential range was not very effectively blocked. In most of the earlier experiments

[1] An identical toxin has been extracted from California salamanders and originally called taricha-toxin (Mosher, Fuhrman, Buchwald, and Fischer, 1964).

(e.g., Narahashi et al., 1964) no equivalent data were available. In one limited set of data on squid axon (Moore, 1965), there appeared to be an equal reduction of the sodium conductance at membrane potentials above and below E_{Na}.

In planning these studies, the following questions arose: (*a*) Does TTX selectively block the sodium conductance; i.e., is its action due to a direct discrimination against this ion per se? (*b*) Or, is its effect a result of its ability to block ions moving in a particular direction; e.g., does it block ions moving inward (most inward current is carried by Na) while sparing ions moving outward (usually K)? (*c*) Or, is its action based on an ability to distinguish between a pathway for the early transient current and one for the late steady current?

These three possibilities may not be mutually exclusive nor all inclusive. For example, the mode of TTX action could be a combination of two or more of these possibilities. However, the experiments were designed to help distinguish between these possible modes of TTX action. Various monovalent cations were substituted for sodium in the bathing sea water and the effect of TTX on voltage-clamped axon membranes observed.

A preliminary report of this work has been given (Moore, Anderson, and Narahashi, 1966).

METHODS

The sucrose-gap voltage-clamp method of measuring ionic currents in squid giant axons was essentially the same as previously described by Moore, Narahashi, and Ulbricht (1964). The only significant change was to rearrange the sucrose flow to obtain more reliable insulation between the central and lateral pools and somewhat more convenient control of the shape and size of the artificial "node."

Leakage currents (observed for small positive and negative potential steps from the holding level) were not constant, but decayed exponentially to a final level of about one-half the initial value. Therefore, we estimated the leakage current by the equation $I_L = (I_0 - I_\infty)e^{-t/\tau} + I_\infty$ where the subscripts represent the leakage current at times $t = 0$ and $t = \infty$ (actually a few milliseconds). The time constant, τ, was in the neighborhood of 0.4 ms. The actual values for I_∞/I_0 and τ were chosen from records for hyperpolarizing potential steps for each experiment and the leakage current subtracted from the total current to obtain the peak sodium and steady potassium currents.

The solutions used to bathe the node are given in Table I. The experiments were done at a temperature of 5–10°C. Three to six consistent and reproducible experiments were deemed sufficient for each experimental situation and a representative example was chosen for illustration.

RESULTS

Natural and/or Artificial Sea Water For purposes of comparison and reference, some characteristic effects of TTX on squid axon membrane ionic

TABLE I

COMPOSITIONS OF EXTERNAL SOLUTIONS

Solution	Na	Li	K	Cs	Rb	Ca	Mg	Cl	SO_4	HCO_3
	mM	mM	mM	mM	mM	mM	mM	mM	mM	mM
Artificial sea water	432.19		9.18			9.46	49.4	504.9	26.0	2.19
Na	526		10			50		636		
Li		526	10			50		636		
K			460			50		560		
Cs			10	526		50		636		
Rb					460	50		560		
½ Na + ½ K	216		216			50		532		

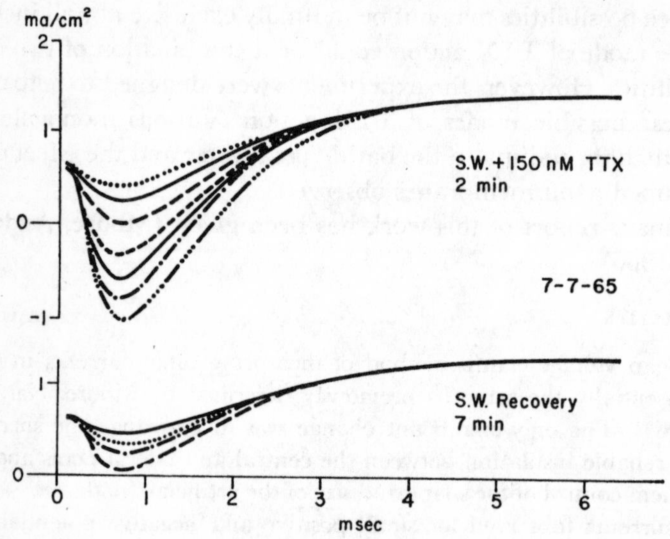

FIGURE 1. Top, selective blockage of inward transient sodium current in voltage-clamped squid axon by tetrodotoxin. The membrane potential was pulsed to a constant potential and the resulting current observed. The lowest curve was the current pattern prior to the addition of 150 nM tetrodotoxin. Successive records were taken at 15 sec intervals over a period of 2 min. The early downward transient (inward current) was reduced while the late steady current was unchanged. The last few curves were almost identical and two have been omitted for clarity. Below, partial recovery of the early transient inward (downward) current over a 7 min washout period.

currents resulting from a voltage-clamp step of membrane potential are shown. Fig. 1 shows the current patterns at a constant potential step observed at 15 sec intervals over a period of 2 min following the addition of 150 nM of TTX to the artificial sea water flowing past the node. While the late steady outward current remains unchanged, the early transient inward Na current is reduced from about 1.5 ma/cm² to nearly zero. Partial recovery of the Na current is seen during a 7 min wash period following the TTX treatment. There is no

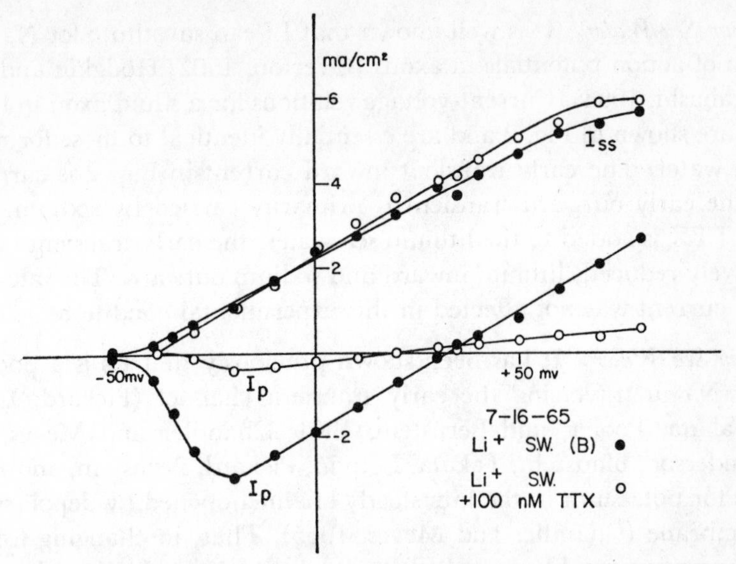

FIGURE 2. Current-voltage relations in an axon membrane bathed in lithium sea water. The branch marked I_{ss} is the late steady outward potassium current. The lower branch is the peak transient inward lithium or outward sodium current. A few minutes after the addition of 100 nM tetrodotoxin to the lithium sea water, the peak transient currents are selectively blocked.

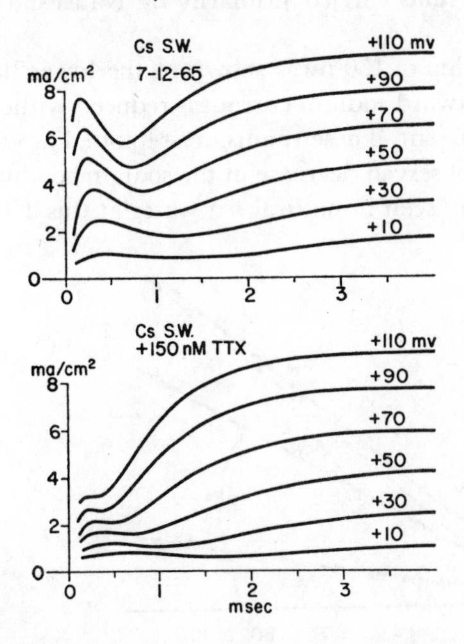

FIGURE 3. Families of current-time curves in an axon membrane bathed in cesium sea water. The absolute value of the voltage-clamped membrane potential is shown at the right of each curve. The lower records were taken 2-3 min after addition of 150 nM tetrodotoxin.

change in the time to the peak transient current throughout the experiment. This illustrates previous observations that there is a reduction of the transient current without a concurrent change in its kinetics (Takata et al., 1966).

Lithium Sea Water It is well known that Li can substitute for Na in the generation of action potentials in axons (Overton, 1902; Hodgkin and Katz, 1949; Narahashi, 1963). Current-voltage relations for a squid axon in lithium sea water are shown in Fig. 2 and are essentially identical to those for natural or Na sea water. The early transient inward current in Fig. 2 is carried by lithium; the early outward transient is primarily carried by sodium. When 100 nM of TTX is added to the lithium sea water, the early transient currents are selectively reduced, lithium inward and sodium outward. The late steady potassium current was not affected in this experimental condition.

Cesium Sea Water It has been shown previously that Cs is a poor substitute for Na in traversing the early transient channel (Pickard, Lettvin, Moore, Takata, Pooler, and Bernstein, 1964; Chandler and Meves, 1965; Moore, Anderson, Blaustein, Takata, Lettvin, Pickard, Bernstein, and Pooler, 1966) and for potassium in the late steady channel opened by depolarization of the membrane (Chandler and Meves, 1965). Thus, in changing from Na to Cs sea water one would expect little or no change in the late steady current carried outwardly by K ions. However, because Cs cannot substitute for the external Na it replaced, the sodium equilibrium potential is shifted by some 75 mv toward a more negative inside potential. Thus, at all potentials there are large outward early transient currents carried primarily by Na as shown in the upper part of Fig. 3.

The effect of TTX at a concentration of 150 nM is shown in the lower half of Fig. 3. It is clear that the early outward sodium current is reduced without effect of the late steady potassium current. These results are replotted as current-voltage relations in Fig. 4. The observed decrease in the sodium conductance may be less marked than usually seen in natural sea water at this TTX

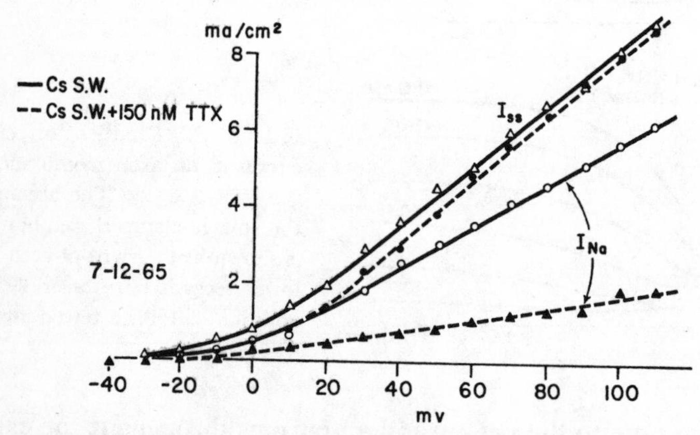

FIGURE 4. Current-voltage relations plotted from Fig. 3. The late steady outward current, I_{ss}, is essentially unaffected by tetrodotoxin in cesium sea water while the early outward transient sodium current is largely blocked.

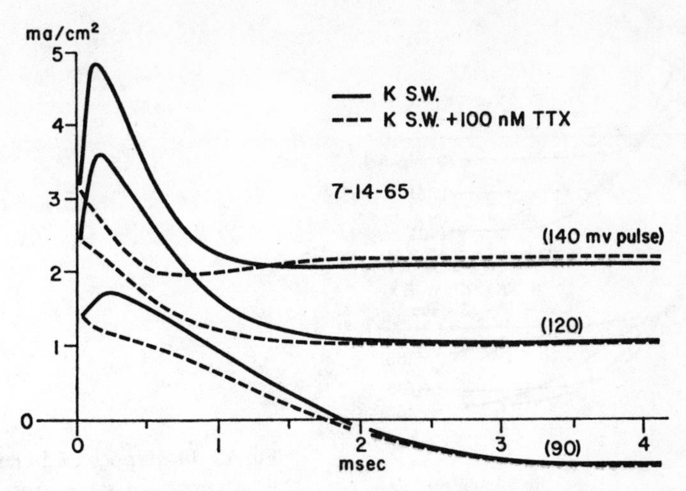

FIGURE 5. Current-time curves in an axon membrane bathed in potassium sea water and a few minutes after the addition of 100 nM tetrodotoxin. In this experiment the base line changed during the course of the experiment and the curves are marked with the applied step of potential change.

concentration but there is no doubt about the fact that the flow of outwardly directed early sodium current is drastically and selectively blocked.

Potassium Sea Water When the axon is bathed in potassium sea water, the internal and external concentrations of potassium are nearly equal resulting in a potassium equilibrium potential near zero. Thus it is possible to have late steady inward potassium currents over a range of negative potentials limited because of the exponentially decreasing conductance in the neighborhood of the normal resting potential (Moore, 1959).

In potassium sea water, there is an early outward transient current through the axon membrane carried primarily by sodium leaving the axoplasm. In Fig. 5 it can be seen that this early outward transient current is blocked by 100 nM of TTX while the late steady potassium current inward or outward is not affected.

Additional records of larger inward potassium currents, taken at slower sweep speeds and more negative potentials, are shown in Fig. 6. Because of a large shift of the zero base line during the experiment on axon 7-14-65 the absolute membrane potential was uncertain. Therefore we have labeled these curves according to the size of the applied step or pulse. It is again evident from the middle section of Fig. 6 that there is no blockade of inward potassium current by addition of TTX to the K sea water.

Rubidium Sea Water Rubidium can substitute for potassium in the external bathing solution (Pickard et al., 1964) and carry late steady inward current although not always as well (Moore et al., 1966). In the lower part

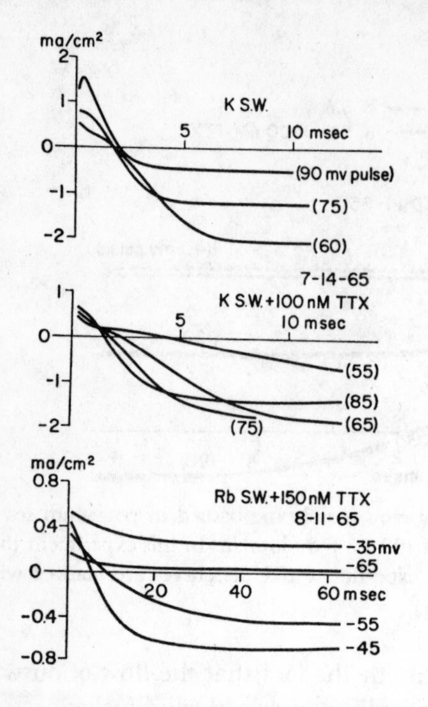

FIGURE 6. Upper and center, current-time curves on same axon with experimental conditions as in Fig. 5 but at slower sweep speeds. Lower, current-time curves for an axon membrane in rubidium sea water containing 150 nM tetrodotoxin.

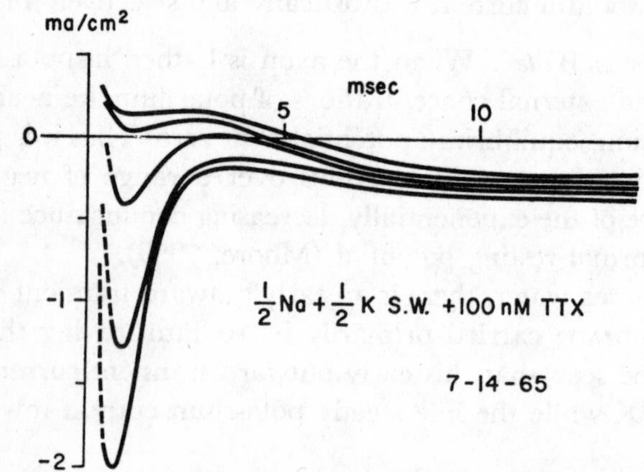

FIGURE 7. Selective effect of tetrodotoxin added to a mixture of sodium and potassium sea waters. The lowest curve is the voltage-clamped axon membrane current before the addition of 100 nM tetrodotoxin. The early transient inward current was rapidly reduced (curves taken at about 30 sec intervals) while the late inward current was only slightly affected.

of Fig. 6 it can be seen that Rb currents can still flow inward in the presence of 150 nM TTX.

One-Half Na + One-Half K Sea Water When using a bathing solution consisting of equal amounts of Na and K, it is possible to find a potential region in which both the early transient current and the late steady current are inward. Fig. 7 shows the essentially complete blockade of the early transient while the late maintained current is only slightly reduced if at all. We attribute most of this reduction in the late current to a drift in the area of the artificial node because of slight changes in the pattern of solution flow past the nerve. The small early current in the latest and uppermost record occurring at about 0.5 msec is of the shape usually attributed to leakage and is the mirror image of the early current pattern seen upon hyperpolarization.

DISCUSSION

These data on the effects of TTX on monovalent cation currents provide some answers to the questions set forth in the Introduction. (*a*) The fact that TTX blocks the transient influx of Li as well as the transient influx of Na shows that the action of TTX is not based on a selective effect on the Na ion per se. (*b*) The early transient outward current in Cs, K, and Rb sea waters was inhibited by TTX, while late steady-state influx of K and Rb was not affected. Thus the selectivity of TTX cannot be based on a discrimination between the inward and outward flow of positive ions. (*c*) In every experiment the early transient current, regardless of the direction of net flow or the ion carrying this current, was depressed by TTX. The late steady current, on the other hand, was unaffected by TTX, irrespective of the direction of net flow and the ion involved. The results are therefore consistent with a mechanism of TTX action based on its ability to distinguish between an early transient conductance path and a late steady conductance path, and to block only the former.

Support for this hypothesis comes from some recent studies of Chandler and Meves (1965). It has been shown that in axons internally perfused with KCl solutions, when the inside is made strongly positive, there is an early outward transient current, presumably carried by K. This has been interpreted as evidence that the membrane is one-twelfth as permeable to K as it is to Na during the early transient period. Recent experiments by Chandler and Meves (personal communication) show that this early outward current is blocked in the presence of external TTX. This therefore substantiates our observations that TTX does not appear to block the early transient by discriminating against the Na ion per se, but rather that it appears to block any ion carrying the early transient current.

A further test of this hypothesis stems from the data on various amines which may act as sodium substitutes in nerve. Studies on frog nerve (Larramendi, Lorente de Nó, and Vidal, 1956; Lorente de Nó, Vidal, and Larramendi, 1957) and on the perfused, isolated squid axon (Tasaki, Singer, and Watanabe, 1965) demonstrate that action potentials may be elicited when NH_3^+, guanidinium$^+$, or hydrazinium$^+$ ions are added to sodium-free external media. In the squid axon, the action potentials obtained in the presence of these ions are all effectively blocked by TTX (Tasaki and Singer, 1966; Tasaki, Singer, and Watanabe, 1966). Voltage-clamp studies by Binstock (personal communication) on NH_3^+, and by Chandler and Meves (personal communication) on guanidinium$^+$, show that the early transient current may be carried by these ions, although they are somewhat less effective than Na^+. Furthermore, the early transient current in a guanidinium solution is blocked by TTX (Tasaki, Singer, and Watanabe, personal communication). Taken together, these results seem to further confirm the notion that it is neither the Na nor even the Li ion per se, which is blocked by TTX, but rather any ion which carries the early transient current.

There is, however, a recurring observation by a number of investigators that the apparent sodium equilibrium potential (that potential at which there is neither an inward nor an outward early transient current) reversibly decreases when TTX is applied (Nakamura et al., 1965; Takata et al., 1966). As noted in the Introduction one of us (MPB) in unpublished experiments on lobster axons has observed a change in the early transient "equilibrium potential" from an initial value of +57 mv to +30 mv after 3.5 min in 90 nM TTX; there was an early outward transient current at +57 mv in the presence of the TTX. The shift in the equilibrium potential is especially apparent in experiments in which the external Ca^{++} concentration is high (Takata et al., 1966; MPB, unpublished data).

The notion that TTX may not block the potassium as well as the sodium in the early transient channel may be an explanation for this recurring observation. On the basis of the Chandler and Meves observation of the potassium to sodium permeability ratio of one-twelfth during the early transient, a preferential block of sodium would lead to a decrease in the apparent Na equilibrium potential. In the results section (cesium), it was noted that TTX was possibly somewhat less effective in blocking the early transient outward current (than in blocking the early transient inward current in other experiments). This might be a reflection of such a preferential block of sodium, the residual current being carried by potassium and leading to an apparent shift in E_{Na}. Alternatively, if there was a slight directional difference in the TTX blockage of the flow of sodium, one might expect to see some rectification rather than a shift in the equilibrium potential.

Although our experiments clearly show that 100 to 150 nM TTX blocks the inward early transient movement of Na or Li we cannot be certain about its ability to block early transient K movements as effectively. The shift in E_{Na} may be interpreted as caused by a differential effect on the ions traversing the early channel. However, the Chandler and Meves experiments (with a somewhat higher concentration of TTX) give rather direct evidence that K movement is blocked. Pertinent to this problem, it must be noted that the precision with which the equilibrium potential can be determined is a function of the relative magnitudes of the early transient and leakage currents. This discrimination becomes particularly difficult when the transient current is drastically reduced, as in the case when TTX is used externally. Another difficulty in the resolution of the early sodium current has been pointed out by Nakamura et al. (1965). If TTX effectively blocks all the transient current (including that fraction carried by K^+), the relative importance of the small fraction of the build-up of the late steady (primarily K^+) current may be magnified and account for the apparent shift in E_{Na}.

On the basis of the data presented in this paper, as well as previous data on the effect of TTX on the voltage-clamped axon, it is tempting to speculate that the early transient and late steady conductance pathways are spatially separated in the nerve membrane. However, it should be clear that the results above do not per se provide any information about the geometry or spatial location of the two channels. Rather, the channels are shown to be operationally different.

The opinions in this paper are those of the authors and do not necessarily reflect the views of the Navy Department or the naval service at large.

We are indebted to Mr. Edward M. Harris for construction of much of the electronic instrumentation, and to Mr. Rodger Solomon for construction of the nerve chamber.
We appreciate the helpful comments on the manuscript by Mr. A. F. Huxley and Dr. Knox Chandler.
This investigation was supported by the National Institutes of Health grant NB 03437, and in part by research task MR 005.08-0020.02, Bureau of Medicine and Surgery, Navy Department.

Received for publication 13 July 1966.

REFERENCES

Chandler, W. K., and H. Meves. 1965. Voltage clamp experiments on internally perfused giant axons. *J. Physiol.*, (*London*). **180**:788.

Hodgkin, A. L., and B. Katz. 1949. The effect of sodium ions on the electrical activity of the giant axon of the squid. *J. Physiol.*, (*London*). **108**:37.

Larramendi, L. M. H., R. Lorente de Nó, and F. Vidal. 1956. Restoration of sodium-deficient frog nerve fibres by an isotonic solution of guanidinium chloride. *Nature*. **178**:316.

Lorente de Nó, R., F. Vidal, and L. M. H. Larramendi. 1957. Restoration of sodium deficient frog nerve fibres by onium ions. *Nature*. **179**:737.

Moore, J. W. 1959. Excitation of the squid axon membrane in isosmotic potassium chloride. *Nature*. **183**:265.

Moore, J. W. 1965. Voltage clamp studies on internally perfused axons. *J. Gen. Physiol.* **48** (5, Pt 2): 11.

Moore, J. W., N. Anderson, M. Blaustein, M. Takata, J. Y. Lettvin, W. F. Pickard, T. Bernstein, and J. Pooler. 1966. Alkali cation selectivity of squid axon membrane. *Ann. N. Y. Acad. Sci.* **137**:818.

MOORE, J. W., N. ANDERSON, and T. NARAHASHI. 1966. Tetrodotoxin blocking: early conductance channel or sodium? *Federation Proc.* **25**:569.

MOORE, J. W., T. NARAHASHI, and W. ULBRICHT. 1964. Sodium conductance shift in an axon internally perfused with a sucrose and low potassium solution. *J. Physiol.*, (*London*). **172**:163.

MOSHER, H. S., F. A. FUHRMAN, H. D. BUCHWALD, and H. G. FISCHER. 1964. Tarichatoxin-Tetrodotoxin: A potent neurotoxin. *Science.* **144**:1100.

NAKAMURA, Y., S. NAKAJIMA, and H. GRUNDFEST. 1965. The action of tetrodotoxin on electrogenic components of squid giant axons. *J. Gen. Physiol.* **48**:985.

NARAHASHI, T. 1963. The properties of insect axons. *Adv. Insect Physiol.* **1**:175.

NARAHASHI, T., J. W. MOORE, and W. R. SCOTT. 1964. Tetrodotoxin blockage of sodium conductance increase in lobster giant axons. *J. Gen. Physiol.* **47**:965.

OVERTON, E. 1902. Beiträge zur allgemeinen Muskel und Nervenphysiologie. *Arch. Ges. Physiol.* **92**:346.

PICKARD, W. F., J. Y. LETTVIN, J. W. MOORE, M. TAKATA, J. POOLER, and T. BERNSTEIN. 1964. Caesium ions do not pass the membrane of the giant axon. *Proc. Nat. Acad. Sci. U. S.* **52**:1177.

TAKATA, M., J. W. MOORE, C. Y. KAO, and F. A. FUHRMAN. 1966. Blockage of sodium conductance increase in lobster giant axon by tarichatoxin (tetrodotoxin). *J. Gen. Physiol.* **49**:977.

TASAKI, I., and I. SINGER. 1966. Membrane macromolecules and nerve excitability. *Ann. N. Y. Acad. Sci.* **137**:792.

TASAKI, I., I. SINGER, and A. WATANABE. 1965. Excitation of internally perfused squid giant axons in sodium free media. *Proc. Nat. Acad. Sci. U. S.* **54**:763.

TASAKI, I., I. SINGER, and A. WATANABE. 1966. Excitation of squid giant axons in sodium-free external media. *Am. J. Physiol.* **211**:746.

II. Synaptic Transmission

Synaptic Transmission

"Synaptic transmission" refers to the process by which a nerve cell influences a contiguous neuron or effector cell over a short period in response to membrane potential changes in the presynaptic cell. Influences of neurons occurring over periods of weeks or months are referred to as trophic effects.

Synapses often occur at axon endings, but they are also found between most combinations of neuron processes and somata. Golgi's hypothesis that every neuron is a distinct anatomical entity (the neuron doctrine) has in general been confirmed. But there are exceptions such as fusion of axons in invertebrate giant fibers.

Membrane changes are transmitted between cells by release and diffusion of a chemical or by direct electrical spread. Correlation of electron-microscopic and physiological observations makes anatomical recognition of some general types of synapse reasonably certain.

Chemically transmitting synapses are thought to correspond to points where two adjacent cells have parallel, thickened, double membranes separated by a gap of the order of 250 Å (synaptic cleft). One side characteristically shows an accumulation of circular "vesicles." Physiological observation shows that transmission at chemical synapses is in one direction and also that transmitter chemical is released in multimolecular packets (quanta). Hence, the vesicle-containing structure is considered to be the presynaptic side of the synapse.

A number of examples of electrical transmission or electrotonic connection between cells have been studied in the last decade. Cases accompanied by electron-microscopic studies usually have revealed either special structures to prevent current escape in the synaptic cleft or areas of close membrane apposition referred to as "tight junctions" or "gap junctions" (see Brightman and Reese, 1969).

Although a number of electrical and chemical synapses seem to transmit nerve impulses without alteration, many others transmit influences to the postsynaptic neuron which do not individually cause it to reach threshold for an action potential. At such synapses, presynaptic impulses individually produce subthreshold excitatory or inhibitory changes in the postsynaptic membrane potential that are integrated in the postsynaptic neuron. The state of the postsynaptic membrane will depend on the pattern and frequency of presynaptic excitation. The previous history of pre- and postsynaptic activity can have subtle effects on further transmission.

These concepts suggest that the complex integrative abilities of nervous systems do not result simply from the pattern of functional connections between neurons (the wiring diagram). The integrative capacities of individual neurons also result from their shape, passive (cable) electrical behavior, the responsiveness of the postsynaptic membrane to specific chemicals, and the previous activity of the system.

To complicate further the study of synaptic transmission, nearly every synapse that has been carefully studied has been proven to have unique features. Here we have chosen papers which develop or further general hypotheses through detailed quantitative analysis and which use techniques of continuing importance in the investigation of synaptic transmission. Important mechanisms by which neurons influence other cells (e.g., neurosecretion) are not represented in this collection and new mechanisms are being described as investigation continues. Eccles (1964) has written one of the most extensive reviews available.

This section begins with an early demonstration of the nature of an electrical synapse. Next, pioneering papers concerned with the identification of neurotransmitters are contrasted with a modern series. Finally, quantitative studies of the biophysical mechanism of excitatory and inhibitory synaptic transmission are presented.

A. AN ELECTRICAL SYNAPSE

An electrical synapse is one at which the flow of current in the presynaptic element directly changes the voltage across the membrane of the postsynaptic cell. Passive electrical (cable) properties suggest that if the presynaptic element is to be capable of more than a subthreshold influence on the postsynaptic cell, it must be of the same size or larger than the postsynaptic element. In addition, "efficient" electrical synapses have structural modifications (as well as the "tight junctions" that have become almost diagnostic at the ultrastructural level) such as ensheathing capsules to prevent current escape between the pre- and postsynaptic elements.

Diagnostic physiological features of electrical synaptic transmission are simultaneity of the electrical events in the pre- and postsynaptic structures and, in most cases, passage of electrical current from one cell to the other in either direction. Transmission is largely determined by passive electrical properties. Its characteristics, therefore, are highly stable and not subject to modification by the immediate past history of activity (for a review, see Bennett, 1966).

Electrical synapses have been found in both vertebrate and invertebrate animals to mediate responses in systems in which there is apparent advantage in great speed and reliability (e.g., Watanabe and Grundfest, 1961; **Furshpan and Potter, 1959**). In a number of cases, the synchronous activity of a distinct population of neurons is achieved by electrical synapses (electrotonic connections) between the cells in the population (e.g., Auerbach and Bennett, 1969). There are cases in which, rather than synchronous discharge of impulses, a more subtle modification of excitability (i.e., tendency to respond to other synaptic influences) of a population of neurons results from electrotonic connections between them (e.g., Hagiwara, Watanabe, and Saito, 1959; Hagiwara and Morita, 1962; Eckert, 1963; Grinnell, 1966; Watanabe, Obara, Akiyama and Yumato, 1967).

Electrically mediated inhibition has been carefully documented. The inhibition modifies the escape reflex mediated by the Mauthner cells of goldfish (Furukawa and Furshpan, 1963).

II. Synaptic Transmission

Both electrically and chemically mediated mechanisms occur in the chick ciliary ganglion. MARTIN AND PILAR (1963a, b) provide instructive examples of the differences in properties of the two synaptic mechanisms and in the methods by which they can be characterized experimentally.

We reproduce here one of the first detailed studies of an electrically transmitting synapse (**Furshpan and Potter, 1959**). This synapse is not typical of electrical synapses because it transmits in one direction only. Therefore, proof that the synapse was indeed operating by spread of electrical current had to be extremely detailed. Preferred current spread in one direction, or rectification, has recently been observed in another electrical synapse (Auerbach and Bennett, 1969).

J. Physiol. (1959) 145, 289–325

TRANSMISSION AT THE GIANT MOTOR SYNAPSES OF THE CRAYFISH

By E. J. FURSHPAN* and D. D. POTTER*

From the Biophysics Department, University College London

(*Received* 30 *July* 1958)

A number of studies have recently been made on the mechanism of impulse transmission across (*a*) the neuromuscular junction (del Castillo & Katz, 1956), (*b*) certain synapses on the motoneurone (Eccles, 1957) and (*c*) the nerve–electroplaque junction (Grundfest, 1957). In each case it seems likely that local circuit action stops at the junction and that the prejunctional action currents contribute very little to the observed post-synaptic potentials. At all these synapses, however, the prejunctional terminals are too small to be impaled with micro-electrodes and it has not been possible to determine quantitatively the resistance of the junction to local current flow (see, however, del Castillo & Katz, 1954). Therefore it seemed of interest to study a synapse in which both elements were amenable to the use of intracellular electrodes. Several such junctions, with large pre- and post-synaptic units, are known in invertebrate nervous systems (Johnson, 1924; Stough, 1926; Young, 1939). The particular ones chosen were the giant motor synapses (GMS's) of the abdominal nerve cord of the crayfish. The anatomy of the GMS's, which will be discussed in greater detail below, has been studied by Johnson (1924) with the light microscope and by Robertson (1952, 1953, 1955) with the electron microscope. Wiersma (1947) has recorded the electrical responses of the synaptic elements using extracellular electrodes. He found that in preparations in good condition, a single presynaptic impulse evoked one conducted post-synaptic spike and that transmission took place in only one direction.

One interesting aspect of the GMS's is that the 'pre-fibres' (presynaptic fibres) are even larger than the 'post-fibres' (post-synaptic fibres). The possibility was, therefore, considered from the beginning that the presynaptic action currents would stimulate the post-fibre, i.e. that an 'electrical' mechanism of transmission operated (Fatt, 1954). Strong evidence has been obtained to support this view. It has also been found that the 'synaptic

* U.S. Public Health Service Fellows. Present address: Wilmer Institute, The Johns Hopkins Medical School, Baltimore 5, Maryland, U.S.A.

membrane' has the properties of an electric rectifier, which explains the one-way character of the transmission. A preliminary account of this work has been published (Furshpan & Potter, 1957).

METHODS

The anatomy of the synaptic elements

The following description is based mainly on Johnson (1924).

Pre-fibres. Two pairs of giant axons constitute the pre-fibres of the crayfish GMS's. The medial pair extend unbranched and without interruption through almost the entire length of the nerve cord. Each member of the lateral pair, on the other hand, consists of a longitudinal chain of separate axons. The abdominal portion of a lateral giant axon is formed by six such axon segments, each joined to the next by a junction which will be referred to as a 'segmental' synapse.

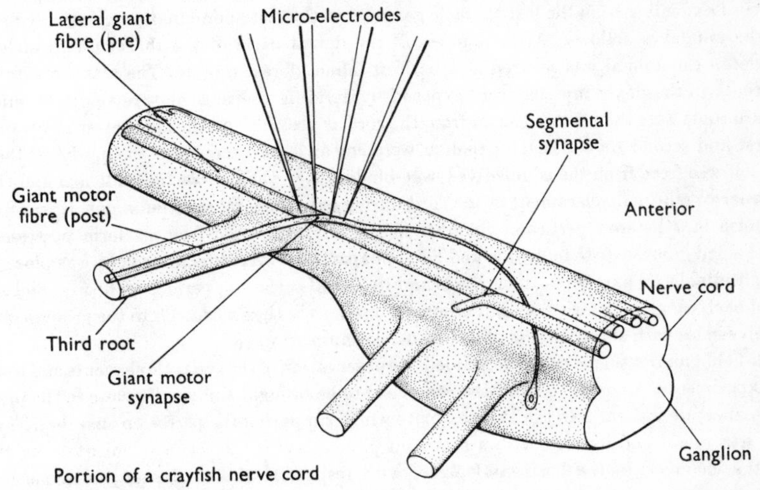

Text-fig. 1. Semidiagrammatic drawing of a portion of a crayfish abdominal nerve cord, containing one ganglion. The course of one motor giant axon is shown from its cell body in the ventral part of the ganglion (Hardy, 1894; Johnson, 1924) until it leaves the third ganglionic root on the opposite side of the cord. Only its junction with the lateral giant pre-fibre is shown; but its synapses with the two medial giant fibres are located just centrally, where the latter fibres cross the motor axon. A synapse between two segments of the lateral giant fibre is also shown. The ventral continuation of the anterior segment presumably goes to the cell body and also provides the collateral, which makes synaptic connexion with that of the contralateral segment (not shown).

Wiersma (1947) has shown that the nerve impulse crosses these junctions in either direction. In preliminary experiments using intracellular electrodes we have found that electrotonic currents flow freely across them in either direction (cf. Kao & Grundfest, 1956); but they seem to be regions of low safety factor at which spikes, in depressed preparations, may be blocked. Text-fig. 1 shows parts of two segments of a lateral giant axon with a synapse between them.

Post-fibres. The post-fibres are giant motor axons which innervate the flexor musculature of the tail (abdomen). There is one such fibre on each side of the first five abdominal ganglia, the sixth ganglion having two pairs. The cell body lies close to the ventral surface of the ganglion (Text-fig. 1). From there the axon extends dorsally, making synaptic connexion with a number of fibres before finally emerging from the cord in the third (ganglionic) root of the opposite side.

The motor giant axons will be designated by the side of the cord from which they emerge rather than the side on which the cell body lies. Histological evidence (Johnson, 1924; Wiersma, 1947) suggests that the motor giant axon forms synapses with the following fibres: (1) the ipsilateral segmental (lateral) giant axon, (2) the ipsilateral medial giant axon, (3) the contralateral medial giant axon, (4) the collateral (see Results, p. 297) of the contralateral segmental giant axon, and (5) the contralateral motor giant axon at the point at which both cross the mid line. Wiersma (1947) has shown that one-to-one transmission can occur at synapses (1), (2) and (3); but there is as yet no physiological evidence for the presence of synapses (4) and (5) (but see below, p. 303). In the experiments to be described synapse (1) has been used almost exclusively and it is the the only one shown in Text-fig. 1. Its structure is described in the Discussion (p. 319). The presence of yet other synapses on the giant motor fibre will be considered in the following paper (Furshpan & Potter, 1959).

Experimental procedure

The preparation. The crayfish *Astacus fluviatilis* was used in all experiments. The part of the nerve cord extending from the last thoracic ganglion to the last abdominal ganglion was dissected from the animal as follows. After removal of the dorsal parts of the thoracic and abdominal exoskeleton the animal was secured in a bath of saline, dorsal side up. The attachments of the third roots to the flexor muscles were exposed by dividing the musculature along the mid line. The third roots were then severed as far from the cord as possible and the muscles entirely removed. The first and second roots of each ganglion were cut at least 1 mm from the cord and then the cord itself was freed from the connective tissue binding it to the floor of the abdomen and thorax.

In order to reduce displacement of the cord during experimental manipulations, it was fixed to a platform in a Perspex perfusion chamber in the following way. The platform possessed two parallel-sided grooves (0·45 mm wide and 2·5 mm apart) between which the cord was placed with its longitudinal axis parallel to the grooves and its dorsal surface uppermost. The first and second roots of each ganglion, which were not otherwise used, were then wedged into the grooves on their respective sides with short lengths of 26 S.W.G. platinum wire.

Dark-field illumination considerably aided the observation of the synaptic elements and was used in all experiments. In order to allow the condenser to be brought sufficiently close to the preparation, an aperture was cut into the Perspex plate which supported the perfusion chamber. The condenser was raised and lowered with a rack and pinion device. It was convenient to be able to move the chamber, in the horizontal plane, with respect to the condenser and a mechanical stage was used to fix the chamber to the supporting plate.

In order to expose the axons for impaling, the sheath which encloses the cord was removed from the dorsal surface of the ganglion and of the region of cord immediately posterior to it. It was especially important to perform this operation with care to avoid serious injury to the synaptic elements.

Both pre- and post-fibres could be stimulated with external electrodes. In the case of the post-fibre the third root was carefully drawn into a glass capillary with the aid of a small hypodermic syringe. The stimulating voltage was then applied between the inside of the capillary and the bath. A similar method was used for the pre-fibre; but instead of drawing the axon into the capillary, the latter was placed on the surface of the cord over the axon. The slight negative pressure supplied by the syringe produced a partial sealing of the capillary to the fibre and reduced the shunting of the stimulating current.

Impaling the axons. The fibres forming the GMS lie very close to the dorsal surface of the cord, and after removal of the sheath they are accessible for the insertion of the micro-electrodes. With dark-field illumination the giant pre-fibres were almost always clearly visible as broad dark bands contrasting with the rest of the cord, which appeared bright. The giant motor fibre usually could not be clearly distinguished, but its position could almost always be inferred from slight visual clues and the fact that it is the most posterior of the large fibres entering the third root (see also Hardy, 1894). In any case, the criteria for accepting a successful entry into a motor giant axon were entirely physiological (see below, p. 298).

In spite of the large number of preparations which were made, each containing ten potentially usable giant motor synapses, the number of successful experiments has been relatively small. The chief impediment has been the difficulty experienced in the insertion of the micro-electrodes. It was usually easier to impale the lateral giant axons than the medial ones, and the former were used as pre-fibres in almost all experiments. The giant motor axons were usually the most difficult of the fibres to enter. The pre-fibre electrodes were inserted first, for once inside the cell they were not easily dislodged by further manipulation.

In some experiments two micro-electrodes were inserted into each fibre; in other experiments only one. When four electrodes were used, the two within a single fibre were separated from one another by 0·05–0·3 mm (usually about 0·15 mm). The distance from the nearest pre-fibre electrode to the edge of the post-fibre (at the synapse) was usually within the same range. At least one, and sometimes both, of the post-fibre electrodes were inserted in the region of fibre-crossing.

Before experience had been gained in locating the motor giant axon, entries were often made, inadvertently, into the smaller axons which also leave the cord in the third root (Pl. 1). These fibres cross over the lateral giant axon in the same general region as does the motor giant axon and there is physiological evidence that at least some of them make synaptic connexion with the lateral giant axon. The position of the junctions, however, does not seem to be at the place of crossing, but at some other site, probably deeper within the cord. The analysis of the GMS which is given below cannot be applied to these synapses, for such analysis requires either that the internal recording electrodes be close to the junction, or at least that their distance from the synapse, as well as the characteristic lengths of the junctional elements, be known. Text-fig. 2 shows the response recorded from one of the smaller third root fibres following stimulation of the lateral giant axon. The post-synaptic potential is much slower and the synaptic delay longer than in the case of a GMS (see Text-fig. 7*b* for comparison), but without knowing the position of the junction, little can be deduced about the mechanism of the transmission process. The prolonged time course and delay of the post-synaptic response might equally well be due to chemical transmitter action, to electrotonic decrement within the post-fibre or to a combination of both factors. It was, therefore, very important to know which of the third-root fibres the post-synaptic electrodes had entered, for only the giant motor axon is known to make a synapse with the lateral pre-fibre at the place of crossing it. The criteria used for recognizing an entry into a motor giant will be described below (see Results, p. 298).

Micro-electrodes. Especially in the case of post-fibres, only micro-pipettes with very fine tip diameters could be successfully inserted; but it was also necessary that the electrodes should pass relatively large currents. It was found that micropipettes of 15–20 MΩ resistance, pulled from Pyrex glass tubing with an outside diameter of about 2 mm and wall thickness about 0·35 mm, usually met the above specifications without being too fragile. Selection of the glass tubing for the appropriate wall thickness was done with the aid of a microscope and ocular micrometer.

The four micro-electrodes were held by separate micromanipulators and the tips of all electrodes could be manœuvred within the same small field. The usual techniques for intracellular potential recording and passing of current across fibre membranes were used (Fatt & Katz, 1951), except for the method of inserting the current-passing electrodes. Each micropipette could be used to pass current or to measure voltage and was used in the latter way to register resting potential during insertion. The probe unit of each of the four cathode-followers contained a pair of micro-switches operated simultaneously by a single lever. In one position of the switches, the Ag–AgCl wire from the micro-electrode was connected to the grid of the cathode-follower probe valve (see Text-fig. 3); and the metal shielding surrounding (*a*) the probe valve, (*b*) the Ag–AgCl wire; and (*c*) the upper part of the micro-electrode, was connected to the cathode of the same valve (Nastuk & Hodgkin, 1950). With the alternative switch position, the connexion of the Ag–AgCl wire was changed from the grid to a lead to the current generator and all the shielding associated with that channel was switched from cathode to earth. Also, when current was being passed, the grid of the reference valve of that cathode-follower was switched from the bath electrode to earth.

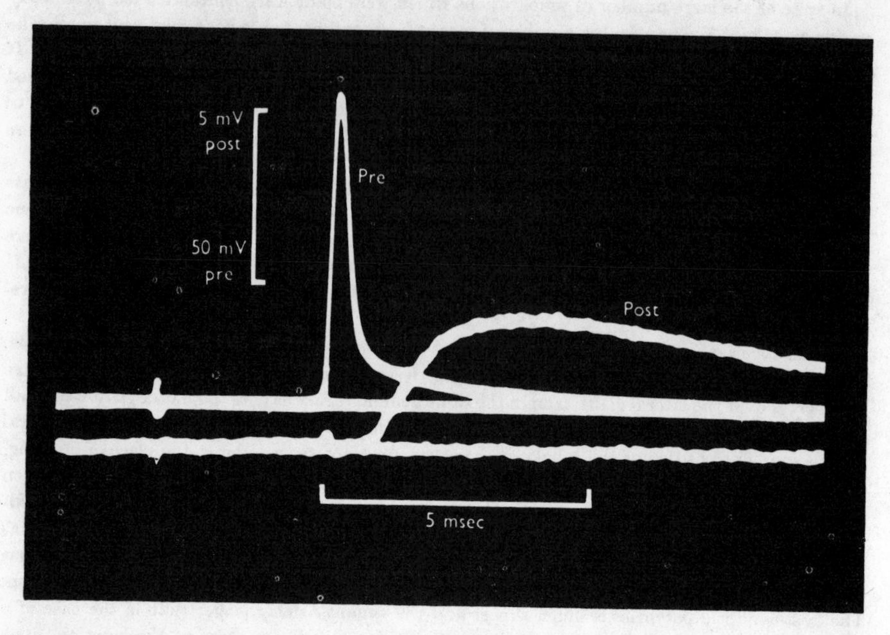

Text-fig. 2. The post-synaptic response, of one of the smaller fibres of the third root, to stimulation of the lateral giant axon of the same side. The pre-fibre spike is shown above, the p.s.p. below. Compare the delay and the time course of the p.s.p. with that of a giant motor synapse (Text-fig. 7). Intracellular recording in this and all subsequent records. Voltage calibration: 50 mV, pre-; 5 mV, post-.

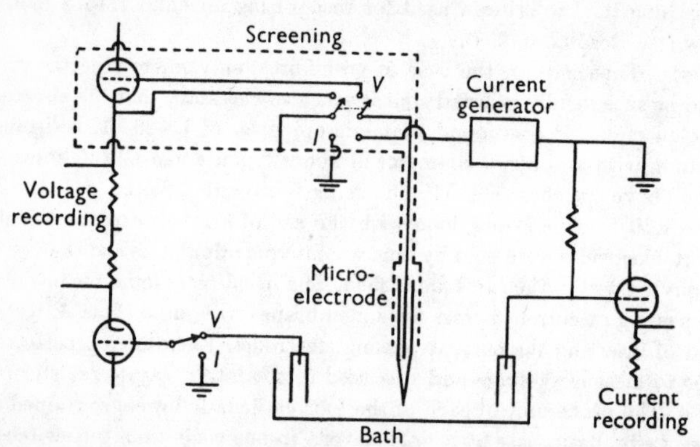

Text-fig. 3. The switching arrangement for alternatively passing current or recording potential with a micro-electrode. There are separate switches and cathode followers for each of the four micro-electrodes, but only one two-channel current generator. Leads from each switch are available, however, for plugging into either channel.

Switching the shielding and the reference-valve grid to earth when current was being passed decreased the capacitative artifact recorded in other channels. The potential across the current-monitoring resistance, as well as the output of any of the cathode-followers, could be led into either of two DC amplifiers, and observed on the screen of a double-beam oscilloscope.

For the experiments to be most useful it was necessary to collect a large amount of data. In order that the measurements could all be made within a short period of time, a recording method similar to that used by Cole & Curtis (1941) was found to be convenient. It is illustrated in Text-fig. 4, which shows the relationship of steady-state current to membrane potential for the lateral giant pre-fibre. The recording was made by switching the connexion of the X-plates of the cathode-ray tube from the time base generator to the second DC amplifier. Then the output of one amplifier moved the beam vertically, while the output of the other gave horizontal deflexions.

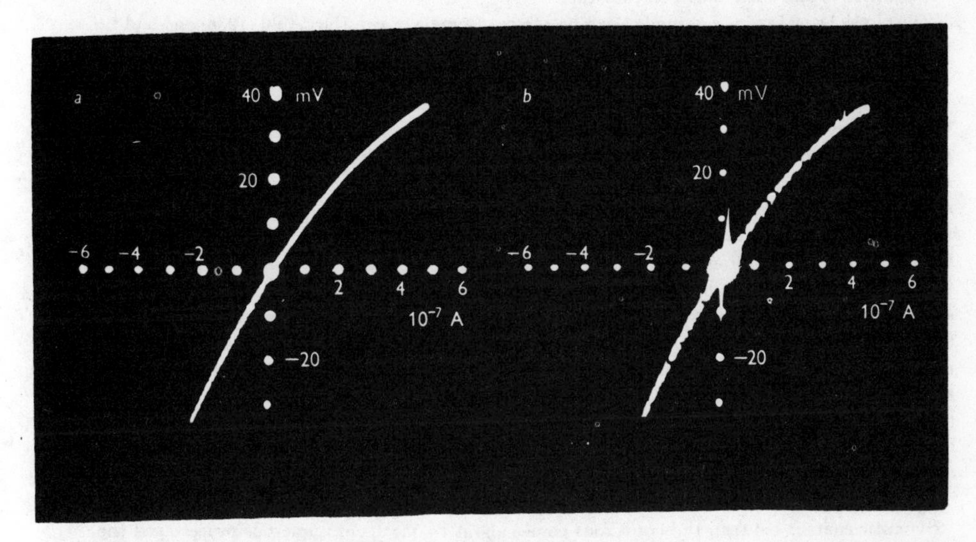

Text-fig. 4. The direct recording from the oscilloscope of current–voltage curves of a latera giant axon. The potential across a monitor resistance, which was proportional to the current applied to the fibre through one micro-electrode, was registered on the abscissa; while the membrane potential of the fibre, recorded by a second nearby intracellular electrode, was recorded on the ordinate. In *a*, the current applied to the fibre was rapidly increased in small increments by manual operation of the wire-wound potentiometer controlling the DC output of the current generator. In *b* rectangular current pulses of varying sizes and 70 msec in duration were applied to the fibre, while recording current and voltage as in *a*. Each pulse produced one of the points on the curve. Records of this kind served as controls for those of the first type.

In Text-fig. 4*a* the applied current was recorded as a horizontal deflexion, while the concomitant changes in the membrane potential were registered as a vertical deflexion. The current, which was varied by manual operation of the output potentiometer of the current generator, was steadily increased during a period of 1–2 sec. It was also possible to pass rectangular (70 msec) pulses of current of various magnitudes. Then a number of points were obtained lying along the same curve as is shown in Text-fig. 4*b*. Both types of record were usually made, the second serving as a control for the first; for there was some question whether the rate of increase of current during manual control of the potentiometer would affect the slope of the current–voltage curve. The error attributable to this factor was usually found to be negligible and it was then preferable to use the first type of record (Text-fig. 4*a*), which allowed more accurate determination of the origin. The points which give the axes in Text-fig. 4 were produced with a decade calibrator. The photographic record, then, is a multiple exposure of calibrating voltage steps of both polarities applied to both channels and the current–voltage relationship of the fibre membrane for both a inward and outward currents. Less than a minute was usually required to complete such record.

When studying the properties of the GMS with only two micro-electrodes, the graphs made were similar, but related membrane potential of one fibre to the current applied to the other. The experiments which gave the most information, however, were those in which two electrodes were inserted into each axon. In these cases the membrane potentials of both fibres were simultaneously recorded on rectangular co-ordinates, while using the remaining internal electrodes to apply current first to one axon and then the other.

The physiological saline was made according to the formula of van Harreveld (1936) and had the following composition: (mM) NaCl 205, $CaCl_2$ 13·5, KCl 5·4, $MgCl_2$ 2·6, $NaHCO_3$ 2·3. Experiments were usually made at about 20° C.

Histology. In order to estimate the diameter of the presynaptic and post-synaptic axons and the area of synaptic contact, histological preparations were made from the nerve cords of five crayfish. The cords were fixed for a few hours in cold vom Rath's fixative (vom Rath, 1895) to which $CaCl_2$ had been added to a final concentration of 1%. It was found that the outside dimensions of the cords were decreased by about 20% during the histological preparation and the measurements of axon diameter were increased in the same proportion. The diameters of the lateral, medial and motor giant axons were variable along the course of the fibres, even within a single segment. It is possible that, in the case of the pre-fibres, this variation was largely an artifact of histological preparation. The diameter of the lateral giant axons was found to be $91·7 \pm 3·0 \mu$ (mean ± S.E.), that of the medial giant axons, $59·1 \pm 2·8 \mu$. The motor axons showed a more or less abrupt increase in diameter, about twofold or more, at the point at which they crossed the mid line and formed synapses with the motor fibre of the opposite side. They usually reached their largest diameter while passing under the ipsilateral medial giant axon, and then gradually became thinner as they crossed over the lateral pre-fibre. The diameter of the motor axons at the point at which they formed synapses with the lateral fibres was $38·5 \pm 1·6 \mu$. All the above diameters are the means of about twenty measurements from seven different ganglia.

Synaptic contact between the two axons comes about by means of numerous processes of the post-fibre which penetrate intervening sheaths and come into close association with pre-fibre membrane (Robertson, 1953, 1955; see Discussion). In several of our preparations it was possible to see that these post-fibre processes expanded before terminating, thus providing an enlarged area of contact between the two membranes. In one of these cords the terminal expansions could be seen sufficiently clearly to be measured in serial sections and thus an estimate of the total area of synaptic contact could be made. The value found was 6×10^{-5} cm², with a range of uncertainty of $3–8 \times 10^{-5}$ cm². This cord, which was not used for measuring axon diameters, was taken from a smaller animal than were those on which the electrical recordings were made; and in calculating the resistance of a unit area of the 'synaptic membrane' (see Discussion) a value of 10^{-4} cm² was used.

RESULTS

Identification of the impaled axons

Lateral giant pre-fibres. With few exceptions the micro-electrode could be placed into the lateral giant axon at the first attempt. Examples of the action potentials of these fibres are shown in Text-fig. 5. The largest part of the falling phase was complete in 1–1·5 msec after the start of the spike and was followed by a more prolonged 'tail'. The small deflexion several milliseconds after the spike, in *a* (indicated by the arrow), was observed in a number of preparations and was found to coincide with the firing of the lateral giant segment on the opposite side of the cord. Text-fig. 5*c* shows simultaneous intracellular recordings from the right and left lateral giant axons, at the same level of the cord. The right segment was stimulated by means of a third internal electrode and the resulting spike arose from the local electrotonic pulse. Accompanying this

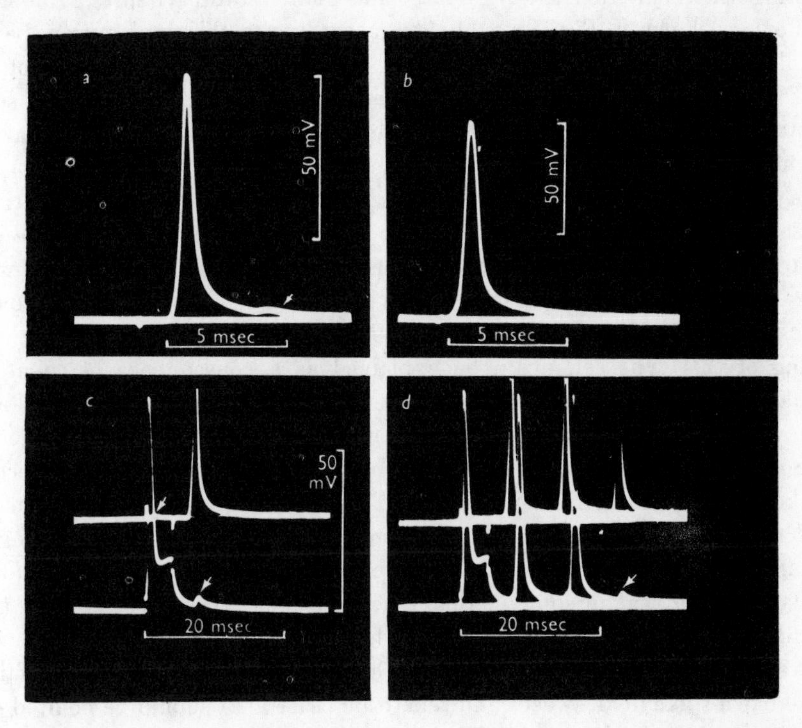

Text-fig. 5. Action potentials recorded intracellularly from lateral giant pre-fibres. Stimulation in *a* and *b* was through external, and in *c* and *d* internal, electrodes. The arrows indicate the small deflexions (collateral potentials) associated with firing of the other lateral giant. In *c* and *d* simultaneous recordings were made from the two lateral giant segments of the same ganglion; a spike in one was accompanied by a small deflexion in the other. When a supraliminal current pulse was applied to the fibre on the right side (lower trace) the resulting spike failed to cross to the other side immediately, but gave rise to firing in that segment about 7·5 msec later. This delayed spike was presumably due to the impulse crossing to the left side in some distant ganglion. In *c* the late left-segment spike failed to cross back to the right side; but in *d* it did so and the cycle was then repeated twice.

spike a small deflexion (upper arrow) was recorded from the left fibre, followed 5·5 msec later by a conducted impulse; and this delayed spike was similarly accompanied by a small deflexion in the right segment (lower arrow). The small deflexions and the late firing on the left side can be attributed to the known cross-connexions between left and right lateral giants. In each ganglion the segmental neurone of one side sends a relatively small collateral branch toward the mid line, where it makes synaptic connexion with a similar branch from the other side (Johnson, 1924; also see Methods). Thus the lateral giant system is a ladder-shaped network, with synapses not only between the segments of the 'uprights' (segmental synapses) but also in the centre of each 'rung'. These last junctions will be referred to as 'collateral' synapses to distinguish them from the segmental and giant motor synapses. The small deflexions shown in Fig. 5 would then, according to this explanation, be the 'collateral' post-synaptic potentials (p.s.p.'s) and the delayed firing of the left segment due to impulse transmission at a collateral synapse in some distant ganglion. The presence of collateral p.s.p.'s were an additional aid in identifying the lateral giant axons.

Another related property of the lateral fibre system which was sometimes observed, and which also helped in its identification, was a repetitive discharge following a single short stimulus. The records of Text-fig. 5*d* were taken from the same preparation and with the same electrode disposition as in *c*, but stimulation had been preceded by a longer period of rest. Rhythmic firing of this type can again be explained as a consequence of collateral synapses. Whereas the collateral p.s.p. in the right segment in *c* (lower arrow) failed to excite, it can be seen that in *d* a conducted spike appeared in the place of the p.s.p. and was presumably evoked by the latter. This second spike in the right segment was followed by a second action potential in the left segment, again after a delay due presumably to the impulse crossing at some distant ganglion. A number of sweeps had been superimposed in *d* and it can be seen that in some of them the cycle was repeated once more (the third spikes in each segment) but in others the second collateral p.s.p. did not succeed in reaching threshold. The third p.s.p. in the right segment failed to evoke a spike in all sweeps and repetition ceased. It should be pointed out that for an impulse to circulate in this way, around a loop of the lateral giant system, at least one of the collateral synapses must show effective one-way transmission. Otherwise, the impulse would cross at the first synapse and the two spikes would traverse the two segmental chains almost simultaneously, cancelling each other in the collateral branches. It can be seen that unidirectional transmission must have occurred in *d*, for the first spike on the right side was not followed by a contralateral impulse immediately, whereas the first left-segment spike did succeed in crossing with short delay. It seems highly probable that such one-way transmission in a system which appears to be anatomically symmetrical is due to injury and that the circuitous repetitive discharge is anomalous.

When two micro-electrodes were inserted into the same segment of a lateral giant axon, the relationship could be obtained between steady current applied through one electrode and membrane potential recorded with the other. Such current–voltage curves were required for the analysis of the GMS described below. Examples are shown in Text-figs. 4 and 13*c*. Their shape and slope at the origin were characteristic of the lateral giant axons and contrasted with a typical curve obtained from a giant motor fibre (Text-fig. 13*d*). Provided that the distance between the two micro-electrodes was small in comparison to the length constant of the fibre, the slope at the origin of such a curve gives a measure of the input resistance, namely, the resistance between a point inside the fibre and the outside. The value of this slope for thirty lateral giant segments, each from a different preparation, was $1{\cdot}03 \pm 0{\cdot}05 \times 10^5\ \Omega$ (mean ± S.E.). The thirty experiments selected were those in which the distance between the current and voltage micro-electrodes was 0·2 mm or less, the mean distance being 0·14 mm. In eleven preparations a third electrode insertion was made and a rough measurement of the length constant (λ) obtained. The mean value (± S.E.) was $1{\cdot}1 \pm 0{\cdot}1$ mm. It was useful to know this quantity in order to correct for errors due to the finite distance between current-passing and voltage-recording electrodes. For example, the above measurement of 'input resistance' should be increased by about 12% to compensate for the mean interelectrode distance of 0·14 mm.

The quantity λ is not strictly applicable to the lateral giant axon, which is segmented and part of a ladder-shaped network rather than a uniform cylinder of indefinite length. The error introduced by ignoring these features, however, is probably not large: (i) the measurements were usually made with the micro-electrodes more than a millimetre from the nearest segmental synapse (i.e. about one 'length constant' away); (ii), the electrical resistance of the segmental synapse appears to be low (unpublished observations); (iii) the nearest collateral branch was always beyond the segmental synapse (see Text-fig. 1) and its 'input resistance' probably high compared to the main part of the segment. Even if the lateral giant axon did not approximate to a linear cable, however, the apparent 'length constant' would still have been useful; for a similar disposition of the electrodes was used in obtaining 'λ' and in making the other measurements to which the correction for spatial decrement was applied.

The properties of the post-fibres. The post-fibre could be distinguished from any of the other fibres in the immediate neighbourhood by a number of physiological criteria. The first test was low-voltage stimulation of the third ganglionic root; and if the impaled axon fired, it was accepted as a third-root fibre. Text-fig. 6*a* shows such a direct response of an axon which was subsubsequently found to be the motor giant axon. In many experiments the impaled post-fibres were incapable of conducting nerve impulses and only graded local responses were obtained. But in either case it was assumed that a third-root fibre had been impaled and the next step was to test its response to stimulation of the nerve cord. If the axon was the motor giant axon, characteristic post-synaptic responses of the type shown in Text-fig. 6*b* were obtained. The two records, taken from different experiments, are multiple exposures and in b_1 the stimulating voltage was varied between sweeps. Provided that the shock intensity was sufficiently high, the early rapid deflexions were always followed by a slow potential (usually consisting of several summed components).

The early responses, as will be shown below, were evoked directly by the giant pre-fibres and were thus the p.s.p.'s of the giant motor synapses. They had an excitatory function and when supraliminal led to firing of the post-fibre. An example of such excitation is shown in *c*, in which cord stimulation intensity was varied while recording from the motor giant. The p.s.p. appeared in two steps, presumably associated with two different giant pre-fibres. In all but one of the sweeps, the summed p.s.p. evoked a post-fibre spike.

The size of the 'giant' p.s.p.'s varied, in different preparations, over a wide range from a few to over 40 mV. But in those cases in which they seemed to be larger than 25–30 mV there was probably a component of local response;

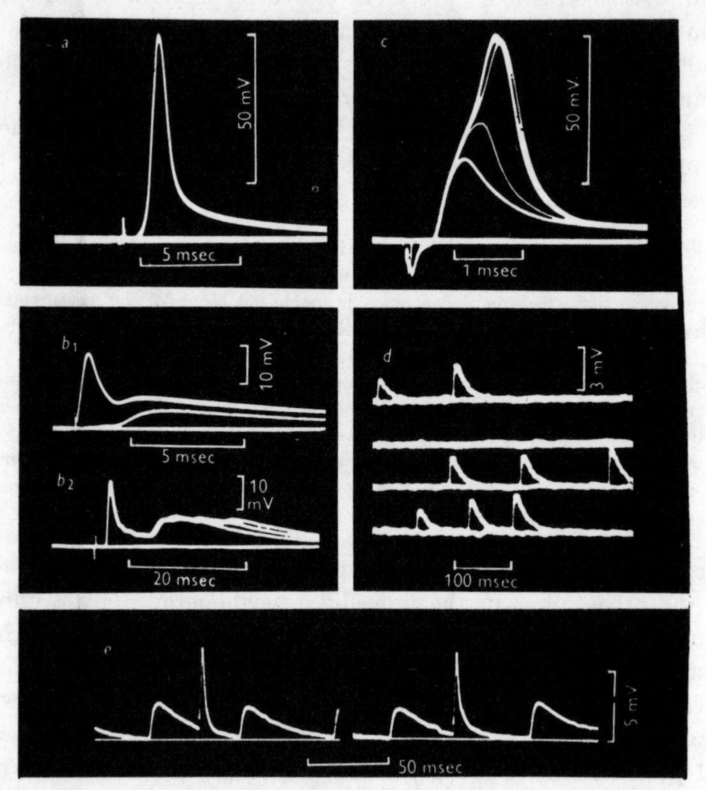

Text-fig. 6. Responses of the giant motor fibre. *a*, a directly evoked spike; the stimulus was applied to the third root by means of the capillary suction electrode. *b*, giant and slow p.s.p.'s following stimulation of the dorsal surface of the nerve cord. Multiple exposures. In b_1 the stimulus intensity was increased between successive sweeps. The pre-fibres for the slow response had the lowest threshold and only a slow potential is seen on the first sweep. *c*, a multiple exposure in which cord stimulation intensity was increased between sweeps. Two giant p.s.p.'s summed to evoke firing of the motor fibre. The record was retouched. *d*, slow potentials in the absence of stimulation. *e*, 'spontaneous' slow potentials and giant p.s.p.'s. The faster response in the centre of each trace was evoked by stimulation of the lateral giant pre-fibre. This 'giant' p.s.p. was among the slowest recorded. The base lines have been drawn.

for in many preparations the post-fibre was unable to conduct action potentials and only showed graded, local spikes. Nevertheless, at some synapses a single p.s.p. could cause firing of the post-fibre and thus effect one-to-one transmission. It seems likely that when the p.s.p.'s were small, this was the result of experimental damage (see Discussion).

TABLE 1. Post-synaptic potentials at the giant motor synapses

	Number of p.s.p.'s	Size (mV)	Rise time (msec)	Time from onset to half-decline (msec)
Lateral pre-fibres	28	17·9 ± 1·4 (7–35)	0·45 ± 0·02 (0·31–0·91)	1·4 ± 0·1 (0·83–3·2)
Medial pre-fibres	8	13·4 ± 1·6 (6–20)	0·41 ± 0·03 (0·31–0·55)	1·2 ± 0·1 (0·80–1·7)

The initial rate of rise of these p.s.p.'s was usually rather slow. In order to obtain a more consistent measure of the time of rise, the approximately linear portion of the rising phase was extrapolated to the base line and this point taken as the onset. Values have been determined separately for p.s.p.'s evoked by lateral and other (medial) giant pre-fibres. The range is given in brackets beneath the mean and S.E. The upper limit of the range of rise times (0·91 msec) for lateral giant p.s.p.'s was considerably larger than the next-to-highest value, which was 0·58 msec.

The time course of the 'giant' p.s.p.'s is summarized in Table 1. P.s.p.'s which were not associated with firing of the lateral giant axon were assumed to be due to either of the medial pre-fibres (see Methods). Although p.s.p.'s can also be evoked by the contralateral segmental axon, such responses can be recognized by their small size and long latency (see below) and they have not been included in the table. The time courses of the two groups of p.s.p.'s (Table 1) were not significantly different. In both cases the rising phase was usually complete in less than 0·5 msec and the decline from peak to half-peak occurred, in most cases, within 1 msec. By contrast, the p.s.p.'s recorded from some of the smaller fibres of the third root were considerably slower. The mean rise-time for twelve such responses was 1·5 msec (1·0–2·5) and the time from onset to half-decline was 7·3 msec (3·5–15·6).

The other characteristic response of the motor giant, the late slow potentials shown in Text-fig. 6*b*, is considered in detail in the following paper (Furshpan & Potter, 1959). Evidence is presented that these potentials are associated with inhibitory effects, despite the fact that they almost always appeared as depolarizations. The difference in their physiological effect, as well as that in time course and latency, makes the slow potentials easily distinguishable from the 'giant' p.s.p.'s.

In addition to these two types of response following cord stimulation, intermittent slow deflexions were seen in most giant post-fibres in the absence of any applied stimulus (Text-fig. 6*d*). Observations discussed in the following paper (Furshpan & Potter, 1959) indicate that the spontaneous potentials are due to firing of the same pre-fibres which evoke the late, slow responses. In Text-fig. 6b_2 the variability in the slow potential is attributable to enhanced firing of the 'spontaneous' potentials following cord stimulation. Text-fig. 6*e* allows a comparison between the time courses of a 'giant' p.s.p. and the spontaneous potentials. Cord stimulation was adjusted so that the only pre-fibre to fire was the ipsilateral segmental giant fibre and the resulting post-fibre response appears in about the centre of each trace. The smaller, slow deflexions are the spontaneous p.s.p.'s. The contrast between the time courses of the two potentials was less than that usually observed. The main point to be emphasized about the above observations, however, is that the spontaneous potentials and the typical pattern of fast and slow responses to cord stimulation were singularly characteristic of the giant motor fibre and allowed it to be distinguished from any of the other axons that were impaled in these experiments.

Following the insertion of a second micro-electrode into the motor giant axon, the current–voltage relationship, as described above, could be obtained. A typical curve (Text-fig. 13*d*) differed from that for the pre-fibre in several respects. (1) The post-fibre curve appeared to deviate more markedly from linearity for similar changes of membrane potential. (2) With inward membrane current (the lower left quadrant) the curvature was opposite for the two fibres. (3) The slope at the origin was steeper for the post-fibre, the mean value in eleven experiments being $6{\cdot}0 \times 10^5\,\Omega$. A large fraction of the higher 'input resistance' is accounted for by the smaller size of the post-fibre and at least part of the opposite curvature found for inward currents (see (2) above) is due to the presence of the GMS, as will be explained below. If these factors are taken into account the discrepancies between the curves for pre- and post-fibres are considerably reduced. Their observed shapes are, however, characteristic and additional means of distinguishing the fibres. Because of the difficulty of impaling the motor giants, a measurement of characteristic length, which requires at least three insertions, was infrequently made and then probably involved considerable error due to damage of the fibre.

Orthodromic nerve-impulse transmission

When micro-electrodes were inserted into both pre- and post-fibres, the succession of potential changes in the two axons could be recorded during an action potential in either. Examples of transmission from pre- to post-fibres (orthodromic) are shown in Text-fig. 7, in which pre-fibre membrane potential always appears on the upper trace of each pair. In a few experiments one-to-one transmission occurred (Text-fig. 7*a*), but in most only a p.s.p. was seen in the post-fibre (*b*). In *a*, the inflected rising phase of the post-fibre spike is similar to that seen in action potentials recorded from other post-junctional regions (e.g. the motor end-plate (Fatt & Katz, 1951)).

The most striking aspect of these records is the brevity of the delay between the foot of the pre-fibre spike and the start of the p.s.p. The initially low rate of rise of both pre- and post-fibre potentials makes the assessment of this delay difficult; in some cases the two potentials seemed to arise at about the same time but at different rates. In order to obtain a measure of delay, the approximately linear parts of the rising phases pre-fibre spike and p.s.p. were extrapolated to the base line. The time between the two points of intersection, in thirty-one experiments (all on lateral GMS's), was $0{\cdot}10 \pm 0{\cdot}01$ msec (mean ± S.E.). Taking pre-fibre spike conduction velocity as 5 m/sec (mean of four experiments) a correction was made for the conduction time between pre-fibre micro-electrode and synapse. The corrected delay was 0·12 ± 0·01 msec (range, 0·07–0·02 msec).

In both *a* and *b* the lateral giant axon had been stimulated by means of an external capillary electrode applied to the surface of the cord. The measure-

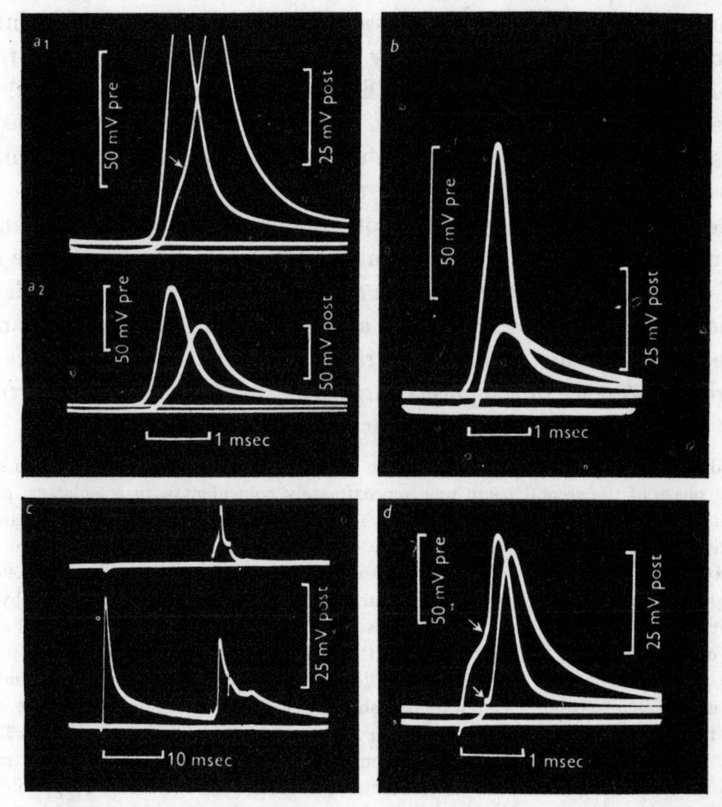

Text-fig. 7. Orthodromic nerve-impulse transmission at the GMS. Simultaneous intracellular recordings from pre- and post-fibres. In each case pre-fibre potential was recorded on the upper trace. The second p.s.p. in *c* and that in *d* were evoked by direct stimulation of the pre-fibre with an intracellular electrode; otherwise the stimulus was applied externally to the dorsal surface of the cord. a_1 and a_2 were recorded from the same synapse at different amplifications. The p.s.p., at about the point indicated by the arrow in a_1, exceeded threshold and evoked a spike. The upper part of the rising phase of the prespike was retouched. Sub-threshold p.s.p.'s are shown in *b*–*d*. In *c*, the first p.s.p., evoked by external stimulation, was apparently due to firing of one of the medial giant fibres, for no spike was seen in the impaled lateral giant fibre (upper trace). The second p.s.p. accompanied a lateral giant spike (85 mV) evoked by intracellular stimulation. The small hump on its falling phase was probably associated with firing of the other lateral giant fibre. *b* and *c* were recorded from the same synapse, but the polarity of the external stimulating pulse was opposite in the two cases. In *d* the arrows indicate the end of the depolarizing current pulse applied to the pre-fibre.

ment of synaptic delay would be invalid if a giant pre-fibre, other than the impaled one, were to fire first and give rise to an earlier p.s.p. To eliminate this possibility, the stimulus strength was varied in order to confirm that both pre- and post-fibre responses had identical thresholds. This was indeed the case in Text-fig. 7*a* and *b*. Nevertheless, the possibility remained that the thresholds of two of the giant pre-fibres were too close to be distinguished. A more decisive procedure was to stimulate the pre-fibre selectively with a second internal electrode. Fig. 7*c* illustrates a situation in which the internal stimulating electrode was particularly useful. A stimulus to the cord just strong enough to evoke a giant p.s.p. (first response on the lower trace) failed to excite the impaled lateral giant axon (upper trace). This post-fibre response was presumably evoked by one of the medial giant axons. Later during the same sweep the lateral giant axon was stimulated by means of an intracellular electrode, and the resulting action potential was accompanied by another p.s.p. In this case one can feel confident that the second p.s.p. was due to the firing of the impaled lateral giant axon and that the short synaptic delay was real. The record of Text-fig. 7*d*, taken at higher sweep speed, shows another example of transmission following intracellular stimulation of the pre-fibre. Although the presence of the pulse complicates the estimation of the synaptic delay, it is clear that the delay was very small.

An incidental observation is also illustrated in Text-fig. 7*c*. The small, late deflexion seen on the falling phase of the second p.s.p. was approximately coincident with a collateral p.s.p. in the lateral giant axon. This has also been observed in three other experiments and suggests that the late deflexion is associated with the contralateral segment of the lateral giant axon. In one of these experiments simultaneous recordings were made from a motor giant axon and the contralateral segmental giant axon. When the ipsilateral segmental giant axon was stimulated by means of an internal electrode, an action potential subsequently appeared in the contralateral segment (as in the case illustrated in Text-fig. 5). It was then seen that this contralateral action potential was approximately coincident with the late deflexion in the motor fibre. Such a coincidence, however, is only suggestive of a causal relationship. If such a relationship exists there are two synapses, reported in histological studies, which might account for the late deflexion. The first (Johnson, 1924) is between the motor giant axon and the collateral branch of the contralateral segmental giant axon, while the second (Robertson, 1955) is between the two opposite motor giant axons. In the second case, for example, the complete pathway would be from ipsilateral to contralateral segmental giant axon, from there to the contralateral motor giant axon and only then to the ipsilateral motor giant axon.

Tests for the presence of antidromic transmission

Wiersma (1947) found that in *Cambarus* stimulation of the third ganglionic root did not evoke firing of any of the giant pre-fibres. A similar test for antidromic transmission has been made in the present experiments but using intracellular recording. Text-fig. 8*a* illustrates the type of result obtained in the six experiments in which internal electrodes had been inserted on either side of a GMS, and in which the post-fibre showed an all-or-nothing action potential. The motor fibre was stimulated directly by means of the capillary electrode containing the third root. In the case shown in Text-fig. 8*a* the magnitude of the post-fibre action potential was 77 mV, while the size of the small accompanying deflexion in the lateral pre-fibre was less than 0·3 mV. The post-fibre spike was thus 250–300 times as large as the consequent pre-fibre potential change. It will be convenient to refer to this ratio as the attenuation factor. Text-fig. 8*b* shows a record of orthodromic transmission at the same synapse.

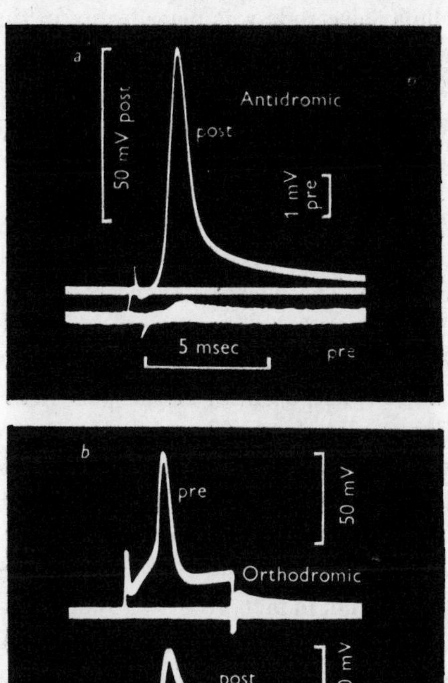

Text-fig. 8. A comparison of the effects of orthodromic and antidromic nerve impulses at the same synapse. Simultaneous intracellular recording from both synaptic elements following stimulation of the post-fibre in *a* and the lateral giant pre-fibre in *b*. The pre-fibre stimulus was applied through a third intracellular electrode.

The lateral giant axon was selectively stimulated by means of a second internal electrode and a pre-fibre spike of 83 mV gave rise to a p.s.p. of 18·5 mV. In this case the attenuation factor was about 4·5. The transmission mechanism was, therefore, about 60 times as effective in the orthodromic as in the antidromic direction. In two other experiments of this type the ratio of antidromic to orthodromic attenuation factors was 210:3·9 = 54 and 350:2·8 = 125. In another two cases the pre-fibre deflexions were too small to be measured at the amplifications used, but were less than 0·3 mV. Thus in five of these six experiments the synapse was highly 'directional', the ratio of attenuation factors being greater than fifty. In the remaining case an appreciable antidromic effect was seen and the attenuation factor ratio was about three; but this synapse was atypical in another important respect, the consideration of which will be deferred to the Discussion. That there is usually no significant antidromic transfer has also been confirmed in twelve additional experiments in which an intracellular electrode was placed only in the pre-fibre. In every case relatively large stimuli applied to the third root resulted in pre-fibre deflexions of less than 1 mV; but in the absence of intracellular recording from the post-fibre there could be no assurance that a conducted, antidromic spike had been evoked.

Tests for the transfer of electrotonic potentials across the synapses

It has been seen that an action potential in the pre-fibre could give rise to a p.s.p. of 25 mV or more. The question then arose whether smaller, or even subthreshold, depolarizations of the pre-fibre would also bring about an appreciable transynaptic effect (i.e. in the post-fibre); and whether local electrotonic alterations in the membrane potential of the post-fibre would have as little effect on the pre-fibre as did the antidromic nerve impulse. In order to test these questions, experiments were made initially with one internal electrode in each synaptic element. Either micro-electrode could be used to pass pulses of current while the other recorded any changes in the membrane potential of the transynaptic fibre. It was found that even subthreshold current pulses applied to the pre-fibre could, in fact, give rise to considerable potential changes in the giant motor fibre and the transynaptic effect had the appearance of an electrotonic potential (Text-fig. 9*a*). A marked effect was only present, however, when the direction of the current across the pre-fibre membrane was outward (i.e. causing depolarization). Any post-synaptic effects produced by inward (hyperpolarizing) currents were very much smaller (Text-fig. 9*b*).

The experimental situation was then reversed, the post-fibre electrode being used to pass current while the pre-fibre electrode recorded potential changes. Now it was found that depolarizing currents applied to the motor giant axon had very little effect on the pre-fibre potential (Text-fig. 9*c*); but, most unexpectedly, hyperpolarizing currents were seen to produce marked transynaptic effects, which had the appearance of anelectrotonic potentials (Text-fig. 9*d*). To summarize, transynaptic membrane potential changes were observed only if depolarizing current was applied directly to the pre-fibre or hyperpolarizing current to the post-fibre. In order to have convenient terms with which to distinguish the experimental situations considered above, the words 'orthodromic' and 'antidromic' will be used, in analogy with nerve-impulse transmission. The orthodromic situation is that in which current is passed with the pre-fibre electrode and the electrotonic potentials appear to cross the junction from pre- to post-fibre. In the antidromic case the current-passing electrode is, of course, in the post-fibre. One can now restate the above observations as follows. Only depolarizations cross the synapse orthodromically in appreciable amounts while only hyperpolarizations do so antidromically.

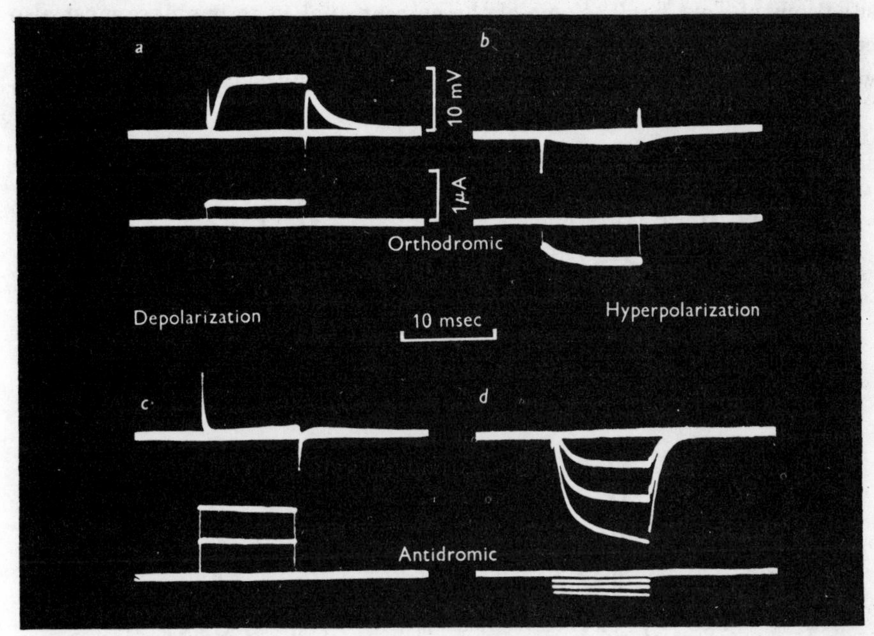

Text-fig. 9. Tests for transynaptic effects of applied current pulses. Inward and outward currents were successively applied to each fibre while recording the membrane potential of the other. The current was applied to the pre-fibre (orthodromic situation) in *a* and *b*, and to the post-fibre in *c* and *d*. The current pulses are shown on the lower, membrane potential on the upper trace of each pair. Outward current and depolarization appear as upward deflexions. The depolarizing pulse in *a* was just below pre-fibre threshold. The calibrations apply to all four cases. The records have been retouched.

Confirmation of these results has been obtained in thirty out of thirty-two experiments in which at least one micro-electrode had been inserted on either side of the junction, and in which both orthodromic and antidromic pulse transmission was tested. In the two experiments which did not conform to this pattern, potential changes of the appropriate sign were seen in one synaptic fibre when current of either direction was applied to the other. One of these two was the experiment mentioned above, in which the ratio of antidromic to orthodromic attenuation factors was very low, about 3.

The synaptic–rectifier hypothesis

The seemingly diverse results of Text-fig. 9 can all be explained by a single mechanism embodied in the following hypothesis: The junction operates by means of 'electrical' transmission, the transynaptic effects seen in Text-fig. 9*a* and *d* resulting from a portion of the applied current crossing the junction; and it behaves like an electrical rectifier, accounting for the negligible transfer of the pulses in *b* and *c*. The hypothesis is illustrated in Text-fig. 10, which shows the direction of flow of current associated with the four experimental situations of Text-fig. 9. Some of the current applied to one fibre must also cross the membrane of the other, by way of the synapse, the amount varying with the 'synaptic resistance'. It is apparent that in the two situations (*a*, *d*),

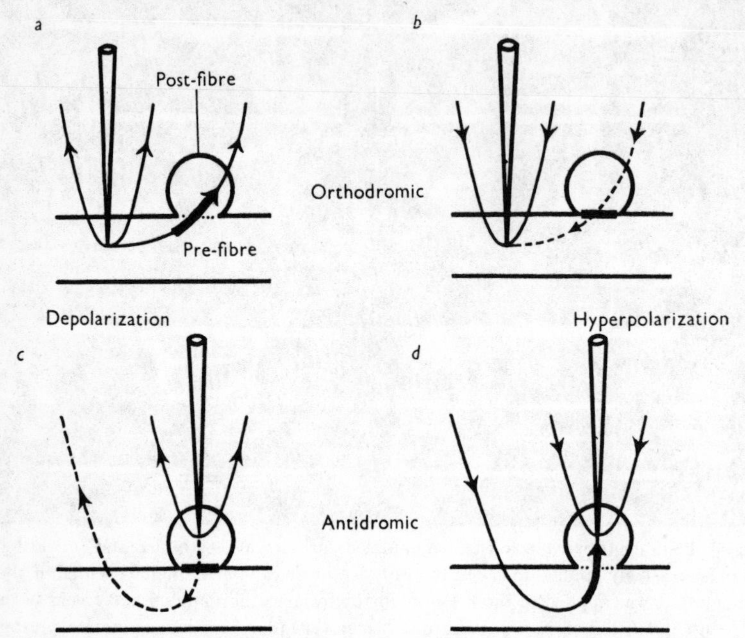

Text-fig. 10. The synaptic-rectifier hypothesis. The diagrams correspond, in the same order, to the four experimental situations of Text-fig. 9. The post-fibre is shown in transverse section at the point at which it crosses over the pre-fibre; the junction is indicated by a dotted line or a heavy bar, representing a low or high synaptic resistance, respectively. The arrows give the direction of (positive) current entering or leaving the current-passing micro-electrode; dashed lines indicate negligible current flow due to high synaptic resistance. Diagrams *a* and *d*, corresponding to the two situations in which transynaptic effects were observed, show that in both cases current would cross the junction in the same direction (indicated by the heavy arrows.)

in which there were considerable transynaptic effects, the direction of any synaptic current was the same, and opposite to that in the other two cases. Therefore, if the synapse were an electrical rectifier and only allowed appreciable (positive) current to cross it in the direction shown by the heavy arrows (Text-fig. 10*a*, *d*) the above results would be explained. The dashed lines (*b*, *c*) indicate that the amount of current flow would be negligible, thus accounting for the absence of transynaptic potential changes in those cases.

An alternative statement of the hypothesis is given in the circuit diagram of Text-fig. 11. For simplicity the two synaptic elements are shown as if they were in line and terminated at the point of juncture, rather than crossing each other (as shown in Text-fig. 1). R_{pre} and R_{post} represent the 'input resistances' of the two fibres. The insides of the axons are separated by the synaptic resistance, S, which is shown as a rectifier. It is apparent that any change in potential across R_{pre} would also alter the p.d. across R_{post}, and vice versa, provided that S is not extremely large. To explain the results of the above experiments, S is assumed to be low only when the inside of the pre-fibre becomes more positive (depolarization) or when the post-fibre becomes more internally negative (hyperpolarization). That is, both situations give rise to a p.d., across the synapse, of the same sign; and the rectifier is assumed to have a low resistance for p.d.'s of this direction (indicated by the plus and minus signs Text-fig. 11.).

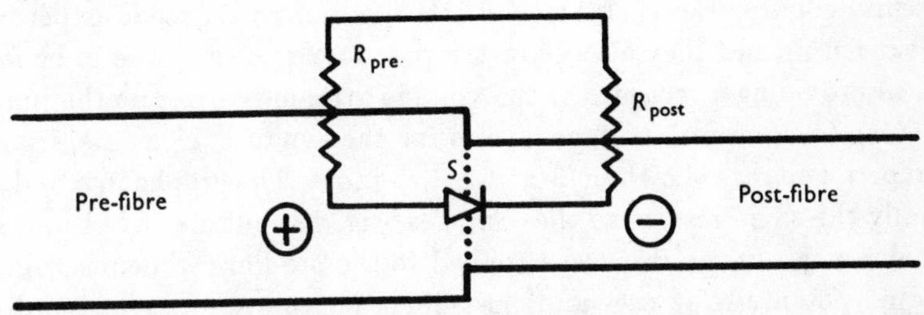

Text-fig. 11. The proposed equivalent circuit for the GMS. Pre- and post-fibres are shown as though terminating end to end (see Text-fig. 1, however). The 'input resistances' of the two fibres are represented by the elements, R_{pre} and R_{post}; and the synaptic resistance by S. The plus and minus signs indicate the direction of the p.d. across the junction for which S is small. Inasmuch as R_{pre} and R_{post} are shown as constant resistive elements, the diagram only holds for small steady potential changes.

Tests of the synaptic–rectifier hypothesis

The current–voltage relationship of the synapse. If the junction behaves like a simple rectifier in an electrical circuit, and the observed transynaptic effects are due entirely to passive current flow across it, the intensity of this synaptic current should depend only on the p.d. across the rectifier, but not on the way in which the p.d. is brought about. In other words, the current–voltage characteristic of the junctional rectifier should account equally well for the transynaptic effects seen in the orthodromic and antidromic situations. In the absence of a direct measure of the current traversing the synapse, however, the rectifier characteristic cannot be directly determined. Instead, we have separately calculated two experimental current–voltage curves for the synapse, one for the orthodromic situation and one for the antidromic. Each characteristic was constructed so that it would exactly account for the transynaptic effects seen in that situation. Comparison of the two curves then provided a critical test of the hypothesis; for if they were the same, it would mean that a synaptic rectifier with that current–voltage characteristic would quantitatively account for the diverse results of the current-passing experiments.

The following data were used in making the required calculations: recordings of membrane potential from both fibres during the application of various currents to each, in turn; and the current–voltage relationships of the two fibres. The most satisfactory method for obtaining this information was to insert a pair of micro-electrodes, for passing current and recording potential, into each synaptic element, thus avoiding the necessity of withdrawing or inserting electrodes during the course of the measurements. Text-fig. 12 shows part of the required data from one experiment, namely simultaneous records of membrane potential from the two fibres of a GMS. The potential changes in *b* and *d* are the transynaptic effects accompanying the directly evoked electrotonic potentials on the other side of the junction (*a* and *c*). The results confirm those obtained in the two-micro-electrode experiments (e.g. Text-fig. 9); but they also allow the p.d. across the synapse to be found, as well as providing a measure of the voltage attenuation across the junction (see below). The current–voltage curves for the synaptic elements, from the same experiment, are shown in Text-fig. 13*c* and *d*. The graphs in *a* and *b* are essentially the same results as those in Text-fig. 12, but displayed in a more convenient way. In *a* current was applied to the pre-fibre while recording the membrane potentials of pre- and post-fibres on the abscissa and ordinate, respectively; *b* shows the analogous results for the antidromic situation. The four curves of Text-fig. 13 provide the necessary data for making the two separate determinations of the apparent current–voltage characteristic of the synaptic rectifier.

The following symbols have been used. V_s and I_s represent changes from the resting values respectively of the p.d. across the junction and of the apparent current traversing it. V_{pre} and V_{post} denote changes in the internal potentials, with respect to the bath, of the two synaptic elements. (Changes in potential and current were used because the resting values were not always known.) The voltage across the synapse is equal to the difference between the local internal potentials of pre- and post-fibres in the immediate vicinity of the junction; and the sign of V_s is taken so that $V_s = V_{pre} - V_{post}$. Since V_{pre} and V_{post} are positive for depolarizations of the fibres, V_s is positive when directed as in Text-fig. 11.

The method of analysis is illustrated in Text-fig. 14. In this example V_s and I_s will be determined for an orthodromic case in which the pre-fibre was depolarized by 30 mV. *a* and *b* are tracings of Text-fig. 13*a* and *d*. It can be seen in *a* that for $V_{pre} = 30\,\text{mV}$, $V_{post} = 10{\cdot}6\,\text{mV}$; and $V_s = V_{pre} - V_{post} = 19{\cdot}4\,\text{mV}$. In determining I_s it was assumed that the transynaptic potential change (in this case V_{post}) was the result of the synaptic current I_s flowing across the 'input resistance' of the post-fibre (R_{post}; see Text-fig. 11). Thus, $I_s = V_{post}/R_{post} = I_{post}$; and the relevant value of I_{post} can be found from the current–voltage curve for the post-fibre. This has been done in Text-fig. 14*b*, giving a value for I_s of $2{\cdot}2 \times 10^{-8}$ A. Perhaps a simpler way of describing the determination of I_s is the following: the amount of current which must be injected into the post-fibre at a point, in order to reproduce the transynaptic potential change (10·6 mV), was found from the current–voltage curve of the fibre to be $2{\cdot}2 \times 10^{-8}$ A. The assumption was made that the two situations, namely, of current entering the post-fibre through the synapse or through a current-passing micro-electrode, are equivalent; and that all of the current crossing the junction entered the post-fibre. The determination of V_s and I_s was then repeated for pre-fibre depolarizations of different sizes. The filled-in

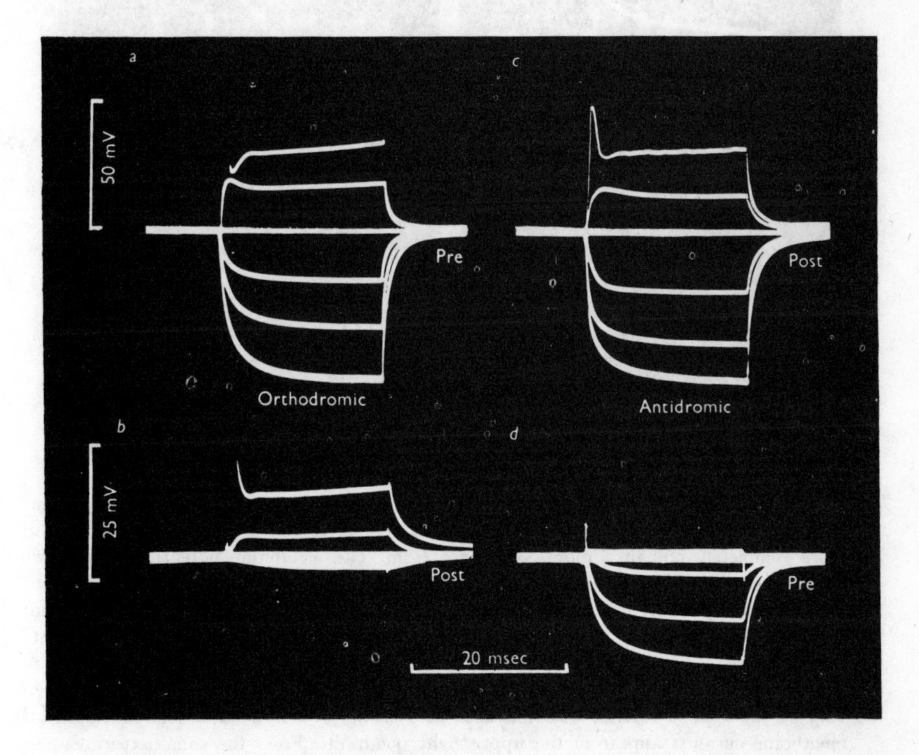

Text-fig. 12. Tests for the transmission of electrotonic potentials across the GMS. The membrane potentials of pre- and post-fibres were simultaneously recorded while applying steady pulses of current, by means of another internal electrode in each axon, to the pre-fibre (*a* and *b*) or post-fibre (*c* and *d*). The direct electrotonic potentials, *a* and *c*, are shown above the corresponding transynaptic effects, *b* and *d*. Deflexions above the base line are depolarizations. The larger catelectrotonic potential in *a* was above threshold, but most of the resulting spike and the accompanying p.s.p. (in *b*) are not visible in the records. The response in *c* was a graded, local spike. Voltage calibrations in *a* and *b* also apply, respectively, to *c* and *d*; the time calibration applies to all records.

circles in Text-fig. 15 give the values obtained for changes in V_{pre} in steps of 5 mV, over the range 0–50 mV. The various points were determined using the upper right quadrants of the curves in *a* and *d* of Text-fig. 13 and the point indicated by the arrow is the one derived in the above example. The open circles in Fig. 15 were obtained from the same experiment for hyperpolarizations in the antidromic situation. V_s and I_s were determined in a way analagous to that just described, but using the lower left quadrants of graphs *b* and

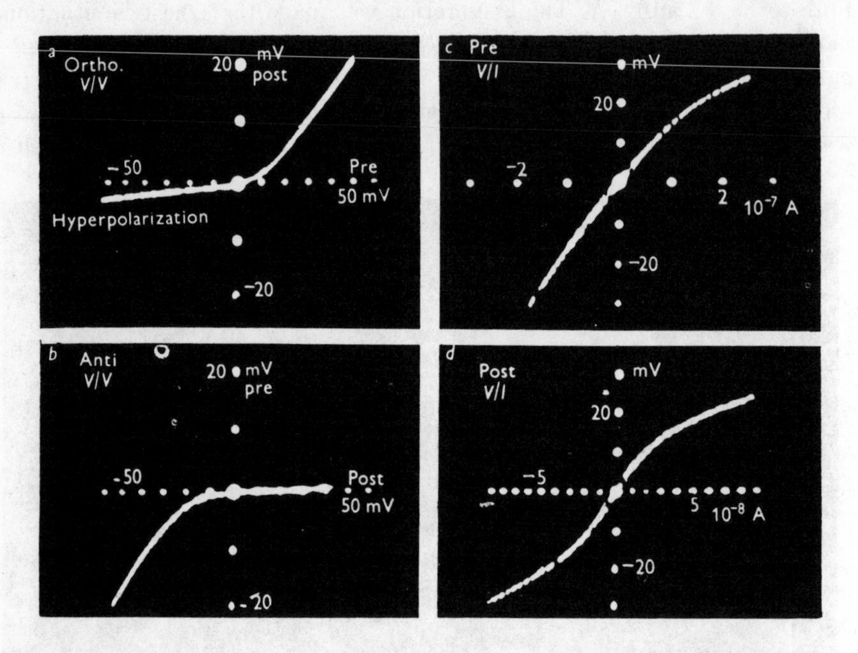

Text-fig. 13. The data necessary for determining the current–voltage characteristic of the synapse in two independent ways. *a* and *b* provide the same type of information as Text-fig. 12, but more completely and in a more convenient form. The membrane potentials of the two fibres were recorded, one on each co-ordinate axis of the oscilloscope, while varying the current applied to the pre-fibre (*a*) or post-fibre (*b*). The potential of the fibre to which the current was directly applied is on the abscissa. *c* and *d* are the nominal (see Discussion) current–voltage curves of pre- and post-fibre. The current was applied with an internal electrode and membrane potential measured nearby, also intracellularly. In all records the method of varying the applied current was that used in Text-fig. 4*a*. Depolarizations and outward membrane currents appear in the upper right quadrant. From the same experiment as Text-fig. 12.

c of Text-fig. 13. The points are those found for alterations in V_{post}, in steps of -5 mV, from 0 to -50 mV. The almost complete agreement between the two relations leaves little doubt of the correctness of the synaptic–rectifier hypothesis, and confirms that the currents that were assumed to cross the synapse, in order to account for the transynaptic effects, were in fact doing so. On any other hypothesis it would be very difficult to understand why the apparent synaptic currents should be identical, during pre-fibre depolarizations and post-fibre hyperpolarizations, whenever the two happened to bring about the same potential difference across the synapse. This is, of course, just the result to be expected if the junction behaved like an electrical circuit element of the kind represented in Text-fig. 11.

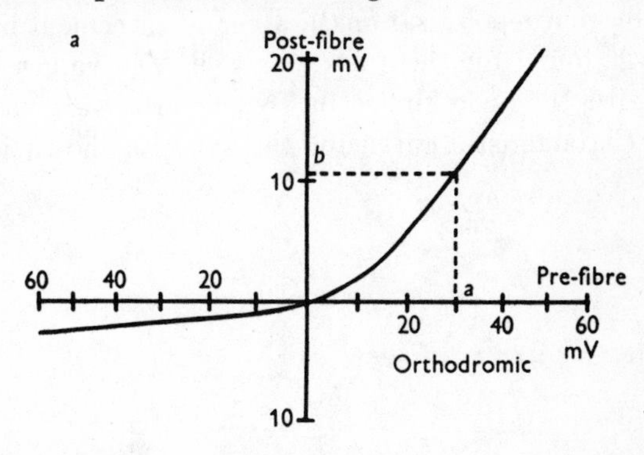

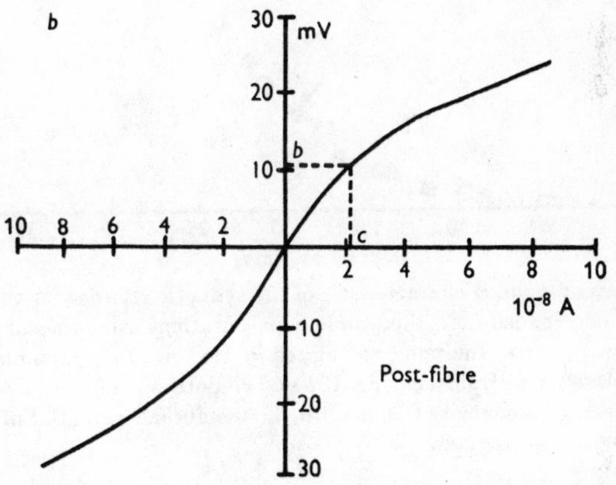

Text-fig. 14. The determination of V_s and I_s. *a* and *b* were traced, respectively, from Fig. 13*a* and *d*. $V_s = V_{pre} - V_{post}$; for $V_{pre} = 30$ mV, $V_{post} = 10{\cdot}6$ mV and thus $V_s = 19{\cdot}4$ mV. I_s was found by a substitution method: that current which duplicated the transynaptic effect, when applied directly to the post-fibre through a micro-electrode, was taken as equal to I_s. The value found from *b* was $2{\cdot}2 \times 10^{-8}$ A.

Curves relating V_{pre} to V_{post} (Text-fig. 13*a*, *b*) do not, by themselves, give an accurate measure of synaptic rectification; for their shape also depends on the membrane properties of the transynaptic fibre. But in deriving the I_s/V_s curve the non-linearities of the individual fibre membranes were eliminated, and Text-fig. 15 represents rather accurately the rectifier properties of the synapse itself. The figure shows only the characteristic for positive values of V_s, so that the resistance of the junction in the forward and reverse directions cannot be compared; but some measure of the degree of 'rectification' is that the 'slope conductance' (dI_s/dV_s) at $V_s = 30$ mV is approximately fifty times that at the origin.

Points relating V_s to I_s were also determined in the same experiment, for the two situations in which V_s was negative (i.e. 'orthodromic hyperpolarizations' and 'antidromic depolarizations'). General agreement between the two sets of points was found for these cases as well. The values of $-I_s$ against $-V_s$, together with the open circles in Text-fig. 15, were used to draw the more complete characteristic curve for the synapse shown in Text-fig. 16.

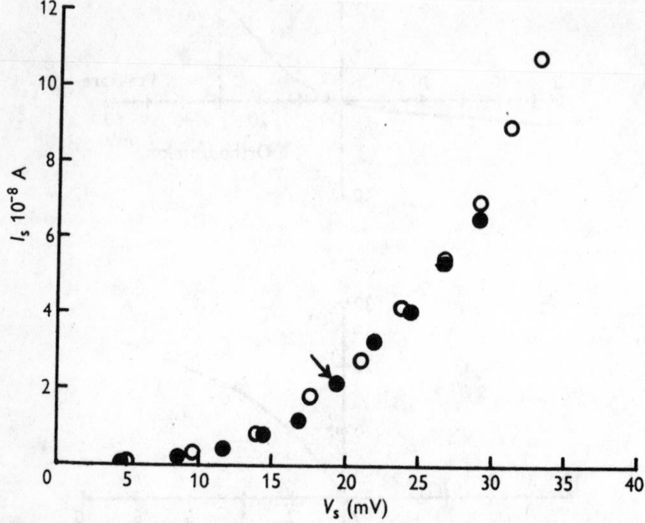

Text-fig. 15. The current–voltage characteristic of the synaptic rectifier for the forward direction only. ●, points determined for orthodromic depolarizations using the curves of Text-fig. 13*a* and *d*. The arrow indicates the point calculated in Text-fig. 14. ○, points obtained for antidromic hyperpolarizations from Text-fig. 13*b* and *c*. Both sets of points were found for increments of 5 mV, up to a change of 50 mV, in the membrane potential of the fibre to which the current was directly applied.

The ratio of the 'slope conductance' at $V_s = -30$ mV to that at the origin is one-third to one-quarter, so that the ratio of slopes compared at $V_s = 30$ and -30 mV was 150 or more. The ratio of 'chord conductances', I_s/V_s, at the same two values of V_s was about 35. Too much emphasis must not, however, be placed on the negative part of the curve in Text-fig. 16. There was considerable uncertainty in estimating such small synaptic currents, although the true values of I were probably not larger than those shown in the figure.

Of the thirty-two experiments mentioned above, fourteen were made with two micro-electrodes in each fibre and these provided sufficient information to construct the two synaptic–rectifier curves. General agreement between I_s against V_s for 'orthodromic depolarizations' and 'antidromic hyperpolarizations' was found in all fourteen cases, but in most of them the correspondence was not as precise as that illustrated in Text-fig. 15. Probably the most important variable influencing such agreement was the spatial decrement of electrotonic potentials from their site of origin (the synapse or one of the current-passing micro-electrodes) to the point at which they were measured. Usually these distances were 10–20% of the 'characteristic lengths' of the fibres. Depending on the positions of the micro-electrodes, however, the discrepancy between the orthodromic and antidromic situations could either be reduced, owing to equal errors in the two cases, or magnified by one or both of

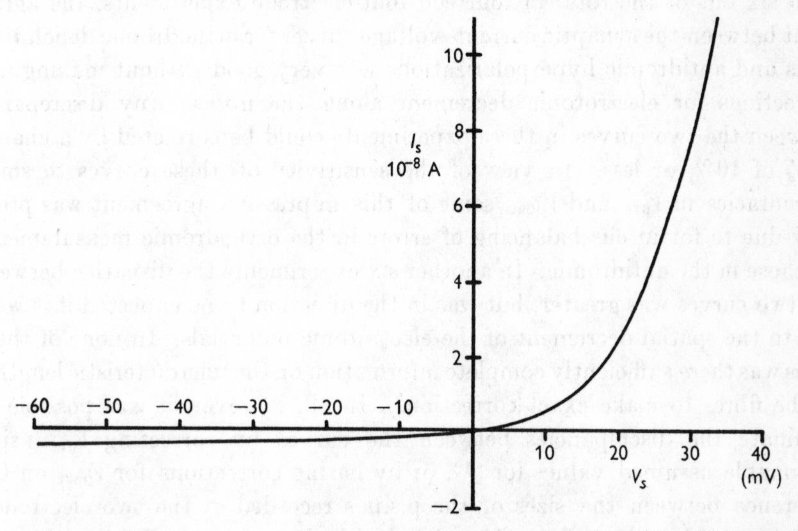

Text-fig. 16. The complete current–voltage characteristic of the synaptic rectifier. Positive values of V_s signify that the pre-fibre side of the junction was electrically positive with respect to the post-fibre side. Taken from the same experiment as Text-figs. 12–15.

the following factors. (1) The membrane resistance, and thus also the spatial attenuation, alters with the level of membrane potential of the fibres; an appreciable decrease in membrane resistance and an increase in spatial attenuation occur with either large depolarizations of the pre-fibre or large hyperpolarizations of the post-fibre (see Text-fig. 13*d*). (2) Since $V_s = V_{pre} - V_{post}$, an error in V_{pre} in the orthodromic case, for example, will give rise to a still larger percentage error in V_s. This point may be illustrated using the experiment of Text-fig. 14. A depolarization of the pre-fibre of 50 mV gave rise to a change in post-fibre potential of 21 mV. The apparent value of V_s was therefore 29 mV. Suppose that the measured value of V_{pre} had been 15% too large, owing to decrement of the catelectrotonus between voltage-recording electrode and synapse; then the true value of V_s would have been 22·5 mV (43·5–21 mV) rather than 29 mV, and the above estimation of V_s would have been 29% too large. This exaggeration of the error in calculating V_s is less for smaller depolarizations, for then the transynaptic potentials are relatively smaller. This type of error was probably less important for the antidromic case, for the post-fibre electrodes were usually closer to the synaptic region. The errors in the current–voltage curves of each fibre, caused by spatial attenuation, were less serious. There was no amplification of the inaccuracy and the percentage error tended to be about the same for both fibres, so that both the orthodromic and antidromic situations were affected to about the same extent and in the same direction.

In six out of the total of fourteen four-electrode experiments, the agreement between the synaptic current–voltage curves for orthodromic depolarizations and antidromic hyperpolarizations was very good without making any corrections for electrotonic decrement along the fibres. Any discrepancy between the two curves in these experiments could be corrected by a change in V_s of 10% or less. In view of the sensitivity of these curves to small inaccuracies in V_{pre} and V_{post}, some of this impressive agreement was probably due to fortuitous balancing of errors in the orthodromic measurements by those in the antidromic. In another six experiments the disparity between the two curves was greater, but was in the direction to be expected if it were due to the spatial decrement of the electrotonic potentials. In none of these cases was there sufficiently complete information on the 'characteristic lengths' of the fibres to make exact corrections. In all, however, it was possible to eliminate the discrepancies between the curves by correcting V_{pre} using reasonable assumed values for 'λ', or by basing corrections for V_{post} on the difference between the sizes of the p.s.p.'s recorded at the two electrodes. The magnitudes of pre-fibre 'λ' which had to be assumed all fell within the range of values found in other experiments.

Of the remaining two experiments, one could not be satisfactorily corrected by the above procedure and the other was one of those in which the synapse showed almost no rectification. In the first of these two experiments, even after assuming a pre-fibre 'λ' as low as 0·5 mm and correcting V_{post} for the difference in the p.s.p.'s recorded at the two electrodes, there remained a disparity of about 20% between the two values of V_s at a given I_s. The difference between the p.s.p.'s recorded at the two post-fibre electrodes was rather large, one being 40% greater than the other for an electrode separation of 0·14 mm; and it seems likely that one of these electrodes was not properly inserted. Despite this disparity, the general shapes of the orthodromic and antidromic curves were quite similar.

In the remaining experiment the almost complete absence of synaptic rectification introduced an error which precluded an accurate analysis of I_s against V_s without additional information. The point is worth considering, for it illustrates how the presence of the synaptic rectifier in the other experiments simplified the investigation. The apparent current–voltage relationship of each fibre was determined with the micro-electrodes in the junctional region, so that the applied current took two parallel pathways. In the case of the post-fibre, for example, some of the applied current, I, flowed across the synapse as well as across the post-fibre membrane (i.e. $I = I_s + I_{post}$). Graphs of the type shown in Text-fig. 13*d* relate V_{post} to I; but for the above analysis the relationship between V_{post} and I_{post} was required. The size of this error ($I - I_{post}$) depends on the extent of shunting by the synapse (i.e. on I_s). Since I_s was very small during pre-fibre hyperpolarizations and post-fibre depolarizations, these parts of the current–voltage curves were not much affected by the proximity of the junction. For example, in the experiment of Text-fig. 13*d*, $6{\cdot}2 \times 10^{-8}$ A of outward current applied to the post-fibre gave a depolarization of 20 mV. The fraction of this current which left the post-fibre by way of the synapse can be estimated as above. Referring to Text-fig. 13*b*, when V_{post} was 20 mV, V_{pre} was less than 0·3 mV. From the pre-fibre current–voltage curve, Text-fig. 13*c* (slope at the origin $= 1{\cdot}5 \times 10^5\ \Omega$), it was found that an applied current of about $0{\cdot}2 \times 10^{-8}$ A was needed to reproduce $V_{pre} = 0{\cdot}3$ mV. This does not give an exact measurement of I_s, however, since the pre-fibre curve is also in error owing to shunting by the synapse; but the direction of the error is such that the true value of I_s would be smaller than that found. Thus, $0{\cdot}2 \times 10^{-8}$ A gives the upper limit of the shunt current and it can be concluded that when V_{post} was 20 mV, the maximum error in I_{post} was 3·2 %. In contrast, similar calculations for the hyperpolarizing part of Text-fig. 13*d* show that, except for very small values of $-V_{post}$, large errors were attributable to shunting by the synapse (compare Text-fig. 10*c*, *d*). At $V_{post} = -20$ mV, up to 35 % of the applied current entered the post-fibre by way of the junction. These large errors did not, however, affect the analysis of the synaptic rectifier described above; for in those calculations only the depolarizing and initial hyperpolarizing parts of the post-fibre curve were used.

The pre-fibre current–voltage curve was less affected by the nearness of the micro-electrodes to the synapse. Because of the lower 'input resistance' of the lateral giant axons, a smaller fraction of the applied current crossed the junction. Thus even for a pre-fibre depolarization of 25 mV, during which the resistance of the synapse was almost as low as for a hyperpolarization of the post-fibre of 20 mV, the error in I_{pre} was not greater than 6 %. For small depolarizations and hyperpolarizations it was negligible; and these were the only parts of the pre-fibre curve used in constructing the characteristic curve of the synapse.

The rectifier in nerve-impulse transmission. The question now arises whether the above results can explain the transmission of the action potential. Are the synaptic currents that accompany a spike in the pre-fibre adequate to account entirely for the p.s.p.'s? The direction of current flow is correct, for the pre-fibre spike represents a large positive increase in V_{pre}; and as long as V_{pre} exceeds V_{post}, V_s and I_s are also positive ($V_s = V_{pre} - V_{post}$). In attempting to make a quantitative test of this question it has not been possible to apply directly the above analysis of the synaptic rectifier, for spikes and the experimental electrotonic pulses differed considerably in amplitude and time course. The largest electrotonic change in V_{pre} was only about 65 % of the pre-fibre action potential. This precludes a direct comparison of the two situations, since the resistance of the synapse is different at different values of V_{pre}. The relatively short time course of the pre-fibre spikes also introduces a difficulty; for the spikes would be subject to additional attenuation, over that found for steady depolarizations, due to the capacitative properties of the post-fibre membrane. Another factor bringing about an increased attenuation of transients would be a delay in the change in synaptic resistance following a change in V_s (i.e. delayed rectification). The available results are, however, insufficient to determine whether a significant delay in synaptic rectification was present. But despite these difficulties it seems likely from indirect evidence that the same mechanism accounts for transmission of electrotonic pulses and action potentials.

The two situations are compared in Table 2. The spikes were more attenuated than the catelectrotonic potentials in every experiment, despite their being larger and the fact that S decreases as V_{pre} increases. The mean difference between the two attenuation factors was about 40% and it seems probable that this difference would have been even greater if the electrotonic potentials had been larger. The greater attenuation of spikes was to be expected, if they were transmitted by the same mechanism as the steady potential changes, for the reasons given above.

TABLE 2. A comparison between the orthodromic attenuation factors for steady catelectrotonic potentials (α_c) and nerve impulses (α_s)

	Steady depolarizations			Spikes			
Expt.	V_{pre} (mV)	V_{post} (mV)	α_c	Spike (mV)	p.s.p. (mV)	α_s	α_s/α_c
1	60	13·5	4·5	100	18	5·6	1·25
2	45	15·9	2·8	80	20	4·0	1·43
3	30	6·7	4·5	72	12	6·0	1·34
4	20	9·5	2·1	70	24·5	2·9	1·38
5	35	9·3	3·8	80	16	5·0	1·33
6	30	10·0	3·0	78	20·7	3·8	1·27
7	50	22·5	2·2	80	28	2·9	1·32
8	50	26·5	1·9	80	30	2·7	1·42
9	45	11·3	4·0	90	13	6·9	1·73
10	55	10·0	5·5	85	9·8	8·7	1·58
Mean	42	13·5	3·4	81·5	19·2	4·9	1·41

In determining α_c, steady pulses of outward membrane current were applied to the pre-fibre while recording pre- and post-fibre membrane potentials (Text-fig. 12*a*, *b*). The maximum alteration in pre-fibre potential is given in the first column (V_{pre}), and the accompanying change in post-fibre potential (V_{post}) is shown in the second column. In the third column, $\alpha_c = V_{pre}/V_{post}$. The next three columns show the analagous findings for pre-fibre spikes and their accompanying p.s.p.'s, α_s being the quotient of spike divided by p.s.p. The ratio of these two attenuation factors is given in the last column.

The variability in α_c and α_s in Table 2 was considerable; but within any particular experiment the two factors tended to vary in the same way. The range of variation for α_s/α_c was only about one-eighth that for either of the attenuation factors by itself. A calculation of the coefficient of correlation between α_c and α_s for the ten pairs of values in Table 2 gave $r = 0{\cdot}95$, suggesting that the same variables determined the effectiveness of transmission in both cases. For the circuit of Text-fig. 11, in the orthodromic situation, $\alpha = 1 + S/R_{post}$, so that variations in either S or R_{post} would equally affect attenuation; and the question arises whether the high correlation between the two factors was due mainly to their dependence on R_{post}. The correlations of α_s with R_{post} and with S, were separately determined; for α_s and R_{post}, $r = -0{\cdot}11$

and for α_s and S, $r = 0{\cdot}82$. Thus the high correlation between α_c and α_s is compatible with the synaptic resistance being the main variable controlling the size of the p.s.p.

DISCUSSION

The experiments described above provide very strong evidence for a type of 'electrical' transmission; that is, a situation in which the pre-fibre action potential, itself, is the electromotive force for the current of the p.s.p. The diverse transynaptic effects obtained by applying depolarizing or hyperpolarizing currents to pre- or post-fibre are simply explained by the synaptic-rectifier hypothesis, while mechanisms based upon the release of transmitter substances do not provide an acceptable alternative. For example, it is difficult to see how the experimentally produced transfer of hyperpolarizations from post- to pre-fibre could be explained in terms of a chemical transmitter mechanism. On the other hand, on an 'electrical' hypothesis this result follows quantitatively from the rectifier properties of the synapse as determined in an entirely different experimental situation, namely by orthodromic depolarizations.

The fact that transmission at the GMS's is usually strictly one-way, an antidromic spike giving rise to a negligible effect in the pre-fibre, is accounted for by the rectifier properties of the synapse. A spike in the post-fibre represents a large, positive increase in V_{post}, so that V_s becomes negative ($V_s = V_{pre} - V_{post}$) and the intensity of the synaptic current is very low (see Text-fig. 16). While it is clear that the rectifier ensures one-way transmission, the fact that the pre-fibre is larger than the post-fibre must also contribute to the 'asymmetry' of transmission. In the absence of synaptic rectification the ratio of the attenuation factors (α_0 and α_a) in the orthodromic and antidromic cases is given by,

$$\alpha_a/\alpha_0 = \{(R_{pre} + S)/R_{pre}\}/\{(R_{post} + S)/R_{post}\}. \quad (1)$$

In a typical situation, in which $R_{post}/R_{pre} = 5$ and taking $S = R_{post}$, $\alpha_a/\alpha_0 = 3$. This was approximately the value found for one of the six cases considered above in the section on antidromic nerve-impulse transmission, and in this case there was no apparent synaptic rectification. At the other five junctions, at which the rectifier was operating, α_a/α_0 was greater than fifty. It should be pointed out that while equation (1) is appropriate for steady potential changes at a non-rectifying synapse, it takes no account of differences in the membrane time constants of the two fibres. Nevertheless, this comparison gives a very rough measure of the importance of the rectifier in blocking antidromic effects. In its presence the pre-fibre potentials accompanying an antidromic spike in the post-fibre are negligible, but if the current–voltage characteristic for the synapse were linear, potential changes of the order of 10 mV (i.e. about one-third of the orthodromic p.s.p.) might be expected. Although it is unlikely that one-to-one antidromic transmission would occur in this situation, rectification does prevent the transmission of appreciable local potential changes in the 'wrong' direction. The preceding discussion concerned synapses at which lateral giant axons were the pre-fibres. At the medial GMS's (i.e. between medial and motor giant fibres) the pre- and post-fibres are more nearly of equal size. In the absence of the synaptic rectifier, it seems likely that antidromic effects would be larger than at the lateral GMS's; but no experimental evidence on this point has been obtained.

An alternative way of discussing the functional significance of the rectifier is to consider it as a mechanism for providing a low synaptic resistance to current flow in the forward direction, rather than as a device for blocking antidromic effects. For example, in the experiment of Fig. 15 the resistance of the synapse was about $5 \cdot 5 \times 10^5 \, \Omega$ when V_s was 25 mV ($S = V_s/I_s$). If the area of the 'synaptic membrane' is taken as 10^{-4} cm^2 (see Methods) then the resistance × unit area of this membrane was 55 Ω cm^2 at this value of 'membrane potential'. In the same experiment the resistance × unit area of the pre-fibre membrane, for a uniform depolarization of 25 mV, was found to be 390 Ω cm^2, using the method of Cole & Curtis (1941). Thus at this value of 'membrane' potential the resistance of the 'synaptic membrane' was about one-seventh that of a similar area of axon membrane.

Although the 'synaptic membrane' seems to be composed of contributions from the two axons, it behaves like a unit structure to the extent that the current traversing it appears to depend only on the p.d. across it. In the above model of the synapse (Figs. 10 and 11) it was assumed that there was no separation between the component membranes; or, if any existed, that the shunt resistance, R_e, between this space and extracellular fluid was high. This was the simplest model and the easiest to analyse. It must be noted, however, that R_e can be given relatively low values, approaching the order of R_{post}, without changing conditions seriously. For example, if R_e is given the same value as R_{post} and is considered to connect the centre of the synaptic resistance to the external fluid (and taking $R_{post}/R_{pre} = 4$), then for a given V_s the apparent synaptic current in the orthdromic situation would be about 20% less than that calculated for the antidromic. In several of the four-micro-electrode experiments, discrepancies of at least this size and direction were found. While the disparities were assumed to arise from errors due to spatial decrement of the electrotonic potentials, it is also possible that a finite value of R_e was a contributory factor. The presence of R_e would reduce the efficiency of transmission, but the effect could readily be compensated by a reduction in the synaptic-membrane resistance, S.

Studies of the structure of the GMS's of the crayfish (*Cambarus*) have been made histologically and with the electron microscope (Robertson, 1952, 1953, 1955) and also provide support for the idea that little separation may exist between the components of the 'synaptic membrane'. Despite both pre- and post-fibres having Schwann-cell and connective-tissue sheaths, the membranes bounding the axoplasms come very close to one another at numerous places in the junctional region. The juxtaposition is brought about by short processes or tubular extensions of post-fibre membrane which penetrate the sheaths of both fibres. The ends of these processes spread out in the space between Schwann cell and pre-fibre axon-membrane, thus making patches in which the two fibre membranes are close together. One of these processes is shown in Pl. 1, a phase-contrast photomicrograph of a transverse section of *Astacus* abdominal nerve cord. The giant motor fibre, labelled 'post', is seen in oblique section crossing over the lateral giant pre-fibre. The arrow indicates a process, about 2–3 μ in diameter, extending from post- to pre-fibre; but the terminal expansion of the process cannot be seen. The section also includes the medial giant fibre of that side and three of the smaller motor axons which accompany the post-fibre in the third root. The disparity between the sizes of lateral and medial pre-fibres is greater than usual.

Electron micrographs show that the juxtaposition of the two fibre membranes can be a very close one. In Fig. 1 of Robertson (1955) the axoplasms of a medial giant pre-fibre and of a post-fibre process are shown separated by a structure which varies from about 150–300 Å in thickness, and which consists of the axon-membranes of the two fibres. Since the individual components of this double membrane are each about 75 Å thick (Robertson, personal communication) the space between them must vary in width from a negligible amount to about 150 Å. It is not known whether the negligible separation of the components, at places in the synaptic membrane, is an artifact.

The synapses in the stellate ganglion of the squid, between second- and third-order giant fibres, morphologically resemble the crayfish GMS's. They were the first example of this type of junction in which giant axons are synaptically connected by processes sent out from post- to pre-fibre (Young, 1939). Despite the anatomical resemblance the mechanism of transmission at the two synapses is apparently different. Bullock & Hagiwara (1957) and Hagiwara & Tasaki (1958) have concluded that transmission across the squid synapse is not brought about by local-circuit current crossing the junction, but by means of a special transmitter substance. Among other evidence supporting this conclusion was the presence of a long synaptic delay, so that the pre-fibre spike could be completed before the onset of the p.s.p. This is in contrast to the very short delays which have been observed at the crayfish GMS and which are compatible with the 'electrical' theory. The fact that the two types of synapse seem to differ in their mechanism of transmission, despite their anatomical similarity, suggests that any structural difference occurs at a finer level of organization. For example, one would like to know more exactly the extent of the space between pre- and post-fibre membranes at the two junctions. For maximum efficiency, at an 'electrical' synapse, the separation would be negligible; but at a 'chemical' junction the space would have to have a relatively low resistance to the external medium. In his earlier studies, Robertson (1952, 1953) also made electron micrographs of the squid giant synapses and confirmed that processes from the post-fibre brought post-synaptic membrane in close apposition to pre-fibre membrane. With the techniques then available, however, it was not possible to determine accurately the size of this space and additional work on this point is needed.

It will be recalled that each post-fibre also makes synaptic connexion with the two medial giant axons (see Methods). Although in most of our experiments the lateral giant pre-fibres were used, in a few cases the GMS's between motor and medial giant axons have also been studied. These junctions have been found to resemble the lateral giant synapses, for electrotonic potentials traversed them but, again, only appreciably when the pre-fibre was depolarized or the post-fibre hyperpolarized. A complete analysis (as in Text-fig. 15) has not been made; but in experiments on five ipsilateral and one contralateral medial GMS's results analogous to those of Text-fig. 9 have been found. In addition, at one of the ipsilateral junctions, a second micro-electrode was inserted into the post-fibre and a curve similar to Text-fig. 13*b* was obtained.

The effects of various pharmacological agents on the GMS of *Cambarus* have been studied by Wiersma & Shalleck (1947, 1948; Schalleck & Wiersma, 1948). Nicotine and related compoounds were among the few substances which had an effect in relatively low concentrations. Facilitation of transmission usually occurred with 10^{-6} g/ml., inhibition with more concentrated solutions. We have made a few preliminary tests, recording the p.s.p. intracellularly, to see if these results could also be obtained with *Astacus*; but only small and variable effects have so far been observed, using nicotine concentrations up to 10^{-4} g/ml. There are a number of ways in which a drug might affect transmission. A reduction in the size of the pre-fibre spike, a lowering of the post-fibre membrane resistance, an increase in post-fibre threshold, or a hyperpolarization or large depolarization of the post-fibre, as well as a direct effect on the 'synaptic membrane', could all serve to bring the p.s.p. below firing level. In addition, nicotine might produce its action by altering the spontaneous activity at other synapses on the post-fibre (Furshpan & Potter, 1959). It is clear that further work is needed to determine the site of action of nicotine in *Cambarus* and to confirm its lack of effect in *Astacus*.

In most of our experiments, the post-fibres were considerably affected by the experimental procedure. Many of them failed to show all-or-nothing action potentials, but gave only graded, local responses to supraliminal depolarizations (e.g. Text-fig. 12*d*). While the absence of a conducted post-fibre spike prevented testing for antidromic impulse transmission, it does not seem to have affected the main results obtained with steady current pulses. In those experiments (e.g. Text-fig. 7*a*) in which the post-fibre did conduct action potentials, the results were essentially the same as those obtained from less satisfactory preparations. Furthermore, the condition of the synapse could apparently vary independently of the condition of the post-fibre. For example, in the experiment of Text-figs. 12–16 the post-fibre gave only local, graded spikes yet the synapse appeared to work well. A pre-fibre spike (90 mV) gave rise to a potential change in the post-fibre greater than 40 mV. Since the membrane potential level at which an active response first appeared was about 30 mV, the post-synaptic potential change must have consisted partly of local response; and one-to-one transmission would presumably have taken place if the action potential mechanism had been operating properly.

The GMS's have several properties which result directly from the presence of the synaptic rectifier mechanism, but are also found at junctions with 'chemical' transmission. These properties cannot, therefore, be used to distinguish this type of 'electrical' synapse from the 'chemical' junctions: such are (1) the almost complete absence of pre-fibre potential change during an antidromic post-fibre spike; (2) an apparent decrease in post-fibre membrane resistance during transmission; (3) monophasic p.s.p.'s; and (4) a high degree of dependence of the size of the p.s.p. on the level of post-fibre membrane potential. The first point has already been discussed. The second, the apparent fall of R_{post} during a p.s.p., has not been demonstrated directly, but its presence is a necessary concomitant of the decrease in synaptic resistance. $S+R_{pre}$ is a shunt in parallel with R_{post} (see Text-fig. 11). Since S decreases to the order of R_{post} during transmission and R_{pre} is very small during pre-fibre activity, $S+R_{pre}$ will also be of the same order as R_{post}. The apparent value of R_{post} measured in the junctional region during transmission probably falls, therefore, to about one half of its resting level. The monophasic p.s.p.'s at the GMS come about because the pre-fibre spike, which provides the e.m.f., is monophasic. In this respect the GMS differs (i) from the situation of electrical interaction between adjacent axons in which the excitability changes in the 'post-fibre' are triphasic (Katz & Schmitt, 1940) and (ii) from the now discarded models for electrical transmission in which the p.s.p.'s were considered to be diphasic (e.g. Eccles, 1946). A more appropriate model for the GMS is a blocked nerve fibre in which the potential changes beyond the block are monophasic (Hodgkin, 1937). Concerning the fourth point, it has been observed in several experiments that the p.s.p. amplitude can be varied over a wide range by displacing post-fibre membrane potential. Depolarizations reduced the size of the p.s.p.'s while hyperpolarizations increased it. In one experiment the p.s.p., which was 17 mV when the steady level of post-fibre membrane potential was at the resting value ($V_{post}=0$), varied from 5 to 28 mV as V_{post} was altered over the range $+37$ to -40 mV. This effect results from the dependence of V_s and S on V_{post} ($V_s=V_{pre}-V_{post}$ and $S=f(V_s)$).

During post-fibre depolarizations, for example, V_s becomes less positive and, owing to the synaptic rectifier characteristic, S increases. Both factors thus cause a reduction in I_s ($=V_s/S$) and, therefore, in the p.s.p.'s. With this type of mechanism it would not be possible, however, to reverse the sign of the p.s.p. and in this respect the GMS differs from 'chemical' synapses (e.g. del Castillo & Katz, 1955; Burke & Ginsborg, 1956; also see Furshpan & Potter, 1959).

The above experiments have shown that pre-fibre depolarizations and post-fibre hyperpolarizations give rise to equal synaptic current provided the same p.d. across the junction is brought about. If the synapse behaves like a passive circuit element, this dependence of I_s on V_s should also hold when a given value of V_s is established at various absolute levels of the fibre membrane potentials. This has been tested in a few preliminary experiments, simultaneously changing V_{pre} and V_{post} by combined application of current to both fibres; and the expected relationship between I_s and V_s was found.

SUMMARY

1. The mechanism of impulse transmission has been studied at the giant motor synapses of the crayfish, by inserting one or two micro-electrodes into both pre- and post-synaptic axons. These fibres could be readily recognized by their distinctive physiological characteristics. The distance of the pre-fibre electrodes from the synapse was usually 10–20% of the characteristic length of that fibre. At least one of the post-fibre electrodes was usually in the immediate region of the junction.

2. With a recording electrode in each fibre it was found, during orthodromic nerve-impulse transmission, that the delay between the foot of the pre-fibre spike and the beginning of the p.s.p. was very small (usually about 0·1 msec).

3. An antidromic nerve impulse in the post-fibre gave rise to only a minute potential change in the pre-fibre (usually less than 0·5 mV).

4. Tests were made as to whether electrotonic potentials could cross the synapse, and it was found that even subthreshold depolarizations of the pre-fibre were accompanied by appreciable, but smaller, depolarizations of the post-fibre. Pre-fibre hyperpolarizations, however, gave rise to negligible changes in post-fibre potential. When the pulses of current were applied to the post-fibre, the results were apparently reversed; for now hyperpolarizations of the post-fibre gave rise to appreciable pre-fibre hyperpolarizations, whereas depolarizations were accompanied by only negligible transynaptic potential changes.

5. These seemingly diverse results could be explained by a simple hypothesis: electrotonic current readily flows across the junction, but the 'synaptic

membrane' is a rectifier allowing positive current to cross only in the direction from pre- to post-fibre.

6. In order to test this hypothesis, the current–voltage characteristic of the synaptic rectifier needed to account for the observed results was constructed. In fact, two such curves were made, one to account for depolarizations traversing the junction from pre- to post-fibre, and the other for hyperpolarizations crossing in the opposite direction. The two curves were found to be identical and the two situations were thus accounted for by the same mechanism, namely, the 'synaptic rectifier'. The results also showed that current flow across the synapse is determined simply by the p.d. across it (i.e. the difference between the internal potentials of pre- and post-fibre). Alternatively, 'chemical' mechanisms cannot adequately account for the transfer of hyperpolarizations from post- to pre-fibre.

7. The 'synaptic rectifier' is oriented in the right direction to allow local currents associated with a pre-fibre action potential to stimulate the post-fibre. It blocks oppositely directed currents, accompanying a post-fibre spike, and thus accounts for the absence of antidromic transmission. There is some evidence that the synaptic current that accompanies a pre-fibre spike is adequate to account for normal transmission.

8. The functional significance and structure of the synapse was discussed. As well as blocking antidromic effects, the 'synaptic rectifier' provides a low resistance for transmission in the orthodromic direction. In one case the resistance of the 'synaptic membrane' was estimated to be about one-seventh that of a comparable area of pre-fibre membrane with the same p.d. across it.

It is a pleasure to thank Professor B. Katz for suggesting the crayfish synapses to us as an object of study and for continual advice and encouragement during the course of the work. We also wish to acknowledge our debt to Professor S. W. Kuffler, Dr B. Ginsborg, and Dr J. del Castillo for their helpful suggestions and comments; and to thank J. L. Parkinson, Audrey M. Paintin and K. S. Copeland for frequent assistance. Financial support was received from the National Science Foundation and the U.S. Public Health Service.

REFERENCES

Bullock, T. H. & Hagiwara, S. (1957). Intracellular recording from the giant synapse of the squid. *J. gen. Physiol.* **20**, 565–578.

Burke, W. & Ginsborg, B. L. (1956). The action of the neuromuscular transmitter on the slow fibre membrane. *J. Physiol.* **132**, 599–610.

Cole, K. S. & Curtis, H. J. (1941). Membrane potential of the squid giant axon during current flow. *J. gen. Physiol.* **24**, 551–563.

del Castillo, J. & Katz, B. (1954). Changes in end-plate activity produced by pre-synaptic polarization. *J. Physiol.* **124**, 586–604.

del Castillo, J. & Katz, B. (1955). Local activity at a depolarized nerve-muscle junction. *J. Physiol.* **128**, 396–411.

del Castillo, J. & Katz, B. (1956). Biophysical aspects of neuro-muscular transmission. *Progr. Biophys.* **6**, 121–170.

Eccles, J. C. (1946). An electrical hypothesis of synaptic and neuro-muscular transmission. *Ann. N.Y. Acad. Sci.* **47**, 429–455.

Eccles, J. C. (1957). *The Physiology of Nerve Cells.* Baltimore: Johns Hopkins.

FATT, P. (1954). Biophysics of junctional transmission. *Physiol. Rev.* **34**, 674–710.

FATT, P. & KATZ, B. (1951). An analysis of the end-plate potential recorded with an intracellular electrode. *J. Physiol.* **115**, 320–370.

FURSHPAN, E. J. & POTTER, D. D. (1957). Mechanism of nerve-impulse transmission at a crayfish synapse. *Nature, Lond.*, **180**, 342–343.

FURSHPAN, E. J. & POTTER, D. D. (1959). Slow post-synaptic potentials recorded from the giant motor fibre of the crayfish. *J. Physiol.* **145**, 326–335.

GRUNDFEST, H. (1957). The mechanisms of discharge of the electric organs in relation to general and comparative physiology. *Progr. Biophys.* **7**, 1–85.

HAGIWARA, S. & TASAKI, I. (1958). A study of the mechanism of impulse transmission across the giant synapse of the squid. *J. Physiol.* **143**, 114–137.

HARDY, W. B. (1894). On some histological features and physiological properties of the post-oesophageal nerve cord of the Crustacea. *Phil. Trans.* B, Part I. **185**, 83–117.

HODGKIN, A. L. (1937). Evidence for electrical transmission in nerve. Part I. *J. Physiol.* **90**, 183–210.

JOHNSON, G. E. (1924). Giant nerve fibres in crustaceans with special reference to *Cambarus* and *Palaemonetes*. *J. comp. Neurol.* **36**, 323–373.

KAO, C. Y. & GRUNDFEST, H. (1956). Conductile and integrative functions of crayfish giant axons. *Fed. Proc.* **15**, 104.

KATZ, B. & SCHMITT, O. H. (1940). Electrical interaction between two adjacent nerve fibres. *J. Physiol.* **97**, 471–488.

NASTUK, W. L. & HODGKIN, A. L. (1950). The electrical activity of single muscle fibres. *J. cell. comp. Physiol.* **35**, 39–73.

ROBERTSON, J. D. (1952). Ultrastructure of an invertebrate synapse. *Thesis*, Massachusetts Institute of Technology: Cambridge, Mass., U.S.A.

ROBERTSON, J. D. (1953). Ultrastructure of two invertebrate synapses. *Proc. Soc. exp. Biol., N.Y.*, **82**, 219–223.

ROBERTSON, J. D. (1955). Recent electron microscope observations on the ultrastructure of the crayfish median-to-motor giant synapse. *Exp. Cell Res.* **8**, 226–229.

SCHALLECK, W. & WIERSMA, C. A. G. (1948). The influence of various drugs on a crustacean synapse. *J. cell. comp. Physiol.* **31**, 35–47.

STOUGH, H. B. (1926). Giant nerve fibres of the earthworm. *J. comp. Neurol.* **40**, 409–443.

VAN HARREVELD, A. (1936). A physiological solution for fresh-water crustaceans. *Proc. Soc. exp. Biol., N.Y.*, **34**, 428–432.

VOM RATH, O. (1895). Zur Conservirungstechnik. *Anat. Anzeig.* **11**, 280–288.

WIERSMA, C. A. G. (1947). Giant nerve fibre system of the crayfish. A contribution to comparative physiology of synapse. *J. Neurophysiol.* **10**, 23–38.

WIERSMA, C. A. G. & SCHALLECK, W. (1947). Potentials from motor roots of the crustacean centra nervous system. *J. Neurophysiol.* **10**, 323–329.

WIERSMA, C. A. G. & SCHALLECK, W. (1948). Influence of drugs on response of a crustacean synapse to pre-ganglionic stimulation. *J. Neurophysiol.* **11**, 491–496.

YOUNG, J. Z. (1939). Fused neurons and synaptic contacts in the giant nerve fibres of cephalopods. *Phil. Trans.* **229**, 465–503.

EXPLANATION OF PLATE

A transverse section of crayfish nerve cord through the region of a lateral GMS. Phase-contrast. The arrow (proc) indicates a post-fibre process which extends to the lateral giant pre-fibre. The medial giant fibre (medial) of the same side is seen on the right, with three of the smaller fibres of the third root (s.m.f.) on the left. Modified vom Rath's (1895) fixative.

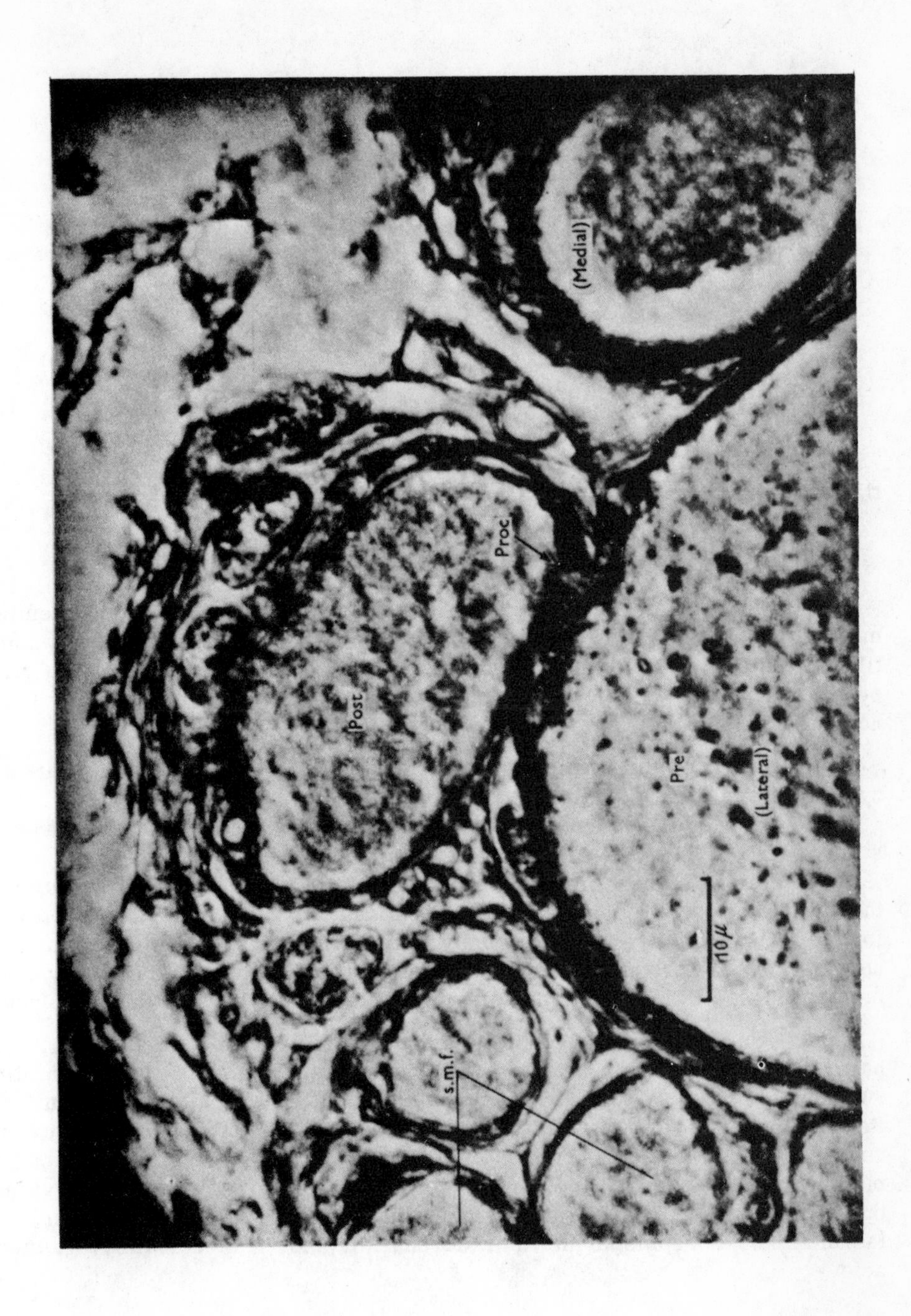

(*Facing p.* 325)

B. IDENTIFICATION OF NEUROTRANSMITTERS

Although the majority of synapses studied seem to operate by chemical mechanisms, in very few has the transmitter chemical been rigorously identified. Criteria establishing that a given chemical is a transmitter in a given synapse have been developed gradually in response to new techniques. These now include the following:

1. Presence of the chemical in the presynaptic neuron together with enzyme systems for its synthesis.
2. Release of the chemical from the presynaptic neurons in response to stimulation.
3. Imitation of the effect on the postsynaptic structure of stimulation of the presynaptic neuron by the pure proposed transmitter applied in physiological concentrations to the postsynaptic structure. Such imitation most notably includes producing identical permeability changes in the postsynaptic membrane (see Part C) and similar interactions with pharmacological agents.

The second criterion is often the most difficult to demonstrate in practice because most synapses have efficient means for removing the transmitter substance, such as hydrolyzing enzymes or active absorption mechanisms. Release of suspected transmitters in response to stimulation is, however, necessary for decisive proof. The chemicals which have come close to meeting these criteria are limited to acetylcholine (ACh), gamma-aminobutyric acid (GABA), 5-hydroxytryptamine (5-HT, serotonin), epinephrine (adrenaline), norepinephrine (nor-adrenalin), and, less certainly, glutamic acid.

1. ACETYLCHOLINE

From the enormous literature concerning the role of acetylcholine (ACh) in transmission we include three historic papers addressed specifically to the point of identifying acetylcholine (or *Vagusstoff*) as a neurotransmitter and collecting it after stimulation of nerves. The series of papers by Otto Loewi, of which the first (**Loewi, 1921**) and tenth (**Loewi and Navratil, 1926**) are reproduced here, constitute the first proof of chemical transmission and the first evidence that acetylcholine is an inhibitory transmitter at the vertebrate heart (see Hutter and Trautwein, 1956, for a more recent account of the mode of action of acetylcholine at the heart). **Dale, Feldberg, and Vogt (1936)** first published direct evidence that acetylcholine is the excitatory transmitter at vertebrate nerve-striated muscle junctions (synapses).

Whether acetylcholine has an inhibitory or excitatory effect depends on the postsynaptic response. In the nervous system of *Aplysia*, a mollusc, there is a well-documented example of an identified neuron which excites a second identified neuron and inhibits a third one. Acetylcholine is the transmitter in each case (Kandel, Frazier, and Coggeshall, 1967).

16

Pflügers Arch. (1921), 189, 239–242

(Aus dem pharmakologischen Institut der Universität Graz.)

Über humorale Übertragbarkeit der Herznervenwirkung.

I. Mitteilung.

Von

O. Loewi.

(Ausgeführt mit Unterstützung der Fürst Liechtenstein-Spende.)

(Mit 5 Textabbildungen.)

(*Eingegangen am 20. März 1921.*)

Der Mechanismus der Wirkung der Nervenreizung ist unbekannt. Mit Rücksicht darauf, daß gewisse Pharmaka fast identisch wirken wie die Reizung bestimmter Nerven, liegt die Möglichkeit vor, daß unter dem Einfluß der Nervreizung Stoffe gebildet werden, die ihrerseits erst den Reizerfolg herbeiführen. Unter den Bedingungen, wie sie beim Arbeiten am ganzen Tier gegeben sind, ist es wohl aussichtslos, diese Frage zu entscheiden. Die einzige Möglichkeit gibt hier der Versuch am isolierten Organ. Von Arbeiten in dieser Richtung liegt eigentlich nur die von Howell vor, wonach die Vaguswirkung durch eine Abscheidung von Kalium während der Reizung bedingt sein soll, doch wurden seine Versuchsergebnisse widerlegt.

Methode.

Ich wählte das Kaltblüterherz, weil hier bei entsprechender Versuchsanordnung die Möglichkeit gegeben ist, infolge Reizung allenfalls entstehende Stoffe in einer geringen Menge von Füllflüssigkeit sich anreichern zu lassen und so nachweisbar zu machen.

Als Methode wählte ich die bekannte Herzkanülenmethode nach Straub mit der Modifikation, daß der linke Vagus mit herauspräpariert am Sinus hängen gelassen und über eine Elektrode gebrückt wurde. Wird der Nerv feucht gehalten und die Reizung mitunter, wenn auch nur kurz unterbrochen, so bleibt er oft viele Stunden lang erregbar.

Die Ringersche Flüssigkeit enthielt 0,6% NaCl, 0,01% KCl, 0,02% $CaCl_2 + 6\ H_2O$, 0,05% $NaHCO_3$. Es wurde dauernd Sauerstoff eingeleitet. Die Versuche wurden an meist frischgefangenen Esculenten (10 Versuche), Temporarien (4 Versuche) und gemeinen Kröten (4 Versuche) im Februar und März ausgeführt.

Versuche.

Sämtliche Versuche fielen gleichsinnig aus.

1. Versuche mit hemmendem Vagusreizerfolg.

Nachdem das Herz zur Entfernung der Blutreste einigemale mit Ringer war ausgespült worden, wurde eine bestimmte Zeit hindurch der Ringer nicht gewechselt und am Ende dieser Periode (Normalperiode) abpipettiert und aufbewahrt. Dann wurde während einer gleichlangen Periode der Vagus mit kurzen Unterbrechungen faradisch gereizt. Der Erfolg war bei den Fröschen der bekannte negativ ino- und chronotrope. Auch die Füllung der Vagusreizperiode wurde abpipettiert und aufgehoben. Nachdem das Herz von der Vagusreizung sich wieder erholt hatte, wurde es abwechselnd mit den Füllungen der beiden Perioden beschickt: die Füllung der Normalperiode wirkte nicht anders als wie frischer Ringer, war also ohne irgend einen Einfluß (s. Abb. 1). Wurde aber der Ringer der Vagusreizperiode eingefüllt,

Abb. 1. Esculenta. 1. Ringer. 2. Ringer aus 15' Vagusreizperiode. 3. Ringer aus 15' Normalperiode. 4. +0,1 mg Atropin.

so trat regelmäßig eine deutliche negativ inotrope (Abb. 1 u. 2) mitunter dazu noch eine negativ chronotrope (Abb. 2) Wirkung ein. Letztere war kaum zu erwarten gewesen, da der Sinus bei der gewählten Versuchsanordnung mit der Füllflüssigkeit kaum in Berührung kommt. Abb. 1 zeigt, daß die Wirkung durch Atropin prompt behoben wird.

2. Versuche mit förderndem Vagusreizerfolg.

Der naheliegende Gedanke, zu prüfen, ob etwa am atropinisierten Froschherzen Vagusreizung infolge der Beimischung von durch Atropin nicht lähmbaren Acceleransfasern zur Abscheidung von „Förderungssubstanz" führe, konnte bislang mangels Froschmaterials nicht durchgeführt werden. So war ich auf die Verwendung von Kröten angewiesen. Diese reagieren in der jetzigen Jahreszeit von vorneherein auf Vagusreizung mit hochgradiger Steigerung der Pulsgröße und -Frequenz (Abb. 3a). Die Versuche wurden ausgeführt wie die oben beschriebenen. Abb. 3b. zeigt, daß während der Inhalt der Normalperiode ganz einflußlos ist, der der Acceleransreizperiode zu einer hochgradigen Vergrößerung des Pulsvolums führt. Dabei ist sehr erwähnenswert, daß der Inhalt der Acceleransreizperiode gewonnen wurde $3^1/_2$ Stunden nach

Abb. 3a. Kröte I. 1. Ringer. 2. Vagusreizung.

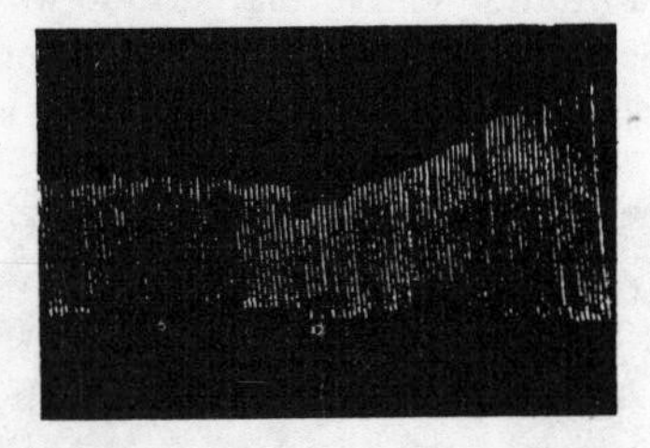

Abb. 3b. 1. Ringer. 2. Ringer aus 25' Normalperiode. 3. Ringer aus 25' Vagusreizperiode.

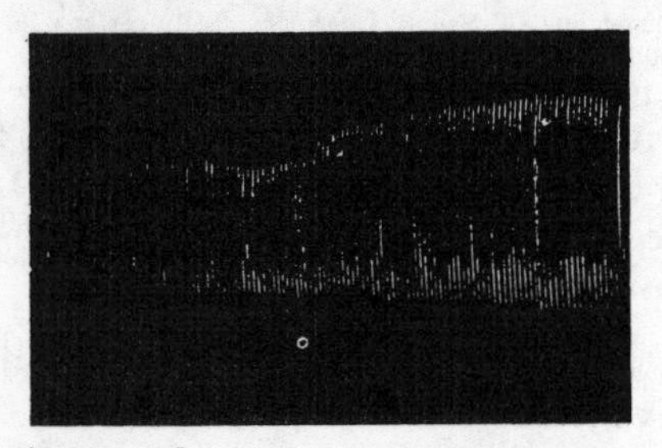

Abb. 3c. Kröte II. 1. Ringer aus 25' Normalperiode von Kröte I. 2. Ringer aus 25' Vagusreizperiode von Kröte I.

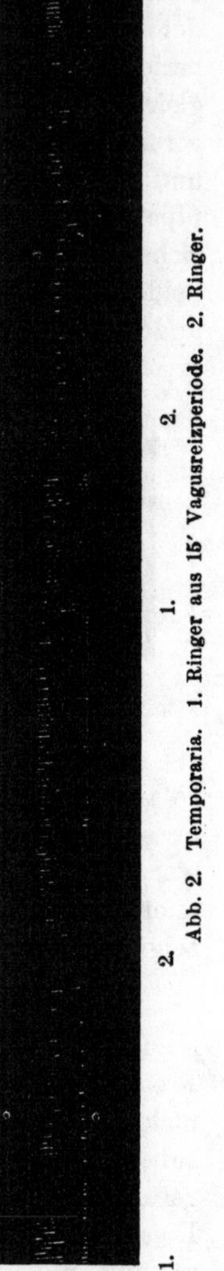

Abb. 2. Temporaria. 1. Ringer aus 15' Vagusreizperiode. 2. Ringer.

Versuchsbeginn, nachdem also das Herz unzählige Male ausgewaschen und der Accelerans 1 Stunde schon gereizt worden war. Mit Rücksicht auf diese lange Vorgeschichte des Herzens schien es wünschenswert, den Inhalt auch an einem ganz frischen Krötenherz auf seine Wirksamkeit zu prüfen. Abb. 3c zeigt, daß sie beim frischen Herzen annähernd die gleiche ist, wie beim vorbehandelten.

Besprechung der Ergebnisse.

Die Versuche lehren, daß unter dem Einfluß der Reizung der herzhemmenden und -fördernden Nerven Stoffe vom gleichen Wirkungscharakter, wie er der Nervreizung eignet, in der Füllflüssigkeit des Herzens nachweisbar werden. Es werden also unter dem Einfluß der Nervenreizung diese Stoffe gebildet oder abgespalten oder sie waren vorgebildet und die Zellen werden erst dafür durchgängig. Was die Bedeutung dieser Stoffe anbetrifft, so liegen zwei Möglichkeiten vor: einmal könnten sie unabhängig von der Art der mechanischen Herztätigkeit direkt unter dem Einfluß der Nervenreizung entstehen und ihrerseits die spezifische Reaktion des Herzens auf den Nervenreiz, der danach nur mittelbar wirksam wäre, auslösen. Wenn ihre Wirkung bei der gewählten Versuchsanordnung quantitativ hinter der der Nervenreizung zurückbleibt, so darf das nicht Wunder nehmen, da anzunehmen ist, daß nur ein geringfügiger Anteil der in oder an der Zelle gebildeten bez. abgespaltnen Stoffe in die Flüssigkeit übergeht, andererseits diese eine hochgradige Verdünnung bewirkt. Zum anderen liegt die Möglichkeit vor, daß die Stoffe nur Produkte der durch die Nervenreizung ausgelösten besonderen Art der Herztätigkeit sind; in diesem Fall würden sie also gewissermaßen nur zufällig identisch wirken wie die Nervenreizung.

Was die Frage nach dem Charakter der Stoffe anbetrifft, so ist bis jetzt nur auszuschließen, daß es sich bei dem Vagusreizprodukt um Kalium handelt, da gesteigerte Kaliumwirkung durch das in unseren Versuchen wirksame Atropin sich nicht beheben läßt.

Sobald ich über das entsprechende Tiermaterial verfüge, beabsichtige ich sowohl die Frage nach der Art der Stoffe als andere, die sich im Anschluß an die mitgeteilten Versuche in großer Zahl aufdrängen, zu bearbeiten.

Pflügers Arch. (1921) 189, 239–242

(From the Pharmacological Institute of Graz University)

ON THE HUMORAL PROPAGATION OF CARDIAC NERVE ACTION

Communication I

by O. Loewi

(Supported by a grant from the Prince Lichtenstein Foundation)
(With 5 figures)
(Released March 20, 1921)

The mechanism by which neural stimulation works its effect is unknown. However, certain chemical agents mimic very closely the effects which follow the stimulation of certain nerves. In view of this fact, it is possible that when a nerve is stimulated substances are formed, which in turn are responsible for the observed effect of stimulation. This question cannot possibly be solved by studying the whole animal. One must instead focus experimentation on an isolated organ. The only study hitherto done in this fashion is by Howell, who claimed that the vagus effect is due to the release of potassium during stimulation. However, his results have been refuted.

Method

I chose to use the hearts of cold-blooded animals. In such hearts, given the proper experimental set-up, there is the following possibility. Any substances which might be formed as a result of stimulation can be concentrated within the small included volume of liquid, thereby rendering them detectable.

I used Straub's well-known heart canula method, with the modification that the left vagus was also dissected. It was left hanging from the sinus and placed over an electrode. The vagus remains excitable for many hours if it is kept humid and if stimulation is occasionally interrupted, even if only briefly.

The Ringer's solution contained 0.6% NaCl, 0.01% KCl, 0.02% $CaCl_2 + 6\ H_2O$, 0.05% $NaHCO_3$. Oxygen was constantly supplied. The experiments were carried out in the months of February and March on generally freshly caught frogs, *Rana esculenta* (10 experiments), *R. temporaria* (4 experiments) and common toads (4 experiments).

Experiments

All experiments yielded consistent results.

1. *Experiments in Which the Effect of Vagus Stimulation Was Inhibitory*

The heart was rinsed several times with Ringer's solution in order to remove blood remnants. Then it was left to stand in a fresh Ringer's solution for a fixed period of time (the normal period). At the end of this period the solution was removed by pipette and stored. The heart was now bathed in fresh Ringer's solution. Then, for a length of time equal to the above-mentioned normal period, the vagus was given intermittent Faradic stimulation. In frogs this resulted in the well-known negative inotropic and chronotropic effect. After the period of vagus stimulation the filling solution was removed by pipette and kept. The heart was allowed to recover from the vagus stimulation. Then the effect upon the heart of the filling solutions from the two periods was compared. The filling solution from the normal period had precisely the same effect as fresh Ringer's solution. That is, it had no effect whatsoever (see Fig. 1.). In contrast, when the Ringer's from the period of vagus stimulation was introduced, a clear negative inotropic reaction regularly occurred (Figs. 1 and 2). Moreover, now and then a negative chronotropic reaction also occurred (Fig. 2). This latter reaction was hardly expected, for in our experimental set-up the sinus barely comes into contact with the solution. Fig. 1 shows that the effect is promptly eliminated by atropine.

2. *Experiments in Which the Effect of Vagus Stimulation Was Excitatory*

The next step in the investigation was to examine the effect of vagus stimulation upon an atropinized heart. Would it bring about the release of an "enhancing substance" because of the admixture of unparalyzed accelerator nerve fibers in the vagus? Since frog material was in short supply I had to use toads. At this time of year toads react similarly to vagus stimulation. It markedly increases both the amplitude and the frequency of their pulse (Fig. 3a). The experiments were carried out in the same manner as those described above. Fig. 3b shows that the filling solution of the normal period has no effect. The filling solution from the period of accelerator nerve stimulation greatly increased the pulse volume. Now it should be stressed that in this experiment the filling solution from the stimulation period was obtained $3\frac{1}{2}$ hours after the beginning of the experiment. The heart had been washed numerous times and stimulated for a full hour. For this reason it seemed wise to test the stimulation period solution with a fresh toad heart. As Fig. 3c shows, the solution had roughly the same effect upon a fresh heart.

Discussion

The investigation shows that after stimulating heart-inhibitory and heart-excitatory nerves, substances were detected in the heart-filling liquid which have the same effect upon the heart as the nerves themselves. These substances must have been either formed or released under the influence of nerve stimulation, or they may have been preexisting and the heart cells were rendered permeable to them. As for the functional significance of these substances, two possibilities come to mind. They could be formed directly under the influence of nerve stimulation, independently of the type of mechanical activity of the heart, and in turn trigger the specific reaction of the heart to nerve stimulation, which afterwards would indirectly have an effect. It is hardly surprising that in our experimental set-up the effect of these substances upon the heart was quantitatively less than direct stimulation of the appropriate nerves. One can assume that only a small portion of the substances formed or released in or at the cell reaches the filling solution, where great dilution occurs. The second possibility is that the substances were merely by-products of a special kind of heart activity caused by nerve stimulation. In this case it would only be to some extent by chance that the substances found in the filling liquid had the same effect upon the heart as nerve stimulation.

As for the nature of these substances, at this point one can only rule out potassium as the product of vagus stimulation; for atropine, which was effective in our experiments, cannot eliminate an increase in activity due to potassium.

As soon as I have the necessary animal material I intend to investigate the nature of these substances as well as to investigate a great number of other questions which have arisen in the course of these studies.

FIGURE LEGENDS

Fig. 1. Esculenta. 1. Ringer's. 2. Ringer's from 15 minute period of vagus stimulation. 3. Ringer's from 15 minute normal period. 4. +0.1 mg atropine.

Fig. 2. Temporaria. 1. Ringer's from 15 minute period of vagus stimulation. 2. Ringer's.

Fig. 3a. Toad #1. 1. Ringer's. 2. Vagus stimulation.

Fig. 3b. 1. Ringer's. 2. Ringer's from 25 minute normal period. 3. Ringer's from 25 minute period of vagus stimulation.

Fig. 3c. Toad #2. 1. Ringer's from 25 minute normal period of toad #1. 2. Ringer's from 25 minute period of vagus stimulation of toad #1.

17

Pflügers Arch. (1926) **214**, 678–688

(Aus dem Pharmakologischen Institut der Universität Graz.)

Über humorale Übertragbarkeit der Herznervenwirkung.

X. Mitteilung.

Über das Schicksal des Vagusstoffs.

Von

O. Loewi und E. Navratil.

Mit 11 Textabbildungen.

(Eingegangen am 29. Juli 1926.)

Bekanntlich hört im Gegensatz zur Wirkung der Acceleransreizung die Wirkung der Vagusreizung mit oder kurz nach Schluß der letzteren auf. In völliger Analogie hierzu steht die von uns bereits in der 7. Mitteilung[1]) gebrachte Beobachtung, wonach die Wirkung ins Herz gebrachten Vagusstoffes[2]) sehr rasch vorübergeht[3]), während die eingebrachten Acceleransstoffes wesentlich länger dauert. Die vorliegende Mitteilung beschäftigt sich mit der Ursache des raschen Abklingens der Vagusstoff- und damit der Vagusreizwirkung.

I. Über Unwirksamwerden von Vagusstoff durch Behandlung mit Herzextrakt.

Am nächsten lag der Gedanke, daß der Vagusstoff vom Herzen rasch zerstört werde. Nachdem die Füllung aus Vagusreizperioden auch nach 24 Stunden noch wirksam ist, also kein zerstörender Stoff aus dem Herzen in sie übergeht, waren wir genötigt zu untersuchen, ob etwa ein Extrakt aus dem Herzen selbst Vagusstoff zerstöre.

Zu diesem Behuf gingen wir bei unseren von Oktober bis April durchgeführten Versuchen folgendermaßen vor: Herzen von Esculenten, mitunter auch Kröten, wurden durch kurzes Waschen mit bicarbonatfreiem Ringer von Blut befreit, in der Reibschale mit Quarzsand zerrieben und mit bicarbonatfreier Ringerlösung aufgenommen. In der Regel benutzten wir für je ein im Durchschnitt 100—200 mg wiegendes Herz 2—5 ccm bicarbonatfreien Ringer. Dann wurde filtriert und das Filtrat auf seine Fähigkeit, Vagusstoff zu zerstören, geprüft. Zu diesem Behuf

---

[1]) Pflügers Arch. f. d. ges. Physiol. **206**, 135. 1924.

[2]) Mit Vagusstoff = V. St. bezeichnen wir der Kürze halber die vagusartig wirksamen Diffusate bzw. Extrakte aus dem Herzen.

[3]) Natürlich besteht eine Abhängigkeit von der Größe der eingebrachten Menge.

hatten wir nichts anderes zu tun, als das Filtrat einerseits unmittelbar nach seiner Herstellung, andererseits nach längerem Stehen bei Zimmertemperatur bzw. bei 37° ins Herz einzubringen und die Wirkung beider Proben miteinander zu vergleichen. Die Begründung dieser einfachen Versuchsanordnung ist darin gegeben, daß das Filtrat, wie wir bereits früher nachwiesen[1]), vagusstoffhaltig ist: es wirkt negativ inotrop; diese Wirkung ist atropinbehebbar[2]), sie ist durch Vagusstoff und nicht durch Cholin bedingt, denn einerseits ist dies nur in völlig unwirksamer Konzentration im Filtrat enthalten, andererseits ergibt die Acetylierung des Filtrats wesentlich weniger Cholin, als einer Cholinmenge von gleich starker Wirksamkeit wie der des Filtrats entsprechen würde[3]).

Das Ergebnis war ganz regelmäßig, daß die längere Zeit gestandene Probe wesentlich weniger negativ inotrop wirkt als die sofort nach der Herstellung untersuchte. Um nun Näheres über den Modus der Abnahme der Wirksamkeit zu erfahren, gingen wir in weiteren Versuchen folgendermaßen vor: das Filtrat wurde in zwei gleiche Teile geteilt; der eine wurde sogleich auf 56° gebracht und 20 Min. dabei gehalten[4]). Dann wurden beide Teile einige Zeit — zwischen $^1/_2$ und 3 Stunden — bei Zimmertemperatur bzw. bei 38° stehengelassen und nunmehr die Wirksamkeit beider Teile am Herzen vergleichend geprüft. Für das Ergebnis war es gleichgültig, ob wir die Filtrate selbst oder deren auf früher beschriebene Weise gewonnene alkoholische Extrakte prüften. Das Ergebnis war, daß in sämtlichen Versuchen die nicht erhitzte Probe durch das Stehenlassen ganz wesentlich an Wirksamkeit einbüßte.

Abb. 1 gibt das Ergebnis eines derartigen Versuches wieder. Hier war ein Extrakt (1 Herz : 30 Ringer) 3 Stunden bei 18° stehengelassen worden. Die Prüfung ergab, daß dieser Extrakt bei weiterer Verdünnung auf das 5fache (Abb. 1 bei 3) nicht mehr negativ inotrop wirkte[5]), im Gegensatz zu dem durch Erhitzen von vornherein inaktivierten (Abb. 1 bei 2).

Um festzustellen, ob der Unterschied in der Wirksamkeit der erhitzten und der nicht erhitzten Probe etwa bedingt sei durch eine durch das Erhitzen gesetzte physikalisch-chemische Änderung des Extraktes

---

[1]) Pflügers Arch. f. d. ges. Physiol. **208**, 694. 1925.

[2]) Nach Atropinisierung wirkt das Filtrat, woferne es genügend konzentriert ist, infolge der Anwesenheit des ebenfalls mit Wasser extrahierbaren Acceleransstoffes meist positiv inotrop. Vgl. hierzu S. 685.

[3]) Diese Versuchsanordnung ist nur im Winter durchführbar, da nur in dieser Jahreszeit das Filtrat regelmäßig und intensiv negativ inotrop wirkt. Ab Mai wirkt es in der Regel positiv inotrop; auf die Ursache dieser Saisondifferenz wird in einer späteren Mitteilung eingegangen werden.

[4]) In eigenen Versuchen stellten wir fest, daß dabei der anfänglich vorhanden gewesene Vagusstoff völlig erhalten bleibt.

[5]) Natürlich besagt das nicht, daß aller Vagusstoff zerstört worden sol, vielmehr gilt der Nachweis der völligen Zerstörung nur für den geprüften Verdünnungsgrad.

und dadurch veranlaßten Unterschied in der Nachweisbarkeit des Vagusstoffes, etwa infolge Änderung der Adsorptionsverhältnisse oder dergleichen, haben wir in weiteren Versuchen die ursprünglich nicht erhitzte Probe nach dem Stehenlassen ebenfalls 20 Min. auf 56° gebracht; die danach erfolgende vergleichende Prüfung führte zum gleichen Resultat wie die oben angeführten Versuche: die erst nach dem Stehenlassen erhitzte Probe war wesentlich weniger wirksam wie die sofort erhitzte. War auch von vornherein anzunehmen, daß das Unwirksamwerden durch das Stehenlassen mit Extrakt sich nicht nur auf den in ihm gerade enthaltenen Vagusstoff beschränken werde, haben wir doch in vielen Versuchen dem Extrakt von vornherein mittels Alkoholextraktion aus anderen Herzen gewonnenen Vagusstoff zugesetzt. Abb. 2 zeigt, daß nicht nur der im Extrakt vorhandene (Abb. 2a bei 2a), sondern auch der zugesetzte Vagusstoff (Abb. 2b bei 3a) durch das Stehenlassen mit dem Extrakt unwirksam geworden ist.

Abb. 1. Esculentenherz. 1. = Ringer. 2. = Herzextrakt inaktiviert. 3. = dasselbe nicht inaktiviert.

Deutet schon die Tatsache, daß durch Erhitzen auf 56° die Zerstörung des Vagusstoffes durch den Extrakt hintangehalten wird, mit großer

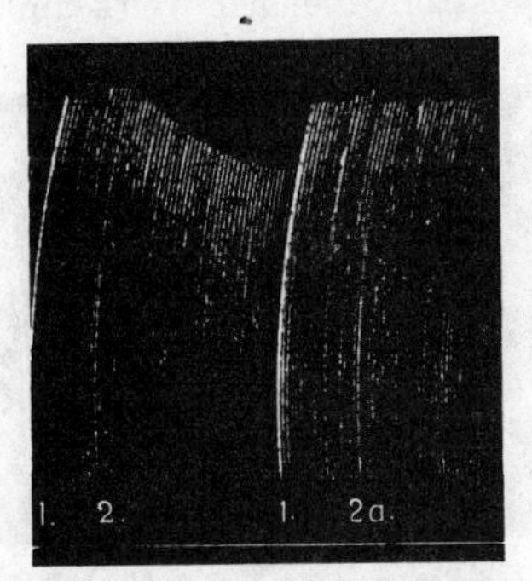

Abb. 2a. 1. = Ringer. 2. = Herzextrakt inaktiviert. 2a. = dasselbe nicht inaktiviert. Nach 2 Stunden Stehen bei 18° C.

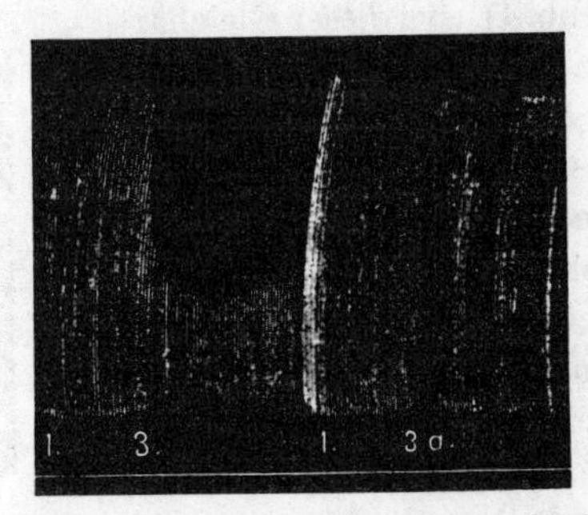

Abb. 2b. 1. = Ringer. 3. = Herzextrakt (aus Abb. 2a) inaktiviert + V.St. 3a. = dasselbe nicht inaktiviert. Nach 3 Stunden Stehen bei 18° C.

Wahrscheinlichkeit darauf hin, daß die Zerstörung auf fermentativem Weg vor sich geht, so wird diese Annahme noch weiter gestützt durch die Beobachtung, daß nicht nur durch Erhitzen auf 56°, sondern auch durch die Behandlung mit ultraviolettem[1]) bzw. Fluorescenzlicht[2]),

[1]) 10 Minuten lang.

[2]) Belichtung der mit Eosin 1 : 100 000 versetzten Probe durch 1 Std. hindurch.

Eingriffe, die die Wirksamkeit des Vagusstoffes selbst nicht beeinflussen, die vagusstoffzerstörende Wirkung des Extraktes hintangehalten wird. Als Illustration für ersteres diene Abb. 3.

Durch die vorliegenden Versuche haben wir den bereits vorhandenen Erkennungszeichen des Vagusstoffes — Atropinbehebbarkeit seiner Wirkung und geringerem Cholingehalt als einer gleichwirksamen Cholinmenge entspricht — ein weiteres hinzugefügt: nämlich Zerstörbarkeit durch Stehenlassen mit Herzextrakt.

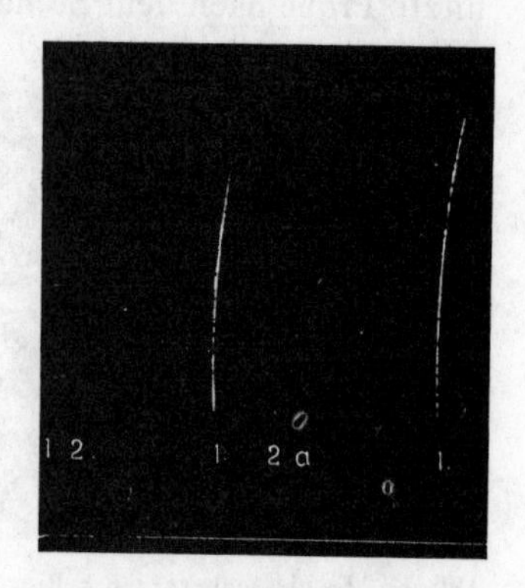

Abb. 3. 1. = Ringer. 2. = Herzextrakt (1:30) 30 Min. ultraviolettem Licht ausgesetzt. 2a = dasselbe nicht belichtet. 2. und 2a. standen 90 Min. bei 17° C.

II. Über Unwirksamwerden von Acetylcholin durch Behandlung mit Herzextrakt.

Wir haben bereits in der II. Mitteilung an die Möglichkeit gedacht, daß der Vagusstoff, von dem wir nachweisen konnten, daß er nicht Cholin selbst ist, ein sehr wirksamer Cholinester sei. Diese Annahme erfährt eine wesentliche Stütze durch folgendes: im Gegensatz zur Wirkung von ins Herz gebrachtem Cholin oder Muscarin[1]), die bekanntlich lange anhält[2]), geht die Wirkung von Acetylcholin gleich der des Vagusstoffes sehr schnell vorüber, und zwar in genau der gleichen Zeit wie die von Vagusstoff gleicher Wirksamkeit (Abb. 4). Bei dieser Gleichartigkeit des Wirkungsablaufes war es natürlich sehr wichtig festzustellen, ob auch die Ursache die gleiche sei. Tatsächlich zeigte sich, daß während unter der gleichen Bedingung Cholin- und Muscarinwirkung unverändert bleiben, das Acetylcholin ebenso wie der Vagusstoff seine Wirksamkeit verliert, wenn es längere Zeit mit nativem Herzextrakt steht, nicht dagegen, wenn letzterer durch Erhitzen, Fluorescenz- oder ultraviolettes Licht vorgängig inaktiviert worden war[3]). Zur

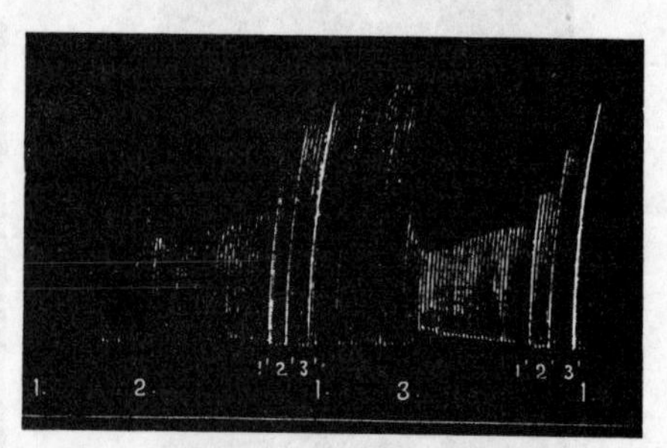

Abb. 4. 1. = Ringer. 2. = V.St. (Alkoholextrakt aus Herz.) 3. = Acetylcholin (1 : 100 Millionen).

[1]) Muscarin synthetic. Grubler.

[2]) Vgl. hierzu Abb. 2 der folgenden Mitteilung: Pflügers Arch. f. d. ges. Physiol. **214**, 690. 1926.

[3]) In den Versuchen mit Muscarin und Cholin wurde, um eine Trübung des Ergebnisses durch den im Extrakt vorhandenen V. St. auszuschalten, das Extrakt zunächst durch Stehenlassen in der Wärme völlig vagusstofffrei gemacht, und

Illustration diene Abb. 5; Abb. 5a zeigt die Abnahme der Wirkung nach 1stündigem Stehen der nicht inaktivierten Probe gegenüber der inaktivierten (20 Min. bei 56°); Abb. 5b gibt einen Ausschnitt aus dem zugehörigen Titrationsversuch wieder, der die absolute Größe der Wirkungsabnahme feststellen sollte; er zeigt, daß bereits nach einer halben Stunde die nichtinaktivierte Probe erst in einer Konzentration von $^1/_{150}$ so stark wirkte wie die inaktivierte in einer Konzentration von $^1/_{1000}$: es hatte also die nichtinaktivierte Probe nach einer halben Stunde um 25%, nach eineinhalb Stunden um 85% an Wirksamkeit eingebüßt.

Es sind also Wirkungsbild, Wirkungsablauf und die Ursache des letzteren bei Vagusstoff und Acetylcholin nicht nur qualitativ, sondern auch quantitativ völlig identisch.

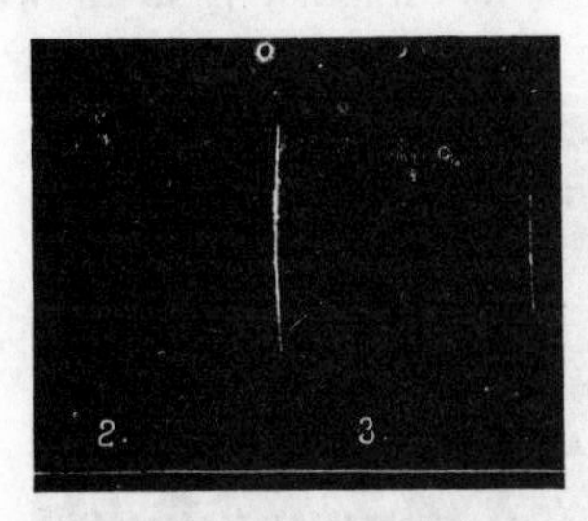

Abb. 5a. Ausgangsmaterial: H.E. (1:20) + Acetylcholin 1:100000. 2. = Obiges inaktiviert. Geprüft 1:1000. 3. = Dasselbe nicht inaktiviert.

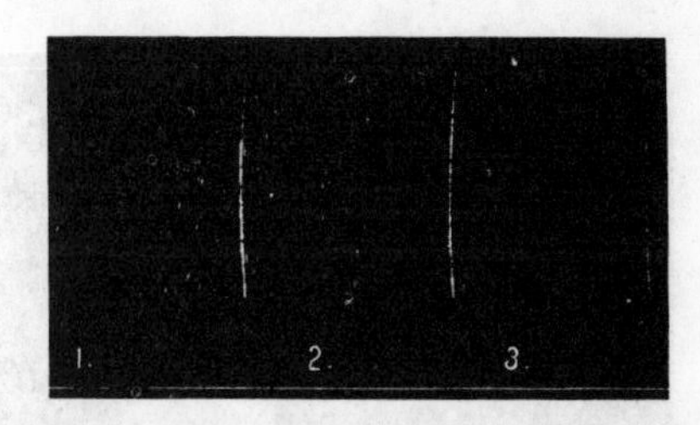

Abb. 5b. Ausgangsmaterial aus Abb. 5a. 1. = nichtinaktiviert nach 30 Min. Stehen. Verd. = 1 : 750. 2. = Inaktiviert nach 90 Min. Stehen. Verd. = 1 : 1000 (s. Abb. 5a). 3. = Nichtinaktiviert nach 90 Min. Stehen. Verd. = 1 : 150.

III. Über die Art der Zerstörung von Vagusstoff und Acetylcholin.

Was die Art der Wirkung des Herzextraktes auf Vagusstoff und Acetylcholin anbetrifft, so liegt es mit Rücksicht auf die Tatsache, daß Esterasen im Organismus vielfach nachgewiesen sind, am nächsten an eine Verseifung des Vagusstoffes und des Acetylcholins zu denken[1]). Der Beweis für das Statthaben einer Verseifung kann nur durch die Feststellung der Art der Spaltprodukte erbracht werden. Im vorliegenden Fall mußte also mindestens festgestellt werden, ob beim fermentativen Unwirksamwerden Cholin entsteht. Diese Frage konnten wir durch Benutzung von Vagusstoff als Ausgangsmaterial nicht eindeutig entscheiden, da zwar in früheren Versuchen nach Vagusreizung eine Zu-

erst jetzt der Muscarin- bzw. Cholinversuch angesetzt. Bei den Versuchen mit Acetylcholin erübrigt sich dies Vorgehen, da in den zur Prüfung der Acetylcholinwirkung nötigen Verdünnungen der vorhandene V. St. völlig unwirksam ist.

[1]) Der Glycerinextrakt des Herzens (1 Herz, 3 Glycerin) ist außerordentlich wirksam.

nahme von Cholin im Herzextrakt bzw. Füllung gefunden wurde, aber der zwingende Beweis bis heute aussteht, daß dies Mehr an Cholin dem Vagusstoff angehört, da uns bis jetzt die Trennung des Cholins vom Vagusstoff noch nicht gelungen ist. Da es aber sehr unwahrscheinlich ist, daß durch Vagusreizung mehr Cholin unabhängig von der Vagusstoffbildung und zwar annähernd im gleichen Verhältnis zu dieser freigemacht wird, haben wir trotzdem zunächst untersucht, ob durch fermentative Zerstörung von Vagusstoff der Cholingehalt sich ändert oder nicht: die Acetylierung des zerstörten und des nicht zerstörten, im Herzextrakt von vornherein vorhandenen und zugesetzten Vagusstoffes ergab identische Cholinmengen.

Dieser Befund erhält seine Bedeutung erst durch den folgenden: in zahlreichen Versuchen haben wir festgestellt, daß die Zerstörung von Acetylcholin durch Herzextrakt eine Verseifung ist[1]); denn wir

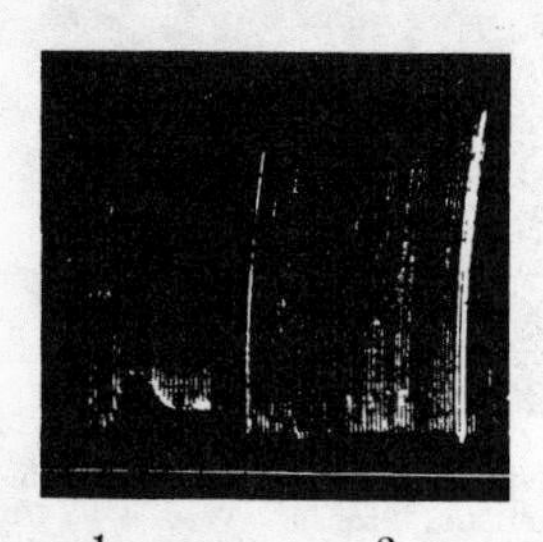

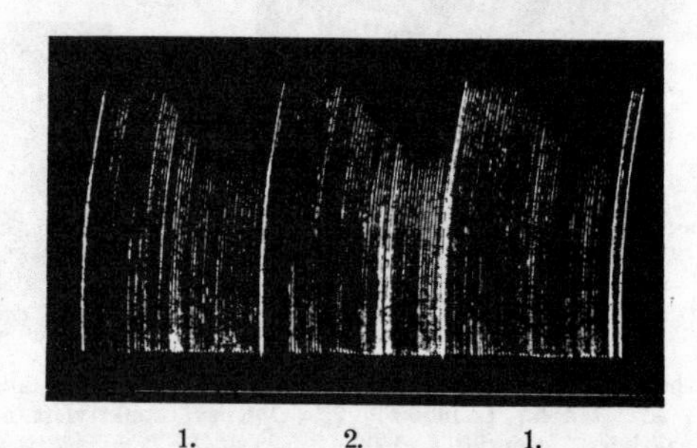

Abb. 6a. H. E. + Acetylcholin (1 : 300 000) nach 60 Minuten Stehen (38° C). 1. = Inaktiviert (geprüft 1 : 100). 2. = Nicht inaktiviert (geprüft 1 : 25).

Abb. 6b. Desgleichen nach Acetylierung (geprüft 1 : 1000).

fanden nach völliger Zerstörung des Acetylcholins durch Herzextrakt mittels Acetylierung das gesamte in diesem enthaltene Cholin wieder (Abb. 6). Abb. 6a gibt den Unterschied der inaktivierten und nichtaktivierten Probe vor, Abb. 6b den Ausgleich des Unterschiedes nach Acetylierung wieder. Nachdem wir nun sahen, daß Wirkungsbild und -ablauf, ferner der Ablauf der Zerstörung von Vagusstoff und Acetylcholin der gleiche ist, ferner nach der Spaltung von Vagusstoff und Acetylcholin das gesamte Cholin wiedergefunden wird, ist es in hohem Maße wahrscheinlich, daß der Vagusstoff, wie wir dies schon früher annahmen, ein Cholinester ist und sein Unwirksamwerden durch Herzextrakt, sowie nachgewiesenermaßen beim Acetylcholin, Folge seiner Verseifung.

---

[1]) Schon *Dale* (Journ. of pharmacol. a. exp. therapeut. **6**, 147. 1914) hat an diese Art der Zerstörung als Ursache des raschen Vorübergehens der Acetylcholinwirkung im Organismus gedacht.

IV. Über Bedingungen des Umfanges der Spaltung durch Herzextrakt.

In mehreren Versuchen wurde der Einfluß der Wasserstoffionenkonzentration auf den Umfang der Spaltung geprüft, und zwar derart, daß gleiche Mengen Herzextrakt in bicarbonatfreiem Ringer mit gleichen Mengen verschiedener froschisotonischer Phosphatmischungen[1]) und dann mit Acetylcholin versetzt stehen gelassen wurden. Abb. 7 zeigt, daß,

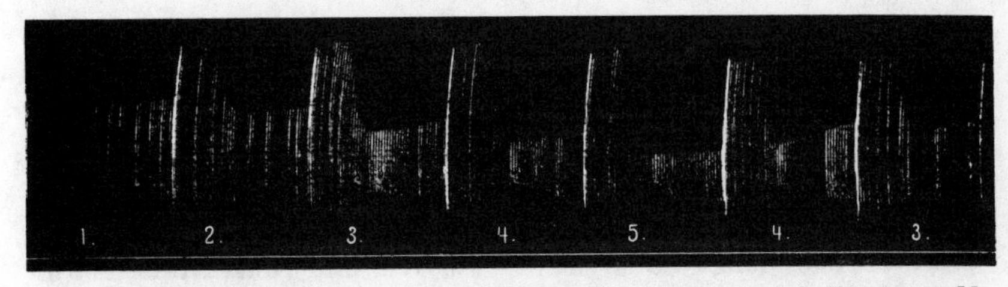

Abb. 7. 1.—4. = H.E. + Acetylcholin (1 : 400 000): geprüft in Verdünnung 1 : 100. 1. = p_H 7,5. 2. = p_H 7,0. 3. = p_H 6,5. 4. = p_H 6. 5. = Acetylcholin in bicarbonatfreiem Ringer ohne H.E.: gleiche Verdünnung.

verglichen mit der Kontrolle (bei 5) bei sämtlichen untersuchten p_H (von p_H 6 bis p_H 7,6) Acetylcholin gespalten wurde, und zwar mit fallender, p_H in steigendem Maße.

In weiteren Versuchen wurde der Einfluß der Temperatur untersucht. Abb. 8 illustriert das Prüfungsergebnis am Herzen. Es zeigt, daß bei 0° keine Zerstörung stattgefunden hat, wohl aber, wie regelmäßig, eine hochgradige bei 38°.

Schließlich untersuchten wir den Einfluß der Einwirkungsdauer und der Menge des Extraktes auf den Umfang der Zerstörung. Daß mit der Länge der Einwirkung die Zerstörung zunimmt, geht schon aus der bereits mitgeteilten Abb. 5b hervor.

Abb. 8. H.E. + Acetylcholin (1 : 300 000) p_H 7,6; 60 Min. gestanden. 1. = H.E. inaktiviert. 2. = H.E. nicht inaktiviert bei 0° C. 3. = H.E. nicht inaktiviert bei 38° C.

Das gleiche zeigt der folgende Versuch, der hauptsächlich den Einfluß der Extraktmenge auf den Grad der Zerstörung dartun sollte. Der Versuch wurde folgendermaßen angestellt: Mit Phosphatmischung p_H 7,5 gepufferte verschiedene Extraktmengen von 0,1—0,7 ccm wurden mit je 0,2 ccm Acetylcholin 1 : 100 000 versetzt und mit bicarbonatfreiem Ringer auf 1 ccm gebracht. Als Kontrolle diente eine Probe ohne Herzextrakt. Die Prüfung fand nach $1^1/_2$- bzw. $2^1/_2$stündigem Stehen bei einer Verdünnung 1 : 200 statt. Die Ergebnisse zeigt die Tabelle.

[1]) Es wurde regelmäßig 1 Teil Phosphatmischung auf 10 Teile Gesamtmischung angewandt.

Stunden	Zerstört in % der Kontrollprobe bei Zusatz von Herzextrakt ccm				
	0	0,1	0,2	0,4	0,7
$1^1/_2$	0	60	85	100	100
$2^1/_2$	0	80	98	[1])	[1])

Die Tabelle zeigt, daß mit steigenden Mengen Extraktes der Umfang der Zerstörung zunimmt.

V. Untersuchung des Einflusses von Extrakten anderer Organe auf Vagusstoff und Acetylcholin.

Außer dem Einfluß des Extraktes aus Herzen haben wir vorläufig nur den der Extrakte aus Darm, Leber und quergestreiftem Muskel (Gastrocnemius) des Frosches untersucht. Wir fanden, daß bei ganz gleicher Anordnung wie in den Versuchen mit Herzextrakt Darm- und Leberextrakt auch quantitativ annähernd ebenso wirken wie Herzextrakt, und nur der Muskelextrakt Vagusstoff und Acetylcholin wesentlich langsamer zerstört. Dieser Befund steht einmal in guter Übereinstimmung mit der nachträglich von uns gefundenen Angabe *Oppenheimers*[2]), wonach im Muskel auch andere esterspaltende Fermente zu fehlen scheinen, zum andern mit der Beobachtung *Riessers*[3]), wonach die Acetylcholinwirkung am Muskel — im Gegensatz zu der am Herzen — von außerordentlich langer Dauer ist. Von Interesse in diesem Zusammenhang ist auch der im hiesigen Institut erhobene Befund *Witanowskis*[4]), wonach im Muskel als im einzigen der untersuchten Organe nur Cholin, kein Vagusstoff vorkommt.

VI. Beobachtungen über das Verhalten des Acceleransstoffes.

In Versuchen, in denen wir nach lange dauerndem Stehenlassen des Herzextraktes die Prüfung auf das Ausmaß der Spaltung des im Extrakt vorhandenen Vagusstoffes in nicht oder wenig verdünnten Extraktlösungen, namentlich solchen aus Krötenherzen, die besonders reich an Acceleransstoff sind, durchgeführt haben, beobachteten wir oft keine negativ inotrope Wirkung mehr, sondern eine positiv inotrope[5]). Als Illustration diene Abb. 9a. Nachdem der Acceleransstoff — wenn auch vielleicht nicht als alleiniger positiv inotrop wirksamer Stoff — ebenso wie der Vagusstoff in das wässerige Extrakt übergeht, ist die positiv inotrope Wirkung wahrscheinlich die Folge davon, daß der

[1]) Sogar unwirksam bei Verdünnung 1 : 10.
[2]) Die Fermente. Lieferung 3, S. 402. Leipzig 1924.
[3]) Arch. f. exp. Pathol. u. Pharmakol. **91**, 342. 1921.
[4]) Pflügers Arch. f. d. ges. Physiol. **208**, 694. 1925.
[5]) Wir haben oben bereits erwähnt, daß auch der frische Extrakt nach Atropinisierung regelmäßig positiv inotrop wirkt.

Acceleransstoff und etwa andre positiv inotrop wirksame Stoffe durch das Stehenlassen des Herzextraktes nicht zerstört worden waren wie der Vagusstoff und in dem Maße wie letzterer unwirksam geworden war, seine bzw. ihre Wirkung sich immer mehr äußert.

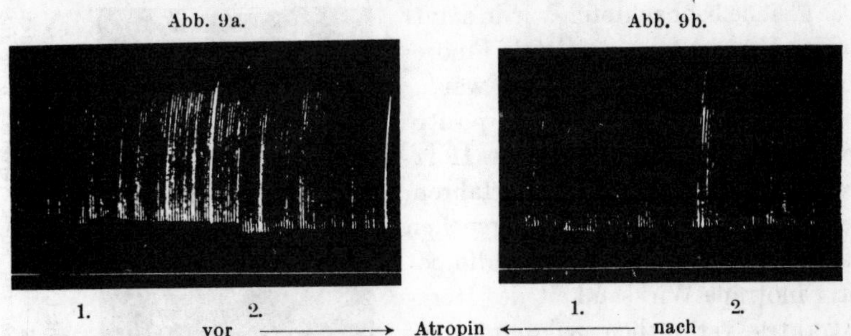

1. = Krötenherzextrakt (1:20) bei 56° C (20 Min.) inaktiviert, 1:5 geprüft. 2. = Dasselbe nicht inaktiviert nach 120 Min. Stehen geprüft.

Nachdem das Schicksal des Acceleransstoffes von nicht geringerem Interesse ist als das des Vagusstoffes, haben wir zunächst geprüft, ob die positiv inotrope Wirkung des Herzextraktes, die mindestens teilweise auf Acceleransstoff zurückzuführen ist, beim Stehenlassen des Extraktes, wenn auch nicht schwindet, so doch abnimmt. Zu diesem Behuf haben wir die Wirksamkeit des durch Erhitzen auf 56° sofort inaktivierten und nativen Herzextraktes nach längerem Stehen am atropinisierten Testherzen miteinander verglichen. Abb. 9 b zeigt, daß die Wirkung identisch ist: es ist also kein positiv inotrop wirksamer, also auch kein Acceleransstoff zerstört worden.

Abb. 10. *Atropinisiertes* Krötenherz. 1. = Ringer. 2. = H. E. (1:30) 30 Min. mit ultraviolettem Licht belichtet. 3. = Dasselbe nicht belichtet. 2. und 3. standen 1 Std. bei 18° C. Prüfung bei Verdünnung 1:5.

Bei Gelegenheit der Prüfung, ob außer durch Erhitzen des Herzextraktes auf 56° die Zerstörung des Vagusstoffes auch durch Behandlung des Herzextraktes mit Fluorescenz- bzw. ultraviolettem Licht hintanzuhalten sei, beobachteten wir einen wesentlichen Unterschied im Verhalten der positiv inotrop wirksamen Substanz des Herzens. Während dieselbe, wie wir soeben sahen, bei Erhitzen auf 56° erhalten bleibt, wird sie durch die Belichtungen zerstört. Abb. 10 stammt aus einem Versuch, wo der eine Teil des Herzextraktes mit ultraviolettem Licht behandelt, der andere

60 Min. bei Zimmertemperatur stehengelassen war; während ersterer am atropinisierten Herzen die übliche positiv inotrope Wirkung zeigt (bei 3), fehlt diese bei der belichteten Probe (bei 2) völlig; der positiv inotrop wirksame Stoff ist also zerstört worden. Das gleiche beobachteten wir in sämtlichen Versuchen, wo mittels Fluorescenzlicht inaktiviert worden war.

Um über die Natur der positiv inotrop wirksamen Substanz des Herzextraktes etwas Genaueres zu erfahren, behandelten wir in weiteren Versuchen diejenigen Substanzen, denen die positiv inotrope Wirksamkeit des Herzextraktes vermutlich zukommt, gesondert mit Fluorescenzlicht: das sind einmal die alkohol-ätherlöslichen Lipoide[1]), zum andern der Acceleransstoff, d. h. die Füllung aus Acceleransreizperioden. Aus Abb. 11a geht hervor, daß durch Belichtung die Wirksamkeit von Alkohol-Ätherextrakt, die in langsam einsetzender schwach inotroper Wirkung besteht, nicht geändert wird, aus Abb. 11b, daß nicht durch Beimischung von Eosin ohne Belichtung (5), wohl aber mit Belichtung die Wirkung des Acceleransstoffes aufgehoben wird. Damit ist mindestens *eine* besondere Eigenschaft des Acceleransstoffes festgelegt. Nicht ohne Interesse ist, daß, wie ebenfalls Abb. 11b zeigt, auch die Wirkung von Adrenalin durch Fluorescenzlicht aufgehoben wird.

a) 1. 1′ 2′ 3′ 2. 1′ 2′ 3′

b) 1. 2. 1. 3. 4. 5.

Abb. 11 a u. b. a) 1. = Alkoholätherextrakt des Herzens mit Eosin versetzt. 1 Stunde belichtet. 2. = Dasselbe nicht belichtet. b) 1. = Acceleransfüllung mit Eosin versetzt; nicht belichtet. 2. = Acceleransfüllung mit Eosin versetzt; belichtet. 3. = Adrenalin 1 : 1000000 mit Eosin versetzt; belichtet. 4. = Adrenalin 1 : 1000000 *nicht* mit Eosin versetzt; belichtet. 5 = Adrenalin 1 : 1000000 mit Eosin versetzt; *nicht* belichtet.

Schlußfolgerung.

Ausgangspunkt für die vorliegende Untersuchung war die bereits früher gemachte Feststellung, daß ebenso wie der Erfolg der Vagusreizung auch der ins Herz gebrachten Vagusstoffe und des Acetylcholins rasch schwindet. Nachdem wir in der vor-

[1]) Vgl. *Clark*, Journ. of physiol. **47**, 66. 1913, und *Loewi*, Pflügers Arch. f. d. ges. Physiol. **170**, 677. 1918.

liegenden Untersuchung feststellen konnten, daß Wasser- und Glycerinextrakt des Herzens Vagusstoff und Acetylcholin esteraseartig verseift, ist der Rückschluß gestattet, daß diese Verseifung auch in vivo vor sich geht und die Ursache des raschen Wirkungsschwundes der Vagusreizung, des Vagusstoffes und des Acetylcholins bildet. Andererseits wird durch den Befund, daß Wirkungsbild, Wirkungsablauf sowie Art und Ausmaß der Wirkungsaufhebung durch die Fermenttätigkeit des Herzextrakts bei Vagusstoff und Acetylcholin völlig gleichartig sind, die schon früher von uns geäußerte Vermutung, in hohem Maße gestützt, daß der Vagusstoff ein Cholinester sei[1]).

Ergebnisse.

1. Die Wirkung von Vagusstoff nimmt bei Gegenwart von wässrigem Herzextrakt nach längerem Stehen bei Zimmer- oder Brutschranktemperatur ab.

2. Diese Abnahme kommt nicht zustande, wenn der Herzextrakt durch Erhitzen auf 56°, Fluorescenz- oder Ultraviolettlicht vorgängig inaktiviert war.

3. In ganz gleicher Weise nimmt die Wirkung von Acetylcholin bei längerem Stehen mit Herzextrakt ab, während die von Muscarin und Cholin völlig erhalten bleibt.

4. Die Abnahme der Wirksamkeit ist die Folge einer fermentativen Verseifung des Acetylcholins; denn die unwirksam gewordenen Proben ergeben bei Reacetylirung den identischen Cholingehalt wie die infolge vorgängiger Inaktivierung des Herzextraktes wirksam gebliebenen.

5. Die Spaltung von Vagusstoff und Acetylcholin geht bei p_H 6 bis 7,5 vor sich und zwar in steigendem Maße mit fallender Wasserstoffionenkonzentration.

6. Bei 0° findet keine Spaltung statt.

7. Der Umfang der Spaltung nimmt zu entsprechend der Einwirkungsdauer und der Menge des Herzextraktes.

8. Darm- und Leberextrakt spalten Vagusstoff und Acetylcholin in ganz der gleichen Weise wie Herzextrakt, Muskelextrakt dagegen spaltet viel langsamer.

9. Der positiv inotrop wirksame Bestandteil des frischen Herzextraktes, dessen Wirkung sich nach Atropinisierung des Testherzens zeigt, wird bei Inaktivierung des Herzextraktes durch Erhitzen nicht zerstört, wohl aber bei Inaktivierung durch Fluorescenz- bzw. Ultraviolettlicht.

10. Ebenso wird der Inhalt aus Acceleransreizperioden sowie Adrenalin durch Fluorescenz- und ultraviolettes Licht zerstört.

[1]) Bei der außerordentlich starken Wirksamkeit des Vagusstoffes könnte sogar daran gedacht werden, daß er Acetylcholin ist.

Pflügers Arch. (1926) **214**, 678–688

(From the Pharmacological Institute of Graz University)

ON THE HUMORAL PROPAGATION OF CARDIAC NERVE ACTION

Communication X

The Fate of the Vagus Substance

by O. Loewi and E. Navratil

(With 11 figures)

(Received July 29, 1926)

In contrast to the effect of accelerator nerve stimulation, the effect of vagus stimulation ceases with or shortly after the last stimulus. Similarly, the effect of the applied vagus substance[1] very quickly wears off,[2] whereas the effect of the introduced accelerator substance lasts much longer—an observation which we reported in our seventh communication.[3] In this paper we consider the cause of the rapid disappearance of the vagus substance and with it of the effect of vagus stimulation.

I. *Inactivation of the Vagus Substance by Treatment with Heart Extract*

One possibility was that the vagus substance was rapidly destroyed by the heart. Since the filling solution from periods of vagus stimulation is still active after 24 hours, it is clear that no "destroying substance" enters the fluid from the heart. What we now had to determine was whether an extract from the heart itself might destroy the vagus substance.

The experiments which we conducted from October to April sought to answer this question. They were carried out in the following manner. Hearts of esculents (or occasionally toads) were cleaned of blood remnants by briefly washing them in a bicarbonate-free Ringer's solution. They were then mashed in a mortar with quartz sand and suspended in a bicarbonate-free Ringer's solution. Each heart weighed on the average 100–200 grams and, as a rule, required 2 to 5 cc of Ringer's solution. The extract was then filtered and the filtrate tested for its ability to destroy the vagus substance. In order to do this we merely compared the effect of two different samples of the filtrate upon a heart. One sample was used immediately after it had been obtained. The other was used after it had stood for a long time at room temperature or at 37°C. We used this simple procedure for the following reasons. The vagus filtrate, as we demonstrated earlier,[4] contains vagus substance, i.e., it has a negative inotropic effect. This effect can be eliminated by atropine[5] and is caused by the vagus substance and not by choline. Choline is present in the filtrate only in a totally inactive concentration. Moreover, acetylation of the filtrate yields much less choline than would be expected from the activity of the filtrate.[6]

The results consistently showed that the sample which had been left standing for some time had a considerably smaller negative inotropic effect than the sample which was tested immediately after it was obtained. In order to gain more insight into the cause of the decrease in activity, we ran some more experiments. They were carried out in the following manner. The filtrate was divided into two equal parts; one was immediately heated to 56°C and kept at this temperature for 20 minutes.[7] Afterwards both parts were left standing at room temperature or 38°C for some time—between ½ hour and 3 hours. Then the action of both samples on the heart was compared. For the results, it made no difference whether we tested the filtrates proper or their alcohol extracts which had been obtained from them as described earlier. The results were the same. In every case the sample which was not heated lost a great deal of its activity through having been left standing.

Fig. 1 gives the results of one such test. An extract (1 heart:30 Ringer's) was left standing for 3 hours at 18°C. When this extract was further diluted five times (Fig. 1.3), it no longer had a negative inotropic effect.[8] In contrast, an extract which had been inactivated by heat at the beginning of the experiment still did (Fig. 1.2).

We now sought to determine why the heated and unheated samples differed in activity. Did heating cause a physicochemical change in the extract which altered the detectability of the vagus substance, perhaps because of a change in absorption conditions? In order to find out we took a sample which had first been allowed to stand and heated it to 56°C for 20 minutes. The ensuing comparison yielded the same results as the one just described. The sample which was heated after it had been allowed to stand was much less active than the sample which had been immediately heated. On *a priori* grounds one would not expect the inactivation to be limited to the vagus substance which is contained in the heart extract under study. And, in fact, in many experiments we have added some vagus substance obtained by alcohol extraction from other hearts and found the same effect. Fig. 2 shows that both the vagus substance from the extract (Fig. 2a.2a) and the added vagus substance (Fig. 2b.3a) are inactivated when the extract is allowed to stand.

It is quite likely that the destruction of the vagus substance takes place in a fermentative manner. The mere fact that heating the extract to 56°C retards the process strongly suggests that this is the case. Moreover, exposure to ultraviolet[9] or fluorescent[10] light, which does not affect the activity of the vagus substance, also reduces the activity of the extract. Fig. 3 shows the effect of ultraviolet light.

We have already shown in previous experiments that atropine neutralizes the vagus substance and that the vagus substance does not contain enough choline to account for its activity. The results of these experiments have enabled us to establish another characteristic of the substance: namely, that it can be destroyed by letting it stand with heart extract.

II. *Inactivation of Acetylcholine by Treatment with Heart Extract*

In our second communication we raised the possibility that the vagus substance, which we had shown was not choline, might be a very active choline ester. Strong support for this hypothesis comes from the following observations. First, the effect of acetylcholine on the heart rapidly wears off, and so does the effect of the vagus substance. In contrast, that of choline and muscarin[11] is long-lasting.[12] Second, acetylcholine activity decays at the same rate as that of vagus substance with the same initial level of activity (Fig. 4). Clearly, it is quite important to determine whether this similarity in activity pattern springs from a common cause. Indeed, acetylcholine, like vagus substance, loses its activity when it is left to stand for a long time with heart extract, but not, however, if the extract has first been rendered inactive by heating, fluorescent light, or ultraviolet light. Given the same conditions of exposure to active heart extract, choline and muscarin activity is unaffected.[13] Fig. 5 serves to illustrate this. Fig. 5a compares the activity of acetylcholine added to active heart extract with that of acetylcholine added to inactivated heart extract (kept at 56°C for 20 minutes) after both mixtures had been allowed to stand for 1 hour. The activity of the acetylcholine in the active heart extract is much lower. Fig. 5b shows a section of the titration experiment run on the material in Fig. 5a in order to determine the absolute magnitude of the decrease in activity. After $\frac{1}{2}$ hour the acetylcholine sample containing active heart extract had lost 25% of its activity. At a concentration of 1/150 it was as active as the other sample at a concentration of 1/1000. After $1\frac{1}{2}$ hours it had lost 85% of its activity.

There are, then, recordings and experimental observations in which the results for the vagus substance and acetylcholine are not only qualitatively but also quantitatively entirely identical.

III. *Mode of Destruction of Vagus Substance and Acetylcholine*

We turn now to the way heart extract affects vagus substance and acetylcholine. Since esterases have often been found in organisms, the most obvious possibility is that there is a saponification of vagus substance and acetylcholine.[14] Now the occurrence of a saponification can only be established indirectly through its end products. In the present case this means the following. We must at the very least show that choline is produced when acetylcholine or vagus substance is rendered inactive by fermentation. We could not unequivocally answer this question by using vagus substance. To be sure, in earlier experiments an increase in choline content was found in heart extract or in the perfusate after vagus stimulation. However, we cannot say for certain that the choline comes from inactivated vagus substance. For we have not yet been able to separate choline from vagus substance ourselves. Although it is very unlikely that, as a result of nerve stimulation, choline is released independently of the release of vagus substance and in approximately the same proportion, we nevertheless first examined whether, through fermentative destruction of vagus substance, the choline content changes or not. The acetylation of inactivated vagus substance or of non-inactivated vagus substance (present from the start or added) revealed identical quantities of choline.

These findings can best be understood in the light of the following observation. We have found in numerous experiments that the destruction of acetylcholine by heart extract is a saponification.[15] After total destruction of acetylcholine by heart extract extract, we recover, upon reacetylation, the total choline contained in the mixture (Fig. 6). Fig. 6a shows the difference existing between the inactivated and the non-inactivated samples, and Fig. 6b shows the elimination of this difference after acetylation.

We have seen that both the vagus substance and acetylcholine have the same patterns of action and destruction. Moreover, we have seen that their decomposition yields choline. For all these reasons it seems quite likely that our earlier assumption was correct. The vagus substance appears to be a choline ester and, like acetylcholine, it seems to be degraded by the heart extract by means of a saponification.

IV. *Conditions Affecting the Extent of Degradation by Heart Extract*

Many experiments examined the influence of hydrogen ion concentration on the magnitude of the degradation. Equal amounts of heart extract were placed in a bicarbonate-free Ringer's solution to which had been added equal amounts of different frog isotonic phosphate mixtures.[15] Then acetylcholine was added and the mixture left standing. Fig. 7 shows that, compared to a control (Fig. 7.5), acetylcholine was split at all pH levels (from pH 6 to pH 7.6). Moreover, the lower the pH the more acetylcholine was split.

Other experiments examined the influence of temperature. Fig. 8 shows the results of the tests on the hearts. At 0°C there was no destruction. But, as always, there was a great deal at 38°C.

Finally, we examined the influence of the length of exposure and the amount of the extract upon the magnitude of destruction. Figure 5b shows that destruction increases with the length of exposure time.

This is also shown by the following experiment whose main purpose was to investigate the influence of the amount of extract upon the degree of destruction. The experiment was carried out in the following manner. Various amounts of extract, from 0.1 to 0.7 cc, were buffered with a phosphate mixture of pH 7.5. Each sample was mixed with 0.2 cc of acetylcholine 1:100,000. Bicarbonate-free Ringer's solution was then added to bring the volume up to 1 cc. The control was a sample without heart extract. The samples were tested at a dilution of 1:200 after having been left to stand for $1\frac{1}{2}$ and $2\frac{1}{2}$ hours. The results appear in the following table.

Amount of Acetylcholine Destroyed by Heart Extract (as Per Cent of Control)

Hours	cc of Heart Extract				
	0	0.1	0.2	0.4	0.7
$1\frac{1}{2}$	0	60	85	100[a]	100[a]
$2\frac{1}{2}$	0	80	98	–	–

[a]See footnote 17.

The table shows that the greater the quantity of extract, the greater the magnitude of the destruction.

V. *The Effect of Other Organ Extracts on Vagus Substance and Acetylcholine*

Besides heart extract, we have so far investigated only intestine, liver, and frog transversely striated muscle (gastrocnemius). Using the same set-up as in our heart experiments, we found that intestine and liver extracts have approximately the same quantitative effect as heart extract. In contrast, muscle extract destroys vagus substance and acetylcholine much more slowly. These findings are consistent with Oppenheimer's observation,[18] which we only came across subsequently, that muscle also seems to lack other ester-splitting ferments. They are also consistent with Riesser's observation[19] that, in contrast to the heart, the effect of acetylcholine on muscle is extraordinarily long-lived. In this context we consider Witanowski's finding,[20] which was made at this institute, particularly interesting. He found that the muscle is the only one of the examined organs to contain only choline and no vagus substance.

VI. *Observations on the Behavior of the Accelerator Substance*

In some experiments we studied the degradation of vagus substance initially present in the heart extract by letting full strength or slightly diluted extract stand for a long time and then determining the magnitude of degradation. Often, instead of a negative inotropic effect, we observed a positive one (see Fig. 9a).[21] This was especially the case when we used toad hearts, which are particularly rich in accelerator substance. This positive effect is probably due to the following. Vagus substance is not the only compound present in aqueous extract. The accelerator substance, which moreover, may not be the only positive inotropic agent, is also present. Probably the accelerator substance and any other positive inotropic compounds are not destroyed by letting heart extract stand. Then, as the vagus substance is destroyed, the effect of the positive agent(s) is progressively more apparent.

Since the fate of the accelerator substance is of no less interest than that of the vagus substance, we pursued the matter further. We first asked if the heart extract's positive inotropic effect (for which the accelerator substance is at least partly responsible) would, if not disappear, at least decrease after the extract had been left to stand. Some heart extract was immediately inactivated by heating to 56°C. Some was left untreated. Both extracts were then allowed to stand for a longer period. Then their effect on an atropinized heart was compared. Fig. 9b shows that they possessed the same effect. Hence, no substance with a positive inotropic effect, such as the accelerator substance, was destroyed.

Earlier we had examined the effect on vagus substance destruction of both heating the heart extract to 56°C and also treating it with fluorescent or ultraviolet light. In the course of this study we observed an important difference in the behavior of the positive inotropic heart substance. As we have recently seen, it remained intact when heated to 56°C, but it was destroyed by exposure to light. Fig. 10 is taken from such an experiment. One part of the extract was treated with ultraviolet light. The other part was left to stand for 60 minutes at room temperature. This had the usual positive inotropic effect on an atropinized heart (Fig. 10.3). The sample exposed to light did not (Fig. 10.2): the positive inotropic effect of the substance had been destroyed. Similar results were observed in all the experiments in which fluorescent light had been used for inactivation.

In other experiments the most likely candidates for the heart extract's positive inotropic activity were treated with fluorescent light. These were the alcohol-ether soluble lipoids[22] and the accelerator substance, that is, the filling solution from the period of accelerator nerve stimulation. Fig. 11a shows that the activity of the alcohol-ether extract, with its slow onset and weak inotropic effect, is not changed by exposure to light. In Fig. 11b the addition of eosine in the absence of light does not affect the activity of the accelerator substance. However, eosine plus light does. This establishes at least *one* property of the accelerator substance. It is of some interest that, as Fig. 11b also shows, fluorescent light also eliminated the effect of adrenaline.

Conclusion

The point of departure of the present study was provided by an observation which we had made earlier: like vagus stimulation, the effect of the vagus substance introduced into the heart and of acetylcholine rapidly wears off. In the preceding study we established that water and glycerine extracts of the heart have an esterase-like saponification effect on vagus substance and acetylcholine. Having shown this, one may now be permitted to conclude that this saponification also occurs *in vivo*, and is responsible for the rapid disappearance of the effects of vagus stimulation, vagus substance, and acetylcholine. Moreover, our earlier assumption that the vagus substance is a choline ester is strongly supported by the evidence. On the one hand, acetylcholine and vagus substance have identical patterns of action. On the other hand, the heart extract's ferment activity affects them in exactly the same way.[23]

Results

1. After standing for a long time at room or incubator temperature, the activity of vagus substance in an aqueous heart extract decreases.

2. This decrease does not occur when the heart extract has first been rendered inactive either by heating to 56°C or by exposure to fluorescent or ultraviolet light.

3. In a similar fashion, the activity of acetylcholine decreases after standing in heart extract for a long time. In contrast, that of muscarin and choline remains completely intact.

4. The decrease in activity is the consequence of a fermentative saponification of the acetylcholine: upon reacetylation the inactivated samples show as much choline as samples which had remained active because they had been placed in previously inactivated heart extract.

5. The degradation of vagus substance and acetylcholine occurs between pH6—7.5. It increases with decreasing hydrogen ion concentration.

6. At 0°C there is no degradation.

7. The magnitude of the degradation increases with the length of exposure and the amount of heart extract.

8. Intestine and liver extracts degrade vagus substance and acetylcholine in exactly the same manner as heart extract. In contrast, muscle extract splits them much more slowly.

9. The positive inotropic component of fresh heart extract, whose effect is apparent after atropinization of the test heart, is not destroyed when the extract is rendered inactive by heating. It is, however, destroyed when the extract is inactivated by fluorescent or ultraviolet light.

10. Similarly, and like adrenaline, the substance from the period of accelerator nerve stimulation is also destroyed by fluorescent and ultraviolet light.

FOOTNOTES

[1]By vagus substance we will mean the diffusates or extracts from the heart which have a vagus-like effect.

[2]Naturally this is a function of how much is used.

[3]Pflügers Arch. f. d. ges. Physiol. 206, 135. 1924.

[4]Pflügers Arch. f. d. ges. Physiol. 208, 694. 1925.

[5]Providing that it is sufficiently concentrated, the filtrate from an atropinized heart generally has a positive inotropic effect. The effect is due to the presence of the accelerator substance, which is also extractable with water. See Section VI, page 482.

[6]This type of experiment can only be done in the winter. Only then does the filtrate possess a consistent and intensely negative inotropic effect. As a rule, it is possitively inotropic from the month of May onward. We will discuss the reason for this seasonal difference in a later communication.

[7]We established in our experiments that all of the original vagus substance remained intact.

[8]Of course, this does not prove that all vagus substance has been destroyed. Rather, there is only evidence for its total destruction at the tested degree of dilution.

[9]For 10 minutes.

[10]The sample mixed with eosine (1:100,000) was exposed for 1 hour.

[11]Muscarin synthetic (Grubler).

[12]See below. Fig. 2 of the following communication: Pflügers Arch. f. d. ges. Physiol. 214, 690. 1926.

[13]In the experiments with muscarin and choline the extract was first completely depleted of any vagus substance. It was left to stand for a while at a warm temperature. This was done in order to avoid any blurring of the result by vagus substance present in the extract. Then and only then was the muscarin or choline experiment begun. This procedure was not necessary in the experiments with acetylcholine. The vagus substance is totally inactive at the dilutions which are necessary for testing acetylcholine.

[14]The glycerine extract of the heart (1 heart, 3 glycerine) is extraordinarily active.

[15]Dale has already suggested that this type of destruction is the cause of the rapid disappearance of the acetylcholine effect in the organism (Journ. of Pharmacol. a. exp. Therapeut. 6, 147. 1914.)

[16]As a rule, 1 part phosphate mixture to 10 parts total mixture was used.

[17]Even inactive at a dilution of 1:10.

[18]Die Fermente. Lieferung 3, S.402. Leipzig 1924.

[19]Arch. f. exp. Pathol. u. Pharmakol. 91, 342. 1921.

[20]Pflügers Arch. f. d. ges. Physiol. 208, 694. 1925.

[21]As we mentioned above, the fresh extract of an atropinized heart also possesses a consistently positive inotropic effect.

[22]See Clark, Journ. of physiol. 47, 66. 1913; and Loewi, Pflügers Arch. f. d. ges. Physiol. 170, 677. 1918.

[23]In view of the extraordinarily strong activity of the vagus substance, one might even entertain the possibility that it is acetylcholine.

FIGURE LEGENDS

Fig. 1. Esculent heart. 1. Ringer's. 2. Inactivated heart extract. 3. The same, not inactivated.

Fig. 2a. 1. Ringer's. 2. Inactivated heart extract. 2a. The same, not inactivated, left to stand at 18°C for 2 hours.

Fig. 2b. 1. Ringer's. 3. Heart extract (from Fig. 2a) inactivated + vagus substance. 3a. The same, not inactivated, left to stand at 18°C for 3 hours.

Fig. 3. 1. Ringer's. 2. Heart extract (1:30) exposed for 30 min. to ultraviolet light. 2a. The same, not exposed. 2. and 2a. were left to stand at 17°C for 90 min.

Fig. 4. 1. Ringer's. 2. Vagus substance (alcohol extract from heart). 3. Acetylcholine (1:100 million).

Fig. 5a. Starting material: heart extract (1:20) + acetylcholine 1:100,000. 2. Above mentioned, inactivated. Tested at 1:1000. 3. The same, not inactivated.

Fig. 5b. Starting material from Fig. 5a. 1. Not inactivated, left to stand for 30 min. Dilution 1:750. 2. Inactivated, left to stand for 90 min. Dilution 1:1000 (see Fig. 5a). 3. Not inactivated, left to stand for 90 min. Dilution 1:150.

Fig. 6a. Heart extract + acetylcholine (1:300,000) left to stand for 60 min. (38°C). 1. Inactivated (tested at 1:100). 2. Not inactivated (tested at 1:25).

Fig. 6b. The same solutions after acetylation (tested at 1:1000).

Fig. 7. 1.-4. Heart extract + acetylcholine (1:400,000); tested at a dilution of 1:100. 1. pH 7.5. 2. pH 7.0. 3. pH 6.5. 4. pH 6. 5. Acetylcholine in bicarbonate-free Ringer's without heart extract: same dilution.

Fig. 8. Heart extract + acetylcholine (1:300,000) pH 7.6; left standing 60 min. 1. Inactivated heart extract. 2. Heart extract, not inactivated, kept at 0°C. 3. Heart extract, not inactivated, kept at 38°C.

Fig. 9. 1. Toad heart extract (1:20), inactivated, kept at 56°C (20 min.), tested at 1:5. 2. The same, not inactivated, tested after standing for 120 min. Fig. 9a. before, Fig. 9b. after atropine.

Fig. 10. *Atropinized* toad heart. 1. Ringer's. 2. Heart extract (1:30) exposed to ultraviolet light for 30 min. 3. The same, not exposed. 2. and 3. were left to stand at 18°C for 1 hr. Tested at a dilution of 1:5.

Fig. 11a and b. a) 1. Alcohol-ether extract of hearts mixed with eosin. Exposed to light for 1 hour. 2. The same, not exposed. b) 1. Accelerator filling solution mixed with eosin; not exposed to light. 2. Accelerator filling solution mixed with eosin; exposed. 3. Adrenalin 1:1,000,000 mixed with eosin; exposed to light. 4. Adrenalin 1:1,000,000 *not* mixed with eosin; exposed. 5. Adrenalin 1:1,000,000 mixed with eosin; *not* exposed to light.

J. Physiol. (1936) 86, 353-380

RELEASE OF ACETYLCHOLINE AT VOLUNTARY MOTOR NERVE ENDINGS

BY H. H. DALE, W. FELDBERG[1] AND M. VOGT[1]

(*From the National Institute for Medical Research, London, N.W. 3*)

(*Received January* 16, 1936)

IN a note published some time ago [Dale and Feldberg, 1934], two of us gave a preliminary description of experiments which indicated that something having the properties of acetylcholine (ACh.) is liberated, when impulses in motor nerve fibres excite contraction of a voluntary, striated muscle. Several earlier observers had recorded observations of this kind, but their significance had not been clear. Geiger and Loewi [1922] estimated the choline present in extracts from frog's voluntary muscle, and observed an apparent large increase (five- to tenfold) after prolonged direct and indirect stimulation. Plattner and Krannich [1932] and Plattner [1932, 1933] found that the substance present in such extracts was rapidly inactivated by fresh blood, like acetylcholine, and that the quantity present had a general correspondence to the wide differences in sensitiveness of different muscles to the stimulating action of acetylcholine. Faradic stimulation of the nerve increased the yield; but Plattner associated the apparent presence of the acetylcholine in the muscle, and its increase on mixed nerve stimulation, with a "parasympathetic" innervation of the blood vessels. In the tongue, excised from a cat treated with eserine, and divided longitudinally into halves, he found that stimulation of the chorda-lingual nerve caused increase of acetylcholine in the extract from one half, while stimulation of the hypoglossal nerve did not significantly increase the yield of the other.

Hess [1923], Brinkman and Ruiter [1924, 1925] and Shimidzu [1926], all perfused the muscles of a frog's hindlimbs and tested the effluent Ringer's solution on isolated preparations sensitive to acetylcholine, with some variations of method. All found evidence of the liberation, when the nerves supplying the muscles were stimulated, of

[1] Rockefeller Foundation Fellow.

something acting like acetylcholine. Brinkman and Ruiter observed that the liberation still occurred, when curare was present in sufficient amount to prevent the motor impulses from causing contractions.

It will be seen that these different observations, though suggestive, are not clear in significance. Something like acetylcholine was apparently liberated, when the mixed nerves to voluntary muscles were stimulated; but the identification of the substance was incomplete, and there was little to indicate what kind of nerve fibres were responsible for its liberation. There is no real evidence, indeed, for a secondary "parasympathetic" nerve supply to the voluntary muscle fibres themselves, responsible for maintenance and variation of tone, such as Frank and his co-workers have assumed [Frank and Katz, 1921; Frank, Nothmann and Hirsch-Kauffmann, 1922]. There is evidence, however, that the sympathetic nerve supply to the blood vessels of the leg muscles contains a cholinergic component in the cat [Hinsey and Cutting, 1933] and in the dog [Bülbring and Burn, 1935]; and the same might be the case in the frog. As regards the vaso-dilator action of sensory axon branches, the mode of its chemical transmission is still in doubt; it is still possible, though in the light of recent evidence no longer probable, that it may also be cholinergic. It is clear, in any case, that the liberation of acetylcholine in a voluntary muscle, as the result of stimulating a mixed nerve containing sensory and sympathetic as well as motor fibres, could not be regarded as having any necessary connection with motor nerve impulses, and the transmission of their excitatory action to the voluntary muscle fibres.

Our object has been to discover whether stimulation of the motor nerve fibres innervating voluntary muscle fibres, to the complete exclusion of the autonomic or sensory fibres running with them in a mixed nerve, causes the liberation of acetylcholine in appreciable quantities; and, if so, to endeavour to obtain evidence as to the site of such liberation.

Such an enquiry formed a natural sequel to the experiments which, during the past few years, have produced evidence of the liberation of acetylcholine when a nerve impulse reaches the ending of a preganglionic fibre of the autonomic system, whether that ending makes contact with a cell of the suprarenal medulla [Feldberg, Minz and Tsudzimura, 1934] or with a nerve cell in a ganglion [Feldberg and Gaddum, 1934; Feldberg and Vartiainen, 1934; Barsoum, Gaddum and Khayyal, 1934]. The acetylcholine is, in these cases, liberated in contact with cells which are responsive to that aspect of its activity in which it resembles nicotine [Dale, 1914]. The direct stimulant action of acetylcholine on certain voluntary muscles after degeneration of their motor nerve supply, also belongs to its "nicotine" action, and not to its "muscarine" action [Dale and Gasser, 1926]. It had further been shown long ago, by Langley and Anderson, that voluntary motor nerve fibres and preganglionic autonomic fibres are functionally interchangeable in crossed regeneration; so that, as evidence for a cholinergic function of preganglionic fibres accumulated, the presumption of a cholinergic function for the motor fibres to voluntary muscle increased [cf. Dale, 1935*a*, *b*). On the other hand, it was clear that the task of demonstrating the liberation of acetylcholine by nerve impulses reaching the endings of motor nerve fibres in a voluntary muscle, would be attended with difficulties of a different kind from those involved in the experiments on a ganglion. For the small substance of the ganglion is closely packed with the synaptic endings of preganglionic fibres; so that if acetylcholine were liberated by impulses arriving at the synapses, it might be expected to appear in relatively high concentration in the slow-dropping venous effluent; and this expectation had been realized in experiment. In a voluntary muscle, on the other hand, the bulk of the tissue, requiring effective perfusion to maintain its functional activity, is relatively enormous in relation to the motor nerve endings. If acetylcholine were liberated by the arrival of impulses at these endings, it could not, therefore, be expected to appear in the perfusion fluid in more than a very low concentration. This expectation has again been realized in our experiments, but we have, nevertheless, been able with regularity to detect the appearance of a substance having the recognizable characters of acetylcholine, when the purely motor nerve supply to a voluntary muscle has been stimulated under the conditions of our experiments.

Most of our experiments have been made on the muscles of cats and dogs. These mammalian muscles are usually regarded as completely insensitive to the action of acetylcholine when their motor nerve supply is intact. Recent evidence, to be discussed later, shows that they are not, in fact, indifferent to acetylcholine in relatively large doses, applied through the circulation or directly; but their response, under such conditions, is by twitches or fibrillation, and they do not exhibit the slow contracture with which many muscles of the frog and other lower vertebrates respond to acetylcholine in low dilutions. It was accordingly of special importance to discover whether acetylcholine was liberated when motor impulses, causing only quick contractions, passed down the motor nerve fibres to such normal, mammalian muscles. A few experiments were also made on frog's muscles, with stimulation of motor fibres separately from the other components of the mixed nerve.

Methods

All our successful experiments on mammalian muscles have been made by perfusing them with Locke's solution at 37° C., pre-oxygenated to saturation, and containing 1 part of eserine in 5×10^5. Perfusion was carried out with the Dale-Schuster pump. In a few experiments the attempt was made to demonstrate the liberation of acetylcholine during stimulation of motor nerve fibres to a muscle with normal circulation, eserine being given to the whole animal with sufficient atropine to prevent excessive circulatory depression. Feldberg [1933*a*], who under such conditions had readily detected acetylcholine in blood of the lingual vein of the dog when the vaso-dilator chorda-lingual nerve was stimulated, had failed to find any when he stimulated the motor hypoglossal nerve. We had a like failure in experiments with natural circulation on the tongue and on the gastrocnemius, stimulating only motor fibres. Franel [1935], in experiments apparently suggested by our preliminary account, also failed in most cases to detect acetylcholine in venous blood from a leg, the muscles of which were thrown into contraction by stimulating the whole limb. The same author, had, indeed, no greater proportion of success when eserinized Locke's solution was used for perfusion. The conditions of these experiments, however, so differed from those of our own that we cannot profitably discuss the cause of their mostly negative results. Our own success in perfusion experiments, with Locke's solution containing eserine, was regular, in contrast to our failure in the few experiments made with normal circulation.

It may be that the relatively low concentration of acetylcholine released by motor stimulation cannot be protected from the blood esterase by eserine in doses insufficient, by themselves, to cause general twitchings of the voluntary muscles. It is possible, on the other hand, that the artificial conditions of saline perfusion in some other way facilitate the escape of the acetylcholine from the site of its origin into the blood vessels. Whatever the reason, artificial perfusion with an eserine-containing saline fluid has been necessary for success. This has caused a limitation of the period of an experiment in which success was possible. Perfusion must be continued long enough to wash the blood thoroughly out of the vessels; on the other hand, it if is continued too long, the muscle becomes more quickly insensitive to motor nerve impulses. In our few experiments on the frog, the skinned hindlimbs were perfused with Ringer's solution containing eserine at the room temperature. Mammals, after preliminary anæsthesia with ether, were given a stable anæsthesia with chloralose, administered intravenously, before the dissection was begun.

Cat's tongue. All branches of both common carotid arteries and all tributaries of both external jugular veins were tied, excepting only the lingual arteries and veins respectively. The transverse vein connecting the jugulars at the hyoid level was left open, small tributaries to it being tied, so that the effluent from both lingual veins could later be collected from one jugular, the other being then tied. Both hypoglossal nerves were tied as near as possible to their exits from the skull, cut and dissected free from the accompanying lingual arteries up to the point of their entry into the tongue muscles. As soon as these dissections had been completed, an injection of "Chlorazol Fast Pink" was given intravenously, to render the blood incoagulable, and thus facilitate its complete removal by perfusion. Cannulæ, united by a Y-junction to the tube leading from the perfusion pump, were then tied into the peripheral ends of both common carotid arteries, the perfusion was begun, and the fluid collected from one jugular vein, as above described. It is, of course, impossible to isolate the vascular system of the tongue from anastomatic communications. While the heart is still beating the perfusion fluid does not become free from blood, unless the pump is so adjusted as to give a comparatively high perfusion pressure and rapid perfusion rate. Our procedure was to cut down the throw of the pump until the venous outflow was about 1–3 c.c. per minute, and then to kill the animal by incising the heart. The last traces of blood then quickly disappeared from the venous fluid. Only an uncertain part of the perfusion fluid pumped into the lingual arteries was thus recovered from the outflow through the lingual veins, a substantial proportion passing by arterial and venous anastomoses and flowing from the open heart cavities into the chest. This unavoidable and unmeasured loss prevented any calculation of the total quantity of acetylcholine liberated in the tongue muscles during stimulation, and restricted observation to a comparison of the acetylcholine contents of the fluid from the lingual veins, when the tongue was at rest and when it was stimulated through its motor nerve. In most of the experiments on the cat's tongue the hypoglossal nerves had been freed from their sympathetic component by aseptic removal of both superior cervical ganglia, under ether anæsthesia, a few weeks before the experiment was made.

Gastrocnemius. In experiments on the gastrocnemius muscle of the cat and dog, the stimulation was in most cases through ventral spinal roots. In the animal under chloralose the sympathetic chain, on the side chosen for experiment, was first removed, through an abdominal incision, from the fourth lumbar to the first sacral ganglion. The muscles covering the lumbar vertebræ were then dissected away, and the neural arches removed from as many vertebræ as necessary, for later exposure of the required roots, with careful hæmostasis at each stage. The popliteal artery and vein were then prepared, all branches except those to and from the gastrocnemius being tied. The Achilles tendon was isolated and a strong ligature passed under it, round the whole of the other tissues of the leg above the ankle. The saphena veins were separately tied. Mass ligatures were also tied round the thigh muscles above the knee, leaving the main vessels and the sciatic nerve free. The crural nerve was cut at the groin. A stout steel rod was pushed through a hole drilled in the lower end of the femur, enabling that bone to be fixed by a clamp, so that muscular contractions would not drag on the popliteal vessels and interfere with the perfusion. The roots to be stimulated were now exposed by opening the *dura mater*. The dorsal roots were separated, and, unless required for control stimulation, were completely removed, so as to leave good lengths of ventral roots for stimulation, without danger of stimulating dorsal roots by escape of current. The ventral roots of the side not under experiment were similarly removed. The last lumbar and first two sacral roots on the experimental side were then tied with fine ligatures close to the cord. These roots were separately tested with short faradic stimuli, those being kept which caused contractions of the gastrocnemius. These were tied together for stimulation, or, in some cases, left attached to a segment of the cord, which could be raised with them for stimulation. The cannulæ were then tied into the popliteal vessels and the perfusion begun. If the exclusion of collateral circulation had not been sufficient to render the venous effluent free from blood, this was effected by lowering the blood-pressure by bleeding. It was not considered desirable, in this case, to deprive the whole nervous pathway, from the roots to the gastrocnemius, completely of its blood supply by killing the animal. In a few experiments, however, in which the whole sciatic nerve was stimulated, or in which the gastrocnemius muscle was stimulated directly, the preparation was isolated completely after the perfusion had been started, by killing the animal and cutting through the femur.

Quadriceps extensor femoris. To exclude by degeneration the sympathetic nerve endings on the blood vessels, which could not easily be effected with the gastrocnemius, a few experiments were made on the quadriceps extensor mass of the dog's thigh. By aseptic operation under ether, the sympathetic chain, from the third lumbar to the first or second sacral ganglion on the experimental side, together with the first sacral ganglion of the other side, was removed some weeks beforehand. At the experiment the quadriceps was isolated by dividing all the other muscles of the thigh, the ilio-psoas and the glutæal muscles between ligatures. The leg was amputated at the knee joint. The perfusion cannulæ were tied into the femoral artery and vein, all branches except those to the quadriceps being tied. Further to minimize collateral circulation, a ligature was tied round the quadriceps mass above the entry of the vessels and nerves supplying it. Stimulation was through the appropriate lumbar ventral roots, or, in some cases, through the crural trunk.

Frog's legs. Large specimens of Hungarian *R. esculenta* were used. The frog was killed by decapitation and skinned. The abdominal sympathetic chains of both sides were removed. Perfusion was through the abdominal aorta and the anterior abdominal vein, the renal portal veins being tied. The roots of the lumbo-sacral plexus were exposed in the spinal canal, dorsal and ventral roots being tied separately and prepared for stimulation. The perfusion fluid was oxygenated Ringer's solution, containing 1 part of eserine in $3\text{–}5 \times 10^5$. A perfusion pressure of 25–35 c.c. of solution was found to give a convenient rate of outflow of about 1 c.c. per minute.

Stimulation. For stimulation the nerves or roots were laid across a suitable pair of chlorided silver electrodes. Direct stimulation of muscles was made through electrodes in the forms of chlorided silver pins, one impaling the muscle near its tendon of insertion, and the other, of several pins, penetrating the muscle near its origin. The stimuli used were maximal break shocks from a secondary coil, Lewis's rotating interruptor being used so as to give from 5 to 15 shocks per second. This method, causing a rhythmic series of twitches, was chosen in preference to tetanization, so as to avoid impediment to the perfusion during the activity of the muscle.

Rate of perfusion. As explained above, the unknown and irregular escape of some of the venous fluid by collateral channels made impossible a strict adjustment of the rate of flow to the size of the muscle perfused. A rough adjustment, however, had to be made; and it has already been indicated that venous fluid was collected from a cat's tongue at about 1–3 c.c., and from the hind legs of a large frog at about 1 c.c.per minute. For a cat's gastrocnemius the rate of collection was 3–8 c.c., and for the quadriceps femoris of a large dog, sometimes as high as 25 c.c. per minute. An estimate of total output rate being impracticable, our main concern was to ensure that an increased liberation of acetylcholine was not masked, on the one hand, by accelerated perfusion, or simulated, on the other, by retarded perfusion during the period of stimulation. Failing adjustment, the rate of perfusion was, in fact, always accelerated during a period of the rhythmic stimulation employed, especially during its latter part. This acceleration was approximately corrected by an assistant, who watched the rate of dropping from the vein cannula, and reduced the throw of the pump as required, so as to ensure that the rate was not unduly accelerated during stimulation, but not, in any case, to retard it below the control rate.

Tests for acetylcholine. Tests were made as usual on the preparation of leech muscle sensitized by eserine. The preparations used usually responded well to acetylcholine in a dilution of 1 in 5×10^8. The tonicity of the mammalian Locke's solution had to be adjusted to that used for the leech by adding distilled water, and was then further diluted, if necessary, in accordance with its activity. The samples as collected were placed on ice and tested with as little delay as possible. In a number of experiments tests were also made on the blood-pressure of cats under chloralose, the sensitiveness of which to small doses of acetylcholine was, when necessary, increased by injecting eserine, and by restricting the circulation volume by removal of the abdominal viscera. For such tests the samples as collected were made acid to congo red by adding a drop of HCl solution, and then kept cold till the perfusion experiment was finished. The stability of the active substance could be tested by either method, and the test on the cat further enabled its sensitiveness to atropine to be demonstrated.

Results

(1) *Stimulation of purely motor fibres to voluntary muscles*

(*a*) *Tongue of the cat.*

The perfusion being started, the rate adjusted, and a bloodless effluent obtained, control samples were tested on the leech. The earlier samples, after the rather prolonged dissection and manipulation, usually showed detectable amounts of a substance acting like acetylcholine, as Hess [1923] observed in his perfusions of frog muscle. The activity might be such as to correspond to an ACh. content of 1 in 4×10^8 of the undiluted fluid. This activity diminished rapidly, however, with successive samples, and, after perfusion for about half an hour, was no longer perceptible by the very delicate test. The hypoglossal nerves were then stimulated rhythmically, the tongue muscles responding with rhythmic contractions. The perfused muscles show rapid fatigue to the effects of such stimuli, the initially vigorous contractions rapidly becoming weaker, until, at the end of a stimulation period of 2–3 min. they have practically ceased. The stimulation was not continued when the muscles showed signs of this failure. The fluid collected during the stimulation showed a pronounced activity, in comparison with the inactive control. The contrast is illustrated in Fig. 1 A and C. All the fluids in this experiment were tested in 50 p.c. dilutions. The control fluid, at A, has no significant action. At C the fluid collected during a first period of stimulation at 5 shocks per sec. was applied and produced an effect closely similar to that of a control solution of ACh. 1 in 10^8, applied at B; so that the undiluted effluent during stimulation would correspond in activity to ACh. 1 in 5×10^7. The rate of outflow being 3 c.c. per min., the amount of ACh.

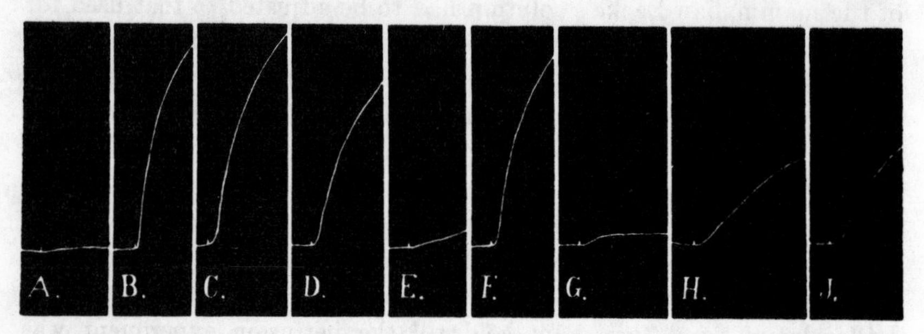

Fig. 1. Leech preparation; eserine 1 in 3×10^5 (same in Figs. 3, 4 and 5). A, C, D, E, F, G, H, venous fluids from cat's tongue, 50 p.c. dilutions (details in text). B and J, control ACh. dilutions; B, 1 in 10^8, J, 1 in 4×10^8.

actually collected in 1 min., apart from what escaped by anastomotic channels, was $0{\cdot}06\gamma$. The active substance was still present in fluid collected after the stimulation period, but in rapidly diminishing concentration in successive samples. Fig. 1 D shows the effect of fluid collected in the 2 min. after the end of stimulation, and E that of fluid collected 20 min. later. A second period of stimulation was then given, again causing vigorous contractions of the tongue at its commencement but with a more rapid onset of fatigue. The fluid collected during this second period, applied at F, was still highly active; and again the stimulating substance had almost disappeared from a further sample, applied at G, which was collected 20 min. after the end of the stimulation. Further periods of stimulation, after the second, were usually progressively ineffective, as regards both the contractile activity evoked in the tongue muscle and the concentration of stimulant substance in the venous fluid. Fig. 1 H shows the effect of the venous fluid collected during a third and less effective period of stimulation. The 50 p.c. dilution at this stage is rather less active than ACh. 1 in 4×10^8, applied as a control at J.

In the later period of the experiment, the tongue gradually becomes œdematous with the prolonged perfusion; and this not only lowers the activity of the fluid collected during stimulation, but causes the activity to disappear more slowly from successive subsequent samples. Eventually, after four, five or more periods of stimulation, the tongue muscles fail to respond further, and the "stimulation" fluid is then not perceptibly more active than the control.

Identity of substance released. The action on the leech muscle, though closely similar to that of acetylcholine, would not by itself identify it. In several cases, when the concentration in the stimulation effluent was sufficiently high, it was directly tested on the blood-pressure of the cat under chloralose, and the activity again matched against known doses of acetylcholine. In every case the match so obtained was identical with that obtained in the comparison made on the leech muscle. In every case, also, the effect on the blood-pressure was completely annulled by a small dose of atropine. In one experiment a larger volume of the effluent was collected, during several periods of stimulation, stabilized by acidification, and evaporated to complete dryness under reduced pressure. The residue was extracted with dry alcohol, and this was evaporated again to dryness, its residue being finally taken up in saline. This solution was then accurately matched against acetylcholine on the eserinized leech preparation, and a solution of acetylcholine corresponding with it in activity, as so determined, was prepared, the actual concentration required being 1 in 2×10^7. These solutions (*S* and *AC*) were then compared against one another on the blood-pressure of a cat under chloralose, the record of the comparison being shown in Fig. 2. It will be seen that in doses of 1 and 0·5 c.c. the two solutions gave indistinguishable effects. The cat was then given 0·15 mg. of eserine per kg. intravenously, and after 15 min. the comparison of the two solutions was repeated (Fig. 2*b*). It will be seen that 0·5 c.c. of the concentrated effluent again accurately matches 0·5 c.c. of the ACh. solution, the effect of both being intensified to the same extent, in comparison with their earlier effects. Finally, after 1 mg. of atropine, the effects of both solutions were annulled (Fig. 2*c*).

If the effluent was made alkaline by adding one-tenth of its volume of *N*/10 NaOH, allowed to stand for 20 min. at the room temperature and then reneutralized with HCl, its actions on the leech and the blood-pressure were found to have disappeared. Since the perfusion fluid contained eserine, it was not possible to test directly the sensitiveness to esterase of the active substance after collection. Indirectly, however, its liability to esterase was demonstrated by an experiment in which the tongue was perfused with Locke's solution containing no eserine. Under these conditions the muscles responded as usual to hypoglossal stimulation, but no trace of stimulant action on the previously eserinized leech muscle was acquired by the effluent. The stimulation was repeated several times, with a similarly negative result, until the tongue muscle ceased to respond. The perfusion fluid remained throughout free from any trace of a substance acting like acetylcholine. Eserine was then added to the perfusion fluid in the usual concentration of 1 in 5×10^5. In the subsequent period of stimulation there was an obvious renewed response of the tongue muscles to the nerve impulses, and the venous fluid now acquired the

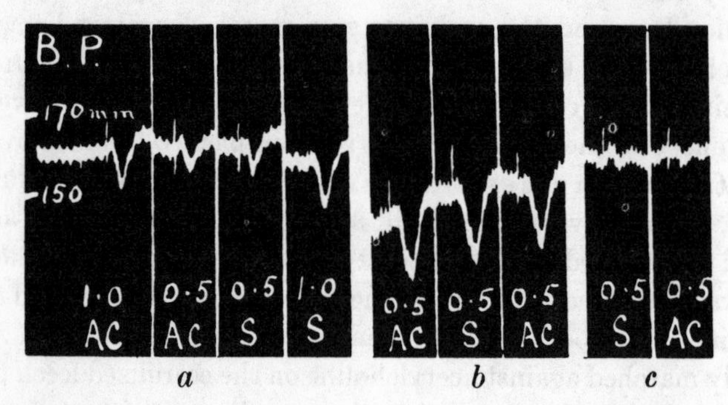

Fig. 2. Carotid blood-pressure of cat under chloralose. Injections (i.v.) of S = concentrated venous effluent from tongue (see text), and AC = acetylcholine in dilution matching S in the leech test. *a*, before eserine; *b*, after eserine; *c*, after atropine.

usual stimulant action on the leech muscle. The revival of muscular response is of interest, but it would not be proper to emphasize its significance on the basis of one observation. For our present purpose the point of importance is the appearance of the stimulant substance in the venous effluent, whereas without eserine it had been consistently absent, even when the muscle contracted well. The substance is normally destroyed on the way from the site of its liberation to the perfusion fluid, but is protected from this destruction by eserine. We are dealing, then, with a substance stable in acid and rapidly destroyed in dilute alkali; protected by eserine from destruction in the tissues; equivalent to the same doses of acetylcholine when tested on the eserinized leech muscle and on the cat's blood-pressure, before and after eserine; and having its action on the cat's blood-pressure annulled by atropine in parallel with that of acetylcholine. There can be no real ground for doubting that it is acetylcholine itself.

Nerve fibres concerned. The tongue muscles and the hypoglossal nerve were chosen, in the first instance, because of the readiness with which this motor nerve could be freed, by degeneration, from sympathetic fibres joining it from the superior cervical ganglion. In most of the experiments above described, both ganglia had been aseptically removed some weeks before the experiment. The hypoglossal nerve has the further advantage that the sensory supply to the tongue runs separately in the lingual nerve, and is not stimulated. Langworthy [1924], has shown, however, that the cat's hypoglossal nerve often contains a vestigial sensory ganglion, in the form of a few cells of sensory type embedded among its fibres. There is no evidence that these are connected with sensory fibres from the tongue; the available evidence, indeed, connects them rather with fibres from the infrahyoid muscle,[1] not included in our perfusion. There was little likelihood, in any case, that this trivial and inconstant sensory component of the hypoglossal nerve would be responsible for the regular liberation of acetylcholine from the tongue, in the significant amounts which we obtained on stimulating the nerve. The exclusion of all sensory elements, however, was even more complete in the later experiments on leg muscles. Another possibility which gave us some concern was that of a mechanical stimulation, by the muscular contractions, of the fibres and endings of the cholinergic chorda tympani, in connection with the blood vessels of the tongue and the glands in its mucous membrane. In several experiments, accordingly, the chorda tympani was also cut by aseptic operation, at a point before its junction with the lingual nerve, and allowed to degenerate completely before the experiment, without affecting the result; stimulation of the hypoglossal, freed from sympathetic fibres, still caused liberation of acetylcholine from the parasympathetically denervated tongue. In another experiment the whole chorda-lingual nerve was cut on one side by previous operation and allowed to degenerate. The margin of the insensitive half of the tongue had been indented by biting before the experiment was performed. The two halves of the tongue were perfused separately on this occasion, and there was no significant difference in their yields of acetylcholine in response to stimulation of their respective hypoglossal nerves.

We shall see later that the mechanical stimulation of chorda-lingual fibres, together with other possibilities, can be more easily excluded by the use of curarine. At the present stage we had established a very strong

[1] Personal communication from Dr D. H. Barron.

presumption that the observed liberation of acetylcholine was due to the impulses passing along motor nerve fibres to excite voluntary muscle fibres. For the further testing of this presumption we turned to other muscles.

(*b*) *Gastrocnemius.*

Cat. The experiments on the cat's gastrocnemius were less successful than those on the tongue. After the rather long preliminary preparation, and the further period required to obtain a bloodless perfusate, we found, in several cases, that the muscle failed to respond to maximal stimulation of the ventral roots, or responded very weakly. In such cases we failed to observe any liberation of acetylcholine during the stimulation. The sciatic nerve was then isolated in the thigh, cut and stimulated peripherally, and the resulting contractions of the gastrocnemius were accompanied by the appearance of acetylcholine in the venous fluid. This latter form of stimulation, however, involved the sympathetic and sensory fibres running in the nerve, and the result was not beyond criticism for our purpose. We may confine attention, therefore, to two experiments in which the roots remained satisfactorily sensitive, the muscle responding when they were stimulated. As in the tongue perfusion, the first samples of perfusion fluid, after it had become free from blood, still contained perceptible traces of acetylcholine. These, however, rapidly disappeared with continued perfusion. When blank controls had been obtained, the roots were stimulated for a period, and with the contractions of the gastrocnemius acetylcholine appeared in the venous fluid, rapidly disappearing again with continued perfusion after the stimulation was stopped. The concentration reached during the stimulation was never high, being about 1 in 4×10^8; on the other hand, the response of this muscle to the root stimulation was never vigorous, even in the most successful experiments.

Dog. The sensitiveness of the dog's roots, as shown by the response of the muscle to their stimulation, survived the conditions of the experiment much better. The first samples of clear venous fluid showed a weak activity on the leech muscle, but practically inactive controls were soon obtained with continued perfusion. Before the motor roots were stimulated, in some experiments, a further control sample was collected during a period of stimulation of the corresponding sensory roots. Fig. 3 shows the record from such an experiment. In all cases the venous fluid was applied in 75 p.c. dilution. A shows the negative effect of a blank control, B that of the fluid collected during stimulation of the sensory roots, and

C the response of the leech muscle to ACh. 1 in 10^9, given for calibration. At D a further resting control sample was applied, and at E the venous fluid collected during a period of stimulation of the motor roots, causing contractions of the gastrocnemius. At F a sample collected 2 min. after the end of stimulation was tested, and at G a sample collected after a further 20 min. perfusion, by which time the effluent was again practically inactive. The sequence could be repeated with a second period of stimulation, but with later periods the response of the muscle, and the concurrent output of acetylcholine, declined together. As with the tongue, if perfusion was continued till the muscle had become visibly œdematous, the appearance in the effluent, of the acetylcholine liberated during stimulation, was delayed and prolonged. Results like the above could be obtained regularly in experiments on the dog's gastrocnemius, stimulated

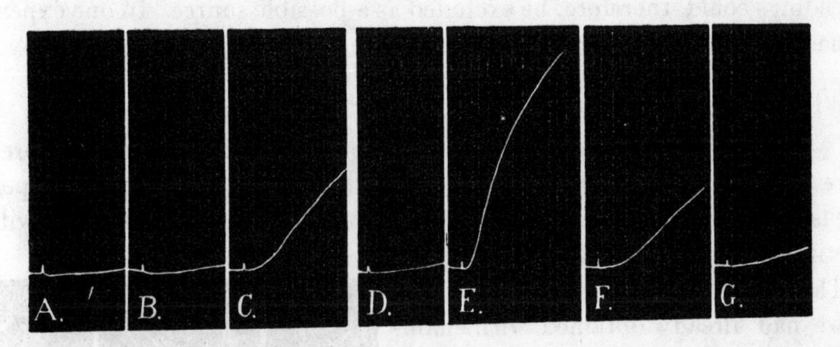

Fig. 3. A, B, D, E, F, G, venous fluids from dog's gastrocnemius, in 75 p.c. dilution (see text). C, ACh. 1 in 10^9.

through the motor roots. The concentration of acetylcholine in the stimulation effluent did not rise above about 1 in 2×10^8, and was accordingly lower than that observed in successful experiments on the tongue. On the assumption that the acetylcholine is liberated by the arrival of impulses at the motor nerve endings, for which evidence will be given later, it may be suggested that the leg muscles, with their relatively long fibres and consequently large mass perfused in relation to the number of nerve endings, might be expected to yield a lower concentration in the venous fluid than the comparatively short-fibred muscles of the tongue. The crude adjustment of the perfusion rate to the size of the muscle, however, forbids any attempts at calculation.

Identification. The substance from the gastrocnemius, like that from the tongue, behaved like acetylcholine in its instability to alkali, and its action on the cat's arterial pressure, abolished by atropine. It was further shown, in one experiment, that an effluent, active on the leech muscle previously sensitized by eserine, had no immediate action on a control strip which had not been so treated.

(*c*) *Quadriceps extensor femoris* (*dog*).

As mentioned earlier, a few experiments were performed on this muscle mass because it could be freed from sympathetic nerve endings, by degeneration following removal of the abdominal sympathetic chain at a preliminary operation. The results obtained by stimulating the appropriate ventral nerve roots were closely similar to those obtained with the gastrocnemius, acetylcholine in 1 in 4×10^8 appearing in the venous effluent, during contractions of the muscle evoked by such stimulation. The mechanical stimulation of cholinergic sympathetic fibres and endings could, therefore, be excluded as a possible source. In one experiment the dorsal roots were also stimulated, with negative result.

(*d*) *Hind leg muscles of the frog.*

The object of these experiments was to identify the nerve fibres, responsible for the output of something like acetylcholine from the perfused frog's muscles, which Hess [1923] and others had observed with stimulation of the mixed nerve supply. The results were very similar to those we had already obtained with mammalian muscles. At the beginning of the perfusion, perceptible amounts of acetylcholine appeared in the venous fluid from the unstimulated muscles, and these disappeared as the perfusion was continued. Stimulation of the ventral roots, the sympathetic chain being extirpated, then regularly caused the appearance of acetylcholine in the venous fluid. The concentrations were low, 1 in 2×10^8 being the maximum obtained. The activity of the fluid, as in the mammalian experiments, slowly disappeared with continued perfusion, and reappeared on renewed stimulation. In two experiments the sympathetic chain was preserved and directly stimulated, causing vaso-constriction, as shown by retarded perfusion; but no acetylcholine appeared. In several experiments the dorsal roots were stimulated with completely negative results. In one experiment only, a perceptible

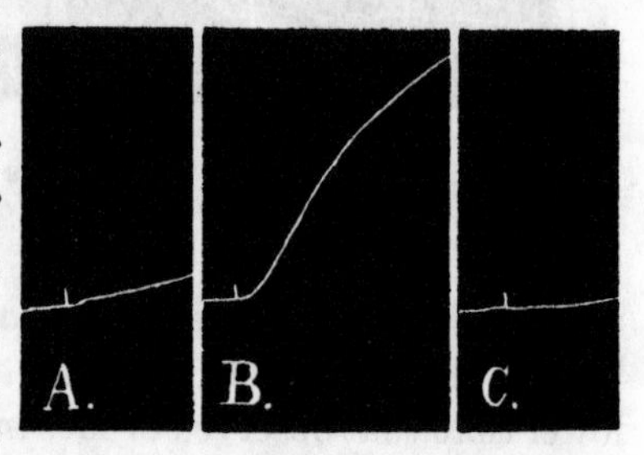

Fig. 4. Venous fluids from skinned hind-limbs of frog. A, before; B, during stimulation of motor roots; C, same fluid as B, after treatment with alkali.

quantity of something acting on the leech like acetylcholine appeared in the venous fluid at an interval of 10 min. after a period of sensory root stimulation. This result cannot properly be regarded as related to the stimulation, and it is only recorded as a presumably accidental, and obviously doubtful exception to otherwise completely negative results. Fig. 4 shows tests, on the eserine treated leech preparation, of a control fluid at A, fluid collected during effective motor root stimulation at B, and a portion of the same fluid after standing with alkali and reneutralization, at C. These experiments show that the substance detected by previous observers, as liberated during the stimulation of the mixed nerve supply, is produced by stimulation of the motor fibres to voluntary muscle, and has the instability to alkali of acetylcholine.

(2) *Direct stimulation of muscles*

We have seen that, when stimulation of the motor fibres failed to produce contractions of the muscle, acetylcholine was no longer liberated. It was possible, therefore, that it might come from the muscle fibres themselves, as a by-product of the contractile process. It was further possible that fluid collecting in inadequately perfused areas, or in the tissue between the muscle fibres, might acquire acetylcholine from some source, and that contractions might mechanically press some of it into the perfusion stream. To test these possibilities we first studied the effects of producing contractions of the muscle by direct stimulation. A normally innervated muscle cannot be effectively stimulated without stimulating, at the same time, the branches and endings of motor nerve fibres in its substance. We accordingly made comparative experiments on normal muscles, and on corresponding muscles denervated by degeneration. The muscles chosen were again the gastrocnemius of the cat and the dog and the quadriceps extensor femoris of the dog, the sciatic or crural nerve on one side having been divided aseptically under ether 10 days previously.

The results with direct stimulation of the normally innervated muscles were not different from those produced by stimulating a similar muscle through its motor nerve supply. The first samples of venous effluent contained some acetylcholine, which disappeared with further perfusion. Stimulation of the muscle then caused acetylcholine to appear, as with motor nerve stimulation, and it disappeared as usual with further perfusion. With the denervated muscle the results were entirely different. In the first place, even the earliest samples of venous fluid, after it had become free from blood, showed no significant activity on the most sensitive leech preparation. Further, though the muscle contracted powerfully in response to direct stimulation, no trace of activity was shown by the fluid collected during or after the stimulation period. A comparison between the results obtained with a normal cat's gastrocnemius and the denervated gastrocnemius of the other leg, similarly perfused and tested in succession, is illustrated in Fig. 5. The venous fluids were tested in 75 p.c. dilutions. At A the control fluid from the denervated muscle was tested, and at B the fluid collected while the muscle was contracting vigorously in response to direct stimulation. In neither case is any trace of activity to be detected. (The small and brief rises of the writing point are accidental mechanical effects of emptying

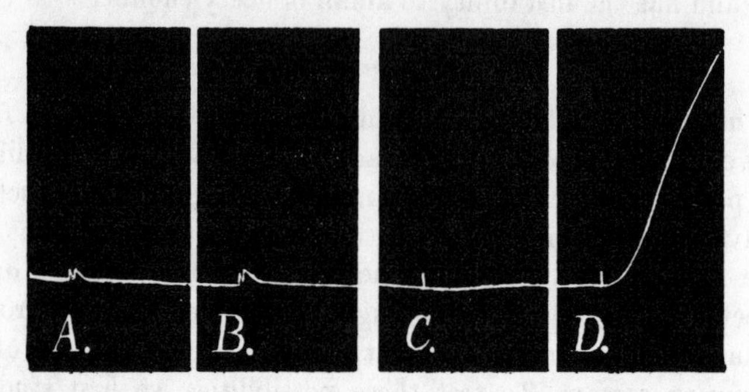

Fig. 5. Venous fluids from gastrocnemius muscles of cat, in 75 p.c. dilution. A and B from denervated muscle, A resting, B during stimulation; C and D from normal muscle, C resting, D during stimulation.

the testing bath and refilling with the test fluid.) At C the control fluid from the normal muscle was tested, and at D that collected during direct stimulation of that muscle.

An experiment on a dog's quadriceps, deprived of its sympathetic supply by degeneration but otherwise normally innervated, showed that it gave the usual yield of acetylcholine to direct stimulation. Another experiment on a sympathetically denervated quadriceps gave a result of special interest. This muscle had initially responded normally by contractions, with output of acetylcholine, to motor nerve stimulation. With continued perfusion successive periods of such stimulation had been progressively less effective, till finally the muscle no longer contracted or yielded acetylcholine, with renewed stimulation of the nerve. Direct stimulation being now applied, the muscle contracted vigorously, but no trace of acetylcholine appeared in the venous fluid. The mechanism concerned with transmission of the excitatory process from the nerve to the muscle fibres being exhausted, acetylcholine was no longer liberated, though the muscle contracted well. The muscle, in this respect, behaved now like one which had been denervated.

(3) *Curarine*

Some of the results described in the foregoing section might still find a conceivable explanation in the passage of acetylcholine from some source into stagnant fluid in the muscle, and the mechanical expression of some of this into the circulation by the contractions. It would be necessary then to assume, indeed, that this source disappeared with degeneration of the nerves and their endings, and that it could be exhausted with repeated stimulation of the nerve to the perfused muscle. It was important, however, to find conditions under which the effects of motor nerve stimulation could be tested, in the complete absence of muscular response. The use of curarine provided the required condition, and gave a decisive result.

Experiments were made on the perfused cat's tongue, stimulated through the hypoglossal nerves, deprived by degeneration of their sympathetic fibres. A little difficulty was experienced in adjusting the dose of curarine, in the necessary presence of eserine in the perfusion fluid, since eserine is well known to be an antagonist of the paralytic action of curare. With our usual concentration of eserine (1 in 5×10^5) curarine even in tenfold concentration (1 in 5×10^4) did not always produce an immediate complete paralysis to the onset of a series of maximal break shocks, applied to the hypoglossal nerves. The experimental stimulation, however, for collection of the venous sample was not carried out until all trace of such initial response had disappeared, the tongue remaining completely passive during the whole period of stimulation. A venous sample was collected, as usual, during a control period, and another during the ineffective stimulation. The presence of curarine made it difficult to test the fluids accurately on the leech muscle, the response of which to acetylcholine is slowly paralysed by curarine. Their activity on the blood-pressure of a cat under chloralose, however, could be readily determined with some accuracy, in comparison with those of known concentrations of acetylcholine. The Locke's solution, containing eserine and curarine in the indicated concentrations, had by itself no perceptible effect on the blood-pressure. Fig. 6 shows a comparison of the depressor effects of the venous effluents collected in such an experiment, with those of two dilutions of acetylcholine. The volume injected was in each case 2 c.c. At A the control venous fluid, collected just before stimulation, was injected; at B the fluid collected during hypoglossal stimulation; at C and D acetylcholine in dilutions of 1 in 2×10^8 and 1 in 10^8 respectively. It will be seen that the stimulation effluent is slightly less active than ACh. 1 in 10^8 and much more active than ACh. 1 in 2×10^8. That is to say, its activity in terms of acetylcholine is not different from that of a venous effluent obtained during hypoglossal stimulation from the uncurarized and actively contracting tongue, in an average, successful experiment. Between D and E 1 mg. of atropine was given, at E acetylcholine 1 in 10^8 and at F the stimulation effluent.

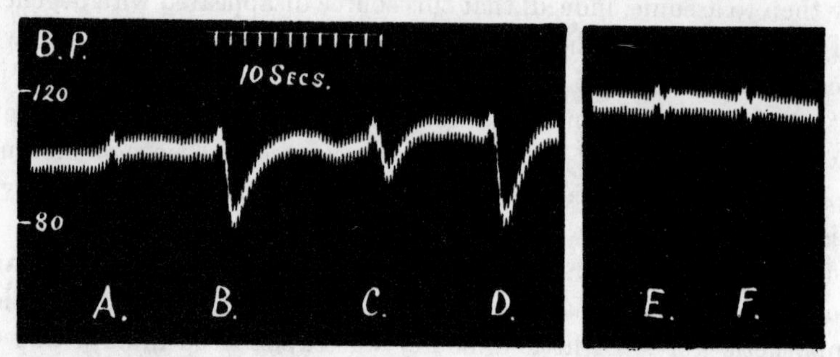

Fig. 6. Carotid blood-pressure of cat, eviscerated under chloralose. Eserine. A and B, 2 c.c. (i.v.) of venous fluids from cat's tongue under curarine. A before, B during motor nerve stimulation. C and D, 2 c.c. of ACh. dilutions; C, 1 in 2×10^8, D, 1 in 10^8. E and F, same as at B and D, after 1·75 mg. atropine.

The leech muscle is not rendered immediately insensitive to acetylcholine by curarine, so that, although no comparison can be obtained between different fluids containing curarine applied in succession to the same preparation, a qualitative contrast was easily demonstrated by applying them to two symmetrical strips of leech, previously sensitized by eserine. A preliminary test with acetylcholine showed that one strip was slightly more sensitive than the other. The control venous fluid from the tongue before stimulation was then tested on the more sensitive strip, and produced no perceptible effect, while the fluid collected during the period of hypoglossal stimulation caused prompt contraction of the other, less sensitive strip. A further application of the same fluid, however, was ineffective, the muscle being now affected by the curarine.

Brinkman and Ruiter [1924] had already shown that curare did not prevent the liberation from the frog's muscles, during nerve stimulation, of the substance stimulating the plain muscle of the cloaca. They stimulated the mixed nerves, however, and it seemed desirable to repeat the observation with stimulation of voluntary motor fibres only, through the ventral roots, and with a different physiological test for acetylcholine. The fluids were therefore compared on symmetrical leech strips, as above described, the hindlimbs of the frog being perfused with Ringer's solution containing both eserine and curarine. As with the tongue, the fluid collected before stimulation was inactive on one strip, while that collected during motor root stimulation, without contraction of the leg muscles, caused the usual contraction of the other leech preparation.

The experiments on the curarized tongue gave opportunity for a passing observation on another point. In the experiments with eserine but without curarine, we had always observed a conspicuous acceleration of the venous outflow when the hypoglossal nerves, freed from sympathetic fibres, were rhythmically stimulated. We had suspected that vasodilatation due to acetylcholine had some part in this effect; but we could not exclude mechanical action of the rhythmically contracting muscles, or the action of products of the contractile metabolism. In the experiments with curarine in addition to eserine contractions were abolished, but the outflow was still accelerated, though only to the extent of 10 p.c., when the hypoglossal nerve was stimulated. When Locke's solution containing curarine, without eserine, was used for perfusion, no vasodilatation was caused by hypoglossal stimulation, just as we had earlier found that, if eserine was not present, no acetylcholine appeared in the venous effluent. We may safely attribute this residual vaso-dilator effect, therefore, to acetylcholine leaking from the motor nerve endings and, if eserine is present, reaching the blood vessels, where it causes arterial dilatation and diffuses into the fluid passing through the capillaries. It should be pointed out that, in the absence of heavy doses of eserine, this action could play no part in the vaso-dilatation which accompanies the contraction of a muscle with normal circulation.

These experiments with curarine show quite definitely that the appearance of acetylcholine in the perfusion fluid is not directly or indirectly connected with contraction of the muscle fibres. Feldberg and Vartiainen, in their experiments on the vagus and sympathetic ganglia, failed to find evidence for the liberation of acetylcholine by impulses passing along uninjured nerve fibres in continuity. We have had the privilege of reading in advance a forthcoming paper by Gaddum and Khayyal, who have observed the liberation of a small quantity of a substance acting on the leech like ACh., when faradic stimulation is applied to the end of a longer stretch of vagosympathetic (preganglionic) nerve. Without anticipating the details of this publication, we may say that, if the intramuscular portions of motor nerves liberated acetylcholine at the maximum rate observed by Gaddum and Khayyal, it would not make a significant contribution to the amount which we have obtained from the tongue during motor nerve stimulation. The only supposition which accords with our facts is that a motor nerve impulse, on reaching the nerve ending, there liberates a small charge of acetylcholine, in close proximity to the motor end plate or other structure immediately subjacent to the nerve ending; and that this liberation is not affected by curarine, in a dose sufficient to prevent the response of the muscle fibre to excitation by the nerve impulse.

Discussion

There is an obvious analogy between the release of acetylcholine by impulses arriving at motor nerve endings in voluntary muscle, for which evidence has been here presented, and its release by impulses reaching the endings of preganglionic fibres in ganglionic synapses. In both cases the chief interest of the phenomenon centres in the question of its relation to the transmission of the excitatory process, with very little delay, across the anatomical discontinuity usually regarded as existing between the nerve terminal, on the one hand, and the ganglion cell or muscle fibre on the other. It may be noted that Samojloff [1925] concluded, from its high temperature coefficient (Av. = 2·37), that the conduction of excitation from motor nerve ending to voluntary muscle involved a chemical process of some kind. He even suggested that a stimulant substance might be liberated at the nerve endings, and drew an analogy from Loewi's observations on the nervous control of the heart.

In the case of the sympathetic ganglion, the direct excitatory effect upon the cells exhibited by acetylcholine with artificial application is of the "nicotine" type [Dale, 1914], being annulled by the secondary, depressant effect of nicotine itself, or by curarine in sufficient concentration [Brown and Feldberg, 1936], but resistant to atropine. It has been shown [Feldberg and Gaddum, 1934; Feldberg and Vartiainen, 1934] that acetylcholine may escape into the fluid perfusing a ganglion, during preganglionic stimulation, in a concentration which excites the cells to the output of impulses, when it is artificially injected into the perfused ganglion; and it has been argued that this indicates its release at the synapses in a concentration which cannot be without excitatory effect. In the case of the voluntary muscle the position is less clear, owing to the relatively low concentration of acetylcholine in the perfusion effluent during motor stimulation, the reason for which has been already discussed, and to the more complex nature of the stimulant actions of acetylcholine, when artificially applied to various types of voluntary muscle. Striated, involuntary muscle fibres can, indeed, be found, such as the outer, striated muscle coat of the intestine in certain fishes [Méhes and Wolsky, 1932], and the sphincter of the pupil in birds, innervated by parasympathetic nerve fibres and responding to the application of acetylcholine with a type of contraction closely simulating the effects of impulses in those nerves; and in such cases the effects of acetylcholine and of parasympathetic impulses differ from those on analogous layers of plain muscle, only in the quickness of the contractile response and in its suppression by curare instead of by atropine.

The reactions to acetylcholine shown by the voluntary striated muscles of different vertebrates, on the other hand, are complex and variable, and it is necessary to consider them in some detail. In the first place, certain normal muscles of the frog, the tortoise and the bird exhibit a prolonged type of contracture, of low tension, in response to nicotine and to various bases which resemble it in action. Riesser and Neuschlosz [1921] first showed that acetylcholine produced this type of effect in relatively low concentrations. It was subsequently shown [Sommerkamp, 1928; Wachholder and Ledebur, 1930] that there are wide differences in the sensitiveness of different muscles of the frog and the tortoise to this action of acetylcholine, corresponding to differences in the prominence of contracture in their natural functions; and, according to Plattner and Krannich [1932], the same muscles show corresponding differences in the amounts of acetylcholine which they yield to artificial extraction. This contracture is relatively resistant to curare, and somewhat sensitive to atropine [Riesser and Neuschlosz, 1921]. Although it seems likely that this type of response has some relation to the normal, functional contractures exhibited by such muscles, it would be difficult to make a case for a participation of acetylcholine in the excitation of voluntary muscle fibres to quick contractions by motor nerve impulses, if this were the only detectable type of reaction shown by skeletal muscle to its artificial application. There is, however, another type of reaction, which the predominant interest given to muscles showing the contracture has been apt to obscure.

Langley [1907], in his experiments with nicotine on the frog's sartorius, showed that it first produced twitches, and a quick type of contraction resembling a short tetanus, prior to the slowly developing contracture. He made the significant observation that, with punctiform application of the alkaloid in low concentrations, the quick reactions could only be elicited from the neighbourhood of the nerve endings, from which the excitation was apparently propagated, whereas the contracture, while it could be produced in any part of the muscle fibres, usually remained localized to the region of application. Sommerkamp [1928] found that some frog's muscles give only quick contractions when immersed in acetylcholine, while others respond mainly by contracture. In the ilio-fibularis muscle the contracture was limited to a small part of the muscle, separable by dissection; the remainder, freed from it, gave only a quick, evanescent contraction, when immersed in ACh. 1 in 10^5. The normally innervated voluntary muscles of the mammal, with the exception of the small muscles moving the eyeball [Duke-Elder, 1930], are often regarded as insensitive to acetylcholine. They do, indeed show a striking contrast in this respect, under ordinary experimental conditions, to muscles denervated by motor fibre degeneration, which respond to very small injections of acetylcholine with a slow type of contraction. Feldberg and Minz [1931] and Feldberg [1933*b*], however, observed quick contractions and fibrillation of normal mammalian muscles when acetylcholine was injected in moderate doses (0·01–0·2 mg.) into the arteries supplying them. According to the Simonarts [1935*a*] the response of normal mammalian muscles to acetylcholine is very readily depressed by ether. In rabbits and cats in the early stages of anæsthesia by a barbiturate, or in spinal preparations freed from the preliminary ether by prolonged ventilation, he observed quick contractions of the muscles, unaffected by fresh section of the nerve supply, when doses of acetylcholine of the order of 1 mg. were injected intravenously. The reaction was unaffected by atropine, which, indeed, had to be given in advance, to eliminate the effects of such doses of acetylcholine on the heart and the blood vessels. On the other hand, the Simonarts found them to be suppressed by a quaternary ammonium salt having a curare action. Acetylcholine solutions applied to the bared surface of a muscle, or injected by a fine needle into its substance, caused brisk and fugitive contractions of fibres, or of whole bundles, localized to the neighbourhood of the application. In a more recent paper A. Simonart [1935*b*] describes results obtained with the more effective method of arterial injection. As was to be expected, the threshold dose of ACh. for normal cat's muscle was much lower by this method, and the tensions recorded with larger doses were of the order of those obtained with maximal indirect faradization. Curare readily abolished these effects. We hope to deal further with this aspect of our problem in a later paper.

Admittedly there are points still requiring further investigation. The nature of the reaction of denervated mammalian muscle to the quicker but less sensitive reaction of normal mammalian muscle, on the one hand, and to the persistent contracture of low tension shown by certain frog muscles on the other, does not seem to be adequately defined by the present evidence. In the case of the mammalian muscles, the facts still suggest that the presence of the normal nerve ending in some way hinders the access, to the sensitive point on the muscle fibre, of acetylcholine applied from without; just as we found that, with normal circulation, there appears to be some hindrance to the escape of acetylcholine from the neighbourhood of the motor nerve endings, where it is liberated by nerve impulses. It may be remarked that such restriction would not be unexpected, in a muscle consisting of fibres which can act as independent physiological units. However that may be, the point requiring emphasis for our purpose is the distinction between the quick contractions elicited by acetylcholine, apparently, on Langley's evidence, by excitation of the structures immediately subjacent to the motor nerve endings, and the persistent contractures produced in certain amphibian and other muscles.

As in the case of the ganglion synapses, the failure of eserine under normal conditions to facilitate or to prolong the excitatory effect of a motor nerve impulse, appears to have been regarded as evidence against the participation of acetylcholine in the transmission of the excitatory process [Kruta, 1935]. In neither case does the argument appear to us well founded. The motor nerve ending is in such immediate contact with the nucleated end plate or other structure, from which the excitatory process in the muscle fibre must start, that there is no room for destruction of acetylcholine during diffusion to its points of action, from which eserine might protect it. On the other hand, when acetylcholine has to reach muscle fibres by diffusion from a solution in which the whole muscle is immersed, the effect of low dilutions in producing contracture should be enhanced by eserine, as Kruta found it to be; just as we have found that eserine is necessary to protect acetylcholine during diffusion from the points of its release into the blood vessels.

The transmission of the effects of nerve impulses by a chemical substance, reaching the effector cells by diffusion, is now an accepted fact in the case of simpler types of contractile cells and tissues, usually displaying an automatic activity which the nerve impulses may modify in either direction. Such transmission by a diffusible stimulant can now be traced in the nervous control of most involuntary muscle, including, as we have seen, some which is striated and relatively quick in contraction. The question which here concerns us is whether in voluntary striated muscle, specialized for the quick contraction of individual fibres in response to nerve impulses, and normally at rest in their absence, this more primitive, chemical method of transmission has been superseded by an entirely different one, in which the chemico-physical disturbance constituting the nerve impulse passes, by continuous propagation, on to the muscle fibre; or whether, on the other hand, the required specialization has been effected by concentrating the release and the action of the chemical stimulant at the point of immediate contact of the nerve ending with the muscle fibre. An analogous question has already arisen in connection with the response to nerve impulses of more than one plain muscle structure. Henderson and Roepke [1934] find that stimulation of the pelvic nerve causes two kinds of reaction of the plain muscle of the urinary bladder—a quick contraction and a maintained tonus. They find that atropine readily abolishes the maintained tonus, just as it abolishes the response to acetylcholine applied from without, leaving the quick contraction. Similarly Bacq and Monnier [1935] distinguish a quick and a slow component in the response to stimulation of the cervical sympathetic nerve of the plain muscle retracting the nictitating membrane. They find that a synthetic substance "F. 933" depresses the slow component, and with it the response to adrenaline, leaving the quick reaction unaffected, or even apparently enhanced. In both cases the observers suggest that the slower reaction is due to chemical transmission, but that this form of control is supplemented by the presence of certain nerve fibres ending directly in plain muscle cells, the propagated change constituting the nerve impulse being directly continued from these to the plain muscle, without chemical intervention. The number of nerve endings in such sheets of plain muscle is known to be small, in relation to the number of muscle cells. It appears to us that the quick reactions, in both these cases, may equally well be explained by the liberation of the transmitter in high concentration in immediate relation to, possibly within the limiting membranes of, the directly innervated cells; the slow reaction being then evoked by its escape and secondary diffusion on to other cells, in a manner analogous to its artificial application through the blood stream or from the surface. In the case of the vaso-dilator effect of the chorda tympani on the blood vessels of the tongue, which is typically resistant to atropine, there is direct evidence of such escape by diffusion of a chemical transmitter with all the properties of acetylcholine, causing contracture of adjacent voluntary muscle fibres if these have been denervated [Bremer and Rylant, 1924; Dale and Gaddum, 1930], and appearing in the blood or perfusion fluid flowing from the tongue [Feldberg, 1933*a*; Bain, 1933].

If Henderson and Roepke, Bacq and Monnier were right in postulating a supplementary, direct transmission of nerve impulses to plain muscle, when quick contraction is required, we should expect it to supersede entirely the chemical method of transmission, for nerve impulses causing excitation of the very rapid and individually reacting fibres of skeletal muscle, giving a single twitch, with minimal transmission delay, in response to each impulse. Similarly we should expect such a direct method of transmission at a ganglionic synapse, where each preganglionic impulse can evoke a single postganglionic impulse, again with minimal transmission delay. Such a conception, however, leaves us with no explanation of the release of acetylcholine at the preganglionic and the motor nerve endings. This can hardly be the survival of an archaic form of transmission, no longer having any function. In the ganglion acetylcholine has been shown to be liberated in a concentration which effectively stimulates ganglion cells; while in the muscle we have shown that, when the liberation of acetylcholine fails by exhaustion, the excitation of the muscle no longer occurs. There seem to be two possibilities.

(1) That the propagated disturbance in the nerve fibre is directly transmitted to the effector cell, but that the latter cannot accept it for further propagation unless sensitized by the action of the acetylcholine, which appears with its arrival at the nerve ending. Such an hypothesis might be stated in terms of Lapicque's well-known conception, by supposing that the action of acetylcholine shortens the chronaxie of the nerve cell, or of the motor end plate of the muscle fibre, so that it is momentarily attuned to that of the nerve. H. Fredericq [1924] has observed, indeed, a shortening of the chronaxie of heart muscle by acetylcholine.

(2) That the acetylcholine, in these as in other cases, acts as the direct stimulant of nerve cell or muscle end plate, releasing an essentially new propagated wave of excitation in postganglionic nerve or muscle fibre, which, however, may so resemble that in the preganglionic or motor nerve fibre as to simulate an unbroken propagation. On this view there is no introduction of a new form of transmission, in evolution from the slowest and most primitive to the most rapid and specialized. The required rapidity of transmission is attained by concentrating the release of the chemical transmitter on the actual surface of the responsive structure.

Of the two possibilities, the latter appears to us to be more easily reconciled with the facts yet available concerning transmission at ganglionic synapses. The former would provide an explanation, alternative to that which we have considered earlier, for the apparently low sensitiveness of some normal muscles to stimulation by acetylcholine. The shortness of the delay in transmission appears to cause no greater difficulty for one conception than the other. The action of curare is explicable, in either case, as rendering the receptive element resistant to the action of acetylcholine, whether this be merely to sensitize or directly to stimulate. On the existing evidence we favour the second conception, while admitting that further facts are required for the exclusion or the establishment of either.

As in the case of transmission at the ganglionic synapse [cf. Feldberg and Vartiainen, 1934], either of the above conceptions of the function of acetylcholine, in the transmission of excitation to the voluntary muscle fibre, would require not only its liberation in immediate relation to the excitable structure, but presumably its very rapid disappearance when the excitatory wave in the muscle had been started. The known extreme liability of acetylcholine to the action of an esterase naturally comes to mind in that connection; but there is no direct experimental evidence to justify an assumption that this esterase is, in fact, responsible for removing acetylcholine from the site of its action in this case, and that eserine would, therefore, increase its persistence at that site. Nor can we predict the effect of such persistence on the transmission under particular conditions, if it could be proved to occur. All that our evidence shows is, that acetylcholine which has escaped from the sites of its liberation requires protection by eserine to enable it to diffuse into a fluid perfused through the blood vessels.

SUMMARY

1. Stimulation of the motor nerve fibres to perfused voluntary muscle causes the appearance of acetylcholine in the venous fluid.

2. Direct stimulation of a normal muscle, or of one deprived only of its autonomic nerve supply, has a similar result; but when the muscle is completely denervated no acetylcholine appears in response to effective stimulation.

3. When transmission of excitation from the nerve to the perfused muscle is prevented by curarine, stimulation of the motor nerve fibres causes the usual release of acetylcholine.

4. When conduction from motor nerve fibres to perfused muscle fails from exhaustion, after repeated stimulation, acetylcholine is no longer released by stimulation of either nerve or muscle.

5. The function of acetylcholine in the transmission of excitation from nerve to voluntary muscle is discussed.

We are indebted to Dr H. King for the supply of pure curarine, and to Dr H. Schriever for kind assistance in certain experiments.

REFERENCES

Bacq, Z. M. and Monnier, A. M. (1935). *Arch int. Physiol.* **40**, 467, 485.
Bain, W. A. (1933). *Quart. J. exp. Physiol.* **23**, 381.
Barsoum, G., Gaddum, J. H. and Khayyal, M. A. (1934). *J. Physiol.* **82**, 9*P*.
Bremer, F. and Rylant, P. (1924). *C. R. Soc. Biol.*, Paris, **90**, 982.
Brinkman, R. and Ruiter, M. (1924). *Pflügers Arch.* **204**, 766.
Brinkman, R. and Ruiter, M. (1925). *Ibid.* **208**, 58.
Brown, G. L. and Feldberg, W. (1936). *J. Physiol.* **86**, 10*P*.
Bülbring, E. and Burn, J. H. (1935). *Ibid.* **83**, 483.
Dale, H. H. (1914). *J. Pharmacol.*, Baltimore, **6**, 147.
Dale, H. H. (1935*a*). *Proc. R. Soc. Med.* **28**, 15.
Dale, H. H. (1935*b*). *Nothnagelvorträge*, Nr. 4. Wien: Urban u. Schwarzenberg.
Dale, H. H. and Feldberg, W. (1934). *J. Physiol.* **81**, 39*P*.
Dale, H. H. and Gaddum, J. H. (1930). *Ibid.* **70**, 109.
Dale, H. H. and Gasser, H. S. (1926). *J. Pharmacol.*, Baltimore, **29**, 53.
Duke-Elder, W. S. and Duke-Elder, P. M. (1930). *Proc. Roy. Soc.* **107**, 332.
Feldberg, W. (1933*a*). *Pflügers Arch.* **232**, 88.
Feldberg, W. (1933*b*). *Ibid.* **232**, 75.
Feldberg, W. and Gaddum, J. H. (1934). *J. Physiol.* **81**, 305.
Feldberg, W. and Minz, B. (1931). *Arch. exp. Path. Pharmak.* **163**, 66.
Feldberg, W., Minz, B. and Tsudzimura, H. (1934). *J. Physiol.* **81**, 286.
Feldberg, W. and Vartiainen, A. (1934). *Ibid.* **83**, 103.
Franel, L. (1935). *Arch. int. Physiol.* **41**, 256.
Frank, E. and Katz, R. A. (1921). *Arch. exp. Path. Pharmak.* **90**, 149.
Frank, E., Nothmann, M. and Hirsch-Kauffmann, H. (1922). *Pflügers Arch.* **197**, 270.
Fredericq, H. (1924). *C. R. Soc. Biol.*, Paris, **91**, 1171.
Geiger, E. and Loewi, O. (1922). *Biochem. Z.* **127**, 66.
Henderson, V. E. and Roepke, M. H. (1934). *J. Pharmacol.*, Baltimore, **51**, 97.
Hess, W. R. (1923). *Quart. J. exp. Physiol.* (Suppl.), p. 144.
Hinsey, J. C. and Cutting, C. C. (1933). *Amer. J. Physiol.* **105**, 535.
Kruta, V. (1935). *Arch. int. Physiol.* **41**, 187.
Langley, J. N. (1907). *J. Physiol.* **36**, 347.
Langworthy, O. R. (1924). *Johns Hopk. Hosp. Bull.* **35**, 239.
Méhes, J. and Wolsky, A. (1932). *Arb. Ung. Biol. Forsch.* **5**, 139.
Plattner, F. (1932). *Pflügers Arch.* **130**, 705.
Plattner, F. (1933). *Ibid.* **232**, 342.
Plattner, F. and Krannich, E. (1932). *Ibid.* **229**, 730; **230**, 356.
Riesser, O. and Neuschlosz, S. M. (1921). *Arch. exp. Path. Pharmak.* **91**, 342.
Samojloff, A. (1925). *Pflügers Arch.* **208**, 508.
Shimidzu, K. (1926). *Ibid.* **211**, 403.
Simonart, A. and Simonart, E. F. (1935*a*). *Arch. int. Pharmacodyn.* **49**, 302.
Simonart, A. (1935*b*). *Ibid.* **51**, 381.
Sommerkamp, H. (1928). *Arch. exp. Path. Pharmak.* **128**, 99.
Wachholder, K. and Ledebur, J. (1930). *Pflügers Arch.* **225**, 627.

2. GAMMA-AMINOBUTYRIC ACID: An Inhibitory Transmitter

Sherrington postulated the existence of inhibitory synapses in the vertebrate central nervous system from his studies of spinal reflexes (1906, 1940). Years later his pupil, J. C. Eccles, demonstrated them (e.g., **Coombs, Eccles, and Fatt, 1955**). In the meantime van Harreveld had demonstrated inhibition at the crayfish nerve-muscle junction (see **Fatt and Katz, 1953**). Gamma-aminobutyric acid (GABA) was identified as the major inhibitory compound in extracts of beef brain (Bazemore, Elliott, and Florey, 1957) by its effect on the crayfish stretch receptor. In the vertebrate central nervous system there is now convincing but incomplete evidence for an inhibitory transmitter role at synapses on certain cortical neurons (e.g., Krnjević and Schwartz, 1967) and at Purkinje axon synapses on Deiters neurons of the cerebellum (Obata, Ito, Ochi, and Sato, 1967). The evidence that GABA is the inhibitory transmitter at crustacean neuromuscular junctions is now quite complete.

The three papers by Kravitz and co-workers on GABA reproduced here provide an example of a "modern" solution to the biochemical problems of identifying a neurotransmitter and an interesting contrast with the older ACh work of Loewi and Dale. The criteria for proof of a transmitter are fulfilled as follows. **Kravitz, Kuffler, and Potter (1963)** show the localization of GABA in the presynaptic axons; **Kravitz, Molinoff, and Hall** (1965) show the existence in the presynaptic axons of the metabolic machinery for producing the transmitter; and **Otsuka**, Iversen, **Hall, and Kravitz (1966)** demonstrate release of the transmitter, GABA, by stimulation of the presynaptic axons. These papers do not consider the effects of artificially applied transmitter on the postsynaptic structures. TAKEUCHI AND TAKEUCHI (1965) give such evidence in the case of crayfish nerve-muscle junctions and KUFFLER AND EDWARDS (1958) do so in the case of crayfish stretch receptor inhibitory synapses.

19

J. Neurophysiol. (1963) **26**, 739–751

GAMMA-AMINOBUTYRIC ACID AND OTHER BLOCKING COMPOUNDS IN CRUSTACEA

III. THEIR RELATIVE CONCENTRATIONS IN SEPARATED MOTOR AND INHIBITORY AXONS[1]

E. A. KRAVITZ,[2] S. W. KUFFLER, AND D. D. POTTER

Neurophysiology Laboratory, Department of Pharmacology, Harvard Medical School, Boston, Massachusetts

(Received for publication February 1, 1963)

INTRODUCTION

IN THE TWO PRECEDING PAPERS it was reported that gamma-aminobutyric acid (GABA) was the most active of ten blocking substances extracted from the nervous systems of lobsters and crabs. The concentrations of GABA in several peripheral nerves were measured and found to be highest in a nerve that contained only one motor and one inhibitory axon. It was natural to wonder if GABA was specifically concentrated in the inhibitory axon. In the present study, therefore, we have isolated individual axons and found, within the limits of sensitivity of our enzymic assay, that motor fibers contained no GABA while inhibitory neurons contained surprisingly high concentrations. Other blocking compounds were also extracted from separated motor and inhibitory axons, but in contrast to GABA, these are present in both neuron types, as is the precursor of GABA, glutamic acid.

The current studies strongly suggest that GABA has a specific physiological role confined to inhibitory neurons.

METHODS

Two types of experiments were performed. The first was to compare enzymically the GABA contents of isolated motor and inhibitory axons and the second was to compare chromatographically the concentrations of other blocking compounds.

Isolation of single motor and inhibitory axons. Long stretches of isolated fibers were most easily obtained from a small nerve bundle in the meropodite (Fig. 1*A* in ref. 18). This bundle contains two prominent axons, one motor and one inhibitory, running side by side unbranched for 90–110 mm. (in 8- to 12-lb. lobsters) with a group of smaller sensory fibers. The motor fiber innervates the stretcher of the carpopodite and the opener (abductor) of the dactyl in the walking leg. The inhibitory axon also innervates the stretcher of the carpopodite but beyond that muscle it separates from the motor axon to supply the closer (adductor) of the dactyl (28). The axons are of similar and uniform diameter (50–60 μ) during their course in the meropodite. The tissue around them and the accompanying smaller nerve fibers were removed and the two axons separated and identified by electrical stimulation through fluid electrodes, noting excitation or inhibition of the innervated

[1] This research was supported by Public Health Service Grants NB-02253-04, NB-03813-02, and NB-K3-7833.

[2] Special Research Fellow of the National Institute of Neurological Diseases and Blindness.

muscles (8). During dissection and separation most of the adhering connective tissue was removed (Fig. 1). Each dissection took about 3 hours and was performed at 6–12° C. in unbuffered saline containing (mM/liter) NaCl 462; KCl 15.6; and $CaCl_2$ 25.9. Several lengths, each up to 120 mm., of the two fiber types were pooled to obtain sufficient material for the analyses (Table 1). Our emphasis lies in comparing the relative amounts of certain materials contained in fibers of similar diameter and of similar length. As far as possible the handling and chemical treatment of fibers were identical.

For GABA assay, the separated fibers were homogenized in ice-cold 5% trichloroacetic acid or 0.5 M acetic acid and assayed as described below. For blocking compounds other than GABA the axons were homogenized in 0.5 M acetic acid, and after centrifugation the supernatant fluid was applied to a continuous flow electrophoresis apparatus as described in the preceding paper (18). The resulting physiologically active regions, B_1, B_2, and B_3 (18) were fractionated further by ascending paper chromatography in the solvent butanol-acetic acid-water (60-15-25). Region E was chromatographed in the solvent ethanol-acetic acid-water (160-15-25) at 4° and two strips, one containing betaine, the other aspartic acid plus glutamine, were eluted, the first for physiological assay, the second for further fractionation (paper chromatography in butanol-acetic acid-water, and phenol-water (70-30): for further details on the methods, see first paper, 7).

GABA assays were also done on the pairs of motor and inhibitory fibers which innervate the opener muscle of the dactyl of the walking leg of lobsters (Fig. 1*B*, ref. 18). No other fibers are known to accompany these two axons in their course over the inner surface of the muscle. The axons were separated for a few millimeters and then stimulated with a fluid electrode. In some experiments recordings were made from a muscle fiber with an intracellular electrode to observe the resulting excitatory or inhibitory junctional potentials. In other experiments the excitatory nerve alone was identified simply by observing muscular contractions. The two fibers were then pulled apart with whatever connective tissue adhered to them, extracted, and treated in the same way as the fibers from the meropodite. There were a number of uncertainties in these dissections. There was a possibility of false identification through escape of the stimulus from the fluid electrode to the fiber lying nearby. Furthermore, when many short lengths of axon had to be pooled for the spectrophotometric enzymic assay (Table 2, exp. 1–4), the chances were increased that segments of inhibitory axon were mistakenly placed in the excitatory pool. Moreover, branches of the two fibers sometimes ran in the main connective tissue sheath for a distance (Fig. 2, ref. 18), and there was a chance that they did not separate properly with the parent stems. Separation was also more difficult than in the meropodite due to the relatively small size and tapering diameters towards the periphery; the range of diameters of isolated fibers in various sizes of lobsters was 40–15 μ. The method was improved by increasing the sensitivity of the enzymic assay until single axon segments of 30–40 mm. could be used. But, in contrast to the meropodite dissection, the possibility of contamination of one fiber with fragments of the other remained.

Enzyme assays for GABA. When sufficient material was available, assays were performed spectrophotometrically as described in the preceding paper (18). If less than 10^{-9} moles of GABA was expected, a fluorimetric procedure was used. The enzyme preparation was the same as that in the spectrophotometric method, but the reduced nicotinamide adenine dinucleotide phosphate (NADPH) was measured by chemical conversion to its oxidized form (NADP) which was measured fluorometrically. The techniques were based on the procedures of Lowry and co-workers (22, 23) and were similar to the GABA assay described by Hirsch and Robins (17). Assay tubes contained enzyme, NADP (9×10^{-4} μmoles), β-mercaptoethanol (0.09 μmole), α-ketoglutarate (9×10^{-3} μmoles), tris buffer, pH 7.9 (1.8 μmoles), and experimental samples in 12-μl. volumes. Zero-time controls (to which α-ketoglutarate had not been added) were run with experimental samples to establish tissue blank levels of fluorescence, and five GABA standards (within the range 2.5×10^{-11} moles to 1.25×10^{-10} moles) were run with each assay. Incubations were for 45 min. at room temperature except for zero-time controls. The reactions were ended, and excess NADP destroyed, by pipetting 10 μl. of the samples into 50 μl. of 0.25 M Na_3PO_4:0.35 M K_2HPO_4 buffer, and heating at 60° for 15 min. One hundred microliters of 10 N NaOH containing 0.03% H_2O_2 was added and the tubes were heated at 60° for 10 min. Finally, 1 ml. of water was added and the fluorescence measured in a Farrand fluorometer, model A, with a Corning 5860 primary filter (365 mμ) and a Baird-Atomic B-1 type interference filter (460 mμ peak transmission) as the secondary filter.

Results

GABA content of motor and inhibitory fibers. The results of the analyses of isolated axons from the meropodite region are presented in Table 1. The total material was over 5 m. of axon length. In each case we have compared the GABA content of paired inhibitory and excitatory neurons; these were of similar diameter and were treated in the same way.

No GABA was found in the meropodite excitatory axons; if it was present at all, it was in amounts too small to be detected by the enzymic assay.

Table 1. GABA content of isolated motor (M) and inhibitory (I) axons from the meropodite

Nerve	Length, mm.	Separation Technique*	GABA, μg/100 mm.	Total GABA		
				Moles	Threshold,† moles	I/M_T‡
M-I	200	Columns	1.2	2.35×10^{-8}		
I	370	Columns	1.0	3.61×10^{-8}		90
M	350		0	0	4×10^{-10}	
I	500	Columns	2.4	1.2×10^{-7}		300
M	500		0	0	4×10^{-10}	
I	500	Electrophoresis	1.1	5.3×10^{-8}		130
M	500		0	0	4×10^{-10}	
I	1,085	Electrophoresis	1.25	1.32×10^{-7}		1,200
M	1,085		0	0	1.1×10^{-10}	

* Separation techniques: columns refers to the Dowex-50-H^+ techniques described in preceding paper (18); electrophoresis means crude extracts were separated by electrophoresis and GABA regions were pooled and assayed for GABA. † The minimum amounts of GABA which could be detected were 10^{-10} moles spectrophotometrically and 8×10^{-12} moles fluorimetrically. The threshold values in the table are higher than this because only part of the sample was used for the assay. ‡ No GABA was detected in the M fiber. This calculation uses the threshold value for GABA (M_T) and represents a minimum ratio of GABA contents.

Therefore, no reliable value can be given to the ratio of GABA concentrations in the two types of fibers. However, it was possible to calculate how much GABA could have been present in the motor fibers and yet have remained undetected by the assay ("Threshold" in Table 1); this amount was dependent on the sensitivity of the assay and the quantity of material used. The threshold value was then compared to the measured GABA content of the corresponding inhibitory fiber to give the minimum value of the ratio of GABA concentrations in the two types of fibers (I/M_T of Table 1). In our most striking experiment, in which the sensitivity of the assay was highest, if GABA was present at all in the motor axons it could not have been more than 1 part in 1,200 of that in the inhibitory fibers. In the other tests, smaller amounts of tissue were used, the assay was less sensitive and therefore the I/M_T values were lower.

In one experiment of this series we also assayed combined motor and inhibitory fibers (no. 1). These contained 1.2 μg. of GABA/100 mm., an amount per unit length similar to the average of inhibitory axons alone.

The GABA contents of isolated axons from the surface of the opener muscle were compared in seven of the eight experiments of Table 2. Assays in the first four experiments were done spectrophotometrically and to obtain

Table 2. GABA content of isolated M and I axons from the opener muscle surface

Nerve	Length, mm.	Separation Technique*	GABA, μg/100 mm.	Total GABA			
				Moles	I/M	Threshold,† moles	I/M_T‡
M-I	320	Columns	0.11	3.4×10^{-9}			
I	350	Columns	0.14	4.6×10^{-9}	11.5		
M	340		0.012	4×10^{-10}			
I	1,010	Columns	0.19	1.91×10^{-8}	9.9		
M	1,010		0.02	1.96×10^{-9}			
I	300	Columns	0.37	1.07×10^{-8}			
M	275		0	0		10^{-9}	10
Connective tissue	§	Columns	0	0		10^{-9}	
I	30	No separation	0.16	4.8×10^{-10}			
M	30		0	0		2×10^{-11}	24
I	36	No separation	0.14	4.8×10^{-10}			24
M	36		0	0		2×10^{-10}	
I	30	No separation	0.15	4.3×10^{-10}	6.2		
M	30		0.023	6.9×10^{-11}			
I	27	No separation	0.13	3.3×10^{-10}	4.8		
M	27		0.026	6.9×10^{-11}			

* No separation, GABA assays were performed directly on crude extracts; columns refers to the Dowex-50-H$^+$ techniques described in preceding paper (18). † See Table 1 for explanation of threshold. ‡ See Table 1 for explanation of I/M_T. § The volume of this tissue was greater than that of the axons.

sufficient material many short fiber segments (8–20 mm. each from 2-lb. lobsters) were pooled. In the last four experiments with single pieces of individual axons from large lobsters, the fluorometric assay was used. In three cases no GABA was found in motor fibers and the I/M_T values were 10 and 24. On the other hand, traces of GABA were found in four experiments, including two (no. 7 and no. 8) in which single fibers were assayed. In experiment no. 4 the connective tissue that was removed from the axons was also analyzed but no GABA could be detected. It should also be noted that while the GABA content of the inhibitory axon of the opener was only 0.1–0.2 μg/100 mm. length, approximately one-tenth that of the meropodite axons, the fiber diameter was smaller; in addition, as mentioned earlier (METHODS), a different inhibitory neuron was involved while the motor axon to the opener was the same as the one analyzed in the meropodite.

The results on the opener fibers raise the unexpected possibility that while GABA is absent in a motor axon along its course, it appears toward the periphery of the same cell. However, such a change in the chemistry of a motor axon in respect to GABA should not be considered seriously on the basis of our results because the experimental conditions were not satisfactory.

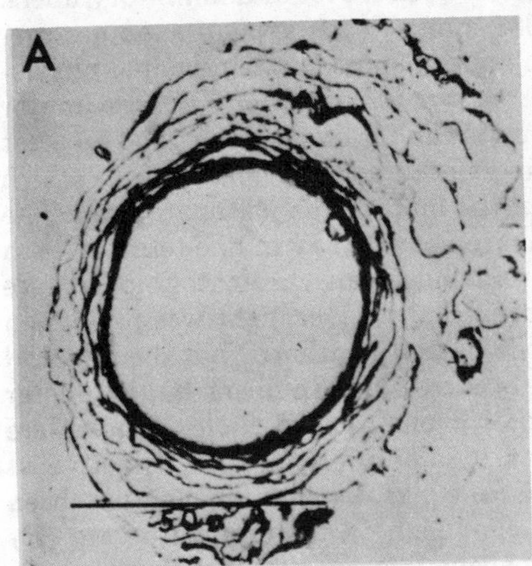

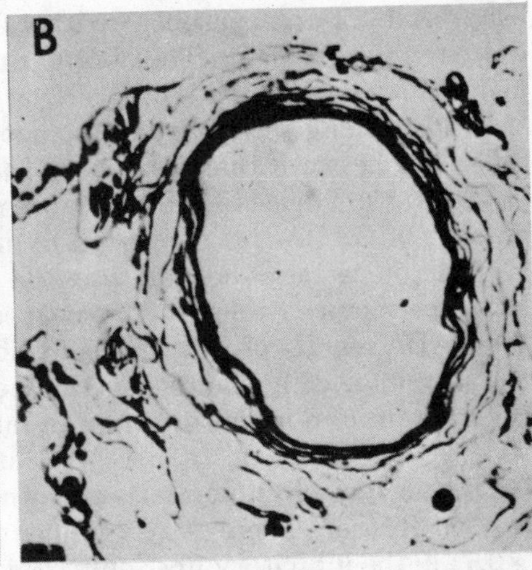

FIG. 1. Cross sections of isolated motor (*A*) and inhibitory (*B*) axons from the meropodite of the walking leg of a large (10 lb.) lobster. The axons were fixed in Millonig's solution after isolation, embedded in Araldite and stained with toluidine blue. In this sample the motor axon is somewhat better cleaned of connective tissue. The scale is the same for the two fibers. Note that diameter of the motor axon is greater in the meropodite than in its peripheral portion (Fig. 2, ref. 18). The sections were kindly provided by Dr. W. Fahrenbach.

There was a good chance of contamination of the excitatory axons by sections of the inhibitory axons, as already discussed (METHODS). It is hoped that with improved techniques of separation and even more sensitive assay methods the opener nerves may be studied more profitably.

One should note that the isolated fibers we analyzed were still a complex tissue because the dissections left varying amounts of connective tissue and the Schwann cell layer around the neuron membrane as indicated in the histological section of isolated axons in Fig. 1. It seems quite unlikely that connective tissue around motor fibers is different from that surrounding inhibitory axons. As far as Schwann cells are concerned there is no evidence at present that their chemical composition differs around various axon types. We therefore attribute the differences in GABA content to the axoplasm. From the data of Table 1 one obtains a GABA content of about 0.5 g. of GABA/100 g. of wet wt. of inhibitory axon. The following figures were used for this calculation: average fiber diameter 60 μ; GABA content 1.4 μg/100 mm.; density of axoplasm, 1.0.

Distribution of other blocking compounds in motor and inhibitory fibers. We did not have specific sensitive enzymic assays for the blocking compounds other than GABA (18). Therefore the relative amounts of these compounds were estimated by comparing the density of Ninhydrin-stained spots on paper chromatograms, or, in the case of betaine, by elution from the paper chromatograms and bio-assay. While these techniques are not quantitative, they easily detected marked differences between motor and inhibitory fibers. The results of an experiment with 1,085 mm. of each type of axon are presented in Figs. 2 and 3. The Ninhydrin-stained spots in the principal physiologically active regions (B_1, B_3, E) are shown for inhibitory (I) and motor (M) fibers along with standard compounds run for comparison. The greatest difference between the two types of fibers occurs in the region B_1 to which GABA runs. There is no GABA spot in the motor axon chromatogram. The absence of GABA was confirmed by fluorometric assay of one-tenth of each B_1 fraction set aside at the time the remainder was chromatographed (see last experiment, Table 1). A spot-absorbing ultraviolet light was present in the GABA regions of I as well as M fibers (lightly outlined) but the material was physiologically inactive. β-Alanine occurred in both fibers, but in greater concentration in inhibitory than in motor axons (dotted circle). Since there was far less β-alanine than GABA and it is only about 1/50 as active as GABA in our physiological assay, this unequal distribution has not been studied further. There is a prominent spot high on the chromatogram (R_F $= 0.7$) in the inhibitory fiber only. We believe this compound to be the cyclic amide (lactam) of GABA, for it has the R_F value and staining characteristics (yellow, slowly turning blue) reported for GABA lactam (27). Presumably it was formed during spotting for chromatography. If this is the case, it is not surprising that we found it only in the inhibitor. The experiment illustrated in Fig. 2 was the only one in which we have seen this fast-moving spot, a fact which strengthens our suspicion that it was an artifact. One of the lower spots in region B_1 is the blocking substance "unknown A" (18), a very weakly active guanido compound which is present equally in the two fibers.

The blocking region, B_3, was subdivided; the bulk of the taurine was in one fraction, (on right in Fig. 2) the remaining taurine and homarine in the other (left). These blocking compounds were present in about equal concentration in the two types of fibers. The position and relative amounts of homarine on the chromatogram were determined with ultraviolet light (254 mμ). The excitatory regions, E, of M and I fibers were not spotted on the same chromatogram, but streaked along the base lines of two chromatograms for 25 cm. Narrow side strips of these chromatograms were stained with Ninhydrin and are shown in Fig. 3. Glutamate was present in both fibers but more prominent in the motor axon. Aspartate and glutamine (no standard

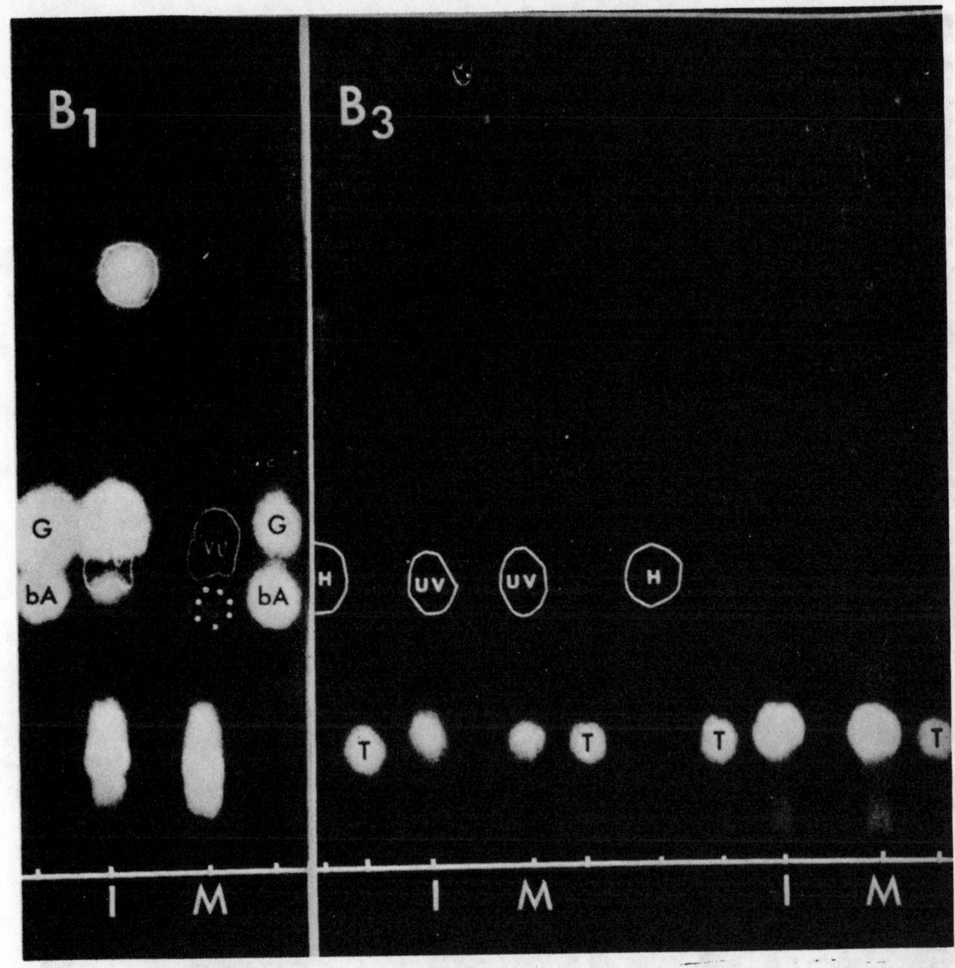

FIG. 2. Comparison of the Ninhydrin-positive components of regions B_1 and B_3 (see also Fig. 3, ref. 18) from electrophoretic fractionation of isolated inhibitory (I) and motor (M) axons; ascending paper chromatography in butanol-acetic acid-water (60-15-25). Standard components were not labeled at the base line but were marked on the spots as follows: G, GABA; bA, beta-alanine; H, homarine; T, taurine; UV, absorption of ultraviolet light. Taurine was present in both fractions of B_3, homarine in one (left). Areas of ultraviolet light absorption and homarine are outlined. In B_1, the outline of the beta-alanine spot in the motor fiber was faint and was dotted in. Solvent fronts and origins are retouched. Note that the large GABA spot appearing in the inhibitor is absent in the motor axon. Both neurons had an UV-absorbing region in GABA area (lightly outlined). The fast-moving spot in the inhibitory fibers is believed to be the cyclic amide of GABA (see text).

illustrated) run together in this solvent, but after elution from the chromatograms they were separated in another solvent and found to be uniformly distributed between the two fibers (not shown). Betaine is not revealed by Ninhydrin but was eluted from the chromatograms and bio-assayed; weak activity was found in both types of fibers. The weak blocking substance

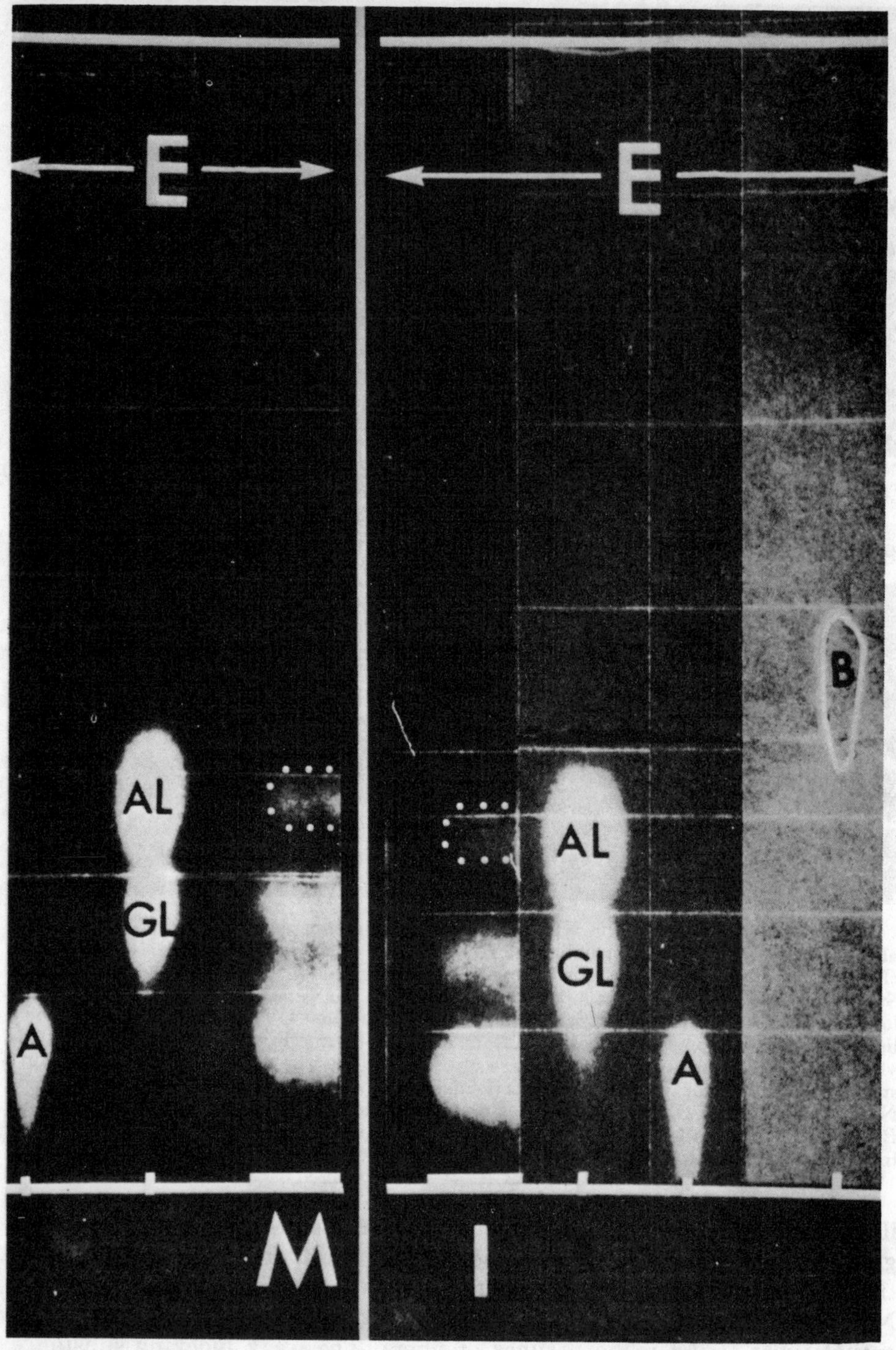

(*Legend on next page*)

alanine (dotted in) may have been present in higher concentration in the motor than in the inhibitory fiber. A large part of the alanine was in region B_2 and is not shown. Ninhydrin-staining of the other regions of the electrophoretically separated extracts failed to reveal any other differences between motor and inhibitory axons.

Therefore, of the ten blocking compounds found in peripheral nerve (see 18) GABA may be the only one which is present exclusively in inhibitory fibers. The other nine are more or less evenly distributed between motor and inhibitory axons, with the possible exception of β-alanine, whose distribution has not been settled with the present material.

Discussion

Widespread interest in GABA was stimulated by the discovery that it occurred in the mammalian CNS in surprisingly high concentrations (for a review see ref. 2). It has been suggested that GABA functions as a key metabolic intermediate in a pathway of glutamate (or α-ketoglutarate) metabolism, bypassing the oxidative decarboxylation of α-ketoglutarate to succinyl coenzyme A. The sequence of enzyme reactions in this pathway is shown in Fig. 4. The enzymes catalyzing these reactions have been demonstrated in extracts prepared from brains of various species (2).

The possible reasons for GABA accumulation in cells have been extensively discussed in the literature (e.g., 17, 25). The relative rates of enzymic synthesis and destruction of GABA and factors which influence these enzymes are considered to be explanations for differences in GABA content of cells. The mechanisms of GABA accumulation are of special interest in the case of crustacean motor and inhibitory axons. At present, the distribution of glutamic decarboxylase, the enzyme which synthesizes GABA, is under study in separated axons.

Physiological interest in GABA was largely sparked by the important findings of Bazemore, Elliott, and Florey (1) that GABA was the principle compound in an extract of the mammalian CNS that blocked the discharges in crustacean stretch receptors. There followed studies by many workers on the pharmacology of GABA and of allied compounds on the vertebrate and invertebrate nervous systems (for reviews of the extensive literature see 4, 9, 14, 21). The physiological aspects of GABA action at synapses could be

FIG. 3. Comparison of the Ninhydrin-stained substances contained in the E region (see Fig. 3, ref. 18) from electrophoretic fractionation of isolated motor (M) and inhibitory (I) axons: two ascending paper chromatograms in ethanol-acetic acid-water (160-15-25) at 4°. Standard compounds (labeled on spots) were indicated as follows: A, aspartate; Al, alanine; Gl, glutamate; B, betaine. Faint horizontal lines mark subdivisions of paper which were eluted for physiological assay. Since tissue betaine was not stained by Ninhydrin, the horizontal strips opposite betaine standards (stained and identified with Dragendorff's reagent) were eluted for bio-asssay. The lowest horizontal strips of I and M contained both aspartate and glutamine; they were also eluted for further chromatography. Solvent fronts are indicated by white lines at top (see *Methods*, ref. 7 for details). Alanine spots were faint and were dotted in.

analyzed most thoroughly in invertebrates. In the CNS of crayfish (13) and in peripheral tissues of Crustacea such as the stretch receptor (20, 16) and the neuromuscular junction (14, 15, 19) GABA imitates the postsynaptic action of the inhibitory transmitter. Recently this analogy was extended to presynaptic inhibition in the crayfish (8) and an effect on the impulse spread in motor nerve terminals has been demonstrated (6). The results of experiments with GABA in the mammalian CNS have been less favorable for an assumption that this compound has a specific synaptic action. For instance,

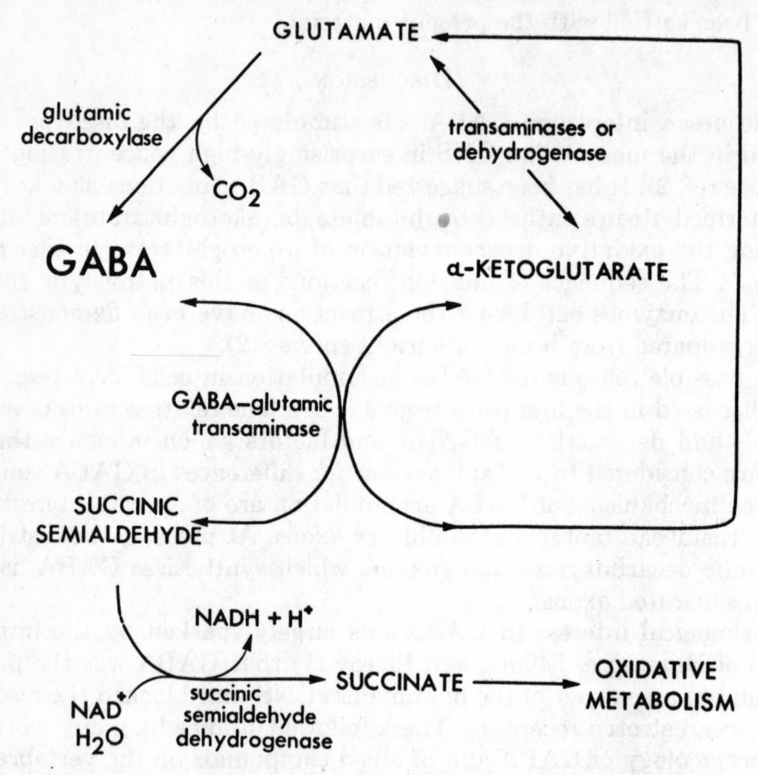

FIG. 4. Enzymic formation and further metabolism of GABA.

Curtis and his colleagues (3, 4), using electrophoretic application from micropipettes on spinal motoneurons, suggested that GABA as well as many of its analogues acted only as general depressants. In recent studies, however, the effect of GABA on the Mauthner cell, a large neuron in the goldfish medulla, was more striking and apparently more specific. Its action was similar to that of the neural transmitter substance and was confined to certain regions of the Mauthner neuron (5).

It has been proposed by many that GABA is an inhibitory transmitter. This hypothesis had the following weakness: the basic requirement that it be present in crustacean tissues, where it had been extensively studied and where its mechanism was best known, had not been met. In fact, it was emphatically stated (12) that there was no GABA in the crustacean nervous system. In the mammalian nervous system, on the other hand, where the presence of GABA had been known for a long time, the evidence was far from convincing that GABA imitated the action of the neurally released transmitter. It was natural, therefore, that in several recent symposia and reviews (e.g., 24, 10, 26, 4) the concensus developed that GABA was not a candidate for a role as an inhibitory transmitter.

Some studies that in several respects are similar to ours were made by Florey and co-workers (11, 12). From motor, sensory, and inhibitory axons they obtained extracts which they tested by physiological assay for blocking activity. It is interesting that in inhibitory neuron extracts they found strong blocking activity but none in sensory and motor fibers. One of the major discrepancies between our findings and theirs is that they could not detect any GABA. Further, their inhibitory material (11) was believed to be more active than GABA. They were led to this conclusion because, in order to explain the great activity of their extract, they would have had to assume that "up to" 3% of wet wt. of their inhibitory neurons was GABA. So far we have not been able to find substances more effective than GABA and in fact there may be no need for them to explain the calculated physiological blocking activity in extracts from crabs (11). Our measured GABA content in lobster inhibitory fibers was about 0.5% of the wet wt. GABA actually accounted for only 30–50% of the total blocking activity in our tissues, the rest being contributed largely by taurine and betaine. If these latter substances, as well as GABA, were present in Florey and Biederman's (11) extracts, much of their blocking activity would be accounted for. In view of the inherent inaccuracies in calculation of fiber weight and in the quantitation by bio-assays, and the fact that different species were used, it is uncertain whether any substantial discrepancies remain. The absence of blocking effects in their motor and sensory extracts (11) was presumably due to the small amount of tissue that was used for the assays and the admixture of excitatory substances.

The present series of studies have a direct bearing on the possible physiological role of GABA in Crustacea only. The following conclusions can be drawn: *1*) GABA is a constituent of the peripheral as well as the central nervous system (7, 18). *2*) GABA is present in surprisingly high concentration in isolated inhibitory axons. It makes up about 0.5% of the wet wt. along the unbranched course of this neuron. We have no way at present to determine the GABA concentration in the nerve terminal regions. *3*) GABA may be confined to inhibitory neurons. On this important point the results are not complete. Principally, we know little about sensory neurons in which a small amount of GABA was found (18). Further, because of technical difficulties, we cannot exclude the presence of GABA in the peripheral portions of motor fibers.

The above conclusions indicate that GABA has a specific role linked with the function of inhibitory neurons. The crucial test that could establish it as a transmitter has not been made. It would have to be shown that it is released from the inhibitory terminals by the inhibitory nerve impulse in adequate amounts at the appropriate time during the process of transmission. In respect to the other blocking compounds all we can conclude is that their function is not uniquely related to inhibitory neurons since they are present in motor and inhibitory fibers.

As far as the vertebrate nervous system is concerned, an extension of single cell analyses, such as those of Hirsch and Robins (17), to known inhibitory and excitatory neurons would be very interesting.

Summary

1. Two efferent axons of similar diameter, one inhibitory, the other excitatory, run side by side in the leg of the lobster (*Homarus americanus*). Long unbranched stretches of these neurons were removed and separated; the isolated axons were analyzed for their content of gamma-aminobutyric acid (GABA) and nine other synaptic blocking compounds that had been previously found in the crustacean nervous system (18). The GABA contents were compared enzymically; the contents of the other blocking substances were compared chromatographically or by physiological assay.

2. The GABA content along the course of the inhibitory axon was about 0.5% of its wet wt., while no GABA was detected in the accompanying excitatory fiber. GABA may therefore be confined to inhibitory nerves. The other blocking compounds, with the possible exception of β-alanine, were found in both neuron types. The distribution of β-alanine (a much weaker blocking substance than GABA) cannot be stated with confidence.

3. The findings indicate that GABA has a function specifically related to inhibitory neurons.

ACKNOWLEDGMENTS

We express our thanks to Dr. Nico van Gelder for valuable assistance in some of the experiments and Dr. W. Fahrenbach who kindly provided histological material for illustration. We also acknowledge continued technical help by R. B. Bosler and Peter Lockwood. Mrs. Marcia Feinlieb and Miss Star Martin contributed greatly in the biochemical laboratory.

REFERENCES

1. Bazemore, A., Elliott, K. A. C., and Florey, E. Isolation of Factor I. *J. Neurochem.*, 1957, *1:* 334–339.
2. Baxter, C. F. and Roberts, E. Gamma-aminobutyric acid and cerebral metabolism. In: *The Neurochemistry of Nucleotides and Amino Acids*, edited by R. O. Brady and D. B. Tower. New York, Wiley, 1960, pp. 127–145.
3. Curtis, D. R., Phillis, J. W., and Watkins, J. C. The depression of spinal neurones by γ-amino-n-butyric acid and β-alanine. *J. Physiol.*, 1959, *146:* 185–203.
4. Curtis, D. R. and Watkins, J. C. The excitation and depression of spinal neurones by structurally related amino acids. *J. Neurochem.*, 1960, *6:* 117–141.
5. Diamond, J. Variation in the sensitivity to GABA of different regions of the Mauthner neurone. *Nature*. In press.
6. Dudel, J. Effect of inhibition on the presynaptic nerve terminal in the neuromuscular junction of the crayfish. *Nature*, 1962, *193:* 587–588.
7. Dudel, J., Gryder, R., Kaji, A., Kuffler, S. W., and Potter, D. D. Gamma-

aminobutyric acid and other blocking compounds in Crustacea. I. Central nervous system. *J. Neurophysiol.*, 1963, *26:* 721–728.
8. DUDEL, J. AND KUFFLER, S. W. Presynaptic inhibition at the crayfish neuromuscular junction. *J. Physiol.*, 1961, *155:* 543–562.
9. ELLIOTT, K. A. C. AND JASPER, H. H. Gamma-aminobutyric acid. *Physiol. Rev.*, 1959, *39:* 383–406.
10. FLOREY, E. Comparative physiology: transmitter substances. *Annu. Rev. Physiol.*, 1961, *23:* 501–528.
11. FLOREY, E. AND BIEDERMAN, M. A. Studies on the distribuiton of Factor I and acetylcholine in crustacean peripheral nerve. *J. gen. Physiol.*, 1960, *43:* 509–522.
12. FLOREY, E. AND CHAPMAN, D. D. The non-identity of the transmitter substance of crustacean inhibitory neurons and gamma-aminobutyric acid. *Comp. Biochem. Physiol.*, 1961, *3:* 92–98.
13. FURSHPAN, E. J. AND POTTER, D. D. Slow post-synaptic potentials recorded from the giant motor fibre of the crayfish. *J. Physiol.*, 1959, *145:* 326–335.
14. GRUNDFEST, H. AND REUBEN, J. P. Neuromuscular synaptic activity in lobster. In: *Nervous Inhibition*, edited by E. Florey. New York, Pergamon, 1961, pp. 92–104.
15. GRUNDFEST, H., REUBEN, J. P., AND RICKLES, W. H., JR. The electrophysiology and pharmacology of lobster neuromuscular synapses. *J. gen. Physiol.*, 1959, *42:* 1301–1323.
16. HAGIWARA, S., KUSANO, K., AND SAITO, S. Membrane changes in crayfish stretch receptor neuron during synaptic inhibition and under action of gamma-aminobutyric acid. *J. Neurophysiol.*, 1960, *23:* 505–515.
17. HIRSCH, H. E. AND ROBINS, E. Distribution of γ-aminobutyric acid in the layers of the cerebral and cerebellar cortex. Implications for its physiological role. *J. Neurochem.*, 1962, *9:* 63–70.
18. KRAVITZ, E. A., KUFFLER, S. W., POTTER, D. D., AND VAN GELDER, N. M. Gamma-aminobutyric acid and other blocking compounds in Crustacea. II. Peripheral nervous system. *J. Neurophysiol.*, 1963, *26:* 729–738.
19. KUFFLER, S. W. Excitation and inhibition in single nerve cells. *Harvey Lectures.* New York, Academic, 1960, pp. 176–218.
20. KUFFLER, S. W. AND EDWARDS, C. Mechanism of gamma-aminobutyric acid (GABA) action and its relation to synaptic inhibition. *J. Neurophysiol.*, 1958, *21:* 589–610.
21. LISSAK, E., ENDRÖCZI, E., AND VINCZE, E. Further observations concerning the inhibitory substance extracted from brain. In: *Nervous Inhibition*, edited by E. Florey. New York, Pergamon, 1961, pp. 369–375.
22. LOWRY, O. H., ROBERTS, N. R., AND KAPPHAHN, J. L. The fluorometric measurement of pyridine nucleotides. *J. Biol. Chem.*, 1957, *224:* 1047–1064.
23. LOWRY, O. H., ROBERTS, N. R., SCHULZ, D. W., CLOW, J. E., AND CLARK, J. R. Quantitative histochemistry of retina. II. Enzymes of glucose metabolism. *J. Biol. Chem.*, 1961, *236:* 2813–2820.
24. MCLENNAN, H. Inhibitory transmitters—A review. In: *Nervous Inhibition*, edited by E. Florey. New York, Pergamon 1961, pp. 350–368.
25. ROBERTS, E. Free amino acids of nervous tissue: some aspects of metabolism of gamma-aminobutyric acid. In: *Inhibition in the Nervous System and Gamma-aminobutyric Acid*, edited by E. Roberts. New York, Pergamon, 1960, pp. 144–158.
26. ROBERTS, E. (ed.). *Inhibition in the Nervous System and Gamma-aminobutyric Acid.* New York, Pergamon, 1960, 591 pp.
27. VAN DER HORST, C. J. G. Artifacts of ornithine and α-γ-diaminobutyric acid found during chromatographic investigation of Rumen liquid. *Nature*, 1962, *196:* 147–148.
28. WIERSMA, C. A. G. AND RIPLEY, S. H. Innervation patterns of crustacean limbs. *Physiol. Comparata et Oecol.*, 1952, *2:* 391–405.

20

Proc. Nat. Acad. Sci. (1965) **54**, 778–782

*A COMPARISON OF THE ENZYMES AND SUBSTRATES OF GAMMA-AMINOBUTYRIC ACID METABOLISM IN LOBSTER EXCITATORY AND INHIBITORY AXONS**

BY EDWARD A. KRAVITZ, PERRY B. MOLINOFF, AND ZACH W. HALL

NEUROPHYSIOLOGICAL LABORATORY, DEPARTMENT OF PHARMACOLOGY, HARVARD MEDICAL SCHOOL, BOSTON

Communicated by Stephen W. Kuffler, July 23, 1965

Inhibitory axons in the lobster nervous system contain about 100 times more gamma-aminobutyric acid (GABA) than excitatory axons, the respective concentrations being about 0.1 *M* and 0.001 *M*.[1] In an attempt to account for this concentration difference, we have begun a study of the enzymes and substrates of GABA metabolism in single excitatory and inhibitory axons. The work was carried out in two phases; first, the pathway of formation and destruction of GABA in the lobster nervous system was established and a study was made of the properties of the enzymes involved; and second, sensitive microassays for the relevant enzymes and substrates were devised and these components were measured in single cells.

Materials and Methods.—Animals: Live lobsters (*Homarus americanus*) were obtained from a local shipper. Generally, 1.5-kg animals were used.

Preparation of central nervous system (CNS) enzymes: Ganglia and connectives of the CNS were homogenized in Kontes Duall tubes in 0.02 *M* potassium phosphate, pH 7.2, 0.01 *M* β-mercaptoethanol, and 10^{-4} *M* pyridoxal phosphate (this solution is hereafter called buffer). This and all subsequent steps were carried out at 4°. The homogenate was centrifuged at 1000 × *g*, the supernatant fluid collected and centrifuged again at 40,000 × *g*. The supernatant solution was then fractionated with solid $(NH_4)_2SO_4$. The fraction precipitating between 35 and 60 per cent saturation was collected, dissolved in a small volume of buffer, and dialyzed against 500 vol of buffer overnight.

Enzyme assays: Glutamic decarboxylase activity was measured by collecting $C^{14}O_2$ released from glutamate-U-C^{14} by a micromodification of published methods.[2, 3] GABA-glutamic transaminase activity was measured by conversion of GABA-U-C^{14} to radioactive succinic semialdehyde and succinate. The method is similar to the microassay procedure described in the legend of Table 1. Succinic semialdehyde dehydrogenase activity was measured either by conversion of radioactive succinic semialdehyde to radioactive succinate or spectrophotometrically by measuring NADH production. The details of the enzyme assay procedures and kinetic studies will be published shortly.

Single fiber studies: The methods of obtaining and identifying single nerve fibers have already been published.[1] The experimental procedures for measuring enzymes and substrates in single axons are described in the legend of Table 1.

Results and Discussion.—The pathway of GABA metabolism and the properties of the enzymes in the pathway were determined with soluble enzyme extracts prepared from the lobster CNS. The pathway was the same as in the mammalian CNS,[4] GABA being formed from glutamic acid by decarboxylation and metabolized through succinic semialdehyde to succinate by transamination and dehydrogenation (Fig. 1). Kinetic properties and conditions for optimum activity were determined

TABLE 1

ENZYMES AND SUBSTRATES OF GABA METABOLISM IN EXCITATORY AND INHIBITORY AXONS

	Excitatory	Inhibitory	I/E
	$\mu\mu$Moles/(cm) (hr)		
(*A*) Enzymes			
Decarboxylase	9	101	11.2
Decarboxylase + 0.1 *M* GABA	—	34	—
Transaminase	20	33	1.65
	Moles/liter axoplasm		
(*B*) Substrates			
Glutamate	0.09	0.07	0.8
α-Ketoglutarate	0.0003	0.0003	1
GABA	0.0006	0.1	170

(*A*) Two excitatory and two inhibitory axons were homogenized in 4-μl volumes of a buffer mixture containing 0.1 *M* potassium phosphate, pH 7.5, 0.03 *M* β-mercaptoethanol, 2.5×10^{-4} *M* pyridoxal phosphate and 0.003 *M* GABA (excitor only—there is sufficient GABA in the inhibitory nerve to bring the extract to the same concentration). The final volumes were adjusted with buffer so that 3-μl aliquots could be withdrawn for the transaminase and decarboxylase assays. The decarboxylase reaction was started by adding glutamate-U-C^{14} (0.3 μmole, 23 μc/μmole) and the final incubation volume was 5 μl. At the end of 1 hr (25°) the reaction was stopped by heating and the material applied to a Dowex-1-acetate column. The product was recovered by washing with 5 ml of water. For the transaminase assay NAD^+ (60 mμmoles), α-ketoglutarate (40 mμmoles), succinate (20 mμmoles), and GABA-U-C^{14} (60 mμmoles, 19 μc/μmole) were added to the tubes and the final incubation volume was 20 μl. After 1 hr trichloroacetic acid was added (final concentration 5%) and the suspensions poured over Dowex-50-H^+ columns. Again the product was obtained by washing with 5 ml H_2O. Unlabeled succinate was added to transaminase incubations and GABA to decarboxylase incubations to protect any radioactive product formed from further metabolism. The amounts of succinate and GABA added did not inhibit the enzymes. (*B*) Assays for α-ketoglutarate and GABA were carried out by the enzymic procedure of Jakoby and Scott.[6] The detailed description of the method for GABA measurement has already been published.[1] Glutamate was converted to GABA with a bacterial decarboxylase and measured as above.

for each of the enzymes. During the course of these studies it was observed that glutamic decarboxylase is markedly inhibited by its product, GABA. This is illustrated in Figure 2 where the effects of two GABA concentrations on enzyme velocity are graphed by the method of Lineweaver and Burk.[5] In the presence of GABA, curves are obtained rather than straight lines, indicating a deviation from classical competitive inhibition kinetics. The enzyme is being further purified to investigate this anomalous behavior.

Next, microassays were devised for the decarboxylase and transaminase so that the activities could be measured in single axons. Succinic semialdehyde dehydro-

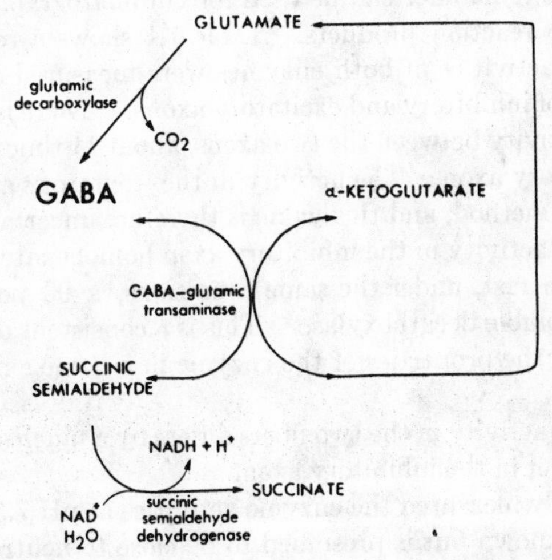

FIG. 1.—The pathway of GABA metabolism in the lobster nervous system.

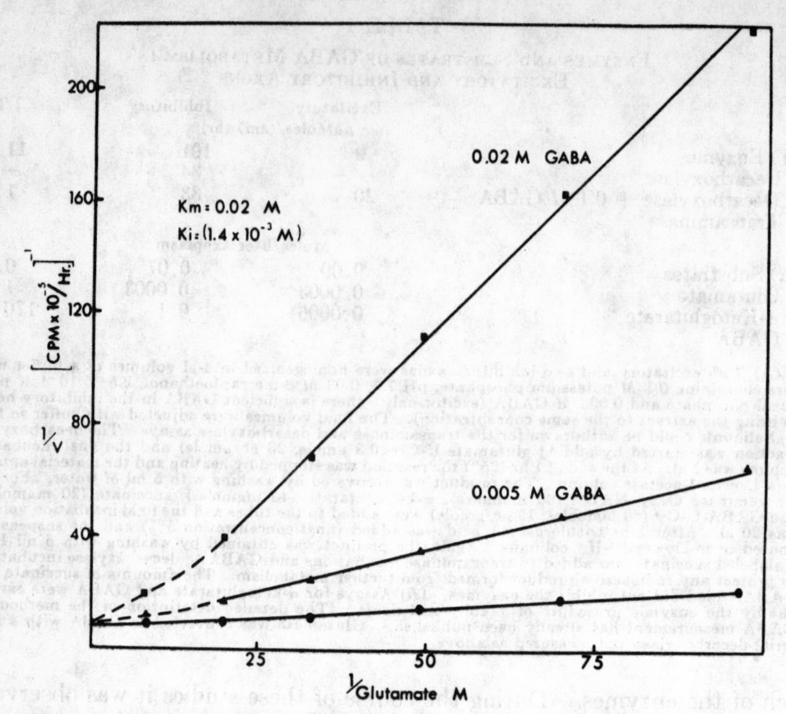

FIG. 2.—GABA inhibition of the glutamic decarboxylase.

genase was not measured since experiments showed that it was not rate-limiting for the removal of GABA. The microassay procedures for the decarboxylase and transaminase are similar. In both assays high specific activity C^{14} substrates are incubated with homogenates of single excitatory and inhibitory axons. At the end of the incubation periods, the radioactive products are separated from the starting materials with ion exchange columns. The radioactivity in a portion of the product is measured and the remainder can be used for chromatographic analysis to verify the identity of the reaction products. Table 1*A* shows a representative experiment in which the activities of both enzymes were measured on aliquots from the same homogenates of inhibitory and excitatory axons. There is a marked difference in decarboxylase activity between the two axons, about 11 times more activity being found in the inhibitory axon. The activity in the excitatory axon is near the limit of sensitivity of the method, and this value is therefore uncertain. In the presence of 0.1 *M* GABA the activity in the inhibitory axon homogenate is reduced by about 65 per cent. In contrast, under the same conditions, a 90 per cent inhibition is observed with the soluble decarboxylase. This is a consistent difference which may reflect differences in the properties of the enzyme in a homogenate and in a soluble dialyzed extract.

The transaminase activity in the two fibers differs by a much smaller factor, being about 1.6 times higher in the inhibitory axon.

We have arbitrarily measured the enzyme activities at pH 7.5. The internal pH of the axons is not known but is presumed to be close to neutrality. At pH 7 the decarboxylase activity is about the same as at pH 7.5. The variation of transaminase activity with pH depends on both the GABA and α-ketoglutarate concentrations. At the concentrations of these compounds measured in the inhibitory axon, there is little change of activity with pH over the range 7.0–7.5, while at the GABA and α-ketoglutarate concentrations found in the excitatory axon, the activity at pH 7.0 is about 2/3 that found at pH 7.5. Thus in the excitatory axon the ratio of decarboxylase to transaminase activity is a function of pH.

Substrate analyses (Table 1*B*) show the glutamate concentration to be slightly higher in the excitatory axon, and α-ketoglutarate concentration to be the same in both fibers. GABA is, as previously reported, present in much higher concentrations in the inhibitory axon. It should be noted that at the GABA and glutamate concentrations existing in inhibitory axons, the decarboxylase is markedly inhibited (90% in soluble extracts, 65% in single cell homogenates), while at the concentrations that exist in the excitatory axons there is only slight ($< 10\%$) inhibition of enzyme activity.

A point of major interest in considering the metabolism of GABA is the question of whether this compound is found in particles in the axoplasm. A bound form of GABA has been described in the mammalian CNS,[7, 8] but we have no evidence that such a form of GABA is present in the lobster nervous system. Moreover, the data in Table 1 are consistent with the idea that the high level of GABA in inhibitory axons can be maintained without sequestering any of the elements in this metabolic pathway, i.e., that enzymes and substrates are freely accessible to each other. The inhibitory axon has the capacity to synthesize more GABA than it can destroy [100 vs. 33 $\mu\mu$moles/(cm)(hr)]. Synthesis would only balance destruction at a high (0.1 *M*) level of GABA where the decarboxylase would be markedly inhibited [34 $\mu\mu$moles/(cm) (hr)]. If the GABA level in the inhibitory fiber should fall for any reason, the inhibition of the decarboxylase would be diminished and GABA would accumulate until the steady-state concentration was re-established. Therefore the high decarboxylase activity and the GABA inhibition together could provide a means of stabilizing the high GABA level in the inhibitory axon. There are, of course, other possible explanations for the maintenance of a high GABA level in inhibitory axons. Regardless of whether the above hypothesis for GABA accumulation proves to be correct or not, the principal difference between excitatory and inhibitory axons (other than GABA itself) is in the activity of glutamic decarboxylase. This therefore seems likely to be the key to the difference in GABA concentration between the two axons.

Summary.—Lobster inhibitory axons contain about 100 times more gamma-aminobutyric acid (GABA) than excitatory axons. A study of the metabolism of GABA in isolated axons demonstrated that GABA-glutamic transaminase, α-ketoglutarate, and glutamate were found in similar amounts in the two types of axons. The activity of glutamic decarboxylase was about 11 times higher in inhibitory axons and it was demonstrated that GABA markedly inhibits this enzyme. It is suggested that the high decarboxylase activity and its GABA inhibition could provide a means of stabilizing the high GABA level found in inhibitory axons.

We thank Dr. D. D. Potter, who participated in all the early phases of this work, and Dr. S. W. Kuffler for his continuing advice and support.

* This research was supported by grants NB-02253-04 and NB-03813-04 from the NIH, and by a Cooperative Graduate Fellowship from the NSF.

[1] Kravitz, E. A., and D. D. Potter, *J. Neurochem.*, **12**, 323 (1965).
[2] Kravitz, E. A., *J. Neurochem.*, **9**, 363 (1962).
[3] Roberts, E., and D. G. Simonsen, *Biochem. Pharmacol.*, **12**, 113 (1963).
[4] Baxter, C. F., and E. Roberts, in *The Neurochemistry of Nucleotides and Amino Acids*, ed. R. D. Brady and D. B. Tower (New York: John Wiley, 1960), p. 127.
[5] Lineweaver, H., and D. Burk, *J. Am. Chem. Soc.*, **56**, 658 (1934).
[6] Jakoby, W. B., and E. M. Scott, *J. Biol. Chem.*, **234**, 937 (1959).
[7] Varon, S., H. Weinstein, and E. Roberts, *Biochem. Pharmacol.*, **13**, 269 (1964).
[8] Elliot, K. A. C., and N. M. van Gelder, *J. Physiol.* (*London*), **153**, 423 (1960).

Proc. Nat. Acad. Sci. (1966) **56,** 1110–1115

*RELEASE OF GAMMA-AMINOBUTYRIC ACID FROM INHIBITORY NERVES OF LOBSTER**

By M. Otsuka,† L. L. Iversen,‡ Z. W. Hall,§ and E. A. Kravitz

DEPARTMENT OF NEUROBIOLOGY, HARVARD MEDICAL SCHOOL

Communicated by Stephen W. Kuffler, August 25, 1966

In 1957, Bazemore, Elliott, and Florey[1] showed that γ-aminobutyric acid (GABA) blocked the discharges of crayfish stretch-receptor neurons. This finding focused attention on the possible role of GABA as an inhibitory transmitter compound. At various crustacean synapses GABA was found to mimic the effects of inhibitory (I) nerve stimulation.[2–9] Furthermore, in the lobster nervous system GABA is highly concentrated in I neurons, the ratio of GABA concentrations in I and excitatory (E) axons being more than 100:1.[10–12] Although this evidence is consistent with GABA being an inhibitory transmitter compound, a crucial experiment has hitherto been lacking: the demonstration that GABA is released from I-nerve terminals in response to nerve stimulation.

The results of the present study show that stimulation of the I-nerve innervating various lobster muscles causes a release of GABA into the bathing fluid; the amount released is proportional to the number of stimuli applied.

Materials and Methods.—Saline medium used to wash nerve-muscle preparations had the following composition: 460 mM NaCl, 15.6 mM KCl, 26 mM $CaCl_2$, 8.3 mM $MgSO_4$.

Dowex-50W-X-2, H^+ resin (100–200 mesh) was obtained from Baker Chemical Co. After fine particles were removed by repeated decantation, the resin was cycled through ammonium and hydrogen forms, and was freshly washed with 5-bed vol of 2 *N* hydrochloric acid and 2-bed vol of water immediately before use.

Dowex-1-X2, Cl^- resin (100–200 mesh) was obtained from Bio Rad Laboratories. The resin was thoroughly washed with 2 *N* hydrochloric acid, and then with 2 *N* acetic acid to convert it to the acetate form; resin columns were freshly washed with 5-bed vol of 2 *N* acetic acid and 5-bed vol of water immediately before use.

3-H^3-GABA (specific activity 5 c/mmole) was synthesized from DL-3-H^3-glutamate (New England Nuclear Corp., Boston, Mass.) by decarboxylation with a bacterial glutamate decarboxylase.

Nerve-muscle preparations: Opener muscles from the crusher and cutter claws of 0.5-kg lobsters (*Homarus americanus*) were set up as illustrated in Figure 1. The single E and I axons innervating the muscle lie in separate bundles in the carpopodite segment: these were dissected and stimulated separately with suction electrodes. The closer muscle and overlying connective tissue were carefully removed to expose the opener muscle. The opener muscles remained *in situ* in the exoskeleton which formed a convenient chamber for washing with saline medium. Junctional (synaptic) potentials were recorded intracellularly from muscle fibers with a glass microelectrode filled with 2 *M* potassium citrate. Chilled, oxygenated saline medium was pumped onto the proximal end of the muscle at a rate of approximately 2 ml/min. The perfusion medium percolated through the muscle and dripped out of the cut end of the dactyl (cleaned of internal tissue). The temperature of the effluent medium was 12–14°. I or E nerves were stimulated at frequencies of 5–20 impulses/sec for the first 15 min of the collection period.

Superficial flexor muscles (M. superfic. vent. abdom.) with nerve supply and skeletal attachments intact were dissected from 1.5-kg lobsters and immersed in 3 ml of saline medium in a small chamber (2.5 × 2.5 × 0.6 cm). Chilled saline medium (7–9°) was pumped continuously through the chamber at a rate of approximately 1.5 ml/min, and 30-min fractions were collected. In most cases stimulation at 5/sec for 30–60 min was applied to the entire bundle of axons innervating the muscle and thus included E nerves; however, in three experiments the single I axon was dissected and stimulated.

Isolation and assay of GABA: GABA was isolated from samples of the saline medium by adsorption and selective elution from a cation exchange resin. Samples (50 ml) of the medium were acidified with 2 ml of 1 N hydrochloric acid, and 2×10^{-11} moles of H^3-GABA were added as an internal standard to estimate recovery. Samples were then passed through a column (1.8 cm $\times$ 20 cm) of Dowex-50-H^+ resin. The resin column was washed with 25 ml of water and 20 ml of IN ammonium hydroxide; GABA and other amino acids were then eluted with a further 30 ml of IN ammonium hydroxide. The eluates were taken to dryness in a rotary evaporator; the residue was dissolved in a total of 4 ml of water and passed through a column (0.4 $\times$ 3 cm) of Dowex-1-acetate resin to remove acidic amino acids and fluorescent contaminants eluted from the Dowex-50 resin. The effluent solution was taken to dryness, and the residue dissolved in 0.1 ml of 0.1 M Tris buffer, pH 7.9. GABA was assayed in 20-μl aliquots of this purified sample, using a micromodification of the highly specific enzymic assay of Jakoby and Scott.[11, 13] For each analysis a blank and two assay samples were run; an internal standard of 5×10^{-11} moles of GABA was added to a fourth sample. The over-all recovery of GABA was determined by measuring the amount of H^3-GABA in an aliquot of each purified sample; assay results were corrected for recoveries which ranged from 75 to 95%. The enzymic assay is sensitive to 10^{-11} moles of GABA and is highly specific; the only other substance known to react in this assay is the β-hydroxy derivative of GABA. This substance, however, is not found in the lobster nervous system.[10]

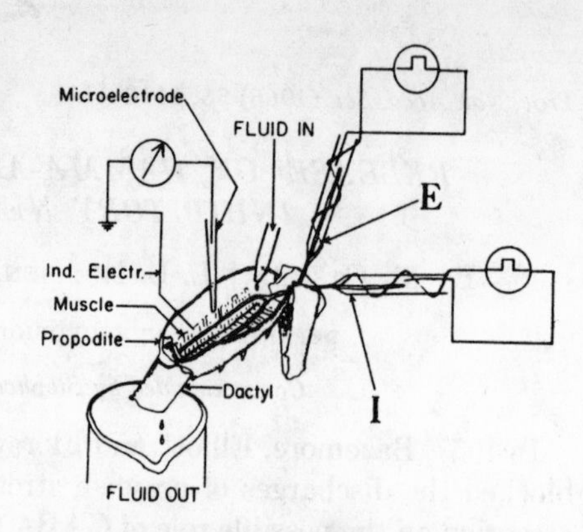

FIG. 1.—Diagrammatic view of opener muscle preparation. E and I are bundles of axons containing the single excitatory and single inhibitory axons which innervate this muscle. The E and I nerve bundles are shown in suction electrodes used for stimulation. The preparation is described in detail in the text.

Results.—Opener muscle of crusher claw: GABA was released in response to I-nerve stimulation in each of 12 experiments with this preparation. It proved necessary to wash the muscles for 3–4 hr before starting an experiment in order to reduce the spontaneous efflux of GABA to a low and stable level. The results of a single experiment are illustrated in Figure 2, which shows that GABA was liberated during I-nerve stimulation but not during E-nerve stimulation. Both periods of I-nerve stimulation released GABA but the amount liberated declined in the course of the experiment. In this particular experiment, stimulation at a frequency of 10 impulses/sec released less than stimulation for the same period at a frequency of 5/sec (75 min earlier). There was a considerable variation in the amount of

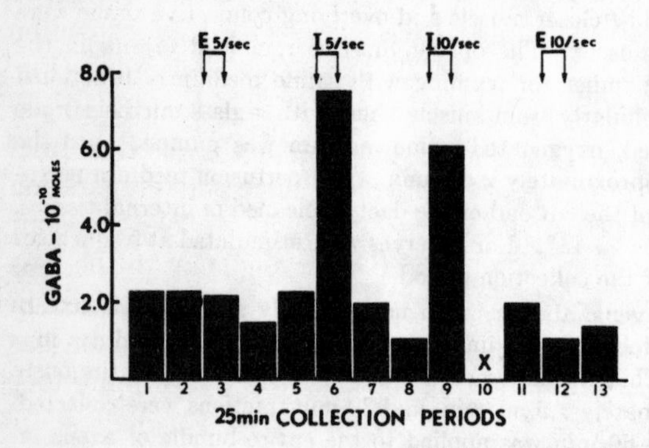

FIG. 2.—Results of a single experiment demonstrating the release of GABA in response to I-nerve stimulation in the opener muscle from the crusher claw. The preparation was washed with saline medium for 4 hr prior to the first resting collection (no. 1). GABA was assayed in 25-min collection samples during rest and during E- and I-nerve stimulation. The values obtained in this experiment for 5/sec are higher than generally observed. In most experiments the assay failed on one or more samples, indicated by X in this figure.

TABLE 1

SUMMARIZED RESULTS FOR GABA RELEASE IN LOBSTER NERVE-MUSCLE PREPARATIONS

Preparation	Duration of collection period (min)	Average GABA Content of Collection Periods—10^{-10} moles: Rest	I Stimulation (5/sec)	I Stimulation (10/sec)	Post I stimulation (rest)	E stimulation (5–10/sec)
Opener of crusher claw	25	1.7 ± 0.12 (46)	3.2 ± 0.76* (8)	4.9 ± 0.61† (5)	2.1 ± 0.28 (10)	1.7 ± 0.26 (10)
Opener of cutter claw	25	1.2 ± 0.24 (15)	—	2.2 ± 0.30* (10)	1.7 ± 0.26 (10)	0.80 ± 0.09 (6)
Flexor muscle						
(*a*) "Positive"	30	2.0 ± 0.24 (19)	5.0 ± 0.66* (13)	—	3.4 ± 0.64‡ (6)	—
(*b*) "Negative"	30	1.8 ± 0.32 (9)	2.0 ± 0.08 (9)	—	2.2 ± 0.38 (5)	—

Average amounts of GABA in rest and stimulation periods in three lobster nerve-muscle preparations. Values are means ± standard errors of means; number of samples indicated in parentheses.
* $P < 0.01$.
† $P < 0.001$ when compared with control rest periods.
‡ $P < 0.05$.

GABA released during I-nerve stimulation. Average values and statistical analyses of the results for the 12 experiments are given in Table 1. During eight I-nerve stimulation periods for 15 min at frequencies of 5/sec, the average amount of GABA released in excess of the spontaneous resting efflux (i.e., net release) was 1.5×10^{-10} moles; during five similar periods of stimulation at a frequency of 10/sec, the average net release was 3.2×10^{-10} moles. The amount of GABA released was thus clearly dependent on the frequency of stimulation, although further experiments will be necessary to define this relationship more precisely. During ten periods of E-nerve stimulation at frequencies of 5/sec or 10/sec for 15 min, there was no detectable release of GABA in excess of the resting levels (Table 1, Fig. 2).

Nerve stimulation was discontinued 10 min before the end of each collection period (e.g., Fig. 2), since studies with C^{14}-sucrose or H^3-GABA demonstrated that the extracellular space of the tissue equilibrated slowly with the external medium under our experimental conditions (half time for washout of extracellular space 13–18 min).

Since at other chemically transmitting synapses calcium ions are known to be necessary for transmitter release,[14–17] we determined the effects of a low-calcium medium on the liberation of GABA. A series of muscles was set up as described above. Intracellular recordings were made of junctional potentials from muscle fibers, and extracellular recordings with a monopolar electrode monitored the action potentials from the E and I nerves on the muscle surface (Fig. 1). Muscles were washed with normal medium for 4 hr before collections were started; after two control collection periods the I nerve was stimulated at 7.5/sec for 15 min; action potentials, inhibitory junctional potentials, and GABA output were recorded (Fig. 3*A*, *1–4*). The muscle was then washed with a medium containing only 10 per cent of the normal calcium content. The low-calcium medium caused a brief period (5–10 min) of high frequency firing in I and E nerves before the preparation became quiescent. This initial burst of spontaneous nerve activity may account for the high GABA content of the first collection period after changing to the low-calcium medium, which is particularly evident in the experiment illustrated in Figure 3*B*, *5*. After being washed for 1 hr in the low-calcium medium, the I axon was stimulated once more at 7.5/sec. I stimulation now failed to produce inhibitory

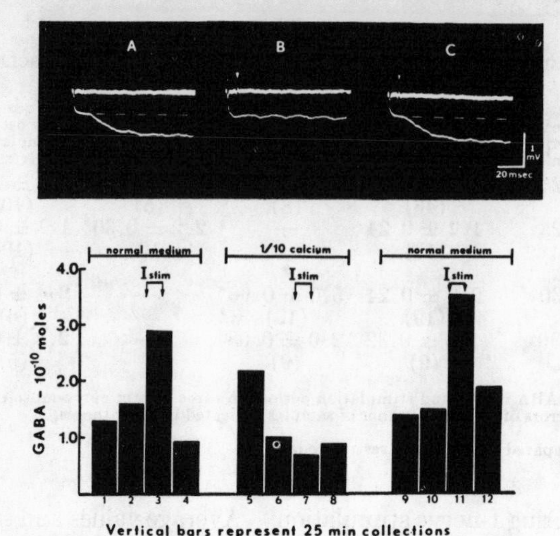

FIG. 3.—Effects of low-calcium medium on inhibitory neuromuscular transmission and GABA release in the opener muscle from the crusher claw. The oscilloscope records in the upper part of the figure show extracellularly recorded action potentials in the nerve (upper trace, *arrows*) and the first part of inhibitory junctional potentials simultaneously recorded intracellularly from a muscle fiber (lower trace). In normal medium (*A*) stimulation of the I nerve evoked inhibitory junctional potentials in the muscle and was accompanied by a release of GABA (lower part of figure). After exposure of the preparation to a medium containing only 10% of the normal calcium (*B*), I stimulation no longer produced inhibitory junctional potentials or GABA release, although action potentials were still recorded (upper trace of *B*). This effect was reversed on return to normal medium (*C*).

junctional potentials in the muscle fibers and no release of GABA was detected, although action potentials were still recorded from the nerve (Fig. 3*B*, *7*). On returning to a normal calcium medium, inhibitory junctional potentials and GABA release could again be demonstrated in response to I-nerve stimulation (Fig. 3*C*, *11*). Average results for four similar experiments are presented in Table 2.

The opener muscle of the crusher claw was the most satisfactory of the three nerve-muscle preparations used in these studies, presumably because it is several

TABLE 2

EFFECT OF LOW-CALCIUM MEDIUM ON GABA RELEASE FROM CRUSHER OPENER DURING I-NERVE STIMULATION

Collection period	No. samples	GABA—10^{-10} moles/25 min
A. Normal medium		
Resting output	11	1.7 ± 0.26
I-nerve stimulation	3	3.2 ± 0.61*
B. Low-calcium medium (2.6 mM)		
Resting output†	7	1.4 ± 0.23
I-nerve stimulation	4	1.7 ± 0.35
C. Normal medium		
Resting output	12	1.9 ± 0.17
I-nerve stimulation	3	3.9 ± 0.61‡

Average results for four expts. with low-calcium medium, similar to that illustrated in Fig. 2. Values are means ± standard errors of the means. Variable number of samples in each group is due to failure of assay in some samples in each experiment.
* $P < 0.05$.
† Resting output for low-calcium medium does not include data obtained in first rest period after exposure to this medium (see Fig. 3); average GABA output for this period was 1.8×10^{-10} moles.
‡ $P < 0.001$ when compared with resting output.

times larger than the other preparations (average wet weight in gm: crusher opener = 1.0; cutter opener = 0.5; superficial flexor = 0.3). The average GABA content of eight nerve-muscle preparations from the crusher claw was 2.2×10^{-7} moles. The amounts of GABA collected in response to I-nerve stimulation in these experiments thus represent only a small proportion of the total GABA of the preparations.

Other muscles: The results obtained from other nerve-muscle preparations were more variable. In 11 experiments with the superficial flexor muscle, a release of GABA in response to nerve stimulation could be demonstrated in only six ("positive") experiments, three with combined E- and I-nerve stimulation, and three with single I-axon stimulation (Table 1). In these experiments stimulation at a frequency of 5/sec was administered during the entire 30-min collection period. Since equilibration of the extracellular space was again slow (half time for washout approximately 15 min), there was a significant release of GABA in the rest collection immediately following I-stimulation periods (Table 1). In some experiments nerve stimulation was continued for up to 60 min. The average net release of GABA (including GABA collected during subsequent 30-min rest period in each case) in 12 separate periods of stimulation for 30 min (9,000 stimuli) was 4.4×10^{-10} moles, and the average net release in four periods of stimulation for 60 min (18,000 stimuli) was 8.5×10^{-10} moles, indicating that GABA release was related to the number of stimuli.

Opener muscles from the cutter claw proved less satisfactory than the larger preparations from the crusher claw. Nevertheless, the average results for six experiments with this preparation demonstrated a significant release of GABA in response to I-nerve stimulation, while no release of GABA was observed during E-nerve stimulation (Table 1).

Discussion.—These results provide the first demonstration of a release of GABA in response to stimulation of an identified I nerve. The amounts of GABA released increased with the number of I impulses, depending on frequency and/or duration of stimulation. No GABA was released by E axons, a result which is consistent with previous findings that E axons contain only small amounts of GABA.[11] In conjunction with findings in previous studies, these results add strong support to the view that GABA is the transmitter substance at inhibitory neuromuscular synapses in the lobster.

In the experiments with a low-calcium medium, nerve conduction along the axons was unimpaired. No recording was made from the terminal region, but in analogy with detailed studies at the frog neuromuscular junction,[17] one may assume that conduction remained intact in the terminals while transmitter release was prevented. The effect of calcium on inhibitory neuromuscular transmission in the lobster thus appears to be similar to that observed at other chemically transmitting synapses.

It should be emphasized that the amounts of GABA collected in the present experiments represent minimum estimates of the actual amounts of GABA released by I nerves during stimulation. A specific transport process accumulates extracellular GABA in lobster nerve-muscle preparations,[18] and this process may serve to inactivate the inhibitory transmitter by removing it from the site of action. The GABA uptake process functions under our experimental conditions and would certainly have removed some of the released GABA from the extracellular space of the tissue. In an attempt to reduce the rate of removal of released GABA by uptake by some other unknown enzymic mechanism, experiments were performed at the lowest practicable temperatures.

In a recent study, Takeuchi and Takeuchi[6] estimated that the amount of GABA required to simulate an inhibitory junctional potential when applied electrophoretically to GABA-sensitive areas of crayfish muscle was 4×10^{-15} moles. In the present experiments the net amount of GABA released per stimulus from the various nerve-muscle preparations was $1\text{–}4 \times 10^{-14}$ moles. In the absence of information concerning the number of inhibitory terminals in the preparations used in the present studies, and since an unknown amount of the released GABA was inactivated in the tissues, it is impossible to make any precise estimate of the amount of GABA released at each nerve terminal.

Summary.—A release of γ-aminobutyric acid (GABA) in response to inhibitory nerve stimulation has been demonstrated at three different neuromuscular junctions of the lobster. The amounts of GABA released were dependent on the frequency of stimulation and the duration of the stimulation period. The average amount of GABA recovered per stimulus ranged from 1 to 4×10^{-14} moles. GABA was not released by stimulation of excitatory nerves, or in response to inhibitory nerve stimulation when neuromuscular transmission was blocked by exposing the tissue to a low-calcium medium.

We are grateful to Drs. S. W. Kuffler and D. D. Potter for their advice and encouragement during these studies.

* This work was supported by U.S. Public Health Service grant NB02253-07.

† Rockefeller Foundation fellow.

‡ Harkness fellow.

§ NSF Cooperative Graduate fellow.

[1] Bazemore, A. W., K. A. C. Elliott, and E. Florey, *J. Neurochem.*, **1**, 334 (1957).

[2] Kuffler, S. W., *Harvey Lectures*, (*1958–1959*), 176 (1960).

[3] Boistel, J., and P. Fatt, *J. Physiol.*, **144**, 176 (1958).

[4] Grundfest, H., J. P. Reuben, and W. H. Rickles, Jr., *J. Gen. Physiol.*, **42**, 1301 (1959).

[5] Dudel, J., *Arch. Ges. Physiol.*, **283**, 104 (1965).

[6] Takeuchi, A., and N. Takeuchi, *J. Physiol.*, **177**, 255 (1965).

[7] Furshpan, E. J., and D. D. Potter, *J. Physiol.*, **145**, 326 (1959).

[8] Dudel, J., and S. W. Kuffler, *J. Physiol.*, **155**, 543 (1961).

[9] Hagiwara, S., K. Kusano, and S. Saito, *J. Neurophysiol.*, **23**, 505 (1960).

[10] Kravitz, E. A., S. W. Kuffler, D. D. Potter, and N. M. van Gelder, *J. Neurophysiol.*, **26**, 729 (1963).

[11] Kravitz, E. A., and D. D. Potter, *J. Neurochem.*, **12**, 323 (1965).

[12] Otsuka, M., E. A. Kravitz, and D. D. Potter, *Federation Proc.*, **24**, 399 (1965).

[13] Jakoby, W. B., and E. M. Scott, *J. Biol. Chem.*, **234**, 937 (1959).

[14] Katz, B., *Proc. Roy. Soc.* (*London*), **B155**, 455 (1962).

[15] Boullin, D. J., *J. Physiol.*, **183**, 76P (1966).

[16] Harvey, A. M., and F. C. MacIntosh, *J. Physiol.*, **97**, 408 (1940).

[17] Katz, B., and R. Miledi, *Proc. Roy. Soc.* (*London*), **B161**, 496 (1965).

[18] Iversen, L. L., and E. A. Kravitz, *Federation Proc.*, **25**, 714 (1966).

C. MECHANISMS OF CHEMICAL TRANSMISSION

The most detailed knowledge of the physiology of chemical synaptic transmission exists for the frog neuromuscular junction and comes from the laboratory of Bernard Katz. The general concepts developed from these studies are confirmed by many excellent studies of other synapses (see Katz, 1962, 1966, 1969 for reviews).

Katz and others postulate that neurotransmitter is stored at presynaptic sites in "packets" containing thousands of molecules; these packets or quanta may in some cases correspond to the electron-microscopically observed vesicles. At random intervals single packets are discharged into the synaptic cleft. Transmitter diffuses across the cleft to the postsynaptic membrane and there produces its characteristic effect–an increase in permeability to one or several ion species; that is, the postsynaptic membrane exhibits increased conductance during transmitter action. If the equilibrium potential resulting from these conductance changes is different from the existing membrane potential, current will flow between the postsynaptic membrane and adjacent membrane areas. The duration of these responses is limited by the duration of transmitter action and special mechanisms are usually present to eliminate the transmitter chemical: enzymes deactivate the chemical; special transport processes resorb it. In the absence of such mechanisms its action decreases as the transmitter diffuses away.

Postsynaptic membrane responses to randomly released transmitter packets are referred to as miniature junctional potentials. The arrival of a nerve impulse in the vicinity of the presynaptic nerve terminal raises the probability of release of such packets by several orders of magnitude. Upon arrival of an impulse at the nerve terminal, relatively large amounts of transmitter are quickly released, resulting in a large postsynaptic response. The increase in probability of transmitter release is due to three things: reduction of the presynaptic membrane potential itself (not some membrane process which this reflects), the duration of this depolarization, and the concentration and proportion of calcium in the bathing fluid.

Fatt and Katz (1951) stated that: "While there is little doubt that acetylcholine is released by the nerve impulse and depolarizes the endplate, the mechanism of these two actions is at present unknown and requires further investigation." Nearly twenty years later we know a great deal more about the physiology of chemical transmission, but we still do not know in a single case how the arriving nerve impulse releases transmitter, nor how the transmitter produces its specific effect on the postsynaptic receptor site.

1. EXCITATION

In 1951, Fatt and Katz presented the first analysis of synaptic transmission utilizing intracellular microelectrode recording techniques. They found that nerve stimulation produces a change at the muscle end-plate (postsynaptic membrane) that cannot be imitated by electrical current passed into the muscle fiber itself. This change involves a marked increase in conductance at the end-plate (a "short-circuit") with a resulting depolarization that tends to a steady state level of − 15 mV (inside relative to outside of the muscle fiber). The cable properties of the muscle influence the spread of the postsynaptic potential away from the end-plate. **Fatt and Katz (1951)** serves as a model for the analysis of the relation of the cable properties of the postsynaptic element to the integration of synaptic events (e.g., on a motor neuron dendrite) in given integrative zones (such as the axon hillock, see, e.g., Coombs, Curtis, and Eccles, 1957).

Observation of miniature junction potentials (**Fatt and Katz, 1952**) provided an important clue to the mechanism of chemical transmission. **Del Castillo and Katz (1954a**, b) relate quantitatively the spontaneously appearing "quanta" to the release of transmitter in response to nerve stimulation and they demonstrate an increase in probability of quantal release during nerve stimulation. The effect of previous activity on synaptic transmission is to change the probability of quantal release (for other examples, see, e.g., Boyd and Martin, 1956; DUDEL AND KUFFLER, 1961a, b; KUNO, 1964a, b).

BITTNER AND HARRISON (1970) have reanalyzed the data of **del Castillo and Katz (1954a**, b), Boyd and Martin (1956), DUDEL AND KUFFLER (1961a, b), and their own data. They show that the nerve impulse may not necessarily cause quantal release of transmitter in a Poisson distribution which is invariant in time. In the case of vertebrate neuromuscular junctions the experimental conditions required to reduce quantal release sufficiently to permit the distinguishing of single events and needed to perform the statistical analyses of "packet" release (low Ca^{++}, high Mg^{++}) are not physiological. It is unproven that the hypothesis remains valid when each impulse releases a large amount of transmitter (see reviews by del Castillo and Katz, 1956; Martin, 1966).

The paper by **Takeuchi and Takeuchi (1960)** exemplifies studies of the permeability changes produced by transmitters. The Takeuchis' conclusion is a modification of the "short-circuit" proposal of **Fatt and Katz (1951)**. They found that a large increase in permeability to cations, but not to anions, results from transmitter action. This study confirms the finding of **Fatt and Katz (1951)** that, in contrast to the permeability changes during the action potential, the conductance change in response to transmitter is independent of membrane potential.

In 1955 DEL CASTILLO AND KATZ showed that the action of acetylcholine is localized at the muscle end-plate. They employed electrophoretic ejection (controlled movement of charged molecules by electrical current) of the transmitter (ACh). That this method has proven a major analytical tool is well illustrated by several of the papers in this collection.

Axelsson and Thesleff (1959) extend important observations from the frog to the mammalian nerve-muscle junction. They show that repeated application of neurotransmitter results in desensitization of the postsynaptic receptor. Otsuka, Endo, and Nonomura (1962) could not reproduce this effect using physiological levels of transmitter. The **Axelsson and Thesleff** paper also introduces the reader to a new category of influences of nerve cells on postsynaptic cells, so-called trophic effects. The paper shows that the presence of the functioning nerve regulates the extent of ACh receptive membrane in the muscle. ACh sensitivity spreads to the entire muscle membrane following denervation. This explains why muscle fibers become supersensitive after denervation, a phenomenon which has been known since the turn of the century. Miledi (1960) at about the same time published a study parallel to that of **Axelsson and Thesleff (1959)** on the frog neuromuscular junction. His study supplies evidence that neither the withdrawal of acetylcholine itself nor the disuse of the muscle resulting from denervation is responsible for the spread of acetylcholine receptors beyond the normal end-plate regions (see review by Miledi, 1962). (Other papers in this collection concerned with long duration changes in synaptic function are **Maturana et al., 1959, and Wiesel and Hubel, 1963b**).

How the arrival of a nerve impulse causes release of transmitter is analyzed in the two studies by **Katz and Miledi (1967a, b)** reproduced here. Katz and Miledi (1965) previously analyzed the delay between the arrival of an impulse at the presynaptic terminal and the postsynaptic effect. They found that diffusion time of transmitter and reaction time at the postsynaptic membrane account for only a few microseconds of the several millisecond delay observed at cold temperatures. Hence the major part of the irreducible delay characteristic of chemical synaptic transmission occurs between the arrival of the impulse at the terminal and the release of transmitter. Katz and Miledi's continuing studies analyze in detail the time course and relationship of presynaptic depolarization and calcium movements to transmitter release. The hypothesis which has emerged is that presynaptic depolarization opens a calcium gate, Ca^{++} enters the cell and increases the probability of the release of transmitter quanta. The major synaptic delay occurs between Ca^{++} entry and transmitter release. Events occurring during this period remain unknown. The squid stellate ganglion has permitted intracellular monitoring and controlled alternation of the membrane potential in the presynaptic terminal while recording the postsynaptic response. Studies of this preparation are providing quantitative evidence for the above release hypothesis (e.g., KATZ AND MILEDI, 1967c, 1969, 1970). The involvement of presynaptic depolarization and Ca^{++} in transmitter · release has thus far proven applicable to all chemical synapses and also to the release of neurohormones from neurosecretory terminals.

J. Physiol. (1951), 115, 320–370

AN ANALYSIS OF THE END-PLATE POTENTIAL RECORDED WITH AN INTRA-CELLULAR ELECTRODE

By P. FATT and B. KATZ

From the Physiology Department and Biophysics Research Unit, University College London

(*Received* 28 *May* 1951)

According to present knowledge, the process of neuromuscular transmission can be described by the following scheme: nerve impulse → acetylcholine → end-plate potential → muscle impulse → contraction. The evidence for this chain has been summarized by Eccles (1948), Hunt & Kuffler (1950) and Rosenblueth (1950), whose reviews may be consulted for further references. While there is little doubt that acetylcholine is released by the nerve impulse and depolarizes the end-plate, the mechanism of these two actions is at present unknown and requires further investigation. The most immediate electrical sign of neuromuscular transmission is the end-plate potential, a local depolarization of the muscle fibre which is presumably due to the direct action of the neuromuscular transmitter. By measuring this electrical change under suitable conditions, some light can be thrown on the preceding steps of the transmission process. The object of this paper is to investigate the properties of the end-plate potential even more closely than has previously been attempted, making use of the method of intra-cellular recording which has been developed by Graham & Gerard (1946), Ling & Gerard (1949) and Nastuk & Hodgkin (1950). This method offers several advantages: resting and action potentials at individual junctions can be recorded in whole muscles, without micro-dissection or even removing the muscle from a Ringer bath, and the measurements do not suffer from uncertainties which are usually associated with the shunting effect of the interstitial fluid.

The immediate concern of the present work is to determine the electric charge which passes through the end-plate membrane during the transmission of one impulse and to throw some light on the mechanism by which the transfer of ions across the end-plate is brought about.

METHODS

The method of intracellular recording was similar to that described by Ling & Gerard (1949) and Nastuk & Hodgkin (1950) except that the same muscle fibre, and often the same spot of the fibre, was used for a series of measurements and thus subjected to repeated insertions and withdrawals of the microelectrode. This introduced an extra risk, for after a number of insertions local damage eventually resulted causing resting and action potentials to decline. It was, therefore, not always possible to complete a set of measurements and, in the non-curarized preparation, active movement of the muscle greatly increased the hazards of the experiment. But in spite of these inherent difficulties the method of repeated local insertions was satisfactory in many cases and gave consistent results which we could not have obtained by other means. For example, a quantitative study of the end-plate potential (e.p.p.) requires that the electric response in the same fibre should be measured at various distances from the junction. Furthermore, the size of the e.p.p. varies in individual muscle fibres much more than their resting potentials or spikes, and it was therefore desirable to compare measurements on the same junction when examining the effect of ions or drugs on the transmission process. Finally, it was only by successive insertion of the electrode at different points along a muscle fibre and so finding the position of maximum e.p.p. (cf. Fatt & Katz, 1950*a*, *b*), that we could be certain of having located an end-plate accurately. (The term 'end-plate' is used here to describe the post-synaptic area of a muscle fibre which is in contact with the motor nerve endings, ignoring the fact that in frog muscle the shape of the junction resembles a 'bush' rather than a 'plate'.)

Preparation. The nerve-sartorius preparation of the frog (*Rana temporaria*) was used and mounted in the chamber shown in Fig. 1. The chamber was moulded from paraffin wax set in a Petri dish and was so arranged that it could be completely drained of fluid from a depression at one side of the

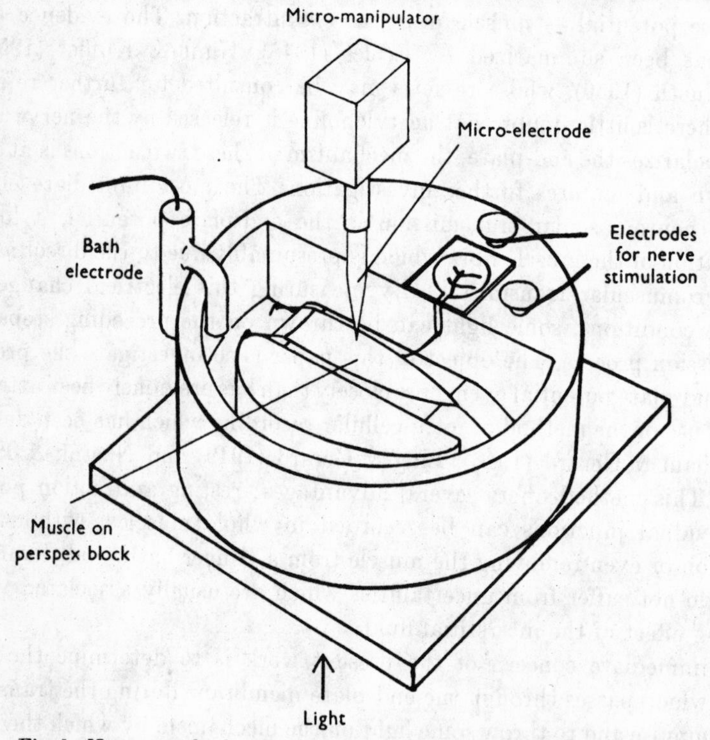

Fig. 1. Nerve-muscle chamber with stimulating and recording electrodes.

central trough. The muscle lay flat, deep surface uppermost, on a transparent Perspex block forming the floor of the central part of the chamber. It was held in this position by threads tied at each end which were looped under silver wire hooks embedded in the paraffin wax. The electrodes for nerve stimulation were situated in a separate moist compartment which was reached by the nerve via a narrow groove. The Petri dish was fixed to a glass plate which was itself carried by a mechanical stage on another large plate. Illumination was provided by light passing up through both glass plates and through the Perspex block. The preparation was viewed with a binocular dissecting microscope of magnification × 39. An eyepiece micrometer served for measurement of short distances along the muscle fibres, while coarser movements were obtained with the mechanical stage and read on the attached vernier.

The depth of fluid in the bath above the muscle was about 3 mm. It was kept at this low value in order to minimize the capacity to earth across the glass wall of the microelectrode.

The bath electrode was in the form of an agar-Ringer solution bridge connecting to a chlorided silver spiral. This led to earth via small series resistances through which steady calibration voltages and square pulses could be applied.

The microelectrode was held by a short piece of rubber tubing which led through an agar-Ringer solution bridge to a chlorided silver ribbon. The microelectrode assembly was carried on a de Fonbrune micromanipulator, the controls for which were placed outside the shielding metal box containing the preparation.

Microelectrode and amplifier. Capillary microelectrodes of external tip diameter less than 0·5 μ. and filled with 3 M-KCl were used. The wire connecting the micro-electrode to the amplifier was shielded, and the shield connected to the cathode of the first valve (RCA 954). A balanced d.c. amplifier with input cathode followers of low grid current and reduced grid-to-earth capacity was

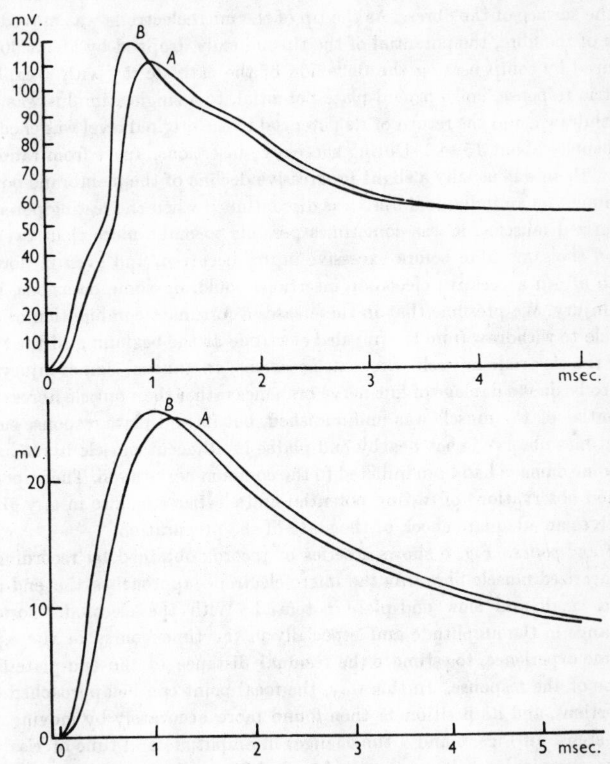

Fig. 2. Effect of amplifier distortion. Curves *A* are tracings of oscillograph records, curves *B* are corrected for high-frequency attenuation. Upper part: end-plate response during normal transmission (the muscle was treated with prostigmine, hence the large residual potential change). Lower part: curarized preparation, showing a pure e.p.p.

used, similar to those described by Nastuk & Hodgkin (1950) and Huxley & Stämfli (1949). The amplifier was calibrated by applying voltages in steps of 10 and 1 mV., and this procedure was also used to balance and measure resting potentials. In addition, a square pulse generator was used to test the time constant of the recording apparatus. This time constant varied with the resistance of the microelectrode which was apt to increase in the process of penetrating a muscle fibre. In some experiments a shielded junction box was inserted between microelectrode and amplifier. The box contained a micro-switch and 20 MΩ. shunt which could be placed across the input when a square pulse signal was applied. In this way it was possible to measure the resistance of the electrode when it was in the recording position. In other experiments a check was kept on the temporal distortion of a square pulse. The voltage wave-form at the amplifier input differed from the square pulse applied between bath and earth, in that it showed two distinct components, an instantaneously rising fraction which can be attributed to the initial displacement of charge at the glass wall of the immersed microelectrode and an exponentially rising portion which gives an indication of the time constant of the recording system. This time constant was usually between 50 and 200 μsec., and in some cases caused appreciable distortion in certain details of the electric response. In Fig. 2 a tracing is shown, together with a correction obtained by 'subtangent analysis' (cf. Rushton, 1937). The difference is rather more pronounced in this than in other experiments. On the whole, it was felt preferable to present the results without such correction, but it should be remembered that most time measurements given below are a little too large, exceeding the true value by about 0·1 msec.

Experimental procedure. In measuring membrane potentials the reference level was the potential of the bath on the surface of the fibres. As the tip of the microelectrode was moved from the bath into the interior of the fibre, the potential of the tip suddenly dropped by about 90 mV., and this drop was measured by compensating the deflexion of the cathode ray with a calibrated voltage input. The electric response, spike or end-plate potential, to a single stimulus was then recorded, the electrode withdrawn, and the return of its potential to the original level was checked, the whole process taking usually about 15 sec. During successive insertions, apart from random variations of a few per cent, there was usually a slight progressive decline of the membrane potential, and as a rule the experiment on an individual fibre was discontinued when the resting potential fell below 80 mV. In curarized muscles, it was sometimes possible to make more than twenty successive measurements on the same fibre before excessive injury occurred, and even in normal twitching muscle twelve to fifteen successive electrode insertions could, on some occasions, be carried out without serious injury. We presume that in these cases a fortunate combination of circumstances allowed the muscle to withdraw from the impaled electrode at the beginning of the twitch without damaging either the electrode or itself. On some occasions, an unusual sign of injury was observed which appeared to be due to damage of fine nerve branches rather than muscle fibres: in these cases, the resting potential of the muscle was undiminished, but its end-plate response suddenly failed, and it was sometimes observed that nearby end-plates in adjacent muscle fibres had also failed, indicating that some damage had been inflicted to the common nerve axon. The important fact was that the continued observations of resting potential and electric response in any given fibre provided by themselves an adequate check of the state of the preparation.

Localization of end-plates. Fig. 5 shows a series of records obtained by recording at different points inside a curarized muscle fibre. As the micro-electrode approaches the end-plate the first sign is always a small and slow end-plate potential. With the electrode closer, there is a characteristic change in the amplitude and especially in the time course of the e.p.p. and it is possible, with some experience, to estimate the residual distance of the end-plate from observations of the shape of the response. In this way, the focal point can be approached quickly, with two or three insertions, and its position is then found more accurately by moving the electrode in 100 or 200 μ. steps. In Figs. 6 and 7 the changes in amplitude and time of rise with distance are shown in two experiments. It was unusual to find fibres which could be followed over long distances along the surface of the sartorius muscle: the outlines of individual fibres often become obscured by adjacent fibres and by nerve branches, blood vessels and connective tissue which tend to run across the surface, especially near the end-plate foci. If a part of the fibre is not perfectly clear in its outlines, therei s a risk that the electrode tip might slip unnoticed into an overlapping neighbour, and we presume that the dotted curve in Fig. 6 arose from such an accident. In the experiment of Fig. 7 no such difficulty was experienced, and there was satisfactory agreement at every observed point between two series of measurements. In the great majority of the experiments, it was not necessary to follow individual fibres for any great length, but it was essential to be certain of the positions of individual end-plates throughout the experiment. For this purpose, the muscle was curarized at the beginning of the experiment, and a suitable number of points of maximum e.p.p. (i.e. 'end-plates') were located in the manner already described. By carefully mapping the microscopic field, noting all outstanding landmarks, it was possible to identify the fibre, and return to the same spot within 50 μ. The reliability of this procedure could be judged only from results; but the electric response at an end-plate differs so much from that of the immediate surroundings that no ambiguities arose, and it was clear that the proportion of failures in identifying previously located end-plates was minute.

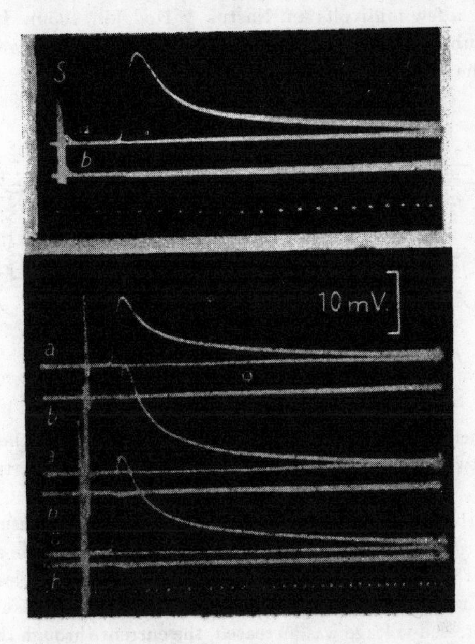

Fig. 3. End-plate potential in curarized muscle. Examples of focal recording, with the micro-electrode (*a*) inside the fibre, (*b*) on the surface of the fibre. *S*: stimulus artifact. Time marks: msec.

Extra-cellular potential changes. The error introduced into measurements of end-plate and action potentials by the existence of a small potential difference outside the fibre had to be considered. The magnitude of the external electric field is proportional to the radial current density at the surface of the fibre and to the specific resistance of the external solution, and it leads to a slight reduction and distortion of the observed membrane potential change. The size of the external potential change was checked in a number of experiments. After recording an end-plate or action potential with the electrode inside the fibre, another record was obtained when the electrode tip had just been withdrawn from the fibre. Examples are shown in Fig. 3. The external potential change varied a good deal: values between less than 1% and 5% of the internal action potential were obtained. Even in low-sodium solutions (4/5 of sodium chloride replaced by osmotically equivalent sucrose), where the effect of external potentials would be greatest, the amplitude of the external potential did not exceed 8% of the internal one, so that only a small correction was required when measurements made in solutions of different conductivities had to be compared. In most experiments of the present paper the conductivity of the bath remained constant (about 90 Ω. cm. at 20° C.), and no correction for external potentials was applied.

Direct stimulation. In some experiments an electric current was sent through the fibre in order to stimulate it directly or to determine its electrotonic 'cable constants' (cf. Hodgkin & Rushton, 1946; Katz, 1948). For this purpose another microelectrode was attached to a second micromanipulator which consisted of a combination of adjustable Palmer blocks and a vertical micrometer drive. The arrangement of stimulating and recording electrodes is shown in the diagram of Fig. 4. The procedure was to insert the recording electrode first, and then to introduce the stimulating microelectrode into the same fibre, either very nearby (at a distance of 20–50 μ.) or 1–2 mm. away. Repetitive subthreshold current pulses were used to indicate the moment when the stimulating electrode penetrated the required fibre, as this coincided with the sudden appearance of an electrotonic potential on the screen of the cathode-ray tube. There was usually also a drop of the resting potential by a few millivolts (cf. Nastuk & Hodgkin, 1950). In some of these experiments, a double-beam tube was used, the membrane potential being recorded by the first, and the current monitored by the second channel (see Fig. 4).

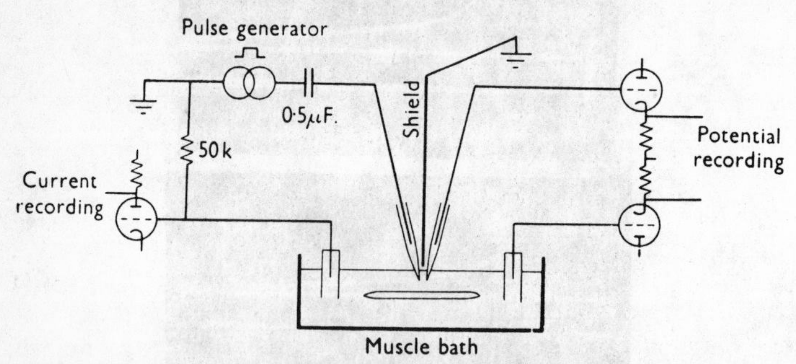

Fig. 4. Arrangement for measuring current and potential across the fibre membrane with two internal electrodes. Note: shield to reduce artifact.

The current passing through electrode and muscle membrane was usually a few tenths of a microampere. This required that several volts be applied to the electrode and gave rise to a high voltage gradient across the wall of the microelectrode tip. Not every electrode was suitable for these experiments, and with some there was evidence of dielectric breakdown occurring at the electrode tip: as the impressed voltage was increased, the current through the electrode would then suddenly rise and by-pass the muscle fibre, failing to produce or maintain a potential change across the membrane.

Temperature. Room temperature was recorded during the experiments, but checks made with a thermocouple indicated that the termperature of the preparation was about 1-1·5° C. lower, evidently because of evaporation occurring from the surface of the shallow portion of the chamber (Fig. 1). To avoid osmotic disturbance of the muscle, the bath was changed at intervals of less than 1 hr.

Solutions. In many experiments, a modified Ringer's solution was used with the composition: 113 mM. sodium, 2·0 mM. potassium, 3·6 mM. calcium, 1 mM. phosphate, 121 mM. chloride. The phosphate buffer maintained the pH at 6·8. This differed from normal Ringer mainly by its higher calcium content (3·6 mM. instead of 1·8 mM.). The advantage of this solution was that it raised the threshold of the muscle fibre, relative to the e.p.p., by about 25% and therefore caused the e.p.p., both in curarized and normal preparations, to become somewhat more conspicuous. In other experiments, normal Ringer was used, either buffered by the addition of 1·0 mM. phosphate or unbuffered. Other solutions containing D-tubocurarine chloride (Burroughs Wellcome), or prostigmine bromide (Roche) were made up as described in the experimental section.

RESULTS

A. *End-plate potential in curarized muscle*

The electric response of the end-plate becomes relatively simple when neuromuscular transmission is blocked by curarine (Eccles, Katz & Kuffler, 1941; Kuffler, 1942*a*). The effect of this drug is to reduce the amplitude of the e.p.p. below the threshold of the muscle fibre, so that no impulse arises and a local subthreshold potential change remains. Its general characteristics have previously been worked out on the whole muscle (Eccles *et al.* 1941). By

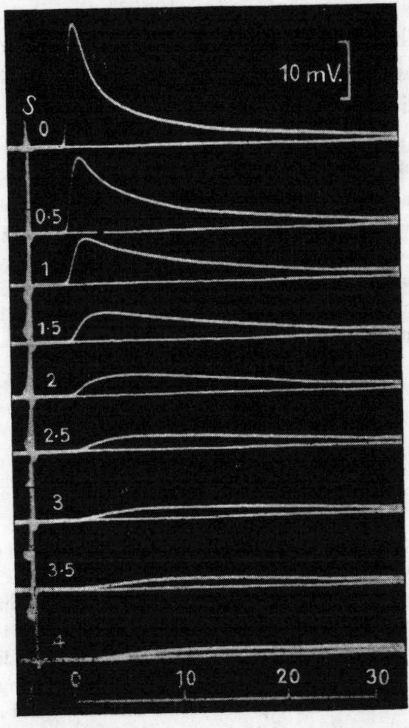

Fig. 5. End-plate potential in a single curarized muscle fibre. The position of the micro-electrode was altered in successive ½ mm. steps. The numbers give the distance from the end-plate focus, in mm. × 0·97. *S*: stimulus artifact. Time in msec.

placing a microelectrode into a curarized muscle fibre the situation becomes further simplified, for the observed response is now confined to that of a single end-plate. The method which was used in approaching the end-plate region of an individual fibre has been described above, and the results of an experiment are illustrated in Fig. 5. The e.p.p. consists of a single monophasic wave, which is rapidly attenuated as it spreads along the fibre. At the centre of the junctional region, the e.p.p. in a completely curarized muscle attains an amplitude of as much as 20–30 mV., but there are large variations of size in different fibres of the same muscle, the observed range of e.p.p. amplitudes with a given dose of curarine being greater than 10-fold. As the e.p.p. represents a graded subthreshold response, there is of course no reason to expect constancy of its amplitude, but it is remarkable that under given experimental conditions a 10- or even 20-fold variation should be found in the size of the end-plate response of different fibres. One may suspect that there are corresponding variations in the size of the junctional contact areas, or in the quantities of acetylcholine ejected from individual nerve endings.

When the e.p.p. is recorded at a 'focal' point, as in Fig. 3, it is found to rise suddenly, reaching its peak in little more than 1 msec. and declining to one-half in another 2 msec. The characteristic features of the 'focal' e.p.p. are listed in Table 1, which again shows that there is large variation in amplitude, but relatively little variation in time course.

TABLE 1. End-plate potential in curarized muscle

(These figures were obtained from ninety-four end-plates with focal recording. Most of these were selected end-plates giving e.p.p.'s of at least 10 mV. in a completely blocked muscle. Figures marked with asterisk*: these are uncorrected figures. To allow for the sytematic errors mentioned in 'Methods', subtract about 1° C. from temperature readings, and 0·1 msec. from time measurements, in this and all subsequent results. Temperature, 20° C. (16–24° C.)*. Calcium concentration, in most cases 3·6 mM. Curarine concentration, $3\text{–}5 \times 10^{-6}$ D-tubocurarine chloride (Burroughs Wellcome).)

Resting potential (mV.)	E.p.p. amplitude (mV.)	Time from onset to peak (msec.) (mean and S.E. of mean)	Time from onset to half-decline (msec.)
90 (75–107)	2·5–29 (usually 10–20)	1·3±0·02 (1·0–1·6)*	3·9±0·08 (2·4–6·0)*

A more complicated picture is sometimes obtained when the electrode is inside a fibre which has a small e.p.p., but next to one with a very large e.p.p. In this case the record becomes seriously distorted by the external field due to the adjacent 'sink', and a combination is recorded of (i) a small true change of membrane potential and (ii) an external p.d. due to the neighbouring fibre. The two changes are of opposite electric sign, and may give rise to a diphasic potential change, starting with a brief downward deflexion (the micro-electrode becoming at first more negative). The diphasic response was seen only under these special conditions, and it should be realized that the initial phase is *not* a 'membrane potential', for it can be seen when the microelectrode is in the bath on the surface of the adjacent fibre. A similar explanation applies to the small diphasic or polyphasic disturbances which were observed in non-curarized muscle (cf. Nastuk & Hodgkin, 1950), and which are due to the external fields of impulses travelling in adjacent fibres.

Effect of curarine concentration on end-plate potential and resting potential

With an increased dose of curarine, the e.p.p. becomes further reduced in size without any other obvious changes. Conversely, it will be seen that the removal of curarine is associated with a large increase in the rate of rise of the e.p.p., but its peak amplitude then becomes obscured by the intervention of the muscle spike. The effect of curarine on the size of the e.p.p. is summarized in Table 2. It is noteworthy that the resting potential of the end-plate membrane is not significantly altered by curarine, the mean values being 90·5 ± 0·5 mV. (S.E. of mean, 176 measurements) for the non-curarized end-plates and 90 ± 0·6 mV. (94) for curarized end-plates. The differences of resting potential

TABLE 2. Effect of curarine on size of end-plate potential

A. Completely blocked muscle. Comparing the peak amplitudes at eight end-plates, with 5×10^{-6} and 2.5×10^{-5} D-tubocurarine chloride. Temperature, 21° C.

	5×10^{-6} curarine (mV.)		2.5×10^{-5} (mV.)		
Fibre	Resting potential	E.p.p.	Resting potential	E.p.p.	E.p.p. reduction
I	93	7·8	90	0·6	0·077
II	93	9·8	92	0·95	0·097
III	86	7·0	90	0·7	0·1
IV	80	6·5	86	0·95	0·146
V	83	6·9	91	0·55	0·08
VI	85	6·6	85	0·8	0·121
VII	83	17·4	81	2·5	0·144
VIII	86	20·7	90	2·7	0·13
Mean	86	10·3	88	1·22	0·112

B. Comparing normal and curarized muscle (4×10^{-6} D-tubocurarine chloride). The e.p.p. was measured at a fixed point of its rising phase, 0·44 msec. after its onset.)

	4×10^{-6} curarine (mV.)		Normal (mV.)		
Fibre	Resting potential	E.p.p. at 0·44 msec.	Resting potential	E.p.p. at 0·44 msec.	E.p.p. reduction
I	95	4·5	96	24·6	0·18
II	89	4·7	94	22·7	0·21
III	90	12·1	97	36·4	0·33
IV	82	8·8	98	34·8	0·25
V	96	6·4	100	32	0·2
VI	89	9·9	97	40	0·25
VII	86	5·4	86	28·4	0·19
VIII	88	12·2	84	36	0·34
IX	86	6·1	84	19·7	0·31
Mean	89	7·8	93	30·5	0·25

in the paired observations of Table 2 are also not significant. In Table 2 A, the mean difference is −2 ± 1·4 mV., the more deeply curarized end-plates having a slightly larger resting potential, while in Table 2 B, the mean difference is +4 ± 2 mV.

An analysis of the spatial spread and decay of the end-plate potential

In two experiments, illustrated in Figs. 6 and 7, a fairly complete series of records was obtained at various distances from the focal point. In Fig. 8 several tracings are superimposed which show the characteristic decline and temporal spreading of the wave-form as the electrode is moved outward from the end-plate. It has been shown by several authors (Eccles *et al.* 1941; Kuffler, 1942*b*; Katz, 1948) that the e.p.p. arises from a rapid initial displacement of electric charge by the neuromuscular transmitter: this active phase appears to subside within a few milliseconds, and thereafter the time course of the e.p.p. is determined by the rate at which the charge spreads along, and leaks across, the muscle membrane. There are certain consequences of this hypothesis which can be subjected to a quantitative test.

The distribution of charge in a resting muscle fibre is described by the classical cable theory (see Hodgkin & Rushton, 1946). According to this theory, the total charge which the transmitter has placed on, or displaced from, the

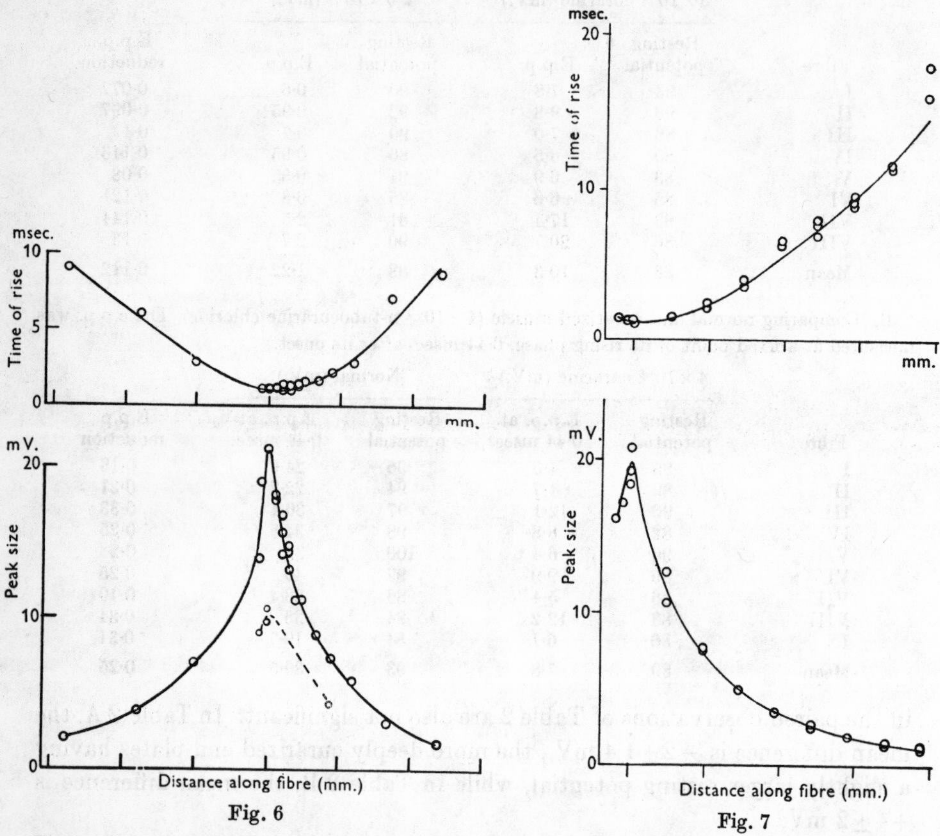

Fig. 6

Fig. 7

Fig. 6. Spread of e.p.p. in a single curarized fibre. Lower part: peak amplitude of e.p.p. is plotted against distance along the fibre (small circles: see text, p. 324). Upper part: time of rise of the e.p.p. is plotted against distance.

Fig. 7. Spread of e.p.p. in another fibre. Same co-ordinates as in preceding figure.

muscle fibre should decay exponentially after the transmitter has subsided. The time constant of the decay of end-plate charge should be the same as the time constant of the muscle membrane as determined by other methods (cf. Katz, 1948). In order to measure the total displacement of charge, the spread of the e.p.p. along the fibre was plotted at various moments (Fig. 9). The area under each curve gives a relative measure of the charge on the fibre surface (or rather of the 'deficit of charge', the fibre surface having been depolarized). The area can be found accurately at short times, up to about 10 msec.; at longer times, an extrapolation is required for distances greater than 4 mm., but this introduces only a slight inaccuracy in the final points. The logarithm of the area is plotted against time in Fig. 10: the end-plate charge is seen to reach a maximum at about 1·5–2 msec. and from then on to decay exponentially with a time constant of 20·6 msec. In another experiment, the maximum was

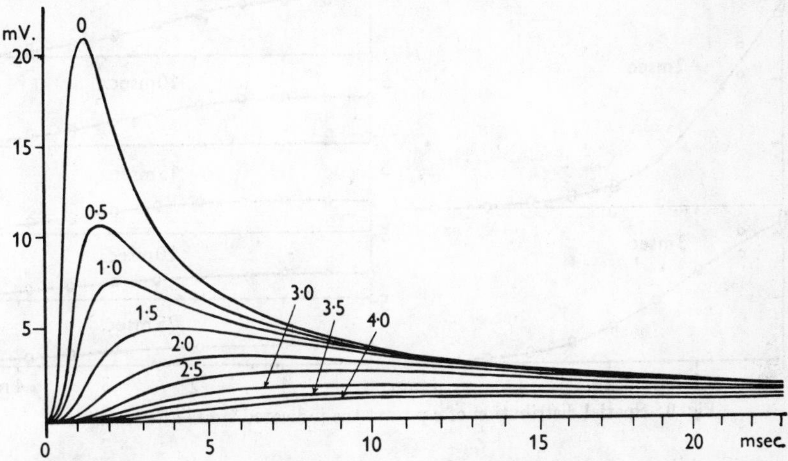

Fig. 8. Tracings of e.p.p.'s at different distances from end-plate focus. In superimposing the records, the stimulus artifact (see Fig. 5) was taken as the common point. The numbers give the distances, in mm. × 0·97, from the end-plate centre.

attained in 2–3 msec., and the charge then subsided with a time constant of 27·4 msec. These time constants are within the range of values previously found for frog muscle (see also Tables 4 and 5), and this confirms the view that, beyond the first 2 or 3 msec., the e.p.p. in curarized muscle is no longer actively maintained, and that its further time course is determined simply by the resistances and capacity of the muscle fibre.

It was pointed out to us by Mr A. L. Hodgkin that the theoretical equations describing the spread of charge along the fibre become greatly simplified if the charge has been applied *instantaneously at a point* of the fibre (see Appendix I). During the e.p.p., charge is placed on (or displaced from) points of the fibre which are usually spread out 50 to 100 μ. on either side of the centre of the nerve endings, and the displacement is nearly complete within 2 msec. These distances and times are small compared with the length and time constant of the fibre and tentatively their finite size may be disregarded, especially when analysing measurements at one or more millimetres from the end-plate focus. According to the simplified theory outlined in Appendix I, the duration T of the rising

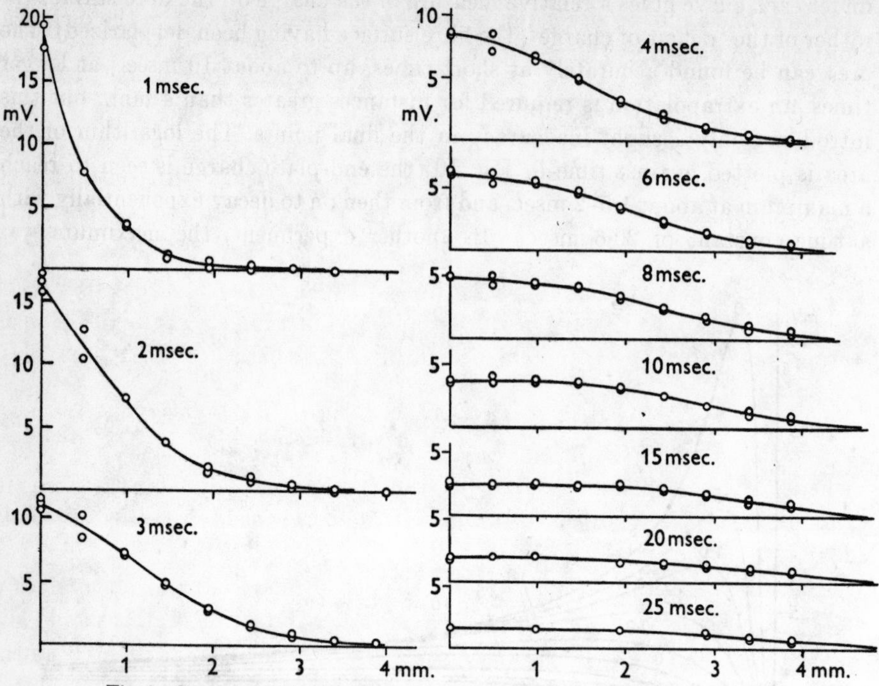

Fig. 9. Spatial distribution of e.p.p. at the indicated times after its start.

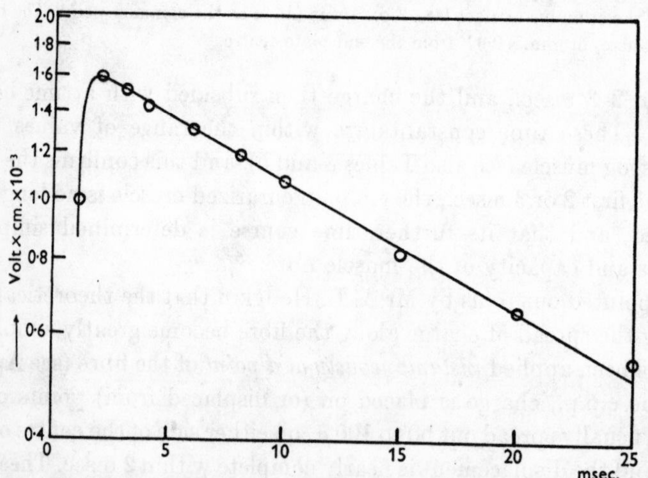

Fig. 10. The exponential decay of the end-plate charge. Ordinates: area ($\int V dx$) of the potential—space curves of Fig. 9, in mV. × cm., on a logarithmic scale. Abscissae: time in msec. Note: to convert ordinate readings into coulombs, multiply by $2c_m$, where c_m is membrane capacity per unit length of fibre, and the factor 2 is required because the e.p.p. spreads in both directions along the fibre.

phase of the e.p.p. at any given point, should then be related to the distance x of this point from the focus by the following equation

$$\frac{x^2}{\lambda^2} = \frac{4T^2}{\tau_m{}^2} + \frac{2T}{\tau_m}, \tag{1}$$

where τ_m is the time constant of the membrane and λ the length constant of the muscle fibre. These constants are related to the resistances and capacity of the fibre as follows:

$$\tau_m = R_m\, C_m,$$

where R_m is the transverse resistance and C_m the capacity of 1 cm.^{2} of membrane, and, when the fibre is immersed in a saline bath,

$$\lambda^2 = \frac{R_m}{R_i} \times \frac{\rho}{2},$$

where R_i is the specific internal resistance and ρ the radius of the fibre (for further details and nomenclature, see Hodgkin & Rushton, 1946, and Katz, 1948).

Having found τ_m, we can now use the curves of Fig. 8 and plot values of $4T^2/\tau_m{}^2 + 2T/\tau_m$ against x^2 (Fig. 11). According to equation (1) this should give a straight line with slope $1/\lambda^2$. The observed relation is approximately linear,

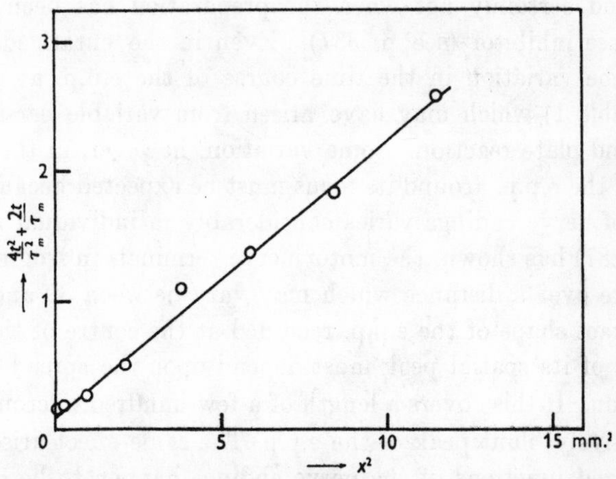

Fig. 11. Analysis of the e.p.p. in curarized muscle (see text and Appendix I). The theoretical relation is linear (slope $= 1/\lambda^2$) and passes through the origin. The divergence from theory at small values of t and x is due to the fact that the rise of the e.p.p. is not instantaneous.

and λ is found to be 2·15 mm. In another experiment, the value of λ determined in the same way was 2·4 mm. This may be compared with the values of λ on p. 335, which were obtained by the 'square-pulse analysis' of the electrotonic potential: the mean value in seven fibres is 2·4 mm. (varying between 2·2 and 2·6 mm.).

To find R_m and C_m, we must make an assumption about the size and internal conductivity of the muscle fibres. The average fibre diameter in a sartorius muscle is about 75–80 μ. (Mayeda, 1890; Katz, 1948), but the fibres on the deep surface of the muscle are larger than average (Hill, 1949). Moreover, in searching for a distinct fibre which can be followed along the muscle, it is probable that one of the largest superficial fibres was chosen. We were inclined to take 100 μ. for fibre diameter and about 250 Ω.cm. for internal resistivity (cf. Bozler & Cole, 1935; Katz, 1948). The more direct experiments described on p. 335 indicate, however, that the fibre diameter was more nearly 140 μ., and this value was therefore adopted. The values calculated for R_m and C_m are 3300–4100 Ω.cm.^{2} and 6–7 μF./cm.^{2}, respectively (see Table 3). As shown in Table 5, these results

TABLE 3. Membrane constants derived from end-plate potential

Fibre	τ_m (msec.)	λ (mm.)	R_m (Ω.cm.^{2})	C_m (μF./cm.^{2})
I	20·6	2·15	3300	6
II	27·4	2·4	4100	7

are within the range of values obtained by more direct methods, and we regard the quantitative agreement as a further confirmation of our premises, namely that the e.p.p. is produced by a brief impulse of transmitter activity.

While this conclusion applies to curarized muscle, it does not hold under all conditions, and certainly not when the preparation has been treated with a cholinesterase inhibitor (see p. 337). Even in the curarized preparation, there was some variation in the time course of the e.p.p. at different end-plates (cf. Table 1) which may have arisen from variable persistence of the transmitter/end-plate reaction. Some variation, however, in the time course and spread of the e.p.p. around its focus must be expected because the spatial distribution of nerve endings varies considerably in individual muscle fibres. As Kühne (1887) has shown, the motor nerve terminals in frog muscle spread along the fibre over a distance which may vary between 30 and as much as 500 μ. The exact shape of the e.p.p. recorded at the centre of this region and the sharpness of its spatial peak must depend upon the spread of the nerve-muscle junction. If this covers a length of a few hundred microns, it will give rise to a relatively blunt peak of the e.p.p. The same effect arises, even with sharply localized junctions, if the nerve endings happen to lie on the buried side of the muscle fibre; the microelectrode cannot then be brought very close, and fine longitudinal adjustment makes little difference. In the course of locating large numbers of end-plates, considerable variations in the sharpness of localization were observed. In some cases, the position of the electrode was more critical even than shown in Fig. 6, while in other cases, a shift of 200 μ., in either direction from the centre, produced little diminution in e.p.p. size. One may surmise that this was associated with an extensive spread of the nerve endings, or their being located on the opposite side of the fibre.

Direct measurement of the membrane constants

It was desirable to determine the resistance and capacity of the muscle fibres, under similar experimental conditions, but in a more direct way than used in the preceding section. For this purpose, the rectangular pulse technique was employed as described by Hodgkin & Rushton (1946) and Katz (1948), except that intracellular electrodes were used to pass current through the membrane, and to record the resulting change of potential across it (see Methods, Fig. 4). The current was an inward directed pulse through the membrane, of about 70 msec. duration and 0·2 μA. intensity, which caused the membrane potential to increase by about 40 mV. The current was delivered by a rectangular pulse generator, but its shape and intensity depended upon the resistance of the microelectrode which was liable to change during the current flow. This showed itself usually in a gradual reduction of current strength, from its initial peak to a more steady level which was reached after some 10–20 msec. The current pulse was examined on a double-beam oscilloscope, and it was ascertained that a period of sufficiently steady current flow, and steady membrane potential, preceded the break of the pulse. Under these conditions, the level of the membrane potential reached at the end of the pulse, and the transient potential changes following its break, could be used to determine the relevant fibre constants.

Applying the cable theory of Hodgkin & Rushton (1946) to the present case, we find that, for a distance x between the two internal electrodes, the steady potential change V recorded at one electrode is related to the steady current I flowing through the other electrode by the following equation:

$$V = \frac{I}{2}\sqrt{(r_m r_i)}\exp\left[-x/\sqrt{(r_m/r_i)}\right], \tag{2}$$

where r_m and r_i are, respectively, the transverse resistance of the membrane times unit length and the longitudinal resistance of the fibre per unit length. The term $\frac{1}{2}\sqrt{(r_m r_i)}$ is the effective resistance between inside and outside, measured at a point far from the tendon, while $\sqrt{(r_m/r_i)}$ is the length constant λ.

Thus the values of r_m and r_i can be found from measurements with two different electrode separations. The time constant of the membrane can be determined from the time course of decay of the membrane potential, for instance by measuring the time of decline to 15% with zero separation, or by comparing half-times at different distances (cf. Hodgkin & Rushton, 1946).

The electrodes were placed into the same muscle fibre about 10 mm. from the pelvic end. The 'polarizing' electrode was left there, while the recording electrode was moved from a position only 20–30 μ. away to a distance of 1·6 mm. and finally back to the original or an intermediate point. The resting potential was measured at the recording electrode, and the current was monitored in every case. The results are shown in Table 4. The values of λ, τ_m, r_m and r_i were obtained directly, while those of d, R_m and C_m are based on an assumption regarding the internal conductivity of the fibre. Its specific

TABLE 4. Membrane constants derived from 'square pulse analysis'

(Temperature 19° C. R_i is assumed to be 250 Ω.cm.)

Fibre	Resting potential (mV.)	λ (mm.)	τ_m (msec.)	$\frac{1}{2}\sqrt{(r_m r_i)}$ (Ω.)	"d" (μ.)	R_m (Ω.cm.^{2})	C_m (μF./cm.^{2})
I	89	2·3	31	210000	132	4000	8
II	91	2·3	37·5	215000	131	4100	9
III	85	2·4	29	135000	168	3400	8
IV	79	2·6	33	215000	139	4900	7
V	90	2·2	33	232000	123	3900	8
VI	89	2·2	34	193000	135	3600	9
VII	82	2·5	44	230000	132	4800	9
Mean	86	2·4	34·5	204000	137	4100	8

resistance R_i was taken as 250 Ω.cm., in accordance with earlier measurements of Bozler & Cole (1935) and Katz (1948). The fibre diameter was then calculated from

$$d = \sqrt{\left(\frac{4}{\pi} \times \frac{R_i}{r_i}\right)}, \tag{3}$$

the mean value of d being 137 μ. This seems rather large, but it is within the known range of fibre diameters of frog's muscle (Mayeda, 1890), and it is likely that during the present experiments the largest fibres have been selected.

The values of R_m and C_m obtained in this series are listed in Table 5, together

TABLE 5. Summary of different measurements of R_m and C_m in muscle

Method and reference	Preparation	R_m (Ω.cm.^{2})	C_m (μF./cm.^{2})
External electrodes (Katz, 1948)	Small bundles and isolated fibres (75 μ.)	1500	6
	Toe muscle (45 μ.)	4000	4·5
Internal electrodes, e.p.p.	Sartorius	3700	6·5
Internal electrodes, 'square pulse'	Sartorius	4100	8

with other measurements on frog muscle. The most notable feature in this table is the large value of the membrane capacity (4·5 − 8 μF./cm.^{2}) which exceeds that of several non-medullated nerve axons by a factor of 5.

The displacement of electric charge at an end-plate by the neuromuscular transmitter

With the use of these figures we can calculate the quantity of electric charge which is removed from the surface of a curarized muscle fibre, during the local action of a nerve impulse. This quantity is of special interest because it gives us an indication of the depolarizing power of the neuromuscular transmitter, and of the minimum number of ions which flow through the active end-plate surface. Presumably, when acetylcholine is released from the nerve endings it reacts with the end-plate so as to form a local 'sink' into which the surrounding muscle membrane discharges. But whatever the mechanism of this action, the discharge of the muscle fibre during the e.p.p. must be brought about by a transfer of ions across the end-plate membrane, and the number of ions which are transported across the end-plate surface must be large enough to provide for the observed displacement of charge.

It might be argued that, even during a subthreshold e.p.p., some regenerative reaction occurs in the surrounding muscle membrane which reinforces the local transfer of ions, quite apart from the primary action at the end-plate itself. But if such a regenerative process were at all important, it would have a noticeable effect on the time course of the e.p.p. and on the membrane constants derived from it. For example, the resistance and time constant so determined should have a larger value than when measured with the usual method of anodic polarization (p. 335). No such difference was observed, and we feel justified in assuming that the muscle membrane is discharged passively into the 'sink' at the motor end-plate.

It was shown in Fig. 10 that during the first 2 msec. of the e.p.p. the displacement of charge from the muscle membrane reaches a maximum and then declines exponentially. The maximum charge amounts to $3\cdot2\times10^{-3}$ V.cm. multiplied by the capacity per unit length of fibre. During normal impulse transmission considerably more charge is transferred across the end-plate: the results of Table 2 indicate a three- to five-fold amplification, following the withdrawal of curarine.

With a membrane capacity of 6 μF./cm.^{2} and a fibre diameter of 135 μ., the capacity of 1 cm. of fibre is $2\cdot45\times10^{-7}$ F., and the transfer of charge during the subthreshold e.p.p. is 8×10^{-10} coulombs. This corresponds to a net transport of at least 8×10^{-15} mol. of univalent cations inward, or anions outward, across the end-plate membrane. In the absence of curarine, the figure increases to $2\text{–}4\times10^{-14}$ mol. This is the *minimum* quantity of ions which the transmitter causes to flow across a single end-plate during one impulse. It is a surprisingly large amount, considering the small size of the end-plate area: it is equivalent, for instance, to the transfer of sodium across 0·8 mm.^{2} of non-medullated axon membrane during a single nerve impulse (Keynes & Lewis, 1950).

The number of ions which contribute to the production of the e.p.p. must, in fact, be larger than this estimate, as it represents only the net transfer of charge, i.e. the excess of cations over anions moving in one direction. Presumably the movement of ions is brought about by a reaction between acetylcholine and its receptors in the end-plate: it may involve a direct entry of acetylcholine ions into the muscle fibre or a permeability change leading to increased flux of other ions across the surface (cf. Fatt, 1950). It would be an important step in the study of this problem if the quantity of acetylcholine released by a nerve impulse at a single end-plate could be compared with the quantity of ions required for the production of an e.p.p.

The local discharge of the fibre surface is opposed, relatively slowly, by the flux across the membrane of potassium, chloride and other 'diffusible' ions which are responsible for the gradual return to the resting level. In the curarized muscle, the initial displacement of charge is much more rapid than subsequent leakage across the membrane, and there is very little overlap between the two phases. A very different situation arises when the action of the transmitter is prolonged, by the use of a cholinesterase inhibitor.

The effect of prostigmine on the end-plate potential

The action of several anti-cholinesterases has been studied carefully by Eccles & MacFarlane (1949), who found that there is invariably a marked lengthening of the active phase during which the e.p.p. is built up. In Fig. 12

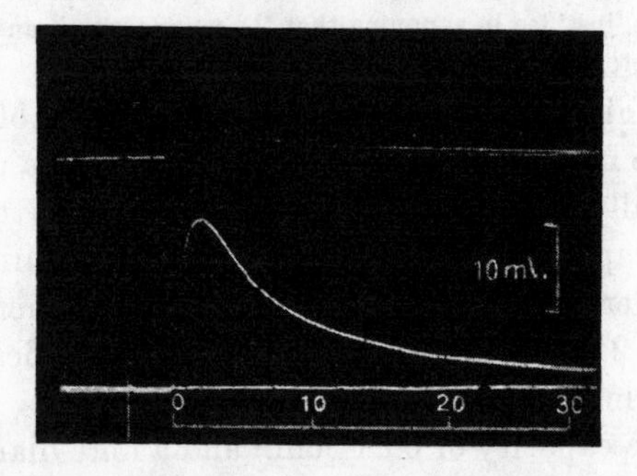

Fig. 12. Effect of prostigmine, in curarized muscle. The lower record was obtained from the same end-plate, after addition of prostigmine bromide (concentration 10^{-6}). Time, msec.

an intracellular record of this effect is shown. In a curarized muscle fibre, (3×10^{-6} D-tubocurarine chloride), an e.p.p. was observed rising to a peak of 7 mV. in 1·1 msec. and falling to one-half in another 2·1 msec. After an addition of 10^{-6} prostigmine bromide, the response at the same end-plate built up to a more rounded peak of 19 mV. in 2·1 msec., and then fell to one-half in another 5·2 msec. The effect is very similar to that previously described on whole muscle, with a moderate dose of eserine (Eccles, Katz & Kuffler, 1942) and other cholinesterase inhibitors (Eccles & MacFarlane, 1949).

It was shown by Eccles *et al.* (1942) that eserine produces a much more dramatic lengthening of the e.p.p. in the uncurarized muscle, though recording becomes then more complicated because of the presence of muscle spikes. We have confirmed their observation under somewhat different conditions. Nerve-muscle transmission can be blocked by lowering the external sodium concentration to one-fifth (cf. Fatt & Katz, 1950*b*) leaving an e.p.p. of similar shape, though usually of somewhat slower rise and fall than in the curarized muscle (see Tables 6 and 1). If prostigmine is added to the solution, a striking

TABLE 6. End-plate potential in sodium-deficient solution

(Mean values of twenty-five experiments at 20° C. Na concentration reduced to one-fifth by substitution of isotonic sucrose.)

Resting potential (mV.)	E.p.p. peak amplitude (mV.)	Time from onset to peak (msec.)	Time from onset to half-decline (msec.)
83	9–28	2·1 (1·1–2·6)	5·9 (3·2–7·8)

change occurs, shown in Fig. 13. The e.p.p. is lengthened enormously, much more than in the experiment of Fig. 12. Instead of passing through a sharp peak, the e.p.p. rises to a plateau which is maintained for some 30–40 msec., and then declines to one-half in 0·1 sec., as compared with 6 msec. (see Fig. 14).

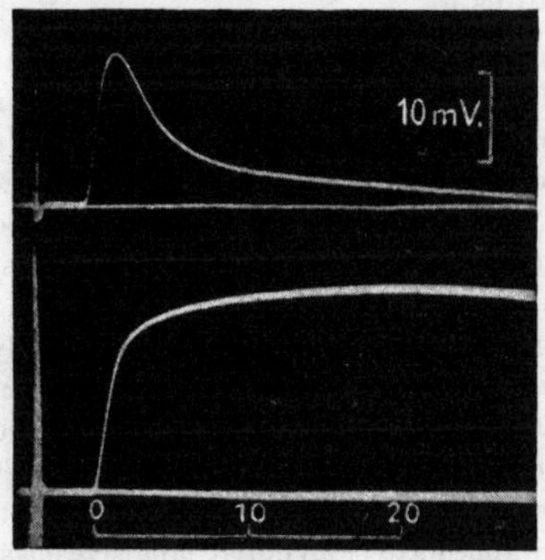

Fig. 13. Effect of prostigmine in a 'low-sodium' muscle. Upper record: E.p.p. in sodium-deficient muscle (4/5 of Na replaced by sucrose). Lower record: after addition of prostigmine bromide (10^{-6}). Time, msec.

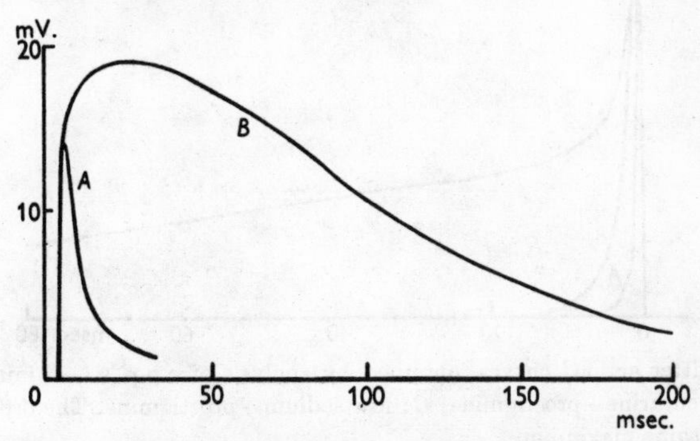

Fig. 14. Superimposed tracings of e.p.p.'s. *A*: low-sodium muscle. *B*: like *A*, but after adding prostigmine bromide (10^{-6}).

There is strong evidence (see Brown, Dale & Feldberg, 1936; Eccles *et al.* 1942; Eccles & MacFarlane, 1949) that this effect is due to the protection of acetylcholine against rapid hydrolysis, and that therefore the amount of acetylcholine, initially released by the nerve impulse, continues to build up the e.p.p. and to maintain it against the simultaneous spread and leakage of charge along and across the surface membrane. The quantity of ions which passes through the end-plate in the prostigmine muscle must greatly exceed the figure given above for the curarized muscle. An estimate of the excess can be obtained by comparing the 'areas', i.e. the time-integrals, of the e.p.p.'s in the two cases: this area is about 50 times larger for the prostigmine-e.p.p. of Fig. 13, than for a 'curarine'-e.p.p. of the same initial rate of rise.

This is an important point in connexion with the alternative modes of acetylcholine action which have been suggested (Katz, 1942; Fatt, 1950). If, for instance, acetylcholine were to depolarize the end-plate by direct penetration, the quantity of ions released by a single impulse must provide not only the electric charge which is placed on the muscle fibre during the ordinary e.p.p. but the much larger quantity which is needed to maintain the e.p.p. in eserine- or prostigmine-treated muscle.

It is of interest to trace the time course of the transmitter/end-plate reaction and its changes under the influence of prostigmine. This can be done approximately by an analysis of the e.p.p. which has been previously described (Katz, 1948, p. 529). The analysis depends upon a knowledge of the time constant of the membrane, and on the assumptions (i) that the time constant is not appreciably affected by prostigmine, and (ii) that the transmitter reaction can be treated as the equivalent of an applied current pulse. There is good evidence that the first assumption holds true (Eccles *et al.* 1942) but the second is oversimplified (cf. Section B below), though not likely to lead to serious error in the present comparison. In Fig. 15 the result of such an analysis is shown for (*a*) the e.p.p. of curarized muscle, (*b*) of curarine-prostigmine-treated muscle and (*c*) of Na-deficient and prostigmine-treated muscle. The curves show, strictly speaking, the time course of three current pulses which, with a membrane time constant of 25 msec., would alter the membrane potential in a manner identical with the three observed types of e.p.p. It will be noted that, even in the presence of a cholinesterase-inhibitor, there appears to be an initial impulsive phase of transmitter action, which is followed by a long period of low-level activity. Similar phenomena have been described and discussed in detail by Eccles *et al.* (1942) and Eccles & MacFarlane (1949).

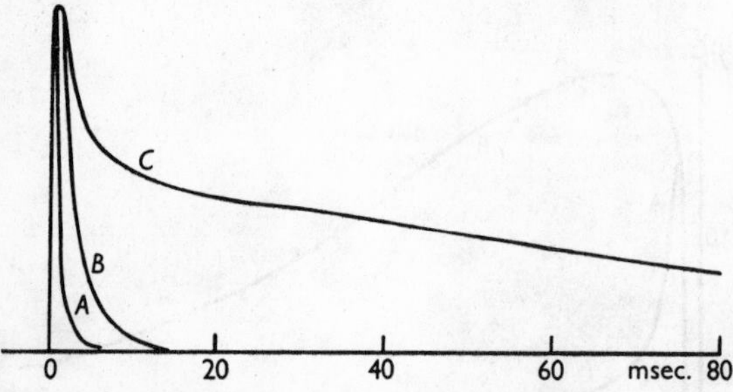

Fig. 15. 'Transmitter action' curves, obtained by analysis of e.p.p.'s. *A*: muscle treated with curarine; *B*: curarine + prostigmine; *C*: low-sodium + prostigmine. The ordinates have been scaled to the same maximum.

B. *The electric response of the normal end-plate membrane*

In the normal muscle fibre, the e.p.p. rises at a much greater rate and leads to a propagating spike and contraction. The electric response at the end-plate differs from a conducted action potential in a characteristic manner (Figs. 16–19, 21). A large e.p.p. invariably precedes the spike and forms a 'step' during the rising phase of the record. After the peak a discrete 'hump' is seen in most cases, indicating a continued action of the transmitter during the falling phase of the potential.

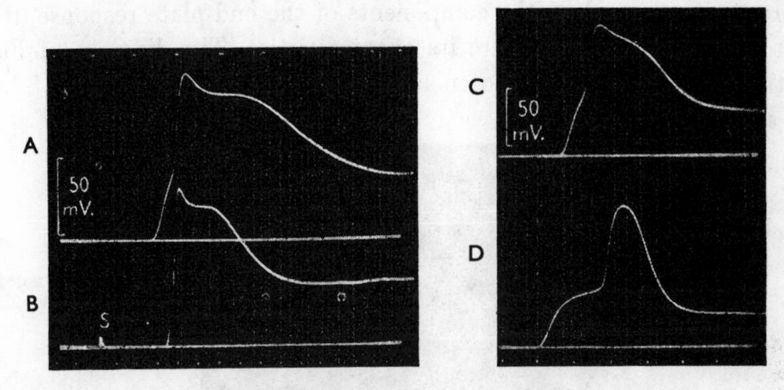

Fig. 16. End-plate responses in normal preparation. Four end-plates, showing step, spike and hump (except in D where the safety margin is low and a delayed spike without hump is seen). *S*, stimulus artifact. Time marks, msec.

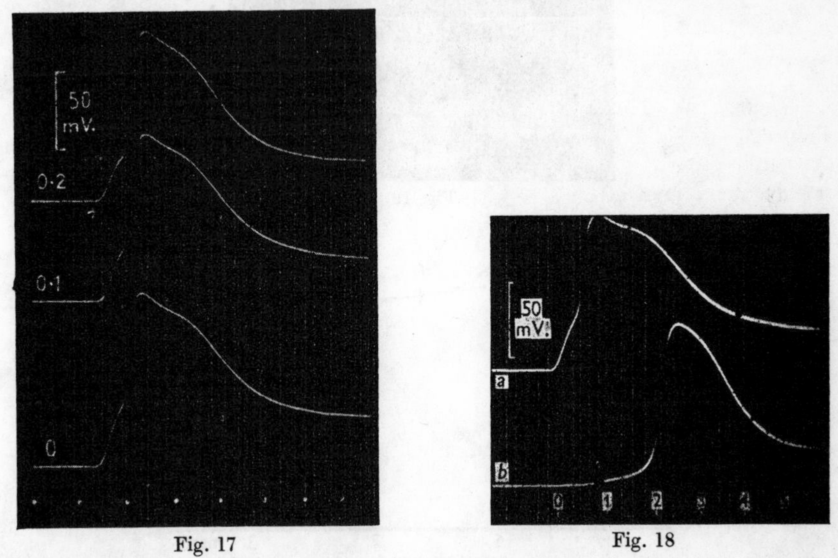

Fig. 17 Fig. 18

Fig. 17. Three records from the same muscle fibre, at distances of 0, 0·1 and 0·2 mm. from the end-plate focus. Time marks, msec.

Fig. 18. Records from the same fibre (*a*) at the end-plate and (*b*) 2·5 mm. away. Time, msec.

The usual procedure was, first, to locate a number of end-plates in a fully curarized muscle and then remove the drug by 30 min. washing in Ringer's solution (see Methods). The same results were obtained in a few cases in which the end-plates were found, by trial recordings, in normal untreated muscle (e.g. Fig. 18). This method naturally involved a considerable wastage of fibres, and was only used as a check to ascertain that the preliminary curarine-treatment had no irreversible effects.

Measurement of 'step' and 'hump'

Before trying to analyse the components of the end-plate response, it is of interest to describe and measure its characteristics. The diagram of Fig. 20 shows the points which were chosen as a convenient measure of 'start', 'step', 'peak' and 'hump'.

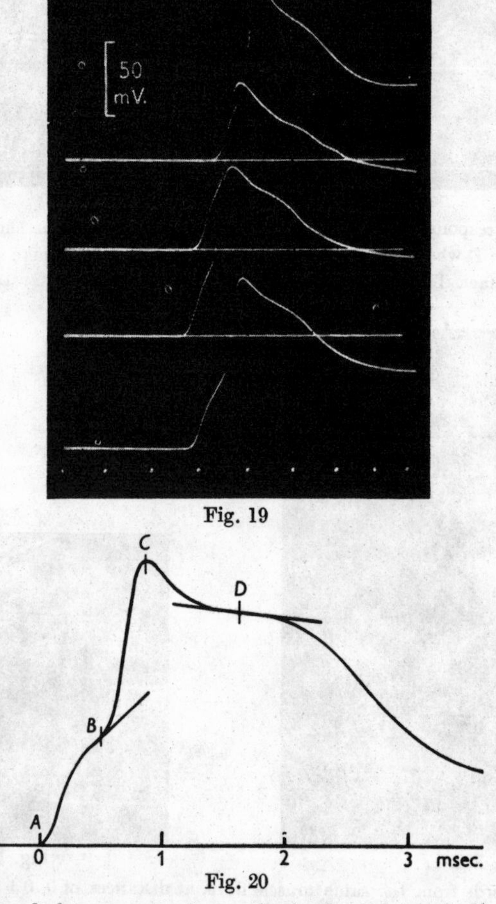

Fig. 19

Fig. 20

Fig. 19. Examples of end-plate responses in prostigmine-treated muscle fibres. Note the larger residual potential change. Time marks, msec.

Fig. 20. Diagram showing 'characteristic points' of the end-plate response.

The 'start' (point A) was taken as the point of just perceptible deflexion (about 0·3 mV., with low amplification). At the end-plate, the response rises sharply above the baseline, and this measurement was accurate within less than 0·1 msec. The 'peak' (point C) provided no difficulty, but the positions of 'step' (B) and 'hump' (D) are subject to some uncertainty. The height of the step was measured near the point of inflexion, at a level at which the spike could be seen to take off. The hump was measured at the mid-point of the flat shoulder on the falling phase. The separation between peak and hump was not always distinct (cf. Fig. 16) and the position of the hump, therefore, not always well defined. We estimate the accuracy of our measurements as being within 2–3 mV. and 0·1 msec. for the step (B), and 5–8 mV. and 0·2–0·3 msec. for the hump (D).

TABLE 7. Electric response at the normal end-plate

(Times are measured from the onset of the potential change. Active-membrane potential: p.d. across active membrane = action potential minus resting potential. Errors are the standard errors of the mean.)

	Temperature (° C.)	Calcium concn. (mM.)	Resting potential (mV.)	Action potential peak (mV.)	Action potential peak Time (msec.)	Active-membrane potential (mV.)	End-plate 'step' (mV.)	End-plate 'step' Time, (msec.)	End-plate 'hump' (mV.)	End-plate 'hump' Time (msec.)
	20 (16–23·5)	3·6 (1·8–9)	91±0·43	113±0·77	1·1 (0·54–2·4)	22±0·67	41±0·6 (25–54)	0·6 (0·31–1·65)	97 (80–117)	1·8 (1·3–2·7)
Number of experiments	—	—	(135)	(134)	(134)	(134)	(135)	(135)	(104)	(104)

Results from 135 end-plates are summarized in Table 7. Most experiments were made with a solution containing 3·6 mM. calcium, i.e. twice the amount normally in Ringer. In Table 7 are included the results of twenty-five experiments in which ordinary Ringer (1·8 mM. calcium) had been used, and thirty experiments in which prostigmine bromide in a concentration of 10^{-6} had been added. These various solutions affected the measurements only in one respect, namely that the height of the initial end-plate step was less with 1·8 mM. calcium (33 mV.) than with 3·6 mM. calcium (41 mV.). The statistical significance of this difference is further shown in Table 8, in which nine 'paired' measurements on the same end-plates are summarized. Prostigmine has an important effect on the membrane potential after the spike (Fig. 19; cf. Eccles *et al.* 1942), but made no appreciable difference to the present results.

TABLE 8. Effect of calcium on end-plate step

	Calcium concn. (mM.)	Resting potential (mV.)	Step height (mV.)	No. of exps.	Step ratio and S.E. of mean
Total measurements	1·8	91	33 (26–44)	25	—
	3·6	91	41 (25–50)	71	—
Paired measurements	1·8	92	32	9	1·25±0·046
	3·6	92	39	9	

As with the curarine experiments (Table 1) a high degree of variability was again encountered in the size of the e.p.p. which differed at individual junctions much more than the resting or action potential of the membrane. In the present measurements, this variability showed itself, not in the level of the e.p.p. at which the muscle impulse takes off—this was relatively constant—but in the *time* needed for the e.p.p. to rise to this threshold level. The variations in the entire muscle must have been greater than is apparent from Table 7, for most of the present results have been obtained from end-plates which had been selected during the preliminary curarine treatment because their e.p.p.'s were found to be large and easy to locate. The differences in the rate of rise of the e.p.p. must mean that even under normal conditions there are large variations in the safety margin of transmission at individual junctions. Such variation has been known for a long time: it was demonstrated by Adrian & Lucas (1912) and by Bremer (1927) who showed that during fatigue or partial curarization a variable number of fibres can be made to respond by varying the interval between two nerve impulses. Another example will be shown on p. 358 below in the variable susceptibility to anodic block at different end-plates.

We did not include in Table 7 the results from a small number of fibres in which the e.p.p. failed to reach the impulse threshold. A delayed spike was then usually recorded coming from a remote junction in the same fibre (cf. Katz & Kuffler, 1941, also Fig. 28*C* below). The local failure was presumably due to some abnormal condition of the muscle, but it was found side by side with end-plates at which transmission did not seem to be impaired and served further to illustrate the high degree of variability in junctional transmission.

The results of Table 7 show a wide dispersion in two other respects: (i) in the latency of the spike peak, and (ii) in the presence of a discrete hump which was clearly discernible in only some of the records. Both variations result from the variable size of the initial e.p.p.: the time to the peak includes the variable duration of the initial step, and a discrete hump could be seen only when the spike took off sufficiently early during the e.p.p. so that a residual transmitter effect, 2 msec. after the start, was not obscured by the spike peak.

In Table 2B, the responses of the same end-plates are compared (i) in fully curarized muscle and (ii) after withdrawal of curarine. The e.p.p. height was measured at a fixed interval, 0·44 msec. after the start. The results indicate that the e.p.p. in the fully curarized muscle was reduced to about one-quarter (with variations between 0·18 and 0·34). In another muscle, the same dose of curarine reduced the e.p.p.'s to about one-eighth. In general, a dose of curarine seemed to depress small e.p.p.'s more than large e.p.p.'s, and the dispersion in e.p.p. sizes, therefore, appeared to be greater in curarized than in normal muscle.

Comparison of the electric response at and off the end-plate

In Fig. 21, an experiment is illustrated in which the action potential was recorded at various distances along the same fibre. As the microelectrode moved away from the end-plate, the complex response (step-peak-hump, Fig. 21, 4–6) changed into a simple conducted spike which travelled in both directions, at a uniform velocity of about 1·4 m./sec. The initial step declined in the manner already shown for the rising phase of the curarized e.p.p., and its sharp ascent was replaced by the gradually increasing 'foot' of the conducted potential wave. The 'hump' was noticeably reduced, a few hundred microns away from the end-plate (e.g. Fig. 17), and vanished as the electrode was moved farther. The shape of the conducted spike varies somewhat from fibre to fibre, and its peak has often an 'angular' appearance as in Fig. 24 (*M*) below, but there is

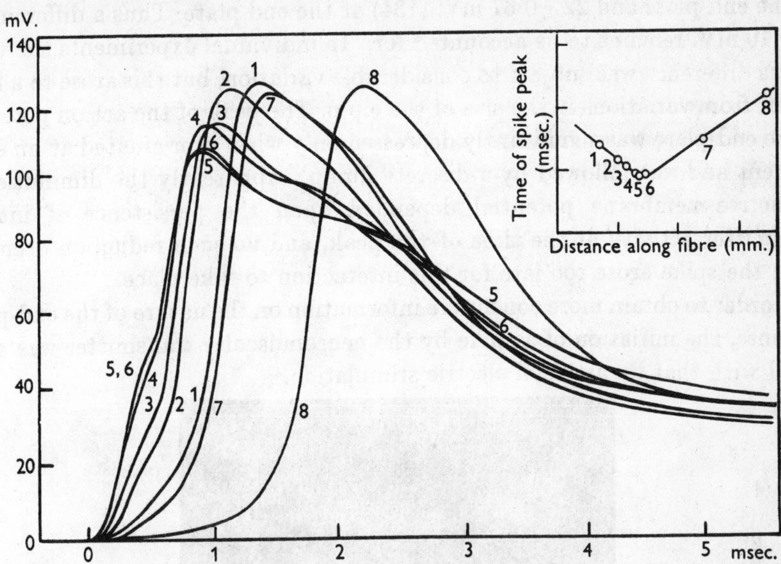

Fig. 21. The transition of electric activity from end-plate to muscle fibre. Calcium concentration, 9 mM. Temp. 17° C. The microelectrode was moved along the fibre, and records were obtained at the following positions (distance from position 1): (1) 0 mm.; (2) 0·3 mm.; (3) 0·45 mm.; (4) 0·6 mm.; (5) 0·65 mm.; (6) 0·75 mm.; (7) 1·75 mm.; (8) 2·75 mm. The resting potential was between 88 and 92 mV. during these records. Note the gradual changes in the shape of the action potential and spike latency. Inset: the time of the spike summit is plotted against distance, showing a propagation velocity of about 1·4 msec. in both directions from positions (5) and (6).

no doubt that the hump-like protrusion is a distinct feature of the end-plate response.

Fig. 21 indicates that the amplitude of the action potential increases by some 10–20 mV. as it is conducted away from the end-plate. In Table 9, the mean values of a large number of measurements, at and off the end-plate, have been listed. The resting potentials do not differ appreciably in the two situations,

TABLE 9. Active-membrane potential *on* and *off* the end-plate

(Mean values and S.E. of means.)

	I Resting potential (mV.)	II Action potential (mV.)	Active membrane potential (II − I)	No. of exps.
End-plate	91 ± 0·43	113 ± 0·77	22 ± 0·67	134
Off the end-plate	88 ± 0·6	123 ± 1	35 ± 1·1	52

Reduction of active-membrane potential at the end-plate: 13 ± 1·3 mV.

but the amplitude of the spike is considerably higher in the nerve-free portion than at the end-plate. The 'active-membrane potential', i.e. the level of the reversed p.d. during the peak, is 35 ± 1·1 mV. (S.E. of mean of 52 experiments) off the end-plate and 22 ± 0·67 mV. (134) at the end-plate. Thus a difference of over 10 mV. remains to be accounted for. In individual experiments the value of this difference was subject to considerable variation, but this arose to a large extent from variations in the size of the e.p.p. The peak of the action potential at the end-plate was significantly depressed only when it originated at an early moment and was followed by a discrete hump. Apparently the diminution of the active-membrane potential depended upon the persistence of intense transmitter activity at the time of the peak, and no such reduction occurred when the spike arose too late for this interaction to take place.

In order to obtain more conclusive information on the nature of the end-plate response, the initiation of a spike by the neuromuscular transmitter was compared with that by artificial electric stimulation.

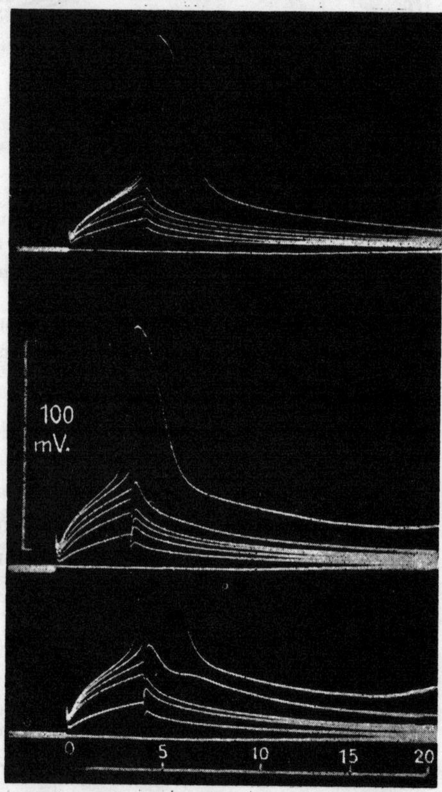

Fig. 22. Membrane potential during direct stimulation. Three different fibres, in which potential changes were recorded, near the cathode, with several subthreshold and one superthreshold current pulse. Note inflexion and local response with subthreshold current pulses. Time, msec.

The end-plate step

Two microelectrodes were placed into the same muscle fibre, less than 50 μ. apart, and one was used as a stimulating electrode by passing an outward current pulse through the membrane, while the change of the membrane potential was recorded by the other electrode. With a sufficient current strength, the membrane potential falls from its resting level, of 85–90 mV., to a point at which a spike is generated. The process is shown in Fig. 22, the time course of the curves being very similar to those previously obtained with external electrodes (Hodgkin & Rushton, 1946; Katz, 1948). The important point is that the height of the step which precedes the action potential is substantially the same as the height of the end-plate step during neuromuscular transmission. The step was measured in the same way as indicated in Fig. 20, and the results are shown in Table 10. In twenty experiments, using 3·6 mM-$CaCl_2$, the mean

TABLE 10. Comparison of initial 'step' with direct and indirect stimulation

	Direct stimulation			End-plate potential		
Calcium concn. (mM.)	Resting potential (mV.)	Step height (mV.)	No. of exps.	Resting potential (mV.)	Step height (mV.)	No. of exps.
3·6	86	39 (31–50)	20	91	41 (25–50)	71
1·8	85	36 (30–41)	7	91	33 (26–44)	25

height of the step was 39 ± 1·2 mV. as compared with 41 ± 0·6 mV. for the e.p.p., while in seven experiments with 1·8 mM-$CaCl_2$, it was 36 mV. (compared with 33 mV. for the e.p.p.). The result was the same whether the stimulus was applied at the nerve-free end of the fibre, or at the end-plate position. The level at which the spike originated seemed to be independent of the time taken to reach it: this time depended upon the current strength, and in the different records varied between 0·3 and 8 msec.

The measurement of the step height is related to the excitation threshold of the muscle fibre, that is to the critical level at which the membrane potential becomes unstable. This level can be found by using a short threshold shock (see Hodgkin, Huxley & Katz, 1949), or by breaking the current at the critical point at which the membrane potential is left 'in the balance', neither rising nor falling for a short time after the break. In practice, the current pulse was increased in small steps, and the largest potential change which just failed to flare up into a spike was taken as an indication of 'threshold' (Fig. 22). Using this method, the threshold depolarization was found to be several millivolts higher than the step, measured in the conventional way adopted above. Measurements of 'step height' and 'threshold level' are shown in Table 11, the means of eight experiments being 38 and 44 mV. respectively. These values are considerably larger than the figure of 15 mV. recently reported for the giant axon of the squid (Hodgkin *et al.* 1949), but the experimental conditions differ in two important respects: (*a*) the threshold of the squid axon was measured by uniform stimulation of a long length of fibre instead of *at one point*, and (*b*) the resting potential of the isolated squid axon is about 30 mV. less than that of

TABLE 11. 'Step' height and threshold level
(Calcium concentration 3·6 mM. in all fibres except VIII where it was 1·8 mM.)

Fibre	I Resting potential, (mV.)	II Step height, (mV.)	III Threshold level (mV.)	Difference (III – II), (mV.)
I	87	39	44	5
II	85	35	37	2
III	82	36	42	6
IV	78	31	36	5
V	81	38	47·5	9·5
VI	91	42·5	50·5	8
VII	90	46	50	4
VIII	88	38	43·5	5·5
Mean	85	38	44	6

frog muscle. If we were to define 'threshold' as a critical *membrane potential*, rather than a critical *depolarization*, the difference between the two sets of measurements would almost vanish, the 'threshold' being at about 45 mV. in either case.

It is safe to conclude from the present experiments that the height of the end-plate step is determined by the threshold of the surrounding muscle membrane, and that the threshold of this region does not differ by more than a few per cent from the threshold of other parts of the muscle fibre.

The end-plate spike

The analogy between an applied current and the neuromuscular transmitter helps us to account for the height of the initial step, but it fails to account for the further course of the end-plate response, for its reduced amplitude and the appearance of a hump on its declining phase.

It might be suggested that the size of the spike would, for some reason, be smaller at the point where it originates than after it has been conducted over a distance, and that this would account for the discrepancy of the active-membrane potentials in Table 9. It was important, therefore, to compare the active-membrane potentials for a locally initiated and a conducted spike. Two successive records were taken from the same point of a muscle fibre, in a nerve-free part: first observing the conducted spike which was elicited some distance away, either via the nerve or by a direct stimulus through another internal electrode. The second microelectrode was then inserted close to the recording point (less than 50 μ. distant) and another, direct, stimulus was applied. An example is shown in Fig. 23 where the peaks of the two action potentials are seen to differ by only 2 mV. The twitch during the first response usually produced some local damage associated with about 10% drop of the resting potential, but evidence will be presented, on p. 354 below, that this did not cause an immediate noticeable change in the active-membrane potentials, whose measurements therefore remained valid.

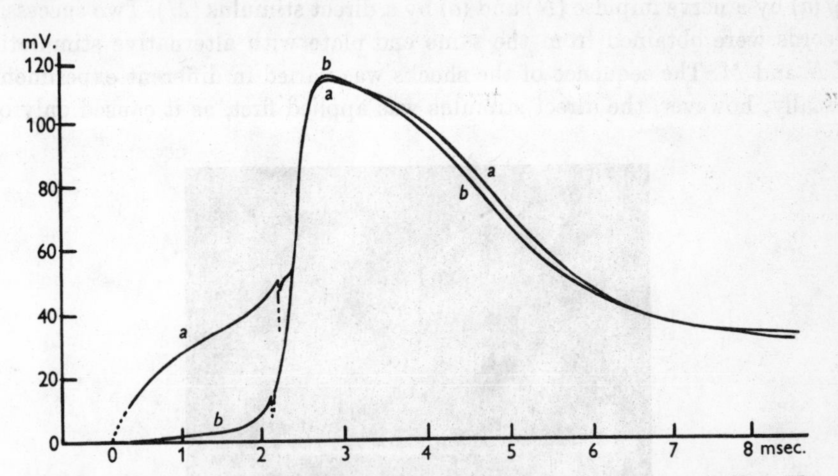

Fig. 23. Comparison of conducted and 'locally originating' spikes in a muscle fibre. The stimulating electrode was (*a*) 40 μ., and (*b*) 2 mm. from the recording electrode.

The effect of local mechanical damage can be appreciated quantitatively in the following manner. In the unstimulated fibre, the resistance across the membrane was on the average 200,000 Ω. (Table 4) and the potential difference 90 mV. Neglecting any initial leakage (cf. Nastuk & Hodgkin, 1950) a 9 mV. (10%) reduction of the resting potential implies that a small leak has sprung around the microelectrode, amounting to a shunt of 1·8 MΩ. (cf. Appendix II). During the spike, the effect of this shunt becomes much less important because the resistance across the membrane has fallen to about 20,000 Ω. (see p. 356 below); the reduction of the active-membrane potential (35 mV.) caused by a leak of 1·8 MΩ. is less than 0·5 mV.

In six experiments in which conducted and locally initiated spikes were compared, the active-membrane potential of the former was on the average 1·5 mV. less than that of the latter, an insignificant difference (S.E. of mean ± 1·3 mV.) and of opposite sign to that required.

The second possibility which had to be examined was that the features of the end-plate response might be imitated by a direct stimulus, if the applied current pulse were maintained throughout the period of electrical membrane activity. When this was done, a number of interesting changes were produced which will be described in the following section, but they bore no resemblance to the end-plate response. On the contrary, under the influence of a maintained outward current, the action potential continued to build up to a higher peak, and there was no indication of a hump during the decline. This effect of the applied current was seen invariably, whether the current was passed through the end-plate or through other regions of the fibre surface.

It might further be suggested that the characteristic features of the end-plate spike depend upon special properties of the muscle fibre at the junction, quite irrespective of release and local action of the transmitter. To decide this question, the action potential must be recorded at the end-plate when it is set up (*a*) by a nerve impulse (*N*) and (*b*) by a direct stimulus (*M*). Two successive records were obtained from the same end-plate with alternative stimulation of *N* and *M*. The sequence of the shocks was varied in different experiments. Usually, however, the direct stimulus was applied first, as it caused only one

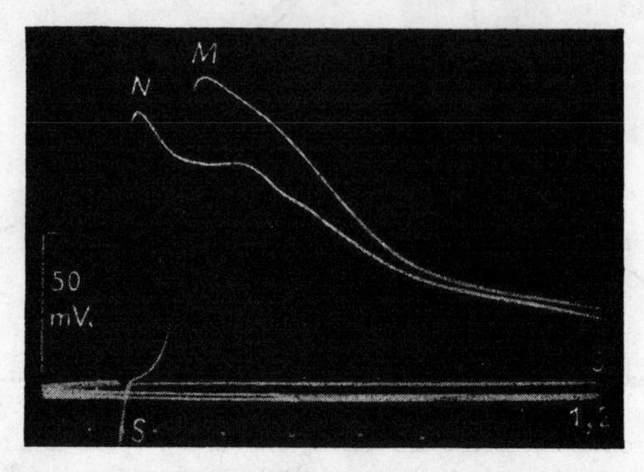

Fig. 24. Response to nerve (*N*) and direct (*M*) stimulation at the end-plate region of a muscle fibre. The direct current pulse was applied about 1·5 mm. away. *S*: stimulus artifact at the end of the direct pulse; 1 and 2: baselines (the lower corresponding to *N* the upper to *M*) showing a small drift of resting potential between the two stimuli, 3: electrotonic potential due to a subthreshold pulse. (The unusual notch in the *N* response following the hump, was due to an external spike potential of adjacent fibres.) Time marks, msec.

fibre to contract. An example is shown in Fig. 24, and for better comparison the two records have been superimposed, by displacing one along the time axis, in the tracings of Fig. 25. It is clear that the muscle spike (*N*) arising from an e.p.p. fails to reach the level which it attains (*M*) in the absence of the e.p.p. The results of fifteen similar experiments are summarized in Table 12. The

Table 12. Comparison of *M* and *N* spikes at the same end-plates

	M			*N*			Difference (*M* − *N*) of active-membrane potential (mV.)
Fibre	Resting potential (mV.)	Action potential (mV.)	Active-membrane potential (mV.)	Resting potential (mV.)	Action potential (mV.)	Active-membrane potential (mV.)	
A. Three selected experiments							
I	88	119	31	85	101	16	15
II	90	121	31	85	97	12	19
III	90	115	25	93	103	10	15
Mean	89	118	29	87	100	13	16
B. Fifteen experiments							
Mean	87	111	24	80	91	11	13±1·3

first part of this table contains the measurements on three fibres, in which little or no local damage occurred, and the resting potential did not change appreciably between the two records. The mean values of all measurements are shown in the second part of Table 12; they include several experiments in which a substantial

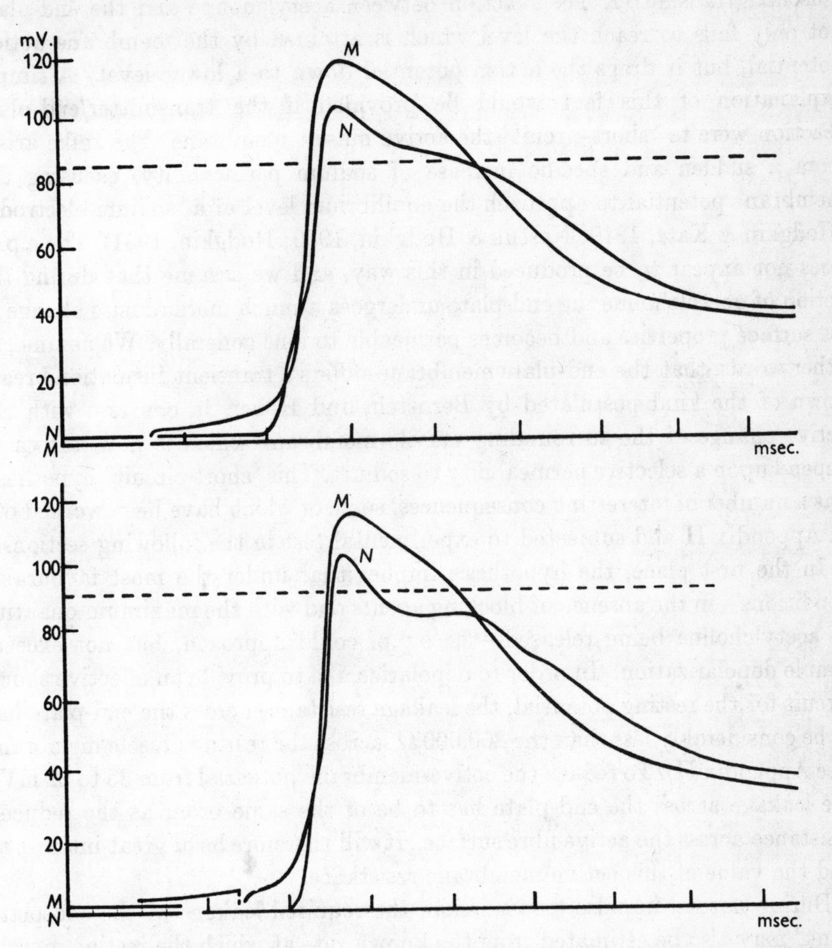

Fig. 25. Tracings of *N* and *M* responses at the end-plate region. Examples from two end-plates. Broken line: zero p.d. across membrane. A small drift of the resting potential between successive stimuli is indicated by the displacement of *M* and *N* baselines.

drop of the resting potential occurred after the first stimulus, but as pointed out above (see p. 354 for further details), the measurement of the active-membrane potential remains valid. Whether we take average values or selected experiments in Table 12, the result is equally conclusive. It shows that the reduction of the active-membrane potential must be attributed to the effect of the neuromuscular transmitter. This reduction amounts to more than 10 mV., and if we compare Table 12 with the previous results of Table 9, it appears that the differences in active-membrane potentials recorded *on* and *off* the end-plates are thus entirely accounted for.

This observation throws some light on the mode of action of the neuromuscular transmitter. The reaction between acetylcholine and the end-plate not only fails to reach the level which is attained by the membrane action potential, but it drags the action potential down to a lower level. A simple explanation of this fact would be provided if the transmitter/end-plate reaction were to 'short-circuit' the active muscle membrane. The spike arises from a sudden and specific increase of sodium permeability, casusing the membrane potential to approach the equilibrium level of a 'sodium electrode' (Hodgkin & Katz, 1949; Nastuk & Hodgkin, 1950; Hodgkin, 1951). The e.p.p. does not appear to be produced in this way, and we assume that during the action of acetylcholine the end-plate undergoes a much more drastic change of its surface properties and becomes permeable to ions generally. We assume, in other words, that the end-plate membrane suffers a transient insulation breakdown of the kind postulated by Bernstein and Höber, in contrast with the active change of the surrounding muscle membrane which is now known to depend upon a selective permeability to sodium. This 'short-circuit' hypothesis has a number of interesting consequences, some of which have been worked out in Appendix II and subjected to experimental test in the following sections.

In the first place, the hypothesis implies that under the most favourable conditions—in the absence of blocking agents and with the maximum quantity of acetylcholine being released—the e.p.p. could approach, but not exceed, simple depolarization. In order to depolarize, i.e. to provide an effective short-circuit for the resting potential, the leakage resistance across the end-plate has to be considerably less than the 200,000 Ω. across the resting muscle membrane (see Appendix II). To reduce the active-membrane potential from 35 to 22 mV., the leakage across the end-plate has to be of the same order as the reduced resistance across the active fibre surface. It will therefore be of great interest to find the value of this active-membrane resistance.

During normal impulse transmission, the required leakage of the end-plate 'sink' may also be estimated from the known rate at which the resting muscle membrane is discharged through this sink (normally 40–50 mV. in about 0·5 msec.) and the two independent estimates should be compared.

In Appendix II, the depolarizing effect of a shunt resistance suddenly placed across the muscle membrane has been calculated and the results indicate that the leakage of the active end-plate is of the order of 20,000–30,000 Ω. during normal transmission. In the following section, experiments are described which indicate that the resistance across the active fibre surface is of the same order of magnitude.

Finally, it follows from the 'short-circuit' hypothesis that the size of the e.p.p.

should be directly proportional to the resting potential. For example, if we were to raise the membrane potential to twice its normal resting level, by 'anodizing' the end-plate region, the same short-circuiting effect of the transmitter should then produce twice as large an e.p.p., and this relation should hold under all conditions, during neuromuscular block as well as in normal impulse transmission.

Resistance changes of the active fibre membrane

Measurements with alternating current have shown that the membrane impedance of nerve and muscle fibres undergoes a rapid diminution during the passage of an impulse (Cole & Curtis, 1939; Katz, 1942). Cole & Curtis concluded that the 'high-frequency conductance' (see Cole, 1949) of the axon membrane increases about 40-fold during the spike. The exact time course of this change was difficult to determine, but it is certain that the membrane conductance reaches its peak very rapidly during the rising phase of the action potential.

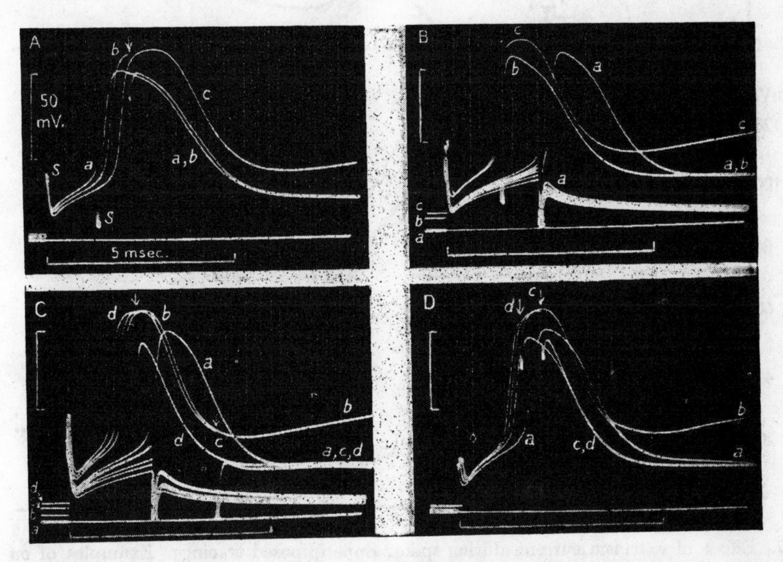

Fig. 26. The effect of an extrinsic current during the muscle spike. Examples from four muscle fibres. A: three responses, with the current (*a*) being broken at the beginning of the spike, (*b*) near its peak (marked by arrow), (*c*) being continued throughout. Note the rapid transition at the arrow, from 'current-on' to 'current-off' curve. B: (*a*) several subthreshold and one superthreshold pulse; (*b*) and (*c*), two responses with current *off* and *on*, respectively. C: showing several subthreshold and four superthreshold stimuli. With the latter, the current was broken either at the beginning (*a*), or during the spike indicated by arrows (*c*), (*d*), or maintained throughout (*b*). In B and C, note drift of resting potential (initial baseline displacement), but little or no change in active-membrane potential. D: four spikes, with the current alternatively *off*, *on*, or broken at moments marked by arrows. Note rapid change at the arrows, from current-*on* to current-*off* type of response. S: stimulus artifacts at make and break of current pulse.

It is possible to obtain an estimate of the changes of membrane resistance by passing a constant current through the fibre membrane and measuring the p.d. which the current adds (henceforth called the 'extrinsic p.d.') during and

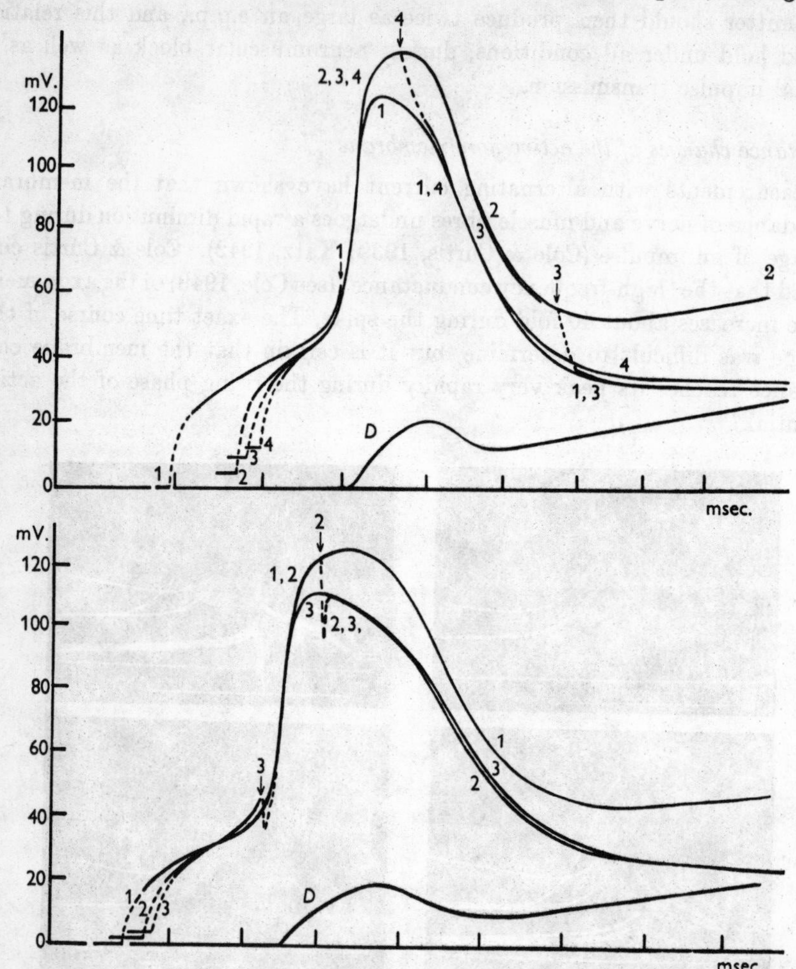

Fig. 27. Effect of extrinsic current during spike. Superimposed tracings. Examples of *on* and *off* curves from two muscle fibres. The 'extrinsic potential' built up by the current pulse during the spike is shown by curve D (which is obtained by subtracting the mean *off* from the *on* curve). Note: although the resting potential fell progressively, as seen by the displacement of initial baselines, the active-membrane potentials showed little change.

after the spike. The procedure was to stimulate with a second microelectrode and to break the applied current at various moments after threshold had been reached (Fig. 26). The experiment depended upon a comparison of successive spike records, with the current on or off, and it was necessary in the first instance to decide whether successive records, obtained usually with progressively falling resting potential, are strictly comparable. It has already been mentioned that a 10 or 20% drop of the resting potential, due to mechanical injury, does not necessarily invalidate the measurements of the active-membrane potential. The justification for this is shown in Fig. 27. In this figure, successive records have been superimposed, by shifting the individual spikes horizontally, until the ascending phases met, but *without* shifting them vertically. Although the resting potentials (initial baselines) differed by several millivolts from one record to the next, the action potentials remained almost exactly superimposable. If the current was maintained beyond the initial subthreshold period, there was:

(i) No noticeable change in the maximum rate of rise of the spike.

(ii) A distinct addition to, and broadening of, the peak.

(iii) An increase in the maximum rate of fall of the spike.

(iv) A later gradual redevelopment of the extrinsic potential difference.

If the current was broken at any moment during this sequence, the membrane potential returned from the 'current on' to the 'current off' curve within a fraction of a millisecond. It would appear from the results in Fig. 27 that we are justified in using the 'off'-curve as a baseline, from which the extra p.d. due to the maintained outward current can be measured. The only region in which reliable measurements could not be made was the steep ascending phase. A slight lateral displacement of the superimposed records would make a considerable difference here. It is unlikely that the applied current produces a large extra p.d. during this phase: (*a*) because the upstrokes of the superimposed curves, as in Fig. 27, cannot be displaced from each other by more than 30 μsec. without noticeably mismatching the later parts of the curves; and (*b*) because the membrane conductance is known to reach its peak during the rising phase, and hence only a small extrinsic p.d. could be expected. It was, indeed, somewhat surprising to find that such a conspicuous potential change is produced by an applied current at the peak of the spike. Previous measurements (Kuffler, 1942*b*) indicated that no such addition occurs and that, on the contrary, the whole of the electrotonic potential collapses during the spike peak. But the discrepancy between Kuffler's and the present results is explained by the fact that we have recorded the action potential within 20–40 μ. of the cathode, while in the previous work an electrode of about 2 mm. width was used which would reduce the observed p.d. effectively to zero. It will be noted from equation 2 (see p. 334) that immediately at the cathode the p.d. produced by an applied current is proportional to $\sqrt{(r_m r_i)}$ so that, even when the membrane resistance has dropped to 1% of its resting value, one-tenth of the previous electrotonic potential should still be observed. But the length constant $\sqrt{(r_m/r_i)}$ is then also reduced to one-tenth so that, 1 or 2 mm. away, the extrinsic p.d. will be effectively abolished.

The time course of the extrinsic p.d. is indicated in Fig. 27 (*D*). As has been pointed out, the initial part of this curve (*D*) coinciding with the steep ascending phase of the spike is uncertain and depends upon the exact point at which the records have been superimposed. There is, however, no such uncertainty about the later course of the extrinsic p.d., measured during and after the peak of the spike. It has a characteristic shape with a maximum 0·5–1·0 msec. after the peak of the spike, and a minimum about 1 msec. later. This was observed in all fifteen experiments of this type, and it suggests that during the spike the muscle membrane undergoes two separate phases of increased ion permeability, one associated with the rise, the other accompanying the fall of the action potential. It will be noted that a relatively large extrinsic p.d. is built up during the slow initial phase of decline (the 'angle'), but it drops to a minimum later when the action potential falls more rapidly. This result seems to be analogous to recent observations by Weidmann (1951) on mammalian heart muscle and has an interesting bearing on the ionic theory of the impulse developed by Hodgkin & Huxley (1950; see also Hodgkin *et al.* 1949). According to this theory, two separate permeability changes occur during the spike: the rising phase is associated with a momentary increase of permeability to sodium ions, but this is a transient change which becomes rapidly exhausted or inactivated. It is followed after a brief delay by a phase of high potassium permeability which leads to a rapid return of the membrane potential to its original level. The present results provide evidence for two separate changes of membrane conductance, a transient increase during the ascending phase, and a second increase during the fall. The two conductance changes may well be associated with the two separate phases of sodium and potassium transfer envisaged by the theory of Hodgkin & Huxley.

In nine experiments, the strength of the outward current was measured simultaneously with the extrinsic p.d. From the ratio of the two values, an approximate estimate of the membrane resistance can be obtained. The estimate depends upon the assumption that the time constant $r_m c_m$ of the active membrane is brief compared with the time course of the extrinsic p.d. and the associated resistance change. Although this is over-simplified, it is approximately true for the relatively slow changes after the peak of the spike, when the time constant of the membrane appears to be of the order of 0·3 msec. (compared with 30 msec. in the resting muscle). In Table 13 the ratio of extrinsic p.d./outward current is given, at the time of the 'dip', in nine experiments. The

TABLE 13. Resistance across fibre membrane during the falling phase of the spike (Measured at the 'dip', cf. Fig. 27, *D*.)

Summary of the values of $\frac{1}{2}\sqrt{(r_m r_i)}$ in nine experiments:

9000 23,000 14,000 24,000 24,000 21,000 42,000 22,000 12,000 Ω.

Mean 21,000 Ω.

mean value is about 20,000 Ω., varying between 9,000 and 42,000 Ω. These values represent the transverse resistance $\frac{1}{2}\sqrt{(r_m r_i)}$ of the active muscle fibre, measured during the falling phase of the spike: they are about one-tenth of the resting value (Table 4) which indicates that the membrane resistance r_m, at that moment, is only about 1% of the resting values, 40 Ω.cm.^{2} instead of 4000 Ω.cm.^{2}. During the rising phase of the spike, the resistance is presumably even lower. At the time of the spike summit, the extrinsic p.d. is of about the same size as during the 'dip', and we may tentatively regard the value of 20,000 Ω. as representing the active-membrane resistance during the peak of the spike as well as later during its falling phase.

The time constant of the active membrane is, therefore, also of the order of 1% of the resting value, about 0·3 msec. compared with 30 msec. This is borne out by the fact that the added extrinsic p.d. disappears within a fraction of a msec. when the current is broken during the spike (Fig. 27).

We suggested that the active end-plate, in spite of its minute size, short-circuits the surrounding active fibre membrane, bringing its potential down, from 35 to 22 mV. In order to produce this effect, it is clear that the active end-plate must itself have a low resistance, of the order of 20,000 Ω. The presence of such a low-resistant sink must have an important influence on the further time course of the action potential, and it is possible to explain the appearance of the end-plate 'hump' without additional assumptions. It can be seen from Fig. 25 that the 'hump' is in reality due to a rapid fall of the active-membrane potential from its peak towards a lower level which is not far from zero, and to a delayed return from this to the resting level. The hump is probably due to a continued short-circuiting of the membrane which not only reduces the 'sodium-potential' of the active fibre membrane, but causes it to discharge quickly when the period of high sodium permeability comes to an end. Similarly, the continued leakage through the end-plate must delay the restitution of the membrane potential, and these effects are probably responsible for the characteristic 'hump' of the end-plate spike.

To summarize, the neuromuscular transmitter not only produces an e.p.p. which gives rise to a muscle spike, but it interacts with the further course of the spike by depolarizing the active membrane and holding its potential close to zero.

The relation between end-plate potential and resting membrane potential

It is possible to change the p.d. across the fibre membrane, by means of electric currents, over a fairly wide range without substantially altering the resistance or capacity of the membrane (Hodgkin & Rushton, 1946; Katz, 1948). This method helps one to distinguish between three conceivable mechanisms by which the e.p.p. may be produced. The e.p.p. might be the result of one of the following processes: (*a*) the transfer across the end-plate of a fixed number of ions (for example, by extrinsic current flow from the motor nerve);

(*b*) a shift of the membrane potential towards a fixed new level (determined, for instance, by a *selective* permeability change to one species of ions); (*c*) the establishment of a *non-selective* ion sink, equivalent to placing a fixed leak resistance across the membrane.

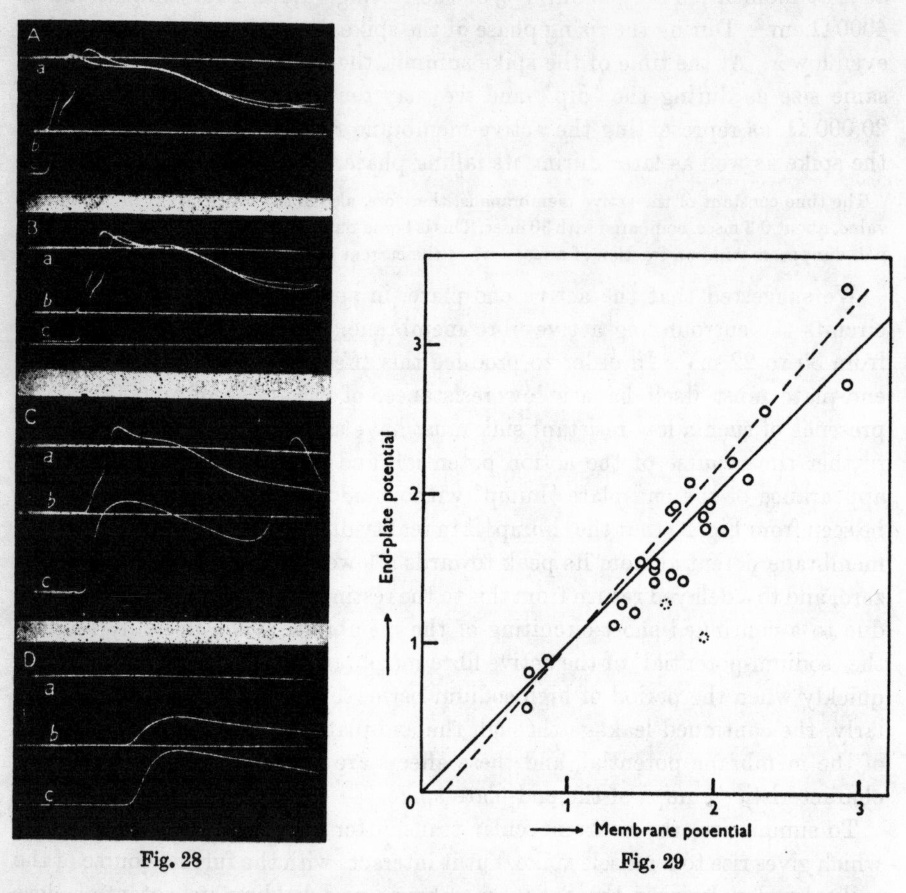

Fig. 28 Fig. 29

Fig. 28. The effect of increased ('anodic') resting potential on the size of the e.p.p. Examples from four end-plates. *a*: zero p.d. across membrane; *b*: normal resting potential; *c*: resting potential has been increased by applied inward current. At C, transmission is blocked by the inward current, while at D transmission had failed at the normal level of the resting potential. Voltage scale: 50 mV., time marks: msec. (Record A(*b*) was taken with too high electrode resistance, giving some amplitude reduction; in all other records no appreciable distortion occurred.)

Fig. 29. Relation between size of end-plate potential and initial membrane potential. Normal resting and end-plate potentials are taken as unity. Normal resting potentials were: 90 mV. mean (65–104 mV.). E.p.p's were measured at a fixed time after their start (before the spike originated or, if transmission was blocked, at the peak of the e.p.p.). The full line indicates direct proportionality; the broken line intersects the horizontal axis at the theoretical junction p.d. between Ringer and myoplasm (see text).

Process (*a*) would give a constant amplitude, and constant charge, of the e.p.p., independent of the initial level of the membrane potential (provided resistance and capacity of the membrane remain constant).

Process (*b*) would result in a variable amplitude, but approximately constant final level of the e.p.p., rather like that attained by the peak of the spike whose level is only slightly affected by changes of resting potential (Fig. 28).

Process (*c*) would reduce the resting potential to a constant fraction and lead to a directly proportional relation between the size of the resting membrane potential and the amplitude of the e.p.p. (see Appendix II).

The experimental procedure was to raise the resting potential to a higher level by subjecting the membrane to an inward current. The e.p.p. was observed at the normal resting potential (mean value 90 mV.) and at the increased level (varying between 118 and 235 mV.), and its size was measured at a fixed interval after the start. Examples are shown in Fig. 28, and the results of twenty-six experiments have been plotted in Fig. 29. The mean increase of resting potential in twenty-one experiments (not including the five observations discussed below) was 87·5%, the corresponding increase of e.p.p. size was 84%. There is little doubt that the size of the e.p.p. is approximately proportional to the value of the initial membrane potential, a result which is consistent with the 'short-circuit' hypothesis (*c*), but not with the other hypotheses stated above.

The results of these experiments were obtained at junctions with widely different safety margins. At some the e.p.p. formed so large a proportion of the resting potential that no anodic block could be produced; at others transmission was readily blocked by an inward current (cf. Katz, 1939), and a pure e.p.p. produced, and finally at some end-plates, the e.p.p. was small and transmission failed even without the application of an inward current. The proportional relation between e.p.p. and membrane potential shown in Fig. 29 was found regardless of the condition of the individual junction.

It may be argued that even when the end-plate has become completely short-circuited, a junction potential of some 14 mV. (Nastuk & Hodgkin, 1950) would remain between the outside bath and the myoplasm. On this basis, the theoretical relation should follow the broken line in Fig. 29 rather than the full, 45°, line.

In Fig. 29, the results of three measurements are included in which the resting potential had dropped below the original level, owing to local injury. Under these conditions, the theoretical relation is no longer exactly linear, but the divergence is too slight to be noticed.

Another point requires comment: not all muscle fibres were able to withstand strong 'anodization', and at times there were signs of dielectric breakdown when the membrane potential had been raised to some 200 mV. This showed itself in a rapid decline of the membrane potential while the inward current was maintained. The two measurements shown in Fig. 29 as dotted circles were made in this unstable condition and should, therefore, be disregarded.

DISCUSSION

A large part of our results is 'descriptive' and deals with the intracellular recording of potential changes at the motor end-plate. These results may be briefly discussed in relation to previous work in which similar techniques or preparations have been employed.

The values of resting and action potentials of the 'muscle fibre', as distinct from its end-plate (Table 9), agree with those reported by Ling & Gerard (1949) and Nastuk & Hodgkin (1950). At the end-plate, the resting potential is the same, but the action potential is reduced, provided excitation occurs via the nerve and the intensity of the transmitter action is high. The general features of the normal end-plate response (the 'step-peak-hump' complex) agree very well with those recently reported by Nastuk (1950).

The composite nature of the end-plate response had previously been demonstrated in an admirable way by Kuffler (1942*a*) who applied an external recording electrode to an isolated nerve-muscle junction. The records obtained by Kuffler differ from our results in some important respects, and these differences require an explanation. Kuffler employed a special technique of 'interface recording' which amounts in effect to the application of a microelectrode to the surface of a muscle fibre in a large volume of saline. This method had certain advantages and was well suited to Kuffler's delicate preparation, but it must be realized that under these conditions the observed potential change follows the time course of the *membrane current*, not that of the *membrane potential* (see Bishop, 1937; Lorente de Nó, 1947; Brooks & Eccles, 1947). This leads to important differences in the shape of the e.p.p. and spike, for the membrane current depends upon d^2E/dx^2, the curvature of the surface potential gradient, and may be directed inward or outward through the membrane. The 'interface' method is, in fact, a differential recording technique which is very sensitive to changes in the local potential gradient, but its results cannot be directly interpreted in terms of the membrane potential. For example, when the interface electrode was moved along the muscle fibre, the response changed very critically, and at 0·5 mm. from the end-plate, the initial e.p.p. deflexion had not only declined, but reversed its sign. This reversal means that the position of the electrode has been moved from the end-plate 'sink' where current flows *into* the fibre, to an adjacent region where the current *leaves* the fibre. At the centre of the junction, a triphasic response was recorded the first phase of which consisted of a 'pure e.p.p.' not superseded by a spike. This naturally led to the supposition that the membrane potential during the e.p.p. may attain the same level as during the spike; but with this conclusion, the present results do not agree. In our opinion, the absence of a spike component in Kuffler's experiment merely implied that the inward current through the end-plate had reached a peak during the initial e.p.p. and begun to decline when the spike originated. To explain this behaviour, we may refer to the results of Fig. 21 and Table 9, where it is shown that the size of the e.p.p. is *greatest*, while the size of the spike is *least* at the centre of the junction. Hence, during the transition from e.p.p. to spike, the electric response at the end-plate changes from the position of a spatial maximum to that of a spatial minimum, and during this process the curvature of the potential gradient, and the membrane current, reverse. The situation may

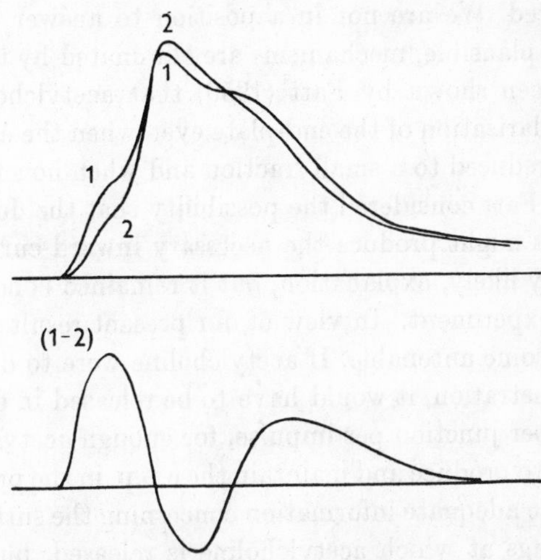

Fig. 30. Diagram explaining the derivation of 'interface' recording (cf. Kuffler, 1942*a*) at the centre of the end-plate. The upper part indicates the changes of membrane potential, (1) at the centre of the end-plate, and (2) a small distance from (1). The lower part illustrates a 'differential' record (1) minus (2), which resembles Kuffler's 'inter-face' recording.

be appreciated more easily by the simplified diagram in Fig. 30 in which Kuffler's relevant record has been reconstructed as differentially recorded between a point at, and slightly off, the centre of the junction.

To summarize, it would appear that the discrepancies between Kuffler's and the present results can be explained by differences in recording technique, remembering that we are concerned with changes of the membrane potential which are not faithfully recorded with an external microelectrode.

In its quantitative aspects the present paper provides strong support for certain views previously presented: for example, it has again been confirmed that the transmitter action at the nerve-muscle junction is a brief impulsive event, and that the characteristic spread and decay of the e.p.p. is largely determined by the resistance and capacity of the resting muscle fibre.

In addition, there have been two pieces of information which invite further comment: first, the determination of the quantity of electric charge which is transferred across the end-plate, and secondly, the fact that the end-plate reaction leads to a smaller potential change, but apparently a larger change of conductance (per unit area), than the normal membrane spike.

It has been pointed out that the electric charge which flows across the normal end-plate during a single impulse requires the net transfer of at least $2–4 \times 10^{-14}$ mol. of univalent cations inward or anions outward and that this quantity becomes multiplied by a factor of about 50 in the presence of a cholinesterase inhibitor. The question arises how such a large flux of ions can be maintained across the presumably minute area of the motor end-plate, and what species of ions are involved. We are not in a position to answer this question, but certain, otherwise plausible, mechanisms are eliminated by the present results. It has recently been shown by Fatt (1950) that acetylcholine ions produce a substantial depolarization of the end-plate even when the external electrolyte content has been reduced to a small fraction and when no sodium is present in the outside fluid. Fatt considered the possibility that the flux of acetylcholine cations themselves might produce the necessary inward current. This did not seem a full, or very likely, explanation, but it remained conceivable under the conditions of his experiment. In view of our present results, we feel that this hypothesis has become untenable. If acetylcholine were to depolarize the end-plate by direct penetration, it would have to be released in quantities of some $1-2 \times 10^{-12}$ mol. per junction per impulse, for enough acetylcholine ions must be made available to produce and maintain the e.p.p. in the prostigmine-treated muscle. We have no adequate information concerning the surface or the volume of the nerve endings at which acetylcholine is released, but it is difficult to believe that they are large enough to contain this amount of acetylcholine. If we take Kühne's (1887) drawings of the terminal arborizations in frog muscle, we are likely to over- rather than under-estimate the size of the nerve endings (see Couteaux, 1947). The surface area of Kühne's nerve-endings tree may be as large as 10^{-4} cm.^{2}, and its volume as much as 2×10^{-8} cm.^{3}. Even if we were to assume that the intracellular cation content of all these structures is made up entirely of acetylcholine, at a concentration of 120 mM., the amount of acetylcholine inside the nerve would be only $2{\cdot}4 \times 10^{-12}$ mol, i.e. barely enough for one or two impulses. Hence, even with such extremely favourable, though unrealistic, assumptions we are led to the absurd result that practically the whole cation store of the nerve endings would have to be exchanged during a single impulse in order to produce an e.p.p.

The only reasonable alternative appears to be that small quantities of acetylcholine alter the end-plate surface in such a way that other ions can be rapidly transferred across it, not only sodium and potassium, but probably all free anions and cations on either side of the membrane. Apparently, we must think in terms of some chemical breakdown of a local ion barrier which occurs as soon as acetylcholine combines with it, and whose extent depends upon the number of reacting molecules.

An explanation of this kind fits the facts reasonably well: it helps us to understand why a depolarization by acetylcholine can still be produced in the absence of external sodium salts, and it explains why the action potential, as well as the resting potential, is short-circuited by the e.p.p. Finally, it satisfies the requirement for a very large amplification of ionic currents which must occur at the point where an impulse is transferred from minute nerve endings to the enormously expanded surface of the muscle fibre.

Our results suggest that the action of acetylcholine 'short-circuits' the muscle fibre at the end-plate and so reduces the active-membrane potential, but this effect does not occur when the muscle fibre is stimulated directly. It appears, then, that the action potential of the muscle fibre, if started elsewhere, sweeps past the end-plate region without stimulating its neuroreceptors, for if they were made to react in the way in which they respond to a nerve impulse, the active-membrane potential would be the same in either case. It is a characteristic property of nerve or muscle membranes to respond to an electric stimulus with a regenerative electrochemical reaction. This reaction is now known to depend upon a selective increase of sodium permeability (Hodgkin & Huxley, 1950), leading to rapid entry of sodium into the fibre with a consequent lowering of its surface potential and reinforcement of the initial electrical alteration. This reaction proceeds towards an equilibrium level which is near the potential of a sodium electrode (Hodgkin & Katz, 1949; Nastuk & Hodgkin, 1950; Hodgkin *et al.* 1949). Our evidence indicates that the end-plate receptors do not behave in this manner: they react to acetylcholine and various other chemical substances, but apparently not to the local currents of the muscle impulse; and if the end-plate does not respond to electric stimulation, then its electrical reaction to acetylcholine cannot be regenerative in the manner of the electric excitation of the surrounding membrane. Thus, it appears that the end-plate, i.e. the neuroreceptive area of the muscle fibre, differs from the surrounding fibre surface not only in its specific sensitivity to chemical stimulants, but in its lack of sensitivity to electric currents.

APPENDIX I

The solution of the problem considered here has been kindly provided by Mr A. L. Hodgkin.

In the special case in which a charge is placed instantaneously, at time $t=0$, on a point along the fibre, at distance $x=0$, the solution of the general differential equation for the leaky capacitative cable without net current, viz.

$$-\lambda\frac{d^2V}{dx^2}+\tau_m\frac{dV}{dt}+V=0$$

takes the form

$$V=\frac{q_0}{2c_m\lambda\sqrt{(\pi t/\tau_m)}}\exp\left(\frac{-x^2\tau_m}{4\lambda^2 t}-\frac{t}{\tau_m}\right), \quad (4)$$

where q_0 is the charge initially on the membrane and c_m is the capacity of the membrane per unit length of fibre. Taking the natural logarithm of equation (4) we obtain:

$$\log_e V=\frac{-x^2\tau_m}{4\lambda^2 t}-\frac{t}{\tau_m}+\log_e\frac{q_0}{2c_m\lambda\sqrt{(\pi t/\tau_m)}}. \quad (5)$$

Since x appears only in the first term on the right of equation (5), if for any given t, $\log_e V$ is plotted against x^2, a straight line will result with slope equal to $-\tau_m/4\lambda^2 t$, from which λ can be obtained. When applied to the curves of Fig. 9, for t greater than 3 msec., this method gives values of λ between 2·35 and 2·1 mm. For the peak of the potential wave at any position x, $dV/dt=0$; so from equation (5) by differentiating and equating to zero, one finds that

$$\frac{x^2}{\lambda^2}=\frac{4t^2}{\tau_m{}^2}+\frac{2t}{\tau_m}.$$

This equation provides a simple means of evaluating λ. Plotting $4T^2/\tau_m{}^2+2T/\tau_m$ against x^2 (T is the time of peak potential at distance x) a straight line is obtained with slope equal to $1/\lambda^2$. This method was used on p. 332, with the results shown in Table 3.

APPENDIX II

The 'short-circuiting' of the end-plate during neuromuscular transmission

It is suggested that acetylcholine short-circuits the end-plate and thereby discharges the surrounding muscle membrane and gives rise to a propagated spike. The short-circuit resistance thus placed across the end-plate surface must be low enough to produce the characteristic features of the normal end-plate response, viz. (i) to depolarize the membrane at an adequate rate and (ii) to shunt the active membrane effectively and reduce its reversed p.d. At the normal nerve-muscle junction, the resting potential is about 90 mV., and the e.p.p. reduces it to one-half in about 0·5 msec. Experiments with anodic block (cf. Fig. 28) indicate that the maximum end-plate depolarization is reached at about 1·2-1·3 msec. (20° C.) and, in the absence of a muscle spike, amounts to about 70% of the resting potential.

The resting muscle fibre can be represented by the electrical cable model shown in Fig. 31. The quantities have been computed from Table 4 ($E_0 = 90$ mV., $r_m = 100{,}000$ Ω.cm., $r_i = 1{\cdot}6$ MΩ./cm., $c_m = 0{\cdot}3$ μF./cm.). If a short-circuit

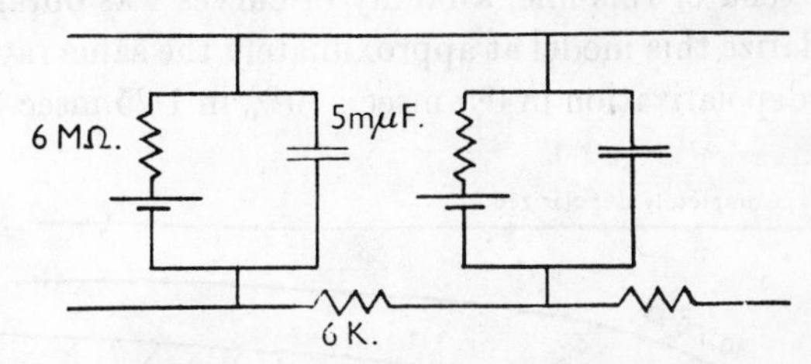

Fig. 31. Two sections of an artificial transmission line representing the passive properties of a muscle fibre. The input resistance at *one* end of the line is approximately equal to the resistance across the midpoint of a muscle fibre; the time constant is 30 msec. and the length constant is represented by thirty-two sections. The e.m.f. was replaced by a dry cell (1·58 V.).

resistance R is placed, at time $t=0$, across the mid-point of this transmission line (neglecting 'liquid junction potentials' and assuming the line to be longer than $8\sqrt{(r_m/r_i)}$) then the potential V at this point changes with time according to the following equation which has been kindly derived for us by Dr E. J. Harris:

$$\frac{V}{E_0} = 1 - a[1 - \exp t/\tau_1 . \operatorname{erfc}\sqrt{(t/\tau_2)}] + b \operatorname{erf}\sqrt{(t/\tau_3)}, \qquad (6)$$

where

$$\tau_1 = R_0^2\, c_m/(r_i - R_0^2/r_m),$$
$$\tau_2 = R_0^2\, c_m/r_i,$$
$$\tau_3 = r_m c_m,$$
$$a = r_i/(r_i - R_0^2/r_m),$$
$$b = R_0\sqrt{(r_i/r_m)}/(r_i - R_0^2/r_m),$$
$$\operatorname{erfc}\sqrt{(t/\tau_2)} = 1 - \operatorname{erf}\sqrt{(t/\tau_2)},$$
$$R_0 = 2R.$$

Using equation (6), the depolarization at $t = 0{\cdot}5$ msec. has been calculated for various values of R (Table 14). Another way of finding the potential changes,

TABLE 14. Relation between short-circuit resistance and 'end-plate potential.'

R (Ω.)	'e.p.p.' at $t=0{\cdot}5$ msec. (mV.)
15,000	65
20,000	56
30,000	46
50,000	32

at various points of the line, consists in constructing an electrical model similar to that of Fig. 31 and recording the changes of potential at the desired point. A line of 100 sections was used, each having the components shown in Fig. 31. This line had the same input impedance and time constant as an average muscle fibre taken from Table 4. The 'characteristic length' (i.e. about 2·5 mm.) was represented by thirty-two sections and the resting potential was replaced by a dry cell (1·58 V.). When a series of different short-circuit resistances was placed across the end of this line, a family of curves was obtained, plotted in Fig. 32. To depolarize this model at approximately the same rate as the normal end-plate (50% depolarization in 0·5 msec., 70% in 1·25 msec.) a short-circuit

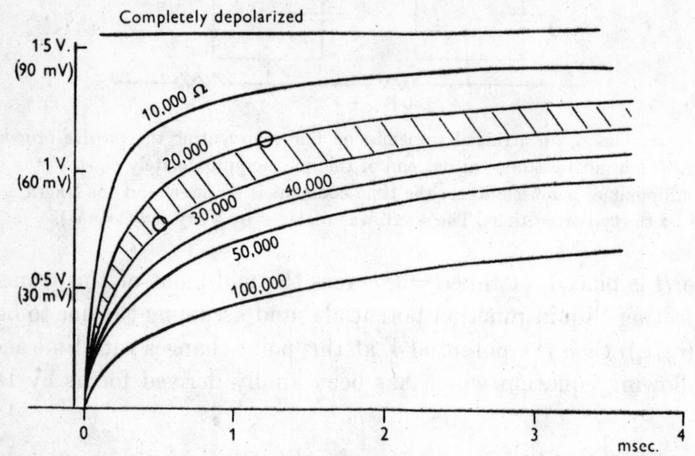

Fig. 32. Depolarization resulting from a short-circuiting of the transmission line. The two circles indicate the depolarization levels observed during the rise of the normal e.p.p. This corresponds to short-circuiting by 20,000–30,000 Ω. (shaded area). Ordinates: depolarization in volts (corresponding values for muscle fibre in brackets). Abscissae: msec.

resistance of 20,000–30,000 Ω. must be used. As the release and decay of the transmitter are gradual processes, one may assume that the resistance of the end-plate membrane falls, during the first msec., to a value rather less than 20,000 Ω. and then gradually recovers, but the average value during the rising phase of the e.p.p. appears to be about 25,000 Ω.

During normal impulse transmission we thus have, very roughly, an end-plate 'sink' with a leak resistance of the order of 25,000 Ω., in parallel with an active muscle membrane which—when *not* short-circuited by the active end-plate—produces a peak potential of 35 mV. and has a resistance of the order of 20,000 Ω. (Table 13). The presence of the end-plate sink reduces the active-membrane potential from 35 mV. to $35 \times 25{,}000/(25{,}000+20{,}000) = 19{\cdot}5$ mV. This drop of 15·5 mV. may be compared with the observed reduction of 13 mV. (Tables 9 and 12) and 16 mV. (Table 12).

An e.p.p. can be imitated in even more realistic fashion by placing a transient short-circuit, e.g. a series combination of resistance and capacity across the artificial line, with the result shown in Fig. 33. It is then a simple matter to reconstruct the experiments of p. 357, where an approximately linear relation

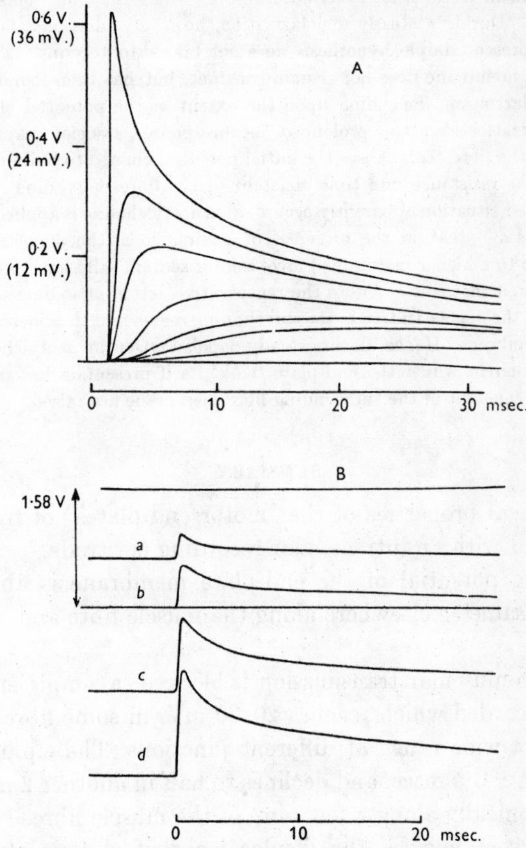

Fig. 33. A: artificial e.p.p.'s. Oscillograph tracings from various points along the artificial line, when a short-circuit of 20,000 Ω., in series with a 0·012 μF. condenser, was placed across it. The distances from the short-circuited points were, successively from above: 0 sections (corresponding to the end-plate centre); 5 sections (corresponding to about 0·4 mm. from the end-plate centre); 10 sections (0·8 mm.); 20 sections (1·6 mm.); 30 sections (2·4 mm.); 40 sections (3·2 mm.); 60 sections (4·8 mm.) Co-ordinates as in Fig. 32. B: relation between artificial e.p.p. and initial voltage level. *b*, 'normal' voltage and e.p.p.; *a*, line voltage reduced by partial short-circuit; *c* and *d*, line voltage increased by applied inward current.

between resting potential and e.p.p. was observed. As during the actual experiment, the resting potential of the model was increased by passing an inward current (through 10 MΩ.) into the line, and it was reduced by a steady shunt imitating the effect of local mechanical injury. Under these conditions, a linear relation betwen e.p.p. and resting potential was obtained over a range from 40 to 240% of the normal level.

According to the present hypothesis, an end-plate which has been depolarized by applied acetylcholine should act as a partial short-circuit to the muscle fibre, and one would expect this to shorten the time course of a superimposed e.p.p. This prediction appears to conflict with experimental observations (e.g Fillenz & Hanafin, 1947) according to which the time course of such an e.p.p. remains unchanged. During a steady depolarization, however, additional factors must be considered which the present simple hypothesis does not take into account. The resistance of the surrounding muscle membrane does not remain constant, but has been found to increase or decrease during depolarization, depending upon the extent of the potential change (Katz, 1948). A moderate depolarization leads to a prolonged 'local response', associated apparently with entry of sodium ions into the fibre. This causes the initial potential change to build up to a higher level, and locally raises the resistance and time constant of the fibre membrane (cf. Hodgkin, 1947; Katz, 1948). A similar situation apparently occurs when acetylcholine is applied, for it has recently been shown (Fatt, 1950) that in the presence of sodium ions, the depolarization around the end-plates builds up to a higher maintained level than if sodium salts have previously been withdrawn. Fatt suggested that this is due to the regenerative action of sodium ions, which tends to spread and reinforce the depolarization in the surrounding region and thus increase the steady state resistance of the membrane. Hence, during steady depolarization by acetylcholine, we may have to consider a situation in which the end-plate 'sink' itself presents a low resistance, while the resistance and time constant of the surrounding fibre membrane are raised.

SUMMARY

1. The electrical properties of the 'motor end-plates' of frog muscle have been investigated with an intracellular recording electrode.

2. The resting potential of the end-plate membrane is about 90 mV. at 20° C.; it is the same as elsewhere along the muscle fibre and is unaffected by curarine.

3. When neuromuscular transmission is blocked, a simple end-plate potential (e.p.p.) is recorded which reaches 20–30 mV. in some fibres, but varies in amplitude over a wide range at different junctions. The e.p.p. rises sharply, reaches a peak in 1–1·5 msec. and declines to half in another 2 msec. The e.p.p. spreads electrotonically along a few mm. of the muscle fibre.

4. In a curarized muscle, the displacement of electric charge from the fibre membrane reaches a maximum at about 2 msec. after the start of the e.p.p., followed by a gradual replacement. The restoration of charge follows an exponential time course, with a time constant of 20–30 msec.

5. An analysis of the distribution of charge indicates that the active phase of neuromuscular transmission is a brief, impulsive, event lasting only a few msec., and that the prolonged spread and decline of the e.p.p. are determined by the resistance and capacity of the resting muscle fibre. The values of the membrane resistance and capacity determined from the properties of the e.p.p. are 4000 Ω.cm.^{2} and 6 μF./cm.^{2} Another series of measurements, using applied inward current and an analysis of the electrotonic potential, gives 4000 Ω.cm.^{2} and 8 μF./cm.^{2} respectively.

6. The net electric charge which is transferred across a curarized end-plate during one impulse is of the order of 8×10^{-10} coulomb, corresponding to 8×10^{-15} mol. of univalent ions.

7. In normal muscle, this value becomes 3–4 times larger. After treating the muscle with a cholinesterase inhibitor, the e.p.p. becomes greatly prolonged and the total amount of charge transferred through the end-plate increases by a factor of up to 50. Under these conditions a charge, equivalent to at least 10^{-12} mol. of univalent ions, passes through the 'end-plate sink', while building up and maintaining the depolarization of the surrounding fibre membrane.

8. During normal impulse transmission, the electric response of the end-plate differs from that of other parts of the muscle fibre in three respects: (i) the response is initiated by a large e.p.p., forming an initial half-millisecond step of about 40 mV. height; (ii) the peak of the spike is *reduced*, the reversed p.d. across the active membrane being about 20 mV., as compared with 35 mV., at other points of the fibre; (iii) during its fall, the action potential passes through a 'hump', at a level of the membrane potential which is not far from zero.

9. The height of the initial step signifies the threshold level at which the potential of the muscle membrane becomes unstable: a step of the same height is seen, when an action potential is set up by passing outward current through a muscle fibre, at or off the end-plate.

10. The subsequent characteristic features of the end-plate response (reduced peak, followed by 'hump') cannot be reproduced by an extrinsic current. Moreover, they are *not* seen when an action potential, produced by *direct* stimulation, is recorded at the end-plate, the active-membrane potential being then about 15 mV. larger than during neuromuscular transmission. Hence, the local action of the transmitter depolarizes not only the resting, but also the active surface of the muscle fibre.

11. A simple hypothesis is put forward to explain the features of the end-plate response, and also certain previous observations concerning the electromotive action of acetylcholine. Assuming that acetylcholine produces a large non-selective increase of ion permeability, i.e. a short-circuit, of the end-plate, then the production of the e.p.p., the diminution of the active-membrane potential, and the hump during the falling phase can all be explained, as well as the fact that acetylcholine depolarizes the end-plate even in the absence of sodium salts (Fatt, 1950).

12. A quantitative estimate, based upon two independent sets of measurements, indicates that the end-plate membrane is converted, during normal impulse transmission, into an ion 'sink' of approximately 20,000 Ω. leak resistance.

13. The size of the e.p.p., at normal or blocked junctions, can be varied over a wide range by increasing the resting membrane potential with anodic polarization. E.p.p. and resting membrane potential are found to be approximately proportional, as would be expected from the above hypothesis.

14. During the muscle spike, the membrane resistance falls to a small fraction, approximately 1 %, of its resting value. The resistance change occurs in two phases, associated respectively with the rise and fall of the action potential, and probably corresponding to the separate phases of increased sodium and potassium permeability (Hodgkin & Huxley, 1950).

We wish to thank Prof. A. V. Hill for the facilities provided in his laboratory and Mr J. L. Parkinson for his invaluable help. This work was carried out with the aid of a grant for scientific assistance made by the Medical Research Council.

REFERENCES

Adrian, E. D. & Lucas, K. (1912). *J. Physiol.* **44**, 68.
Bishop, G. H. (1937). *Arch. int. Physiol.* **45**, 273.
Bozler, E. & Cole, K. S. (1935). *J. cell. comp. Physiol.* **6**, 229.
Bremer, F. (1927). *C.R. Soc. Biol., Paris*, **97**, 1179.
Brooks, C. McC. & Eccles, J. C. (1947). *J. Neurophysiol.* **10**, 251.
Brown, G. L., Dale, H. H. & Feldberg, W. (1936). *J. Physiol.* **87**, 394.
Cole, K. S. (1949). *Arch. Sci. physiol.* **3**, 253.
Cole, K. S. & Curtis, H. J. (1939). *J. gen. Physiol.* **22**, 649.
Couteaux, R. (1947). *Rev. Canad. Biol.* **6**, 563.
Eccles, J. C. (1948). *Ann. Rev. Physiol.* **10**, 93.
Eccles, J. C., Katz, B. & Kuffler, S. W. (1941). *J. Neurophysiol.* **4**, 362.
Eccles, J. C., Katz, B. & Kuffler, S. W. (1942). *J. Neurophysiol.* **5**, 211.
Eccles, J. C. & MacFarlane, W. V. (1949). *J. Neurophysiol.* **12**, 59.
Fatt, P. (1950). *J. Physiol.* **111**, 408.
Fatt, P. & Katz, B. (1950*a*). *Nature, Lond.*, **166**, 597.
Fatt, P. & Katz, B. (1950*b*). *J. Physiol.* **111**, 46*P*.
Fillenz, M. & Hanafin, M. (1947). *J. Neurophysiol.* **10**, 189.
Graham, J. & Gerard, R. W. (1946). *J. cell. comp. Physiol.* **28**, 99.
Hill, A. V. (1949). *Proc. Roy. Soc.* B, **136**, 228.
Hodgkin, A. L. (1947). *J. Physiol.* **106**, 305.
Hodgkin, A. L. (1951). *Biol. Rev.* (in the Press).
Hodgkin, A. L. & Huxley, A. F. (1950). *Abstr. XVIII int. physiol. Congr.* p. 36.
Hodgkin, A. L., Huxley, A. F. & Katz, B. (1949). *Arch. Sci. physiol.* **3**, 129.
Hodgkin, A. L. & Katz, B. (1949). *J. Physiol.* **108**, 37.
Hodgkin, A. L. & Rushton, W. A. H. (1946). *Proc. Roy. Soc.* B, **133**, 444.
Hunt, C. C. & Kuffler, S. W. (1950). *Pharmacol. Rev.* **2**, 96.
Huxley, A. F. & Stämpfli, R. (1949). *J. Physiol.* **108**, 315.
Katz, B. (1939). *J. Physiol.* **95**, 286.
Katz, B. (1942). *J. Neurophysiol.* **5**, 169.
Katz, B. (1948). *Proc. Roy. Soc.* B, **135**, 506.
Katz, B. & Kuffler, S. W. (1941). *J. Neurophysiol.* **4**, 209.
Keynes, R. D. & Lewis, P. R. (1950). *Nature, Lond.*, **165**, 809.
Kuffler, S. W. (1942*a*). *J. Neurophysiol.* **5**, 18.
Kuffler, S. W. (1942*b*). *J. Neurophysiol.* **5**, 309.
Kühne, W. (1887). *Z. Biol.* **23**, 1.
Ling, G. & Gerard, R. W. (1949). *J. cell. comp. Physiol.* **34**, 383.

Lorente de Nó, R. (1947). A study of nerve physiology. *Stud. Rockefeller Inst. med. Res.* **131–132.**
Mayeda, R. (1890). *Z. Biol.* **27**, 119.
Nastuk, W. L. (1950). *Abstr. XVIII int. physiol. Congr.* p. 373.
Nastuk, W. L. & Hodgkin, A. L. (1950). *J. cell. comp. Physiol.* **35**, 39.
Rosenblueth, A. (1950). *The Transmission of Nerve Impulses at Neuro-Effector Junctions and peripheral Synapses*, p. 325. New York.
Rushton, W. A. H. (1937). *Proc. Roy. Soc.* B, **123**, 382.
Weidmann, S. (1951). *J. Physiol.* **115**, 227.

23

J. Physiol. (1952), 117, 109–128

SPONTANEOUS SUBTHRESHOLD ACTIVITY AT MOTOR NERVE ENDINGS

By P. FATT and B. KATZ

From the Biophysics and Physiology Departments, University College, London

(*Received* 5 *December* 1951)

The present study arose from the chance observation that end-plates of resting muscle fibres are the seat of spontaneous electric discharges which have the character of miniature end-plate potentials. The occurrence of spontaneous subthreshold activity at an apparently normal synapse is of some general interest, and a full description will be given here of observations which have been briefly reported elsewhere (Fatt & Katz, 1950*a*).

METHOD

The intracellular recording technique (Ling & Gerard, 1949; Nastuk & Hodgkin, 1950) was used in most experiments, confirmatory evidence being obtained with external recording in a few cases. The adaptation of the microelectrode technique to a study of the motor end-plate has been described in detail in a previous paper (Fatt & Katz, 1951).

The preparation was the m. ext. longus dig. IV and m. sartorius of the frog. In a few experiments, limb and abdominal muscles of lizards (*Lacerta dugesii*) and tortoises (*Testudo graeca*) were used.

RESULTS

Preliminary observations

In the course of some earlier work, while recording from the surface of isolated muscle fibres, we occasionally noticed a spontaneous discharge of small monophasic action potentials. The potentials varied somewhat in size, but had a very consistent time course, rising rapidly in 1–2 msec, and declining more slowly, to one-half in about 3–4 msec. They were localized at one region of the fibre, and in their shape and spatial spread resembled the end-plate potential (e.p.p.) (cf. Eccles, Katz & Kuffler, 1941; Fatt & Katz, 1951). Moreover, the discharge disappeared when a moderate dose of curarine was applied to the muscle.

Not much attention was paid to the phenomenon at the time, and it was suspected to be due to a local injury discharge at the nerve endings, the axon having been cut close to the muscle fibre. In more recent experiments, however, we observed the same phenomenon in nerve-muscle preparations which seemed to be above any such suspicion, and this induced us to examine the nature of the discharge in detail.

Localization of the spontaneous discharge at the end-plate region

The formal resemblance of the spontaneous discharges to the e.p.p. suggests that they originate at the nerve-muscle junction. This is not by itself critical evidence, for any brief disturbance of the muscle membrane—anywhere along the fibre—would be followed by a local potential change similar in time course and spatial spread to the e.p.p. (see Eccles *et al.* 1941; Katz, 1948; Fatt & Katz, 1951). A more decisive point was that the discharge could never be seen at

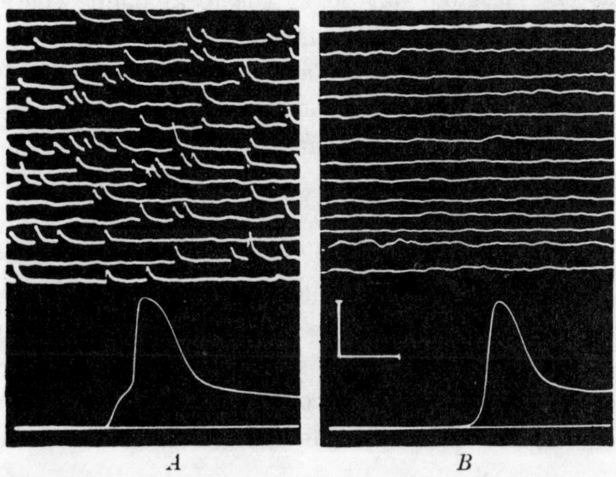

Fig. 1. Localization of spontaneous activity. M. ext. l. dig. IV. Intracellular recording. Part *A* was recorded at the end-plate, part *B* at a distance 2 mm away, in the same muscle fibre. The lower part, taken at high speed and low amplification, shows the response to a nerve stimulus (shock applied at beginning of sweep); the upper part, taken at low speed and high amplification, shows the spontaneous activity at the end-plate. Voltage and time scales: 50 mV and 2 msec for the lower part, and 3·6 mV and 47 msec for the upper part.

nerve-free regions of the muscle fibre, and whenever the test was applied, its place was found to coincide exactly with that of the e.p.p. An example of this is shown in Fig. 1, where the response to a nerve impulse is recorded, together with the subthreshold activity, at two different points of a muscle fibre. The spontaneous discharge was seen at the 'end-plate', where the e.p.p. is large and the muscle spike originates, but not at a distance of 2 mm where the e.p.p. is small and the spike appears after a delay due to muscle conduction. In other experiments a large number of end-plates were mapped out in a curarized muscle, in the manner previously described (Fatt & Katz, 1951), and after removal of curarine the end-plates as well as nerve-free portions of the fibres were explored, by observing the membrane potential for periods of about 10–15 sec with high amplification. Experiments of this kind indicated that the majority of the end-plates in normal, resting, muscle fibres of the frog are the seat of spontaneous miniature e.p.p.'s.

The term 'normal fibres' should perhaps be qualified, as we have studied the phenomenon only in isolated tissue. But it was seen in every (except denervated) muscle, and in fibres whose resting, action and end-plate potentials would be considered as 'normal'.

It might be argued that, by inserting a microelectrode at the end-plate, the nerve endings may be damaged or stimulated mechanically, and that this would lead to the observed local activity. This argument, however, is answered by the following finding: in many muscle fibres the occurrence of a spontaneous discharge was first detected at a distance of 1–1·5 mm from the end-plate, and only later was the electrode placed at the focal point where the potential changes were largest. Clearly, the miniature e.p.p.'s were present before the microelectrode approached the nerve endings sufficiently closely to be able to injure them. Damage to the nerve endings has apparently some potentiating effect (see p. 122 below), but it cannot be regarded as the cause of the phenomenon.

Size and frequency of miniature end-plate potentials

The amplitude of the miniature discharges varied from one fibre to the next, and even at a single end-plate the sizes of individual discharges were scattered over a fairly wide range around the mean (Fig. 13). Their order of magnitude was 0·5 mV (or approximately 1/100 of the size of the normal e.p.p., Fatt & Katz, 1950*a*, 1951). There were some differences between the two muscles

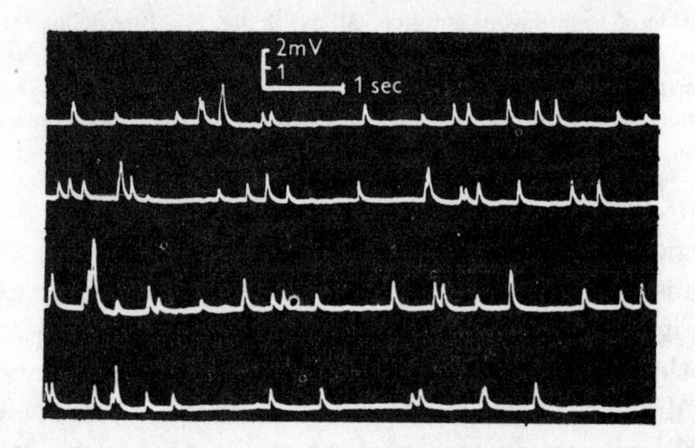

Fig. 2. Example of miniature e.p.p.'s in a muscle treated with 10^{-6} prostigmine bromide.

which we studied most extensively: the potential changes were usually larger in the m. ext. l. dig. IV than in the sartorius, where they sometimes barely exceeded 0·1 mV. The largest individual amplitudes were obtained in prostigmine-treated muscle (Fig. 2, up to 3 mV), and much higher levels occurred during temporal summation of several discharges (cf. Fig. 8).

During any one experiment on a given end-plate, the *mean amplitude* remained fairly constant. The *mean frequency*, however, was unstable and subject to progressive change. In different fibres, often from the same muscles, frequencies varied over a thousand-fold range (between about $\frac{1}{10}$ per sec and 100 per sec, at a temperature of about 20° C). The intervals between successive discharges, at any one end-plate, varied in a random manner which will be analysed in some detail below.

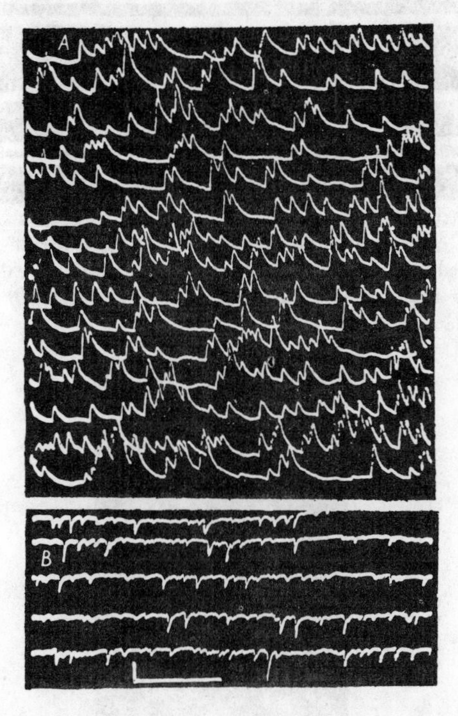

Fig. 3. 'High-frequency' discharge, recorded from the same end-plate as in Fig. 1, during a later stage of the experiment. *A*: microelectrode inside the fibre; *B*: microelectrode on the surface contacting an active patch (see text). Note: the polarity of the deflexion reverses when the electrode is on the outside. Scales: 1 mV and 50 msec.

Active spots in the end-plate surface

The results hitherto described were obtained with an intracellular electrode, with which the p.d. across the muscle fibre membrane was recorded. On a few occasions it was possible to locate one or two discrete spots on the fibre surface from which miniature e.p.p.'s of as much as 1–2 mV amplitude could be recorded externally, i.e. *without* penetrating the fibre and recording the resting potential. The discharges so recorded (e.g. Figs. 3–5) differed in certain respects from the 'internal' potential changes: (i) Their polarity was reversed, the microelectrode becoming negative with respect to the bath (with the electrode

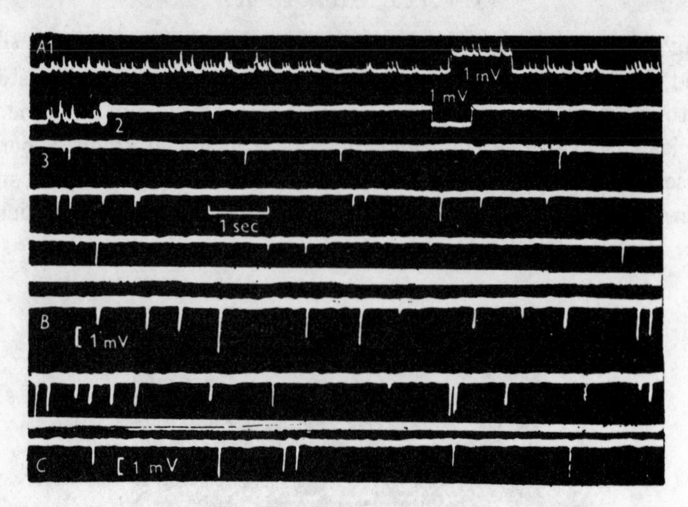

Fig. 4. 'Internal' and 'external' records. *A–C*: examples from three different end-plates (all from prostigmine-treated muscle). *A* 1: internal recording, with a 1 mV calibration step. *A* 2 and 3: external recording at two different spots on the same end-plate. *B* and *C*: external records from other end-plates.

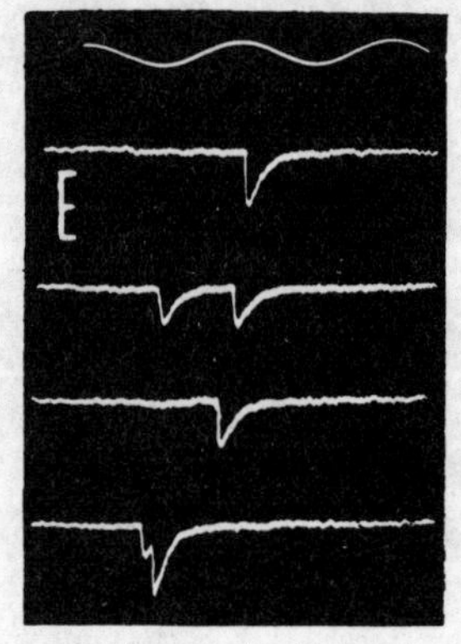

Fig. 5. External records from a single spot, using fast time base. Voltage scale: millivolts. Time: 50 c/s.

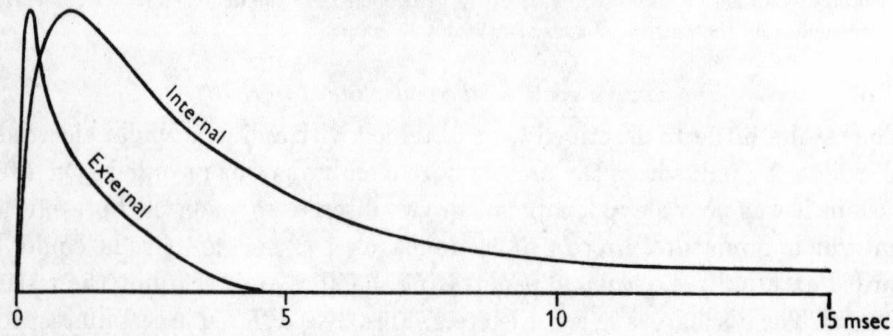

Fig. 6. Time course of internal and external miniature potentials taken from the experiment of Figs. 1 and 3. The two curves have been drawn to the same maximum, disregarding the opposite polarity of the records.

inside the fibre, a steady resting potential of about 90 mV is obtained which diminishes during a discharge, i.e. the internal microelectrode becomes *less* negative). (ii) The localization of the external miniature potentials was extremely critical, the whole phenomenon vanishing when the microelectrode was moved some microns along the fibre surface. (iii) The external miniature e.p.p.'s had a very brief time course. For example, in the experiment of Figs. 3 and 6, the duration of rise and fall of the potentials was only about one-fifth of that internally recorded, and the decline terminated sharply giving the potential a characteristic 'triangular' shape. (iv) The size and frequency pattern of the external potentials was often quite distinct from those recorded internally, at the same end-plate. In Fig. 4, for instance, a few large discharges

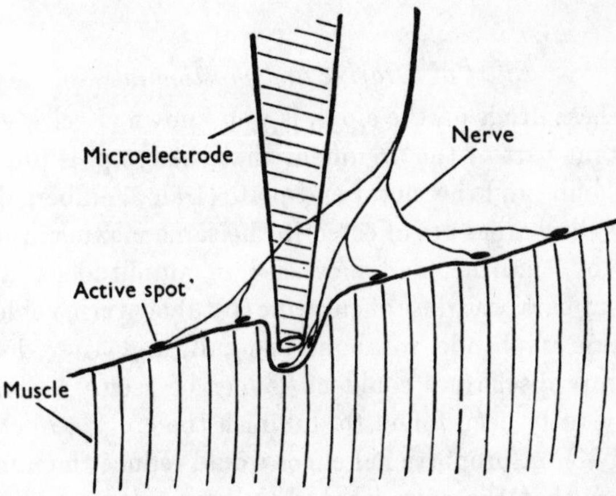

Fig. 7. Diagram to explain the characteristics of the externally recorded discharge.

are seen with the external electrode, their amplitude being more uniform and their recurrence rate much less than observed with the electrode inside the fibre. In other experiments, a few external potentials of large size were found to stand out among the remainder of the discharge which consisted of minute potentials more or less merging into the base-line noise.

It appeared in these cases that, with the microelectrode on the surface of the fibre, the recording of miniature e.p.p.'s became much more 'selective' and that all but a few active points of the end-plate, in closest proximity to the electrode, were 'rejected'. The diagram in Fig. 7 indicates a simple way of explaining the features of the external records. There are presumably a large number of active spots distributed within the end-plate surface at which individual units of the spontaneous discharge originate. With the electrode inside the fibre, the discharges of all these units are recorded without much discrimination, the separation of the active points being much less than the length constant of the fibre. But when the microelectrode is in the outside bath, no electric activity is seen unless the electrode tip happens to be placed directly over one, or a few, of these active patches, in which case a relatively large external p.d. may be recorded from these units alone. This p.d. arises from the flow of current across the end-plate membrane and in the external fluid immediately surrounding the electrode tip, and the time course of these local currents (and therefore of the externally recorded potential) is considerably more rapid than that of the underlying membrane potential change (cf. Lorente de Nó, 1947; Fatt & Katz, 1951).

Finally, it might be questioned whether we are justified in attributing the external records to potential changes in the end-plate membrane (i.e. in the muscle), rather than to 'presynaptic' changes in the nerve endings. The fact that a moderate dose of curarine abolishes 'external' as well as 'internal' miniature potentials, strongly indicates that they both arise 'post-synaptically', in the end-plate.

Effect of curarine and prostigmine

The effect of these drugs on the e.p.p. is well known (Eccles *et al.* 1941) and forms an important part of the argument that the e.p.p. is due to a reaction between acetylcholine and the motor end-plate (Dale, Feldberg & Vogt, 1936). The spontaneous discharges are affected in the same manner: they are greatly reduced in size by curarine, and increased in amplitude and duration by prostigmine. A subparalytic dose of curarine (D-tubocurarine chloride 5×10^{-7}) reduced the mean amplitude to about one-half, and after doses exceeding 10^{-6}, the miniature discharges could no longer be seen. A concentration of 4×10^{-6} has previously been found to diminish the e.p.p. to between $\frac{1}{4}$ and $\frac{1}{8}$ (Fatt & Katz, 1951). A proportional effect would reduce the miniature potentials to little more than the noise level of the apparatus.

The effect of prostigmine is illustrated in Fig. 8. The increase in amplitude and duration resembles qualitatively that observed with the e.p.p. However, the effect shown in this figure is not as striking as the dramatic lengthening of the e.p.p. in the non-curarized state (Eccles, Katz & Kuffler, 1942; Fatt & Katz, 1951).

The prostigmine action had some interesting corollaries: (i) Successive miniature potentials summed to a higher level and occasionally gave rise to muscle impulses. (ii) The miniature e.p.p.'s, being larger and longer than in normal muscle, have a more effective electrotonic spread. This enables one to see the discharges at a distance of a few millimetres from the end-plate and makes their detection much easier.

It might appear from Fig. 8 that prostigmine increased the *frequency* of the discharge, but we were unable to find a statistically significant effect on frequency of either prostigmine or curarine (5×10^{-7}) in several other experiments.

The effect of denervation

The preceding experiments show that the spontaneous discharges are localized at the end-plates of resting muscle fibres. The time course of the individual potential indicates that it is due to a brief, impulsive, disturbance

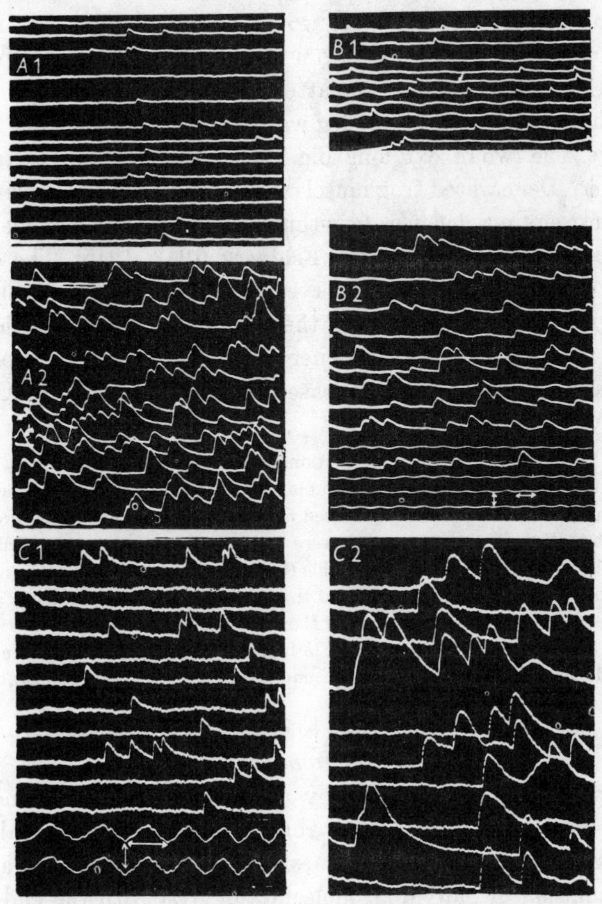

Fig. 8. Effect of prostigmine on the size and time course of the miniature e.p.p. *A–C*: three different end-plates (*A* and *C* from m. ext. l. dig. IV, *B* from sartorius). 1: in normal Ringer; 2: after addition of 10^{-6} prostigmine bromide. (In *C* 2, potential changes of slow rise can be seen, beside those of the usual rapid rate of rise. This was observed in several other experiments (cf. also Fig. 9*B*), and is presumably to be explained by electrotonic spread from an accessory end-plate on the same fibre.) Arrows indicate 1 mV scale and 20 msec (50 c/s).

of the membrane similar to the e.p.p., and the effects of curarine and prostigmine suggest that both arise from the action of ACh. A simple picture would be that at each nerve-muscle junction, the nerve terminals contain a large number of discrete 'active spots' at which ACh is released. The miniature potentials would be due to an asynchronous, spontaneous, activity of individual spots, while the e.p.p. is due to a synchronous discharge of the whole apparatus. The e.p.p. could thus be regarded as the sum of miniature potentials produced by all the units.

While it seems very probable that the miniature e.p.p.'s are due to local discharges of acetylcholine, it may be questioned whether these discharges originate in the nerve endings. It would be conceivable, for instance, that stray amounts of ACh are released from vascular tissues surrounding the muscle fibres. Pertinent evidence on this matter was obtained by studying denervated muscles. The sciatic nerve of one leg was severed, and the frogs kept at room temperature. The two m. ext. long. dig. IV were isolated and tested more than 2 weeks later. Denervated frog muscle is known to become sensitized to ACh, while its nerve endings degenerate within a few weeks and lose their capacity of building up and releasing ACh (cf. Feldberg, 1943, 1945). The experimental result was clear: muscles whose nerves had been cut 2–3 weeks previously showed no miniature e.p.p.'s, while they were readily seen in the innervated companion muscles. The effect of denervation thus strongly supports the view that the discharges are due to a release of small quantities of ACh from functioning nerve terminals.

After more prolonged denervation, fibrillation gradually developed (cf. Reid, 1941). This was seen in only a few fibres of any one muscle, and in some experiments very infrequent subthreshold potential changes were observed in denervated muscle which may have been precursors of fully propagated fibrillation. These phenomena were too infrequent for any detailed study, but we noticed that the amplitude was highly variable (from less than 1 up to 10 mV) and the duration of the local potential change more prolonged than that of the miniature e.p.p.'s on the normal side. Moreover, the local changes as well as the fibrillation in denervated muscle seemed to be unaffected by curarine and prostigmine, and, in further contrast to the miniature e.p.p.'s (p. 120) were abolished by a small increase of calcium concentration (from 1·8 to 3·6 mM).

Can the miniature discharges be attributed to molecular leakage of acetylcholine from nerve endings?

ACh is synthesized continually by cholinergic nerves and nerve endings (Feldberg, 1945), and there is probably a slow continuous leakage of ACh molecules from this reservoir. The question arises whether such a leakage, i.e. a random collision of individual molecules of ACh with the end-plate, could produce the miniature e.p.p.'s. It would be difficult to explain the effect of prostigmine on such a 'single-molecule' hypothesis, but there are two more cogent arguments which indicate that the miniature potentials must be due to the synchronous release of a large number of acetylcholine molecules and cannot be explained by a mechanism of simple molecular diffusion.

(i) The quantity of ACh released by an impulse at one end-plate has been estimated (cf. Acheson, 1948; Rosenblueth, 1950) as being of the order of $1\cdot5 \times 10^{-16}$ g or about 10^6 molecules. This value was derived from perfusion experiments with prolonged stimuli, in which large losses must have occurred, and the true amount released locally during the first impulse was presumably much larger. But even if we accept this estimate, the amount responsible for a miniature e.p.p. would be of the order of 1/1000–1/100 of that for an e.p.p., which is still an aggregate of thousands of ACh molecules.

(ii) If the miniature discharges were to be regarded as the local depolarizing effect of individual ACh molecules then, by the same argument, the application of ACh in solution should greatly increase the frequency of the miniature discharges; in fact, the steady depolarization which is produced by applied ACh would have to be regarded as a fusion of miniature e.p.p.'s, like the fusion of twitches in a tetanus.

This conclusion, however, is contrary to our observations. A moderate concentration of ACh (5×10^{-8}, with 10^{-6} prostigmine bromide) which depolarized the end-plates by a few mV, did not appreciably alter the frequency of the miniature discharges. The only noticeable change was a slight reduction of amplitudes, similar to the reduction of the e.p.p. observed at depolarized end-plates (Nastuk, 1950; Fatt & Katz, 1951). These results seem to us to be incompatible with the suggestion that the miniature potential might be attributed to the action of one (or very few) ACh molecules; if individual molecular collision between ACh and the end-plate builds up a steady depolarization then the molecular units of this depolarization must be much smaller than the recorded miniature e.p.p.

Miniature end-plate potentials and spontaneous excitation at motor nerve terminals

On the basis of the preceding argument it can be said that a miniature e.p.p. is due to a spontaneous momentary release of many molecules of ACh from a small area of the motor nerve endings. The quantity may be estimated as being rather less than 1/100 of that during the transmission of a motor nerve impulse. Successive miniature e.p.p.'s are of uniform shape, though of variable amplitude (Fig. 13) and follow one another in a random sequence (p. 122).

Suppose the motor axon has about 100 branches, or alternatively, that there are within the terminal area some 100 discrete 'patches' concerned with the release of ACh. If these terminal structures have a special tendency to spontaneous excitation, then our observations would be easily understood. It has been pointed out (Fatt & Katz, 1950*a*) that spontaneous excitation might simply be the result of excessive voltage noise across the nerve membrane, and that this noise level is likely to be largest at the smallest nerve endings. This idea, which was originally suggested to us by Mr A. L. Hodgkin, is discussed more fully below. Whether it provides the complete explanation or not, the present experiments make it very likely that spontaneous excitation at individual motor nerve endings does occur and that it is the immediate cause of the miniature e.p.p.'s. In addition to the above results, the following observations are of interest in this connexion.

(i) Agents which are known to abolish electric excitation also extinguish the spontaneous discharge. Thus, the miniature e.p.p.'s cannot be seen in sodium-free solutions (made by substitution of isotonic sucrose), or in Ringer solution to which a small dose of a local anaesthetic (0·01 % novocaine) has been added. Partial withdrawal of sodium (see below), or low concentrations of novocaine (0·001–0·002 %) cause the amplitude of the miniature potentials to diminish. These results are interesting, though not unequivocal, for withdrawal of sodium has, and application of novocaine may have, some curare-like action on the end-plate (Fatt, 1950), quite apart from affecting nerve excitability.

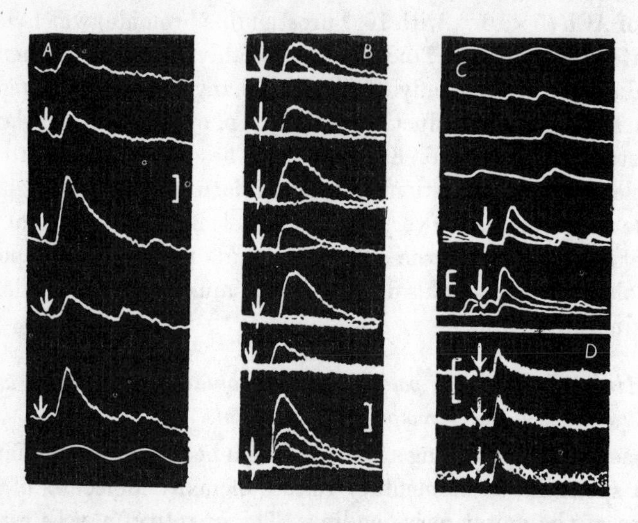

Fig. 9. *Effect of* Ca-*lack*. *A–C*: three different experiments. Muscles soaked in reduced ($\frac{1}{4}$) Ca concentrations; in *B*, prostigmine bromide 10^{-6} was present. Note: the records in *A* and the three top records in *C* were single sweep records. All other records were obtained with multiple sweeps repeated at about 1 per sec, during *each* of which the nerve was stimulated (at the instant marked by an arrow), though there was not always an end-plate response (*B*). The top record in *A*, and the three top records in *C* show spontaneous discharges only. All other records show *e.p.p. responses to nerve stimulation*, varying in step-like manner between zero and a few millivolts (e.g. *B*, bottom record). In some records (in *A* and *C*) spontaneous discharges are seen on the same sweep, immediately before or after the e.p.p. response. For comparison with the effect of Ca-lack, the relative constancy of the e.p.p. response in a *curarized* fibre is shown in *D* (5×10^{-6} D-tubocurarine chloride; three successive records, each with three superimposed sweeps). Volt scale: millivolts. Time: 50 c/s.

(ii) A curious effect was observed when reducing the calcium concentration. This causes the size of the e.p.p. to diminish *without* affecting the size of the miniature potentials. With a sufficiently low calcium level, the response to a motor nerve impulse may be no larger than a single miniature e.p.p. In this condition, successive nerve impulses evoke a random display of minute e.p.p.'s whose sizes vary in a step-like fashion, seemingly corresponding to multiples of a miniature discharge (cf. Fig. 9 *B*, bottom record, where steps of 0, 1, 2 and 3 can be recognized). In the experiment of Fig. 9 *B*, 328 strong shocks were applied to the nerve, 188 of which failed to elicit an end-plate response. Of the observed 140 e.p.p.'s, 100 had an amplitude which was within the range of the spontaneous miniature potentials (mean of 33 spontaneous discharges: 0·87 mV, standard deviation ±0·18 mV), while the residual 40 e.p.p.'s were greater (up to 2·92 mV) and presumably represent responses of two or three miniature units. The contrast between this behaviour and the relatively constant response of a deeply curarized muscle fibre is illustrated in Fig. 9 (*A–C* with low calcium, *D* with curarine). The experiment throws some new light on the action of calcium at the nerve-muscle junction: lack of calcium apparently reduces the e.p.p. in definite 'quanta', as though it blocks individual nerve terminals, or 'active patches' within them, in an all-or-none manner. The normal e.p.p. can be seen, as in Fig. 9 *A–C*, to break down into individual miniature units. Conversely, it may be said that an individual spontaneous discharge does not differ, in appearance, from the response of a terminal unit to the motor nerve impulse.

The effect of sodium and calcium ions

Variations of Na or Ca concentrations have very similar effects on the e.p.p. (Fatt & Katz, 1950*b*; Castillo-Nicolau & Stark, 1951). With both ions the e.p.p. decreases as the concentration falls, the relation being approximately proportional over a certain range.

The effects of these cations on the miniature potentials, however, are different: the size of the miniature e.p.p. falls as the Na concentration is lowered, but it is independent of Ca concentration (see preceding section). For example, a reduction of Na concentration to one-quarter diminished the mean size of the miniature e.p.p.'s in one experiment from 2·28 ± 0·05 to 0·82 ± 0·03 mV, in another experiment from 0·72 ± 0·03 to 0·31 ± 0·01 mV (prostigmine being used throughout). The average reduction of the mean amplitude in several experiments with one-quarter Na was to 0·4, while the resting potential remained constant, at about 90 mV.

A reduction of Ca concentration to one-quarter had no significant effect, the mean size of the miniature potentials increasing, on the average to 1·01 (S.E. of mean of five experiments was ±0·02). A fourfold increase of Ca concentration produced a small diminution of the miniature potential, on the average to 0·89 (S.E. ±0·02, five experiments).

There are reasons for supposing that both species of cations play a specific part in the release of ACh from active nerve endings, and the present observations are likely to be relevant in this connexion. A full discussion of this matter will be deferred until a later paper.

As regards the mean frequency of the spontaneous discharges, Na and Ca have no clear-cut effect. In some experiments lack of Ca reduced the mean frequency appreciably, but the effect was not consistently observed, and in several later experiments, no change was seen. Lowering the Na concentration to one-quarter did not significantly alter the recurrence frequency.

Agents which have a pronounced effect on the recurrence frequency

The changes in the spontaneous discharge which have been described concerned the *size* of the miniature potentials, while little or no effect on the mean *frequency* was found. Because of its gradual uncontrolled variations, the frequency cannot be measured accurately over a long period, and it is only possible to look for gross changes.

Of the various agents studied, changes of temperature and of osmotic pressure were found to produce marked effects on the frequency of the discharge.

Temperature. In these experiments, the bath was replaced at intervals by Ringer's solution of different temperature (varying between 8 and 25° C), and records of spontaneous discharges were obtained from a number of initially located end-plates. The frequency increased with temperature, corresponding to a positive temperature coefficient with a Q_{10} of about 3. It should be noted that we have only studied the *steady* effect; our procedure did not allow us to observe initial transient changes which might conceivably have a different, or even a negative, temperature coefficient (see Sand, 1938).

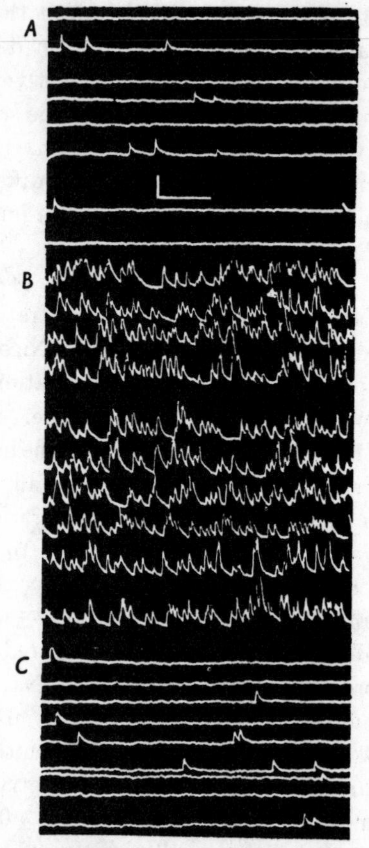

Fig. 10. Effect of osmotic pressure on discharge rate. *A* and *C*: normal Ringer, before and after applying (in *B*) a 50% hypertonic solution. Scales: 1 mV and 0·1 sec.

Osmotic pressure. The most dramatic effect was observed when the osmotic pressure of the bathing fluid was changed (cf. Fig. 10). The addition of a small quantity of sucrose or NaCl to the bath causes a striking increase in the frequency of the discharge, and conversely dilution of the Ringer's fluid reduces the frequency. For example, a 50% increase of tonicity (by addition of sucrose) was followed by a reversible 45-fold increase of discharge frequency (from 2 to 90 per sec). In other experiments, raising the osmotic pressure by 30% caused the frequency to increase from 15 to 150 per sec, and lowering the osmotic pressure by 50% reduced the frequency from 28 to 0·9 per sec.

Finally, there was some indication that damage to the nerve endings is followed by an increase in the frequency of the discharge. In the course of repeated 'focal' insertions of the microelectrode, we observed sometimes that the e.p.p. response to a motor impulse progressively diminished (while the resting potential was well maintained), and at the same time the miniature potentials became progressively more frequent. Stretching a muscle 10–15% beyond its resting length reversibly increased the rate by a factor of 2·5–3, and pulling on the motor nerve sometimes caused a similar 'speeding up' of the discharge. It is possible that these procedures produce some local depolarization and thereby increase the instability of the nerve endings. There are however no means at present of obtaining direct evidence of the state of the nerve terminals and of their membrane potential.

The 'random' nature of the spontaneous discharge

The sequence of miniature potentials appeared to be completely irregular. It is true that, occasionally, this irregularity was broken by a short burst of high-frequency discharges, which might have been due to an extraneous stimulus or more probably to interaction between the different contributing units, e.g. to electrotonic currents from an active nerve ending lowering the threshold of adjoining endings. There were, however, long periods during which no bursts could be seen, and this suggests that any 'coupling' between the various units must be very weak and only rarely and temporarily leads to effective interaction.

The random succession of the discharge was subjected to a statistical test. It is a characteristic property of this type of random time series that the probability of occurrence of any one event does not depend upon past history. For an interval Δt, which is very brief compared with the mean interval T, the probability P of at least one occurrence is simply $\Delta t/T$ (for a complete theory see Feller, 1950, chapter 17). As the interval t becomes greater, P increases exponentially according to the equation $P = 1 - \exp(-t/T)$. Similarly, the intervals between successive discharges, in a very large series of observations (total number N), should be distributed exponentially, the frequency of occurrence n of any interval between t and $t + \Delta t$ being given by $n = N\Delta t/T \exp(-t/T)$. To test the applicability of these equations, a series of 800 miniature potentials, covering a total period of 176·8 sec (i.e. with a mean frequency of 800/176·8 = 4·52 per sec) was chosen. This series was selected because it showed no obvious bursts of synchronized activity, and the mean frequency was suitable for accurate measurement (at too high frequencies, summation and coincidences between several unitary discharges make measurements uncertain; at too low rates, the experiment becomes too long and progressive changes of the mean are likely to occur). To check that the mean frequency had not appreciably altered during the period of observation, the measurements were carried out separately for two series of 500 and 300 discharges respectively, and these agreed satisfactorily.

Successive intervals between individual discharges were divided into groups, as in Fig. 11. The distribution fits a simple exponential curve

$$n = N\Delta t/T \exp(-t/T),$$

where T is the mean interval ($\chi^2 = 27{\cdot}9, f = 21$). If we plot the *total* number of intervals, whose duration is smaller than t, against t, as in Fig. 12, the results are seen to fall closely along the predicted curve $P = 1 - \exp(-t/T)$.

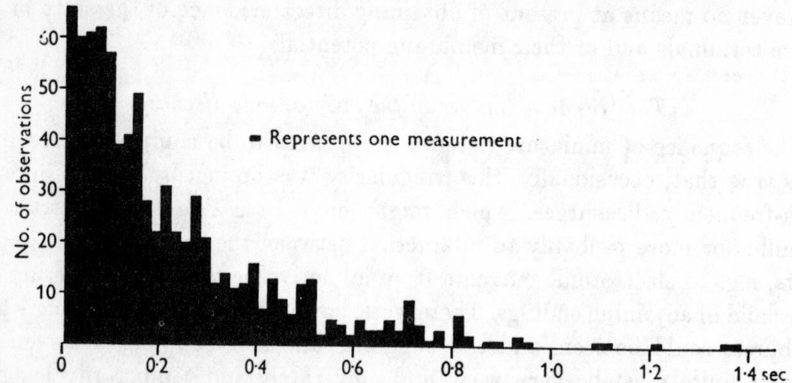

Fig. 11. Distribution of time intervals between discharges, in a series of 800 miniature potentials. The shortest interval which could be resolved in this experiment was 5 msec (the interval between 0 and 0·005 sec has been left blank). The observations were grouped in classes of 20 msec. The mean interval was 0·221 sec.

This random sequence of miniature potentials should, however, be interpreted with some caution. It evidently means that in this particular set of observations there was no noticeable interaction between the various contributing units, but it does not prove that the constituent units themselves discharged in a completely random manner. The record of impulses from a whole sense organ may appear to be chaotic, yet each nerve unit carries impulses of remarkable regularity. Whether such regular rhythms are concealed in the present picture of miniature e.p.p.'s cannot be said, because it has not yet been possible to isolate the constituent units satisfactorily. Selective recording from critical surface spots (cf. p. 112) might provide the answer. Our external records did not show any regular rhythmicity, but they were too few to provide a conclusive test.

Statistical distribution of amplitudes

We have suggested that there are a number of individual units (nerve terminals, or discrete 'ACh-release patches' within them) which independently contribute to the spontaneous discharges, and one might expect to find a grouping of the individual sizes corresponding to such units. In some experiments, this appeared to be the case, especially when external records were obtained (p. 114). However, when a large series of measurements was tested

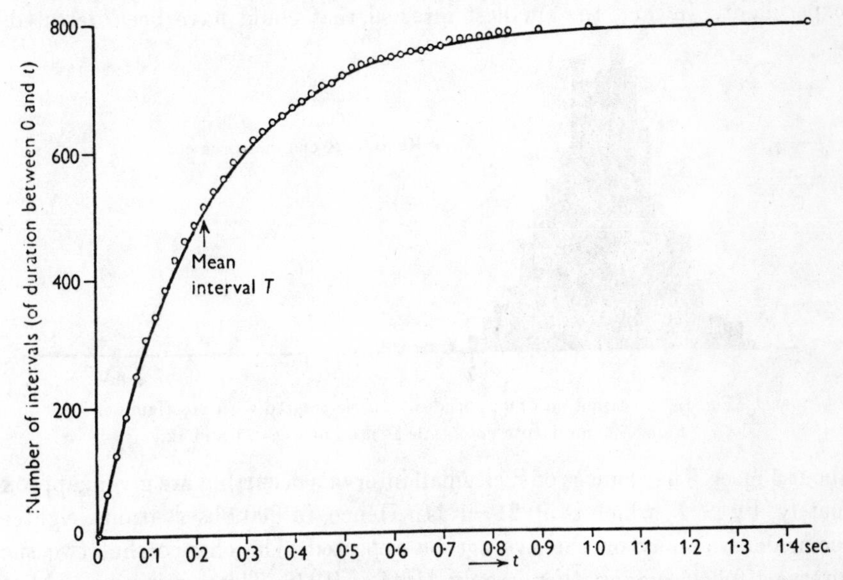

Fig. 12. Total number of observed intervals, of duration less than t (from the experiment of Fig. 11), has been plotted against interval duration t. Circles: observed numbers. Full line: a theoretical curve, $y = N[1 - \exp(-t/T)]$, where N is total number of observations, and T the mean interval, marked by arrow).

as in Fig. 13, the sizes were usually found to be scattered in an approximately normal manner around a simple mean value. In Fig. 13 the coefficient of variation was about 30 %, and there were no outstanding secondary humps or peaks. The explanation of this scatter is probably several fold: (i) the mean amplitude of the individual potentials is about 1/100 of the e.p.p., hence some 100 different units may be involved, and a discrimination of individual sizes would then be impossible; (ii) the size of an individual unitary discharge must itself be subject to some variation, depending upon the interval between discharges (both 'facilitation' and 'depression' may be expected, as with two successive e.p.p.'s; cf. Eccles *et al.* 1941); (iii) the accuracy of individual measurements is limited by a random error of a few per cent, due to the baseline noise. These factors might easily account for the statistical spread of amplitudes shown in Fig. 13.

Perhaps the most convincing evidence for the existence of discrete terminal units is that obtained in calcium-deficient muscle (Fig. 9), where it was shown on p. 119 that successive e.p.p. responses varied in a step-like manner, corresponding to units of miniature e.p.p.'s.

In Fig. 13 there is indication of several discharges of about twice the mean amplitude, and of one isolated discharge of three or four times the mean size. These are very probably examples of a 'coincidence' of two (or three) unitary discharges which could not be resolved with the slow film speed used in this experiment. In fact, the smallest interval that could have been detected is

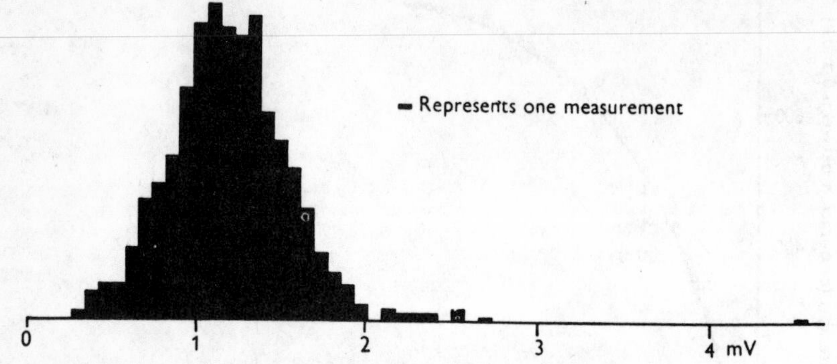

Fig. 13. Distribution of amplitudes. Muscle treated with prostigmine. Same 800 miniature potentials as used in Figs. 11 and 12.

about 5 msec. The chances of such small intervals occurring are given, approximately, by $\Delta t/T$, which is $5/221 = 1/44$. Hence, in 800 observations, eighteen 'coincidences' may on the average be expected. The chance that two such intervals follow one another is only $1/44^2 = 1/1936$. Thus, it is reasonable to suppose that the single observation of 4·6 mV amplitude was a 'triple' coincidence while the thirteen observations between 2 and 3 mV were ordinary 'double' coincidences.

DISCUSSION

The results point to the conclusion that some terminal spots of the motor nerve endings are spontaneously active and release ACh in the same impulsive manner as they do after the arrival of a normal motor nerve impulse.

The question arises why such 'neurogenic' activity does not lead to a back-firing into the main motor axon and, by axon reflex, to a total discharge of the motor unit. The answer is probably that the terminal area which, at any time, is spontaneously active is too small to produce a propagating nerve impulse. It should, however, be noted that in the eserinized mammalian muscle spontaneous 'fibrillation' (fibre activity) and 'fasciculation' (motor unit activity) have both been described (Masland & Wigton, 1940), which indicates that under certain conditions a back-firing into the motor nerve axon does, in fact, occur. While in most of our experiments the level of the spontaneous activity remained below the threshold of either nerve or muscle fibres, in some prostigmine-treated muscles occasional fibrillation was seen which was presumably due to a summation of miniature potentials exceeding the threshold of a muscle fibre. It is likely that the spontaneous twitching in eserinized or tetanus-poisoned mammalian muscle arises from the same phenomenon.

That the occurrence of miniature e.p.p.'s is not an exclusive property of frog's muscle was made evident by a few tests on lizard and tortoise preparations in which the same type of spontaneous activity was found in innervated regions of muscle fibres.

The second problem which confronts us is the origin of the spontaneous excitation at nerve terminals. In a previous note (Fatt & Katz, 1950*a*) it has been suggested that the 'noise' voltage across the axon membrane may become so large, at a sufficiently minute structure, that it may occasionally exceed the threshold level at some point. By 'noise' voltage is meant the random fluctuation of the resting potential due to thermal agitation of ions within the membrane.

This noise level may be calculated from the formula

$$E^2 = 4kT\int_0^{f_1} r\,\mathrm{d}f \tag{1}$$

(see Campbell & Francis, 1946), where

E = r.m.s. value of noise fluctuation (volts);
k = Boltzmann's constant $= 1{\cdot}38 \times 10^{-23}$ joules per degree Kelvin;
T = absolute temperature (degrees Kelvin);
r = resistive component of impedance across nerve ending (ohms);
f_1 = frequency range over which the effective noise energy is distributed (c/s).

The resistive component r at the *end* of a long non-medullated axon is twice as large as in its middle, and is given by

$$r = a\sqrt{(r_i r_m)},$$

where

$$a = \sqrt{\{\tfrac{1}{2}[1/(1+\omega^2\tau_m^2) + 1/\sqrt{(1+\omega^2\tau_m^2)}]\}},$$

$$\omega = 2\pi f,$$

$$\tau_m = r_m c_m.$$

and r_m, r_i and c_m are, respectively, the transverse membrane resistance, the longitudinal axon cylinder resistance and the membrane capacity of one centimetre length of fibre (or terminal branch). These values are related to the more fundamental fibre constants as follows:

$$r_m = R_m/\pi d, \quad r_i = 4R_i/\pi d^2, \quad r_m c_m = R_m C_m,$$

where d is fibre diameter (cm), R_m specific membrane resistance ($\Omega \times \mathrm{cm}^2$), R_i specific axoplasm resistance ($\Omega \times$ cm), and C_m specific membrane capacity ($\mu\mathrm{F/cm}^2$). Equation (1) can, therefore, be rewritten as

$$E^2 = \frac{8kT}{\pi}(R_i R_m)^{0{\cdot}5} d^{-1{\cdot}5}\int_0^{f_1} a\,\mathrm{d}f,$$

from which it is seen that the noise voltage increases as the fibre size diminishes.

The absolute values can only be guessed, as we have no direct information about the size and electrical characteristics of nerve endings. Supposing that the fundamental electrical constants of nerve endings are of the same order of magnitude as in several types of non-medullated axons, we may use the following values: $R_m = 5000\Omega\ \mathrm{cm}^2$; $R_i = 200\Omega$ cm; $C_m = 1\mu\mathrm{F/cm}^2$. In computing $\int a\,\mathrm{d}f$, an arbitrary upper limit of integration (f_1) had to be chosen; we assumed that the upper frequency limit of effective noise components is at about 10 kc/s, and that at frequencies above it the fluctuations of membrane potential would be too rapid to lead to local excitation. On these assumptions, $\int a df$ is calculated to be approximately 800 c/s.

The noise level, for an axon of these properties, depends then only on its diameter, the noise voltage increasing as the fibre size is reduced. For a 100μ axon, for instance, E is about $2\mu V$; for the terminal of an extremely fine fibre (or sufficiently long branch) of, say, $0 \cdot 1\mu$ diameter, E becomes about 0·5 mV. These are r.m.s. values; the peak amplitudes of the fluctuation are several times larger, but they become very infrequent above 3–4 times the r.m.s. value, and in practice will not exceed five times this size.

The result of this calculation remains indecisive: it indicates that at very fine nerve terminals, thermal agitation noise may reach an amplitude of 1 or 2 mV and thereby a physiologically important range; nevertheless, on the present assumptions, there is little chance of the membrane noise approaching the 15–20 mV which is the threshold level of a *Loligo* axon (Hodgkin, Huxley & Katz, 1949). Although thermal agitation of ions may play an important part in the production of the discharge, some other property of the nerve endings, as yet unspecified, may well be involved.

SUMMARY

1. End-plates of many resting muscle fibres are the seat of spontaneous subthreshold electrical activity. It consists of a random succession of miniature end-plate potentials, their amplitude being of the order of 1/100 of the normal end-plate response to a motor nerve impulse.

2. The miniature potentials are greatly reduced in size by a small dose of curarine, and are increased in size and duration by prostigmine. They are abolished by denervation, and by nerve anaesthetics.

3. The frequency of the discharges varies over a wide range. It increases with temperature and, strikingly, with small increases of osmotic pressure.

4. There is evidence that the discharges are due to spontaneous local excitation of individual motor nerve endings, or of even smaller specialized membrane areas which are concerned with the release of acetylcholine.

5. The effect of sodium and calcium ions has been studied. Calcium deficiency is known to reduce the size of the end-plate potential (e.p.p.); this effect appears to take place in 'steps', involving an all-or-none blockage of a variable number of miniature components. The size of individual miniature potentials in contrast to that of the e.p.p. is not affected by calcium-lack. Sodium deficiency, on the other hand, reduces the amplitudes of both e.p.p. and miniature potentials.

6. The origin of the spontaneous discharge is discussed. It is possible that thermal agitation of ions across the nerve membrane plays an important part in their initiation.

We are indebted to Prof. A. V. Hill for the facilities provided in his laboratory, to Mr J. L. Parkinson for frequent help, and to Dr R. B. Makinson, of Sydney University, for solving some of the mathematical problems for us. This work was carried out with the aid of a grant for scientific assistance made by the Medical Research Council.

REFERENCES

Acheson, G. H. (1948). Physiology of neuro-muscular junctions: Chemical aspects. *Fed. Proc.* **7**, 447–457.

Campbell, N. R. & Francis, V. J. (1946). A theory of valve and circuit noise. *J. Instn elect. Engrs*, **93**, pt. 3, 45–52.

del Castillo-Nicolau, J. & Stark, L. (1951). Effect of calcium ions on motor end-plate potentials. *J. Physiol.* **114**, 31*P*.

Dale, H. H., Feldberg, W. & Vogt, M. (1936). Release of acetylcholine at voluntary motor nerve endings. *J. Physiol.* **86**, 353–380.

Eccles, J. C., Katz, B. & Kuffler, S. W. (1941). Nature of the endplate potential in curarized muscle. *J. Neurophysiol.* **4**, 362–387.

Eccles, J. C., Katz, B. & Kuffler, S. W. (1942). Effect of eserine on neuromuscular transmission. *J. Neurophysiol.* **5**, 211–230.

Fatt, P. (1950). The electromotive action of acetylcholine at the motor end-plate. *J. Physiol.* **111**, 408–422.

Fatt, P. & Katz, B. (1950*a*). Some observations on biological noise. *Nature, Lond.*, **166**, 597–598.

Fatt, P. & Katz, B. (1950*b*). Membrane potentials at the motor end-plate. *J. Physiol.* **111**, 46–47*P*.

Fatt, P. & Katz, B. (1951). An analysis of the end-plate potential recorded with an intra-cellular electrode. *J. Physiol.* **115**, 320–370.

Feldberg, W. (1943). Synthesis of acetylcholine in sympathetic ganglia and cholinergic nerve. *J. Physiol.* **101**, 432–445.

Feldberg, W. (1945). Synthesis of acetylcholine by tissue of the central nervous system. *J. Physiol.* **103**, 367–402.

Feller, W. (1950). *Introduction to Probability Theory and its Applications*, vol. 1, p. 419. New York.

Hodgkin, A. L., Huxley, A. F. & Katz, B. (1949). Ionic currents underlying activity in the giant axon of the squid. *Arch. Sci. physiol.* **3**, 129–150.

Katz, B. (1948). The electrical properties of the muscle fibre membrane. *Proc. Roy. Soc.* B, **135**, 506–534.

Ling, G. & Gerard, R. W. (1949). The normal membrane potential of frog sartorius fibers. *J. cell. comp. Physiol.* **34**, 383–396.

Lorente de Nó, R. (1947). A study in nerve physiology, pt. 1, pt. 2 (chap. 16). In *Stud. Rockefeller Inst. med. Res.* **131**, **132**. New York.

Masland, R. L. & Wigton, R. S. (1940). Nerve activity accompanying fasciculation produced by prostigmine. *J. Neurophysiol.* **3**, 269–275.

Nastuk, W. L. (1950). The electrical activity of single muscle fibres at the neuro-muscular junction. *Abstr. XVIII int. physiol. Congr.* pp. 373–374.

Nastuk, W. L. & Hodgkin, A. L. (1950). The electrical activity of single muscle fibres. *J. cell. comp. Physiol.* **35**, 39–74.

Reid, G. (1941). The reaction of muscle to denervation in cold-blooded animals. *Aust. J. exp. Biol.* **19**, 199–206.

Rosenblueth, A. (1950). *The Transmission of Nerve Impulses at Neuro-Effector Junctions and Peripheral Synapses*, p. 325. New York.

Sand, A. (1938). The function of the ampullae Lorenzini, with some observations on the effect of temperature on sensory rhythms. *Proc. Roy. Soc.* B, **125**, 524–553.

24

J. Physiol. (1954) 124, 560–573

QUANTAL COMPONENTS OF THE END-PLATE POTENTIAL

By J. del CASTILLO and B. KATZ

From the Department of Biophysics, University College, London

(*Received* 25 *January* 1954)

In this paper a further study is made of the spontaneous synaptic potentials in frog muscle (Fatt & Katz, 1952*a*), and their relation to the end-plate response. It has been suggested that the end-plate potential (e.p.p.) at a single nerve-muscle junction is built up statistically of small all-or-none units which are identical in size with the spontaneous 'miniature e.p.p.'s'. The latter, therefore, could be regarded as the least unit, or the 'quantum', of end-plate response. A convenient picture of how hundreds of such quanta, each capable of producing a miniature potential of 0·5–1·0 mV, can build up an e.p.p. of, say, 70–80 mV is provided by the hypothesis that separate parcels of acetylcholine (ACh), released from discrete spots of the nerve endings, short-circuit the muscle membrane. The unit changes of membrane conductance produced at many parallel spots summate and lead to an intense depolarization of the muscle fibre.

Although this is a plausible view, there is no direct proof that the normal e.p.p. is made up in this quantal fashion. The evidence comes from experiments in which the 'quantum content' of the e.p.p. had been reduced to a small number by lowering the external calcium concentration (Fatt & Katz, 1952*a*). It was then found that the size of the end-plate response approached that of the spontaneous potential and at the same time exhibited large random fluctuations, apparently involving steps of unit size. Similar observations were made by Castillo & Engbaek (1954) on muscles treated with Mg-rich solutions. The statistical behaviour of the end-plate response under these conditions has been investigated in more detail and subjected to a quantitative analysis.

METHODS

The technique for intracellular recording of e.p.p.'s and miniature potentials has been described in earlier papers (Fatt & Katz, 1951, 1952*a*; Castillo & Katz, 1954). The m. ext. l. dig. IV of English frogs was used, immersed in an isotonic solution containing concentrations of $CaCl_2$ and $MgCl_2$ adjusted so as to reduce the response to any desired level. In most experiments prostigmine (10^{-6}, w/v) was added to increase the amplitude of the potentials (without altering their 'quantum content'). The usual procedure was to locate a suitable spot with the internal electrode and record spontaneous potentials on moving film. Then, a large number of end-plate responses to single or pairs of nerve volleys were recorded, using a swept time-base and one or a few seconds interval between records. Finally, another series of spontaneous potentials was recorded before the micro-electrode was withdrawn from the fibre.

The amplitudes of the potentials were measured and their distribution displayed in a histogram. With low Ca and high Mg concentrations, all-or-none fluctuations are observed in successive records, with frequent total failures of e.p.p. response. Special importance was attached to the counting of 'failures' and 'successes', as their proportions provided a simple and decisive test of our hypothesis. A precaution which had to be taken was to guard against intermittent failures of response due to other causes, e.g. inadequate stimulation or nerve damage leading to block. The first source of trouble was avoided by using a strong shock, the second source could be recognized without much difficulty, because nerve block, if it occurred at all, developed in a rapidly progressive manner and was unrelated to the size of the initial end-plate response.

In several muscle fibres there was evidence of a remote, second, motor nerve supply (cf. Katz & Kuffler, 1941) producing small and slow miniature potentials and e.p.p. response. These were of discrete shape and could be discarded without ambiguity, when counting responses and measuring amplitudes.

The ext. l. dig. IV contains some muscle fibres of the 'slow system' supplied by small motor axons (Kuffler & Vaughan Williams, 1953; Katz, 1949). As most experiments were made below the level of propagated spikes, the question arises whether we may not sometimes have been recording from 'slow' muscle fibres. This is unlikely because the characteristics of the response were those of the e.p.p.'s of 'twitch fibres' (sharp localization and low threshold whenever tested, high resting potential, short latency, monophasic e.p.p. response). Spontaneous discharge of miniature potentials had previously been shown to occur at end-plates of ordinary 'twitch fibres' (Fatt & Katz, 1952*a*).

RESULTS

When a muscle was soaked in a solution containing approximately 10 mM-$MgCl_2$, transmission became blocked and subthreshold e.p.p.'s could be recorded at individual junctions. A characteristic feature of these responses was their random fluctuation in successive records. This is illustrated in Fig. 1 where twelve responses, together with some spontaneous miniature potentials, are shown. If the response was further reduced, by increasing Mg or lowering Ca concentrations, the amplitude fluctuations became even more pronounced and were found to be of discontinuous character. In the experiment of Fig. 2, for instance, the majority of records showed no response at all. On the average only about one out of seven nerve impulses elicited an e.p.p. whose size was of the same order of magnitude as the spontaneous potentials.

This behaviour is characteristic of block by high Mg and low Ca, and very different from curare-block. With increasing doses of curarine the e.p.p., at individual junctions, is progressively reduced in size and may eventually become undetectable, but we have never found the response to be abolished, or to fluctuate, in the quantal manner shown in Fig. 2.

If one proceeds to add Mg or withdraw Ca, a practical limit is reached when the e.p.p. response becomes too infrequent to be distinguished from a spontaneous discharge. There are no differences in amplitude which would enable one to discriminate between the two forms of activity; the distinction depends entirely on the constant latency of the response and random timing of the spontaneous discharges. In a normal frog muscle, at 20° C, the latency of the e.p.p. varies only within a fraction of a millisecond, but in the present experiments we have accepted 1–2 msec as the maximum latency fluctuation, and disallowed as 'response' any potentials which arose outside these limits. In practice, unless the frequency of spontaneous firing was high and the

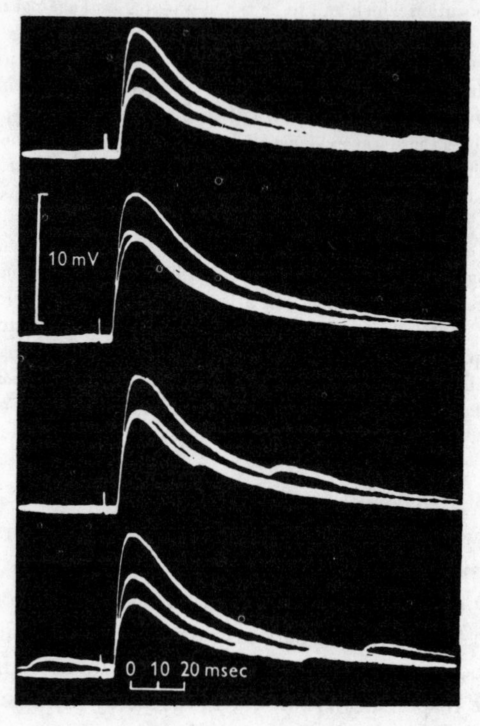

Fig. 1. Fluctuation of e.p.p. response at a single nerve-muscle junction, treated with 10 mM-Mg (Ca concentration was normal: 1·8 mM; prostigmine 10^{-6}). Intracellular recording. In each record, three superimposed responses are seen. Note scattered spontaneous miniature potentials.

frequency of responding very low, there was little chance of confusing a spontaneous potential with an e.p.p.-response: for example, in Fig. 2 (latency of five 'accepted responses' being constant within 1 msec; spontaneous firing rate 2·2 per sec) the chances of one of the 'accepted responses' being 'spontaneous' are about 5 %, and the chances of more than one arising spontaneously are quite negligible.

Most experiments were made at an intermediate level of blocking when the proportion of failures at individual end-plates was of the order of 50 %. The remaining responses were scattered in amplitude over a wide range, as illustrated in Fig. 3. (Responses to pairs of nerve impulses are shown in this figure.) Many e.p.p.'s fall evidently within the range of sizes of the spontaneous potentials. Others are larger and probably represent multiple units of response. It is interesting that the large e.p.p.'s occasionally show a just noticeable inflexion on their rising phase (e.g. Fig. 3, record C_1) indicative of their composite nature and of imperfect synchronization of the contributing units.

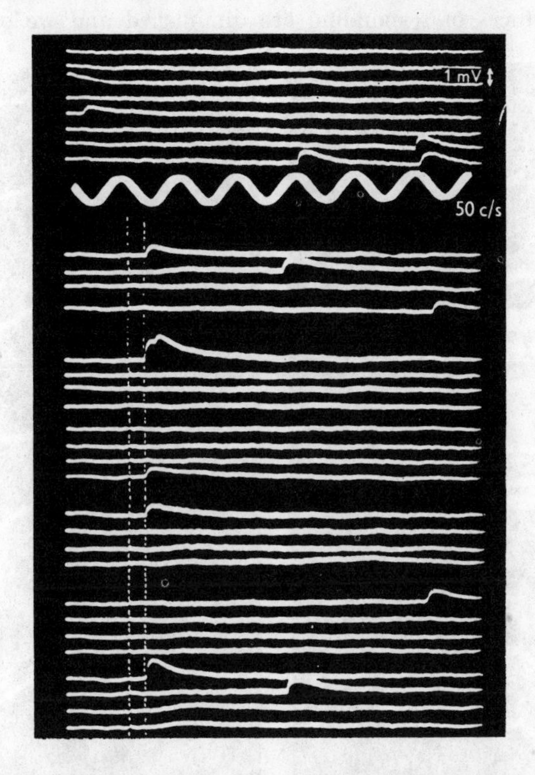

Fig. 2. This muscle was treated with reduced Ca (0·9 mM) and 14 mM-Mg concentration. The top part shows a few spontaneous potentials (traces separated by 1 mV steps). The lower part (below the 50 c/s time signal) shows responses to single nerve impulses. Stimulus artifact and response latency are indicated by a pair of dotted vertical lines. The proportion of failures was very high: there are only five responses to twenty-four impulses.

The experiments of Figs. 1–3, made at different levels of neuromuscular block, have one feature in common, namely a wide fluctuation in e.p.p. amplitudes. In Figs. 4 and 5, the distribution of amplitudes in two experiments is shown, both of spontaneous potentials and response. It is clear that these results cannot be analysed, nor even satisfactorily described, without a statistical treatment.

Suppose we have, at each nerve-muscle junction, a population of n units (cf. Fatt & Katz, 1952a, 1953) capable of responding to a nerve impulse. Suppose, further, that the average probability of responding is $\bar{p}$ (the chances p

may differ greatly for the individual constituents, but are supposed to remain constant during the experiment) then the mean number of units responding to one impulse is $m = n\bar{p}$. Under normal conditions, $\bar{p}$ may be assumed to be relatively large, that is a fairly large part of the synaptic population responds to an impulse. However, as we reduce the Ca and increase the Mg concentration, the chances of responding are diminished and we observe mostly

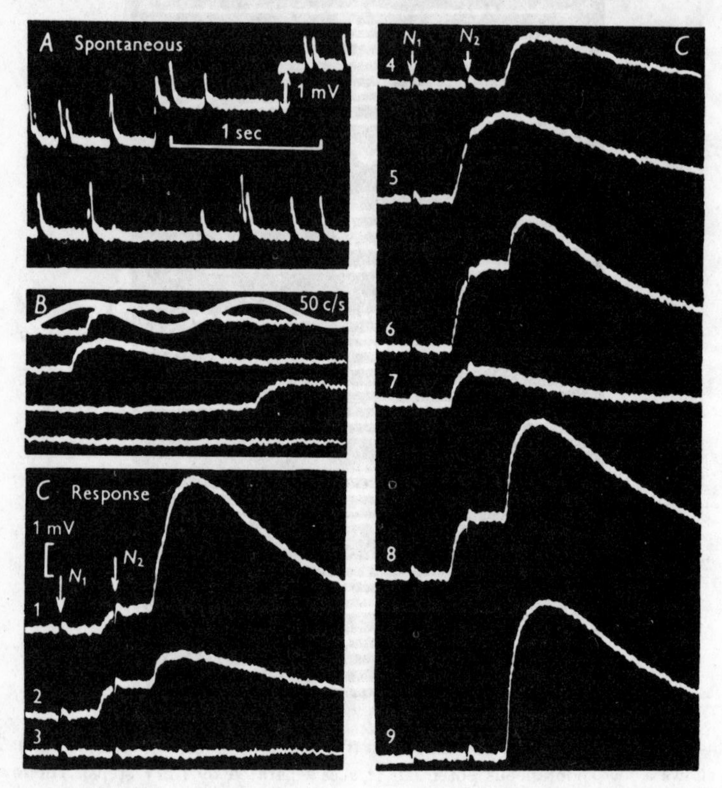

Fig. 3. Muscle was treated with a solution containing 0·45 mM-Ca and 6 mM-Mg. Intracellular recording from single junction. *A* and *B*: spontaneous miniature e.p.p.'s. *C*: examples of responses to paired nerve impulses. Timing of stimuli N_1 and N_2 is indicated by arrows. Failure of response to N_1 in C_4 and C_9, failure to N_2 in C_5 and C_7, double failure in C_3. 50 c/s time signal applies to *B* and *C*. *A* was recorded on slow time base and shows two calibration steps of 1 mV.

complete failures with an occasional response of one or two units. Under these conditions, when p is very small, the number of units x which make up the e.p.p. in a large series of observations should be distributed in the characteristic manner described by Poisson's law (their relative frequencies being given by $\exp(-m)\, m^x/x!$).

To test the applicability of Poisson's law may seem difficult, because all we can do is measure amplitudes of supposedly composite e.p.p.'s; we cannot count the components directly. The task is, however, made easier because the presence of spontaneous activity gives us an independent measure of unit size.

We can obtain the value of m, i.e. the mean number of units responding to one impulse, in two ways: first, from the relation

$$m = \frac{\text{mean amplitude of e.p.p. response}}{\text{mean amplitude of spontaneous potentials}}. \quad (1)$$

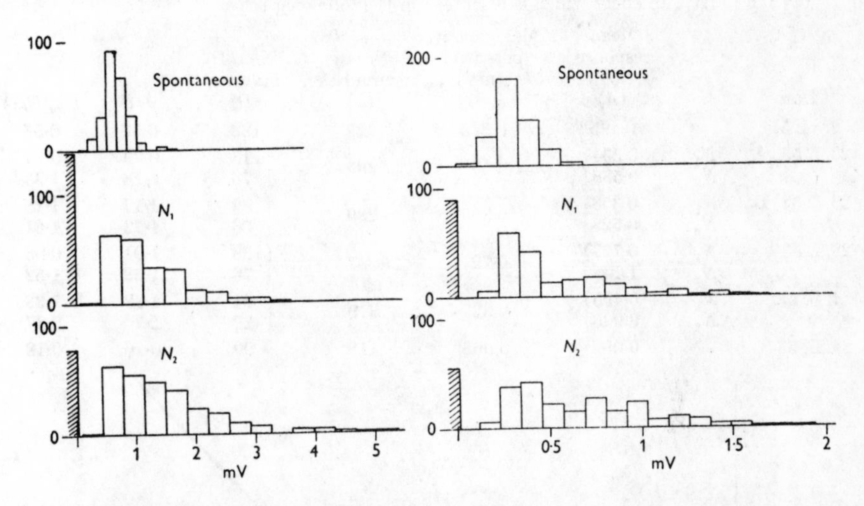

Figs. 4 and 5. Histograms from two end-plates, showing distribution of amplitudes of spontaneous miniature potentials and of the responses to pairs of nerve impulses (7 msec interval between N_1 and N_2). *Failures* are not represented as a 'class', but their number is indicated by the height of the shaded columns.

Equation (1) is a simple re-statement of our hypothesis, namely that the e.p.p. is made up of units of the same size (though not necessarily composed of the same individuals) as the spontaneous miniature potentials. Equation (1) depends on the assumption that there is linear summation of the miniature components of the e.p.p.: this is justified provided the amplitude of the e.p.p. is only a few per cent of the resting potential (cf. Fatt & Katz, 1951), but equation (1) fails to apply to larger responses.

Secondly, we can use the first term of the Poisson series ($\exp(-m)$, for $x=0$) which gives the proportion of failures. Hence

$$m = \log_e \frac{\text{number of nerve impulses}}{\text{number of failures of e.p.p. response}}. \quad (2)$$

Combining (1) and (2) we obtain

$$\frac{\text{mean amplitude of response}}{\text{mean amplitude of spontaneous potentials}} = \log_e \frac{\text{number of impulses}}{\text{number of e.p.p. failures}}. \quad (3)$$

Equation (3) provides a useful test of our hypothesis and depends only on measurements of mean amplitudes and counting of 'failure' and 'success' of e.p.p. response. The results of several experiments in which this test has been applied are shown in Table 1 and Fig. 6. The agreement between the two

TABLE 1. In the last two columns the validity of equation (3) is tested in ten experiments. They include four experiments in which responses to pairs of impulses have been utilized (N_1–N_2 intervals 3·5–11 msec). It will be seen that the value of m (i.e. A/B or $\log_e C/D$) is larger for N_2 than for N_1, an effect which is discussed in the following paper.

Date		Mean response (mV) (*A*)	Mean spont. potential (mV) (*B*)	No. of impulses (*C*)	No. of failures (*D*)	*A/B*	$\log_e C/D$
2. vi. 51		0·495	0·875	328	188	0·57	0·56
23. i. 53, A	N_1	0·334	0·46	289	113	0·73	0·94
	N_2	0·588			76	1·28	1·33
23. i. 53, B	N_1	0·358	0·305	280	89	1·17	1·15
	N_2	0·528			56	1·73	1·61
28. i. 53	N_1	0·727	0·72	357	138	1·01	0·95
	N_2	1·14			78	1·58	1·52
4. ii. 53	N_1	0·495	0·335	319	84	1·48	1·33
	N_2	0·905			27	2·7	2·47
24. ii. 53		0·089	0·565	118	99	0·16	0·18

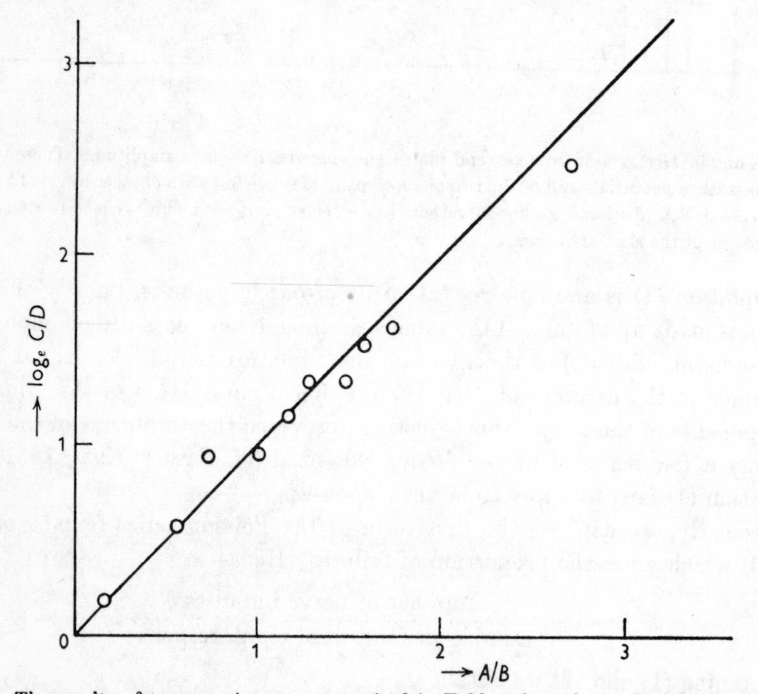

Fig. 6. The results of ten experiments summarized in Table 1 have been plotted, showing the consistency of the two methods of determining the value of m (equations (1) and (2)). Ordinate: $\log_e \frac{\text{number of impulses}}{\text{number of e.p.p. failures}}$. Abscissa: $\frac{\text{mean e.p.p. response}}{\text{mean amplitude of spontaneous potentials}}$. The line corresponds to equality of these two estimates of m.

determinations of m, corresponding to the right and left sides of equation (3), is very satisfactory and may be regarded as a strong support of our initial hypothesis.

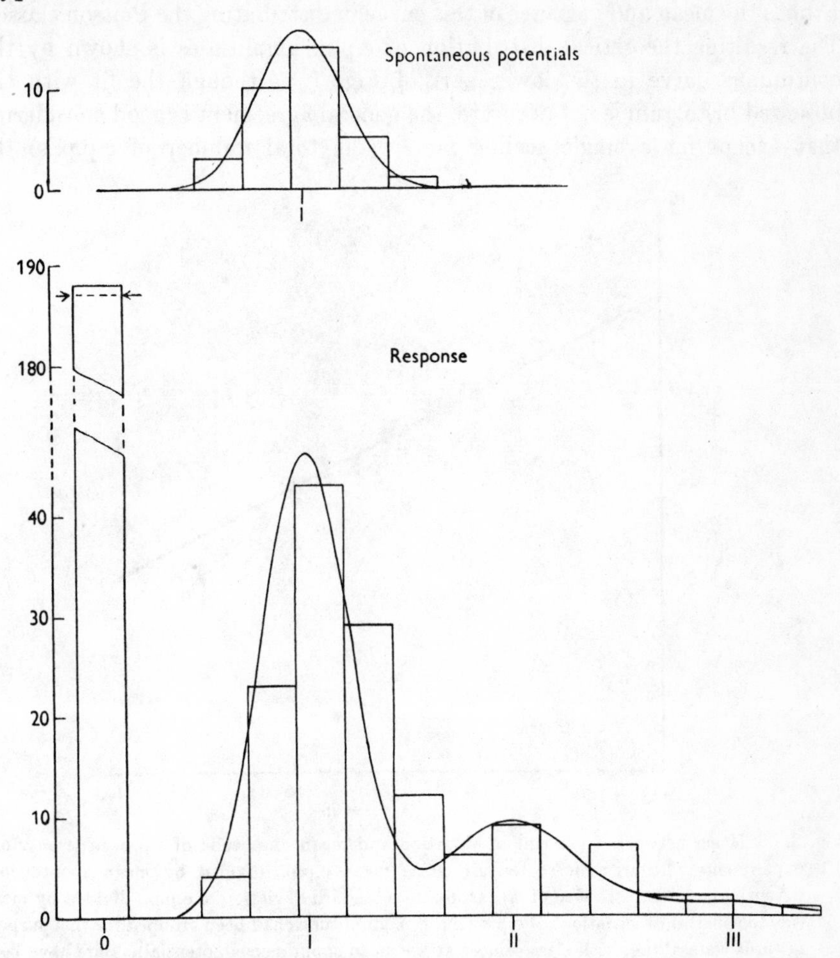

Fig. 7. Histogram showing distribution of amplitudes of spontaneous miniature potentials and end-plate responses at a Ca-deficient junction (experiment of Fatt & Katz, 1952*a*, pp. 119–120). In the lower part, the continuous curve has been calculated on the hypothesis that the responses are built up statistically of units whose mean size and amplitude distribution are identical with those of the spontaneous potentials (see text). Expected number of failures shown by arrows. Abscissae: scale units = mean amplitude of spontaneous potentials (0·875 mV).

The experiment of Fig. 7, the results of which were reported by Fatt & Katz (1952*a*), has been analysed more fully. The value of m was first determined by equation (1), and the expected numbers of the Poisson series were calculated. For $x=0$ (failure of response), there was excellent agreement between calculated and observed values, but for the terms $x > 0$ account had to be taken of the scatter of amplitudes of the 'unitary' spontaneous potentials. This was done by (*a*) fitting a Gaussian curve to the spontaneous potentials, and (*b*) using x times the mean and variance of this curve in distributing the Poisson classes. The resulting theoretical distribution of e.p.p. amplitudes is shown by the continuous curve in the lower part of Fig. 7. Although the fit with the observed histogram is not accurate, the general agreement is good considering that except for a single scaling factor (the total number of e.p.p.'s) the

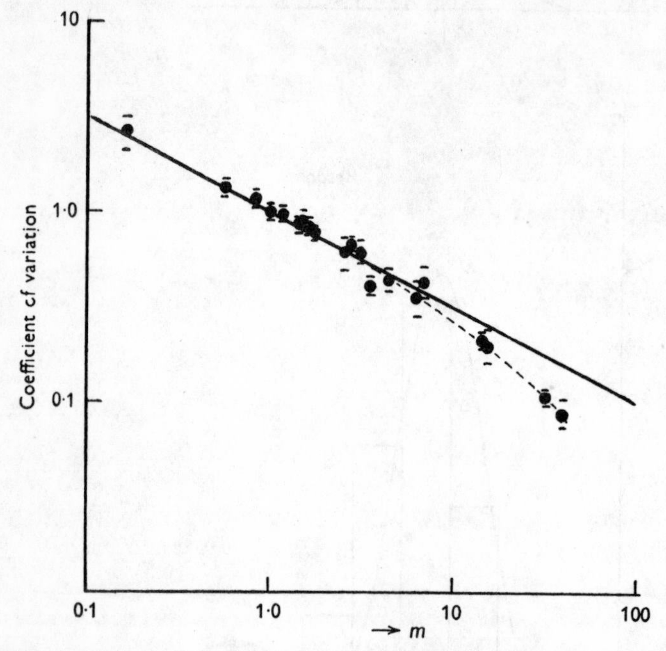

Fig. 8. Relation between coefficient of variation and mean amplitude of e.p.p. in twenty-one experiments. Logarithmic scales. Abscissa: mean e.p.p., divided by mean spontaneous potential (i.e. nominal value of m). Ordinate: standard deviation of e.p.p., divided by mean (i.e. 'coefficient of variation' of e.p.p.). E.p.p. amplitudes had been grouped for this purpose in 'unit classes' (i.e. with class centres at $n \times$ mean spontaneous potential). Bars have been placed at ± 2 S.E. of the 'coefficient of variation'. Full line shows theoretical relation for Poisson-distributions.

constants chosen in calculating the curve had been determined independently. The main discrepancies vanish if the mean size of the unit response is taken to be 7% larger than the mean spontaneous potential, a difference which is probably within limits of experimental error.

In other experiments the e.p.p. amplitudes were grouped more coarsely into classes of unit-width, and a χ^2 test was applied; also, the coefficient of variation of e.p.p. amplitudes (grouped in such unit classes) was determined and compared with the expected coefficients $m^{-0.5}$ (Fig. 8). These tests were less accurate than the preceding analysis, but the results agreed with the view that the responding units are distributed in Poisson-fashion provided the quantum content of the e.p.p. is small (<3). When the tests were extended to larger e.p.p.'s (m exceeding 10), there was a consistent discrepancy, the observed fluctuation of e.p.p. amplitudes covering a smaller range than expected (see Figs. 8 and 9).

DISCUSSION

The most interesting evidence is that shown in Table 1 and Fig. 6 for small values of m. The agreement between the two determinations of m can hardly be fortuitous and supports the view that the spontaneous miniature potential is the least 'quantum of action' at the nerve-muscle junction, the e.p.p. being built up statistically of such quanta. Furthermore, one may conclude that at this reduced level of m, the statistical chances of any one unit responding to a single impulse are very low, and in successive records the responses represent different members of a large, mostly inactive, population.

It is tempting to speculate what the precise probability of the unit response may be. For this, it is not sufficient to know only the value of m; we also require information of the total number of available units n. Moreover, it does not follow from the results that all units have the same chance of responding; a Poisson distribution would be obtained even from a non-uniform population, provided only the probabilities of responding are small and constant for each individual member (Kendall, 1948). If the whole synaptic population consisted of, say, 500 units, and m is unity, then the average chance of any unit responding to one impulse would be 1/500, but individual probabilities may be considerably higher for some and much smaller for many other members of the population.

What happens under more normal conditions when we raise the Ca and lower the Mg concentration? The value of m becomes large and the statistical analysis unsatisfactory. It is clear, however, that the response fluctuates much less than predicted from our equations (Fig. 9). Now suppose the size of the population n remains constant, then the increase of m would be due to an increased probability p. If the population is uniform, the distribution of responses would change from a Poisson to a binomial form. Associated with this one may expect a reduction in statistical spread, for the coefficient of variation for a Poisson series is $\sqrt{(1/m)}$, while that of a binomial distribution is only $\sqrt{\left(\frac{1}{m}-\frac{1}{n}\right)}$. Closer examination, however, shows that this argument is insufficient to account for the observed divergences. We can set a lower limit to the value of n: the normal e.p.p. is about 100 times larger than a miniature potential and must be composed of an even greater number of units because unit increments of the e.p.p. would diminish at high levels of depolarization (cf. Fatt & Katz, 1951). There is also reason to believe that the normal e.p.p.

does not involve the whole population, so that $n=200$ is a conservative estimate. With $m=32$, the coefficient of variation would be $\sqrt{(\frac{1}{32}-\frac{1}{200})}=0{\cdot}162$, compared with 0·177 in a Poisson distribution (when m/n is very small). The observed coefficient, however, is about $0{\cdot}11 \pm 0{\cdot}005$, and a significant discrepancy of this kind remained for all experiments in which m was greater than 10.

There are two other factors which are more likely to provide an explanation. One factor has already been mentioned, viz. a failure of linear summation of miniature potentials, when m becomes large and the amplitude of the total e.p.p. an appreciable fraction of the resting potential. Application

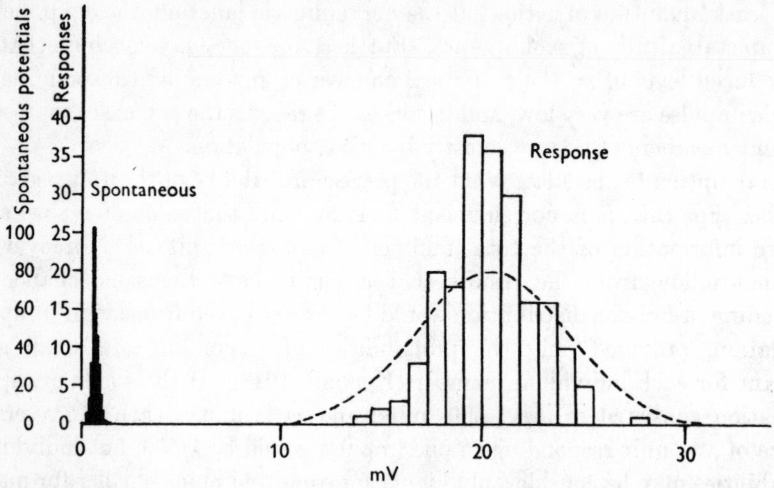

Fig. 9. Histogram from an experiment with large e.p.p. Nominal value of m (using equation (1)) is 32. Dotted curve: expected distribution of e.p.p.'s (modified Gaussian curve allowing for scattered unit size: mean = 20·4 mV), $\sigma = 3{\cdot}7$ mV. Note large discrepancy between observed and expected distribution.

of equation (1) and of the superposition theorem may lead to serious error if the e.p.p. response exceeds a small fraction (5 %) of the resting potential. Suppose each 'transmitter unit' produces a fixed leakage conductance ΔG across the end-plate membrane, then the increment of potential ΔP which it contributes to the e.p.p. becomes less the greater the existing leakage and the lower the membrane potential (cf. Fatt & Katz, 1951). This must have an important effect on the observed coefficient of variation, because (*a*) the actual number m would be greater than that calculated from equation (1), and (*b*) the scale of the amplitude fluctuations would be reduced in proportion to ΔP. We have made only a rough estimate of this effect, but it seems that it may account for a large part, if not the whole, of the observed discrepancy.

The other factor which may be involved is that different members of the population may *not* have the same chances of success, and that for large values of m some individual units have a high probability and respond almost every time, while others have a low probability and contribute to the e.p.p. only occasionally. The presence of some units which respond regularly is bound to diminish the statistical fluctuation of the e.p.p. In general, the coefficient of variation for this case is less than that expected for a binomial distribution (see Table 2).

TABLE 2. Coefficients of variation for different distributions

(From Kendall, 1948; the coefficient of variation is expressed here as a simple fraction, instead of per cent.)

Poisson ($p \ll 1$)	Binomial (var $p=0$)	'Non-uniform population'
$\sqrt{\left(\frac{1}{m}\right)}$	$\sqrt{\left(\frac{1}{m}-\frac{1}{n}\right)}$	$\sqrt{\left(\frac{1}{m}\left(1-\frac{\operatorname{var} p}{\bar{p}}\right)-\frac{1}{n}\right)}$

n = total number of units available at a single junction.
m = mean number of units responding to one impulse.
$\bar{p}=m/n$ = average probability of response (per unit per impulse).
var p = variance of individual probabilities (p being assumed to vary among responding units, but not during successive impulses).

This last factor should not be confused with the case in which probabilities of response vary during the set of observations, e.g. if the value of m suffered a progressive change. In this case the standard deviation of the e.p.p. amplitude would become greater, not less. A small effect of this kind was present in some experiments and could be checked by dividing the observations into groups. The drift of the mean value, however, was not large enough to affect the result seriously.

CONCLUSIONS

The following picture emerges from the present study: transmission at a nerve-muscle junction takes place in all-or-none 'quanta' whose sizes are indicated by the spontaneously occurring miniature discharges. The number of quantal units responding to a nerve impulse fluctuates in a random manner and can be predicted only in statistical terms. The average 'quantum content' of the e.p.p. depends on the probability of response of the individual units, and this varies with the external Ca and Mg concentration (for a more detailed hypothesis, see Castillo & Katz, 1954). It is possible that some synaptic units respond more readily than others, but with a sufficiently high Mg and low Ca level the chances of excitation of all units are so small that a Poisson distribution is obtained.

Under more normal conditions, the e.p.p. is large and the statistical fluctuation small. While the evidence for the quantal composition of low-level e.p.p.'s ($m < 5$) seems conclusive, inferences about the normal behaviour are indirect and can only be made by extrapolating into a range in which the present statistical analysis can give no useful information. There are, however, good reasons for supposing that the normal e.p.p. is built up of a large number of units of the same kind as described here, furthermore that even the normal e.p.p. involves only a fraction of the total synaptic population, the average probability of response apparently being less than unity. This suggestion is based on the finding that the size of the e.p.p. can be increased from nil to well above the 'normal-Ringer' amplitude by raising the Ca concentration, *without* increasing the size of the spontaneous miniature e.p.p. (Fatt & Katz, 1952*a*, *b*; Castillo & Stark, 1952). If one accepts the present results as showing that the miniature e.p.p. is the basic unit of response, then the effect of Ca must be to raise the quantum content m of the e.p.p., either by increasing the size of the population n or its probability of responding $\bar{p}$. We have assumed in our argument that a change of probability, rather than population size, is involved, though the formal distinction between these two modes of action is not very profitable until more is known about the nature of the molecular reaction whose probability we are considering.

SUMMARY

1. The relation between response and spontaneous activity at a single nerve-muscle junction has been studied.

2. By increasing Mg and lowering Ca concentration, the amplitude of the e.p.p. can be reduced to that of a spontaneous 'miniature potential'. At the same time, a large random fluctuation of successive e.p.p. amplitudes is observed.

3. Statistical analysis indicates that the e.p.p. is built up of small all-or-none quanta which are identical in size and shape with the spontaneously occurring miniature potentials.

4. When the average 'quantum content' (m) of the e.p.p. is small ($m < 3$), its amplitude fluctuates in a manner predictable by Poisson's law. At higher levels ($m > 10$), deviations occur which may be due to a reduction in the 'unit-increment' of the e.p.p., or to variation in the probability of response among different synaptic units.

5. The statistical behaviour of the normal nerve-muscle junction and the influence of Ca and Mg ions are discussed.

We are indebted to Mr J. L. Parkinson for his unfailing assistance. This work was supported by a research grant made by the Nuffield Foundation.

REFERENCES

DEL CASTILLO, J. & ENGBAEK, L. (1954). The nature of the neuromuscular block produced by magnesium. *J. Physiol.* **124**, 370–384.

DEL CASTILLO, J. & KATZ, B. (1954). The effect of magnesium on the activity of motor nerve endings. *J. Physiol.* **124**, 553–559.

DEL CASTILLO, J. & STARK, L. (1952). The effect of calcium ions on the motor end-plate potential. *J. Physiol.* **116**, 507–515.

FATT, P. & KATZ, B. (1951). An analysis of the end-plate potential recorded with an intra-cellular electrode. *J. Physiol.* **115**, 320–370.

FATT, P. & KATZ, B. (1952*a*). Spontaneous subthreshold activity at motor nerve-endings. *J. Physiol.* **117**, 109–128.

FATT, P. & KATZ, B. (1952*b*). The effect of sodium ions on neuromuscular transmission. *J. Physiol.* **118**, 73–87.

Fatt, P. & Katz, B. (1953). Chemo-receptor activity at the motor end-plate. *Acta physiol. scand.* **29**, 117–125.

Katz, B. (1949). The efferent regulation of the muscle spindle in the frog. *J. exp. Biol.* **26**, 201–217.

Katz, B. & Kuffler, S. W. (1941). Multiple motor innervation of the frog's sartorius muscle. *J. Neurophysiol.* **4**, 209–223.

Kendall, M. G. (1948). *The Advanced Theory of Statistics.* 4th ed., vol. I, pp. 122 *et seq.* London: Griffin.

Kuffler, S. W. & Vaughan Williams, E. M. (1953). Small-nerve junctional potentials. The distribution of small motor nerves to frog skeletal muscle, and the membrane characteristics of the fibres they innervate. *J. Physiol.* **121**, 289–317.

J. Physiol. (1954) 124, 574–585

STATISTICAL FACTORS INVOLVED IN NEUROMUSCULAR FACILITATION AND DEPRESSION

BY J. DEL CASTILLO AND B. KATZ

From the Department of Biophysics, University College, London

(*Received* 25 *January* 1954)

When a series of impulses arrive at the nerve-muscle junction, the end-plate potentials (e.p.p.) which they produce are not constant but vary in size, depending on the number and frequency of the stimuli. Two main types of phenomena have been observed: (*a*) a progressive increase of the e.p.p. (facilitation, post-tetanic potentiation) and (*b*) a phase of depression (Wedenski-inhibition, junctional fatigue). The present paper is concerned with the stage of neuromuscular transmission at which facilitation and depression occur and with the question whether 'quantal' changes of e.p.p. amplitude are involved.

It has been suggested that the progressive synaptic changes during repetitive stimulation are due to a variation in the output of acetylcholine (ACh) from the motor nerve endings rather than to post-synaptic events (e.g. Feng, 1941*b*; Hutter, 1952; Eccles, 1953, p. 89 *seq.*). In view of the recent evidence indicating that ACh release occurs in discrete quanta, it is of interest to inquire whether a functional change of ACh output takes place at quantal or molecular level, involving either the number or the size of the miniature units of which the end-plate response is composed.

METHODS

The technique was essentially the same as described previously (Fatt & Katz, 1951; Castillo & Katz, 1954*a*, *b*). Intracellular recording electrodes were applied to m. ext. l. dig. IV of the frog, and spontaneous miniature e.p.p.'s, as well as responses to single and repetitive nerve impulses, were recorded.

The majority of the experiments were devoted to a statistical study of facilitation, i.e. of the increase of e.p.p. amplitudes during a series of impulses. As pointed out in the preceding paper, a successful analysis can only be made when the response at a single junction has been reduced to a very small number of unit components.

Measurements of e.p.p. amplitudes are less accurate for the second than for the first of a pair of closely spaced responses, because the second e.p.p. is superimposed on the declining phase of the first and some error is, therefore, likely to occur in estimating its height. More reliance was placed on 'success-and-failure' counts which are free from such uncertainty, but their application means that the experimental conditions have to be further restricted. The most significant change in failure counts may be expected when the proportion of failures is about 50% and the value of m (the mean number of responding units, Castillo & Katz, 1954*b*) between 0·5 and 1·0. If facilitation causes m to increase by a factor c, then the largest change in failure counts is obtained when $m = (\log_e c)/(c-1)$. For values of c ranging from 1·3 to 3, $m_{opt.}$ lies between 0·87 and 0·55. Whether the results obtained under the special conditions required for the statistical analysis can be applied generally will be discussed later.

In a number of 'facilitation' experiments, tetani of up to fifty successive stimuli, at 100 per sec, were used. It was then necessary to start with a very high proportion of failures ($m = 0·1$–$0·2$ for the first impulse) in order to be able to follow their statistical decline during the tetanus. The failure counts for the first, and to a diminishing extent of subsequent, impulses in this series were affected by error due to uncertainty of excluding spontaneous potentials. As explained previously (Castillo & Katz, 1954*b*), the distinction between a single-unit response and a spontaneous miniature potential depended on the constancy of timing of the response. In the tetanus experiments a slow time-base was used, and as much as 3 msec had to be allowed for variations in latency. This means that, over a large number of records, some spontaneous potentials are likely to have been counted as response, their most probable number being $N\tau/T$, where N is the number of records, τ the allowance for latency variations after each stimulus, and T the mean interval of spontaneous potentials in the absence of stimulation. Even under the most adverse conditions, the error was not very serious: thus, in Table 4, of 110 counted responses to 711 N_1 impulses, about ten were probably spontaneous potentials (making the correct proportion of failures 0·86 instead of 0·845). Higher response counts are less affected.

A source of error of opposite direction was that a mechanical contact breaker was used in applying the tetanus, with a small chance of 'cutting' the first shock and making it ineffective. This error, however, was much less important and not likely to have caused an omission of more than one response count.

Solutions. Ringer's solution was used modified by adding $MgCl_2$ and lowering $CaCl_2$ so as to reduce the quantum content of the end-plate response to the required level (see Castillo & Katz, 1954*a*). In most experiments, prostigmine (10^{-6}, w/v) was added to increase the size of the unit response, but a few experiments were made without prostigmine and found to give the same results. The temperature during these experiments was 15–20° C.

PART I. FACILITATION

RESULTS

Double impulses

When two impulses, at an interval of several milliseconds, are sent into a curarized frog muscle and e.p.p.'s are recorded from the whole muscle, the second e.p.p. is found to be larger than the first (Schaefer & Haass, 1939; Eccles, Katz & Kuffler, 1941; Feng, 1941*a*, *b*). A similar, and even somewhat greater, effect has been observed in magnesium-treated muscles (Feng, 1941*b*). When the experiment is made on a muscle fibre blocked with high Mg and low Ca, and the response of a single end-plate is studied with an internal electrode, the amplitude of the response varies, and facilitation effects are not immediately obvious (see preceding paper, fig. 3). The range of fluctuation far exceeds the expected increase in amplitude of the second response. However, when a large number of paired responses are measured, one finds that the average size of the second e.p.p. is significantly larger than that of the first (see Table 1; Castillo & Katz, 1954*b*, figs. 4, 5).

While the response of a single junction, under these conditions, can only be summarized in statistical terms, records taken from a whole muscle with external electrodes are easier to describe. They do not show marked random fluctuations, and in successive responses the second e.p.p. is consistently larger than the first. It is clear, however, that external records, while giving the average response from hundreds of end-plates, conceal the statistical properties of the facilitation process.

The first question which confronts us is whether the increase in the average size of the e.p.p., at each junction, is due to an increased *amplitude* or increased *number* (m) of its unit components. In the former case, the distribution of e.p.p. amplitudes would be the same for N_2 as for N_1, except for an increased voltage scale. The number of failures would remain unchanged. In the latter case, the distributions should be different, corresponding to different mean numbers m_1 and m_2. In particular, the number of failures should be reduced, in a ratio obtainable from equation (1) (Castillo & Katz, 1954*b*).

$$\frac{\text{mean size of response}}{\text{mean size of spontaneous potentials}} = \log_e \frac{\text{number of impulses}}{\text{number of failures}}. \quad (1)$$

The first point to settle was whether the number of failures of e.p.p. response was significantly less for the second of two impulses. Pooling the failure counts of several experiments, with an average N_1 response of approximately one miniature unit, we found 534 failures to N_1 and 292 failures to N_2 (intervals 3·5–11 msec), giving a highly significant difference (242 ± 25). This suggests that facilitation in these experiments did involve an increase in the *number* of responding units.

TABLE 1. Statistical properties of facilitation. If facilitation is entirely due to recruitment of units, i.e. an increase of m, then equation (1) should apply and we obtain $m_2 - m_1 = \log_e (f_1/f_2)$, where $m = A/B$ and f_1 and f_2 are the number of failures to N_1 and N_2 respectively.

Expt.	N_1 N_2 interval (msec)		Mean response (mV) (*A*)	Mean spont. potential (mV) (*B*)	Number of impulses (*C*)	Number of failures (*D*)	*A/B*	$\log_e C/D$	$m_2 - m_1$	$\log_e (f_1/f_2)$
1	10·8	N_1	0·334	0·46	289	113	0·73	0·94	0·55	0·4
		N_2	0·588			76	1·28	1·33		
2	6·7	N_1	0·358	0·305	280	89	1·17	1·15	0·56	0·47
		N_2	0·528			56	1·73	1·61		
3	6·9	N_1	0·727	0·72	357	138	1·01	0·95	0·57	0·57
		N_2	1·14			78	1·58	1·52		
4	3·5	N_1	0·495	0·335	319	84	1·48	1·33	1·22	1·13
		N_2	0·905			27	2·7	2·47		

In four experiments, equation (1) was applied to N_1 and N_2 responses, and satisfactory agreement was obtained for both (Table 1). This indicates that a large part, if not the whole, of the observed increase in e.p.p. amplitude is due to a statistical recruitment of quantal components.

Additional tests were made (χ^2, and coefficient of variation on data grouped in classes of 'unit-width'; see Castillo & Katz, 1954*b*), and these gave confirmatory results.

A recruitment of miniature units could be brought about in different ways: (*a*) It may be that only those units which *responded* to the first impulse are 'facilitated', that is preceding activation may raise the probability of responding to the next impulse. (*b*) Alternatively, facilitation may affect the whole latent population and leave behind a state of greater probability of activation, independent of previous success or failure.

A decision between these possibilities was reached by examining those records in which N_1 had failed to elicit an e.p.p. Clearly, if facilitation depended upon there having been a previous response, the mean size of the *selected* N_2 potentials ($N_1=0$) should show no facilitation and not differ significantly from the mean N_1 potential. On the other hand, if facilitation is independent of previous response or failure, then the mean of the selected N_2 potentials should show the full increase in amplitude over the N_1 response. The results of four experiments are summarized in Table 2. The selected N_2 responses

TABLE 2. Comparison of total and selected ($N_1=0$) responses to N_2
(Numbers of responses shown in brackets)

	Mean of all N_2 responses		Mean of selected N_2 responses ($N_1=0$)	
Expt.	mV	N_2/N_1	mV	N_2/N_1
1	0·59 (288)	1·76	0·635 (113)	1·9
2	0·53 (280)	1·47	0·55 (89)	1·53
3	1·14 (357)	1·57	1·09 (138)	1·5
4	0·905 (319)	1·83	1·1 (84)	2·22
Mean	0·815 (1244)	1·66	0·86 (424)	1·75

TABLE 3. Occurrence of 'double-failures' of e.p.p. response to paired impulses

N: number of paired impulses. F_1 and F_2: observed number of failures to N_1 and N_2, respectively. F_{1+2}: observed number of 'double failures'. The last column shows the number of 'double-failures' to be expected by chance coincidence.

		Observed			Calculated,
Expt.	N	F_1	F_2	F_{1+2}	$F_1 \times F_2/N$
1	289	113	76	32	30
2	280	89	56	16	18
3	357	138	78	30	30
4	319	84	27	9	7
Total	1245	424	237	87	85

were, on the average, about 5% *larger* than the unselected N_2 potentials, a small difference which may be due to sampling or experimental error. In any case, it is clear that the full facilitation effect was obtained even when not a single unit had responded to the first impulse. As a corollary, it was found that the 'double-failures' were no more frequent than one would expect on chance coincidence (see Table 3, columns 5 and 6).

One may conclude that success or failure of response to N_1 did not significantly alter the chances of response to N_2. However, a nerve impulse even though it may fail to produce any end-plate response, leaves behind a state of increased probability of activation of the terminal units. This is of interest because it indicates that facilitation must operate at a very early stage of the transmission process, before the liberation of ACh.

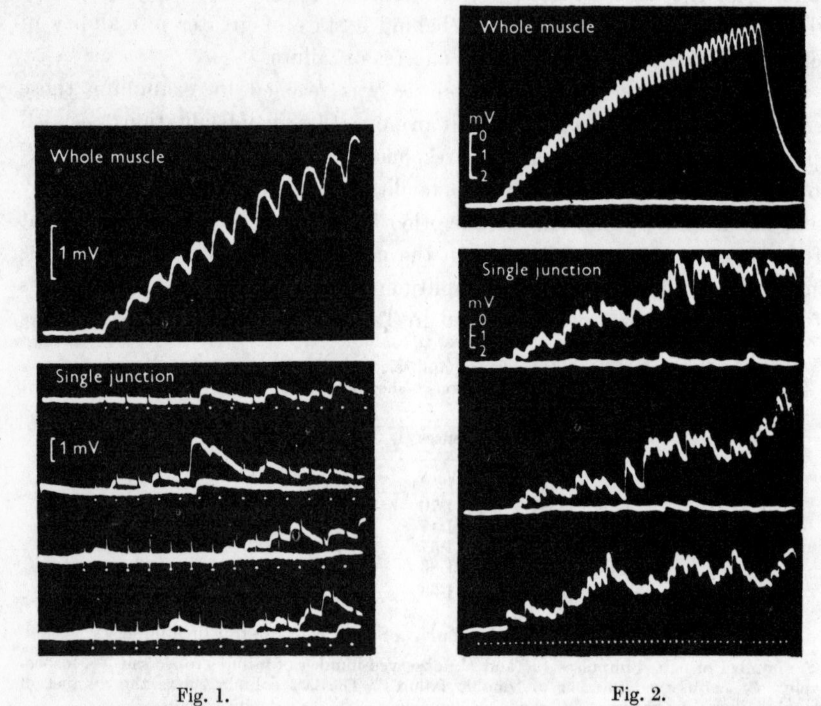

Fig. 1. Fig. 2.

Fig. 1. Recruitment of end-plate potentials during a tetanus. Muscles treated with 14 mM-Mg and reduced Ca (0·9 mM). Nerve stimulated at 100 per sec. Upper part: external recording from focal region of frog's sartorius. Lower part: intracellular recording from single end-plate of m. ext. l. dig. IV. Stimuli indicated by dots. Note spontaneous potentials on the superimposed 'base-lines'.

Fig. 2. Recruitment of e.p.p.'s. From the same experiments as Fig. 1, but on a reduced time scale, showing the progressive building up of response during a $\frac{1}{2}$ sec tetanus. Upper part: external recording from sartorius. Lower part: intracellular recording from m. ext. l. dig. IV. Stimuli indicated by dots. Voltage scale: mV.

Statistical recruitment of units during a tetanus

The quantal character of facilitation is further illustrated in Figs. 1 and 2 (see also Castillo & Katz, 1953) which show the building up of the end-plate responses during a tetanus. In these figures, records from a whole sartorius muscle, representing the average response of hundreds of end-plates, are compared with the responses of a single junction of the m. ext. l. dig. IV. The sartorius records show a smooth progressive increase in the size of the e.p.p. On the other hand, the response of a single junction fluctuates in a random manner, and a smooth effect is only obtained if one averages a large number of *successive* records (cf. Fig. 3).

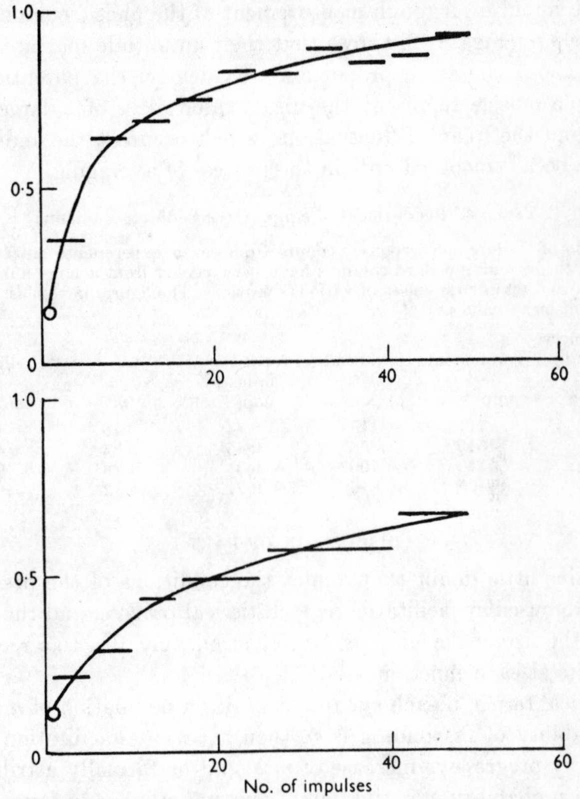

Fig. 3. Statistical increase of response at single junctions. Plots were obtained from two experiments of the kind illustrated in Figs. 1 and 2 (lower parts). 'Successes' and 'failures' of end-plate response were counted in a large number of records from the same junction, and the ratio of responses/impulses was plotted. Ordinate: proportion of responses ('1·0' means that every nerve impulse elicited an e.p.p.). Abscissa: serial number of impulses during a tetanus at 100 per sec. The initial point shows the statistical response to a single nerve impulse; the horizontal bars show the extent of grouping employed during the count.

At a single end-plate the first half a dozen impulses either fail to elicit a response or, occasionally, fire a single 'miniature unit'. As stimulation continues, unit responses become more frequent, occasionally much larger e.p.p.'s occur indicative of multiple unit activity, and failures become rare. In seven experiments, success-and-failure counts were made during tetani of approximately fifty impulses lasting 0·5 sec; examples are shown in Fig. 3 and a summary in Table 4. The progressive diminution of failure counts may be used to calculate the progressive increase of m (eqn. (1)). The result is given in column 5 of Table 4; it suggests that the average number of unit responses increases from 0·18 for the first impulse to about 0·83 for the 10th and 1·72 for the 40th impulse. A rough measurement of the phasic components of the 'external' e.p.p. (Figs. 1, 2) shows that their amplitude increases during the tetanus in about the same proportion. Evidently, the potentials recorded from a whole muscle represent the mean value of m of a large number of junctions, and the quantal fluctuations which occur at the individual end-plates have been 'smoothed out' in the process of averaging.

TABLE 4. Recruitment of e.p.p. response during a tetanus

Mean results of 'failure-and-response' counts from seven experiments on single junctions. The results in the second and third columns have been pooled from a large number of records covering single and tetanic responses of various durations. The figures in the last three columns are independent mean values.

Serial no. of nerve impulses during tetanus at 100 per sec	Total no. of nerve impulses	No. of end-plate responses	Proportion of failures $(=\exp(-m))$	m	'Facilitation factor' (m_n/m_1) (mean and S.E.)
N_1	711	110	0·84	0·18	1·0
N_6-N_{15}	1615	858	0·45	0·83	$4·7 \pm 0·44$
$N_{16}-N_{25}$	1140	799	0·31	1·28	$7·3 \pm 0·7$
$N_{36}-N_{45}$	1105	886	0·21	1·72	$10·0 \pm 1·21$

DISCUSSION OF PART I

There remains little doubt that, under the conditions of the present experiments, neuromuscular facilitation is a statistical process, and the progressive increase in the size of the e.p.p. is, largely or entirely, due to a recruitment of quantal units at each junction.

In statistical terms, if each synapse contains a population of n units whose mean probability of responding is $\bar{p}$, then $m = n\bar{p}$. 'Facilitation', being accompanied by progressive increase of m, could be formally attributed either to increased probability of activation of existing units or to formation of new units by preceding impulses. There is little evidence to choose between these alternatives, but we find it easier to think in terms of an increase in probability.

The basic mechanism of facilitation, that is the reason why a greater number of units respond to successive nerve impulses, is a problem on which our experiments have no direct bearing. It has recently been suggested by Lloyd (1949, 1952) that certain processes of potentiation in the spinal cord are due to increased amplitude of pre-synaptic spikes, arriving during a positive after-potential left by preceding impulses. A mechanism of this kind might possibly explain the increased transmitting power of successive impulses at the nerve-muscle junction, and it is compatible with our conclusion that facilitation operates at a very early stage of the transmission process, even before the release of ACh has occurred. It is of interest in this connexion that the size, and quantum content, of the e.p.p. can be increased by anodic polarization of the pre-synaptic structures (Castillo & Katz, 1954*c*), but it is as yet doubtful whether a single impulse can produce a positive after-potential of the required magnitude.

The present results were obtained under special conditions when the quantum-content of the e.p.p. had been greatly reduced by the use of low calcium and high magnesium concentrations. Can the conclusions from these experiments be applied to cases where the value of m is large (e.g. to facilitation in curarized muscle)? Recruitment of units is conceivable in principle so long as m is smaller than n. Even under 'normal' conditions, when no magnesium is used and with a calcium concentration of 1·8 mM, there is evidence that only a part of the synaptic population responds to a nerve impulse, for with higher Ca concentrations the e.p.p. continues to rise while the spontaneous unit-potentials remain constant or decrease slightly in amplitude (Fatt & Katz, 1952*a*, *b*). The fact that the e.p.p. may be increased more than threefold by calcium excess (Castillo & Stark, 1952) suggests that normally there is still a wide margin for recruitment of additional units.

It is, however, quite possible that other factors play an increasingly important part when the value of m becomes large. In the present experiments, the probability of each unitary response was very low; the value of p may have increased about tenfold during the tetanus, but it remained throughout in the low-level range in which a Poisson distribution applies. Hence the individual units activated by successive impulses were in general not the same, for when p is very small, the chances of the same unit responding twice in succession are negligible. The situation is quite different when p is large: individual units must then be expected to respond more regularly and contribute to many successive e.p.p.'s.

The present experiments give us no information on the recovery of a unit-response after previous activity, and it may well be that its *amplitude* passes through a 'supernormal phase'. We cannot, therefore, exclude the possibility that an increase in unit size, i.e. an increase at the molecular rather than quantal level of ACh output, may contribute to facilitation when m is large.

PART II. NEUROMUSCULAR DEPRESSION

The results in Fig. 2 show that facilitation of the e.p.p. is cumulative and continues throughout the period of stimulation. This behaviour is characteristic of preparations blocked by high Mg and low Ca but differs from that of curarized muscles (Feng, 1941*b*; Eccles & MacFarlane, 1949). During curare block, the increase of the e.p.p. is a transient event, and after a number of impulses the amplitude of the e.p.p.'s declines to a low value. A similar phase of depression probably underlies some of the phenomena known as

Wedenski-inhibition and neuromuscular fatigue in untreated preparations. It is interesting that depression of repetitive e.p.p.'s fails to occur in Mg-treated muscle, when the quantal release of ACh has been reduced to a low level. It seems that depression acts at a later stage than facilitation and only comes into operation after a period of intensive ACh release.

Experiments were made to find out whether this effect also involves a quantal reduction of ACh output. The procedure was to 'fatigue' transmission by prolonged nerve stimulation at low rate (e.g. at 2 per sec for 10 min), and re-insert the recording electrode into a previously selected muscle fibre when twitching had ceased. The main finding was that, as the result of prolonged stimulation, the amplitude of the e.p.p. was reduced to a few millivolts and showed large fluctuations, while the rate of spontaneous miniature potentials was greatly increased. Examples are shown in Figs. 4 and 5. When stimulation was stopped, and replaced by occasional test shocks, the e.p.p. slowly increased in size, and the excessive firing of spontaneous potentials gradually died down. If nerve stimulation was resumed, the response once more declined and could be practically abolished (Fig. 4, *G*) while the random activity flared up again.

The observations were repeated in nine experiments, and qualitatively the result was always the same. There was, however, no close quantitative relation between the reduction in e.p.p. size and increase in random firing rate. Thus, during 'recovery' the random firing may have returned to the low initial rate, or less, long before the e.p.p. had risen to a steady amplitude (Fig. 5). Furthermore, while the frequency of the discharge increased say, tenfold, there was no relation between the depression of the e.p.p. and the *absolute* rate of random firing. In the experiment of Fig. 4 it rose from an initial frequency of 20 per sec to a rate of over 200 per sec, too high to be measured. In another experiment, during the same depression of the response, the frequency changed from about 1 to 10 per sec.

This was of interest because it ruled out any suggestion that the increased random activity might have rendered the nerve endings 'refractory' to impulses and so indirectly caused a reduction of the response. The depression of the e.p.p. still occurred when the firing rates were far too low for this argument to hold.

Size of miniature potentials. At the height of the random activity, no reliable estimates of the size of the potentials could be made, and it is possible that their amplitudes were reduced to, say, one half. When the rate of firing was moderate, no obvious changes in size were observed. There was at no stage a reduction of amplitudes comparable to that of the e.p.p. Taken together with the fluctuating character of the reduced e.p.p., this suggests that the depression is accompanied by a 'quantal' breakdown of the response. The effect resembles, in this respect, the block produced by Mg or low Ca, but differs from it by the increase in the rate of random activity.

A few tests were made to find out which phase of neuromuscular activity was required to produce the increased random firing. 'Post-synaptic' activity was not responsible, for the effect was still observed when the prolonged

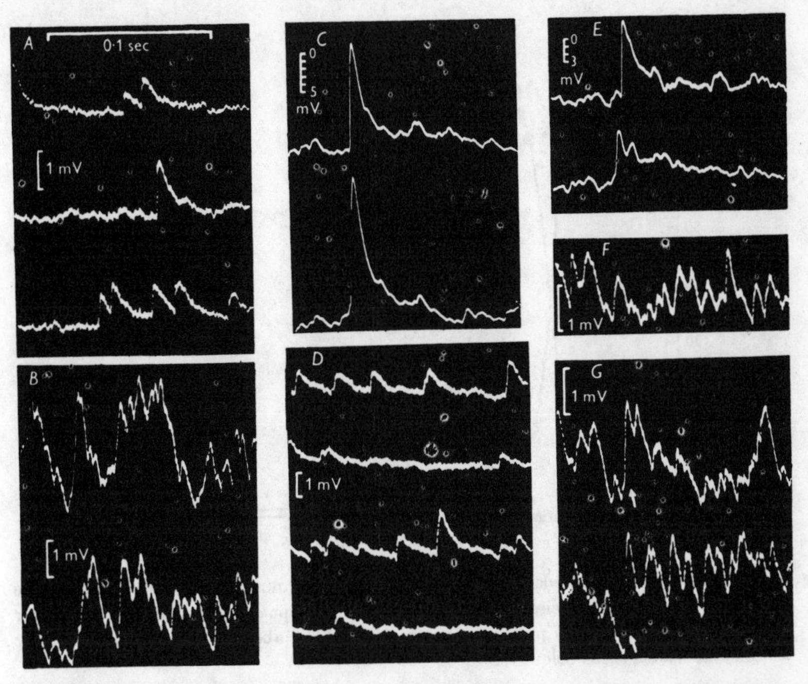

Fig. 4. Effect of prolonged stimulation on end-plate response and random activity. M. ext. l. dig. IV, in normal Ringer. Intracellular recording from single junction. *A*: spontaneous activity in normal muscle. *B*: after several minutes nerve stimulation, at about 2 per sec, showing increased random activity. This was followed immediately by records *C* showing random firing and e.p.p. response at lower amplification. *D*: during partial recovery. Random firing dying down. (Muscle twitches again on nerve stimulation; no response shown.) *E*: after renewed prolonged stimulation of nerve. Random activity has flared up. E.p.p. response again reduced to well below threshold. This was followed by *F* and *G*, at higher amplification, *F* showing random firing and *G*, in addition, two e.p.p. responses (at arrows) now barely distinguishable from the background activity.

stimulation was carried out on a temporarily curarized nerve-muscle preparation. When the drug was removed after the period of stimulation, the random discharge was found to be greatly increased and then gradually subsided.

On the other hand, if the preparation was first treated with a large dose of Mg (17 mM), prolonged low-rate stimulation caused only a very small, statistically insignificant, increase in spontaneous firing rate. This is in keeping with earlier findings (Boyd, Brosnan & Maaske, 1938; Maaske, Boyd & Brosnan, 1938; Luco & Rosenblueth, 1939; Feng, 1941*b*) that neuromuscular 'fatigue' occurs in the curarized muscle, but is small or absent in Mg-treated preparations.

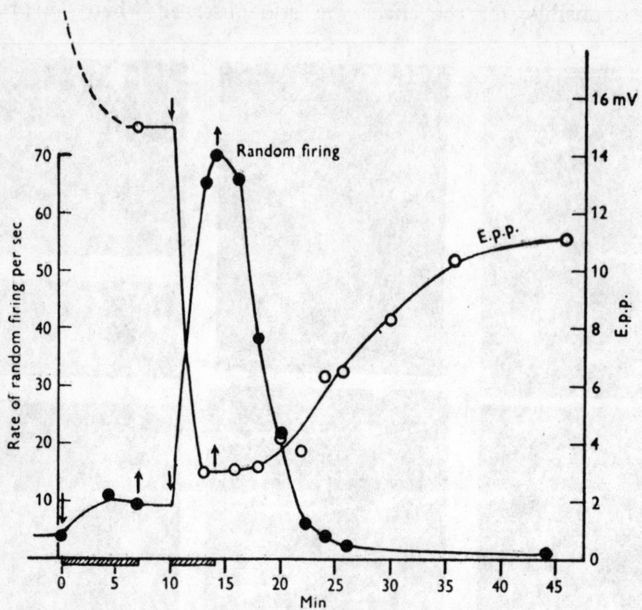

Fig. 5. Effect of prolonged stimulation on e.p.p. and rate of random activity. Full circles: rate of random firing. Hollow circles: amplitude of e.p.p. response. This was recorded after twitching had ceased. Two periods of nerve stimulation (about 5 per sec), indicated by arrows and shaded areas below.

The results of this section provide further evidence for the suggestion that the changes involved in a number of important synaptic events are of a quantal nature. Such changes were previously observed during ionic alterations in the environment, and have now been found to occur during the course of repetitive activity, 'facilitation' being associated with an increase, and 'depression' with a decrease in the number of units of response. Further discussion of these relations will be deferred to a subsequent paper (Castillo & Katz, 1954*c*) in which the effects of electrotonic changes in the motor nerve endings are studied.

SUMMARY

1. The process of neuromuscular facilitation has been studied using intracellular recording in frog muscle fibres.

2. By raising Mg and lowering Ca concentrations, the 'quantum content' of the end-plate potential (e.p.p.) can be reduced to a low level. Under these conditions, the statistical properties of the e.p.p. can be analysed following single and repetitive nerve impulses.

3. Facilitation of the e.p.p. is shown to be a quantal process involving at each junction a progressive recruitment of responding miniature units.

4. The statistical chances of any one quantal unit responding to a nerve impulse appear to be increased by a preceding impulse. The stage of transmission at which this facilitating effect occurs, its possible mechanism, and limits of its application in the normal or curarized muscle are discussed.

5. Observations dealing with neuromuscular 'depression' (during prolonged activity) are described.

6. In a normal (or temporarily curarized) muscle, prolonged nerve stimulation causes a progressive decline of the e.p.p., associated with a greatly increased frequency of random miniature discharges. The depression of the response is largely due to a reduction in the number of quantal units.

We are indebted to Mr J. L. Parkinson for his unfailing assistance. This work was supported by a research grant made by the Nuffield Foundation.

REFERENCES

BOYD, T. E., BROSNAN, J. J. & MAASKE, C. A. (1938). The summation of facilitating and inhibitory effects at the mammalian neuro-muscular junction. *J. Neurophysiol.* **1**, 497–507.

DEL CASTILLO, J. & KATZ, B. (1953). Statistical nature of 'facilitation' at a single nerve-muscle junction. *Nature, Lond.*, **171**, 1016.

DEL CASTILLO, J. & KATZ, B. (1954*a*). The effect of magnesium on the activity of motor nerve endings. *J. Physiol.* **124**, 553–559.

DEL CASTILLO, J. & KATZ, B. (1954*b*). Quantal components of the end-plate potential. *J. Physiol.* **124**, 560–573.

DEL CASTILLO, J. & KATZ, B. (1954*c*). Changes in end-plate activity produced by pre-synaptic polarization. *J. Physiol.* **124**, 586–604.

DEL CASTILLO, J. & STARK, L. (1952). The effect of calcium ions on the motor end-plate potentials. *J. Physiol.* **116**, 507–515.

ECCLES, J. C. (1953). *The Neurophysiological Basis of Mind.* Oxford: Clarendon Press.

ECCLES, J. C., KATZ, B. & KUFFLER, S. W. (1941). Nature of the 'end-plate potential' in curarized muscle. *J. Neurophysiol.* **5**, 362–387.

ECCLES, J. C. & MACFARLANE, W. V. (1949). Actions of anticholinesterases on end-plate potential of frog muscle. *J. Neurophysiol.* **12**, 59–80.

FATT, P. & KATZ, B. (1951). An analysis of the end-plate potential recorded with an intracellular electrode. *J. Physiol.* **115**, 320–370.

FATT, P. & KATZ, B. (1952*a*). Spontaneous subthreshold activity at motor nerve endings. *J. Physiol.* **117**, 109–128.

FATT, P. & KATZ, B. (1952*b*). The effect of sodium ions on neuromuscular transmission. *J. Physiol.* **118**, 73–87.

FENG, T. P. (1941*a*). The local activity around the skeletal *n–m* junctions produced by nerve impulses. *Biol. Symp.* **3**, 121–152.

FENG, T. P. (1941*b*). The changes in the end-plate potential during and after prolonged stimulation. *Chin. J. Physiol.* **16**, 341–372.

HUTTER, O. F. (1952). Post-tetanic restoration of neuromuscular transmission blocked by D-tubocurarine. *J. Physiol.* **118**, 216–227.

LLOYD, D. P. C. (1949). Post-tetanic potentiation of response in the monosynaptic reflex pathway of the spinal cord. *J. gen. Physiol.* **33**, 147–170.

LLOYD, D. P. C. (1952). Electrotonus in dorsal nerve roots. *Cold. Spr. Harb. Symp. quant. Biol.* **17**, 203–219.

LUCO, J. V. & ROSENBLUETH, A. (1939). Neuromuscular 'transmission-fatigue' produced without contraction during curarization. *Amer. J. Physiol.* **126**, 58–65.

MAASKE, C. A., BOYD, T. E. & BROSNAN, J. J. (1938). Inhibition and impulse summation at the mammalian neuromuscular junction. *J. Neurophysiol.* **1**, 332–341.

SCHAEFER, H. & HAASS, P. (1939). Über einen lokalen Erregungsstrom an der motorischen Endplatte. *Pflüg. Arch. ges. Physiol.* **242**, 364–381.

J. Physiol. (1960), **154**, *pp.* 52–67
With 9 text-figures
Printed in Great Britain

ON THE PERMEABILITY OF END-PLATE MEMBRANE DURING THE ACTION OF TRANSMITTER

By A. TAKEUCHI* and N. TAKEUCHI
Department of Physiology, University of Utah College of Medicine, Salt Lake City, Utah, U.S.A.

(*Received* 25 *April* 1960)

Movements of specific ions in excitable tissues have been emphasized by a number of recent experiments. In particular, there have been many quantitative experiments which show that the rising phase of the nerve impulse in invertebrate giant axons is associated with an inflow of sodium and the falling phase with an outflow of potassium (Hodgkin, Huxley & Katz, 1952; Hodgkin & Huxley, 1952) and similar selective permeability changes have been observed accompanying the propagated action potential in muscle fibres and in myelinated nerve fibres (Dodge & Frankenhaeuser, 1959; Hodgkin & Horowicz, 1959*a*).

In addition, the inhibitory potentials observed in some nerve cells (Coombs, Eccles & Fatt, 1955; Edwards & Hagiwara, 1959), heart muscle fibres (Burgen & Terroux, 1953; Hutter & Trautwein, 1956; Trautwein & Dudel, 1958), and crustacean muscle (Fatt & Katz, 1953; Boistel & Fatt, 1958) have been shown to be produced by changes in permeability of the post-synaptic membrane to potassium and/or chloride ions.

On the other hand, it has been considered that at the end-plate the transmitter produces a rapid simultaneous transfer of sodium and potassium, and possibly also of all other free ions on either side of the membrane (Fatt & Katz, 1951; del Castillo & Katz, 1954, 1955, 1956). An approximately linear relationship has been observed between the amplitude of the end-plate potential (e.p.p.) and the membrane potential, the equilibrium potential being about -15 mV (Fatt & Katz, 1951). A similar relationship has also been observed in the amplitude of end-plate current (e.p.c.) obtained when the membrane was clamped at a constant potential during neuromuscular transmission (Takeuchi & Takeuchi, 1959). The membrane potential at which e.p.c. becomes zero may be called provisionally 'e.p.c. equilibrium potential'. In the present experiment, the e.p.c. equilibrium potential was measured in solutions of various ionic composition and the ions which contributed to the e.p.c. were determined. It will be shown

* Fellow of Rockefeller Foundation.

that during the action of transmitter the end-plate becomes permeable mainly to sodium and potassium ions, but probably not to chloride ions.

METHODS

The sartorius muscle was dissected with the sciatic nerve from winter frogs (*Rana pipiens*) and mounted in a chamber made of methacrylate resin (lucite). The nerve was introduced into a wet chamber and stimuli were applied by a pair of silver electrodes. The capacity of the muscle chamber was about 4 ml. When the ionic composition was changed, at least 20 ml. of solution was exchanged by a different solution and the preparation was usually left for 10 min in each solution before measurements were made, except in chloride-deficient solution. When potassium-deficient solution was used, the fluid was kept flowing during the experiment.

The voltage clamp technique was similar to that previously reported (Takeuchi & Takeuchi, 1959), although a slightly modified amplifier for current recording was used. In order to change the clamped membrane potential, a square voltage pulse of about 15 msec duration was applied to the middle stage of the negative feed-back amplifier. A relatively short pulse for changing membrane potential has the advantage of avoiding a possible change in potassium concentration in nearby muscle fibres, which would occur if long pulses or d.c. had been used (Takeuchi & Takeuchi, unpublished). However, the short pulse has the disadvantage that when electrodes were inserted somewhat away from the end-plate, the measured e.p.c. equilibrium potential would tend to assume a lower value than that obtained with longer pulses.

Intracellular electrodes filled with 3 M-KCl were used for recording potential changes and, in most cases, for passing current. In some cases, especially when chloride concentration was changed, 0·6 M potassium-sulphate- or 1·5 M sodium-citrate-filled electrodes were used for current electrodes. In experiments employing electrophoretic injection of sodium ions into muscle fibres, electrodes filled with 3 M-NaCl or 1·5 M sodium citrate were used for passing current.

Electrodes were inserted under microscopic control and the end-plate was located exactly by inserting the electrode at several points along the length of muscle fibre. In order to ensure accurate localization of the electrodes at the end-plate, two criteria were used: (1) the e.p.p. had to show maximal amplitude and shortest rise time at the site of electrode impalement, and (2) the e.p.c.–membrane-potential relationship had to intercept the voltage axis at -10 to -20 mV. If the electrodes were slightly away from an end-plate focus, the measured e.p.c. equilibrium potential tended to approach zero. Generally the current-passing electrode was left in the fibre during the entire experimental series in order to maintain this localization. The recording electrode could then be withdrawn during changing of solutions.

The membrane potential of muscle may be defined as the difference between potentials recorded at points just outside the muscle membrane and inside the membrane. The tip potential may change in different ionic composition, and this might influence the value of the resting potentials measured in various solutions. Relatively low-resistance KCl-filled electrodes were used for potential recording and this served to minimize the change in tip potential in different media (Adrian, 1956). Before and after each series of experiments the recording electrode was withdrawn and the resting potential and the drift of recording system checked. The resting potential tended usually to decrease with time, but during the experiment the membrane potential was clamped within ± 1 mV of the previously determined value.

The normal Ringer's solution used had the following ionic composition (mM): Na^+ 113·6; K^+ 2·5; Ca^{2+} 2·0; Cl^- 117·5; HPO_4^{2-} 1·1; $H_2PO_4^-$ 0·4. In addition, tubocurarine chloride (3×10^{-6} g/ml.) was added to block neuromuscular transmission. Sodium-deficient solutions were obtained by mixing normal Ringer's solution and isotonic sucrose or dextrose solution containing the same concentrations of K^+, Ca^{2+} and phosphate buffer as normal solution. Sodium-rich solutions were made by adding crystalline NaCl. In lower concentrations of sodium (less than about 35 mM) neuromuscular transmission was blocked, so that tubocurarine was omitted. Solutions with sodium concentration less than 20 mM were not used, because at room temperature conduction block of the nerve frequently occurred in these solutions.

The concentration of potassium ions was altered by replacing sodium chloride by potassium sulphate, concentrations of other ions being kept constant. When the concentration of potassium ion in Ringer's solution was increased more than about 8 mM, neuromuscular transmission was frequently blocked, probably owing to conduction block at nerve branches. In lower concentrations an increase in concentration of potassium ions caused an increase in quantum content (Takeuchi & Takeuchi, unpublished) and the concentration of tubocurarine was increased to block the neuromuscular transmission.

In order to reduce the concentration of chloride, sodium chloride in Ringer's solution was substituted by sodium glutamate. In glutamate Ringer's solution containing 2 mM calcium chloride the muscle fibre had a tendency to contract spontaneously and the amplitude of the e.p.c. showed a large variation. At the time the solution was changed from normal Ringer's solution to glutamate solution contraction occurred frequently, damaging the muscle fibre impaled by the current electrode. This contraction may be due in part to the decrease in resting potential caused by the sudden decrease in chloride ions in outside solution (Hodgkin & Horowicz, 1959*b*), and partly may be due to the decrease in concentration of calcium ion (Fatt & Ginsborg, unpublished observation cited in Boistel & Fatt, 1958). In the present experiments with glutamate Ringer's solution calcium chloride was increased to 5 mM to avoid these disadvantages. Thus glutamate solution contained 10 mM-Cl^-.

Experiments were done at room temperature (18–23° C).

RESULTS

Curarization

When the muscle end-plate membrane, soaked in normal Ringer's solution containing about 3×10^{-6} g/ml. tubocurarine, was clamped at the resting potential by a negative feed-back system, a stimulus applied to the nerve produced an inward current. When a square voltage was applied to the feed-back amplifier, the membrane potential at the point where the electrodes were inserted was suddenly changed to a new level. After a rapid transient current due to charging of the capacitive component of nearby muscle membrane, a residual current flowed through the current electrode, charging the membrane somewhat more remote from the electrode. By changing the amplitude of the applied voltage pulse the membrane potential at the end-plate could be clamped at various potential levels. A stimulus applied to the nerve produced an e.p.c. superimposed on the residual current. Figure 1 shows these currents obtained from a curarized end-plate in normal Ringer's solution. The amplitude of the e.p.c. varied with the value of the clamped membrane potential, the e.p.c.–membrane-potential line crossing the abscissa at about 10–20 mV (inside negative). When the movement artifact was reduced by decreasing the concentration of sodium in the Ringer's solution, the linearity held until after the sign of the e.p.c. was reversed by depolarizing the membrane,

It is known that curarine changes the sensitivity of the end-plate to transmitter. The relation between e.p.c. height and membrane potential obtained from the same end-plate in two different concentrations of tubocurarine is shown in Fig. 2. The slopes of the two lines were different, but the points at which the lines cross the abscissa were almost the same. Since the slope of the line represents the conductance change at the peak of e.p.c., this result shows that tubocurarine decreased the magnitude of the conductance change produced by transmitter, but had little or no influence on other factors which determine the e.p.c. equilibrium potential. From this

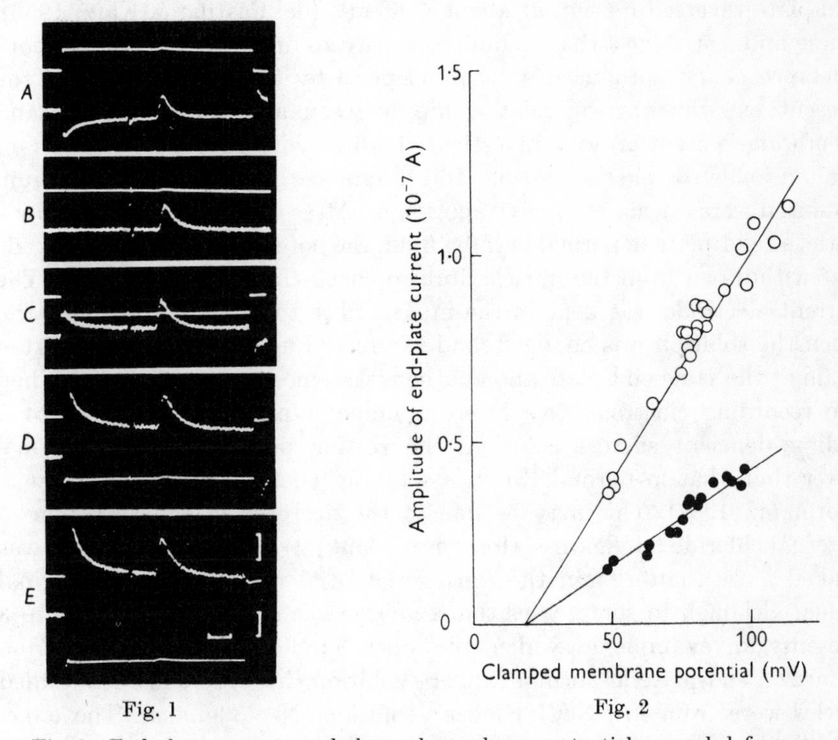

Fig. 1 Fig. 2

Fig. 1. End-plate currents and clamped membrane potentials recorded from a curarized end-plate in normal Ringer's solution. Upper traces represent clamped membrane potentials and lower traces feed-back currents containing the end-plate currents. In *A* a square pulse was applied to the feed-back system to depolarize the end-plate membrane. *B*, end-plate current obtained when the membrane was clamped at the resting potential (85 mV). In *C*, *D* and *E* the membrane potential was hyperpolarized to various values. The end-plate currents are superimposed on the current which maintains the membrane potential at various levels. Upper bar, 1×10^{-7}A; lower bar, 10 mV. Time marker, 2 msec. Temperature 20° C.

Fig. 2. Relationship between amplitude of e.p.c. and membrane potential in different concentrations of tubocurarine. Open circles were obtained from an end-plate in 3×10^{-6} g tubocurarine/ml. and filled circles are from same end-plate in 4×10^{-6} g tubocurarine/ml.

result it may be said that when the ionic composition of Ringer's solution was changed, adequate concentration of tubocurarine to block the neuro-muscular transmission could be used without influence on the e.p.c. equilibrium potential.

Sodium ions

It has been shown that when sodium concentration in Ringer's solution was decreased, the amplitude of e.p.p. was reduced (Fatt & Katz, 1952), and when the NaCl in Ringer's solution was totally replaced by sucrose, the potential change produced by electrophoretic application of ACh at the end-plate reversed its sign at about −60 mV (del Castillo & Katz, 1955). These findings suggest that sodium ions play an important role in the conductance change produced at the end-plate by the transmitter. In the present experiments the relationship between amplitude of e.p.c. and membrane potential was investigated with various concentrations of sodium ions in Ringer's solution, and differences in the e.p.c. equilibrium potential were estimated by extrapolation. After measurements were done with the end-plate in normal bathing fluid, the potential recording electrode was withdrawn from the muscle fibre to check the resting potential. The current electrode was kept in the muscle fibre to identify the end-plate. Then the solution was changed, and the recording electrode was inserted again at the same end-plate and another set of measurements made. When the recording electrode was inserted immediately after changing to a sodium-deficient sucrose solution, the resting potential was 10–20 mV lower than that in normal Ringer's solution (Giebisch, Kraupp, Pillat & Stormann, 1957). This may be due to the decrease in concentration of external chloride (Hodgkin & Horowicz, 1959*b*). When the preparation was soaked more than 10 min the resting potential returned to the original value, although in some cases the recovery was not complete. Figure 3 presents an example in which the open circles were obtained from curarized end-plate in normal Ringer's solution (Na^+ 113·6 mM), and filled circles were from low-NaCl Ringer's solution (Na^+ 33·6 mM). The e.p.c. equilibrium potentials obtained on extrapolation are separated by about 17 mV. It will be shown that chloride ions had little or no effect on the e.p.c. equilibrium potential. Therefore, since potassium and calcium concentrations in the bathing solution were kept constant, the shift of e.p.c. equilibrium potential should be attributed to the change in concentration of sodium ions. The e.p.c. equilibrium potentials obtained in various sodium concentrations are presented in Fig. 4. Although variations of the e.p.c. equilibrium potential are rather great, there is a definite tendency for the e.p.c. equilibrium potential to increase in sodium-deficient solutions.

When an outward current flows from a NaCl-filled electrode, outflow of Na^+ from the electrode will increase whereas that of Cl^- will decrease

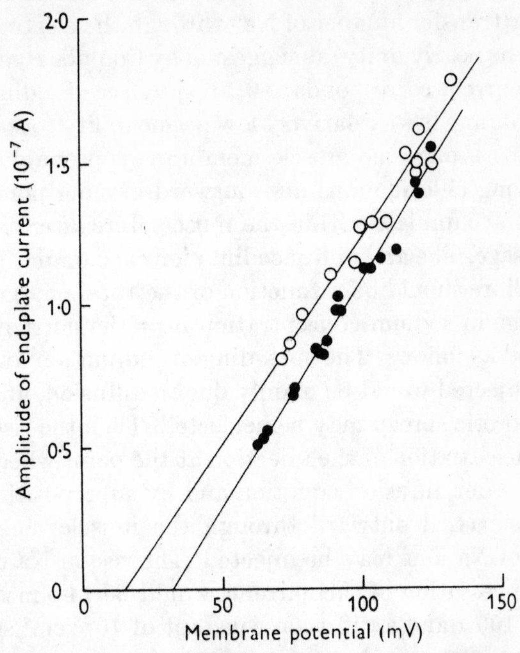

Fig. 3. Relationship between amplitude of e.p.c. and membrane potential obtained from an end-plate in two different concentrations of sodium ion in outside solution. Open circles obtained in normal Ringer's solution (113·6 mM-Na^+) and filled circles in dextrose Ringer's solution containing 33·6 mM-Na^+.

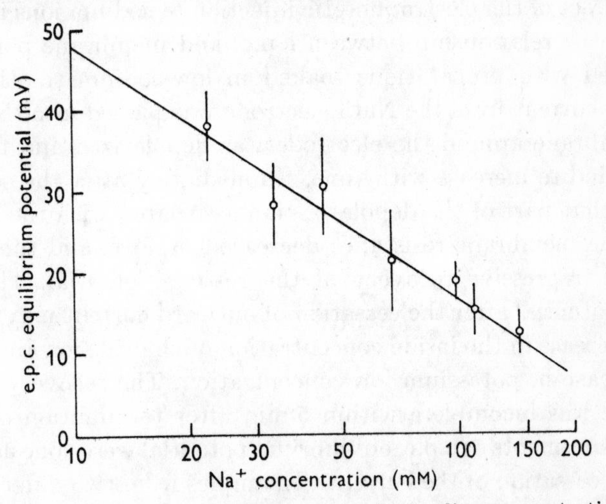

Fig. 4. E.p.c. equilibrium potential plotted against sodium concentration in outside solution on semilogarithmic scale (mean ± standard deviation about the mean). The line is drawn according to $(1/2{\cdot}29)\,\{99 + 1{\cdot}29 \times 58\log_{10}(15{\cdot}5/Na_0)\}$ mV.

relative to the diffusion of these ions from the electrode in the absence of current. The transfer number of Na^+ through the electrode is not clear, but if it is taken as nearly unity, as suggested by Coombs *et al.* (1955), 1×10^{-8} A of outward current corresponds to 0·1 pequiv/sec of sodium ions. The resting muscle membrane has a relatively low permeability to sodium ions and outward current through the muscle membrane is probably mostly carried by inward-moving chloride ions and outward-moving potassium ions. Thus, sodium ions accumulate within the muscle fibre after a period of outward current passage. The amount of sodium ions accumulating at a point along the muscle fibre should be a function of the total charge.

The change in sodium concentration near the current electrode tip can be calculated as follows. The spreading of sodium ion away from the point where it is injected would be mainly due to diffusion, and the contribution of electrophoretic spread may be neglected. Then the rise in Na concentration after the cessation of the injection at the point where it is injected can be obtained from diffusion equation and by superposition theorem. When 5×10^{-8} A is passed outward through the muscle membrane for 7 min, 210 pequiv of Na ions may be injected. The rise in Na concentration, 100 sec after the cessation of the current, would be 14·5 mM, assuming a fibre diameter of 100 μ and a diffusion constant of 10^{-5} cm^2/sec. This change in internal concentration of Na ions might shift the e.p.c. equilibrium potential by 9·4 mV (see Discussion). At the same time the inside concentration of potassium ions may be decreased and that of chloride ions increased to maintain electrical neutrality.

The effect of the electrophoretic injection of sodium ions into the muscle fibre on the relationship between e.p.c. and membrane potential was investigated with preparations soaked in low-sodium solution. While the outward current from the NaCl electrode was passed into the muscle fibre, the membrane around the electrode was depolarized and the depolarization tended to increase with time. Immediately after the cessation of the current that part of the depolarization associated with the potential drop across the membrane resistance decreased rapidly, and this was followed by slow progressive recovery of the resting potential. The decrease in resting potential after the cessation of outward current may be mostly due to the increase in the inside concentration of chloride ion and partly due to the decrease in potassium ion concentration. The recovery of the resting potential was incomplete within 5 min after termination of the current. The measurements of e.p.c. equilibrium potential were done usually 1–2 min after the cessation of the current. Examples from two different end-plates are presented in Fig. 5. The filled circles represent the e.p.c.–membrane-potential relation before passing current and the open circles after passing current (*A*, 5×10^{-8} A for 7 min; *B*, $4{\cdot}7 \times 10^{-8}$ A for 11 min). After passing

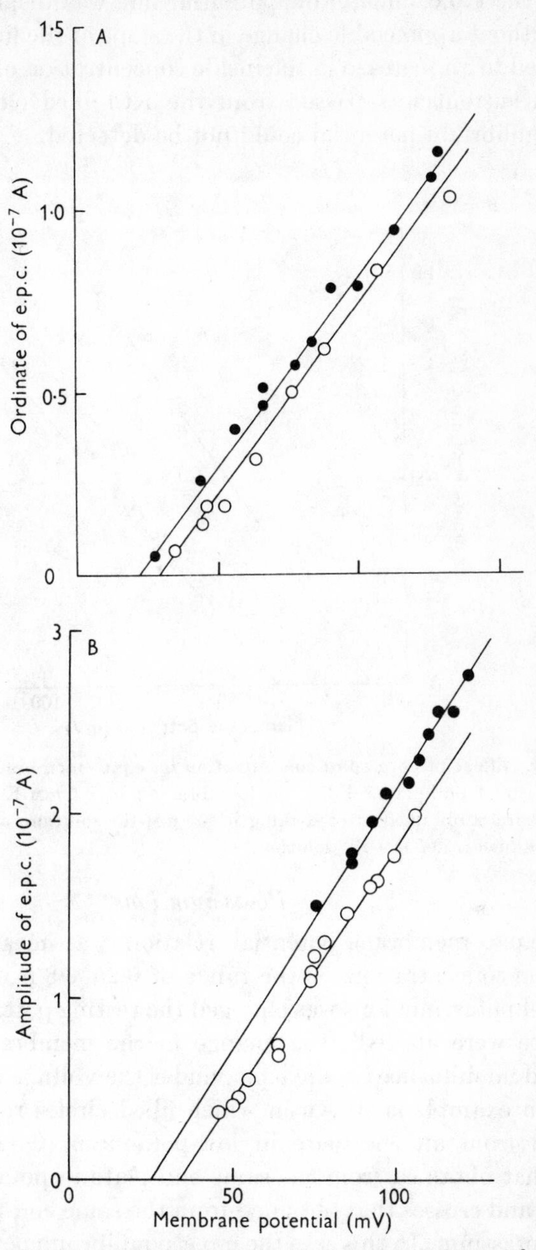

Fig. 5. Effect of sodium injection into muscle fibre at end-plate on the e.p.c.–membrane-potential relation obtained from two different end-plates soaked in Na-deficient Ringer's solution. External Na concentration was 33·6 mM. Filled circles are obtained from end-plates before injection. Open circles in *A* are obtained after electrophoretical injection of sodium ion 5×10^{-8} A for 7 min into muscle fibre, and those in *B* are obtained after injection of $4{\cdot}7 \times 10^{-8}$ A for 11 min.

current the e.p.c.–membrane-potential line was displaced slightly to the right, without appreciable change in the slope of the line. This shift may be attributed to an increase in the inside concentration of sodium, since when outward current was passed from the KCl-filled electrode, the shift in e.p.c. equilibrium potential could not be detected.

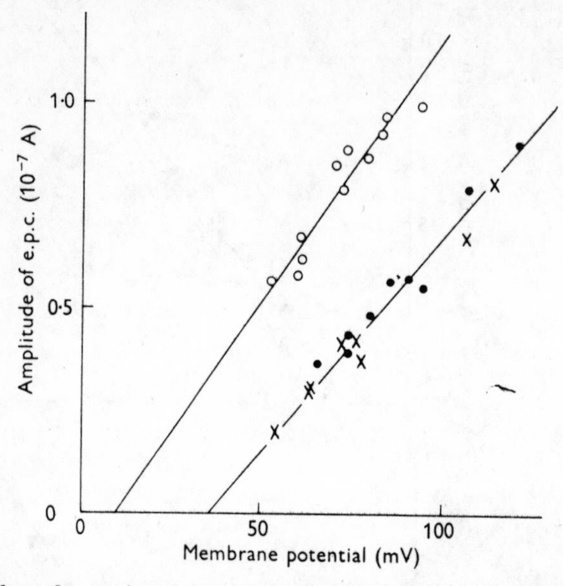

Fig. 6. Effect of potassium concentration on e.p.c.–membrane-potential relation obtained at end-plate. Filled circles obtained in 0·5 mM-K^+ Ringer's solution. Open circles obtained after soaking in 4·5 mM-K^+ solution, and crosses obtained after return to 0·5 mM-K^+ solution.

Potassium ions

The e.p.c.–membrane-potential relation was measured with various potassium concentrations in the range of 0·25–6·8 mM. When the concentration of potassium ions was changed the resting potential and membrane resistance were altered. The change in the membrane resistance, however, had no influence on the e.p.c. under the voltage clamp condition. In Fig. 6 an example is shown in which filled circles represent the relation obtained from an end-plate in low-potassium (0·5 mM) medium, open circles that obtained from the same end-plate in potassium-rich (4·5 mM) medium and crosses that obtained from the same end-plate after returning to low potassium. In this case the e.p.c. equilibrium potential shifted about 28 mV. When the solution was returned to 0·5 mM potassium, the e.p.c. equilibrium potential agreed closely with the initial value. The e.p.c. equilibrium potentials obtained in various potassium concentrations are presented in Fig. 7.

Chloride ions

In order to investigate whether chloride plays some role in the production of e.p.c., chloride was replaced by glutamate, which is probably too large to pass through the membrane, and the e.p.c. equilibrium potential was measured. The resting potential was initially reduced, but recovered gradually to the original value. This decrease in resting potential in low-chloride solutions may be due to the decrease in chloride equilibrium potential and the recovery of the resting potential may be due to restoration of the chloride equilibrium potential as the chloride concentration in

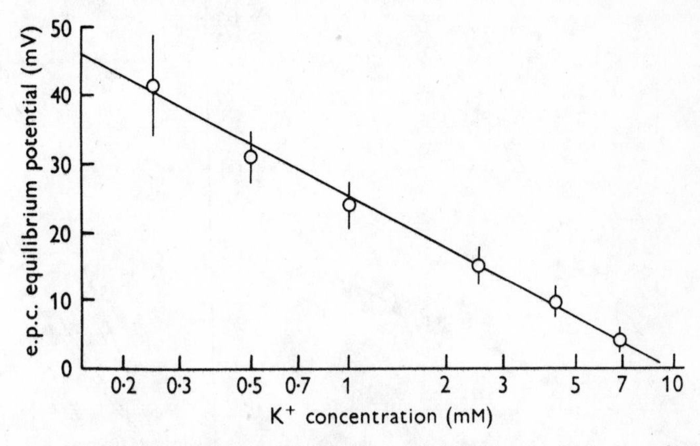

Fig. 7. E.p.c. equilibrium potential plotted against potassium concentration in outside solution on semilogarithmic scale (mean ± S.D.). The line is drawn according to $(1/2{\cdot}29)\{58\log_{10}(126/K_0) - 1{\cdot}29 \times 50\}$ mV.

the muscle fibre decreases. After the muscle is equilibrated with low-chloride solution, and then returned to normal solution, the chloride potential suddenly increases and the resting potential is intermediate between the chloride and potassium potentials. The resting potential then returns to its normal value, as the chloride concentration in the muscle fibre increases (Hodgkin & Horowicz, 1959*b*). In Fig. 8 *A* filled circles represent the relationship in normal solution, while open circles and crosses represent points obtained from the same end-plate 2–3 min and 15 min, respectively, after soaking in glutamate solution. When the outside solution was replaced by glutamate solution the resting potential decreased from 90 to 77 mV, followed by recovery to the original value (86 mV) in about 10 min, indicating a change in the chloride potential; however all lines of the e.p.c.–membrane-potential relation cross the abscissa at about the same point. Similar results were obtained in the muscle equilibrated with low-chloride solution. Open circles in Fig. 8 *B* represent the relationship obtained from an end-plate soaked in glutamate solution for about 30 min, filled circles are those obtained from the same end-plate 2–3 min after returning to normal solution and crosses are those 10 min after returning to normal solution. In this case the resting potential was increased from 69 to 86 mV by bringing the preparation back to normal solution and returned to 77 mV in 10 min, but little or no change in the e.p.c. equilibrium potential was observed. These results suggest that chloride ions contribute

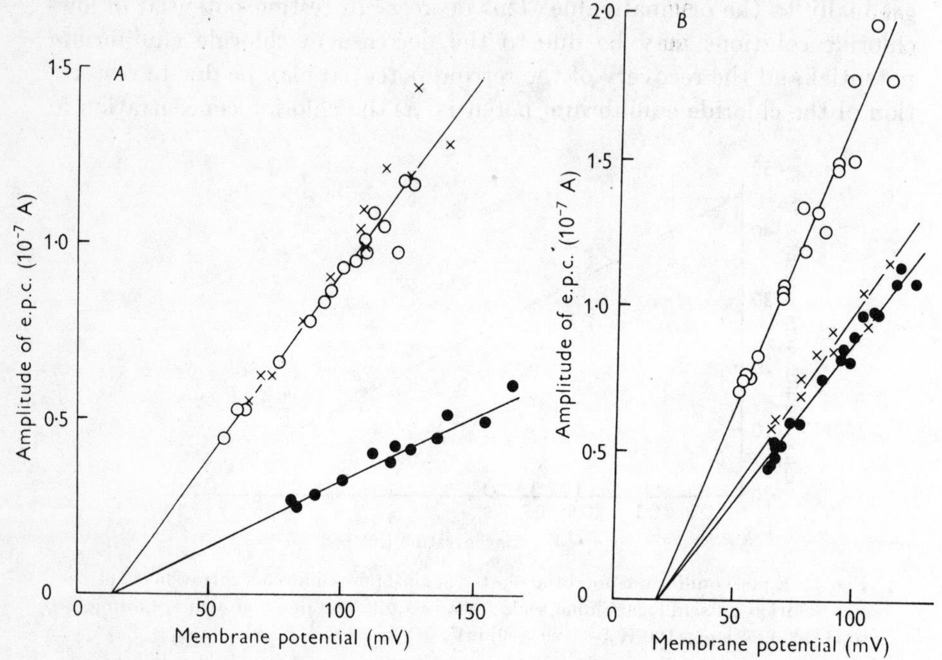

Fig. 8. Effect of chloride concentration on e.p.c.–membrane-potential relation. *A*, filled circles obtained in normal Ringer's solution, open circles show potential after soaking in glutamate Ringer's solution 2–3 min, and crosses after soaking in glutamate solution for 15 min. *B*, open circles obtained from an end-plate equilibrated in glutamate solution, filled circles 2–3 min after bathing in normal solution, crosses 10 min after addition of normal solution. When using glutamate, the concentration of tubocurarine was reduced to $\frac{2}{3}$ or $\frac{3}{4}$ of that in the chloride Ringer, and the calcium concentration increased to 5 mM.

much less to the production of the e.p.c. than do sodium and potassium ions. When chloride ion was replaced by nitrate or sulphate ion, similar results were obtained, i.e. varying the concentrations of these anions had no effect on the e.p.c. equilibrium potential.

DISCUSSION

The potential at which the e.p.c.–membrane-potential line crosses the membrane potential axis is the potential at which the ionic fluxes caused by the transmitter convey no net charge across the membrane. In normal

Ringer's solution this membrane potential was about 15 mV (inside negative). Possible mechanisms of this phenomenon are as follows (del Castillo & Katz, 1956; Katz, 1958): (*a*) the transmitter causes the end-plate membrane to develop a non-selective permeability to free ions on both sides of membrane and there remains the junctional potential between physiological solution and myoplasm (Fatt & Katz, 1951; del Castillo & Katz, 1954), (*b*) the membrane becomes permeable to more than one ion species, e.g. sodium and potassium or sodium and chloride (del Castillo & Katz, 1954) or sodium, potassium, chloride and probably calcium (Nastuk, 1959) and the sum of the ionic fluxes across the membrane becomes zero at −15 mV. In the present experiments the contribution of sodium, potassium and chloride ions to the e.p.c. was tested. The method adopted was to estimate the e.p.c. equilibrium potential by extrapolating the linear relation between e.p.c. amplitude and membrane potential in various solutions. Although this method may have rather large errors because of the extrapolation, a shift of the e.p.c. equilibrium potential could be obtained when the outside concentration of sodium or of potassium was changed. In contrast, no appreciable change could be detected when chloride was replaced by glutamate. When the muscle was soaked in glutamate solution for a long period, the resting potential tended to increase. Although this may suggest that glutamate has some influence on the membrane, e.g. the accumulation of potassium in the cell (Krebs & Eggleston, 1949; Davies & Krebs, 1952), immediately after soaking the preparation in glutamate solution the resting potential decreased, followed by gradual recovery. Thus, immediately after its application, glutamate may be considered as an inert substitute for chloride. In crustacean muscle it is known that glutamate elicits muscle contraction (van Harreveld, 1959) and has a specific action on the muscle membrane at the neuromuscular junction, causing at first depolarization and then desensitization of the membrane to transmitter (van Harreveld & Mendelson, 1959). In frog muscle, however, glutamate had no excitatory action in concentrations as low as that used in crustacean muscle, and the contraction which did occur when most of chloride was substituted by glutamate may be explained by a decrease in chloride equilibrium potential and in calcium concentration. If glutamate neither changes the permeability of the end-plate membrane to other ions nor passes through the end-plate membrane during the action of the transmitter (as is postulated in inhibitory synapses, cf. Shanes (1958)), the present results indicate that as a first approximation the transmitter increases the permeability of the end-plate membrane to sodium and potassium ions but not to chloride ion. There may be other ions to which the end-plate becomes permeable during the action of transmitter, e.g. there is some evidence that, at least in special conditions, permeability of end-plate membrane to calcium may be increased by action of ACh (Takeuchi & Takeuchi, unpublished). However, in physiological conditions the e.p.c. may be composed of sodium and potassium fluxes and other ions may play only a small role in producing the e.p.c.

The electrical behaviour of the end-plate membrane may be represented approximately by the network shown in Fig. 9. Current can be carried through the membrane by movement of ions through the resistances in the end-plate. The ionic current is divided into components carried by sodium and potassium ions. In a previous report (Takeuchi & Takeuchi, 1959) the electrical behaviour of the end-plate membrane was represented by a series

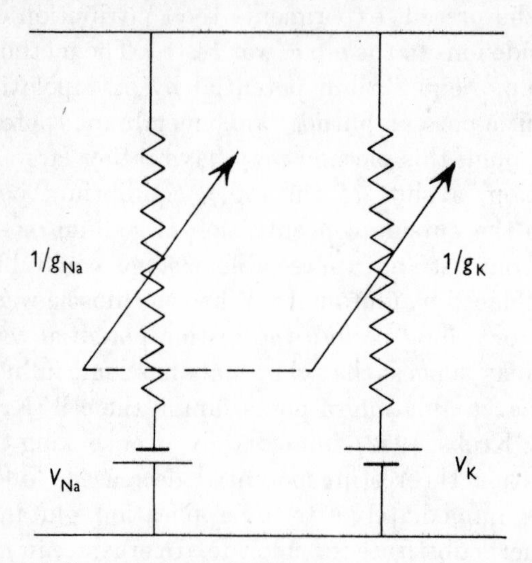

Fig. 9. Electrical diagram of end-plate membrane. For explanation see text.

resistance $R(t)$ and an e.m.f. E where $R(t)$ is a function of time and E is a constant voltage independent of time. Comparing this scheme with that presented in the present report, from the condition that E is independent of time it can be derived that the time course of the changes of resistance in the sodium and potassium channels is the same as that of $R(t)$, i.e. although the resistances in the sodium and potassium channels varied, the ratio of the resistances remains constant.

The results in the present experiments were obtained in the condition that $dV/dt = 0$ during the action of transmitter, where V is membrane potential. Thus the change in the characteristics of the membrane away from the end-plate and the contribution of the membrane capacity may be neglected, since only the change in the end-plate membrane contributes to the change in current. The current carried by potassium and sodium may be simply represented by a conductance and a driving force in each channel.

When the transmitter arrived at the end-plate, the conductances of the sodium and potassium channels increased from their resting values, g°_{Na} and g°_{K}, giving rise to the e.p.c. Thus the e.p.c. may be represented by the following equations:

$$I_{e.p.c.} = \Delta g_{Na}(V - V_{Na}) + \Delta g_K(V - V_K), \quad (1)$$
$$\Delta g_{Na} = g_{Na} - g^{\circ}_{Na},$$
$$\Delta g_K = g_K - g^{\circ}_K,$$

where V_{Na} and V_K are equilibrium potential for sodium and potassium, respectively. On rearrangement

$$I_{e.p.c.} = (\Delta g_{Na} + \Delta g_K)\left(V - \frac{V_K + (\Delta g_{Na}/\Delta g_K)\,V_{Na}}{1 + (\Delta g_{Na}/\Delta g_K)}\right). \quad (2)$$

$(\Delta g_{Na} + \Delta g_K)$ corresponds to $1/R(t)$ of previous notation and

$$\left(V_K + \frac{\Delta g_{Na}}{\Delta g_K} V_{Na}\right)\left(1 + \frac{\Delta g_{Na}}{\Delta g_K}\right)^{-1}$$

is e.p.c. equilibrium potential. If the e.p.c. equilibrium potential in normal Ringer's solution is assumed to be -15 mV and V_K and V_{Na} are taken as -99 and $+50$ mV, respectively (assuming internal concentration of potassium and sodium ions as 126 and 15·5 mM, respectively, from the data of Boyle & Conway (1941)), $\Delta g_{Na}/\Delta g_K$ is 1·29.

If eqn. 2 represents the characteristics of the end-plate response to the transmitter, the equation should predict the changes in e.p.c. equilibrium potential in various ionic conditions. In Figs. 4 and 7 lines are drawn according to the eqn. 2, assuming the ratio $\Delta g_{Na}/\Delta g_K$ and the internal concentrations of potassium and sodium are unchanged by altering the external concentrations. Although the experimental data have rather large variations, the observed values agree roughly with the predicted lines. Possible changes in inside concentrations of ions were minimized by soaking the preparation in test solutions for rather short periods (usually 10–20 min). Although the inside concentrations of sodium and potassium were taken from the data of Boyle & Conway (1941) and it is uncertain whether the values are applicable to the present conditions, the coincidence of the values obtained experimentally with those obtained from eqn. 2 support the above assumption. The assumption indicates that at the membrane potential where $V = V_K$, e.p.c. may be composed of inward sodium current, at a membrane potential $V_{Na} > V > V_K$ e.p.c. may be made of inward sodium and outward potassium fluxes, and at the e.p.c. equilibrium potential the sodium and potassium currents are equal and opposite in direction.

If the transmitter opens new channels to sodium and potassium which are closed in the absence of the transmitter, the e.p.c. equilibrium potential is the potential at which net current through the end-plate is zero. The equation of Goldman (1943), which has been successfully applied to the resting potential (Hodgkin, 1951; Jenerick, 1953) might be applied to the calculation of the e.p.c. equilibrium potential. But this equation could not describe the shift of e.p.c. equilibrium potential in various ionic conditions, if P_{Na}/P_K was taken as constant. This may suggest that e.p.c. equilibrium potential does not satisfy the conditions for which the equation was derived.

In conclusion it may be said from the present results that the characteristics of the conductance change produced by transmitter at the end-plate membrane are (1) the sodium and potassium conductances increase, with the ratio $\Delta g_{Na}/\Delta g_K$ constant, (2) the amount of the conductance change is independent of the membrane potential, and (3) curarine changes the conductance ($\Delta g_{Na}+\Delta g_K$), keeping the ratio $\Delta g_{Na}/\Delta g_K$ constant.

SUMMARY

The relationship between the amplitude of end-plate current (e.p.c.) and the membrane potential was investigated under various ionic conditions.

1. In normal Ringer's solution when the e.p.c.–membrane-potential line was extrapolated, the line cut the voltage axis at the point 10–20 mV (inside negative). This point (e.p.c. equilibrium potential) was not changed by altering the curarine concentration.

2. In sodium-deficient solution e.p.c. equilibrium potential became more negative.

3. In potassium-deficient solution the e.p.c. equilibrium potential became more negative.

4. When chloride was replaced by glutamate, no appreciable change in e.p.c. equilibrium potential could be detected, although the resting potential was changed.

5. It may be concluded that, as a first approximation, the transmitter makes the end-plate more permeable to sodium and potassium, but not to chloride.

We wish to express our sincere thanks to Drs C. C. Hunt, A. R. Martin, and C. Edwards for their constant advice and criticism.

REFERENCES

Adrian, R. H. (1956). The effect of internal and external potassium concentration on the membrane potential of frog muscle. *J. Physiol.* **133**, 631–658.

Boistel, J. & Fatt, P. (1958). Membrane permeability change during inhibitory transmitter action in crustacean muscle. *J. Physiol.* **144**, 176–191.

Boyle, P. J. & Conway, E. J. (1941). Potassium accumulation in muscle and associated changes. *J. Physiol.* **100**, 1–63.

Burgen, A. S. V. & Terroux, K. G. (1953). On the negative inotropic effect in the cat's auricle. *J. Physiol.* **120**, 449–464.

COOMBS, J. S., ECCLES, J. C. & FATT, P. (1955). The specific ionic conductances and the ionic movements across the motoneuronal membrane that produce the inhibitory post-synaptic potential. *J. Physiol.* **130**, 326–373.

DAVIES, R. E. & KREBS, H. A. (1952). Biochemical aspects of the transport of ions by nervous tissue. *Biochem. Soc. Symp.* **8**, 77–92.

DEL CASTILLO, J. & KATZ, B. (1954). The membrane change produced by the neuromuscular transmitter. *J. Physiol.* **125**, 546–565.

DEL CASTILLO, J. & KATZ, B. (1955). Local activity at a depolarized nerve-muscle junction. *J. Physiol.* **128**, 396–411.

DEL CASTILLO, J. & KATZ, B. (1956). Biophysical aspects of neuro-muscular transmission. *Progr. Biophys.* **6**, 21–170.

DODGE, F. A. & FRANKENHAEUSER, B. (1959). Sodium currents in the myelinated nerve fibre of *Xenopus laevis* investigated with the voltage clamp technique. *J. Physiol.* **148**, 188–200.

EDWARDS, C. & HAGIWARA, S. (1959). Potassium ions and the inhibitory process in the crayfish stretch receptor. *J. gen. Physiol.* **43**, 315–321.

FATT, P. & KATZ, B. (1951). An analysis of the end-plate potential recorded with an intra-cellular electrode. *J. Physiol.* **115**, 320–370.

FATT, P. & KATZ, B. (1952). The effect of sodium ions on neuromuscular transmission. *J. Physiol.* **118**, 73–87.

FATT, P. & KATZ, B. (1953). The effect of inhibitory nerve impulses on a crustacean muscle fibre. *J. Physiol.* **121**, 374–389.

GIEBISCH, G., KRAUPP, O., PILLAT, B. & STORMANN, H. (1957). Der Ersatz von extra-cellulärem Natriumchlorid durch Natriumsulfat bzw. Saccharose und seine Wirkung auf die isoliert durchströmte Säugetiermuskulatur. *Pflüg. Arch. ges. Physiol.* **265**, 220–236.

GOLDMAN, D. E. (1943). Potential, impedance and rectification in membranes. *J. gen. Physiol.* **27**, 37–60.

HODGKIN, A. L. (1951). The ionic basis of electrical activity in nerve and muscle. *Biol. Rev.* **26**, 339–409.

HODGKIN, A. L. & HOROWICZ, P. (1959*a*). Movements of Na and K in single muscle fibres. *J. Physiol.* **145**, 405–432.

HODGKIN, A. L. & HOROWICZ, P. (1959*b*). The influence of potassium and chloride ions on the membrane potential of single muscle fibres. *J. Physiol.* **148**, 127–160.

HODGKIN, A. L. & HUXLEY, A. F. (1952). Currents carried by sodium and potassium ions through the membrane of the giant axon of *Loligo*. *J. Physiol.* **116**, 449–472.

HODGKIN, A. L., HUXLEY, A. F. & KATZ, B. (1952). Measurement of current-voltage relations in the membrane of the giant axon of *Loligo*. *J. Physiol.* **116**, 424–448.

HUTTER, O. F. & TRAUTWEIN, W. (1956). Vagal and sympathetic effects on the pacemaker fibres in the sinus venosus of the heart. *J. gen. Physiol.* **39**, 715–733.

JENERICK, H. P. (1953). Muscle membrane potential, resistance and external potassium chloride. *J. cell. comp. Physiol.* **42**, 427–448.

KATZ, B. (1958). Microphysiology of the neuromuscular junction. *Johns Hopkins Hosp. Bull.* **102**, 275–312.

KREBS, H. A. & EGGLESTON, L. V. (1949). An effect of L-glutamate on the loss of potassium ions by brain slices suspended in a saline medium. *Biochem. J.* **44**, vii.

NASTUK, W. L. (1959). Some ionic factors that influence the action of acetylcholine at the muscle end-plate membrane. *Ann. N.Y. Acad. Sci.* **81**, 317–327.

SHANES, A. M. (1958). Electrochemical aspects of physiological and pharmacological action in excitable cells. Part I. The resting cell and its alteration by extrinsic factors. *Pharmacol. Rev.* **10**, 59–164.

TAKEUCHI, A. & TAKEUCHI, N. (1959). Active phase of frog's end-plate potential. *J. Neurophysiol.* **22**, 395–411.

TRAUTWEIN, W. & DUDEL, J. (1958). Zum Mechanismus der Membranwirkung des Acetyl-cholin an der Herzmuskelfaser. *Pflüg. Arch. ges. Physiol.* **266**, 324–334.

VAN HARREVELD, A. (1959). Compounds in brain extracts causing spreading depression of cerebral cortical activity and contraction of crustacean muscle. *J. Neurochem.* **3**, 300–315.

VAN HARREVELD, A. & MENDELSON, M. (1959). Glutamate-induced contractions in crustacean muscle. *J. cell. comp. Physiol.* **54**, 85–94.

27

J. Physiol. (1959) 147, 178–193

A STUDY OF SUPERSENSITIVITY IN DENERVATED MAMMALIAN SKELETAL MUSCLE

By J. AXELSSON and S. THESLEFF

From the Department of Pharmacology, University of Lund, Sweden

(*Received* 26 *January* 1959)

It is well known that organs chronically deprived of their motor nerves develop an increased sensitivity to the neurohumoral transmitter and to other chemical agents. This phenomenon is observed in several types of tissue, e.g. striated muscle, smooth muscle, ganglia and glands, and is known as 'denervation supersensitivity' (Cannon & Rosenblueth, 1949). The mechanism underlying the change has been extensively studied and has been the subject of much speculation, but it has, up to now, remained unknown (Emmelin, 1952).

Supersensitivity is particularly marked in chronically denervated skeletal muscle. Denervated frog muscle is overexcitable to acetylcholine (ACh) by a factor of about one hundred (Nicholls, 1956), and denervated mammalian muscle (cat gastrocnemius) becomes one thousand times more excitable to ACh administered by close arterial injection (Brown, 1937).

The sensitivity in striated muscle to ACh and other chemicals is increased far more than that to electric stimulation, and cannot be explained by changes in the electric properties of the muscle membrane (Nicholls, 1956). It is consequently reasonable to suppose that supersensitivity following denervation results from an increase in the chemical excitability of the muscle, with an enhanced responsiveness of the membrane to the depolarizing action of ACh.

Recent electrical methods for close-range, localized drug application (Nastuk, 1953) are particularly suited for studying drug effects on small and limited parts of the muscle membrane (del Castillo & Katz, 1955*a*). Iontophoretic micro-application of ACh to single muscle fibres is for that reason likely to offer a possibility for investigating the distribution and sensitivity of ACh receptors in denervated muscle. In the present study we have used the technique of localized ACh application with the chronically denervated tenuissimus muscle of the cat. It will be shown that after denervation the entire muscle membrane becomes as sensitive to ACh as the end-plate region, which maintains its original responsiveness to the drug. A preliminary report of some of the observations has been given (Axelsson & Thesleff, 1959).

METHODS

Unless otherwise stated, the experiments were made on the isolated tenuissimus muscle of the cat. This muscle has the advantage of being covered by only a thin layer of connective tissue and can be kept for a long time in oxygenated Tyrode fluid without showing signs of deterioration (Boyd & Martin, 1956*a*); for details of its anatomy see Adrian (1925).

For denervation and removal of the muscle the cats were anaesthetized with Nembutal (pentobarbitone; Abbott Laboratories) 40 mg/kg body wt., injected intraperitoneally. Denervation was carried out 2–21 days before the experiment by the removal of 1 cm of the motor nerve close to its point of entry into the muscle. Generally only the muscle on one side was denervated, while the other served as control. Denervated and control muscles were always examined on the same day.

During the experiment the muscle was mounted around a Perspex rod in a constant-temperature bath kept at 35–37° C. The bathing fluid had the composition used by Liley (1956), and it was oxygenated by bubbling a O_2 95% + CO_2 5% mixture through it immediately before its introduction into the muscle bath. The bath held about 30 ml. of solution, which was changed continuously at the rate of about 500 ml./hr.

The drugs used in the study were released iontophoretically from the tip of a micro-pipette as described by del Castillo & Katz (1955*a*). Drug pipettes had a tip diameter of less than 0·5 μ, and the discharge of the drug was regulated by 'breaking' or 'releasing' voltages (making the interior of the pipette more negative, or positive, respectively). The sensitivity of the muscle membrane to ACh was tested by applying positive voltage pulses of constant intensity and duration (1–10 msec) to the pipette. When the tip of the pipette was close to the receptor structure, the ACh released by a current pulse produced a transient membrane depolarization of a few millivolts amplitude. The position of the tip of the pipette was then adjusted, until the ACh potential reached its shortest rising time and highest peak amplitude. In an innervated muscle the end-plate region was located by pursuing fine superficial nerve twigs and by the recording of miniature end-plate potentials with a rapid time course.

The resting membrane potential and membrane potential changes were recorded with conventional capillary micro-electrodes, as described by Fatt & Katz (1951). The tip of the micro-electrode was inserted into the muscle fibre within 0·1 mm of the point of drug application. The time constant with a 10 MΩ electrode was of the order of 50 μsec. The current passing through the drug pipette was in all instances monitored and recorded on the second beam of the oscilloscope.

The drug pipette and the micro-electrode were carried on a Zeiss slide micromanipulator and their position was viewed with a binocular dissecting microscope at 80 × magnification.

The drugs employed in the study were: acetylcholine and edrophonium chloride (Hoffmann-La Roche), carbachol and tubocurarine chloride (Burroughs Wellcome).

RESULTS

Sensitivity to ACh

In an innervated muscle ACh produces membrane depolarization only when applied to the end-plate region of a muscle fibre (Figs. 1 and 2). The area sensitive to ACh is small and covers not more than 0·1 mm of the fibre length. Within this limited part of the membrane an ACh pulse of 1–10 msec duration produces a transient membrane depolarization, which at critical 'spots' may have a rising phase lasting about 10 msec, as is shown in the lower record of Fig. 3. At such a 'spot', however, the position of the pipette is critical and small movements horizontally or vertically cause the potential to decline to a small fraction and the rising phase to lengthen several times. In a muscle with intact innervation it is not possible by microapplication of ACh to produce membrane depolarization outside the end-plate area (Figs. 1 and 2).

In chronically denervated muscles, however, the entire muscle membrane is sensitive to applied ACh (Fig. 1). Four to five days after denervation, locally released ACh produces depolarization with a slow time course, outside the former end-plate region. After one week of denervation and subsequently throughout our observation period (21 days) the whole surface of the muscle is

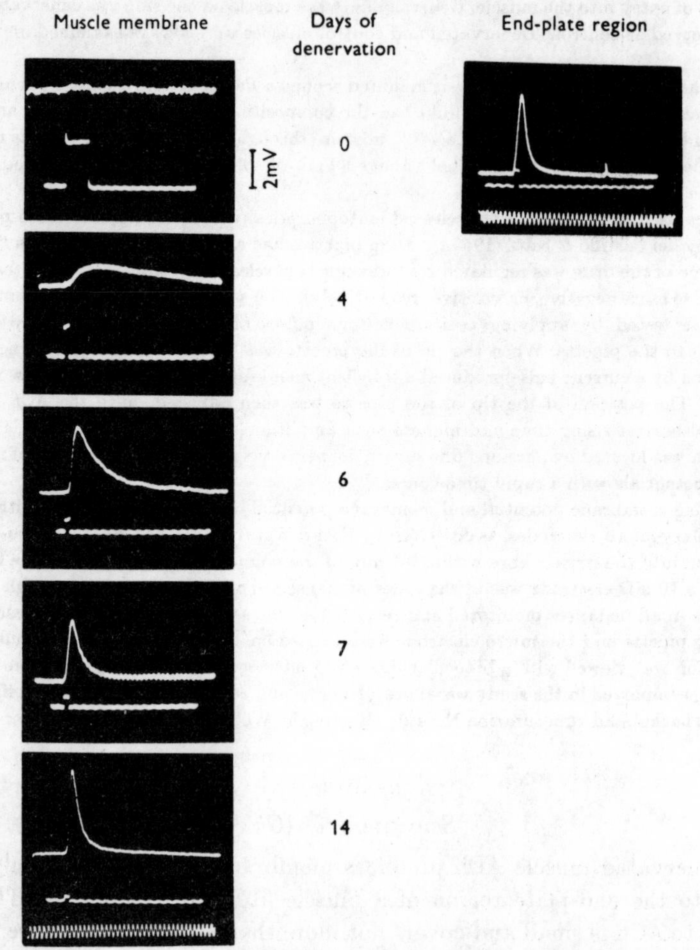

Fig. 1. An illustration of the sensitivity of innervated and denervated muscle fibres to iontophoretically applied ACh. The membrane potential of the fibre is recorded in the upper tracing and the current passing through the pipette in the lower tracing of each record. In an innervated muscle ACh produces membrane depolarization only when applied to the end-plate region (upper right record) but has no effect when released outside this area of the membrane (upper left record). After denervation the whole muscle becomes sensitive to ACh, which wherever applied to the membrane causes a potential change (subsequent records). Time marker, 100 c/s.

uniformly sensitive to ACh. As an example of this, it can be mentioned that in twelve muscles, denervated 1–3 weeks previously, there was among more than two hundred tested spots not a single one that was insensitive to applied ACh. The uniformity of ACh sensitivity in denervated muscles made only vertical movements of the pipette critical for the amplitude and time course of an ACh potential. This is illustrated in Fig. 2, where the sensitivity of a muscle fibre to ACh was determined by moving the tip of the pipette close to the fibre membrane at points separated by distances of 0·5 mm. In a 14-days

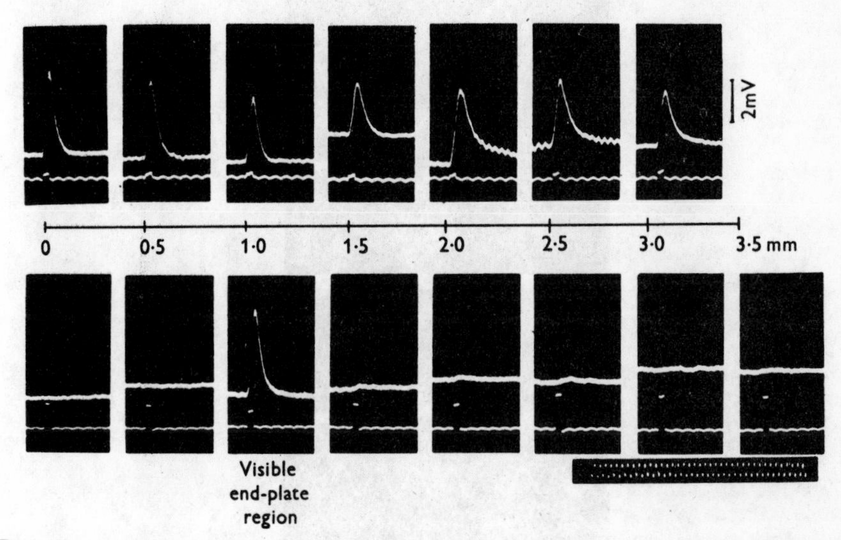

Fig. 2. The sensitivity of a muscle fibre to locally applied ACh was tested by moving the tip of an ACh pipette close to the muscle membrane at points separated by distances of about 0·5 mm. In an innervated fibre (lower records) only the visible end-plate region is sensitive to ACh. In a 14-days denervated fibre a constant pulse of ACh produced at each point of the membrane a potential change, as is shown in upper records. To obtain a maximum response only vertical movements of the drug pipette were used. Time marker, 100 c/s.

denervated muscle (upper record) an ACh pulse of constant intensity and duration produced a membrane depolarization of about similar amplitude and time course each time the pipette approached the membrane, while in an innervated fibre (lower record) ACh was effective only when applied at a visible end-plate region. Because of damage caused to the muscle membrane by the insertion of the recording electrode, only about six points were tested in each muscle fibre.

The time courses of an ACh potential at a sensitive 'spot' in an innervated end-plate region and that observed anywhere in a 2-weeks denervated muscle were similar, as is shown in Fig. 3. Time to peak and amplitude of an ACh potential are related to the distance between the tip of the pipette and the receptor site in a way depending on the principles of diffusion (del Castillo &

Katz, 1955*a*). The formula given by del Castillo & Katz suggests that for ACh potentials with a rising phase of 10 msec duration (Fig. 3), the distance between source and receptor is about 10 μ, and that consequently the spacing of receptors in a denervated muscle membrane corresponds to that of an innervated end-plate. Another method of estimating the 'sensitivity' of ACh receptors is to calculate the number of coulombs responsible for the

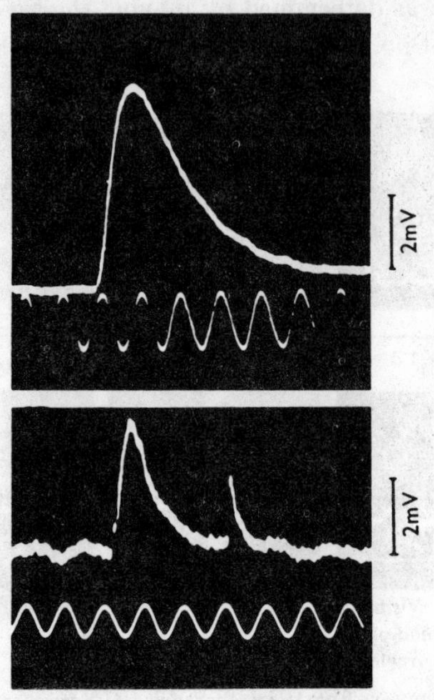

Fig. 3. The time course of an ACh potential produced by a current pulse of 1 msec duration at a sensitive 'spot' in an innervated end-plate (lower record) and anywhere at the membrane of a 14-days denervated muscle fibre (upper record). Note a miniature end-plate potential after the ACh pulse in the lower record. Time marker, 100 c/s.

TABLE 1. The sensitivity of ACh receptors was estimated by calculating the numbers of coulombs required for the release of the quantity of ACh^+ that produced a membrane depolarization of one mV amplitude. The experiments were made on tenuissimus muscles after 7 and 14 days denervation and on the innervated end-plates of corresponding control muscles. The same batch of micropipettes was used in all tests. S.E.'s of the means are shown

Days of denervation	No. of fibres	C/mV
0 (end-plate region)	29	$1{\cdot}4\ (\pm 0{\cdot}13) \times 10^{-10}$
7	16	$1{\cdot}7\ (\pm 0{\cdot}24) \times 10^{-10}$
14	9	$6{\cdot}6\ (\pm 0{\cdot}73) \times 10^{-11}$

release of ACh ions from the tip of the micro-pipette (del Castillo & Katz. 1957). This was done in experiments such as those shown in Fig. 1, and the results are given in Table 1, expressed as C/mV amplitude of the resulting ACh potential. The table shows that the quantity of ACh^+ which produces a given depolarization in a denervated muscle is of the same order of magnitude as that which causes a similar potential change at the end-plate of an innervated fibre.

A few days after denervation miniature end-plate potentials disappear and it is impossible to locate the former end-plate with certainty. However, one may observe degenerating nerve twigs and thereby obtain an idea about the approximate site of this region. In these instances we found that the supposed end-plate had a sensitivity to ACh which corresponded to that of the end-plate region in an innervated fibre. In muscles denervated for 7–21 days no particular region of the membrane was more sensitive to ACh than other parts. Consequently, it is unlikely that the end-plate, after denervation, becomes supersensitive to ACh. Presumably it retains its original responsiveness to ACh throughout the denervation period and is not affected by the process which renders the rest of the cell membrane sensitive to the drug.

Determinations have also been made of the sensitivity of innervated and chronically denervated tenuissimus muscles to more conventional forms of drug application. In these experiments the drug was either applied to the bathing fluid in a constant-temperature bath or administered by close arterial injection. To avoid errors caused by enzymotic and spontaneous hydrolyses of ACh, carbachol was used. The threshold was taken as the concentration or dose of the salt which produced a visible muscle twitch or in denervated muscles an isotonic contracture. With bath application, carbachol in an effective concentration produced an electrically silent contracture in denervated muscles and in innervated muscles fibrillary muscle twitches without a change in tension. In innervated muscles the threshold concentration was of the order of 10^{-5} (w/v), and in muscles denervated for 2 weeks 10^{-6} (w/v). Close arterial injections were made in a volume of 0·2 ml. and corresponding threshold doses were 5 and 0·05 μg.

In addition to the tenuissimus muscle of the cat, the sensitivity of chronically denervated muscles to iontophoretically applied ACh has been studied in the hemidiaphragm and in the extensor digitorum longus muscles of the rat. The results were qualitatively similar to those obtained in the tenuissimus muscle. About 2 weeks after denervation, the entire muscle membrane, with the occasional exception of the very distal end of the muscle fibre, was sensitive to locally applied ACh. The rising phase of the ACh potentials was, however, somewhat slower in a denervated muscle of the rat than it was at the end-plate region of an innervated control muscle.

Properties of ACh receptors in denervated muscle

The length of a muscle fibre in the tenuissimus muscle of the cat is about 2 cm and its diameter 30 μ (Adrian, 1925; Boyd, 1956). Assuming that the ACh-sensitive region at the end-plate covers 0·1 mm of the fibre length, it is obvious that after denervation the size of the receptor area increases by a factor of at least 200. The enormous increase of the receptor area made it desirable to ascertain whether these new ACh receptors have properties identical with those at the end-plate.

At an innervated end-plate ACh raises the Na and K conductance of the membrane and possibly also its permeability to other ions. This general permeability increase tends to push the membrane potential, irrespective of its initial level, towards an equilibrium value of about 10–20 mV, negative inside (cf. del Castillo & Katz, 1954). We have determined the equilibrium

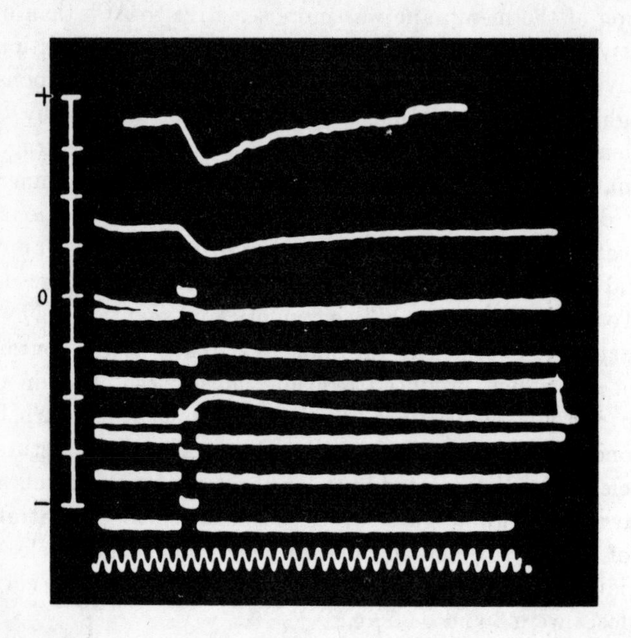

Fig. 4. The equilibrium potential during the action of ACh in a 10-days denervated muscle fibre The membrane potential of the fibre was progressively altered by passing rectangular current pulses through the membrane. At a membrane potential level of about – 10 mV the direction of an ACh pulse reversed. Mechanical movements were reduced by the use of Tyrode solution with a 2 × normal concentration of NaCl. Time marker, 100 c/s; voltage calibration 10 mV steps.

potential during the action of ACh in a 10-days-denervated muscle and established that it is similar to that at a normal end-plate. This is illustrated in Fig. 4, which shows a muscle fibre in which the membrane potential was progressively altered by passing rectangular current pulses through the membrane. (Mechanical responses were reduced by the use of a bathing fluid containing a 2 × normal concentration of NaCl, as described by Howarth, (1958). The reversal of the ACh potential occurred at a membrane potential level of about – 10 mV, which is about the same equilibrium potential as that observed at an innervated end-plate.)

As in the end-plate (del Castillo & Katz, 1955*b*), so also in the denervated muscle membrane, ACh raises the permeability to K^+ in the absence of external Na^+. This is shown in an experiment in which a denervated muscle was immersed in an isotonic solution of K_2SO_4 (Fig. 5). The membrane potential was displaced from zero by the passage of a transverse current and ACh was released from the tip of a micropipette. As a sign of increased membrane permeability the pulse of ACh reduced the potential difference built up across the membrane.

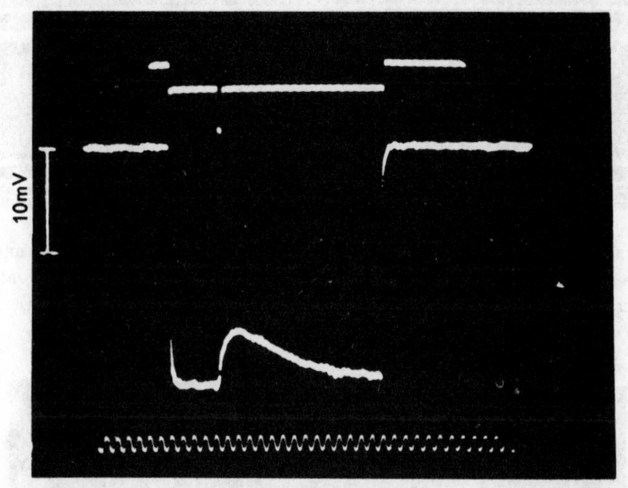

Fig. 5. ACh causes a reduction of the membrane resistance in a 7-days denervated muscle fibre immersed in isotonic K_2SO_4 solution. In this medium the cell membrane is completely depolarized but is electrotonically displaced from zero by the passing of a 'catelectrotonic' current pulse. When ACh is released its effect is to reduce the membrane resistance as is shown by the prompt decline in the electrotonic potential. Time marker, 100 c/s.

In the continued presence of ACh, end-plate receptors become refractory to the depolarizing action of the drug (Thesleff, 1955; Axelsson & Thesleff, 1958). This occurs also in a denervated muscle fibre but the time course of the 'desensitization' process is slower than at an innervated end-plate, as is shown in Fig. 6. The experiments illustrated in Fig. 6 were made with double-barrelled ACh pipettes, as described by Katz & Thesleff (1957*a*). Short 'test' pulses of ACh were released from one barrel while the other barrel was used for a prolonged and steady release of ACh. Receptor 'desensitization' is shown as a reduction in the amplitude of the test response during the continuous output of ACh.

The response of a chronically denervated muscle to some other drugs was found to be similar to that of an innervated end-plate. Carbachol released iontophoretically by a current pulse produced a transient depolarization with a rapid time course, and tubocurarine reversibly reduced the sensitivity of ACh receptors (Fig. 7).

On the other hand, with anticholinesterases a difference was found. When a drug such as edrophonium (Tensilon) is applied to the normal end-plate

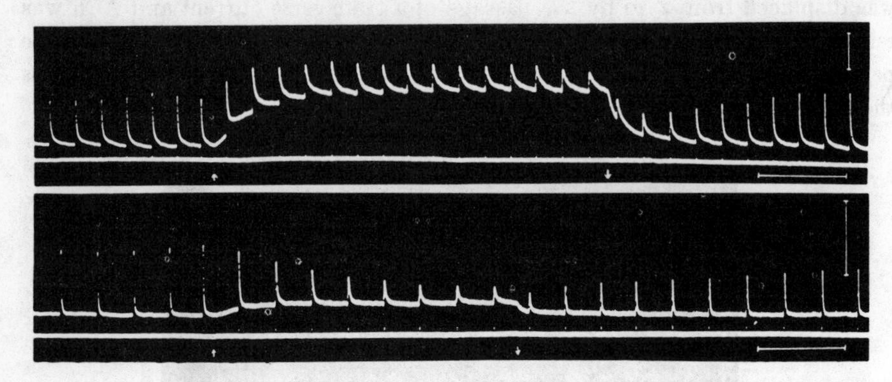

Fig. 6. 'Desensitization' of ACh receptors by the continuous release of ACh (for explanation see text). The lower record illustrates the time course of this process in an innervated end-plate region and the upper record in the membrane of a 14-days denervated muscle fibre. Time marker, 5 sec; voltage calibration 5 mV.

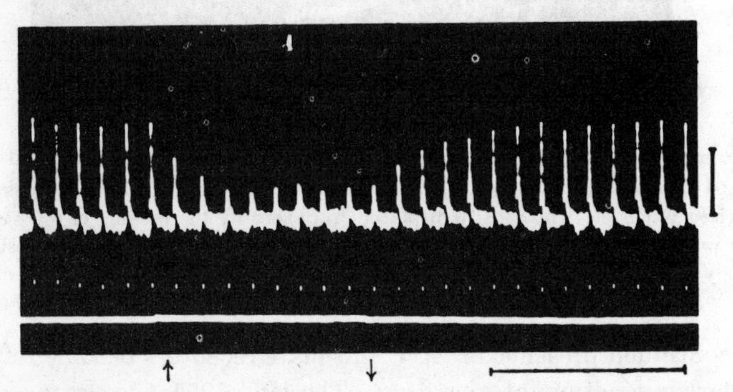

Fig. 7. Carbachol released by current pulses produces in a 14-days denervated muscle fibre transient depolarizations. When tubocurarine is released (between arrows) the sensitivity of membrane receptors is reversibly reduced, as is illustrated by the decline in amplitude of the carbachol pulses. Time marker, 5 sec; voltage calibration 2 mV.

region it potentiates the responses to ACh, as is shown in the lower record of Fig. 8. In a denervated muscle, however, its effect is to inhibit the response to ACh and to produce a slight membrane depolarization (upper record). The general view is that the main action of edrophonium is inhibition of cholinesterase (Smith, Cohen, Pelikan & Unna, 1952; Nastuk & Alexander, 1954; Katz & Thesleff, 1957*b*); hence the absence of ACh potentiation indicates that the membrane is, in a practical sense, devoid of cholinesterase.

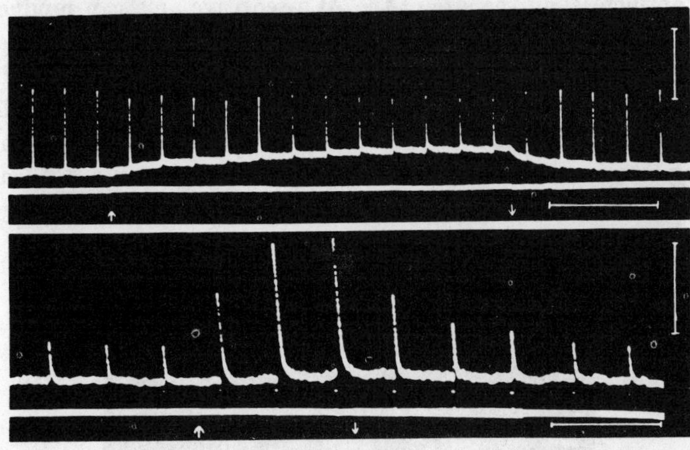

Fig. 8. The anticholinesterase drug edrophonium (Tensilon) potentiates the effect of ACh at the innervated end-plate region (lower record) and inhibits its action on the membrane of a 12-days denerv :ted fibre (upper record). The release of edrophonium is marked by arrows. Time marker, 5 sec; voltage calibration 5 mV.

Electrical excitability

It has been suggested that, unlike the rest of the muscle membrane, the ACh receptors are not excitable by electrical stimulation (del Castillo & Katz, 1954). Grundfest (1957), moreover, has concluded that electrical inexcitability is a common feature of chemically activated post-synaptic membranes. It was therefore of some interest to examine whether a denervated muscle would, despite its high and uniform sensitivity to ACh, respond by a conducted response to electrical stimulation. This was investigated in a number of experiments in which the electrical thresholds of single fibres were determined by the use of two intracellular electrodes, one for passing a stimulating electric current and the other for recording resulting potential changes. During an observation period of 3 weeks after denervation all the fibres were electrically excitable and a depolarizing current gave rise to a propagated action potential (Fig. 9). The critical membrane potential, i.e. the potential difference across the membrane at which excitation occurred, was unaltered by denervation, as is shown in Table 2.

DISCUSSION

The present investigation has shown that in a mammalian muscle about 4 days after denervation the size of the ACh-sensitive area starts to increase and 3–6 days later covers the larger part or all of the muscle membrane. Brown (1937) observed, in cat muscle, that supersensitivity to applied ACh begins 4 days after denervation and is fully developed within 1 week. The parallelism in time course of the two processes, i.e. the increase in the receptor surface and the supersensitivity, suggests that they are closely related.

That an increase in the size of the ACh-sensitive area can produce supersensitivity to the drug is evident when the electric properties of a muscle fibre are considered. Fig. 10 is a diagram of an electric model with cable conductor properties similar to those of a muscle fibre. A potential difference (e.g. the membrane potential) between *A* and *B* can be reduced by short-circuiting either a fraction, or all, of the parallel resistors r_m. According to Ohm's law the current divides itself in inverse proportion to the relative

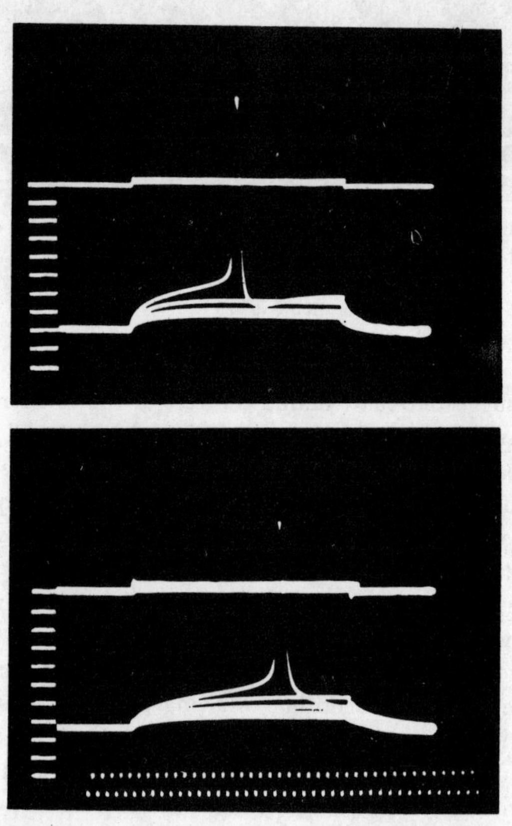

Fig. 9. A stimulating electric current (upper line in each tracing) causes membrane excitation, i.e. an action potential, in a 14-days denervated fibre (lower record). For comparison, the upper record shows a similar experiment in an innervated muscle. Time marker 1000 c/s; voltage calibration 10 mV steps.

TABLE 2. The potential difference across the membrane at which excitation occurs in fibres of the tenuissimus muscle. The S.E.'s of the means are shown

Days of denervation	No. of fibres	Critical membrane potential (mV)
0	34	$48 \pm 0{\cdot}6$
7	39	$47 \pm 0{\cdot}8$
14	62	$51 \pm 0{\cdot}6$

resistance, and it follows that to produce the same potential reduction between A and B the drop in resistance for a localized short circuit has to be greater than for a more generalized one.

In order to produce a depolarization at the innervated end-plate it is necessary to reduce the total effective resistance between the inside and the external medium. For this particular purpose this resistance can be considered as made up of two components in parallel, the resistance of the end-plate membrane and a shunt resistance due to structures outside the end-plate. The magnitude of the latter resistance is a function of the core conductor properties of the muscle fibre. Since normally only the end-plate membrane is affected by ACh it follows that a relatively larger reduction of the resistance in the end-plate membrane is needed in order to produce a given decrease of the total effective resistance.

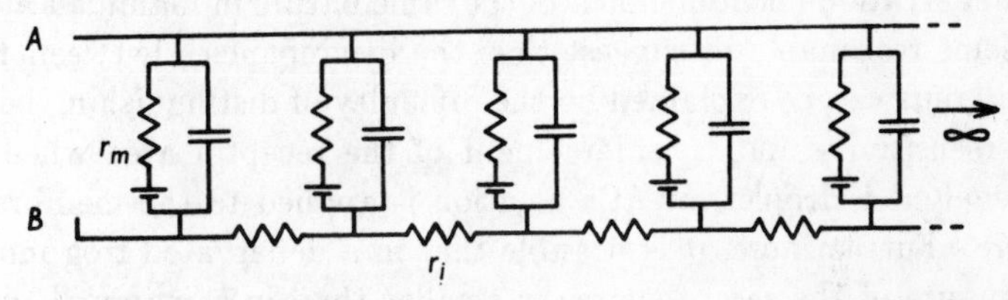

Fig. 10. A diagram of an electric model with cable conductor properties similar to those of a muscle fibre: r_m = membrane resistance; r_i = longitudinal resistance of myoplasm.

It is possible to calculate the degree of supersensitivity that can be expected from this mechanism if we know the membrane resistance, the area of end-plate membrane and the total effective resistance ($\frac{1}{2}/\sqrt{r_m \times r_i}$), where r_m = transverse resistance of the membrane times unit length; r_i = longitudinal resistance of the fibre per unit length). For example, if the total effective resistance is 0·5 MΩ and the resistance of the end-plate membrane 100 MΩ, it is necessary to reduce the resistance of the end-plate membrane to 2 MΩ in order to decrease the total effective resistance by 20%. On the other hand, in the denervated muscle fibre the same effect will result simply when the membrane resistance decreases by 20%. Hence, in this example a supersensitivity of 40 times would result.

The ratio of the decrease in the resistances discussed is an index of supersensitivity only when the dose–response relation of ACh is linear. This is, however, presumably not the case, since it has been shown that the relation between ACh dose and membrane depolarization has an 'S-shaped' rather than linear start (Katz & Thesleff, 1957*a*). Such a non-linear dose–response relation is likely to augment further the difference in sensitivity between innervated and denervated fibres. With the membrane constants obtained for the tenuissimus muscle (Boyd & Martin, 1956*b*) it can be estimated that the sensitivity of this muscle to ACh should, after denervation, increase by a factor of 10–500, depending upon whether 100 or 20 μ is taken as the diameter of the active end-plate membrane.

The possibility that denervation supersensitivity can be explained by an increase of the receptor surface has not previously received attention. The prevailing hypothesis has been that, after denervation, the sensitivity of ACh receptors at the end-plate increased, leaving the rest of the membrane relatively unaffected. This view originated from results obtained by Kuffler (1943). He applied droplets (0·3 μl.) of ACh solution to single frog muscle fibres, and observed that after denervation the sensitivity of the end-plate region to ACh increased 1000 to 100,000 times while that of other parts showed little change. In denervated mammalian muscles we have been unable to observe a selective supersensitivity of end-plate receptors to ACh, but we have not examined chronically denervated frog muscle. Assuming, however, that the denervation phenomenon is of the same nature in mammals and in the frog it seems reasonable to suggest that the discrepancies between Kuffler's results and ours can be explained by the difficulty of distinguishing between a true supersensitivity and an enlargement of the receptor area when a large volume, such as a droplet, of ACh solution is applied to the membrane of a single fibre. Furthermore, it is possible that in a denervated frog muscle the increase in size of the receptor area is smaller than in a mammalian muscle and probably never affects more than a part of the membrane.

It is obvious that when a large part or the whole of a muscle becomes uniformly sensitive to ACh the observed degree of supersensitivity will vary according to the mode of ACh application. Close arterial injections, for instance, cannot give a true estimate of supersensitivity, nor are results obtained by local and by bath application strictly comparable.

As previously mentioned, the observed extension of ACh sensitivity can account for a several hundred times increase in the responsiveness of a denervated mammalian muscle fibre to applied ACh. This process is therefore, in our opinion, an adequate explanation for what is known as denervation supersensitivity. It is, however, likely that other factors, e.g. a lack of cholinesterase and an increased electrical excitability, also contribute to the phenomenon. Whether our results explain denervation supersensitivity in other organs remains to be shown. Considering the close similarity of this process in various types of tissues it would, however, be rather surprising if the mechanism varied markedly.

Brown (1937), and subsequently Rosenblueth & Luco (1937), showed that close arterial injection of ACh in a chronically denervated mammalian muscle produced, after an initial burst of twitches, an electrically silent contracture. In an innervated muscle ACh caused muscle twitches but never an increase in tension. Observations made in the present investigation (see also Zaimis. 1954) offer an explanation for the contracture caused by ACh in denervated muscles. The spread of the receptor area to embrace the whole membrane will allow ACh to depolarize the entire length of a muscle fibre and thereby to produce a contracture unaccompanied by conducted responses. This type of contracture is thus similar to that elicited by direct current or by a high external potassium concentration.

It is of interest that the size of the receptor area in a muscle fibre is determined by external influences. Reduced muscular activity after denervation could be such an external factor, but this is rather improbable, since it has been shown that disuse atrophy of muscle is not in itself accompanied by supersensitivity. An increased sensitivity to the transmitter agent and other chemicals is only observed when there is a lack of a living nerve connexion (Eccles, 1941). Hence, there appears to be a true influence of the nerve on the muscle membrane (cf. nervous control of muscle differentiation, Buller, Eccles & Eccles, 1958). Such an influence is probably exerted by the transmitter agent or by some other chemical substance released from nerve terminals. This is a new and particularly challenging aspect, since it points to the possibility that presynaptic activity can control the size of the chemically sensitive region in post-synaptic membranes.

SUMMARY

1. The sensitivity to acetylcholine (ACh) of chronically denervated mammalian skeletal muscle was studied by iontophoretic micro-application of the agent to single muscle fibres. The membrane potential of the fibres was simultaneously recorded with an intracellular electrode inserted close to the point of ACh application.

2. In an innervated muscle ACh produces membrane depolarization only when applied to the end-plate region of a muscle fibre. The area sensitive to ACh is small and occupies only a very limited part of the muscle membrane.

3. In a denervated muscle, however, the entire muscle membrane becomes sensitive to ACh. One to two weeks after denervation the whole surface of the muscle is about as sensitive to ACh as the end-plate region, which has maintained its original responsiveness to the drug. The muscle is uniformly sensitive to ACh and wherever applied on the membrane ACh produces a depolarization of a similar magnitude and time course as when applied to the end-plate region of an innervated muscle.

4. In a denervated muscle, as at the normal end-plate, ACh causes a non-selective increase in the membrane permeability to Na and K and possibly also to other ions.

5. ACh receptors in a denervated muscle respond to drugs in a manner qualitatively similar to that of the innervated end-plate. Anticholinesterases, on the other hand, do not potentiate the effects of ACh, suggesting that the denervated muscle membrane is practically devoid of cholinesterase.

6. Although the whole membrane in a denervated muscle is sensitive to ACh, the muscle fibres remain electrically excitable.

7. It is concluded that the increase in the size of the receptor area after denervation accounts for the supersensitivity of denervated muscles to ACh and other chemical substances. Furthermore, the extension of the ACh-sensitive region to the whole muscle will allow the drug to depolarize the entire length of a fibre and thereby to produce the ACh contracture observed in denervated mammalian muscles.

The expenses of this investigation were aided by grants from the Swedish Medical Research Council, The Muscular Dystrophy Association of America, Inc. and the Air Research and Development Command, United States Air Force, through its European Office. We are indebted to Professor B. Katz and Dr A. Lundberg for valuable suggestions and criticism. Unfailing technical assistance was provided by Miss E. Adler.

REFERENCES

ADRIAN, E. D. (1925). The spread of activity in the tenuissimus muscle of the cat and in other complex muscles. *J. Physiol.* **60**, 301–315.

AXELSSON, J. & THESLEFF, S. (1958). The 'desensitizing' effect of acetylcholine on the mammalian motor end-plate. *Acta physiol. scand.* **43**, 15–26.

AXELSSON, J. & THESLEFF, S. (1959). A study on supersensitivity in denervated mammalian skeletal muscle. *J. Physiol.* **145**, 48*P*.

BOYD, I. A. (1956). The tenuissimus muscle of the cat. *J. Physiol.* **133**, 35–36*P*.

BOYD, I. A. & MARTIN, A. R. (1956*a*). Spontaneous subthreshold activity at mammalian neuromuscular junctions. *J. Physiol.* **132**, 61–73.

BOYD, I. A. & MARTIN, A. R. (1956*b*). The end-plate potential in mammalian muscle. *J. Physiol.* **132**, 74–91.

BROWN, G. L. (1937). The actions of acetylcholine on denervated mammalian and frog's muscle. *J. Physiol.* **89**, 438–461.

BULLER, A. J., ECCLES, J. C. & ECCLES, R. M. (1958). Controlled differentiation of muscle. *J. Physiol.* **143**, 23–24*P*.

CANNON, W. B. & ROSENBLUETH, A. (1949). *The Supersensitivity of Denervated Structures.* New York: The Macmillan Company.

DEL CASTILLO, J. & KATZ, B. (1954). The membrane change produced by the neuromuscular transmitter. *J. Physiol.* **125**, 546–565.

DEL CASTILLO, J. & KATZ, B. (1955*a*). On the localization of acetylcholine receptors. *J. Physiol.* **128**, 157–181.

DEL CASTILLO, J. & KATZ, B. (1955*b*). Local activity at a depolarized nerve-muscle junction. *J. Physiol.* **128**, 396–411.

DEL CASTILLO, J. & KATZ, B. (1957). A study of curare action with an electrical micro-method. *Proc. Roy. Soc.* B, **146**, 339–356.

ECCLES, J. C. (1941). Changes in muscle produced by nerve degeneration. *Med. J. Aust.* i, 573–575.

EMMELIN, N. (1952). 'Paralytic secretion' of saliva: An example of supersensitivity after denervation. *Physiol. Rev.* **32**, 21–46.

FATT, P. & KATZ, B. (1951). Analysis of the end-plate potential recorded with an intra-cellular electrode. *J. Physiol.* **115**, 320–370.

GRUNDFEST, H. (1957). The mechanisms of discharge of the electric organs in relation to general and comparative electrophysiology. *Progr. Biophys.* **7**, 1–85.

HOWARTH, J. V. (1958). The behaviour of frog muscle in hypertonic solutions. *J. Physiol.* **144**, 167–175.

KATZ, B. & THESLEFF, S. (1957*a*). A study of the 'desensitization' produced by acetylcholine at the motor end-plate. *J. Physiol.* **138**, 63–80.

KATZ, B. & THESLEFF, S. (1957*b*). The interaction between edrophonium (Tensilon) and acetylcholine at the motor end-plate. *Brit. J. Pharmacol.* **12**, 260–264.

KUFFLER, S. W. (1943). Specific excitability of the end-plate region in normal and denervated muscle. *J. Neurophysiol.* **6**, 99–110.

LILEY, A. W. (1956). An investigation of spontaneous activity at the neuromuscular junction of the rat. *J. Physiol.* **132**, 650–666.

NASTUK, W. L. (1953). Membrane potential changes at a single muscle end-plate produced by transitory application of acetylcholine with an electrically controlled microjet. *Fed. Proc.* **12**, 102.

NASTUK, W. L. & ALEXANDER, J. T. (1954). The action of 3-hydroxyphenyldimethylethylammonium (Tensilon) on neuromuscular transmission in the frog. *J. Pharmacol.* **111**, 302–328.

NICHOLLS, J. G. (1956). The electrical properties of denervated skeletal muscle. *J. Physiol.* **131**, 1–12.

ROSENBLUETH, A. & LUCO, J. V. (1937). A study of denervated mammalian skeletal muscle. *Amer. J. Physiol.* **120**, 781–797.

SMITH, C. M., COHEN, H. L., PELIKAN, E. W. & UNNA, K. R. (1952). Mode of action of antagonists to curare. *J. Pharmacol.* **105**, 391–399.

THESLEFF, S. (1955). The mode of neuromuscular block caused by acetylcholine, nicotine, decamethonium and succinylcholine. *Acta physiol. scand.* **34**, 218–231.

ZAIMIS, E. J. (1954). Transmission and block at the motor end-plate and in autonomic ganglia. *Pharmacol. Rev.* **6**, 53–57.

28

Reprinted without change of pagination from the
Proceedings of the Royal Society, B, *volume* 167, *pp.* 23–38, 1967

The release of acetylcholine from nerve endings by graded electric pulses

By B. Katz, F.R.S. and R. Miledi
Department of Biophysics, University College London

(*Received* 29 *June* 1966)

1. The effect of brief depolarizations focally applied to a motor nerve ending was studied. Particular attention was paid to the relation between (i) strength and duration of the pulse and (ii) the size and latency of the resulting end-plate potential.

2. The release of acetylcholine lags behind the depolarization which causes it. If pulses of less than 4 ms duration are used (at 5 °C), the release starts after the end of the pulse.

3. Within a certain range, lengthening the pulse increases the *rate* of the ensuing transmitter release.

4. Unexpectedly, lengthening the depolarizing pulse also increases the latency of the transmitter release. This finding is discussed in detail. It is regarded as evidence suggesting that entry into the axon membrane of a positively charged substance (external Ca^{2+} ions or a calcium compound CaR^{+}) is the first step leading to the release of acetylcholine packets from the terminal.

Introduction

It has been shown in the preceding paper that the use of tetrodotoxin enables one to eliminate the nerve impulse and to study the process of transmitter release by applying electric pulses of graded strength and duration to the motor nerve ending. In the present paper this method will be employed to study certain quantitative aspects in more detail, in particular to examine the relation between the parameters of the applied pulse and the latency and size of the evoked response.

It had previously been found (Katz & Miledi 1965*b*) that there is an appreciable delay between the rise of the action potential wave in the nerve terminal and the release of acetylcholine. The cause of this delay is unknown, and it is of interest to find out whether this latent period can be altered by varying the intensity or duration of the imposed depolarizing pulse.

Methods

Much of the technical detail has been described in previous papers (Katz & Miledi 1965*a*, *e*). For most of the present work the focal method of pulse application (Katz & Miledi 1967*b*, method *B*) was preferred. By placing a microelectrode (a capillary usually of 2 to 4 μm tip diameter, filled with 0·5 or 3 M NaCl) over a junctional area and applying negative-going pulses to the pipette, a minute part of the junction could be depolarized for defined brief periods.

The pulse applied in this way alters the potential of a small surface of both the axon terminal and the underlying muscle fibre. This is represented schematically in figure 1. Suppose a 2 μm length of nerve ending of 1·5 μm diameter is covered by the micropipette and becomes uniformly depolarized by the applied surface-negative pulse. The area of membrane which becomes discharged amounts to approximately 10 μm^2. The corresponding area of the muscle membrane underlying this part of the nerve ending is of the same order of size. The membrane potential change follows the applied current with very little time lag; a 'time constant' can be derived approximately from the product of the capacity of the small membrane area (about 10^{-7} μF) and the input resistance of the fibre which is of the order of 10^5 ohms for the muscle and 10^8 ohms for the nerve terminal. This would give 10^{-2} μs for the muscle and 10 μs for the nerve.

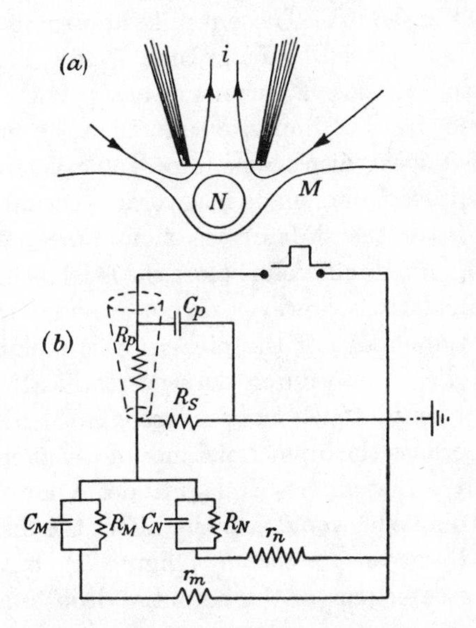

FIGURE 1. Diagram illustrating method of focal depolarization. (*a*) Tip of pipette placed over nerve-muscle junction. *N*, cross section of nerve terminal. *M*, muscle fibre. *i*, current entering tip of pipette. (*b*) Equivalent circuit, with tip of pipette indicated by dashed outline. C_p, capacity of glass wall of pipette; R_p, resistance of tip of pipette. R_s, shunt resistance on surface of nerve-muscle junction. C_m and R_m, capacity and resistance of localized surface of muscle fibre. C_N and R_N, capacity and resistance of local surface of nerve terminal. r_m and r_n, input resistances of muscle fibre and nerve terminal, respectively.

In the steady state most of the potential difference applied across R_s (i.e. between inside and outside of the pipette) will appear across the focal area of the membrane because the transverse resistance of this area is of the order of 10^{10} ohms, i.e. very much higher than the input resistance of the fibre. However, this ceases to be the case if the membrane resistance falls to a very low value or is actually caused to break down under the influence of an excessive electric pulse. This did, in fact, occur quite often, at least in the muscle fibre from which intracellular records were taken. It showed itself by the sudden appearance of an 'electrotonic potential', indicating that a low resistance had suddenly been established between the micropipette and the interior of the muscle fibre (see also Huxley & Taylor 1958). Evidence of local breakdown in the axon terminal was observed less frequently; it showed itself by a sudden prolonged outburst of miniature e.p.ps. When these symptoms occurred the series of observations had to be discontinued and the pipette moved to another spot, or to another fibre.

The chief drawback of the method is that it does not provide adequate information on the amplitude of the pre-synaptic potential change. The best one can do is to obtain an indication of the *local displacement of the post-synaptic membrane potential* by the applied current. This is revealed by the change in size, and ultimately reversal of sign of the miniature e.p.ps which are elicited during the continued flow of the applied current. For example, suppose the membrane potential of the muscle fibre is −90 mV and the average unit e.p.p. (spontaneous, or occurring after a pulse) is 0·5 mV. The e.p.p. is known to reverse sign when the membrane potential is displaced by more than 75 mV, i.e. beyond −15 mV (del Castillo & Katz 1954*c*). Suppose a current of one particular strength elicits unit potentials of the same size, but opposite sign, then the membrane potential has presumably suffered a local displacement of 150 mV, from −90 to +60 mV. Rough estimates of this kind were made in several experiments and indicated that, post-synaptically at least, the shift of the membrane potential could reach an amplitude of several hundred millivolts before the local insulation of the membrane broke down. These calculations, however, do not provide a safe basis for estimating even approximately the change of the pre-synaptic membrane potential. Apart from the inaccuracy of this procedure, 'delayed rectification' may limit the focal displacement of the potential much more severely in the nerve ending than in the muscle fibre. This is because the input resistance of the axon terminal exceeds that of the muscle fibre by a few orders of magnitude. A lowering of the membrane resistance to, say, 10 ohm cm² would scarcely affect the distribution of potentials in the muscle fibre (between r_m and R_m; figure 1), but in the nerve ending this could cause the greater part of the potential drop to appear electrotonically across r_n.

A further difficulty is that a strong negative pulse from the pipette might conceivably subject the surrounding parts of the nerve terminal to a strong electrotonic hyperpolarization and lead to the 'anodal breakdown' effect, i.e. bursts of miniature e.p.p. (del Castillo & Katz 1954*b*; Katz & Miledi 1965*a*), originating *outside* the focal area. Such an event, however, would be easily recognized (*a*) from the abrupt appearance of bursts and long after-discharges following the pulse, (*b*) from the fact that the size and sign of such extra-focal discharges would *not* be altered by prolongation of the applied current. In our experience, with electrodes of 2 μm or larger diameter, the upper limit for the usable current strength was usually set by the occurrence of *post*-synaptic breakdown.

The upshot of these technical complications was that current intensities had to be kept within a limited range in which repeatable and consistent responses were obtained, and that only those problems could be studied which did not depend on an exact knowledge of the amplitude of the pre-synaptic potential change. The frequency of focal stimulation was usually kept low, with intervals between pulses ranging from 1·5 to 10 s. Even at these rates, e.p.p. responses tended to decline, and the statistical occurrence of failures to increase progressively. This was especially marked when the initial failure rate was low (and the value of *m* greater

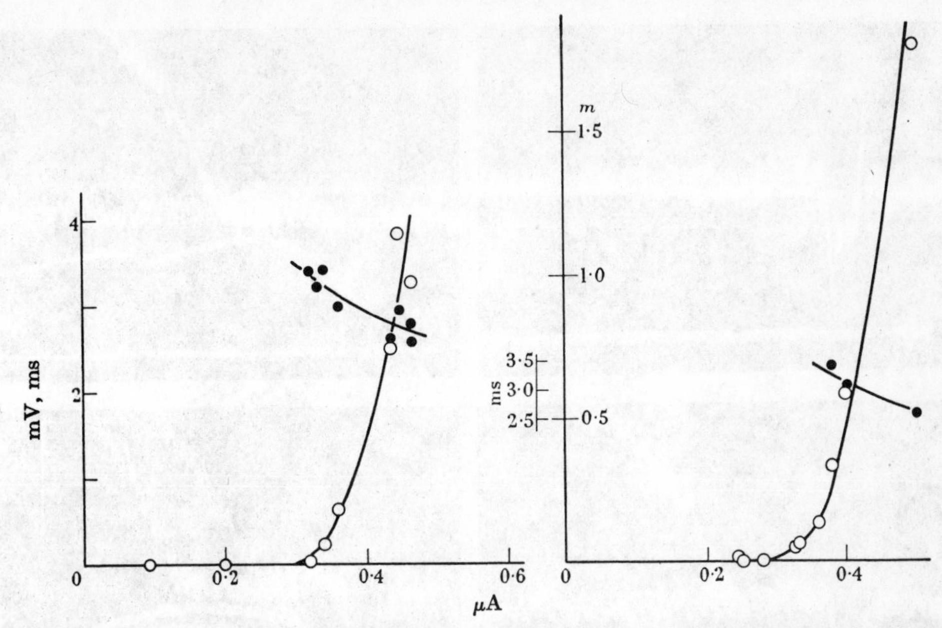

FIGURE 2. Strength-response relations obtained with focal, 1 ms pulses. Temp. 3·5 °C. In the series on the left, hollow circles show average amplitudes, in mV, of evoked e.p.p. On the right (from another series): hollow circles show mean 'quantal content' of evoked response, calculated from the proportion of failures. The mean unit size was approx. 1 mV in both series. Full circles: minimum latencies, in ms, measured from the start of the applied pulse to the beginning of the earliest e.p.p. Note: results shown in all figures were obtained by intracellular recording from muscles paralysed by tetrodotoxin (10^{-6} g/ml.).

than a small fraction of unity). This is not very surprising, for an average release of one unit per pulse from a few microns length of nerve terminal corresponds to nearly the full rate of release by a normal nerve impulse.

Recording

The response was recorded intracellularly from the muscle fibre, with a suitably placed external microelectrode serving as reference lead. The residual artifacts associated with the applied pulse were not sufficiently large to obscure the time course of the evoked e.p.p. The 'artifact' included a small electrotonic potential change in the muscle fibre which arose from the 'convergence field' set up in the Ringer's fluid surrounding the polarizing micropipette. This amounted usually to less than 1 mV, well below 1 % of the focally applied potential change. As before (Katz & Miledi 1967*b*) prostigmine was used in most experiments.

RESULTS

The effect of strength and duration of focally depolarizing pulses on the size of the end-plate response

If one increases the intensity of a short pulse of constant duration, the response rises in a non-linear manner as shown in figure 2. The behaviour resembles the

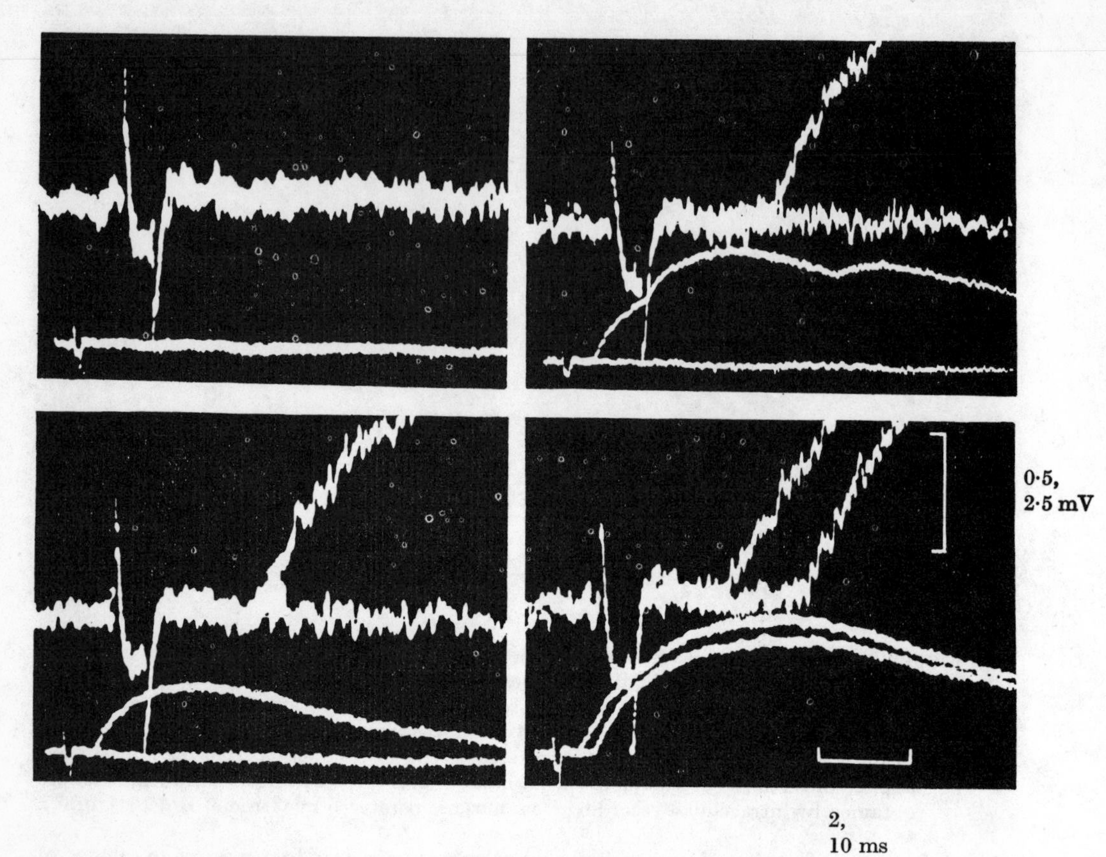

FIGURE 3. Examples of 'quantal' e.p.ps evoked by brief focal pulses. Temp. 4·5 °C. Each block shows two simultaneous recordings, the top record at the higher speed and amplification indicated on the scales. Each recording consists of two superimposed traces. The column on the left was obtained with a slightly weaker pulse. Note: 'all-or-none' appearance of discrete units of response, and variable latencies.

effect already described with the method of electrotonic application. Below a certain intensity, the probability of observing even a single unit discharge is very small; above this apparent threshold, occasional unit e.p.ps are seen (figure 3), and a further increase of current strength causes a steep rise in the amplitude, i.e. the number of quantal units, of the e.p.p. (see Katz & Miledi 1967*b*).

A similar non-linear increase in the size of the e.p.p. is observed if one increases the duration of a pulse of constant intensity from, say, 0·2 ms to a few milliseconds (figure 4).

If one determines the 'threshold' intensity of pulses of different durations, taking the appearance of an occasional unit e.p.p. as threshold index, one obtains a 'strength duration' curve of the kind shown in figure 5 (cf. Katz & Miledi 1965*d*). This has the usual hyperbolic shape indicating that for very brief pulses, the transmitter release depends on the time integral of the depolarization rather than on its amplitude.

The interpretation of this curve differs from that of the *conventional* type of strength-duration relation which is related to the electric time factor of the membrane (see Katz 1939). In the latter case depolarization rises gradually during a current pulse; the shorter the current, the greater must be its strength if

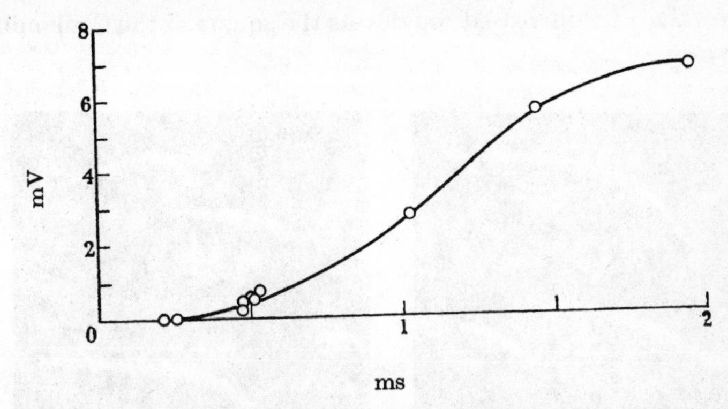

FIGURE 4. Effect of pulse duration on size of evoked response. Temp. 7 °C. Ordinate: average amplitude of evoked e.p.p. Abscissa: duration of applied pulse. Pulse intensity was 3·25 μA.

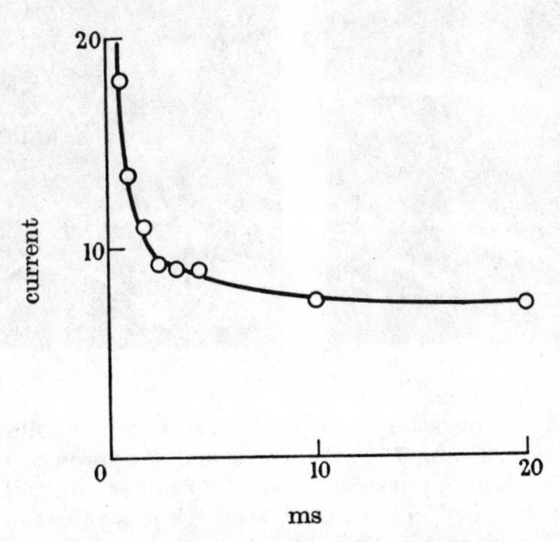

FIGURE 5. 'Strength-duration' curve, showing the relation between duration and intensity of equally effective pulses. Temp. 9·5 °C. The duration (abscissa) was varied, and the current strength (ordinate, relative units) was determined which caused occasional unit responses to appear at the end of the pulse. The 'rheobase' (intensity required for 20 ms pulse) was about 0·15 μA.

a given potential change is to be attained. In the present experiments, in which a focal microelectrode is used, the electric time factor is probably much less than 0·1 ms (see Methods), and the pulses of focal depolarization produced by the currents in figures 4 and 5 were presumably 'square' except for a small fraction of a millisecond at beginning and end.

The disproportionate increase in the response with increasing pulse duration (figure 4) indicates that the *rate* of transmitter release which results from a depolarization of given amplitude, is not determined by that amplitude alone but continues to rise when the depolarization is prolonged up to several milliseconds in duration. Further examples are shown in figure 9 where an increase in pulse length from 0·5 to 2·12 ms caused the average quantal release m (del Castillo & Katz 1954*a*) to increase from 0·07 to 0·97 units per pulse, in spite of a falling pulse intensity. The value of m increased roughly as the square of the Coulomb quantity of the pulse.

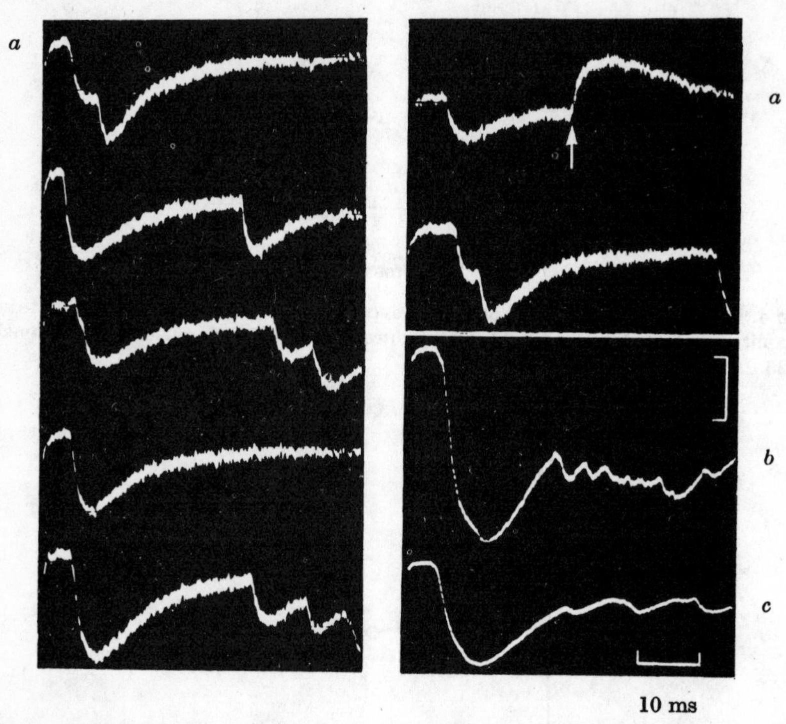

FIGURE 6. Inverted e.p.ps (i.e. negative-going potentials with intracellular recording) *during* focal current application. Temp. 7 °C. Current strength approx. 0·7 μA (varied between 0·57 and 0·76 μA). Duration of pulse 50 ms, starting at beginning of trace. Vertical scale 0·4 mV in *a*, 1 mV in *b*, 2 mV in *c*. Current strength was raised in *b* and *c*. About 150 pulses were given at 5 s intervals. Throughout the series, only one potential (at arrow) had the normal, positive-going, sign; this was presumably a spontaneous min. e.p.p. arising outside the focal depolarized zone.

This applies to *brief* periods of depolarization, say up to 2 or 3 ms at 5 °C. If the duration of the current is lengthened further, the rate of release, measured as the frequency of evoked miniature e.p.ps, undergoes a complex sequence of changes, an example of which is shown in figures 6 and 7. These figures illustrate the temporal distribution of miniature e.p.ps recorded during a series of strong depolarizing pulses of 50 ms duration. Samples of the records from which the histograms in figure 7 were compiled are shown in figure 6. It will be noted that these potential changes are of opposite electric sign to spontaneous potentials and to those recorded after the end of a pulse (e.g. figure 12), for reasons already explained (Methods and

Katz & Miledi 1965*a*). The size of the early peak in the histograms must be an underestimate because, during 'peak times', units were crowded together into nearly synchronous reponses, and not all their miniature components could be resolved. The general features, however, are clear, viz. that during the large depolarizing current the frequency of quantal transmitter release rises after a brief delay and builds up to a peak in several milliseconds, from which it falls to a low level. This is followed by a slow secondary rise. A detailed investigation of these

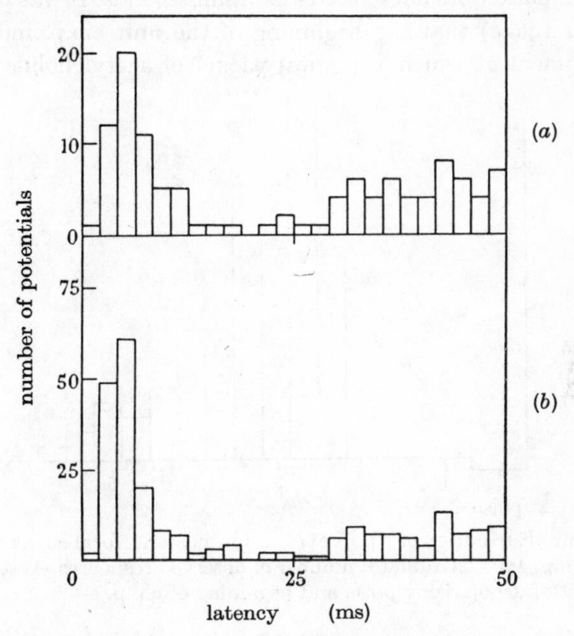

FIGURE 7. Latency histogram of e.p.ps from the experiment illustrated in figure 10. (*a*): Response during the first 50 pulses. (*b*): Histogram of the whole series (145 pulses). There was a gradual decline in the *later* part of the response.

complex changes which are presumably related to the 'refractoriness' and 'facilitation' described in the previous paper must be left to a later study. In some experiments, the phase of depression shown in figure 7 was not evident; this may well depend on such variable factors as the size of the focal pipette and the relative resistances of the nerve ending and its depolarized membrane area (see p. 25).

The present observations indicate that one of the assumptions which was basic to Liley's (1956) hypothesis, namely that the rate of transmitter release is determined by the value of the pre-synaptic membrane potential at that instant, is not valid. The reason why the striking time-dependence of the process has not previously been recognized lies in the facts that earlier experiments were made at a high temperature at which transient phenomena are confined to a much shorter initial period, and that in the absence of tetrodotoxin the currents had to be established gradually to avoid initiation of impulses. This procedure inevitably obscures the phenomena which have been reported here.

The delay between depolarization and quantal release

The results so far described show the relations between size and duration of the depolarizing pulse and the resulting *intensity* of transmitter release. In the present section, the *time course* of the resulting transmitter release will be considered.

The main observations are presented in figures 11 and 12 and the histograms of figures 8 to 10. They contain two important features. First, unit e.p.ps do not begin to appear until well after the end of the depolarizing pulse, provided the duration of this pulse does not exceed a few milliseconds. It was previously shown (Katz & Miledi 1965*b*) that the beginning of the unit e.p.p. indicates, approximately, the moment at which a quantal packet of acetylcholine is released. The

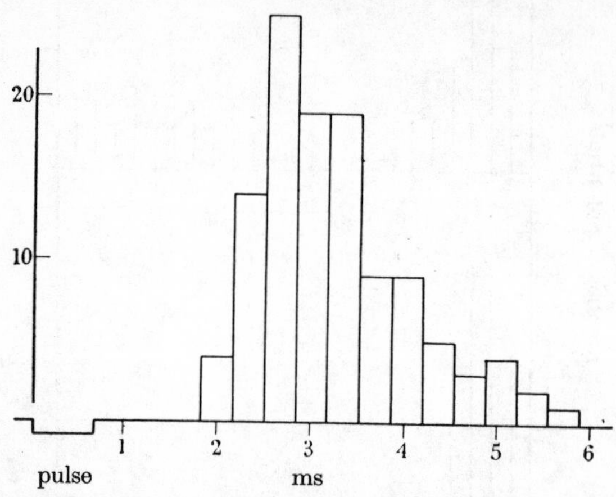

FIGURE 8. Latency distribution of e.p.ps evoked by constant, focal current pulses of 0·68 ms duration. Temp. 4·5 °C. Ordinate: number of observed responses. Abscissa: time interval between start of depolarizing pulse and beginning of e.p.p.

error incurred in assuming coincidence of these two events is probably no more than 0·1 ms (Katz & Miledi 1965*e*). It seems safe to conclude that, with these brief focal pulses of depolarization, the resulting transmitter release does not start until some 1 to 2 ms have elapsed after the depolarization has subsided. This is quite unlike the release of potassium ions (Hodgkin & Huxley 1952) which attains its maximum intensity *during* the depolarization. These observations suggest that there is likely to be a series of intermediate reactions between the membrane potential change and the release of acetycholine.

The second point which deserves attention is the latency shift which was consistently observed when a pulse of given strength was altered in duration.

The 'latency shift'

When the pulse is lengthened, e.g. from 0·5 to 2 ms, the process of transmitter release is influenced in two apparently opposite ways. There is the very large increase in the total amount released which has already been described. Unexpectedly, however, the delay—measured from the start of the depolarizing pulse—is *lengthened*. This effect is small, but consistent. Examples are shown in figures 9 to 12 and table 1.

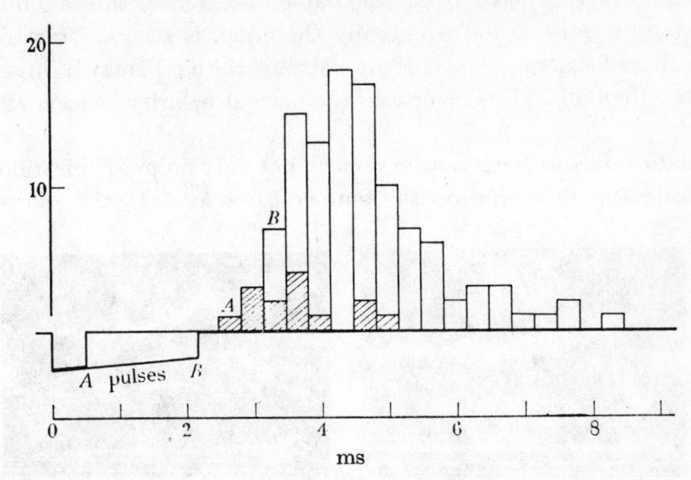

FIGURE 9. Effect of pulse duration on the latency histogram. Two pulses (A and B) were applied. The latencies of evoked e.p.ps are shown, respectively, by shaded and clear columns. Temp. 4·4 °C. The short pulses (A) evoked only 14 e.p.ps in 211 trials; hence $m = \ln 211/197 = 0{\cdot}07$. The longer pulses ($B$) produced 104 responses in 168 trials; hence $m = \ln 168/64 = 0{\cdot}97$.

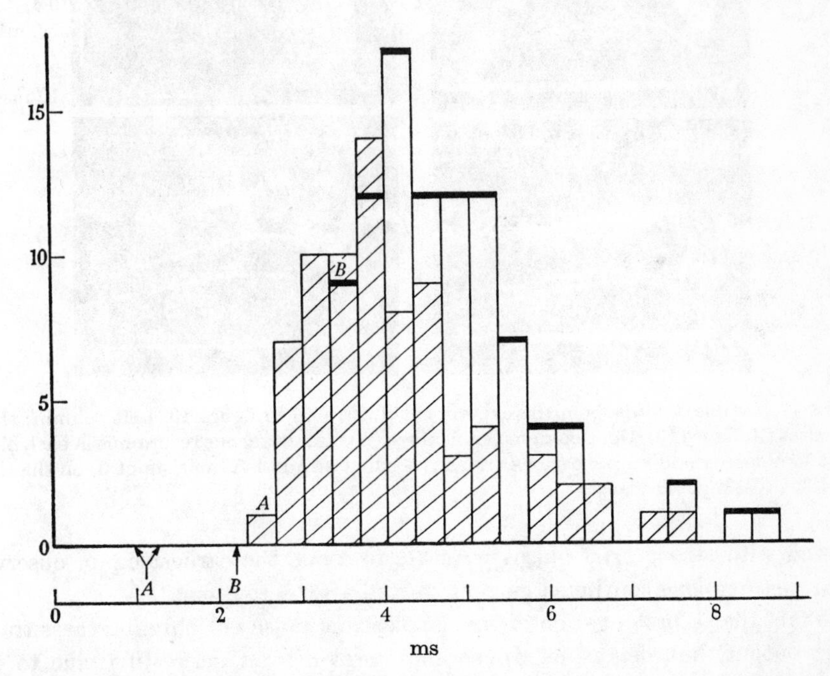

FIGURE 10. Another example of the 'minimum-latency shift' caused by a change in pulse duration at constant pulse intensity. Durations of pulses are indicated by arrows at A and B. The responses to A (424 pulses) are shown by shaded columns, those to B (166 pulses) by heavy crossbars (and, where numbers of B responses exceed those of A, by clear rectangles). Temp. 5 °C.

If one measures the delay from the *end* of the pulse, this diminishes with increasing pulse duration, and eventually the e.p.p. is seen to start during the pulse itself. If the focally applied current is strong, the e.p.p. may be inverted while the pulse lasts (figure 12), but reverts to the normal polarity as soon as the pulse is withdrawn.

The absolute values of the latencies depend not only on pulse duration, but also on its intensity (figure 2) and on the temperature. At 5 °C, the shortest delays

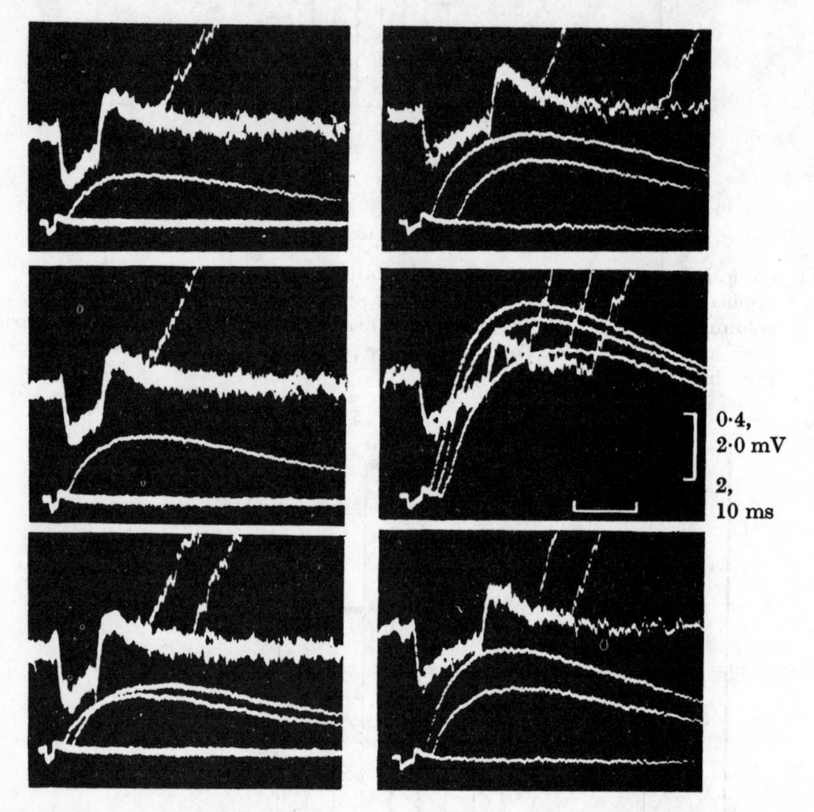

FIGURE 11. Sample records from the experiment illustrated in figure 10. Left column: short pulses (*A*, figure 10); right column; long pulses (*B*). Simultaneous recordings in each block at two speeds and amplifications. Each recording on the left consists of 6, on the right of 3, superimposed traces.

observed with strong *brief* pulses were 1·5 to 2 ms, the earliest e.p.p. observed during a strong *long*-continued current started after 4 to 5 ms.

The fact that a prolongation of the depolarizing pulse not only increases transmitter output, but delays its appearance, provides an interesting clue to the physico-chemical mechanism of release and will be discussed in detail below.

An obvious further experiment was to apply pulses of opposite sign during the latent period, in order to see whether hyperpolarization immediately after the depolarizing pulse would cancel, or at least modify, the transmitter release. The results of this type of experiment, however, were not very revealing. The procedure was to use depolarizing pulses (P), of 0·4 to 2 ms. duration, and follow them after a brief interval (20 to 140 μs) with 'anodic' pulses (A) of the same or somewhat lower intensity. Several series of observations were made; to summarize, the anodic pulses were either ineffective or produced a small reduction in transmitter output, without obvious change in the latency distribution. With very brief intervals between P and A pulses, the suppressing effect of the hyperpolarization, though small, was statistically significant. Thus, with approximately 20 μs separation

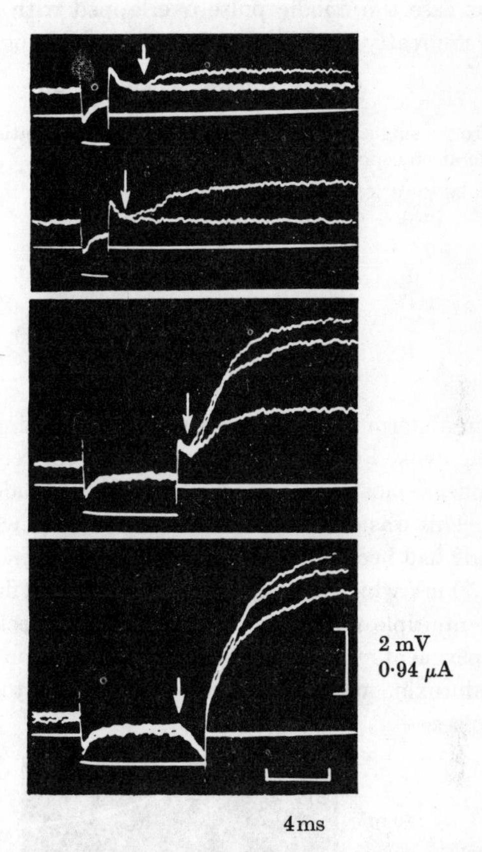

Figure 12. Further example of 'latency shift'. Three pulse durations. Beginning of e.p.ps shown by arrows. Note: inversion of e.p.p. *during* long current pulse (bottom record). Temp. 4·5 °C. The preparation was in a medium of low sodium and calcium content (half sodium replaced by sucrose, 0·2 mM Ca).

257 P pulses evoked 100 responses. The same number of combined $P+A$ pulses produced only 74 responses. When intervals of 115 μs were used, the proportion of responses was 139/419 and 118/400, to P and $P+A$ respectively, a difference which does not pass the usual tests of statistical significance. Similar results were obtained in another experiment, where at 45 μs separation the depolarizing pulses alone evoked 154 responses in 417 trials, while $P+A$ gave 59 responses to 243 pulses. When the intervals were lengthened to 140 μs, no difference between P and $P+A$ effects was seen. Thus, it appears that the anodic pulses given during the

latent period are not very effective in interfering with the release mechanism. This contrasts with the powerful anodic suppression described in a previous paper (Katz & Miledi 1967*a*) where the hyperpolarizing pulse was applied immediately after the negative peak of the focal axon spike, i.e. during the decline of the membrane action potential. The difference presumably arises from the fact that, in the present case, there is no appreciable *direct* interference between opposite membrane potential changes, the anodic pulse being applied *after* the depolarization, while in the previous case the anodic pulse overlapped with and curtailed the action potential, and so greatly reduced its efficacy in releasing transmitter.

TABLE 1. THE 'LATENCY' SHIFT

Results were from a single end-plate at about 5·2 °C; observations with 1·0, 1·27 and 2·2 ms pulses were made at one spot.

pulse duration (ms)	number of applied pulses	number of responses	minimum latency (ms)
0·5	23	3	2·1
1·0	98	14	2·4
1·27	326	61	2·7
2·2	166	96	3·4

Triggered 'giant' responses

Under certain abnormal conditions, the graded relation between the size of the pulse and of the evoked e.p.p. is replaced by an entirely different phenomenon, viz. a huge, almost explosive release which is triggered suddenly at an unpredictable pulse strength. This was observed in an experiment in which 5 mM tetraethylammonium bromide had been added (figures 13 and 14). The response consisted of an e.p.p. of about 50 mV which subsided slowly and showed fluctuating high-frequency components of multiple miniature e.p.ps during its decline. A similar effect was seen in two experiments in which a dose of tubocurarine had been given together with the tetrodotoxin, and the nerve was subjected to a series of very strong depolarizing pulses.

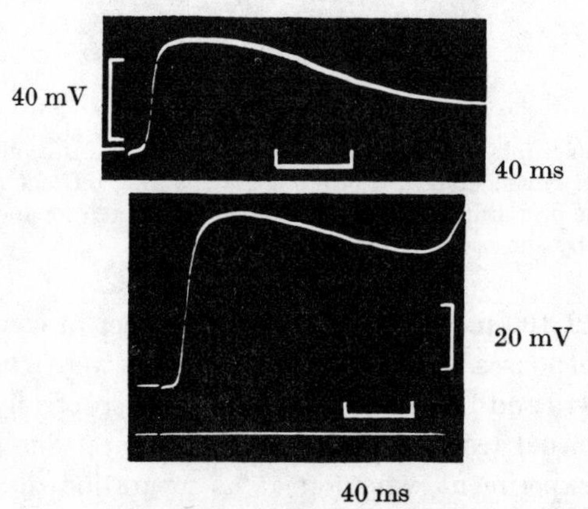

FIGURE 13. Triggered 'giant' e.p.ps Preparation was in a solution containing 5 mM tetraethylammonium bromide. Temp. 3·5 °C. Pulse was applied to motor axon ('electrotonic' method).

The quantal content of the enormous e.p.ps shown in figure 13 must have been well over one thousand, judging from the relatively small sizes of the spontaneous miniature potentials (about 0·15 mV). The mechanism by which these huge responses are triggered is obscure. It was noted that they occurred during a transient state of abnormal instability, which was characterized by a high frequency of spontaneous discharge and which was followed by total failure of any evoked response.

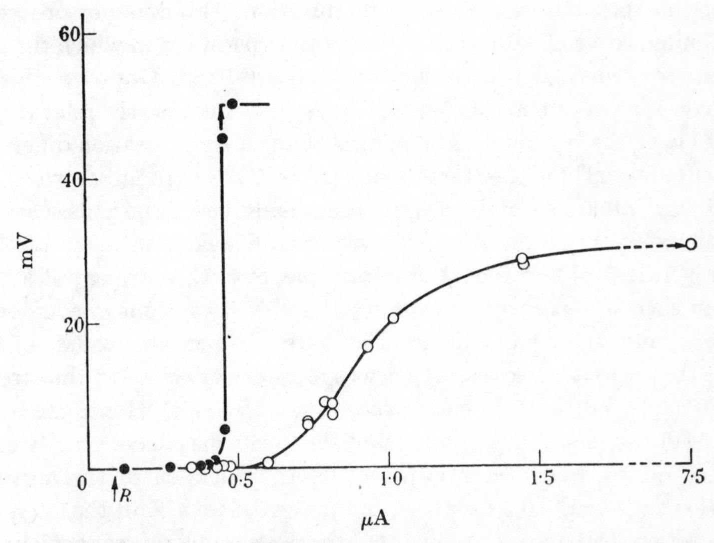

FIGURE 14. Strength-response relations, showing amplitude of e.p.ps (ordinate) evoked by 5·5 ms current pulses applied to the motor axon (method *A*, Katz & Miledi 1967*b*). Abscissa, current strength. Hollow circles: before adding 5 mM tetraethylammonium. Full circles: after adding *T.E.A.* *R*: 'rheobase' determined at beginning of experiment (before applying tetrodotoxin).

DISCUSSION

The central piece of evidence presented in this paper is contained in figures 9 to 12. These show (*a*) that the release of the transmitter is delayed and, in fact, only starts after the end of the brief depolarization which initiated it, (*b*) that lengthening the depolarizing pulse causes a larger amount to be released *after a longer delay*.

In searching for a reasonable interpretation of these findings, it should be recalled that external calcium ions are essential for the depolarization to become effective (Katz & Miledi 1965*c*).

We envisage that the evoked release of quantal packets of acetylcholine proceeds in several stages the first of which is an inward movement through the axon membrane either of Ca^{2+} itself, or of a positively charged calcium compound CaR^{+}. The subsequent steps could be pictured in various ways, e.g. according to the vesicular hypothesis as a reaction on the inside of the membrane by which Ca or Ca*R* causes momentary fusion of vesicular and axon envelopes at their point of collision (del Castillo & Katz 1957; Katz 1962).

We are concerned here only with the first link of this chain. The surface-negative pulse has two opposite effects: it facilitates this process by raising the membrane 'permeability' to Ca^{2+} or CaR^+ ions; but the negative pulses also oppose the inward movement of the positive ions. The former reaction presumably develops and decays with a time lag, while the latter is a direct electrophoretic effect which might well account for the delaying action of the applied current.

On the basis of this hypothesis, one might expect that a sufficiently large negative surface potential would completely prevent inward movement of the postulated Ca-compound, and that transmitter release could in principle be delayed until the end of the pulse regardless of its duration. This was not observed: the delay was limited to 4 to 5 ms at 5 °C, even in an experiment in which the external calcium concentration had been reduced to about 0·1 mM. However, this finding presents no decisive argument against our hypothesis because the relevant quantities (amplitude of the pre-synaptic potential change, concentration difference and 'equilibrium potential' for the transfer of Ca^{2+} or CaR^+) are unknown.

The minimum latencies obtained with intense and very brief pulses are shorter than the minimum 'synaptic delay' between nerve spike and e.p.p. at the same temperature (Katz & Miledi 1965*e*). For example, at 5 °C, a strong pulse of 0·5 ms duration can elicit a unit response with a latency of 1·5 to 2 ms as compared with 2 to 4 ms minimum spike-e.p.p. delay. There are two probable causes of this difference: (*a*) the current pulse was of very high intensity, close to the strength at which postsynaptic breakdown tends to occur (see Methods). Hence the brief local displacement of the membrane potential of the nerve may have greatly exceeded that attained during the normal impulse; (*b*) the duration of the nerve action potential is longer than the 0·5 ms pulse, and that would also shift the latency in the observed direction. It is indeed possible that the large temperature coefficient of the synaptic delay results in part from the large effect of temperature on the falling phase of the action potential, a correlation to which attention has already been drawn in a previous paper (Katz & Miledi 1965*e*).

The 'strength-response' relations shown in figure 2 are of interest in that they help one to assess the relative power of the nerve impulse in releasing transmitter. Only a portion of the focal 'strength-response' relation was obtained, because the usable range of intensities is limited by the current strength at which the muscle membrane breaks down (see p. 24). The length of nerve subjected to the stimulus was of the order of a few microns, approximately 1% of the total length of the junctional branches. Normally (with a calcium concentration of 1·8 mM), a nerve impulse causes the release of about 200 quantal units. In the experiments of figure 2, the size of the unit potentials was approx. 1 mV; the 1 ms pulses were able to release 2 to 4 units from the small area under the electrode, and there was no sign of a 'ceiling' being reached. Moreover, the duration of the pulse was shorter than that of a nerve spike at 3·5 °C. It appears from this experiment that the 'secretory power' of the nerve impulse is well below that of a 'maximum stimulus' though it might approach saturation in high calcium, or during tetanic potentiation.

That the release of transmitter increases more than in direct proportion to the imposed depolarization is known from previous work (del Castillo & Katz 1954*b*; Liley 1956; Takeuchi & Takeuchi 1962; Miledi & Slater 1966). It was not known, however, that there is a similar non-linear increase when the *duration* of the depolarization is lengthened (figure 4). This observation must affect any analysis of the release evoked by a nerve impulse. If one visualizes a depolarization of a few milliseconds duration as being made up of a series of much briefer pulses, then each of them apparently facilitates, or 'potentiates', the effect of its successor. The basis of this process of facilitation is not known. It might be due to accumulation of calcium ions or a calcium compound at the inside of the axon membrane (cf. Katz & Miledi 1965*c*; Gage & Hubbard 1966), an idea which would also fit with recent observations by F. A. Dodge & R. Rahamimoff (unpublished), that transmitter release increases much more than in direct proportion to the external calcium concentration.

References

del Castillo, J. & Katz, B. 1954*a* Quantal components of the end-plate potential. *J. Physiol.* **124**, 560–573.

del Castillo, J. & Katz, B. 1954*b* Changes in end-plate activity produced by pre-synaptic polarization. *J. Physiol.* **124**, 586–604.

del Castillo, J. & Katz, B. 1954*c* The membrane change produced by the neuromuscular transmitter. *J. Physiol.* **125**, 546–565.

del Castillo, J. & Katz, B. 1957 La base 'quantale' de la transmission neuro-musculaire. In 'Microphysiologie comparée des éléments excitables.' *Coll. Internat. C.N.R.S. Paris*, no. 67, 245–258.

Gage, P. W. & Hubbard, J. I. 1966 An investigation of the post-tetanic potentiation of end-plate potentials at a mammalian neuromuscular junction. *J. Physiol.* **184**, 353–375.

Hodgkin, A. L. & Huxley, A. F. 1952 The components of membrane conductance in the giant axon of *Loligo*. *J. Physiol.* **116**, 473–496.

Huxley, A. F. & Taylor, R. E. 1958 Local activation of striated muscle fibres. *J. Physiol.* **144**, 426–441.

Katz, B. 1939 *Electric excitation of nerve* 151 p. London: Oxford University Press.

Katz, B. 1962 The Croonian Lecture. The transmission of impulses from nerve to muscle, and the subcellular unit of synaptic action. *Proc. Roy. Soc.* B **155**, 455–477.

Katz, B. & Miledi, R. 1965*a* Propagation of electric activity in motor nerve terminals. *Proc. Roy. Soc.* B **161**, 453–482.

Katz, B. & Miledi, R. 1965*b* The measurement of synaptic delay, and the time course of acetylcholine release at the neuromuscular junction. *Proc. Roy. Soc.* B **161**, 483–495.

Katz, B. & Miledi, R. 1965*c* The effect of calcium on acetylcholine release from motor nerve terminals. *Proc. Roy. Soc.* B **161**, 496–503.

Katz, B. & Miledi, R. 1965*d* Release of acetylcholine from a nerve terminal by electric pulses of variable strength and duration. *Nature, Lond.* **207**, 1097–1098.

Katz, B. & Miledi, R. 1965*e* The effect of temperature on the synaptic delay at the neuromuscular junction. *J. Physiol.* **181**, 656–670.

Katz, B. & Miledi, R. 1967*a* Modification of transmitter release by electrical interference with motor nerve endings. *Proc. Roy. Soc.* B **167**, 1–7.

Katz, B. & Miledi, R. 1967*b* Tetrodotoxin and neuromuscular transmission. *Proc. Roy. Soc.* B **167**, 8–22.

Liley, A. W. 1956 The effects of presynaptic polarization on the spontaneous activity of the mammalian neuromuscular junction. *J. Physiol.* **134**, 427–443.

Miledi, R. & Slater, C. R. 1966 The action of calcium on neuronal synapses in the squid. *J. Physiol.* **184**, 473–498.

Takeuchi, A & Takeuchi, N. 1962 Electrical changes in pre- and post-synaptic axons of the giant synapse of *Loligo*. *J. gen. Physiol.* **45**, 1181–1193.

PRINTED IN GREAT BRITAIN AT THE UNIVERSITY PRINTING HOUSE, CAMBRIDGE

J. Physiol. (1967), **189**, *pp.* 535–544 535
With 8 text-figures
Printed in Great Britain

THE TIMING OF CALCIUM ACTION DURING NEUROMUSCULAR TRANSMISSION

BY B. KATZ AND R. MILEDI
From the Department of Biophysics, University College London

(*Received* 7 *November* 1966)

SUMMARY

1. When a nerve–muscle preparation is paralysed by tetrodotoxin, brief depolarizing pulses applied to a motor nerve ending cause packets of acetylcholine to be released and evoke end-plate potentials (e.p.p.s), provided calcium ions are present in the extracellular fluid.

2. By ionophoretic discharge from a 1 M-$CaCl_2$ pipette, it is possible to produce a sudden increase in the local calcium concentration at the myoneural junction, at varying times before or after the depolarizing pulse.

3. A brief application of calcium facilitates transmitter release if it occurs immediately before the depolarizing pulse. If the calcium pulse is applied a little later, during the period of the synaptic delay, it is ineffective.

4. It is concluded that the utilization of external calcium ions at the neuromuscular junction is restricted to a brief period which barely outlasts the depolarization of the nerve ending, and which precedes the transmitter release itself.

5. The suppressing effect of magnesium on transmitter release was studied by a similar method, with ionophoretic discharges from a 1 M-$MgCl_2$-filled pipette. The results, though not quite as clear as with calcium, indicate that Mg pulses also are only effective if they precede the depolarizing pulses.

INTRODUCTION

The presence of extracellular calcium ions is known to be essential for neuromuscular transmission. The principal point of calcium action is the process by which the nerve impulse releases acetylcholine from the motor nerve endings (Katz & Miledi, 1965*c*). This process can be studied even when the nerve impulse and its accompanying sodium current have been eliminated by tetrodotoxin; under these conditions a brief depolarizing pulse locally applied to the nerve ending causes 'packets' of acetylcholine to be released, provided calcium ions are present in the extracellular medium (Katz & Miledi, 1967*a*). The time course of the transmitter release could be determined with great accuracy, by measuring the statistically varying intervals between pulse and quantal e.p.p.s in a long series of observations. After a brief pulse (0·5–1 msec), there is a short delay, about 1–2 msec at 5° C, during which the probability of release does not perceptibly exceed the low background level. This is followed by a rapid rise to a peak and a gradual decline of the probability of release which may extend over 10 msec or more.

It was of interest to find out at what stage during this sequence of events the external calcium ions come into play. By close ionophoretic application, it is possible to produce a sudden increase in the local calcium concentration, and to time it fairly accurately in relation to the depolarizing pulse. The question which the present experiments are meant to answer is whether the extracellular calcium is utilized during the period preceding the transmitter release (i.e. during depolarization plus initial latency), or whether it becomes effective during the transmitter release itself.

METHODS

The procedure follows that described in previous papers (Katz & Miledi, 1965*a*, *d*; 1967 *a*, *b*). The muscle (sartorius of *Rana pipiens* or *temporaria*) was placed in a Ringer solution of low calcium (< 0·1 mM) and added magnesium (about 1 mM) content, paralysed by tetrodotoxin (about 10^{-6} g/ml.) and kept at low temperature (about 4° C). The main difference from previous work is that a twin-pipette was employed, containing 1 M solutions of NaCl and $CaCl_2$ respectively in the two barrels. The sodium channel was used to apply depolarizing and various electrical control pulses to the surface of the nerve terminal, while the other channel was used to raise the local calcium concentration at desired moments. Once a calcium effect was seen, the strength and duration of the pulse was reduced and the pipette carefully re-positioned until an optimum effect was obtained. E.p.p.s were recorded with an intracellular electrode. Prostigmine (10^{-6} g/ml.) was used in most experiments. In other experiments, twin- or triple-pipettes were used one of whose barrels contained 1 M-$MgCl_2$ replacing, or in addition to, the calcium pipette.

The use of multiple-barrel pipettes made special precautions necessary. A strong current pulse through one of the barrels was apt to cause a transient resistance change in an adjacent barrel, possibly by dislodging charged particles from the tip. This was checked by monitoring the current intensities, and it was verified that, usually, barrel interactions of this kind were not large enough to vitiate the results.

When applying a strong positive-going pulse to the calcium pipette, the possibility of electrical (as distinct from ionophoretic) effects had to be considered (see Katz & Miledi, 1967 *b*, Methods). Certain control experiments have already been reported (Katz & Miledi, 1967 *b*): depolarizing pulses were followed after intervals of 20–140 μsec by similar hyperpolarizing (positive-going) pulses from the same sodium pipette. It was shown that there is no significant interference by the positive pulse, if it was applied more than 100 μsec after the end of the depolarization, and only a small reduction of the transmitter release when applied within 20–50 μsec. These earlier findings are also relevant to the observations described below.

RESULTS

Figures 1 and 2 illustrate results of two experiments. A series of depolarizing pulses (*P*) was applied via the sodium barrel, at intervals of several seconds. The pulses given by themselves failed to evoke more than very infrequent unit e.p.p.s. If P was preceded by a brief ionophoretic discharge of calcium from the other barrel (positive-going pulse Ca), the failure rate was greatly reduced, and unit responses occurred at varying times as shown in Figs. 1–4. (Italicized symbols (Ca, Mg) refer to ionophoretically applied doses of these substances.)

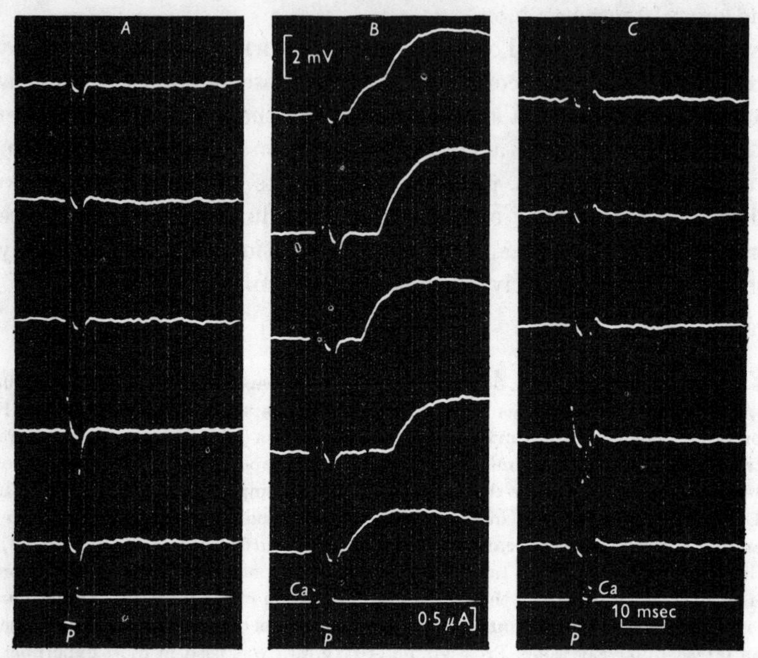

Fig. 1. Effect of ionophoretic pulses of calcium (Ca) on end-plate response. Depolarizing pulses (P), and calcium were applied from a twin-barrel micropipette to a small part of the nerve–muscle junction. Intracellular recording from the end-plate region of a muscle fibre. Bottom traces show current pulses through the pipette. Column *A*. Depolarizing pulse alone. *B*. Calcium pulse precedes depolarizing pulse. *C*. Depolarization precedes calcium pulse. Temperature 4° C.

To obtain a maximum effect, the interval between Ca and P pulses had to be adjusted so as to allow the calcium concentration to reach a peak at the critical site and time. In the case of Fig. 5, the optimum Ca–P interval was about 10–20 msec. But a significant effect could be obtained with much shorter intervals, and this was important for the present study. With careful placing of the pipette, a facilitating action of calcium could be observed when a Ca pulse as brief as 1 msec was applied, separated from the start of the depolarizing pulse by as little as 50–100 μsec.

In the experiments illustrated in Figs. 1 and 2, pairs of $Ca+P$ pulses alternated with pairs in which the time sequence of Ca and P was reversed. The result was unequivocal: Ca pulses given *after* the depolarization were ineffective. They failed to raise the level of response even if they were applied immediately after P, i.e. during the minimum latent period of the recorded response. It should be noted that it was not possible to synchronize the Ca and P pulses, because simultaneous application of the two pulses of opposite sign cancelled some of the depolarization produced by P alone.

One may conclude from these experiments that the utilization of external calcium ions occurs during, and possibly immediately after, the depolarization, that is during a period preceding that of increased probability of transmitter release. It is true that the time resolution of the ionophoretic method is limited because of inevitable diffusion delay. Nevertheless, the difference in timing between effective Ca pulses, immediately before P,

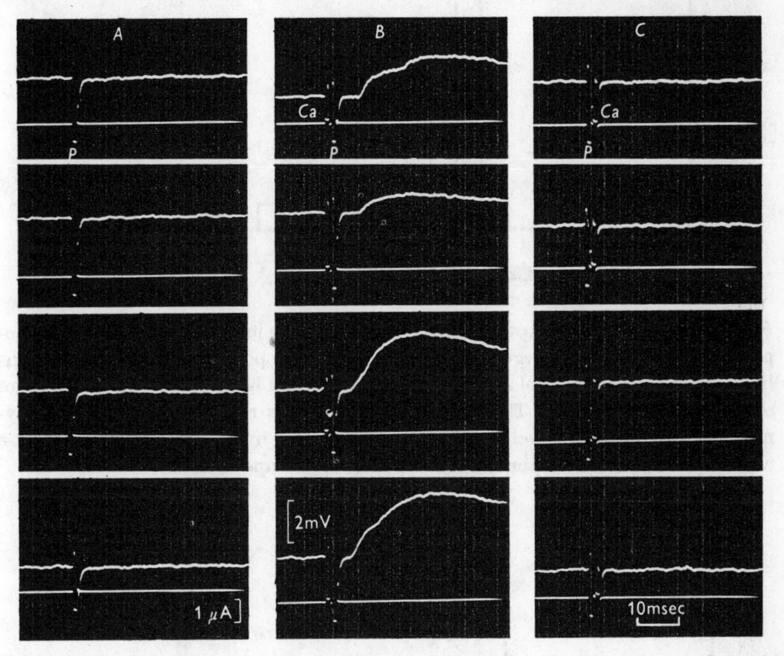

Fig. 2. A, depolarizing pulses (P); B, Ca pulses $+P$; C, $P+Ca$. The currents through the twin-pipette are shown in the bottom trace of each record. Temperature 3° C.

and ineffective Ca pulses, after P, was so small that our interpretation cannot be in doubt. If one were to argue that the responses obtained with $Ca+P$ are due to external calcium ions reaching the critical membrane sites only after the initial latent period, then there would be no explanation for the fact that calcium ions discharged ionophoretically during the latency fail to promote the occurrence of later units of response (Figs. 3, 4).

A number of controls were made to check on any electric interaction between the two barrels. In addition to the observations mentioned in Methods, a test was made to see whether the negative P pulse from the sodium barrel interfered in any way with the discharge of calcium by an immediately following pulse from the other barrel. This point was examined by giving a triple sequence of pulses $P_1 + Ca + P_2$, and comparing the effect with the usual $Ca + P$ action (relating, in each case, the response to that obtained without the *Ca* pulse). The result is shown in Table 1. The calculated values of '*m*' are not at all accurate, but there was clearly no evidence for any substantial reduction of the calcium effect by the immediately preceding negative pulse *P*.

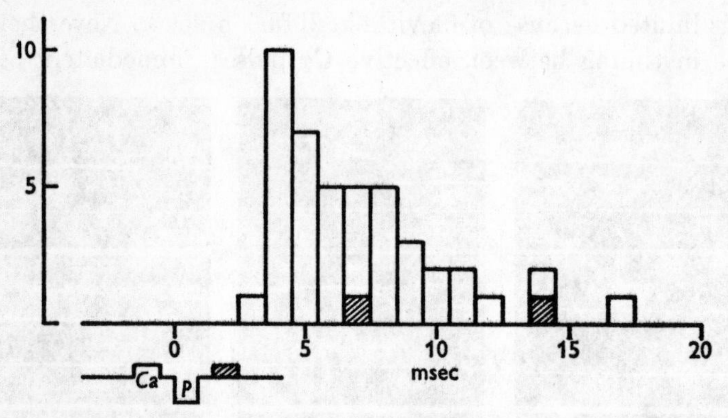

Fig. 3. Histogram of 'synaptic delays'. Abscissa: time interval between start of depolarizing pulse (*P*) and beginning of a unit end-plate potential. The time relations between calcium and depolarizing pulse are indicated below. Ordinate: number of observed unit potentials. The main histogram shows responses evoked by forty-nine ($Ca+P$) pulses. Shaded blocks: responses evoked by 48 ($P+Ca$) pulses. (These were as infrequent as responses due to *P* alone.) Temperature 3° C.

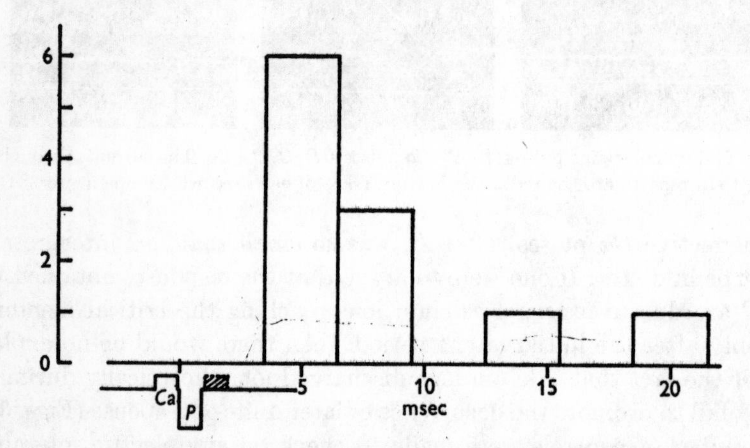

Fig. 4. Another histogram, obtained with very brief separation between *P* and *Ca* pulses. $Ca+P$ gave eleven unit responses (as shown) to twenty-nine pulses. $P+Ca$ (see shaded Ca block below) produced no response in fourteen trials. *P* alone also failed to evoke any response. Temperature 4·5° C.

Another test was to verify that a *Ca* pulse following *P* did not *reduce* the response. This was done simply by raising the steady Ca-efflux from the pipette (see Katz & Miledi, 1965*a*, *c*) and repeating the experiment with an increased initial rate of *P* responses.

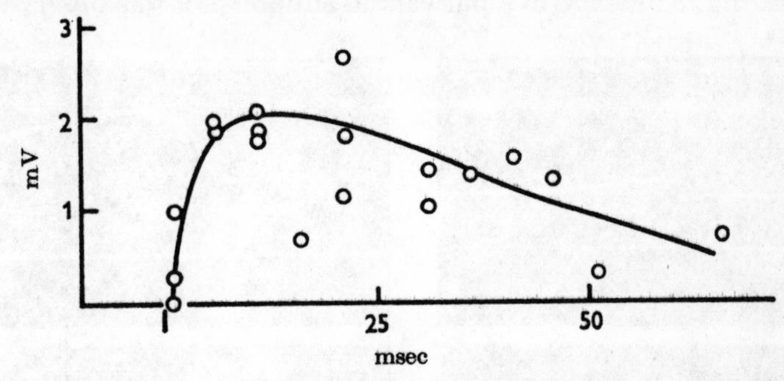

Fig. 5. Effect of time interval between *Ca* and *P* on end-plate response. Abscissa: Interval between start of *Ca* pulse (1 msec duration) and 2·7 msec depolarizing pulse *P*. Ordinates are amplitudes of individual e.p.p.s evoked by *P*. Temperature 3° C.

TABLE 1. Control experiment

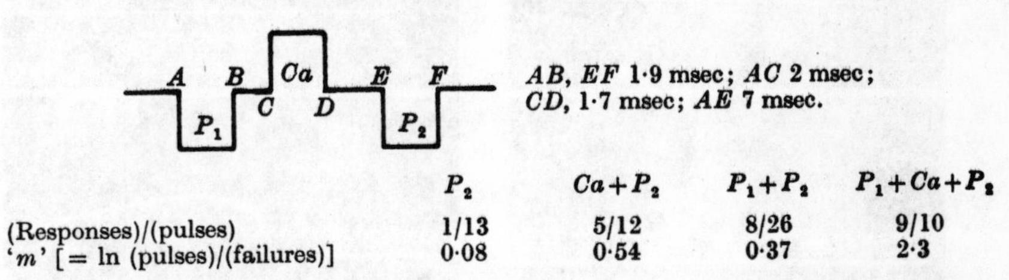

AB, *EF* 1·9 msec; *AC* 2 msec; *CD*, 1·7 msec; *AE* 7 msec.

	P_2	$Ca+P_2$	P_1+P_2	P_1+Ca+P_2
(Responses)/(pulses)	1/13	5/12	8/26	9/10
'*m*' [= ln (pulses)/(failures)]	0·08	0·54	0·37	2·3

In a few experiments, the inhibitory influence of magnesium on transmitter release was studied by ionophoretic application (Figs. 6–8). Magnesium is less potent in equimolar concentration than its antagonist calcium, and as a consequence it was difficult to produce a sufficiently intense suppression by brief ionophoretic discharges from the *Mg*-barrel. Indeed, the positive pulses had to be made so strong that local membrane break-down was a serious risk (see Katz & Miledi, 1967*b*). It was possible, nevertheless, to obtain clear inhibitory effects with *Mg* pulses of 5 msec duration provided they were applied *before* *P*. Figs. 7 and 8 illustrate an experiment with a triple pipette (the three barrels containing respectively, Ca, Na and Mg). A *Ca* pulse was given throughout, preceding *P* by about 15 msec, in order to elicit a background rate of response sufficiently high for testing the suppressing action of *Mg*. Figure 8 combines the results of two complete series in which somewhat different interval settings for *P* and *Mg* pulses were chosen. The bottom part shows the responses to eighty-three *P* pulses in the absence of magnesium. When *Mg* pulses preceded *P*, the average value of m (i.e. ln (number of pulses)/(number of failures)) was reduced from 0·96 to 0·22, and from 0·4 to 0·03 in the two series, respectively. The residual responses (to sixty-one pulses) are represented in the histogram at the top (Fig. 8A). When the Mg pulses *followed* P (middle histogram, Fig. 8B, sixty-five pulses), no suppression was observed.

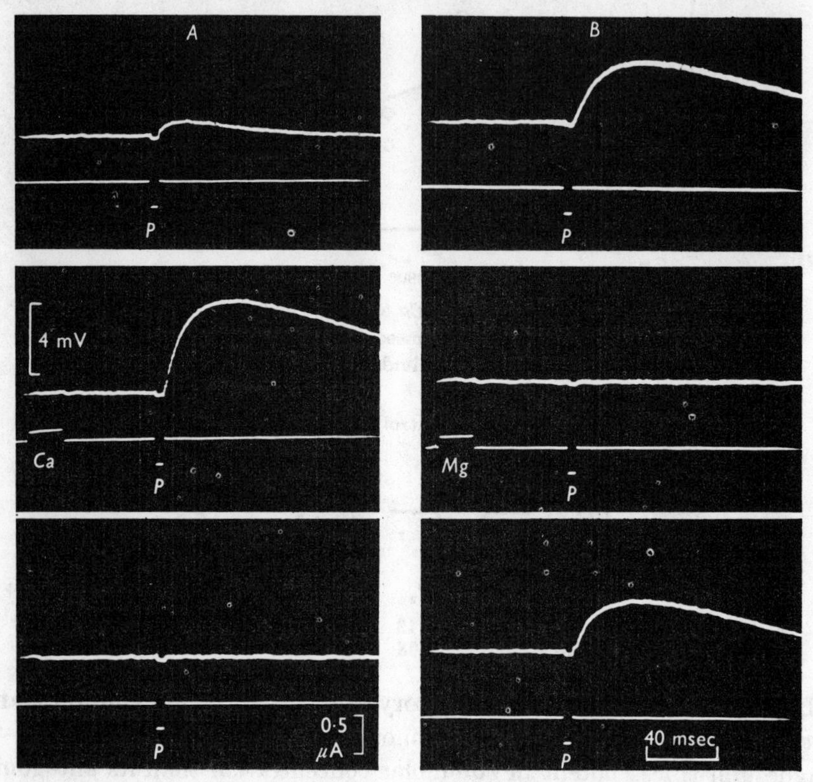

Fig. 6. A triple-barrel pipette containing, respectively, Ca, Na (for depolarization P), and Mg, was used. Column A shows the facilitating action of a calcium pulse; column B shows the inhibiting influence of a magnesium pulse. To demonstrate the inhibitory effect, the bias on the calcium barrel was adjusted between A and B, so as to increase the response to P alone. Temperature 5° C.

Several other experiments of a similar kind were made; in some the result differed from that shown in Fig. 8, in that the Mg pulse reduced the response slightly even when it was applied immediately after P. It is not certain whether this was attributable to the magnesium ion, or to the small suppressing effect which a hyperpolarizing pulse had occasionally been found to produce (Katz & Miledi, 1967b). Apart from this small and inconsistent effect, the results were in line with those of Fig. 8.

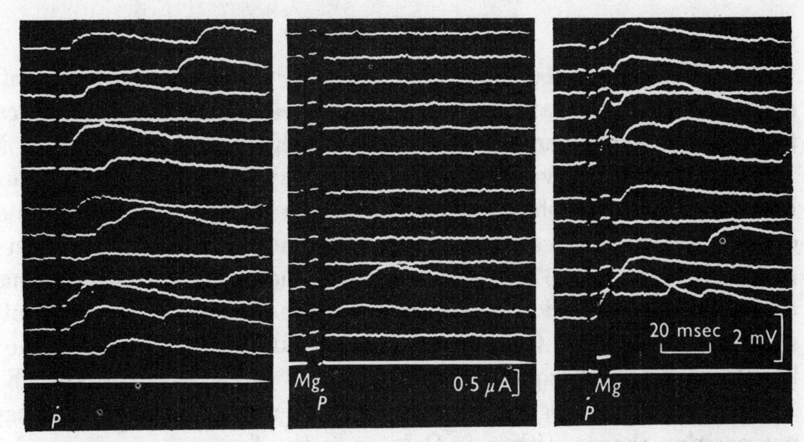

Fig. 7. Effect of *Mg*-pulses on end-plate response. Left column: Responses to depolarizing pulses *P* alone. Middle column: *Mg* pulse precedes *P*. Right column: *Mg* pulse follows *P*. Note: in the right column, the positive-going *Mg* pulse overlapped in time with the rising phase of some of the responses, and the associated focal hyperpolarization caused a transient increase in the amplitude of the e.p.p. (see Katz & Miledi, 1965*a*). Temperature 4° C. In this figure and Fig. 8, a *Ca*-pulse was applied in all trials, preceding *P* by about 15 msec, to raise the level of response sufficiently for inhibition to be demonstrable.

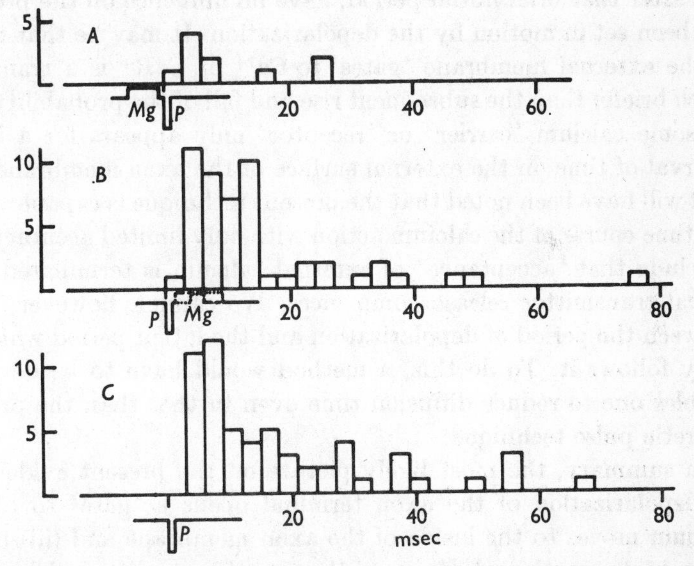

Fig. 8. Histograms of synaptic latencies. Abscissa: time interval between start of depolarizing pulse and start of unit e.p.p. Ordinate: number of observed unit potentials. Top histogram *A* shows distribution of responses to 61 $(Mg+P)$ pulses. *Middle* (*B*): Responses to 65 $(P+Mg)$ pulses. Two series with different *P–Mg* intervals as indicated by the interrupted and dotted outlines of the *Mg* pulses. Bottom histogram *C*: responses to eighty-three depolarizing pulses alone. Temperature 3° C.

DISCUSSION

It has been suggested that the first step by which depolarization of nerve endings leads to quantal release of acetylcholine is influx of calcium through the axon membrane (Hodgkin & Keynes, 1957; Katz & Miledi, 1967*b*). Recent observations have shown that there is a considerable delay between a brief depolarizing pulse and the release. The release may not even commence until sometime after the end of the depolarization (Katz & Miledi, 1965*b*, 1967*b*). The question, therefore, arises what happens during this latent period. Does the inward movement of calcium not begin until sometime after the end of the depolarizing pulse? Or is the delay largely due to subsequent reaction steps? In a recent paper (Katz & Miledi, 1967*b*), we suggested that calcium enters the membrane carrying net positive charge, either as Ca^{2+} or in the form of a compound CaR^{+}. This view was put forward to explain the peculiar 'latency shift', that is an increase in the minimum latency when the depolarizing pulse was lengthened.

The present findings suggest that the postulated entry of external calcium ions into the axon membrane is halted very soon after the end of the depolarizing pulse. Calcium ions which are made available on the outside after that brief initial period, have no influence on the process which has been set in motion by the depolarization. It may be that the opening of the external membrane 'gates' to Ca^{2+} or CaR^{+} is a transient event much briefer than the subsequent rise and fall of the probability of release. Or some calcium 'carrier' or 'receptor' only appears for a brief initial interval of time on the external surface of the axon membrane.

It will have been noted that the present technique is capable of resolving the time course of the calcium action with only limited accuracy. One may conclude that 'acceptance' of external calcium is terminated before the actual transmitter release commences. We cannot, however, distinguish between the period of depolarization and the latent period which immediately follows it. To do this, a method would have to be devised which enables one to reduce diffusion time even further than the present iono-phoretic pulse technique.

In summary, the most likely picture on the present evidence is this: (i) depolarization of the axon terminal opens a 'gate' to calcium; (ii) calcium moves to the inside of the axon membrane and (iii) becomes involved in a reaction which causes the rate of transmitter release to increase and which contributes a large part of the synaptic delay.

REFERENCES

HODGKIN, A. L. & KEYNES, R. D. (1957). Movements of labelled calcium in squid giant axons. *J. Physiol.* **138**, 253–281.

KATZ, B. & MILEDI, R. (1965*a*). Propagation of electric activity in motor nerve terminals. *Proc. R. Soc.* B **161**, 453–482.

KATZ, B. & MILEDI, R. (1965*b*). The measurement of synaptic delay, and the time course of acetylcholine release at the neuromuscular junction. *Proc. R. Soc.* B **161**, 483–495.

KATZ, B. & MILEDI, R. (1965*c*). The effect of calcium on acetylcholine release from motor nerve terminals. *Proc. R. Soc.* B **161**, 496–503.

KATZ, B. & MILEDI, R. (1965*d*). The effect of temperature on the synaptic delay at the neuromuscular junction. *J. Physiol.* **181**, 656–670.

KATZ, B. & MILEDI, R. (1967*a*). Tetrodotoxin and neuromuscular transmission. *Proc. R. Soc.* B **167**, 8–22.

KATZ, B. & MILEDI, R. (1967*b*). The release of acetylcholine from nerve endings by graded electric pulses. *Proc. R. Soc.* B **167**, 23–38.

2. INHIBITION

Inhibitory synaptic influence reduces the effectiveness of excitatory agents in influencing the postsynaptic element or reduces (or stops) spontaneous (pacemaker) activity. Three inhibitory synaptic mechanisms have been detailed. One is electrically mediated inhibition, seen in goldfish Mauthner neurons (Furukawa and Furshpan, 1963). In the other known cases inhibition is chemically mediated and involves specific ionic conductance changes of the postsynaptic membrane or an effect exerted by the transmitter on presynaptic terminals (Dudel and Kuffler, 1961c). The similarities in the mechanisms of excitatory and inhibitory synaptic transmission suggest that the general principles linking release of inhibitory neurotransmitter and nerve impulses are the same as were developed in the studies of excitatory synaptic transmission.

Inhibitory transmitter substances produce an increase of the postsynaptic membrane conductance to ionic species having their equilibrium potential near the resting membrane potential (K^+ and/or Cl^-). Two effects result. First, the increase in conductance makes excitatory sources of current (e.g., excitatory junction potentials, sensory generator potentials) less effective in changing the membrane potential. Second, if excitatory influence depolarizes the membrane potential from the equilibrium level for the inhibitory conductance change, inhibitory synaptic current will flow to restore the membrane potential toward the inhibitory equilibrium potential. These mechanisms are involved to different degrees at different synapses: the conductance increase seems to be the more important one at the crustacean nerve-muscle junction (Fatt and Katz, 1953; Dudel and Kuffler, 1961c), and currents seem more important at the spinal motor neuron (Coombs, Eccles, and Fatt, 1955). Further work along the lines pioneered by Coombs *et al.* to determine the specificity of the inhibitory conductance changes have emphasized the involvement, to differing degrees at different synapses, of both K^+ and Cl^- (e.g., Boistel and Fatt, 1958; Hagiwara, Kusano, and Saito, 1960; Eccles, Eccles, and Ito, 1964. The last of these revises the conclusion that K^+ is the major ion involved reached by Coombs et al., 1955).

Presynaptic inhibition, here described by Dudel and Kuffler (1961c) is an important integrative phenomenon. In the crustacean neuromuscular junction and the mammalian central nervous system (see KUNO, 1964a, b), presynaptic inhibition can be explained as resulting from a decreased probability of quantum release which may be due to decreased size of the action potential in the presynaptic terminal. The presynaptic action potential is thus less effective in releasing excitatory transmitter. A possible anatomical basis for presynaptic inhibition has been found in electron-microscopic studies of crustacean peripheral synapses (Atwood and Jones, 1967) and of vertebrate central nervous system (e.g., Gray, 1962). These studies show axon terminations, with typical synaptic structures, on other axon terminals.

Other chemical synaptic mechanisms than those described here have been suggested. An interesting example is that of PINSKER AND KANDEL (1969). They show that *Aplysia* abdominal ganglion interneuron L10 mediates with acetylcholine a prolonged postsynaptic membrane depolarization (late inhibitory postsynaptic potential, IPSP). The late IPSP does not invert with increases in membrane potential of 80 mV and is not altered by substituting propionate for chloride ion or reducing external potassium ion. Pinsker and Kandel postulate the existence of an acetylcholine activated, metabolically dependent, electrogenic sodium pump in the postsynaptic membrane.

30

THE EFFECT OF INHIBITORY NERVE IMPULSES ON A CRUSTACEAN MUSCLE FIBRE

By P. FATT and B. KATZ

From the Biophysics Department, University College, London

(*Received* 28 *January* 1953)

Synaptic inhibition has recently been studied by Brock, Coombs & Eccles (1952), who found that an inhibitory nerve impulse increases the resting potential of a motor nerve cell, thereby raising its threshold to excitatory impulses. The authors concluded that this was the sole mechanism by which direct inhibition of spinal motoneurones is produced, and it is of interest to inquire to what extent this conclusion may be generalized. A very different preparation for a study of direct synaptic inhibition is the crustacean nerve-muscle system (Biedermann, 1887; Hoffmann, 1914; Marmont & Wiersma, 1938; Kuffler & Katz, 1946) which enables one to stimulate single inhibitor and motor nerve axons and record their effects on individual muscle fibres (see Fatt & Katz, 1953*a*). Previous work indicated that the inhibitory process interferes with the crustacean muscle response at two stages: (*a*) in blocking transmission from the nerve to the muscle membrane; and (*b*) in uncoupling excitation and contraction processes within the muscle fibre. Only the first of these two actions is considered in this paper, which deals particularly with the electrical membrane changes set up in a crustacean muscle fibre by inhibitory nerve impulses.

It will be shown that the main effect of inhibitory impulses is to *attenuate the 'end-plate potentials'*, i.e. to diminish the local depolarization produced by motor impulses. Inhibitory impulses do not by themselves change the resting potential of the muscle fibres, unless this has previously been displaced from its normal level. But even though no potential change may be recorded, the inhibitory impulse was found to have a peculiar effect on the electrical properties of the muscle membrane: it always produces a transient increase of membrane conductance (or 'ion permeability') whose nature will be discussed below.

METHODS

The two nerve-muscle preparations used for most experiments were: the opener of the claw of the hermit crab (*Eupagurus bernhardus*) and the flexor of the dactylopodite of the shore crab (*Carcinus maenas*). The former muscle is supplied by two axons, one motor and one inhibitor, which run in separate nerve bundles (see Wiersma & Harreveld, 1935) and thus can be stimulated separately without requiring isolation. In *Carcinus*, the flexor muscle is supplied by two motor axons which provide a double innervation for the muscle fibres (cf. Fatt & Katz, 1953*b*) and one inhibitor axon. The motor axons are easily found because they are the two outstanding fibres in the thicker nerve bundle and can be isolated without difficulty. Muscle fibres were exposed by carefully removing the part of the shell which covered them.

The preparations were mounted on glass slides under rubber bands; a rubber tube was slipped over the opened tip of the dactylopodite (*Carcinus*) or propodite (*Eupagurus*) and connected to a perfusion bottle. The preparation was then immersed in crab Ringer (Fatt & Katz, 1953*a*) in a large Petri dish which also contained the 'indifferent' stimulating and recording electrodes.

The motor and inhibitor axons were stimulated separately, without lifting them from the saline bath, by applying capillary electrodes close to their surface (cf. Fatt & Katz, 1953*b*). This was a convenient procedure, but sometimes gave rise to rather large shock artifacts (e.g. Figs. 1 and 2). Two pulse stimulators were used to provide excitor (*E*) and inhibitor (*I*) shocks of adjustable intervals and independently variable intensities.

The electric responses of crustacean muscle increase in amplitude with the frequency of nerve impulses (Katz, 1936; Wiersma, 1941; Katz & Kuffler, 1946), and the response produced by a single impulse is generally very small. It was necessary, therefore, to use repetitive stimulation: the frequency was kept at about 30 per sec, adjusted so as to give moderately large potential changes without vigorous mechanical responses. The records shown below were all due to multiple cathode-ray sweeps whose frequency was the same as, or one half of (Fig. 6), that of the stimuli.

Two intracellular electrodes were used, one to record membrane potentials of individual muscle fibres and the other, when required, to pass current through the fibre membrane. The procedures which were followed in recording membrane potentials, displacing the resting potential and measuring the fibre constants have been described in detail in previous papers (Fatt & Katz, 1951, 1953*a*).

RESULTS

The effect of inhibitory impulses on the 'end-plate potential'

When the motor nerve to a crustacean muscle is stimulated at a frequency of about 30 per sec, non-propagated action potentials are set up in the muscle fibres accompanied usually by a weak contraction (cf. Katz & Kuffler, 1946). The action potentials are in many ways analogous to the end-plate potentials (e.p.p.) of vertebrate muscle (Wiersma, 1941; Katz, 1949), and this term has been used to describe them in spite of morphological differences between crustacean and vertebrate nerve-muscle junctions. In Fig. 1, *E*, e.p.p.'s are shown obtained with an intracellular electrode from the opener of the claw of a hermit crab. The record is due to repetitive responses during stimulation at 33 per sec. The size of the e.p.p. varies in different muscles and different fibres, and depends upon the frequency and number of nerve impulses (see Katz & Kuffler, 1946). The time course of the e.p.p. was also found to vary widely: in some fibres the e.p.p. decayed with a time constant of 2–3 msec, in others 10 or 20 times more slowly. This large difference might be due either to a variable persistence of the nerve-muscle transmitter, or to a variable time constant of the muscle membrane. We observed that the longest e.p.p. (decay constant approximately 65 msec) occurred in a fibre of exceptionally large membrane time constant (53 msec) determined by the rectangular pulse technique (Hodgkin & Rushton, 1946; Fatt & Katz, 1951). In a few other experiments in which both time constants were measured, a similarly good agreement was obtained, and it appears therefore that in these experiments the rate of fall of the e.p.p. is controlled by the resistance and capacity of the muscle fibre, rather like that of the e.p.p. in curarized vertebrate muscle.

There are, however, certain exceptions: for example, in some fibres of the flexor muscle of the dactylopodite (*Carcinus maenas*) e.p.p.'s of different rates of decay were produced by stimulating different motor axons (Fatt & Katz, 1953*b*) an effect which cannot be explained on the present simple hypothesis.

Usually, the decline of the e.p.p. in a crustacean muscle fibre, though not strictly exponential, approaches the exponential time course much more closely than does that of the 'focal' e.p.p. in a frog muscle fibre (Fatt & Katz, 1951). The analysis is thus simplified considerably; the reason for the observed exponential decay is almost certainly to be found in the much more uniform spatial spread, along the fibres, of the crustacean e.p.p. (Fatt & Katz, 1953*b*).

Another peculiarity was the large amplitude fluctuation of the response with successive impulses (e.g. Fig. 1, *E*). It could not be attributed to mechanical movement artifacts, for it was observed even when no muscle contraction was seen. It might be due to an intermittent failure of nerve impulses to conduct into all the terminal axon branches, or alternatively to some random event affecting each nerve terminal individually.

When both inhibitor (*I*) and motor (*E*) axons were stimulated, the e.p.p. was reduced in size as previously reported by Marmont & Wiersma (1938) and Kuffler & Katz (1946). The attenuation of the e.p.p. depended upon the time interval between *I* and *E* shocks; it was greatest when *I* slightly preceded *E*. The maximum reduction of the e.p.p. varied considerably; in some fibres it amounted to as much as 90%. Typical results are shown in Figs. 1, 2 and 4. When measuring the e.p.p. size, the 'phasic' amplitude of the responses was used (see Fig. 3), not the total deflexion which is complicated by summation of successive potentials. An 'inhibited' e.p.p. differed from the normal not only in amplitude, but also in time course, its rate of decay being faster. Thus, in the example of Figs. 1 and 2 the decay constant (fall to $1/e$) of the normal e.p.p. was 27 msec, that of the maximally 'inhibited' e.p.p. was 10 msec. When an inhibitory impulse arrived during the falling phase of an e.p.p. it caused the rate of decline of the e.p.p. to increase abruptly, by at least 50% (e.g. Fig. 4, record 1*a*).

Inhibited e.p.p.'s showed even more pronounced amplitude fluctuations than normal e.p.p.'s (cf. Fig. 5). It seems that fluctuations occurring simultaneously at excitatory and inhibitory nerve endings have additive effects; moreover, fluctuations in the time of arrival of *I* and *E* impulses would also accentuate the phenomenon.

In Fig. 5 'inhibition curves' are plotted, showing the attenuation of the e.p.p. at various $I-E$ intervals. The results are similar to those already described by Marmont & Wiersma (1938) and Kuffler & Katz (1946). The curve is of interest because it provides an indication of the time course of inhibitory activity. When the inhibitor impulse arrives at a nerve terminal, it presumably releases some agent which has the power of locally suppressing the subsequent e.p.p.'s. This power gradually disappears so that less and less attenuation is produced as the excitor impulse arrives progressively later. The gradual return of the 'inhibitor curve' from maximum to no attenuation

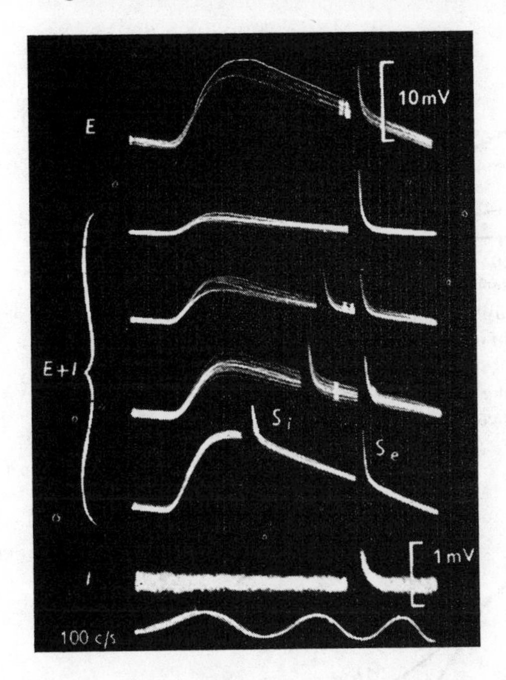

Fig. 1.

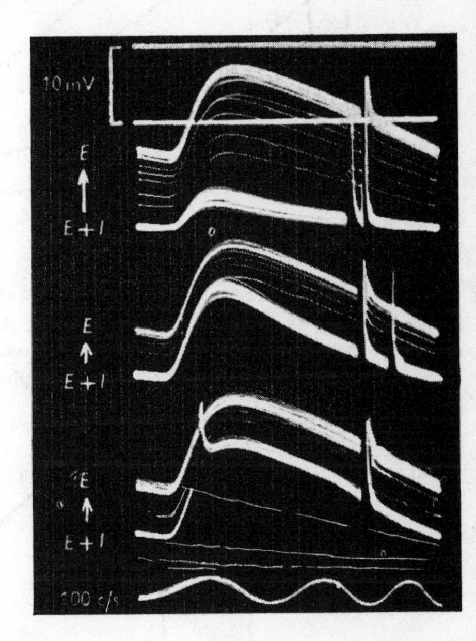

Fig 2.

Fig. 1. Effect of inhibitory impulse on crustacean 'end-plate potential'. Intracellular recording. Muscle fibre of opener of claw of hermit crab. Stimulation at 33/sec. Each record is a photograph of multiple sweeps, the stimulus artifacts (S_e and S_i) appearing on the falling phases of preceding responses. E (low amplification): stimulation of excitor axon, producing e.p.p. I (at higher amplification): stimulation of inhibitor axon alone, producing no potential change. $E+I$ (low amplification): combined stimulation of both axons, producing attenuated e.p.p. E shock was fixed (right hand artifact), the I shock preceding by varying intervals (successively from above: 1·2, 4·2, 6·1 and 11·3 msec).

Fig. 2. Effect of inhibitory impulse on crustacean 'e.p.p.' Repetitive time-base sweeps and stimulation, at 33 per sec, as in Fig. 1. Three recordings at different $I-E$ intervals, showing the response to combined $(E+I)$ stimulation and the transition to a pure E response, when inhibitory impulses are stopped. Note the difference between the 'inhibited' and the first 'uninhibited' e.p.p. in each recording. Time interval between I and E shocks: top records, I precedes E by 1·4 msec; middle, I follows E after 3·8 msec; bottom, I half-way between E shocks (15 msec interval). The decline of the last e.p.p. towards the base-line is also shown in this recording.

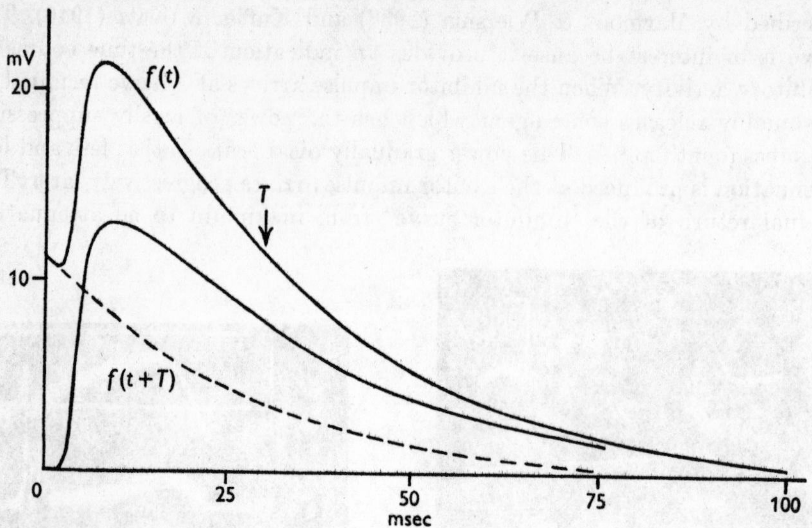

Fig. 3. Procedure used in measuring 'phasic' amplitude of e.p.p. during repetitive stimulation. The top record $f(t)$ shows the last of a series of e.p.p.'s obtained during steady stimulation at 33 per sec. The broken curve $f(t+T)$ shows the remainder of the potential after time T, T being the interval between stimuli. Assuming that the recorded potential is due to algebraic summation of successive e.p.p.'s, the phasic component is given by the difference between $f(t)$ and $f(t+T)$.

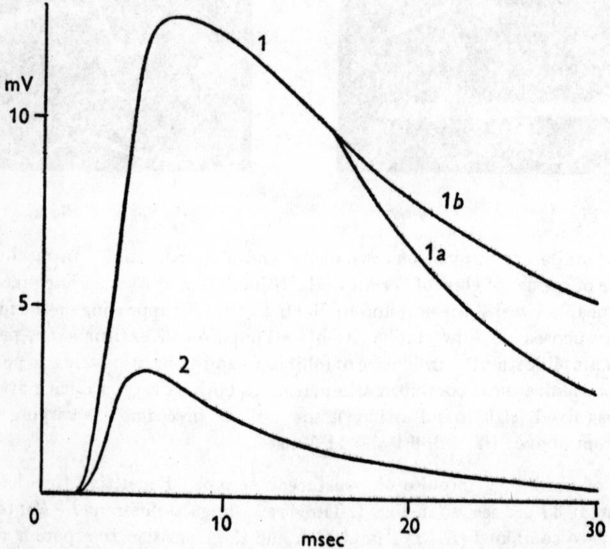

Fig. 4. Superimposed drawings of 'phasic' e.p.p.'s, from the experiment illustrated in Figs. 1 and 2. (These curves are single examples; to appreciate the random fluctuations of successive e.p.p.'s, Figs. 1 and 2 should be consulted.) 1: I shock follows 15 msec after E and accelerates the decay of the e.p.p.; 1 a, during steady $E+I$ stimulation; 1 b, first response after cessation of inhibitor impulses. 2: I precedes E by 1·6 msec, attenuating the amplitude of the e.p.p.

indicates, therefore, the time course of subsidence of the inhibitory transmitter potency. The initial rise of the inhibitor action cannot be deduced from Fig. 5; it may well be faster than the initial falling phase of the 'inhibition

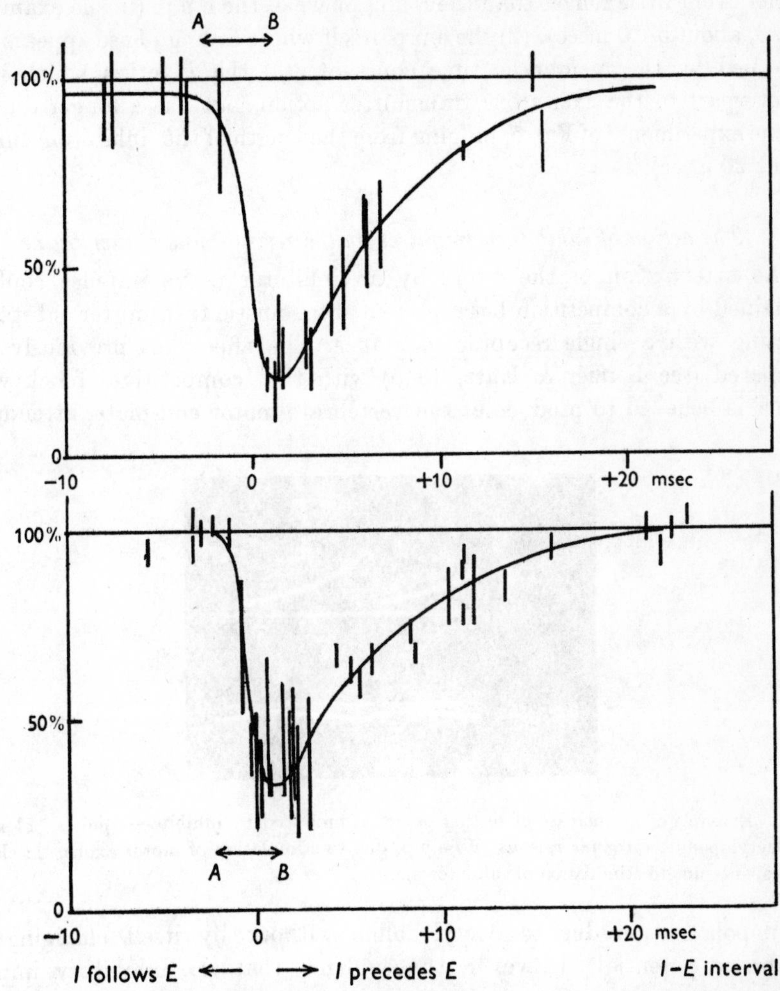

Fig. 5. 'Inhibition curves.' Results of two experiments showing dependence of e.p.p. amplitude on time interval between I and E impulses. Abscissa: interval between inhibitor and excitor shocks, in msec. Ordinate: 'phasic' amplitude of e.p.p. as a percentage of the normal. The vertical bars indicate the range of scatter observed in the recordings (cf. Figs. 1 and 2). Both experiments from opener of claw of hermit crab.

curve' $(A-B)$. When the I impulse arrives *after the beginning of the e.p.p.* it can only stop its further ascent; hence, the initial portion $(A-B)$ of the 'inhibition curve' has no special significance, except that it cannot be shorter than the rising phase of the e.p.p.

In considering the interaction of excitor and inhibitor impulses we have encountered, so far, three events of different time course, (1) the excitatory transmitter process ($E-$action) which gives rise to the e.p.p.: it is probably a brief event little longer than the rising phase of the e.p.p. (in the example of Fig. 3, about 5–10 msec); (2) the e.p.p. itself whose falling phase appears to be governed by the membrane time constant; (3) the I action which is the counterpart to the excitatory transmitter action, but has a longer duration (in the experiments of Fig. 5, judging from the length of the 'inhibition curves', about 20 msec).

The action of inhibitory impulses on the resting muscle membrane

The attenuation of the e.p.p. by an inhibitor nerve impulse could be explained by a competition between two antagonistic transmitter substances, reacting with a single receptor substance. The effect has previously been compared (see Kuffler & Katz, 1946) with the 'competitive' block which curare is believed to produce at the vertebrate motor end-plate, attenuating

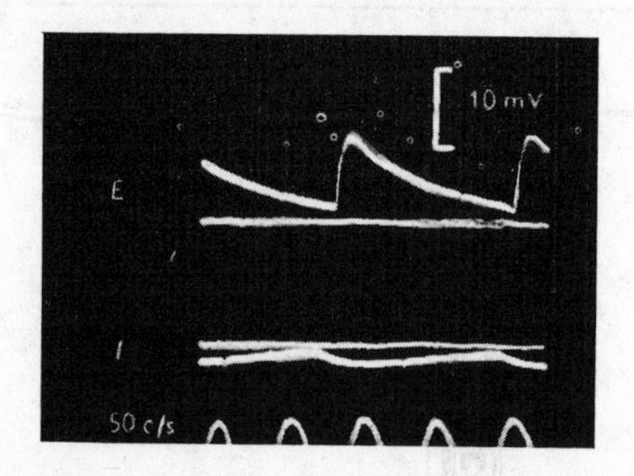

Fig. 6. Example of an increase of resting potential produced by inhibitor impulses. Flexor of dactylopodite, *Carcinus maenas*. E: e.p.p. due to stimulation of motor axons. I: electric response due to stimulation of inhibitor axon.

the depolarization due to acetylcholine without by itself changing the membrane potential. However, the evidence that the inhibitory impulse produces by itself no electrical change in a crustacean muscle fibre is not unequivocal. Biedermann (1887) claimed that inhibitory nerve impulses increase the resting potential of the muscle, while other authors (Wiersma & Harreveld, 1935; Kuffler & Katz, 1946) found no such effect.

In our own experiments, we did observe at times that stimulation of the inhibitory axon produced an increase of the resting membrane potential, but the effect was small and irregular, being only seen in relatively few fibres. An example is shown in Fig. 6 where a small *inverted* e.p.p. (henceforth called 'I-potential') can be seen which is of slower time course than the 'motor e.p.p.'.

The presence or absence of this I-potential had no relation to the reduction of the e.p.p. which was always observed. In the experiment of Fig. 1, for instance, there was no trace of an I-potential, but e.p.p.'s were attenuated by 80%.

What is the explanation of this irregular potential change produced by the inhibitor impulse? It might be thought that in some, possibly damaged, preparations a steady leakage of transmitter substances occurs from the nerve endings. The muscle membrane being thus exposed to a 'steady background concentration' of E substance might be kept slightly depolarized. Under these conditions an inhibitory nerve impulse would indeed produce a transient repolarization (just as an application of curare repolarizes a vertebrate motor end-plate previously exposed to acetylcholine). This suggestion then implies that the I-potential is simply a reduction of a steady 'background e.p.p.', and that the inhibitor impulse affects the membrane potential of the muscle fibre only indirectly, by stopping the E transmitter from depolarizing it. This explanation, however, became untenable in the course of further experiments described below. In these experiments I-potentials were provoked, abolished, or *reversed* by passing conditioning currents directly through the muscle membrane, without involving excitatory nerve endings at all.

The effect of inhibitor impulses on electrically applied potential changes in the muscle fibre. When the membrane potential of the muscle fibre was displaced by an applied current, an interesting and unexpected result was obtained. There was always a unique level of the membrane potential at which inhibitory impulses produced no potential change at all. Often this level was identical with the resting potential, but occasionally several millivolts higher. When the membrane potentials had been reduced below this equilibrium level, I-potentials like that in Fig. 6 were observed, giving a transient *increase* of membrane potential, the increase being approximately proportional to the initial depolarization. Conversely, when the membrane potential had been raised above the equilibrium level, the I-potentials became reversed (e.g. Fig. 7, record Ic), the inhibitory nerve impulse now producing a transient *depolarization.*

Thus, in fibres in which no I-potential was normally seen, it could be provoked by a preliminary depolarization and reversed by a preliminary hyperpolarization of the membrane. In fibres in which a small I-potential was present 'normally', it could be increased in size by an applied outward current, or abolished and reversed by an applied inward current.

Hence, the inhibitory nerve impulse diminishes any displacement of the membrane potential, in either direction, from a certain equilibrium level; the presence, sign and magnitude of the I-potential is merely a manifestation of this effect and depends upon the sign and magnitude of the initial displacement. The situation is illustrated by the circuit diagram of Fig. 8, in which four membrane components are shown, a fixed capacity C_m, a fixed battery V_0 representing the 'equilibrium potential', and two variable conductances G_1

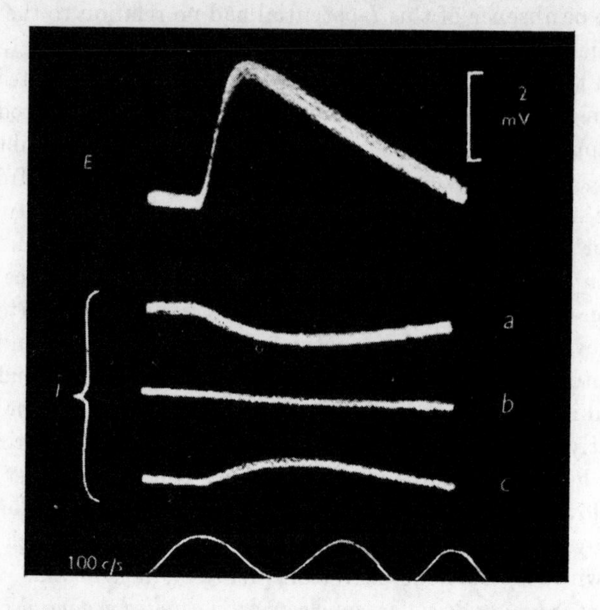

Fig. 7. *I*-potentials in a crustacean muscle fibre. Opener of claw, hermit crab. *E*: e.p.p. due to stimulation of motor axon. *I*: electric potential changes due to inhibitor impulses. The level of the resting membrane potential was lowered to 48 mV (cathodic) in *a*, and raised to 95 mV (anodic) in *c*. The resting potential in the other records was 73 mV. Note the dependence of the *I*-response on the level of the membrane potential: its absence at the normal level (*b*), and its reversal during hyperpolarization (*c*).

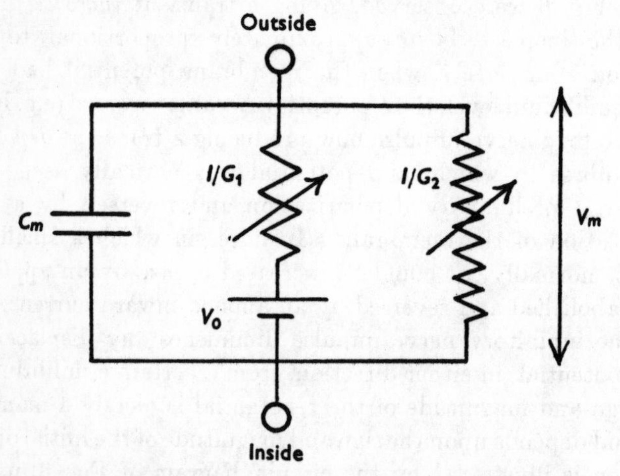

Fig. 8. Circuit model of crustacean muscle membrane. For details see text.

and G_2 which represent the permeability of the membrane to two groups of ions: (1) those whose movement serves to restore the resting 'equilibrium potential' (e.g. K and Cl, represented by the conductance G_1 in series with V_0); (2) those ions whose movements tend to depolarize (e.g. Na, represented by the leakage conductance G_2).

Suppose the inhibitory nerve impulse releases an agent which causes the 'permeability' G_1 to increase. As a result, the membrane potential V_m will tend to approach the equilibrium level V_0. If initially the 'leakage conductance' G_2 was nil, and no net current flowed through the membrane, then the membrane potential V_m was already at the equilibrium level and the inhibitory impulse, though always increasing G_1, produces no potential change.

However, if the value of V_m was initially lowered either by passing outward current through the membrane, or by introducing a leakage through G_2, then the inhibitory impulse would cause the membrane potential to increase towards V_0.

Thus, the *I*-potential could be attributed to an increase of membrane permeability to those ions (e.g. K) whose movement normally serves to maintain and restore the resting potential.

Size and time course of the I-potential. The 'phasic' amplitude of the *I*-potential amounted to only about 5% of the initial displacement of membrane potential; i.e. if the resting potential had been reduced by 20 mV below the 'equilibrium level', the *I*-deflexion after each inhibitory impulse was about 1 mV. This percentage value varied in different preparations between 3 and 10% and, as with the e.p.p., the amplitude depended upon the frequency of impulses. The total deflexion was much greater than the 'phasic', because at frequencies of 20–30 per sec successive *I*-potentials summed to at least twice their individual height.

The time course of the *I*-potential was generally slower than that of the e.p.p. observed in the same fibre. An example is shown in Fig. 9 where the time of rise of the e.p.p. is 5 msec and that of the *I*-potential about 15 msec.

Now, as with the e.p.p., the *I*-potential must be slower than the process which produces it. Suppose, for instance, that the *I*-potential is due to a transient increase of the series conductance G_1. Then even when G_1 has returned to normal, a residual *I*-potential will be left which gradually subsides with the normal time constant of the membrane. Qualitatively, it can be said that the fundamental change must be briefer than the potential which it produces, but at least as long as the period of rise of the *I*-potential.

In Fig. 9 three events, measured on the same fibre, are shown on the same time scale: the e.p.p., the *I*-potential and the 'inhibition curve' (cf. Fig. 5) which, as discussed earlier, indicates the time course of the inhibitory transmitter action. It is noteworthy that the time relations between *I*-potential and 'inhibition curve' are not unlike those to be expected between a membrane potential change and its causative agent. It thus seems possible that a *single* process, or a single reaction between inhibitor substance and membrane substrate, may simultaneously produce two effects: (*a*) attenuation of the e.p.p., and (*b*) an increase of membrane conductance G_1, which then gives rise to the *I*-potential.

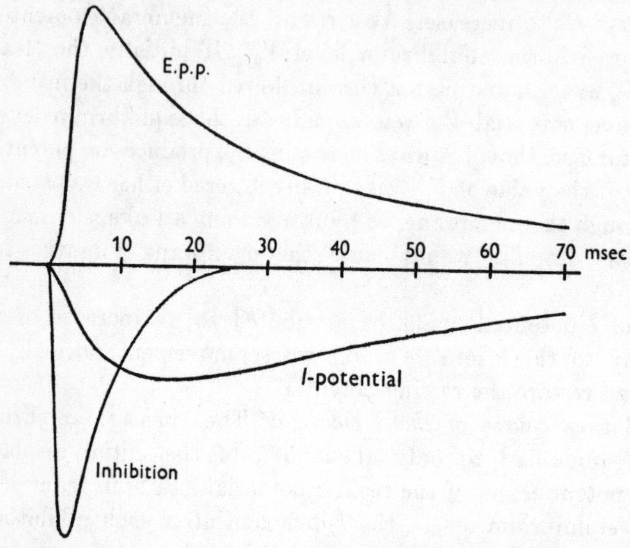

Fig. 9. Time relations of e.p.p., 'inhibition curve' (cf. Fig. 5) and *I*-potential taken from an experiment on a fibre of the opener of the claw (hermit crab). The amplitudes of the three curves have been chosen arbitrarily.

A further point of interest is the *magnitude* of the conductance change. A fairly simple estimate can be obtained in two ways. (*a*) We can make use of the sudden increase in the rate of decay of the e.p.p. which occurs when an inhibitory impulse impinges during its falling phase (e.g. Fig. 4, record 1*a*). Measurements of these rates indicate a transient increase of membrane conductance of about 50%. (*b*) It can be shown that the initial steep rate of rise of the *I*-potential (dp/dt) is related to the conductance change, in the following way:

$$r_m/r_m' = 1 - (\tau_m/p) \times dp/dt,$$

where τ_m is resting time constant of membrane, p displacement of membrane potential ($V_m - V_0$), r_m and r_m' membrane resistances, at rest and immediately following inhibitory impulse, respectively. Measurements of this kind were made in four experiments indicating a 20–50% increase of membrane conductance. Thus, while the peak of the *I*-potential amounts to only 5% reduction of the imposed p.d., the underlying conductance change appears to be several times greater.

The relation between I-potential and electric inhibition. It need hardly be emphasized that the I-potential, as such, has no significance as an inhibitory *agent*: the real inhibitory effect is the reduction of the e.p.p., which was always observed, regardless of presence or absence of the I-potential or of its electric sign.

When the membrane potential was raised by anodic current, the inhibitor impulse still caused a reduction of the e.p.p., but the effect tended to become obscured by the simultaneous appearance of the I-potential which was itself a depolarization summing with the e.p.p. This is illustrated in Fig. 10, where the electric responses to E, I and $E+I$ are shown at two levels of the membrane

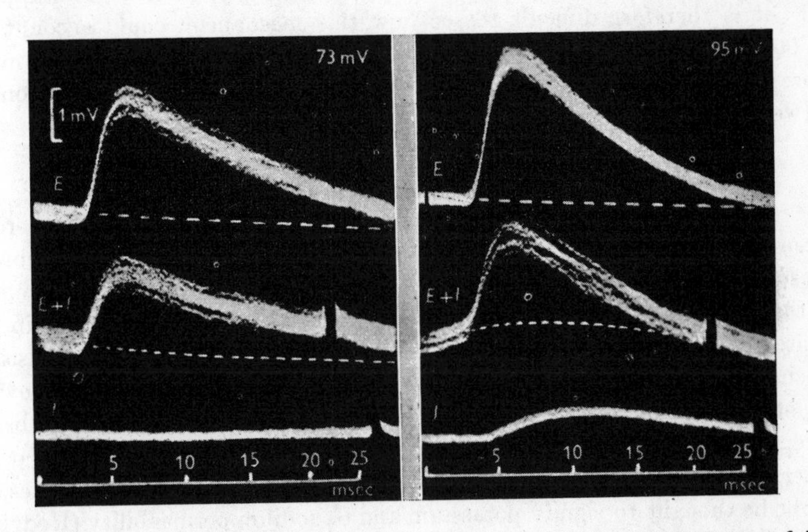

Fig. 10. E.p.p. and inhibitory effects at different levels of membrane potential. Opener of claw, hermit crab. The responses in the left-hand part were obtained at the normal resting potential (73 mV), those on the right at an anodically increased level (95 mV). E: e.p.p. due to stimulation of motor axon. Note increased size of e.p.p. at higher level of membrane potential. I: potential change due to inhibitor impulses. (The gaps are due to inhibitor shock artifacts.) $E+I$: combined stimulation of both axons. The broken lines have been drawn to indicate approximately the 'base-lines' of the phasic e.p.p.'s.

potential, (*a*) at the resting level, 73 mV, and (*b*) 22 mV higher. The 'phasic' amplitude and rate of rise of the e.p.p. were reduced by the inhibitory impulse in both cases; but in the 'hyperpolarized' state, the contribution of the I-potential was so large that the combined deflexion was only slightly less than the uninhibited e.p.p.

It would clearly be difficult to resolve the two components of the $(E+I)$ potential with any accuracy, because mutual interaction must be expected, and it is probable that the I-potential becomes reduced by the action of the excitor impulse, just as the e.p.p. is diminished by the action of the inhibitor.

Although the I-potential appears to be merely a by-product of the inhibitory process, the question arises whether the mechanism which produces it, namely an increase of the membrane conductance G_1, might not directly account for the attenuation of the e.p.p.; for the e.p.p. itself is a displacement of the membrane potential from the equilibrium level V_0. Returning to the scheme of Fig. 8 one might suggest that the excitatory transmitter substance E reacts with the membrane so as to increase the shunt conductance G_2 (e.g. sodium permeability) and thereby produces the observed depolarization. The inhibitory substance I then counteracts this effect by increasing the series conductance G_1. This simple explanation, however, fails on quantitative grounds. The increase of G_1 accounts for only a 5% reduction of an applied p.d.; it is therefore difficult to see how this mechanism could account for 80–90% reduction of an e.p.p. The attenuation of the e.p.p. is clearly a much more powerful effect and probably depends on a more specific antagonism between transmitters E and I.

DISCUSSION

The electrical effects of an inhibitor nerve impulse are twofold: there is attenuation of a subsequent e.p.p. and, *in addition*, a lowering of the membrane resistance (more specifically, of the resistance in series with the resting e.m.f. of the membrane). These two actions appear to be synchronous, within the limited accuracy of our analysis. Yet, they must be regarded as two distinct mechanisms. To explain the dual nature of the inhibitory effect, the following hypothesis may be considered. In Fig. 8, the ion permeability of the membrane is represented by two variable conductances, G_1 and G_2, one in series and the other in parallel with the resting e.m.f. (By analogy with other tissues, G_1 might be thought to signify potassium and G_2 sodium permeability (Hodgkin, 1951), but in crustacean muscle the evidence is not sufficient for such specific conclusions.) We assume that motor and inhibitor nerve impulses affect conductances G_1 and G_2 by way of a competitive reaction with a single receptor substance.

The reactions may be described as $E+R \rightleftharpoons ER$, and $I+R \rightleftharpoons IR$ (hence $ER+I \rightleftharpoons IR+E$), where R is the receptor substance in the muscle fibre, and E and I are transmitter substances released, respectively, at excitor and inhibitor nerve endings. Suppose that the value of G_2 increases with the concentration, or surface density, of ER, and that G_1 increases with IR. A motor nerve impulse will then depolarize the fibre by increasing the shunt conductance G_2; an inhibitor impulse produces two effects: (*a*) by its competitive reaction with the receptor it reduces ER and thereby attenuates the e.p.p., (*b*) its reaction-product increases G_1 and thus can give rise to an I-potential.

Thus, in spite of their apparent complexities, the observed phenomena might all be due simply to interaction of three basic entities: one receptor and two antagonistic transmitter substances.

In connexion with this hypothesis, two observations may be mentioned. (1) When the membrane potential was varied by an applied current, the size of the e.p.p. was found to vary in the same sense (e.g. Fig. 10), as it should if the effect of the motor impulse is to shunt the membrane. The effect was similar to that obtained at the amphibian nerve-muscle junction (Fatt & Katz, 1951), but the results were less accurate and can only be stated qualitatively. (2) The effect of the inhibitory impulse was studied when the muscle was soaked in a potassium-free medium. If the *I*-potential and the underlying change of G_1 were due exclusively to an increase of potassium permeability, then it should be impossible to reverse the *I*-potential in a potassium-free medium, because the theoretical equilibrium potential for potassium has been raised to infinity. However, while the resting potential increased in the potassium-free medium by some 10 mV (cf. Fatt & Katz, 1953*a*) the *I*-potential which was observed could still be reversed by 'anodizing' the membrane and raising its potential by a further 5–10 mV. It would appear, therefore, that potassium is not exclusively or specifically involved in the presumed change of G_1.

The mechanism of synaptic inhibition described here differs substantially from that found by Brock *et al.* (1952) in the cat's spinal cord. A 'hyperpolarization' of the post-synaptic membrane, superficially resembling that described by Brock *et al.*, is occasionally observed in crustacean muscle, but only when the fibres are partly depolarized, and in any case the effect is small and must be regarded as a by-product, rather than an agent, of the inhibitory process.

On the other hand, the attenuation of the e.p.p. is a very powerful effect, reaching in some cases as much as 80–90%. Brock *et al.* (1952) found no such effect in the spinal motoneurone of the cat, and perhaps we are dealing here with a mechanism which is peculiar to the crustacean nerve-muscle system. But it is also possible that a number of different synaptic mechanisms are at play at the different points in the central nervous system where inhibitory effects are produced.

The question of a direct mechanism by which the inhibitory nerve impulse can inactivate muscular contraction (Marmont & Wiersma, 1938; Kuffler & Katz, 1946) has not been investigated in the present experiments, but our results have some bearing on it. In the earlier studies, inhibitory impulses were found to stop contraction even though they arrived too late to reduce the amplitude of the e.p.p. It was suggested, therefore, that the inhibitory impulse can interfere with the contractile process directly, without operating on the fibre membrane. Our present results throw considerable doubt on this interpretation.

In the first place, it has now been shown that the inhibitor impulse does alter the membrane properties even when no potential changes are observed. Secondly, while it is true that the 'phasic' amplitude of successive e.p.p.'s may remain unchanged, if the I impulses arrive after their peaks, the decay of the e.p.p.'s becomes accelerated and their summation less effective. In this way, the mean level of the membrane potential may rise and the 'total' depolarization diminish, possibly below the level needed for contraction. The effect is illustrated in Fig. 2 where the level of the membrane potential is seen to shift by several millivolts even though the 'phasic' amplitude of the e.p.p. remains unaltered. Although this may not explain all cases of mechanical inhibition which have been observed, we feel that this matter needs re-examination.

SUMMARY

1. Experiments were made to study the effect of single excitor and inhibitor nerve fibres on the membrane of single crustacean muscle fibres.

2. An inhibitor impulse reduces the amplitude of the 'end-plate potential' (e.p.p.) due to a subsequent motor nerve impulse. This effect is observed when the interval between the antagonistic impulses is less than about 20 msec.

3. When the inhibitor impulse arrives during the falling phase of the e.p.p., it accelerates the decay of the e.p.p.

4. A 'direct' effect of the inhibitor impulse on the resting membrane potential of the muscle fibre is not usually seen. However, when the membrane potential has been displaced in either direction, by means of an applied current, the inhibitor impulse produces a potential change which is directed towards the normal resting level.

5. This action, and the acceleration of decay of the e.p.p. by the inhibitor impulse, can be explained by a lowering of the membrane resistance of the muscle fibre (more specifically of the resistance *in series* with the resting e.m.f.).

6. All the electrical effects of the inhibitor impulse are compatible with the concept of a single inhibitor/receptor reaction ($I + R \rightleftharpoons IR$), which changes the ion permeability of the fibre membrane and, at the same time, competes with the action of the excitatory transmitter ($E + R \rightleftharpoons ER$) on the common receptor molecule.

We are greatly indebted to Mr J. L. Parkinson for his invaluable help. This work was carried out with the aid of a grant for scientific assistance made by the Medical Research Council.

REFERENCES

BIEDERMANN, W. (1887). Beiträge zur allgemeinen Nerven und Muskelphysiologie. XX. Über die Innervation der Krebsschere. *S.B. Akad. Wiss. Wien*, Abt. 3, **95**, 7–40.

BROCK, L. G., COOMBS, J. S. & ECCLES, J. C. (1952). The recording of potentials from motoneurones with an intracellular electrode. *J. Physiol.* **117**, 431–460.

FATT, P. & KATZ, B. (1951). An analysis of the end-plate potential recorded with an intracellular electrode. *J. Physiol.* **115**, 320–370.

FATT, P. & KATZ, B. (1953*a*). The electrical properties of crustacean muscle fibres. *J. Physiol.* **120**, 171–204.

FATT, P. & KATZ, B. (1953*b*). Distributed 'end-plate potentials' of crustacean muscle fibres. *J. exp. Biol.* (in the Press.)

HODGKIN, A. L. (1951). The ionic basis of electrical activity in nerve and muscle. *Biol. Rev.* **26**, 339–409.

HODGKIN, A. L. & RUSHTON, W. A. H. (1946). The electrical constants of a crustacean nerve fibre. *Proc. Roy. Soc.* B, **133**, 444–479.

HOFFMANN, P. (1914). Über die doppelte Innervation der Krebsmuskeln. Zugleich ein Beitrag zur Kenntnis nervöser Hemmungen. *Z. Biol.* **63**, 411–442.

KATZ, B. (1936). Neuro-muscular transmission in crabs. *J. Physiol.* **87**, 199–221.

KATZ, B. (1949). Neuro-muscular transmission in invertebrates. *Biol. Rev.* **24**, 1–20.

KATZ, B. & KUFFLER, S. W. (1946). Excitation of the nerve-muscle system in crustacea. *Proc. Roy. Soc.* B, **133**, 374–389.

KUFFLER, S. W. & KATZ, B. (1946). Inhibition at the nerve-muscle junction in crustacea. *J. Neurophysiol.* **9**, 337–346.

MARMONT, G. & WIERSMA, C. A. G. (1938). On the mechanism of inhibition and excitation of crayfish muscle. *J. Physiol.* **93**, 173–193.

WIERSMA, C. A. G. (1941). The efferent innervation of muscle. *Biol. Symp.* **3**, 259–289.

WIERSMA, C. A. G. & VAN HARREVELD, A. (1935). On the nerve-muscle system of the hermit crab (*Eupagurus bernhardus*). III. The action currents of the muscles of the claw in contraction and inhibition. *Arch. néerl. Physiol.* **20**, 89–102.

31

J. Physiol. (1955) 130, 326–373

THE SPECIFIC IONIC CONDUCTANCES AND THE IONIC MOVEMENTS ACROSS THE MOTONEURONAL MEMBRANE THAT PRODUCE THE INHIBITORY POST-SYNAPTIC POTENTIAL

BY J. S. COOMBS, J. C. ECCLES AND P. FATT
Department of Physiology, The Australian National University, Canberra, Australia

(*Received* 28 *March* 1955)

In the first account of potentials recorded intracellularly from motoneurones by microelectrodes, it was reported that inhibitory synaptic action evoked a hyperpolarization of the neuronal membrane (Brock, Coombs & Eccles, 1952), which is the reverse potential change from that produced by excitatory synaptic action. This inhibitory post-synaptic potential (i.p.s.p.) of motoneurones appears to be analogous to the hyperpolarizing responses that have been observed during inhibitory action on other tissues, for example, during inhibition of cardiac muscle (Gaskell, 1887; Monnier & Dubuisson, 1934; Burgen & Terroux, 1953) and of crustacean muscle when the membrane potential is abnormally low (Fatt & Katz, 1953).

There is now much evidence that the excitatory responses of nerve and muscle cells are caused by specific increases in the permeability of the surface membrane to ions and the consequent ionic fluxes, and not, for example, by changes in the membrane capacity (cf. Cole & Curtis, 1939; Hodgkin, 1951). On this basis the electrical responses of giant nerve fibres and of muscle fibres have been satisfactorily explained by the movements of Na^+ and K^+ ions across the surface membrane (cf. Hodgkin, 1951; Hodgkin & Huxley, 1952*c*). Also the excitatory response at the neuromuscular junction (the end-plate potential) has been satisfactorily explained by the postulate that the neuromuscular transmitter causes a large transient increase in the permeability of the end-plate membrane to all ions (Fatt & Katz, 1951; Castillo & Katz, 1954). However, there has as yet been no complete explanation of the inhibitory responses of crustacean muscle (Fatt & Katz, 1953) or cardiac muscle (Burgen & Terroux, 1953). There is evidence that an increased permeability to K^+ ions occurs at both junctional regions, but at least with crustacean muscle it had to be postulated that there was also an increased permeability to some other ion species which like K^+ ions would normally be close to electrochemical equilibrium.

The present paper gives an account of investigations designed to discover the ionic movements that are responsible for producing the i.p.s.p. of motoneurones. It has been found that identical ionic mechanisms are concerned in five types of inhibitory synaptic action that are exerted on motoneurones in the spinal cord: direct inhibition of antagonist motoneurones by group I*a* impulses from the annulo-spiral endings of muscle spindles (Lloyd, 1941, 1946; Laporte & Lloyd, 1952; Bradley, Easton & Eccles, 1953); the disynaptic inhibition by group I*b* impulses from Golgi tendon organs (cf. Granit, 1950; Laporte & Lloyd, 1952); the polysynaptic inhibition by group III muscle impulses (Lloyd, 1943); the polysynaptic inhibition by cutaneous impulses (Renshaw, 1942; Hagbarth, 1952); and the disynaptic inhibition by impulses in the motor axon collaterals, which will henceforth be called antidromic inhibition because it can be evoked by antidromic volleys (Renshaw, 1941; Eccles, Fatt & Koketsu, 1954). However, most of the investigation has been concerned with the first and the last of the above series, which give the most satisfactory i.p.s.p.'s for experimental investigation, because with the simple techniques of direct nerve stimulation they may be obtained uncomplicated by excitatory synaptic action. The relation of the results presented here to the inhibitory suppression of the discharges of motoneurones will be considered in a later paper.

A preliminary account of some of these investigations has been published (Coombs, Eccles & Fatt, 1953). The experimental procedures have already been fully described: the operative, and electrical procedures by Brock *et al.* (1952) and by Coombs, Eccles & Fatt (1955*a*); the micromanipulator technique by Eccles, Fatt, Landgren & Winsbury (1954); and the extrinsic current and intracellular injection procedures by Coombs *et al.* (1955*a*). Throughout the investigations all photographic records have been formed by the superposition of many faint traces (usually about 40) in order to reject random noise. This 'noise' was often considerable when large currents were being passed through the microelectrode.

RESULTS

A. *Effect of alterations in membrane potential on the i.p.s.p.*

By passing a direct current through one barrel of a double microelectrode that has been inserted into a motoneurone, the membrane potential may be altered to any desired value within a wide range, as has been fully described in the previous paper (Coombs *et al.* 1955*a*).* The actual change in the membrane potential has been calculated as the potential alteration measured through the other barrel of the microelectrode less the potential drop that the current would cause in the coupling resistance between the two barrels of the microelectrode (cf. Coombs *et al.* 1955*a*). All potential measurements designated membrane potentials have been so corrected.

*See pages 764–766 for "Methods" from Coombs, Eccles and Fatt, 1955 [eds.].

In Fig. 1 (A–G) it is seen that the brief hyperpolarization (the i.p.s.p.) produced in a biceps-semitendinosus motoneurone by a volley in group I*a* quadriceps afferent fibres at the normal membrane potential of −74 mV (D) was greatly modified by relatively small changes in the membrane potential, being increased by depolarizations (A–C) and diminished or even reversed (E–G) by hyperpolarizations. There is much evidence that this volley from the antagonist muscle exerts a purely inhibitory action on the motoneurone (Lloyd, 1946; Laporte & Lloyd, 1952; Bradley *et al.* 1953), hence it appears that an increase in the membrane potential above the resting level actually reversed the sign of the i.p.s.p., i.e. it converted the normal hyperpolarizing i.p.s.p. into a depolarizing i.p.s.p. This is similar to the alteration of inhibitory junctional potentials with change in membrane potential that occurs in the crustacean muscle, although in that case the reversal normally occurs at the resting potential (Fatt & Katz, 1953).

When plotted against the membrane potential, the corresponding peak voltages of the i.p.s.p. are seen to lie on a curve that crosses through zero at a membrane potential which may be called the 'reversal-potential' for the i.p.s.p. and which is approximately −80 mV in Fig. 2A, i.e. with the interior of the motoneurone 80 mV negative to the external indifferent electrode. The simplest explanation of this reversal is that the i.p.s.p. is generated by the net flux of some specific type of ion (or ions) through the post-synaptic membrane, this flux being consequent on a specific increase of the membrane permeability. The 'reversal-potential' would thus represent the membrane potential at which there was equality of the electric charges carried by the fluxes in the two directions across the membrane. With the normal resting membrane potential there must be a net outward passage of charge in order to produce the observed hyperpolarizing i.p.s.p. (cf. Fig. 1D). Hence, for example if the specific ions are anions, the ionic flux must be preponderantly inward, and if cations, it must be outward. A more complex situation will exist if, as seems likely, both anions and cations are included in the specific group (cf. Discussion, § A). The upward convexity of the curve in Fig. 2 is characteristic of our most reliable experiments. The theoretical significance will be considered in the Discussion (§ D).

As shown in Figs. 1H to N, the i.p.s.p. set up by an antidromic volley by means of the axon collaterals and repetitive interneuronal discharges (Eccles *et al.*, 1954) was almost identically affected by changing the membrane potential. This similarity both of reversal-potential and curvature is illustrated in Figs. 2A and B, and in Table 1 the reversal-potentials are seen to be almost the same in the four neurones where both were determined. Furthermore, the mean reversal-potential was about 11·5 mV (extremes 5 to 26) more negative than the mean resting potential in the eight motoneurones where curves similar to that of Fig. 2 were drawn for the i.p.s.p. of

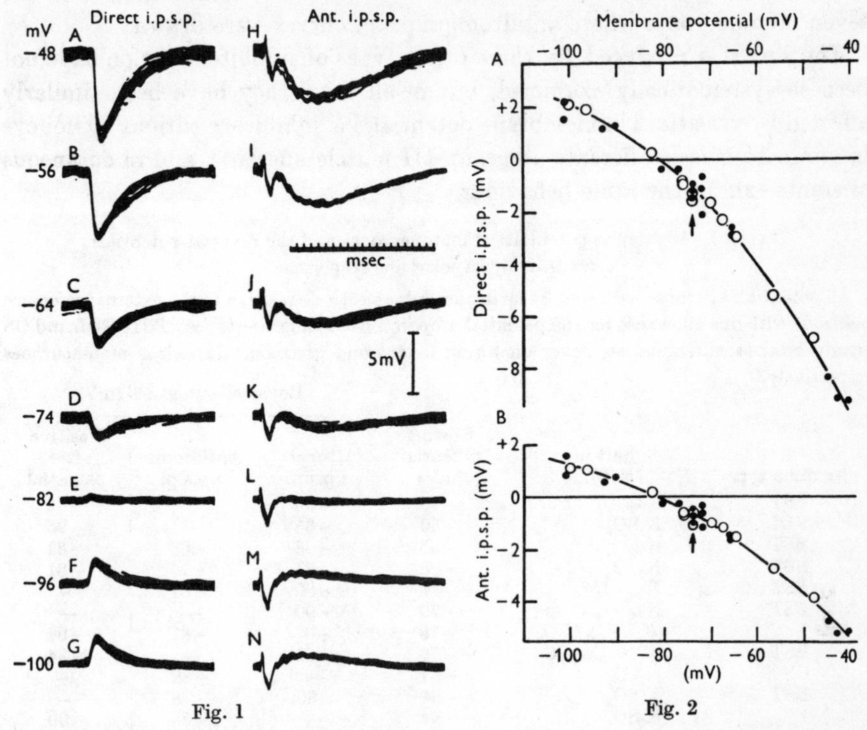

Fig. 1. Potentials recorded intracellularly from a biceps-semitendinosus motoneurone by means of a double-barrelled microelectrode filled with 0·6 M-Na_2SO_4. All records are formed by the superposition of about forty faint traces. A to G show potentials set up by the direct inhibitory action of a group I*a* quadriceps afferent volley, while with H to N the inhibitory action is exerted by an antidromic volley set up in L_7 and S_1 ventral roots by a stimulus just below threshold for the axon of the motoneurone. The initial diphasic spike is the characteristic field potential produced by antidromic invasion of adjacent motoneurones. By means of a steady background current through one barrel of the double microelectrode, the membrane potential has been preset at the voltage indicated on each record. In the absence of such current the membrane potential was − 74 mV. Potential and time scales apply to all records. Positivity of the microelectrode relative to the indifferent external electrode is recorded upwards. All voltages are given relative to this external reference electrode, hence hyperpolarization of the motoneurone is registered by an increasing negativity, i.e. by a downward deflexion, while the diphasic spike in the antidromic record (due to the external field from adjacent motoneurones) appears with an initial upward deflexion, which is the inverse of the usual convention.

Fig. 2. Maximum voltages of the i.p.s.p.'s of the series partly shown in Fig. 1 are plotted as ordinates against the respective membrane potentials as abscissae, A being for the direct i.p.s.p.'s and B for the antidromic i.p.s.p.'s. According to the convention adopted throughout this paper, hyperpolarizing and depolarizing i.p.s.p.'s are plotted respectively as negative and positive voltages. Points designated ● were obtained before, and those designated ○ some 9 min after, a depolarizing current of 5×10^{-8} A for 90 sec. There was no significant change in the resting potential, the mean value being indicated by the arrows.

direct inhibition, and about 14 mV (extremes 7 to 25) more negative in the seven motoneurones where antidromic i.p.s.p. curves were drawn.

The i.p.s.p.'s produced by three other types of inhibitory action have not been so systematically examined, but in all cases they have been similarly affected by variations in membrane potential, i.e. inhibitory actions by volleys in group I*b* muscle afferents, in group III muscle afferents, and in cutaneous afferents exhibit the same behaviour.

TABLE 1. Membrane potentials of motoneurones and the reversal-potentials for their hyperpolarizing responses

All potentials are those indicated by an intracellular electrode relative to the external reference electrode with due allowance for the potential drop in the coupling resistance. BST, FDL and GS signify biceps-semitendinosus, flexor digitorum longus and gastrocnemius-soleus motoneurones respectively.

			Reversal-potentials (mV)		
Neurone type	Salt in electrode	Resting potential (mV)	Direct i.p.s.p.	Antidromic i.p.s.p.	Positive after-potential
BST	Na_2SO_4	-70	-80	—	-95
FDL	K_2SO_4	-60	-65	—	-95
BST	K_2SO_4	-56	-68	-66	-81
BST	K_2SO_4	-68	-77	-77	-81
BST	Na_2SO_4	-74	-80	-81	—
BST	Na_2SO_4	-75	-90	—	—
?	$((CH_3)_4N)_2SO_4$	-70	—	-88	-94
BST	$((CH_3)_4N)_2SO_4$	-76	-81	—	-84
GS	K_2SO_4	-57	—	-82	—
BST	K_2SO_4	-54	-80	-78	—
?	K_2SO_4	-68	—	-88	-90
Mean values	—	-66	-77·5	-80	-88·5

Discussion. The similarity of the reversal-potentials indicates that all types of i.p.s.p. are generated by essentially the same ionic mechanisms. In contrast, the e.p.s.p. (excitatory post-synaptic potential) was reversed, i.e. changed to a hyperpolarizing potential, only when the membrane potential was itself reversed, i.e. changed to internal positivity (cf. Coombs *et al.* 1953 and unpublished observations). Even when the i.p.s.p. has been changed to the depolarizing type, the effect of varying the membrane potential readily distinguished it from the e.p.s.p. (cf. Figs. 3C, D; 9C, D; 13A, B, C, K, L, M) because of the difference in the values of the reversal-potentials, hence a discrimination was still possible when the recorded potential was a mixture of i.p.s.p. and e.p.s.p. Thus, in respect of its reversal-potential the e.p.s.p. resembles the end-plate potential of the neuromuscular junction (Fatt & Katz, 1951; Castillo & Katz, 1954), whereas the i.p.s.p. is similar to the inhibitory potential in crustacean muscle (Fatt & Katz, 1953).

This investigation into the effect of varying the membrane potential leads immediately on to the question: What species of ions are concerned in generating the i.p.s.p.? Or, alternatively, the question could be formulated: What are the species of ions to which the post-synaptic membrane becomes more permeable when it is acted on by inhibitory impulses? In an attempt to answer the question the most direct experimental procedure would be to investigate the changes produced in the i.p.s.p. by altering ionic concentrations on one or other side of the post-synaptic membrane. For example the effect of varying the external potassium concentration has been studied in attempts to discover the ionic mechanisms producing the inhibitory potentials of crustacean muscle (Fatt & Katz, 1953) and mammalian heart (Burgen & Terroux, 1953). Analogous experiments on mammalian motoneurones *in situ* are technically difficult, and hitherto only a few preliminary attempts have been made. On the other hand, on account of the relatively small volume, large changes can readily be produced in the internal ionic composition of the motoneurone. As described in the subsequent sections these changes have been brought about either by diffusion of various salts out of the intracellular microelectrode or by the ionic movements that are produced when a current is passed through the motoneurone either from the indifferent electrode to the microelectrode, or in the opposite direction. Such currents would have respectively a hyperpolarizing and a depolarizing action on the motoneuronal membrane, and they have been designated accordingly.

B. *Effect on the i.p.s.p. of diffusion of ions out of the microelectrode*

When a microelectrode filled with a concentrated chloride solution was inserted into a motoneurone, the i.p.s.p of that motoneurone usually showed a gradual change from a hyperpolarizing to a depolarizing response (Fig. 3 A–C). With a microelectrode having a resistance as low as 5 MΩ, this inversion of the i.p.s.p. occurred very rapidly and after a few minutes a fairly steady state was generally maintained. This effect of diminution and eventual inversion of the i.p.s.p. was observed whether potassium, sodium or choline was the cation in the microelectrode, but did not occur when such anions as sulphate, phosphate or bicarbonate replaced the chloride. However the same effect was observed when the anion was bromide, nitrate (cf. Fig. 9 A–C) or thiocyanate.

Careful inspection shows that the inverted potentials such as those of Fig. 3 B, C are not mirror images of the initial hyperpolarizing i.p.s.p. They are much briefer, particularly on the declining phase, which often passes over into a low hyperpolarization (cf. Fig. 9 E). It may further be noted in passing that i.p.s.p.'s in the form of a depolarization have already been observed with chloride-filled electrodes, though they were not recognized as such (cf. Brock *et al.* 1952, fig. 6 A).

Since the membrane potential did not increase in Fig. 3 A–C, the change in the i.p.s.p. was not comparable with that seen in Fig. 1 D–G. However a diminution of membrane potential by an extrinsically applied current caused the i.p.s.p. to revert to the hyperpolarizing type (Fig. 3 D; cf. also Fig. 9 D). When the curve relating membrane potential to i.p.s.p. was determined as in

Fig. 2, it was found to be shifted to the right, having, for example, a reversal potential of only about −50 mV (curve through ● points in Fig. 4). Thus essentially the effect produced by a chloride-containing electrode was a displacement of the i.p.s.p./membrane potential curve to the right (and also upwards) so that the reversal potential occurred at a lower membrane potential. On account of this effect, microelectrodes filled with salts having such indifferent anions as sulphate or phosphate (cf. Table 1) must be used in determining the normal reversal potential for the i.p.s.p. On the other hand, no significant difference was observed between the three cations (Na^+, K^+ and $(CH_3)_4N^+$) used in the series of Table 1.

Discussion. The simplest explanation of the change in the i.p.s.p. from Fig. 3A to C would postulate that it is due to the diffusion of chloride ions (or anions of like action, cf. Fig. 9A–C) out of the microelectrode into the neurone with a consequent increase in the ratio of internal to external chloride concentration across the motoneuronal membrane. There would be the further provisional postulate that such anions are especially concerned in the ionic fluxes that are causally related to the i.p.s.p., i.e. that under the influence of inhibitory impulses, areas of the motoneuronal membrane become highly permeable to these anions, whereas they remain relatively impermeable to the ineffective anions. The change to a depolarizing i.p.s.p. would thus be caused by the change from a net inward to a net outward flux of the effective anions resulting from the increased intracellular concentration. This explanation receives support from an approximate calculation of the concentrations and fluxes obtaining for chloride ions under such experimental conditions.

The normal intracellular concentration of chloride has been estimated to be about 9 μequiv × cm^{-3} (cf. Discussion, § C). Taking the volume of the motoneurone to be $2{\cdot}3 \times 10^{-7}$ cm^3 (Coombs *et al.* 1955*a*), its total chloride content would be about 2 p-equiv. It has been calculated (cf. Appendix, § A) that there would be a chloride diffusion of about 0·06 p-equiv/sec from a 3 M-KCl-filled electrode that has a resistance of 5 MΩ when immersed in 0·15 M-KCl. Assuming an exponential time constant of 30 sec for the attainment of chloride equilibrium between a motoneurone and its environment (cf. Discussion, § B), this steady chloride injection from the microelectrode at a rate of 0·06 p-equiv/sec would cause the chloride content of the neurone to increase, with an exponential time constant of 30 sec, to a new steady level which is 1·8 p-equiv (30 × 0·06) higher than the initial level. Hence in a few minutes after insertion of a 5 MΩ electrode filled with 3 M-KCl, diffusion of chloride would produce a new steady state with about double the normal intracellular concentration, which is a relative increase of the order required for the above explanation of the observed change in the i.p.s.p. After insertion of the microelectrode the observed rate of change of the i.p.s.p. has often been much slower than the expected rate (time constant of about 30 sec), possibly on account of temporary obstruction of the electrode orifice during the insertion. However, in a later section (D, 2*a*) an indirect method will be described for determining the rate at which the diffusional equilibrium is attained and the comparison will then be made (Discussion, § B).

The briefer time course already noted for the depolarizing type of i.p.s.p. produced by ionic diffusion (cf. Figs. 3 B, C and 9 E) can be explained as follows. The ions injected from the electrode would preponderantly affect the responses of those inhibitory post-synaptic areas of the membrane close to the microelectrode which would consequently give the depolarizing i.p.s.p., while more remote areas would be less affected and still give the hyperpolarizing i.p.s.p. The delay produced by electrotonic spread would cause this hyperpolarization largely to follow the depolarization, hence the diphasic character of the i.p.s.p. An alternative explanation would regard the diphasic type of i.p.s.p. as indicative of a dual composition, being attributable to two independent transmitter and ionic mechanisms. However, by varying the strength of stimulus setting up the inhibitory volley, it has not been possible to obtain a threshold separation between afferent fibres that give one or other type of response. Thus presumably the diphasic type of i.p.s.p. in Fig. 3 B is given by a homogeneous group of afferent impulses (group I*a*) and, after further diffusion it is converted to a pure depolarizing i.p.s.p. (Fig. 3 C, Q–T), which runs a time course virtually identical with the e.p.s.p. (cf. Fig. 16 F, G), except that the latent period is almost 1 msec longer when measured from the time of entry of the respective volleys into the spinal cord (cf. Brock *et al.* 1952; Eccles, Fatt & Landgren, 1954). When fully developed, the depolarizing i.p.s.p. is a mirror image of the hyperpolarizing i.p.s.p. (Figs. 3 A, C and 9). This mirror-image relationship has already been observed for the i.p.s.p.'s recorded when the neuronal membrane was set at potentials far on either side of the normal resting potential (cf. Fig. 1).

This technique of injecting ions by diffusion out of the microelectrode is unsatisfactory because there is no control of the rate of injection out of any given electrode. For example the rate at which cations can be injected by diffusion is too low to give significant changes. If the probable value of 150 mM is assumed for the potassium concentration of the interior of the motoneurone, the total potassium content would be over 30 p-equiv. Moreover, the time constant for attainment of potassium equilibrium across the membrane would probably be briefer than the value of 30 sec determined for chloride ions. Hence the diffusional rate of potassium out of a microelectrode of 5 MΩ resistance, 0·06 p-equiv/sec, would not significantly affect the potassium concentration of a motoneurone. Diffusion experiments are thus valueless as a test of a possible participation of potassium ions in the generation of the i.p.s.p. As described in the two following sections, effective control of the rate of ionic injection from a microelectrode has been achieved by applying a voltage to it so that current passes between the microelectrode and the cell either hyperpolarizing or depolarizing it.

C. *Effect produced on the i.p.s.p. by a hyperpolarizing current*

(1) *Chloride-filled microelectrode*

The membrane potential and the i.p.s.p. responses before, during and after a prolonged hyperpolarizing current ($3{\cdot}2 \times 10^{-8}$ A for 1 min) are shown in the d.c. amplifier records of Fig. 3 E–L, which were obtained after the preliminary diffusional injection of chloride (cf. Fig. 3 A–C). The vertical positions of the records are significant in giving the level of membrane potential. The hyperpolarizing current caused an immediate large increase of the membrane potential, from −59 to −125 mV, and at the same time, as described in § A, there was an immediate large increase in the depolarization produced by the i.p.s.p., 3·6–14 mV (Fig. 3 F). However, these effects did not remain constant during the passage of the hyperpolarizing current, nor did they vary in the same sense. The membrane potential progressively declined down to −111 mV, while the i.p.s.p. response progressively increased to 25·5 mV (Fig. 3 G, H).

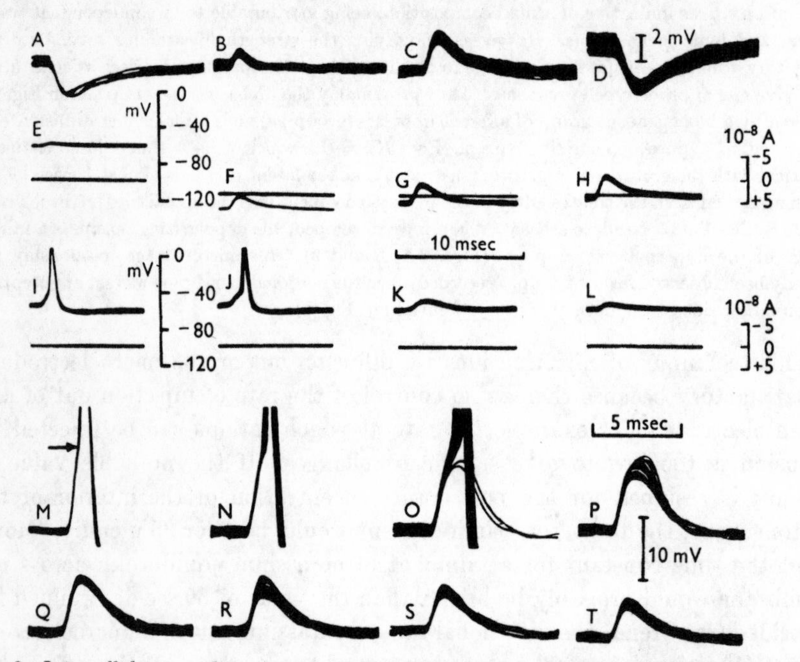

Fig. 3. Intracellular recording through a double-barrelled microelectrode filled with 3 M-KCl, of i.p.s.p.'s generated in a biceps-semitendinosus motoneurone by a quadriceps group I*a* afferent volley. A, B and C show effect of diffusion of Cl^- ions out of the microelectrode in changing the i.p.s.p. recorded at the resting membrane potential (−59 mV), while D shows restoration of the hyperpolarizing i.p.s.p. at a lower membrane potential (−27 mV). E–L show records both of current through one barrel of the electrode (straight lines) and of simultaneous i.p.s.p.'s obtained with a d.c. amplifier. The successive records are so placed that the scales on each side (potential on left, current on right) obtain right across the series, giving the actual levels of membrane potential and current respectively. E is before, F, G, H during and I, J, K, L after the application of $3{\cdot}2 \times 10^{-8}$ A for 60 sec. C and E are the same responses recorded at high and low amplification (note respective potential scales), while M–T cover same range of recovery as I–L, but again at a high amplification, though less than for C (note potential scales). Two time scales are shown, one for E–L, the other for the remaining records. All records formed by superposition of about forty traces. Note effect of wide variation in spike latency in N and O. Spikes in M, N and O are truncated.

On cessation of the current the membrane potential immediately fell to −51 mV, which was well below the initial level, and at the same time the i.p.s.p. was diminished in size as shown by the less steep rising phase (Fig. 3I), though to a value much larger than that initially existing before the hyperpolarization (cf. Fig. 3E with I, and C with M). At this membrane potential the depolarization of the i.p.s.p. was more than adequate to generate an impulse, as may be seen in the early records of the subsequent recovery series taken both at low (Fig. 3J–L) and at high (Fig. 3N–T) amplification. During

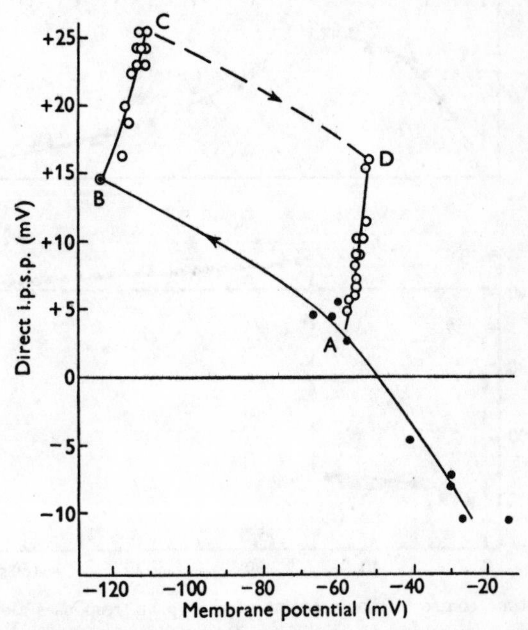

Fig. 4. Points designated ● are from initial records of i.p.s.p.'s at various membrane potentials as in Fig. 2, two such records being shown in Fig. 3C and D. ○ points plotted from B to C were obtained during the passage of the hyperpolarizing current ($3 \cdot 2 \times 10^{-8}$ A, cf. records F to H of Fig. 3, F being shown by ⊙ in Fig. 4), while ○ points from D to A were obtained after cessation of this current (records I to T of Fig. 3). Further description in text.

this recovery, the membrane potential gradually returned towards the initial resting level and the i.p.s.p. response declined so that eventually it resembled the relatively small depolarizing response observed before the hyperpolarization (cf. Fig. 3T with C).

The full course of the changes during and after the hyperpolarizing current of Fig. 3 is plotted in Fig. 4 on the co-ordinates used for the i.p.s.p./membrane potential curve which was determined by preliminary records as in Fig. 2. On hyperpolarization there was an immediate change from the initial condition obtaining at the resting potential, point A (Fig. 3E) to point B (Fig. 3F).

Thereafter, during the hyperpolarization, the gradually changing conditions were defined by the series of points lying approximately along the line BC, there being a progressive decrease in the rate of change. On cessation of the hyperpolarization, there was an immediate change from C to D (Fig. 3I, M) and then again a gradual change back to the original condition, as defined by the series of points lying approximately along the line DA. The generation of spike potentials prevented the direct measurement of the points near to D

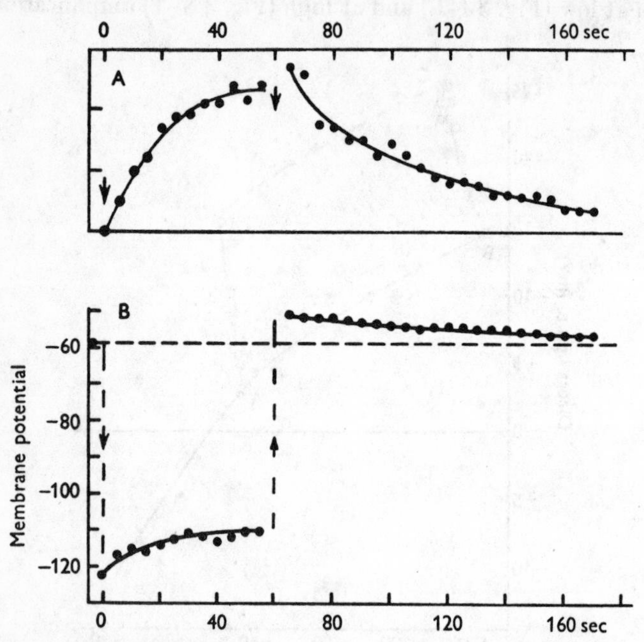

Fig. 5. A. Plot of time course of displacement of the i.p.s.p. responses along the line BC of Fig. 4 during the current and along the line DA after its cessation, as described in text. Ordinates are in arbitrary units measured along the lines BC and DA of Fig. 4. Arrows mark onset and cessation of current. B. Plot of the membrane potentials on same time scale as in A in order to show the similar time courses of the changes during and after the current. Arrows mark onset and cessation of current.

(cf. Fig. 3I, J, M, N). However, the summit-heights of the i.p.s.p.'s were calculated from the slopes of their rising phases on the assumption that similar time courses prevailed throughout.

It would appear that the hyperpolarization has caused the i.p.s.p./membrane potential curve to be displaced progressively upwards and to the right, and during recovery there was the reverse displacement. The time course of these changes may be determined approximately by plotting the displacements along the lines BC and DA against the times of the successive responses. As shown in Fig. 5A, the time-course is approximately exponential with a half time of about 15 sec during the hyperpolarization and about 30 sec after its cessation. Fig. 5B also shows typically that the changes in membrane potential followed approximately the same time course. Usually the membrane potential has been depressed immediately after a hyperpolarizing current as in Fig. 5B, but sometimes there was no change, and a slight increase has even been observed (cf. arrows of Fig. 6).

On cessation of the hyperpolarization the time course of recovery with a chloride electrode has been so rapid that it has not been possible with the present technique to determine a satisfactory curve for i.p.s.p./membrane potential. For example, in Fig. 6A the ● points give the curve as determined before the hyperpolarization, while the + points give the curve at 10–35 sec after the end of a hyperpolarization (6×10^{-8} A for 60 sec). This curve appears to join the initial curve at its lower end, but this effect is at least partly attributable to the progressive recovery during the observations. The upper three

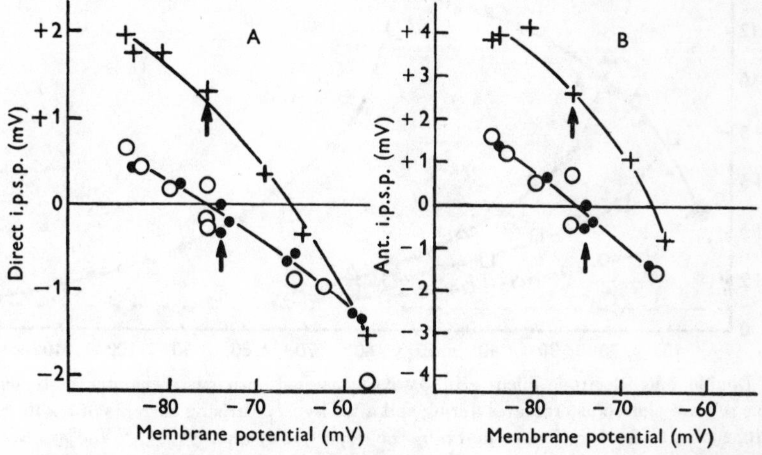

Fig. 6. A. Plot as in Fig. 2 of direct i.p.s.p.'s against membrane potential, but with a double microelectrode filled with 3M-KCl. Points designated ● were obtained before and + at 10–35 sec after passage of a hyperpolarizing current of 6×10^{-8} A for 60 sec. Finally, ○ points plot records at 2–2½ min after the current. Arrows indicate mean resting potentials. B. As in A, but for antidromic i.p.s.p.'s determined simultaneously.

+ points were determined first and the remaining four in serial order therefrom. The points indicated by ○ were obtained at 2–2·5 min after the cessation of the hyperpolarization and agree closely with the initial curve, showing that the recovery was then complete. Fig. 6B gives the simultaneously determined effects on the antidromic i.p.s.p. The series illustrated in Figs. 4 and 6 show typically that, immediately after the passage of a hyperpolarizing current through an electrode filled with chloride, the i.p.s.p./membrane potential curve was displaced upwards and to the right so that the reversal-potential for the i.p.s.p. was lowered, for example from −76 to −66·5 mV in Fig. 6 (cf. also Fig. 17C).

The effects of varying the duration of a hyperpolarizing current are illustrated in Fig. 7, in which the size of the i.p.s.p. is plotted before, during and after the currents. The small changes that occurred in the resting potential have been neglected in plotting the curves, but Fig. 4 B, C shows that the time courses of change of the i.p.s.p. would not be significantly distorted thereby. During the longest hyperpolarizing current (55 sec), the time course resembles that of Fig. 5 A, the curve flattening to a plateau in its later stages. On cessation of the current the consequent fall in membrane potential caused a sudden decrease in the i.p.s.p. (cf. Fig. 4 C, D) and thereafter there was the slow exponential decline as in Fig. 5 A. Similar features were observed for the briefer hyperpolarizing currents, but there was not then time for the full change in the i.p.s.p.

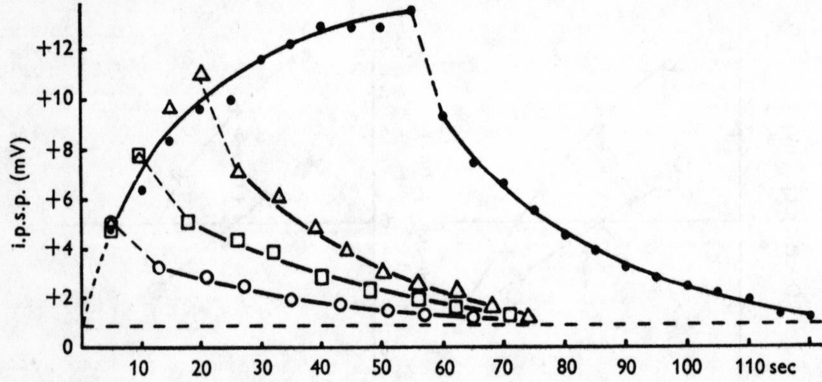

Fig. 7. Double microelectrode filled with 3 M-KCl inserted into gastrocnemius motoneurone. I.p.s.p.'s are plotted as ordinates during and after hyperpolarizing currents of 5×10^{-8} A for 5, 10, 20 and 55 sec duration, as shown by the respective symbols ○, □, △ and ●. Note the increase in the depolarizing i.p.s.p. during the current as in Fig. 5 A, and the approximately exponential decline thereafter. At onset and cessation of the currents the curve is shown as a broken line because it is partly due to the effect of the changed membrane potential on the i.p.s.p. The horizontal broken line corresponds to the small initial depolarizing i.p.s.p.

All types of i.p.s.p. have been similarly affected when the chloride content of the motoneurone was increased. For example, comparison of Fig. 6 A and B shows that there was virtually the same change in the inhibitory potentials produced either by direct inhibition or by antidromic inhibition. The results plotted in Fig. 7 were obtained with a polysynaptic inhibitory potential that was produced in a gastrocnemius motoneurone by an afferent volley in quadriceps nerve that included group I*b* and group II impulses.

The rapid and complete reversibility of the effects illustrated in Figs. 3–7 has not been observed with all motoneurones. In the earliest experiments (cf. Coombs *et al.* 1953) reversibility was rarely observed, there being generally a cumulative effect of successive hyperpolarizing currents, and often, too, a continuous drift to more depolarizing i.p.s.p. responses even when no current was applied. In our more recent experiments such effects were observed only with badly injured cells that were rapidly deteriorating. One can be certain that the cells giving the reversible reactions of Figs. 3–7 were in a much better physiological condition than cells whose reactions to a comparable disturbance were irreversible. It seems likely that the improved physiological condition of the cells in our more recent experiments is attributable to the technical improvements which have reduced mechanical damage (Eccles *et al.* 1954).

Discussion. It would appear that the chloride which was injected into the cell by the hyperpolarizing current has caused the i.p.s.p./membrane potential curve to be displaced so that the reversal potential was at a lower level of membrane potential, though its more general features were retained. This result resembles that produced by diffusion, but may be made much larger and can be graded in intensity (cf. Fig. 7). In general, these effects may be explained in the same way as the comparable effects that have been attributed to diffusion of chloride ions. However, it is necessary to determine the effects which other ions have on the i.p.s.p. before attempting to develop any quantitative explanations of the changes produced in the i.p.s.p. by increased Cl^- ion concentrations. Nevertheless, the time course of the changes occurring during and after the passage of a hyperpolarizing current (Figs. 5 and 7) may profitably be considered at this stage, though a full treatment must await the final Discussion (§ B).

During the passage of the hyperpolarizing current it has been shown in Appendix B that the intracellular Cl^- concentration will increase exponentially to a level which would be much higher than the initial level. For example, a hyperpolarizing current of 5×10^{-8} A would be expected to increase the concentration by about 20 mM, the increase occurring with a time constant probably briefer than 20 sec. Changes of this magnitude and time course would account satisfactorily for the observed changes in the i.p.s.p. in Figs. 5 and 7. Likewise, on cessation of the hyperpolarizing current, the recovery towards the initial i.p.s.p. response may be explained by considerations of ion flux. The high intracellular concentration of Cl^- ion would be expected to cause a net outward flux of Cl^- ions across the membrane. Provided that the rate of diffusion out of the microelectrode has remained constant, and that the cell has remained in an unchanged condition, there should be a complete recovery to the initial i.p.s.p. response, as is indicated in Figs. 5 and 7.

(2) *Sulphate- or phosphate-filled microelectrode*

When even a large hyperpolarizing current was applied through a sulphate-filled electrode, there was no significant change in the i.p.s.p. (Fig. 8A, C). The e.p.s.p. was also virtually unaffected (Fig. 8B, D), while the resting membrane potential was a little depressed (−84 to −75 mV) and later recovered. A more complete test is illustrated in Fig. 8E, where the i.p.s.p. was determined over a wide range of membrane potentials before and after a hyperpolarizing current of 5×10^{-8} A for 90 sec. Points ⊕ determined as early as 5 and 10 sec after the end of the hyperpolarizing current are seen to lie close to the curve determined before the current. If there was any change, it was in the reverse direction from that produced by the monovalent anions considered above. The membrane potential fell from -64 to -61 mV, and there was no recovery. A similar absence of significant change in the i.p.s.p. has been observed after each of the eleven hyperpolarizing currents that have been passed into six different cells. Currents of 4 to 8×10^{-8} A have usually

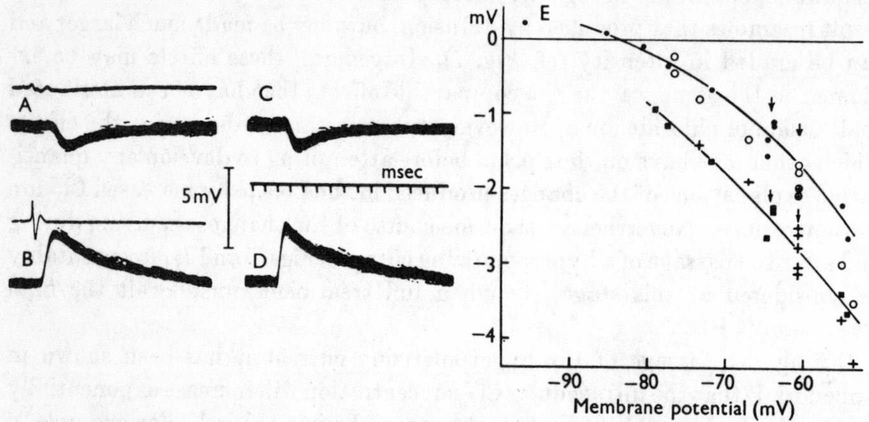

Fig. 8. A and B show respectively the direct i.p.s.p. and the monosynaptic e.p.s.p. set up in a biceps-semitendinosus motoneurone by the appropriate afferent volleys. Surface records from L 7 dorsal root are also shown, negativity being downwards. C and D show the respective responses after passage of a hyperpolarizing current (12×10^{-8} A for 90 sec) through an electrode filled with 1·6 M-$[(CH_3)_4N]_2SO_4$. A small spike-like local response is superimposed on the e.p.s.p. of record D, being produced probably on account of the fall in resting potential. E. Points designated ● show the i.p.s.p.'s at various membrane potentials as in Fig. 2A, but with a double microelectrode filled with 0·6 M-K_2SO_4. Points designated ○ show that there was little change after the passage of a hyperpolarizing current (5×10^{-8} A for 90 sec). The first two records after the current are designated ⊕. Similarly, there was little change in the positive after-potential, points designated ■ being before and + after the same hyperpolarizing current. Arrows indicate mean resting potentials which were 3 mV lower after current. Curves are drawn through the initial series of records, ● and ■ respectively.

been applied for 1–3 min. Usually there has been a small fall in resting potential as in Fig. 8, and sometimes recovery occurred. It has been immaterial whether the cation in the electrode has been K^+ as in Fig. 8E or $(CH_3)_4N^+$ as in Fig. 8A, C. Fig. 8E also shows that after the current there was no change in the hyperpolarizing after-potential over a wide range of membrane potentials.

Similarly, when hyperpolarizing currents have been applied through an electrode filled with equimolar concentrations of K_2HPO_4 and KH_2PO_4 (seven applications of current to four cells), there has been subsequently no significant change in the i.p.s.p. (cf. Coombs *et al.* 1953, fig. 1I, J), or in the resting potential and the e.p.s.p. Currents of 2 to 10×10^{-8} A have usually been applied for 1 min.

Discussion. This invariable failure of these polyvalent anions to affect the i.p.s.p. when injected by a hyperpolarizing current is paralleled by the absence of any spontaneous change in the i.p.s.p. immediately after the insertion of the microelectrode. For this reason sulphate (or phosphate) electrodes are preferred, and are essential when determining the reversal-potentials for the i.p.s.p. (Figs. 1, 2), and comparing them with the resting membrane potentials and the reversal-potentials for the hyperpolarizing after-potential as in Table 1. The absence of any effect by a hyperpolarizing current applied through an electrode filled with sulphate or phosphate is further of importance because it shows that the changes in the i.p.s.p. described in § C 1 are not consequential on the hyperpolarizing current itself or on osmotic swelling (cf. Appendix, § B), but presumably are due to the injection of chloride.

(3) *Electrodes filled with salts of various monovalent anions*

Among the several anions that have been tested, it has been found that, when hyperpolarizing currents were applied through electrodes filled with potassium nitrate, bromide or thiocyanate, the subsequent changes in the i.p.s.p. were indistinguishable from those observed with a chloride-filled electrode. Though the observations with chloride were of more significance because it is present in high concentration in the environment of the neurone and so might be expected to take part normally in the production of the i.p.s.p., tests with a number of other anion species were of interest in an attempt to investigate further the postulated permeability change in the inhibitory post-synaptic membrane. Instead of reporting the intensity and duration of the hyperpolarizing currents, it will be convenient henceforth to report merely the quantity of anions so injected (6 p-equiv being injected by 10^{-8} A flowing for 60 sec), for it is now justifiable to assume that hyperpolarizing currents produce the observed changes in the i.p.s.p. by means of anion injection. With the readily diffusible anions such as Cl^-, Br^-, NO_3^-, SCN^-, much of the quantity injected leaks out across the surface membrane: so that, after currents have flowed for 60 sec, they do little more than maintain the raised intracellular concentration (cf. Appendix, § B). This complication seriously limits the significance of the specification of the quantity injected.

As shown typically in Fig. 9 E, F, after the injection of about 18 p-equiv of NO_3^- ions by passage of a hyperpolarizing current through a KNO_3-filled electrode, the i.p.s.p. was transformed from a small depolarizing to a large depolarizing response which evoked the discharge of an impulse at just the same threshold depolarization (15 mV) as the e.p.s.p. (compare Fig. 9 F and G with L). It may be noted in passing that the injection caused no significant change in the e.p.s.p. (Fig. 9 L), or in the threshold at which it set up an impulse.

The series from F to K shows the partial recovery of the i.p.s.p. towards the initial response (compare Fig. 9K with E). The resting membrane potential showed the characteristic depression immediately after the hyperpolarizing current and a slow recovery thereafter, much as in Fig. 5B. The time course of the recovery from the large depolarizing response (Fig. 10A) closely resembled that observed for chloride (Fig. 5A). Altogether there were fifteen injections of NO_3^- ions into five motoneurones, and virtually the same result was observed in all. The usual injection rates varied from 18 to 42 p-equiv a minute and were

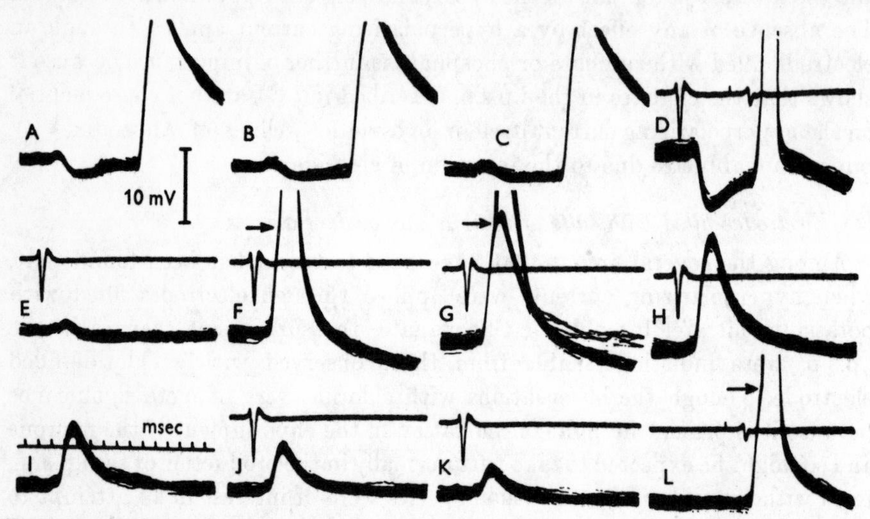

Fig. 9. Intracellularly recorded potentials from biceps-semitendinosus motoneurone with a single microelectrode filled with 3 M-KNO_3. In A to D a quadriceps group I*a* afferent volley set up an i.p.s.p. and a later biceps-semitendinosus volley set up an e.p.s.p. Note in D–H and J–L the surface-recorded potentials from the L6 dorsal root, negativity being downwards. With D the membrane potential was diminished from 79 to 60 mV, resulting in a large hyperpolarizing i.p.s.p., and the generation of a spike by the e.p.s.p. All spikes are truncated. E shows the i.p.s.p. before and F immediately after the passage of a hyperpolarizing current of 3×10^{-8} A for 60 sec. Note that the large depolarizing i.p.s.p. sets up a spike in F and G, the threshold shown by arrow being about the same as the threshold for the e.p.s.p. (arrow in L). Records G–K show progressive recovery of the i.p.s.p. towards the initial low depolarizing response (E). The potential and time scales apply to all records.

continued for 1–2 min. The half times of recovery varied from 23 to 35 sec (cf. Fig. 10A), and the resting membrane potential was always temporarily depressed, usually by about 5 mV.

Fig. 10B shows typically the similar effects produced by an injection of about 24 p-equiv of Br^- ions by a hyperpolarizing current. The large depolarizing i.p.s.p. returned virtually to the initial value, while the membrane potential was slightly depressed from -70 to -69 mV after the current and then recovered to -75 mV. In contrast there was no appreciable change in the e.p.s.p. The recovery closely followed an exponential curve with a half-time of about 14 sec. Altogether the effect of Br^- ions on the i.p.s.p. was tested with nine injections into four cells, usually at a rate of 18–24 p-equiv per min for 1–1½ min. With the exception of one cell, into which two injections were made, the effect of injection was reversible as in Fig. 10B, the half-times of recovery varying from 10 to 20 sec.

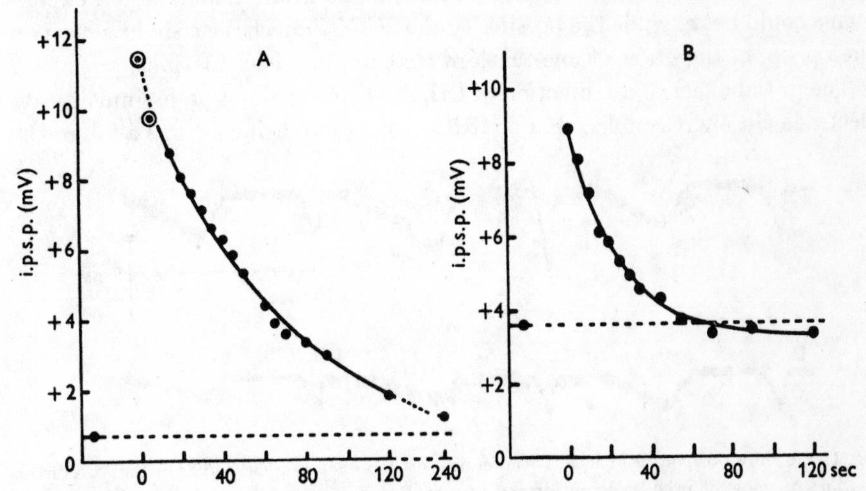

Fig. 10. A. Plot of heights of i.p.s.p.'s, showing time course of recovery after a hyperpolarizing current for series partly shown in Fig. 9. The two points designated ⊙ are the first two responses (cf. Fig. 9F) where the spike obscured the summit, the points being calculated from the slopes of the rising phases. The initial point on the horizontal broken line gives the initial response (Fig. 9E). B. Plot of i.p.s.p.'s as in A, but during recovery after a hyperpolarizing current (4×10^{-8} A for 60 sec) had been applied through an electrode filled with 4 M-KBr.

The injection of SCN^- ions also produced similar changes in the i.p.s.p. Altogether ten injections were made into six cells, and a reversible change to the depolarizing i.p.s.p. was observed in all but one cell into which three injections were made. The membrane potential was always depressed by the injection, and recovery occurred with the five cells showing recovery of the i.p.s.p. The half-times of the recovery of the i.p.s.p. fell within the range of 15–20 sec.

As shown in Fig. 11, the injection of even a large quantity of HCO_3^- ions from a $KHCO_3$-filled electrode into a motoneurone did not change the i.p.s.p. to the depolarizing type of response. There was in Fig. 11B, after injection of about 30 p-equiv, a large increase in both types of i.p.s.p. in the hyperpolarizing direction, but this could have been caused by the depression of the resting potential (−66 to −59 mV), which was the usual effect of injections of HCO_3^- ions. As the resting potential recovered, the i.p.s.p.'s returned towards their initial size (cf. Fig. 11C and D with A). The complicating effects of changes in resting potential were minimal after a second injection of HCO_3^- ions (60 p-equiv given in 1 min) into this cell. The resting potential was even slightly increased, an unusual finding, and correspondingly there was a diminution of both i.p.s.p.'s (cf. Fig. 11E, F with D). Altogether there were twelve injections of HCO_3^- ions (ranging from 15 to 60 p-equiv) into six motoneurones, and in every case the changes produced in the i.p.s.p. were negligible when allowance was made for the effect of the depression of resting potential. However, our experiments could not exclude the possibility that HCO_3^- ions have a slight action on the i.p.s.p. of the order of one-tenth of that produced by Cl^- ions.

The initial effect of an injection of CH_3COO^- ions from a potassium acetate electrode closely resembled that of HCO_3^- ions, there being always a fall in the

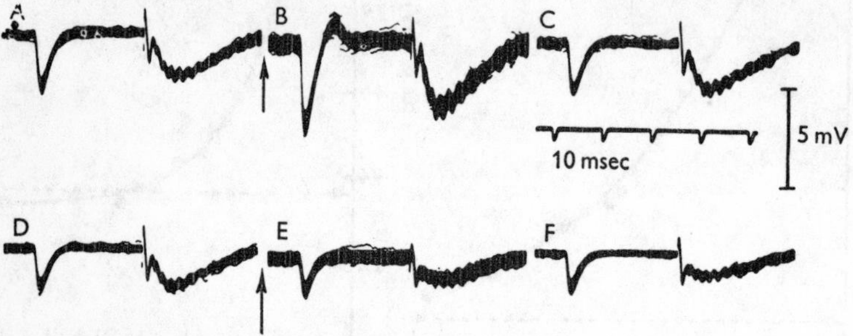

Fig. 11. A shows responses of biceps-semitendinosus motoneurone elicited as in Fig. 1, a direct i.p.s.p. being followed by an antidromic i.p.s.p. B shows increase in the hyperpolarization of the i.p.s.p.'s after the passage of a hyperpolarizing current (5×10^{-8} A for 60 sec) through a microelectrode filled with 2M-$KHCO_3$. C shows recovery about 1 min later. Later still the responses of D were slightly diminished after the passage of a hyperpolarizing current of 10×10^{-8} A for 60 sec (responses E and F). Injections are indicated by arrows.

resting potential and an increase in the hyperpolarization of the i.p.s.p. as in Fig. 11B. Membrane potential/i.p.s.p. curves were determined before and after several injections. Fig. 12 shows typically that immediately after the injection of about 60 p-equiv of CH_3COO^- ion there was no significant displacement from the curve obtaining before injection (the two points $\oplus$). However, with the later progressive fall in the resting potential there was also a displacement of the curve upwards and to the right, i.e. in the same direction as was observed after an injection of Cl^- ions. Discrimination is easy, for the acetate displacement developed late and was irreversible, being apparently associated with the progressive fall in resting potential, which indicated a specific injurious action either of acetate ions or of acetic acid. A further injurious action is shown by the blockage of axon-soma transmission that invariably developed after injection of acetate ions and from which recovery was never observed. Altogether eight injections of acetate ions have been made into five cells, and all have conformed to the above pattern, recovery never being observed. It seems that acetate ions do not have any appreciable action on the i.p.s.p., other than that attributable to a change in membrane potential or to a slowly developing irreversible injury.

In contrast to acetate, motoneurones tolerated injections of glutamate ions very well, the microelectrode being filled with 4 M-mono-potassium glutamate. The resting potential was changed little if at all after any of our seven injections into four cells, for example being depressed from −83 to −80 mV after injection of about 100 p-equiv. Also axon-soma transmission of antidromic impulses and the e.p.s.p. were not affected even by this large dosage. There was always a small and irreversible diminution of the i.p.s.p. by each dosage of glutamate ions. Since the resting potential was usually unchanged, this effect would appear to resemble that of Cl^- ions, but even with a large dosage (100 p-equiv) it was not possible to halve the hyperpolarization of the

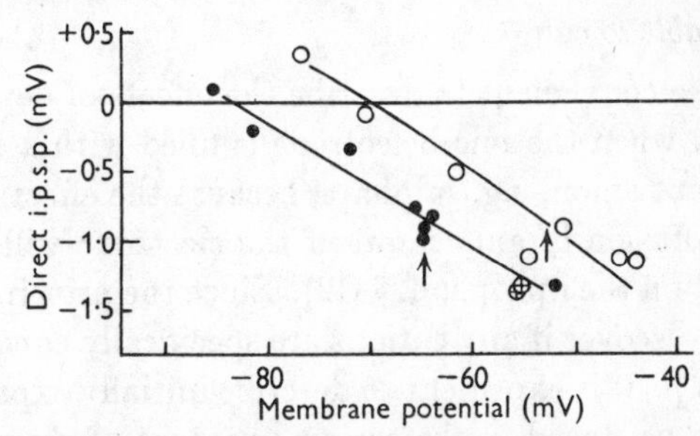

Fig. 12. Points designated ● plot i.p.s.p.'s against membrane potential with a double microelectrode filled with 5 M-potassium acetate. Points designated ○ give records after the passage of a hyperpolarizing current of 4×10^{-8} A for 90 sec. The two records immediately after the current are shown by ⊕. Mean resting potentials before and after the current are shown by the arrows.

i.p.s.p. It may be concluded that the glutamate ion probably shares to a very slight extent the property of changing the i.p.s.p. in the manner of Cl^-, Br^-, NO_3^- and SCN^- ions.

Discussion. Some generalizations may be made at this stage concerning the physico-chemical features which discriminate the anions which affect the i.p.s.p./membrane potential curve from those that do not. Chloride, bromide nitrate and thiocyanate apparently belong to the former group. In fact they are indistinguishable even quantitatively in their effects on the i.p.s.p., though comparison is not precise because it can be made only between the effects produced by injecting different ion species into different motoneurones. These ions are comparable in being monovalent and in having similar mobilities in aqueous solution, which presumably means that the hydrated ions are of similar size. In contrast polyvalent anions, SO_4^{2-} and HPO_4^{2-}, do not have any effect on the i.p.s.p. The individual ions in solution, besides carrying more than one negative charge, are also larger than those in the former group. When anions are injected from the potassium phosphate electrode with a hyperpolarizing current, they will consist in part of singly charged $H_2PO_4^-$ ions. Nevertheless, no effect was obtained from such an injection. Furthermore, injections of HCO_3^-, CH_3COO^- and glutamate$^-$ were also without effect or had only a very slight effect. As judged by their mobilities these anions are distinct from the Cl^-, Br^-, NO_3^-, SCN^- group in having considerably greater hydrated size. Thus a common criterion determining whether a given type of anion will modify the i.p.s.p. appears to be ion size (cf. Discussion, § A). Before discussing further the significance of ion size, it is necessary to discover if any cations are also specifically concerned in producing the i.p.s.p. Injections of cations into motoneurones have been effected by depolarizing currents.

D. *Effects produced on the i.p.s.p. by a depolarizing current*

(1) *Effects attributable to cations*

Initially it will be convenient to describe the effects of depolarizing current flow on the i.p.s.p. when the microelectrode is filled with a salt solution containing an indifferent anion, e.g. sulphate, because the effect of the current on the background diffusion of anions out of the electrode will not then have a complicating effect on the i.p.s.p. (cf. § C2). Since the principal interest of this investigation is to discover if any cations are specifically concerned in the production of the i.p.s.p., it is expedient to describe initially experiments in which large changes were produced in the cationic content of the motoneurone, i.e. when the depolarizing current was passed through electrodes filled with sodium sulphate or tetramethylammonium sulphate. Little cationic change can be expected when the depolarizing current is passed through an electrode filled with potassium sulphate (cf. Appendix C). However, the results obtained with such an electrode are important since they provide a control for the effects produced by the ionic flux carrying the current outwards across the neuronal membrane. Necessarily, with our present technique, these important control observations are very imperfect because they are made with different electrodes inserted into different cells.

(*a*) *Electrode filled with sodium sulphate.* Comparison of Fig. 13 with Fig. 1 shows the change produced in both the direct and antidromic i.p.s.p.'s when a depolarizing current of 5×10^{-8} A was passed for 90 sec out of a microelectrode filled with 0·6 M-Na_2SO_4. The resting membrane potential was lowered from -74 to -57 mV, and at the same time the i.p.s.p. was converted from a hyperpolarizing into a depolarizing type (compare Fig. 1D with Fig. 13D), i.e. there was, as shown by the plotted curve (+ points, Fig. 14A), a large displacement upwards and to the right of the i.p.s.p./membrane potential curve, the reversal-potential changing from -80 mV to -35 mV. The effects produced by the depolarizing current thus resembled those arising from injections of Cl^- ions (compare Fig. 14A with Fig. 6A). The change in i.p.s.p. after a depolarizing current was always transient, regression to the initial condition occurring within a few minutes. For example, the family of curves of Fig. 14A (cf. the i.p.s.p. records of Fig. 13A–E and K–O) and the plotted reversal potentials (▲ points of Fig. 14D) show the approximate time course of this regression, complete recovery having occurred in about 9 min. The antidromic i.p.s.p. was similarly affected and recovered similarly (Figs. 13F–J and P–T; 14B). After a much longer depolarizing current (5×10^{-8} A for 300 sec), there was a much larger change in the i.p.s.p., and regression followed a slower time course; yet,

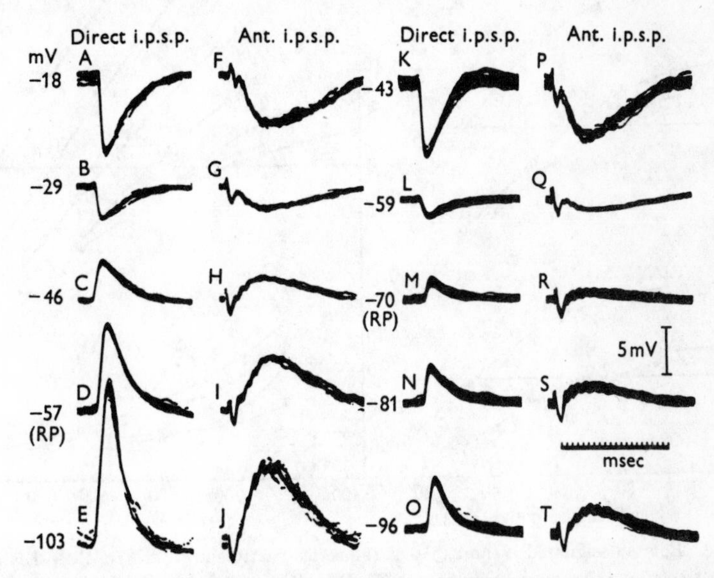

Fig. 13. A–E show records at the indicated membrane potentials of the same i.p.s.p. response in the same motoneurone as in Fig. 1A–G, but immediately (5–40 sec) after the passage of a depolarizing current, 5×10^{-8} A for 90 sec, the resting potential being −57 mV. Records F–J show simultaneously recorded antidromic i.p.s.p.'s (cf. Fig. 1H–N). K–O and P–T as in A–E and F–J respectively, but at different membrane potentials as indicated, and showing partial recovery at 180–230 sec, the resting potential being then −70 mV. The potential and time scales apply to all records. Electrode filled with 0·6 M-Na_2SO_4.

nevertheless, there was again complete recovery within 15 min, as revealed by the family of curves of Fig. 14C, and the plotted reversal-potentials (○ points of Fig. 14D). Similar behaviour was observed with all of the other eight neurones into which a depolarizing current was passed from a Na_2SO_4-filled microelectrode, there being in all seventeen applications of depolarizing current through this type of electrode. Half-recovery times have varied from about 100 to 250 sec for moderate doses of current as in Fig. 14A and B.

The depression of the resting membrane potential after such depolarizing currents is satisfactorily explained by the reduction which would occur in the intracellular potassium (Coombs *et al.* 1955*a*). The recovery of resting potential followed much the same time course as the reversal-potential for the i.p.s.p. (■ and ● points, Fig. 14D). However this reversal-potential was diminished more than the resting potential; hence, immediately after cessation of the depolarizing current (Fig. 13D) there was a depolarizing i.p.s.p. at the resting

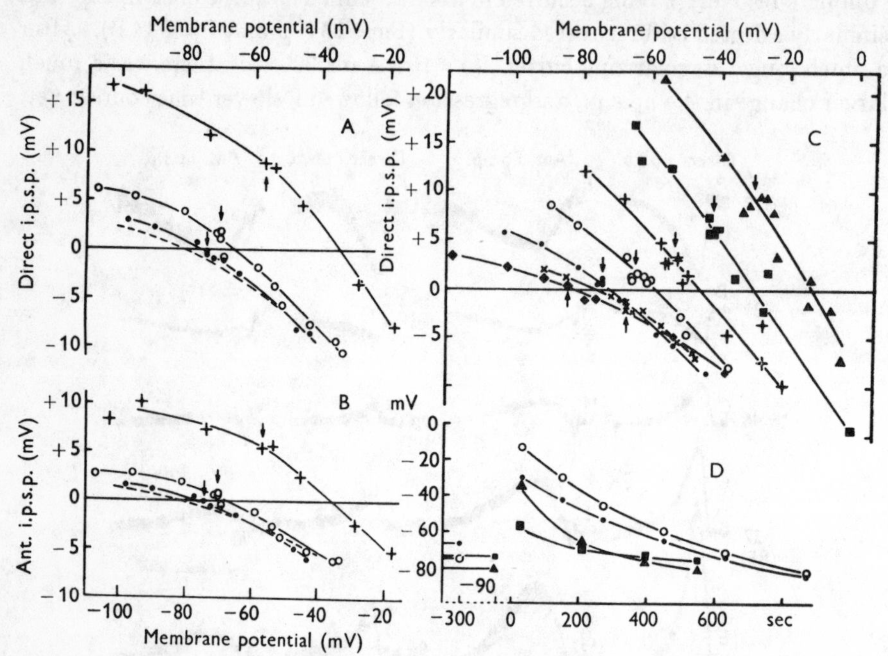

Fig. 14. A. Points designated + and ○ plot the series partly shown in Fig. 13 A–E and K–O, respectively, on the same co-ordinates as Fig. 2A. Points ● show a further stage of recovery at 360 to 430 sec after the passage of the depolarizing current. The broken line shows the curve of Fig. 2A which was obtained for this motoneurone both before the depolarizing current and after complete recovery at 510–580 sec. Note that compared with Fig. 2A the ordinate scale is halved relative to the abscissal scale. Arrows indicate respective resting potentials. B. As in Fig. 14A, but for antidromic inhibitory potentials that are partly illustrated in Fig. 13F–J and P–T. The broken line shows curve of Fig. 2B, but at half the ordinate scale as in Fig. 14A. C. Same motoneurone and i.p.s.p.'s as in Fig. 14A, but showing recovery after the passage of a much longer depolarizing current (5×10^{-8} A for 300 sec). The curve through the × points is the control curve before the current with the arrow showing resting potential of −67 mV, while immediately after the current (5–60 sec) the curve indicated by ▲ points was obtained. Progressive recovery is shown by the successive groups of records at 120–180 (■), 240–300 (+), 420–480 (○), 600–660 (●) and 840–900 (◆) sec respectively. Note that as shown by the arrows the resting potential had increased at 14–15 min to a value (−79 mV) considerably in excess of the initial value (−67 mV). D. Plot of recovery after depolarizing currents of responses of Fig. 14A and C. Zero on abscissal time scale gives instant of cessation of the current, the points to the left thereof giving the initial control values before the currents of 90 and 300 sec duration (note scale readings of −90 and −300 sec). Points indicated by ▲ and ■ plot respectively, the reversal potentials and the resting potentials for the series of Fig. 14A after a 90 sec current, while ○ and ● give the corresponding values for the series of Fig. 14C after a 300 sec current.

membrane potential then obtaining. After 10 min the resting potential had recovered to a value considerably above its initial level (Fig. 14C, D), as already described by Coombs *et al.* (1955*a*); hence the i.p.s.p. at the resting potential continued to be a depolarizing response even after 15 min (cf. Fig. 14C).

(*b*) *Electrode filled with tetramethylammonium sulphate.* With tetramethylammonium sulphate electrodes we have observed that a depolarizing current caused the same change in the i.p.s.p. as with Na_2SO_4-filled electrodes, as is shown by the family of curves plotted in Fig. 15A. The curves have been shifted to the right and become steeper with each successive current, and the magnitude of the shift has been of the same order as for Na_2SO_4-filled electrodes.

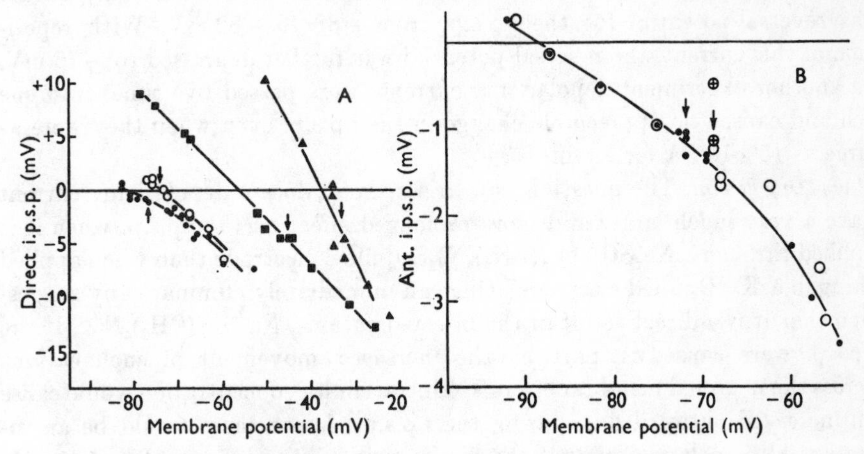

Fig. 15. A. A double microelectrode filled with 1·6M-tetramethylammonium sulphate was inserted into a biceps-semitendinosus motoneurone. Points shown as ● plot direct i.p.s.p. against membrane potential as in Fig. 2A. After a depolarizing current of 3×10^{-8} A for 60 sec the points are shown as ○, while ■ points give records after a repetition of this current and ▲ after a still further repetition. Arrows indicate mean resting potentials. B. A double microelectrode filled with 0·6M-K_2SO_4 was inserted into an unidentified motoneurone. Points shown as ● plot antidromic i.p.s.p. against membrane potential as in Fig. 2B. After a depolarizing current of 10×10^{-8} A for 90 sec the records are plotted as ○. The two records immediately following the current are shown by ⊕, the first being high above the control curve and the second showing a considerable recovery at about 15 sec. Arrows marking the resting potentials show that it was depressed by about 3 mV after the current.

However, an important difference has been that recovery towards the initial condition either failed altogether or at best was slow and incomplete. It has already been reported that the membrane potential has been depressed after each depolarizing current with a tetramethylammonium electrode and has recovered very slowly, if at all, the effects of successive currents being cumulative as shown by the arrows in Fig. 15A (Coombs *et al.* 1955*a*). This was attributed to the difficulty of eliminating the injected $(CH_3)_4N^+$ ion, and hence of recovering the lost potassium. With the tetramethylammonium injection the neurone also finds it difficult to reverse the ionic shifts across the membrane that displace the i.p.s.p. response in the depolarizing direction. Thus it seems likely that both effects are attributable to the same ionic shifts.

(c) *Electrode filled with potassium sulphate.* When the microelectrode was securely implanted in the cell, the passage of a depolarizing current through a potassium sulphate electrode caused little or no change in the resting potential (cf. Coombs *et al.* 1955*a*). As a rule there was a brief displacement of the i.p.s.p. in the depolarizing direction, but this displacement was much smaller than with the sodium sulphate electrode and the time course of recovery was much faster, the half-time being usually less than 20 sec (cf. Fig. 15B). As a further example, a relatively large current of 7×10^{-8} A for 90 sec only shifted the reversal-potential for the i.p.s.p. from -88 to -82 mV. With repetition of this current, the reversal-potential was further depressed to -76 mV. In another experiment depolarizing currents were passed five times into one cell and caused no appreciable change in the i.p.s.p. even when they were as large as 15×10^{-8} A for 2 min.

(d) *Discussion.* The question now arises: Why does a depolarizing current have a very much larger and more prolonged effect on the i.p.s.p. when it is applied through a Na_2SO_4- or $((CH_3)_4N)_2SO_4$-filled electrode than when applied through a K_2SO_4-filled electrode? One can immediately eliminate any suggestion that it is a direct result of the injected cations, Na^+ or $(CH_3)_4N^+$. If the i.p.s.p. were caused in part by the increased movement of such cations across the neuronal membrane, a raised intracellular concentration would cause an increased outward flux during the i.p.s.p.; hence there would be an increased hyperpolarization, not the diminution or reversal actually observed. On the contrary the observed change in the i.p.s.p. could be caused by diminution of the concentration of a cation species that was specially concerned in the generation of the i.p.s.p. Thus the effect of the Na^+ and $(CH_3)_4N^+$ injections on the i.p.s.p. would be due not to any specific effect of these ions, but rather to the diminution of K^+ ions, which is an inevitable consequence of such injections (cf. Appendix, § C). It is therefore postulated that, under the influence of inhibitory impulses, the inhibitory post-synaptic membrane becomes highly permeable to K^+ ions as well as to the anions of similar small size (cf. Results, § C).

In part, the postulated permeability change of the membrane resembles that postulated to occur during the positive after-potential that follows a spike (cf. Coombs *et al.* 1955*a*). However, in this latter condition the high permeability was restricted to K^+ ions, whereas in the former it would be shared with the small anions. As a consequence of this complication, careful controls have to be made with regard to the effect on the i.p.s.p. that would be produced by the influx of Cl^- ions across the neuronal membrane. A considerable part (up to 40%) of the depolarizing current used to inject the cations from the microelectrode is probably carried across the neuronal membrane by the influx of Cl^- ions. Experiments in which a depolarizing current is passed through a K_2SO_4-filled electrode provide a partial control for the effect of this Cl^- influx. With such an electrode a depolarizing current will also produce a considerable increase in the intracellular Cl^- ion concentration, but there will be virtually no change in the intracellular K^+ ion concentration (cf. Appendix, § C). However, the control is defective in that the Cl^- increase is likely to be much smaller than with the passage of the same current through Na_2SO_4- or $((CH_3)_4N)_2SO_4$-filled electrodes, which causes the readily penetrating K^+ ions to be replaced by less penetrating cations, with the consequence that a greater proportion of the current through the membrane is carried by the inward passage of Cl^- ions (cf. Appendix, § C).

On the basis of these expected differences it is possible to account for the respective changes observed in the i.p.s.p. in terms of the hypothesis that the i.p.s.p. is caused by the ionic fluxes consequent on a greatly increased permeability of the membrane to K^+ ions as well as to Cl^- ions. A hyperpolarizing i.p.s.p. would be produced by an inward movement of Cl^- ions and/or an outward movement of K^+ ions, while the inverse movements would give a depolarizing i.p.s.p. With the depolarizing current applied through Na_2SO_4- and $((CH_3)_4N)_2SO_4$-filled electrodes, both the increase in Cl^- and the decrease in K^+ ion concentration would contribute to the displacement of the i.p.s.p. in the depolarizing direction (Figs. 14, 15 A). On the other hand, with the K_2SO_4-filled electrode the displacement would be much smaller (cf. Fig. 15 B) because there would be less increase in the Cl^- and virtually no change in the K^+ ion concentration. With the Na_2SO_4-filled electrode the recovery to the initial condition would be attributable both to the outward pumping of Na^+ and consequent replenishment of K^+ ions, which in part at least is coupled with the Na^+ efflux (cf. Hodgkin & Keynes, 1954; Keynes, 1954), and to the net loss of Cl^- ions by diffusion outwards along their electro-chemical gradient. The slow and incomplete recovery with the $((CH_3)_4N)_2SO_4$-filled electrode presumably would be attributable to the difficulty in eliminating the $(CH_3)_4N^+$ ion from the cell interior and hence in regaining the lost K^+ ions. Some recovery would be expected while the raised Cl^- ion concentration was being lowered by outward leakage across the membrane. In contrast, the recovery of the i.p.s.p. after passage of a depolarizing current through a K_2SO_4-filled electrode was very rapid (cf. Fig. 15 B), which is to be expected because it involves only the outward diffusion of Cl^- ions.

Despite the sufficiency of these explanations the results on cation replacement are not so convincing as those involving anion injection. To a considerable extent the difficulty in investigating this problem arises from the fact that initially there is a high internal K^+ concentration. It is thus much easier to produce a *relative* change in chloride concentration than a relative change in potassium concentration, and furthermore the osmotic influx of water frustrates any attempt to increase the intracellular potassium concentration (Appendix, § C). In principle, a better approach would be to change the extracellular, rather than the intracellular, potassium concentration, as has been done for the isolated crustacean neuromuscular and mammalian cardiac muscle preparations, where inhibition also occurs (Fatt & Katz, 1953; Burgen & Terroux, 1953). However, on account of technical difficulties, the few preliminary experiments have not yielded any significant results.

(2) *Effects attributable to anions*

When a depolarizing current was passed through a microelectrode filled with a potassium salt, there would be virtually no change in the cation content of the motoneurone (Appendix, § C). Hence any change in the i.p.s.p. would presumably be caused by changes in the anions. Experiments with the three potassium salts, KCl, KBr, and KNO_3 gave indistinguishable results, hence they will be considered together. The effects of depolarizing currents have not been investigated with a KSCN-filled electrode.

(*a*) *Electrodes filled with potassium chloride, nitrate and bromide.* In most of our experiments with these salts (sixteen injections into six cells) the i.p.s.p. had already been transformed into the depolarizing type by diffusion (cf. § B). After the passage of the depolarizing current there was then invariably a large diminution in the depolarizing i.p.s.p. and even a momentary reversal to a hyperpolarizing type (Fig. 16A, B), but the effect was always transient, recovery rapidly occurring in the direction of the initial depolarizing i.p.s.p. (Fig. 16C–F). The e.p.s.p. was unaffected (Fig. 16G, H). As shown in Figs. 16I, 17A, B, the recovery curves for the i.p.s.p. following depolarizing currents were exponential in type and approximately mirror images of the recovery curves already described following hyperpolarizing currents (cf. Figs. 7, 10). The half-times of recovery have varied from 10 to 26 sec. At times there has been a small 'overshooting' of the recovery curve (cf. Fig. 16I, 17A), i.e. the i.p.s.p. was finally displaced further in the depolarizing direction than it had been initially. The resting potential was usually increased just after the current. It was never decreased. The maximum increase was 10 mV, but usually it was much less, about 2 mV. It returned with much the same time course as the change in i.p.s.p. Thus, for a constant membrane potential, the change in the i.p.s.p. would have been somewhat greater than appears in Figs. 16 and 17.

When diffusion out of the microelectrode was so small that the i.p.s.p. was not transformed into the depolarizing type, a depolarizing current invariably caused very little change in the i.p.s.p. (five injections into three cells). For example, there was possibly a slight shift in the hyperpolarizing direction with the records of Fig. 17C (○ points) immediately after a depolarizing current of 6×10^{-8} A for 60 sec, but recovery had occurred in a few seconds, as may also be seen with the ● points of Fig. 17D. This contrasts with the effect seen after passing a hyperpolarizing current through the electrode (points designated ■ in Fig. 17C and ○ in Fig. 17D). The displacement of the i.p.s.p. in the hyperpolarizing direction is seen to be negligible when compared with the opposite displacement by a hyperpolarizing current of equal strength and duration. Fig. 17C and D thus contrast with Figs. 16I and 17A, B. Furthermore, in the experiments illustrated by Fig. 17C and D the resting potential was increased very little, if at all, just after the depolarizing current.

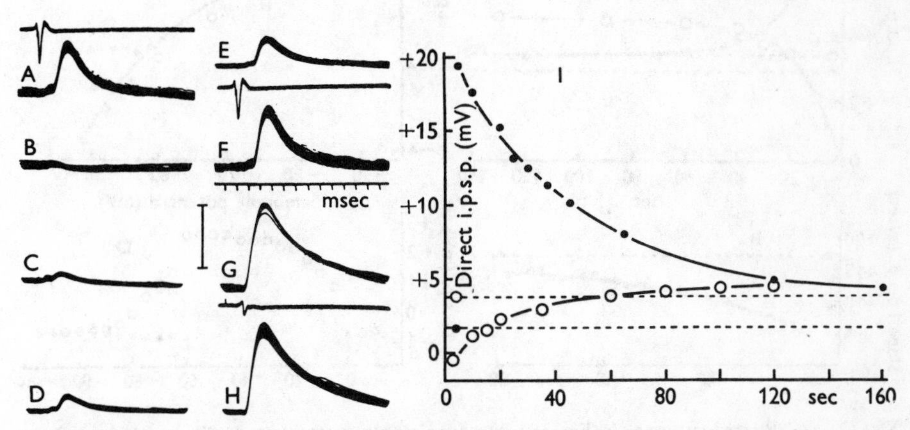

Fig. 16. A–H. A single microelectrode filled with 3M-KNO_3 was inserted into a biceps-semitendinosus motoneurone. A shows the depolarizing i.p.s.p. set up by a quadriceps group I*a* volley. After the passage of a depolarizing current (4×10^{-8} A for 90 sec) there was a momentary reversal of the i.p.s.p. to the hyperpolarizing type (B) followed by recovery to the original response (C–F). G and H show e.p.s.p.'s set up by monosynaptic excitation before and immediately after the same depolarizing current. Records of the quadriceps afferent volley in L6 dorsal root are shown in A and F, while the biceps-semitendinosus afferent volley similarly recorded is seen between G and H. The resting potential was −79 mV before passage of current and −82 mV immediately after. Potential scale gives 5 mV for records A and F, and 10 mV for all other records. Time scale applies to all records. I. Points shown as ○ plot recovery of i.p.s.p.'s for series partly illustrated in Fig. 16B–F, the initial depolarizing i.p.s.p. (Fig. 16A) being shown by ○ on the broken line, which thus represents a base-line. Time is measured from cessation of current. Points shown as ● represent recovery series for the i.p.s.p. after the earlier passage of a hyperpolarizing current (3×10^{-8} A for 90 sec), the initial value for the i.p.s.p. again having a base-line drawn through it (cf. Fig. 10).

(*b*) *Electrodes filled with sodium chloride.* With a NaCl-filled microelectrode a depolarizing current has always produced the same changes in the i.p.s.p., and the resting potential, as with a Na_2SO_4-filled electrode. There was a reversal of the i.p.s.p. to the depolarizing type, if it had not already been reversed by the diffusion of Cl^- ions out of the electrode (§ B), or, in the event of prior reversal, a further increase in the size of the depolarizing i.p.s.p. After the passage of the current there has often been recovery towards the initial condition.

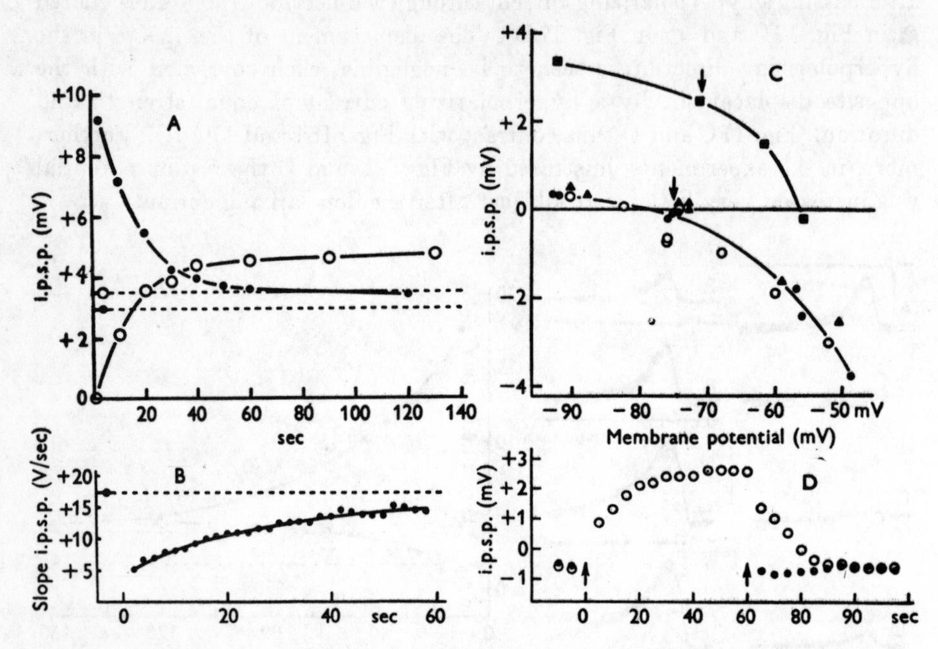

Fig. 17. A. Plotted curves as in Fig. 16I, but for a single microelectrode filled with 4 M-KBr in a biceps-semitendinosus motoneurone. Points shown as ● and ○ plot respectively the recovery of i.p.s.p.'s after a hyperpolarizing and a depolarizing current (both 4×10^{-8} A for 60 sec). The initial responses and base-lines are shown as in Fig. 16I. B. Plotted curve as in Fig. 16I, but for a double microelectrode filled with 3 M-KCl and inserted into a biceps-semitendinosus motoneurone. Zero time corresponds to cessation of a depolarizing current of 6×10^{-8} A for 60 sec. On the ordinate scale are plotted the steepest slopes of rising phases of the direct i.p.s.p.'s. C. A double microelectrode filled with 3 M-KCl was inserted into a gastrocnemius motoneurone. Points shown as ● plot the antidromic i.p.s.p.'s against the membrane potentials as in Fig. 2B. ■ points plot four records obtained within 15 sec after cessation of a hyperpolarizing current (6×10^{-8} A for 60 sec). The arrows indicate the mean resting potential and show a small depression after the current. Points designated ▲ show that recovery was almost complete at about 60 sec after the current. After a further 60 sec, in order to ensure complete recovery, a depolarizing current (6×10^{-8} A for 60 sec) was passed through the electrode and the points designated as ○ plot the records just after this current. There was little change in the curve and no detectable change in the resting potential. D. A double microelectrode filled with 3 M-KCl was inserted into a biceps-semitendinosus motoneurone. Points designated ○ plot the antidromic i.p.s.p.'s before, during and after a hyperpolarizing current of 5×10^{-8} A for 60 sec plotted as in Fig. 7, but for an i.p.s.p. that was initially of hyperpolarizing type. Points designated ● give the antidromic i.p.s.p.'s before and after a depolarizing current of 5×10^{-8} A for 60 sec. No records were possible during this current because the antidromic impulse then invaded the motoneurone. Arrows mark the onset and cessation of the currents.

(*c*) *Discussion.* The effects produced when a depolarizing current is passed through an electrode filled with KCl, KBr or KNO_3 solution can be satisfactorily explained by the diminution or suppression of the anionic diffusion out of the electrode. A steady state will prevail before the application of the current, because the anion concentration in the cell has been raised until the diffusion rate outwards across the membrane equals the steady diffusion rate from the electrode into the cell (cf. § B). During the flow of the depolarizing current there will be diminution or suppression of the diffusion of anions out of the electrode (cf. Appendix, § C); hence the anion concentration of the cell will decline towards its normal level. On cessation of the current there would be an immediate restoration of the anion diffusion out of the electrode, consequently the anion concentration of the cell would increase again towards the steady-state concentration that prevailed before the current. Thus the changes occurring after cessation of the depolarizing current should resemble those occurring initially on insertion of the microelectrode into the cell.

In this manner a satisfactory explanation is provided for the general observation that the faster the diffusion rate out of the electrode (as revealed by the magnitude of the displacement of the i.p.s.p. to the depolarizing type), the larger is the restorative effect produced during a depolarizing current (Figs. 16, 17). Furthermore, it leads to the expectation that, on cessation of the current, recovery to the initial steady state, i.e. towards the depolarizing type of i.p.s.p., would occur with approximately the same exponential time constant that obtains for recovery in the reverse direction after a hyperpolarizing current (cf. Figs. 7, 10, 16, 17). The above explanation takes no account of the Cl^- influx which would occur across the membrane during the depolarizing current (cf. Appendix, § C). Presumably in our experiments this influx has never been larger than the depression of the anion diffusion out of the electrode. When the depolarizing current is passed through an electrode filled with a potassium salt, there will be no significant change in the intracellular potassium concentration. However, with a NaCl-filled electrode, this current will cause, in addition to the suppression of Cl^- diffusion out of the electrode, a depletion of intracellular potassium, and also a larger Cl^- influx across the membrane (cf. Appendix, § C). The observed displacement of the i.p.s.p. in the depolarizing direction indicates that the latter two effects are predominant.

DISCUSSION

A. *Ionic permeabilities of the membrane during the inhibitory post-synaptic potential*

Experimental investigation of the i.p.s.p. at varying membrane potentials in § A led to the assumption that the i.p.s.p. is due to a great increase in the flux of some particular ion or ions across the post-synaptic membrane. The normal

23-2

level of the reversal-potential for the i.p.s.p. shows that the total ionic flux conveys no net charge across a membrane which is about 10 mV more hyperpolarized than the resting membrane (Table 1). This means that, if the particular ion species is an anion, its concentration in the neurone will be lower than the concentration giving equilibrium with the external concentration at the resting membrane potential, while, if a cation, it will be higher than this equilibrium concentration. Finally, if more than one ion species is involved, the i.p.s.p. will be the integrated result of the various ionic fluxes across the membrane, and the ratio of extracellular to intracellular concentration for any one species will probably give an equilibrium potential different from that derived for the i.p.s.p. Since the equilibrium potential for the i.p.s.p.

TABLE 2. Ion diameters in aqueous solution as derived from limiting ion conductances and expressed relative to $K^+ = 1{\cdot}00$. Values derived from Landolt-Börnstein (1936).

The horizontal broken line gives the division between the small ions that pass readily through the post-synaptic inhibitory membrane and the larger ions that pass with much greater difficulty or not at all.

Cations		Anions	
K^+	1·00	Br^-	0·94
		Cl^-	0·96
		NO_3^-	1·03
		SCN^-	1·11
— — —	— — —	— — —	— — —
Na^+	1·47	HCO_3^-	1·65
$N(CH_3)_4^+$	1·60	$CH_3CO_2^-$	1·80
		SO_4^{2-}	1·84
		$H_2PO_4^-$	2·04
		HPO_4^{2-}	2·58

is at a high level of internal negativity (about −80 mV), the ion species concerned must be so distributed that if they are anions there is a high ratio of extracellular to intracellular concentration, while if cations there would be an inverse distribution.

In order to explain the changes produced in the i.p.s.p. by the injection of various ions, it has been postulated (§§ C1, C3, D1*d*) that, when an impulse reaches an inhibitory synapse, it acts on the post-synaptic membrane (presumably by an inhibitory transmitter substance) and greatly increases its permeability to the monovalent anions, Cl^-, Br^-, NO_3^-, SCN^-, and also to the cation, K^+. On the other hand, there must be little or no change in the permeability to the anions SO_4^{2-}, HPO_4^{2-}, $H_2PO_4^-$, HCO_3^-, $CH_3CO_2^-$ and glutamate, and to the cations, Na^+ and $(CH_3)_4N^+$. In Table 2 these anions and cations have been arranged in order of their diameter in aqueous solution, as calculated according to Stokes's law from the limiting ion conductances (cf. Boyle & Conway, 1941). It will be seen that a greatly increased permeability has been postulated for all ions having diameters not exceeding 1·11 times that of K^+ ions, irrespective of their charge, while little or no increase in permeability has been postulated for the larger ions ranging from Na^+ (1·47 times the diameter of K^+ ions) upwards. Thus the production of the i.p.s.p. may simply be due to the inhibitory transmitter substance acting on the post-synaptic membrane to make it a very selective ionic sieve, all ions smaller than a critical size passing through readily, i.e. those above the horizontal line in Table 2, while larger ions virtually fail to pass. In particular the hyperpolarization of the i.p.s.p. could be produced only if there is a very effective blocking of Na^+ ions. Further experiments with ions of relative diameters ranging from 1·1 to 1·6 are necessary in order to find the critical size.

This postulated sieve-like action of the cell membrane towards ions in aqueous solution resembles that proposed by Boyle & Conway (1941) for the surface membrane of muscle fibres, in that the permeability is determined by the size of the hydrated ion and not by its charge. As stated in § D 1*d* the evidence for high potassium permeability during the i.p.s.p. is not so convincing as for the high anion permeability; however, the close analogy with Boyle & Conway's sieve theory supports our interpretation of the experimental evidence. Further indirect support is provided by the evidence that there is a selective increase in the membrane permeability to K^+ ions during inhibitory responses of crustacean muscle (Fatt & Katz, 1953) and of mammalian cardiac muscle (Burgen & Terroux, 1953). It is probable that during these inhibitory responses there is also an increase in chloride permeability, because, at least with crustacean muscle, the flux of some ion other than potassium must have been responsible for the depolarizing inhibitory potentials that could still be evoked in a potassium-free medium (Fatt & Katz, 1953).

B. *Ionic equilibrium between a motoneurone and its environment*

It has been shown in §§ C 1 and C 3 that the injection of the order of 20–50 p-equiv of Cl^- (or of the comparable anions NO_3^-, Br^- or SCN^-) caused the i.p.s.p./membrane potential curve to shift so that for a given membrane potential the i.p.s.p. was changed to a smaller hyperpolarization or greater depolarization (Figs. 3, 4, 6, 7, 9 and 17 C). Subsequently, the i.p.s.p. curve returned toward its original position with a half-time of about 20 sec (range from 10 to 35 sec for the whole group of anions, Figs. 5, 7, 10, 16 I and 17 A). This regression is attributable to a fall of the Cl^- (or other anion) concentration immediately under the surface membrane of the motoneurone. The most likely explanation of this falling concentration is that the high anion concentration built up within the cell by the injection is reduced as the ion diffuses outward across the membrane.

However an alternative explanation would be that the hyperpolarizing current used for the injection has caused a gradient of Cl^- (or other anion) concentration to be set up within the cell with the concentration highest immediately under the membrane, and that the decrease in concentration indicated by the change of the i.p.s.p. would be due secondarily to diffusion from this region under the membrane to re-establish a uniform concentration within the neurone. This postulated excess of anion concentration immediately under the membrane is likely to occur, since the transference number for the anion (the fraction of current due to a movement of this ion) will probably be less in the membrane than in the intracellular space. Nevertheless, the changes that were observed to last some tens of seconds cannot be due to this factor. Thus for a sphere of 70 μ diameter and a substance with a diffusion coefficient of $1 \cdot 2 \times 10^{-5}$ cm^2/sec, which are probable values for this case, it can be shown that any initial concentration difference between the surface layer and the remainder of the interior will rapidly diminish, so that within 0·3 sec there is a variation of no more than 10% of the initial difference (cf. Carslaw & Jaeger, 1947, for similar problems on the conduction of heat in solids).

Similar considerations apply to the situation following a depolarizing current as described in § D 2*a*. The i.p.s.p. was temporarily converted towards the hyperpolarizing type from which recovery to the initial condition occurred with approximately the same time constant as recovery in the reverse direction after a hyperpolarizing current (cf. Figs. 16 I, 17 A, B). Thus it appears that, after an applied current has caused either an increase or decrease in the intracellular concentration of Cl^-, NO_3^-, or Br^- ions, diffusion processes across the surface membrane cause a restoration of the initial concentration. Because of the high ionic concentration in the microelectrode there would be in the absence of current a steady output of anions (and cations) by diffusion. The initial steady-state condition would be restored as the net outward anionic flux across the surface membrane falls or rises to equal the output from the electrode. Thus it would be expected that the time constant of recovery, whose mean value is about 30 sec (derived from the mean half-time of about 20 sec), is dependent on the properties of the cell and not of the electrode, and that it would be the same in both directions, as is actually observed (Figs. 16 I, 17 A, B). For present purposes, no serious error is introduced by assuming that, when the microelectrode is filled with a potassium salt, these curves for the i.p.s.p.'s also express changes in the intracellular concentration of the anion species under investigation.

One may now profitably develop theoretical equations to describe the changes in intracellular concentration of such an anion species after the passage of a current through a microelectrode containing the appropriate potassium salt. During the initial steady state before the passage of the current, the anionic flux out of the microelectrode will equal the net outward flux across the surface membrane of the neurone. Suppose now that a hyperpolarizing or depolarizing current is passed through the microelectrode, so increasing or decreasing the quantity of that anion species within the cell. After this current the rate of change of the amount of anion within the cell may be expressed as

$$v \frac{dC}{dt} = -kA\,(C - C_i), \tag{1}$$

where v is the volume of the cell, A is the area of its surface membrane and k is the proportionality constant that at the resting membrane potential relates the ionic efflux across the membrane to the intracellular concentration, C, the initial steady-state value of which is C_i. k is not a true permeability constant, in that it includes the effect of the electrical potential gradient across the membrane as well as the membrane permeability as strictly defined.

Integrating and rearranging eqn. (1):

$$C = C_i + (C_0 - C_i)\, e^{-kAt/v}, \tag{2}$$

where C_0 is the concentration at zero time, i.e. at the cessation of the hyperpolarizing or depolarizing current, t being measured from this instant. This expression shows that the concentration will fall or rise from C_0 to C_i with an exponential time constant of v/kA, which is the time constant that has a duration of about 30 sec in our experiments.

A more complex situation exists *during* injection of anions by a hyperpolarizing current, because the membrane potential is then displaced far from the initial resting value, and as a consequence there is a large alteration in the value of k, when the conditions are expressed as in eqn. (1). For example, when the interior of the cell is made more negative by a hyperpolarizing current, there will be a large increase in the value of k because there is an increase in the membrane potential, and hence in the driving force for the outward movements of anions. Thus, though still expressed as v/kA, the time constant for establishment of equilibrium during the passage of a hyperpolarizing current would be expected to be much briefer than the time constant for recovery after the current. The expected difference may be seen between the rising and falling phases of the curve in Fig. 5A, where the respective time constants are about 20 and 40 sec. On the other hand, no large error has been introduced by assuming a constant value for k after cessation of the applied current, because the change in membrane potential has been small throughout the recovery process (cf. Fig. 5B).

Having determined the value for k (as above defined), it is possible to calculate the total Cl^- ion efflux from the motoneurone under normal steady-state conditions. Thus, if M_{Cl} denotes the outward flux of Cl^- ions, the total efflux will be given by two expressions, which may therefore be equated:

$$AM_{Cl} = -kA[Cl]_i, \tag{3}$$

where $[Cl]_i$ is the normal intracellular concentration of Cl^- ions. Since at the resting membrane potential the time constant, v/kA, has been found to be about 30 sec for Cl^- ions,

$$AM_{Cl} = \frac{-v[Cl]_i}{30} = -0{\cdot}07 \text{ pmole/sec}$$

if $[Cl]_i = 9$ mM (cf. § C, below) and $v = 2{\cdot}3 \times 10^{-7}$ cm^3 (Coombs *et al.* 1955*a*). Since this chloride efflux occurs at the normal steady state, and since it has not been necessary to postulate a chloride pump (cf. § C, below), there will be the same diffusional influx. It thus is possible to calculate the membrane conductance for chloride ions, G_{Cl}, from the formula derived by Hodgkin (1951),

$$G_{Cl}=\frac{F^2}{RT}M_{Cl},$$

where F, R and T have the usual connotation. The value so calculated for AG_{Cl}, $0{\cdot}25\times10^{-6}$ mho, represents the chloride conductance for the whole neuronal surface, and hence is to be compared with the approximate measured value, 10^{-6} mho, for the total conductance (Coombs *et al.* 1955*a*). This proportion of 20–30% of the total membrane conductance is in good agreement with the proportions calculated for the chloride conductance of giant axons and striated muscle fibres (Hodgkin & Katz, 1949; Hodgkin, 1951; Hodgkin & Huxley, 1952*b*; Shanes, Grundfest & Freygang, 1953). The most unreliable estimate used in the above calculation was the value for the volume of the motoneurone. It has been suggested that this estimate could be too low by a factor of two (Coombs *et al.* 1955*a*). By assuming double the value for v, G_{Cl} is doubled in size and so would account for about half of the total conductance. If an area of 5×10^{-4} cm² is assumed for A, i.e. for the effective surface area of the motoneurone (Coombs *et al.* 1955*a*), a value of 0·07 pmole/sec for the whole neuronal surface corresponds to a value for M_{Cl} of 140 pmole cm^{-2} sec^{-1}, which is several times greater than the largest ionic fluxes observed for giant axons or muscle fibres (cf. Hodgkin & Keynes, 1954).

C. *The equilibrium potential for the ionic fluxes during the i.p.s.p.*

The inhibitory response of the motoneurone may be considered to be due to an increased ionic conductance in those parts of the motoneuronal membrane under the inhibitory synapses. The consequent ionic fluxes would tend to change the potential of these inhibitory patches to a new equilibrium value, but local currents through the capacity and resistance of the remainder of the membrane would limit the recorded change. The actual potential attained by the i.p.s.p. results from this in-parallel relationship of the inhibitory patches and the rest of the membrane. These local currents are eliminated when, by a background of applied current, the membrane potential is adjusted so that it is not changed by the i.p.s.p., which is precisely the reversal-potential for the i.p.s.p. as reported in §§ A, C and D of the Results (cf. Table 1). If it be assumed that the applied current has not modified the action of the inhibitory transmitter on the post-synaptic membrane, the reversal-potential for the i.p.s.p. would give the value for the inhibitory equilibrium potential under any given condition of ionic concentrations.

If the assumption be made that there is a uniform electric field through the membrane, the net flux outwards across unit area of the membrane will be expressed for any cation species by

$$M_c = P_c \frac{EF}{RT}\left(\frac{C_{co} - C_{ci}\,e^{EF/RT}}{1 - e^{EF/RT}}\right), \tag{4}$$

and for any anions species by

$$M_a = -P_a \frac{EF}{RT}\left(\frac{C_{ai} - C_{ao}\,e^{EF/RT}}{1 - e^{EF/RT}}\right), \tag{5}$$

where E is the membrane potential, inside with respect to outside, R, T, and F have the usual connotations, P_c and P_a are the permeabilities for the cation and anion species considered, while C_{co} and C_{ci} are the respective external and internal concentrations for the cations, and C_{ao} and C_{ai} for the anions (Goldman, 1943; Hodgkin & Katz, 1949). The i.p.s.p. will be generated by the current due to the fluxes of all the ion species involved. The current density,

$$I_{\text{i.p.s.p.}} = F\Sigma(M_c - M_a). \tag{6}$$

At the reversal or equilibrium potential for the i.p.s.p., $F\Sigma(M_c - M_a) = 0$.

It must be assumed that the permeability to Na^+ ions is not significantly increased by the inhibitory transmitter, otherwise it would not be possible to account for the experimental observations on the i.p.s.p., for example, its hyperpolarizing character at the normal resting potential. Hence we may regard the fluxes of K^+ and Cl^- ions as alone producing the i.p.s.p. Since the inhibitory transmitter appears to act by converting the inhibitory areas into a sieve selective towards the size of ions, but indifferent to their charge (§ A) it may be postulated, as a first approximation, that during the inhibitory process the inhibitory areas are equally permeable to ions as similar in size as K^+ and Cl^- ions (cf. Table 2); hence it will be assumed that $P_K = P_{Cl}$ and at the inhibitory equilibrium potential that $M_K - M_{Cl} = 0$. Substituting Eqns. (4) and (5) for M_K and M_{Cl} respectively and cancelling common factors,

$$[K]_o - [K]_i \exp(E_{\text{i.p.s.p.}}F/RT) + [Cl]_i - [Cl]_o \exp(E_{\text{i.p.s.p.}}F/RT) = 0, \tag{7}$$

where $[K]_i$, $[K]_o$, $[Cl]_i$ and $[Cl]_o$ are the intracellular and extracellular concentrations of K^+ and Cl^- ions and $E_{\text{i.p.s.p.}}$ is the equilibrium potential for the i.p.s.p. This may be expressed by

$$E_{\text{i.p.s.p.}} = -\frac{RT}{F}\ln\left(\frac{[K]_i + [Cl]_o}{[K]_o + [Cl]_i}\right). \tag{8}$$

The external ionic concentrations may be assumed to be those of an ultrafiltrate of cat's blood plasma, i.e. $[K]_o = 5{\cdot}5$ mM and $[Cl]_o = 125$ mM (mean values derived from Davson, Duke-Elder & Benham (1936), D'Silva (1936) and Krogh (1946) for cat's plasma potassium and from Davson, Duke-Elder & Maurice (1949), Eggleton (1937), Davson *et al* (1936) for cat's plasma chloride, the ratios of ultrafiltrate to plasma concentration being taken as 0·93 and 1·015 for potassium and chloride respectively (Manery, 1954)). There is good reason for regarding the equilibrium potential for the positive after-potential (mean value −88·5 mV in Table 1 column 6) as the equilibrium potential for K^+ ions, i.e. as E_K, (Coombs *et al.* 1955*a*); hence it is possible to calculate $[K]_i$ according to the formula

$$[K]_i = [K]_o \exp(-E_K F/RT) = 5{\cdot}5 \exp(88{\cdot}5/26{\cdot}8) = 151 \text{ mM}.$$

If the mean equilibrium potential for the i.p.s.p. ($E_{i.p.s.p.}$) is taken as −79 mV (cf. columns 4 and 5, Table 1), $[Cl]_i$ can be directly calculated from eqn. 8 as having the value of 8·9 mM.

Given $[Cl]_o$ as 125 mM, the equilibrium potential for Cl^- ions may be determined from the Nernst equation as being −71 mV, which is in close agreement with the normal resting potential across the motoneuronal membrane (cf. Brock *et al.* 1952; Coombs *et al.* 1955*a*; Table 1, column 3). Thus it appears that normally Cl^- ions are in electro-chemical equilibrium across the membrane. This conclusion agrees with that derived from investigations on other membranes, e.g. of muscle fibres and giant axons (Hodgkin, 1951) and of erythrocytes (Krogh, 1946). However, the significance of our calculation must be evaluated against the assumption on which it was based, viz. that the inhibitory membrane is equally permeable to K^+ and Cl^- ions.

D. *The ionic fluxes generating the i.p.s.p.*

It is now possible to construct a diagrammatic representation of the explanation that has been proposed for the generation of the i.p.s.p. On the left side of Fig. 18A there is shown an element of an inhibitory patch of the motoneuronal membrane with K^+ and Cl^- resistance-voltage components in parallel and on the right side there is an ordinary element of the membrane. The battery on the right side represents the resting membrane potential with a voltage of −70 mV. For present purposes it is sufficient to operate the inhibitory components in an all-or-nothing manner by a ganged switch as shown. The actual time course of operation will be considered in a later paper. In accord with the above considerations the K^+ and Cl^- components are shown with identical resistances and with batteries of −90 and −70 mV respectively. It is to be understood that the separation between K^+ and Cl^- elements is for diagrammatic purposes only. Presumably the respective ionic fluxes occur through the same channels in the sieve-like membrane. With the switch open at the resting membrane potential, there will be no current through the ordinary membrane element. Closure of the inhibitory switch will cause an inward current to flow through the K^+ component, so hyperpolarizing the membrane condenser. The Cl^- component and the ordinary membrane elements will in part shunt this effect, i.e. it is postulated that, under the conditions of normal resting potential and ionic composition, the hyperpolarizing i.p.s.p. is generated solely by the flux of K^+ ions. By applying an extrinsic current the membrane potential can be preset at any desired level (Results, § A). If it is preset at -80 mV, on closing the inhibitory switch in Fig. 18A, the inward current through the K^+ component just equals the outward current through the Cl^- component, i.e. the inhibitory element provides no current for the remainder of the membrane. This is precisely the condition obtaining at the equilibrium potential for the i.p.s.p.

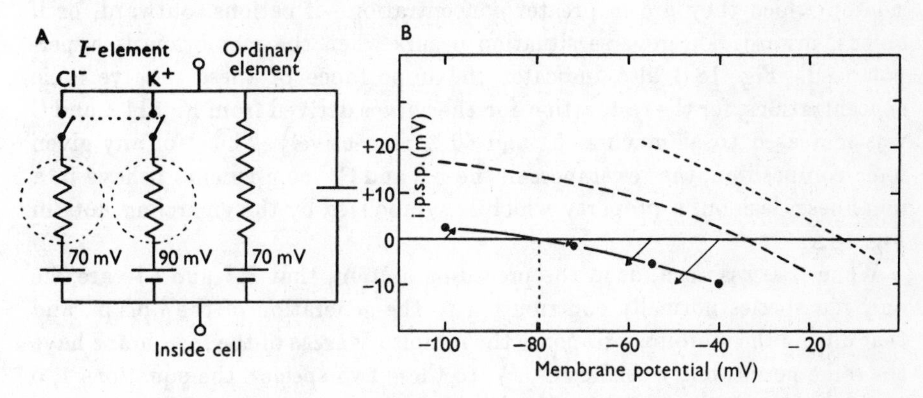

Fig. 18. A. Diagrammatic representation of the electrical properties of an ordinary element of the neuronal membrane and of an inhibitory element with K^+ and Cl^- ion components in parallel. Further description in text. B. I.p.s.p./membrane potential curves drawn on co-ordinates having equal values as described in the text. The arrows give the actual potential loci for i.p.s.p. responses corresponding to the four points designated ● on the continuous line. Further description in text.

When the membrane potential is displaced from this equilibrium potential, Fig. 2 shows that the i.p.s.p. is changed in the direction that counteracts this initial displacement, i.e. it can be considered as effecting a partial restoration. The mechanism of this restoration can be appreciated by reference to Fig. 18A. Closure of the *I*-switch will cause current to flow through the *I*-element in the direction which will displace the membrane potential back towards the equilibrium potential of -80 mV. The effectiveness of this restorative action is illustrated in Fig. 18B, where the curves of Figs. 2A, 14A (+) and 14C (▲) are drawn on equi-valued potential co-ordinates. For the curve of Fig. 2A the potential loci during the i.p.s.p.'s are represented by arrows for a series of initial membrane potentials. A background of extrinsic current sets the level of each membrane potential and continues throughout the response. The loci are drawn at 45° because, when the i.p.s.p. attains any potential, there is necessarily an equivalent alteration of membrane potential. If there was 100% restoration at the summit of the i.p.s.p., the arrows would end on the vertical line through the equilibrium potential of -80 mV. It will be seen that the restoration was as large as 25% for the depolarized membrane, whereas it was only 16% at the normal resting potential. This variation is of course also indicated by the increasing steepness of the i.p.s.p./membrane potential curve as the membrane becomes more depolarized (cf. Figs. 2, 4, 6, 8E, 14 and 16B). Fig. 18A is defective in that it does not account for this non-linear behaviour. In order to do so, account must be taken of the Cl^- and K^+ ion concentrations on the two sides of the membrane. The more depolarized the membrane the less effectively will it impede the movement of ions from the side on which they are in greater concentration—if cations, outward, or if anions, inward. The reverse situation occurs when the membrane is hyperpolarized. Fig. 18B also indicates the importance of these relative ionic concentrations, for the restoration for the curves derived from Fig. 14A and C was increased to as much as 51 and 60% respectively. Thus, for any given ionic composition, the resistances in the K^+ and Cl^- components behave in a non-linear fashion, a property which is symbolized by the encircling dots in Fig. 18A.

When it is assumed, as in the preceding section, that K^+ and Cl^- are the only ion species normally contributing to the generation of the i.p.s.p., and that during the inhibitory response the inhibitory areas of the membrane have the same permeability coefficient, P_I, to these two species, the equations 4, 5 and 6 may be arranged in the form,

$$\frac{I_{\text{i.p.s.p.}}}{FP_I}=\frac{EF}{RT}\left(\frac{[\mathrm{K}]_o+[\mathrm{Cl}]_i-([\mathrm{K}]_i+[\mathrm{Cl}]_o)\,\mathrm{e}^{EF/RT}}{1-\mathrm{e}^{EF/RT}}\right) \tag{9}$$

In the application of this equation a fundamental assumption is that the permeability factor P_I is not changed in value by changes in the membrane potential. This is reasonable in view of the chemical transmission hypothesis, according to which the peculiar changes in the post-junctional membrane which are elicited by the junctional transmitter cannot be elicited by electrical means. In plotting eqn. 9 (Fig. 19A) the values for $I_{\text{i.p.s.p.}}/FP_I$ have been calculated for various values of the membrane potential (E) from −140 to 0 mV when the assumed values for $[\mathrm{K}]_o$, $[\mathrm{K}]_i$, $[\mathrm{Cl}]_o$, $[\mathrm{Cl}]_i$ are 5·5, 150, 125 and 9 mM respectively, i.e. approximately at normal levels (cf. § C). No further assumption is made in plotting the curve. The two other curves drawn as broken lines show the curves for the two components,

$$\frac{EF}{RT}\left[\frac{[\mathrm{K}]_o+[\mathrm{Cl}]_i}{1-\mathrm{e}^{EF/RT}}\right] \quad \text{and} \quad \frac{EF}{RT}\left[\frac{([\mathrm{K}]_i+[\mathrm{Cl}]_o)\,\mathrm{e}^{EF/RT}}{1-\mathrm{e}^{EF/RT}}\right]$$

of the right side of eqn. 9, which plot respectively ionic fluxes giving internal positivity (the upper broken line) and internal negativity (the lower broken line). The continuous line represents the sum of these two curves.

Since P_I has been assumed to be constant for any particular inhibitory response, the values for the ordinates of Fig. 19A, i.e. for $I_{\text{i.p.s.p.}}/FP_I$, would be proportional to the ionic current densities ($I_{\text{i.p.s.p.}}$) at the respective membrane potentials. Furthermore, since the membrane capacity may be assumed to be unaltered, these values would be approximately proportional to the maximum rates of rise of the recorded i.p.s.p.'s, the additional assumption being made that the intracellular microelectrode picks up a constant fraction of the i.p.s.p. that is generated by the currents through the inhibitory patches. This assumption appears to be justified because the membrane conductance is not greatly altered over the range of membrane potentials employed in our experiments (Coombs *et al.* 1955*a*). Hence, with suitable scaling of the ordinates,

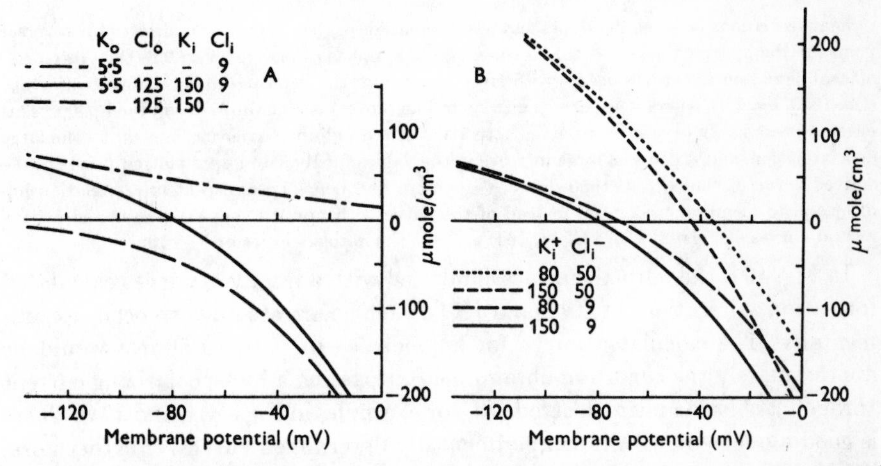

Fig. 19. A. The continuous line gives the curve when computed values for $I_{\text{i.p.s.p.}}/FP_I$ at the assumed normal intracellular ionic concentrations are plotted against membrane potential as described in the text (cf. eqn. 9). Since the ordinate scaling is for the calculated net ionic flux rates divided by the permeability constant of the membrane it has the dimensions of a concentration as shown. The derivation of the two broken lines is described in the text. B. The continuous line is plotted as in Fig. 18A, while the other lines represent the computed curves when the intracellular ionic concentrations are altered as described in the text, and indicated in the figure.

the curve shown by the continuous line in Fig. 19A can be regarded as plotting the theoretical curve for the effect of variations in the membrane potential on the maximum slope of an i.p.s.p. However, our experimental curves have been drawn, not for the maximum slopes of the i.p.s.p.'s, but for their peak amplitudes, which on two counts are less satisfactory for comparison with Fig. 19A: first, the potentials attained by the peak amplitudes are much more modified by decay due to current flow through the general neuronal membrane than are the maximum slopes; secondly, there must necessarily be a considerable change in membrane potential, i.e. in abscissal scaling, at the peak amplitude (cf. arrows in Fig. 18B), whereas the maximum slope of the i.p.s.p. occurs when there is a relatively small change in membrane potential. However, the time courses of the i.p.s.p.'s are approximately constant over our experimental ranges of membrane potential (cf. Fig. 1), so the forms of the experimental curves for i.p.s.p. summits should not deviate widely from those for slopes. As expected, good agreement has been observed when both curves were plotted for i.p.s.p.'s that were recorded with a much faster time-base than those of Figs. 1, 3 and 13. It is therefore significant that the experimental curves plotting the peak potentials of the i.p.s.p. against membrane potentials (Figs. 2, 4, 6, 8E, 14 and 16B) agree closely with the theoretical curve of Fig. 19A, having approximately the same reversal potential and upward convexity.

Small deviations between the slopes and peak potentials of the i.p.s.p. occur close to the reversal point for the i.p.s.p., where the time course shows a diphasic form (cf. Fig. 9B, C). Also a significant deviation sometimes occurs with large i.p.s.p.'s in the depolarizing direction (cf. Coombs *et al.* 1953, fig. 2B), where it appears that a spike-like process is superimposed on the i.p.s.p. This effect is presumably caused by a partial activation of the sodium-carrier mechanism by the large depolarization, and it causes a large increase in the values for the two upper plotted points of the derived curve (Coombs *et al.* 1953, fig. 3, ○ points), and hence an apparent linear relationship of i.p.s.p. to membrane potential instead of the curve with upward convexity as in the experimental curves of Figs. 2, 4, 6, 8E, 14, 16B and the theoretical curves of Fig. 19.

In Fig. 19B the normal curve is combined with a family of curves calculated for various alterations of $[Cl]_i$ and $[K]_i$, such as are assumed to occur experimentally. The calculated curve for an increase in $[Cl]_i$ to 50 mM would be approximately the condition obtaining after passing a hyperpolarizing current through a chloride microelectrode as, for example, in Figs. 4, 6 and 17C. There is good agreement with the experimentally determined curves. Furthermore, if it be assumed that during the i.p.s.p. there is the same increase in permeability towards Br^-, NO_3^- and SCN^- ions, the internal concentrations of such ions would be merely added to the $[Cl]_i$ value and a similar curve would be obtained by calculation, which again agrees with experiment.

With the passage of a depolarizing current through a Na_2SO_4- or $((CH_3)_4N)_2SO_4$-filled microelectrode it has been postulated that both the decreased $[K]_i$ and increased $[Cl]_i$ would displace the i.p.s.p. in the depolarizing direction. The calculated curves of Fig. 19B reveal that an assumed diminution of $[K]_i$ from 150 to 80 mM has a much smaller effect than an assumed increase of $[Cl]_i$ from 9 to 50 mM for all membrane potentials beyond the very low value of -14 mV, though the actual increase in Cl^- ions was less than 60% of the decrease in K^+ ions. The combined effect of $[Cl^-]_i$ increase and $[K^+]_i$ decrease, however, gives a curve which is in reasonable agreement with the curve plotted in Fig. 14A (+ points) after a depolarizing current that would cause approximately this depletion of $[K^+]_i$ ions, but probably a smaller increase in $[Cl^-]_i$ ions.

CONCLUSIONS

A simple hypothesis has been developed which provides satisfactory explanations of the effects produced on the inhibitory post-synaptic potential of motoneurones by variations of their membrane potential and of their ionic composition. According to this hypothesis the inhibitory transmitter substance that is liberated from the inhibitory presynaptic terminals acts on the inhibitory patches of the post-synaptic membrane and greatly increases their permeability to all ions below a critical size, i.e. to K^+, Cl^-, NO_3^-, Br^- and SCN^-. Under ordinary physiological conditions K^+ and Cl^- ions are the important permeable ions, the membrane being about equally permeable to them, while the relative impermeability of the membrane to Na^+ ions is unchanged. The equilibrium potential for the inhibitory post-synaptic potential is about -80 mV, i.e. about 10 mV more negative than the normal resting potential, while the equilibrium potential for the positive after-potential, i.e. for K^+ ions only, is about -90 mV (Coombs *et al.* 1955*a*). From these values and the external K^+ and Cl^- ion concentrations, the internal Cl^- concentration has been calculated to be about 9 mM which indicates that there is approximately chloride equilibrium at the normal resting potential. Thus, according to the hypothesis, at the resting membrane potential the hyperpolarizing inhibitory potential would be entirely due to the flux of K^+ ions, the flux of Cl^- ions actually limiting its size. However, when the membrane is depolarized, as would occur during an excitatory post-synaptic potential, the Cl^- ion flux will operate in the same way as the K^+ ion flux to counteract the depolarization. The effectiveness of the conjoint K^+ and Cl^- fluxes in preventing the generation of an impulse by the excitatory post-synaptic potential will be discussed in a later paper (Coombs, Eccles & Fatt, 1955*b*), where it will be shown that it is possible in this way to account satisfactorily for the observed inhibitory effects. That paper will also deal with the time course of the raised ionic permeability produced by a single inhibitory volley.

SUMMARY

1. Two new techniques have been applied to motoneurones in an investigation of the ionic mechanisms concerned in producing the inhibitory response known as the inhibitory post-synaptic potential, (i.p.s.p.), which normally is a hyperpolarization of the neuronal membrane.

(A) Double-barrelled microelectrodes have been inserted into motoneurones of anaesthetized cats so that the inhibitory potentials can be recorded through one barrel when the membrane potential is preset at any desired value above or below the resting potential by current through the other barrel.

(B) By passing through the intracellular microelectrode a measured current for a measured period, the ionic composition of the neurone has been altered by injecting known amounts of ions. Currents that hyperpolarize the neuronal membrane are almost entirely carried out of the microelectrode by anions, which are thus injected from the microelectrode into the neurone. Similarly, cations are injected by the reverse current. Complications arise because there is a background diffusion of ions from the electrode into the neurone, and also on account of the ionic flux which carries the applied current across the neuronal membrane. It has been possible experimentally to make approximate estimates of the effects produced by these complications.

2. Using technique (A) it has been possible to construct curves expressing the relationship of the i.p.s.p. to the membrane potential. Normally the i.p.s.p. is reversed to a depolarizing response when the membrane is hyperpolarized beyond about 10 mV above the resting level.

3. Using technique (B) it has been shown that the i.p.s.p. is converted to a depolarizing response by increase in the intracellular concentration of some monovalent anions, Cl^-, Br^-, NO_3^- and SCN^-, which appear about equipotent in this respect. The larger monovalent anions, HCO_3^-, CH_3COO^- and glutamate have little or no effect and the large polyvalent anions SO_4^{2-} and HPO_4^{2-} are quite ineffective. During any injecting current the changes develop exponentially towards an equilibrium with a time constant of about 20 sec, and on cessation of the current there is usually a fairly complete recovery along an exponential time course, with a rather longer time constant, about 30 sec.

4. When a depolarizing current is passed through the microelectrode, it appears that the changes in the i.p.s.p. are largely attributable to the ionic flux that conveys the current across the neuronal membrane, the neurone suffering both a loss of potassium and a gain of chloride. Comparison of experiments with potassium, sodium and tetramethylammonium salts in the electrode has indicated that the displacement of the i.p.s.p. in the depolarizing direction is attributable not only to the gain in chloride, but also to the loss of potassium. As indicated by the changes in the i.p.s.p., when the potassium is replaced by tetramethylammonium, the lost potassium of the neurone is recovered very slowly if at all. Presumably the neurone has difficulty in eliminating the tetramethylammonium. When it is replaced by sodium, there is always complete recovery with a time constant of about 5 min, which may be regarded as the time-constant for elimination of the excess sodium by the sodium pump.

5. The depolarizing current also appears to act by depressing the diffusion of anions (Cl^-, Br^- or NO_3^-) from the microelectrode into the neurone. As would be expected, on cessation of the current the diffusion from the microelectrode again increases the anion content with the same time constant of about 30 sec that governs diffusional exchange between the cell and its environment.

6. There is a brief theoretical treatment of the diffusional relationships between the motoneurone and its environment. The calculated resting rate of chloride diffusion across the membrane is considerably higher than the values that have been given for corresponding areas of giant fibre membranes and it accounts for a considerable fraction (up to half) of the membrane conductance.

7. It is postulated that the inhibitory transmitter substance greatly increases the permeability of the inhibitory patches of the post-synaptic membrane to some ions (K^+, Cl^-, NO_3^-, Br^-, SCN^-) and not at all or very slightly to others (Na^+, $(CH_3)_4N^+$, SO_4^{2-}, HPO_4^{2-}, HCO_3^-, CH_3COO^-). On the basis of this hypothesis explanations have been offered for all the changes that are produced in the i.p.s.p. by varying the membrane potential and by the various changes in intracellular ionic composition. Since all permeable ions have diameters (calculated from limiting ion conductances) smaller than all the non-permeable, it seems that the inhibitory transmitter acts by converting the membrane into a sieve with pores that are small enough to block the passage of sodium and all larger ions.

APPENDIX

Changes in intracellular ionic concentrations produced by injection from a microelectrode

(A) *Ionic injection by diffusion*

Since the intracellular microelectrode has in our experiments been filled with a concentrated salt solution, at least 1·2 equiv/l., the ionic movements from the electrode into the cell would always have been much greater than from the relatively low concentration in the cell, about 0·15 equiv/l., into the electrode. When the tip of a microelectrode is immersed in a solution identical with that filling it, the ionic movements into and out of the orifice will be equal, and, if the ionic conductance for any one ion (G_n) is known, the corresponding flux for that ion (M_n) can be calculated from the relationship given by Hodgkin (1951),

$$M_n = \frac{RT}{z^2F^2} G_n,$$

where R, T, z and F have the usual connotation. For example, when an electrode filled with 3M-KCl has a resistance of $3{\cdot}5 \times 10^6\,\Omega$ when immersed in 3M-KCl, it may be assumed that, on account of the similar mobilities of the K^+ and Cl^- ions, the conductance for each will be $1{\cdot}4 \times 10^{-7}$ mho and hence the efflux of each ion species out of the electrode can be calculated from the above equation to be about 0·04 pmole/sec. When such an electrode is withdrawn from the 3M-KCl and similarly immersed in a 0·15M-KCl solution, i.e. in a solution approximately isotonic with the interior of a cell, the resistance rises within a second or so to a new steady level almost three times the value measured in a 3M-KCl solution; for example, for the electrode considered above, the resistance rises to about 10 MΩ. A ratio of this order has been calculated on theoretical grounds and is also implicit in the calculated values reported by Nastuk & Hodgkin (1950). This increase in resistance is attributed to the dilution of the KCl solution in the terminal segment of the microelectrode due to the outward diffusion of KCl at the tip, where there is initially a very high concentration gradient. It should be noted that, on immersion in the lower concentration of KCl, almost all of the raised resistance occurs on account of dilution within the electrode, for only about 200,000 Ω could occur on account of the increased specific resistance of the external medium.

When immersed in 0·15 M-KCl the efflux of ions from the electrode cannot be calculated from the above equation because there will no longer be equality of efflux and influx. The reduction in concentration of KCl within the microelectrode near the orifice will diminish the efflux below the level calculated for the immersion in 3 M-KCl. On the other hand, the efflux would be considerably higher than the value calculated according to the above equation from the conductance (10^{-7} mho) of the electrode in the 0·15 M solution. For present purposes it is sufficient to regard 0·04 pmole/sec as being an upper limit for the efflux of K^+ and Cl^- ions from an electrode filled with 3 M-KCl and having a resistance of 10 MΩ, either when it is immersed in a solution of 0·15 M-KCl, or when it is implanted in a cell where the solution is of about the same concentration. The ionic efflux varies inversely with the resistance of the microelectrode. For example, with a KCl-filled electrode of 5 MΩ resistance, which is the lowest resistance ordinarily used, an approximate estimate of the ionic efflux would be 0·06 pmole/sec, which is the value quoted in Results, § B, of the text.

If the potential of the intracellular microelectrode is altered with respect to the indifferent electrode, current will flow from it into the cell or vice versa, and as a consequence there will be modification of the ionic exchanges which are occurring by diffusion across the two interfaces between the cell and its environment, i.e. between the cell and the microelectrode and between the cell and the surrounding medium. The changes so produced in the ionic composition of a cell by current in either direction are complex, and not simply attributable to the injection by the current of anions or cations from the microelectrode. The ionic injections with the two types of current are considered in the next two sections. The effects on the intracellular ionic concentrations are dealt with in § B of the Discussion.

(B) *Ionic injection by hyperpolarizing currents*

When the microelectrode is made electrically negative to the indifferent electrode, current will flow from the indifferent electrode through the cell membrane into the neurone and then into the microelectrode, i.e. it will pass through the cell membrane in the direction that will hyperpolarize. The potential gradient giving such a hyperpolarizing current will modify the ionic diffusion obtaining between the microelectrode and the cell, increasing the anion movements into the cell and decreasing the cation movements. In fact the current is carried from the cell into the electrode by the sum of the two changes in these ionic currents, $-i = F\,(m_c - m_a)$, where i is the current in amperes, the negativity denoting hyperpolarizing current, m_c and m_a are the respective changes in the diffusional transport of cations and anions measured in equivalents per sec, and F is the Faraday. With the largest size of KCl-filled microelectrode that we have used (5 MΩ resistance) the rates of diffusion of K^+ and Cl^- ions into the cell have been calculated to be each about 0·06 p-equiv/sec (Appendix, § A), while the rates of diffusion in the reverse direction (cell to microelectrode) would be so much less that they may be neglected in the present approximate calculations. A very small hyperpolarizing current would be carried in about equal proportions by the increase of the anion movement into the cell and by the decrease of the cation movement, e.g. with $0{\cdot}5 \times 10^{-8}$ A, the Cl^- efflux from the microelectrode would be increased from 0·06 to about 0·085 p-equiv/sec, while the K^+ efflux would be decreased from 0·06 to about 0·035 p-equiv/sec. However, the current applied through the microelectrode has been much larger, often about ten times. Assuming complete suppression of the background K^+ efflux, which would carry a current of about $0{\cdot}6 \times 10^{-8}$ A, the Cl^- efflux would have to be increased from 0·06 to 0·5 p-equiv/sec in order to carry a hyperpolarizing current of 5×10^{-8} A. This increase in Cl^- efflux is overestimated by the small amount of K^+ influx from the cell into the microelectrode. Usually the microelectrode has had a resistance of 10–20 MΩ, so the effects of background ionic diffusion from the microelectrode were usually much smaller than in this numerical illustration. When the microelectrode is filled with other salts, there must be a similar rate of injection of other anions by a hyperpolarizing current (approximately 0·1 p-equiv/sec for each 10^{-8} A of current).

However, it cannot be assumed that the injected anions simply accumulate progressively in the cell. For example, chloride ions (and also bromide, nitrate and thiocyanate ions) diffuse freely across the cell membrane (Discussion, § B), and their outward flux across the membrane will be increased as the internal concentration rises and also on account of the larger driving force exerted by the increased potential gradient which the current produces in the membrane. Since the internal mobile anions would be initially in low concentration, the inward membrane current (i.e. the hyperpolarizing current) would be carried largely by the influx of K^+ and Na^+ ions, a likely proportion being about 30% for the anionic share of the current (cf. Discussion, § B). With increasing concentration of internal anions, however, the outward anion flux would contribute progressively more to this current. A further mechanism tending to limit the progressive increase in the concentration of internal anions is provided by the osmotic influx of water across the cell membrane which would occur on account of the increase in salt content produced in the cell due to the Na^+ and K^+ ions coming from outside and anions from the electrode. For example, when passed through a chloride-filled electrode, a hyperpolarizing current of 5×10^{-8} A injects about 0·5 p-equiv/sec of Cl^- ions into the cell. If the exponential time constant for Cl^- equilibrium between the cell and its environment is 30 sec (Discussion, § B), a steady state is reached during the passage of the current when the Cl^- content of the cell is increased by 15 p-equiv. However, due to the increased membrane potential the time constant may be decreased even to one-half (cf. Fig. 5A), in which case the Cl^- content of the cell would be increased by only 7·5 p-equiv. Nevertheless, there would still be a considerable increase in the concentration of Cl^- ions, more than 30 mM, if the cell volume is assumed to be $2{\cdot}3 \times 10^{-7}$ cm^3 (cf. Coombs *et al.* 1955*a*). This represents a large relative change, for the normal concentration has been calculated to be only about 9 mM (Discussion, § C). In making this calculation the osmotic influx of water across the membrane has been neglected, but it can be shown to have a relatively small effect for such injections of anions. For example, an increase in both the anion and cation concentrations by 30 mM increases the osmotic pressure of the cell by about 20%. Thus the consequent influx of water would cause a 20% increase in the water content of the cell, and hence the increase in anion concentration would be reduced from 30 to 24 mM. However, the osmotic influx of water would be much more important when relatively impermeable anions such as sulphate or phosphate were injected. Probably osmotic influx of water then provides the only significant check against a progressive increase in anion concentration during a hyperpolarizing current.

(C) *Ionic injection by depolarizing currents*

When the microelectrode is made electrically positive to the indifferent electrode, i.e. when a depolarizing current is passed, the ionic diffusion between the microelectrode and the cell will be affected in the direction opposite to that for a hyperpolarizing current, the anion movements into the cell being decreased or even suppressed, while the cation movements are increased. It can be assumed, likewise, that the cation flux out of the microelectrode will inject about 0·1 p-equiv/sec of cations into the cell for each 10^{-8} A of current. On analogy with the membranes of other excitable cells (Hodgkin & Huxley, 1952*a*, *b*), it may be assumed that normally more than half (perhaps about 0·7) of the depolarizing current is carried across the membrane by the net outward flux of K^+ ions, with the inward flux of Cl^- ions carrying most of the remainder (cf. Discussion, § B).

With a K_2SO_4-filled electrode there would be virtually no progressive change in the proportional carriage of current across the membrane. Since current would be carried out of the electrode almost exclusively by K^+ ions, and the transference number for K^+ ions is probably about 0·7 for the outward current across the membrane, there would tend to be an increase in intracellular K^+ ion concentration. Since the remaining 0·3 of the membrane current would be carried largely by the inward flux of Cl^- ions, for each 10^{-8} A of depolarizing current there would be an addition of about 0·03 p-equiv/sec of K^+ and Cl^- ions to the cell. This increase in salt concentration would cause an osmotic influx of water across the surface membrane which would tend to restore the osmotic pressure to normal, i.e. to bring the K^+ ion concentration down to 150 mM. An approximate value for this rate of osmotic change may be derived from measurements on the giant axons of cephalopods (Hill, 1950). Making due allowance for the different volume to surface ratios, the time constant for osmotic equilibration would lie between 4 and 8 sec for the motoneurone. Probably the value would be still shorter, for the motoneuronal membrane has been calculated to be several times more permeable to ions than the giant fibre membrane (Discussion, § B), and hence presumably it is also more permeable to water. Thus the time constant for osmotic equilibration might be well less than 10% of that for ionic equilibration, which is about 30 sec for the small anions. Under such conditions, during the passage of a depolarizing current through a K_2SO_4 electrode, there would be a negligible increase in the intracellular K^+ concentration above its normal value of 150 mM, for the influx of water across the cell membrane would virtually parallel the net gain of K^+ ions produced by the current flow; for example, the usual intensity of current, 5×10^{-8} A, would cause an increase of only 4% in the intracellular K^+ ion concentration. As with a hyperpolarizing current, this osmotic influx of water would not seriously depress the relative increase in the internal Cl^- concentration by the influx across the membrane, for, in contrast to K^+, the initial Cl^- concentration is very low. However, the respective transference numbers that have assumed for the hyperpolarizing and depolarizing currents indicate that the rate of addition of Cl^- by a depolarizing current would be less than half of that produced by a similar hyperpolarizing current.

Very different conditions prevail when the depolarizing current is applied through a Na_2SO_4-filled electrode. Throughout there would be a sodium transference number of nearly unity for the current from electrode to cell. As above, the current would be carried initially across the membrane by the outward flux of K^+ ions and the inward flux of Cl^- ions with respective transference numbers probably of about 0·7 and 0·3. Thus there will be a progressive loss of K^+ ions and gain of Na^+ and Cl^- ions. The consequent rapid depletion of intracellular K^+ ions (initially at about 0·07 p-equiv/sec for every 10^{-8} A) will result in a progressive change in the transference numbers across the membrane: the potassium efflux will fall progressively, while the chloride influx will rise (perhaps even to double the initial value), and to a smaller extent there will be an increase in the net outward flux of Na^+ ions. Thus the principal effects of the current on the intracellular ionic composition will be a substitution of Na^+ ions for K^+ ions and an increase in Cl^- ions. This latter increase will be considerably larger than when the current is passed through a K_2SO_4-filled electrode. Similar changes in the K^+ and Cl^- ion concentrations would be expected when the depolarizing current is passed through a $((CH_3)_4N)_2SO_4$-filled electrode.

REFERENCES

BOYLE, P. J. & CONWAY, E. J. (1941). Potassium accumulation in muscle and associated changes. *J. Physiol.* **100**, 1–63.

BRADLEY, K., EASTON, D. M. & ECCLES, J. C. (1953). An investigation of primary or direct inhibition. *J. Physiol.* **122**, 474–488.

BROCK, L. G., COOMBS, J. S. & ECCLES, J. C. (1952). The recording of potentials from motoneurones with an intracellular electrode. *J. Physiol.* **117**, 431–462.

BURGEN, A. S. V. & TERROUX, K. G. (1953). On the negative inotropic effect in the cat's auricle. *J. Physiol.* **120**, 449–464.

CARSLAW, H. S. & JAEGER, J. C. (1947). *Conduction of Heat in Solids.* Oxford: Clarendon Press.

CASTILLO, J. DEL & KATZ, B. (1954). The membrane change produced by the neuro-muscular transmitter. *J. Physiol.* **125**, 546–565.

COLE, K. S. & CURTIS, H. J. (1939). Electric impedance of the squid giant axon during activity. *J. gen. Physiol.* **22**, 649–670.

COOMBS, J. S., ECCLES, J. C. & FATT, P. (1953). The action of the inhibitory synaptic transmitter. *Aust. J. Sci.* **16**, 1–5.

COOMBS, J. S., ECCLES, J. C. & FATT, P. (1955*a*). The electrical properties of the motoneuronal membrane. *J. Physiol.* **130**, 291–325.

COOMBS, J. S., ECCLES, J. C. & FATT, P. (1955*b*). The inhibitory suppression of reflex discharges from motoneurones. *J. Physiol.* **130**, 396–413.

DAVSON, H., DUKE-ELDER, W. S. & BENHAM, G. H. (1936). The ionic equilibrium between the aqueous humour and blood plasma of cats. *Biochem. J.* **30**, 773–775.

DAVSON, H., DUKE-ELDER, W. S. & MAURICE, D. M. (1949). Changes in ionic distribution following dialysis of aqueous humour against plasma. *J. Physiol.* **109**, 32–40.

D'SILVA, J. L. (1936). Adrenaline and potassium in serum. *J. Physiol.* **86**, 219–228.

ECCLES, J. C., FATT, P. & KOKETSU, K. (1954). Cholinergic and inhibitory synapses in a pathway from motor-axon collaterals to motoneurones. *J. Physiol.* **126**, 524–562.

Eccles, J. C., Fatt, P. & Landgren, S. (1954). The 'direct' inhibitory pathway in the spinal cord. *Aust. J. Sci.* **16**, 130–134.

Eccles, J. C., Fatt, P., Landgren, S. & Winsbury, G. J. (1954). Spinal cord potentials generated by volleys in the large muscle afferents. *J. Physiol.* **125**, 590–606.

Eggleton, M. G. (1937). The behaviour of muscle following the injection of water into the body. *J. Physiol.* **90**, 465–477.

Fatt, P. & Katz, B. (1951). An analysis of the end-plate potential recorded with an intracellular electrode. *J. Physiol.* **115**, 320–370.

Fatt, P. & Katz, B. (1953). The effect of inhibitory nerve impulses on a crustacean muscle fibre. *J. Physiol.* **121**, 374–389.

Gaskell, W. H. (1887). On the action of muscarine upon the heart, and on the electrical changes in the non-beating cardiac muscle brought about by stimulation of the inhibitory and augmentor nerves. *J. Physiol.* **8**, 404–415.

Goldman, D. E. (1943). Potential, impedance, and rectification in membranes. *J. gen. Physiol.* **27**, 37–60.

Granit, R. (1950). Reflex self-regulation of muscle contraction and autogenetic inhibition. *J. Neurophysiol.* **13**, 351–372.

Hagbarth, K. E. (1952). Excitatory and inhibitory skin areas for flexor and extensor motoneurones. *Acta physiol. scand.* **26**, Suppl. **94**, 1–58.

Hill, D. K. (1950). The volume change resulting from stimulation of a giant nerve fibre. *J. Physiol.* **111**, 304–327.

Hodgkin, A. L. (1951). The ionic basis of electrical activity in nerve and muscle. *Biol. Rev.* **26**, 339–409.

Hodgkin, A. L. & Huxley, A. F. (1952*a*). Currents carried by sodium and potassium ions through the membrane of the giant axon of *Loligo*. *J. Physiol.* **116**, 424–448.

Hodgkin, A. L. & Huxley, A. F. (1952*b*). The components of membrane conductance of the giant axon of *Loligo*. *J. Physiol.* **116**, 449–472.

Hodgkin, A. L. & Huxley, A. F. (1952*c*). A quantitative description of membrane current and its application to conduction and excitation in nerve. *J. Physiol.* **117**, 500–544.

Hodgkin, A. L. & Katz, B. (1949). The effect of sodium ions on the electrical activity of the giant axon of the squid. *J. Physiol.* **108**, 37–77.

Hodgkin, A. L. & Keynes, R. D. (1954). Movements of cations during recovery in nerve. *Symp. Soc. exp. Biol.* **8**, 423–437.

Keynes, R. D. (1954). The ionic fluxes in frog muscle. *Proc. Roy. Soc.* B, **142**, 359–382.

Krogh, A. (1946). The active and passive exchanges of inorganic ions through the surfaces of living cells and through living membranes generally. *Proc. Roy. Soc.* B, **133**, 140–200.

Landolt-Börnstein, H. H. (1936). *Physikalisch-chemische Tabellen*, 5th ed., Book III, Part 3, Berlin: Springer.

Laporte, Y. & Lloyd, D. P. C. (1952). Nature and significance of the reflex connections established by large afferent fibres of muscular origin. *Amer. J. Physiol.* **169**, 609–621.

Lloyd, D. P. C. (1941). A direct central inhibitory action of dromically conducted impulses. *J. Neurophysiol.* **4**, 184–190.

Lloyd, D. P. C. (1943). Neuron patterns controlling transmission of ipsilateral hind limb reflexes in cat. *J. Neurophysiol.* **6**, 293–315.

Lloyd, D. P. C. (1946). Facilitation and inhibition of spinal motoneurones. *J. Neurophysiol.* **9**, 421–438.

Manery, J. F. (1954). Water and electrolyte metabolism. *Physiol. Rev.* **34**, 334–417.

Monnier, A. M. & Dubuisson, M. (1934). L'action des nerfs extrinsèques du cœur considérée comme phénomène de subordination. I. Étude des variations de polarisation du myocarde sous l'action du vague (effet Gaskell). *Arch. int. Physiol.* **38**, 180–206.

Nastuk, W. L. & Hodgkin, A. L. (1950). The electrical activity of single muscle fibres. *J. cell. comp. Physiol.* **35**, 39–74.

Renshaw, B. (1941). Influence of discharge of motoneurones upon excitation of neighbouring motoneurones. *J. Neurophysiol.* **4**, 167–183.

Renshaw, B. (1942). Reflex discharges in branches of the crural nerve. *J. Neurophysiol.* **5**, 487–498.

Shanes, A. M., Grundfest, H. & Freygang, W. (1953). Low level impedance changes following the spike in the squid giant axon before and after treatment with 'veratrine' alkaloids. *J. gen. Physiol.* **37**, 39–51.

J. Physiol. (1955) **130:** 291-326

METHODS FROM

ELECTRICAL PROPERTIES OF THE MOTONEURONE MEMBRANE

J. S. Coombs, J. C. Eccles and P. Fatt

Experiments were performed on the lower lumbar region of the cat's spinal cord under pentobarbital anaesthesia. The technique of micromanipulation was the same as that described by Eccles, Fatt, Landgren & Winsbury (1954).

Microelectrodes. Capillary microelectrodes for intracellular usage were drawn from hard glass tubing and filled with a concentrated aqueous solution of an electrolyte. Filling with the desired solution was accomplished by immersing the microelectrode in the boiling solution for a period up to ½ hr or, alternatively, by immersing it in boiling distilled water for this period and, after cooling, placing it in a solution from which the electrolyte would pass into the water-filled microelectrode. Connexion of the microelectrode with the electrical equipment was made through Ag–AgCl junctions. The indifferent lead from the cat was made through an Ag–AgCl junction and a saline-soaked cotton gauze pad which made a low resistance contact with the surface of the lumbar musculature.

In order to record potentials from the interior of a cell at the same time as a current was being applied, a double-barrelled microelectrode was devised. This was fabricated by placing a flat glass partition longitudinally within a glass tube. One end of the tube was sealed and two side arms were attached there, one leading from each side of the partition. The tube was then drawn into a microelectrode and filled in the usual manner (cf. sketch of electrode in Fig. 1). On examination under the microscope, using a water-immersion objective of n.a. 1·20, the separate barrels were seen to end together at the tip, which still had a circular or oval cross-section. The overall diameter of the tip was not different from that found on single electrodes, i.e. it ranged to below 0·5 μ. The tips of these electrodes were more liable to break when subjected to mechanical stress than single-barrelled electrodes, and it was a common occurrence that the double-barrelled electrode was damaged on being driven through the spinal cord, as was inferred from alterations of its electrical properties. Because of electrical artifacts inherent in the technique to which the double-barrelled electrode was applied, it was not usually possible to use electrodes with a tip diameter of less than about 0·7 μ. On microscopic examination after use, successful electrodes were seen to have tip diameters of 0·7–2·0 μ, and occasionally one barrel would open a few microns short of the other.

Electrical artifacts. Both the recording of potential and the passage of current individually presented the usual technical difficulties, such as have been described in various papers where single-barrelled electrodes have been used for these two operations. In addition, special problems arise with the use of a double-barrelled electrode, due to electrical coupling between the barrels. Typical values of the relevant electrical characteristics are indicated in Fig. 1. The capacity between the barrels (of the order of 20 pF) caused such a large artifact to appear in the recording system that it was impossible to follow in detail potential changes occurring in a cell within 2–3 msec after a change in applied current. When a voltage step is applied to the free end of the 100 MΩ resistance represented in Fig. 1, a current develops through the microelectrode tip, rising toward a plateau with a time constant of 0·025 msec (5 pF coupled to 10, 10 and 100 MΩ in parallel). Initially, the current will be carried via both barrels because of the large capacity coupling the two. In the recording system, a large voltage pulse will appear, which is entirely an artifact, as it does not represent a potential change at the microelectrode tip. It rises with a time constant of 0·025 msec toward a potential equal to $\frac{1}{21}$ of the voltage step produced by the current-generating device; and it decays with a time constant of 0·4 msec (20 pF discharging through 10 and 10 MΩ in series). Transient artifacts having this origin appear in Figs. 4, 5 and 13. It will be appreciated that there is little possibility of making accurate measurements until this artifact has very largely subsided.

The large resistance (100 MΩ) was placed between the current-applying device and the microelectrode in order to minimize distortion in the recording of potential changes that arise as the response of the cell. If the barrel for applying current were directly connected to the current-generating apparatus, which was of comparatively low impedance, the capacity between the barrels would cause an attenuation in the recording of rapid potential changes at the tip of the

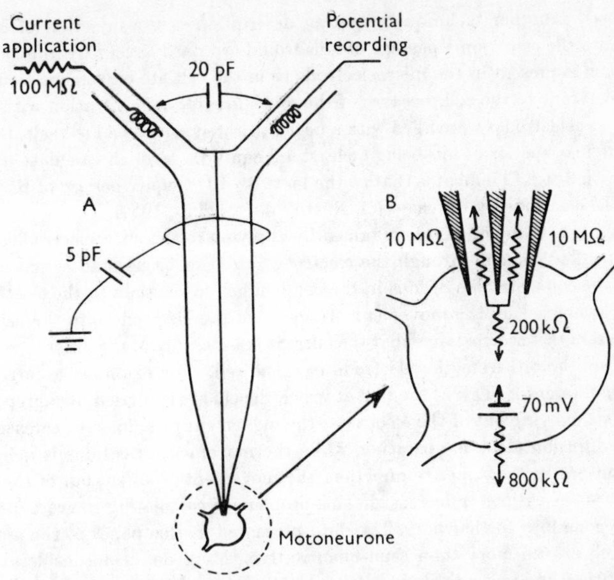

Fig. 1. A. Double-barrelled microelectrode and its immediate connexions. Typical values are given of the several electrical characteristics which are significant in the use of the electrode. B. Enlarged view of the microelectrode tip in the motoneurone. The motoneurone properties represented are the potential and resistance (ignoring the reactance) between the inside and outside of the inactive cell, as determined in this paper. For diagrammatic purposes the microelectrode tip is shown greatly magnified relative to the motoneurone.

microelectrode (the time constant of the circuit attenuating the high frequencies would be given by 20 pF × 10 MΩ). With the introduction of the 100 MΩ resistance, the whole of the microelectrode was free to follow transient potentials developed in the cell, while the steady polarizing currents were only slightly affected by these potentials.

A serious problem in the measurement of membrane potential changes produced by extrinsic current arises from the resistive coupling between the barrels. With the electrode dipping into physiological saline the application of a steady current through one barrel caused a potential change to be recorded from the other. This was due to the resistance immediately around the electrode tip shared by the current-applying and potential-recording systems. The greater the tip diameter the less was the coupling resistance. It was, furthermore, much reduced if the electrode was fractured so that one barrel opened a few microns short of the other. The coupling resistance was liable to fluctuate considerably on pushing the microelectrode through the spinal cord. It was therefore necessary to make frequent checks of its value. The values subtracted from the intracellular determinations were usually those obtained immediately after withdrawing from the cell, this being based on the assumption that the coupling resistance exclusive of the cell membrane was the same inside the cell as outside. The changes in coupling resistance could be explained by breakage of the tip decreasing its value, or by joint blockage of the barrels by some material of high resistivity increasing its value. Since the resistance between the interior of a motoneurone cell body and the external conducting medium was, as eventually determined, of the order of 800 KΩ, reliance could not be placed on measurements made with electrodes with coupling resistance greater than 300 KΩ. It was this consideration that compelled the rejection of electrodes with tip diameters less than about 0·7 μ. Usable electrodes had coupling resistance ranging between 60 and 300 KΩ.

Injection of ions. Another technique requiring description is the injection of ions from the microelectrode into the cell. Some passage of ions would ordinarily occur by diffusion. When the same species of ion is present in the microelectrode as in the cell, its movement would be mainly from the microelectrode to the cell, because of the much greater concentration within the microelectrode. The microelectrodes were filled with a near-saturated solution of the salt, the concentration of any particular species of ion being at least 1·2 equiv/l. A rough calculation made for an electrode filled with 3 M-KCl indicates that at the most 6×10^{-14} equiv per sec of K^+ and Cl^- ions will be leaving the electrode by diffusion (cf. Nastuk & Hodgkin, 1950).

Control over the rate of injection was obtained by applying to the microelectrode a potential of up to a few volts. The current through the electrode can then be used to estimate the rate of injection. Since the concentration of ions in the cell is much lower than in the electrode, most of the current will be carried by the movement of ions from the electrode into the cell. However, a problem arises from the circumstance that the current is a measure of the difference in the movement of cations and anions from the electrode into the cell. For example, a current from the electrode to the indifferent lead gives the rate at which the charge is carried by a preponderance of cations over anions, moving out of the electrode; the movement of anions is decreased from that prevailing when diffusion alone is operating, while the movement of cations is increased. Conversely, with a current in the opposite direction, the movement of anions out of the electrode is increased, while that of cations is decreased. This problem of calculating the rate of ionic movement will not be gone into further here. The data presented in this paper of the amount of ion injected into a cell are no more than semi-quantitative. More detailed consideration of ionic movement by diffusion and under the influence of electric fields is given in another paper (Coombs, Eccles & Fatt, 1955).

The ion species which have been injected into motoneurones are the following:

Cations: potassium, sodium, tetramethylammonium, choline.

Anions: chloride, bromide, nitrate, thiocyanate, bicarbonate, glutamate, sulphate, phosphate.

In electrodes used for investigating the effect of a given cation, the accompanying anion was usually sulphate. Correspondingly, for investigating the effects of a given anion, the accompanying cation was usually potassium.

32

J. Physiol. (1961), **155**, *pp.* 543–562
With 10 *text-figures*
Printed in Great Britain

PRESYNAPTIC INHIBITION AT THE CRAYFISH NEUROMUSCULAR JUNCTION

BY J. DUDEL* AND S. W. KUFFLER

From the Neurophysiology Laboratory, Department of Pharmacology, Harvard Medical School, Boston 15, *Massachusetts, U.S.A.*

(*Received* 24 *October* 1960)

There are three possible ways in which synaptic inhibition may reduce excitation. (1) Inhibition may reduce the amount of excitatory transmitter that is released from nerve terminals when they are activated by a nerve impulse; this is presynaptic inhibition. (2) The reaction of the post-synaptic receptors to a constant amount of released transmitter may be reduced, for instance by competition. (3) The post-synaptic membrane may be altered in such a way that the depolarizing action of the excitatory transmitter is opposed.

There exists good evidence for the last alternative only, namely, that the membrane conductance of the post-synaptic region is increased, with the membrane potential staying near the resting potential. This inhibitory mechanism of a specific post-synaptic permeability increase (to K^+ and/or Cl^-) seems to be widespread in different species and has been demonstrated in the vertebrate heart, the mammalian central nervous system, in crustacea at neuromuscular junctions and nerve cell synapses, as well as in various other preparations (Fatt & Katz, 1953; Coombs, Eccles & Fatt, 1955; Kuffler & Eyzaguirre, 1955; Trautwein & Dudel, 1958; for a recent review see Kuffler, 1960).

In this study on the crayfish neuromuscular junction we present evidence for a presynaptic mechanism. The inhibitory nerve impulse acts on the excitatory nerve terminals and decreases the probability of release of quanta of excitatory transmitter. A preliminary report has appeared (Dudel & Kuffler, 1960).

METHODS

As previously described, the abductor muscle of the dactylopodite in the crayfish was used (Dudel & Kuffler, 1961*a*).† For altering the muscle membrane potential the current-passing electrode was filled with saturated potassium citrate solution (Boistel & Fatt, 1958). These electrodes were preferable because they carried more readily maintained currents of up to 10^{-6} A.

* Present address: Physiologisches Institut, Akademiestr. 3, Heidelberg, Germany.

† Methods appear on page 787.

RESULTS

Conductance increase during inhibition

Some features of the conductance-increase mechanism, first demonstrated by Fatt & Katz (1953), are presented here largely to make clear the distinction from presynaptic inhibition. Inhibitory stimulation reduces the potential change that is produced when a current is applied across the junctional membrane, i.e. the resistance of the membrane has decreased

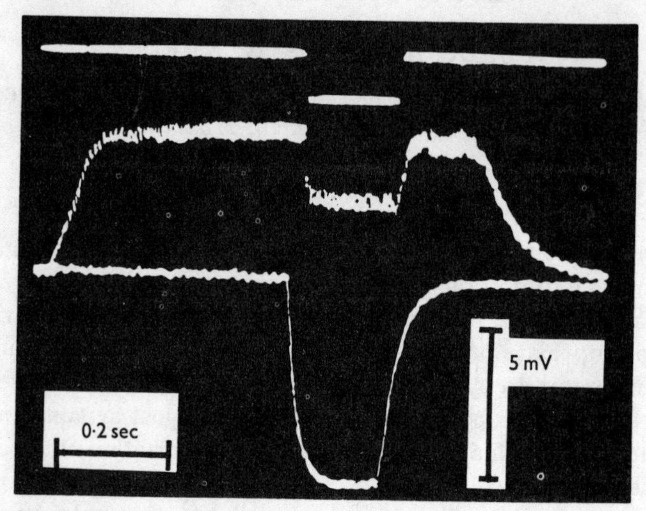

Fig. 1. Conductance increase during depolarizing inhibitory junctional potential in abductor of the dactyl. One recording and one current-passing electrode inserted into a muscle fibre. Current pulses of $1{\cdot}5 \times 10^{-8}$ A monitored in upper sweep were passed across the membrane at rest (lower sweep) and during inhibitory potential. Stimulation frequency of inhibitory nerve was 150/sec. Note reduction of the electrotonic potential from 7 to 2 mV by inhibition.

during inhibition. This is seen in Fig. 1, where two sweeps are superimposed. The lower record gives the electrotonic potential in the resting state of the membrane caused by the current pulse (upper trace). In the middle trace the inhibitory axon was stimulated by a short train of inhibitory impulses at 150/sec and a small depolarization of the fibre resulted. The same current pulse, if applied during the plateau of inhibitory junctional potentials (i.j.p.s), set up a much reduced electrotonic potential with a more rapid time course. The calculated membrane conductance during inhibition, as measured by this method, was frequently increased ten times above the resting value.

The most characteristic feature of the conductance type mechanism is the reversal of the i.j.p. near the resting potential. The method of determining the reversal potential was used routinely in the present studies.

An example is shown in Fig. 2. The membrane was depolarized by passing current through a second intracellular micro-electrode. Starting with a resting potential of 80 mV the i.j.p.s became negligibly small at 72 mV and reversed their direction with additional depolarization. Arrows mark the reversal potential.

The reversal potential obtained in Fig. 2 may not indicate exactly the true inhibitory equilibrium potential, being slightly too positive, because the current electrode does not depolarize the whole fibre uniformly (see Burke & Ginsborg, 1956, for discussion of an analogous problem). Here, however, this introduces only a small error. The reversal or equilibrium

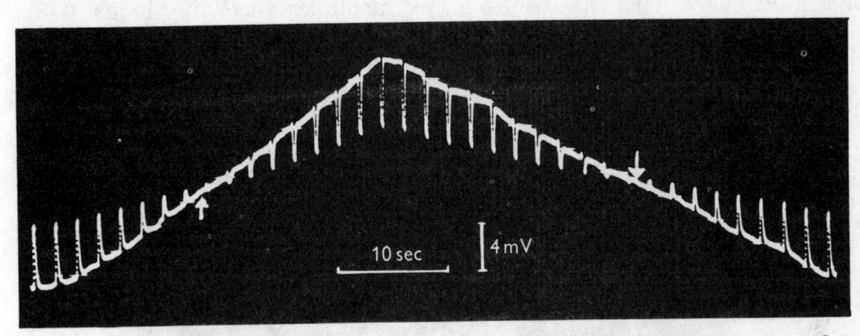

Fig. 2. Trains of inhibitory junctional potentials (150/sec for 0·2 sec) repeated every 2 sec. Resting potential at start 80 mV shifted to 61 mV by passing current through a second intracellular micro-electrode. Arrows mark the reversal potential of inhibitory potentials at 72 mV.

potential for i.j.p.s was also determined by depolarizing the fibre uniformly by excess K^+, instead of by current, and no significant difference from the result of Fig. 2 (arrows) was noted.

An additional mechanism for inhibition

In their original work Fatt & Katz (1953) thought that the reduction of excitatory junctional potentials (e.j.p.s) was too great to be brought about by a conductance increase alone, and they suspected an additional mechanism (cf. Kuffler & Katz, 1946). Experimental conditions have now been found in which the post-synaptic conductance increase mechanism cannot account for any of the observed reduction of the e.j.p. In Fig. 3 single excitatory stimuli set up excitatory junctional potentials of 2 mV (Fig. 3*A*) Single inhibitory stimuli caused a small depolarization of about 0·2 mV (Fig. 3*C*). The inhibitory equilibrium level, however, as found by the method shown in Fig. 2, and further determined by K depolarization (above), was at the dotted line 6 mV more positive than the resting potential. In this situation in which the *excitatory* junctional potential does not reach the *inhibitory* reversal level, the inhibitory conductance mechanism cannot reduce the e.j.p. If the i.j.p. is timed to fall on the peak of the e.j.p., it should still depolarize, since the i.j.p. is still 'below' (more negative than) its reversal potential. In fact, it should add to the e.j.p. This seemed to be the case in Fig. 3*B*, where the combined potential, if changed at all, was somewhat bigger. To our surprise, however, if the i.j.p. was timed to precede the e.j.p. by 3 msec, the e.j.p. was greatly reduced (Fig. 3*D*). This cannot be explained by a mechanism of inhibitory conductance increase, which can shift the membrane potential only toward the inhibitory reversal level. Therefore, an additional mechanism had to be postulated. Still further evidence for an additional mechanism will be presented below with the results on extracellular recording (Figs. 6–8).

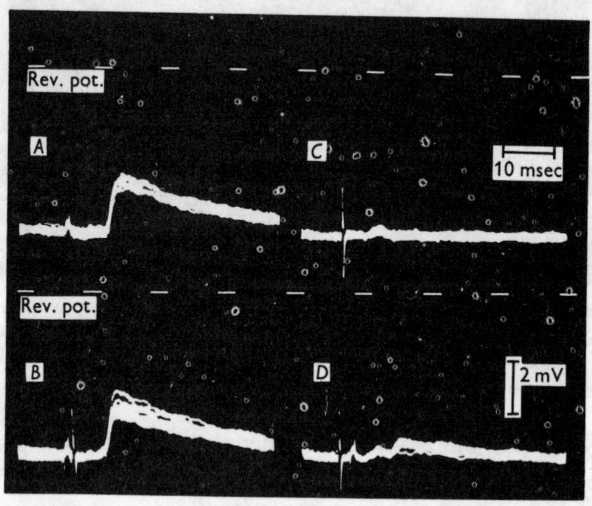

Fig. 3. Intracellular records from abductor muscle fibre. Resting potential 86 mV, inhibitory reversal level (rev. pot.) 6 mV more positive (interrupted line), K^+ in solution reduced to 2·7 mM. *A*: single excitatory junctional potential (e.j.p.) of 2 mV. *C*: single inhibitory junctional potential (i.j.p.) depolarizing 0·2 mV. *B*: inhibitory axon stimulated 1·5 msec after excitatory axon; i.j.p. coincides with peak of e.j.p. *D*: inhibitory axon stimulated 3 msec before excitatory axon; i.j.p. seen before e.j.p. which is greatly reduced.

A close examination of Fig. 3*D* reveals that the rising phase of the small i.j.p. was over when the e.j.p. started. The maximum of the inhibitory conductance increase occurs during the rise of the i.j.p., and the residual conductance increase during the falling phase of the i.j.p. must have been relatively small at a time when the reduction of the subsequent e.j.p. was seen to be maximal.

To obtain the above type of reduction of e.j.p.s the timing of inhibitory stimuli in relation to the excitatory ones was quite critical. If the inhibitory stimulus preceded the excitatory one by 1–3 msec the effect was maximal and then declined over a further 3–5 msec. The importance of timing of the excitatory–inhibitory stimuli for reducing excitatory potentials, as is shown in Fig. 4, has been known for a long time and similar curves were obtained by Marmont & Wiersma (1938), Kuffler & Katz (1946) and Fatt & Katz (1953). The significance of the timing effect for our experiments is that it revealed a type of inhibition that could not be accounted for by the conductance-increase mechanism.

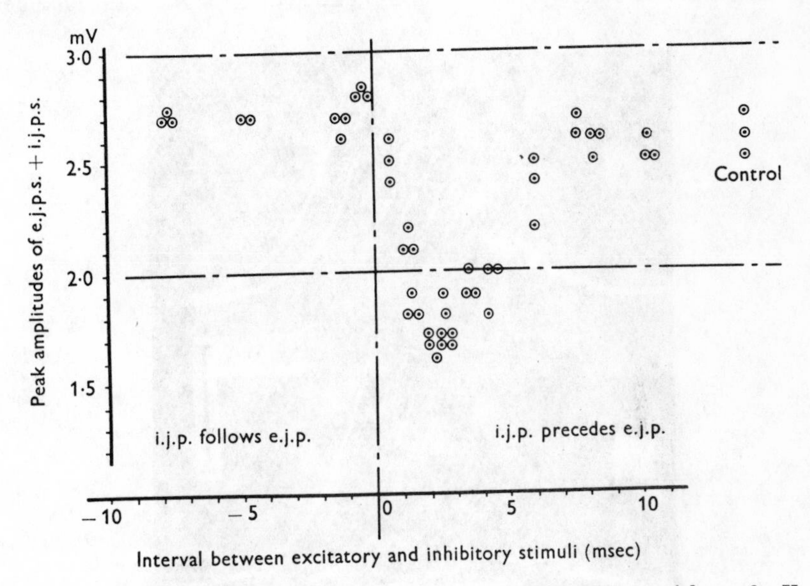

Fig. 4. Experiment as in Fig. 3. The inhibitory equilibrium potential was 6 mV more positive than resting potential. Control gives size of e.j.p. without any inhibitory stimulation.

Another way of showing the influence of timing of impulses is illustrated in Fig. 5. Inhibitory and excitatory nerves were stimulated at rates of 100/sec for 120 msec. In Fig. 5*B* each inhibitory nerve stimulus was delivered 2·5 msec before the excitatory one, while in Fig. 5*A* it was given 8·0 msec before each excitatory stimulus. The e.j.p.s were much more reduced in Fig. 5*B*, where the timing was maximally effective.

Effect of inhibition on extracellularly recorded e.j.p.s

The existence of a second type of inhibition in addition to the postsynaptic conductance-increase mechanism can also be shown by recording extracellularly from individual junctional areas with micro-electrodes. This recording method has the advantage that the quantum content of the synaptic potentials can be determined statistically (Dudel & Kuffler,

35-2

1961*a*, *b*), and more specific results on the nature of the second type of inhibition can be obtained.

In the lower sweep of Fig. 6*A* are shown extracellular records from a single junctional area. The e.j.p.s are negative, in contrast to the intracellular junctional potentials which are recorded simultaneously (upper sweep). The latter are made up of activity in the entire muscle fibre. If properly timed inhibitory impulses were added in Fig. 6*B*, the intracellular e.j.p. heights were reduced, as one would expect. The significant change in

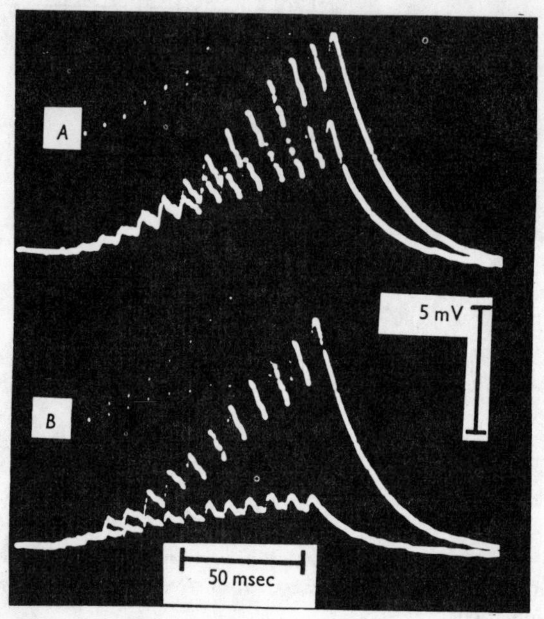

Fig. 5. Minimally and maximally effective timing for inhibition. Intracellular recording, resting potential near 75 mV. *A*: upper sweep, excitatory stimuli at 100/sec for 120 msec; lower sweep, inhibitory stimuli precede excitatory stimuli by 8 msec. *B*: same excitatory and inhibitory trains as in *A*, but in lower sweep the inhibitory stimulus precedes each excitatory one by 2·5 msec, resulting in maximally effective reduction of the e.j.p.s.

the extracellular records from the single junction was the increased number of failures (arrows), i.e. the excitatory impulse failed to release packets of transmitter. At the same time, the mean amplitude of the 'hits' was reduced in size (see below, Table 1).

In Fig. 7 more detail is included in a fast sweep. Three consecutive sweeps were superimposed, repeated at a rate of 5/sec. Fig. 7*A* shows excitation alone, with the stimulus artifacts in the lower record followed by the motor nerve impulse (arrow) arriving near the junction. Then two negative extracellular e.j.p.s appeared on the lower sweep, which lasted a little longer than the rising phase of the intracellular e.j.p.s in the upper sweep. There was, in this sequence of three stimuli, one failure of transmission at the single junctional area. The intracellular e.j.p.s did not fluctuate greatly. In Fig. 7*B* an inhibitory stimulus (first artifact)

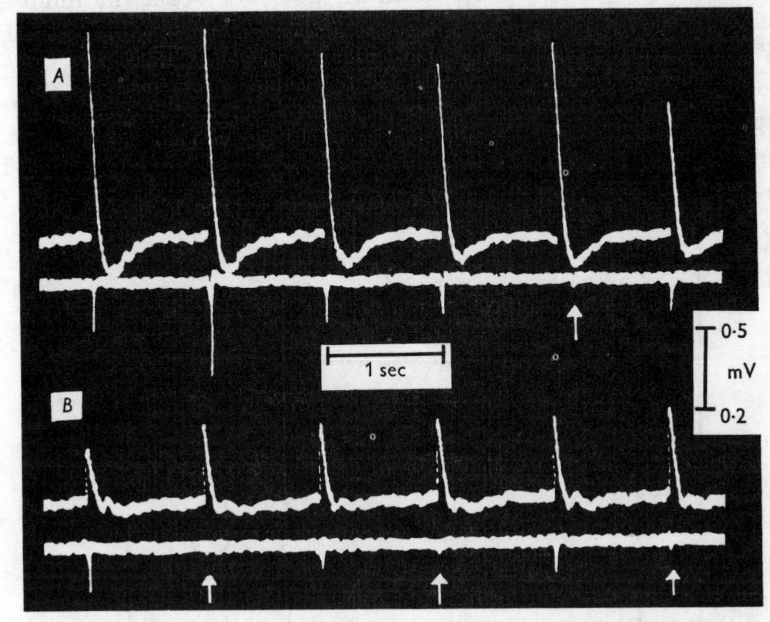

Fig. 6. Intracellular records (upper sweeps) from a muscle fibre and simultaneous extracellular measurements from single junctional area (lower sweep) at stimulation rate of 1/sec. *A*: excitatory stimulation alone. *B*: an inhibitory stimulus is added, preceding the excitatory one by 2 msec; e.j.p.s are reduced and number of transmission failures in single junctional area (arrows) increased. The diphasic portion in the intracellular records is due to AC amplifier.

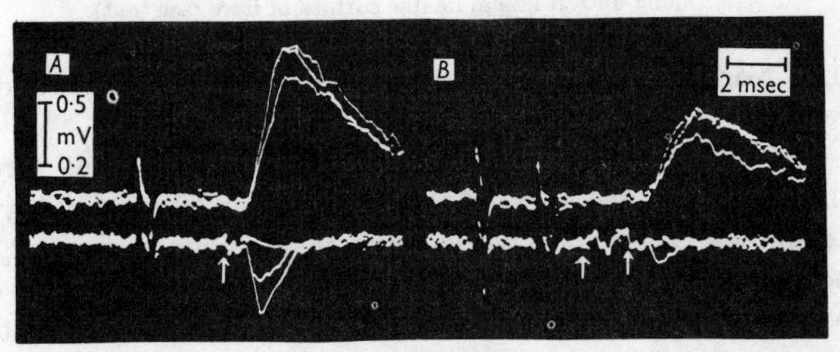

Fig. 7. Simultaneous intra- and extracellular records, as in Fig. 6, three consecutive junctional potentials superimposed at rate of 1/sec. *A*: e.j.p.s alone; failure of transmission occurs once in extracellular records; arrow marks extracellularly recorded motor nerve impulse near the junction. *B*: inhibitory stimulus preceding excitatory one by 2 msec, reducing the intracellular e.j.p.'s and resulting in two failures of transmission at the single junctional area; arrows mark nerve impulses.

preceded the excitatory one by 2 msec. On the lower beam we now see two extracellular nerve impulses (arrows), the first being inhibitory. They are 1·5 msec apart, a little less than the stimulus interval, presumably owing to different conduction distances and speeds of propagation. Again, all the intracellular e.j.p.s were reduced and in the lower beam the number of transmission failures was increased from 1 to 2. Only one e.j.p. of about 50 μV remains (see below).

Figures 6 and 7 show that inhibition can considerably reduce the average extracellular response. This reduction of the extracellular e.j.p. cannot be brought about by an increased membrance conductance. This

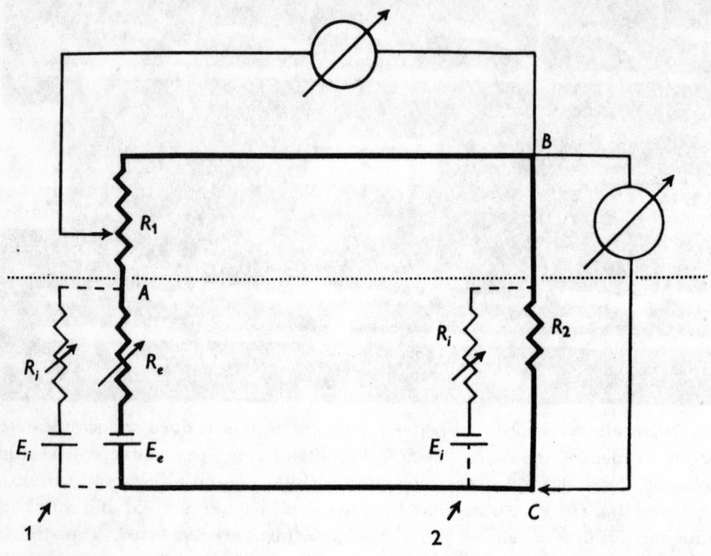

Fig. 8. 'Equivalent circuit' diagram showing external and internal recording conditions. Long dotted line indicates surface of fibre (see text).

point can be readily appreciated if the extracellular recording conditions are analysed.

Figure 8 gives an 'equivalent circuit', modified from that given by del Castillo & Katz (1956, Fig. 17*b*) for recording extracellularly from vertebrate end-plates. *A* is the active excitatory junctional area, R_e is the membrane resistance of this area and E_e is the driving force for the e.j.p.: R_1 is the resistance of the small extracellular fluid volume outside the junction, where the synaptic current density is very high. *B* represents the distant indifferent electrode. R_2 is the 'input resistance' (i.e. between a point inside the fibre (*C*) and the extracellular fluid (*B*)). The intracellular electrode records between *C* and *B*, the extracellular electrode between *B* and a part of R_1. There are two alternative locations for an inhibitory membrane conductance increase to act on this circuit. First, the inhibitory synaptic membrane could be so close to the excitatory one that R_e and R_i are in parallel and the inhibitory and excitatory synaptic currents flow through the same restricted extracellular fluid volume with a resistance R_1. The second alternative is for the inhibitory synaptic area to be some distance away from A, so that it is in parallel with R_2. It can be shown that in both cases the reduction of R_i (inhibitory conductance increase) does not diminish the extracellularly recorded e.j.p. If in alternative (1) an appreciable current were produced, extracellular potentials should be recorded on inhibitory stimulation. We have not seen such potentials. In alternative (2) the inhibitory resistance decrease works essentially as a shunt to R_2. R_2 is small compared with R_1+R_e, as is seen from the fact that the intracellular junctional potential is much smaller than its driving force E_e, as pointed out by del Castillo & Katz (1956). Thus shunting R_2 cannot diminish current flow through R_1; it could only slightly increase it, producing a larger extracellular e.j.p.

Therefore, the reduction of the extracellularly recorded e.j.p.s during optimally timed inhibition provides another demonstration of an inhibitory mechanism which cannot be accounted for by an increase in membrane conductance. By themselves the above results do not allow us to distinguish the site of action of this other inhibitory mechanism. The site may be presynaptic, with less transmitter released per impulse, or it may be post-synaptic, for instance, with competition taking place between excitatory and inhibitory transmitters. By the use of a statistical analysis, however, it was possible to show that less transmitter was released during inhibition and that the site of action must therefore be presynaptic. That is, presynaptic inhibition should reduce the number rather than the size of excitatory transmitter quanta. Competitive inhibition would have the converse effect. Information about quantum size and number can be obtained by statistical treatment of our extracellular records from single synaptic areas.

Statistical treatment

Figure 9 gives the size distribution of extracellular e.j.p.s from a single junctional area with and without inhibition. Each graph contains 1000 counts. At a frequency of 5/sec there are relatively few transmission failures in the uninhibited junction. With inhibition the failures increase from 72 to 570 and there are fewer large e.j.p.s. For instance, in the uninhibited junction there are 32 e.j.p.s of 150 μV, while inhibition reduces them to 2. The whole distribution is shifted to the left.

The inhibited e.j.p.s were sharply grouped around a peak of about 40 μV. They must have a very low quantum content, because more than half the responses were failures. A fitting Poisson curve could, therefore, be determined readily for these potentials by the method previously used (Dudel & Kuffler, 1961*a*). The theoretical curve (interrupted line) fits the histogram well. The unit size derived from this Poisson distribution was $E_1 = 40\ \mu V$ and the number of quanta released by one stimulus was $m = 0{\cdot}56$.

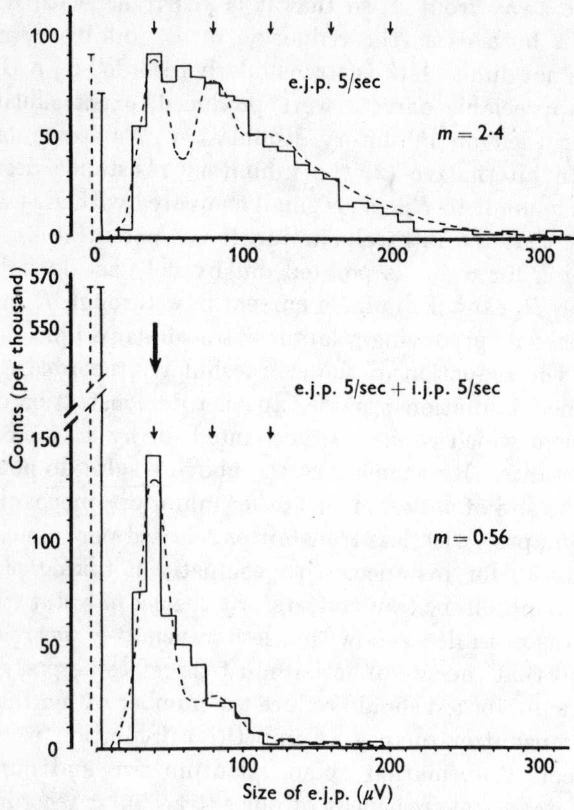

Fig. 9. Histograms giving size distribution of extracellularly recorded e.j.p.s from a single junctional area at stimulation rate of 5/sec. Upper graph, junction not inhibited. Lower graph, inhibitory impulses precede excitatory ones by 2 msec; note increased number of failures from 72 to 570. Interrupted line is calculated theoretical (Poisson) distribution for unit quantum size E_1 (40 μV) with standard deviation σ_1 (10 μV), and an average quantum content per stimulus m. Small arrows indicate multiples of quantum size, heavy arrow gives average size of spontaneous miniature potentials.

The question now was whether the uninhibited e.j.p.s had the same quantum size as the inhibited ones. This could be tested by calculating a Poisson curve for the uninhibited e.j.p.'s, using only the 40 μV unit size obtained above, and the average size of the uninhibited e.j.p.s (see Dudel & Kuffler, 1961*b*). The calculated curve was a reasonably good fit.

Thus it was shown that inhibition of the e.j.p.s could be described as a reduction of the average quantum content from 2·4 to 0·6 while the size of the unit remained unchanged.

An independent control of the size of the quantum E_1 is derived from the average size of the spontaneous miniature potentials. Fourteen spontaneous potentials were observed during the experiment of Fig. 9, with an average size of $E_s = 41\ \mu V$ and a standard deviation of $\sigma_1 = 11\ \mu V$. E_s thus agrees well with the quantum size E_1.

The possibility could be considered that the inhibitory effect on the release of excitatory transmitter fluctuates over a wide range, possibly in a quantal way. This would result in a large fluctuation of the probability for the release of quanta of excitatory transmitter. However, the Poisson theorem assumes that the probability of release of quanta is the same for each stimulus. A wide fluctuation of this probability would distort the Poisson distribution considerably. Since we were able to fit the inhibited e.j.p.s by a Poisson distribution, the inhibition of the excitatory transmitter release should be relatively constant with each inhibitory stimulus. If the inhibitory transmitter is released in quanta, the quantum content of the inhibitory potentials would have to be relatively high even at low frequencies. Such an assumption would be in line with the observation that at low stimulation frequencies (around 1/sec) the amplitude of the intracellular i.j.p.s fluctuated less than that of the intracellular e.j.p.s.

A simpler statistical treatment consists in using a relation derived from Poisson's theorem. It relates the number of trials (n) and failures (n_0) of transmission n/n_0 to mean quantum content m. m is given by the average size $\bar{E}$ of e.j.p.s divided by the unit size E_1, as treated in the previous papers (Dudel & Kuffler, 1961*a*, *b*).

$$\log_e \frac{n}{n_0} = \frac{\bar{E}}{E_1} = m. \quad (1)$$

All values except E_1 can be determined directly and so E_1 can be calculated.

Table 1 gives a comparison of e.j.p.s with and without inhibition in six experiments. The calculated values for E_1 (unit size) agree well for inhibited and uninhibited junctions. In the same experiments the average size of spontaneous potentials (E_s) was measured and, as expected, was found to be in close agreement with E_1 (Dudel & Kuffler, 1961*a*). The last column (E_i) gives the intracellular e.j.p.s, i.e. the integrated activity from the whole fibre. A comparison of the intracellular e.j.p. reduction during inhibition with the reduction of extracellular e.j.p.s gives an interesting picture of the variability of activity in individual junctions. The differences in reduction can be quite striking. In experiment No. 3 the average extracellular e.j.p. size is radically reduced to less than 1/20 by inhibition, while the intracellular e.j.p. is only halved. This must mean that the effectiveness of inhibition at individual junctional areas varies greatly.

In the preceding study (Dudel & Kuffler, 1961*b*) it was shown that failure by a nerve impulse to set up an e.j.p. at a single junctional area did not alter the average size of the next e.j.p. and did not affect its facilitation. This was in agreement with the basic assumption made for the Poisson treatment, that each stimulus has the same probability of releasing quanta, i.e. each stimulus reaches the junction. The rather noncommittal term 'stimulus' is used here because we do not know whether the impulse seen in the axon actually travels right to the terminals. It was found that also in groups of extracellular e.j.p.s which were inhibited by optimally timed inhibition, failures of transmission did not affect the size of succeeding

TABLE 1

Expt. no.	Stimulation frequency	n	m	$\bar{E}$ (μV)	E_1 (μV)	E_s (μV)	E_i (mV)
1	5/sec	961	1·3	85	65	71	1·47
	5/sec + inhib.	1375	0·72	50	69	—	1·17
2	5/sec	728	2·8	94	34	41	1·2
	5/sec + inhib.	699	0·57	24	42	—	0·45
3	5/sec	228	1·3	35	30	30	1·7
	5/sec + inhib.	308	0·03	1·2	40	—	0·8
4	2/sec	200	0·60	30	49	51	0·12
	2/sec + inhib.	136	0·28	14	48	—	0·08
5	5/sec	264	0·79	34	43	—	0·25
	5/sec + inhib.	167	0·31	13	42	—	0·11
6	5/sec	570	0·74	46	62	55	0·09
	5/sec + inhib.	409	0·50	31	62	—	0·07

n = number of stimuli; m = average quantum content of extracellular e.j.p. determined from $m = \log_e n/n_0$; n_0 = number of transmission failures; $\bar{E}$ = average size of extracellular e.j.p.; E_1 = size of quantum determined from $E_1 = \bar{E}/m$; E_s = average size of spontaneous extracellular potential; E_i = average size of intracellular e.j.p.

e.j.p.s. The average size of the e.j.p.s preceded by a failure was the same as the average size of the whole group. This finding further justifies the treatment of inhibited e.j.p.s by the Poisson analysis (see Discussion).

The statistical treatment has shown that inhibition can be described as a reduction of the number of released quanta, while the quanta themselves are not changed. Thus, optimally timed inhibitory stimuli act presynaptically on the excitatory system.

The effect of inhibition on spontaneous miniature potentials

Having shown that appropriately timed inhibitory impulses act presynaptically, their effects on the post-tetanic increase of the frequency of spontaneous miniature potentials (Dudel & Kuffler, 1961*b*) are of interest. The experiment was done in the following way: Trains of excitatory stimuli, for instance, at 75/sec for 200 msec each, were repeated every 2 sec. The miniature frequency was measured in the intervals between the trains and was found increased above the rate seen in the non-stimulated muscle. The excitatory trains were then repeated with optimally (as

Fig. 5*B*) and minimally (Fig. 5*A*) effective inhibitory timing. A count of spontaneous miniature potentials between the trains showed that they were reduced when the excitatory–inhibitory timing was set for maximal reduction of the e.j.p.s. A tabulation of the results from two experiments is shown in Table 2. The decrease of miniature frequency was significant at the 99·9 % level. Since miniature potentials are released presynaptically, anything that changes the frequency of release presumably acts presynaptically. This effect therefore shows once more that inhibition has a presynaptic effect.

TABLE 2

Expt. no.	Conditioning train	Inhibition timing	Measurements *n*	Miniature frequency	S.E.
1	50/sec	min.	106	1·44/sec	0·06
	50/sec	optimal	159	1·13/sec	0·06
2	75/sec	min.	148	2·61/sec	0·09
	75/sec	optimal	94	1·98/sec	0·10

One more observation, in which inhibition is likely to affect the presynaptic terminals, concerns post-tetanic facilitation. A train of conditioning excitatory stimuli, for instance, at 50/sec for 1 sec, was followed by a facilitation of single test stimuli for 10 sec or longer. If in the conditioning train inhibitory stimuli were set to precede each excitatory one by 2 msec (as in Fig. 5*B*), the subsequent post-tetanic facilitation was significantly reduced. If, however, the same inhibitory impulses were timed to arrive at the junction after the excitatory ones, there was no noticeable effect on post-tetanic facilitation.

Relationship of inhibition to facilitation

In a formal sense the reduction of the probability of release by inhibition is the opposite of facilitation. Thus presynaptic inhibition could be interpreted as a removal or prevention of facilitation. If this were true, presynaptic inhibition would reduce e.j.p.s to their non-facilitated size. As a consequence, greatly facilitated, large e.j.p.s would be reduced by a greater factor than the smaller, less facilitated ones. This, however, is not so, because it was found that if e.j.p.s were set up at 10 or 20/sec they were reduced by optimally timed presynaptic inhibition by the same proportion as the much less facilitated e.j.p.s at 1/sec. An illustration is given in Table 3. The percentage of e.j.p. reduction seen in the last column clearly does not depend on the frequency of excitatory stimulation. This reduction by the same proportion, independent of e.j.p. size, seems remarkable (see Discussion). To obtain it, however, one has to take the precaution in these experiments of keeping the inhibitory facilitation constant. For instance, the inhibitory frequency was left unchanged at 10/sec in the above tests. The change in excitatory stimulation rate from 10/sec to 1/sec was simply made by cutting out nine of the ten stimuli by an electronic gate (decade counter). Only the inhibitory i.j.p. which precedes the e.j.p. by 2 msec will have a presynaptic inhibitory action. The others precede the e.j.p. by more than 100 msec and have no inhibitory effect (cf. Fig. 4); they only serve to keep facilitation of the i.j.p.s constant.

The above experiments had a very different outcome if both excitatory and inhibitory frequencies were reduced together. Then there was much less reduction of the e.j.p. at a lower frequency. An example is given in experiment No. 3, Table 3. With inhibitory stimulation at 2/sec there was

TABLE 3

Expt. no.	e.j.p. frequency	e.j.p. size (mV)	i.j.p. frequency	e.j.p. + i.j.p. size (mV)	Reduction of e.j.p. (%)
1	20/sec	16·0	20/sec	7·0	56
	2/sec	1·6	20/sec	0·7	56
	10/sec	4·5	10/sec	2·4	47
	1/sec	0·7	10/sec	0·37	47
2	5/sec	8·5	5/sec	5·8	32
	0·5/sec	3·6	5/sec	2·8	22
	10/sec	11·0	10/sec	7·0	36
	1/sec	4·4	10/sec	2·8	36
3	2/sec	2·0	2/sec	1·9	5
	20/sec	15·2	20/sec	10·2	33
	2/sec	2·4	20/sec	1·6	33

practically no reduction of e.j.p.s (5 %). This experiment demonstrates that presynaptic inhibition also is facilitated like the post-synaptic inhibitory potential (Fig. 1*B*, Dudel & Kuffler, 1961*b*).

The effect of gamma-aminobutyric acid (GABA) and presynaptic inhibition

In search of a mechanism for presynaptic inhibition one has to examine whether the inhibitory nerve acts on the excitatory terminals electrically, i.e. by current spread, or chemically, by the action of a transmitter. Our only available evidence indicates that there are chemoreceptor sites on the excitatory fibre endings. We were able to mimic presynaptic inhibition by GABA.

It is known that GABA increases the post-synaptic membrane conductance in a manner similar to inhibitory nerve stimulation. Accordingly it shifts the membrane potential to the inhibitory equilibrium level, i.e. the potential change due to GABA reverses at the same potential at which inhibitory potentials reverse (Kuffler, 1960). For the present test one needs conditions in which the known inhibitory post-synaptic conductance increase by GABA can be excluded, as it was excluded for neural inhibition in the experiment of Fig. 3. These requirements were met in the following manner. In the preparation of Fig. 10 the inhibitory reversal potential (rev. pot.) was 7 mV more positive than the resting potential (interrupted line), determined as in Fig. 2. Excitatory stimuli at 10/sec set up e.j.p.s, which after a period of facilitation reached a fairly steady height near 5·5 mV. If, for instance, 5×10^{-5} M or a stronger concentration of GABA was applied (not illustrated), the resting potential became depolarized toward the inhibitory reversal potential (cf. Kuffler, 1960, Fig. 11) and the e.j.p.s further added to the depolarization, their peaks exceeding the reversal potential. At certain concentrations of GABA

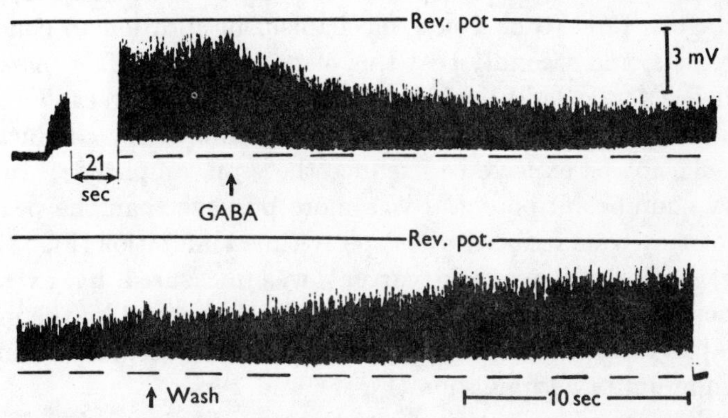

Fig. 10. Second 'inhibitory' action of GABA. Intracellular records of e.j.p.s at stimulation rate of 10/sec. Peaks of e.j.p.s do not reach the inhibitory reversal potential (rev. pot.), which was 7 mV more positive than the resting potential (interrupted line). 10^{-5} M GABA reduced e.j.p.s moving their peaks away from the reversal potential. The reversal potential for GABA coincided with the inhibitory reversal potential.

(10^{-5} M in the preparation of Fig. 10) no depolarization of the membrane was noted while the e.j.p. peaks were reduced, *away* from the inhibitory equilibrium potential. (Note a small hyperpolarization, due to the non-specific 'wash effect' of the applied solution.)

In interpreting the result of Fig. 10 one needs to point out that the known conductance increase of the post-synaptic membrane by GABA can shift the potential only toward the GABA equilibrium potential. As long as the e.j.p. peaks are *below* the inhibitory equilibrium level, their diminution away from this potential cannot be accounted for by an inhibitory conductance increase. Therefore, GABA has, in addition to its post-synaptic conductance increase, a second inhibitory action on the e.j.p. On the basis of the analogy with presynaptic inhibition described in this study, the effect of GABA is probably due to its action on the motor endings.

DISCUSSION

Our results concerning the conductance increase in crustacean muscle fibres during inhibition are in full agreement with those of other workers (Fatt & Katz, 1953; Boistel & Fatt, 1958; Grundfest, Reuben & Rickles, 1959; Hagiwara, Kusano & Saito, 1960). The conductance increase in crayfish muscle during inhibition is predominantly an increase in the permeability to Cl^-, a conclusion confirmed once more by our finding that changing the membrane potential by excess K^+ did not significantly alter the inhibitory equilibrium potential (Fig. 2).

An examination of the evidence for presynaptic inhibition may be made in two stages. First, that a new mechanism in addition to conductance increase exists, and secondly that this mechanism is, in fact, presynaptic. The evidence for an additional mechanism is quite direct. Two experimental situations were presented in which an inhibitory conductance increase could not be expected to reduce the e.j.p. amplitude. In one the inhibitory equilibrium potential was more positive than the peak of the e.j.p.s, yet they were reduced by properly timed inhibition (Fig. 3). In the other the excitatory synaptic current was measured by extracellular micro-electrodes. This synaptic current was not changed by the inhibitory post-synaptic conductance increase. We found, however, very strong reductions with inhibitory stimulation (Figs. 6 and 7).

The positive evidence for the presynaptic mechanism is less direct than the evidence which excludes the post-synaptic conductance increase in specific instances. The principal new result reported here is that during inhibition the size of the released quanta was unchanged, while their number was reduced, i.e. less transmitter was released. This alone requires that the process be presynaptic provided the quanta are released presynaptically. The reasons for the latter assumption were presented in the first two papers in this series (Dudel & Kuffler, 1961*a*, *b*). It should be recalled that these papers did not contain direct evidence for the presynaptic release of quanta but showed the close and detailed analogy with the quantal release in the vertebrate junction; there the evidence for quantal release from the presynaptic terminals was complete (Katz, 1958).

There are numerous questions left about the details of the presynaptic events. Not even the structural relationship between the excitatory and inhibitory terminals is known. Under the light microscope the axons run together and split up together into fine arborizations on the muscle surface (for early references see D'Ancona, 1923; Tiegs, 1924; van Harreveld, 1939).

Physiologically there is the problem whether the inhibitory axon exerts its effect on the motor terminals electrically, by current flow, or chemically, by means of a transmitter. The motor endings appear to have chemoreceptor sites because the presynaptic neural inhibition could be reproduced by applying GABA. This substance mimics, in all known respects, the effects of the inhibitory transmitter (Kuffler, 1960). Another point in favour of chemical interaction is the timing relationship of impulses. The inhibitory impulse has to arrive near the junction more than 1 msec before the excitatory impulse for a maximal presynaptic inhibitory effect. Synaptic delays of 1 msec are usually seen in our nerve–muscle junctions. If the inhibitory transmitter acts on the excitatory nerve with the same delay, its effect would appear just at the time when the motor impulse arrives at that region. For an electrical interaction, on the other hand, a 1 msec minimal delay seems rather long. In addition, one can detect some presynaptic inhibition even when the inhibitory impulse precedes the excitatory one by 5–6 msec (Fig. 4). This fits much better if the interaction is chemical, and makes electrical transmission unlikely.

In respect to the mechanisms of the reduction of the probability of release by inhibition, very few facts are available. We have evidence that under normal conditions the nerve impulse, or perhaps its attenuated electrotonic extension, does get to the terminals even when no quanta are released (Dudel & Kuffler, 1961*b*). The chemical steps between the membrane potential change and the release of transmitter are unknown. Thus, when we measure the probability of release we are dealing with the end-product of an unknown chain of events. Therefore, any discussion about the control of transmitter release during facilitation or inhibition remains speculative (Curtis & Eccles, 1960). The principal facts for which any eventual theory will have to account are these: inhibition can reduce post-tetanic facilitation or the release of spontaneous miniature potentials only if the inhibitory impulses precede the excitatory ones by a critical time; and further, the inhibited e.j.p.s still fit the Poisson distribution curve, only the probability of release is reduced. Significantly, even during inhibition, the failure to release a quantum after a stimulus does not influence the average quantum content of the e.j.p. set up by the next stimulus. This excludes inhibitory action being an intermittent conduction block at such distance from the junction that no effective membrane potential change would reach the terminal.

Although many possible mechanisms might explain presynaptic inhibition, we would point out that the main known action of the inhibitory transmitter can be used for a consistent explanation. One may assume that the inhibitory nerve impulse releases in its terminal region a transmitter which acts on the neighbouring motor nerve fibre by increasing the conductance for a few milliseconds, as it does in the post-synaptic muscle membrane. The increased conductance, in turn, will decrease the electrical potential change which serves as a 'stimulus' for the terminal. If this change were electrotonic (see below) one could explain the fact that the inhibitor reduces the e.j.p. independently of its size (Table 3). The size of the stimulus probably controls the average amount of quanta which are released and could also determine the extent of post-tetanic facilitation. That the size of a nerve impulse can control transmitter output has been suggested by the experiments of del Castillo & Katz (1954), who hyperpolarized the terminals and obtained larger end-plate potentials. Hagiwara & Tasaki (1958) also were able to reduce or increase synaptic potentials in the squid stellate ganglion, by changing the presynaptic membrane potential.

The advantage of the above suggestions is the possibility of an experimental test. Attempts to record changes in the size of the motor nerve impulse in the terminals have not yet been successful. We have also tried to determine the 'excitability' of the terminals (Wall, 1958) by stimulating them at the spots where the extracellular junctional potentials were recorded. It was not possible to set up impulses in the motor axon by passing current through this terminal area. This may lead one to suspect that the fine terminal stretch actually does not carry all-or-none impulses as the axon does, but is reached by electrotonic potentials only.

There are some suggestions for a presynaptic mechanism of inhibition elsewhere in the nervous system. Of special interest is the observation by Frank & Fuortes (1957), who saw inhibition of monosynaptic excitation in the spinal cord of cats without any appreciable change in the excitability of the post-synaptic cell. This 'remote inhibition' of Frank & Fuortes may well be presynaptic. A similar report comes from J. C. Eccles and his co-workers (personal communication). Therefore a definite possibility exists that the mammalian nervous system may also make use of a presynaptic mechanism.

SUMMARY

1. At crustacean neuromuscular junctions, as well as in numerous synapses elsewhere, the inhibitory transmitter acts by increasing the post-synaptic permeability to certain ions while the membrane potential remains near the resting level.

2. If inhibitory impulses are timed to arrive at the neuromuscular junction 1–6 msec before the excitatory impulse, they reduce the excitatory junctional potentials (e.j.p.s). It is demonstrated that in specific instances this reduction of e.j.p.s cannot be brought about solely by an increase in a post-synaptic inhibitory conductance.

3. The average number and size of quanta released by an excitatory nerve stimulus from nerve terminals were determined at individual junctional areas with extracellular micro-electrodes. Properly timed inhibition

reduced the number of released quanta while their size remained unchanged. Therefore, this type of inhibition acts at a presynaptic site on the excitatory nerve terminals, reducing their output of transmitter. These results do not support as mechanisms of inhibition either a post-synaptic decrease of sensitivity to the transmitter or a conduction block of the excitatory nerve.

4. Applied gamma-aminobutyric acid has an inhibitory effect in addition to its post-synaptic inhibitory conductance increase. This and a minimal synaptic delay of about 1 msec suggest chemical transmission from the inhibitory to the excitatory nerve terminals.

5. Presynaptic inhibition reduces the post-excitatory facilitation of e.j.p.s as well as the post-excitatory frequency increase of spontaneous miniature potentials. Presynaptic inhibition itself shows a process of facilitation, much like the inhibitory junctional potential.

We wish to thank Dr David Potter who participated in some of the experiments, Dr Edwin Furshpan for helpful discussions, and Mr R. B. Bosler for his unfailing assistance during the course of this study. Financial support by the United States Public Health Service and by the Mallinckrodt Foundation is acknowledged.

REFERENCES

BOISTEL, J. & FATT, P. (1958). Membrane permeability change during inhibitory transmitter action in crustacean muscle. *J. Physiol.* **144**, 176–191.

BURKE, W. & GINSBORG, B. L. (1956). The action of the neuromuscular transmitter on the slow fibre membrane. *J. Physiol.* **132**, 599–610.

COOMBS, J. S., ECCLES, J. C. & FATT, P. (1955). The specific ionic conductance and the ionic movements across the motoneuronal membrane that produce the inhibitory post-synaptic potential. *J. Physiol.* **130**, 326–373.

CURTIS, D. R. & ECCLES, J. C. (1960). Synaptic action during and after repetitive stimulation. *J. Physiol.* **150**, 374–398.

D'ANCONA, V. (1923). Per la miglior conoscensa delle terminosioni nervose nei muscoli somatici dei crosticei decopodi. *Trab. Lab. Invest. biol. Univ. Madr.* **23**, 393–423.

DEL CASTILLO, J. & KATZ, B. (1954). Changes in end-plate activity produced by presynaptic polarization. *J. Physiol.* **124**, 586–604.

DEL CASTILLO, J. & KATZ, B. (1956). Localization of active spots within the neuromuscular junction of the frog. *J. Physiol.* **132**, 630–649.

DUDEL, J. & KUFFLER, S. W. (1960). A second mechanism of inhibition at the crayfish neuromuscular junction. *Nature, Lond.*, **187**, 247–248.

DUDEL, J. & KUFFLER, S. W. (1961*a*). The quantal nature of transmission and spontaneous miniature potentials at the crayfish neuromuscular junction. *J. Physiol.* **155**, 514–529.

DUDEL, J. & KUFFLER, S. W. (1961*b*). Mechanism of facilitation at the crayfish neuromuscular junction. *J. Physiol.* **155**, 530–542.

FATT, P. & KATZ, B. (1953). The effect of inhibitory nerve impulses on a crustacean muscle fibre. *J. Physiol.* **121**, 374–389.

FRANK, K. & FUORTES, M. G. F. (1957). Presynaptic and post-synaptic inhibition of monosynaptic reflexes. *Fed. Proc.* **16**, 39–40.

GRUNDFEST, H., REUBEN, J. P. & RICKLES, W. H., Jr. (1959). The electrophysiology and pharmacology of lobster neuromuscular synapses. *J. gen. Physiol.* **42**, 1301–1323.

HAGIWARA, S. & TASAKI, I. (1958). A study of the mechanism of impulse transmission across the giant synapse of the squid. *J. Physiol.* **143**, 114–137.

HAGIWARA, S., KUSANO, K. & SAITO, S. (1960). Membrane changes in crayfish stretch receptor neuron during synaptic inhibition and under action of gamma-aminobutyric acid. *J. Neurophysiol.* **23**, 505–515.

KATZ, B. (1958). Microphysiology of the neuro-muscular junction. A physiological 'quantum of action' at the myoneural junction. *Johns Hopkins Hosp. Bull.* **102**, 275–295.

KUFFLER, S. W. (1960). Excitation and inhibition in single nerve cells. *Harvey Lect.* 1958–1959. New York: Academic Press.

KUFFLER, S. W. & EYZAGUIRRE, C. (1955). Synaptic inhibition in an isolated nerve cell. *J. gen. Physiol.* **39**, 155–184.

KUFFLER, S. W. & KATZ, B. (1946). Inhibition at the nerve-muscle junction in crustacea. *J. Neurophysiol.* **9**, 337–346.

MARMONT, G. & WIERSMA, C. A. G. (1938). On the mechanism of inhibition and excitation of crayfish muscle. *J. Physiol.* **93**, 173–193.

TIEGS, O. W. (1924). On the mechanisms of muscular action. *Aust. J. exp. Biol. med. Sci.* **1**, 11–29.

TRAUTWEIN, W. & DUDEL, J. (1958). Zum Mechanismus der Membranwirkung des Acetylcholin an der Herzmuskelfaser. *Pflüg. Arch. ges. Physiol.* **266**, 324–334.

VAN HARREVELD, A. (1939). The nerve supply of doubly and triply innervated crayfish muscles related to their function. *J. comp. Neurol.* **70**, 267–284.

WALL, P. D. (1958). Excitability changes in afferent fibre terminations and their relation to slow potentials. *J. Physiol.* **142**, 1–21.

J. Physiol. (1961) **155**: 514-529

METHODS FROM

THE QUANTAL NATURE OF TRANSMISSION AND SPONTANEOUS MINIATURE POTENTIALS AT THE CRAYFISH NEUROMUSCULAR JUNCTION

J. Dudel and S. W. Kuffler

METHODS

Almost all experiments in the current studies were done on the abductor muscle of the dactyl (corresponding to the opener of the claw) in the 1st or 2nd walking leg of the crayfish *Orconectes virilis* (Steinhilberi). The muscle is relatively thin, consisting of several layers of short fibres which originate on the exoskeleton and are inserted on a central tendon. Most of the fibres which were used were 200–300 μ in diameter and about 2–3 mm long in the relatively small animals. The adductor of the dactyl was removed, leaving the flat inner surface of the abductor exposed. The intact exoskeleton around the muscle formed a natural chamber with a volume of not more than 0·1 ml. The leg was placed in a Lucite chamber, with the dactyl and the carpopodite fixed (Fig. 1). The inhibitory and excitatory axons to this muscle are contained in separate bundles which were exposed in the meropodite. The bundles were kept submerged in the second compartment of the chamber and were stimulated with fluid electrodes. These consist of glass tubes with a small opening, large enough for a nerve bundle to go through, connected to a syringe by fine tubing. Various lengths of nerve can be pulled up into the tube, together with physiological solution. The stimulus is then applied between the inside and outside of the glass tube.

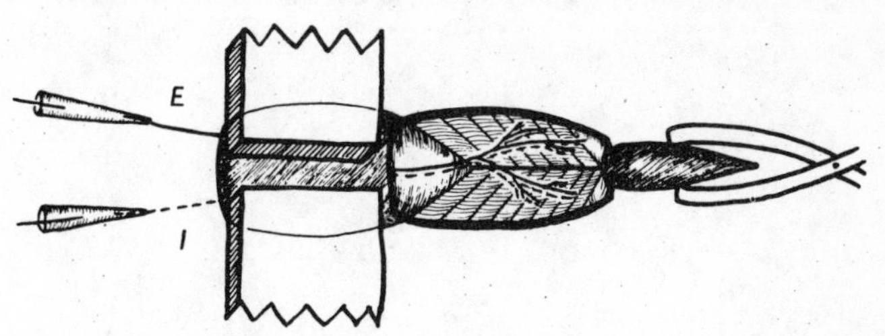

Fig. 1. Scheme of the preparation of the abductor of the dactyl in the crayfish, viewed from above. Adductor muscle removed. E = excitatory axon, I = inhibitory axon.

The composition of van Harreveld's solution (1936) is (mM) NaCl 195, KCl 5·4, $CaCl_2$ 13·5 and $MgCl_2$ 2·6. The pH was kept near 7·5 with 10 mM tris maleate buffer. Most experiments were done at 21–23° C. The muscle fibres were kept covered by physiological solution and the fluid was periodically replaced by adding a few drops of fresh solution. Alternatively a constant stream of fluid was kept flowing past the muscle fibres.

The recording system consisted of conventional DC and AC amplifiers. The micropipettes were generally filled with 3 M-KCl and had resistances of 7–20 MΩ. Their tip diameters were 1 μ or smaller. For exploration of the potentials along the fibre surface, larger pipettes (1–5 MΩ) were sometimes used, filled with saturated NaCl solution and having tips of up to several micra diameter.

Muscle fibres were viewed under dark-field illumination, which enabled one to see the course of nerve branches. The finer ramifications near the junctions were not visible. The excitatory and inhibitory axons run together and also branch in unison innervating fibres at multiple spots (van Harreveld, 1939; Fatt & Katz, 1953).

III. Physiology of Sensory Neurons

EXAMPLES FROM VISUAL SYSTEMS

It was in the study of sensory neurons that the analytical advantages of working with "single units" (advantages to which all the papers in this collection attest) were first exploited. Among these early studies might be mentioned those by Adrian and Zotterman (1926) and by Matthews (1931) on vertebrate muscle spindles, and one of particular relevance to this collection, the study of HARTLINE AND GRAHAM (1932) of *Limulus* optic nerve fiber responses. Elegant and quantitative studies, using sophisticated electrophysiological techniques, have now been made of primary sensory neurons responding to a great variety of specific kinds of stimuli in a great many different animals. From such studies a number of generalizations have emerged about the physiology of sensory neurons. These are reviewed briefly below.

We have chosen a series of papers from studies of visual systems that both illustrate points of the general picture and provide specific background for the papers in the following chapter on neural integration. The physiology of vision has been studied more intensively than that of any other sensory modality and provides a fine example of the benefits derived from approaching a problem from the varied points of view of different disciplines, in this case, psychophysics, biochemistry, morphology (including electron microscopy), and electrophysiology. The physiology of two visual systems, the vertebrate, camera-type eye, and the faceted, compound eye of *Limulus*, the horseshoe crab, are emphasized. In both systems there is not only relatively complete knowledge of the physiology of individual primary receptors but detailed knowledge of how the individual receptor responses are further integrated toward the extraction of information relevant to the behavior of the animal.

Specific sensitivity. Most sensory neurons respond best to a particular kind of exciting energy referred to as the "adequate stimulus." The degree to which the adequate stimulus is specified is sometimes extraordinary (e.g., the odorant molecule which excites male moth antennal chemoreceptors, or frequency for response of a vibration receptor). Often specificity is the result of non-neural structures associated with the sensory neuron that collect or reject energy, channel, transform, and perhaps amplify it to a form appropriate for the excitation of the sensory neuron (a well-studied example is the vertebrate ear). In a number of cases sensory systems achieve detection of a specific signal close to the theoretical limit for its detection from random nonspecific energy.

In most sensory neurons the earliest recordable sign of the conversion of an adequate stimulus into a neural event is a depolarization which results from an increase in conductance of the neural membrane. This conductance change usually takes place on a process of the neuron, the sensory dendrite. It is a curiosity that such dendrites often bear ultrastructural similarity to cilia. Both the outer segment of the vertebrate rod and the dendrite of the eccentric cell of the *Limulus* eye which are discussed in papers reproduced here show this similarity. Vertebrate rods and cones, however, show hyperpolarization as the first response to light (see the following). In the case of *Limulus*, Smith, Stell, and Brown (1969a, b) considered the hypothesis that the effect of light is not due to a conductance increase mechanism but rather results from light-induced alterations in a constant current generator, here an electrogenic sodium pump.

Hecht, Schlaer, and Pirenne (1942) seek to define in absolute terms the light energy necessary to produce a visual perception. Their psychophysical findings define quantitatively the problem to be solved in biophysical terms: how do 5-14 quanta of green light arriving at the sensory receptors produce a response by the subject? Or, at the level of the primary sensory neuron, how does one photon excite a rod? It is noteworthy that this very basic work was accomplished on unanesthetized human subjects.

Contributions of biochemical studies to the solution of the problem of visual excitation are summarized in the article by Wald, Brown, and Gibbons (1963). This is the only review included in this collection. It provides a concise description of the photochemistry of visual pigments (findings which are proving remarkably similar for photopigments of diverse animals). It also states clearly two of the unsolved problems: that of the understanding of structural design in molecular dimensions and that of how photochemical events result in a change in the membrane properties of the receptor neuron. Indeed, there is no sensory neuron for which there is an explanation of the mechanism by which energy (photochemical, chemical, or mechanical) transducing mechanisms affect the neuron membrane conductance. This is, however, merely one facet of a major unsolved problem of cellular neurophysiology: what is the mechanism of membrane conductance changes'

A recent finding may lead to new understanding of the link between the photochemical and membrane events. Cone, Hagins, and others (e.g., see Cone, 1968) have recorded very rapid electrical potentials recordable from living and from nonliving, fixed visual receptors, provided there is no structural disarray or denaturation of the photopigments. Application of a very intense, brief flash results in an electrical potential whose time course, polarity, and wave length required for maximum amplitude can be related to the characteristics of biochemical intermediates resulting from rhodospin bleaching. It is not the potentials themselves which have importance for neuronal mechanisms, but the clue which they give to structural organization and light induced changes in that organization at molecular dimensions.

The observations presented by **Hagins, Zonana, and Adams (1962)** provide a very clear demonstration of a local conductance increase produced by a stimulus on a sensory neuron. These authors show that the origin of extracellularly recorded currents is precisely at the point where a small stimulating spot of light shines on the squid rod.

In vertebrate photoreceptors (rods), Penn and Hagins (1969) have shown a very different mechanism of electrical response. In the dark, the rod produces a large, metabolically driven current flowing between inner and outer segments. This is analogous to (if not the same as) the hyperpolarizing current resulting from the activity of the Na^+ pump in some neurons (e.g., the crayfish stretch receptor neuron, Nakajima and Takahashi, 1966). The response to light recorded in rod inner segments is a hyperpolarization (Tomita, Kaneko, Murakami, and Pautler, 1967; **Werblin and Dowling, 1969**). This appears to be the result of an increase in the resistance of the rod membrane (perhaps a reduction in Na^+ conductance) in response to light. The pump is thus more effective in altering (hyperpolarizing) the rod membrane potential in the light than in the dark (Hagins, Colloquium, Harvard University, 1970).

Generator potential. The conductance change in the sensory dendrite in most receptors results in movement of ions along preestablished concentration gradients and thus to a depolarizing current. The resulting change in membrane potential has been termed the generator potential (Granit, 1955) or receptor potential. Its magnitude is a function (sometimes logarithmic, often complex) of the stimulus intensity. The generator potential persists even when the regenerative conductance changes that give rise to action potentials are blocked by, for example, tetrodotoxin (Lowenstein, Terzuolo, and Washizu, 1963) or procaine (KATZ, 1950b). The generator potential is a graded representation of the stimulus which is not regenerative; the conductance changes which give rise to it involve mechanisms different from those giving rise to action potentials.

Impulse initiation. When the sensory generator potential is sufficiently large it triggers action potentials which are propagated along the sensory axon. The site of impulse initiation is often separate and distinct from the place at which the generator potential originates. This site is often interpretable in relation to the neuronal architecture by a consideration of the cable properties. For example, impulse initiation frequently occurs where the axon leaves the soma and there is a sharp reduction in membrane surface relative to volume (see e.g., Edwards and Ottoson, 1958). The location and the properties of the spike initiating zone have a profound influence on the relation between the stimulus energy arriving at the neuron and the pattern of nerve impulses conducted toward the central nervous system. The difference between a tonic receptor (one that continues to fire impulses for the duration of a maintained stimulus) and a phasic receptor (one which fires impulses at the onset of a stimulus or during rapid increases in stimulus strength) has in some cases been shown to lie entirely in the spike initiation mechanism (Nakajima and Onodera, 1969a, b). The bases for these differences are largely unknown.

The first direct demonstration of a sensory generator potential and of its relation to impulse frequency was provided by KATZ (1950a, b) for a tonic receptor, a frog muscle spindle. He found a linear relation between the magnitude of the generator potential and the frequency of impulses. Such a relation has now been demonstrated in a number of tonic receptors for the steady firing rate during a maintained stimulus.

Adaptation. Most receptors, whether tonic or phasic, decrease their rate of impulse firing immediately after a maintained stimulus is applied; many respond with progressively fewer impulses when an identical stimulus is given repeatedly. Such decreases in responsiveness are termed adaptation. The mechanisms responsible for adaptation are not clearly known. Changes in the non-neural apparatus responsible for channeling and filtering the stimulus energy to the receptor neuron are sometimes involved. In such cases these changes may be under control of efferents from the central nervous system. However, adaptation occurs within isolated sensory neurons. The spike-triggering mechanism may play a role, as indicated above. In addition, Na^+ entering the cell as a result of action potentials and of the sensory generator current may stimulate a Na^+ pump whose activity results in a hyperpolarizing current which opposes the generator potential (Nakajima and Takahashi, 1966). In photoreceptors the physical bleaching of the light-absorbing pigments by the light can play an important role (**Wald, Brown, and Gibbons, 1963**; Dowling, 1963).

The Limulus eye. A number of the generalizations about the relations between stimulus, generator potential and impulse initiation are illustrated in the observations presented by **Fuortes and Poggio (1963)** in their study of the *Limulus* lateral eye. This eye has been studied with both extracellular and intracellular recording techniques. In each ommatidium approximately a dozen neurons, the retinula cells, are arranged like segments of an orange (Miller, 1957). These contribute to a highly ordered array of membrane folds, the rhabdom, on which the photopigment is arranged. A single, larger cell, the eccentric cell, sends a distal process up the center of the ommatidium (filling the core of the orange, as it were). Although all the retinula cells have axons, only the axon of the eccentric cell carries impulses from the ommatidium toward the brain. The cells of an ommatidium are electrotonically coupled so that changes of potential occurring in any part can be recorded intracellularly from other elements of an ommatidium (Behrens and Wulff, 1965). Thus, although the eccentric cell is strictly speaking a second-order neuron, integration here most probably involves only simple summation.

Recording with an electrode intracellular to the eccentric cell, **Fuortes and Poggio (1963)** can detect both the generator potential in response to light stimuli and the impulses which arise in the axon at some distance from the cell. The depolarization, after an initial larger transient, is proportional to the logarithm of the light intensity and accompanies a conductance increase. The frequency of impulse firing is directly proportional to the magnitude of the generator potential or to the amount of current passed through the recording electrode. This suggests the equivalence, with respect to the impulse triggering mechanism, of current arising from the photochemically mediated conductance change and of an artificially imposed current. The slope relating impulse frequency to generator potential size or to current intensity becomes less steep as measurements are taken at longer times after the beginning of a stimulus. Such a decay in firing frequency during a maintained stimulus is an example of adaptation.

Further studies of the *Limulus* eye have suggested a direct relation between the absorption of light by the photopigment and the generator potential. In dim light and sometimes even in complete darkness, spontaneous depolarizations can be detected which have a superficial resemblance to the miniature potentials recorded at neuromuscular junctions. Analysis has shown that they follow Poisson statistics for independent, random events. Their average frequency is proportional to light intensity for short durations of dim light (Adolf, 1968). Statistical analysis allows the suggestion that absorption of one quantum of light results in one depolarizing event (Fuortes and Yeandle, 1964). It seems that these sometimes spontaneously occurring depolarizations represent units (termed quanta by analogy to transmitter quanta) from which, in bright light, the receptor or generator potential is built up. Similar spontaneous, miniature receptor potentials have been recorded from locust ommatidia and from the leech eye, suggesting that the phenomenon may be a general one in photoreceptors.

The vertebrate eye. The paper by **Werblin and Dowling (1969)** in Section IV could have been placed here equally well. It reports intracellular recordings from cells of the amphibian retina. In contrast to the depolarization seen in invertebrate photoreceptors the response to light recorded from the inner segments of vertebrate rods and cones (the primary photoreceptors) is a hyperpolarization which is graded with the intensity of the light (see also Tomita *et al.*, 1967). Evidence concerning the mechanism of this hyperpolarizing response has been obtained by Penn and Hagins (1969) and has been mentioned previously.

In the vertebrate retina transmission and integration of primary sensory cell responses take place through several interneurons without the mediation of conducted impulses.

Further reading. The physiology of crustacean abdominal stretch receptor neurons has been more rigorously studied than that of any other sensory neuron. Two papers by EYZAGUIRRE AND KUFFLER (1955a, b) set forth the range of responses seen in the isolated receptor organs. Edwards and Ottoson (1958) define the site of impulse initiation as the initial segment of the axon. Nakajima and Takahashi (1966) provide evidence which suggests that activity stimulates an electrogenic Na^+ pump which may play a role in adaptation. Obara (1968) has shown that the conductance change responsible for the generator potential involves an increase in permeability to monovalent cations, but not to divalent cations or to anions. A detailed analysis of the sites at which adaptation occurs in both tonic and phasic receptors of the stretch receptor organ is described in two papers by Nakajima and Onodera (1969a, b).

A symposium on sensory receptors (Frisch, Ed., 1965) provides accounts of varied research approaches to problems of sensory physiology. Mellon (1968) has written a recent review of the physiology of sense organs.

33

[Reprinted from THE JOURNAL OF GENERAL PHYSIOLOGY, July 20, 1942, Vol. 25, No. 6, pp. 819–840]

ENERGY, QUANTA, AND VISION*

BY SELIG HECHT, SIMON SHLAER, AND MAURICE HENRI PIRENNE‡

(*From the Laboratory of Biophysics, Columbia University, New York*)

(Received for publication, March 30, 1942)

I

Threshold Energies for Vision

The minimum energy required to produce a visual effect achieves its significance by virtue of the quantum nature of light. Like all radiation, light is emitted and absorbed in discrete units or quanta, whose energy content is equal to its frequency ν multiplied by Planck's constant h. At the threshold of vision these quanta are used for the photodecomposition of visual purple, and in conformity with Einstein's equivalence law each absorbed quantum transforms one molecule of visual purple (Dartnall, Goodeve, and Lythgoe, 1938). Since even the earliest measurements show that only a small number of quanta is required for a threshold stimulus, it follows that only a small number of primary molecular transformations is enough to supply the initial impetus for a visual act. The precise number of these molecular changes becomes of obvious importance in understanding the visual receptor process, and it is this which has led us to the present investigation.

The first measurements of the energy at the visual threshold were made by Langley (1889) with the bolometer he invented for such purposes (Langley, 1881). He found the energy to be 3×10^{-9} ergs for light of 550 mμ. Langley worked before the physiology of vision was understood, so that he used the wrong light and took none of the precautions now known to be necessary; even so, his results are too high only by a factor of 10.

In the fifty years since Langley there have been eleven efforts to redetermine the minimum energy for vision. We have carefully studied all these accounts and have done our best to evaluate the measurements. Unfortunately, many of them contain serious errors which invalidate them. Most of them involved no direct energy determinations; instead, the investigators relied on previously measured energy distributions in standard sources and made elaborate computations from them. Only a few can be considered as reliable.

After Langley, the earliest paper is by Grijns and Noyons (1905). Their data differ widely from all other measurements and cannot be accepted even

* A preliminary report of these measurements was published in *Science* (Hecht, Shlaer, and Pirenne, 1941), and presented to the Optical Society in October, 1941 (Hecht, 1942).

‡ Fellow of the Belgian American Educational Foundation.

though it is hard to discover their precise errors because the description is too obscure. Zwaardemaker (1905), in whose laboratory their measurements were made, reports some of his own rough determinations, which turn out to be near Langley's. Neither Grijns and Noyons nor Zwaardemaker actually measured the energies involved, but relied on Ångström's (1903) determinations of the energy distribution in the Hefner lamp.

The best of the early efforts is by von Kries and Eyster (1907); and though the results involve many calculations, they come very close to the most careful of modern measurements. Von Kries and Eyster made no direct energy determinations; they measured brightnesses, durations, and areas. The conversion of these factors into final energies requires skill and care in the evaluation of absorptions, reflections, lens factors, and the like, and it is gratifying to see the admirable way in which von Kries accomplished this task.

TABLE I

Minimum Energy for Vision

Wavelength	Energy	No. of quanta	Source
mμ	*ergs*		
505	$0.66\text{–}1.17 \times 10^{-10}$	17–30*	Chariton and Lea (1929)
507	$1.3\text{–}2.6 \times 10^{-10}$	34–68	von Kries and Eyster (1907)
530	$1.5\text{–}3.3 \times 10^{-10}$	40–90	Barnes and Czerny (1932)

* For inexperienced observers.

Computations from star magnitudes were made by Ives (1916) and by Russell (1917). However, neither they nor Reeves (1917) and Buisson (1917), who both reproduced star observations in the laboratory, employed the best physiological conditions for the measurements. Moreover, none of them took consideration of the different luminosity curves for rod vision and cone vision, and used the latter as standard in the computations.

Direct energy measurements were made by du Noüy (1921), but his work involves serious physical errors, and his results are too low by a factor of more than 100—so low indeed as to seem impossible.

The most recent determinations are by Chariton and Lea (1929), by Wentworth (1930), and by Barnes and Czerny (1932), all of whom agree in the order of magnitude of their results. Wentworth's exposures were too long to yield minimal values; otherwise her work is excellent. She measured the energies involved, which Barnes and Czerny also did, but not as directly.

From these twelve researches, we have chosen the three sets of measurements which are free from what can now be recognized as obvious error. These are given in Table I. Even though they differ by a factor of about 3, these data can be considered as roughly confirming one another. However, since for our purposes a factor of 3 cannot be ignored, we undertook to make the measurements again, but under the best physical and physiological conditions.

II

Visual Conditions

The circumstances which will yield the maximum retinal sensibility have been adequately known for years. They involve dark adaptation, peripheral vision, small test fields, short exposures, and selected portions of the spectrum.

Complete dark adaptation means a stay of at least 30 minutes in the dark before measurements can be begun (Piper, 1903; Hecht, Haig, and Chase, 1937). After thorough dark adaptation the periphery of the retina is much more sensitive than its center. The greatest density of rod elements begins at about 18° out (Østerberg, 1935), and exploration shows that between 20 and 30° from the center there is a region of maximum sensibility to light (Wentworth, 1930). The variation within this region is not large, and for convenience we chose a retinal area situated 20° temporally on the horizontal axis.

In visual threshold measurements it has been established that the larger the test area, the smaller need the intensity be for its recognition (*cf.* summary by Wald, 1938 *a*). This reciprocal relation is exact only for small areas. Our preliminary experiments, as well as the work of other investigators, show a minimum for the product of area and intensity for fields of the order of 10 minutes diameter. We therefore chose a circular retinal area of 10 minutes diameter for the test field.

The energy required to pass over the visual threshold involves an approximately reciprocal relationship between intensity and time of exposure. For exposures shorter than 0.01 second, the reciprocal relation holds perfectly (Graham and Margaria, 1935). To be sure of falling within this most efficient range, our exposures were 0.001 second long.

Finally, from the measurements of the scotopic luminosity curve (Hecht and Williams, 1922), it is known that for dim vision the eye is most sensitive to a wavelength of 510 mμ, and this is the light which we used for making the measurements.

III

Apparatus and Calibrations

The physical arrangements may be seen in Fig. 1. The light source *L* is a ribbon filament lamp run on constant current obtained from storage cells and measured potentiometrically. By means of a lens, it is focussed on the slit of a double monochromator M_1M_2 and finally on the artificial pupil *P*. The subject, who sits in a dark cabinet in the dark room, has his head in a fixed position by keeping his teeth in a "bite" or hard impression of his upper jaw. He has his left eye next to the pupil *P*, and on looking at the red fixation point *FP* he sees the field lens *FL*. The light intensity of this uniformly illuminated field is varied in large steps by the neutral filters F, and in a gradual way by the neutral wedge and balancer W. The size of the field is controlled by the diaphragms D. Its exposure is fixed by the shutter S, and is initiated by the subject.

For the record it is necessary to describe the apparatus and calibrations in detail. The double monochromator is made of two individual constant deviation monochromators, M_1 and M_2, which are arranged for zero dispersion by means of the reversing prism RP. In this way, all the light passes through an equal thickness of glass, and assures a uniform brightness of the field lens FL. The exit slit of M_1 has been removed, and the entrance slit of M_2 serves as the middle slit of the combined double monochromator. The entrance and exit slits of the combination are kept at 1.2 mm., which corresponds to a band width of 10 mμ centered at 510 mμ. The middle slit, before which the shutter is placed, is kept at 0.1 mm.

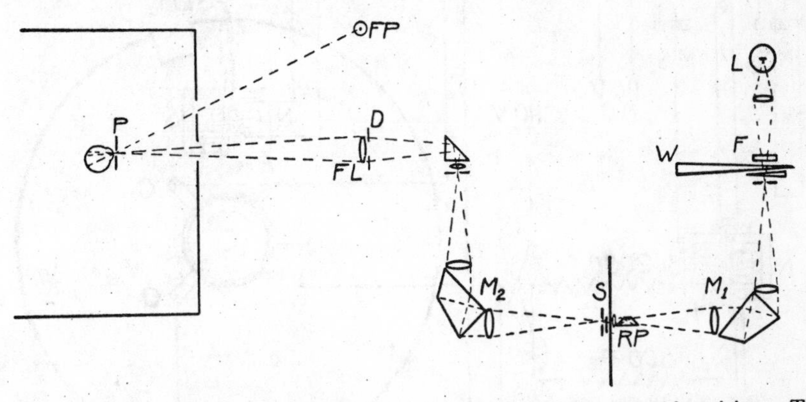

FIG. 1. Optical system for measuring minimum energies necessary for vision. The eye at the pupil P fixates the red point FP and observes the test field formed by the lens FL and the diaphragm D. The light for this field comes from the lamp L through the neutral filter F and wedge W, through the double monochromator M_1M_2 and is controlled by the shutter S.

The field lens FL magnifies the exit slit by a factor of 2, and thus yields an image of it 2.4 mm. wide and over 10 mm. high at the pupil P. The image is sufficient to cover uniformly not only the pupil P, but also the linear thermopile used for the energy calibration. The pupil mount at P and the field lens FL are connected by a carefully diaphragmed and blackened tube. The 2 mm. circular pupil P used for the visual measurements can be replaced by a slit 2 mm. wide and 10 mm. high behind which is the receiving surface of the thermopile for energy measurements.

S is a precision shutter made of two parts. One part is a thin circular aluminum disc with a small sector of 10.8° removed and properly balanced. It is run at 1800 R.P.M. by means of a synchronous motor, and therefore permits light to pass through the middle slit for 1/1000 second during each revolution. The other part is a polar relay shutter, which, by means of a phasing commutator on the shaft of the synchronous motor, is opened for only one passage of the rotating disc aperture whenever the subject releases a push button.

The essentials of the shutter are shown in diagrammatic detail in Fig. 2. On the same shaft with the disc there is mounted a commutator having a "live" sector, which together with the brush occupies somewhat less than 90°. Two brushes are arranged on this commutator 90° apart, and are so phased with the A.C. line voltage that one of these brushes receives only a positive impulse while the other receives only a negative impulse. These impulses control a polar relay PR_2, which then actuates a pair of single pole, double throw micro switches, MS_1 and MS_2. These are arranged with their springs in opposition in such a manner that the switches are in equilibrium

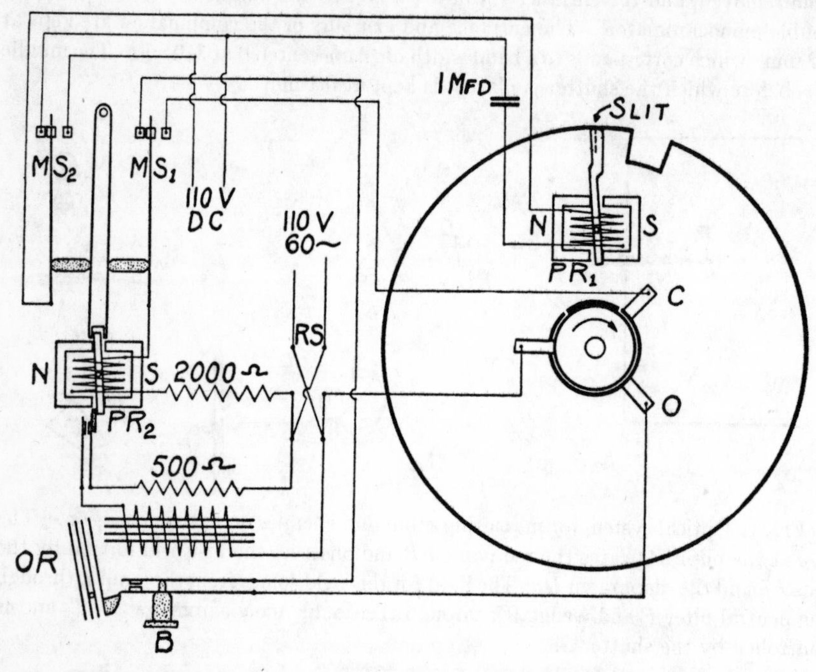

FIG. 2. Shutter for obtaining a single exposure of 1/1000 second. The details are described in the text.

at either of their two positions, and require but a small force and movement to kick them over to their other positions. Micro switch MS_1 is in series with the winding of PR_2, and in one position connects with the opening brush O and in the other position with closing brush C. The other micro switch, MS_2, charges and discharges a 1 μfd. condenser from the 110 volt, D.C. line through the polar relay PR_1. These impulses in and out of the condenser actuate PR_1 whose armature movement then uncovers and covers the middle slit.

The operation is seen by following a single cycle of operation of circuit and shutter. Fig. 2 shows the apparatus during its rest or closed period. The 110 volt, 60 cycle power enters through a pole-reversing switch, RS, to the neutral brush on the commutator. The impulse through the closing brush C is blocked, since it enters an open contact in MS_1. The impulse going to the opening brush O is blocked at the secondary contacts of the overload relay OR, the push button of which, B, is controlled by the observer. When B is released so that contact is made, the next impulse which leaves the opening brush O goes through the left hand contact of MS_1, through the winding of PR_2, and through a 2,000 ohm resistance to the other side of the power line. This impulse through PR_2 is adequate to throw its armature to the other position, thus switching over both MS_1 and MS_2, and closing the power circuit through the primary of the overload relay OR. The activation of OR closes its armature, whose movement opens the secondary contacts attached to it, thereby breaking the circuit from the opening brush O so that the cycle does not repeat itself. The switching of MS_2 to its right contact charges the 1 μfd. condenser through PR_1, which moves its armature and thereby exposes the slit. The switching of MS_1 to its right contact sets the circuit for the very next impulse through the closing brush C to PR_2. This closing impulse comes exactly three half-cycles or 3/120 second after the original opening impulse, and causes PR_2 to return to its original position. Now MS_2 discharges the 1 μfd. condenser, which actuates PR_1 so that its armature moves to cover the slit and terminate the cycle.

The pole-reversing switch RS enables one to select the correct polarity for the operation of this circuit. It needs to be set only at the beginning of an experiment when the synchronous motor is first started.

PR_1 and PR_2 are old Baldwin speaker units in which all the spring tension restraint of the armature has been removed; they thus act as very fast polar relays. An oscillographic study of PR_2, which is essentially unloaded, shows that the micro switches are thrown to the right contacts before the end of the half cycle which actuates it. However, PR_1, due to the loading of the shutter vane attached to the armature, is not nearly so fast, but opens in less than 3/120 second and closes in less than 4/120 second, which are the limits required for its operation. MS_1 and MS_2 are a pair of micro switches, type Z,—BZ-R, selected for near equality of spring tension. They are mounted plunger to plunger with a loose bar between them. This bar has a fulcrum at one end, and a fork at the other. Inside the fork is located the armature of PR_2. The fork width is so adjusted that it offers no resistance to the movement of the armature except at the very end of its motion when the impact of the armature is sufficient to kick over both micro switches.

It was necessary to calibrate the neutral filters, the wedge and balancer, the diaphragm openings, and the energy at the pupil P. The filters and the wedge and balancer were measured with our photoelectric spectrophotometer (Shlaer, 1938) at the same wavelength used in the experiments, and in an analogous optical position in front of the entrance slit of the first monochromator. We first used filters and wedges made of gelatin; later they were replaced with neutral glass. The smaller diaphragms were calibrated under the microscope with a filar micrometer by measuring several diameters for each opening; the larger ones were similarly measured with a comparator.

The energy density at the pupil P was measured with a Hilger linear thermopile and a Paschen galvanometer. The thermopile was first standardized against a standard carbon filament lamp of known energy radiation. To do this we used the tube holding the pupil and the field lens, first removing the field lens and substituting the slit for the pupil, and fixing the thermopile immediately behind the slit. This assembly of tube, slit, and thermopile was then mounted on an optical bench so that the standard lamp was at the specified distance of 2 meters from the receiver of the thermopile. The thermopile and its end of the tube was then covered with a thermos flask and allowed to reach thermal equilibrium. Between the source and the opening of the tube was mounted a triple leafed shutter with about 20 cm. spacing between the leaves. The surfaces facing the thermopile were blackened while those facing the source were shiny. This shutter was used to open and close the radiation to the thermopile.

The thermopile was connected to a Paschen galvanometer, which is a moving magnet type of very high sensitivity (about 2×10^{-9} volts per mm. at a meter). In series with the thermopile and galvanometer was a resistance of about 0.1 ohm, across which known potentials could be inserted to counterbalance the potential generated by the thermopile, thus using the galvanometer as a null point instrument. The radiation was first permitted to fall on the thermopile, and the galvanometer brought back to zero by means of measured counter-potentials. The radiation was then occluded and the counter-potential switched off to check the zero of the galvanometer. In this way we could measure large potentials corresponding to galvanometer swings of several meters without actually using such scale distances. The thermopile was calibrated as potential *vs.* radiant energy density incident upon its receivers for three different energy densities which covered a range of about 3 to 1, and included the actual energy density delivered by the ribbon filament lamp and the monochromators.

For calibrating the energy density through the monochromators, the field lens was replaced in the tube and the tube placed in its correct position in the apparatus. Diaphragm D was removed, the middle slit of the monochromator was opened to 1.5 mm., and the wedge and balancers were removed. The energy was then measured with the same thermopile and the same electrical system. With the lamp current at 19 amperes, the energy density at the pupil P was 27.5 microwatts per square centimeter; with the current at 18 amperes, it was 18.3 microwatts per square centimeter. In the early visual determinations we used the lamp at 19 amperes; in the later determinations at 18 amperes.

In order to convert these measurements into values of the energy at the pupil during the visual determinations, it is necessary to reduce the measured energy density by factors corresponding (*a*) to the change of the middle slit from 1.5 to 0.1 mm., (*b*) to the change in aperture of the field lens from its largest opening of 25.9 mm. diameter to the sizes of the particular diaphragms used, and (*c*) to the insertion of the wedge and balancer. All these factors were known from previous separate measurements, but we calibrated them again in their places in the apparatus by means of a sensitive dry-disc photocell in place of the thermopile behind the thermopile slit. The results merely confirmed the previous calibrations. By applying these reduction factors for the wedge at its thinnest place, the middle slit at 0.1 mm., the 10 minute diaphragm at the field lens, and the 2 mm. pupil at P, we found that the energy density through the pupil is 3.4×10^{-4} ergs per second when the ribbon filament lamp is running at 18 amperes. The energy calibrations were run through twice several months apart and agreed almost perfectly.

IV

Visual Measurements

From the subject's point of view, an experiment involves the report of whether or not he has seen a flash of light after he has opened the shutter for an exposure. Fixation of the red point need not be continuous, a circumstance which avoids undue fatigue. The observer is told by the operator that conditions are set and that he should try a flash when he is ready. He fixates the red point, and at the moment which he considers propitious, he exposes the light to his eye. The operator changes the position of the wedge, or removes or introduces a filter until he is satisfied with the precision of the measurements.

TABLE II

Minimum Energy for Vision

Each datum is the result of many measurements during a single experimental period, and is the energy which can be seen with 60 per cent frequency. $\lambda = 510\ m\mu$; $h\nu = 3.84 \times 10^{-12}$ ergs.

Observer	Energy	No. of quanta	Observer	Energy	No. of quanta
	ergs $\times 10^{10}$			*ergs* $\times 10^{10}$	
S. H.	4.83	126	C. D. H.	2.50	65
	5.18	135		2.92	76
	4.11	107		2.23	58
	3.34	87		2.23	58
	3.03	79			
	4.72	123	M. S.	3.31	81
	5.68	148		4.30	112
S. S.	3.03	79	S. R. F.	4.61	120
	2.07	54			
	2.15	56	A. F. B.	3.19	83
	2.38	62			
	3.69	96	M. H. P.	3.03	79
	3.80	99		3.19	83
	3.99	104		5.30	138

In the early measurements we considered that the threshold had been reached when the observer saw a flash of light at a given intensity six times out of ten presentations. Later the measurements were made somewhat more elaborately. Each of a series of intensities was presented many times and the frequency of seeing the flash was determined for each. From the resulting plot of frequency against intensity we chose the threshold as that amount of light which could be seen with a frequency of 60 per cent.

During 1940 and 1941 we measured the threshold for seven subjects. With four we made several determinations each, extending over a year and a half; one subject we measured on two occasions 3 months apart; and two we measured only once. For all these observers the minimum energy necessary for vision ranges between 2.1 and 5.7×10^{-10} ergs at the cornea. These small energies represent between 54 and 148 quanta of blue-green light. The results for the individual subjects are in Table II, and are given as energy and as the number of quanta required.

It is to be noticed that these values are of the same order of magnitude as those of von Kries and Eyster, and of Barnes and Czerny, but almost twice as large. Because of the fairly wide ranges, these previous measurements and our own overlap to some extent, and it is conceivable, though not probable, that their observers may actually have needed somewhat smaller energies than ours. Chariton and Lea's results, however, are much too small. Actually their value of 17 $h\nu$ is an extrapolation to zero frequency of seeing; if we take as threshold a 60 per cent frequency, their data come more nearly to 25 $h\nu$. This is still too small a value, and is probably in error, as will be apparent in later sections of our paper.

V

Reflections and Absorptions

The values in Table II, as well as those of previous investigators, are the energies incident at the cornea. Nevertheless the tacit supposition has generally been made that they represent the actual energies necessary to initiate a visual act. It is important to recognize that this assumption is incorrect. Before one can know how many quanta are required to start the visual process, one must apply at least three corrections to the measurements.

The first is reflection from the cornea. This is about 4 per cent and is obviously of not much importance. The second involves loss by the ocular media between the outer surface of the cornea and the retina. It has been common opinion that this loss is small. However, the measurements of Roggenbau and Wetthauer (1927) on cattle eyes, as well as the recent measurements of Ludvigh and McCarthy (1938) on human eyes, have shown that this loss is large. From the values of Ludvigh and McCarthy it appears that at 510 mμ the ocular media transmit almost exactly 50 per cent of the light entering the cornea of a young person, and less of an older one.

The next correction is much more difficult to evaluate with precision and involves the percentage of the energy absorbed by the retinal elements themselves. Since visual purple is the photosensitive substance concerned in this particular act, light which is not absorbed by it is visually useless. One cannot assume that visual purple absorbs all the light incident on the retinal cells. The fraction which it does absorb must be found by experiment.

Koenig (1894) determined the absorption of the total amount of visual purple which can be extracted from the human eye. If this amount of visual purple is spread evenly over the whole retina, his data show that it will absorb only 4 per cent of light of 510 mμ. This is a small value. Nevertheless, it is about the same as the 4 per cent and the 13 per cent recently found by Wald (1938 *b*) with a similar method for the absorption of the visual purple of the rabbit and rat retinas respectively.

These figures are probably too low, first because it is unlikely that all of the visual purple in the eye has been extracted, and second, because visual purple is not evenly distributed over the retina. It is lacking in the fovea; and even in the periphery the density of the rods is known to vary in a definite way. However, these absorptions may be considered as lower limiting values.

VI

Visual Purple Absorption

We have estimated the absorption of visual purple in the retina in a completely independent manner by comparing the percentage absorption spectrum of different concentrations of visual purple with the scotopic (rod) luminosity curve of the eye measured at the retina. The comparison rests on the fact that the shape and width of the percentage absorption spectrum of a substance varies with its concentration, and that the luminosity curve must represent the percentage absorption curve of a particular concentration of visual purple in the retina.

Fig. 3 shows the absorption spectrum of frog's visual purple as determined by Chase and Haig (1938) in our laboratory, by Lythgoe (1937) in London, and by Wald (1938 *b*) at Harvard. The agreement of the data is obvious, and shows that the absorption spectrum of visual purple may be considered as well established. Table III gives the average of these three series of measurements computed so that the maximum density at 500 mμ has a value of 1.

From these data in Table III we may prepare a series of percentage absorption spectra for different concentrations of visual purple. Since we are not interested in the absolute concentration of visual purple, but rather in its absorption capacities, we can deal with the series of percentage absorption spectra entirely in terms of maximum absorption. It will be recalled that the photometric density d is related to the transmission I_t by the equation $d = \log (1/I_t)$, and since the absorption $I_a = 1 - I_t$, it is a simple computation to find the percentage absorption corresponding to any density value, or the reverse.

We have made such computations for a variety of visual purple densities, and Fig. 4 shows the resulting percentage absorption curves for the different maximal absorptions of visual purple. For comparisons among the curves in Fig. 4 the maxima have all been made equal to 1, but their actual values are indicated in the figure. It is clear that the width of the curves increases as the concentration of visual purple increases.

The scotopic luminosity curve, as measured experimentally, records the reciprocal of the relative energy in different parts of the spectrum required for the production of a constant and very low brightness in the eye (Hecht and Williams, 1922). If this is to be compared with the absorption spectrum of

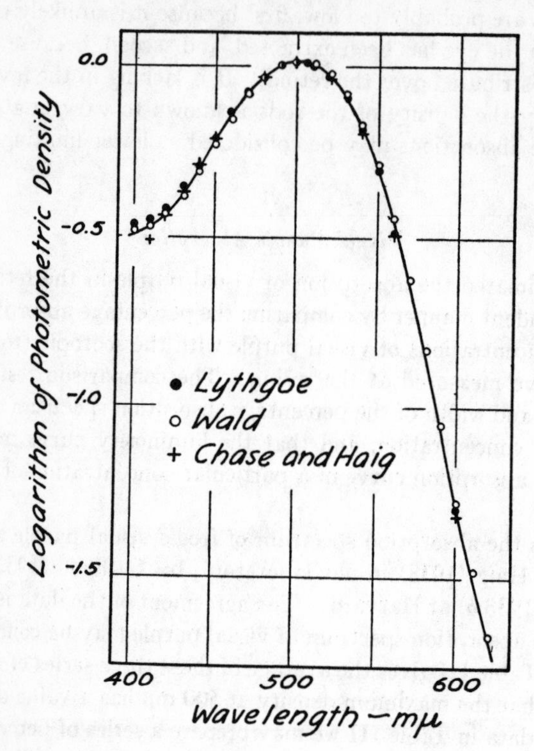

FIG. 3. Absorption spectrum of frog's visual purple. The data from the three sources have been made equal at 500 mμ.

visual purple, it must be converted into a quantum luminosity curve instead of an energy luminosity curve, because it is the number of quanta which determines the photochemical effectiveness of light and not just its energy content (Dartnall and Goodeve, 1937). Moreover, since our interest lies in retinal comparisons, the luminosity curve must be corrected for ocular media absorption in terms of the data of Ludvigh and McCarthy.

The scotopic luminosity data have been corrected in these two ways; the computed values are given in Table IV and shown as circles in Fig. 5. Included in the same figure are two percentage absorption spectra of visual purple; the upper curve represents 20 per cent maximal absorption, while the lower curve is 5 per cent maximal absorption.

TABLE III

Absorption Spectrum of Visual Purple

Average of data from Chase and Haig (1938), Wald (1938 *b*), and Lythgoe (1937).

λ – mμ	Density	λ – mμ	Density	λ – mμ	Density
400	0.306	480	0.900	560	0.321
410	0.317	490	0.967	570	0.207
420	0.353	500	1.000	580	0.131
430	0.408	510	0.973	590	0.0805
440	0.485	520	0.900	600	0.0473
450	0.581	530	0.780	610	0.0269
460	0.691	540	0.628	620	0.0150
470	0.811	550	0.465		

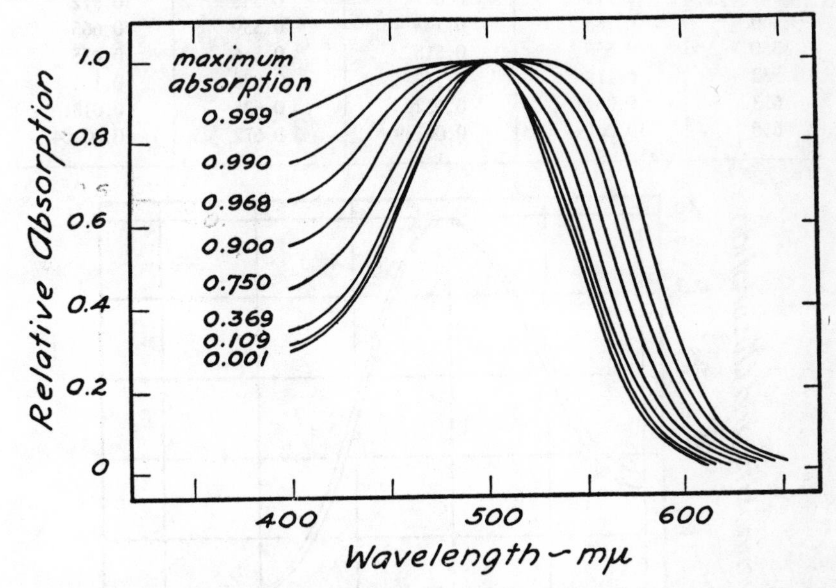

FIG. 4. Percentage absorption spectra of various concentrations of visual purple. For convenience in comparing the shapes of the curves, their maxima have all been equated to 1 and superimposed. The actual fraction absorbed at the maximum is shown for each curve. It is apparent that with increasing concentration the absorption curve steadily increases in width.

For comparing the luminosity and absorption data, it is well to confine our attention mostly to the long wave half of the luminosity curve because of the larger number of points involved. From the comparison it is apparent that the 5 per cent maximum absorption curve describes the points quite well, but

TABLE IV

Rod Luminosity Distribution in Spectrum

The original energy luminosity data of Hecht and Williams (1922) in column 2, when divided by the corresponding wavelengths in column 1, yield the quantum luminosity values in column 3 after being multiplied by a factor so that the maximum at 511 mμ equals 1. When these values in column 3 are divided by the ocular media transmission data in column 4 from Ludvigh and McCarthy (1938), they yield the spectral luminosity distribution at the retina given in column 5 after multiplication by a factor so that the maximum at 502 mμ is 1.

λ − mμ	Energy luminosity at cornea	Quantum luminosity at cornea	Ocular transmission	Quantum luminosity at retina
412	0.0632	0.0779	0.116	0.336
455	0.399	0.447	0.410	0.545
486	0.834	0.874	0.472	0.926
496	0.939	0.964	0.490	0.984
507	0.993	0.998	0.506	0.986
518	0.973	0.957	0.519	0.921
529	0.911	0.877	0.540	0.812
540	0.788	0.743	0.559	0.665
550	0.556	0.515	0.566	0.455
582	0.178	0.155	0.596	0.131
613	0.0272	0.0226	0.625	0.0181
666	0.00181	0.00139	0.672	0.00104

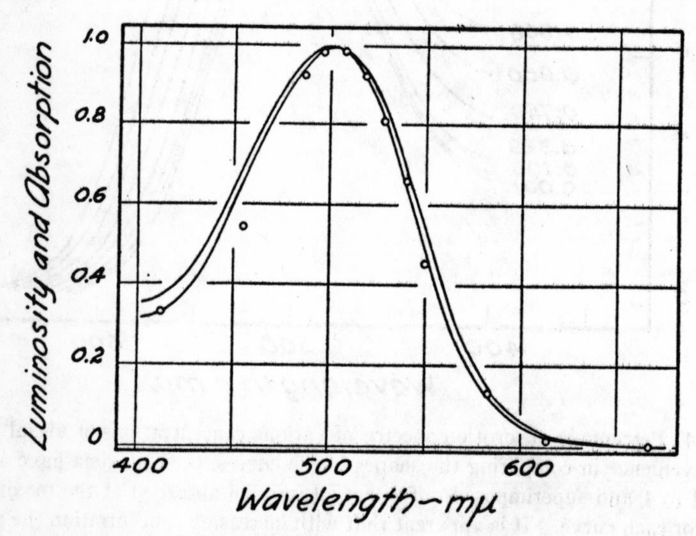

FIG. 5. Comparison of scoptopic luminosity at the retina with visual purple absorption. The points are the data of Hecht and Williams corrected for quantum effectiveness and ocular media transmission. The curves are the percentage absorption spectra of visual purple; the upper curve represents 20 per cent maximal absorption, and the lower one 5 per cent maximal absorption. All curves have been made equal to 1 at the maximum, 500 mμ, for ease in comparison.

that the 20 per cent curve is definitely excluded, because its absorption on both sides is just too high. The 10 per cent absorption curve, not shown in the figure, is perhaps slightly better than the 5 per cent one; it cuts through more points. In any case, both values are of the same order of magnitude as those found by Koenig and by Wald. However, to be quite safe, we may take 20 per cent as the upper limit for the absorption of 510 mμ by the visual purple in the human retina after complete dark adaptation.

VII

Energy Absorbed by the Rods

It is clear now why the 54 to 148 quanta required at the cornea cannot represent the energy actually employed in vision. About 4 per cent of this incident light is reflected by the cornea; almost precisely 50 per cent is absorbed by the lens and other ocular media; and of the rest, at least 80 per cent passes through the retina without being absorbed. If corrections are made for these factors, the range of 54 to 148 quanta at the cornea becomes as an upper limit 5 to 14 quanta absorbed by the visual purple of the retina.

Visual purple is in the terminal segments of the rods, and the 10 minute circular visual field contains about 500 rods (Østerberg, 1935). Since the number of absorbed quanta is so small, it is very unlikely that any one rod will take up more than one quantum. In fact, the simplest statistical considerations show that if 7 quanta are absorbed by 500 rods, there is only a 4 per cent probability that 2 quanta will be taken up by a single rod. We may therefore conclude that in order for us to see, it is necessary for only 1 quantum of light to be absorbed by each of 5 to 14 retinal rods.[1]

It is very likely that the photodecomposition of visual purple in solution has a quantum efficiency of 1 (Dartnall, Goodeve, and Lythgoe, 1938). Our data then mean that 1 molecule of visual purple needs to be changed simultaneously in each of 5 to 14 rods, in order to produce a visual effect. This is indeed a small number of chemical events, but by virtue of its very smallness, its reality may be tested in an entirely independent manner.

VIII

Poisson Distributions

The energy calibration of the light gives merely the average number of quanta per flash. This is in the nature of the measurement, because the

[1] These data disprove the supposition made by Granit, Holmberg, and Zewi (1938) that most of the visual purple in the retina is inert as sensory substance, and that sensory impulses from the rods are "initiated by the bleaching of a thin surface film, which had to contain only an immeasurably small fraction of the total quantity present" (Granit, Munsterhjelm, and Zewi, 1939). Since the maximum visual purple concentration which the retina can achieve is able to absorb only 5 to 14 quanta at the threshold of vision, a very small fraction of the total visual purple would absorb much less than one quantum and would be ineffective for visual purposes.

thermopile records only the energy density, which is the number of quanta per second from a continuously incident light. Each flash, however, will not always deliver this average number. Sometimes the flash will yield fewer, sometimes more, quanta.

Since absorption of this group of quanta by the retina represents discrete and independent events which occur individually and collectively at random, the actual number of such retinal events which any given flash provides will vary according to a Poisson probability distribution (Fry, 1928). Let n be the number of quanta which it is necessary for the retina to absorb in order for us

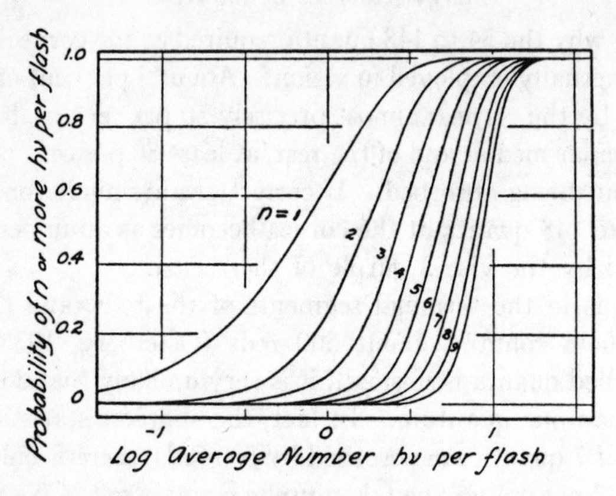

FIG. 6. Poisson probability distributions. For any average number of quanta ($h\nu$) per flash, the ordinates give the probabilities that the flash will deliver to the retina n or more quanta, depending on the value assumed for n.

to see a flash of light. Let a be the average number of quanta which any flash yields to the retina. Then the Poisson distribution states that

$$P_n = a^n/e^a n!$$

in which P_n is the probability that the flash will yield the necessary n quanta, and e is the base of natural logarithms. A special virtue of the Poisson distribution is that it has only one parameter, and is thus determined when the average number a is set. The values of P_n for various values of a and n are available in printed tables (*e.g.* Fry, 1928).

Since for us to see a flash of light the retina must absorb n quanta, we shall also see when the retina absorbs more than n quanta. From the published Poisson distributions, one can then compute the probability that n or more quanta will be delivered to the retina in a given flash when the average number of quanta delivered by that flash is known. The values computed in this way for different values of a and n are shown in Fig. 6.

There are two significant features of Fig. 6. One is that the shape of the distributions is fixed and different for every value of n. The curve becomes steeper as n increases. It follows from this that if the probability distribution could be determined by experiment, its shape would automatically reveal the value of n corresponding to it.

Another and equally important feature of Fig. 6 is that the relationship is expressed in terms of the logarithm of the average number of quanta per flash. Therefore, for comparison with the distributions in Fig. 6, the experiments need not employ the absolute values of the average number of quanta delivered per flash, but merely their relative values.

The experiments may then be made quite simply. On many repetitions of a flash of given average energy content, the frequency with which the flash is seen will depend on the probability with which it yields n or more quanta to the retina. When this frequency is measured for each of several intensities, a distribution is secured whose shape, when plotted against the logarithm of the average energy content, should correspond to one of the probability distributions in Fig. 6, and should thus show what the value of n has been.

IX

Frequency of Seeing

We have made determinations of this kind. The experimenter varies the intensity of the light by placing the wedge in specific positions unknown to the observer. The observer then elicits the flash whenever he is ready, and merely reports whether he has seen it or not. The intensities are presented in a deliberately random sequence, each for a specific number of times, usually 50. The procedure is simplified for the operator by a series of accurately made stops against which the wedge may be rapidly set in predetermined positions. A complete series in which six intensities are used requires about $1\frac{1}{2}$ hours of continuous experimentation composed of two or three periods of intensive work.

The comfort of the observer is of great importance and this must be at a maximum. It is equally important that fixation should not be rigidly continuous because this is fatiguing. Above all, the observer must be on guard to record any subjective feelings of fatigue the moment they become apparent. The experiment is much facilitated by the fact that the observer controls the occurrence of the flash, and can set it off only when he is thoroughly fixated and ready for an observation.

The data for the three observers who engaged in this experiment are given in Table V. One experiment for each observer is plotted in Fig. 7. The points in the figure record the percentage frequency with which a flash of light is seen for flashes of average quantum content shown in the abscissas. Comparison with the curves in Fig. 6 shows that the measurements are best fitted by Poisson distributions in which n is 5, 6, and 7 quanta per flash. For the two other experiments in Table IV, n is 7 and 8. No special statistical methods are necessary to determine which curve fits the data, since smaller and larger values of n are easily excluded by the simplest visual comparison.

TABLE V

Energy and Frequency of Seeing

Relation between the average number of quanta per flash at the cornea and the frequency with which the flash is seen. Each frequency represents 50 flashes, except for S. H. for whom there were 35 and 40 for the first and second series respectively.

S. H.		S. H.		S. S.		S. S.		M. H. P.	
No. of quanta	Frequency	No. of quanta	Frequency	No. of quanta	Frequency	No. of quanta	Frequency	No. of quanta	Frequency
	per cent		*per cent*		*per cent*		*per cent*		*per cent*
46.9	0.0	37.1	0.0	24.1	0.0	23.5	0.0	37.6	6.0
73.1	9.4	58.5	7.5	37.6	4.0	37.1	0.0	58.6	6.0
113.8	33.3	92.9	40.0	58.6	18.0	58.5	12.0	91.0	24.0
177.4	73.5	148.6	80.0	91.0	54.0	92.9	44.0	141.9	66.0
276.1	100.0	239.3	97.5	141.9	94.0	148.6	94.0	221.3	88.0
421.7	100.0	386.4	100.0	221.3	100.0	239.3	100.0	342.8	100.0

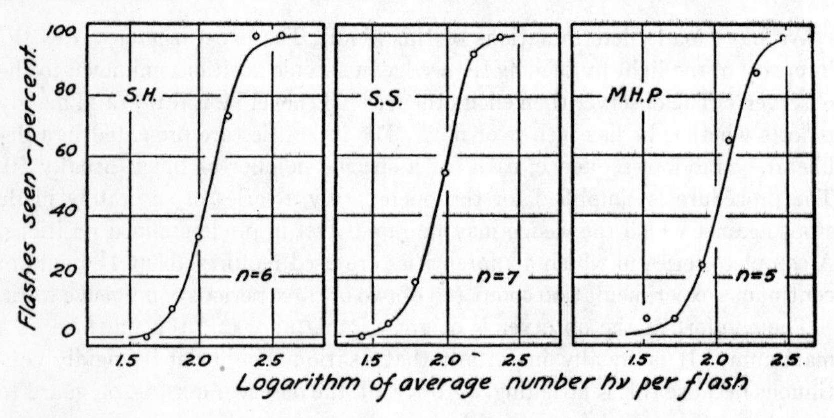

FIG. 7. Relation between the average energy content of a flash of light (in number of $h\nu$) and the frequency with which it is seen by three observers. Each point represents 50 flashes, except for S.H. where the number is 35. The curves are the Poisson distributions of Fig. 6 for n values of 5, 6, and 7.

From these measurements it is apparent that the number of critical events in the retina required to produce a visual effect lies between 5 and 8. These values are in such good agreement with the results determined by the straightforward physical measurements already described that we must consider them as the actual number of quanta absorbed by the retina.

X

Physical Fluctuation and Biological Variation

It is unimportant that the number of quanta delivered to the cornea is very much higher than the number finally involved in vision according to these measurements. This is because most of the light incident on the cornea is wasted and does not contribute to the initiation of a visual act. The amount falling on the cornea could be greatly increased by any arrangement in the eye which would act as a filter. Thus, the cornea and the lens might be pigmented, and this probably contributes to the fact that the oldest investigator (S.H.) actually requires the highest number of quanta incident on the cornea. Indeed, one might even put a filter immediately in front of the eye since the precise position of the filter in the optical system is immaterial. Nevertheless, the probability distributions would still remain the same, and by their shape would yield the magnitude of the number of events involved in the visual act.

It is necessary to amplify this point somewhat. Fluctuations are part of all physical systems, but they become significantly large only when the number of individual events, in the modern physical sense, is small. The general phenomenon is known as the shot-effect and has been studied extensively in electron emission, though it has wide application in the problem of measurements (Schottky, 1922; Barnes and Czerny, 1932). As a rough approximation, one may say that the range of variation is proportional to the square root of the number of individual events involved in the process.

In the optical system of our apparatus, the light from the ribbon filament lamp varies in intensity from moment to moment, but because the number of quanta emitted is enormous, the variation is almost too small to be measured. However, when the light intensity has been reduced first by the filters and wedge, then by the monochromators, then by the shutter, then by the ocular media, and finally by the retina itself, it has become so low that it represents only a few quanta per flash, and is therefore subject to great variation.

Barnes and Czerny (1932), and following them Brumberg and Vavilov (1933) realized that fluctuations must occur in the energy necessary for vision, and both groups of investigators looked for them. But they both missed the point of where the source of the fluctuations is and supposed it to be the energy deposited at the cornea. Brumberg and Vavilov even expected differences in the fluctuations for different wavelengths because of the greater energy required for seeing red light, for example, than blue-green light in conformity with the scotopic visibility curve of Fig. 5. However, the comparisons in Fig. 5 show that the differences in number of quanta required for vision in different parts of the spectrum record merely their relative absorption by visual purple. The number of absorbed quanta for an ultimate effect is the same regardless of wavelength and it is this number which sets the magnitude of the physical fluctuation encountered.

In deriving the curves of Fig. 6 for the quantitative statement of this physical fluctuation in terms of the Poisson probability distribution, we have made the single assumption that a constant number of quanta n must be absorbed by the retina in order for us to see a flash of light. Since it is conceivable, in view of the variability of an organism from moment to moment, that this value n is not constant, we have considered the consequences of assuming that the number n varies from time to time. The results show that biological variation is a factor of no great importance.

The situation may be best made clear by an example. Suppose that instead of n being constant, it varies between 4 and 8 quanta per visual act, and that the frequency with which 4, 5, 6, 7, and 8 quanta are necessary is distributed in terms of an ordinary probability distribution. The curves in Fig. 6 representing the frequency distributions for various values of n may then be weighted in this way and averaged. The average curve which is then secured is practically the same as the original Poisson distributions in Fig. 6, and may be fitted by the curves for $n = 4$ or 5.

Thus, when biological variation is imposed upon the physical variation, there is no change in the essential characteristics of the physical distribution. Instead, the value of n merely falls below the average of the biological distribution, and is never below the lowest value in the distribution. This tells us that when, as in Fig. 7, the measurements yield n values of 5, 6, or 7, these numbers represent lower limiting values for the physical number of quanta. In other words, the only effect which biological variation has on the physical variation is to decrease the slope of the curves in Fig. 7 and thus make the apparent number of quanta smaller than the real number.

These considerations serve for understanding the meaning of the fluctuations shown by an organism in its response to a stimulus. It has generally been assumed that a constant stimulus, when presented frequently, remains constant, and that the fluctuations in response are an expression of the variations undergone by the organism. Indeed, this is one of the tenets of psychological measurements, and an elaborate structure of psychometrics has grown up on it as a basis (*cf.* Guilford, 1936).

The present evaluation of our measurements shows, however, that at the threshold the emphasis has been in the wrong place. At the threshold where only a few quanta of energy are involved, it is the stimulus which is variable, and the very nature of this physical variability determines the variation encountered between response and stimulus. Moreover, even when biological variation is introduced, it is the physical variation which essentially dominates the relationship.

This is at the absolute threshold. One may wonder, however, whether a differential threshold at any level of intensity may also involve a small number of events which determines the differentiation, and which may therefore be subject to a similar physical variation as at the absolute threshold itself. Only experiment can decide this.

The fact that for the absolute visual threshold the number of quanta is small makes one realize the limitation set on vision by the quantum structure of light. Obviously the amount of energy required to stimulate any eye must be large enough to supply at least one quantum to the photosensitive material. No eye need be so sensitive as this. But it is a tribute to the excellence of natural selection that our own eye comes so remarkably close to the lowest limit.

SUMMARY

1. Direct measurements of the minimum energy required for threshold vision under optimal physiological conditions yield values between 2.1 and 5.7×10^{-10} ergs at the cornea, which correspond to between 54 and 148 quanta of blue-green light.

2. These values are at the cornea. To yield physiologically significant data they must be corrected for corneal reflection, which is 4 per cent; for ocular media absorption, which is almost precisely 50 per cent; and for retinal transmission, which is at least 80 per cent. Retinal transmission is derived from previous direct measurements and from new comparisons between the percentage absorption spectrum of visual purple with the dim-vision luminosity function. With these three corrections, the range of 54 to 148 quanta at the cornea becomes as an upper limit 5 to 14 quanta actually absorbed by the retinal rods.

3. This small number of quanta, in comparison with the large number of rods (500) involved, precludes any significant two quantum absorptions per rod, and means that in order to produce a visual effect, one quantum must be absorbed by each of 5 to 14 rods in the retina.

4. Because this number of individual events is so small, it may be derived from an independent statistical study of the relation between the intensity of a light flash and the frequency with which it is seen. Such experiments give values of 5 to 8 for the number of critical events involved at the threshold of vision. Biological variation does not alter these numbers essentially, and the agreement between the values measured directly and those derived from statistical considerations is therefore significant.

5. The results clarify the nature of the fluctuations shown by an organism in response to a stimulus. The general assumption has been that the stimulus is constant and the organism variable. The present considerations show, however, that at the threshold it is the stimulus which is variable, and that the properties of its variation determine the fluctuations found between response and stimulus.

BIBLIOGRAPHY

Ångström, K., Energy in the visible spectrum of the Hefner standard, *Physic. Rev.*, 1903, **17**, 302.

Barnes, R. B., and Czerny, M., Lässt sich ein Schroteffekt der Photonen mit dem Auge beobachten?, *Z. Physik.*, 1932, **79,** 436.

Brumberg, E., and Vavilov, S., Visuelle Messungen der statistischen Photonenschwankungen, *Bull. Acad. Sc. U.R.S.S.*, 1933, 919.

Buisson, H., The minimum radiation visually perceptible, *Astrophys. J.*, 1917, **46,** 296.

Chariton, J., and Lea, C. A., Some experiments concerning the counting of scintillations produced by alpha particles. Part I, *Proc. Roy. Soc. London, Series A*, 1929, **122,** 304.

Chase, A. M., and Haig, C., The absorption spectrum of visual purple, *J. Gen. Physiol.*, 1938, **21,** 411.

Dartnall, H. J. A., and Goodeve, C. F., Scotopic luminosity curve and the absorption spectrum of visual purple, *Nature*, 1937, **139,** 409.

Dartnall, H. J. A., Goodeve, C. F., and Lythgoe, R. J., The effect of temperature on the photochemical bleaching of visual purple solutions, *Proc. Roy. Soc. London, Series A*, 1938, **164,** 216.

du Noüy, P. Lecomte, Energy and vision, *J. Gen. Physiol.*, 1921, **3,** 743.

Fry, T. C., Probability and its engineering uses, New York, Van Nostrand, 1928, 476.

Graham, C. H., and Margaria, R., Area and the intensity-time relation in the peripheral retina, *Am. J. Physiol.*, 1935, **113,** 299.

Granit, R., Holmberg, T., and Zewi, M., On the mode of action of visual purple on the rod cell, *J. Physiol.*, 1938, **94,** 430.

Granit, R., Munsterhjelm, A., and Zewi, M., The relation between concentration of visual purple and retinal sensitivity to light during dark adaptation, *J. Physiol.*, 1939, **96,** 31.

Grijns, G., and Noyons, A. K., Ueber die absolute Empfindlichkeit des Auges für licht, *Arch. Anat. u. Physiol., Physiol. Abt.*, 1905, 25.

Guilford, J. P., Psychometric methods, New York, McGraw-Hill, 1936.

Hecht, S., The quantum relations of vision, *J. Opt. Soc. America*, 1942, **32,** 42.

Hecht, S., Haig, C., and Chase, A. M., The influence of light adaptation on subsequent dark adaptation of the eye, *J. Gen. Physiol.*, 1937, **20,** 831.

Hecht, S., Shlaer, S., and Pirenne, M. H., Energy at the threshold of vision, *Science*, 1941, **93,** 585.

Hecht, S., and Williams, R. E., The visibility of monochromatic radiation and the absorption spectrum of visual purple, *J. Gen. Physiol.*, 1922, **5,** 1.

Ives, H. E., The minimum radiation visually perceptible, *Astrophys. J.*, 1916, **44,** 124.

Koenig, A., Ueber den menschlichen Sehpurpur und seine Bedeuting für das Sehen, *Sitzungsber. k. Akad. Wissensch.*, Berlin, 1894, 577.

von Kries, J., and Eyster, J. A. E., Über die zur Erregung des Sehorgans erforderlichen Energiemenzen, *Z. Sinnesphysiol.*, 1907, **41,** 394.

Langley, S. P., The bolometer and radiant energy, *Proc. Am. Acad. Sc.*, 1881, **16,** 342.

Langley, S. P., Energy and vision, *Phil. Mag.*, 1889, **27,** series 5, 1.

Ludvigh, E., and McCarthy, E. F., Absorption of visible light by the refractive media of the human eye, *Arch. Ophth.*, Chicago, 1938, **20,** 37.

Lythgoe, R. J., The absorption spectra of visual purple and of indicator yellow, *J. Physiol.*, 1937, **89,** 331.

Østerberg, G., Topography of the layer of rods and cones in the human retina, *Acta Ophth., Copenhagen*, 1935, suppl. 6, 106 pp.

Piper, H., Über Dunkeladaptation, *Z. Psychol. u. Physiol. Sinnesorgane*, 1903, **31**, 161.

Reeves, P., The minimum radiation visually perceptible, *Astrophys. J.*, 1917, **46**, 167.

Roggenbau, C., and Wetthauer, A., Über die Durchlässigkeit der brechenden Augenmedien für langwelliges Licht nach Untersuchungen am Rindsauge, *Klin. Monatsbl. Augenheilk.*, 1927, **79**, 456.

Russell, H. N., The minimum radiation visually perceptible, *Astrophys. J.*, 1917, **45**, 60.

Shottky, W., Zur Berechnung und Beurteilung des Schroteffektes, *Ann. Physik.*, 1922, **68**, 157.

Shlaer, S., A photoelectric transmission spectrophotometer for the measurement of photosensitive solutions, *J. Opt. Soc. America*, 1938, **28**, 18.

Wald, G., Area and visual threshold, *J. Gen. Physiol.*, 1938 *a*, **21**, 269.

Wald, G., On rhodopsin in solution, *J. Gen. Physiol.*, 1938 *b*, **21**, 795.

Wentworth, H. A., A quantitative study of achromatic and chromatic sensitivity from center to periphery of the visual field, Psychological Monographs, No. 183, Princeton, New Jersey, and Albany, New York, Psychological Review Co., 1930, **40,** 189 pp.

Zwaardemaker, H., Die physiologisch wahrehmbaren Energiewanderungen, *Ergebn. Physiol.*, 1905, **4**, 423.

JOURNAL OF THE OPTICAL SOCIETY OF AMERICA VOLUME 53, NUMBER 1 JANUARY 1963

The Problem of Visual Excitation*

GEORGE WALD, PAUL K. BROWN, AND IAN R. GIBBONS†
Biological Laboratories of Harvard University, Cambridge

This paper attempts to come to grips with the problem, how the action of light on a visual pigment results in a nervous excitation. The only action of light in vision is to isomerize retinene, the chromophore of the visual pigments, from the 11-cis to the all-trans configuration. This change triggers a progressive opening-up of the protein structure, exposing new reactive groups. Since the absorption of one photon by one molecule of visual pigment may stimulate a dark-adapted rod, some large amplification process is needed between the act of absorption and the response. This may be an enzymatic catalysis, or the consequence of puncturing a critical membrane. A microspectrophotometric study of retinas and single rods shows the outer segment to have a quasi-crystalline structure, in which the visual pigments are almost perfectly oriented, and even "free" molecules capable of diffusion maintain a degree of orientation. Examination of mud puppy retinas in the electron microscope has revealed several new aspects of structure: (1) A system of cytoplasmic filaments ("dendrites") springing from the inner segments of the rods and cones and standing like palisades around the outer segments. These may facilitate exchanges of material between the inner and outer segments. (2) Systems of particles in the membranes of the dendrites and pigment epithelium processes, which may be involved in interchanges of material with the outer segments. (3) A system of particles in crystalline array in the rod lamellae, which may contain the visual pigment. If so, measurements of the visual pigment *in situ* show that each particle should contain about 50 molecules of pigment. Such typically solid-state processes as exciton migration and photoconduction probably have at most very limited scope in the outer segments of rods and cones. The seat of excitation is probably the plasma membrane which envelops rod outer segments and composes also the lamellae in cones.

PHOTOCHEMISTRY OF VISUAL PIGMENTS

THE past years have taught us something of the chemistry of the visual pigments and the action of light upon them. The great problem now is to try to understand how these processes result in a nervous excitation. Such a question is meaningless in the context of extracts and solutions. Its answer must be sought in the retina itself, and in the relationships of the visual pigments to the highly organized structures of which they form part.

All the known visual pigments share a common type of structure and response to light.[1] Each is composed of a chromophore, 11-*cis* retinene$_1$ or retinene$_2$, the aldehyde of 11-*cis* vitamin A_1 or A_2, joined to a specific type of protein, rod or cone opsin:

$$\text{Vitamin } A_1 \underset{\text{DPN}\cdot\text{H}}{\overset{\text{DPN}^+}{\rightleftharpoons}} \text{Retinene}_1 \begin{cases} + \text{rod opsin} \overset{\text{light}}{\rightleftharpoons} \text{Rhodopsin} & 500 \\ + \text{cone opsin} \overset{\text{light}}{\rightleftharpoons} \text{Iodopsin} & 562 \end{cases}$$

(alcohol dehydrogenase)

$$\text{Vitamin } A_2 \underset{\text{DPN}\cdot\text{H}}{\overset{\text{DPN}^+}{\rightleftharpoons}} \text{Retinene}_2 \begin{cases} + \text{rod opsin} \overset{\text{light}}{\rightleftharpoons} \text{Porphyropsin} & 522 \\ + \text{cone opsin} \overset{\text{light}}{\rightleftharpoons} \text{Cyanopsin} & 620 \end{cases}$$

(Last column: Approx λ_{max} ($m\mu$))

The structural formulas of all-*trans* and 11-*cis* vitamin A and retinene follow. Vitamin A_2 and retinene$_2$ differ only in possessing an added double bond in the 3,4 position in the ring:

all-*trans* Vitamin A: ring C1–C6 (H_2C_4, H_2C_3, C2 H_2, C1 with two CH_3, C5 with CH_3), side chain C7 (H), C8 (H), C9 (CH_3), C10 (H), C11 (H), C12 (H), C13 (CH_3), C14 (H), C15 CH_2OH

Retinene: C14 (H)–C15 (H)=O

* Invited paper presented at the Symposium on Physiological Optics, Joint Session of the Armed Forces–NRC Committee on Vision, the Inter-Society Color Council, and the Optical Society of America, 14–15 March 1962, Washington, D. C.

† These investigations were supported in part with funds from the Rockefeller Foundation, the U. S. Office of Naval Research, the National Science Foundation, and the Public Health Service.

[1] Recent discussions of the chemistry of visual processes and their bearing on visual excitation can be found in H. J. A. Dartnall, *The Visual Pigments* (Methuen and Company, London, 1957); G. S. Brindley, *Physiology of the Retina and the Visual Pathway* (Edward Arnold, London, 1960); R. A. Morton and G. A. J. Pitt, Progr. in Chem. Org. Nat. Prods. **14**, 244 (1957); W. A. H. Rushton Progr. in Biophys. and Biophys. Chem. **9**, 239 (1959); G. Wald and R. Hubbard in *The Enzymes*, edited by P. D. Boyer, H. Lardy, and K. Myrbäck (Academic Press Inc., New York, 1960), 2nd ed., Vol. 3, Chap. 19, p. 369; and G. Wald, in *Life and Light*, edited by W. D. McElroy and B. Glass, (Johns Hopkins Press, Baltimore, 1961), p. 724; also Vitamins and Hormones **18**, 417 (1961). Detailed references to the literature will be found in these books and papers.

```
  CH3  H   CH3  H
  |    |    |   |
  C    C    C   C    H
 / \\ / \\ / \\ / \\ /
H2C  C    C    C    C      11-cis (neo-b)
 |   |    |    |    |
 |   |    H    H    |
H2C  C—CH3          C    H
  \ /           //  \\ /
   C  CH3     CH3    C
   |                 |
   H2               CH2OH
```

The absorption of light by a visual pigment has the immediate effect of changing the configuration of the chromophore (i.e., isomerising it) from 11-*cis* to all-*trans*. *The only action of light in vision is this isomerization.* Everything else that happens, physicochemical and physiological, represents "dark" consequences of this initial reaction.[2]

The isomerization of the chromophore is followed immediately by a spontaneous opening-up of the structure of opsin. Some years ago we identified what we thought to be the first step in this rearrangement, which we called lumirhodopsin, by irradiating rhodopsin in a glycerol-water glass at temperatures of −50° to −100°C, at which lumirhodopsin is stable.[3]

Recently, however, Yoshizawa and his collaborators at Osaka University began to find evidence of a still

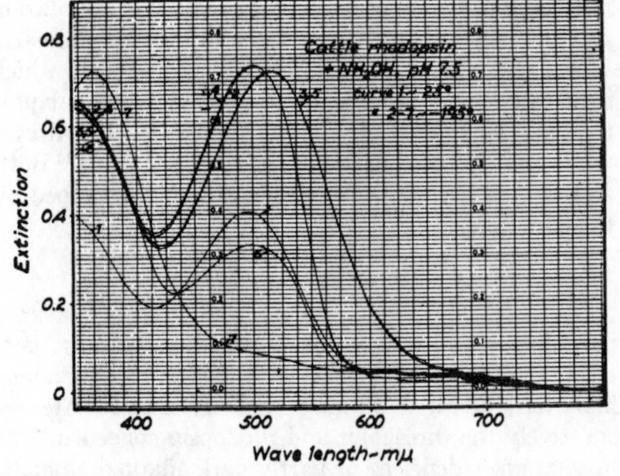

FIG. 1. Photoreversibility of rhodopsin and pre-lumirhodopsin at liquid-nitrogen temperature. Cattle rhodopsin in a glycerol-water mixture (55:45) containing 0.1M hydroxylamine. (1) Rhodopsin at 25°C. (2) Cooled to −195°C. (3) Irradiated at 440 mμ to a steady-state mixture of rhodopsin and pre-lumirhodopsin (i.e., chromophores respectively 11-*cis* and all-*trans*). (4) Irradiated at 600 mμ; the pre-lumirhodopsin is photo-isomerized back to rhodopsin. (5) Reirradiated at 440 mμ, reproducing the same steady-state mixture as in 3. (6) Warmed to 25°C in the dark, so allowing the pre-lumirhodopsin in this mixture to bleach to a mixture of retinene oxime and opsin, leaving the rhodopsin as a residue. Then recooled to −195°C to yield the spectrum shown. This is then the spectrum of the rhodopsin present in the steady-state Mixture 5. Subtracting it from 5 yields the spectrum of pre-lumirhodopsin itself, with λ_{max} about 540 mμ. (7) Finally the solution is warmed again to 25°C, bleached to completion, and recooled. (From Yoshizawa and Wald.[6])

[2] R. Hubbard and A. Kropf, Proc. Natl. Acad. Sci. U. S. **44**, 130 (1958); Ann. N. Y. Acad. Sci. **81**, 388 (1959).

[3] G. Wald, J. Durell and R. C. C. St. George, Science **111**, 179 (1950).

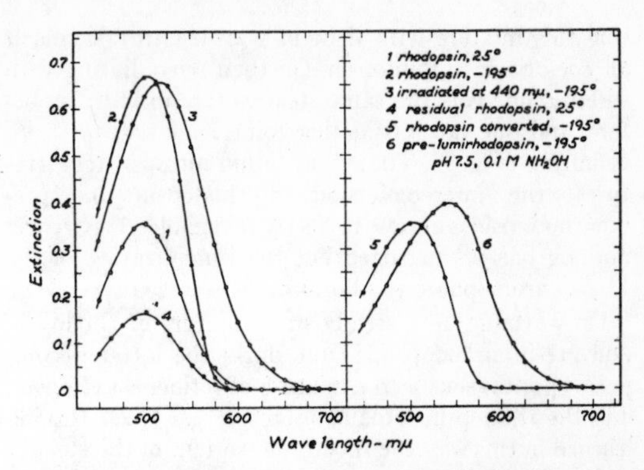

FIG. 2. Conversion of rhodopsin to pre-lumirhodopsin at liquid nitrogen temperature. Cattle rhodopsin in a glycerol-water mixture (55:45) containing 0.1M hydroxylamine. At the end of this experiment, the preparation was bleached completely at room temperature to a mixture of retinene oxime, opsin, and whatever impurities the preparation contained, comparable to Curve 7 in Fig. 1. Since retinene oxime does not absorb light appreciably above about 440 mμ, the absorption of this product above that wavelength is due entirely to impurities. This spectrum has been subtracted from all the others as recorded to yield the spectra of the pure pigments, shown here. (1) Rhodopsin at 25°C (λ_{max} 498 mμ). (2) Cooled to −195°C (λ_{max} about 506 mμ). (3) Irradiated at 440 mμ to yield a steady-state mixture of rhodopsin and pre-lumirhodopsin (λ_{max} about 518 mμ). (4) Warmed to 25°C in the dark, the pre-lumirhodopsin in 3 bleaches spontaneously to retinene oxime and opsin, leaving rhodopsin. This shows, therefore, the fraction of the original rhodopsin present in the steady-state Mixture 3. Subtracting this from 3 yields the spectrum of pre-lumirhodopsin 6 with λ_{max} about 543 mμ [i.e., *6=3−(4/1×2)*]. The difference (*1-4*) represents the rhodopsin converted to pre-lumirhodopsin, the spectrum of which at −195°C (Curve 5) is obtained by taking this proportion of Curve 2—i.e. [*5=2×(1-4)/1.*] A number of such measurements show that the maximal extinction (E_{max}) of pre-lumirhodopsin is on the average 1.13 times as high as that of rhodopsin at −195°C. (From Yoshizawa and Wald.[6])

earlier product, which we now call pre-lumirhodopsin. This is formed on irradiating rhodopsin at liquid-nitrogen temperatures (about −190°C).[4] Pre-lumirhodopsin is stable up to about −140°; above this temperature it goes rapidly to lumirhodopsin. Grellmann, Livingston, and Pratt[5] have extended these observations and have studied the kinetics of this transformation by flash-photolytic procedures. Simultaneously Yoshizawa, now at our laboratory, has carried out a detailed spectrophotometric study of these reactions (cf. Figs. 1 and 2).[6]

The irradiation of rhodopsin in a glycerol-water mixture (2:1) at −195°C, with light of wavelength 440 mμ, isomerizes the rhodopsin's 11-*cis* chromophore to a steady-state mixture of 11-*cis* and all-*trans* retinene. The 11-*cis* fraction (about 43%) is rhodopsin: the all-*trans* fraction (57%) is pre-lumirhodopsin. On irradia-

[4] T. Yoshizawa and Y. Kito, Nature **182**, 1604 (1958); T. Yoshizawa, Y. Kito and M. Ishigami, Biochim. et Biophys. Acta **43**, 329 (1960); Y. Kito, M. Ishigami and T. Yoshizawa, *ibid.* **48**, 287 (1961).

[5] K.-H. Grellmann, R. Livingston, and D. Pratt, Nature **193**, 1258 (1962).

[6] T. Yoshizawa and G. Wald, (to be published).

ting this mixture with light of wavelength 600 mμ, it all goes back to rhodopsin. On then reirradiating with 440 mμ, one gets the same steady-state mixture as before, and one can repeat this going back and forth indefinitely (Fig. 1). That is, at liquid-nitrogen temperatures, the interconversion of rhodopsin and pre-lumirhodopsin is almost perfectly reversible. Apparently nothing has yet occurred but the isomerization of the 11-*cis* chromophore to all-*trans*.

On warming up a steady-state mixture of rhodopsin and pre-lumirhodopsin in the dark, the latter decomposes spontaneously to a mixture of retinene and opsin, and the rhodopsin remains intact (Figs. 1, 2). Having learned in this way the rhodopsin content of the steady-state mixture, one can by subtraction obtain the spectrum of pre-lumirhodopsin. This has λ_{max} 543 mμ (Fig. 2).

At temperatures above −140°C, pre-lumirhodopsin goes over in the dark to lumirhodopsin; and above −20°C this in turn yields metarhodopsin. These intermediates appear to represent progressive stages in the opening-up of opsin structure. By the time metarhodopsin has formed, two new sulfhydryl (−SH) groups and one H^+−binding group with pK about 6.6 (imidazole?) have been exposed. By this time excitation has also occurred. Metarhodopsin then slowly hydrolyzes—except in some invertebrate retinas—to a mixture of all-*trans* retinene and opsin. These relationships are summarized diagrammatically in Fig. 3.

From the point of view of visual excitation, the critical event seems to be the exposure of new and chemically active groupings on opsin, as the result of the spontaneous opening-up of the opsin structure that follows the isomerization of the retinene chromophore by light. The critical product is lumi- or metarhodopsin. Bleaching—the hydrolysis to retinene and opsin—has not yet occurred, and even if it does occur later, is much too slow a process to have a direct part in excitation.

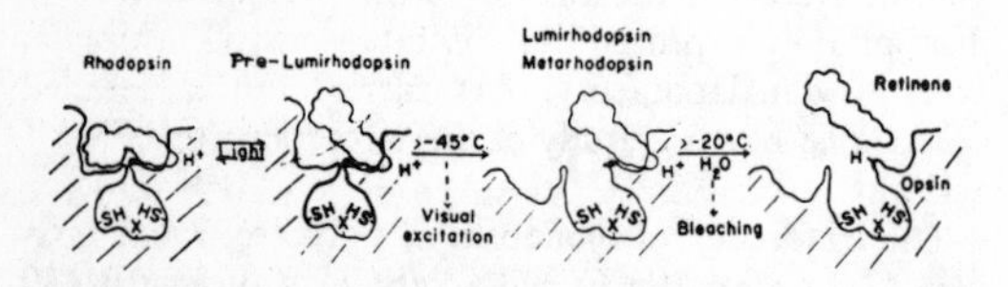

FIG. 3. The action of light on rhodopsin. The absorption of light by rhodopsin isomerizes its 11-*cis* chromophore to the all-*trans* configuration, yielding as a first product the all-*trans* chromoprotein, pre-lumirhodopsin. At temperatures above −140°C, this goes over to lumirhodopsin, and above −45°C the latter yields metarhodopsin, both products representing progressive stages in the opening-up of structure of the protein opsin. This unfolding of structure exposes two sulfhydryl (−SH) groups and one H^+–binding group, and may be responsible for triggering visual excitation. Vertebrate metarhodopsins are unstable, and above −20°C hydrolyze to all-*trans* retinene and opsin. The hydrolysis process is responsible for the major loss of color (i.e., bleaching) of the visual pigments. (Modified from Hubbard and Kropf.[2])

VISUAL THRESHOLD VS VISUAL PIGMENT CONCENTRATION

A fundamental relationship now casts a bridge between the chemistry and the physiology of vision. This is the relationship that connects the visual threshold with the concentration of visual pigment. Selig Hecht attempted over many years to arrive at such a relationship analytically. For a time he entertained the view that visual-pigment concentration might be proportional to 1/log threshold,[7] but this proved unsatisfactory, and later he abandoned it. He stated his final position as follows: "In general, human visual dark adaptation runs roughly parallel with the accumulation of visual purple in the dark adapting animal retina. Efforts to study this parallelism experimentally have not been successful. . . . In fact, even the sensitivity data of human dark adaptation, though very precise, are still without adequate theoretical treatment in terms of visual purple concentration changes."[8]

It has recently been shown that the visual threshold is connected with the concentration of visual pigment by a very simple relationship, at least to a first approximation, and over a wide middle range. The logarithm of the visual sensitivity (i.e., log 1/threshold or −log threshold) is linear with rhodopsin concentration. This relationship was first proposed on the basis of qualitative comparisons between the course of dark adaptation and of visual-pigment synthesis in various animals.[9] It was then demonstrated directly in the rat, in which it holds equally throughout the course of dark adaptation, and for the rise of visual threshold that constitutes night blindness in vitamin A deficiency (Fig. 4).[10] Both sets of measurements are reasonably well described by the equation

$$1-(R_t/R_0)=0.28\log(I_t/I_0),$$

in which I_0 and R_0 are constants representing respectively the threshold and rhodopsin concentration in normal, dark-adapted animals; and I_t and R_t are respectively the thresholds and rhodopsin concentrations in vitamin A deficient or partly dark-adapted animals.

This relationship (with, of course, other constants) was later confirmed by Rushton in a study of human rod dark adaptation,[11] and Rushton's measurements of the regeneration of cone pigments in the human fovea

[7] S. Hecht, in *Handbook of General Experimental Psychology*, edited by C. Murchison (Clark University Press, Worcester, 1934), p. 728.

[8] S. Hecht, Ann. Rev. Biochem. **11**, 465 (1942).

[9] G. Wald, in "Visual Problems of Colour," National Physical Laboratory Symposium No. 8 (Her Majesty's Stationery Office, London, 1957), Vol. 1, p. 7.

[10] J. E. Dowling and G. Wald, Proc. Natl. Acad. Sci. U. S. **46**, 587 (1960); **44**, 648 (1958) [also in "Vitamin Metabolism," *Proceedings of the Fourth International Congress of Biochemistry* (Pergammon Press, London, 1959), p. 185]; J. E. Dowling, Nature **188**, 114 (1960).

[11] W. A. H. Rushton, in *Light and Life*, edited by W. D. McElroy and B. Glass, (Johns Hopkins Press, Baltimore, 1961), p. 706.

would lead one to suppose that the same type of relationship also holds in cone vision.[9]

It should be noted at once that the linearity of this relationship is only an approximation that holds reasonably well over the main body of the function. As visual concentration falls to zero, the threshold must rise to infinity, whereas this equation has it rise only to a modest limiting value. A further adjustment may also be needed in the region near 100% rhodopsin, in accord with evidence that the bleaching of the first few percent of rhodopsin causes a disproportionately large rise of threshold,[12] almost surely more evident in the light-adapting retina, in which neural activity contributes to the rise of threshold,[9] than in the dark-adapting or adapted retina, which is relatively quiet. The true relationship is very likely a flattened reversed S, concave to the origin at high concentrations of rhodopsin and convex to the origin at low concentrations (Fig. 4, broken line). It is significant that in Fig. 4 almost all the points at the left lie above the straight line as drawn, and almost all those to the right lie below it. The S-shaped curve fits these points better than the straight line. Useful as the linear relationship is, therefore, it does not represent the final solution of this problem.

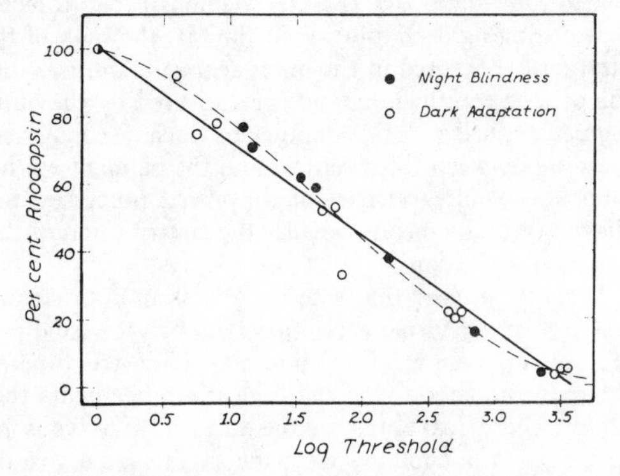

FIG. 4. Relation between the rhodopsin content of the rat retina and the logarithm of the visual (ERG) threshold, in animals becoming night blind owing to vitamin A deficiency, and in normal animals during dark adaptation. The relationship is almost identical in both cases, the log threshold varying almost linearly over a wide middle range with the retinal content of rhodopsin (solid line). A flat, S-shaped curve (broken line) fits the measurements more accurately; and the rise of threshold to infinity as rhodopsin falls to zero is, of course, essential. (Modified from Dowling and Wald.[10])

ONE PHOTON TO EXCITE

A dominant consideration in the excitation problem is that the absorption of a single quantum of light can excite a dark-adapted rod. We owe this realization initially to Hecht, Shlaer and Pirenne,[13] and it is now generally conceded on various grounds. Perhaps it is demonstrated most simply by such a computation as in Table I, of the number of photons absorbed by single rods at various levels of brightness. This shows that at the absolute threshold of human vision—about 10^{-6} mL—a single rod absorbs a photon on the average about once in 40 min. Even at 0.01 mL a rod absorbs a photon on the average less than once per 1/4 sec. Above this, luminance cones begin to dominate vision. It is clear that if the rods were not single-photon detectors, they would be of little value in vision.

[12] W. A. H. Rushton and R. D. Cohen, Nature **173**, 301 (1954); G. Wald, Science, **119**, 887 (1954).

[13] S. Hecht, S. Shlaer, and M. H. Pirenne, J. Gen. Physiol. **25**, 819 (1942).

MECHANISMS OF AMPLIFICATION

The absorption of a single quantum of light by a rod can excite only one molecule of rhodopsin, thereby isomerizing a single retinene chromophore from 11-*cis* to all-*trans*, and exposing, as said, two sulfhydryl groups and one hydrogen ion-binding group. Clearly this is all too little in itself to account for excitation or for any of the electrical signs of response. One must assume that a large amplification process intervenes between the action of light on a visual pigment and the response of a receptor cell.[14]

Two ways have been suggested in which such an amplification might occur.[15] The first is particularly attractive from a biochemical point of view. Most digestive enzymes are liberated in an inactive form, as proenzymes or zymogens (pepsinogen, trypsinogen, etc.), which are converted to active enzymes always by the same process, the splitting-off from the zymogen of a small fragment of its structure, so uncovering the active catalytic center. This process is sometimes called "uncorking." We have already noted that the effect of light on a visual pigment, by isomerizing retinene, exposes new groups on opsin. It is possible that the visual pigment is a zymogen, "corked" by the close attachment of the retinene chromophore. The absorption of a photon, by opening this attachment, may expose a catalytic center, converting the molecule to an active enzyme. An enzyme can turn over many times its equivalent of sub-

[14] The amount of amplification needed is of some interest. A hint of this is contained in the following remarks of Sir John Eccles, written to G. Wald (January 30, 1959) in reply to a question concerning the minimum ion transfer that accompanies the excitation of a nerve fiber or cell. Eccles wrote: "I actually gave one such calculation in my book on the Neurophysiological Basis of Mind (p. 77). There you will see that the actual ionic movement across the end-plate membrane produced by a single impulse would be as high as 3×10^{-14} mole. With a large nerve cell, such as a motoneurone, the approximate figure I can give for a powerful synaptic activation is 10^{-16} mole. That is still an enormous number of ions. On the same basis I can answer your question by assuming a capacity of 1 μF/cm^2 for your rod membrane. The total capacity of 20 square micra (i.e., the approximate area of one transverse membrane in a frog rod outer segment) would then be 2×10^{-13}F. If we were to assume under the most favorable conditions that for generating an impulse the voltage change need be only 5 millivolts, the quantity of electricity to cross the membrane would be 10^{-20} mole of univalent ions. So my answer is that you fail by a factor of some thousands. . . ."

[15] G. Wald, in *Enzymes: Units of Biological Structure and Function*, edited by O. Gaebler) (Academic Press Inc., New York, 1956), p. 355.

Table I. Absorption of light by rhodopsin in human rods at various luminances.

The first two columns show the flux density at the retina, corrected for the opening of the natural pupil, and allowing for an ocular transmission of 0.5. Converted to quanta per sec on the basis that for scotopic vision at 507 mμ, 1 lumen = 5.72×10^{-4} W = 1.37×10^{-4} cal/sec = 1.47×10^{15} quanta/sec. Cross-sectional area of human rod outer segment, 2.54×10^{-6} mm². The outer segment contains about 4×10^{7} molecules of rhodopsin when dark adapted, on the assumption that it absorbs about 30% of incident light of wavelength 507 mμ [W. A. Rushton, J. Physiol. 25, 819 (1942)]. In the last column, no allowance is made for the bleaching of rhodopsin, which at higher luminances would lower the numbers of quanta absorbed.

Field luminance (millilamberts)	Flux density at retina: Lumens per sq mm	Flux density at retina: Quanta per sec per sq mm	Quanta per sec per rod: Incident	Quanta per sec per rod: Absorbed
10^{-6}	3.6×10^{-13}	510	0.0013	0.00043
10^{-5}	3.5×10^{-12}	5000	0.0127	0.0043
10^{-3}	3.2×10^{-10}	4.6×10^{5}	1.17	0.39
10^{-2}	2.9×10^{-9}	4.18×10^{6}	10.6	3.6
0.1	2.4×10^{-8}	3.4×10^{7}	87	29
1.0	1.7×10^{-7}	2.38×10^{8}	605	202
10	9.5×10^{-7}	1.4×10^{9}	3430	1144
100	4.8×10^{-6}	7×10^{9}	17,750	5917
1000	2.8×10^{-5}	4×10^{10}	103,000	34,334

strate. It is the biochemical equivalent of an amplifier. If the product of its catalysis were a second enzyme, that would constitute a second stage of amplification.

This would be a happier thought if one knew how to go on with it. If rhodopsin is a zymogen, and metarhodopsin an active enzyme, what process does metarhodopsin catalyze? If we had some idea what we would like such an enzyme to do, one might test for its activity, but we have as yet no such idea.

An alternative possibility rests on the realization that the transverse membranes that constitute the microstructure of the outer segments are made in large part of visual pigment. The attack of light on a molecule of visual pigment might, in effect, punch a unimolecular hole in such a membrane. That might permit a flow of ions, resulting in a local depolarization or loss of resistance sufficient to excite.

NATURE OF THE RECEPTOR RESPONSE

These considerations raise a further problem. How does a rod or cone respond to excitation? What is one trying to explain? Is the response limited to a graded, maintained change of potential, or does it go on to yield an all-or-nothing spike? We have as yet no definitive answer to this question. All-or-nothing spikes are certainly picked up in the vertebrate retina at the level of the ganglion cells. Brindley[1] has reported finding spikes perhaps 40% of the time in the inner nuclear layer of the frog retina, and perhaps 20% of the time at the outer ends of the rods and cones. Brown and Wiesel[16] have reported spikes from the inner nuclear layer in the cat retina. On the other hand, we now have extensive records of slow-graded potentials evoked from the region of synaptic connections between cones and bipolar cells (the outer reticular layer) in fish retinas.[17] It would be important to know at what level in the visual pathway spikes first arise, and particularly whether the responses of the rods and cones include spikes, or are confined to slow, graded changes of potential. Until this issue is decided, one cannot hope to come to clear conclusions regarding the excitation process.

ROLE OF THE PIGMENT EPITHELIUM

A final element in the excitation problem involves the role of the pigment epithelium.[18] This is a true retinal layer, that originates embryonically from cells adjacent to, and of the same kind, as those that form the rods and cones. Dowling[19] has recently demonstrated directly that during light adaptation in the rat, the bulk of the vitamin A liberated in the outer segments diffuses into the pigment epithelium, and is recaptured by the outer segments during dark adaptation. Such exchanges of material between the receptors and the pigment epithelium are an integral part of the visual processes, but they may by no means exhaust the contribution of this tissue to excitation.

It must be clear that a primary task in approaching the problem of visual excitation is to fit the visual pigments and their reactions into the microstructures of the rods and cones—the anatomical arrangements that exploit those reactions—in the form of a nervous response. In what follows we come at this from two directions: (a) the spectrophotometric examination of the visual pigments and their products *in situ*; and (b) an electron-microscope study of the ultrastructures of the rods, cones, and pigment epithelium. Such a visual pigment as rhodopsin, if spherical, is 40–50 Å in diameter. This is also the thickness of a typical transverse membrane in an outer segment. At this level of analysis the chemistry fuses with the anatomy, and one finds oneself engaged in what may be characterized as the "chemo-anatomical" study of visual excitation.

VISUAL PIGMENTS *IN SITU*[20]

In these experiments we follow closely in the footsteps of Eric Denton, who, with simple and ingenious

[16] K. T. Brown and T. N. Wiesel, Am. J. Ophthalmol **46**, No. 3, 91 (1958).

[17] G. Svaetichin, Acta Physiol. Scand. **39**, Suppl. 134, 17 (1956); T. Tomita, Japan, J. Physiol. **7**, 80 (1957); E. F. MacNichol, M. L. Wolbarsht and H. G. Wagner, in *Light and Life*, edited by W. D. McElroy and B. Glass (Johns Hopkins Press, Baltimore, 1961), p. 795.

[18] Cf. G. Wald, Exptl. Cell Research Suppl. 5, 389 (1958).

[19] J. E. Dowling, Nature **188**, 114 (1960).

[20] These measurements have already been described at a Symposium on Receptor Mechanisms held in Birmingham, England, in September 1961 in a paper by the present authors in *Biological Receptor Mechanisms*, edited by J. K. Grant and D. J. Bell (Cambridge University Press, New York, 1962), p. 32. They will be reported in detail elsewhere by P. K. Brown and G. Wald. Cf. also the paper by P. A. Liebman, Biophys. J. 2, 161 (1962).

procedures, has shown that the absorption spectra of visual pigments measured in whole retinas rival in accuracy the best spectra obtained in solution; and he has also explored the molecular orientation of the pigments in the outer segments.[21]

For these measurements one of us designed a microspectrophotometer which permits the accurate and rapid recording of absorption spectra in small areas of tissue.[22] A special compartment built into the light path of a Cary 14 recording spectrophotometer holds a low-magnification quartz microscope ("macroscope") with which measurements can be made between 300 and 700 mμ in fields 0.1–1 mm in diameter; or this can be replaced by a conventional compound microscope, with which spectra can be measured from 350–650 mμ in fields as small as 4 μ in diameter. With the macroscope we measure the spectra of small areas of retina, and with the microscope the spectra of single rods.

The latter have been measured with the light passing through them axially and transversely. The outer segment of a frog rod is cylindrical, and on the average is about 6 μ wide and 50 μ long. The field of measurement is delimited by masking down the greatly magnified projected image of the rod at the level of the photocell compartment of the instrument. Transverse measurements are performed on isolated outer segments lying on their sides in suspension. For these an oblong mask is used that delimits an area 4×40 μ at the level of the specimen. For axial measurements a piece of whole retina is mounted on a slide with the outer segments pointing directly upward, and a circular mask is positioned within the boundaries of a single rod so as to delimit an area 4 μ across at the level of the specimen.

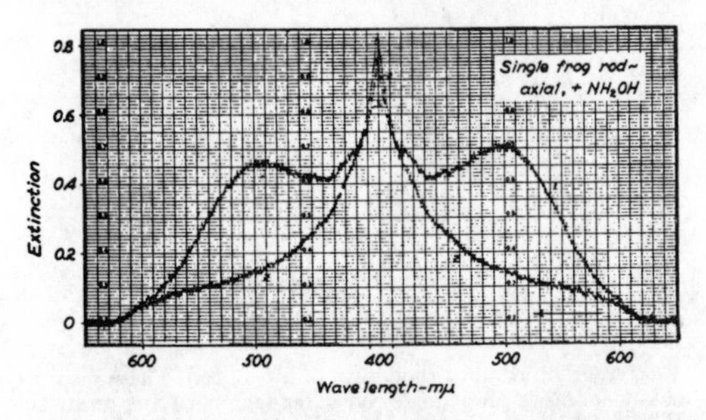

FIG. 5. Absorption spectra of a single frog rod, recorded with the light passing down the long axis of the outer segment. A piece of dark-adapted frog retina was mounted in 55% glycerol in Ringer solution containing 0.1M hydroxylamine. The field of measurement was masked down to 4 μ at the level of the specimen, so as to fall within the boundary of a single rod. Since some rhodopsin bleaches during the measurement, Spectrum 1 was recorded from the red to the violet and immediately back to the red (see arrow). Then the retina was bleached, and Spectrum 2 recorded. The subtraction of 2 from 1 yields a pair of difference spectra, each slightly distorted in opposed senses. These were averaged to yield the result shown in Fig. 6.

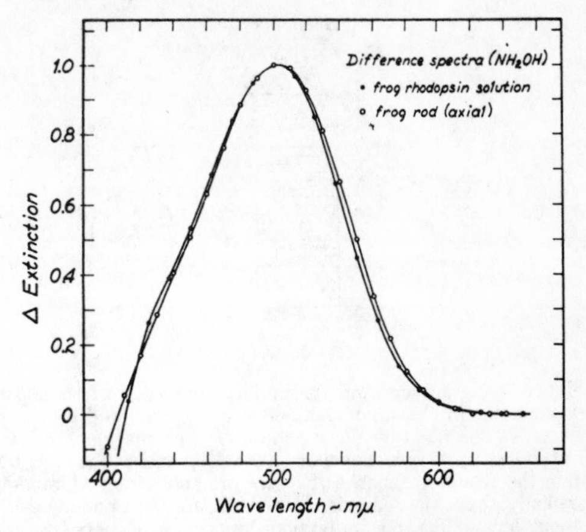

FIG. 6. Comparison of the averaged difference spectrum of a single rod obtained from Fig. 5 (λ_{max} 504 mμ) with the difference spectrum of frog rhodopsin in 2% digitonin solution in the presence of 0.1M hydroxylamine (λ_{max} 502 mμ).

Further details of the instrument and procedure and sample measurements have already been described.[22] We attempt no more here than to summarize some of the more interesting results.

We have tried, throughout this work, to maintain the standards of good spectrophotometry in solution. The first thing to say of our measurements is that they contain no surprises. The absorption spectra of rhodopsin and porphyropsin measured in single rods are usually identical with those measured in large patches of retina, and differ from the spectra of these pigments in solution only by being displaced 2–3 mμ toward the red (Figs. 5, 6).

It has sometimes been thought that the individual receptor cells may vary widely in the pigments they contain and hence in their spectral properties, and that measurements on whole retinas or extracts may represent no more than statistical summaries, in which large individual differences have been obliterated by averaging. Our measurements, at least so far as the rods of frogs and mud puppies are concerned, lend no support to this view.

We have also examined the orientation of rhodopsin in the outer segments of frog rods. The classic observations in this field were performed by Schmidt of Giessen[23] with the polarizing microscope on the same material, and have since been confirmed and considerably extended by Denton.[21] One can conclude from these experiments that to a first approximation the molecules of rhodopsin are oriented in the structure of the outer segment so that their chromophores lie transversely, i.e., perpendicularly to the long axis of the rod (Fig. 7). They can be thought of as lying within a series of transverse planes, comparable with the transverse layers of which the outer segment is constructed anatomically (Fig. 10 and below).

[21] E. J. Denton, Proc. Roy. Soc. (London) **B150**, 78 (1959).
[22] P. K. Brown, J. Opt. Soc. Am. **51**, 1000 (1961).
[23] W. J. Schmidt, Kolloid Z. **85**, 137 (1938).

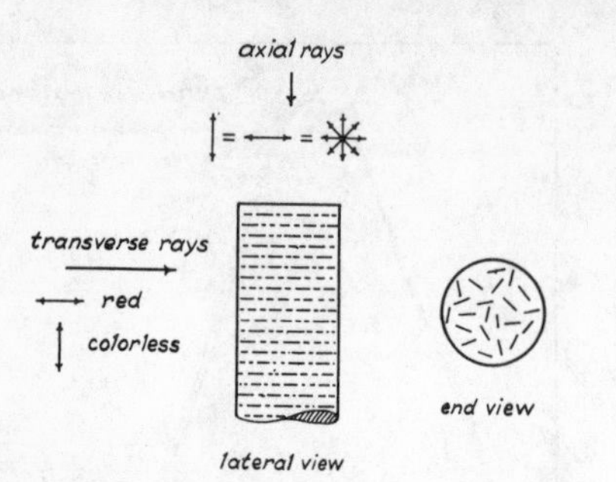

FIG. 7. Diagram showing the major orientation of rhodopsin chromophores in the outer segment of a frog rod, and its consequences, as described by W. J. Schmidt.[23] The chromophores lie, for the most part, perpendicular to the long axis of the rod; but within the transverse planes they are oriented either at random or radially about the rod axis. As a result, for light passing down the long axis of the rod, its normal direction in the eye, the rod displays the same color in all planes of polarized light, and in unpolarized light. In plane-polarized light passing through its side, however, the rod looks red when the electric vector of polarization is in the transverse direction, and colorless when it is axial.

This arrangement has important consequences. All organic pigments owe their color to the presence of conjugated systems of alternate single and double bonds. A pigment absorbs light to the degree that its conjugated system lies perpendicular to the light ray, and parallel with the electric vector of the wave motion of the light. In the eye, light normally passes down the long axis of the rods. The transverse orientation of the visual-pigment chromophores thus spreads them out in the position most favorable for absorption. Within the transverse planes in which all the chromophores may be imagined to lie, there are no preferred directions. Either they lie randomly, or are arranged radially about the rod axis. As a result, as Schmidt observed, a dark-adapted rod displays the same depth of color for all orientations of plane-polarized light passing down its long axis (Fig. 7).

On the other hand, for plane-polarized light passing through the side of the rod, the chromophores are in position to absorb maximally the light vibrating in the transverse direction, and minimally that vibrating parallel to the rod axis. Schmidt found that in such lateral illumination, dark-adapted rods look red in light polarized transversely, as regards its electric vector, and colorless in light polarized in the axial direction. All these relationships are summarized in Fig. 7.

We have carried out two kinds of measurements that bear upon these relationships. On the one hand, we have measured the absorption spectra of single dark-adapted rod outer segments, in plane-polarized and in unpolarized light, passing through them laterally (Fig. 8). When the plane-polarized light was oriented with its electric vector perpendicular to the rod axis, one recorded maximal extinction (curve A). If the chromophores of the visual pigments lay wholly in the transverse plane, the extinction measured in this way should be twice that recorded in unpolarized light (B), since the component of the latter, parallel to the rod axis (that is, half the light), should not be absorbed at all. In place of this ideal factor 2, the ratio of extinctions observed in this experiment (A/B in Fig. 8) is 1.8.

The degree of orientation of the chromophores in this rod is therefore high, though not perfect. If it were perfect, we should record no rhodopsin absorption for light polarized parallel to the rod axis; instead we record the extinction 0.017 (D in Fig. 8). For the same reason, the difference between the extinctions measured in the transverse and vertical planes, instead of being equal to A (0.072), is only 0.057, 81% as high (C). Since the extinction in the transverse plane A is 0.072, and that in the axial plane D is 0.017, 81% of the total extinction is in the transverse plane.[24]

Similar measurements made in 0.2 mm fields, through several thicknesses or rod outer segments pressed down flat on the retina with a cover slip, have yielded much the same result. For plane-polarized light passing through the sides of the rods, the ratio of extinctions of

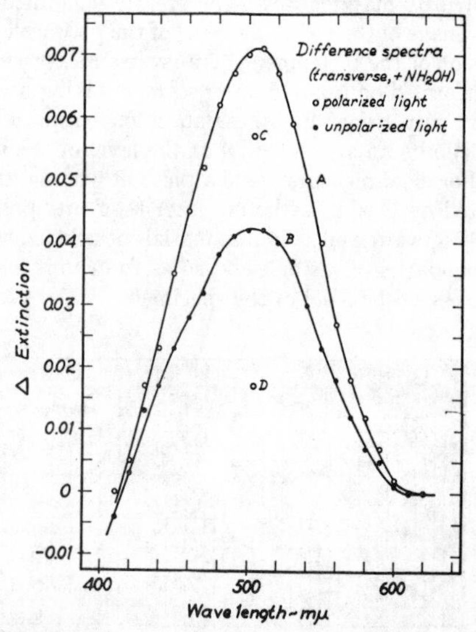

FIG. 8. Orientation of rhodopsin in a frog rod. These measurements were made on a single rod outer segment lying on its side in 30% methyl cellulose solution in the presence of 0.1M hydroxylamine. Four types of absorption spectrum were recorded: (1) Dark-adapted outer segment in plane-polarized light, the electric vector of which was oriented at right angles to the rod axis; (2) the same, but with the plane of polarization turned through 90°, so as to be parallel with the rod axis; (3) the same, in unpolarized light; and (4) after wholly bleaching the rhodopsin. The curves and points shown in the figure are: A: (1)–(4); B: (3)–(4); C: (1)–(2)—only the maximum extinction is shown; and D: the same, (2)–(4). Were the rhodopsin chromophores oriented ideally in the transverse plane, D would be zero, and the ratio of maximum extinctions A/B would be 2.0; here it is 1.8. About 80% of the rhodopsin extinction is concentrated in the transverse plane in this instance.

[24] Compare P. A. Liebman, Biophys. J. 2, 161 (1962).

rhodopsin measured in the transverse and axial planes is about 82:18.

It is important to distinguish such a ratio of extinctions from the ratio of chromophore orientations that it implies. Orientations in the transverse plane can be resolved into two coordinates of space at right angles to each other, only one of which should be effective for absorbing light passing through the side of the rod; whereas the axial component of orientation, involving a single coordinate of space, is always effective in absorbing light passing laterally through the rod. Hence, for this direction of incidence, the axial component of orientation is twice as effective in absorption as is the transverse component. A ratio of extinctions in the two directions of 18:82 therefore implies a ratio of orientation vectors of 9:82; i.e., the vectors of orientation of the individual rhodopsin chromophores are as 0.1 axial to 0.9 transverse.

An independent method for estimating the orientation of retinal pigments is illustrated in Fig. 9. In this method relatively large fields can be used, and the light passes through the outer segments axially, as in the intact eye.

Consider a pigment oriented randomly in the rod, as in true solution. For simplicity, let us resolve its orientation into three Cartesian coordinates, one parallel to the rod axis, the other two at right angles to each other in the transverse plane. Then *unpolarized* light passing down the axis of the rod will find 2/3 of the chromophores in position to absorb—the 2/3 lying in the transverse plane.

We have just seen, however, that, in fact, the chromophores of rhodopsin lie about 0.9 in the transverse plane. For this reason the absorption by rhodopsin of *unpolarized* light passing down the axis of the rods should be 0.9/0.67 or 1.35 times as great as if the rhodopsin were randomly oriented.

The further working out of a procedure can best be understood with reference to Fig. 9. A dark-adapted frog retina was mounted in 43% sucrose solution containing 0.1*M* hydroxylamine, and the absorption spectrum recorded in unpolarized light passing down the rod axes. Then the retina was bleached, and the spectrum again recorded. The difference between these spectra (dark-bleached) is shown with a solid line in Fig. 9. The dashed line shows a similar difference spectrum obtained with frog rhodopsin in digitonin *solution* in the presence of hydroxylamine.

The maximal extinctions at 500 mμ, due wholly to rhodopsin, have both been set arbitrarily at 1.0. When this is done, the minimum at about 370 mμ, which marks the formation of retinene oxime as the product of bleaching, is much larger in the case of the bleached solution than the bleached retina. The reason for this is that in solution both rhodopsin and the retinene oxime that results from its bleaching are randomly oriented. However in the retina the rhodopsin because of its primarily transverse orientation, has its extinction increased 1.35 times, whereas the retinene oxime, having lost this advantageous orientation, displays an abnormally small extinction relative to the rhodopsin.

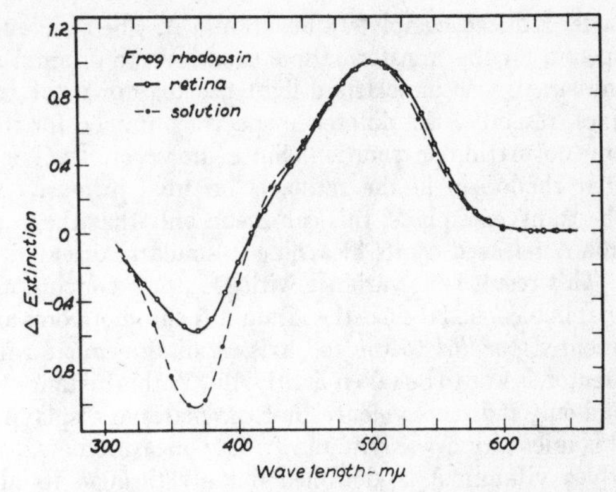

FIG. 9. Orientation of rhodopsin chromophores in rod outer segments. *Solid line:* Difference spectrum of rhodopsin measured *with unpolarized light* in a frog retina suspended in 43% sucrose +0.1*M* hydroxylamine. A circular area of retina was measured, about 0.2 mm in diameter, with the light passing down the long axes of the rods. The wavelength of maximum absorption (λ_{max}) was 504 mμ, and the extinction at this wavelength (E_{max}) was 0.546. *Dashed line:* difference spectrum of a solution of frog rhodopsin in 2% digitonin with hydroxylamine; in this case λ_{max} was 502 mμ and E_{max} 0.475. For comparison, both spectra have been brought arbitrarily to E_{max} of 1.0. Under these circumstances the minimum at about 367 mμ, owing to the formation of retinene oxime as the product of bleaching, is much larger in the solution than in the retina. The reason is that the transverse orientation of the rhodopsin chromophores increases their extinction in unpolarized light penetrating axially, by a factor of about 1.35; but this advantage is lost on bleaching, since the retinene oxime has a partly reversed orientation. The comparison of these curves shows that about 90% of the chromophore extinction is concentrated in planes perpendicular to the long axis of the rods.

When rhodopsin is bleached to retinene oxime in solution, the ratio of extinctions at 370 and 500 mμ in the difference spectrum is about −1.01 (cf. Fig. 9, dashed line). If the transversely oriented rhodopsin in the retina had bleached to *randomly* oriented retinene oxime, this ratio should have been −0.67. In the experiment of Fig. 6 it is −0.58, and our best present average value in such experiments is 0.47. This must mean that, as opposed to the transverse orientation of the rhodopsin chromophore, retinene oxime goes beyond random orientation to some special degree of axial orientation.

Direct measurements made with plane-polarized light passing through the sides of the outer segments confirm this observation. The retinene oxime extinction is about twice as great for light polarized in the axial plane as for transversely polarized light. If one recognizes as above that the axial component of *orientation* is about twice as effective in absorption as the transverse component, this result implies that the orientation vectors of the individual retinene oxime chromophores are about equal in the axial and transverse directions.

We have carried out similar experiments in which rhodopsin in the retina is bleached to retinene and the latter reduced completely to vitamin A. The difference spectra for the transformation, rhodopsin to vitamin A, measured with unpolarized light passing down the rod axes, resemble the difference spectra obtained for this transformation in solution. Since, however, we know that rhodopsin in the retina is oriented primarily in the transverse plane, this can mean only that the vitamin A released by its bleaching is similarly oriented.

This result is at variance with Denton's conclusion[21] that in bleached retinas the vitamin A chromophores are oriented *parallel* to the rod axis. A disagreement with Denton is not to be taken lightly, but in this instance he had only indirect evidence that the material displaying this orientation was vitamin A. Our measurements, in which vitamin A is identified reliably through its absorption spectrum, and its formation in the retina is followed in detail, show clearly that its major orientation is transverse, like that of the rhodopsin chromophore.

What do these orientations mean? In the case of rhodopsin, the chromophore is covalently bound to protein; the whole molecule is fixed in place, and makes a large contribution to the organized structure of the outer segment. Neither retinene oxime nor vitamin A, however, is attached to other molecules. Vitamin A can be shown directly to be "free," in that it is readily extracted from the tissues with petroleum ether, and readily moves back and forth between the retina and the pigment epithelium in the intact eye.[19] (Of course this measure of freedom does not exclude transitory attachments to other molecules on the basis of hydrogen bonds and van der Waals forces.) Retinene oxime is not only similarly "free," but is an artificial structure with no business in the retina. Yet both molecules are oriented.

The point is that in such a highly organized structure as the outer segment, even "free" molecules become part of the organization. We had supposed originally that our measurements might distinguish bound from free molecules by finding the former oriented, the latter disoriented. It turns out, on the contrary, that in such a quasi-crystalline structure even "free" molecules must find their place in the molecular lattices, and so maintain some degree of orientation.

The relatively small volume of ground substance in the outer segment appears homogeneous, and may in part approximate true solution. This also represents the closest approach to an aqueous phase in the outer segment. On the other hand, the transverse layers which compose the bulk of the outer segment are formed of highly oriented molecules. Since Schmidt[23] it has been generally assumed that these are arranged in alternate layers, mainly protein and mainly phospholipid in composition. Phospholipid accounts for 35–40% of the dry weight of the outer segment, protein for most of the remainder. The protein molecules in the layers appear to lie transversely, the phospholipid molecules axially.[23] The more axial orientation of retinene oxime may imply a greater tendency to enter the phospholipid layers, the more transverse orientation of vitamin A a greater seeking-out of the protein lamellae.

In either case these molecules find place and diffuse in what are essentially two-dimensional areas, rather than three-dimensional volumes. Diffusion within such surfaces should be considerably more rapid than in bulk solution, probably 1.5 times as rapid, since it is confined within two of the three planes of space. To say this more simply, a molecule confined to such a transverse layer does all its diffusing laterally, and wastes no time wandering upward or downward. This may be an important factor physiologically in promoting the transport of vitamin A between the interior of the outer segment and its periphery where exchanges with the pigment epithelium occur.

The orientation of rhodopsin in the outer segments is remarkably precise. It should not be assumed that our finding a small axial component of orientation in addition to the main transverse component need represent a lapse from perfect orientation. For the chromophores of rhodopsin to lie and absorb light wholly in one plane, they should have to be planar themselves; and we know that they are not. Oroshnik *et al.*[25] pointed out some years ago that in all such structures steric hindrance causes a twist between the β-ionone ring and side chain; and the 11-*cis* isomer of retinene that constitutes the chromophore of all these visual pigments has an additional twist owing to steric hindrance at the 11-*cis* linkage. This chromophore, therefore, is distinctly nonplanar, and we have every reason to expect that even with perfect orientation, though the largest absorption vector lies transversely in the outer segment, this must be accompanied by a small absorption vector in the axial direction. Our finding 90% transverse and 10% axial orientation may well represent this division of absorption vectors in what are, in fact, perfectly oriented rhodopsin molecules.

Molecular orientation is an approach to crystallinity, and hence the solid state. In the outer segment we find a well-nigh perfect orientation of rhodopsin molecules. The outer segment is to this degree a crystal. If it were all rhodopsin, it would be a crystal of rhodopsin. But even in a frog rod, rhodospin accounts for only about 40% of the dry weight, and perhaps 60% of the protein, and these are unusually high proportions of visual pigment among rods and cones generally. The rhodopsin molecules must be set among other molecules (themselves probably oriented since rhodopsin is oriented). Such sections of mixed crystallinity must be accompanied by other sections of quite different material, for example, the phospholipids that account for almost 40% of the dry weight of the outer segment, and form their own oriented structures. Other, though minor, sections of the rod structure may be amorphous. The multilayered ultrastructure of the outer segment is a visible

[25] W. Oroshnik, G. Karmas, and A. D. Mebane, J. Am. Chem. Soc. 74, 295 (1952).

sign both of its quasi-crystallinity and of its heterogeneity of composition.

To a degree then, a rod outer segment approaches crystallinity—we have called it "quasi-crystalline"—and the solid state; yet only mixed and aperiodic crystallinity, probably very much broken-up in this regard. Were it more homogeneous in structure, we should assume at once that solid-state phenomena play a large role in its function: exciton migration, semiconduction, photoconduction, and so on. Its heterogeneity of structure throws all this into question. Such phenomena may indeed play a role in the function of the outer segments, but this cannot be assumed; it will have to be demonstrated in each instance.

ULTRASTRUCTURE OF VISUAL RECEPTORS[26]

We have examined in the electron microscope the microstructure of rods, cones, double cones, and pigment epithelium cells in the mud puppy, *Necturus*. Taken together with microspectrophotometric measurements of the visual pigments *in situ*, these observations have consequences for the theory of visual excitation. We can do no more than summarize these observations here.

The general features of ultrastructure of the retinal cells are by now sufficiently familiar so that we can go at once to the special properties of these tissues in *Necturus*, and whatever may be new in our observations.

The *Necturus* retina contains rods, cones, and double cones in the approximate proportions 35:28:10.[27] The light microscopy of the visual cells has been very faithfully described by Howard.[28] The visual cells are among the largest known, even among amphibia. (In what follows, all dimensions are for fixed, dehydrated and embedded preparations. The fresh dimensions are about 25% larger.) The rod outer segments are about 12 μ wide and 30 μ long; cone and double-cone outer segments are about 24 μ long, tapering from about 9 μ at the base to about 5 μ at the tip.

Whole eyes were fixed for 1 hr in 2% osmium tetroxide, buffered to *p*H 7.8 with veronal acetate, containing 45 mg/ml of sucrose and 0.002*M* calcium chloride. After fixation, the eyes were dehydrated in graded acetone–water mixtures, and embedded in Araldite epoxy resin. Thin sections were cut with a Porter–Blum microtome and stained with saturated uranyl acetate in 50% ethyl alcohol. They were examined in an RCA EMU-3D electron microscope, operated at 100 kV.

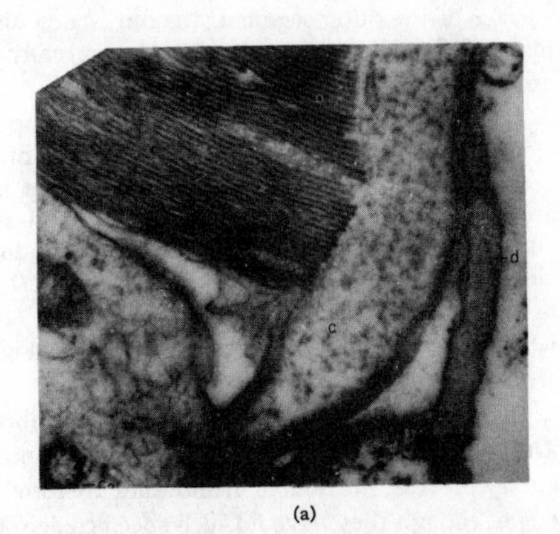

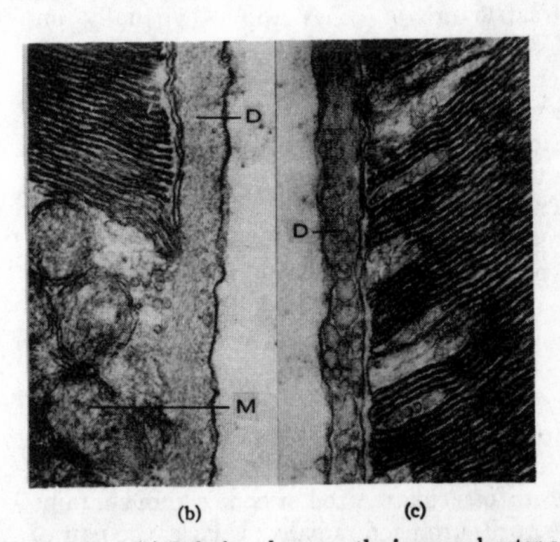

(a) (b) (c)

FIG. 10. Longitudinal sections of portions of two rods and a cone in *Necturus*. (a) Relations between the inner and outer segment in a rod. A heavy ciliary process C projects into the outer segment from the basal body formed from one centriole of the inner segment c_1. The second centriole c_2 lies nearby, characteristically at right angles to the first. A dendrite D, a prolongation of the cytoplasm of the inner segment, runs up beside the ciliary process. The pile of double-membrane discs that forms the bulk of the outer segment lies in the same ground substance as fills the ciliary process; and the plasma membrane of the inner segment is extended so as to envelop the outer segment and the ciliary process, and the dendrites. (b) Longitudinal section of a rod, showing the mitochondria M clustered in the inner segment at the base of the outer segment, and a dendrite D continuous with the cytoplasm of the inner segment in this region. This section also shows plainly the two thicknesses of plasma membrane that separate the outer segment and the dendrite, with one thickness contributed by each of these structures. (c) Longitudinal section of a cone outer segment, at the side across from the ciliary process. At this side the outer segment lacks a separate plasma membrane, so that only the plasma membrane of the dendrite D is present. Also at this side the pairs of membranes that form the transverse lamellae, instead of meeting at the edge, part company, each single membrane coming around to form one of the next pair. At several points this arrangement is disturbed, in that a wider loop of membrane encloses 1–3 closed pairs.

[26] Much of this material has been reported by the present authors in the Symposium on Receptor Mechanisms (1961) already cited in Footnote 20. This work will be reported in detail by P. K. Brown, I. A. Gibbons and G. Wald in the J. Cell Biol., 1963 (to be published).
[27] S. C. Palmer, J. Comp. Neurol. 22, 405 (1912).
[28] A. D. Howard, J. Morphol. 19, 561 (1908).

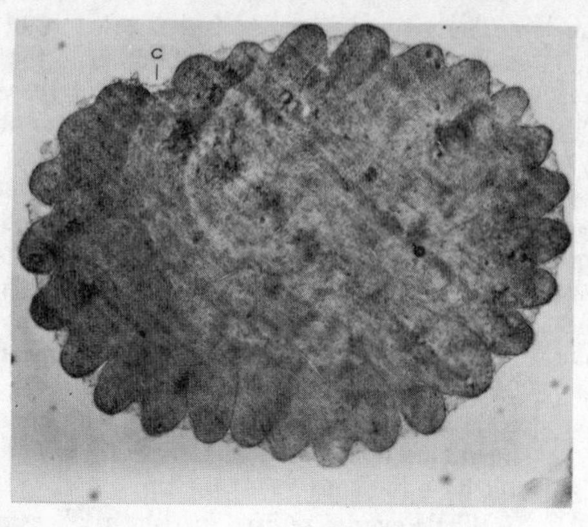

(a)

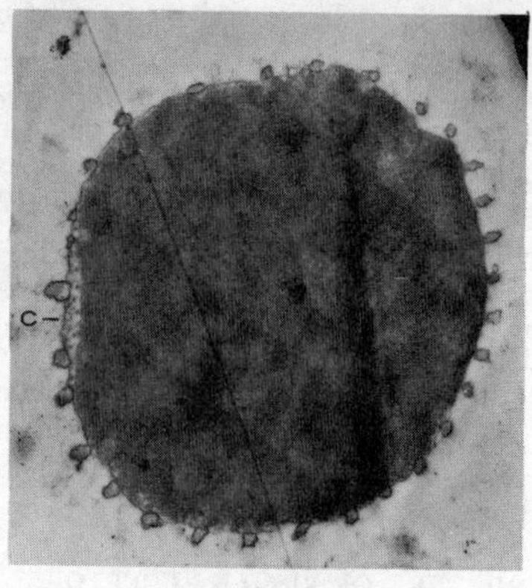

(b)

FIG. 11. Cross sections of outer segments of a rod (a) and cone (b) in *Necturus*. (a) The rod cross section has a scalloped appearance, owing to the presence of 27 deep fissures which run longitudinally. A dendrite appears in cross section in the mouth of each fissure; and the flattened ciliary process C contains single fibrils, no longer in radial array. Large numbers of regularly spaced, deeply staining particles are distributed widely over the cross section. It is possible that the visual pigment, in this case porphyropsin, is concentrated in these lamellar particles. (b) The cone cross section is smooth in outline, since cone outer segments lack fissures. Thirty dendrites are seen in cross section standing about it, as is the flattened ciliary process C, containing single fibrils. A plasma membrane lies over the ciliary process and extends to both sides, part way around the outer segment; but the side of the outer segment across from the ciliary process lacks a separate plasma membrane. The sparsity of processes from the pigment epithelium shows that these sections cut the outer segments toward their proximal ends.

Some general features of rod and cone structure can be appreciated from Fig. 10. Whatever we have to say of the ultrastructure of cones applies virtually unchanged to double cones.

The outer segment is completely enclosed in a plasma membrane which is continuous with that of the inner segment. Embryologically and in regeneration the outer segment is derived from a cilium, and what had been the plasma membrane of the cilium becomes the plasma membrane surrounding the outer segment.[29] Much of the cilium persists, forming the backbone structure of the mature rod or cone outer segment [Fig. 10(a)]. The cilium springs from a typical basal body, derived from one of the centrioles of the inner segment. The second centriole typically lies at right angles to it. The centrioles of *Necturus* are composed of triple filaments, having the typical form and arrangement found in the basal bodies of protozoan cilia.[30] A rod or cone cilium contains nine double fibrils arranged radially; it lacks the pair of central fibrils characteristic of motile cilia [Fig. 10(a)]. The cilium has previously been described as ending shortly after entering the outer segment.[31,32] Actually our cross sections show that it extends at least halfway up the rod outer segments, and almost to the tips of the cone outer segments. Distally the ciliary processes become progressively thinner, and flatten out over the surface of the outer segment; the fibrils, now single, are losing their radial arrangement, and eventually petering out [Fig. 11(a), (b); Fig. 13].

The great bulk of the outer segment is composed of a stack of transversely oriented double membranes. In rod outer segments each double membrane is sealed around its circumference by a special, differentiated edge structure, which stains more deeply, so forming a closed "double-membrane disk" (Sjöstrand). The stack of double-membrane disks lies in a ground substance which also surrounds the ciliary process [Fig. 10(a), (b)].

In cones and double cones the double membranes are arranged differently. On the side that contains the ciliary process, the double membranes are fused at their edges, though they have no such specific edge structure as found in the rod disks. At the opposite side, however, the membranes of each pair part company at the edge, each membrane coming around to form one of the adjoining pair (Fig. 10). Also, this side of the outer segment is not covered by a plasma membrane [Fig. 10(c), 11(b)]. It is as though the pile of transverse membranes in the cone were formed by the repeated infolding of the plasma membrane at the side opposite from the ciliary process.[31,32] These relationships are shown diagrammatically in Fig. 13.

This otherwise regular structure is occasionally interrupted, as rather frequently in Fig. 10(c), by the appearance of wide loops of membrane enclosing one or

[29] E. De Robertis, J. Gen. Physiol. **43**, Suppl. 2, 1 (1960); J. E. Dowling and I. R. Gibbons, in *The Structure of the Eye*, edited by G. K. Smelser (Academic Press Inc., New York, 1961), p. 85.

[30] I. R. Gibbons, Nature **190**, 1128 (1961).

[31] F. S. Sjöstrand, in *The Structure of the Eye*, edited by G. K. Smelser (Academic Press Inc., New York, 1961), p. 1.

[32] F. D. Sjöstrand, Ergeb. Biol. **21**, 128 (1959); Revs. Modern Physics **31**, 301 (1959). E. De Robertis and A. Lasansky, in *The Structure of the Eye*, edited by G. K. Smelser (Academic Press Inc., New York, 1961), p. 29.

more double membranes within them. Cohen[33] has observed similar formations in monkey cones. Such loops-within-loops probably occur through the local fragmentation of double membranes into vesicles. Wherever a double membrane breaks into vesicles toward its edge, that leaves the remainder of the double membrane closed off by a new edge-loop lying within the larger loop formed by the adjoining membranes. Almost everywhere that loops-within-loops appear in Fig. 10(c), one also sees vesicles separating the inner loops from the outer.

Each rod outer segment contains on the average 1100 double-membrane disks. Within a disk each membrane is about 50 Å thick, and the space between the membranes ranges from 0–40 Å. In *Necturus* the double-membrane disks are about 150 Å apart, making a repeating unit (disk+interdiskal space) of about 270 Å.

The outer segments of cones and double cones contain a total of about 750 paired membranes, or 1500 single membranes, each about 50 Å thick. The membranes of a pair are about 0–40 Å apart, and the pairs about 200 Å apart, making a repeating unit of about 320 Å.

In *Necturus* the rod outer segments are cut by a series of deep longitudinal grooves or fissures into an array of lobules, each as long as the outer segment, arranged radially about a more or less solid center [Fig. 11(a), 12]. There are 22–31 (average 27) such fissures in a single rod. It has been suggested earlier that they may represent a device for increasing the surface of these extraordinarily broad outer segments, so facilitating interchanges of material with contiguous tissues.[18] The plasma membrane lines the mouths of the fissures, though it does not penetrate into the fissures themselves, so giving the surface of the outer segment a fluted appearance, and making the rod in cross section appear scalloped [Fig. 11(a), 12]. The cones do not possess such fissures, and therefore present a smooth outer surface [Fig. 11(b)].

We now should like to discuss a structure that is widely distributed among retinas, yet has received little previous attention. This is an array of processes, continuous with and extending upward from the cytoplasm of the inner segment, and standing like a palisade about the outer segments of the rods and cones. These processes can be seen in longitudinal section in Fig. 10(a), (b), and (c), and in cross section in Fig. 11(a) and (b), and 12. Clearly they correspond with the "cytoplasmic prolongations" that Carasso[34] found in visual cells of frog tadpoles, and with the structure that Cohen[33] calls a "calyx or cup" enveloping the bases of the outer segments of rods and cones in the Rhesus monkey. Since these structures represent neuronal processes which extend from the cell body on the side from which excitation is received, we speak of them as rod and cone

[33] A. I. Cohen, in *The Structure of the Eye*, edited by G. K. Smelser (Academic Press Inc., New York, 1961), p. 151.

[34] N. Carasso, Compt. rend. **247**, 527 (1958).

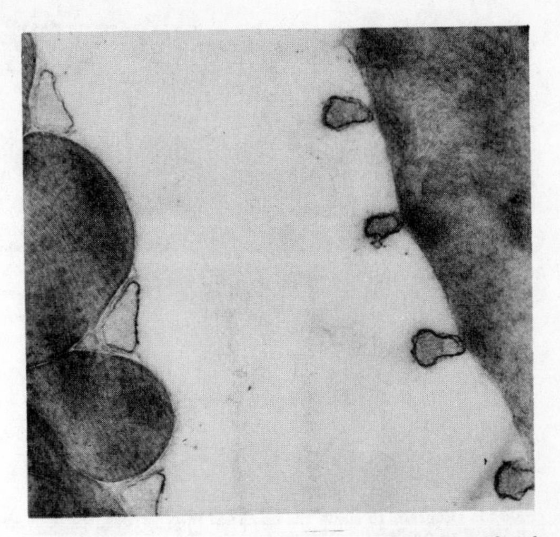

Fig. 12. Portions of cross sections of an adjoining rod and a cone, showing the dendrites and the rod-lamellar particles. It should be noted that in the rod each dendrite is separated from the outer segment by two thicknesses of plasma membrane—one that of the dendrite itself, the other that of the outer segment. In the cone only the plasma membrane of the dendrites is evident; this is the side of the cone opposite the cilium, where no separate plasma membrane appears. The rod section also displays a system of deeply staining particles, so regularly spaced as to approach crystalline array. The cone section lacks such particles.

dendrites, though without wishing to imply thereby that they are in any way concerned with the conduction of excitation.

In rod outer segments one such dendrite ordinarily stands at the mouth of each fissure; on the average, therefore, there are about 27 dendrites in all. Since about half our rod cross sections, which by other signs we know to represent relatively distal portions of the outer segment, lack dendrites, as also any recognizable ciliary processes, we assume tentatively that both the ciliary process and the dendrites extend about halfway up the length of the outer segment, both ending at about the same level. About 30 dendrites on the average are found standing around the outer segment of a cone, and since almost all our cone cross sections contain both dendrites and the ciliary process, we assume that in the cones both these structures extend nearly to the tip. The same plasma membrane of the inner segment that is prolonged to envelop the outer segment is also reflected over the dendrites, so that each of the dendrites is separated from a rod outer segment by two thicknesses of this membrane [Fig. 10(b), 12]. In cones the same two thicknesses of plasma membrane appear on the side that contains the ciliary process, whereas on the opposite side one sees only the single plasma membrane that covers the dendrite [Figs. 10(c), 11(b), 12; cf. also Fig. 13].

The outer segments are surrounded not only by this system of cytoplasmic filaments extending upward from

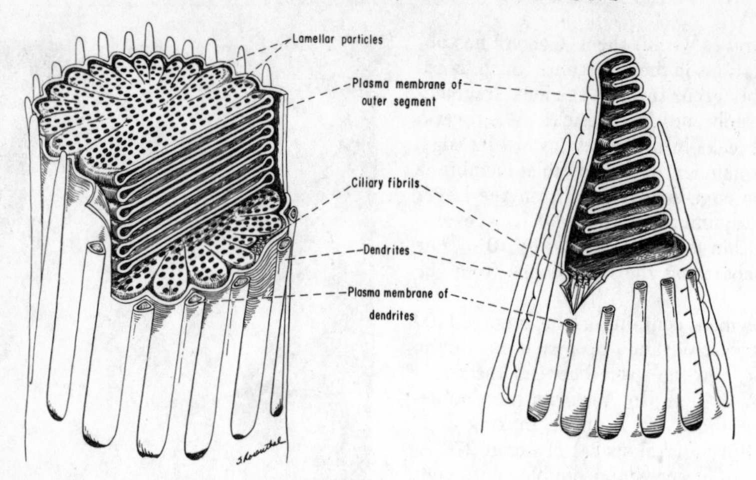

FIG. 13. Diagram to show the essential microstructures of *Necturus* rod and cone outer segments. The outer segment of a rod is constructed of a pile of double-membrane discs, each sealed off by a differentiated rim structure. These lamellae are cut radially by a system of deep fissures, and bear a system of deeply staining particles in regular array. The whole is enclosed in a plasma membrane continuous with that of the inner segment. The same plasma membrane is also reflected over the dendrites which stand like a palisade around the outer segment, one in the mouth of each fissure. Running up the outer segment is the residue of the primitive cilium from which the outer segment is derived embryonically. In the cone, the lamellae are formed differently, by the repeated infolding of the plasma membrane. No fissures are present, nor do the lamellae display the deeply staining particles characteristic of *Necturus* rods.

the inner segments, but also by a second system of cytoplasmic processes extending downward from the pigment epithelium (Fig. 14). These are relatively sparse and distant toward the proximal ends of the outer segments (hence almost absent from Figs. 11 and 12), but become increasingly dense and more and more intimately associated with the outer segments as the latter approach the pigment epithelium. Conversely, a point is reached at which the rod outer segments have lost their dendrites, though the cones may still possess them (Fig. 14). At their distal ends both rod and cone outer segments are embedded in the pigment epithelium cells and their processes.

We have observed systems of small, densely staining particles, not described previously, on both the dendrites and the pigment epithelium processes. The dendrite particles are button-shaped, about 17–25 mμ in diameter, and seem to lie on the outer surface of the plasma membrane. The pigment epithelium particles appear as little lens-shaped thickenings of the membranes of the pigment epithelium processes, which are also about 25 mμ in diameter. Such particles appear very prominently also in pigment-epithelial processes of the rat retina,[35] so they may represent a general type of structure.

Probably the most significant observation that we have to report involves a system of deeply staining particles in the transverse membranes or lamellae of the rod outer segments. These can be seen clearly on close examination of the rod outer segment shown in Fig. 11(a) and are shown at higher magnification in Figs. 12 and 15. We shall refer to them hereafter as rod-lamellar particles. They are not apparent in cross sections of cone outer segments [Figs. 11(b), 12].

The most notable feature of these particles is that they are distributed so evenly over the surfaces of the rod lamellae as to approximate a crystalline array. The particles are circular, and are about 30 mμ in diameter. If, as we suppose, they lie in the lamellar membranes, they can at most be as thick as those membranes, about 5 mμ. We assume tentatively that the particles are disk-shaped, and are about 30 mμ across and 5 mμ thick. In that case, the volume of each particle is about 3500 $m\mu^3$, and the volume of one mole of such particles is about 2.1 million cm^3. If one assumes a typical protein density of 1.3, one mole of such particles would weigh about 2.7 million grams.

There are about 140 such particles per μ^2 of membrane; and since each membrane in the rod outer segment has an area of about 113 μ^2, each membrane contains about 16 000 particles, and each double-membrane disk therefore contains about 32 000 particles. The 1100 disks that compose a rod outer segment therefore contain about 35 million particles.

We should like to consider the possibility that the visual pigment of *Necturus* rods—which is primarily the retinene$_2$ pigment porphyropsin—is concentrated in the lamellar particles. Something can be said for this possibility; though one of the great weaknesses of present-day electron microscopy is that it tells one so little of the composition of the structures it makes visible. The only highly organized structures that appear in our preparations of rod outer segments are the mem-

[35] J. E. Dowling and I. A. Gibbons (personal communication).

branes that form the double-membrane disks. Rods of other animals sometimes display wider and more constant spaces between the membranes of each disk, filled with material that might well be highly organized; but in *Necturus* this space is so narrow and variable in width as to discourage the thought that it is highly organized. The spaces between and around the disks appear to be filled with the ordinary intracellular ground substance. It is at least reasonable to expect that porphyropsin, particularly since it has proved to be highly oriented as is rhodopsin in frog rods, is located in the membranes. Furthermore, as a protein carrying a highly unsaturated chromophore, porphyropsin may be expected to stain deeply, as do the membranes, and particularly their lamellar particles.

We have measured the difference spectrum of porphyropsin in the *Necturus* retina in the microspectrophotometer described above [Fig. 16(a)]. A number of such measurements reveals an average extinction at 522 mμ of 0.07. The molar extinction of porphyropsin at this wavelength (the extinction of a solution containing 1 mole porphyropsin per liter, measured in a layer 1 cm in depth) is about 30 000. Measurements in polarized light passing through the sides of the rods show that porphyropsin is oriented so as to lie predominantly in the transverse plane. We will assume tentatively that because of this orientation its extinction for unpolarized light passing down the rod axes is increased by a factor of 1.35, as is that of rhodopsin in frog rods. We make no allowance for light passing between the rods, because owing to the outer segments acting as guides, the great bulk of the light seems to be conducted through them [Fig. 16(b)]. On this basis one can calculate that a single rod outer segment in *Necturus* contains about 1800 million molecules of porphyropsin.

If therefore the porphyropsin is located with the lamellar particles, each of the approximately 35 million particles must contain about 50 molecules of porphyropsin. Since one mole of lamellar particles may weigh about 2.7 million grams, if the particle were made wholly of porphyropsin, the molecular weight of the latter would be $(2.7\times10^6)/50$, or 54 000, a figure not far from the molecular weight of cattle rhodopsin (about 40 000). It is at least possible, therefore, that the lamellar particles consist primarily, even perhaps wholly, of porphyropsin.

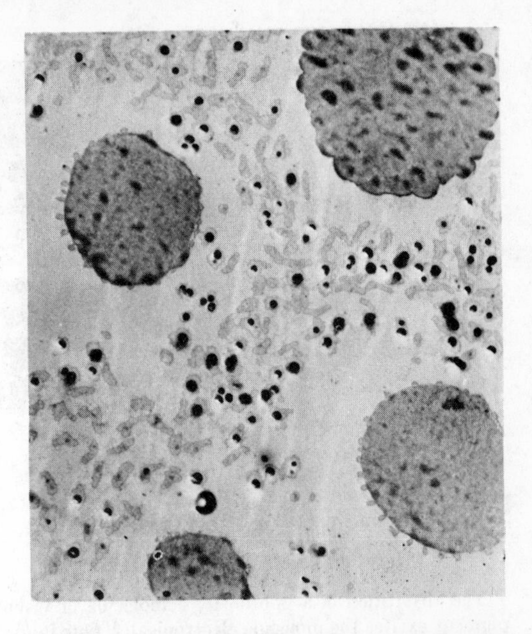

FIG. 14. Tangential section showing three cone outer segments and one rod (scalloped border, upper right), and the arrangement of pigment-epithelium processes about them. The large, black pigment granules in the processes are ovoid in long section, but all are seen here in circular cross sections, since the processes are so narrow that all the granules lie in them lengthwise. The large number of these processes indicates that the section lies near the pigment epithelium, i.e., toward the distal ends of the outer segments. At this level, though the cones still possess dendrites and ciliary processes, the rods have lost both.

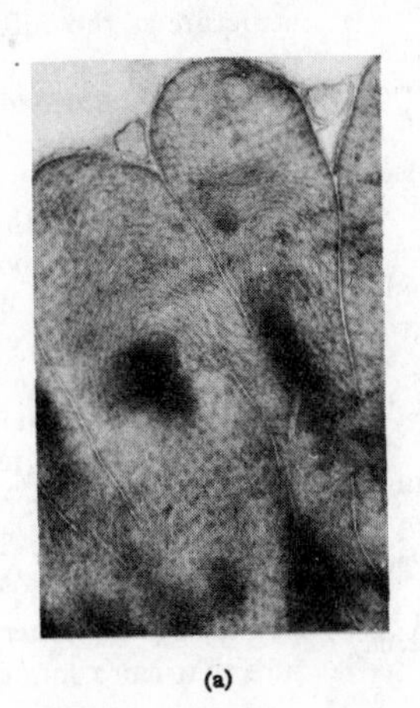

(a)

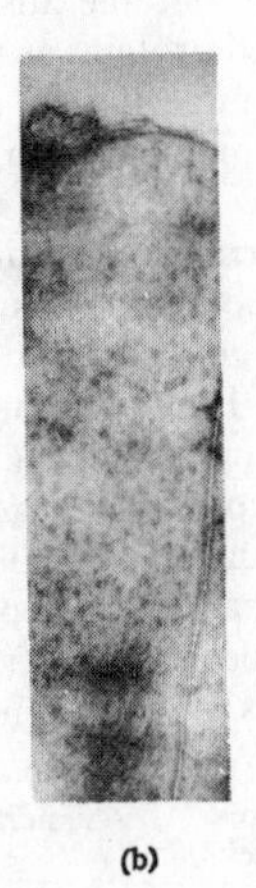

(b)

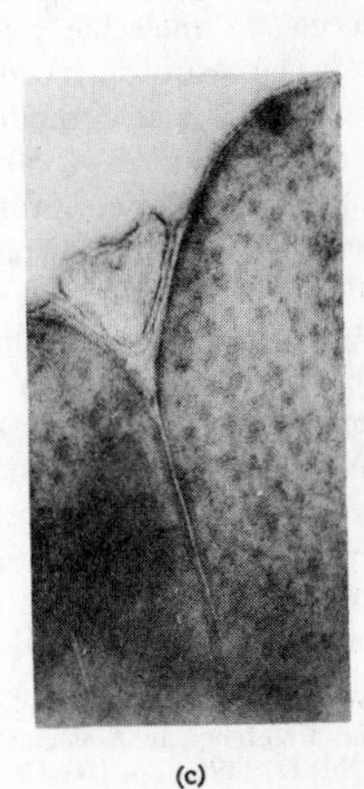

(c)

FIG. 15. Portions of cross sections of rod outer segments, principally to show the arrangement of rod-lamellar particles. Sections a and b were photographed at a magnification of 39 000 diameters, c at 61 200 diameters. The deeply staining particles are about 30 mμ in diameter, and so regularly spaced as to be in virtually crystalline array.

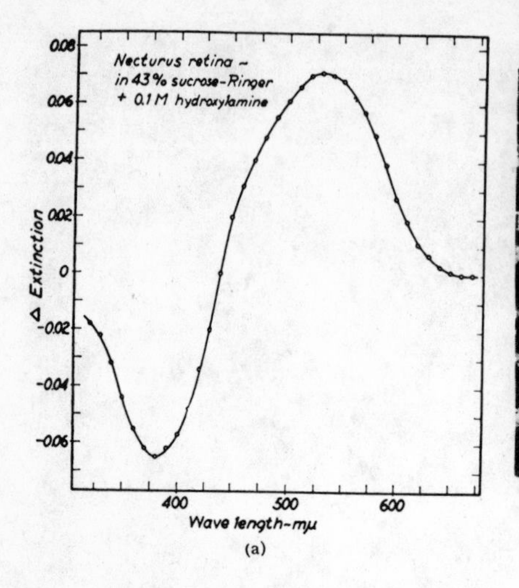

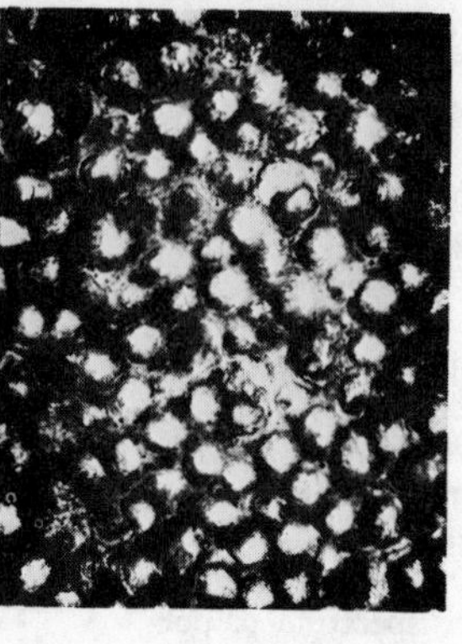

FIG. 16. (a) The spectrum of porphyropsin *in situ*. Difference spectrum of porphyropsin (λ_{max} 530 mμ), measured in a circular area of *Necturus* retina about 1.17 mm in diameter. The retina was mounted in 43% sucrose in Ringer solution containing 0.1M hydroxylamine. (b) Photomicrograph of a portion of *Necturus* retina mounted in Ringer solution for such a measurement, showing the ends of the rods and cones brilliantly lighted, whereas little light comes through the spaces among them. The outer segments act as light pipes, conveying most of the transmitted radiation, and being primarily responsible therefore for the absorption measured in the retina as a whole.

PROBLEM OF EXCITATION

Migration of Molecular Excitation

The absorption of a photon by a molecule of visual pigment excites the molecule electronically. Our fundamental problem is to understand how this molecular excitation leads eventually to a nervous excitation. As one element in that problem, the possibility must be considered that the molecular excitation migrates from the site of absorption of light to some other point in the outer segment, where it exercises its effect.

Two types of processes have to be considered: the transfer of electronic excitation from molecule to molecule without emission or absorption of light (inductive resonance, radiationless transfer, exciton migration); and the migration of electrons or positive "holes" (semi- or photoconduction).

Ordinarily exciton migration is thought to occur only over distances of the order of a chromophore diameter, and not exceeding, at best, about 100 Å. It must be said at once that the prospects of such migration from molecule to molecule over considerable distances is poor in the case of such conjugated proteins as the visual pigments. A single rhodopsin molecule, if spherical, has a diameter of 40–50 Å, which is almost wholly occupied by colorless protein, so that even were such molecules packed in crystalline array, the chromophores would be separated from one another by distances already approaching the limit over which radiationless transfer of excitation can occur. By clustering a group of molecules with their chromophores together, the chances of inductive resonance would be improved, but such a group could not include many molecules.

Hagins and Jennings[36] have examined the possibility of exciton migration in a frog rod in which rhodopsin is unusually concentrated. They could find no experimental evidence that this process, if it occurs at all, can convey electronic excitation over distances comparable with the dimensions of the outer segment. In *Necturus*, the much lower content of visual pigment in the rods, together with the possibility that it is concentrated in the lamellar particles, makes such radiationless transfers still less likely, unless they are within a single lamellar particle. Since adjacent particles are about 55 mμ apart—edge-to-edge—no such transfers could occur between them. Transfers between the membranes of a single double-membrane disk might be more possible, but here on *a priori* grounds one would think that such transfers would occur relatively rarely. To summarize a much more extensive argument that there is no space for here, we think the chances of exciton migration in the rod outer segment are poor, unless perhaps within the boundaries of a single lamellar particle, and at most a single double-membrane disk.

It is more difficult to evaluate the possibility of semiconduction and photoconduction. We should like, nevertheless, to point out that in such a highly heterogeneous structure as the outer segment, the conditions for such processes are far from ideal. At most, perhaps one might hope that photoconduction might occur within a single double-membrane disk, if indeed it can occur that widely.

Altogether, pending direct evidence to the contrary, we think there is good reason to assume that molecular excitation does not migrate far from the site of absorption of a photon by a molecule of visual pigment. In terms of the problem of nervous excitation, this probably means that the latter also has its source close to the site of absorption, and most probably at the same level in the structure of the outer segment at which the absorption has occurred. That would mean that excitation must take effect in some structure that runs up or beside the outer segment, i.e., some axially oriented structure that can communicate the excitation to the inner

[36] W. A. Hagins and W. H. Jennings, in "Energy Transfer with Special Reference to Biological Systems," Discussions Faraday Soc. No. 27, (1960), p. 180. Cf. also Liebman.[24]

segment. We probably should think, therefore, in terms of a "Principle of Axial Pick-Up."

We have discussed four such axial structures: the plasma membrane, the ciliary process, and the dendrites, all of which originate in and are continuous with structures in the inner segment; and the processes from the pigment epithelium. Three of them—the plasma membrane, cilium, and pigment-epithelium processes—appear to be universally associated with outer segments. The dendrites, though distributed widely (*Necturus*, frogs, monkeys), may not be universal; they have not yet been seen, for example, in the rat.[35] This consideration, therefore, weighs against the dendrites as the conductors of excitation to the inner segment. Furthermore, if the excitation originates in the transverse lamellae of the outer segments, it should have to cross the plasma membrane of the rod before encountering a further extension of the same plasma membrane covering the dendrite; and if the point is to depolarize that membrane, why not the first time it is encountered, the membrane of the outer segment itself? As for cones, if it is true that their transverse lamellae are infoldings of the plasma membrane, one sees little reason to go beyond this to excite the extension of the same plasma membrane that covers a dendrite. It should be recalled also that in the rods the dendrites probably reach only about halfway up the length of the outer segments. For these and for other reasons we doubt that the dendrites play a primary role as conductors of excitation.

Sjöstrand suggested some years ago that the cilium may conduct excitation from the outer to the inner segment. This is a serious possibility, indeed made more likely by our observation that the ciliary process extends much further into the outer segment than has generally been supposed. Yet we find it extending only about halfway up a rod outer segment, and if it were true that the cilium must pick up the excitation at the level of light absorption, that would mean wasting half the outer segment. Furthermore, the cilium is known to have other uses connected with the embryogenesis of the outer segment and with its regeneration in the adult animal.[29] It seems to exercise an organizer function in both these developments; and when, in vitamin A deficient rats maintained on vitamin A acid, the rods have regressed to the point of losing their ciliary structures (basal body, root), they appear to be incapable of regenerating an outer segment.[10]

This exceedingly condensed and incomplete presentation of the argument leads to a rather orthodox conclusion: that on many counts the best present candidate as the source of nervous excitation is the plasma membrane of the outer segment. This is present universally, encompasses the entire outer segment, and is continuous with the plasma membrane of the inner segment. Presumably the excitation takes effect at about the same level in the outer segment at which the absorption of light occurs. In rods it may involve an interaction between the membrane and the edge of a double-membrane disk. In cones, as already said, the lamellae themselves may be portions of the plasma membrane.

We should like, in these terms, to summarize our present concept of the major functions of the structures discussed above. Certainly one function, and perhaps the main business, of the pigment-epithelium processes is to facilitate exchanges of material between the outer segments and the pigment epithelium. Similarly, the main business of the dendrites may be to facilitate exchanges of material between the outer segments and the inner segments. The extraordinarily high concentration of mitochondria that constitutes the so-called ellipsoid, lying in the inner segment just below the outer segment, occupies just the region of the inner segment from which the dendrites spring [Fig. 10(b)]. It is an attractive thought that the dendrites may carry metabolites from the mitochondria, notably perhaps ATP, to the outer segment.

The cilium, as already said, seems to play an essential role in organizing both the embryonic development and the regeneration of outer segments.[29] That leaves to the plasma membrane its traditional role in excitatory structures: the maintenance of a potential difference between the intracellular and extracellular fluids, which on depolarization of the membrane which separates them, results in nervous excitation.

(*Reprinted from Nature, Vol.* 194, *No.* 4831, *pp.* 844–847, *June* 2, 1962)

LOCAL MEMBRANE CURRENT IN THE OUTER SEGMENTS OF SQUID PHOTORECEPTORS

By W. A. HAGINS, H. V. ZONANA and R. G. ADAMS

National Institute of Arthritis and Metabolic Diseases, National Institutes of Health, Bethesda, Maryland; Naval Medical Research Institute, Bethesda, Maryland; and Johns Hopkins Medical School, Baltimore, Maryland

WHEN a quantum of light is absorbed by any of the millions of rhodopsin molecules in a dark-adapted human retinal rod there is a high probability that the cell will become excited[1]. Other photoreceptors may be equally sensitive to single photons[2]. The mechanism linking each of such a large number of photopigment molecules to the sensory output of one cell is not yet known; but two physical processes have been suggested as playing a part in conveying information about absorbed photons from the light-absorbing organelles of photoreceptors to the distal neurones of the visual system. The first is diffusion of light-created electrons and holes through the ordered structures in which photopigments are usually concentrated. It has been shown that such migration in the form of excitons (coupled electron-hole pairs) is improbable[3]; but photoconductivity may exist in rods under some conditions[4] and might contribute to the spread of excitation within outer segments (or to the flow of current through them). Its importance in living cells is still to be shown. A second suggestion is that light might release locally a chemical transmitter substance which diffuses to the dendrites of nearby neurones[5]. But the slow speed of molecular diffusion limits its range to a few microns during the short latency of vision, while the distal neurons of the visual system are often separated from the sites of light absorption by the much greater lengths of the receptor cells. Therefore, some additional mechanism seems required to account for signal transmission in long photoreceptors. A likely possibility is displacement of the cell membrane potential by light-induced flow of local membrane current, for such local currents are believed to produce the graded receptor potentials often seen in other kinds of sensory cells[6] and, in one

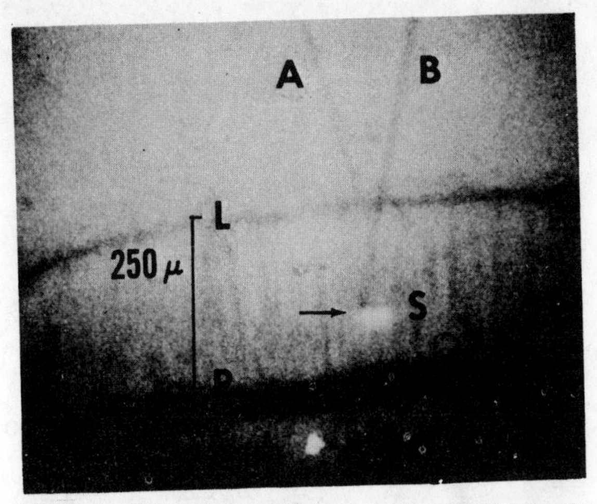

Fig. 1. Slice of living squid retina photographed with light of wave-lengths 900–1,000 mμ. Electrode *A* at inner limiting membrane *L*. Tip of electrode *B* at arrow in layer of outer segments just above pigmented zone *P*. Stimulus spot *S* consisted of wave-lengths 630–650 mμ

case, to account for spread of excitation from sites where stimuli act to regions of the receptors where nerve impulses are generated or transmitter substances are released[7]. There is, in fact, evidence that some photoreceptor cells do depolarize when illuminated, although the source of the depolarizing current was not located[8].

This article reports some experiments which suggest that such locally originating receptor currents are capable of transmitting signals along both cell bodies and outer segments of squid photoreceptors.

The squid retina, like that of the octopus, is a packed array of long cells of which the photopigment-bearing processes (here called 'outer segments' by analogy with vertebrate rods and cones) are covered with piles of microvilli[9]. The outer segments are attached to nucleated cell bodies which lie just outside a dense layer of black pigment. From the cell bodies axones are given off which pass through a plexus and enter the optic lobe situated behind the eye[10]. Apart from the photoreceptors, only a few very small glial and pigment-bearing cells are present.

When live retinæ were isolated from dark-adapted animals (*Loligo pealii*) and sliced into strips about 100μ thick by cuts parallel with the long axes of the receptors, the strips survived and gave good electroretinograms for several hours in cold moist air

2

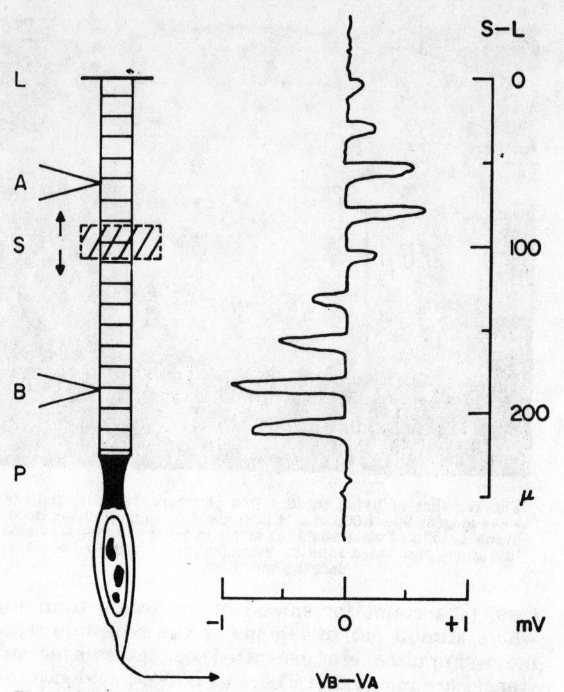

Fig. 2. Extracellularly recorded potentials in a retinal slice with stimuli at various depths in the layer of outer segments. Ordinate, distance of stimulus *S* from internal limiting membrane *L*. Abscissa, potential difference between electro tips. Stimulus intensity equivalent to 3×10^{11} photons cm.$^{-2}$sec.$^{-1}$ at 500 mμ. See legends of Figs. 1 and 3 for explanation of symbols

(5°–10° C.) on the stage of an inverted microscope. In order to avoid light adaptation of the slices, all manipulations and observations were made in infra-red radiation of wave-lengths 0·9–1·0μ rendered visible with the aid of photoelectric image converters. Micropipettes were inserted into the slices and zones in the layer of outer segments were stimulated with movable spots of red light 30–40μ in diameter. The positions of receptors, stimulus, and electrodes were recorded continuously with a motion picture camera photographing the screen of an image converter attached to the microscope. One frame from a record is shown in Fig. 1. The slice lies with the long axes of the receptors in the plane of the page. The tip of pipette *A* lies at the inner limiting membrane *L* while that of *B* has been thrust into the retina so that its tip (arrow) is surrounded by the proximal parts of the same group of outer segments which end

3

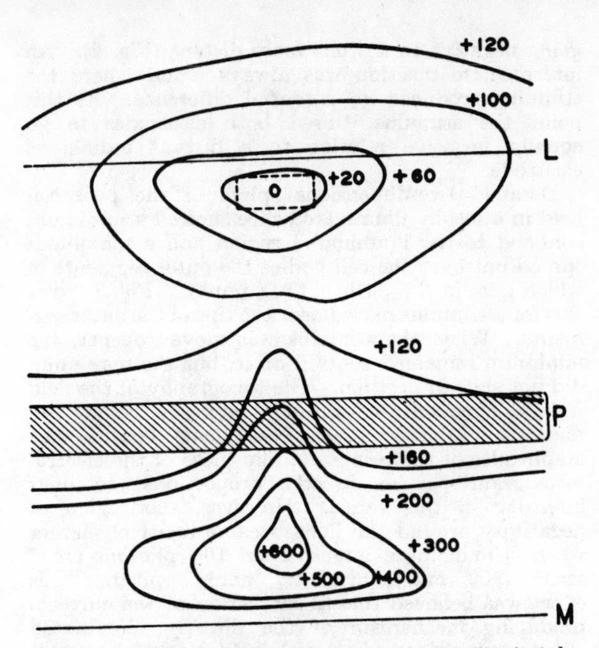

Fig. 3. Isopotential contours in a retinal slice illuminated near the tips of the outer segments. Internal limiting membrane at L, stimulus indicated approximately by dashed square. Cell bodies of receptors lie between pigment layer P and glial membrane M. Contours indicated in microvolts relative to potential at centre of stimulus. Plot was made by inserting a microelectrode into the slice in the direction from L to M and recording responses relative to an electrode at L as the stimulus was swept along a line parallel to L. Stimulus intensity equivalent to 6×10^{10} photons cm.$^{-2}$ sec.$^{-1}$ at 500 mμ

at A. The cell bodies and axons lie below the pigmented layer P. The stimulus S enters the slice perpendicularly to the plane of the page, passing through parts of the outer segments of about 200 receptor cells.

When two pipettes filled with sea-water and with tips 1–5μ in diameter were placed as shown in Fig. 1, no resting potential difference was recorded between them. This was not surprising, since the receptor outer segments are not more than 4–8μ in cross-section and the electrodes would be expected to create new extracellular spaces or enlarge existing ones by destroying some photoreceptors during penetration of the slice. However, when an intermittent light stimulus was moved along a line connecting the two tips, small graded slow potentials were seen which resembled the externally recorded electroretinogram[11]. The sign and amplitude of the potentials depended on the position of the spot, the tip nearest the light

4

going negative to the one more distant (Fig. 2). An intermediate position was always found where the stimulus produced no potential difference. At this point the stimulus caused both electrodes to go equally negative relative to a distant indifferent electrode.

Detailed two-dimensional plots of the potential field in a locally illuminated slice showed a minimum confined to the illuminated region and a maximum spread out over the cell bodies the outer segments of which were in the path of the stimulus. Fig. 3 shows this for a stimulus placed near the tips of the outer segments. When the stimulus was moved deeper, the minimum remained centred on it, but the maximum did not shift in position. The topography of the field was independent of stimulus intensity below 10^{11} photons cm.$^{-2}$ sec.$^{-1}$ (500μ equivalent) because the amplitudes of the responses, like those of the electroretinogram[11b], were directly proportional to light intensity in this range. However, the region of negativity around the light became relatively larger when stimuli more intense than 10^{13} photons cm.$^{-2}$ sec.$^{-1}$ (500 mμ equivalent) were applied. This effect was believed due to saturation of the current-producing mechanism of the directly illuminated outer segments combined with excitation of surrounding cells by stray light scattered from the primary beam.

The fields observed were those to be expected if positive membrane current arose in the cell bodies of the illuminated group of receptors, flowed through the intercellular spaces and entered the outer segments where photons were actually being absorbed. Such a current should change the membrane potentials of the cells which produce it. Attempts were therefore made to record these changes directly with intracellular electrodes.

If a pipette with tip diameter less than 0·2μ and filled with potassium chloride solution (resistance 50–100 megohms in 3 M potassium chloride) was substituted for electrode B of Fig. 1 and carefully tapped into a slice, many regions were encountered in all retinal layers which had negative resting potentials of 10–50 mV. When an electrode was in such a region, illumination of the outer segments in line with its tip produced graded potentials of the usual form but much larger than those recorded with large pipettes filled with sea-water and always positive in sign. In bright light, the responses nearly abolished the resting potential but did not reverse it. Within 2 min. after penetration, the resting potential and light response usually disappeared, declining together until only a small but typical extracellular response from surrounding cells remained. Since the electrode resistance never increased more than 5 megohms when it was inserted into a region giving large responses, it is likely that the tip was in a photoreceptor, because their interiors formed the largest and most numerous spaces present. On this assumption, the space constants of the outer segments were estimated by measuring the amplitudes of the large responses elicited by stimuli at varying distances along the cells from the electrode tip. Values of 400–600μ were usually obtained. No intracellularly recorded action potentials were seen, perhaps because the photoreceptor axons were too severely depolarized by the injury associated with dissection and penetration; but it seems clear that light does produce local currents which markedly depolarize both cell bodies and outer segments and would be capable of initiating or modifying trains of spikes at the origins of the receptor axons.

The preceding results suggest that the action of light is to produce some local electrical change which allows current to flow into the outer segments within a few microns of the point where the photons are absorbed, but they say little else about what the change might be. However, some further information was obtained by light-adapting zones in the outer segments. When the tips of the outer segments the responses of which are shown in Fig. 2 were exposed to a flash containing $7{\cdot}4 \times 10^{14}$ photons $cm.^{-2}$ of wavelength 550 ± 10 mμ, the responses of the tips to subsequent test flashes were initially reduced more than tenfold, recovering their original responsiveness gradually over a period of 15 min. The unilluminated deeper parts of the same outer segments were not detectably affected by the adapting light. Adaptation of deeper zones produced corresponding effects. Intracellular recordings from the adapted cells showed no prolonged changes in their resting potentials, so that it is unlikely that their internal ionic composition was much affected by the light. Moreover, the adapting exposures were not great enough to activate more than 5 per cent of the rhodopsin in the illuminated zones, since the extinction coefficient of the squid photopigment is less than 20,000 $cm.^2$ $m.mol.^{-1}$ at 550 mμ (ref. 12). This result is reminiscent of the drastic reduction in responsiveness of vertebrate retinæ by adapting exposures whose photochemical effects are minor[13].

Two alternative explanations can be offered for the ability of light to adapt zones in outer segments without depolarizing the receptors or destroying their photopigment. First, adaptation might alter the intercellular concentration of some ion which carries current into the receptors. The intercellular spaces seem very small in electron micrographs[9d] and their ionic composition could be greatly affected by a flow of charge that would leave the cytoplasm of the receptors almost unchanged. The 15-min. period required for an adapted zone to recover its sensitivity would then be due to diffusion of ions into the illuminated zones from neighbouring intercellular spaces. However, the diffusion of most small ions in water seems to be nearly two orders of magnitude too fast for this, and it would be necessary to suppose that diffusion is greatly retarded in the spaces between the outer segments. Local adaptation can also be explained in a second more abstract way. If the photopigment molecules were organized into functional units in which single absorbed photons inactivated at least 20 chromophores for a time, the observed adaptation would result. Further work will be necessary to assess the value of these suggestions; but it is clear, nevertheless, that the flow of the receptor current depends to a remarkable degree on highly localized changes induced by light in the outer segments of the receptors. The roles of electron-hole migration and diffusion of transmitter substances thus seem restricted to signal transmission through distances of only a few microns from the sites of light absorption, if indeed they play any part at all in visual excitation in the squid retina.

A detailed account of these experiments and of the electrochemical basis of the receptor current of the squid retina will be published separately. We thank the staff of the Marine Biological Laboratory, Woods Hole, Massachusetts, for their help during this work, and Mr. M. W. Klein of the Engineering Research Development Laboratories, U.S. Army Corps of Engineers, Fort Belvoir, Virginia, for the loan of infrared optical equipment. The opinions expressed herein are ours and are not to be construed as representing those of the U.S. Navy or the naval service in general.

[1] Hecht, S., Shlaer, S., and Pirenne, M. H., *J. Gen. Physiol.*, **25**, 819 (1941–42).

[2] (*a*) Barlow, H. B., Fitzhugh, R., and Kuffler, S. W., *J, Physiol.*, **137**, 327 (1957). (*b*) Yeandle, S., *Amer. J. Ophthal.*, **46**, 82 (1958).

[3] Hagins, W. A., and Jennings, W. H., *Disc. Faraday Soc.*, **27**, 180 (1959).

[4] Rosenberg, B., Orlando, R., and Orlando, J., *Arch. Biochem. Biophys.*, **93**, 395 (1961).

[5] Fuortes, M. G. F., *J. Physiol.*, **148**, 14 (1959).

[6] (*a*) Gray, J. A. B., *Handbook of Physiology*, **1**, 123 (Amer. Physiol. Soc., Washington, 1959). (*b*) Granit, R., *Receptors and Sensory Perception* (Yale, New Haven, 1955). (*c*) Davis, H., *Physiol. Revs.*, **41**, 391 (1961).

[7] Loewenstein, W. R., *Nature*, **183**, 1724 (1959).

[8] Naka, K., *J. Gen. Physiol.*, **44**, 571 (1961).

[9] (a) Wolken, J. J., *J. Biophys. Biochem. Cytol.*, **4**, 835 (1958). (b) Wald, G., and Philpott, D. E., *Exp. Cell. Res.*, Supp. 5, 389 (1958). (c) Moody, M. F., and Robertson, J. D., *J. Biophys. Biochem. Cytol.*, **7**, 87 (1960). (d) Zonana, H., *Johns Hopkins Med. Bull.*, **109**, 185 (1961).

[10] Young, J. Z., *Nature*, **186**, 836 (1960).

[11] (a) Frohlich, H., *Z. Psychol. Physiol. Sinnesorgane*, **48**, 28 (1913). (b) Hagins, W. A., Adams, R. G., and Wagner, H. G., *Biol. Bull.*, **119**, 317 (1960).

[12] Hubbard, R., and St. George, R. C. C., *J. Gen. Physiol.*, **41**, 501 (1957–58).

[13] (a) Dowling, J. E., *Nature*, **188**, 114 (1960). (b) Rushton, W. A. H., *J. Physiol.*, **156**, 193 (1961).

36

Reprinted from *The Journal of General Physiology*, January, 1963
Volume 46, Number 3, pp. 435–452
Printed in United States of America

Transient Responses to Sudden Illumination in Cells of the Eye of *Limulus*

M. G. F. FUORTES and G. F. POGGIO

From the Ophthalmology Branch, National Institute of Neurological Diseases and Blindness, National Institutes of Health, Public Health Service, United States Department of Health, Education, and Welfare, Bethesda; Marine Biological Laboratory, Woods Hole; and Department of Physiology, Johns Hopkins Medical School, Baltimore

ABSTRACT Responses recorded from visual cells of *Limulus* (presumably eccentric cells) following abrupt and maintained illumination consist of depolarization with superimposed spikes. Both the depolarization and the frequency of firing are greater at the beginning of the response than later on. Frequency of firing decreases with time also during stimulation with constant currents, but the decay is then less than it is during constant illumination. Early and steady-state responses do not increase in the same proportion following illumination at different intensities. Membrane conductance is higher during the early peak of the response than in steady state. Early and late potential changes appear to tend to the same equilibrium value. The results support the assumptions that: (*a*) discharge of impulses is the consequence of depolarization of a specialized "pacemaker region" in the axon; (*b*) depolarization induced by light is the consequence of increase of membrane conductance. The major conductance changes occurring during constant illumination may be due to corresponding changes of the "stimulus" supplied by the photoreceptor or to changes of sensitivity of the eccentric cell's membrane to this stimulus. Some accessory phenomena may be the consequence of regenerative properties of the nerve cell itself.

Many receptor organs respond to a suddenly applied and steadily maintained stimulus (a step function) with a discharge of impulses which decrease in frequency with time. It seems important to inquire about the causes of this decay (usually called adaptation), not only because its clarification is a necessary preliminary to the study of other problems, but also because its understanding may further our insight into the basic functions of receptor organs.

The present article is an analysis of the changes occurring with time in the responses evoked by a step of light in cells of the eye of *Limulus*. Findings obtained in previous work (Fuortes, 1959), in which the action of light was studied several seconds after the beginning of the illumination (when the responses had reached conditions approaching steady state), were interpreted on the basis of the model reproduced in Fig. 1. It is assumed in this model that the primary effect of illumination upon the eccentric cell is to increase

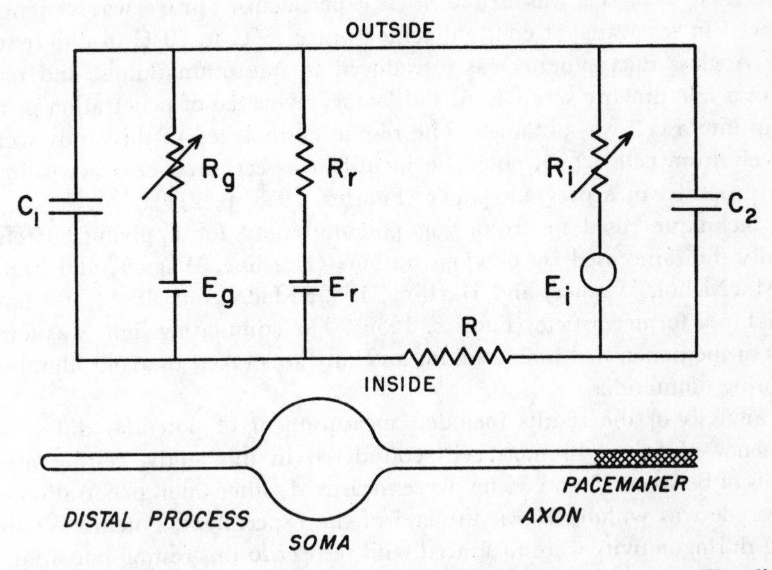

FIGURE 1. Electrical circuit to illustrate properties of eccentric cells. The diagram is essentially identical with those of Figs. 11 and 12 in Fuortes, 1959. A resistance R has been explicitly included to represent separation of pacemaker region from soma. R_g and E_g represent resistance and electromotive force of those parts of the eccentric cell's membrane which are responsible for production of generator potentials; R_r and E_r, of the parts which maintain resting values; and R_i and E_i, of the membrane areas subserving generation of impulses. R_g is supposed to decrease with light, possibly owing to the action of a chemical transmitter. R_g and R_r are thought to be insensitive to the voltage across the membrane, except under special circumstances briefly mentioned under Discussion (electrically inexcitable membrane according to Grundfest, 1959). R_i and E_i are represented as single elements for simplicity but should be imagined to include all mechanisms required for impulse production. Their value is, therefore, supposed to change as a function of the voltage across the membrane.

conductance of part of its membrane: the increase of conductance brings about depolarization and this depolarization is the cause of firing. This same model will be applied to the interpretation of the results obtained in the present study. It will be seen that the initial response to a step of light, as well as the transition from early to "steady state" response, is a disappointingly complex process. Some of these complications should be expected if responses to light are a consequence of the processes implied in the model proposed above; others seem to depend upon features which are not explicitly included in the original model.

METHODS

The lateral eye of *Limulus* was used for these experiments. The eye was sectioned in two and placed in sea water at constant temperature (6°C to 20°C in different experiments). A glass micropipette was introduced in one ommatidium and moved by means of a micrometric screw until satisfactory evidence of penetration of the electrode tip into a cell was obtained. The responses analyzed in this study were taken exclusively from cells which could be identified as eccentric cells according to the criteria proposed in a previous paper (Fuortes, 1958, p. 219).

The techniques used for recording potentials and for applying currents were essentially the same used by previous authors (Hartline, Wagner, and MacNichol, 1952; MacNichol, Wagner, and Hartline, 1953; MacNichol, 1956) and have been described in a former article (Fuortes, 1959). The stimulating light was focused on the lens of the penetrated ommatidium and care was taken to avoid illumination of neighboring ommatidia.

The analysis of the results included measurement of potential differences and of frequency of firing. In most cells considered in this study, resting membrane potentials of between 45 and 55 mv were measured either upon penetration or when the electrode was withdrawn, at the end of the experiment. Potential changes developing during activity were measured with respect to this resting potential. As was done in previous studies (Fuortes, 1958, 1959), in the presence of firing of impulses, the measurement of membrane potential was taken in the intervals between successive spikes.

In the former work, in which only steady-state conditions were considered, frequencies were measured by counting the number of impulses discharged in a convenient time unit, but in the present study, dealing with rapidly changing responses, frequencies were measured as the inverse of the intervals between successive impulses. These intervals were recorded on analog magnetic tape, measured by means of a time interval digitizer, and punched on paper tape by means of a high speed perforator. Since the data analyzed here were not recorded originally on magnetic tape, photographic records were transferred onto tape by moving the film at uniform speed and projecting the enlarged records on a photocell provided with a black screen which left only a slit of about 0.5 mm exposed to light. When the shadow of a spike passed the slit, a sharp electrical transient was generated and recorded on the magnetic tape. The results obtained with this method were checked on occasion by direct measurements on the film record, and were found to be satisfactory.

RESULTS

Responses to Lights of Different Brightness A series of responses to lights of different intensities is shown in Fig. 2. The voltage change underlying the

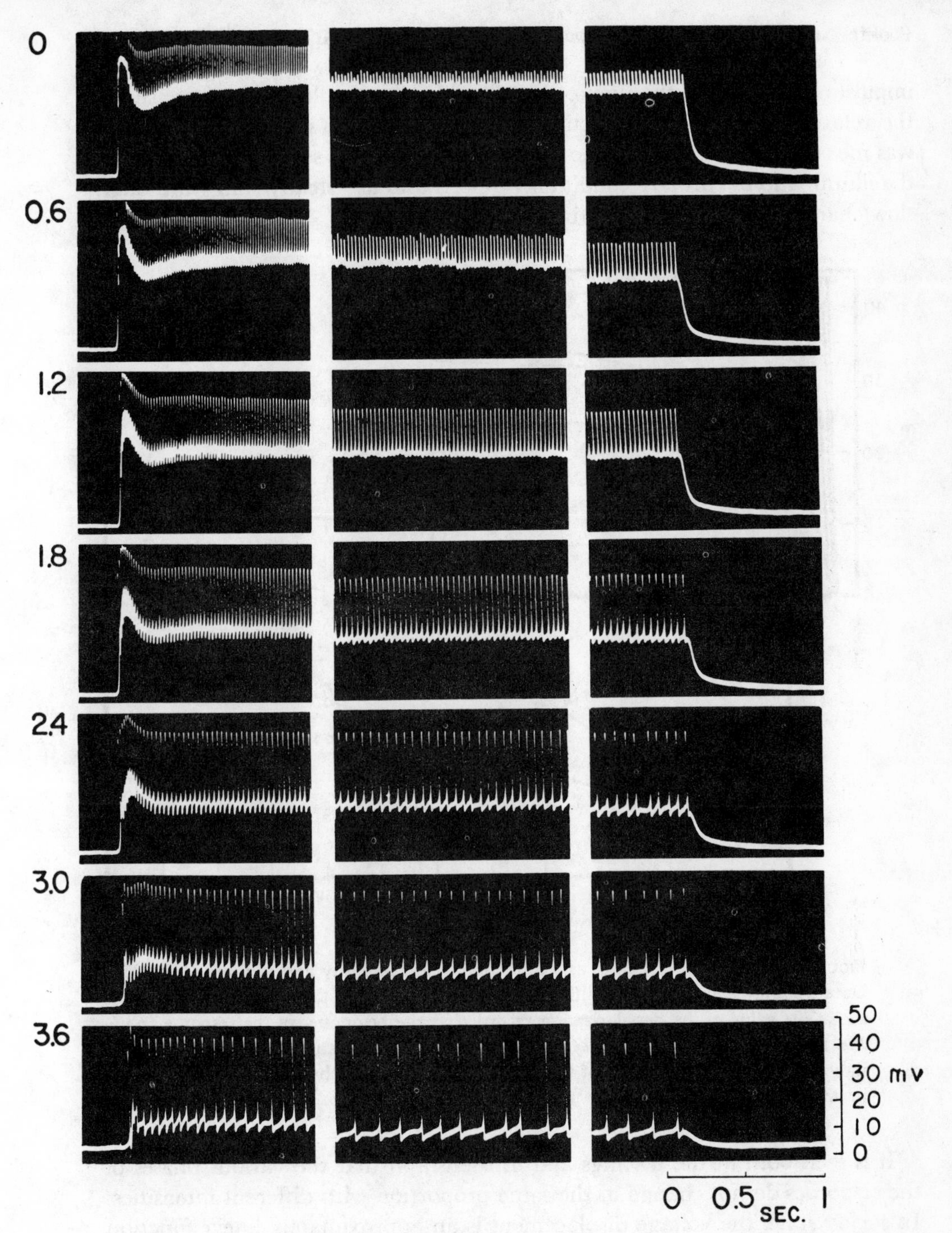

FIGURE 2. Responses to steps of light. Unit 3358 (same cell as in Figs. 3 and 6 through 10). Figures at left indicate attenuation of light intensity in logarithmic scale. Steps of light lasted 20 seconds. Gaps between the three records in each row are 8.4 seconds each. The stimuli were applied in order of increasing intensity and were separated by 20 second intervals. Temperature, 18°C.

impulses is retraced in the inset of Fig. 3, while the graph in this figure shows the relation between voltage change and intensity of stimulation. Voltage was measured at the peak of the early response and 20 seconds after onset of the illumination. This late condition is referred to as "steady state" although slow changes still occur at this time.

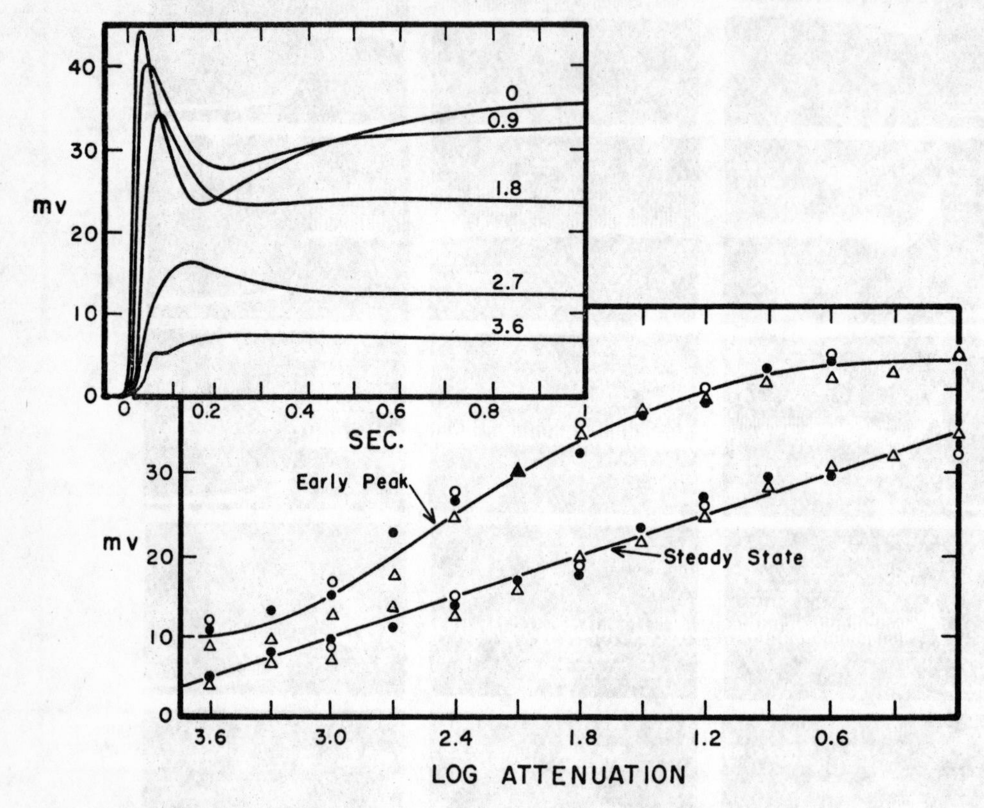

FIGURE 3. Height of responses as a function of light intensity and of time. Unit 3358. Data from three experiments as illustrated in Fig. 2, all from the same unit. The graph shows the value of the depolarization recorded at the beginning of the response (*Early Peak*) and after 20 seconds (*Steady State*), for different attenuations of light intensity, as indicated in abscissa. Tracings of the voltage changes evoked by light are superimposed in the inset.

It is seen both in the tracings and in the graph that the various phases of the responses do not change in the same proportion with different intensities. In steady state, the voltage displacement is an approximately linear function of the logarithm of the intensity of illumination (MacNichol, 1956; Fuortes, 1958), but an S-shaped relation is found for the early peak, so that the difference between peak and steady-state potentials is largest for responses to moderate light intensities.

The course of decay from early to steady-state responses is in general quite complex, although it may fit a simple exponential function for responses to stimuli of moderate intensity (*e.g.* 2.7 in the inset of Fig. 3).

Some of these features may be explained assuming that "saturation" of

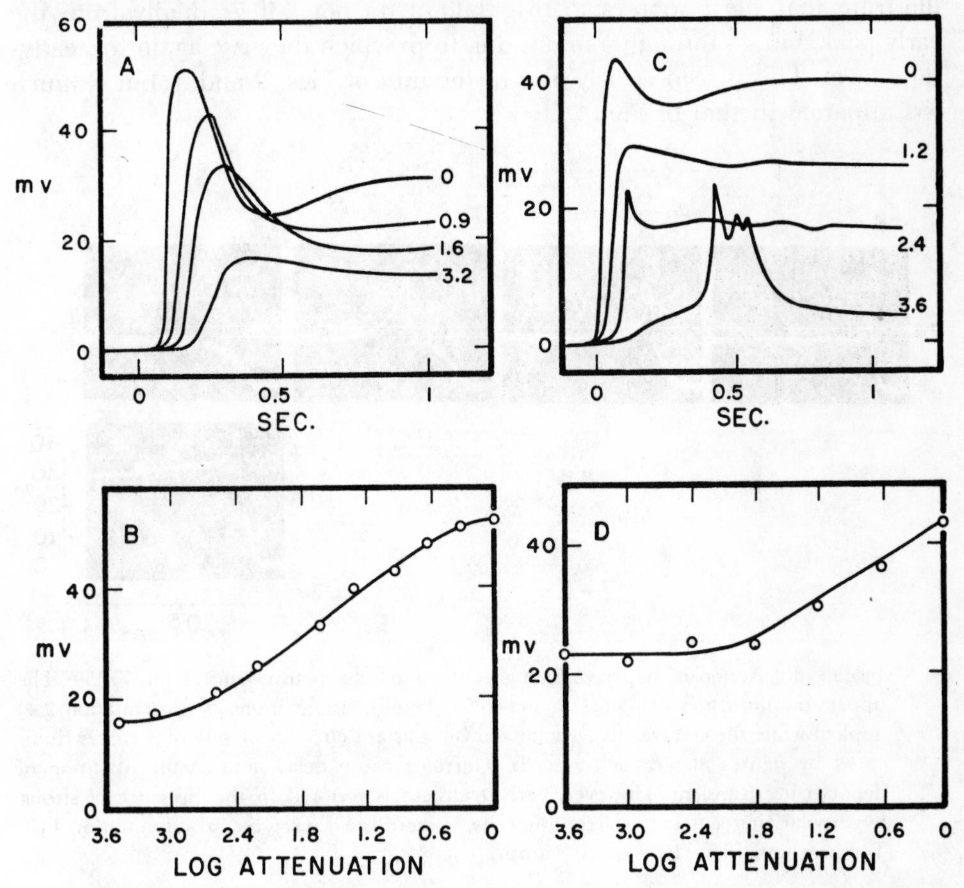

FIGURE 4. Responses to light in two different cells. Tracings and plots like those of Fig. 3, but omitting graph of steady-state responses. The cell of A and B (unit 11157) gave results similar to those of Fig. 3. The responses of the cell of C and D (unit 42358) showed instead sharp transient waves of depolarization which, in the lowest tracing, appeared to originate abruptly from a slower response. In both cases, temperature was about 20°C.

one of the processes leading to depolarization of the eccentric cell occurs following stimulation with bright lights, but early responses may be complicated also by other processes, as revealed by findings to be mentioned in the following section.

Oscillations and Sharp Transients in Early Responses to Steps of Light Fig. 4 shows tracings and plots of the early potential changes recorded from two different cells, following stimulation with steps of light of different intensity.

In the unit of Fig. 4 A and B, the potential change follows a smooth course for all light intensities used, and the height of the early peak grows progressively with increasing light intensity. The superimposed tracings of Fig. 4 A illustrate that the responses to bright light do not fall gradually from the early peak, but go through a minimum, from which they rise again to steady-state level. This is seen clearly also in the unit of Figs. 2 and 3, but is much less apparent in that of Fig. 4 C.

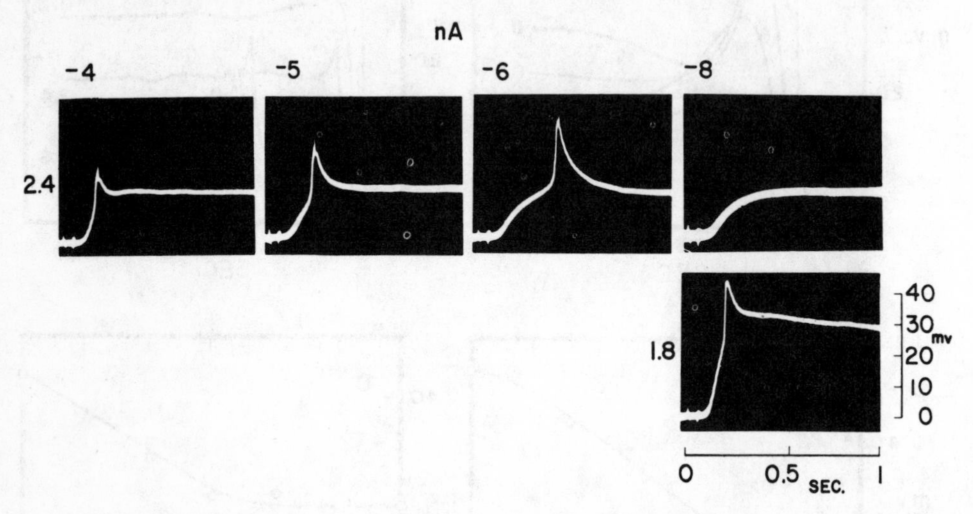

FIGURE 5. Action of hyperpolarizing currents on sharp transients. Unit 71359. The upper row illustrates responses to steps of light of constant intensity (attenuation 2.4) applied while the cell was hyperpolarized by constant currents of different values (indicated by figures above each record). Currents cause delay and finally abolition of depolarizing transient. However, early transient is restored in the presence of strong hyperpolarizing current if light intensity is increased (lower record, attenuation 1.8). Temperature, 18°C. 1 nA = 10^{-9} amp.

In the cell of Fig. 4 C and D, the response to a light attenuated by 3.6 logarithmic units shows sudden sharp transients superimposed upon a more gradually developing response. The sharp initial peak of the response to illumination attenuated by 2.4 resembles the first delayed transient of the bottom tracing and might be due to the same process. The graph of peak height as a function of light intensity (Fig. 4 D) shows that peak height does not change appreciably for attenuations between 3.6 and 1.8. These observations suggest that the early response to a step of light may include two components. The process giving rise to the fast transient does not appear to increase systematically with increasing light intensity and is predominant in the responses to weak lights. The slower component grows as light intensity is increased and becomes predominant with bright lights. When the sharp transient is small or absent, the shape of the response will be controlled by the slower process and results such as are illustrated in Fig. 4 A and B will be obtained.

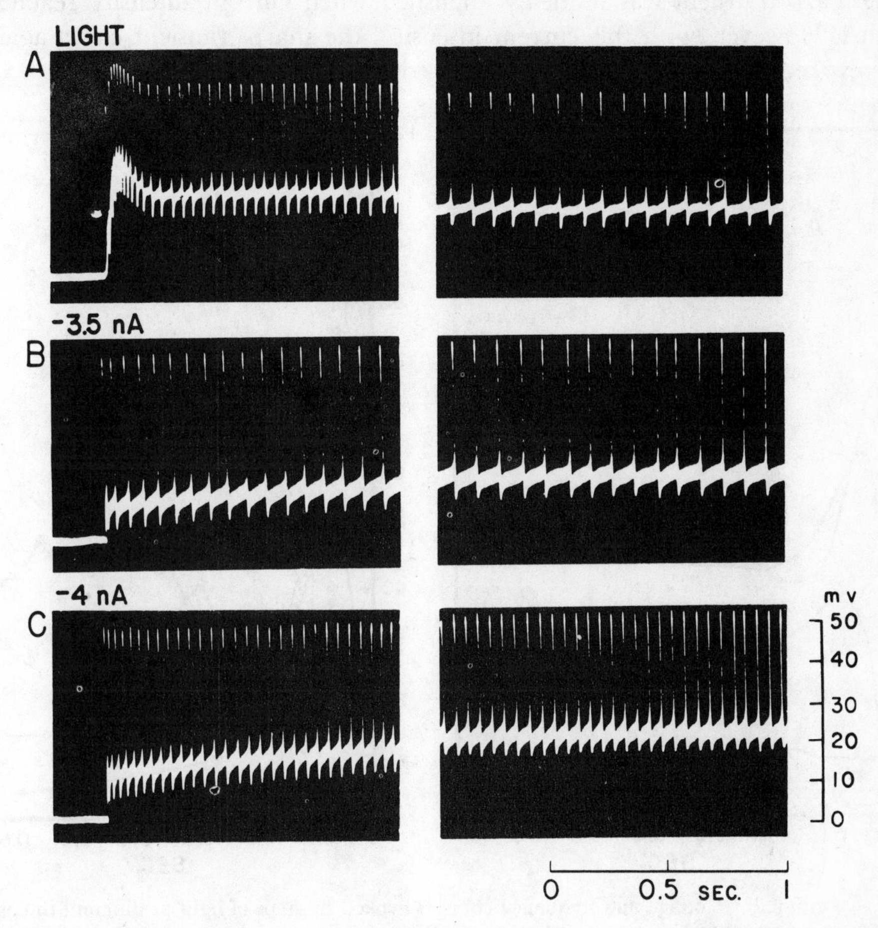

FIGURE 6. Frequency of firing in responses evoked by light or by currents. Unit 3358. A: response to a step of light of moderate intensity. B and C: responses to currents of intensity as indicated. Left and right hand records are separated by an interval of 8.5 seconds. The slow drift of the baseline in B and C is probably due to changes of electrode resistance and does not indicate changes of membrane resistance or membrane potential (see Frank and Fuortes, 1956). Temperature, 18°C.

Effects of Currents on the Sharp Transients Evoked by Steps of Light Fast transients can be separated from the slower component of the early response to a step of light by means of hyperpolarizing currents through the microelectrode. Fig. 5 shows responses recorded when a step of light of moderate intensity (attenuation 2.4 log units) was applied while constant currents of different intensity were passed through the cell's membrane. The hyperpolarizing current increased both the size and the delay of the sharp transient, and separation of the two components of the early response became more and more obvious as current intensity was increased. In the experiment considered, the sharp transient was suddenly abolished when current intensity reached 8 nA. However, with this current intensity, the sharp transient could again be evoked if light intensity was increased (attenuation 1.8).

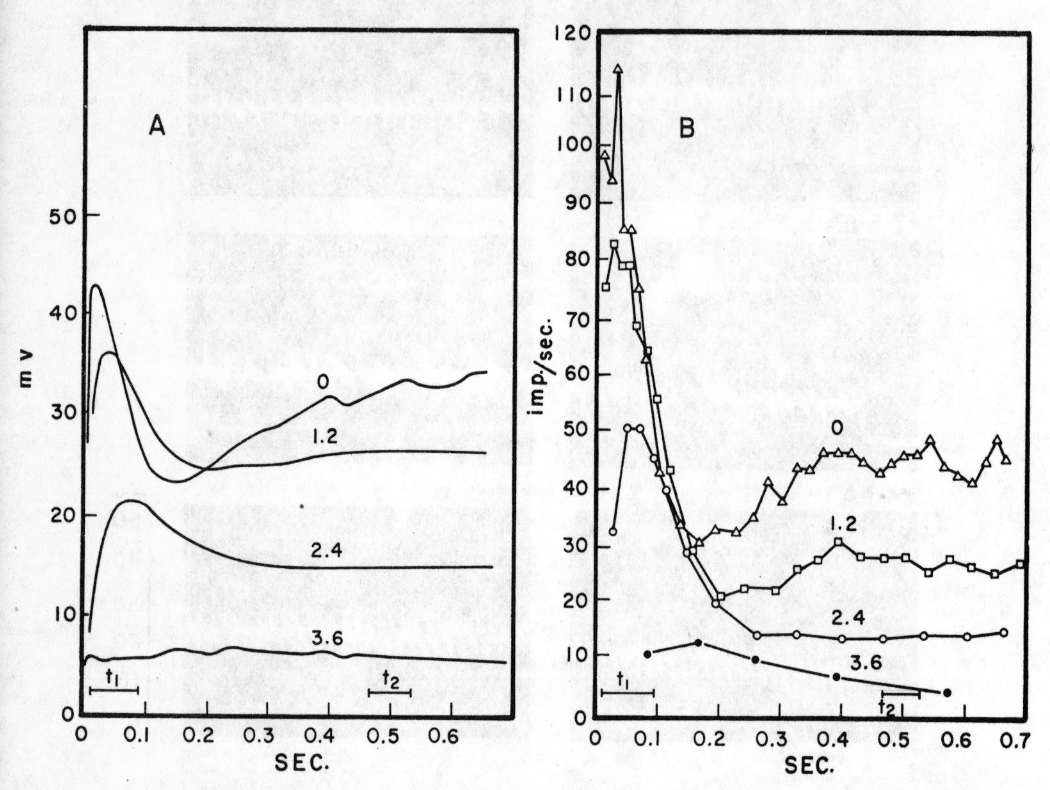

FIGURE 7. Voltage and frequency changes evoked by steps of light at different intensities. Unit 3358. A: tracings of voltage changes as in inset of Fig. 3. Numbers near tracings indicate light attenuation. B: Plots of reciprocals of intervals between successive impulses in responses traced in A. Both in A and B, zero in the abscissa indicates time of firing of first impulse. The points in B are placed at the end of each interval so that they indicate time of firing of following impulses.

In certain experimental conditions (different light intensities applied during flow of constant current), these sharp transients were generated when a critical value of membrane voltage was reached. However, under other conditions it could be shown that these transients do not necessarily originate when the same critical membrane voltage is attained. For instance, only very rarely could sharp transients be evoked when membrane voltage was displaced by currents in the absence of illumination.

Frequency of Firing in Responses to Constant Lights or Constant Currents Hartline, Coulter, and Wagner (1952) stated that frequency of impulses decays in similar manner in responses to steps of light or to steps of depolarizing current applied with external electrodes. This was observed, in the present investigation, only for weak stimuli, when the initial transient of responses to light is

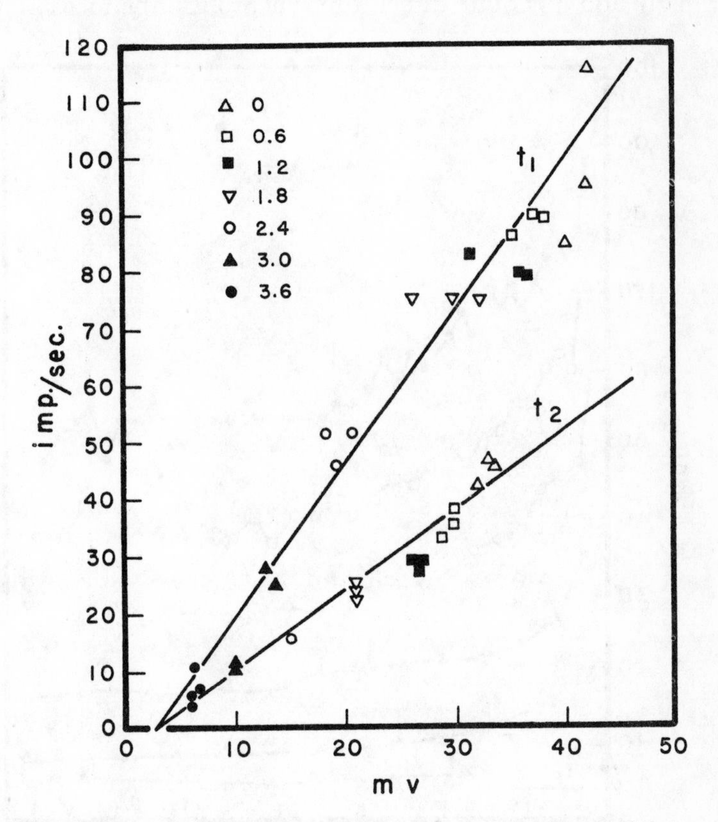

FIGURE 8. Relation between voltage and frequency of firing at different times. Unit 3358. Data taken from experiment of Fig. 7 but including other measurements from same cell. One to three intervals were measured within the period t_1 (indicated by the short horizontal line in Fig. 7) and corresponding voltages were measured from the tracings at the mid-point of each interval. Three measurements within the period t_2 were taken graphically from plots such as those of Fig. 7 B. The different symbols refer to measurements taken at different light intensities, as indicated.

small (see bottom tracings of Figs. 2, 4 A, and 7 A). In general, however, frequency decays more rapidly during constant illumination than during stimulation with constant currents, as illustrated in Fig. 6.

Fig. 7 shows the time course of the voltage displacement (A) and of the rate of discharge (B) evoked in one cell by lights of different intensities. The curves in the two figures present a general resemblance, but they cannot be superimposed by scaling because frequency decays more sharply with time than voltage does.

In Fig. 8, the same data are plotted to show the instantaneous relation between voltage and frequency at selected times (t_1 and t_2 in Fig. 7). These measurements are subject to large errors since both voltage and frequency change rapidly and it is quite difficult to synchronize the measurements with

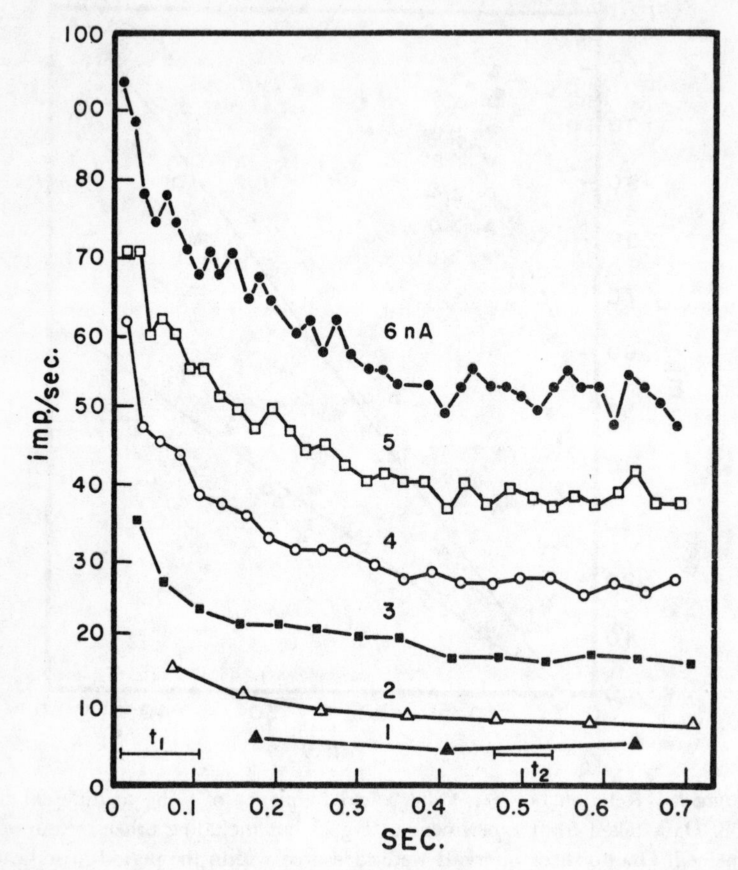

FIGURE 9. Decay of frequency during stimulation with constant currents. Unit 3358. Plots of reciprocals of intervals between successive impulses in responses elicited by depolarizing current steps of intensity as indicated.

accuracy. Still, the data taken at t_1 and t_2 could both be roughly fitted by straight lines, and there is little doubt that the slope of the line was steeper for the measurements taken at the earlier time.

The same type of analysis was then performed for the responses elicited in the same cell by steps of current, and the results illustrated in Figs. 9 and 10 were obtained. It is seen in these plots that frequency decays smoothly with time during flow of a constant current. Comparison of the frequency-intensity plots of Fig. 10 with the frequency-voltage plots of Fig. 8 shows that the slope changes with time in the same manner. This would be expected if the stimulus responsible for generation of impulses were constant during constant current stimulation but changed as the voltage recorded across the soma membrane during illumination.

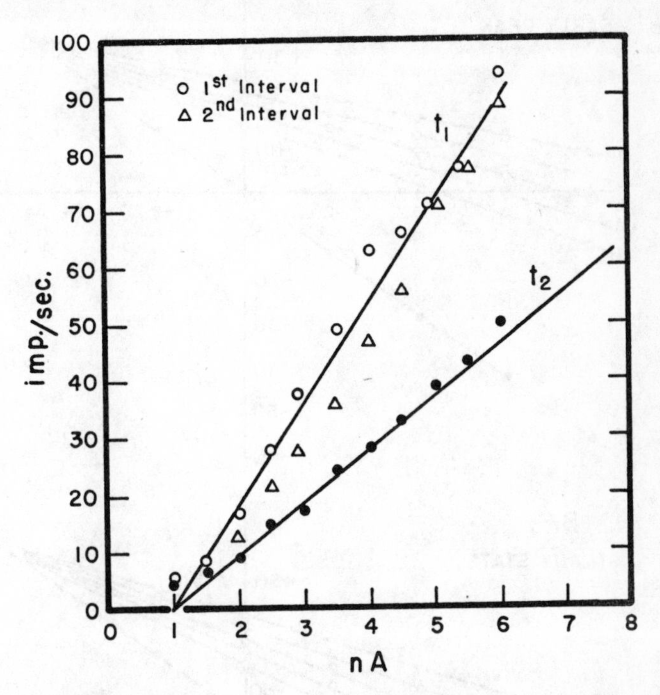

FIGURE 10. Relation between current intensity and frequency at different times. Unit 3358. Reciprocal of intervals measured at an early time t_1 or at a later time t_2 (see short horizontal lines in Fig. 9) after onset of stimulation, plotted as a function of intensity of the stimulating current. Points around the line t_1 measure inverse of first and second intervals in the trains of impulses elicited by each current intensity used. Points around the line t_2 are the average of as many measurements as could be taken within the period t_2, as shown in Fig. 9. The straight lines t_1 and t_2 have slopes in the same ratio as those of the corresponding lines in Fig. 8.

Membrane Resistance during Illumination A relation was found in a previous paper between voltage displacement evoked by light and conductance of the eccentric cell's membrane. This relation was the basis for the model proposed to interpret the mechanisms of responses to light in the eye of *Limulus* (Fig. 1). The measurements performed in the older work in "steady state" conditions (Fuortes, 1959) have been applied in this research to the early phases of the response. Steady currents were passed through the microelectrode, and for each current intensity used, responses to steps of light of different intensities were recorded. The potential drop recorded through the

Wheatstone bridge in each condition was measured at the peak of the initial response and 1.5 seconds after start of the illumination. The results of this experiment are illustrated in Fig. 11. Membrane resistance in darkness was

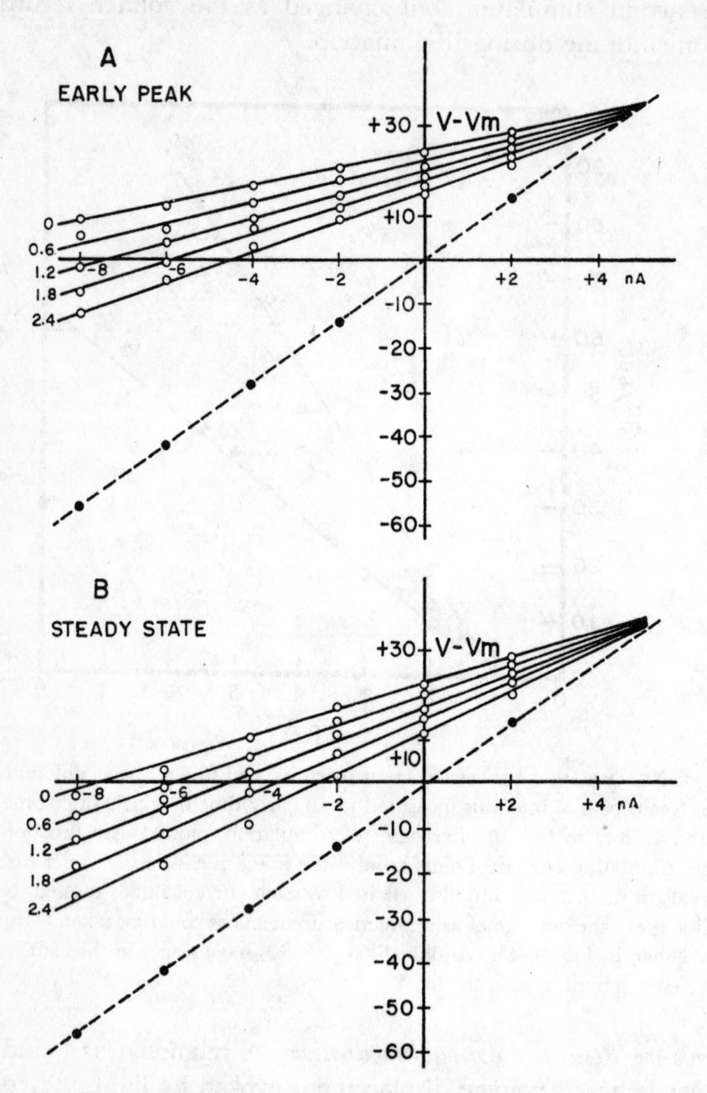

FIGURE 11. Potentials resulting from the combined action of lights and currents. Unit 72059. The dashed lines were drawn to have a slope corresponding to a resistance of 7 MΩ, presumed to measure membrane resistance in darkness (see text for explanation). The vertical distances between dashed lines and open circles measure the potential changes evoked by lights of five different intensities (light attenuation indicated by figures at left) applied during flow of constant currents of the intensities shown in abscissa. Hyperpolarizing and depolarizing currents are indicated by minus and plus signs respectively. Ordinate measures the difference between internal potential and resting potential ($V - Vm$).

determined using the postulation proposed in a former article (Fuortes, 1959) that frequency of firing is determined exclusively by membrane voltage. This resistance was calculated to be 7 MΩ. The voltage displacement evoked by currents could not be measured in the records, because the bridge was balanced for currents applied in darkness, but if the resistance value calculated indirectly is correct, then currents of −8, −6, −4, −2, 0, and +2 nA will change membrane potential by −56, −42, −28, −14, 0, and +14 mv respectively. These values are marked in the plot by solid circles joined by the dashed line. The potential changes evoked by illumination are then measured in the records and are plotted starting from these levels of membrane potential. The open circles in the upper graph (A) in Fig. 11 measure the height of the transient phase of the response to light; the lower graph (B) measures the height of the response 1.5 seconds after start of illumination. It is seen that in either case, the measurements obtained with a given illumination fall on a straight line (see Fuortes, 1959, Figs. 6 and 7). Under some assumptions (mentioned in previous work), the slopes of the lines in this plot measure membrane resistance in darkness (dashed line) and at different intensities of illumination. It follows that membrane resistance is decreased by light and that the resistance change is greater at the beginning of the response than later on.

The lines traced through the experimental points of both plot A and plot B in Fig. 11 converge around a point corresponding to about +35 mv. This value is about 10 mv negative with respect to the outside potential and is approximately the same value obtained in previous experiments (Fuortes, 1959; Rushton, 1959). According to the model presented in Fig. 1, this value represents the equilibrium potential for the changes evoked by light. The results indicate, therefore, that this equilibrium point is the same at the beginning and at a later stage of the response, and that the decay of potential is due to a decrease of membrane conductance rather than to a change in the electromotive force.

DISCUSSION

According to the interpretation advanced in previous work, the essential processes leading to generation of impulses following photic stimulation in the eye of *Limulus* are as follows: (*a*) light is absorbed in a specialized portion of the retinula cells, adjacent to the distal process of the eccentric cell; (*b*) following absorption of light, a chemical (transmitter) substance is liberated by the retinula cells and diffuses to the eccentric cell; (*c*) the transmitter substance combines with the membrane of the eccentric cell process, evoking increase of its conductance and decrease of its potential;[1] (*d*) depolarization of the distal process is accompanied by a current which tends to depolarize soma and proximal portions of the axon of the eccentric cell; impulses are generated in a localized region of the axon (pacemaker region, see Tomita, 1957) as a consequence of this depolarizing current. These suggested mechanisms are largely analogous to those which are supposed to operate in transmission of excitation across the central synapses and at the neuromuscular junctions.

If these are the processes which operate in the transformation of light into nerve impulses, it is clear that a large number of factors may influence the relations between the original stimulus (light) and the final response (discharge of impulses).

The processes controlling the relations between depolarizing currents and frequency of firing can be best analyzed by examining responses to currents applied through the microelectrodes. It was stated in an earlier paper (Fuortes, 1959) that resistance of the eccentric cell's membrane does not change appreciably owing to flow of current, except when impulses are discharged. Thus, a constant stimulating current will produce (in the interval between impulses) a constant potential drop across the soma membrane and an approximately constant current through the pacemaker region of the axon, where impulses are generated. To this constant current, the pacemaker region responds with decreasing frequency of firing, owing to changes occurring there in the stimulus-response relations (Fuortes and Mantegazzini, 1962). These changes occur also when firing is evoked by light, but in addition, constant illumination produces a decaying potential drop across the soma membrane and thus a decaying current through the pacemaker region. The decrease of frequency occurring during constant illumination can be explained as being due to the decrease of voltage at the soma membrane and to changes occurring at the pacemaker region, presumed to be situated in the axon. (See Tomita, 1956, 1957; Fuortes, 1960.)

But while analysis of responses to currents may be useful for understanding the features of firing, it cannot explain the features of the potential changes evoked by light, since these are presumably the consequence of a chemical action.

[1] The assumption of a chemical transmission between rhabdome and eccentric cell was proposed to explain conductance changes following illumination but not following application of currents (Fuortes, 1959; Rushton, 1959). Dr. H. K. Hartline has pointed out to us that conductance changes can be explained without invoking a chemical transmitter, if one assumes (*a*) that the membrane of the rhabdome is very close to the membrane of the eccentric cell, and (*b*) that the membrane of the rhabdome changes conductance with illumination. An analogous interpretation had been advanced a few years ago by Dr. Jerry Lettvin as an alternative explanation of results obtained on spinal motoneurones. Both suggestions seem to fit equally well the data obtained so far, and a choice can be postponed until the need arises.

Experiments such as those described here were performed with the hope that the electrical changes recorded from eccentric cells would furnish rather direct clues for identifying the processes initiated by light in photoreceptors, but it appears that such inference may be instead quite indirect. Clearly, the course of the nerve cell's response should not be expected to reproduce the course of the "stimulus" supplied by the photoreceptor changes, since the ability of the nerve cell's membrane to respond to the stimulus may change with time. For instance, the decay of voltage recorded from eccentric cells during constant illumination might be due to decreased concentration of the postulated transmitter agent, or to decreased sensitivity of the eccentric cell to the transmitter. Such decrease of response due to "desensitization" has been observed in the neuromuscular junction by Katz and Thesleff (1957).

Moreover, it is possible that voltage changes across the eccentric cell's membrane may develop independently both of transmitter concentration and of chemical sensitivity of its membrane, and there are reasons for considering that the sharp, transient potential waves illustrated in Figs. 3 C and 5 may be the consequence of regenerative properties of the eccentric cell's membrane. It has been seen that production of these transients is affected by currents through the eccentric cell's membrane. Depolarization of the eccentric cell does not seem to be sufficient to evoke them, because only rarely could they be produced by means of depolarizing currents in the absence of light, but depolarization to a certain level appears to be required for their production, since they are retarded or abolished by hyperpolarizing currents.

It will be noted that the sharp transients described here have features similar to those of the transient potential changes which sometime occur in dark-adapted preparations, following illumination with dim lights (Yeandle, 1958). To explain production of Yeandle's potentials, it has been suggested that the "transmitter substance" may be liberated in the form of discrete packages, and the effects of currents on Yeandle's potentials were tentatively interpreted assuming that the transmitter carries a negative charge (Fuortes, 1960). It appears difficult, however, to apply these suggestions to the present results. One of the major difficulties is raised by the observation that currents which drastically affect the sharp transients do not influence other components of the generator potential in a comparable way. It is possible, therefore, that the transients described here are due not to discontinuities of transmission but to some process originating in the nerve cell itself when a certain membrane voltage is reached, provided that some other condition (perhaps a certain membrane conductance) is also attained.

A further complication is revealed by the "undershoot" which follows the early peak of responses elicited by bright lights (see Figs. 2, 3, and 4 A). This phenomenon has been ascribed by Jones, Green, and Pinter (1962) to an autocatalytic process, but it is not clear what reactions may be involved. Inhibition due to scattered light falling upon ommatidia adjacent to the one impaled by the microelectrode (Hartline, Wagner, and Ratliff, 1956) may contribute to the decrease of frequency; however, it is unlikely that it may explain the fall of voltage, since only minor voltage changes are usually associated with this type of inhibition (Fuortes, 1960; Hartline, Ratliff, and Miller, 1961, Fig. 10).

It must be concluded that several unknown processes affect the early responses to a step of light and may be responsible for the finding that the early response is not proportional to the response recorded in steady state.

Despite the complex nature of early responses evoked by a step of light, the relation between voltage displacement and conductance was found to be very similar at the peak of the early transient and in steady state.

According to the results of Fig. 11, illustrating the changes of response induced by currents through the eccentric cell's membrane, the equilibrium potential (represented by the value of the battery, Eg in Fig. 1) is approximately the same for early and steady-state responses. Apparently, the early voltage change is large because conductance is high, and decays later because conductance decreases.

Benolken (1961) has recently described results showing that the membrane potential of cells in the eye of *Limulus* may be reversed during the early phase of responses to bright lights and has proposed that the model represented in Fig. 1 should be modified to take this reversal of potential into account. It should be pointed out, however, that the early reversal of membrane potential has been observed by Benolken in cells which could not be classified as eccentric cells according to the criteria proposed in previous work (Fuortes, 1958). In these cells (presumed to be retinula cells), an early reversal of potential had already been reported (Fuortes, 1958, Fig. 2), but the model discussed in this and in preceding papers was proposed to explain the features of responses of presumed eccentric cells, in which potential reversal has not been observed so far.

The results described in this paper on the early phases of responses to steps of light are consistent with the model proposed in a previous article (Fuortes, 1959) and reproduced in Fig. 1. It appears that both the voltage changes occurring during the early phase of visual responses and those occurring in steady state can be ascribed to changes of the resistance represented by R_g in Fig. 1. But the model says nothing about the processes controlling the resistance R_g, and the results described here indicate only that the changes of the value of R_g follow a complex course, being probably controlled by a variety of factors which remain largely unknown at present.

Received for publication, May 1, 1962.

BIBLIOGRAPHY

Benolken, R. M., Reversal of photoreceptor polarity recorded during the graded receptor potential response to light in the eye of *Limulus*, *Biophys. J.*, 1961, **1**, 551.

Frank, K., and Fuortes, M. G. F., Stimulation of spinal motoneurones with intracellular electrodes, *J. Physiol.*, 1956, **134**, 451.

Fuortes, M. G. F., Electrical activity of cells in the eye of *Limulus*, *Am. J. Ophth.*, 1958, **46**, 210.

Fuortes, M. G. F., Initiation of impulses in visual cells of *Limulus*, *J. Physiol.*, 1959, **148**, 14.

Fuortes, M. G. F., Inhibition in *Limulus* eye, *in* Inhibition of the Nervous System and γ-Aminobutyric Acid, (E. Roberts, editor), Oxford, Pergamon Press, 1960, 418.

Fuortes, M. G. F., and Mantegazzini, F., Interpretation of the repetitive firing of nerve cells, *J. Gen. Physiol.*, 1962, **45**, 1163.

Grundfest, H., Synaptic and ephaptic transmission, *in* Handbook of Physiology. Neurophysiology I, (J. Field, ed., editor), Washington, American Physiological Society, 1959, 147.

Hartline, H. K., Coulter, N. A., Jr., and Wagner, H. G., Effects of electric current on responses of single photoreceptor units in the eye of *Limulus*, *Fed. Proc.*, 1952, **11**, 65.

Hartline, H. K., Ratliff, F., and Miller, W. H., Inhibitory interaction in the the retina and its significance in vision, *in* Nervous Inhibition, (E. Florey, editor), Oxford, Pergamon Press, 1961, 241.

Hartline, H. K., Wagner, H. G., and MacNichol, E. F., The peripheral origin of nervous activity in the visual system, *Cold Spring Harbor Symp. Quant. Biol.*, 1952, **17**, 125.

Hartline, H. K., Wagner, H. G., and Ratliff, F., Inhibition in the eye of *Limulus*, *J. Gen. Physiol.*, 1956, **39**, 651.

Jones, R. W., Green, D. G., and Pinter, R. B., Mathematical simulation of certain receptor and effector organs, *Fed. Proc.*, 1962, **21**, 97.

Katz, B., and Thesleff, S., A study of the "desensitization" produced by acetylcholine at the motor end-plate, *J. Physiol.*, 1957, **138**, 63.

MacNichol, E. F., Jr., Visual receptors as biological transducers, *in* Molecular Structure and Functional Activity of Nerve Cells, (R. G. Grenell and L. J. Mullins, editors), Washington, American Institute of Biological Sciences, 1956, 34.

MacNichol, E. F., Wagner, H. G., and Hartline, H. K., Electrical activity recorded within single ommatidia in the eye of *Limulus*, *in* Abstracts 19th International Physiological Congress, Montreal, 1953, 582.

Rushton, W. A. H., A theoretical treatment of Fuortes' observations upon eccentric cell activity in *Limulus*, *J. Physiol.*, 1959, **148**, 29.

Tomita, T., The nature of action potentials in the lateral eye of horseshoe crab as revealed by simultaneous intra- and extracellular recording, *Jap. J. Physiol.*, 1956, **6**, 327.

Tomita, T., Peripheral mechanisms of nervous activity in lateral eye of horseshoe crab, *J. Neurophysiol.*, 1957, **20**, 245.

Yeandle, S., Electrophysiology of the visual system, discussion, *Am. J. Ophth.*, 1958, **46**, 82.

[illegible] and Holmes [illegible] 1949, 38: 105–135.

Grundfest, H. [illegible] Chicago: [illegible], 1950: [illegible].

Hodgkin, A. L. [illegible] 1950, 8: 141–219.

[illegible] and [illegible] for measurement [illegible] of the [illegible] 1954, 2: 40–57.

[illegible] J. Physiol., 1953, 120: [illegible].

[illegible] and [illegible]. J. Gen. Physiol., 1941, 24: 349–371.

Tomita, T. [illegible] 1957, 55: [illegible].

Yoshida, [illegible] 1953, 4: 139–162.

Hursh, [illegible] 1939, 127: 131–139.

Lorente de Nó, R. A study of nerve physiology. [illegible] 1947, 131, 132.

[illegible]

[illegible]

IV. Neuronal Integration

STUDIES IN VISUAL SYSTEMS

The term "integration" applied to a neuron refers to the process by which excitatory and inhibitory inputs to a neuron, together with the neuron's own spontaneity (if any), are summed to produce conducted nerve impulses or alter ongoing impulse activity. Integration by a system of neurons refers to the orderly processing of inputs (information) to produce a reordered or abstracted output. An ultimate goal of studies of neural integration is the explanation of behavioral responses to specific stimuli in terms of the activity of individual neurons.

Studies such as those presented in the preceding sections form an essential background to the study of neuronal integration. The integrative capabilities of a neuron depend first on its shape, passive (cable) electrical characteristics, and the topographical relation of inputs (synapses) to specialized regions such as the impulse initiating zone. They further depend on the characteristics of the neuron's active responses, for example, its tendency to fire spontaneously and the degree to which it fires repetitively to sustained depolarization. In considering the transformation of inputs between neurons, the characteristics of synaptic transmission are fundamental. Not only can the effect be inhibitory or excitatory, both pre- and postsynaptically, but systematic variations in efficacy can occur as a result of immediate (on the order of milliseconds) past history, as seen in processes such as facilitation, post-tetanic potentiation, or fatigue. Very probably changes also result from the history of use over hours, weeks, and months. These changes may involve metabolic, hormonal, or trophic (growth effecting influences (see the last two papers of this section and also **Axelsson and Thesleff, 1959**). Obviously in deriving mechanisms of neuronal integration one must know or deduce the pattern of anatomical connections among the elements, the "wiring diagram."

In the study of the wiring diagram certain (usually invertebrate) nervous systems offer special advantages because reduced cell numbers or increased cell size allow the repeated identification, from animal to animal, of individual neurons. (See, e.g., Kandel, Frazier, Washizu, and Coggeshall, 1968, on *Aplysia*, and Baylor and Nicholls, 1969, on leech). The development of a dye (procion yellow) which can be injected from microelectrodes and will diffuse through the neuron to its smallest processes (Stretton and Kravitz, 1968) has provided a powerful technique for studying the relation between neuronal form and integrative capabilities, as well as for mapping the complex pattern of processes in neuropiles (areas within nervous systems containing exclusively dendrites and axons. See, e.g., Kennedy, Selverston, and Remler, 1969; Davis, 1970; Kaneko, 1970).

Neuronal integration has mainly been studied by the most accessible routes, either tracing the processing of a sensory stimulus from the periphery toward higher order neurons of the nervous system or starting from a muscular movement or behavioral act and tracing the neural commands centrally.

The explanation of specific behavioral movements in terms of the activity of the neurons commanding them is possible in crustaceans such as crayfish because few neurons are involved and because final integration occurs at the muscle (see **Fatt and Katz, 1953; Dudel and Kuffler, 1961c**). Kennedy and his collaborators provided evidence that motor neurons are driven by a hierarchy of interneurons (Kennedy, Evoy, and Hanawalt, 1966; Kennedy, Selverston, and Remler, 1969), and showed that stimulation of a single "command fiber" can call forth a complex behavioral sequence. The existence of command fibers has also been demonstrated in molluscs (Willows, 1968).

It has not been possible to study integration in terms of identified, individual neurons in the nervous system of vertebrates. But generalizations about morphologically or functionally distinct classes of neurons and their role in neuronal integration have been derived. An example is the description of events in motor neurons of a particular muscle in relation to the activity of the muscle during a reflex movement. Recent studies of the cerebellum show how morphological and physiological observations can be combined to deduce the integrative contributions of each morphological class of neurons (Eccles, Ito, and Szentágothai, 1967).

Hartline and his co-workers began analysis of integration in the visual system at the receptor organ itself. Hartline, Wagner, and Ratliff (1956) exploited the relative simplicity of the compound eye of *Limulus* to show that stimulation in one receptor unit (ommatidium) inhibits activity in neighboring units. Such lateral inhibition alters the threshold for firing or decreases the rate of firing under steady illumination. Neighboring ommatidia inhibit one another mutually and this effect is abolished by cutting nerve connections between them. **Hartline and Ratliff (1957)** show quantitatively that the amount of inhibition a *Limulus* receptor exerts on a neighbor can be predicted by two simple simultaneous equations. The magnitude of inhibition is a function of the level of activity in the neuronal axon and this in turn depends on both the stimulus to the receptor and the level of inhibition to which the unit itself is subject. Hartline and Ratliff describe the partial release from inhibition of a neuron when the receptor unit which inhibits it is in turn inhibited by a third unit which does not affect the first cell. They call this effect disinhibition. Lateral inhibition, mutual inhibition, and disinhibition have the effect of increasing output in ommatidia near units in which there is a discontinuity in illumination. These effects increase the response by the mosaic of receptors to edges, borders, and shadows. Disinhibition may be used (speculatively) to explain the optical illusion seen in the Hering's grid on the cover of this book. The importance of such antagonistic relations between receptors becomes clear in the next four articles presented. Such mechanisms may explain some highly complex abstractions of visual perception.

Hartline earlier originated systematic study of discharge patterns in isolated vertebrate (frog) optic nerve fibers (see review by Hartline, 1940). He defined the receptive field of a single optic nerve fiber as the area of retina which must be illuminated to produce an active response in that unit (Hartline, 1938). His studies began a continuing school of investigation of the sequential processing of visual stimuli (input, information) along the vertebrate visual pathway by recording responses of individual cells to specific stimuli. In the last few years, responses of virtually all neuron types in the vertebrate retina have been analyzed with intracellular techniques.

DOWLING AND WERBLIN (1969) provided a detailed anatomical study of synaptic connections among cells of the amphibian retina. The large size of the retinal neurons permitted **Werblin and Dowling (1969)** to record from elements of each layer and correlate these recordings, their anatomical findings, and a variety of stimuli. They confirmed Granit's (1955) notion that the vertebrate retina is a visual system in itself. They demonstrated the occurrence of several integrative events performed by retinal neurons before the first action potential is seen. Kaneko (1970) has presented similar findings in goldfish (see also Tomita *et al.*, 1967, for recordings from fish cones). Grundfest (1961) suggests that axon conduction is not necessary over such short distances.

Kuffler (1953) provided a detailed description of the functional organization and discharge pattern of cat retinal ganglion cells years before the beginning of the current blossoming of studies of retinal organization. The ganglion cells are the first (in the progression centrally from the receptors) to send their axons (and propagated impulses) from the retina to the brain (lateral geniculate in cats). Kuffler broadens the concept of receptive field to include all those areas of retina which affect responses of the cell under study. His analysis has provided a model for study of single unit function in the visual pathway and has been extended by others to the squirrel, rabbit, and monkey, to rods and cones, and so forth (see, e.g., Michael, 1968; and Gouras, 1968). Kuffler describes a general organization common to retinal ganglion cell receptive fields. There is a central region, illumination of which produces a response which is opposite to that observed when a surrounding annulus is illuminated; and there is also an intermediate zone. Illumination of the center portion of the receptive field may increase firing in the unit (an on-center unit) or decrease it (an off-center unit). Illumination with an annulus will inhibit the on-center or excite the off-center units. Hence the cell responds to a balance of influences from the differing portions of its receptive field to produce an "integrated" impulse pattern in its axon. Kuffler emphasizes that sudden changes (e.g., motion) of light produce the most marked effects.

Lettvin, Maturana, McCulloch, and Pitts (1959) suggest that use of small spots of light may exclude recognition of some kinds of integration. Using a large variety of stimuli, in the frog instead of the cat, they identified optic nerve fibers which they call sustained contrast, net convexity, moving edge, and net dimming detectors. **Barlow and Levick (1965)** analyze the mechanism of similar moving-edge (direction selective) units found in the rabbit retina and suggest the presence of a mechanism involving lateral inhibition at the level of the horizontal cells.

Hubel and Wiesel examined sequentially the neuronal relays of the vertebrate visual pathway including the lateral geniculate, striate cortex, and nonstriate areas in cats and monkeys (see HUBEL AND WIESEL, 1968). In the study reproduced here (**Hubel and Wiesel, 1962**) they identify several types of receptive fields with respect to the distribution of on and off centers and surrounds. They analyze these with respect to the kinds of stimuli likely to produce maximal response (e.g., edges, moving bars in various orientations). They found a hierarchy of cells with respect to the complexity of their responses to linear stimuli (lines, bars, edges). When responses of the unit to forms of light were predictable from the excitatory and inhibitory regions mapped out with small spots of light, the unit was called simple. For so-called complex cells it was not possible to map out distinct excitatory and inhibitory regions, and responses to small spots of light were weak and inconsistent. Both simple and complex types are, however, responsive to linear stimuli in a particular orientation, but complex cells respond to the properly oriented stimulus placed anywhere in the receptive field. The hierarchy of complexity has since been extended to hypercomplex and hyper-hypercomplex cells (e.g., HUBEL AND WIESEL, 1968), each representing a further abstraction of some property of the stimulus. **Hubel and Wiesel (1962)** also show that some cells respond only to binocular stimuli (binocular synergy). Finally they confirm the presence of topographic representation of the visual field in the cortex. They find that the cortex can be divided into cell columns, extending from surface to white matter, which respond to similar orientation and types of stimuli. Such columnar organization was shown previously in somatosensory cortex of cats by MOUNTCASTLE (1957). In the monkey (HUBEL AND WIESEL, 1968) two independent sorts of columns are found. The first groups cells with common receptive field orientations and the other contains cells of similar eye preference. Additionally, Hubel and Wiesel show that horizontal layering with respect to complexity of abstraction occurs.

How functional neuronal connections of the complexity just described become established remains an open question. Sperry proposed that (in frogs) neuronal specificity develops during maturation and that the correct cells of the retina and optic tectum connect during growth (e.g., Attardi and Sperry, 1963). This theory excludes any role for visual function in establishing the "wiring diagram." The brief paper of **Maturana, Lettvin, McCulloch, and Pitts** (1959) is included as the briefest introduction to these unresolved problems that we could find. It supports Sperry's view (see Sperry and Hibbard, (1968), by showing electrophysiologically the specific regrowth of cut optic nerve fibers to form functional synapses at the proper region and depth of the tectum in the frog. Gaze, Keating, Szekely, and Beazley (1970) present further experiments which indicate that normal activity of the eye may be needed during regrowth to establish some normal connections.

Wiesel and Hubel (1963b) provide further evidence of the importance of functional interaction in neuronal development. They first established the presence of functioning cortical cells in kittens which had not yet used their eyes. Thus they established that the basic "wiring diagram" was established without previous visual experience. However, in kittens in which one eye had been sutured shut at birth there was marked atrophy in the lateral geniculate projections of the deprived eye when examined three months after birth Wiesel and Hubel, 1963a). In the companion paper reproduced here (Wiesel and Hubel, 1963b) they demonstrate failure of most cortical cells to respond to stimulation in the previously deprived eye. Thus previously functional synapses have been "lost." In recent work they have shown that there is a relatively brief period (perhaps as short as three or four days) during which the visual system is susceptible to deprivation. This suggests a physiological basis for the explanation of observations of critical periods in development of behavior (Hubel and Wiesel, 1970).

The study of binocularly blinded kittens (Wiesel and Hubel, 1965) produced a further observation which was surprising. Fewer, rather than more, cells were abnormal. Hence the functional development or degeneration of visual pathways may depend more on interactions between afferent components than on the mere quantity of input.

37

J. Gen. Physiol. (1957) **40**, 1357-376.

INHIBITORY INTERACTION OF RECEPTOR UNITS IN THE EYE OF LIMULUS*

By H. K. HARTLINE and FLOYD RATLIFF

(From The Rockefeller Institute for Medical Research)

(Received for publication, July 30, 1956)

In the lateral eye of the horseshoe crab, *Limulus*, the visual receptor units exert an inhibitory influence mutually upon one another. The discharge of impulses in any one optic nerve fiber, generated in the sensory structure of the particular ommatidium from which that fiber arises, is determined principally by the intensity of the light stimulus to the ommatidium and the state of adaptation of this receptor unit. However, the ability of an ommatidium to discharge impulses is reduced by illumination of the ommatidia in neighboring regions of the eye: its threshold to light is raised, and the frequency of the discharge that it can maintain in response to steady suprathreshold illumination is decreased. This inhibitory action is exerted reciprocally between any two ommatidia in the eye that are separated by no more than a few millimeters. As a result of inhibitory interaction among neighboring receptors, patterns of optic nerve activity are generated which are not direct copies of the patterns of external stimulation, but are modified by this integrative action that takes place in the eye itself.

These basic features of inhibition in the eye of *Limulus* have been described in detail in a recent paper (Hartline, Wagner, and Ratliff, 1956). In that paper it was shown that the anatomical basis for the inhibitory interaction is a plexus of nerve fibers lying just back of the layer of ommatidia, connecting them together. Furthermore, a direct experiment demonstrated mutual inhibitory action between two ommatidia whose respective optic nerve fibers were placed on the recording electrodes together. It is the purpose of the present paper to analyze the inhibitory interaction of receptor units in the eye of *Limulus*, and to describe quantitative properties of receptor activity that arise as a consequence of this interaction.

Method

The experiments reported in the present study are based on the measurement of the frequency of the discharge of nerve impulses from two receptor units simultane-

* This investigation was supported by a research grant (B864) from the National Institute of Neurological Diseases and Blindness, Public Health Service, and by Contract Nonr 1442(00) with the Office of Naval Research.

ously, enabling the exact description of their interaction. In each experiment, a lateral eye of an adult *Limulus* was excised with 1 to 2 cm. of optic nerve and mounted in a moist chamber (maintained at 18°C.). Two small strands were dissected from the optic nerve and each placed over a pair of wick electrodes connected to its own separate amplifier and recording system. In some experiments we dissected each strand until only a single fiber remained, as evidenced by the uniformity and regularity of the action potential spikes elicited in response to illumination of the eye. In other experiments, bundles containing many active fibers were used and the isolation of single units was accomplished by coating the eye with opaque wax (a heavy suspension of lampblack in paraffin wax) and then removing the coating carefully from a very small region, exposing the corneal facet of just that one ommatidium from which it was desired to record impulses. The black wax evidently prevents internal reflections inside the cornea of the eye, for by this method perfect isolation of single units can often be obtained,—a result rarely achieved merely by focussing a small spot of light on the facet by means of a lens.

The receptors of the eye were illuminated by the same optical system that was described in the paper mentioned above. Small spots of light were projected on the eye, their sizes controlled by diaphragms and their intensities by neutral wedges. The direction of incidence of each beam could be adjusted for maximal effectiveness by a system of mirrors. A separate system was employed for each spot of light, to avoid scatter in a common optical path.

Oscillograms of the amplified action potentials in the nerve fibers were recorded photographically in some experiments. Often, however, it was preferable to measure directly the frequency with which impulses were discharged in each nerve fiber. This was done by leading the output of each amplifier through a pulse-shaper into an electronic counter. The threshold of the pulse-shaper was calibrated, and could be set to discriminate between the action potential spikes and amplifier noise with perfect reliability (in *Limulus* optic fibers, spikes can usually be obtained that are many times greater than any fluctuations of potential due to noise); the uniform output of the pulse-shaper insured perfect operation of the counting circuits. Each counter was "gated" by an electronic timer, so that only those nerve impulses occurring within a specified interval of time (usually several seconds) were registered. The gating timer was activated by a delaying timer, which permitted the counting period to be started at any desired time (usually 1 or 2 seconds) after the onset of illumination to the eye.

To obtain maximum precision in the measurement of frequency of discharge, we displayed the gating voltages and the pulses to the counter from both recording channels on a dual trace oscilloscope. We then estimated for each channel that fraction of an interval between impulses that occupied the time between the onset of the counting period and the occurrence of the first counted impulse, at the beginning of the counting period, and the corresponding fraction between the last counted impulse and the cessation of the counting period. These fractions were added to the total number of intervals registered. For greater convenience in some experiments the delaying timer, instead of activating the counter gate directly, was arranged to sensitize an electronic "trigger" which, upon the occurrence of the next nerve impulse, activated the gate to the counter. Thus the counting period always started at the occurrence of an impulse, and only the fractional interval at the end of the gated counting period needed to be estimated. The precision gained by these methods was necessary when the counting period was short (1 or 2 seconds), and was desirable for the longer periods usually employed (7 to 10 seconds). Measurement to within about one-quarter interval was warranted by the regularity of the discharge found in many preparations, and the reproducibility of the frequencies observed.

In the experiments reported in this paper, we have confined our attention to the frequency of the discharge of impulses that is maintained at a more or less steady level during steady prolonged illumination of the eye. The transient changes in frequency that occur during the first second or two after light is turned on or off were excluded from the measurements. Exposures were fixed in duration (usually less than 10 seconds), and were repeated at fixed intervals of 2 to 5 minutes, with longer periods of rest interspersed, to minimize cumulative effects of light adaptation.

RESULTS

In the recent paper to which we have referred, direct evidence was given that receptor units in the *Limulus* eye may inhibit one another mutually. Recording the action potentials in a nerve strand containing two active fibers, it was shown that the discharge of impulses in either of the fibers was slowed when the other was brought into activity by illuminating the ommatidium from which it arose. A similar experiment is illustrated in Fig. 1, in which two strands dissected from the optic nerve, each containing a single active fiber, were placed on separate recording electrodes so that the action potentials of each of them were recorded separately. A small spot of light was centered on each of the ommatidia (designated "A" and "B") in which the fibers originated. The oscillograms show the effects of illuminating each of these small regions of the eye separately and together. The steady frequency of discharge of impulses in each fiber was less when both receptor units were active than when they were stimulated singly.

In the experiment shown in Fig. 1 the spots of light centered on the respective ommatidia were each made large enough to illuminate several receptors immediately adjacent to them in order to make the slowing of the discharges large enough to be apparent at a glance. Strictly mutual inhibition of the individual units, however, was exerted by these receptors, for similar (though less pronounced) slowing of the discharge was produced when each was illuminated by a spot so small as to be confined to the facet of its ommatidium, except for slight amounts of light that may have been scattered in the eye. In many of the experiments to be reported below, we took precautions to ensure that we were dealing with the mutual interaction of only two ommatidia by using opaque wax to effect complete optical isolation, as described in the section on Method.

From our previous study, we know that the inhibition of a receptor unit, measured by the decrease in the frequency of its discharge, is greater the

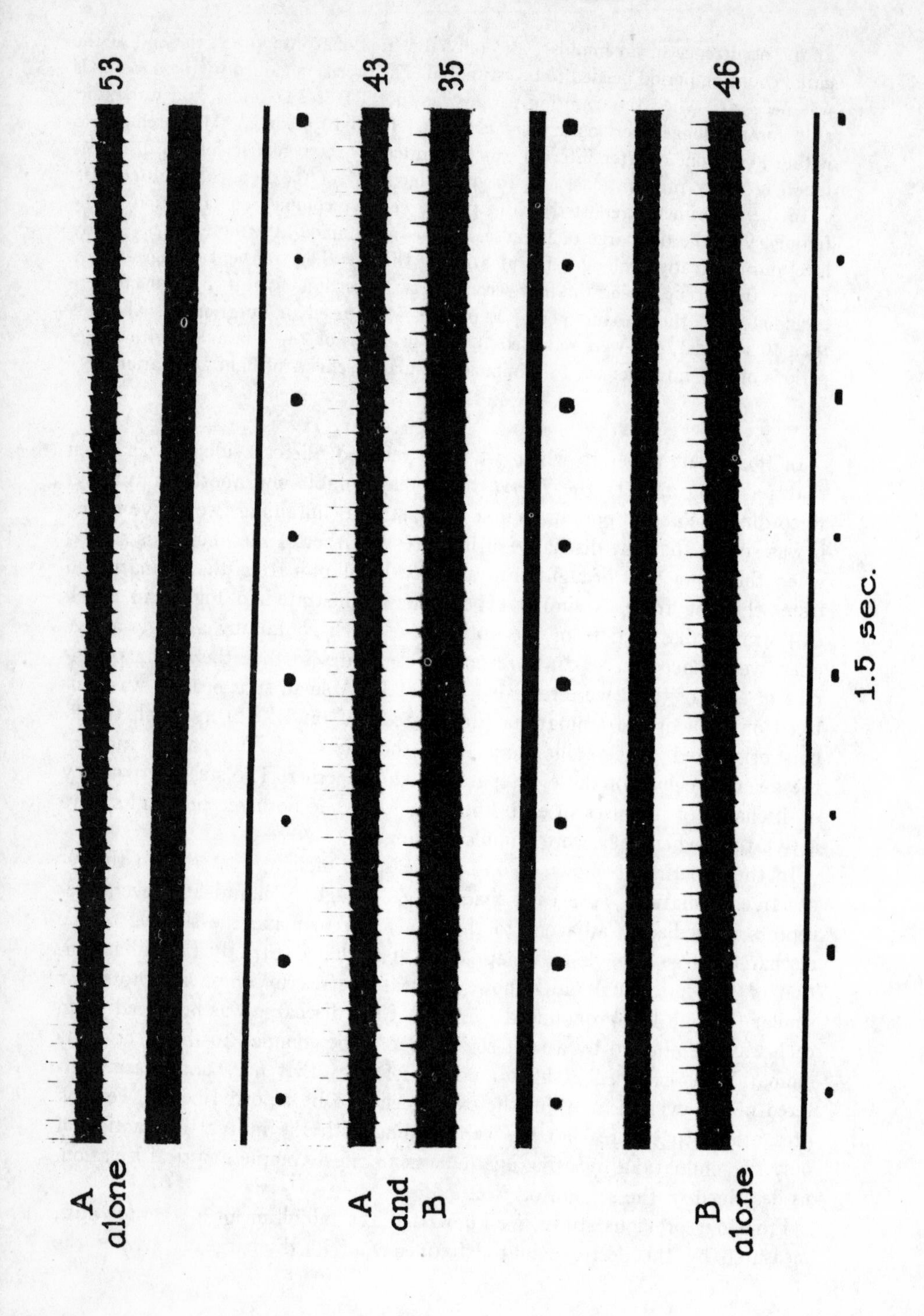
A alone
53
A and B
43
35
B alone
46
1.5 sec.

Fig. 1. Oscillograms of action potentials recorded simultaneously from two optic nerve fibers of a lateral eye of *Limulus*, showing the discharge of nerve impulses when the respective ommatidia in which these fibers originated were illuminated singly and together. In the top record, one ommatidium ("A," nerve fiber activity recorded by upper oscillographic trace) was illuminated by itself at an intensity that elicited the discharge of 53 impulses (as indicated at the right) in the period of 1.5 seconds covered by the records. In the bottom record, the other ommatidium ("B," activity recorded by lower trace) was illuminated by itself at an intensity that elicited the discharge of 46 impulses in 1.5 seconds. In the middle record, both ommatidia were illuminated together, each at the same intensity as before; ommatidium A discharged 43 impulses, ommatidium B 35 impulses, in 1.5 seconds. For A, the decrease in frequency of 10 impulses per 1.5 seconds is taken as the magnitude of the inhibition exerted upon it while B was discharging at the rate of 35 impulses per 1.5 seconds; for B, the decrease of 11 impulses per 1.5 seconds measures the inhibition exerted upon it while A was discharging at the rate of 43 impulses per 1.5 seconds. Two separate optical systems were used, each focusing a small spot of light (approximately 0.5 mm. in diameter) on the eye, one centered on ommatidium A, the other centered on ommatidium B. The spots were 1 mm. apart, center to center. Each spot illuminated about 5 ommatidia in addition to A and B. For each record, the light had been turned on 7 seconds before the start of that portion of the record shown in the figure. Time marked in one-fifth seconds; black bands above time marks are the signals of the stimulating illumination.

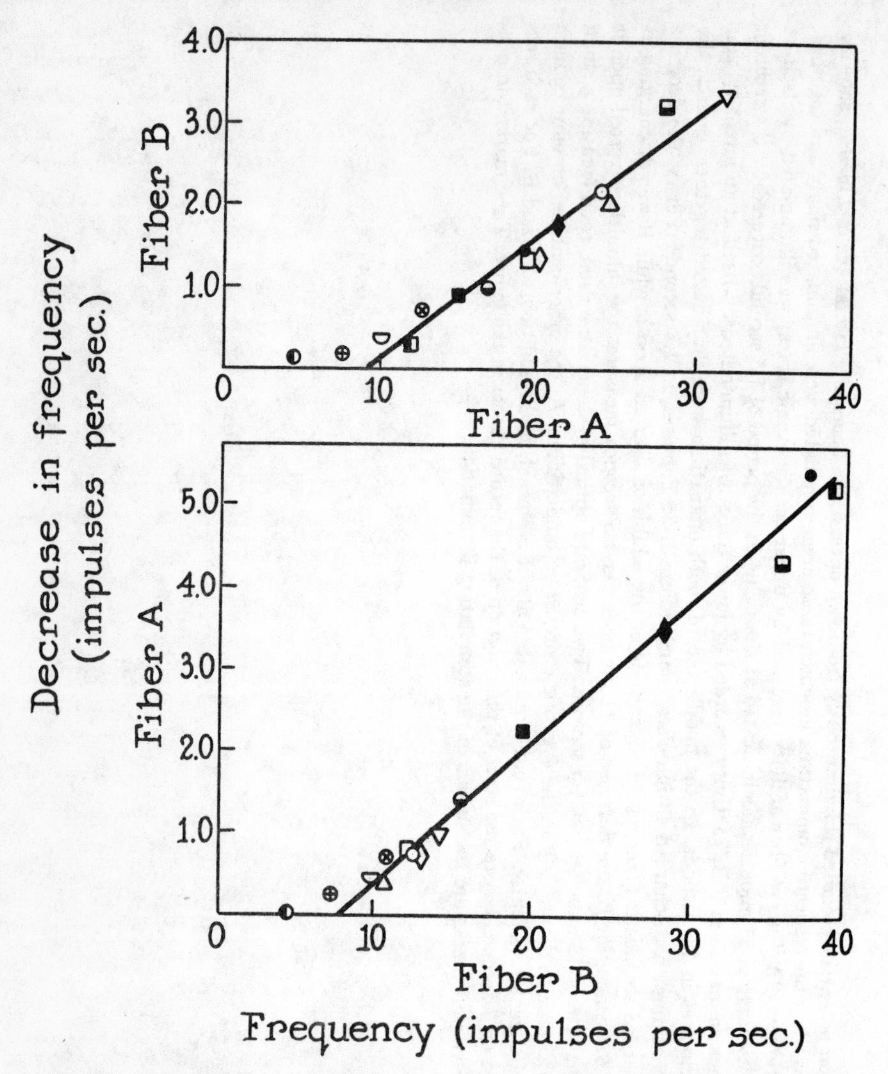

FIG. 2. Graphs showing mutual inhibition of two receptor units in the lateral eye of *Limulus*. In each graph the magnitude of the inhibition of one of the ommatidia is plotted (ordinate) as a function of the degree of concurrent activity of the other (abscissa). Sets of optic nerve fiber responses similar to those shown in Fig. 1 were analyzed as explained in the legend of that figure, each set yielding a point in the upper graph (inhibition of ommatidium B by ommatidium A) and a corresponding point (designated by the same symbol) in the lower graph (inhibition of ommatidium A by ommatidium B). The different points were obtained by using various intensities of illumination on ommatidia A and B, in various combinations.

Frequencies were determined by counting the number of impulse intervals during stronger the stimulus to receptors that inhibit it. Experiments similar to that illustrated in Fig. 1 make it possible to show quantitatively how the amount of inhibition exerted on a receptor varies with the degree of activity of a nearby receptor unit that exerts this inhibition; the mutual interaction of two receptors can be analyzed by stimulating each of them at different intensities singly and in combination. The result of such an experiment is shown in Fig. 2. In this experiment, the frequencies of discharge of each of two ommatidia were measured, for various intensities of illumination, when each was illuminated alone and when both were illuminated together. The decrease in the frequency of discharge of each has been plotted as ordinate against the frequency of the concurrent discharge of the other as abscissa. The upper graph shows the amount of inhibition exerted upon ommatidium B by ommatidium A, as a function of the degree of activity of A; the lower graph shows the converse effect upon A of the activity of B. Both sets of points are adequately fitted by straight lines. In each case there was a fairly distinct threshold for the inhibition; each ommatidium had to be brought to a level of activity of 8 or 9 impulses per second before it began to affect the discharge of the other. Above this threshold, the frequency of discharge of B was decreased by 0.15 impulse per second for each increment of 1 impulse per second in the level of activity of A; the corresponding coefficient of the inhibitory action in the reverse direction (A acted on by B) was 0.17.

We have performed many similar experiments. Six of them, including the one just described, were done with "optical isolation," employing large nerve bundles that had exhibited activity of many fibers before the application of

---

the last 5 seconds of a 7 second exposure to light (so that only the steady discharge was measured). To obtain each pair of points in the two graphs two such counts were made for each of the following conditions of illumination: A alone, B alone, A and B together, presented in an order designed to minimize systematic errors. The averages of such duplicate determinations of the magnitude of the inhibition are the values plotted in the graph. From the distribution of the differences between the individual measurements in each duplicate determination the standard error of the points in the graph was calculated to be 0.12 impulse per second. The straight lines were fitted by the method of least squares. In the upper graph the line has a slope of 0.15, which is the value of the "inhibitory coefficient" $K_{B,A}$ (effect of A on B); in the lower graph the slope is 0.17 ($= K_{A,B}$, the coefficient of the effect of B on A). The intercept of the line on the axis of abscissae is 9.3 impulses per second for the upper graph, 7.8 for the lower. Disregarding a possible "toe" at the bottom of each plot, these give the values, respectively, of the thresholds of the inhibitory effect of A acting on B, designated later in the text as r_A^0, and of B acting on A (r_B^0).

In this experiment illumination was restricted to the two ommatidia from which activity was recorded by masking the rest of the eye with opaque wax (see text). These ommatidia were 1 mm. apart.

opaque wax to the eye to mask all but those two receptor units singled out for observation. In such experiments we could be quite certain that not any of the receptors adjacent to those under observation were excited by scattered light (since no nerve impulses from them were observed). In these experiments, therefore, the observed inhibitory effects were entirely those exerted mutually by the two receptor units upon one another. In other experiments we could be less certain about possible contributions from adjacent receptors excited by scattered light, although the scattered light was never very strong, and its effects were probably below threshold in most cases. All these experiments have shown features similar to those exhibited in Fig. 2. All showed a linear relation between the magnitude of the inhibition of one receptor (measured by the decrease in its frequency) and the degree of concurrent activity of the other (measured by its frequency). Nearly all experiments showed a "threshold" frequency below which no inhibitory effect was detected. The threshold was usually about as distinct as that shown in Fig. 2—a slight "toe" at the bottom of the curve was often noted. Although the values of the two thresholds were nearly identical in the experiment of Fig. 2, in other experiments they were not always the same for both members of an interacting pair. Likewise the slopes of the two curves often differed more than was the case in the experiment we have figured, sometimes by as much as a factor of 2.

The key to the analysis of the mutual interaction in the eye of *Limulus* lies in the correlation between the magnitude of the inhibition of a receptor and the degree of concurrent activity of the receptors that inhibit it. The degree of activity of any one of these receptors, however, depends not only on the stimulus to it but also on whatever inhibitory influences it may be subjected to in turn. It is the resultant level of activity of a receptor unit that determines the strength of the inhibition it exerts on a neighboring receptor. We have direct experimental evidence for this. Fig. 3 shows a small portion of the upper graph of Fig. 2; the points plotted as open symbols are measurements of the inhibition of ommatidium B produced by illuminating ommatidium A at two different intensities (two points at each intensity), with B illuminated at a low intensity. At the higher of the two intensities on A, which elicited a discharge in fiber A of approximately 24 impulses per second, the response of B was decreased by a little more than 2.0 impulses per second; at the lower intensity (A discharging at the rate of approximately 20 impulses per second) the discharge of B was reduced by about 1.3 impulses per second, following the trend of the solid line, which is a portion of that drawn through all the experimental points in the upper graph of Fig. 2. For these points plotted as open symbols, the activity of B itself was small (11 to 12 impulses per second); consequently the inhibition that B exerted back on A was also small (a little less than 1 impulse per second). This is indicated by the short dotted arrows; the "tails" of these arrows are plotted at the abscissae that represent the values of the frequency obtained when ommatidium A was illuminated alone. The two points marked by the solid symbols, on the other hand, were obtained with B illuminated at higher in-

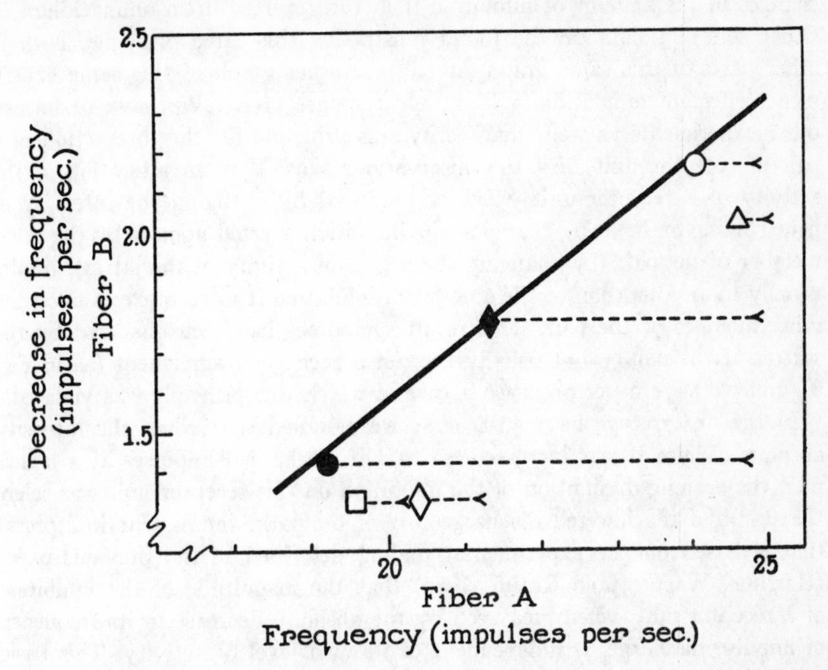

FIG. 3. Portion of the upper graph of Fig. 2. Inhibition exerted on ommatidium B is correlated with the degree of activity of ommatidium A. For the open symbols, B was illuminated at low intensity and exerted very little inhibition back on A. This is shown by the short lengths of the dotted lines, the right hand ends of which are plotted at the abscissae which give the frequency of A when it was illuminated alone. For the solid symbols, B was illuminated at high intensity and exerted strong inhibition back on A, as shown by the long lengths of the dotted lines associated with these points. For the solid symbols, ommatidium A was illuminated at the higher of the two intensities used for the open symbols. The solid line is a portion of that plotted in Fig. 2. The symbols are the same as those used for these same points in Fig. 2.

tensities. As a consequence of the resulting higher levels of activity of B (28 and 37 impulses per second), the discharge rates of ommatidium A (which for these points was illuminated at the higher of the two intensities used before) were much reduced, as can be seen by the lengths of the dotted arrows associated with these points. Corresponding to the reduced activity of A, the magnitude of the inhibition it exerted on B was smaller. This also followed the trend of the solid line. Thus any change in the frequency of A, whether brought about directly by changing the intensity of its stimulating light, or indirectly by changing the amount of inhibition exerted upon it as a consequence of altering the level of activity of B, resulted in comparable changes in the amount of inhibition it in turn exerted upon ommatidium B. Other sets of points can be found illustrating this same principle, both in other parts of this same graph, and in the other graph of this same experiment (effect of ommatidium B on the response of A). We have performed other experiments as well, that verify this principle for the interaction of a pair of receptor units. All the observations show that an alteration in the activity of a receptor unit, whether produced by changing the intensity of light shining on it, or by changing the inhibition exerted upon it by the other member of the pair (by changing the degree of activity of the latter), results equally in an alteration of the amount of inhibition it in turn exerts upon the other member of the pair. This result sometimes has been obscured by the scatter of the points, but usually there has been good agreement (as in Fig. 3), and we have never observed a case in which this principle was violated.

In the analysis we have just made we assumed that when the intensity on ommatidium B was increased, so that it discharged impulses at a higher rate, the ensuing diminution of the inhibition on this receptor unit was solely the result of the lowered discharge rate of ommatidium A. Our interpretation is based on the experimental finding described in the previous paper (Hartline, Wagner, and Ratliff, 1956) that the magnitude of the inhibition of a receptor unit, when measured by the absolute decrease in its frequency of impulse discharge, is independent of its own level of activity. This basic result, however, was established only as an approximation; indeed, it was noted in that paper that as the level of excitation of a "test" receptor was raised, the reduction in its frequency resulting from a fixed illumination of nearby ommatidia did in fact decrease slightly but significantly, in most experiments. This was attributed to an appreciable inhibition of the nearby ommatidia by the test receptor, just as we have done in the present case. But it might be argued alternatively that the measure of inhibition we have adopted has the inherent property that it yields a smaller value as the frequency of discharge of the test receptor is increased, and that the quantitative correlation of this measure of inhibition of one receptor unit with degree of activity of the other is only fortuitous in the present experiments. Independent experimental evidence is required to establish our interpretation more firmly.

Such independent evidence is furnished by experiments in which a third spot of light has been introduced to provide additional inhibitory influences that could be controlled independently of the two interacting receptor units whose activity was being measured. We have made use of the fact that the

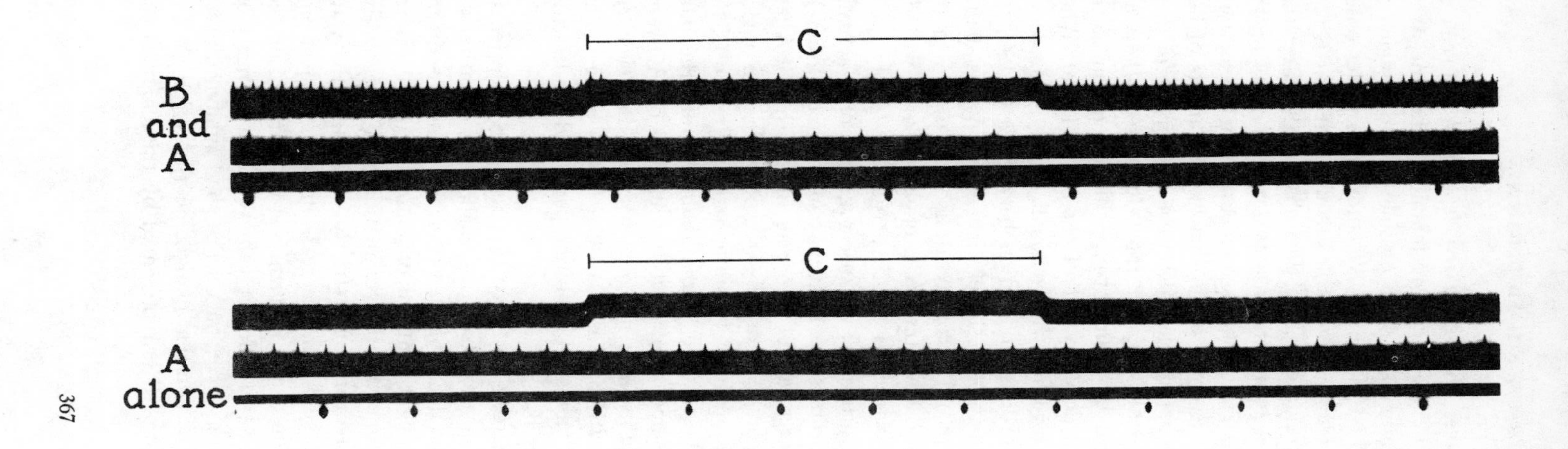

FIG. 4. Oscillograms of the electrical activity of two optic nerve fibers, showing disinhibition. In each record, the lower oscillographic trace records the discharge of impulses from ommatidium A, stimulated by a spot of light 0.1 mm. in diameter confined to its facet. The upper trace records the activity of ommatidium B, located 3 mm. from A, stimulated by a spot of light 1 mm. in diameter, centered on the facet of B, but that also illuminated approximately 8 to 10 ommatidia in addition to B. A third spot of light ("C"), 2 mm. in diameter, was directed onto a region of the eye centered 2.5 mm. from B and 5.5 mm. from A (B approximately midway between A and C); exposure of C was signalled by the upward offset of the upper trace. Lower record; activity of ommatidium A in the absence of illumination on B, showing that illumination of C had no perceptible effect under this condition. Upper record; activity of ommatidia A and B together, showing (1) lower frequency of discharge of A (as compared with lower record) resulting from activity of B, and (2) effect of illumination of C, causing a drop in the frequency of discharge of B and concomitantly an increase in the frequency of discharge of A, as A was partially released from the inhibition exerted by B.

Time marked in one-fifth seconds. The black band above the time marks is the signal of the illumination of A and B, thin when A was shining alone, thick when A and B were shining together.

inhibitory influence becomes weaker with increased separation between an affected receptor and the region of the eye used to inhibit it (Hartline, Wagner, and Ratliff, 1956). Consequently, it is often possible to find a region on the eye that is too far from the first of the two receptor units under observation to affect that one directly by an appreciable amount, but that is near enough to the second to inhibit it markedly. We then observe the effect that the altered frequency of discharge from this inhibited receptor has on the response of the first ommatidium. Fig. 4 shows oscillograms of the activity recorded simultaneously from two receptor units, showing the effects of illuminating regions of the eye in the manner just described. When one of these receptor units (A) was illuminated alone (lower trace, lower record) its activity was not appreciably affected by illuminating a distant region of the eye (C) (signalled by the upward displacement of the upper trace). When ommatidium A was illuminated together with a small region centered on ommatidium B, which was intermediate in position between A and C, the discharge of impulses by A was markedly slower than when A was illuminated alone (lower trace, upper record). This result is attributable to the vigorous activity of ommatidium B and the receptors stimulated with it, evidenced by the discharge of impulses in B's optic nerve fiber (upper trace, upper record). Then when C was turned on, the discharge rate of A actually increased, concomitantly with a decrease in frequency of discharge from B. When C was turned off, the discharge rate of B rose again and that of A fell. We interpret this result to mean that as the receptors in the region that included ommatidium B were inhibited by illumination of region C, the decrease in their activity partially released ommatidium A from the inhibition they exerted upon it. The amount of inhibition exerted on A by region B, measured by the difference in frequency of A between the lower record (A alone) and the upper (A with B) was less when C was being illuminated than when it was not; this diminished inhibition paralleled the lessened degree of activity recorded in fiber B.

The parallelism between the degree of activity of a receptor subjected to inhibition and the inhibition it in turn exerts on its neighbors is quantitative. This is shown in Fig. 5, drawn from data obtained from the same experiment as Fig. 4, except that the spot of light centered on ommatidium B was reduced in size, so that it was confined to that ommatidium. Consequently, the inhibition exerted by B may be correlated strictly with the activity recorded in its axon. In Fig. 5, the inhibition (decrease in frequency) of ommatidium A is plotted as a function of the frequency of discharge of ommatidium B; the open symbols are for two different values of light intensity on ommatidium B with no illumination on region C. The solid symbols are for a high intensity on B, but with the addition of light on the region C. The effect that C had on the discharge of B is represented by the length of the

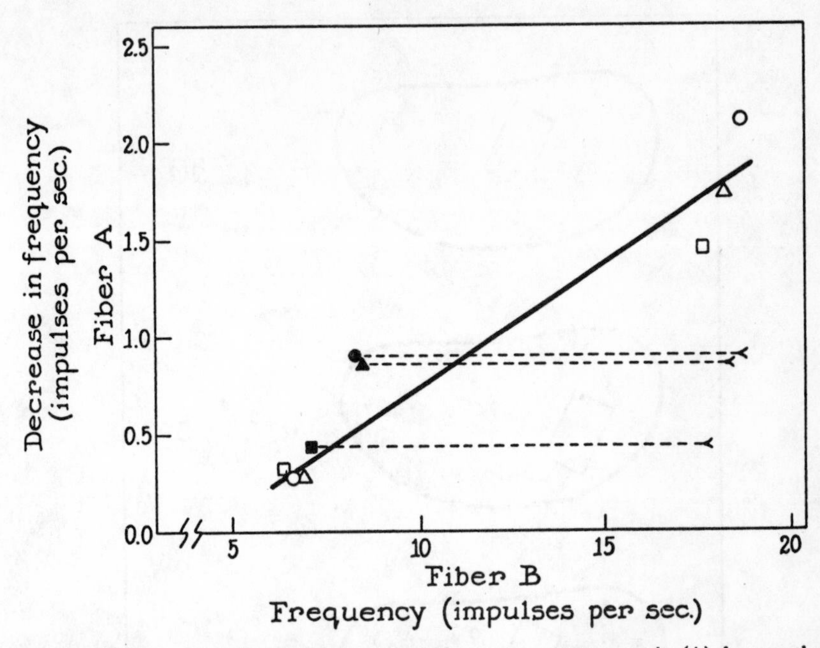

FIG. 5. Decrease of the inhibition exerted on one receptor unit (A) by another (B), as a result of inhibiting the activity of the second by illuminating a region of the eye (C) close to it. From the same experiment as that of Fig. 4 (see legend), but with the spot of light on ommatidium B reduced to 0.2 mm. diameter. The inhibition of ommatidium A was measured by the difference between its frequency when illuminated alone and when it was illuminated together with ommatidium B; this has been plotted as ordinate against the frequency of B as abscissa (as in Figs. 2 and 3). Three points were determined for a high intensity and three for a low intensity on B, when there was no illumination on C (open symbols). Three points were similarly determined for a high intensity on B when the nearby region of the eye, C, was illuminated (see legend of Fig. 4). These points are designated by the solid symbols. The lengths of the dotted lines associated with these points show the amount of reduction in the frequency of discharge of ommatidium B, as a result of the inhibition exerted upon it by C. Corresponding to this reduction in the activity of B, the inhibitory effect it in turn exerted on A was reduced, by an amount that is in quantitative agreement with the reduction obtained by lowering the intensity on B, as given by the solid line drawn through the open symbols. Illumination of the region C with no light shining on ommatidium B had very little effect on the activity of A: a reduction in frequency of 0.3 impulse per second was the maximum observed (the region C must have contributed even less than this amount to the total inhibition, since the receptors in it were also subject to inhibition by the activity of B).

Determination of the frequencies was made as described in the legend of Fig. 2.

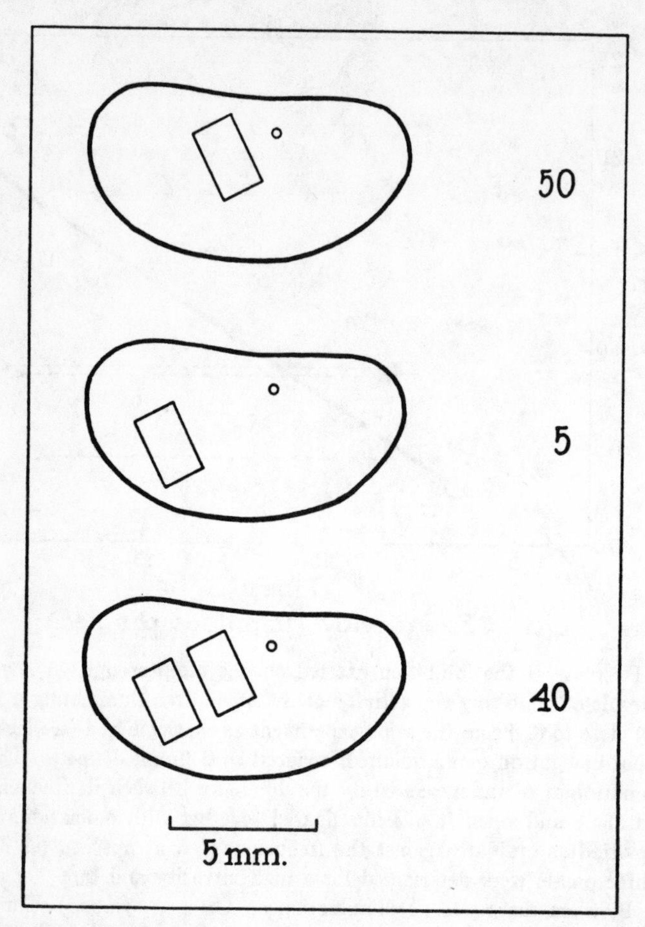

FIG. 6. A schematic diagram of patterns of light on a lateral eye of *Limulus* in an experiment illustrating disinhibition. The heavy lines are sketches of the outer margins of the eye. A small spot of light marked "o" was centered on the facet of a "test" ommatidium whose activity was measured by recording the action potentials in the optic nerve fiber arising from it. This spot was small enough to illuminate only the ommatidium on which it was centered. For each measurement the small spot of illumination was turned on for 12 seconds at a constant intensity. The number of impulses discharged by the test ommatidium in the last 10 seconds of such exposure was determined when the ommatidium was illuminated alone, and again when it was illuminated together with rectangular patches of light on other regions of the eye, as shown in the three sketches. The difference in the counts (decrease in frequency, in impulses per 10 seconds) measures the magnitude of the inhibition exerted on the test ommatidium by the receptors in the regions illuminated by the rectangular patches of light; these differences are given at the right for the respective parts of the experiment. Upper sketch, a rectangular patch of light near the test omma-

dotted lines. This experiment shows that the diminution in activity of ommatidium B had the same effect in reducing the inhibition exerted on A, whether that diminution was the result of inhibition of B by illumination of region C, or the result of reduction in the intensity of the light stimulus to B. Thus the degree of activity of a receptor unit does indeed determine, quantitatively, the strength of the inhibition it exerts on another receptor unit. Our analysis of the interaction between two receptor units illuminated together is therefore substantiated.

The release of a receptor unit from the inhibiting effects of others, by causing those others to be inhibited by yet a third group of receptors, is interesting physiologically. Such "disinhibition" is not difficult to obtain, though it may require some pains to show a strong effect. We have performed one other experiment similar to that of Figs. 4 and 5, recording from two fibers and using a third spot of light to inhibit one and disinhibit the other. It is considerably easier to show disinhibition when recording from only one receptor, for then it is possible to choose a favorable combination of locations for the spots of light that serve to inhibit and to disinhibit this test receptor. We have done many such experiments. An example is given in Fig. 6; the experiment is explained in its legend. Instead of focussing spots of light in various locations on the surface of the eye, disinhibition can also be demonstrated by using sources of light in various places in the external visual field, where the directional sensitivity of the ommatidia determines the location in the eye of the groups of receptors stimulated by the respective light sources. Dr. William Miller, in our laboratory, has also demonstrated disinhibition of receptors in the median eye of *Limulus* (a simple eye), using light sources in the external visual field.

Disinhibition simulates facilitation: illumination of a distant region of the eye results in an increase in the activity of the test receptor. In the *Limulus* eye the dependence of such an effect on the stimulation of receptors in an intermediate region of the eye (to produce the original inhibition) makes it

---

tidium produced a decrement of 50 impulses in 10 seconds. Middle sketch, a similar rectangular patch of light farther away from the test ommatidium produced a decrement of 5 impulses in 10 seconds. Lower sketch, both patches of light shining together produced a decrement of only 40 impulses in 10 seconds.

Thus in the last case the distant patch, rather than adding to the inhibition exerted by the near one, caused a disinhibition of 10 impulses per 10 seconds. As established by the experiments of Figs. 4 and 5, this was the result of the inhibition of the receptors in the near patch by the activity of those in the distant one, with the consequence that they in turn exerted less inhibition on the test ommatidium. This release of the test ommatidium from the inhibition exerted by the receptors in the near patch was greater than the slight inhibitory effect exerted directly on the test ommatidium by the receptors in the distant patch.

easy to recognize the mechanism involved. But if such a group of intermediate inhibiting elements were active spontaneously, or through uncontrolled influences, it might be difficult to recognize the true nature of a disinhibiting action. Perhaps the most significant aspect of these experiments showing disinhibition is the principle that they reveal, that indirect effects may extend considerably beyond the limit of the direct inhibitory connections among the receptors of the eye. Indeed, no member of the population of receptors is completely independent, under every condition, of the activity in any part of the eye. This is a direct consequence of the principle of interaction that we have established: the inhibiting influence exerted by a receptor depends on its activity, which is the resultant of the excitatory stimulus to it and whatever inhibitory influences may in turn be exerted upon it.

The principles that we have established experimentally may be conveniently summarized in a simple algebraic expression. The activity of a receptor unit—its response (r)—is to be measured in the present case by the frequency of the discharge of impulses in its axon. This response is determined by the excitation (e) supplied by the external stimulus to the receptor, diminished by whatever inhibitory influences may be acting upon the receptor as a result of the activity of neighboring receptors. The excitation of a given receptor is to be measured by its response when it is illuminated by itself, thus lumping together the physical parameters of the stimulus and the characteristics of the photoexcitatory mechanism of the receptor. Each of two interacting receptor units inhibits the other to a degree that depends (linearly) on its own activity. The responses of two such units are therefore given by a pair of simultaneous equations:

$$r_A = e_A - K_{A,B}(r_B - r_B^0)$$

$$r_B = e_B - K_{B,A}(r_A - r_A^0)$$

in which the subscripts are used to label the respective receptor units. In each of these equations, the magnitude of the inhibitory influence is given by the last term, written in accordance with the experimental findings as a simple linear expression. The "threshold" frequency that must be exceeded before a receptor can exert any inhibition is represented by r^0. The "inhibitory coefficient," K, in each equation is labelled to identify the direction of the action: $K_{A,B}$ is the coefficient of the action of receptor B on receptor A; $K_{B,A}$ *vice versa*. It is to be clearly understood that the equations do not apply in the ranges of responses for $r < r^0$ in either case: negative values of inhibition must be excluded since they are never observed. Also, r and e, by their nature (being measured by frequencies), cannot be negative. Appropriate changes must be made in the equations in those ranges of the variables where these restrictions apply: if, for example, $r_B < r_B^0$, the first equation must be replaced by $r_A = e_A$; if, to choose another example, r_B is

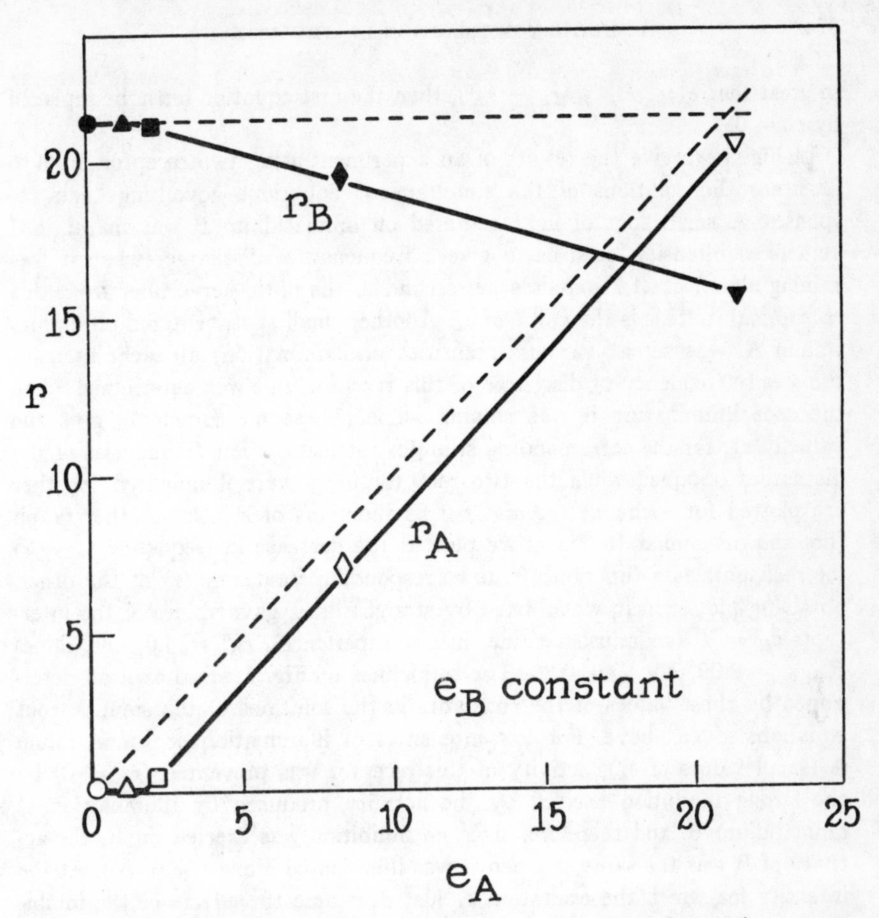

FIG. 7. Graph showing the relation of the responses of two interacting receptor units. One ommatidium (B) was illuminated at a fixed intensity; the intensity on the other (A) was varied. Responses (r) were measured by the frequency of the steady discharge of nerve impulses (last 7 seconds of a 10 second exposure to light). When both A and B were illuminated together, r_A refers to the response of ommatidium A (open symbols), r_B refers to the response of ommatidium B (solid symbols). The excitation of A, designated by e_A, is measured by the response of this ommatidium when it was illuminated alone; the excitation of B, designated by e_B, is measured by the response of B when it was illuminated alone. For each set of exposures (A alone, B alone, A and B together) with a given intensity for ommatidium A, the values of r_A and r_B obtained have been plotted (ordinates) against the value of e_A (abscissa). Values obtained for e_B were consistent within 0.1 impulse per second for all the sets of exposures; their average (21.2 impulses per second, shown by the horizontal dotted line) has been used in the calculations. The solid lines are the solutions of the simultaneous equations given in the text; the calculations and values of the constants are given there. The vertical distance from each dotted line ($r = e_A$, and $r = e_B$) to the corresponding solid line beneath it shows the amount of the inhibition for each receptor at that value of e_A.

so great that $e_A < K_{A,B}(r_B - r_B^0)$, then the first equation must be replaced by $r_A = 0$.

In Fig. 7 we give the results of an experiment using two receptor units to illustrate the solutions of the simultaneous equations governing their responses. A small spot of light centered on ommatidium B was maintained at a fixed intensity, such as to give a frequency of discharge (when it was shining alone) of 21.2 impulses per second in the optic nerve fiber from this receptor unit. This is the value of e_B. Another small spot, centered on ommatidium A, was set at various intensities of illumination; at each intensity the steady frequency of discharge of this receptor unit was determined when the spot illuminating it was shining alone. These measurements give the value of e_A, for the corresponding stimulus intensities. The frequencies of the discharges obtained when the two receptor units were illuminated together are plotted for each one (r_A and r_B) as functions of e_A. On another graph (not shown) similar to Fig. 2 we plotted the decrease in frequency ($e - r$) for each unit as a function of the corresponding frequency (r) of the other, obtaining plots which, when fitted by straight lines, gave values of the intercepts $r_A^0 = 0$ (an unusual value, in our experience), $r_B^0 = 4.0$, and slopes $K_{A,B} = 0.09$, $K_{B,A} = 0.26$. The solid lines of Fig. 7 are drawn as determined by these values of the constants in the solutions of the simultaneous equations given above. For low intensities of illumination on ommatidium A (small values of e_A), activity of this receptor was prevented ($r_A = 0$) by the strong inhibition exerted by the activity produced by illumination of ommatidium B, and therefore, since no inhibition was exerted on B, the activity of B was the same as when it was illuminated alone ($r_B = e_B$). At the intensity for which the excitation e_A just overcame the effects of the inhibition exerted by B ($e_A = K_{A,B}[e_B - r_B^0] = 1.6$), receptor A began to respond; as e_A increased, the frequency of its discharge increased linearly with a slope of 1.024 $\left(=\frac{1}{1 - K_{A,B}K_{B,A}}\right)$. At this same value of e_A (since $r_A^0 = 0$), the inhibition by A on B began to be exerted, and r_B decreased linearly with increasing e_A, the slope being -0.27 $\left(=\frac{-K_{B,A}}{1 - K_{A,B}K_{B,A}}\right)$. (The dotted lines show the form the graphs would have taken had there been no inhibition.) In this experiment each spot of light actually illuminated about 8 or 9 other ommatidia in the immediate vicinity of the one on which it was centered and from which activity was recorded. For purposes of illustration, it is permissible to treat the results as though individual units were interacting, although in actuality it was each small group. That the principles involved hold rigorously when only two single receptor units are actually used is inherent in the treatment, for these principles were derived from the experiment of Fig. 2 (and those like it), in which strict optical isolation of individual ommatidia was employed.

The mutual interdependence of two receptor units responding to steady illumination is thus concisely and accurately described by a pair of simultaneous equations. Similar equations hold for the responses of any two ommatidia that are close enough together in the eye to interact. When more than two interacting elements are activated, similar relations apply simultaneously to the responses of all of them. In addition, however, each receptor unit is subjected to inhibitory influences from all the others, and the degree to which its response is decreased is known to be greater, the greater the number of neighboring ommatidia that are stimulated (Hartline, Wagner, and Ratliff, 1956). The simultaneous equations governing the responses of more than two ommatidia therefore must contain terms expressing the inhibition contributed from all the active elements, combined according to some law of spatial summation. Experiment shows that simple arithmetic addition of such terms is adequate to describe spatial summation of inhibitory influences in the eye of *Limulus*. In a paper that will follow (see also Hartline and Ratliff, 1954 and Ratliff and Hartline, 1956) we will describe the experiments that establish this law of spatial summation and will illustrate some of the effects that are obtained when more than two receptor units in the eye inhibit one another mutually.

SUMMARY

The inhibition that is exerted mutually among the receptor units (ommatidia) in the lateral eye of *Limulus* has been analyzed by recording oscillographically the discharge of nerve impulses in single optic nerve fibers. The discharges from two ommatidia were recorded simultaneously by connecting the bundles containing their optic nerve fibers to separate amplifiers and recording systems. Ommatidia were chosen that were separated by no more than a few millimeters in the eye; they were illuminated independently by separate optical systems.

The frequency of the maintained discharge of impulses from each of two ommatidia illuminated steadily is lower when both are illuminated together than when each is illuminated by itself. When only two ommatidia are illuminated, the magnitude of the inhibition of each one depends only on the degree of activity of the other; the activity of each, in turn, is the resultant of the excitation from its respective light stimulus and the inhibition exerted on it by the other.

When additional receptors are illuminated in the vicinity of an interacting pair too far from one ommatidium to affect it directly, but near enough to the second to inhibit it, the frequency of discharge of the first increases as it is partially released from the inhibition exerted on it by the second (disinhibition).

Disinhibition simulates facilitation; it is an example of indirect effects of interaction taking place over greater distances in the eye than are covered by direct inhibitory interconnections.

When only two interacting ommatidia are illuminated, the inhibition exerted on each (decrease of its frequency of discharge) is a linear function of the degree of activity (frequency of discharge) of the other. Below a certain frequency (often different for different receptors) no inhibition is exerted by a receptor. Above this threshold, the rate of increase of inhibition of one receptor with increasing frequency of discharge of the other is constant, and may be at least as high as 0.2 impulse inhibited in one receptor per impulse discharged by the other. For a given pair of interacting receptors, the inhibitory coefficients are not always the same in the two directions of action. The responses to steady illumination of two receptor units that inhibit each other mutually are described quantitatively by two simultaneous linear equations that express concisely all the features discussed above. These equations may be extended and their number supplemented to describe the responses of more than two interacting elements.

BIBLIOGRAPHY

Hartline, H. K., and Ratliff, F., Spatial summation of inhibitory influences in the eye of *Limulus*, *Science*, 1954 **120,** 781 (abstract).

Hartline, H. K., Wagner, H. G., and Ratliff, F., Inhibition in the eye of *Limulus*, *J. Gen. Physiol.*, 1956, **39,** 651.

Ratliff, F., and Hartline, H. K., Inhibitory interaction in the eye of *Limulus*, *Fed. Proc.*, 1956, **15,** 148 (abstract).

38

J. Neurophysiol. (1969) **32**, 339-355

Organization of the Retina of the Mudpuppy, *Necturus maculosus.* II. Intracellular Recording

FRANK S. WERBLIN AND JOHN E. DOWLING

The Wilmer Institute, The Johns Hopkins University School of Medicine, Baltimore, Maryland 21205

FROM ELECTRICAL RECORDINGS at the level of the optic nerve, it has been possible to specify many of the functions performed by the vertebrate retina. These include brightness detection (16, 18, 27), center-surround contrast detection (1, 8, 20, 23), and motion detection (3, 4, 26–28). It has not been possible, however, to determine how the retina organizes the visual message recorded at the optic nerve, primarily because intracellular recording from single cells distal to the ganglion cells has been difficult.

Detailed structural studies of the vertebrate retina, such as the one in the preceding paper (14), provide a framework within which the functional organization of the retina can be described. The anatomical studies show a limited number of clearly defined synaptic structures at which interaction between specific neurons can take place. In this paper we shall describe the intracellularly recorded response characteristics of each type of neuron in a vertebrate retina, and then relate the response of each neuron to the responses of those neurons to which it is synaptically coupled. By following the responses through the synaptic pathways, we can begin to describe how information from the visual field is abstracted and encoded in the retina.

Intracellular recording throughout most retinas has been difficult because even the finest available micropipettes fail to penetrate the small retinal neurons consistently without damage. Bortoff (5–7) showed that this difficulty could be overcome by recording in an animal with larger retinal neurons: the mudpuppy, *Necturus maculosus.* As described in the preceding paper (14), the retinal neurons in *Necturus* are about 30 μ in diameter, as compared to diameters of less than 10 μ for many cells in the retinas of frogs or mammals. This represents a difference in cell volume of 3^3, more than an order of magnitude. Bortoff was able to record intracellularly from cells throughout the retina of *Necturus* and to stain these cells for later identification. He described several response types in the retina, primarily those of the more distal neurons, which give slow, graded potentials. Bortoff used diffuse illumination in his experiments, and did not study the spatial organization of the receptive fields for each neuron. It will be shown here that study of spatial organization is crucial for interpreting the functional organization of the retina.

In other vertebrate retinas, intracellular recording and staining have been possible in only a few cell types. Fish, for example, have relatively large horizontal cells that can be penetrated easily, and these cells have been extensively studied. The potentials recorded from within these cells—the luminosity- or L-type S potentials—are sustained and graded with illumination over a limited range of intensity and are always hyperpolarizing (15, 24, 25, 27, 29, 30, 32–35, 37–39, 42). Recently, Tomita (40) and Kaneko and Hashimoto (21) obtained intracellular responses from fish cones and have shown that these receptor cells also respond with sustained, hyperpolarizing potentials that are graded over a limited range of intensity.

The finding of graded, hyperpolarizing potentials in both receptors and horizontal cells has been somewhat surprising. Svaetichin and his co-workers (24, 30, 37) have

Received for publication October 15, 1968.

suggested that the horizontal cells are not neurons but glia-like cells. They have shown that horizontal cells are more easily affected by metabolic poisons than classical neurons and under these adverse conditions tend to polarize in the direction opposite to that of classical neurons. They have offered the hypothesis that such "glia" cells control transmission of information through the retina. It has been shown anatomically, however, that horizontal cells make synaptic connections similar to those of classical neurons, which indicates that horizontal cells should be considered as neurons (see 13, 14). Grundfest (17) has pointed out that since such cells as horizontal cells or receptors need not communicate over long distances, no axonal (conductile) process is necessary. Thus, it is not necessary that the cell be depolarized for initiation of impulse activity, and so the direction of polarization during activation may not be significant. Thus, excitation could be signaled with either hyperpolarizing or depolarizing potentials.

Intracellular recording from neurons in the inner nuclear layer other than horizontal cells has seldom been achieved. The few studies that have been made indicate that some units show impulse activity while others do not (5, 9, 22, 39, 41). These findings suggest that it is within the inner nuclear layer that the transition from slow-potential generators to spike generators takes place.

METHODS

In all experiments adult *Necturus*, about 12 inches long, were used. The animals were trapped in the rivers of Wisconsin, flown to Baltimore, and kept in a tank of cold spring water in a dark room until the time of the experiment. The animal was decapitated, and the anterior part of one eye dissected away. The head was placed in a contoured holder so that the eye could be properly positioned with respect to the stimulus and the electrode. During the dissection no attempt was made to remove the vitreous humor. Under these conditions the retina remained functional for as long as 6 hr. Drying seemed to be the primary cause of functional deterioration.

The micropipette electrodes were made with a Livingston-type spring puller, modified so that the pull was initiated only after a fixed heating interval. Corning, type 7740 capillary tubing with 0.8-mm outside diameter and 0.4-mm inside diameter was used. All electrodes were filled by heating to about 70 C and then boiling under reduced pressure for about 10 min. The electrodes, when filled with 2 M potassium chloride, had tip resistances of 100–150 megohms measured in Ringer solution. Similar electrodes filled with the staining solution of 4% Niagara blue (21), had resistances of about 700 megohms. Electrodes were used only on the day they were filled. Those that were stored for longer periods were less effective in penetrating cells.

To eject Niagara blue from the electrodes for staining, a negative potential of 400 v was applied across the electrode tip through a current-limiting resistance of 100 megohms (Fig. 1). Within a few seconds after the voltage was applied, the first few microns of the electrode tip were destroyed and stain began to enter the cell through the widened tip. The current flowing through the electrode tip was monitored via a loudspeaker. With practice it was possible to deliver the proper quantity of stain by "listening" to the flow of stain through the tip.

After the stain had been ejected, the eyecup was fixed for about 15 min in 6% glutaraldehyde and 2.5% potassium dichromate. This solution had a pH of 4, which kept the stain insoluble. The retina was then firm enough to be removed from the eyecup without tearing or folding. It was transilluminated, and the blue spot located with the aid of a dissecting microscope. A 2 x 2 mm square of tissue containing the spot was cut out, dehydrated, and embedded in soft plastic. Sections were cut at about 10 μ on a glass knife until the stained cell was located.

The first stage of the d-c recording amplifier consisted of a field effect transistor with an input impedance of 10^{13} ohms and an input capacitance of about 10^{-12} farads. Although negative capacitance was used to increase the bandwidth, the system was always band-limited by the high distributed capacitance and resistance at the pipette tip. With the very high impedance electrodes used in these experiments, action potentials were recorded only when the pipette tip was within the cytoplasm of a cell. A probable explanation for this is that the tip capacitance acts in parallel with the tip resistance when the electrode is in the cytoplasm, thus increasing the bandwidth; but the tip capacitance acts as a shunt when the tip of the electrode is outside the cell membrane, thus decreasing the bandwidth. Even the intracellularly recorded action potentials were

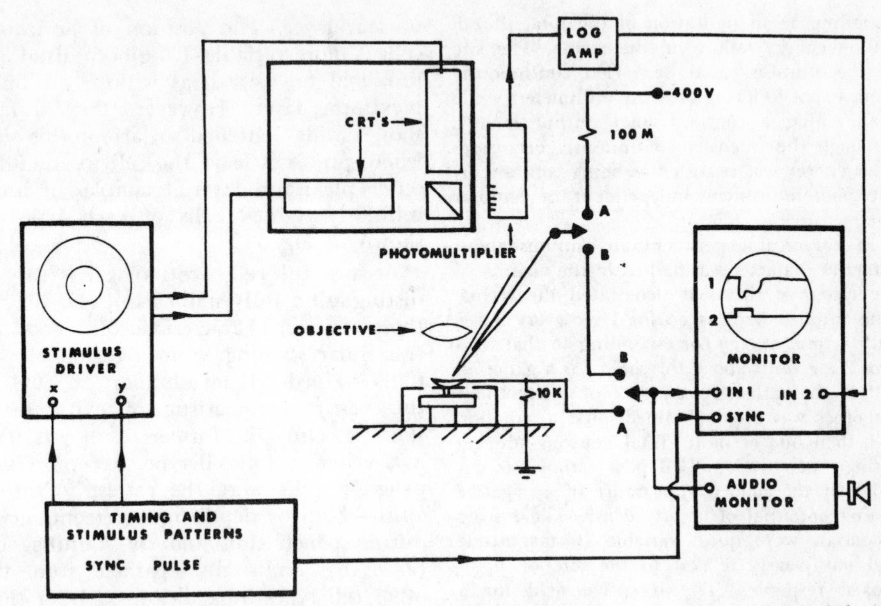

FIG. 1. Apparatus for stimulation, recording, and staining. Patterns generated on the faces of the two cathode-ray tubes (CRTs) are combined in the beam-splitting prism and focused on the retina. The other output from the prism illuminates the face of a photomultiplier. Output from the photomultiplier is linear with intensity throughout the range used here. This linear signal is logarithmically compressed and appears as the lower trace, 2, in all recordings presented in the paper. One log unit of intensity is represented by one vertical division. The electrode is coupled, either to the d-c recording amplifier and then to the oscilloscope (B) or to the negative 400-v staining potential (A), by the relay system. When the stain is ejected, current flows through the electrode, the preparation, and the 10-kilohm resistor to ground. The potential produced by the staining current across the 10-kilohm resistor is monitored by the audio amplifier. Timing is controlled so that the oscilloscope sweep begins in all recordings 200 msec before the retina is stimulated. In this way, timing for all recordings can be estimated from the initial segment of the trace.

attenuated by the limited bandwidth of the recording electrode. The recorded signal was amplified, displayed on an oscilloscope, and recorded on tape.

The retina was stimulated by patterns of illumination produced on the faces of two 1-inch cathode-ray tubes as shown in Fig. 1. The patterns were combined through a beam-splitting prism and focused on the retina. Each cathode-ray tube was driven by a master oscilloscope on which the stimulus pattern also appeared. A 1-cm pattern on the master oscilloscope corresponded to a 0.5-mm replica focused on the surface of the retina. The stimulus patterns could be monitored as they were moved over the surface of the retina and the receptive fields for units were plotted on the face of the master oscilloscope. The patterns in these experiments were simple. consisting of spots and annuli (Lissajous patterns) of continuously variable diameter and intensity. The width of the spot or line width of the annulus was 100 μ when focused on the retina. At high intensities the width of the spot or line width of the annulus increased somewhat, but receptive fields were plotted at lower intensities. The patterns were focused on the retina by viewing through a dissecting microscope; they were not noticeably distorted in passing through the vitreous or retinal tissue. The stimulus intensity was limited to a maximum of about 3.5 log units above threshold for units in the *Necturus* retina. The cathode-ray tube stimulators (RCA type 1EP1) are available with numerous phosphors having narrow emission spectra over the visible range. For most of these experiments phosphors which had emission peaks at 525 nm were used. When color-coded responses were looked for, tubes with phosphors with peak spectral emission at 425 nm or 600 nm were employed.

The light from the beam splitter at right angles to that focused on the retina fell on a photomultiplier tube (RCA type 1P21). The output from the photomultiplier was logarithmically compressed and appeared with each recording as an indication of the total illumination energy falling on the retina. The size of the annulus could be varied continuously from a spot to a ring of many millimeters while maintaining a constant total output energy, although the intensity continuously decreased. This energy was recorded as being constant by the photomultiplier, independent of annular size.

In a typical experiment an annulus about 2 mm in diameter was flashed in the vicinity of the electrode tip as it penetrated the retina. The annulus was flashed for 1 sec every 5 sec, with a total energy corresponding to that of a spot 2 log units above threshold for a ganglion cell in the retina. A sequence of intracellular responses was often obtained—first a ganglion cell, then one or more distal neurons—during a single penetration. Cell penetration was signaled by the sudden appearance of a negative resting potential of about 30 mv. The resting potentials were quite variable in magnitude and not clearly related to the size of light-evoked responses. The receptive field for a penetrated unit was then determined, either in the dark or with the center of the field or the entire retina illuminated. Cells could often be recorded from for more than 15 min.

The term "receptive field" is used here to indicate the extent of the retinal surface over which a unit could be influenced by illumination, whether this illumination evoked or antagonized a response. In units with impulses, "threshold" is defined as the intensity of illumination at which impulses occurred. For the slow-potential generators, "threshold" refers to the intensity at which a response could be seen above the electrical noise.

RESULTS

Intracellular staining

To identify the type of cell generating a particular response, it was necessary to stain many cells intracellularly after recording from them. Stained cells, corresponding to different response types, tended to group at characteristically different depths within the retina, and thus it was possible in most cases to identify with confidence the intracellularly stained cells by their shape and depth (see 14). For example, bipolar cells almost always lie within the outer half of the inner nuclear layer and have radially elongated cell bodies. The amacrine cells almost always lie along the inner margin of the inner nuclear layer. The position of horizontal cells is more variable, lying both distal and proximal to, as well as within, the outer plexiform layer. However, they usually show a wide, flattened surface on one side, where processes leave the cell to enter the outer plexiform layer. Examples of intracellularly stained cells of each type are shown in Fig. 2.

Because the cells are large, it is easy to distinguish a fully stained cell (ca. 40% of those stained) characteristic of a good intracellular staining, from a region of partially stained cell membranes (ca. 60% of the cases) characteristic of extracellular staining. Only the former result was used as a criterion for cell type. Receptors and ganglion cells were the easiest to distinguish—both by depth measurements made during penetration and by staining—because these cells are separated from the other cell types by a plexiform layer (Fig. 2 *a, e*). About 5 of each of these 2 types of cell were stained, but at least 50 of each type were recorded from during these experiments.

Cells that stained consistently along the inner margin of the inner nuclear layer were identified as amacrine cells (Fig. 2*d*). Twenty stained amacrine cells were recovered, and more than 100 were recorded from. Cells identified as bipolar cells were usually found within the outer half of the inner nuclear layer. Fourteen such cells were stained and more than 100 recorded from. The stained bipolar cell shown in Fig. 2*c* is typically oval, and shows staining of its Landolt club process. The horizontal cell shown in Fig. 2*b* is located at the outer margin of the inner nuclear layer, and shows a flattened apical surface, typical of many horizontal cells. Twenty-five horizontal cells were stained and recovered successfully, but many more were recorded from during the course of these experiments. As noted above, it was often possible to obtain a sequence of intracellular recordings while penetrating the retina with a single pipette. The sequence of responses was always consistent with that predictable from the position of the stained cells. This added further support to the identification of cell types.

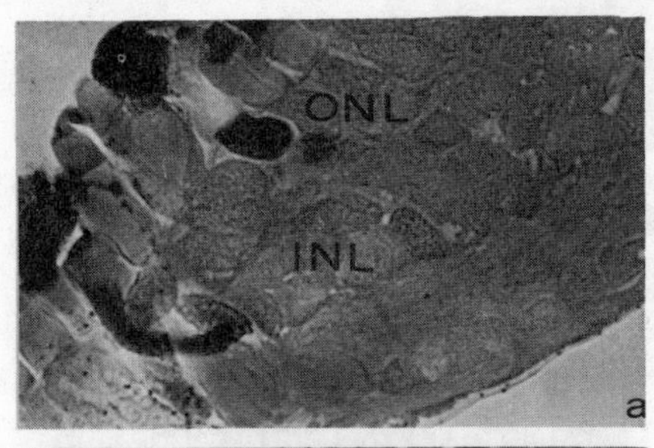

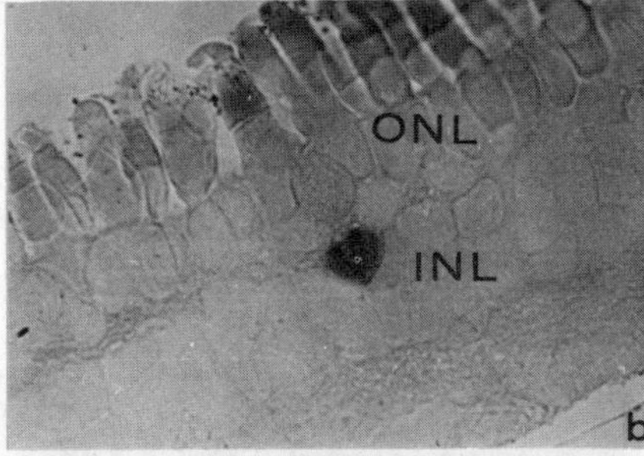

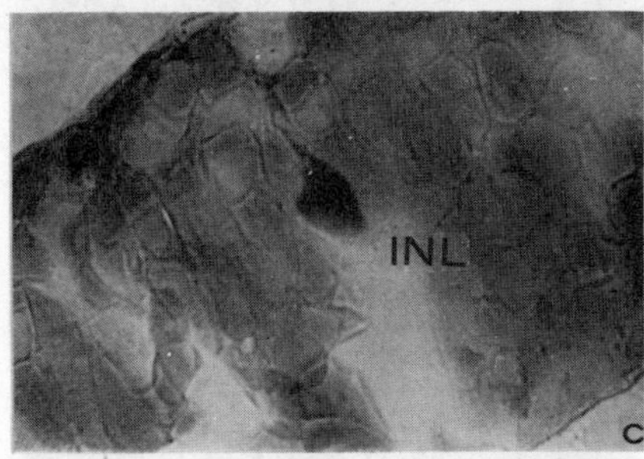

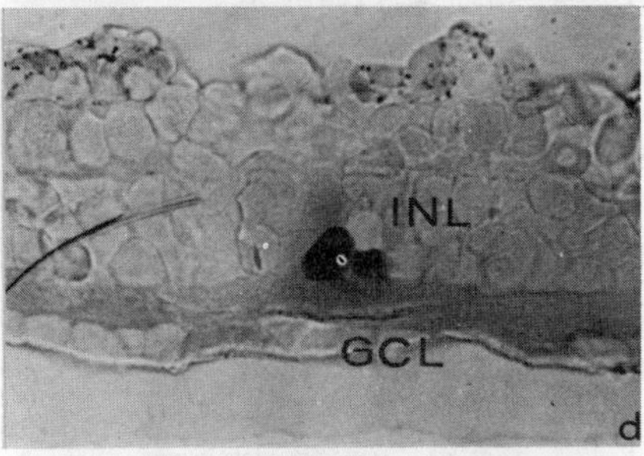

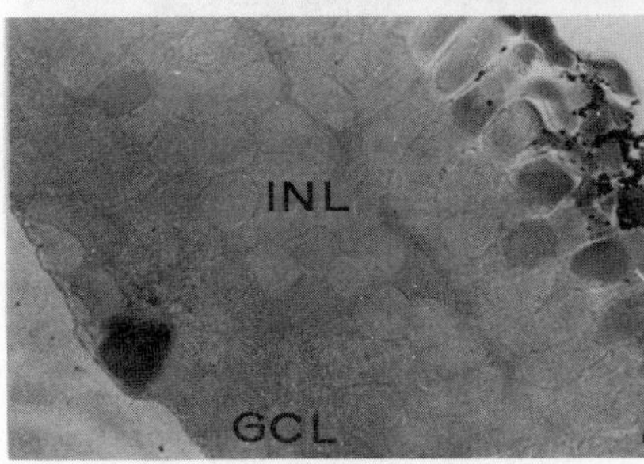

FIG. 2. Intracellularly stained cells such as these were used to establish the identity of recorded responses. The receptor cells (*a*) always show stain in the inner segment or nucleus. The horizontal cell (*b*) shown here is located along the outer margin of the inner nuclear layer and has a flattened distal surface where processes extend into the plexiform layer. Bipolar cells (*c*) are located in the outer half of the inner nuclear layer. The bipolar cell morphology of an elongated cell body is represented in this section, although the overall preservation of this retina is poor. Amacrine cells (*d*) stain typically along the inner margin of the inner nuclear layer. In the example shown here there is a faint blue staining of an adjacent cell, probably due to leakage of the stain. The deeply stained cell is almost certainly the cell recorded from, but in any case the layering of the stain along the inner margin of the inner nuclear layer is unequivocal. Ganglion cells (*e*) are easily identified in the single row of cells proximal to the inner plexiform layer. ONL, outer nuclear layer; INL, inner nuclear layer; GCL, ganglion cell layer.

Intracellular recording

Intracellularly recorded responses from each of the types of retinal neurons in *Necturus* are shown in Fig. 3. Neurons in the distal retina respond with slow, graded, mostly hyperpolarizing potentials, as shown in the upper half of the figure. In the proximal retina (lower half of the figure) most neurons respond with transient, depolarizing potentials on which impulses are superimposed.

For each type of neuron the response both to focal (left side of figure) and to annular (right side of figure) illumination centered on the receptive field was recorded. These two measurements provided sufficient information to characterize each response type. For example, the receptor has a narrow receptive field such that spot illumination evokes a much larger response than annular illumination. The horizontal cell has a broad, uniform receptive field, so that both spot and annular illumination evoke a sizable hyperpolarizing response. The bipolar cell responds with a sustained polarization when the center of its receptive field is illuminated. The sustained response is reduced when illumination is added at the periphery of the receptive field (right column). The units proximal to the bipolar cell reflect the antagonism between center and periphery established at the level of the bipolar cell, but these units are depolarizing and spike-generating neurons.

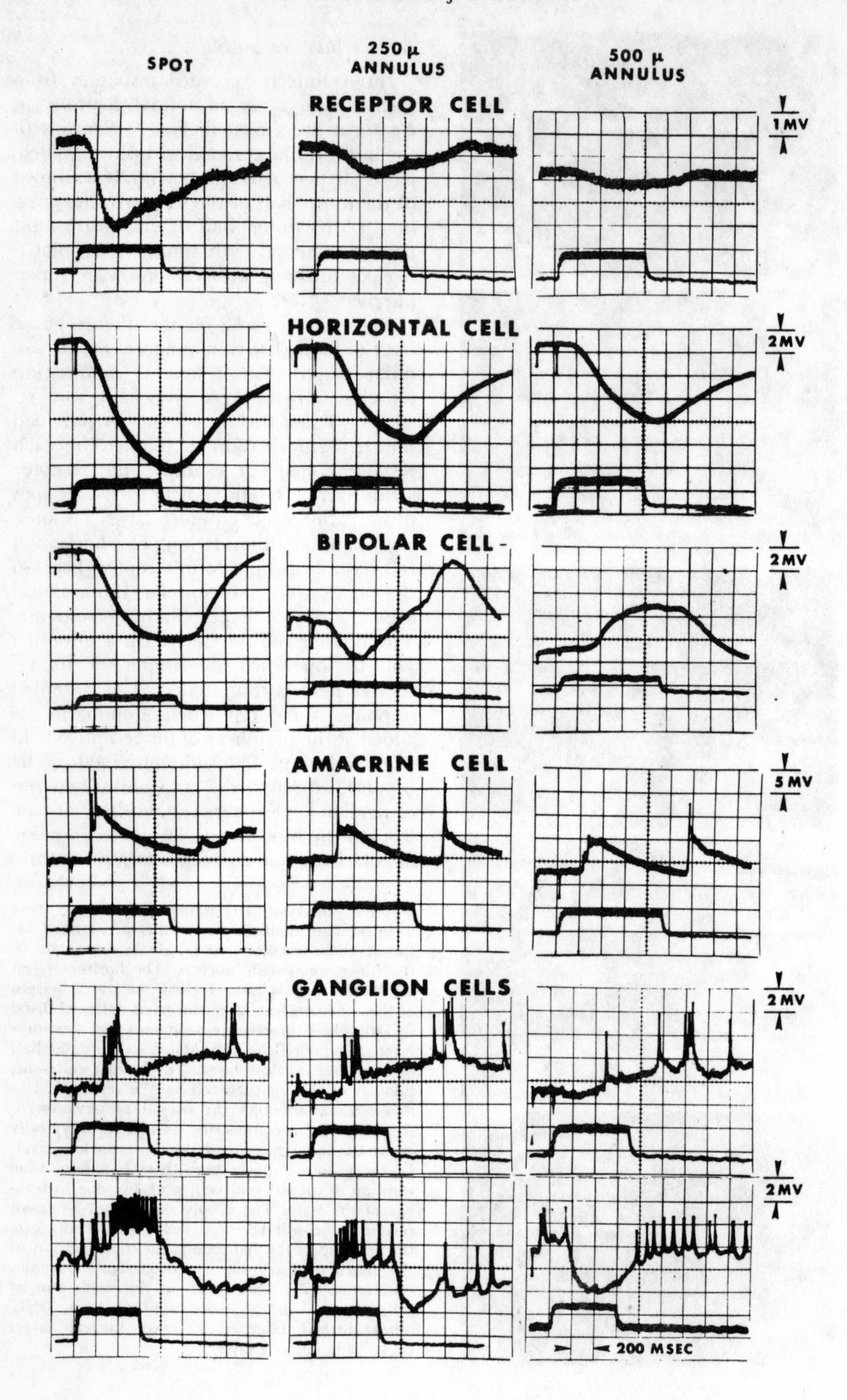
SPOT
250 μ ANNULUS
500 μ ANNULUS
RECEPTOR CELL
1 MV
HORIZONTAL CELL
2 MV
BIPOLAR CELL
2 MV
AMACRINE CELL
5 MV
GANGLION CELLS
2 MV
2 MV
200 MSEC

The characteristics of each type of response will be discussed in detail in the following sections. The recordings in Fig. 3 are typical examples of about 80% of the recordings made in *Necturus*. The other 20% (not discussed here) represent either rare response types, damaged units, or those that, for unknown reasons, are not easily or consistently penetrated.

RECEPTOR RESPONSE. The receptor response is a hyperpolarization, consisting of an initial transient that decays to a steady level and is graded with stimulus intensity. When illumination is removed the potential returns very slowly to the base line, with a small "off" transient. Receptor responses are illustrated in Figs. 3, 4, and 5.

The response occurs after a latency of about 50 msec and has a time to peak of 50 msec when elicited with a stimulus about 1 log unit above threshold. Resting potentials are quite variable in the receptors, but are typically 30 mv. The response is only about 5 mv in magnitude. It is graded over a relatively narrow range of intensities —less than 2 log units from threshold. At higher intensities the latency and time to peak decrease, but the magnitude of the response remains unchanged.

Figure 5 shows an experiment in which a receptor was recorded from while a 100-μ spot, centered on the receptor, was flashed. This was done first with the retina in the dark and again while steady annular illumination (250 μ in radius, centered on the receptor, and of equal total energy as the stimulating spot) illuminated the retina. The two superimposed traces indicate that the receptor response was not altered by the annular illumination, so that under these conditions there is no evidence of a surround effect. In the more proximal retina it will be shown that under the same conditions a dramatic surround effect is observed.

The response from cones in fish has been recorded intracellularly by Tomita (40). He also reported a hyperpolarizing response that is sustained and graded with intensity. He has seen no appreciable difference in response when stimulating with a spot or disk of similar illumination centered on the receptor, indicating that the receptor is not susceptible to a "surround effect." The lack of a surround effect is a characteristic feature of the receptor and is illustrated also in Fig. 3. The significance of this will be discussed below with respect to the more proximal neurons.

Recordings from the receptors were probably always obtained from the inner segments or nuclei of the cells, since this is where the stain was located. The outer segments were often lost during the preparation of the tissue for histologic study following staining, so that it has not been possible to distinguish the responses of rods and cones. The inner segments of all re-

---

FIG. 3. Recordings show the major response types in the *Necturus* retina and the difference in response of a given cell type to a spot and to annuli of 250- and 500-μ radius. Receptors have relatively narrow receptive fields, so that annular stimulation evokes very little response. Small potentials recorded upon annular stimulation were probably due to scattered light. The horizontal cell responds over a broader region of the retina, so that annular illumination with the same total energy as the spot (left column) does not reduce the response significantly (right columns). The bipolar cell responds by hyperpolarization when the center of its receptive field is illuminated (left column). With central illumination maintained (right trace; note lowered base line of the recording and the elevated base line of the stimulus trace in the records) annular illumination antagonizes the sustained polarization elicited by central illumination, and a response of opposite polarity is observed. In the middle column the annulus was so small that it stimulated the center and periphery of the field simultaneously. The amacrine cell was stimulated under the same conditions as the bipolar cell, and gave transient responses at both the onset and cessation of illumination. Its receptive field was somewhat concentrically organized, giving a larger "on" response to spot illumination, and a larger "off" response to annular illumination of 500-μ radius. With an annulus of 250-μ radius, the cell responded with large responses at both on and off. The ganglion cell shown in the upper row was of the transient type and gave bursts of impulses at both on and off. Its receptive-field organization was similar to the amacrine cell illustrated above. The ganglion cell shown in the lower row was of the sustained type. It gave a maintained discharge of impulses with spot illumination. With central illumination maintained, large annular illumination (right column) inhibited impulse firing for the duration of the stimulus. The smaller annulus (middle column) elicited a brief depolarization and discharge of impulses at on, and a brief hyperpolarization and inhibition of impluses at off.

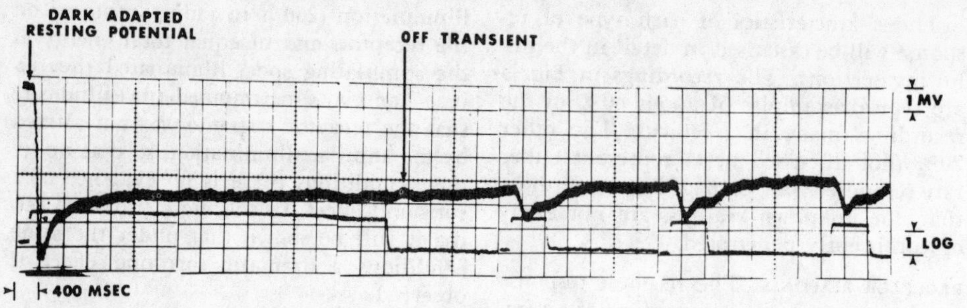

FIG. 4. Time course of response for a receptor. Response of the dark-adapted receptor begins with a large initial transient which decays to a steady hyperpolarization that is graded with the intensity of stimulus. When the stimulus is terminated, the response begins to decay to the original resting potential, with a surprisingly small off-transient. Additional stimuli of the same intensity, presented before the cell fully dark adapts, evoke responses with small initial transients that decay to the same steady hyperpolarized level that is independent of the state of adaptation.

ceptors in *Necturus* are of comparable size, so there is no apparent reason to suppose that the electrode should have penetrated either rods or cones preferentially.

HORIZONTAL CELL RESPONSE. The intracellularly recorded horizontal cell response, like that of the receptor, is hyperpolarizing and sustained with intensity. Figure 3 shows, however, that the horizontal cell response is slower, with a latency of about 100 msec and a time to peak of over 300 msec. At the cessation of illumination, the horizontal cell response decays with a similar slow time course, lasting about 300 msec.

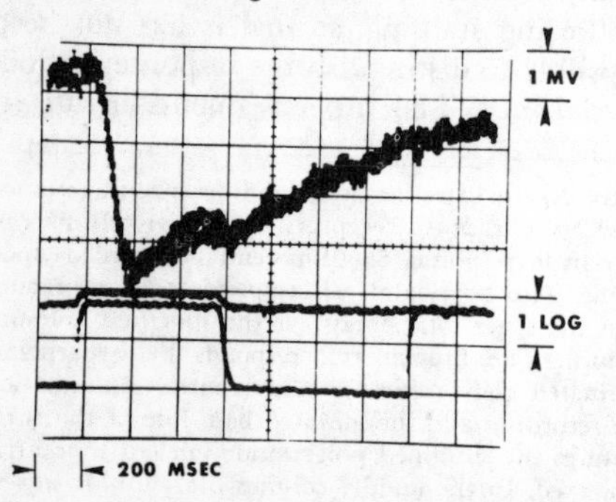

FIG. 5. The autonomy of the receptor response is illustrated by these two superimposed records. One record was obtained with spot illumination flashed on the receptor with the retina in the dark. For the second record, an annulus (of 250-μ radius) around the electrode was maintained on (elevated base line in stimulus recording), and then the receptor was illuminated with the same spot as in the first case. The two records are almost identical, indicating that under these conditions illumination of the surround does not affect the response of the receptor.

The magnitude of the response is graded with intensity over about 3 log units. However, the shape of the intensity-response curve depends on the configuration of the stimulus. Figure 6 shows that the response

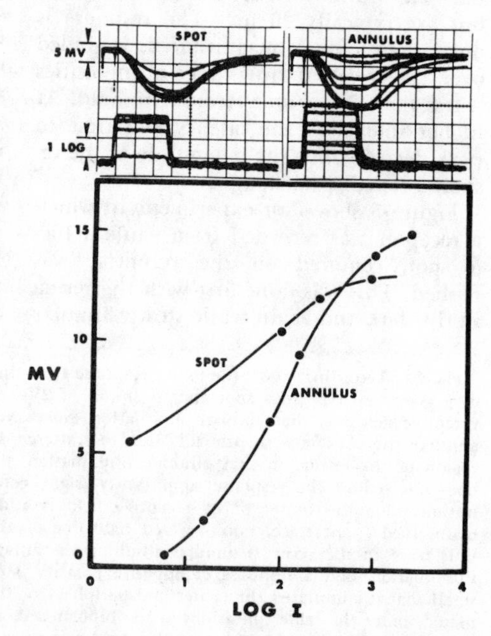

FIG. 6. Summation properties of the horizontal cell. The intensity-response relations of the horizontal cell were measured with stimuli of two different configurations. In one experiment all of the energy is confined to a spot of illumination. In the other the stimulus energy is distributed in an annulus with a radius of 0.5 mm. The response is greater, and saturates at a higher total energy level, when the stimulus is diffuse than when it is concentrated.

is greater and saturates at a higher absolute intensity when annular rather than spot stimulation is used. This suggests that the potential recorded in the horizontal cell is formed through the weighted summation of potentials from many sites, each of which can be saturated. Figure 3 shows that the horizontal cell summates over a wide area (compared to the receptor), since the response is not greatly reduced for stimulation with an annulus of 250- to 500-μ radius.

The potentials recorded from horizontal cells resemble the non-color-coded, or luminosity-type S potentials recorded in fish (15, 25, 34, 39, 42). The resting potential is typically 30 mv, with a response potential as large as 20 mv. The units summate over a wide area of the retina and are graded and sustained with intensity. No units corresponding to the chromaticity or C-type S potentials, which change their polarity of response with wavelength, have been observed in *Necturus* (25, 33).

BIPOLAR CELL RESPONSE. Like receptors and horizontal cells, bipolar cells generate only slow, graded potentials in response to illumination of the retina. However, unlike receptors and horizontal cells, the receptive field of the bipolar cell is concentrically organized into two antagonistic zones. The polarization of the cell in response to illumination of the center of the receptive field is antagonized by additional illumination falling on the periphery of the field (Fig. 3). About half of the units hyperpolarize to central illumination, while the other half depolarize. Representative recordings of each type of cell are shown in Fig. 7. In both cases additional illumination of the retina at the periphery of the receptive field antagonized the response to central illumination. Illumination of the periphery alone, however, did not polarize the cell in the opposite direction to that of the central illumination. Peripheral illumination appears only to turn off the bipolar responses to central illumination.

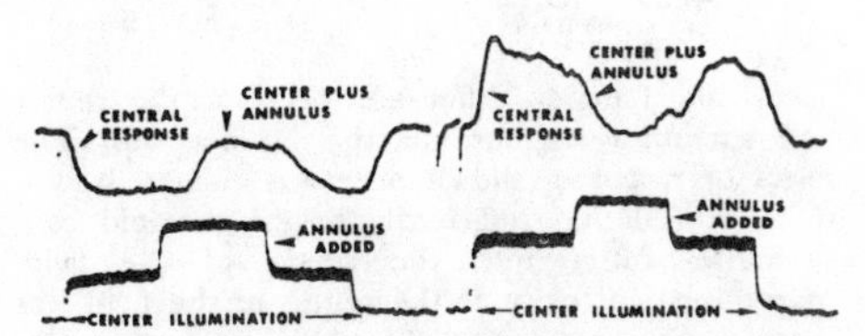

FIG. 7. The antagonistic effect of annular illumination on the bipolar cell response. In each experiment the center illumination of the receptive field for the bipolar cell was maintained on while an annulus of 250-μ radius was flashed. The bipolar cell polarization produced by central illumination was antagonized by the annular illumination. This was true for both the hyperpolarizing type (left) and the depolarizing type (right) of bipolar cell.

The bipolar responses bear a superficial similarity to the color-coded C-type S potentials (25, 33) in that it is possible to polarize the cell in opposite directions by the addition of appropriate stimuli. We have, however, been unable to find any evidence for color coding in the reversal of the bipolar response.

The magnitude of the bipolar cell response is typically 10 mv, starting from a resting potential of about 30 mv. The latency of the response to central illumination is about 100 msec at the intensity levels used in these experiments, which were within 3 log units of threshold. The center of the receptive field is about 100 μ in diameter, whereas an annulus with a radius of 250 μ is most effective in eliciting the antagonistic surround response. Smaller annuli scattered considerable light into the center of the field, whereas larger annuli fell outside the most efficient antagonistic peripheral zone.

Figure 8 shows intensity-response relations for a typical bipolar cell. These curves show how the response to central illumination increased with increasing intensity for three different levels of fixed illumination at the periphery of the receptive field. The curves demonstrate that the potential of the bipolar cell can be differentially controlled by varying the ratio of total flux energy falling at the center and periphery of its receptive field. To illustrate this more clearly, the points on the curves in Fig. 8 have been replotted in Fig. 9, with polarization potential rather than annular energy as the parameter. These curves show that under steady-state conditions the bipolar cell behaves as a center-surround contrast detector. The magnitude of the bipolar response can be held

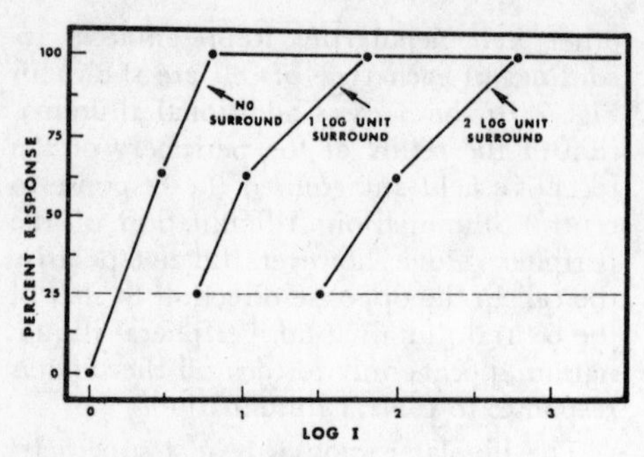

FIG. 8. Intensity-response relations for the bipolar cell. The bipolar response is determined by two parameters: energy falling in the center and energy falling in the periphery of the unit's receptive field. The three curves plotted here for a typical bipolar cell show how the polarization increases with increasing central illumination for three different values of peripheral illumination. The level of saturation is the same in all cases, but the dynamic range for the response is shifted along the intensity axis as the surround illumination is increased. Dotted line above the resting potential indicates that in this cell the surround could not turn off the polarization completely.

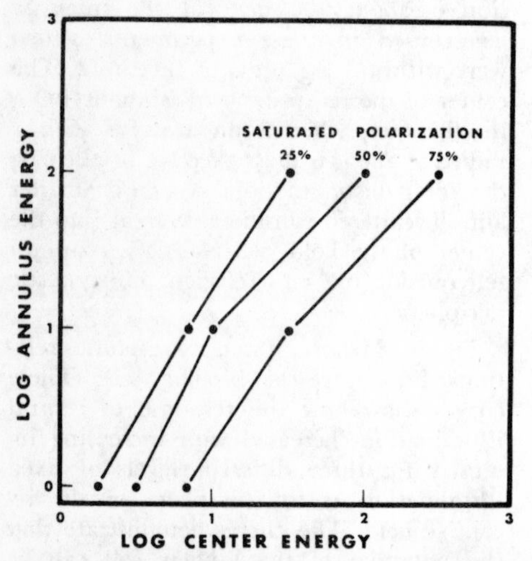

FIG. 9. Contrast detection properties of the bipolar cell. The abscissa and ordinate in this figure represent the total energies of stimulation falling in the center and periphery of receptive field for the bipolar cell. Each curve represents the relative quantities of each energy required to keep the polarization in the bipolar cell fixed. These curves indicate that the polarization is maintained by a fixed ratio of energy falling at the center versus the surround of the field. The curves suggest that the difference function is computed before saturation of either component. If this were not the case it would be impossible to increase the polarization of the unit after it had once been saturated.

constant for a fixed ratio of center-to-surround illumination over a wide range of absolute intensities.

The response to central illumination always precedes the antagonism evoked by peripheral illumination, regardless of the relative intensities of central and peripheral illumination. This is illustrated in Fig. 10. For these recordings, an annulus with radius of 250 μ was used throughout. The annulus was sufficiently bright so that scatter into the center of the receptive field generated an initial center response (here a hyperpolarization). The effect of the scatter was reduced in each successive (lower) trace by increasing sustained illumination at the center of the receptive field. In the lowermost record of Fig. 10 the central illumination was so great that the unit was saturated and no center response was evoked by the annulus. In all cases except the last the center hyperpolarizing response preceded the antagonism by more than 100 msec. Because of this latency difference there is in the bipolar cell a transient center response to any change in illumination even if there is not a change in contrast. The response is maintained only if the center-surround contrast is altered.

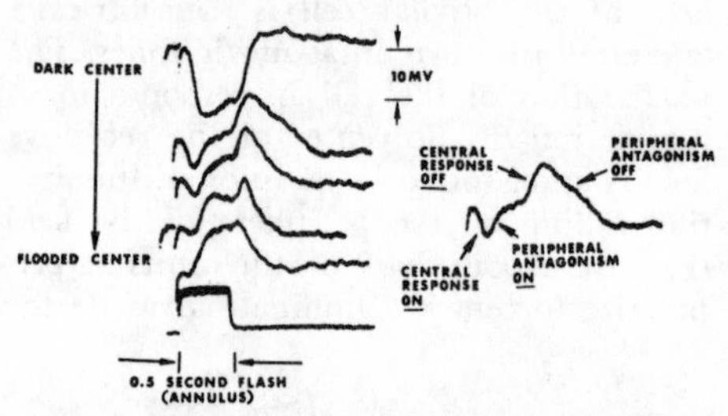

FIG. 10. Latency difference between the center and surround response of the bipolar cell. The series of responses shown here was elicited by annular stimulation sufficiently bright to yield considerable scatter into the center of the field. Maintained intensity at the center of the field was gradually increased to reduce the effect of scatter from the flashing annulus. The effective ratio of stimulation of center and surround is thereby gradually shifted to favor the surround. However, for any ratio of central and surround stimulation, the central response always occurs first. This is shown by the transient hyperpolarization in all traces except the lowermost trace. No center response is observed in the lowermost trace because the maintained central illumination was saturating.

AMACRINE CELL RESPONSE. At the level of the inner plexiform layer in the retina, a transition occurs from units that generate slow, sustained, graded, mostly hyperpolarizing potentials in response to steady illumination, to units that generate transient, depolarizing, regenerative potentials in response to changes in illumination. The transition from the bipolar cell response to the amacrine cell response is the first example of this.

Intracellularly recorded amacrine cell responses are shown in Fig. 3. The resting potential of these cells is typically 30–40 mv, whereas light-evoked responses are up to 30 mv in magnitude. The light-evoked responses consist typically of one or two spikes superimposed on a large, transient, depolarizing slow potential that lasts for about 250 msec. Such transient responses are seen at both the onset and cessation of illumination. Although the magnitude of the regenerative portion of the response is fixed, the latency of the spike varies dramatically with intensity of illumination, as illustrated in Fig. 11. Latencies of over 600 msec at threshold, and less than 200 msec at intensities 3 log units above threshold, were seen in these units.

The threshold for the amacrine response is easy to measure because of the nature of the all-or-none regenerative potentials. Thresholds are at least as low as the measured thresholds for receptors and bipolar cells. This suggests that we do not see the true threshold for the slow-potential units, which is probably buried in the noise of the recording system. This suggests that visual information can be carried by the slow-potential generators in the distal retina with little sign of electrical activity recorded intracellularly in these units.

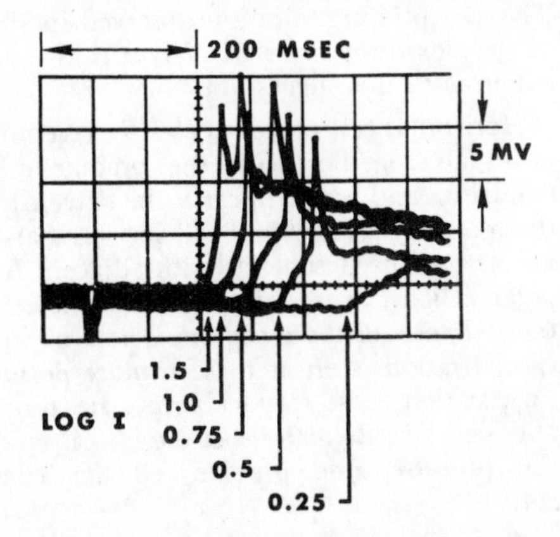

FIG. 11. Latencies of the amacrine response as a function of stimulus intensity. Five responses obtained with different intensities of stimulation were superimposed. Recordings show how the latency decreases with increasing stimulus intensity although the form of the response—one or two spikes superimposed on a transient slow potential—remains relatively unchanged.

Threshold measurements for amacrine cells were made to square pulses of light stimulation, because amacrine cells respond best to change in illumination. If the stimulus intensity was varied slowly enough, it was possible to increase the intensity by many orders of magnitude above threshold without evoking a regenerative response in an amacrine cell.

The dimensions of the receptive fields for the amacrine cells have been difficult to determine accurately. Some units had very broad, uniformly sensitive fields and responded at both "on" and "off" to illumination of any area of the receptive field. Others had narrow centers (measuring 100–200 μ) and larger surrounds like the receptive fields of the bipolar cells (Fig. 3). These units responded at on to central illumination, and at off to peripheral illumination. With diffuse illumination they responded at both on and off.

GANGLION CELL RESPONSES. The response of most of the ganglion cells in the retina of *Necturus* was transient, consisting of a brief burst of impulses superimposed on a small membrane depolarization. Such responses are illustrated in Fig. 3. The rate of impulse firing was roughly proportional to the membrane polarization, thus distinguishing this response from that of the amacrine cell, which consists of only one or two spikes superimposed on a large depolarization (Fig. 3 and Fig. 11). Resting potentials were quite variable and low in the ganglion cells, never exceeding 40 mv. Many of the units recorded from appeared to be spontaneously active, but this may be because they were often damaged upon penetration. The micropipette had to puncture through the internal limiting membrane just before it penetrated a ganglion cell.

Some ganglion cells responded at on, others responded at off, and some responded at both on and off. In addition to these ganglion cells that responded transiently with illumination, some cells responded with a sustained discharge of impulses when the center of the cell receptive field was illuminated. In such units the sustained discharge elicited by central illumination was inhibited by additional illumination at the periphery of the receptive field, as shown in Fig. 3. The receptive-field organization of these units was very much like that of the bipolar cells. The similarity between these sustained, spike-generating units and the bipolar response is illustrated in Fig. 3 where the membrane potentials for both cell types, although opposite in polarity, follow the same time course for all three conditions of stimulation. The dimensions of the receptive fields for the spike-generating cells were roughly similar to the dimensions for the bipolar cell receptive field, having centers no greater than 200 μ.

A small number of the sustained-type units have been stained, and they were located in the inner nuclear layer. We refer to these cells as ganglion cells because *a*) they show large spikes superimposed on a small membrane depolarization as in the normally situated ganglion cells, *b*) there are sustained-type units in the optic nerve of *Necturus* (19), and *c*) the discharge of these units resembles the responses of ganglion cells found in the frog (19), cat (23), rabbit (3), ground squirrel (28), and primates (20). This suggests that these cells may be the so-called displaced ganglion cells, or Dogiel cells, known to occur in many retinas (11). It will be most interesting if further experiments show that the displaced ganglion cells in the retina have different properties from those of cells located in the ganglion cell layer, as the results here suggest.

DISCUSSION

Outer plexiform layer

The mechanisms by which slow-potential generators in the retina communicate at their synaptic contacts are not understood. However, by comparing the dimensions of receptive fields and cell processes it is possible to infer the neural pathways in the outer plexiform layer and to suggest how the receptive field of the bipolar cell is formed.

The central part of the bipolar cell receptive field, measured physiologically, is roughly 100 μ in width. It is surrounded by an antagonistic region, which is best stimulated with an annulus of 250-μ radius. The dimensions of the central region of the bipolar cell receptive field correspond closely to the width of the dendritic fields of the bipolar cells, as measured in Golgi-stained material (14). No bipolar cell so far observed has a dendritic spread large enough to encompass both center and surround regions of the bipolar cell receptive field. However, horizontal cell processes extend laterally in the outer plexiform layer over distances of 200–400 μ, approximately the dimensions of the surround. Thus, the activity produced by an annulus of 250-μ radius could reach bipolar cell dendrites via the horizontal cell processes. This requires that the horizontal cell be driven by receptors in the periphery and modulate receptor-to-bipolar activity at the center of the bipolar cell receptive field. The synaptic organization observed in the outer plexiform layer of *Necturus* is consistent with this suggestion.

Horizontal cell processes contact receptor terminals together with the bipolar cell dendrites, and both appear to be driven by the receptors. Horizontal cell processes also make specialized contacts with adjacent bipolar cell dendrites, often close to the ribbon synapses of the receptors. The synaptic specializations seen at these contact points suggest that these contact points are probable sites of interaction between horizontal cell processes and bipolar cell dendrites (14).

Thus, anatomical evidence shows that the processes of horizontal cells extend far enough to mediate the receptive-field surround, and that the processes are both postsynaptic to receptors and presynaptic to bipolar cell dendrites. The physiology shows that the effect of the surround is to antagonize the bipolar cell central response. This suggests that the bipolar cell is polarized by receptors directly at the center of its receptive field and this polarization is antagonized by surrounding receptors acting through horizontal cells, so that differences in levels of illumination in these two areas control bipolar cell polarization. Figures 8 and 9 show that it is the ratio of energy falling in each of these areas that determines that polarization.

There is no direct physiological evidence that the horizontal cell exerts a controlling effect on the bipolar cell, but there have been some suggestive experiments. Byzov (10) recorded the local ERG in the frog while passing current through a micropipette which was located inside an S-potential unit, probably a horizontal cell. He was able to control the magnitude of the local ERG in response to a flash of fixed intensity by passing current through the S-potential unit via the electrode. Since the horizontal cell is situated between the receptors and the presumed site of ERG generation, Byzov's experiment suggests that the horizontal cell can regulate the effect of transmission from receptors to the more proximal retina.

To explain his results, Byzov (10) has postulated that horizontal cells exert their effect presynaptically back onto the receptors. However, we have seen no suggestion of a surround antagonism in the receptors under the same conditions which evoke the surround effect in the bipolar cells, and there is no anatomical suggestion for synapses back onto receptors (14). Tomita (40) also has found no effect of surround illumination in the intracellularly recorded receptor response. Although lack of an apparent surround effect in receptors recorded intracellularly does not prove its absence, both the anatomical and physiological evidence suggest that the horizontal cell acts proximal to the receptors, on the bipolar cell dendrites. Thus it would seem most likely that horizontal cells mediate the surround of the bipolar cell receptive field by antagonizing the effect of direct receptor-to-bipolar transmission.

The results indicate that one of the functions of the outer plexiform layer, in which neurons showing sustained, graded responses interact, is to register sustained differences in intensity as graded polarizations in the bipolar cell. In this way, contrasting boundaries can be accentuated in the visual information brought to the inner plexiform layer by the bipolar cell.

Inner plexiform layer

The neurons proximal to the bipolar cell—the amacrine and ganglion cells—respond to changes in illumination by depolarizing and initiating impulse activity. Ganglion cells showing this behavior have been studied for over 30 years, but amacrine cell activity has not been previously identified although it has probably been recorded before (22, 31, 36, 39).

In the experiments reported in this paper the units we identify as amacrines, generating one or two spikes superimposed on a large membrane depolarization, were consistently found located along the inner margin of the inner nuclear layer. The histology in *Necturus* shows that this region is almost always populated by amacrine cells, and almost never by ganglion cells or bipolar cells (14). In the frog, similar spike-generating units which show a larger membrane depolarization associated with impulse activity than do ganglion cells were also found in the inner nuclear layer (31, 39). These units have not been identified but it was demonstrated that such units cannot be driven when the optic nerve is stimulated antidromically, indicating that they are not ganglion cells (22).

A comparison of the responses recorded under the same stimulus conditions in the bipolar and amacrine cells shows that the amacrine cell is active only during the initial part of the transient in the bipolar cell response, whether this transient is positive-going or negative-going, or at on or off. This raises the question of how the relatively sustained bipolar response is converted to a transient response across the bipolar-to-amacrine synapse. An anatomical observation may have bearing on the question. Adjacent to each ribbon synapse of the bipolar cell terminal, at which there is a bipolar-to-amacrine synapse, there is a return synapse from amacrine to bipolar terminal. If the return synapse were inhibitory, the amacrine cell could turn off its own excitation in some way. However, the solution to this problem requires further investigation.

The activity of the ganglion cells in the *Necturus* retina can be interpreted in terms of the intracellular activity of the neurons that provide their synaptic input. The anatomy suggests two types of synaptic input. Ganglion cells can be driven directly by bipolar cells at the synaptic ribbon or by amacrine cells through conventional synapses, and it has been postulated that some ganglion cells may be driven primarily by bipolar cells while others may receive their major input from the amacrine cells (12, 14). The physiology shows that there are two distinct forms of ganglion cell response: some ganglion cells respond like bipolars, having concentrically organized receptive fields and showing sustained responses which can be antagonized. However, most ganglion cells in the *Necturus* retina behave more like amacrine cells, giving transient responses at on or off or both. Thus the suggested anatomical dichotomy seems to have a physiological expression.

The sustained type of ganglion cell having a concentrically organized, antagonistic receptive field has been described for many vertebrates (2, 3, 19, 23). Some of the special properties of these units, described previously, are consistent with the behavior of bipolar cells in *Necturus* as well as ganglion cells. For example, Barlow, Fitzhugh, and Kuffler (2) in the cat and Barlow and Levick (4) in the rabbit showed that the antagonistic surround is best elicited when the retina is light adapted. This behavior applies to bipolar cells in *Necturus* because the surround can only "turn off" the polarization resulting from central stimulation (Figs. 3 and 7). Barlow and Levick (4) also found in the rabbit that the central response always precedes the peripheral response, regardless of the relative intensities of the two. Figure 10 shows this phenomenon clearly for the bipolar cell in *Necturus,* and Fig. 3 for the ganglion cell.

The on-off units in *Necturus* seem to follow more closely the activity of amacrine cells (see Fig. 3) from which cells the anatomy suggests they receive their primary synaptic input (14). Anatomical studies in many vertebrates (12, 14, 27) suggest that the more complex processing in the retina is carried out in the inner plexiform layer, probably by amacrine cells. The electrophysiology of the neurons in *Necturus* shows that amacrine cells are primarily responsive to temporal changes in illumination, and that their activity is particularly suitable to accentuate dynamic properties of the visual world such as motion. Further work should serve to elucidate the mechanisms by which the complex retinal functions are performed in the inner plexiform layer.

In conclusion, the data presented in this paper show that two transformations are made in tandem in the retina at each of the plexiform layers. Figure 12 shows schematically how these transformations are accomplished. At the outer plexiform layer a contrast detection is effected so that the final bipolar cell polarization reflects the ratio of energy falling in the center of its receptive field to the energy falling over a wider peripheral area. This transformation appears to be accomplished primarily at the base of the receptor. Here processes from the bipolar and horizontal cells receive input from the receptors, whereas the horizontal cell processes feed across to adjacent bipolar dendrites. It is suggested that the effect of the horizontal cell contacts is to antagonize the effect of the receptor-to-bipolar response. Because horizontal cells have a much wider dendritic spread than do bipolars, a center-surround antagonism is formed. All elements involved in the synaptic interaction are slow-potential generators, mostly hyperpolarizing, and the responses are all graded and sustained with illumination. Some ganglion cells with response properties similar to those of the bipolars carry the transformation effected at the outer plexiform layer to higher centers.

The inner plexiform layer operates on the transformed information coming to it via the bipolar cells. Here the dynamic qualities of the visual image—temporal changes in either the intensity or configuration—are detected and amplified by the amacrine cell system. Amacrine cells impinge on ganglion cells and drive them

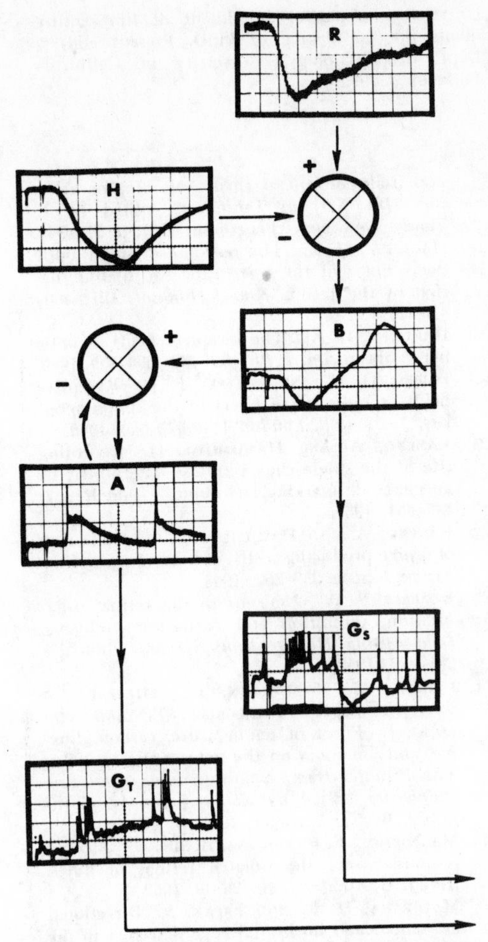

FIG. 12. Summary diagram of synaptic organization of the retina. Transformations taking place in each plexiform layer are represented here by summing junctions. At the outer plexiform layer the direct input to the bipolar cell from the receptor is modified by input from the horizontal cells. At the inner plexiform layer the bipolar cell drives some ganglion cells directly. These ganglion cells generate a sustained response to central illumination, which is inhibited by additional annular illumination; and thus these cells follow the slow, sustained changes in the bipolar cells' response. Bipolar cells also drive amacrine cells, and the diagram suggests that it is the amacrine-to-bipolar feedback synapse that converts the sustained bipolar response to a transient polarization in the amacrine cell. These amacrine cells then drive ganglion cells which, following the amacrine cell input, respond transiently.

transiently. The ganglion cells so driven carry information about the changes in the characteristics of the visual field. The optic nerve, which contains axons from both types of ganglion cells, thus carries information both about relative intensities in the visual field and temporal changes in these intensities.

SUMMARY

The responses of neurons throughout the retina of the mudpuppy have been studied by intracellular recording with micropipettes. These neurons were subsequently identified by intracellular staining. The results of the electrophysiological studies were correlated with the synaptic organization of the retina of *Necturus,* and through this combined study of structure and function it has been possible to begin a description of the functional organization in the inner and outer plexiform layers of the retina.

Neurons that form synapses at the outer plexiform layer—the receptors, horizontal cells, and bipolar cells—respond with slow, graded, sustained potentials to illumination. Receptors and horizontal cells always respond by hyperpolarizing, whereas bipolars may either hyperpolarize or depolarize in response to illumination. The polarization of a bipolar cell, produced by illumination at the center of its receptive field, is antagonized when the area surrounding the receptive field is illuminated.

In the inner plexiform layer, the neurons driven by the bipolars—the amacrine and ganglion cells—respond to illumination by depolarizing. Depolarization beyond a threshold level in these cells results in classical regenerative "spike" activity. Amacrine cells respond transiently to either increasing or decreasing intensity of illumination, provided the change is rapid enough. Most ganglion cells respond transiently to changes in illumination. A few ganglion cells respond tonically to steady illumination within their receptive fields.

ACKNOWLEDGMENTS

B. DeHaven provided excellent technical assistance. Publication No. 21 from the Augustus P. Long Laboratories of the Alan C. Woods Building, The Wilmer Institute.

This research was supported in part by National Institutes of Health Grant NB-05336 and by Research to Prevent Blindness, Inc.

F. S. Werblin was the recipient of a National Institutes of Health Special Fellowship, 1967–68. The experiments reported here were described in a thesis presented by F. S. Werblin to the Department of Biomedical Engineering, Johns Hopkins University in partial fulfillment of the requirements for the degree of Ph.D. Present address: Dept. of Physiology, University of California, Berkeley, Calif.

REFERENCES

1. Barlow, H. B. Summation and inhibition in the frog's retina. *J. Physiol., London* 119: 69–88, 1953.
2. Barlow, H. B., Fitzhugh, R., and Kuffler, S. W. Change of organization in the receptive fields of the cat's retina during dark adaptation. *J. Physiol., London* 137: 338–354, 1957.
3. Barlow, H. B., Hill, R. M., and Levick, W. R. Retinal ganglion cells responding selectively to direction and speed of image motion in the rabbit. *J. Physiol., London* 173: 377–407, 1964.
4. Barlow, H. B. and Levick, W. R. The mechanism of directionally selective units in rabbit's retina. *J. Physiol., London* 178: 477–504, 1965.
5. Bortoff, A. Localization of slow potential responses in the *Necturus* retina. *Vision Res.* 4: 626–627, 1964.
6. Bortoff, A. and Norton, A. Positive and negative potential responses associated with vertebrate photoreceptor cells. *Nature* 206: 626–627, 1965.
7. Bortoff, A. and Norton, A. Simultaneous recording of photoreceptor potentials and the P-III component of the ERG. *Vision Res.* 5: 527–533, 1965.
8. Brown, J. E. and Rojas, J. A. Rat retinal ganglion cells: receptor field organization and maintained activity. *J. Neurophysiol.* 28: 1073–1090, 1965.
9. Byzov, A. L. Functional properties of different cells in the retina of cold-blooded vertebrates. *Cold Spring Harbor Symp. Quant. Biol.* 30: 547–558, 1966.
10. Byzov, A. L. Horizontal cells of the retina as regulators of synaptic transmission. *Fiziol. Zh. SSSR* 53: 1115–1124, 1967; and *Neurosci. Transl.* No. 3: 268–276, 1967–1968.
11. Cajal, Ramón y, S. *Die Retina der Werbelthiere,* translated by Greeff. Wiesbaden: Bergmann, 1894.
12. Dowling, J. E. Synaptic organization of the frog retina: an electron miscroscopic analysis comparing the retinas of frogs and primates. *Proc. Roy. Soc., London, Ser. B* 170: 205–222, 1968.
13. Dowling, J. E., Brown, J. E., and Major, D. Synapses of horizontal cells in rabbit and cat retinas. *Science* 153: 1639–1641, 1966.
14. Dowling, J. E. and Werblin, F. S. Organization of the retina of the mudpuppy, *Necturus maculosus.* I. Synaptic structure. *J. Neurophysiol.* 32: 315–338, 1969.
15. Gouras, P. Graded potentials of bream retina. *J. Physiol., London* 152: 487–505, 1960.
16. Granit, R. *Sensory Mechanisms of the Retina.* London: Oxford Univ. Press, 1947.
17. Grundfest, H. Excitation by hyperpolarizing potentials. A general theory of receptor activities. In: *Nervous Inhibition,* edited by E. Florey. New York: Pergamon, 1961, p. 326–341.
18. Hartline, H. K. The response of single optic nerve fibers of the vertebrate eye to illumination of the retina. *Am. J. Physiol.* 121: 400–415, 1938.
19. Hartline, H. K. The receptive fields of optic nerve fibers. *Am. J. Physiol.* 130: 690–699, 1940.
20. Hubel, D. H. and Wiesel, T. N. Receptive fields of optic nerve fibers in the spider monkey. *J. Physiol., London* 154: 572–580, 1960.
21. Kaneko, A. and Hashimoto, H. Recording site of the single cone response determined by an electrode marking technique. *Vision Res.* 7: 847–851, 1967.
22. Kaneko, A. and Hashimoto, H. Localization of spike-producing cells in the frog retina. *Vision Res.* 8: 259–262, 1968.
23. Kuffler, S. W. Neurons in the retina: organization, inhibition and excitation problems. *Cold Spring Harbor Symp. Quant. Biol.* 17: 281–292, 1952.
24. Laufer, M., Svaetichin, G., Mitarai, G., Fatehchand, R., Vallecalle, E., and Villegas, J. Effect of temperature, carbon dioxide, and ammonia on the neuron-glia unit. In: *The Visual System: Neurophysiology and Psychophysics,* edited by R. Jung and H. Kornhuber. Berlin: Springer, 1961, p. 457–463.
25. MacNichol, E. F. and Svaetichin, G. Electric responses from the isolated retinas of fishes. *Am. J. Ophthalmol.* 46: 26–40, 1958.
26. Maturana, H. R. and Frenk, S. Directional movement and horizontal edge detectors in the pigeon retina. *Science* 142: 977–979, 1963.
27. Maturana, H. R., Lettvin, J. Y., McCulloch, W. S., and Pitts, W. H. Anatomy and physiology of vision in the frog. *J. Gen. Physiol.* 43 (No. 6, part 2): 129–171, 1960.
28. Michael, C. R. Receptive fields of single optic nerve fibers in a mammal with an all-cone retina. *J. Neurophysiol.* 31: 249–282, 1968.
29. Mitarai, G. Determination of ultramicroelectrode tip potential in the retina in relation to S potential. *J. Gen. Physiol.* 43: 95–99, 1960.
30. Mitarai, G., Svaetichin, G., Vallecalle, E., Fatehchand, R., Villegas, J., and Laufer, M. Glia-neuron interactions and adaptational mechanisms of the retina. In: *The Visual System: Neurophysiology and Psychophysics,* edited by R. Jung and H. Kornhuber. Berlin: Springer, 1961, p. 463–481.
31. Naka, K., Inoma, S., Kosugi, Y., and Tong, C. Recording of action potentials from single cells in the frog retina. *Japan. J. Physiol.* 10: 436–442, 1960.
32. Naka, K. and Kishida, K. Simultaneous re-

cording of S and spike potentials from the fish retina. *Nature* 214: 1117–1118, 1966.

33. Naka, K. and Rushton, W. A. H. S-potentials from color units in the retina of fish (*Cyprinidae*). *J. Physiol., London* 185: 536–555, 1966.
34. Naka, K. and Rushton, W. A. H. The generation and spread of S-potentials in fish (*Cyprinidae*). *J. Physiol., London* 192: 437–561, 1967.
35. Orlov, O. Yu. and Maksimova, E. M. S-potential sources as excitation pools. *Vision Res.* 5: 573–582, 1965.
36. Svaetichin, G. Horizontal and amacrine cells of the retina: properties and mechanisms of their control upon bipolar and ganglion cells. *Acta Cient. Venezolana* 8: Suppl. 3, 254–276, 1967.
37. Svaetichin, G., Laufer, M., Mitarai, G., Fatehchand, R., Vallecalle, E., and Villegas, J. Glial control of neuronal networks and receptors. In: *The Visual System: Neurophysiology and Psychophysics,* edited by R. Jung and H. Kornhuber. Berlin: Springer, 1961, p. 445–456.
38. Svaetichin, C. and MacNichol, E. F. Retinal mechanisms for chromatic and achromatic vision. *Ann. N. Y. Acad. Sci.* 74 (part 2): 385-404, 1958.
39. Tomita, T. Electrical activity in the vertebrate retina. *J. Opt. Soc. Am.* 53: 49–57, 1963.
40. Tomita, T. Mechanisms subserving color coding. *Cold Spring Harbor Symp. Quant. Biol.* 30: 559–566, 1966.
41. Tomita, T., Murakami, M., Hashimoto, Y., and Sasaki, Y. Electrical activity of single neurons in the frog retina. In:*The Visual System: Neurophysiology and Psychophysics,* edited by R. Jung and H. Kornhuber, Berlin: Springer, 1961, p. 463–481.
42. Witkovsky, P. A comparison of ganglion cell and S-potential response properties in carp retina. *J. Neurophysiol.* 30: 546–561, 1967.

39

Reprinted from
J. Neurophysiol., 1953 *16*: 37-68

DISCHARGE PATTERNS AND FUNCTIONAL ORGANIZATION OF MAMMALIAN RETINA*

STEPHEN W. KUFFLER
The Wilmer Institute, Johns Hopkins Hospital and University
Baltimore, Maryland

(Received for publication December 11, 1951)

INTRODUCTION

THE DISCHARGES carried in the optic nerve fibers contain all the information which the central nervous system receives from the retina. A correct interpretation of discharge patterns therefore constitutes an important step in the analysis of visual events. Further, investigations of nervous activity arising in the eye reveal many aspects of the functional organization of the neural elements within the retina itself.

Following studies of discharges in the optic nerve of the eel's eye by Adrian and Matthews (2, 3), Hartline and his colleagues described the discharge pattern in the eye of the Limulus in a series of important and lucid papers (for a summary see 20). In the Limulus the relationship between the stimulus to the primary receptor cell and the nerve discharges proved relatively simple, apparently because the connection between sense cell and nerve fiber was a direct one. Thus, when stimulation is confined to one receptor the discharge in a single Limulus nerve fiber will provide a good indication of excitatory events which take place as a result of photochemical processes. Discharges last for the duration of illumination and their frequency is a measure of stimulus strength. Lately, however, it was shown by Hartline *et al.* (22) that inhibitory interactions may be revealed when several receptors are excited. On the whole, the Limulus preparation shows many features which are similar to other simple sense organs, for instance, stretch receptors. In the latter, however, instead of photochemical events, stretch-deformation acts as the adequate stimulus on sensory terminals and is translated into a characteristic discharge pattern.

The discharge from the cold-blooded vertebrate retina (mainly frogs) proved much more complex. Hartline found three main types when recording from single optic nerve fibers: (i) "on" discharges, similar to those in the Limulus, firing for the duration of the light stimulus, (ii) "off" discharges appearing when a light stimulus was withdrawn, and (iii) "on-off" discharges, a combination of the former two, with activity confined mainly to onset and cessation of illumination. The mammalian discharge patterns were studied in a number of species by Granit and his co-workers in the course of their extensive work on the physiology of the visual system (summaries in 13, 15). On the whole, they did not observe any fundamental differences between frog and mammalian discharge types (see later).

* This investigation was supported by a research grant from the National Institutes of Health, U. S. Public Health Service.

The present studies were begun several years ago with the intention of examining the retinal organization and particularly processes of excitation and inhibition. As a first step, the discharge patterns were re-examined. It was assumed, in line with other workers, that the deviations in vertebrate eyes from the simple Limulus, or "on" discharge type, are due to the nervous structures and to their interconnections between the rod and cone layers on the one hand and ganglion cells on the other. Therefore, an extension of such studies should shed further light on the functional organization of the retina.

A preparation was used which approached fairly satisfactorily the "normal" state of the cat's eye. The discharge patterns reported by Hartline and those extensively studied by Granit were readily obtained. Single receptive fields—areas which must be illuminated to cause a ganglion cell to discharge—were explored with small spots of light and thereby some new aspects of retinal organization were detected. Specific receptive subdivisions, arranged in a characteristic fashion and connected to the common ganglion cells, seem to exist within each receptive field. This finding made it possible to study in detail some of the factors which normally contribute to the changing discharge pattern during vision. The present set-up also furnishes a relatively simple preparation in which the neural organization resembles the spinal cord and probably many higher centers of the nervous system. Many analogies have been found with discharge patterns in the spinal cord which are currently under study.

METHOD

The experimental arrangement, particularly the details of the optical system, has been described in full in a preceding paper (31). The main instrument, the "Multibeam Ophthalmoscope," consisted of a base which carried a holder in which the cat's head was rigidly fixed. Above the head, and also carried by the base, was the viewing-stimulating apparatus, which could be freely rotated and tilted. It contained three light sources with independent controls. This optical system was aligned with the cat's eye which thus was in the center of a spherical coordinate system and the eye's ordinary channels were used for illumination of the retina. One light provided adjustable background illumination and thereby determined the level of light adaptation. It was also used as a source for observation of the retinal structures. A maximal visual magnification of about 40 was obtained. The background illumination covered a circle of not less than 4 mm. (16° for the cat) in diameter, centered on the recording electrode. Two Sylvania glow-modulator tubes were used for stimulation of restricted areas of the retina. They illuminated patterns, mostly circles of varying diameter, which were imaged on the retina. The smallest light spots were 0.1 mm. in diameter on the retina. Thus, two images could be projected and their size and location varied independently on the retina. All three light sources used a common optical path, led into the eye through a pupil maximally dilated by Atropine or Neo-synepherine.

Complications from clouding of the cornea were prevented by the use of a glass contact lens, while the rest of the eye's optical system, lens and vitreous, remained intact. The circulation of the retina was under direct minute observation and whenever the general condition of the animal deteriorated this was readily noticed. The eye, as judged by its circulation and its discharge patterns, remained in good condition for the duration of the experiments which frequently lasted for 15–18 hours. Dial-urethane (Ciba) anesthesia (0.5 cc./kg.) or decerebration was used. The effect of anesthesia on the discharges is discussed later.

The eyeball was fixed by sutures to a ring which was part of the microelectrode manipulator. This fixation was generally satisfactory and breathing or minor body movement did not disturb the electrode position on ganglion cells. Sudden movements, however, such as coughing, jerking, etc., prevented continuous recording from single units. Occasionally a persistent slow nystagmus developed and, in order to abolish this movement, the tendons of extraocular muscles were severed at their insertions.

Microelectrodes were introduced into the eye protected by a short length of #19 hypodermic tubing which served to penetrate the scleral wall near the limbus. The unprotected electrode shaft, less than 1 mm. thick, then traversed the vitreous and made contact with the retinal surface and toward the tip it was drawn to a fine taper. The shadow of the electrode thus covered only a small portion of the receptive field. If hit by the narrow light beams the electrode shaft caused scattering. All these phenomena and the positions and imagery of the stimulating beams or patterns were directly observed during the experiments and thereby a subjective evaluation of illumination conditions could be formed.

Electrical contact with the retinal cells was made by 10–15 μ Platinum-Iridium wires which were pushed to the tip of the glass tubes. The metal was either flush with the surrounding glass jacket which was sealed around it, or it protruded several micra. The configuration of the electrode tip was purposely varied a good deal, especially when the ganglion discharge was to be blocked by pressure. The potentials varied in size, and the largest were around 0.6 mV. The position of the indifferent electrode could be anywhere on the cat's body. In technically satisfactory preparations no difficulty was encountered in finding ganglion cells in quick succession and individual units could be observed for many hours (see later).

The second beam of the oscilloscope was used as an indicator of the current flow through the Sylvania glow-modulator tubes. The current was proportional to the light output but the spectral distribution of the light varied with different current strength. Therefore, Wratten neutral filters were used when the white light of the stimulators had to be attenuated. For the purpose of the present experiments the wave length variation which occurred played no significant role in those cases where intensities were varied by current flow adjustment (see Figs. 7*III*, 8, 10). The accurate electronic control of the stimulating light sources made an adjustment of flash durations quick and convenient. The time base was also recorded on the second sweep by intensity modulation through a square-wave oscillator.

Illumination values are given in meter candles; the calibration was made for flux reaching the corneal surface above the pupil and calculated for the area which it covered on the retina. Losses within the eye's media are neglected. The maximal available background illumination was about 6000 meter candles at the retina and could be attenuated to any desired extent. Since 1 m.c. at the retina corresponds to 10 mL external brightness (see 31) the samples illustrated here were taken well within the photopic range. Discharge patterns were, however, also studied in the absence of background illumination. In most experiments the exploring spot's intensity was approximately 100 m.c.

Results

1. *Some characteristics of single unit discharge*

Differentiation between ganglion and axon potentials. As a recording electrode of 5 15 μ diameter at the tip made contact with the surface of the retina, a mass of potentials was usually recorded on illumination of the eye. Very light touch of the retinal surface rarely yielded differentiated single unit potentials. The latter could, however, be obtained with a slight further advance of the electrode, still without marked pressure against the tissue. Different degrees of "touch" and pressure were easily differentiated under close direct observation (see Method). The most common and most easily recorded potential seen in the retina was a polyphasic spike, starting with an initial positive deflection, similar to that shown in Figure 1*B*. Such potentials are generally set up by a small spot of light at some distance from the recording electrode. From this observation it follows that conduction to the recording lead has taken place and that the potential is derived from a nerve fiber. The polyphasic shape is typical of conducted potentials in a volume conductor. Similar potentials are familiar from recordings in other parts of the nervous sytem where microelectrodes are employed. The propagated potentials in nerve fibers could be used in the present studies, but they were small and could not be kept under the electrodes for prolonged periods. In contrast, potentials were recorded which always originated under the electrode tip (Fig. 1*A*). These were simpler and larger and usually started with

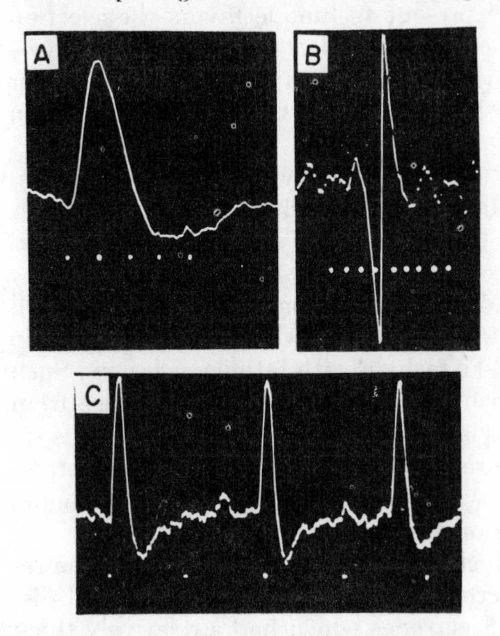

Fig. 1. Potentials from different retinal elements recorded with microelectrode. A: Ganglion cell discharge, caused by stimulation of retina in proximity of recording electrode. B: Nerve impulse in an axon, set up by retinal stimulation some distance from electrode. C: Three ganglion cell potentials from middle portion of a high-frequency discharge which is illustrated in Fig. 9c; potentials become progressively smaller at this rate. Negative deflexion of A and C 0.4 mV. and 0.1 mV. in B. Time intervals in A and B 0.1 msec. and in C 1.0 msec. Note that ganglion potential can also start with small positive inflexion if recording electrode is somewhat shifted.

a sharp negative inflection which was followed by a relatively smaller positivity. The potentials were generally about 0.3–0.4 msec. at the base, and their negative phase was of longer duration than in the potentials of Fig. 1*B*, where the whole triphasic complex is of a similar duration. The distortion of the real potential time course is due to the smallness of the effective interelectrode distance with the present electrodes. In a volume conductor the potentials which arise close to or under the electrodes start with a sharp negative inflection, as in Figure 1*A*. On such grounds this potential is likely to be a ganglion cell potential which lies in the vicinity of the electrode contact. Physiological tests furnish convincing evidence for such a conclusion. The area of the retina, which on illumination caused discharges in the "ganglion" cells, was found to be in the immediate neighborhood of the electrode tip; this also was the place where the lowest intensity light spot was effective in setting up discharges. As an exploring spot was moved further from the tip of the electrode, stronger stimuli were needed. The active unit lay in the approximate center of the "receptive field" (see later) and excitation apparently reached it through converging pathways from its immediate neighborhood. Such an arrangement is typical of ganglion cells. Figure 1*C* illustrates three ganglion potentials which form part of the high-frequency discharge series of Figure 9*c*; the impulses follow at intervals of 2.0 and 1.7 msec. At these rates the potential heights decline.

It follows from the relationship between receptive area and recording electrode that one can distinguish between conducted potentials in axons and those arising from ganglion cells. The latter may, however, also show a more complex shape, presumably when the recording electrode is some distance away from the cell body. The present technique favors the selection of larger ganglion cells but the extent of this selection is not known (see Discussion). The findings agree with those of Rushton (29), who by different methods showed the large single retinal discharges to arise from ganglion cells.

The potentials can also be easily distinguished by listening to their discharge in the loudspeaker. The ganglion potentials, which arise in the center of the receptive fields, have a lower pitch, apparently because of less high-frequency components than in the axon spike.

Evidence for single cell discharges. The conventional criteria of single cell discharges are usually potentials of uniform size which arise at a sharp threshold and do not vary in a step-like fashion with fatigue or injury. Such criteria are generally sufficient to insure that potentials do not arise from several cells which fire in unison. In view of later findings, however, it is especially important to know that one really deals with single cell discharges.

The following procedures, which were incidental to many experiments, gave additional convincing evidence on this point.

(i) During *progressive pressure* which was obtained by advancing the recording electrode by means of a micrometer control, the ganglion cell discharge could frequently be blocked. Electrodes which had a relatively thick jacket flush with the platinum tip, were most convenient for such pressure blocks. By these procedures, potentials could be separated into two components. The first component was variable over a very wide range and its height depended on the amount of pressure, while the other varied much less. The small potential had the characteristics of a local potential which precedes propagated spikes as described by Katz (25) and Hodgkin (23). Accordingly, whenever such a "prepotential" was sufficiently reduced the spike disappeared abruptly. These events leading up to pressure block are illustrated in Figure 2. Pressure itself frequently stimulated the ganglion cells and the ensuing activity was usually photographed by exposing a fast recurring sweep until a required number of impulses was obtained. In Figure 2*a*, at the beginning of pressure, an inflexion marked by arrows is seen on the upper half of the rising phase. In *b* the two phases are more marked, the spike taking off from the beginning of the falling phase of the prepotential. In *c* a critical level is reached and at the first arrow a pure prepotential appears. The second arrow indicates two potentials which, by chance, were accurately superimposed; in one case the prepotential causes a spike, in the other it just fails to do so. In *d* the prepotential alone is seen. It should be noted that the time course of the potentials under pressure is slower than under normal (Fig. 1) conditions. This applies especially to the prepotential. While the microelectrodes give a distorted (shortened) time course of potentials, the difference between spike and prepotential seems significant. Decreasing the pressure restored the prepotential size and when it reached a critical height the spike suddenly reappeared; the process could then be repeated. With excessive pressure, however, the whole potential disappeared irreversibly. The constancy of the spike under such conditions of block, and recovery from block, confirm the assumption that it is derived from one ganglion cell only. It is unlikely that two cells should be so located in the vicinity of the electrode tip as to be affected in a quite similar and simultaneous manner. The origin of the variable prepotential was not studied in detail. It probably also originates in the ganglion cell, and such potentials may be set up there by the bipolars. It resembles some of the potentials obtained by Svaetichin (29) in spinal ganglion cells. Similar potential sequences are also seen at neuromuscular junctions or ganglionic synapses with curare or fatigue blocks.

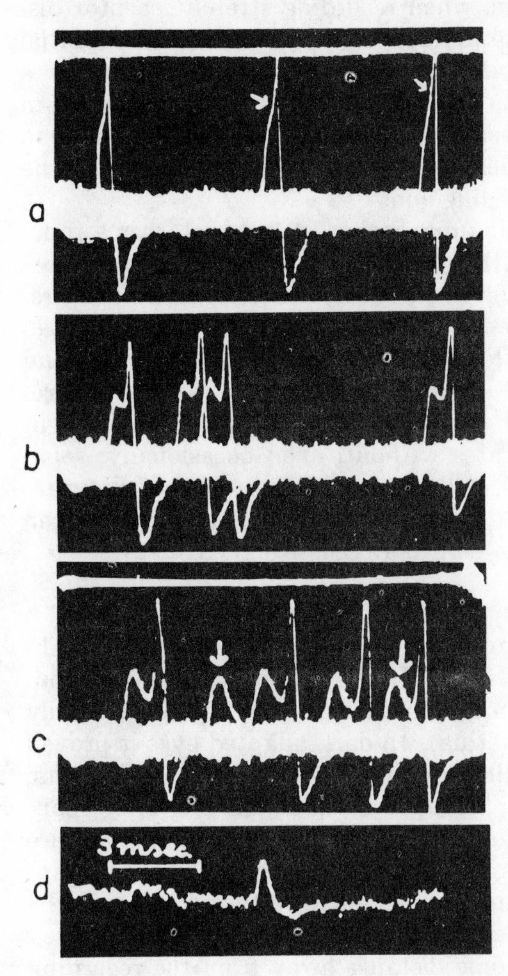

FIG. 2. Progressive pressure block of ganglion cell discharge. Exposures made with sweep recurring at high frequency. Four successive stages of pressure block. *a:* An initial inflexion (arrow) appears on upper portion of rising phase. *b:* A more discrete "prepotential" is seen. *c:* Prepotential is further reduced and occasionally (arrows) no spike appears. At second arrow a chance superposition of two apparently identical prepotentials occurred; one sets up spike, other fails to do so. *d:* spike is completely blocked, prepotential alone recorded on single sweep. Potential size in *a* is 0.3 mV.; note that also spike diminishes. Under progressive pressure potentials are of longer duration than normal (see Fig. 1).

(ii) The *potential size* of impulses at high frequencies is further evidence that single cell discharges are recorded. In the eye discharge, frequencies of 200–700/sec. and more are quite common. During these high-frequency bursts the potential size may decline, sometimes to about half of its original size. The decline is generally smooth in its progression and therefore cannot be due to one or two units dropping out during the discharge (Figs. 9, 10). If one cell ceased to fire the potential should abruptly decrease. Alternatively it could hardly be assumed that several units should be so closely coupled. Variability of potential size in single peripheral nerve fibers has been observed at frequencies around 500/sec. when recording stretch receptor discharges (24). There seem to be some differences, however, in the potential height changes between axons and ganglion cells. The latter tend to show a fall in height at lower frequencies, a fact already studied by Renshaw in spinal motoneurons (27). In the present instances (*e.g.*, Fig. 9) the ganglion cell probably fires near its physiological limit, each impulse following in the relative refractory period of the preceding one.

The most convincing test of single unit discharge, however, was a functional one revealed in the mapping of the receptive fields. As will be shown below, discharge patterns are distributed in a characteristic fashion within receptive fields (*e.g.*, Fig. 6). That more than one ganglion cell should happen to have identical receptive fields with such a great regularity as was found in the present experiments would be a difficult assumption to make. Moreover, one would have to postulate that the cells always gave coupled high-frequency discharges at near-limit rates without, even occasionally, separating. Further, interaction, such as will be seen in the series of Figure 8, where regular mutual suppression of discharges occurs, could hardly happen if one recorded simultaneously from two or more cells.

2. *Spontaneous retinal activity*

Spontaneous activity in the mammalian retina has been regularly observed by Granit in dark-adapted cats (13). In the present preparation con siderable background discharge was a dominant feature especially in dimly illuminated retinae (1–5 m.c. at the retina). In dark-adapted eyes it proved very difficult to investigate the detailed discharge patterns of single units, since they fired frequently at "resting" rates of about 20–30/sec. The "spontaneous" activity in the absence of illumination seems to be a normal feature for the following reasons: discharges due to injury of nerve fibers or ganglion cells under the recording lead, due to movement and pressure, could be excluded; spontaneous activity in many isolated units could be suppressed by illuminating the receptive fields some distance away from the recording lead (see also later); similarly, an electrode with a tip of 10–15 μ, if gently placed near the middle of the optic disc, recorded massed spontaneous discharges which originated elsewhere, since illumination of the whole eye suppressed a great portion of the discharge; injury discharges along nerve fibers could not be expected to be modified by illumination in such a fashion.

Spontaneous activity was particularly pronounced in decerebrate animals, but was also regularly seen under Dial-urethane anesthesia. The latter seemed to reduce the activity. Similarly, intravenous Nembutal, in amounts such as 20 per cent of the anesthetic dose, had an immediate and prolonged effect in arresting or diminishing discharges from the retina. A similar effect with a slower onset was seen with intraperitoneal injections.

Since a great part of the present studies was done on cats under Dial-urethane the effect of the anesthetic will influence the findings to an unknown degree. All essential observations, however, were also repeated in decerebrate preparations.

As indicated above, spontaneous activity, when recorded from isolated dark-adapted units, was generally suppressed or decreased for varying periods after application of increased background illumination. In the course of light adaptation discharges usually returned gradually, or the slowed rates increased again. However, once a unit discharges in the light-adapted state, it is not possible to say how "spontaneous" the activity is.

Of particular interest are those discharges which were apparently not due to injury and were not appreciably modified by general illumination of the eye. No detailed study of their nature could be made since they were never recorded in complete isolation. It is possible that during a steady increased background illumination many units appear which have previously not discharged, while others drop out. Such switching of active units may make it impossible to decide whether certain units have been continuously active or not. This important aspect of retinal activity has yet to be explored. In many cats grouped discharges in numerous nerve fibers were seen. They could usually be suppressed by illumination of the eye, but again their origin was not studied.

While most features of "spontaneous" activity remain to be investigated, it is a noteworthy phenomenon, since it is upon such a high level of background activity that patterns of many visual events are superimposed. Rhythmic and "spontaneous" activity is common to the central nervous system in mammals and has also been observed in a variety of other visual systems (1, 4, 7).

3. Extent of receptive fields of cat's retina

The receptive field of a single unit was defined by Hartline as the area of the retina which must receive illumination in order to cause a discharge in a particular ganglion cell or nerve fiber. Hartline (17, 18) was the first to study the physiological characteristics of receptive fields of single optic nerve fibers in frogs in a precise and thorough manner, by exploring the area with a small spot of light. Since the retina is composed of a group of overlapping receptive fields, the extent of these is of obvious interest. By charting the boundaries of an area over which a spot of light sets up impulses in a ganglion cell or in its nerve fiber, one will obtain the configuration of the receptive field. The field size depends on stimulus strength, the size of the exploring spot and the state of dark adaptation. The latter will largely determine the level of sensitivity of the area. For instance, if an exploring spot is made smaller, or if the level of background illumination is increased, the intensity of the spot has also to be increased in order to set up responses over as large an area as previously. The problems of determining receptive field sizes have been discussed in detail by Hartline (18), and his results on frogs were found to apply equally to the mammalian retina.

The receptive field definition may be enlarged to include all areas in *functional* connection with a ganglion cell. In this respect only can the field size change. The anatomical configuration of a receptive field—all the receptors actually connected to a ganglion cell by some nervous pathways—is, of course, assumed to be fixed. As will be seen below, not only the areas from which responses can actually be set up by retinal illumination may be included in a definition of the receptive field but also all areas which show a functional connection, by an inhibitory or excitatory effect on a ganglion cell. This may well involve areas which are somewhat remote from a ganglion cell and by themselves do not set up discharges.

The optical conditions in the mammal present additional difficulties for mapping of receptive fields, as contrasted to those in the opened frog's eye. Because of the imperfections of the optical system, an appreciable amount of light scattering occurs and the images will be less sharply focussed. The most advantageous situation for the full exploration of the receptive fields, which approximates the anatomical receptive field boundaries, is complete dark adaptation. During this state, however, most units discharge spontaneously, making threshold determination or detection of changes in response patterns difficult. The mapping was mostly carried out in different states of light adaptation, and even under such conditions a "steady" state cannot be maintained. As implied in the term "adaptation," thresholds change, drifting towards a lower value, and discharge patterns may also vary correspondingly. Such changes seem to be part of normal events in the eye. In spite of these factors some relevant data of the size of receptive fields can be obtained.

Figure 3 illustrates a chart of a retinal region which contains receptors with connections which converge upon one ganglion cell and cause it to discharge. The exploring spot was 0.2 mm. in diameter and the background illumination approximately 10 m.c. The smallest inner area was obtained by an exploring spot, about five times threshold for a position near the electrode tip. If the spot was moved outside this area (5×), no discharges were set up. If the spot intensity was increased 10 times, by removing a Wratten neutral filter, and thus making it 50 times threshold, it caused discharges within the larger area (50×). Further increase in the stimulus strength to 500 times threshold expanded the receptive field on three sides (500×) while the demarcation line on the left remained practically unchanged. This may indicate that light scattering was not a very great factor in this particular mapping. Otherwise such a fixed portion of the boundary, in spite of an increase in stimulus intensity, could hardly be obtained. Frequently a receptive field as shown here was charted and then the exploring spot was further increased in strength. The field suddenly expanded several times and then generally no distinct boundary demarcation was obtained. It is thought that this was clearly due to scatter of light since a reduction of the stimulating spot size again resulted in a definite limit of the receptive field.

The present technique, using small exploring spots, is suited to detect relatively dense concentrations of receptors which feed into a single ganglion cell, and therefore provides only an approximate estimate of the actual anatomical receptor distribution. Evidence suggests that the density of receptors beyond the receptive field limit (Fig. 3) may be insufficient to produce more than subthreshold effects on a ganglion cell (see Discussion). Stimulation with larger spots may overcome the difficulties and extend the recep-

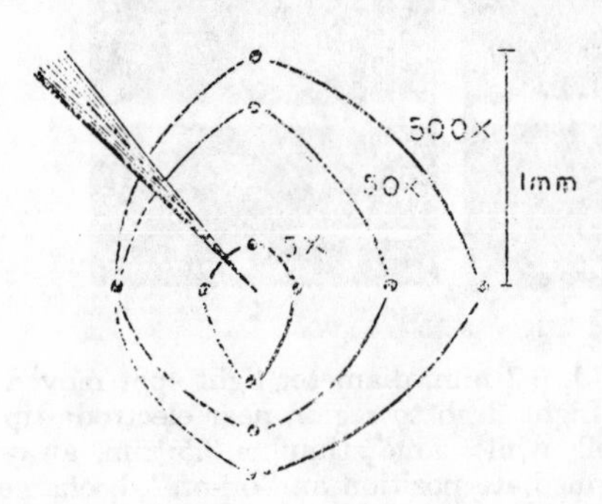

FIG. 3. Extent of receptive field obtained with exploring beam of 0.2 mm. in diameter at three different intensities. Electrode (shaded) on ganglion cell. Background illumination about 10 m.c. Inner line encloses retinal region within which light spot, about 5 × threshold at electrode tip, sets up discharges. Other boundaries of field were mapped at intensity 50 × and 500 × threshold. Note that on left, receptive field does not expand appreciably as stimulating spot intensity is increased.

tive field into areas where the receptor concentration is low. By increasing the spot size, in fact, receptive fields apparently 3–4 mm. in diameter were found, but scatter of light makes those findings unreliable. The experiment should be done by the use of a great variety of illumination patterns near threshold intensities which would allow a more exclusive excitation of the "surround," while the central region is not illuminated. Most determinations in the present study were made in the region of the cat's tapetum, a highly reflecting region where the anatomical features of the retina can be observed with greater accuracy through the optical system. Further, the tip of the recording electrode can be seen, the stimulating spot can be followed, and in this way conditions can be checked by direct observation, provided the background illumination is sufficiently bright. The receptive field diameters varied between 0.8 and 2.0 mm. with the present method. Small ganglion cells may have fields of different extent. No determinations have been made in the periphery of the retina (see Discussion).

4. Stimulation of subdivisions of receptive fields

(*a*) *Specific areas within receptive fields.* In Hartline's (17) experiments stimulation anywhere within a receptive field of the frog caused essentially the same discharge pattern in a given fiber; *i.e.*, either "on," "on-off" or pure "off" responses resulted. Accordingly the discharge type from the frog's receptive field seems relatively fixed (see, however, Discussion). This question was investigated in the present study.

It was found that the discharge patterns from ganglion cells whose receptive fields were explored varied with the specific subdivisions which were illuminated. Figure 4 illustrates such findings. A light spot, 0.2 mm. in diameter, was moved to different positions, all within an area of 1 mm. in diameter. In Figure 4*a* a discharge appeared during illumination; this "on" response was of a transient nature and although stimulation was continued at the same intensity, the discharge ceased within less than one second (see Section 6). In Figure 4*b* when the light spot was moved 0.5 mm. from the

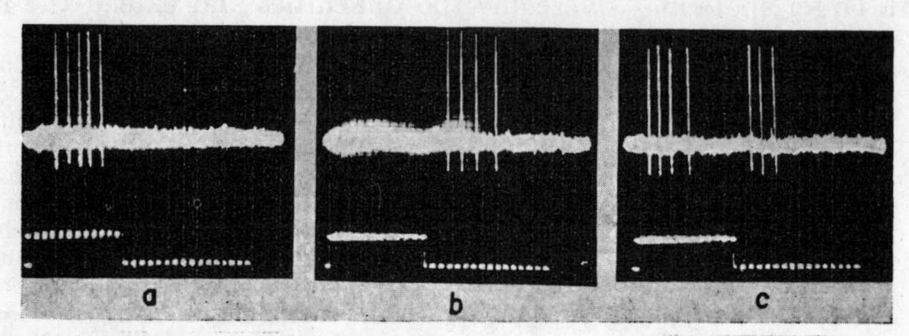

FIG. 4. Specific regions within receptive field. 0.2 mm. diameter light spot moved to three different positions within receptive field. Light flash to region near electrode tip in (a) causes only "on" discharges in ganglion cell, while same stimulus 0.5 mm. away is followed by "off" responses (b) and in an intermediate position an "on-off" discharge is set up (c). In this and subsequent records second beam signals intensity and duration of light flash; intensity modulation of 50/sec. gives time base. Impulses 0.5 mV.

first position no "on" discharge at all appeared and the response was of the pure "off" type, *i.e.*, discharges occurred after the cessation of illumination. At an intermediary position of the exploring spot, a combination of the first two responses resulted, and an "on-off" discharge is seen (*c*). All transitions in discharge patterns from those here shown were seen when the light spot of fixed intensity was moved to a number of positions within the receptive field, while the background illumination of the eye remained constant. Other illustrations of changes in discharge patterns with illumination of different areas within the receptive field are seen in Figures 7 and 8. Thus, the ratio and number of "on" or "off" discharges varied with the specific area which was illuminated. The changes in discharge type, caused by merely shifting an exploring spot, were not always striking in all units. To obtain the varied discharge patterns it was frequently necessary to change, in addition, the state of light adaptation, the stimulus intensity, or area of the stimulating light (see below).

It is concluded that within the receptive fields of single ganglion cells (or nerve fibers) there exist areas which can contribute differing discharge patterns. The discharge, as seen with stimulation of the whole receptive field, is the resultant of the contribution and interaction of all of these areas.

(*b*) *Distribution of discharge patterns in receptive fields*. All units had a central area of greatest sensitivity in which either the "on" or the "off" component predominated in the discharge pattern. Flashes of 0.5–1.0 sec. duration, for instance, to subdivisions of an area of perhaps 0.5 mm. in diameter around the ganglion cell would give "on" responses only. Within this area the "on" frequency decreased as the spot was shifted away from the most sensitive region in the center. This is shown in Figure 5. A spot 0.1 mm. in radius was projected onto the retinal region around the tip of the recording electrode which was placed on a ganglion cell. In this and nearly all other ex-

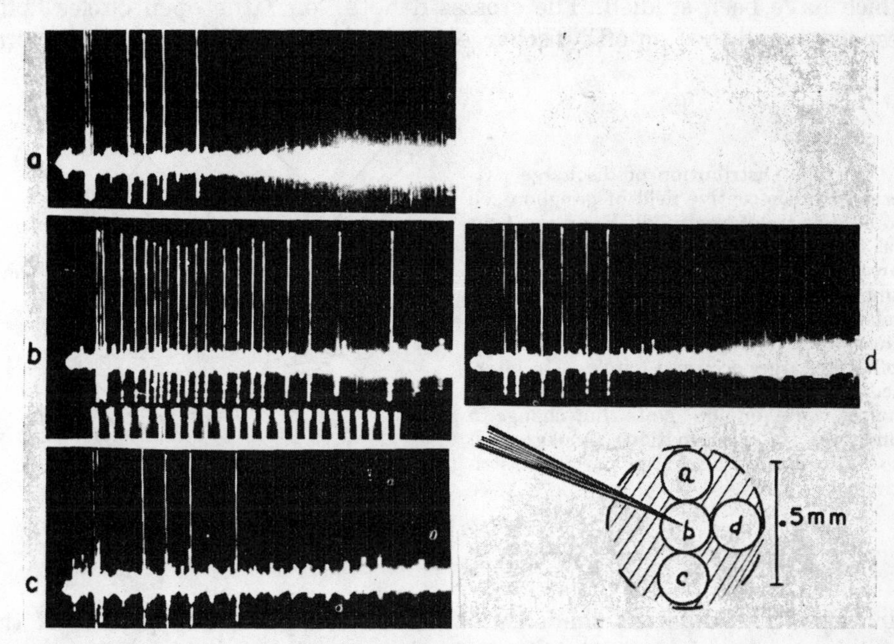

FIG. 5. Center portion of receptive field. Ganglion cell activity caused by circular light spot 0.2 mm. in diameter, 3–5 times threshold. Background illumination was about 30 m.c. Positions of light spot indicated in diagram. In *b* an "on" discharge persists for duration of flash. Intensity modulation at 20/sec. Movement of spot to positions *a*, *c*, and *d* causes lower frequency discharge which is not maintained for duration of light stimulus. Movement of spot beyond shaded area fails to set up impulses (see, however, extent of receptive field in similar unit with stronger stimuli in Fig. 6). Potentials 0.5 mV.

periments the region of electrode contact proved to be the most light-sensitive part of the receptive field. The area of lowest threshold and the geographical center of the receptive fields usually coincide. If the stimulating light spot was made 3–4 times threshold for the central location it evoked there a vigorous "on" response for the duration of illumination (Fig. 5*b*). A shift of the light spot, as illustrated in the scheme included in Figure 5, made it much less effective. The "on" discharges set up by the same stimulus became shorter and of lower frequency, and with further movement away from the center no discharges at all were set up. The boundaries of the receptive field with this relatively weak stimulus strength at a background of 30 m.c. are indicated by the broken circle.

The records of Figure 5 show only a central area of a receptive field similar to the one which is within the inner circle of Figure 3. If the small exploring spot is made 100–1000 times threshold, a more complete picture of the discharge pattern distribution in receptive fields can be formed. The chart of Figure 6 was obtained from a unit under a background illumination of about 25 m.c. It is characteristic in a general way of the majority of units which have been studied. The crosses denote "on," the open circles "off" responses, and the "on-off" discharges are indicated by the cross-circle combinations. The different shaded areas give an approximate picture of the predominant areal organization within the receptive field, *i.e.*, of receptors and neural connections (see Discussion). The center-surround relationship may be the converse in other units, with the "off" responses predominating in the center; the area ratio between center and surround also fluctuates greatly. Further, the discharge pattern distribution shifts with changing conditions of illumination (see below).

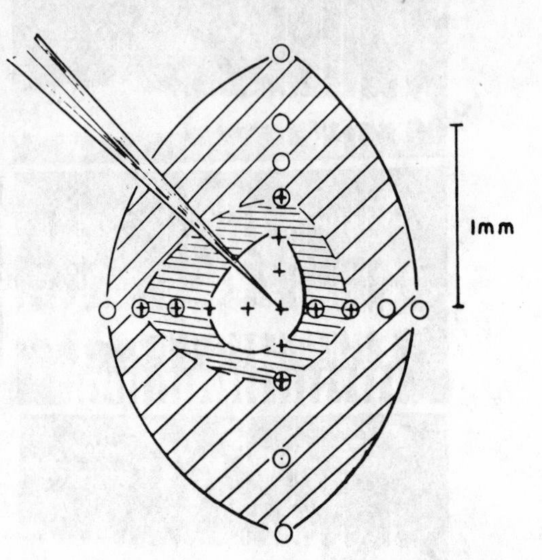

FIG. 6. Distribution of discharge patterns within receptive field of ganglion cell (located at tip of electrode). Exploring spot was 0.2 mm. in diameter, about 100 times threshold at center of field. Background illumination approximately 25 m.c. In central region (crosses) "on" discharges were found, while in diagonally hatched part only "off" discharges occurred (circles). In intermediary zone (horizontally hatched) discharges were "on-off." Note that change in conditions of illumination (background, etc.) also altered discharge pattern distribution (see text).

Not in all units was the field laid out in a regular concentric manner as in Figure 6. The areas were frequently irregular. In some instances there appeared "gaps" between regions; *i.e.*, isolated spots in the periphery seemed to be functionally connected to a ganglion cell.

(*c*) *Factors modifying discharge patterns and size of receptive fields.* As indicated above, the discharge patterns arising in single receptive fields may vary, if conditions of illumination are altered. The four upper records of Figure 7 show "on" discharges produced by a 0.2 mm. diameter light spot. In the lower records is seen a corresponding series of "on-off" discharges which were obtained from the same unit by changing different parameters of illumination. In *I* the area of the stimulating spot was increased so as to include the whole receptive field and thereby the "on" was converted into an "on-off" discharge. In *II* the same effect was obtained by decreasing the background illumination while leaving all other conditions unchanged. In *III* merely the intensity of the testing spot was increased, while in *IV* the spot was moved to another portion of the receptive field, without altering its intensity or area. It follows from these observations that a modification of any of these variables of light stimulation, alone or in combination, will in turn lead to modifications of the discharge pattern. In addition to the factors illustrated in Figure 7, the duration of stimulation also plays a role. The direction of the changes can usually be predicted. If, in a composite discharge pattern, one of the components—for instance, the "on" portion—predominates strongly, a reduction of stimulus strength will cause the relatively weak

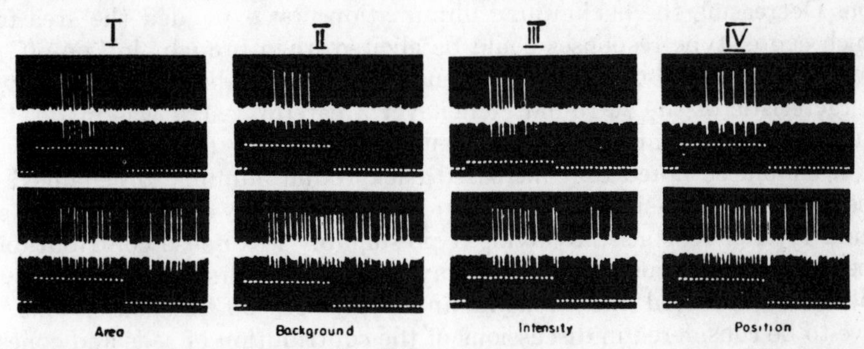

FIG. 7. Change in discharge pattern from "on" response (upper records) in single ganglion discharge into an "on-off" response (lower records). In I stimulating spot of 0.2 mm. diameter in central region of receptive field set up "on" discharge. Increasing spot diameter to 3 mm. set up more "on" impulses and brought in an "off" component. Same result was obtained in II by merely decreasing background illumination from 19 m.c. to 4 m.c. and in III by increasing stimulus spot intensity (intensity scale in III different). In IV exploring spot was shifted by about 0.4 mm. from central into more peripheral part of receptive field. Intensity modulation 50 p.s.

"off" fraction to disappear, while the "on" may be only little affected. The same result can generally be obtained by merely increasing the background illumination or reducing the area of the stimulating spot. Conversely, a combination of a weak "on" and a strong high-frequency "off" component can be changed into a pure "off" response by reducing the stimulus strength or increasing the background illumination intensity. Discharge patterns can frequently be altered by variation of background and stimulating light intensities even when the whole receptive field is illuminated. However, results are usually not as clear-cut as with fractional activation of the receptive field.

The *effect of background illumination* deserves more detailed analysis since it is one of the most potent factors in altering discharge conditions. As the background illumination is increased, the boundaries of the receptive fields "contract" and also the discharge pattern distributions change. The response type which is characteristic of the surround (diagonally hatched area of Fig. 6) tends to disappear and the pattern of the center (non-hatched region) will predominate. In fact, some units even with careful exploration, using small 0.1 0.2 mm. light spots under photopic conditions, gave only pure "on" or "off" responses within the limits of the receptive field which might be only 0.5 mm. in diameter. If the area of the stimulating spot was increased without changing its intensity for instance, by illuminating a retinal patch 1 mm. in diameter then the stimulus occasionally brought in an additional weak response which was characteristic of the "fringe" or surround. Thus, an "on" type of response would be converted into an "on-off" as the spot size was increased (see also Fig. 7*I*). The characteristic response of the surround could always be made evident by using a dim background, or after a short period (several minutes) of complete dark adaptation. Decreasing the background illumination first expanded the area from which center-type responses could be elicited, then brought in "on-off" responses around its boundary and eventually disclosed discharges which were characteristic of the surround. Whenever a careful search was made, both "on" and "off" components were seen in all receptive fields.

It should be noted that increased background illumination changed the receptive field in a similar manner in all units which were studied. The surround type of response, involving a presumably less dense contribution of receptors (see Discussion), was always suppressed first, independently of whether it consisted of a predominantly "on" or "off" response. This will have to be considered in discussions of the contribution of rods and cones to discharge patterns.

The great range of flexibility at the level of the single unit discharge is of particular interest, since all the factors which were found to affect the discharges play a role under normal conditions of vision.

5. *Interaction of different areas within receptive field*

It may be presumed that one of the basic contributions of interneurons within the retina (cells between the photoreceptors and the ganglion cells which give rise to the optic nerve fibers) consists in modifying the pattern of discharges which are set up by excitation of rods and cones. The impulses emerging through the optic nerve show the result of a complex series of events which have taken place in the retina, such as spatial interaction and processes of facilitation and inhibition. These problems have already been considered by Adrian and Matthews in their classical investigations on the eel's eye (2, 3) and by Hartline in the early studies of the organization of the receptive field (17, 18, 19). A wealth of data on the functional organization of the retina has also emerged from Granit's laboratory (13, 14, 15).

An additional approach is made possible by the present findings that certain areas within a single receptive field make a predominant "on" or "off" contribution to the discharge pattern. Two spots of light were projected onto the retina; each came from a separate light source, and the location, size, brightness and duration of illumination of both were controlled independently. The two light beams could be shifted on the retina in relation to each other and their temporal sequence was controlled electronically. There are numerous possible variants under which the experiments could be done. The first and simplest is illustrated in Figure 8. One of the exploring spots, (A), with a radius of 0.1 mm. was placed near the tip of the recording electrode in the center of the receptive field, and it caused a high-frequency "on" response during illumination. The other spot, (B), twice the diameter of the

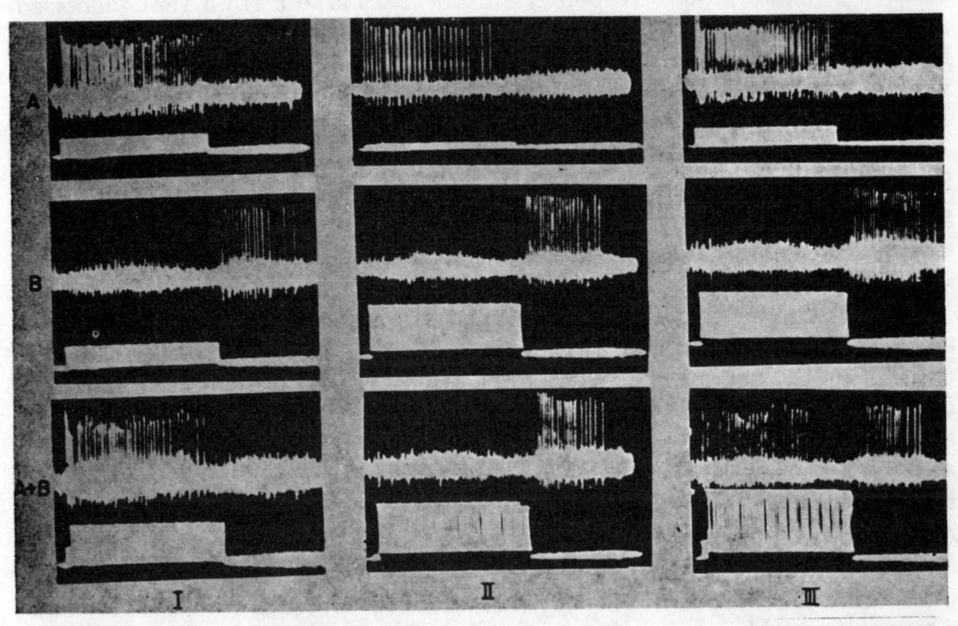

FIG. 8. Interaction of two separate light spots. Single ganglion cell discharge during background illumination of 20 m.c. Spot A, 0.1 mm. in radius, was placed in center of receptive field at tip of recording electrode. Spot B, 0.2 mm. in radius, was 0.6 mm. away in surround. Flashed separately they set up "on" (A) and "off" (B) responses. With a simultaneous flash, A+B in column I, "off" response was suppressed and at same time number of "on" discharges in A+B is slightly reduced as compared with A. In II, intensity of spot A was reduced, while spot B was increased (note flash strength indication on second beam). As a consequence B suppressed "on" discharge of A. In III, both spots were "strong." When flashed together (A+B) they reduced each others' discharges. Flash duration was 0.33 sec., potentials were 0.3 mV.

first, with its center 0.6 mm. from the ganglion cell and the recording electrode, was in the surround and set up a simple "off" discharge in this unit. When both spots were flashed on the retina simultaneously, the "off" response was suppressed (Fig. 8*I*, A+B). At the same time the number of "on" impulses was somewhat reduced as compared with the control response to stimulation of spot (A). Such situations could be produced regularly with two spatially separated light spots within a receptive field, *i.e.*, illumination of one area could suppress discharges arising from stimulation of another. The reverse situation from Figure 8*I* could be produced in the same unit as is shown in Figure 8*II*. Spot (A) was made less intense while (B) in the surround was made stronger. When these stimuli were given together (A+B), the "on" effect of (A) was completely suppressed, while the off discharge was but little affected. An intermediate situation between Figure 8*I* and *II* could also be created by altering the intensities so as to make the effects from spots (A) and (B) equally "strong." When flashed simultaneously in Figure 8*III* (A+B), they simply reduced each other's effect, setting up a relatively weak "on-off" response. In order to make certain that increased scatter of light with two spots was not responsible for the effects, the two light beams were superimposed. In such cases their effect on the discharge was simply additive. Figure 8 illustrates only a few of the possible variants in discharge which can be produced by two interacting light patches. Instead of changing the intensities of the stimulating spots, results similar to Figure 8 could also be obtained by merely varying the areas of spots (A) and (B) so as to produce the required amount of "on" or "off" discharge. Alternately, shifting the location of the light stimuli or altering the background illumination would balance the "on" and "off" relationship in any required direction.

In many experiments one light spot was fixed and the other was moved around it in the manner of a satellite. In this way a systematic study was made of the interacting regions within a receptive field. As might be expected from the above results, one could produce all combinations of response types and variants of the "on-off" ratio. Once the receptive field with its boundaries and discharge patterns within that area was plotted (see Fig. 6), the result of interaction of two spots could usually be predicted. It is worthy of note that in the present experiments not only the excitatory result of a light stimulus, such as an "on" discharge, could be inhibited, but also the "off" discharge—itself a consequence of inhibitory processes—could be suppressed. As a rule, then, when two light stimuli within the receptive field interact, *both* become modified, but if the effect of one is much "stronger" than the other, its discharge may not be appreciably affected.

Suppression of "off" responses could also be seen some time after stimulation of an "on" area. The time course of this inhibitory effect, presumably caused by persistent excitation after previous illumination, could be studied in the following manner. In units similar to that shown in Figure 8*I* the duration of the stimulus to the "on" area (A) could be progressively shortened while (B) was kept constant. It was found that beam (A) could suppress (B) for varying periods after (A) had been turned off. The time course of the inhibitory after-effect depended on the duration and intensity of (A). There was a transition from complete suppression of the "off" discharge to partial suppression and to a mere delay in the onset of the "off" discharge.

In these investigations it was surprising that frequently a ganglion cell, which gave an "off" effect, was largely unresponsive to stimulation of an "on" area during the period of the "off" discharge. Further, in the tests where the interaction of two "on" areas was studied, lack of addition of excitatory influences frequently developed. Since these observations on interaction phenomena have a bearing on functional organization of the retina a more thorough analysis will be presented in a separate publication.* Particularly the combination of spatial and temporal effects opens up some further approaches. These instances are mentioned here because they present a wider picture of factors which play a role in the production of discharge patterns. Further, they tend to explain some "anomalous" observations, such as lengthening of latent periods with stronger stimuli, or increased discharge frequencies with weaker ones (Figs. 11, 12).

6. *Characteristics of "on" discharge*

(*a*) *Transient and maintained "on" response.* From analogies with the Limulus eye there may be reason to suspect that the maintained "on" response in mammals, which keeps discharging for long periods during illumination, is set up in receptors which have a fairly "direct" connection from photoreceptors to bipolars and to ganglion cells. On the other hand, the "on" which is part of the frequently occurring "on-off" type may be set up in units where the receptive field has different neuroanatomical connections.

In the preparations studied there were units which gave only the Limulus type of "on" response when the whole retina was stimulated under photopic or scotopic conditions. Under careful scrutiny, when restricted subdivisions of the receptive field were stimulated, with dim background illumination, it was always observed that these "on" units also received "off" contributions from the periphery (see Section 4). More frequent were those units which gave a transient "on" response lasting about 1–2 sec. with diffuse maintained retinal stimulation. These were generally followed by an "off" response, depending on the background illumination (see above). The most frequent units were those with "on-off" discharges, the "on" lasting 1 sec. or less. The following modifications of the transient "on" responses were of special interest because they revealed some further aspects of receptive field organization: (i) When the central portion of some receptive fields was illuminated by a spot of 0.1–0.2 mm. in diameter an "on" discharge resulted lasting for seconds or, in several instances, even minutes. Either increasing or decreasing the stimulus strength frequently shortened the duration of the "on" discharge. (ii) Moving the stimulating spot as little as 0.1–0.2 mm. from the center of the receptive field greatly shortened the discharge and at the same time the onset of the discharges could be delayed (Fig. 10b). Further, units were observed which gave a maintained "on" response at the center, "off" responses in the periphery and transient "on" responses coupled with "off" discharges in intermediate regions of the receptive field. This required the selection of an appropriate background illumination, stimulating intensity and size of the exploring spot. (iii) In some units a small central spot gave maintained "on" responses and, as the area of the illuminating patch was enlarged to include the surround, the discharge became of the transient type (see also 17). (iv) One isolated instance in which, however,

* These questions are discussed more fully in *Cold Spr. Harb. Symp. quant. Biol.*, 1952, *17*.

the unit gave easily repeatable responses for several hours deserves mentioning. Under a background illumination of 10 20 m.c. the unit showed an "on" response which could not be maintained for longer than 1 2 sec. at any available intensity of the stimulating spot which was 0.2 mm. in diameter and directed onto the central region. When the background illumination was increased (60–100 m.c.) this discharge was converted into a maintained "on" type although the stimulus was of the *same* intensity as that which gave the shorter "on" response before. This situation was the reverse of the more common one since, by increasing the background illumination, a given stimulating intensity usually becomes less effective. One may surmise that in this unit the background illumination preferentially suppressed inhibitory influences from the "off" areas.

The above findings suggest the following interpretation: the maintained "on" discharge is converted into the transient type by activation of elements which converge onto the same ganglion cell from the periphery of the receptive field. Accordingly a unit which is so organized that it has a strong "on" center and a weak "off" surround will tend to give a well-maintained discharge even with illumination of the whole eye. The discharge will shorten in proportion to the peripheral "off" contribution. Such a view is also supported by the interaction experiments in which a simultaneous second spot in the surround weakens and shortens the discharge set up by the central one. The duration of the "on" discharge then will depend on how many "off" pathways to a ganglion cell are active in relation to the "on" fraction. It is realized that the inhibitory "off" action starts approximately simultaneously with the "on" action. Therefore, if both continued simultaneously at the same strength, one would expect merely a reduction of the "on" discharge frequency scale and not a shortening when a certain "off" component is added. Such a reduction of an "on" discharge is seen in Figure 8*III*. However, the "on" discharges which are generally observed start at a relatively high frequency which subsequently tends to decrease. With reduction of the stimulus strength producing such an "on" discharge, the initial high frequency will be reduced while the later discharge of lower frequency may drop out completely (see also 18). Therefore a similar "weakening" of an "on" discharge by an inhibitory action may lead to a shortened "on" response. Further, the suppressing effect from "off" zones does not necessarily start with its full force, but may increase with prolonged stimulation as can frequently be seen in its action of stopping "off" discharges. The presence of inhibitory contributions in many pure "on" elements has already been shown by Donner and Willmer (9).

(*b*) *Latent period and discharge frequency of "on" responses*. Generally one can cause increased excitation, as measured by frequency of response and shortened latent period, by (i) increased stimulus intensity, (ii) increase in stimulated area, (iii) decrease in background illumination (or increased dark adaptation), (iv) moving the stimulating spot toward the center of the receptive field.

A fairly typical effect of stimulus strength on the latent period and discharge frequency is seen in Figure 9. A spot 0.2 mm. in diameter was flashed at four different intensities onto the "on" center of a receptive field, increasing in steps of 10 from *a* to *d*, with the eye under a background illumination of about 2 m.c. This illustration is of particular interest because it shows how short the latent period can be and how high the discharge rate can become in the cat's retina even with moderate intensities of stimulation. In *a* the stimulus is near threshold and the latent period is 93 msec. In *b* the latency is 36 msec. and the average discharge rate for the first 8 impulses is about 180/sec. In *c* the discharge frequency is 300/sec. for the first 13 discharges and the latency is 22 msec. and in *d* a peak frequency of over 800/sec. is reached between the 4th and 10th impulse, the latency being 15 msec. This rate of discharge is much higher than is customarily obtained from nervous structures under physiological conditions. A pause as in *d* is common, both after "on" or "off" bursts. Increasing the area of stimulation within the center of the receptive field, starting with a relatively weak stimulus, also caused higher discharge frequency and latency shortening in this unit (see, however, below). The latent period of 15 msec. in Figure 9*d* is shorter than hitherto seen in mammals, presumably due to restriction of the stimulus to a predominantly "on" area (see below).

Figure 10 illustrates a unit which gave an "on" discharge lasting several seconds with illumination of the whole eye and a somewhat longer one with illumination confined to its "on" center. In *a* it showed a high-frequency initial burst of 575/sec. for the first 8 impulses with the potential size sharply declining (followed by a pause). The latent period of 15 msec. in *a* was lengthened and the discharge frequency and duration was reduced in the subsequent three records by the following: (i) in *b* the light spot of the same intensity as in *a* was moved from the center of the field by 0.1 to 0.2 mm.; (ii) in *c* with the light spot in the center again, the background illumination was increased; and (iii) in *d*, the stimulus intensity was reduced. Reducing the stimulating area or shortening the duration of the light flash (not illustrated) had a similar result as shown in *b–d*. The findings of Figures 9 and 10 are in general agreement with the early work of Adrian and Matthews (2), Hartline (17) and Granit (13).

Some notable exceptions to the general "rules" as discussed above were also observed—and, in fact, could frequently be produced by appropriately arranging the conditions of the experiment. Thus, in contrast to the usual results, the latent period of discharge was actually prolonged in the unit of Figure 11 when the area of stimulation on the retina was changed from a patch 0.2 mm. in diameter (a) to one of 3 mm. (b). Similarly, increasing the light intensity could have the same effect. One may assume that stimulation of the larger area brought in a strong "off" component from the surround, causing a delay in the "on" response. Such a situation could actually be produced frequently by stimulation of two separate small "off" and "on" areas. Another "exception" is seen in Figure 12 where an "on-off" response is con-

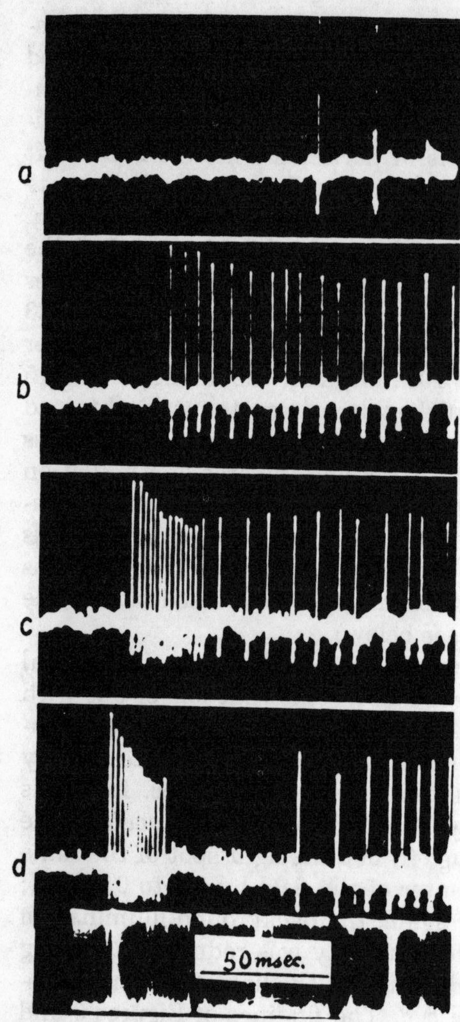

FIG. 9. Effect of stimulus strength on latent period and discharge frequency. 0.2 mm. diameter spot projected onto "on" center of a receptive field at illumination background of about 2 m.c. Between *a* and *d* stimulus was increased in steps of 10. Latent periods were 93, 36, 22 and 15 msec. Peak frequency in *d* was over 800 sec. Transient pause after a high-frequency burst occurred regularly, as did decline of potential size. Impulses 0.4 mV.

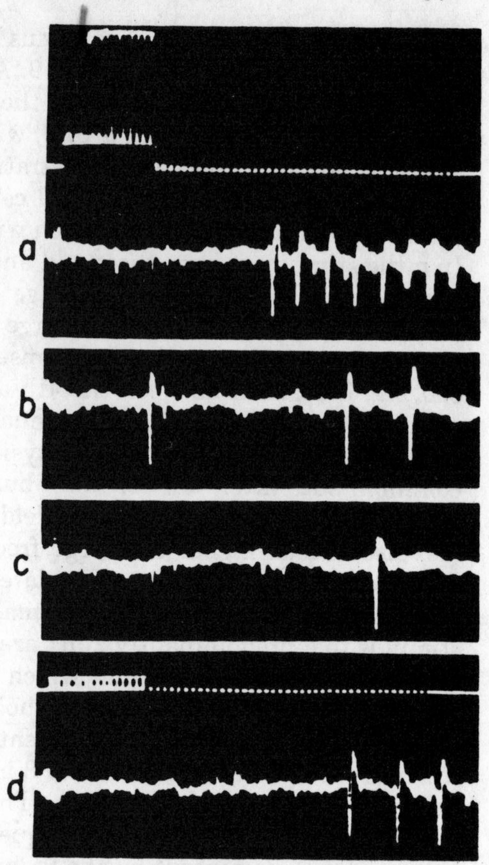

FIG. 10. Ganglion discharge with spot (0.2 mm. diam.) illumination. *a*: flash of 6.5 msec. in duration to center of receptive field set up response with initial frequency of 575 sec. and latent period of 15 msec. Prolonging illumination did not change latent period but caused an "on" response for 2 3 sec. *b*: same flash; image moved 0.1 0.2 mm. from central position. Latent period 21 msec., only two impulses set up. First impulse on sweep was "spontaneous" and not related to flash. *c*: same flash as *a*, but background illumination increased. Only one impulse set up. *d*: conditions as in *a*, but stimulus intensity decreased. Latent period 22 msec., discharge burst shorter. Effects seen in *b–d* were also obtained by shortening flash or reducing spot size. Intensity modulation 2000/sec.

verted into an "off" by reducing the stimulus strength. Surprisingly, however, the latent period of the "off" response is shorter and the number of impulses is greater with this weaker illumination (see also 9). Again, an explanation can be sought in the antagonism of "on" and "off" influences. The weaker stimulus, by failing to excite the "on" fraction, caused less inhibition

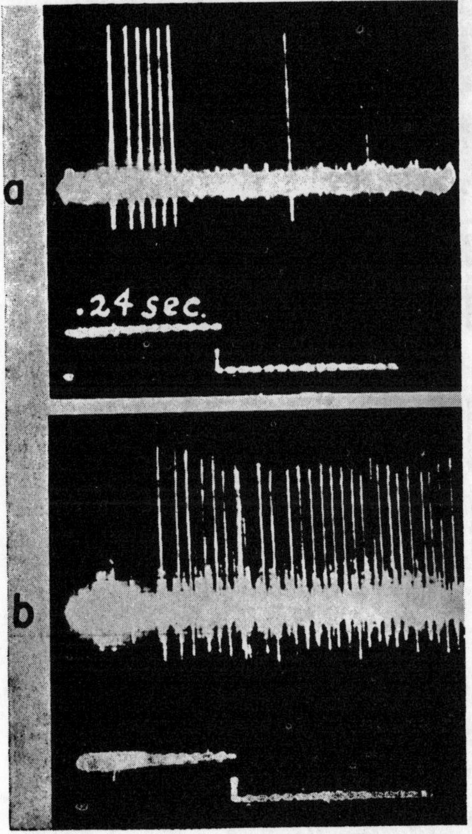

FIG. 11. "Anomalous" effect of change in stimulus area on latent period. *a:* ganglion discharge set up by 0.2 mm. diam. spot within central region of receptive field. *b:* spot size increased so as to include whole field. Note the greatly prolonged latent period of "on" component. Potentials 0.6 mV.

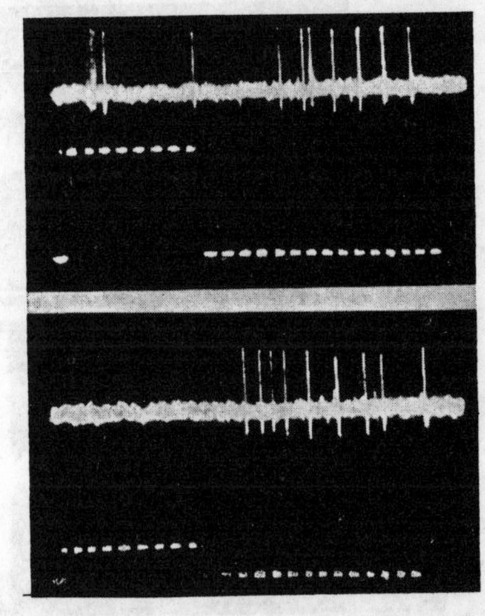

FIG. 12. "Anomalous" effect of change in stimulus intensity on discharge. Upper record: "on-off" ganglion discharge. Below: with stimulus intensity decrease "on" component drops out. Note, however, the shorter latent period and increased number of impulses in "off" discharge (see text). Frequency 50/sec.

of the "off" component. In all these "anomalous" instances it must be noted that a non-homogeneous population of receptors is activated and the discharge pattern depends on the proportion of "off"- and of "on"-oriented receptors which are excited.

7. *"Off" response*

As appears from Section 4, no pure "off" units were found when the receptive fields were explored with small spots of light and suitable background illumination. Those units which gave an "off" response alone with illumination of the whole eye were always found to have an "off" center and "on" surround, while units giving "on-off" responses could have either type of center. The "off" activity of an area could be tested by the ability of a light stimulus to set up impulses when its intensity was reduced or the light turned off, or by the suppression of spontaneous activity.

The interaction between separate stimuli to "on" and "off" areas was shown in Figure 8; in Figure 13 a similar experiment is illustrated with both

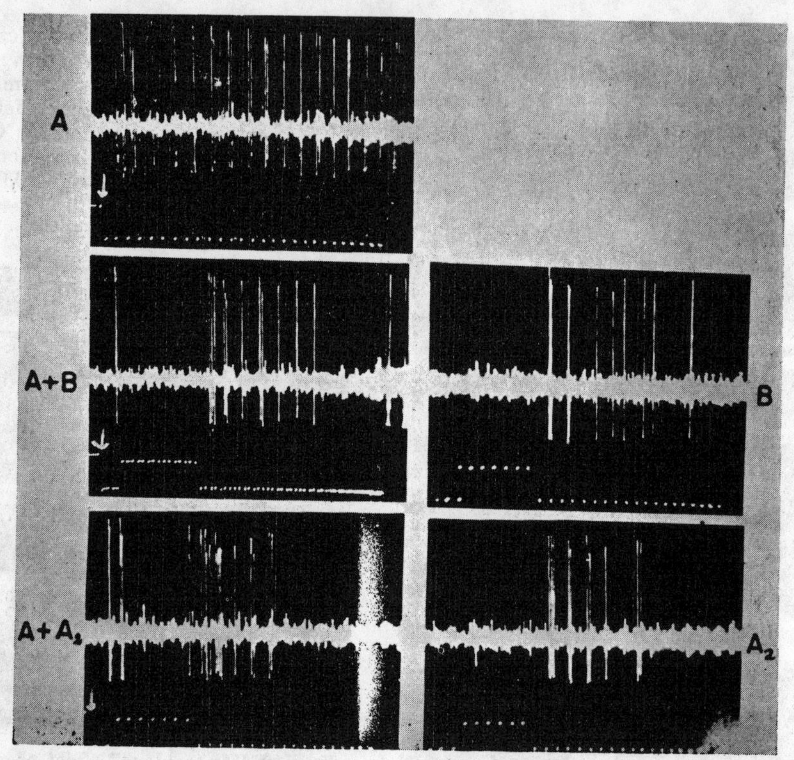

FIG. 13. Inhibitory action of light on "off" response. Light beams A and B projected onto separate areas, each 0.2 mm. in diameter, in central region of a receptive field near tip of recording electrode. Both regions give "off" responses only. Background 18 m.c A: "Off" discharge produced following termination (arrow), at beginning of sweep, of stimulation by beam A. A + B: Beam B, applied during "off" discharge, suppresses impulses. B: spot B alone. A + A_2: Stimulus to spot A ceases near beginning of sweep, as above, but same area is re-illuminated by second flash. Not only is there suppression of "off" discharges during flash of A_2, but also subsequent "off" response duration is reduced as compared with A. A_2: second flash of beam A alone. Note that "off" discharges set up in one region of receptive field can be suppressed by stimulation of another "off" region, or by restimulation of same area. The grouped discharges occurred in many units of this experiment. Time base 100/sec. in A + B, 50/sec. in all other records. Potentials 0.3 mV.

light stimuli to an "off" region. A Spot caused a strong "off" response by illumination of an area 0.2 mm. in diameter in the central portion of a receptive field, just about 0.1 mm. away from the area of lowest threshold at the electrode tip. The illumination was started before the sweep and only the cessation of the light signal appears on the record (marked by arrow). Grouped discharges similar to those in this figure were frequently seen and have been also noted by others during the "off" effect (16, 17). Spot B was the same distance from the electrode tip as A, but on the opposite side. This stimulus was shorter and by itself caused a briefer discharge (Fig. 13B). When B followed A, it suppressed the "off" discharge for the duration of its flash (Fig. 13A+B). When both stimuli were given to spot A in succession, the second (A_2) also suppressed the "off" impulses. A_2, being a shorter flash than the preceding A, set up a shorter "off" response than A alone. It is noteworthy, however, that the "off" discharges of A were not reinforced at the end of flash A_2. In this unit it seems that A_2 during its flash not only suppressed the impulses, but also the processes which "survived" after cessation of A. The inhibitory action of light on the "off" discharge by reillumination of the whole eye is well known from the work of Granit and Therman (16) and Hartline (17). Suppression of "off" discharges, set up in one region of the receptive field, by subsequent excitation of another "off" area is to be expected from the experiments on interaction (Section 5) and have also been seen by Hartline (20a).

The duration of the latent period of the "off" responses was studied, since it is a measure of the processes which have preceded the discharge. It is known that the latent period shortens and the discharge frequency increases as a function of the intensity and duration of the preceding illumination (13, 20). Again, however, exceptions to this rule occur. In some experiments latent periods as short as 10–15 msec., similar to the shortest periods for "on" discharges, could be seen. Another indication concerning the processes which are involved in inhibitory activity can be obtained from a determination of the time which is taken up between stimulation of the receptors and the first sign of suppression of activity at the ganglion cell level. Some conclusion may then be drawn regarding the mechanism of excitation spread within the retina. The speed and mode of this spread will be important in the competitive situation when both "on" and "off" areas are excited simultaneously, as must occur normally in the eye when stimulation is not confined to subdivisions of a receptive field. Such latent periods of inhibitory action are obtained by measuring the time it takes for a second flash (B or A_2 in Fig. 13) to suppress a discharge. The time between the onset of the "off" flash and the first suppressed impulse would clearly be the most accurate determination. This method will be most precise if the suppression is tested and measured on a well-maintained and regular high-frequency discharge. By such determinations the shortest latent periods of inhibition were around 10 msec. These times may, in fact, be too long since they do not indicate the actual onset of inhibitory action at the ganglion cell. The processes may start acting well before they become evident by their action of suppressing a discharge. Further information in this connection will be presented in a study of the inhibitory and excitatory pathways which converge on ganglion cells.

Discussion

Sampling of units within retina. An advantage of the present technique is the ease of recording retinal activity and the intactness of the eyeball which enables the normal optical channels to be used for illumination and observation. The method, however, will tend to select the potentials from the larger ganglion cells. On the other hand, since one generally can find suitable cells for recording within any small area of the retina, such as 1 square mm., it is quite likely that these cells can be smaller than the "giant" cells described by Rushton (27). Further, nearly all the work was done on cells within a radius of about 5–8 mm. from the optic disc, particularly in the two quadrants above the disc within the highly reflecting region of the tapetum. No positive evidence has been found that within these areas there are specific subsections which give different discharge patterns. The cat has no fovea but there exists a region on the visual axis of the eye, called centralis (6), about 1–2 mm. temporal from the disc, which has an especially dense representation in the visual cortex (30). This region was included in the present studies and found to show no qualitative differences from other areas. No activity of bipolars has been recorded and therefore all the discharge patterns which are described, while derived from ganglion cells, represent also the discharges in the optic nerve fibers.

Since in each preparation the discharges from numerous units can be observed in quick succession, *e.g.*, 30–40 within an hour, it is possible to collect statistical data on discharge types. It was, however, found more informative to obtain detailed results from a relatively small number of units and frequently these were kept on the electrodes for 5–6 hours. Only those cases are presented which, at the present stage, seem more representative or important. The great majority of experiments were done well within photopic levels, with the background illumination between 1 and 50 m.c. All the essential features of discharge pattern behavior, however, were also present under scotopic conditions in the absence of background illumination. It should also be noted that in this study relatively short flashes were used and no "equilibrium" conditions were attained.

Fluidity of discharge patterns. The most outstanding feature in the present analysis is the flexibility and fluidity of the discharge patterns arising in each receptive field. Stability of discharge type can be obtained in the present preparations in units under certain conditions, especially with a relatively strong background illumination, when the surround is suppressed. A constant "on" or "off" response may then be seen even with spot illumination. Such stability, however, disappears when one or more of several parameters, such as the adaptation level, stimulus intensity, and area of illumination, are changed singly or in combination. In the absence of a fixed pattern from the whole receptive field, it does not appear accurate enough to speak of "on," "on-off" or "off" fibers in the cat's retina. The difference in retinal discharge pattern distribution between frog and cat is worthy of note, particularly since the analyses in frog were made by Hartline (18) with a well-controlled and accurate technique. Although he reported the discharge patterns in receptive fields fixed, he points out many exceptions and reports occasional units in which a change in discharge patterns did occur. He also presents data which may be interpreted to indicate the presence of inhibitory surrounds, such as a decline of discharge frequencies with strong stimuli or with large areas of excitation (17, 19). The difference between cat and frog may turn out to be largely a quantitative one. A less flexible system of discharge in frogs may perhaps not be surprising. By using a different approach, such as pharmacological techniques (12) and passing current through the eye or varying the wave-length and intensity of the stimulating light, shifts in discharge patterns have already been observed by Granit and his colleagues. They also repeatedly pointed out the lability of certain portions of the discharges, particularly in connection with work concerning the on-off ratios (8, 14, 15). Donner and Willmer (9), working with dark-adapted cats and stimulation of the whole retina, also observed a great range of variability in ganglion response patterns during stimulation at different intensities. They have shown that visual-purple-dependent receptors can give rise to both "on" and "off" discharge components.

Functional organization of receptive fields. There seems to exist a very great variability between individual receptive fields and therefore a detailed classification cannot be made at present. Some features, however, of general organization were found common to all. In all fields there exists a central region giving a discharge pattern which is the opposite from that obtained in the periphery. The center may be either predominantly "off," the surround "on," or vice versa. A transitional zone is in between (see Fig. 6). The essential character of discharge within the centers cannot be changed by altering any of the parameters of illumination, *i.e.*, an "off" center cannot be converted into an "on" center. It must not be inferred, however, that the centers are quite uniform and receive no contribution which is characteristic of the surround. In view of the fixed nature of the center discharge, it may be convenient to classify receptive fields into "on" center and "off" center fields. In line with this the respective elements may be similarly designated as "on" center or "off" center units. No accurate record of distribution has been made in hundreds of units which were investigated. The "off" center units seemed to occur more frequently. Functionally the center and surround regions are opposed, the one tending to suppress the other. The ganglion cell is subjected to multiple influences from its receptive field and its discharge will express the balance between these opposing and interacting contributions. In view of the relative ease with which the peripheral receptive field contribution can be altered (see later), and thereby the balance within the unit changed, the discharge pattern fluidity is readily appreciated.

From a functional point of view, then, the important aspect of the present findings is not that one unit can give under special conditions either "on" or "off" responses but that there exists a mixture of contributing receptors, perhaps with their specific pathways (below). In proportion to their activation they can produce all shades of transitions from one response pattern to another. In any event, illumination of the whole receptive field will always produce a push-pull action as the opposing components are thrown into activity.

Specific neural pathways. The nervous organization of all the elements functionally connected to a ganglion cell constitutes an example of the complexity of the central nervous system, well known from the studies of Cajal (5) and lately especially of Polyak (26). It is natural that a specific organization should be suggested for excitatory and inhibitory pathways for which there is physiological evidence. Experiments seem to show (Fig. 6) that excitation of a certain number of receptors by restricted illumination causes one type of response only. Presumably a given pathway is utilized by a given group of receptors. The principal reason for a change in response type seems to be that either additional receptors have been brought in or receptors have been eliminated (see below). Suggestions as to the specific neural connections, based on present evidence, are clearly speculative and grossly simplified. One may think of a neural arrangement which parallels the roughly concentric functional pattern, with relatively uniform connection types between receptors in the center and the ganglion cell and a differing pathway set-up from the surround receptors to the ganglion cell. The in-between region may present the zone where the pathway types are more mixed than anywhere else. A correlation of greater significance between neural pathways and discharge patterns may perhaps be obtained from a study of animals with a fovea, *e.g.*, monkey, where the neuroanatomical connections are simpler and better known. It may be predicted, accordingly, that the foveal paths are associated with specific discharge behavior.

Receptor density in receptive fields. Any given small area of the retina which has been studied presumably has a dense and fairly uniform receptor population (26). Histological data also show that adjoining receptors, or even the same receptors, may have connections to different ganglion cells. Further, we know that the same receptor may connect to different ganglion cells in differing ways. This is the neuroanatomical basis for overlapping receptive fields. The present study gives some information about the density distribution of receptors which are *functionally* connected to one ganglion cell. The following type of experiment supports the assumption that the central portion of receptive fields hold a denser population of receptors per unit area than do the peripheral regions: units which have an "on" center and "off" surround, when tested with stimuli 100–1000 times threshold (Fig. 6), may be excited in their most sensitive central region by a 0.2 mm. diameter spot of near-threshold intensity. At this strength "on" discharges of quite short duration are evoked within a small central area (as Fig. 5). Such a small spot is well below threshold for the outlying portions of the receptive field. Placing a ground glass in front of the eye, thereby reducing, and in addition scattering, the light beam, will produce "on-off" or pure "off" responses. This suggests that receptors dispersed in peripheral regions have been reached and summation has occurred in their pathways leading to the ganglion cell. The experiment also shows that the receptors which contribute the "off" component do not have a lower threshold than those in the central "on" region. Threshold differences, however, within the receptor population are not contra-indicated by such results. Presumably because of the density of receptors, the center is found to be more sensitive when tested with small beams.

The scatter of receptors in the periphery makes obvious the difficulties of receptive field mapping with small light beams, since one has to assume that a sufficient number of receptors must be activated to evoke a ganglion cell discharge. Receptors, functionally connected but located in the periphery, will be missed and an error in underestimating the field size is likely to be made. When mapping is done in units with an "off" surround, while they show spontaneous activity, the field periphery can be delineated by the area over which a small spot will produce slowing or stoppage of firing. This method is more sensitive than the one described in Section 3 and receptive fields extending over 3–4 mm. (12–16° in cat), could occasionally be obtained. The effect of light scatter, however, could not be estimated closely enough to make these findings reliable.

The low density of receptors in the surround also makes readily understandable the observed shrinkage of receptive fields with augmented background illumination. If tested with a *small spot*, the dropping out of receptors by raised thresholds in the field surround will be of more consequence than in the dense central region, since the outer receptor family, being scattered, operates nearer to the margin for firing the ganglion cell. Even with illumination of the *whole field* the peripheral contribution, if it depends more on facilitation and summation, should be more affected if a portion of component pathways is put out of action. The background changes should affect an "on" or "off" surround equally, in line with present observations.

Changes in receptive field contribution to ganglion cells. A special nervous organization of receptive fields alone could not account for all the observed discharge pattern changes under diverse conditions of illumination. There is a great body of evidence for a diversity of receptor properties in respect to thresholds, adaptation, wave-lengths, etc. For instance, in view of the changes in receptive field size at low or high levels of light adaptation, it is clear that under such changing conditions a largely differing set of receptors will be thrown into activity with a given stimulus. Hence, this alone will bring a different set of active connections with a ganglion cell into being. The differing connections, in turn, are likely to cause changes in discharge patterns. Since steady states cannot be attained, a shift in the active receptor population is likely to go on continuously even in the dark-adapted eye, as indicated by the background activity, unless the latter is entirely due to spontaneous rhythms in the neural elements.

Psychophysical aspects. A transference of information about discharge patterns, as obtained here, to psychophysical data is obviously based on speculation. However, the data must be used with all their imperfections since they provide the components which form the basis of the message content reaching the higher centers. The most potent stimuli, those causing the greatest nervous activity, are relatively sudden changes. These may be either changes of the general illumination level or such changes as occur during movement of images (see also 17). In the latter case the antagonistic arrangement of central and peripheral areas within receptive fields seems important since the smallest shift can cause a great change in the discharge pattern. This should be advantageous in the perception of contrast and in acuity. In view of this the importance of small eye displacements, as would occur in any scanning movement, is clear. It should be noted that zonal gradients within fields, between center and surround, will change with different levels of adaptation as the receptive field shrinks or expands. It may also be tempting to consider in this connection well-known suppression phenomena like lateral inhibition which has been studied by many investigators. Particularly the interaction of two light patches with facilitation and inhibition as observed in humans over small distances on the retina may be considered in view of the extent of the cat's receptive field (10, 11). It is clear, however, that even the smallest light beams used in the present experiments do excite a great number of ganglion cells through their overlapping receptive fields. It is not known how the latter are functionally related to each other. For instance, it would be of interest to know whether the same receptor can be connected to one ganglion cell through an excitatory and to another cell through an inhibitory pathway.

Regarding the information content carried by a single ganglion cell or its connected nerve fiber, the following two phenomena may be briefly considered: (i) If a ganglion cell can discharge at one time during illumination and then be converted into one which signals only when a light stimulus is withdrawn, one may assume that it does not carry merely the information which is suggested by a "simple" interpretation of the discharge pattern. The higher centers may receive identical impulse patterns in both cases. (ii) A unit giving "on-off" discharges in a given situation, according to a "simple" interpretation, sends information first about an increased and then about a decreased level of luminosity. The identical discharge pattern may, however, be evoked by turning a light on, and then instead of turning it off, one may further increase its intensity. Such fibers then merely signal change, but not necessarily the direction of change, such as brightness or darkness.

In view of the massive continued nervous activity in eyes "at rest" or during illumination it is difficult to think of information content in terms of single unit contributions. One may rather have to consider that groups of fibers modulate activity levels and patterns by superposition or subtraction. The latter—for instance, transient cessation or diminution—is likely to be as meaningful as the opposite in terms of message content. Further, similar discharge patterns at different background illuminations may convey a different meaning, since they are superimposed on a different background activity. These examples merely illustrate some of the difficulties inherent in this type of analysis. They indicate that on the basis of the single unit discharge a 1:1 agreement between discharge patterns and information should not always be sought. At the same time it should be recalled that there is agreement between psychophysical measurements such as the visibility curves and the analogous curves, obtained in different mammals (13), or the Limulus (20), from nervous discharges.

In this study the influence of transient light changes on discharge patterns has been emphasized; in view of the importance of background activity the effect of steady levels of illumination on discharge behavior must be analyzed in great detail.

Summary

Discharge patterns from the unopened cat's eye have been studied by recording from single cells in the retina. Small electrodes, inserted behind the limbus, traversed the vitreous and made light contact with different regions of the retina. The normal optics of the eye were used for stimulation by two independently controlled light beams. Circular stimuli of various dimensions, duration and intensities were applied to different areas of the retina. A third light source provided the background illumination, determining the adaptation level, and also served for simultaneous direct observation of the fundus.

1. The discharges arising in nerve fibers and ganglion cells can be readily distinguished through differences in their time course and the location of their respective receptive fields. The present study was done on ganglion cell activity.

2. Ganglion cells can be blocked reversibly by pressure and the potentials can be split into "prepotentials" and "spikes."

3. Under dim background illumination and during dark adaptation the cat's retina is dominated by generalized spontaneous activity. The latter is reduced by illumination and anesthetics such as Dial or Nembutal. Certain discharges do not seem to be influenced appreciably by illumination. These observations are in general agreement with Granit's findings.

4. The configuration of receptive fields—those areas of the retina which must be illuminated to cause a discharge in a ganglion cell—were studied by exploration with small spots of light. The fields are usually concentric, covering an area of 1–2 mm., or possibly more, in diameter. The boundaries and extent of receptive fields cannot be delineated accurately. They shrink under high background illumination and expand during dark adaptation.

5. The discharge pattern from individual ganglion cells is not fixed. "On," "off" or "on-off" discharges can be obtained from one ganglion cell if specific zones within its receptive field are stimulated by small spots of light. The discharge pattern from a ganglion cell depends, amongst others, on the following factors: Background illumination and state of adaptation, intensity and duration of stimulation, extent and location of area which is stimulated within a receptive field. Each of these parameters can alter the discharge pattern by itself or in conjunction with the others.

6. The general functional organization of each receptive field is the following: There exists a central area of low threshold as tested by a small spot

of light. The discharge pattern of the central region is the opposite of that found in the periphery or surround. The center may give predominantly "off," the surround "on" discharges, or the reverse. An intermediary region gives "on-off" discharges. The units which carry discharges from "center on" or "center off" receptive fields may accordingly be classified as "on" center or "off" center units. A conversion of one type into another by changing conditions of illumination has not been possible.

Experiments indicate that receptors in the periphery of receptive fields are less dense per unit area than in the central regions.

7. Interaction of different regions within single receptive fields was studied by simultaneous excitation by two small beams of light. Depending on a number of factors, "off" areas may suppress the discharge from "on" regions, or vice versa. All degrees of mutual modification can be obtained. It is assumed that specific areas give rise to predominantly inhibitory or excitatory pathways to a given ganglion cell.

8. The discharge pattern of a ganglion cell, set up by illumination of an entire receptive field, depends on the summed effects of interacting pathways converging on the cell. The ratio of functionally opposing center and surround regions varies greatly in different receptive fields. Under diverse conditions of illumination the balance of active inhibitory and excitatory contributions changes within the same receptive field. This seems to be responsible for the varied discharge patterns.

9. The character of "on" components in discharge patterns was also studied. The maintained "on" response discharges for the duration of illumination while the transient "on" adapts quickly. Transitions between these two "on" types were found in the same receptive fields. It is suggested that the transient "on" discharges are the result of various amounts of addition by the "off" surround to the "on" center.

10. The latent periods of "on" or "off" responses were found to be shorter than hitherto observed, presumably because of more exclusive stimulation of their specific receptive areas. High discharge frequencies of 200–800/sec. were found to be within the normal range in the cat's eye.

11. Some "anomalous" observations, such as lengthened latent periods with increased stimulus intensities or higher frequency discharges with weaker stimuli, are interpreted in the light of receptive field organization.

ACKNOWLEDGMENT

I am grateful to Dr. S. A. Talbot for his help, particularly in the design of the optical and electronic instruments, which made this study possible. My thanks are also due to Mr. Albert Goebel for constructing the optical apparatus.

REFERENCES

1. Adrian, E. D. Synchronized reactions in the optic ganglion of Dytiscus. *J. Physiol.*, 1937, *91*: 66–89.
2. Adrian, E. D. and Matthews, R. The action of light on the eye. Part I. The discharge of impulses in the optic nerve and its relation to the electric changes in the retina. *J. Physiol.*, 1926, *63*: 378–414.

3. ADRIAN, E. D. AND MATTHEWS, R. The action of light on the eye. Part III. The interaction of retinal neurones. *J. Physiol.*, 1928, *65*: 273–298.
4. BERNHARD, C. G. Contributions to the neurophysiology of the optic pathway. *Acta physiol. scand.*, 1940, *1*: Suppl. 1.
5. CAJAL, S. RAMÓN Y. La rétine des vertébrés. *Trab. Lab. Invest. biol. Madr.*, 1933, Suppl. *28*.
6. CHIEVITZ, J. H. Über das Vorkommen der Area centralis retinae in den vier höheren Wirbelthierklassen. *Anat. entw.-gesch. Monogr.*, 1891, pp. 311–334.
7. CRESCITELLI, F. AND JAHN, T. L. The effect of temperature on the electrical response of the grasshopper eye. *J. cell. comp. Physiol.*, 1939, *14*: 13–27.
8. DONNER, K. O. AND GRANIT, R. The effect of illumination upon the sensitivity of isolated retinal elements to polarization. *Acta physiol. scand.*, 1949, *18*: 113–120.
9. DONNER, K. O. AND WILLMER, E. N. An analysis of the response from single visual-purple-dependent elements in the retina of the cat. *J. Physiol.*, 1950, *111*: 160–173.
10. FRY, G. A. Depression of the activity aroused by a flash of light by applying a second flash immediately afterwards to adjacent areas of the retina. *Amer. J. Physiol.*, 1934, *108*: 701–707.
11. GRANIT, R. Comparative studies on the peripheral and central retina. I. On interaction between distant areas in the human eye. *Amer. J. Physiol.*, 1930, *94*: 41–50.
12. GRANIT, R. Some properties of post-excitatory inhibition studied in the optic nerve with micro-electrodes. *K. svenska VetenskAkad. Arkiv. Zool.*, 1945, *36*: 1–8.
13. GRANIT, R. *Sensory mechanisms of the retina.* London, Oxford Univ. Press, 1947, 412 pp.
14. GRANIT, R. Neural organization of the retinal elements, as revealed by polarization. *J. Neurophysiol.*, 1948, *11*: 239–253.
15. GRANIT, R. The organization of the vertebrate retinal elements. *Ergebn. Physiol.*, 1950, *46*: 31–70.
16. GRANIT, R. AND THERMAN, P. O. Excitation and inhibition in the retina and in the optic nerve. *J. Physiol.*, 1935, *83*: 359–381.
17. HARTLINE, H. K. The response of single optic nerve fibers of the vertebrate eye to illumination of the retina. *Amer. J. Physiol.*, 1938, *121*: 400–415.
18. HARTLINE, H. K. The receptive field of the optic nerve fibers. *Amer. J. Physiol.*, 1940, *130*: 690–699.
19. HARTLINE, H. K. The effects of spatial summation in the retina on the excitation of the fibers of the optic nerve. *Amer. J. Physiol.*, 1940, *130*: 700–711.
20. HARTLINE, H. K. The nerve messages in the fibers of the visual pathway. *J. opt. Soc. Amer.*, 1940, *30*: 239–247.
20a. HARTLINE, H. K. The neural mechanisms of vision. *Harvey Lect.*, 1941, *37*: 39–68.
21. HARTLINE, H. K. AND GRAHAM, C. H. Nerve impulses from single receptors in the eye. *J. cell. comp. Physiol.*, 1932, *1*: 277–295.
22. HARTLINE, H. K., WAGNER, H. G., AND MCNICHOL, E. C. The peripheral origin of nervous activity in the visual system. *Cold Spr. Harb. Symp. quant. Biol.*, 1952, *17*.
23. HODGKIN, L. A. The subthreshold potentials in a crustacean nerve fiber. *Proc. Roy. Soc.*, 1938, *B126*: 67–121.
24. HUNT, C. C. AND KUFFLER, S. W. Stretch receptor discharges during muscle contraction. *J. Physiol.*, 1951, *113*: 298–315.
25. KATZ, B. Experimental evidence for a non-conducted response of nerve to subthreshold stimulation. *Proc. Roy. Soc.*, 1937, *B124*: 244–276.
26. POLYAK, S. L. *The retina.* Chicago, Univ. of Chicago Press, 1941, 607 pp.
27. RENSHAW, B. Effects of presynaptic volleys on spread of impulses over the soma of the motoneuron. *J. Neurophysiol.*, 1942, *5*: 235–243.
28. RUSHTON, W. A. H. The structure responsible for action potential spikes in the cat's retina. *Nature*, 1949, *164*: 743–744.
29. SVAETICHIN, G. Analysis of action potentials from single spinal ganglion cells. *Acta physiol. scand.*, 1951, *24*: Suppl. 86.
30. TALBOT, S. A. (personal communication).
31. TALBOT, S. A. AND KUFFLER, S. W. A multibeam opthalmoscope for the study of retinal physiology. *J. opt. Soc. Amer.*, Dec. 1952 (in press).

J. Physiol. (1965), 178, pp. 477–504
With 11 text-figures
Printed in Great Britain

THE MECHANISM OF DIRECTIONALLY SELECTIVE UNITS IN RABBIT'S RETINA

By H. B. BARLOW* and W. R. LEVICK†
From the Physiological Laboratory, University of Cambridge and the Neurosensory Laboratory, School of Optometry, University of California, Berkeley 4, U.S.A.

(*Received* 29 *September* 1964)

Directionally selective single units have recently been found in the cerebral cortex of cats (Hubel, 1959; Hubel & Wiesel, 1959, 1962), the optic tectum of frogs and pigeons (Lettvin, Maturana, McCulloch & Pitts, 1959; Maturana & Frenk, 1963), and the retinae of rabbits (Barlow & Hill, 1963; Barlow, Hill & Levick, 1964). The term 'directionally selective' means that a unit gives a vigorous discharge of impulses when a stimulus object is moved through its receptive field in one direction (called the preferred direction), whereas motion in the reverse direction (called null) evokes little or no response. The preferred direction differs in different units, and the activity of a set of such units signals the direction of movement of objects in the visual field.

In the rabbit the preferred and null directions cannot be predicted from a map of the receptive field showing the regions yielding on or off-responses to stationary spots. Furthermore, the preferred direction is unchanged by changing the stimulus; in particular, reversing the contrast of a spot or a black-white border does not reverse the preferred direction. Hubel & Wiesel (1962) thought that the directional selectivity of the cat's cortical neurons could be explained by the asymmetrical arrangement of on and off zones in the receptive field, and the simple interaction of effects summated over these zones, but the foregoing results rule out this explanation, at least in the rabbit's retina (Barlow & Hill, 1963).

In the present paper we go on from this point to describe experiments which show, first, that directional selectivity is not due to optical aberrations of some kind and, secondly, that it is not a simple matter of the latency of response varying systematically across the receptive field. After these negative results we describe experiments upon the organization of directional selectivity within the receptive field, and upon its mechanism. These lead us to the conclusion that the ganglion cells responding to a

* Present address: Neurosensory Laboratory, School of Optometry, University of California, Berkeley 4.
† C. J. Martin Travelling Fellow on leave of absence from the University of Sydney.

particular direction of motion are fed by a subset of bipolar cells that respond to the corresponding sequence of excitation of two neighbouring retinal regions with which they connect. Furthermore there is evidence that this sequence-discrimination is brought about by a laterally connecting inhibitory element from one of these regions, and this seems a likely function for the horizontal cells to perform. At this stage identification of the elements concerned is obviously tentative, but we were pleasantly surprised to find well known histological structures already at hand to fill the roles that the functional organization seemed to require.

All the experiments described here were performed upon on–off directionally selective units. We have reason to believe that the mechanism is different for the rare, on-type, directionally selective units and also for the centrifugal and centripetal motion sensitivity of the ordinary concentric type of units.

METHODS

Action potentials were recorded from the unopened eyes of rabbits. As described elsewhere (Barlow *et al.* 1964), it proved most effective to use fine tungsten electrodes, decerebrate or lightly anaesthetized (urethane and chloralose mixture) animals, and immobilization by continuous infusion of gallamine triethiodide (Flaxedil). Periodically the animal was allowed to recover from paralysis by using an infusion fluid without relaxant. One could thus ensure that the level of anaesthesia was neither too deep nor too light at the rate of anaesthetic infusion employed.

When a good on–off, directionally selective unit was isolated the first step was always to map out its receptive field on a plotting board 57 cm from the eye. A stationary spot was turned on and off; in most cases this was $\frac{3}{4}°$ diameter at an intensity of 12 cd/m^2 and was superimposed on a background of 0·6 cd/m^2. Directional selectivity was tested for and null and preferred directions determined. Various techniques were used to provide the temporal and spatial patterns of light stimuli. For the two-spot experiment we initially used two glow modulator tubes controlled by pulse generators, but we later found that black and white cards moved behind apertures in grey paper provided a more flexible means of delivering the required stimuli. This method also has disadvantages (see p. 486) which we finally overcame by illuminating the apertures from behind with thin Perspex light pipes lit by low-current torch bulbs turned on and off manually. The changes of luminance occurring within the receptive field of the unit were monitored by a photocell whose output was recorded with the action potentials.

RESULTS

The results are presented in four sections. These are: (1) controls and negative results which rule out various preliminary hypotheses on the mechanism; (2) experiments which lead to the conclusion that the directional selectivity of the ganglion cell results from the sequence-discriminating activity of subunits—probably bipolar cells; (3) observations and experiments which show that sequence-discrimination is achieved by an inhibitory mechanism that prevents responses to sequences in the null direction; (4) observations showing that inhibition also occurs with stimulation of the surround of the receptive field.

Controls and negative results

Optical controls. It might be thought that the unequal responses to motion in the null and preferred directions were the result of a peculiarity of the light distribution in the retinal image caused by aberrations of the optical system. Although none of the schemes suggested to us, and none we could imagine, seemed at all promising, we considered this possibility, but were forced to abandon it at an early stage. The most direct disproof is given by the observation that two units recorded simultaneously, or within a short period of time, in the same retinal region can have their preferred axes opposite or at right angles to each other. An optical explanation of the phenomenon requires that there be some asymmetry in the light distribution to cause the asymmetry in the responses, and two different asymmetrical light distributions cannot co-exist in the same region.

A second disproof is provided by observing how little the phenomenon is affected by deliberately introduced optical aberrations. Figure 1 shows that clear-cut directional selectivity persists with spherical supplementary lenses causing more than 10 dioptres of refractive error in either direction. This was for the normal pupil diameter of our preparation—6 mm or more. Furthermore, reducing the aperture of the rabbit's optical system to 3 or 1·25 mm with an artificial pupil must greatly improve retinal image quality (since diffraction is most unlikely to be a limiting factor); yet we found that such stopping down merely extended the range of refractive error over which the directionally selective property was shown. It did not even improve acuity as judged by the finest grating giving any response to movement.

The best spherical correction was determined in each preparation, judging this opthalmoscopically and sometimes retinoscopically. In all the preparations the cornea and media were free from clouding, except occasionally at the very end of a 2-day experiment. Optical effects from the shank of the electrode can be excluded, for directional selectivity is often observed when recording from a nerve fibre. In such cases, no portion of the electrode intercepts light reaching the retina from the receptive field projection.

Finally, we should point out that for most of the tests used here the human eye's performance is superior by a factor of 10. The demands made of the rabbit's optical system are thus not at all severe, but of course the blur of the retinal image should be taken into account when interpreting quantitatively the results of our later experiments.

Latencies. The first idea about mechanism was that the latency of the response might be shorter at successive positions along a path through the receptive field traced in the preferred direction. It was thought that when the image of an object moved in this direction the excitation from successive positions would arrive synchronously at the ganglion cell, and thus might be more effective than when movement was in the null direction and it was dispersed in time. An alternative scheme can be devised in which the temporally dispersed sequence is more effective because it avoids

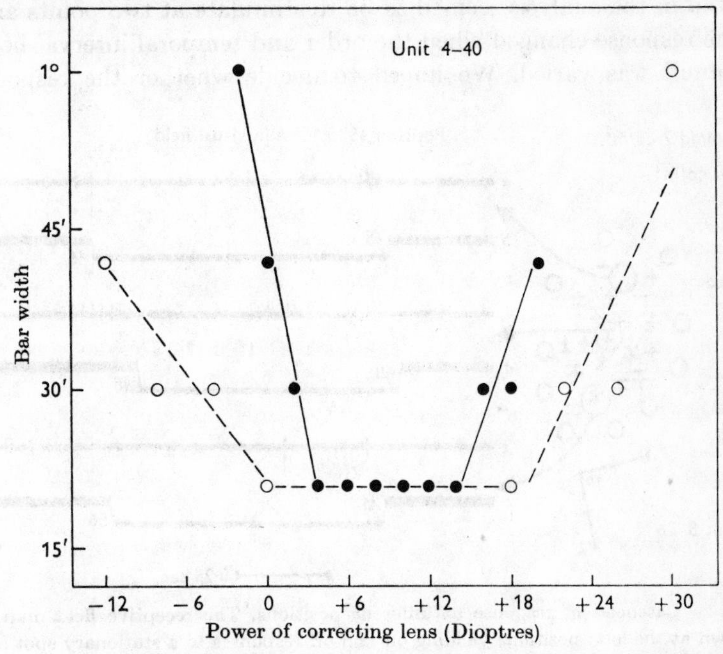

Fig. 1. The effect of refractive error and pupil diameter upon the acuity for a moving grating. Pupil dimensions were 6 × 8 mm (●) and 3 mm diameter (○); luminance of the white bars was 2·5 cd/m² for the large, 25 cd/m² for the small pupil. The points show the finest grating for which there was a distinctly greater response to movement in the preferred direction. Paradoxical responses, greater in the null direction, were not observed. A small discharge with movement was detectable in most cases for the next finer grating (71 % of the period), but this was not obviously different in the two directions. Notice that directional selectivity persists over a large range of refractive error, and that the finest resolvable grating is not affected by pupil diameter: optical aberration is not likely to be the cause of directional selectivity, and probably does not limit the optimum acuity of this preparation.

refractoriness. Thomson's (1953) work on the rabbit's retina, together with our own findings on the different latencies of centre and surround in concentric units (Barlow *et al.* 1964) lent some plausibility to the suggestion that the latencies might vary with position, but the result shown in Fig. 2 is entirely negative. In this and other experiments 'on' and 'off' latencies showed no sign of changing systematically with position in the receptive field.

Unsuccessful two-spot experiments. One sees from the records of Fig. 2 that excitation of a point in the field causes a response at on and off: yet if a spot of light is moved through the field in the null direction no response occurs, even though each point crossed by the spot must have received an 'on' followed by an 'off' stimulus as the spot passed over it. The obvious next step in the analysis seemed to be to stimulate at two points and see how the response changed when the order and temporal interval between the stimuli was varied. We hoped to decide whether the response to

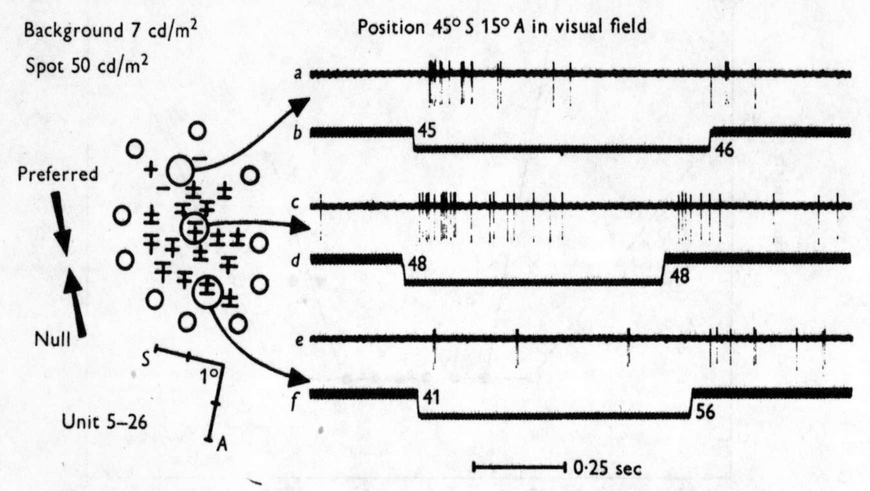

Fig. 2. Latencies of response at different positions. The receptive field map is shown at the left; positions yielding on and off responses to a stationary spot are marked + and −; positions yielding both are marked ∓ if off was greater, ± if on was greater; no responses were obtained on or outside the ring of O's. Records *a, c, e,* are the responses to a light turned first off, then on, at three successive positions along a line through the receptive field in the preferred-null axis, as shown to the left (negativity is upwards). Records *b, d, f,* are from a photomultiplier observing the receptive field (decreased light is downward), and the numbers show the latency of response in msec (one impulse was ignored in *a*). There is no significant trend in latency as one moves across the receptive field.

motion in the preferred direction was greater than the sum of responses to excitation of separate points along the path, or whether the unequal responses to motion in the two directions resulted from inhibition occurring when the motion was in the null direction.

The question seemed clear-cut, but the results were not. In the first place it was not nearly as easy to obtain unequal responses for the two sequences as it was to obtain unequal responses with real moving objects. Secondly, when we did get evidence of sequence-dependent responses, the result seemed highly variable and we were unable to decide whether summation or inhibition or both were occurring. This failure forced us to realize that we did not know whereabouts in the receptive field we should put the spots, nor how far apart they should be. There was in fact a prior question to answer before the two-spot type of experiment could be performed and interpreted. The question, put in a form that avoids implications as to mechanism, is this: does the ganglion cell respond selectively to one direction of motion over all parts of its receptive field, or is there some critical zone or line which must be crossed? The following observations show that there is no such line and the directionally selective property is distributed over the receptive field.

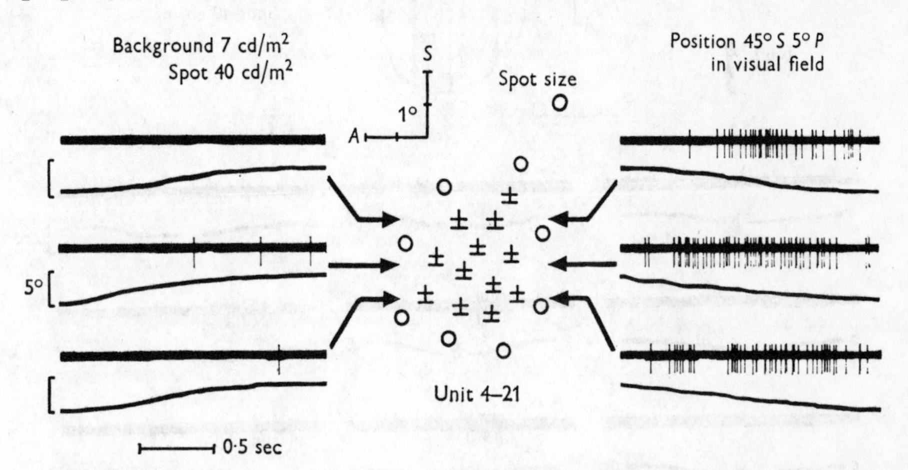

Fig. 3. Responses to motion along three different paths through the receptive field. The map in the centre shows the field and the paths through it; symbols as in Fig. 2. The records of the responses to traverses in the null direction are to the left, those for the preferred direction to the right. The lower trace of each pair is from a potentiometer and shows the position of the spot as it moved through the field (calibration at left). Top, middle and bottom parts of the receptive field all show the same directional selectivity.

Sequence-discrimination by subunits

Distribution of directional selectivity. Figure 3 shows the responses obtained when a spot of light was moved across the receptive field and back along three parallel lines. These were separated by more than the breadth of the spot, and therefore different receptors were covered by the geometric image of the moving spot in each case. It will be seen that the selectivity clearly exists along all these three pathways.

Figure 4 shows typical responses obtained when the spot was moved several times from one position to the next and back, as marked. It is clearly not necessary for the spot to cross any definite line in order to obtain different responses for the two directions of motion. If the experiment is repeated using a black spot the same result is obtained; direction of motion, independent of contrast, can be picked out in a large number of widely separated regions of the receptive field. However, there is an interesting exception to the rule that all regions of the receptive field have the capacity to distinguish between null and preferred sequences of excitation of the receptors they contain. There is a zone adjacent to the edge of the field that is first crossed when motion is in the preferred

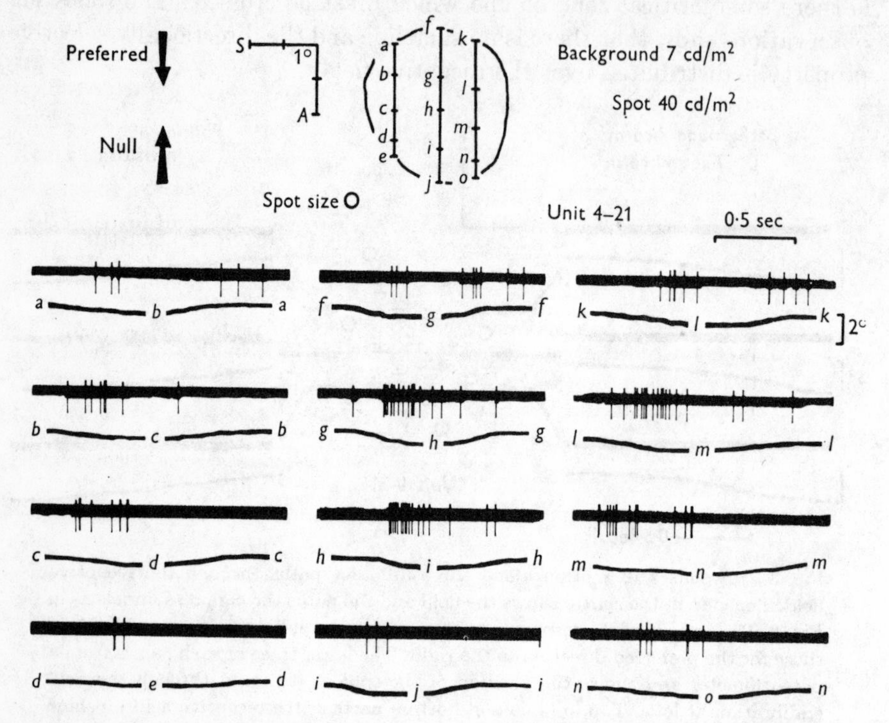

Fig. 4. Back and forth motion in different parts of the receptive field. The edge of the receptive field is mapped at the top, and the positions *a*, *b*, *c*, . . ., *o*, within it are indicated. The spot was moved back and forth several times between *a* and *b*, then between *b* and *c*, and so on. The records are samples of these back and forth motions. The lower trace of each pair shows the position of the spot in the field: downward movement of the trace corresponds to movement of the spot in the preferred direction. Marked asymmetry of response for the opposite directions holds in most positions in the field. Its absence in the top row of records is expected in the inhibitory scheme (see Fig. 7 and p. 490).

direction where this capacity is lacking: motion in either direction causes a response. This is shown in the top line of records in Fig. 4, and a possible explanation for the effect is given later (see p. 490).

Smallest region giving directional selectivity. Figure 4 shows clear directional selectivity when a spot of light is moved to and fro through 1° in a receptive field whose total diameter is about 3°. What then is the smallest distance over which responses to motion in the two directions differ?

To answer this question we moved a strip of white card with a section painted black behind an aperture of variable width in a sheet of grey paper. The aperture breadth was 20 mm (subtending 2°) and the width measured in the direction in which the card moved varied from 1 mm (6′) to 11 mm (1° 06′). When the card was moved there was no change until the border of the black section entered the aperture, it then moved through a variable distance before it disappeared behind the other edge, after which there was again no further change and the aperture was wholly black. The sequence described above was called 'off' stimulation since with black trailing the receptors were successively exposed to a reduction of illumination. On the other hand, when the border of the white portion appeared in

TABLE 1. Single slit experiment

A black–white border was moved through the receptive field, but the view of the motion was restricted by a fixed rectangular slit in a grey card placed immediately in front of the moving border. The width of this slit, measured in the direction of motion, was varied. Movement of the border causing the slit to change from white to black was called 'off' stimulation, movement causing it to fill with white was called 'on' stimulation. Each of these stimuli was applied with the edge moving in both null and preferred directions. We graded the unit's response from 0 to 6 after listening to several repetitions of these stimuli on the loudspeaker. This unit could distinguish the preferred from the null direction with slit widths down to about 17′ for both 'on' and 'off' stimuli.

Slit width	'Off' stimulus		'On' stimulus	
	Preferred	Null	Preferred	Null
1° 36′	6	1	4	0
1° 06′	6	2	4	1
48′	5	2	4	0
34′	4	2	3	1
24′	4	2	3	1
17′	3	2	3	2
12′	3	3	3	2
8′	2	2	2	2

the aperture and moved along it they were successively exposed to 'on' stimulation. Clearly 'off' and 'on' stimulation could be applied in any direction and at any velocity, but we confined our attention to the preferred and null directions, and moved the cards by hand at velocities chosen to give optimum discrimination between the two directions—in most cases about 5°/sec. The variable studied was the width of the aperture, and the feature of the response attended to was the existence of a clear-cut difference between responses to stimulation in the preferred and null directions.

The amplitude of the response was graded subjectively from 0 to 6, and Table 1 gives a typical result. The subjective grading was a crude but convenient way of quantification: records of typical responses are shown in Fig. 5. Altogether 10 units have been studied in this way, and the threshold aperture for directional discrimination varied from 6′ to 24′.

The straightforward interpretation of this result is that the complete mechanism for directional selectivity is contained within a subunit of the receptive field extending not much more than $\frac{1}{4}$° in the preferred–null axis. Since the result does not depend critically upon the position of the slit within the receptive field, it looks, again taking the straightforward view, as

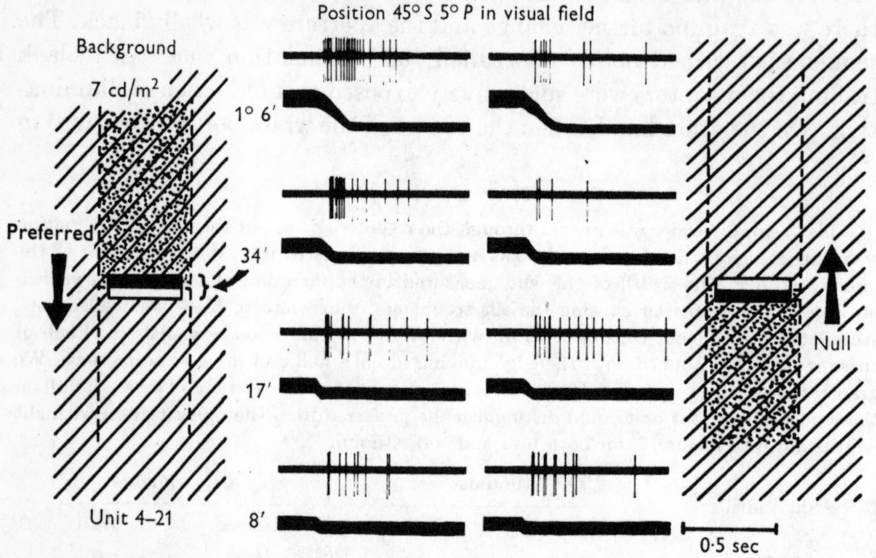

Fig. 5. Responses to motion of a black edge across slits of various widths (breadth was 2° in all cases). Results are shown for four slit widths, from 1° 06′ to 8′ measured in the preferred-null axis. The records to the left were obtained when the black border advanced across the slit in the preferred direction, those to the right when it moved in the null direction. The lower trace of each pair is the response of a photomultiplier aimed at the field. Notice that the differences between preferred and null motions are obvious in the top two records, when the distances through which the edge could be seen moving were 1° 06′ and 34′; the difference is only just detectable at 17′ and has vanished at 8′. The directionally selective mechanism appears to be contained within a retinal region subtending $\frac{1}{4}$° to $\frac{1}{2}$°, whereas the whole receptive field subtended $4\frac{1}{2}$°.

if the sequence-discriminating mechanism must be reduplicated perhaps a dozen or more times to cover the whole receptive field. Would there be any escape from this conclusion if the image was of very poor quality so that, even for the small slits, it was diffused over a large part of the receptive field?

If the optics are poor, then the time course of light intensity changes at the two edges of the receptive field will be slightly different for the two directions of motion, and one can set up somewhat elaborate schemes in which these differences form the basis of directional selectivity. The schemes have to be made even more complicated to account for the fact that black or white edges advancing through the slit show the same directional preference. Now it will be shown in the next section that the interaction responsible for directional selectivity occurs better at small separations of the stimuli than at large separations. We can think of no way in which neighbouring retinal regions would show more interaction than widely separated ones unless the subunit responsible for directional selectivity is itself a small compact one. In this way the experiment about to be described is a useful control confirming the conclusion reached from the single slit experiments.

Discrimination of sequence. The fact that movements within a region of the receptive field subtending less than $\frac{1}{4}^\circ$ are sufficient to give directionally selective responses suggests a possible explanation for our failure to get clear-cut sequence-dependent results when we first attempted to excite with a pair of static stimuli at various temporal intervals. In these experiments spots closer together than about 1° had never been tried and we therefore designed new apparatus so that two strips of light each subtending $0{\cdot}1^\circ \times 2^\circ$ could be brought to within $0{\cdot}1^\circ$ of each other and turned on and off in either sequence (see Methods).

At small separations the experiment gave definite evidence of sequence-dependence, as shown in Fig. 6. The response is much greater in the sequence corresponding to movement in the preferred direction than for null sequences, and this is true for both 'on' and 'off' stimuli. However, these differences become less when the separation of the slits is increased to over 1° even though both slits remain inside the receptive field.

We have done similar two-slit experiments in which moving cards were used to provide 'on' or 'off' excitation at the two slits. These also indicated that sequence discrimination occurs at small separations but is reduced at large separations as shown in Table 2. It was not easy to judge the responses accurately but the difference between null and preferred sequences faded out for separations greater than $\frac{1}{2}^\circ$ and other units showed a similar reduction for large separations. There is, however, a defect in these experiments which was not always fully controlled. In order to provide a vigorous stimulus with each slit it was made $0{\cdot}1^\circ$ wide. This is below the threshold for directional selectivity in most preparations, hence in these cases each slit by itself was equally effective for null and preferred sequences. However, this was not always true, and some of our results lack the necessary controls. In addition it might be held that there are effects of movement which are subthreshold for each slit by itself but become suprathreshold with the pair, and that it is the summation of these subliminal effects that produces the asymmetry for the two sequences. A control observation (made on a unit in which directional selectivity occurred across the 0·1° slit) meets this criticism and is worth reporting because it reinforces the conclusion that sequence as such is effective.

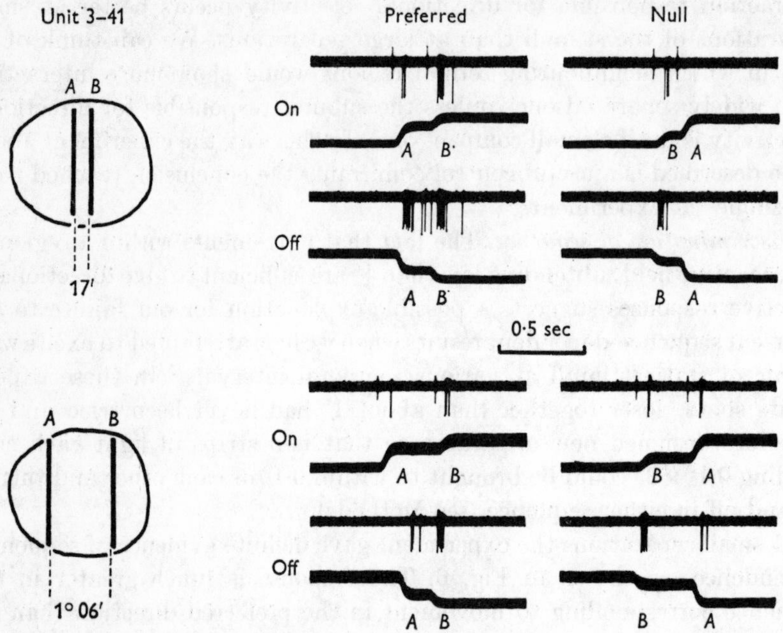

Fig. 6. Responses to different temporal sequences of two static stimuli. On the left the positions of the pair of stimuli are shown within the outline of the receptive field. Records for the small separation are shown above, those for the large separation below. Within each half, records for on are above those for off. The lower trace of each pair is the photomultiplier output: increasing light moves the trace upwards, and slit *A* was arranged to give the bigger step in every case, even though it was not brighter. Preferred sequences are on the left, null on the right. Notice that the preferred sequence yields more spikes than the null at the small spatial separation, but this difference ceases to be clearly visible when the separation of the slits is increased.

In this experiment the card behind the pair of slits had white and black regions arranged so that when the card was moved in one direction the sequence of lightening or darkening at the two slits corresponded to motion in the opposite direction. Under these conditions each slit by itself gave a greater response for movement of the card in the preferred direction. However, when both slits were used there was a greater response for movement of the card in the null direction: that is, the sequence of activation of the slits was overriding the effects of motion in the opposite direction within each slit.

It must be pointed out that the time interval between the two stimuli is an important variable that we have not yet studied systematically. In timing these stimuli manually we have varied the interval over as wide a range as possible, but in spite of this we never found as strong an interaction at large as at small spatial separations of the slits. Some further results with this type of experiment are given in Table 3 (see later) and it is hoped that a more systematic exploration will be presented in the future.

TABLE 2. Two-slit experiment

Same unit as in Table 1. The stimulus was again a black–white border moved through the receptive field in the null or the preferred direction, but the view of the motion was now restricted by a pair of narrow slits in a grey card placed immediately in front of the moving border. Each slit subtended only 6′, and responses to preferred and null directions were indistinguishable for each slit by itself. However, when the pair of slits was darkened or lightened in sequence, the strength of the response was found to depend upon the order in which the slits changed. The separation of the two slits, measured in the direction of motion, was varied and the unit's response graded as in Table 1. The effect of the order of stimulation was greatest at small separations, but null and preferred sequences were still distinguishable up to 24′ separation.

Slit separation	'Off' stimulus		'On' stimulus	
	Preferred	Null	Preferred	Null
1° 36′	2	2	2	2
1° 06′	3	3	3	3
48′	3	3	2	2
34′	3	3	2	2
24′	4	3	3	2
17′	3	2	3	1
12′	4	2	3	1
8′	3	1	3	1
6′	3	1	3	2

We think the results already given are sufficient to establish that directional selectivity may be based upon the discrimination of the sequence of excitation of only a pair of regions. Even though the image of a moving object falls on a long succession of receptors in a continuous succession of time intervals it is unnecessary to postulate the interaction of more than two regions to account for the directionally selective property.

Responses to gratings. The results so far reported suggest that sequence-discrimination is performed by subunits of the receptive field. Figure 1 shows that a directionally selective unit can discriminate the direction of motion of the bars of a grating subtending 15′ (period 30′). It is hard to see how this discrimination could be performed if the bars of the grating were small compared to the size over which each subunit integrates or averages the light, and in fact the resolvable grating size fits, to a first approximation, the size of subunit suggested by the preceding tests. Another result that may also fall into line is the shape of the curves found when determining threshold as a function of area (Fig. 5 of Barlow *et al.*

1964; see also Barlow, 1953). Complete summation (that is, threshold $\propto$ 1/area) does not hold out to the full diameter of the receptive field in the directionally selective units of the rabbit, and it is tempting to identify the limit to which it does hold (approx. 20′) as the integrating area of the subunits.

One negative result in the grating tests is worth comment. Hassenstein (1951) found paradoxical optokinetic movement responses in beetles: when a grating of period slightly greater than the angular separation of the axes of the ommatidia was moved in one direction, the beetles responded as for the opposite direction of movement. Nothing of this sort was seen in the present tests: when motion of a grating caused a response, this was never greater in the null direction than in the preferred direction. This is not too surprising, for the occurrence of paradoxical movement responses in the beetle must depend upon the regular spacing of its ommatidia.

Mechanism of sequence-discrimination

The foregoing experiments show that the directional selectivity of ganglion cells is based upon sequence-discrimination within subunits of their receptive fields, but they tell us nothing about the mechanism whereby a pair of stimuli causes a greater discharge in one sequence than in reverse. Figure 7 shows two schemes in which the preferred sequence, corresponding to motion in the preferred direction, elicits a greater response than the null sequence. These are intended to exemplify two broad alternatives, not to make exact specifications.

The left-hand scheme works by detecting a specific conjunction of excitations: activity aroused by increase or decrease of illumination in region *A* is delayed and arrives at the 'and' gate in the next layer synchronously with activity aroused when the image moves on to region *B*. Activity from *B* passes to the 'and' gate below it, and is also passed laterally to interact with activity from *C*. The sequence *ABC* is the preferred sequence, and the gates only respond when their respective conjunctions 'both *B* and delayed *A*', or 'both *C* and delayed *B*' occur.

Instead of selecting the preferred stimulus by a logical conjunction, the right hand scheme rejects the null stimulus by veto. Activity aroused at 'on' or 'off' in region *B* or *C* is again passed laterally and acts after a delay, but in this case it inhibits the next unit. As before, *CBA* is the null sequence, and when it occurs the inhibition from *C* prevents the response that would have resulted from *B* alone, and inhibition from *B* likewise vetoes *A*'s response. On the other hand if the sequence is the preferred one, *ABC*, then the inhibition from *B* does not arrive until the excitation from *A* has already got through, and likewise *C* is unsuccessful in vetoing *B*. It will be observed that this scheme only requires that inhibition persists longer than excitation; a definite delay when it is passed laterally is not strictly necessary.

Some evidence favouring the right-hand, inhibitory, scheme has already been given. (1) As shown in Fig. 2 a stationary spot turned on and off elicits a response. If the excitatory conjunction scheme was modified to account for this it would probably still predict a considerably lower threshold for a moving than for a stationary spot. As shown in Fig. 5 of Barlow *et al.* (1964), the thresholds for spots of various sizes moving in the preferred direction differ by small and inconstant amounts from those for the same spot turned on or off. (2) The most striking feature of these directional units is the absence of any impulses when movement is in the null direction. This prompts one to look for a mechanism that inhibits unwanted responses. (3) When testing for directional selectivity in

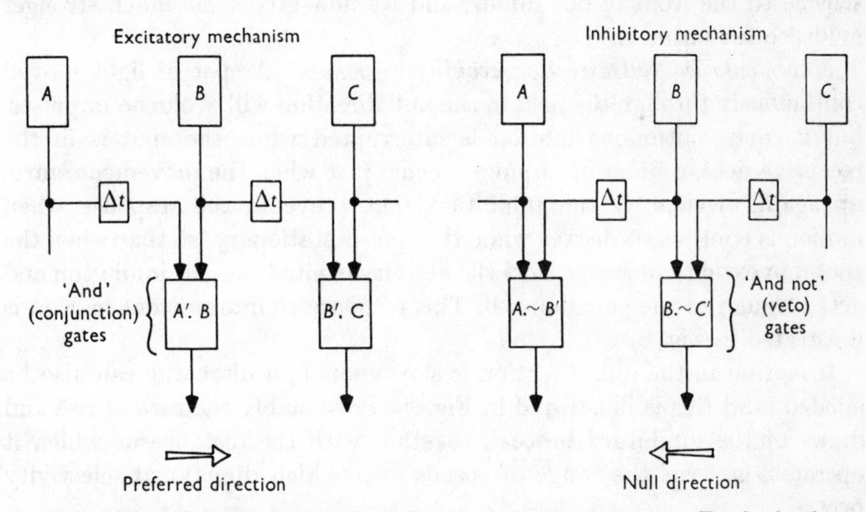

Fig. 7. Two hypothetical methods for discriminating sequence. For both, the preferred direction would be from left to right, null from right to left. In the excitatory scheme activity from the groups of receptors *A* and *B* is delayed before it is passed laterally in the preferred direction to the 'and' (conjunction) gates. If motion is in the preferred direction *A′* (delayed *A*) occurs synchronously with *B*, *B′* occurs synchronously with *C*, and these conjunctions cause the units in the next layer to fire. In the scheme on the right the activity spreads laterally, but in the null direction, from the groups of receptors *B* and *C*, and it has an inhibitory action at the units in the next layer; hence these act as 'and not' (veto) gates. The inhibition prevents activity from *A* and *B* passing through these gates if motion is in the null direction, but arrives too late to have an effect if motion is in the preferred direction. Notice that a special delay unit is not really necessary, for this scheme works if inhibition simply persists longer than excitation and can thus continue to be effective after a lapse of time. The excitatory scheme works by picking out those stimuli with the desired property, whereas the inhibitory scheme works by vetoing responses to unwanted stimuli; the latter is the one favoured by the experimental evidence.

different parts of the receptive field we found that the results obtained at one edge were anomalous in that movements in both null and preferred directions gave responses. Such responses are illustrated in the top row of records in Fig. 4. On the inhibitory scheme it may be possible to resolve this anomaly along the following lines. Responses from the last points crossed by a spot moving in the null direction are normally prevented by inhibition coming from the penultimate regions that have just been crossed. If the spot is moved to and fro solely in the rim, this penultimate region is avoided, and consequently it never inhibits the responses coming from the rim. Measurements of this 'inhibition-free' zone at the rim suggest that it may extend for as much as 1° inwards from the extreme edge of the receptive field.

These observations are not decisive, but they brought the inhibitory scheme to the front of our minds, and we now give some much stronger evidence favouring it.

Movements in null direction evoking responses. A spot of light moved continuously through the field in the null direction will evoke no impulses, but if such continuous motion is interrupted while the spot is in the receptive field, a burst of impulses occurs just when the movement starts up again. Evidently the inhibition that prevents the response when motion is continuous decays while the spot is stationary, so that when the spot moves on to new receptors the activity excited escapes inhibition and gets through to the ganglion cell. This response to intermittent motion is illustrated in Fig. 8.

If motion in the null direction is slow enough, a discharge can also be elicited, and this is illustrated in Fig. 9. Presumably the rate of rise and decay of the inhibitory process, together with the distance at which it operates, governs the range of speeds over which directional selectivity occurs.

Responses to slits singly and in sequence. What was thought to be a crucial test of the inhibition hypothesis was devised. Two slits were placed close to each other and the responses to each in isolation were recorded several times at on and off. The slits were then presented in null or preferred sequence, and several responses again recorded. The records were analysed by counting the impulses that occurred within $\frac{1}{2}$ sec of stimulation, and the averages of 4 to 7 responses are presented in Table 3.

First compare the figures in the last two columns, and notice that the result confirms what has already been said. Preferred sequences are more effective stimuli than null sequences at all separations studied, but the difference is most pronounced at small separations and decreases at the separations greater than 17′. Now compare the figures in the 'Null' column with those in the '$A+B$' column. In every case the 'Null' has the lower figure, so that inhibition certainly occurs: when the sequence is in the null direction fewer impulses occur than when each region is excited separately. Finally, compare the figures in the 'Preferred' column with those in the '$A+B$' column. Here there is an excess in the 'Preferred' column at small separations, but not at large separations.

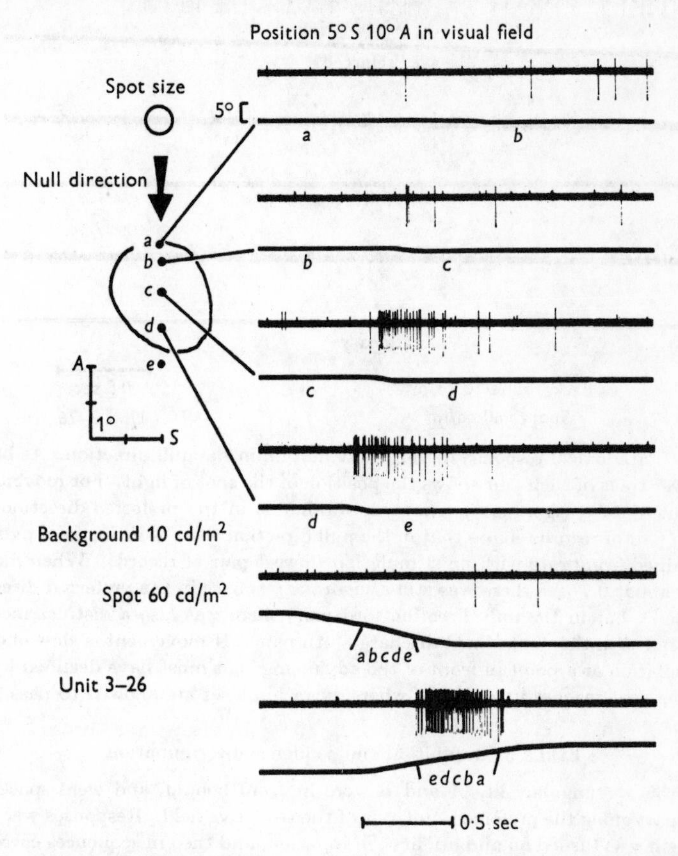

Fig. 8. Escape of impulses with intermittent movement in the null direction. Five positions are marked in relation to the outline of the receptive field shown on the left. The lowest two pairs of records show the effect of sweeping continuously through these positions in the null (*abcde*) and preferred (*edcba*) directions. In the upper four pairs the spot was moved discontinuously, first from *a* to *b*, then from *b* to *c*, then from *c* to *d*, then from *d* to *e* just outside the field. The lower trace of each pair shows the position of the spot in the field. As an example of the escape phenomenon notice that no impulses occur when movement from *c* to *d* is part of a continuous sweep (5th pair of records), but they do occur when this movement is made in isolation (3rd pair). The suggested interpretation is that 'on' or 'off' stimulation at any point inhibits 'on' or 'off' excitation of the next point in the null direction, but this inhibition decays with time. When the spot pauses at *c*, off excitation from *c*, and on excitation of the next point, occur after inhibition has decayed and impulses therefore escape.

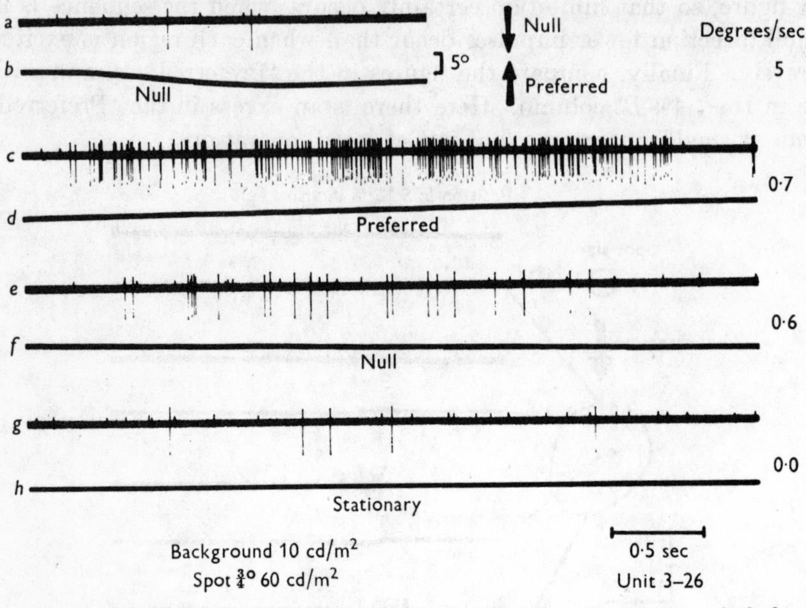

Fig. 9. Paradoxical response to very slow motion in the null direction. As before the lower trace of each pair shows the position of the spot of light. For movements at about 5°/sec a vigorous response was obtainable in the preferred direction, but the top pair of records shows that in the null direction there is no increase over the maintained firing rate with no stimulation (lowest pair of records). When motion was at about 0·7°/sec there was still a vigorous response in the preferred direction (2nd pair), but in the null direction (3rd pair) there was also a distinct increase compared with the maintained discharge (4th pair). If movement is slow enough, the inhibition at a point in front of the advancing spot must have declined by the time the spot reaches it to a level where extra impulses are allowed to pass.

TABLE 3. Inhibition and sequence-discrimination

Two narrow rectangular slits *A* and *B* were lit from behind, and were spaced various distances apart along the preferred-null axis of the receptive field. Responses were recorded when each slit was turned on and off, first, in isolation, and then in sequences corresponding to the null (*BA*) and preferred (*AB*) directions. The figures are the average numbers of spikes that occurred within ½ sec of stimulation (4–7 responses averaged). For the null sequence there was always a deficit of spikes compared with the sum of the spikes produced by the two slits separately.

Slit separation	Stimulus	*A*	*B*	*A* + *B*	Null *BA*	Preferred *AB*
1° 06′	On	2·6	1·7	4·3	2·0	1·8
	Off	9·3	4·7	14·0	5·9	8·0
34′	On	5·2	3·2	8·4	3·4	6·3
	Off	10·2	6·2	16·4	4·6	14·9
17′	On	5·1	3·2	8·3	1·6	13·9
	Off	9·1	4·1	13·2	3·2	19·8
8′	On	5·0	4·0	9·0	2·0	13·0
	Off	8·5	5·5	14·0	1·8	17·7

Individual responses are highly variable and the experimental situation needs systematic exploration with averaging techniques. At this stage we can say that the experiment obviously supports the idea that null sequences are ineffective stimuli because of inhibition. It also indicates, however, that there is some degree of facilitation for preferred sequences, though it is fair to add that this seems a less important effect than the inhibition.

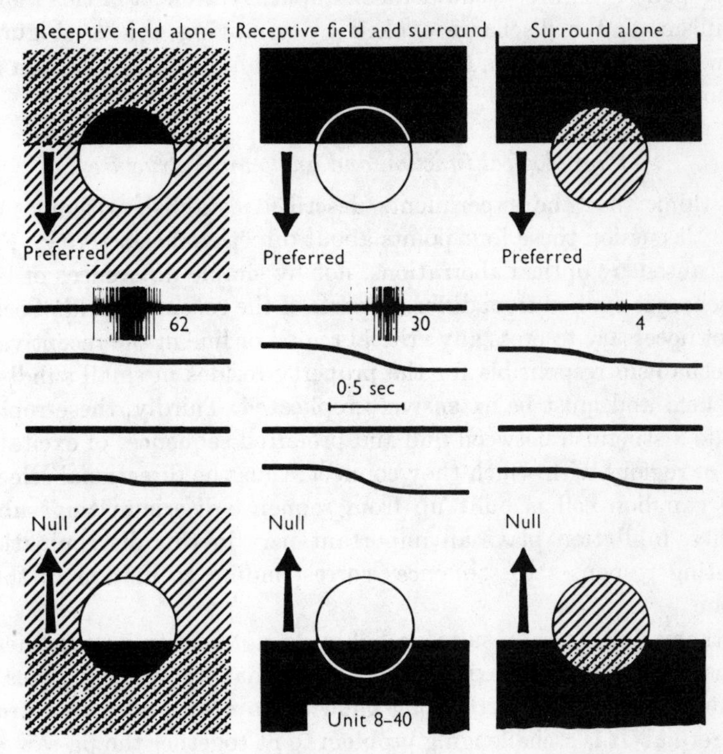

Fig. 10. Lateral inhibition and responses to movement. A black edge was moved behind a mask of grey paper (cross-hatched) so that the advancing border crossed a 4° hole exposing the receptive field alone with the surround masked off (left), the surround alone with the receptive field masked off (right), or it crossed both together, with no mask (centre). The records show the responses; the lower trace of each pair came from a photocell aimed at the receptive field. No impulses were obtained when motion was in the null direction (lower pair of records). In the preferred direction some were obtained in each case, but the response was much greater with the surround masked off than when it crossed surround and centre together (62 instead of 30 impulses). Motion in the surround inhibits the response to motion in the centre, just as light going on or off in the surround inhibits on or off responses from the centre.

Inhibition from outside the receptive field

The type of inhibition postulated to account for directional selectivity, and shown up in the experiment of Table 3, comes from within the receptive field—that is, from within the region where light can evoke impulses. There is also an inhibitory mechanism acting from outside the receptive field—that is, from the surrounding region where light stimuli evoke little or no response. Figure 10 shows an example of the effect of this inhibitory mechanism on the discharge evoked by a moving object. Figure 5 of Barlow *et al.* (1964) shows the effect of this inhibitory mechanism on the threshold.

DISCUSSION

Physiological function and anatomical structure

We think that the experiments described establish without need of further discussion these four points about directional selectivity. First, it is not caused by optical aberrations, nor by simple differences of latency for discharges evoked from different parts of the receptive field. Secondly, it is not necessary to cross any critical region or line in the receptive field: the mechanism responsible for the property resides in small subdivisions of the field and must be extensively replicated. Thirdly, these replicated subunits distinguish between null and preferred sequences of excitation of a pair of regions with which they connect; thus the directional selectivity of the ganglion cell is built up from sequence-discriminating subunits. Fourthly, inhibition plays an important part in this discrimination by preventing responses to sequences corresponding to motion in the null direction.

By themselves these results probably do not justify any further conclusions, but the complexity of function that they have revealed is beginning to match up to the long-known complexity of neural structure in the retina. It is a challenging problem to fit together the jig-saw puzzle of anatomical elements in the hope of revealing the picture of physiological function, and a tentative solution is shown in Fig. 11. It is certainly incomplete, for it does not specify the connexions of the concentric type of ganglion cell, nor of those selectively responsive to fast and slow movement (Barlow *et al.* 1964). Furthermore, we assume that there is a duplicate set of bipolar and horizontal cells that are activated at 'off'. We have some evidence, to be presented elsewhere, that 'on' and 'off' systems do not interact with each other at this level, and therefore for simplicity we have omitted the 'off' system. Because of the diversity of types of bipolar and horizontal cells (on and off for at least four different directions) one can see why a very large number of bipolar cells are required to handle the input from a group of receptors.

At various points there are alternatives to our scheme that are not excluded by the evidence at present available. On the other hand the roles of the anatomical elements and their postulated connexions are not as arbitrarily assigned as a naive reader is liable to suppose. For discussion, take what is perhaps the most controversial and interesting feature of the scheme—the assignment to horizontal cells of the role of inhibitory elements that prevent bipolar cells responding to null sequences. There are two main questions to be answered: why place the inhibitory element in the inner nuclear layer? And why postulate that the horizontal cell

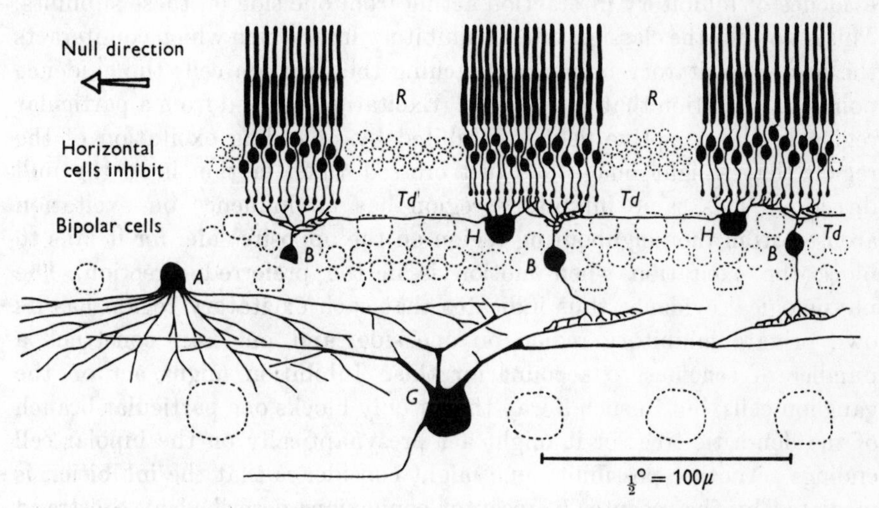

Fig. 11. Suggested functional connexions of the retinal elements concerned with directional selectivity. The elements are freely adapted from Cajal (1893), and are assembled in accordance with the functional organization suggested in this paper. The scale of the diagram is approximate and a posterior nodal distance of 11·5 mm has been assumed. The pathway of excitation is from receptors (R), through bipolars (B), to the ganglion cell (G), but activity in this direct pathway is modified by the associational cells. The horizontal cells (H) pick up from receptors, conduct laterally in the null direction through a teledendron (Td), and inhibit bipolars in the neighbouring region. This prevents responses when an image moves in the null direction, but has no effect when motion is in the preferred direction. Horizontal cells have the function of the laterally conducting elements in the inhibitory scheme shown in Fig. 7. The amacrine cells (A) are thought to pick up from bipolar endings in the inner plexiform layer and to conduct activity throughout their axo-dendritic ramifications; they are assumed to make synaptic connexion with the ganglion cells and inhibit them, thus mediating lateral inhibition of the type illustrated in Fig. 5 of Barlow *et al.* (1964) and Fig. 10 of this paper. The off-responding mechanism is not illustrated, but seems to require duplicate horizontal cells and bipolar cells. Notice that the ganglion cell must connect selectively to those particular bipolars which respond selectively to the sequences for one particular direction. Its response is specific for this pattern of stimulation but is invariant with respect to contrast and position in the receptive field. It may be said to achieve some degree of 'stimulus generalization'.

connects from receptors to bipolar cells, rather than, for instance, from bipolars to bipolars?

The strength of the proposed scheme arises from the fact that a function can naturally be assigned to the neural elements that are known to exist, without making esoteric or revolutionary assumptions about how they work. Sequence-discrimination is assigned to bipolar cells because the ganglion cell appears to pick up from subunits that are replicated in different parts of the receptive field, and bipolar cells are the replicated anatomical elements that feed ganglion cells. There is physiological evidence of inhibitory interaction acting from one side on these subunits. This is not like the classical lateral inhibitory interaction which counteracts the pooled excitatory influences reaching the ganglion cell: the evidence points to inhibition that acts locally. Excitation aroused from a particular region of the receptive field is inhibited by preceding excitation of the region that a light image has just crossed when motion is in the null direction. This same inhibitory region has no influence on excitation aroused from the neighbouring region on the opposite side, for it fails to block the excitation when motion is in the preferred direction. The physiological evidence thus indicates that each excitatory region has its own private inhibitory region on one side, and one can construct a number of schemes to account for this. Inhibition might act on the ganglion cells, but in such a way that it only blocks one particular branch of the dendritic tree: or it might act presynaptically on the bipolar cell endings. Another possibility one might consider is that the inhibition is mediated by the receptor-to-receptor connexions described by Sjöstrand (1958) in the guinea-pig. The distance over which the inhibitory effects have to be passed may be a difficulty here, and this notion shares the difficulty described below for other forms of inhibition which act on the receptors. Since the horizontal cells are known to have processes conducting laterally the natural starting hypothesis is that they are the cells carrying this inhibition from one region to another.

If this is granted there is still scope for argument as to where this inhibition is picked up from, and where it feeds to. Might it not inhibit receptors rather than bipolar cells? Might it not even pick up from bipolar cells and feed back to receptors? The key observation here is that a region which has itself been inhibited from its own private inhibitory zone on one side can none the less inhibit activity aroused in the neighbouring zone on the other side. Motion through the receptive field in the null direction may elicit no impulses whatever. Consider what is happening half way through such a traverse: one sees that excitation of the receptors at the mid-point prevents the discharge from the next group of receptors the spot is going to cross, even though no activity is transmitted centrally from the receptors at the mid-point. This indicates that the inhibitory connexion runs from an early point on the path from the inhibition-arousing zone to a later point on the path from the zone that is inhibited. Presumably then it runs from receptors to bipolar cells, and in that case the inhibition can act in the ordinary way by stabilizing the membrane potential of the bipolar cells. However there is clearly a point here open to histological investigation. Do the horizontal cells make this pattern of connexion in the rabbit? Polyak (1941) describes the horizontal cells of the monkey as making receptor-to-receptor connexions.

No further information has come to light on the pathway mediating ordinary lateral inhibition of the type shown in Fig. 10. This probably acts on the ganglion cells, and amacrine cells remain the most plausible guess.

It is clear that our allocation of functions to particular structures must be regarded as provisional, but we were pleased to find how well the physiological organization seems to fit in with the anatomical structure.

Other proposed anatomical correlates in other species. Maturana, Lettvin, McCulloch & Pitts (1960) and Lettvin, Maturana, Pitts & McCulloch (1961) have also attempted to relate structure and function in the retina, in their case in the frog. Their discussion has something in common with ours, but they place greater emphasis on the concept that the ganglion cell's properties are determined by the shape and size of its dendritic tree. They believe that the different strata of the inner plexiform layer carry information as to different properties of the pattern of light falling on the receptors; the ganglion cell is then thought to pick up the appropriate combination of these properties by ramifying in the various layers. This may explain how a ganglion cell is able to make connexion with a specific subset of bipolar cells, and their notion does not contradict ours. Where we feel that our scheme goes further is in showing how the complex task of signalling direction of movement can be broken down into simpler tasks that can be performed by elements making simple excitatory and inhibitory connexions.

Maturana & Frenk (1963) have described directionally selective units in the pigeon's retina. These obviously have much in common with the units in the rabbit, for they show the same directional selectivity independent of the path through the field and the contrast of the moving object. Furthermore, they made an interesting observation which led them to the conclusion that an inhibitory mechanism is involved in directional selectivity. They turned a spot of light on and off in one place in the receptive field, eliciting responses in the usual way. While the light was off they moved it to another position in the field displaced in the null direction from the first position, and turned it on and off again. No responses were obtained, whereas, if the spot had been displaced in the preferred direction, responses were obtained as usual. Clearly this is similar to the two-spot experiment described here, but it seems from their brief description that inhibition must persist for a long time in the pigeon. They do not attempt to make detailed suggestions about which anatomical structures are responsible for the specificity of the stimuli that generate responses in a particular unit, but they give the impression that they believe it is achieved by the ganglion cell. In our view the specificity originates with the bipolars, and the ganglion cell generalizes for position and contrast by picking up only from those bipolars that respond to sequences of on or off stimuli corresponding to one particular direction of motion.

Grüsser-Cornehls, Grüsser & Bullock (1964) tested movement-sensitive units in the frog with various stimuli, and came to the conclusion that movement detection was really 'change-of-position' detection. Their experiment suggests that they are distinguishing between continuous and discontinuous change of position, and in that case our conclusions are not too far apart: discontinuous change of position, as in the two-spot experiment, can activate the directionally selective mechanism. However, they were not dealing with units responding selectively to the direction of motion, for unlike Maturana *et al.* (1960) they failed to find such units in the frog, although they confirmed many of these authors' other findings.

Directional system in insects. Reichardt (1957, 1961*a*, *b*) has proposed a mechanism capable of explaining the responses of insects to movement in their visual field. This seems at first sight very different from the one we have arrived at, for his scheme depends upon evaluating the cross-correlation between the signal from an ommatidium and that from its neighbour modified by passage through a low-pass filter. This is closer to the excitatory-conjunction scheme that we rejected than it is to the inhibitory scheme. However, one should probably regard Reichardt's proposal as the simplest physical system with a performance specification similar to the beetle's eye, and one should not be too surprised if the realization of a system in 'biological hardware' is different from what it would be in physical hardware, even if the operation performed is very similar.

Pattern recognition, trigger features, and stimulus generalization

Maturana & Frenk (1963) suggest that an understanding of the type of behaviour they describe in the ganglion cells of the pigeon retina clarifies certain problems of pattern recognition. We think there are two aspects of recent work on the visual pathway that are interesting in this respect. The first is the specificity of the features that are effective in triggering the activity of sensory neurones. Examples of this are provided by the 'fly detectors' (Barlow, 1953) and 'convexity detectors' (Lettvin *et al.* 1959) of the frog's retina, the linear elements of the cat's cortex (Hubel & Wiesel,

1959, 1962), the 'horizontal edge detectors' of the pigeon retina (Maturana & Frenk, 1963), and the directionally selective elements found in all these preparations as well as in the rabbit's retina. Now in pattern recognition by machines, Grimsdale, Sumner, Tunis & Kilburn (1959) broke the task into two stages by first detecting the presence of certain key features of the patterns to be discriminated and then looking for the characteristic combinations of these features. Most of the successful systems for recognizing printed or handwritten characters make use of a similar scheme (Selfridge & Neisser, 1960; Uhr & Vossler, 1961; Frishkopf & Harmon, 1961; Kamentsky & Liu, 1963), and it is interesting to see why it is necessary for the computer to view its text through these 'feature filters'. It is because even the largest computor cannot recognize letter *A* by comparing the input with a complete list of all members of the class of *A*'s. Such an approach would require the separate representation of each of the 2^n possible states of the n binary inputs and this becomes unmanageable for values of n that are very small by biological standards. Presumably the trigger features of the visual system likewise enable the input states to be classified in an effective way without requiring a googolian number of separate representations.

The second aspect we want to draw attention to is the detailed manner in which the specific and general properties of these trigger features are picked out. This discussion will be based upon our suggested mechanism for directional selectivity, and we shall introduce certain simplifications which, though not entirely justifiable, make it easier to compare the neural process with artificial pattern recognition.

According to our analysis the operation of abstracting direction of movement is done in two stages, each with the same two steps. The first step in each case is the summation or pooling of selected excitatory influences, and the second step is the inhibitory interaction of another element that has, as it were, the power of veto. The first step loses information, for the bipolar cell which pools inputs from a number of receptors does not reflect in its output which particular ones were active. As pointed out by Reichardt (1961*a*, *b*) the inhibitory step could in principle regain this lost information, but in the case of bipolar cells it does not do this; instead it makes the response more selective by bringing in new information. Without this inhibitory interaction a bipolar cell would simply say, when it became active, 'Light fell in this region'; with the inhibition it says 'Light fell in this region and was not preceded by light falling in that region'. Compared with the receptors in the preceding layer, the bipolars have lost some information about the exact position of the stimulus, but they have extracted some information about the presence of a particular sequential pattern in the stimulus.

The same two steps are taken in the next stage, occurring in the next layer. Here a ganglion cell does not pool from all the bipolars in the receptive field, but it picks up selectively from all those which respond when a stimulus moves in a particular direction, irrespective of the location of the bipolar cell or whether it belongs to the 'on' class or the 'off' class. It thus discards the information as to the contrast of the stimulus object and whereabouts in the receptive field it was; it 'generalizes' by grouping together activity resulting from movement in a particular direction, regardless of contrast and exact position. This is followed by inhibitory interaction which again makes the response more specific. Light going on or off in the surround (Fig. 5 of Barlow *et al.* 1964), or movement in the surround (Fig. 10, this paper) reduces or prevents the response, so that when activity occurs it implies that changes were not occurring in the surrounding retina at the time they occurred within the receptive field.

Let us now express the logical pattern of these repeated operations symbolically. The pooling or generalizing operation is equivalent in some ways to the formation of a logical union (inclusive 'or', symbolized by $\vee$), and the inhibitory or veto operation is equivalent to 'and not...' (symbolized by $.\sim$). If B is the class of inputs to which a bipolar cell responds, and R_a, R_b, etc., are the inputs causing activity in the receptors a, b, etc., then

$$B = (R_a \vee R_b \vee R_c \ldots) . \sim (R_r \vee R_s \vee R_t \ldots).$$

Likewise the class G of inputs causing activity in a ganglion cell is expressed in terms of B_a, B_b, etc., the inputs which activate the selection of bipolars it connects with; thus

$$G = (B_a \vee B_b \vee B_c \ldots) . \sim (B_r \vee B_s \vee B_t \ldots).$$

If we symbolize by E^ψ the class of inputs which is effective in exciting a particular element after ψ synapses, and by $E^{\psi+1}$ the class effective for an element after one more synapse, then $E^{\psi+1}$ is given by

$$E^{\psi+1} = (E^\psi_a \vee E^\psi_b \vee E^\psi_c \ldots) . \sim (E^\psi_r \vee E^\psi_s \vee E^\psi_t \ldots).$$

Notice that only a small proportion of the possible logical functions of the E^ψ can be expressed in this form, and it is therefore not at all a trivial restriction.

We are suggesting that the classification system at one level in the nervous system is built out of the classification at the preceding level by a combination of pooling or union, and inhibition or veto (and not...). Can we regard the proposed mechanism for directionally selective units as a paradigm of the neural mechanisms responsible for the classification system imposed on our sensory input? Is pooling analogous to 'stimulus generalization', and is greater specificity of response always achieved by the veto of an associational neurone, an interposed inhibitory element? These are intriguing questions.

SUMMARY

1. The mechanism of directional selectivity has been investigated in retinal ganglion cells of decerebrate or lightly anaesthetized rabbits.

2. The property of responding to one direction of motion (preferred) but not to the opposite (null) direction occurs in on–off units, but the responses to movement cannot be predicted from the map of the receptive field obtained with static stimuli; the property cannot be explained by optical aberrations (see Fig. 1), nor by progressive changes of latency across the field (see Fig. 2).

3. There is no critical line or region that must be crossed to produce unequal responses to preferred and null motion (Fig. 4): small subsections of the receptive field possess the property (Fig. 5).

4. The response to successive stimulation of two small regions depends upon whether the order corresponds to motion in the preferred or null direction (Fig. 8). This effect is strong when the two regions are within about $\frac{1}{2}°$ of each other, but declines at greater separations.

5. This is thought to indicate that directional selectivity results from the discrimination of sequence. Normal movement excites many points in a long succession, but the mechanism works by discriminating the sequence of individual pairs of regions.

6. When two stimuli are presented in the null sequence the number of impulses elicited is much less than the sum of the numbers elicited from each stimulus in isolation (Table 3). There is a small excess of impulses over this sum when the stimuli are presented in the preferred sequence.

7. From this and other findings it is concluded that sequence-discrimination results primarily from an inhibitory mechanism that vetoes the response to null sequences, rather than from the detection of the conjunction of excitation from two regions with an appropriate delay (see Fig. 7).

8. If the image of a moving object spreads outside the receptive field on to its surround there are fewer impulses than when it is confined to the receptive field alone (Fig. 10). This must be the inhibitory mechanism that elevates the threshold for large compared with small spots, and it is presumably different from the inhibition responsible for sequence-discrimination.

9. The functional organization is discussed in relation to the anatomical organization (Fig. 11). It is suggested that horizontal cells conduct

laterally and inhibit the bipolars on one side, thus preventing them from responding to null sequences; the ganglion cells then pick up from the bipolars responsive to like sequences and it is thought that the inhibition from the surround may be mediated by amacrine cells.

10. The ability to abstract direction of motion irrespective of the position in the receptive field and the contrast of the moving object has elements in common with much more complex feats of pattern recognition. The two steps—inhibition by associational neurones and selective pooling—may also play a part in these more complex feats.

We wish to thank W. A. H. Rushton, P. A. Merton, P. E. K. Donaldson and G. Westheimer for the loan of apparatus, and W. Hall, C. Hood, R. Rumble and P. Starling for help in construction and photography. This work was supported in part by Grants NB 05215 and NB 03154 from the U.S. Public Health Service. The micromanipulator for the Cambridge experiments was constructed in the Department of Physiology, University of Sydney to the design of P. O. Bishop and W. Kozak.

REFERENCES

Barlow, H. B. (1953). Summation and inhibition in the frog's retina. *J. Physiol.* **119**, 69–88.

Barlow, H. B. & Hill, R. M. (1963). Selective sensitivity to direction of motion in ganglion cells of the rabbit's retina. *Science*, **139**, 412–414.

Barlow, H. B., Hill, R. M. & Levick, W. R. (1964). Retinal ganglion cells responding selectively to direction and speed of image motion in the rabbit. *J. Physiol.* **173**, 377–407.

Cajal, S. Ramon y (1893). La rétine des vertébrés. In *La Cellule*, **9**, 119–257, ed. Carnoy, J. B., Gilson, G. & Denys, J. Lierre: J. van In and Co.; Louvain: A. Uystpruyst, Libraire.

Frishkopf, L. S. & Harmon, L. D. (1961). Machine reading of cursive script. In *Proc. 4th Lond. Symp. on Information Theory*, pp. 300–316, ed. C. Cherry. London: Butterworth.

Grimsdale, R. L., Sumner, F. H., Tunis, C. J. & Kilburn, T. (1959). A system for the automatic recognition of patterns. *Proc. Inst. elect. Engs*, B, **106**, 210–221.

Grüsser-Cornehls, U., Grüsser, O. J. & Bullock, T. H. (1964). Unit responses in the frog's tectum to moving and non-moving stimuli. *Science*, **141**, 820–822.

Hassenstein, B. (1951). Ommatidienraster und afferente Bewegungsintegration. *Z. vergl. Physiol.* **33**, 301–326.

Hubel, D. H. (1959). Single unit activity in striate cortex of unrestrained cats. *J. Physiol.* **147**, 226–238.

Hubel, D. H. & Wiesel, T. N. (1959). Receptive fields of single neurones in the cat's striate cortex. *J. Physiol.* **148**, 574–591.

Hubel, D. H. & Wiesel, T. N. (1962). Receptive fields, binocular interaction and functional architecture in the cat's visual cortex. *J. Physiol.* **160**, 106–154.

Kamentsky, L. A. & Liu, C. N. (1963). Computer-automated design of multifont print recognition logic. *I.B.M. Journal of Research and Development*, **7**, 2–13.

Lettvin, J. Y., Maturana, H. R., McCulloch, W. S. & Pitts, W. H. (1959). What the frog's eye tells the frog's brain. *Proc. Inst. Radio Engrs, N.Y.*, **47**, 1940–1951.

Lettvin, J. Y., Maturana, H. R., Pitts, W. H. & McCulloch, W. S. (1961). Two remarks on the visual system of the frog. In *Sensory Communication*, pp. 757–776, ed. W. Rosenblith. M.I.T. Press; New York: John Wiley.

Maturana, H. R., Lettvin, J. Y., McCulloch, W. S. & Pitts, W. H. (1960). Anatomy and physiology of vision in the frog (*Rana pipiens*). *J. gen. Physiol.* **43**, suppl. 2, Mechanisms of Vision, 129–171.

Maturana, H. R. & Frenk, S. (1963). Directional movement and horizontal edge detectors in pigeon retina. *Science*, **142**, 977–979.

Polyak, S. L. (1941). *The Retina.* Chicago: University of Chicago Press.

REICHARDT, W. (1957). Autokorrelationsauswertung als Funktionsprinzip des Zentralnervensystems. *Z. Naturf.* **12**, 447–457.

REICHARDT, W. (1961*a*). Autocorrelation, a principle for the evaluation of sensory information by the central nervous system. In *Sensory Communication*, pp. 303–317, ed. W. ROSENBLITH. M.I.T. Press; New York: John Wiley.

REICHARDT, W. (1961*b*). Über das optische Auflösungsvermögen der Facettenaugen von Limulus. *Kybernetik.* **1**, 57–69.

SELFRIDGE, O. & NEISSER, U. (1960). Pattern recognition by machine. *Sci. Amer.* **203**, 60–68.

SJÖSTRAND, F. S. (1958). Ultrastructure of retinal rod synapses of the guinea pig eye as revealed by three-dimensional reconstructions from serial sections. *J. Ultrastr. Res.* **2**, 122–170.

THOMSON, L. C. (1953). The localisation of function in the rabbit's retina. *J. Physiol.* **119**, 191–209.

UHR, L. & VOSSLER, C. (1961). A pattern-recognition program that generates, evaluates and adjusts its own operators. *Proceedings of the Western Joint Computer Conference*, **19**, 555–570.

106 J. Physiol. (1962), **160**, pp. 106–154
With 2 plates and 20 text-figures
Printed in Great Britain

RECEPTIVE FIELDS, BINOCULAR INTERACTION AND FUNCTIONAL ARCHITECTURE IN THE CAT'S VISUAL CORTEX

BY D. H. HUBEL AND T. N. WIESEL
From the Neurophysiology Laboratory, Department of Pharmacology Harvard Medical School, Boston, Massachusetts, U.S.A.

(*Received* 31 *July* 1961)

What chiefly distinguishes cerebral cortex from other parts of the central nervous system is the great diversity of its cell types and interconnexions. It would be astonishing if such a structure did not profoundly modify the response patterns of fibres coming into it. In the cat's visual cortex, the receptive field arrangements of single cells suggest that there is indeed a degree of complexity far exceeding anything yet seen at lower levels in the visual system.

In a previous paper we described receptive fields of single cortical cells, observing responses to spots of light shone on one or both retinas (Hubel & Wiesel, 1959). In the present work this method is used to examine receptive fields of a more complex type (Part I) and to make additional observations on binocular interaction (Part II).

This approach is necessary in order to understand the behaviour of individual cells, but it fails to deal with the problem of the relationship of one cell to its neighbours. In the past, the technique of recording evoked slow waves has been used with great success in studies of functional anatomy. It was employed by Talbot & Marshall (1941) and by Thompson, Woolsey & Talbot (1950) for mapping out the visual cortex in the rabbit, cat, and monkey. Daniel & Whitteridge (1959) have recently extended this work in the primate. Most of our present knowledge of retinotopic projections, binocular overlap, and the second visual area is based on these investigations. Yet the method of evoked potentials is valuable mainly for detecting behaviour common to large populations of neighbouring cells; it cannot differentiate functionally between areas of cortex smaller than about 1 mm^2. To overcome this difficulty a method has in recent years been developed for studying cells separately or in small groups during long micro-electrode penetrations through nervous tissue. Responses are correlated with cell location by reconstructing the electrode tracks from histological material. These techniques have been applied to the somatic sensory cortex of the cat and monkey in a remarkable series of studies by Mountcastle (1957) and Powell & Mountcastle (1959). Their results show that the approach is a powerful one, capable of revealing systems of organization not hinted at by the known morphology. In Part III of the present paper we use this method in studying the functional architecture of the visual cortex. It helped us attempt to explain on anatomical grounds how cortical receptive fields are built up.

METHODS

Recordings were made from forty acutely prepared cats, anaesthetized with thiopental sodium, and maintained in light sleep with additional doses by observing the electrocorticogram. Animals were paralysed with succinylcholine to stabilize the eyes. Pupils were dilated with atropine. Details of stimulating and recording methods are given in previous papers (Hubel, 1959; Hubel & Wiesel, 1959, 1960). The animal faced a wide tangent screen at a distance of 1·5 m, and various patterns of white light were shone on the screen by a tungsten-filament projector. All recordings were made in the light-adapted state. Background illumination varied from $-1{\cdot}0$ to $+1{\cdot}0$ $\log_{10}$ cd/m^2. Stimuli were from 0·2 to 2·0 log. units brighter than the background. For each cell receptive fields were mapped out separately for the two eyes on sheets of paper, and these were kept as permanent records.

Points on the screen corresponding to the area centralis and the optic disk of the two eyes were determined by a projection method (Hubel & Wiesel, 1960). The position of each receptive field was measured with respect to these points. Because of the muscle relaxant the eyes usually diverged slightly, so that points corresponding to the two centres of gaze were not necessarily superimposed. In stimulating the two eyes simultaneously it was therefore often necessary to use two spots placed in corresponding parts of the two visual fields. Moreover, at times the two eyes were slightly rotated in an inward direction in the plane of their equators. This rotation was estimated by (1) photographing the cat before and during the experiment, and comparing the angles of inclination of the slit-shaped pupils, or (2) by noting the inclination to the horizontal of a line joining the area centralis with the optic disk, which in the normal position of the eye was estimated, by the first method, to average about 25°. The combined inward rotations of the two eyes seldom exceeded 10°. Since the receptive fields in this study were usually centrally rather than peripherally placed on the retina, the rotations did not lead to any appreciable linear displacement. Angular displacements of receptive fields occasionally required correction, as they led to an apparent difference in the orientation of the two receptive-field axes of a binocularly driven unit. The direction and magnitude of this difference were always consistent with the estimated inward rotation of the two eyes. Moreover, in a given experiment the difference was constant, even though the axis orientation varied from cell to cell.

The diagram of Text-fig. 1 shows the points of entry into the cortex of all 45 microelectrode penetrations. Most electrode tracks went no deeper than 3 or 4 mm, so that explorations were mainly limited to the apical segments of the lateral and post-lateral gyri (LG and PLG) and a few millimetres down along the adjoining medial and lateral folds. The extent of the territory covered is indicated roughly by Text-figs. 13–15. Although the lateral boundary of the striate cortex is not always sharply defined in Nissl-stained or myelin-stained material, most penetrations were well within the region generally accepted as 'striate' (O'Leary, 1941). Most penetrations were made from the cortical region receiving projections from in or near the area centralis; this cortical region is shown in Text-fig. 1 as the area between the interrupted lines.

Tungsten micro-electrodes were advanced by a hydraulic micro-electrode positioner (Hubel, 1957, 1959). In searching for single cortical units the retina was continually stimulated with stationary and moving forms while the electrode was advanced. The unresolved background activity (see p. 129) served as a guide for determining the optimum stimulus. This procedure increased the number of cells observed in a penetration, since the sampling was not limited to spontaneously active units.

In each penetration electrolytic lesions were made at one or more points. When only one lesion was made, it was generally at the end of an electrode track. Brains were fixed in 10% formalin, embedded in celloidin, sectioned at 20 μ, and stained with cresyl violet. Lesions were 50–100 μ in diameter, which was small enough to indicate the position of the electrode tip to the nearest cortical layer. The positions of other units encountered in a cortical penetration were determined by calculating the distance back from the lesion along the track,

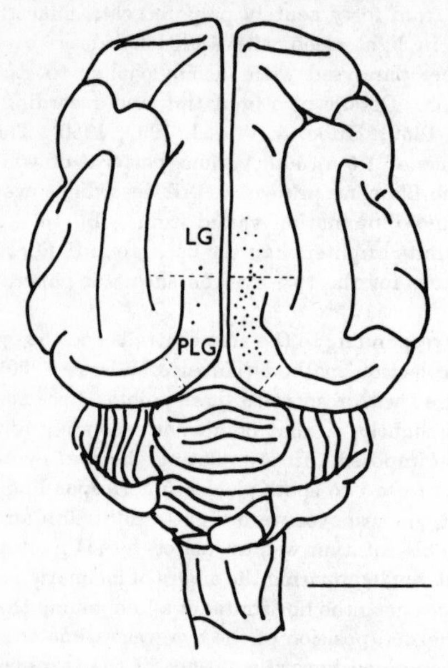

Text-fig. 1. Diagram of dorsal aspect of cat's brain, to show entry points of 45 micro-electrode penetrations. The penetrations between the interrupted lines are those in which cells had their receptive fields in or near area centralis. LG, lateral gyrus; PLG, post-lateral gyrus. Scale, 1 cm.

using depth readings corresponding to the unit and the lesion. A correction was made for brain shrinkage, which was estimated by comparing the distance between two lesions, measured under the microscope, with the distance calculated from depths at which the two lesions were made. From brain to brain this shrinkage was not constant, so that it was not possible to apply an average correction for shrinkage to all brains. For tracks marked by only one lesion it was assumed that the first unit activity was recorded at the boundary of the first and second layers; any error resulting from this was probably small, since in a number of penetrations a lesion was made at the point where the first units were encountered, and these were in the lower first or the upper second layers, or else at the very boundary. The absence of cell-body records and unresolved background activity as the electrode passed through subcortical white matter (see Text-fig. 13 and Pl. 1) was also helpful in confirming the accuracy of the track reconstructions.

PART I

ORGANIZATION OF RECEPTIVE FIELDS IN CAT'S VISUAL CORTEX: PROPERTIES OF 'SIMPLE' AND 'COMPLEX' FIELDS

The receptive field of a cell in the visual system may be defined as the region of retina (or visual field) over which one can influence the firing of that cell. In the cat's retina one can distinguish two types of ganglion cells, those with 'on'-centre receptive fields and those with 'off'-centre fields (Kuffler, 1953). The lateral geniculate body also has cells of these two types; so far no others have been found (Hubel & Wiesel, 1961). In contrast, the visual cortex contains a large number of functionally different cell types; yet with the exception of afferent fibres from the lateral geniculate body we have found no units with concentric 'on'-centre or 'off'-centre fields.

When stimulated with stationary or moving patterns of light, cells in the visual cortex gave responses that could be interpreted in terms of the arrangements of excitatory and inhibitory regions in their receptive fields (Hubel & Wiesel, 1959). Not all cells behaved so simply, however; some responded in a complex manner which bore little obvious relationship to the receptive fields mapped with small spots. It has become increasingly apparent to us that cortical cells differ in the complexity of their receptive fields. The great majority of fields seem to fall naturally into two groups, which we have termed 'simple' and 'complex'. Although the fields to be described represent the commonest subtypes of these groups, new varieties are continually appearing, and it is unlikely that the ones we have listed give anything like a complete picture of the striate cortex. We have therefore avoided a rigid system of classification, and have designated receptive fields by letters or numbers only for convenience in referring to the figures. We shall concentrate especially on features common to simple fields and on those common to complex fields, emphasizing differences between the two groups, and also between cortical fields and lateral geniculate fields.

RESULTS

Simple receptive fields

The receptive fields of 233 of the 303 cortical cells in the present series were classified as 'simple'. Like retinal ganglion and geniculate cells, cortical cells with simple fields possessed distinct excitatory and inhibitory subdivisions. Illumination of part or all of an excitatory region increased the maintained firing of the cell, whereas a light shone in the inhibitory region suppressed the firing and evoked a discharge at 'off'. A large spot confined to either area produced a greater change in rate of firing than a small spot, indicating summation within either region. On the other hand, the two types of region within a receptive field were mutually antagonistic. This was most forcefully shown by the absence or near absence of a response to simultaneous illumination of both regions, for example, with diffuse light. From the arrangement of excitatory and inhibitory regions it was usually possible to predict in a qualitative way the responses to any shape of stimulus, stationary or moving. Spots having the approximate shape of one or other region were the most effective stationary stimuli; smaller spots failed to take full advantage of summation within a region, while larger ones were likely to invade opposing regions, so reducing the response. To summarize: these fields were termed 'simple' because like retinal and geniculate fields (1) they were subdivided into distinct excitatory and inhibitory regions; (2) there was summation within the separate excitatory and inhibitory parts; (3) there was antagonism between excitatory and inhibitory regions; and (4) it was possible to predict responses to stationary or moving spots of various shapes from a map of the excitatory and inhibitory areas.

While simple cortical receptive fields were similar to those of retinal ganglion cells and geniculate cells in possessing excitatory and inhibitory subdivisions, they differed profoundly in the spatial arrangements of these regions. The receptive fields of all retinal ganglion and geniculate cells had one or other of the concentric forms shown in Text-fig. 2*A*, *B*. (Excitatory areas are indicated by crosses, inhibitory areas by triangles.) In contrast, simple cortical fields all had a side-to-side arrangement of excitatory and inhibitory areas with separation of the areas by parallel straight-line boundaries rather than circular ones. There were several varieties of fields, differing in the number of subdivisions and the relative area occupied by each subdivision. The commonest arrangements are illustrated in Text-fig. 2*C*–*G*: Table 1 gives the number of cells observed in each category. The departure of these fields from circular symmetry introduces a new variable, namely, the orientation of the boundaries separating the field subdivisions. This orientation is a characteristic of each cortical cell, and may be vertical, horizontal, or oblique. There was no indication that any one orientation was more common than the others. We shall use the term *receptive-field axis* to indicate a line through the centre of a field, parallel to the boundaries separating excitatory and inhibitory regions. The *axis orientation* will then refer to the orientation of these boundaries, either on the retina or in the visual field. Axes are shown in Text-fig. 2 by continuous lines.

Two common types of fields, shown in Text-fig. 2*C*, *D*, each consisted of a narrow elongated area, excitatory or inhibitory, flanked on either side by two regions of the opposite type. In these fields the two flanking regions were symmetrical, i.e. they were about equal in area and the responses obtained from them were of about the same magnitude. In addition there were fields with long narrow centres (excitatory or inhibitory) and asymmetrical flanks. An example of an asymmetrical field with an inhibitory centre is shown in Text-fig. 2*E*. The most effective stationary stimulus for all of these cells was a long narrow rectangle ('slit') of light just large

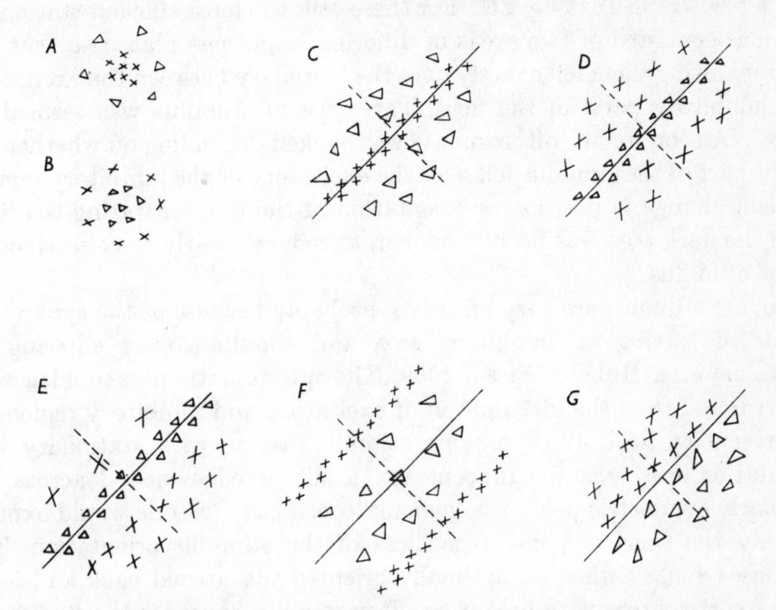

Text-fig. 2. Common arrangements of lateral geniculate and cortical receptive fields. *A*. 'On'-centre geniculate receptive field. *B*. 'Off'-centre geniculate receptive field. *C–G*. Various arrangements of simple cortical receptive fields. ×, areas giving excitatory responses ('on' responses); △, areas giving inhibitory responses ('off' responses). Receptive-field axes are shown by continuous lines through field centres; in the figure these are all oblique, but each arrangement occurs in all orientations.

enough to cover the central region without invading either flank. For maximum centre response the orientation of the slit was critical; changing the orientation by more than 5–10° was usually enough to reduce a response greatly or even abolish it. Illuminating both flanks usually evoked a strong response. If a slit having the same size as the receptive-field centre was shone in either flanking area it evoked only a weak response, since it covered only part of one flank. Diffuse light was ineffective, or at most evoked only a very weak response, indicating that the excitatory and inhibitory parts of the receptive field were very nearly balanced.

In these fields the equivalent but opposite-type regions occupied retinal areas that were far from equal; the centre portion was small and concentrated whereas the flanks were widely dispersed. A similar inequality was found in fields of type *F*, Text-fig. 2, but here the excitatory flanks were elongated and concentrated, while the centre was relatively large and diffuse. The optimum response was evoked by simultaneously illuminating the two flanks with two parallel slits (see Hubel & Wiesel, 1959, Fig. 9).

Some cells had fields in which only two regions were discernible, arranged side by side as in Text-fig. 2*G*. For these cells the most efficient stationary stimulus consisted of two areas of differing brightness placed so that the line separating them fell exactly over the boundary between the excitatory and inhibitory parts of the field. This type of stimulus was termed an 'edge'. An 'on' or an 'off' response was evoked depending on whether the bright part of the stimulus fell over the excitatory or the inhibitory region. A slight change in position or orientation of the line separating the light from the dark area was usually enough to reduce greatly the effectiveness of the stimulus.

Moving stimuli were very effective, probably because of the synergistic effects of leaving an inhibitory area and simultaneously entering an excitatory area (Hubel & Wiesel, 1959). The optimum stimulus could usually be predicted from the distribution of excitatory and inhibitory regions of the receptive field. With moving stimuli, just as with stationary, the orientation was critical. In contrast, a slit or edge moved across the circularly symmetric field of a geniculate cell gave (as one would expect) roughly the same response regardless of the stimulus orientation. The responses evoked when an optimally oriented slit crossed back and forth over a cortical receptive field were often roughly equal for the two directions of crossing. This was true of fields like those shown in Text-fig. 2*C*, *D* and *F*. For many cells, however, the responses to two diametrically opposite movements were different, and some only responded to one of the two movements. The inequalities could usually be accounted for by an asymmetry in flanking regions, of the type shown in Text-fig. 2*E* (see also Hubel & Wiesel, 1959, Fig. 7). In fields that had only two discernible regions arranged side by side (Text-fig. 2*G*), the difference in the responses to a moving slit or edge was especially pronounced.

Optimum rates of movement varied from one cell to another. On several occasions two cells were recorded together, one of which responded only to a slow-moving stimulus (1°/sec or lower) the other to a rapid one (10°/sec or more). For cells with fields of type *F*, Text-fig. 2, the time elapsing between the two discharges to a moving stimulus was a measure of the rate of movement (see Hubel & Wiesel, 1959, Fig. 5).

If responses to movement were predictable from arrangements of excitatory and inhibitory regions, the reverse was to some extent also true.

The axis orientation of a field, for example, was given by the most effective orientation of a moving slit or edge. If an optimally oriented slit produced a brief discharge on crossing from one region to another, one could predict that the first region was inhibitory and the second excitatory. Brief responses to crossing a very confined region were characteristic of cells with simple cortical fields, whereas the complex cells to be described below gave sustained responses to movement over much wider areas.

TABLE 1. Simple cortical fields

	Text-fig.	No. of cells
(a) Narrow concentrated centres		
(i) Symmetrical flanks		
Excitatory centres	2*C*	23
Inhibitory centres	2*D*	17
(ii) Asymmetrical flanks		
Excitatory centres	—	28
Inhibitory centres	2*E*	10
(b) Large centres; concentrated flanks	2*F*	21
(c) One excitatory region and one inhibitory	2*G*	17
(d) Uncategorized	—	117
Total number of simple fields		233

Movement was used extensively as a stimulus in experiments in which the main object was to determine axis orientation and ocular dominance for a large number of cells in a single penetration, and when it was not practical, because of time limitations, to map out every field completely. Because movement was generally a very powerful stimulus, it was also used in studying cells that gave little or no response to stationary patterns. In all, 117 of the 233 simple cells were studied mainly by moving stimuli. In Table 1 these have been kept separate from the other groups since the distribution of their excitatory and inhibitory regions is not known with the same degree of certainty. It is also possible that with further study, some of these fields would have revealed complex properties.

Complex receptive fields

Intermixed with cells having simple fields, and present in most penetrations of the striate cortex, were cells with far more intricate and elaborate properties. The receptive fields of these cells were termed 'complex'. Unlike cells with simple fields, these responded to variously-shaped stationary or moving forms in a way that could not be predicted from maps made with small circular spots. Often such maps could not be made, since small round spots were either ineffective or evoked only mixed ('on-off') responses throughout the receptive field. When separate 'on' and 'off' regions could be discerned, the principles of summation and mutual antagonism, so helpful in interpreting simple fields, did not generally hold. Nevertheless, there were some important features common to the two types of cells. In the following examples, four types of complex fields will be illustrated. The numbers observed of each type are given in Table 2.

TABLE 2. Complex cortical receptive fields

	Text-fig.	No. of cells
(*a*) Activated by slit—non-uniform field	3	11
(*b*) Activated by slit—uniform field	4	39
(*c*) Activated by edge	5–6	14
(*d*) Activated by dark bar	7–8	6
Total number of complex fields		70

The cell of Text-fig. 3 failed to respond to round spots of light, whether small or large. By trial and error with many shapes of stimulus it was discovered that the cell's firing could be influenced by a horizontally oriented slit $\frac{1}{8}°$ wide and 3° long. Provided the slit was horizontal its exact

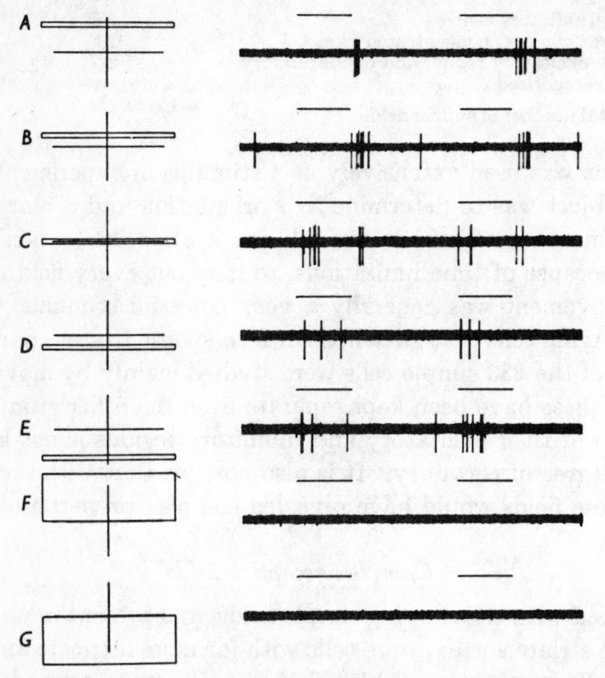

Text-fig. 3. Responses of a cell with a complex receptive field to stimulation of the left (contralateral) eye. Receptive field located in area centralis. The diagrams to the left of each record indicate the position of a horizontal rectangular light stimulus with respect to the receptive field, marked by a cross. In each record the upper line indicates when the stimulus is on. *A–E*, stimulus $\frac{1}{8} \times 3°$, *F–G*, stimulus $1\frac{1}{2} \times 3°$ (4° is equivalent to 1 mm on the cat retina). For background illumination and stimulus intensity see Methods. Cell was activated in the same way from right eye, but less vigorously (ocular-dominance group 2, see Part II). An electrolytic lesion made while recording from this cell was found near the border of layers 5 and 6, in the apical segment of the post-lateral gyrus. Positive deflexions upward; duration of each stimulus 1 sec.

positioning within the 3°-diameter receptive field was not critical. When it was shone anywhere above the centre of the receptive field (the horizontal line of Text-fig. 3) an 'off' response was obtained; 'on' responses were evoked throughout the lower half. In an intermediate position (Text-fig. 3*C*) the cell responded at both 'on' and 'off'. From experience with simpler receptive fields one might have expected wider slits to give increasingly better responses owing to summation within the upper or lower part of the field, and that illumination of either half by itself might be the most effective stimulus of all. The result was just the opposite: responses fell off rapidly as the stimulus was widened beyond about $\frac{1}{8}°$, and large rectangles covering the entire lower or upper halves of the receptive field were quite ineffective (Text-fig. 3*F*, *G*). On the other hand, summation could easily be demonstrated in a horizontal direction, since a slit $\frac{1}{8}°$ wide but extending only across part of the field was less effective than a longer one covering the entire width. One might also have expected the orientation of the slit to be unimportant as long as the stimulus was wholly confined to the region above the horizontal line or the region below. On the contrary, the orientation was critical, since a tilt of even a few degrees from the horizontal markedly reduced the response, even though the slit did not cross the boundary separating the upper and lower halves of the field.

In preferring a slit specific in width and orientation this cell resembled certain cells with simple fields. When stimulated in the upper part of its field it behaved in many respects like cells with 'off'-centre fields of type *D*, Text-fig. 2; in the lower part it responded like 'on'-centre fields of Text-fig. 2*C*. But for this cell the strict requirements for shape and orientation of the stimulus were in marked contrast to the relatively large leeway of the stimulus in its ordinate position on the retina. Cells with simple fields, on the other hand, showed very little latitude in the positioning of an optimally oriented stimulus.

The upper part of this receptive field may be considered inhibitory and the lower part excitatory, even though in either area summation only occurred in a horizontal direction. Such subdivisions were occasionally found in complex fields, but more often the fields were uniform in this respect. This was true for the other complex fields to be described in this section.

Responses of a second complex unit are shown in Text-fig. 4. In many ways the receptive field of this cell was similar to the one just described. A slit was the most potent stimulus, and the most effective width was again $\frac{1}{8}°$. Once more the orientation was an important stimulus variable, since the slit was effective anywhere in the field as long as it was placed in a 10 o'clock–4 o'clock orientation (Text-fig. 4*A*–*D*). A change in orientation of more than 5–10° in either direction produced a marked reduction in the response (Text-fig. 4*E–G*). As usual, diffuse light had no influence on the firing. This cell responded especially well if the slit, oriented as in *A–D*, was moved steadily across the receptive field. Sustained discharges were evoked over the entire length of the field. The optimum rate of movement was about 1°/sec. If movement was interrupted the discharge stopped, and when it was resumed the firing recommenced. Continuous firing could be maintained indefinitely by small side-

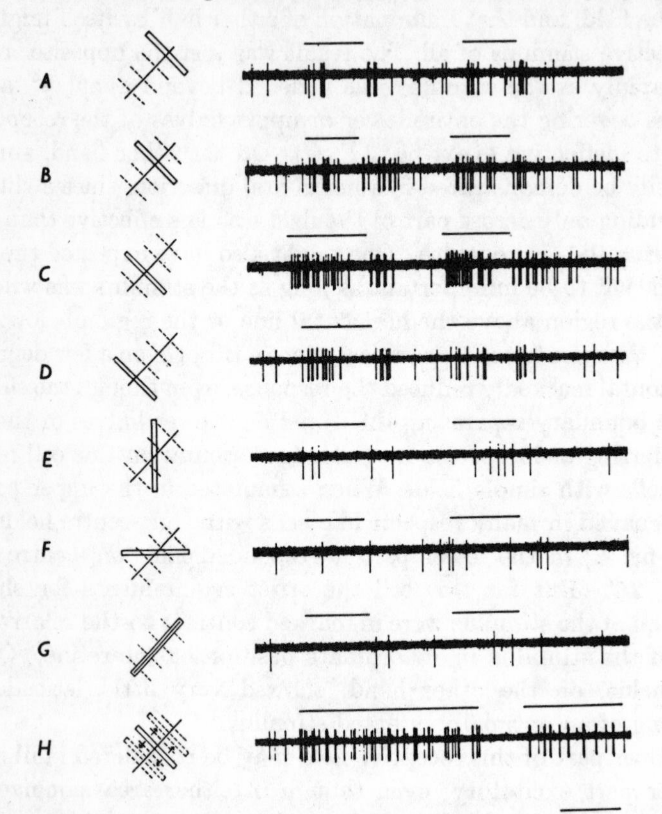

Text-fig. 4. Responses of a cell with a complex field to stimulation of the left (contralateral) eye with a slit $\frac{1}{8} \times 2\frac{1}{2}°$. Receptive field was in the area centralis and was about 2 × 3° in size. *A–D*, $\frac{1}{8}°$ wide slit oriented parallel to receptive field axis. *E–G*, slit oriented at 45 and 90° to receptive-field axis. *H*, slit oriented as in *A–D*, is on throughout the record and is moved rapidly from side to side where indicated by upper beam. Responses from left eye slightly more marked than those from right (Group 3, see Part II). Time 1 sec.

to-side movements of a stimulus within the receptive field (Text-fig. 4*H*). The pattern of firing was one characteristic of many complex cells, especially those responding well to moving stimuli. It consisted of a series of short high-frequency repetitive discharges each containing 5–10 spikes. The bursts occurred at irregular intervals, at frequencies up to about 20/sec. For this cell, movement of an optimally oriented slit was about equally effective in either of the two opposite directions. This was not true of all complex units, as will be seen in some of the examples given below.

Like the cell of Text-fig. 3 this cell may be thought of as having a counterpart in simple fields of the type shown in Text-fig. 2*C–E*. It shares with these simpler fields the attribute of responding well to properly oriented slit stimuli. Once more the distinction lies in the permissible variation in position of the optimally oriented stimulus. The variation is small (relative to the size of the receptive field) in the simple fields, large in the complex. Though resembling the cell of Text-fig. 3 in requiring a slit for a stimulus, this cell differed in that its responses to a properly oriented slit were mixed ('on-off') in type. This was not unusual for cells with complex fields. In contrast, cortical cells with simple fields, like retinal ganglion cells and lateral geniculate cells, responded to optimum restricted stimuli either with excitatory ('on') responses or inhibitory ('off') responses. When a stimulus covered opposing regions, the effects normally tended to cancel, though sometimes mixed discharges were obtained, the 'on' and 'off' components both being weak. For these simpler fields 'on-off' responses were thus an indication that the stimulus was not optimum. Yet some cells with complex fields responded with mixed discharges even to the most effective stationary stimuli we could find. Among the stimuli tried were curved objects, dark stripes, and still more complicated patterns, as well as monochromatic spots and slits.

A third type of complex field is illustrated in Text-figs. 5 and 6. There were no responses to small circular spots or to slits, but an edge was very effective if oriented vertically. Excitatory or inhibitory responses were produced depending on whether the brighter area was to the left or the right (Text-fig. 5*A*, *E*). So far, these are just the responses one would expect from a cell with a vertically oriented simple field of the type shown in Text-fig. 2*G*. In such a field the stimulus placement for optimum response is generally very critical. On the contrary, the complex unit responded to vertical edges over an unusually large region about 16° in length (Text-fig. 6). 'On' responses were obtained with light to the left (*A–D*), and 'off' responses with light to the right (*E–H*), regardless of the position of the line separating light from darkness. When the entire receptive field was illuminated diffusely (*I*) no response was evoked. As with all complex fields, we are unable to account for these responses by any simple spatial arrangement of excitatory and inhibitory regions.

Like the complex units already described, this cell was apparently more concerned with the orientation of a stimulus than with its exact position in the receptive field. It differed in responding well to edges but poorly or not at all to slits, whether narrow or wide. It is interesting in this connexion that exchanging an edge for its mirror equivalent reversed the response, i.e. replaced an excitatory response by an inhibitory and vice versa. The ineffectiveness of a slit might therefore be explained by supposing that the opposite effects of its two edges tended to cancel each other.

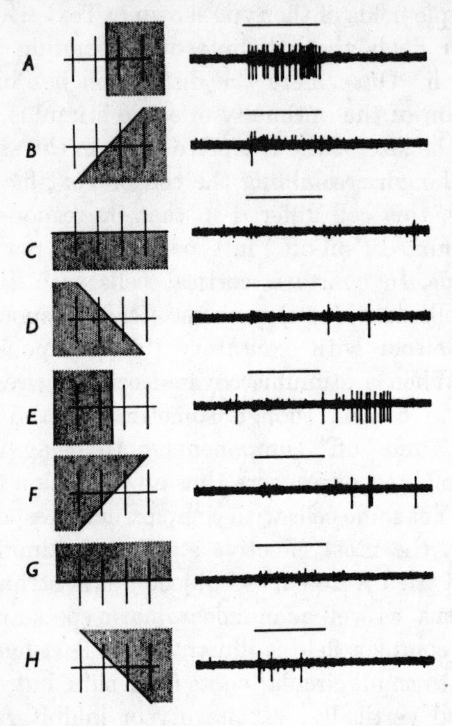

Text-fig. 5. Responses of a cell with a large ($8 \times 16°$) complex receptive field to an edge projected on the ipsilateral retina so as to cross the receptive field in various directions. (The screen is illuminated by a diffuse background light, at 0·0 $\log_{10}$ cd/m^2. At the time of stimulus, shown by upper line of each record, half the screen, to one side of the variable boundary, is illuminated at 1·0 $\log_{10}$ cd/m^2, while the other half is kept constant.) *A*, vertical edge with light area to left, darker area to right. *B–H*, various other orientations of edge. Position of receptive field 20° below and to the left of the area centralis. Responses from ipsilateral eye stronger than those from contralateral eye (group 5, see Part II). Time 1 sec.

As shown in Text-fig. 6, the responses of the cell to a given vertical edge were consistent in type, being either 'on' or 'off' for all positions of the edge within the receptive field. In being uniform in its response-type it resembled the cell of Text-fig. 4. A few other cells of the same general category showed a similar preference for edges, but lacked this uniformity. Their receptive fields resembled the field of Text-fig. 3, in that a given edge evoked responses of one type over half the field, and the opposite type over the other half. These fields were divided into two halves by a line parallel to the receptive-field axis: an edge oriented parallel to the axis gave 'on' responses throughout one of the halves and 'off' responses through the other. In either half, replacing the edge by its mirror image reversed the response-type. Even cells, which were uniform in their response-types, like those in Text-fig. 4–6, varied to some extent in the magnitude of their responses, depending on the position of the stimulus. Moreover, as with most cortical cells, there was some variation in responses to identical stimuli.

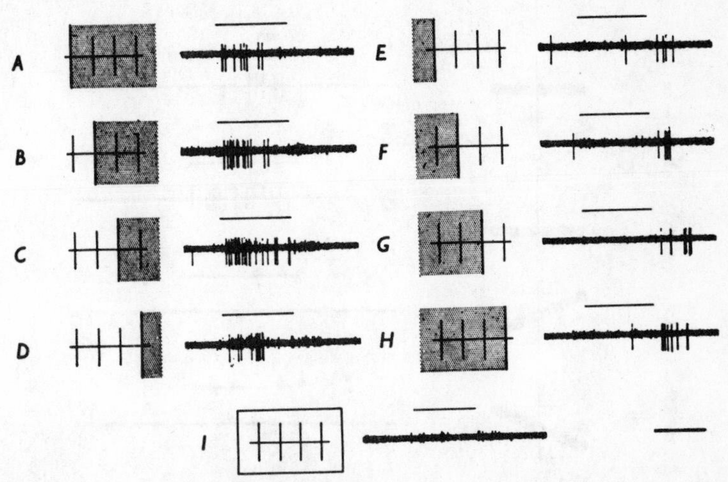

Text-fig. 6. Same cell as in Text-fig. 5. *A–H*, responses to a vertical edge in various parts of the receptive field: *A–D*, brighter light to the left; *E–H*, brighter light to the right; *I*, large rectangle, 10 × 20°, covering entire receptive field. Time, 1 sec.

A final example is given to illustrate the wide range of variation in the organization of complex receptive fields. The cell of Text-figs. 7 and 8 was not strongly influenced by any form projected upon the screen; it gave only weak, unsustained 'on' responses to a dark horizontal rectangle against a light background, and to other forms it was unresponsive. A strong discharge was evoked, however, if a black rectangular object (for example, a piece of black tape) was placed against the brightly illuminated screen. The receptive field of the cell was about 5 × 5°, and the most effective stimulus width was about $\frac{1}{3}$°. Vigorous firing was obtained regardless of the position of the rectangle, as long as it was horizontal and within the receptive field. If it was tipped more than 10° in either direction no discharge was evoked (Text-fig. 7*D*, *E*). We have recorded several complex fields which resembled this one in that they responded best to black rectangles against a bright background. Presumably it is important to have good contrast between the narrow black rectangle and the background; this is technically difficult with a projector because of scattered light.

Slow downward movement of the dark rectangle evoked a strong discharge throughout the entire 5° of the receptive field (Text-fig. 8*A*). If the movement was halted the cell continued to fire, but less vigorously.

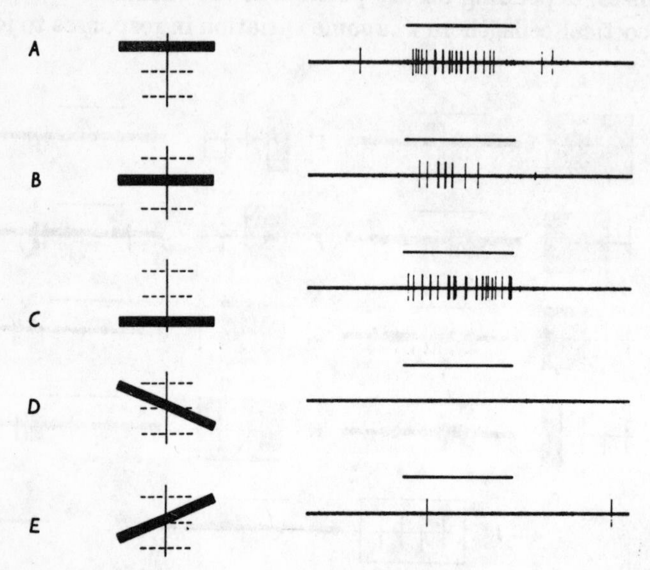

Text-fig. 7. Cell activated only by left (contralateral) eye over a field approximately $5 \times 5°$, situated 10° above and to the left of the area centralis. The cell responded best to a black horizontal rectangle, $\frac{1}{3} \times 6°$, placed anywhere in the receptive field (*A–C*). Tilting the stimulus rendered it ineffective (*D–E*). The black bar was introduced against a light background during periods of 1 sec, indicated by the upper line in each record. Luminance of white background, 1·0 $\log_{10}$ cd/m^2; luminance of black part, 0·0 $\log_{10}$ cd/m^2. A lesion, made while recording from the cell, was found in layer 2 of apical segment of post-lateral gyrus.

Upward movement gave only weak, inconsistent responses, and left–right movement (Text-fig. 8*B*) gave no responses. Discharges of highest frequency were evoked by relatively slow rates of downward movement (about 5–10 sec to cross the entire field); rapid movement in either direction gave only very weak responses.

Despite its unusual features this cell exhibited several properties typical of complex units, particularly the lack of summation (except in a horizontal sense), and the wide area over which the dark bar was effective. One may think of the field as having a counterpart in simple fields of type *D*, Text-fig. 2. In such fields a dark bar would evoke discharges, but only if it fell within the inhibitory region. Moreover, downward movement of the bar would also evoke brisker discharges than upward, provided the upper flanking region were stronger than the lower one.

In describing simple fields it has already been noted that moving stimuli were often more effective than stationary ones. This was also true of cells with complex fields. Depending on the cell, slits, edges, or dark bars were most effective. As with simple fields, orientation of a stimulus was always critical, responses varied with rate of movement, and directional asymmetrics of the type seen in Text-fig. 8 were common. Only once have we seen activation of a cell for one direction of movement and suppression of

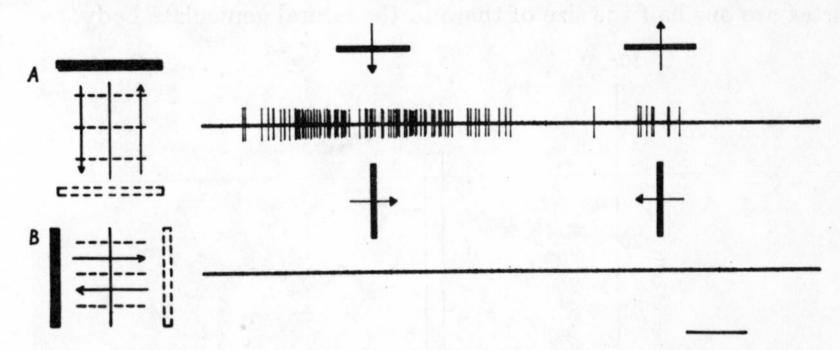

Text-fig. 8. Same cell as in Text-fig. 7. Movement of black rectangle $\frac{1}{3} \times 6°$ back and forth across the receptive field: *A*, horizontally oriented (parallel to receptive-field axis); *B*, vertically oriented. Time required to move across the field, 5 sec. Time, 1 sec.

maintained firing for the opposite direction. In their responses to movement, cells with complex fields differed from their simple counterparts chiefly in responding with sustained firing over substantial regions, usually the entire receptive field, instead of over a very narrow boundary separating excitatory and inhibitory regions.

Receptive-field dimensions

Over-all field dimensions were measured for 119 cells. A cell was included only if its field was mapped completely, and if it was situated in the area of central vision (see p. 135). Fields varied greatly in size from one cell to the next, even for cells recorded in a single penetration (see Text-fig. 15). In Text-fig. 9 the distribution of cells according to field area is given separately for simple and complex fields. The histogram illustrates the variation in size, and shows that on the average complex fields were larger than simple ones.

Widths of the narrow subdivisions of simple fields (the centres of types *C*, *D* and *E* or the flanks of type *F*, Text-fig. 2) also varied greatly: the smallest were 10–15 minutes of arc, which is roughly the diameter of the smallest field centres we have found for geniculate cells. For some cells with complex fields the widths of the most effective slits or dark bars were also of this order, indicating that despite the greater overall field size these cells were able to convey detailed information. We wish to emphasize that in both geniculate and cortex the field dimensions tend to increase with distance from the area centralis, and that they differ even for a given location in the retina. It is consequently not possible to compare field sizes in the geniculate and cortex unless these variations are taken into account. This may explain the discrepancy between our results and the findings of Baumgartner (see Jung, 1960), that 'field centres' in the cortex are one half the size of those in the lateral geniculate body.

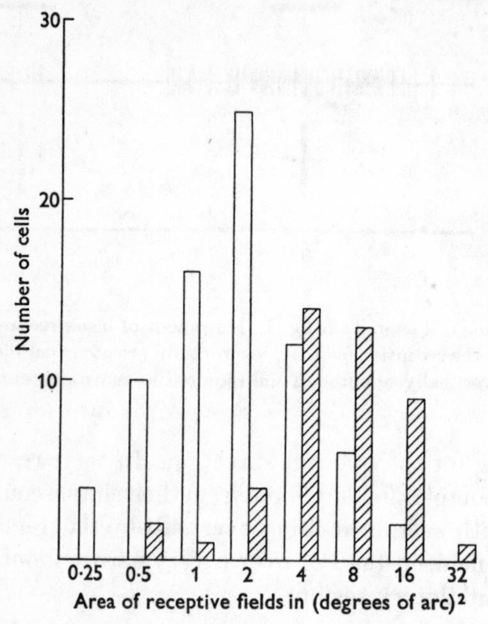

Text-fig. 9. Distribution of 119 cells in the visual cortex with respect to the approximate area of their receptive fields. White columns indicate cells with simple receptive fields; shaded columns, cells with complex fields. Abscissa: area of receptive fields. Ordinate: number of cells.

Responsiveness of cortical cells

Simple and complex fields together account for all of the cells we have recorded in the visual cortex. We have not observed cells with concentric fields. Except for clearly injured cells (showing extreme spike deformation or prolonged high-frequency bursts of impulses) all units have responded to visual stimulation, though it has occasionally taken several hours to find the retinal region containing the receptive field and to work out the optimum stimuli. Some cells responded only to stimuli which were optimum in their retinal position and in their form, orientation and rate of movement. A few even required stimulation of both eyes before a response could be elicited (see Part II). But there is no indication from our studies that the striate cortex contains nerve cells that are unresponsive to visual stimuli.

Most of the cells of this series were observed for 1 or 2 hr, and some were studied for up to 9 hr. Over these periods of time there were no qualitative changes in the characteristics of receptive fields: their complexity, arrangements of excitatory and inhibitory areas, axis orientation and position all remained the same, as did the ocular dominance. With deepening anaesthesia a cell became less responsive, so that stimuli that had formerly been weak tended to become even weaker or ineffective, while those that had evoked brisk responses now evoked only weak ones. The last thing to disappear with very deep anaesthesia was usually the response to a moving form. As long as any responses remained the cell retained the same specific requirements as to stimulus form, orientation and rate of movement, suggesting that however the drug exerted its effects, it did not to any important extent functionally disrupt the specific visual connexions. A comparison of visual responses in the anaesthetized animal with those in the unanaesthetized, unrestrained preparation (Hubel, 1959) shows that the main differences lie in the frequency and firing patterns of the maintained activity and in the vigour of responses, rather than in the basic receptive-field organization. It should be emphasized, however, that even in light anaesthesia or in the attentive state diffuse light remains relatively ineffective; thus the balance between excitatory and inhibitory influences is apparently maintained in the waking state.

PART II

BINOCULAR INTERACTION AND OCULAR DOMINANCE

Recording from single cells at various levels in the visual system offers a direct means of determining the site of convergence of impulses from the two eyes. In the lateral geniculate body, the first point at which convergence is at all likely, binocularly influenced cells have been observed, but it would seem that these constitute at most a small minority of the total population of geniculate cells (Erulkar & Fillenz, 1958, 1960; Bishop, Burke & Davis, 1959; Grüsser & Sauer, 1960; Hubel & Wiesel, 1961). Silver-degeneration studies show that in each layer of the geniculate the terminals of fibres from a single eye are grouped together, with only minor overlap in the interlaminar regions (Silva, 1956; Hayhow, 1958). The anatomical and physiological findings are thus in good agreement.

It has long been recognized that the greater part of the cat's primary visual cortex receives projections from the two eyes. The anatomical evidence rests largely on the observation that cells in all three lateral geniculate layers degenerate following a localized lesion in the striate area (Minkowski, 1913). Physiological confirmation was obtained by Talbot & Marshall (1941) who stimulated the visual fields of the separate eyes with small spots of light, and mapped the evoked cortical slow waves. Still unsettled, however, was the question of whether individual cortical cells receive projections from both eyes, or whether the cortex contains a mixture of cells, some activated by one eye, some by the other. We have recently shown that many cells in the visual cortex can be influenced by both eyes (Hubel & Wiesel, 1959). The present section contains further observations on binocular interaction. We have been particularly interested in learning whether the eyes work in synergy or in opposition, how the relative influence of the two eyes varies from cell to cell, and whether, on the average, one eye exerts more influence than the other on the cells of a given hemisphere.

RESULTS

In agreement with previous findings (Hubel & Wiesel, 1959) the receptive fields of all binocularly influenced cortical cells occupied corresponding positions on the two retinas, and were strikingly similar in their organization. For simple fields the spatial arrangements of excitatory and inhibitory regions were the same; for complex fields the stimuli that excited or inhibited the cell through one eye had similar effects through the other. Axis orientations of the two receptive fields were the same. Indeed, the only differences ever seen between the two fields were related to eye dominance: identical stimuli to the two eyes did not necessarily evoke equally strong responses from a given cell. For some cells the responses were equal or almost so; for others one eye tended to dominate. Whenever the two retinas were stimulated in identical fashion in corresponding regions, their effects summed, i.e. they worked in synergy. On the other hand, if antagonistic regions in the two eyes were stimulated so that one eye had an excitatory effect and the other an inhibitory one, then the responses tended to cancel (Hubel & Wiesel, 1959, Fig. 10*A*).

Some units did not respond to stimulation of either eye alone but could be activated only by simultaneous stimulation of the two eyes. Text-figure 10 shows an example of this, and also illustrates ordinary binocular synergy. Two simultaneously recorded cells both responded best to transverse movement of a rectangle oriented in a 1 o'clock–7 o'clock direction (Text-fig. 10*A*, *B*). For one of the cells movement down and to the right was more effective than movement up and to the left. Responses from the individual eyes were roughly equal. On simultaneous stimulation of the two eyes both units responded much more vigorously. Now a third cell was also activated.

The threshold of this third unit was apparently so high that, at least under these experimental conditions, stimulation of either eye alone failed to evoke any response.

A second example of synergy is seen in Text-fig. 11. The most effective stimulus was a vertically oriented rectangle moved across the receptive field from left to right. Here the use of both eyes not only enhanced the response already observed with a single eye, but brought into the open a tendency that was formerly unsuspected. Each eye mediated a weak

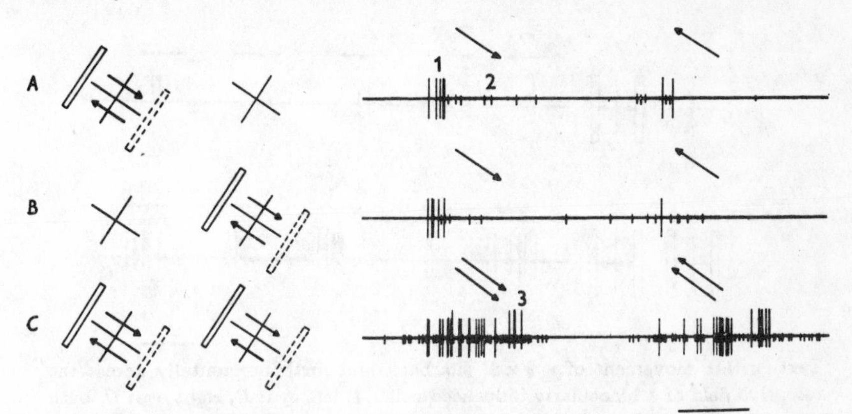

Text-fig. 10. Examples of binocular synergy in a simultaneous recording of three cells (spikes of the three cells are labelled 1–3). Each of the cells had receptive fields in the two eyes; in each eye the three fields overlapped and were situated 2° below and to the left of the area centralis. The crosses to the left of each record indicate the positions of the receptive fields in the two eyes. The stimulus was $\frac{1}{8} \times 2°$ slit oriented obliquely and moved slowly across the receptive fields as shown; *A*, in the left eye; *B*, in the right eye; *C*, in the two eyes simultaneously. Since the responses in the two eyes were about equally strong, these two cells were classed in ocular-dominance group 4 (see Text-fig. 12). Time, 1 sec.

response (Text-fig. 11*A*, *B*) which was greatly strengthened when both eyes were used in parallel (*C*). Now, in addition, the cell gave a weak response to leftward movement, indicating that this had an excitatory effect rather than an inhibitory one. Binocular synergy was often a useful means of bringing out additional information about a receptive field.

In our previous study of forty-five cortical cells (Hubel & Wiesel, 1959) there was clear evidence of convergence of influences from the two eyes in only one fifth of the cells. In the present series 84 % of the cells fell into this category. The difference is undoubtedly related to the improved precision in technique of binocular stimulation. A field was first mapped in the dominant eye and the most effective type of stimulus determined. That stimulus was then applied in the corresponding region in the other eye. Finally, even if no response was obtained from the non-dominant eye, the two eyes were stimulated together in parallel to see if their effects were synergistic. With these methods, an influence was frequently observed from the non-dominant eye that might otherwise have been overlooked.

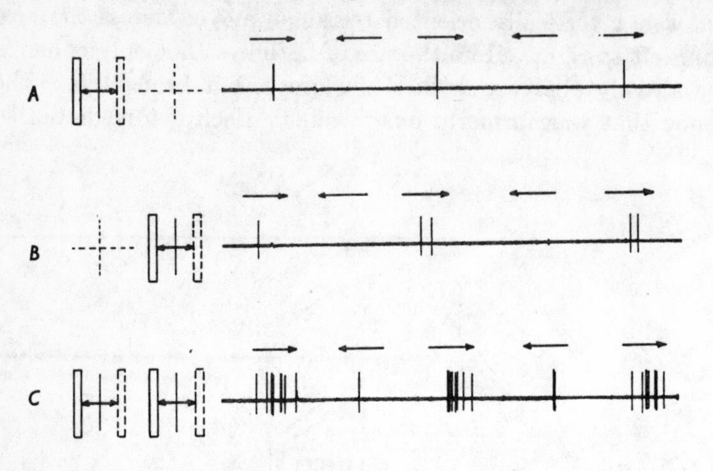

Text-fig. 11. Movement of a $\frac{1}{4} \times 2^\circ$ slit back and forth horizontally across the receptive field of a binocularly influenced cell. *A*, left eye; *B*, right eye; *C*, both eyes. The cell clearly preferred left-to-right movement, but when both eyes were stimulated together it responded also to the reverse direction. Field diameter, 2°, situated 5° from the area centralis. Time, 1 sec.

A comparison of the influence of the two eyes was made for 223 of the 303 cells in the present series. The remaining cells were either not sufficiently studied, or they belonged to the small group of cells which were only activated if both eyes were simultaneously stimulated. The fields of all cells were in or near the area centralis. The 223 cells were subdivided into seven groups, as follows:

Group	Ocular dominance
1	Exclusively contralateral
2*	Contralateral eye much more effective than ipsilateral eye
3	Contralateral eye slightly more effective than ipsilateral
4	No obvious difference in the effects exerted by the two eyes
5	Ipsilateral eye slightly more effective
6*	Ipsilateral eye much more effective
7	Exclusively ipsilateral

* These groups include cells in which the non-dominant eye, ineffective by itself, could influence the response to stimulation of the dominant eye.

A histogram showing the distribution of cells among these seven groups is given in Text-fig. 12. Assignment of a unit to a particular group was to some extent arbitrary, but it is unlikely that many cells were misplaced by more than one group. Perhaps the most interesting feature of the histogram is its lack of symmetry: many more cells were dominated by the contralateral than by the ipsilateral eye (106 vs. 62). We conclude that in the part of the cat's striate cortex representing central vision the great majority of cells are influenced by both eyes, and that despite wide variation in relative ocular dominance from one cell to the next, the contralateral eye is, on the average, more influential. As the shaded portion

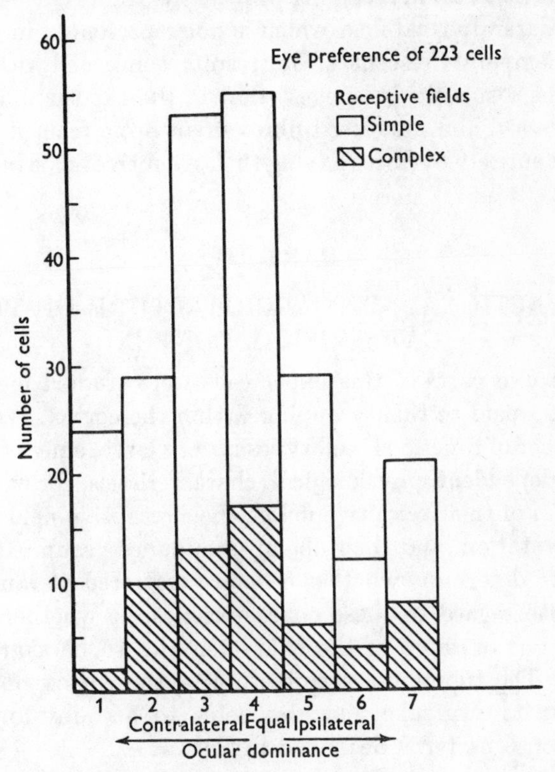

Text-fig. 12. Distribution of 223 cells recorded from the visual cortex, according to ocular dominance. Histogram includes cells with simple fields and cells with complex fields. The shaded region shows the distribution of cells with complex receptive fields. Cells of group 1 were driven only by the contralateral eye; for cells of group 2 there was marked dominance of the contralateral eye, for group 3, slight dominance. For cells in group 4 there was no obvious difference between the two eyes. In group 5 the ipsilateral eye dominated slightly, in group 6, markedly; and in group 7 the cells were driven only by the ipsilateral eye.

of Text-fig. 12 shows, there is no indication that the distribution among the various dominance groups of cells having complex receptive fields differs from the distribution of the population as a whole.

A cortical bias in favour of the contralateral eye may perhaps be related to the preponderance of crossed over uncrossed fibres in the cat's optic tract (Polyak, 1957, p. 788). The numerical inequality between crossed and uncrossed tract fibres is generally thought to be related to an inequality in size of the nasal and temporal half-fields, since both inequalities are most marked in lower mammals with laterally placed eyes, and become progressively less important in higher mammals, primates and man. Thompson *et al.* (1950) showed that in the rabbit, for example, there is a substantial cortical region receiving projections from that part of the peripheral contralateral visual field which is not represented in the ipsilateral retina (the 'Temporal Crescent'). Our results, concerned with more central portions of the visual fields, suggest that in the cat the difference in the number of crossed and uncrossed fibres in an optic tract is probably not accounted for entirely by fibres having their receptive fields in the temporal-field crescents.

PART III

FUNCTIONAL CYTOARCHITECTURE OF THE CAT'S VISUAL CORTEX

In the first two parts of this paper cells were studied individually, no attention being paid to their grouping within the cortex. We have shown that the number of functional cell types is very large, since cells may differ in several independent physiological characteristics, for example, in the retinal position of their receptive fields, their receptive-field organization, their axis orientation, and their ocular-dominance group. In this section we shall try to determine whether cells are scattered at random through the cortex with regard to these characteristics, or whether there is any tendency for one or more of the characteristics to be shared by neighbouring cells. The functional architecture of the cortex not only seems interesting for its own sake, but also helps to account for the various complex response patterns described in Part I.

RESULTS

Functional architecture of the cortex was studied by three methods. These had different merits and limitations, and were to some extent complementary.

(1) *Cells recorded in sequence.* The most useful and convenient procedure was to gather as much information as possible about each of a long succession of cells encountered in a micro-electrode penetration through the cortex, and to reconstruct the electrode track from serial histological sections. One could then determine how a physiological characteristic (such as receptive-field position, organization, axis orientation or ocular dominance) varied with cortical location. The success of this method in

delineating regions of constant physiological characteristics depends on the possibility of examining a number of units as the electrode passes through each region. Regions may escape detection if they are so small that the electrode is able to resolve only one or two cells in each. The fewer the cells resolved, the larger the regions must be in order to be detected at all.

(2) *Unresolved background activity.* To some extent the spaces between isolated units were bridged by studying unresolved background activity audible over the monitor as a crackling noise, and assumed to originate largely from action potentials of a number of cells. It was concluded that cells, rather than fibres, gave rise to this activity, since it ceased abruptly when the electrode left the grey matter and entered subcortical white matter. Furthermore, diffuse light evoked no change in activity, compared to the marked increase caused by an optimally oriented slit. This suggested that terminal arborizations of afferent fibres contributed little to the background, since most geniculate cells respond actively to diffuse light (Hubel, 1960). In most penetrations unresolved background activity was present continuously as the electrode passed through layers 2–6 of the cortical grey matter.

Background activity had many uses. It indicated when the cells within range of the electrode tip had a common receptive-field axis orientation. Similarly, one could use it to tell whether the cells in the neighbourhood were driven exclusively by one eye (group 1 or group 7). When the background activity was influenced by both eyes, one could not distinguish between a mixture of cells belonging to the two monocular groups (1 and 7) and a population in which each cell was driven from both eyes. But even here one could at least assess the relative influence of the two eyes upon the group of cells in the immediate neighbourhood of the electrode.

(3) *Multiple recordings.* In the series of 303 cells, 78 were recorded in groups of two and 12 in groups of three. Records were not regarded as multiple unless the spikes of the different cells showed distinct differences in amplitude, and unless each unit fulfilled the criteria required of a single-unit record, namely that the amplitude and wave shape be relatively constant for a given electrode position.

In such multiple recordings one could be confident that the cells were close neighbours and that uniform stimulus conditions prevailed, since the cells could be stimulated and observed together. One thus avoided some of the difficulties in evaluating a succession of recordings made over a long period of time span, where absolute constancy of eye position, anaesthetic level, and preparation condition were sometimes hard to guarantee.

Regional variations of several physiological characteristics were examined by the three methods just outlined. Of particular interest for the present study were the receptive-field axis orientation, position of receptive fields on the retina, receptive-field organization, and relative ocular dominance. These will be described separately in the following paragraphs.

Orientation of receptive-field axis

The orientation of a receptive-field axis was determined in several ways. When the field was simple the borders between excitatory and inhibitory regions were sufficient to establish the axis directly. For both simple and complex fields the axis could always be determined from the orientation of the most effective stimulus. For most fields, when the slit or edge was placed at right angles to the optimum position there was no response. The receptive-field axis orientation was checked by varying the stimulus orientation from this null position in order to find the two orientations at which a response was only just elicited, and by bisecting the angle between them. By one or other of these procedures the receptive-field orientation could usually be determined to within 5 or 10°.

One of the first indications that the orientation of a receptive-field axis was an important variable came from multiple recordings. Invariably the axes of receptive fields mapped together had the same orientations. An example of a 3-unit recording has already been given in Text-fig. 10. Cells with common axis orientation were therefore not scattered at random through the cortex, but tended to be grouped together. The size and shape of the regions containing these cell groups were investigated by comparing the fields of cells mapped in sequence. It was at once apparent that successively recorded cells also tended to have identical axis orientations and that each penetration consisted of several sequences of cells, each sequence having a common axis orientation. Any undifferentiated units in the background responded best to the stimulus orientation that was most effective in activating the cell under study. After traversing a distance that varied greatly from one penetration to the next, the electrode would enter an area where there was no longer any single optimum orientation for driving background activity. A very slight advance of the electrode would bring it into a region where a new orientation was the most effective, and the succeeding cells would all have receptive fields with that orientation. The change in angle from one region to another was unpredictable; sometimes it was barely detectable, at other times large (45–90°).

Text-figure 13 shows a camera lucida tracing of a frontal section through the post-lateral gyrus. The electrode track entered normal to the surface, passed through the apical segment in a direction parallel to the fibre bundles, then through the white matter beneath, and finally obliquely through half the thickness of the mesial segment. A lesion was made at the termination of the penetration. A composite photomicrograph (Pl. 1) shows the lesion

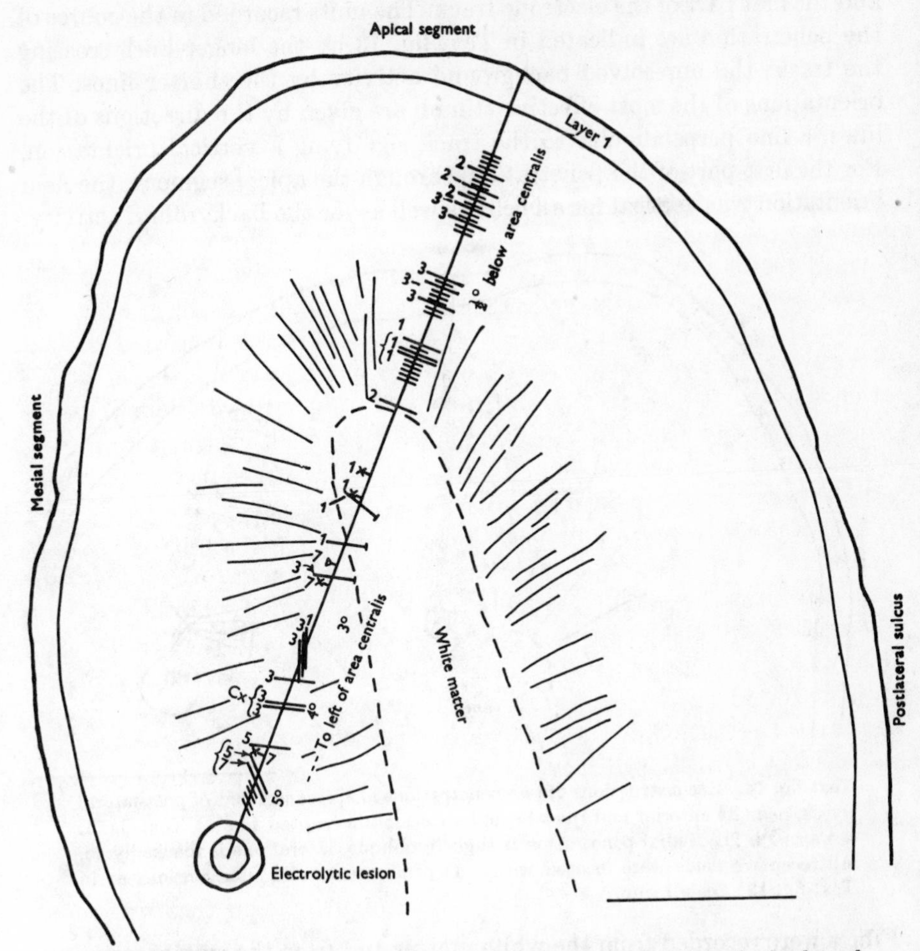

Text-fig. 13. Reconstruction of micro-electrode penetration through the lateral gyrus (see also Pl. 1). Electrode entered apical segment normal to the surface, and remained parallel to the deep fibre bundles (indicated by radial lines) until reaching white matter; in grey matter of mesial segment the electrode's course was oblique. Longer lines represent cortical cells. Axons of cortical cells are indicated by a cross-bar at right-hand end of line. Field-axis orientation is shown by the direction of each line; lines perpendicular to track represent vertical orientation. Brace-brackets show simultaneously recorded units. Complex receptive fields are indicated by 'Cx'. Afferent fibres from the lateral geniculate body indicated by ×, for 'on' centre; Δ, for 'off' centre. Approximate positions of receptive fields on the retina are shown to the right of the penetration. Shorter lines show regions in which unresolved background activity was observed. Numbers to the left of the penetration refer to ocular-dominance group (see Part II). Scale 1 mm.

and the first part of the electrode track. The units recorded in the course of the penetration are indicated in Text-fig. 13 by the longer lines crossing the track; the unresolved background activity by the shorter lines. The orientations of the most effective stimuli are given by the directions of the lines, a line perpendicular to the track signifying a vertical orientation. For the first part of the penetration, through the apical segment, the field orientation was vertical for all cells as well as for the background activity.

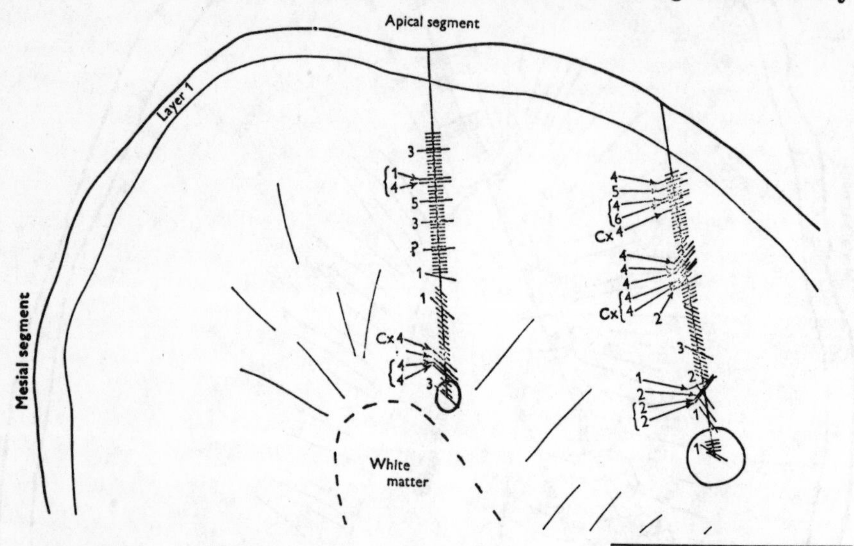

Text-fig. 14. Reconstructions of two penetrations in apical segment of post-lateral gyrus, near its anterior end (just behind anterior interrupted line in Text-fig. 1, see also Pl. 2). Medial penetration is slightly oblique, lateral one is markedly so. All receptive fields were located within 1° of area centralis. Conventions as in Text-fig. 13. Scale 1 mm.

Fibres were recorded from the white matter and from the grey matter just beyond it. Three of these fibres were axons of cortical cells having fields of various oblique orientations; four were afferent fibres from the lateral geniculate body. In the mesial segment three short sequences were encountered, each with a different common field orientation. These sequences together occupied a distance smaller than the full thickness of the apical segment.

In another experiment, illustrated in Text-fig. 14 and in Pl. 2, two penetrations were made, both in the apical segment of the post-lateral gyrus. The medial penetration (at left in the figure) was at the outset almost normal to the cortex, but deviated more and more from the direction of the deep fibre bundles. In this penetration there were three different axis orientations, of which the first and third persisted through long sequences.

In the lateral track there were nine orientations. From the beginning this track was more oblique, and it became increasingly so as it progressed.

As illustrated by the examples of Text-figs. 13 and 14, there was a marked tendency for shifts in orientation to increase in frequency as the angle between electrode and direction of fibre bundles (or apical dendrites) became greater. The extreme curvature of the lateral and post-lateral gyri in their apical segments made normal penetrations very difficult to obtain; nevertheless, four penetrations were normal or almost so. In none of these were there any shifts of axis orientation. On the other hand there were several shifts of field orientation in all oblique penetrations. As illustrated by Text-fig. 14, most penetrations that began nearly normal to the surface became more and more oblique with increasing depth. Here the distance traversed by the electrode without shifts in receptive-field orientation tended to become less and less as the penetration advanced.

It can be concluded that the striate cortex is divided into discrete regions within which the cells have a common receptive-field axis orientation. Some of the regions extend from the surface of the cortex to the white matter; it is difficult to be certain whether they all do. Some idea of their shapes may be obtained by measuring distances between shifts in receptive-field orientation. From these measurements it seems likely that the general shape is columnar, distorted no doubt by any curvature of the gyrus, which would tend to make the end at the surface broader than that at the white matter; deep in a sulcus the effect would be the reverse. The cross-sectional size and shape of the columns at the surface can be estimated only roughly. Most of our information concerns their width in the coronal plane, since it is in this plane that oblique penetrations were made. At the surface this width is probably of the order of 0·5 mm. We have very little information about the cross-sectional dimension in a direction parallel to the long axis of the gyrus. Preliminary mapping of the cortical surface suggests that the cross-sectional shape of the columns may be very irregular.

Position of receptive fields on the retina

Gross topography. That there is a systematic representation of the retina on the striate cortex of the cat was established anatomically by Minkowski (1913) and with physiological methods by Talbot & Marshall (1941). Although in the present study no attempt has been made to map topographically all parts of the striate cortex, the few penetrations made in cortical areas representing peripheral parts of the retina confirm these findings. Cells recorded in front of the anterior interrupted lines of Text-fig. 1 had receptive fields in the superior retinas; those in the one penetration behind the posterior line had fields that were well below the horizontal meridian of the retina. (No recordings were made from cortical regions receiving projections from the deeply pigmented non-tapetal part of the inferior retinas.) In several penetrations extending far down the mesial (interhemispheric) segment of the lateral gyrus, receptive fields moved further and further out into the ipsilateral half of each retina as the electrode advanced (Text-fig. 13). In these penetrations the movement of fields into the retinal periphery occurred more and more rapidly as the electrode advanced. In three penetrations extending far down the lateral segment of the post-lateral gyrus (medial bank of the post-lateral sulcus) there was likewise a clear progressive shift of receptive-field positions as the electrode advanced. Here also the movement was along the horizontal meridian, again into the *ipsilateral* halves of both retinas. This therefore confirms the findings of Talbot & Marshall (1941) and Talbot (1942), that in each hemisphere there is a second laterally placed representation of the contralateral half-field of vision. The subject of Visual Area II will not be dealt with further in this paper.

Cells within the large cortical region lying between the interrupted lines of Text-fig. 1, and extending over on to the mesial segment and into the lateral sulcus for a distance of 2–3 mm, had their receptive fields in the area of central vision. By this we mean the area centralis, which is about 5° in diameter, and a region surrounding it by about 2–3°. The receptive fields of the great majority of cells were confined to the ipsilateral halves of the two retinas. Often a receptive field covering several degrees on the retina stopped short in the area centralis right at the vertical meridian. Only rarely did a receptive field appear to spill over into the contralateral half-retina; when it did, it was only by 2–3°, a distance comparable to the possible error in determining the area centralis in some cats.

Because of the large cortical representation of the area centralis, one would expect only a very slow change in receptive-field position as the electrode advanced obliquely (Text-fig. 13). Indeed, in penetrations through the apex of the post-lateral gyrus and extending 1–2 mm down either bank there was usually no detectable progressive displacement of receptive fields. In penetrations made 1–3 mm apart, either along a parasagittal line or in the same coronal plane (Text-fig. 14) receptive fields again had almost identical retinal positions.

Retinal representation of neighbouring cells. A question of some interest was to determine whether this detailed topographic representation of the retina held right down to the cellular level. From the results just described one might imagine that receptive fields of neighbouring cortical cells should have very nearly the same retinal position. In a sequence of cells recorded in a normal penetration through the cortex the receptive fields should be superimposed, and for oblique penetrations any detectable changes in field positions should be systematic. In the following paragraphs we shall consider the relative retinal positions of the receptive fields of neighbouring cells, especially cells within a column.

In all multiple recordings the receptive fields of cells observed simultaneously were situated in the same general region of the retina. As a rule the fields overlapped, but it was unusual for them to be precisely superimposed. For example, fields were often staggered along a line perpendicular to their axes. Similarly, the successive receptive fields observed during a long cortical penetration varied somewhat in position, often in an apparently random manner. Text-figure 15 illustrates a sequence of twelve cells recorded in the early part of a penetration through the cortex. One lesion was made while observing the first cell in the sequence and another at the end of the penetration; they are indicated in the drawing of cortex to the right of the figure. In the centre of the figure the position of each receptive field is shown relative to the area centralis (marked with a cross); each field was several degrees below and to the left of the area centralis. It will be seen that all fields in the sequence except the last had the same axis orientation; the first eleven cells therefore occupied the same column. All but the first three and the last (cell 12) were simple in arrangement. Cells 5 and 6 were recorded together, as were 8 and 9.

In the left-hand part of the figure the approximate boundaries of all these receptive fields are shown superimposed, in order to indicate the degree of overlap. From cell to cell there is no obvious systematic change in receptive-field position. The variation in position is about equal to the area occupied by the largest fields of the sequence. This variation is undoubtedly real, and not an artifact produced by eye movements occurring between recordings of successive cells. The stability of the eyes was checked while studying each cell, and any tendency to eye movements would have easily been detected by an apparent movement of the receptive field under observation. Furthermore, the field positions of simultaneously recorded cells 5 and 6, and also of cells 8 and 9, are clearly different; here the question of eye movements is not pertinent.

Text-figure 15 illustrates a consistent and somewhat surprising finding, that within a column defined by common field-axis orientation there was no apparent progression in field positions along the retina as the electrode advanced. This was so even though the electrode often crossed through the column obliquely, entering one side and leaving the other. If there was any detailed topographical representation within columns it was obscured by the superimposed, apparently random staggering of field positions. We conclude that at this microscopic level the retinotopic representation no longer strictly holds.

Receptive-field organization

Multiple recordings. The receptive fields of cells observed together in multiple recordings were always of similar complexity, i.e. they were either all simple or all complex in their organization. In about one third of the multiple recordings the cells had the same detailed field organization; if simple, they had similar distributions of excitatory and inhibitory

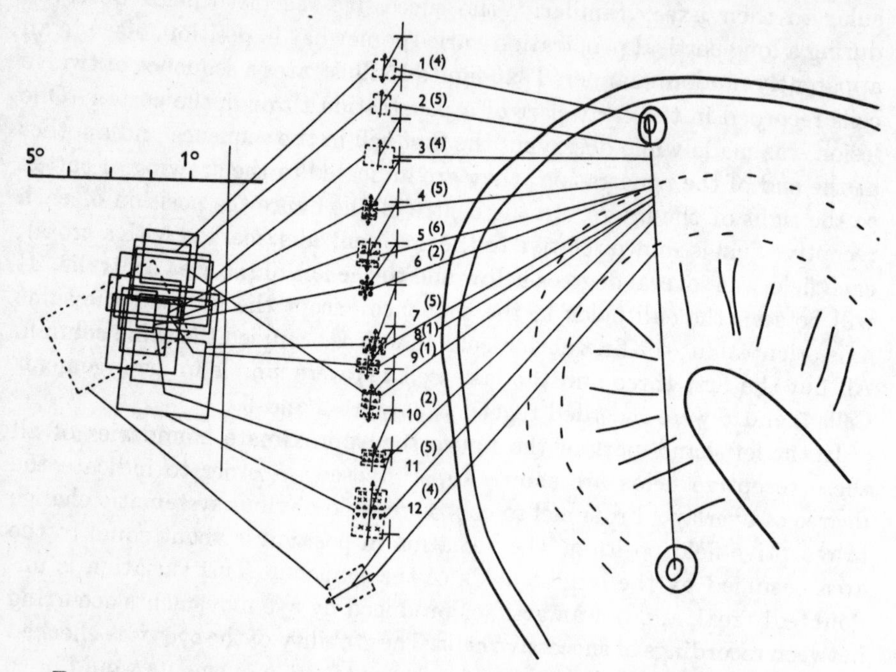

Text-fig. 15. Reconstruction of part of an electrode track through apical and mesial segments of post-lateral gyrus near its anterior end. Two lesions were made, the first after recording from the first unit, the second at the end of the penetration. Only the first twelve cells are represented. Interrupted lines show boundaries of layer 4.

In the centre part of the figure the position of each receptive field, outlined with interrupted lines, is given with respect to the area centralis, shown by a cross. Cells are numbered in sequence, 1–12. Numbers in parentheses refer to ocular-dominance group (see Part II). Units 5 and 6, 8 and 9 were observed simultaneously. The first three fields and the last were complex in organization; the remainder were simple. ×, areas giving excitation; △, areas giving inhibitory effects. Note that all receptive fields except the last have the same axis orientation (9.30–3.30 o'clock). The arrows show the preferred direction of movement of a slit oriented parallel to the receptive-field axis.

In the left part of the figure all of the receptive fields are superimposed, to indicate the overlap and variation in size. The vertical and horizontal lines represent meridia, crossing at the area centralis. Scale on horizontal meridian, 1° for each subdivision.

areas; if complex, they required identical stimuli for their activation. As a rule these fields did not have exactly the same retinal position, but were staggered as described above. In two thirds of the multiple recordings the cells differed to varying degrees in their receptive field arrangements. Two types of multiple recordings in which field arrangements differed seem interesting enough to merit a separate description.

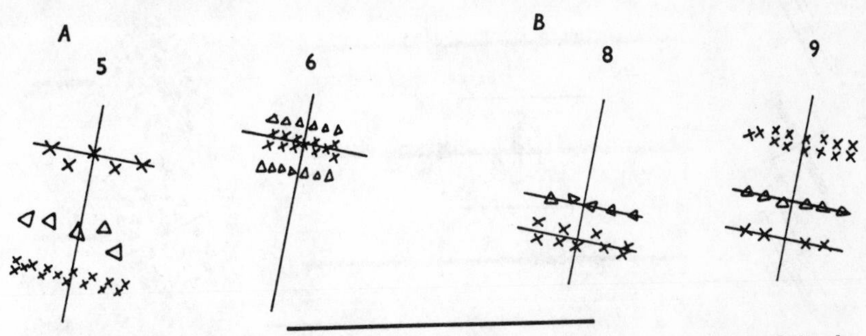

Text-fig. 16. Detailed arrangements of the receptive fields of two pairs of simultaneously recorded cells (nos. 5 and 6, and 8 and 9, of Text-fig. 15). The crosses of diagrams 5 and 6 are superimposed as are the double crosses of 8 and 9. Note that the upper excitatory region of 5 is superimposed upon the excitatory region of 6; and that both regions of 8 are superimposed on the inhibitory and lower excitatory regions of 9. Scale, 1°.

In several multiple recordings the receptive fields overlapped in such a way that one or more excitatory or inhibitory portions were superimposed. Two examples are supplied by cell-pairs 5 and 6, and 8 and 9 of Text-fig. 15. Their fields are redrawn in Text-fig. 16. The fields of cells 5 and 6 are drawn separately (Text-fig. 16*A*) but they actually overlapped so that the reference lines are to be imagined as superimposed. Thus the 'on' centre of cell 6 fell directly over the upper 'on' flank of 5 and the two cells tended to fire together to suitably placed stimuli. A similar situation existed for cells 8 and 9 (Text-fig. 16*B*). The field of 9 was placed so that its 'off' region and the lower, weaker 'on' region were superimposed on the two regions of 8. Again the two cells tended to fire together. Such examples suggest that neighbouring cells may have some of their inputs in common.

Cells responded reciprocally to a light stimulus in eight of the forty-three multiple recordings. An example of two cells responding reciprocally to stationary spots is shown in Text-fig. 17. In each eye the two receptive fields were almost superimposed. The fields consisted of elongated obliquely oriented central regions, inhibitory for one cell, excitatory for the other, flanked on either side by regions of the opposite type. Instead of firing together in response to an optimally oriented stationary slit, like the cells in Text-fig. 16, these cells gave opposite-type responses, one inhibitory and the other excitatory. Some cell pairs responded reciprocally to to-and-fro movements of a slit or edge. Examples have been given elsewhere (Hubel, 1958, Fig. 9; 1959, Text-fig. 6). The fields of these cell pairs usually differed only in the balance of the asymmetrical flanking regions.

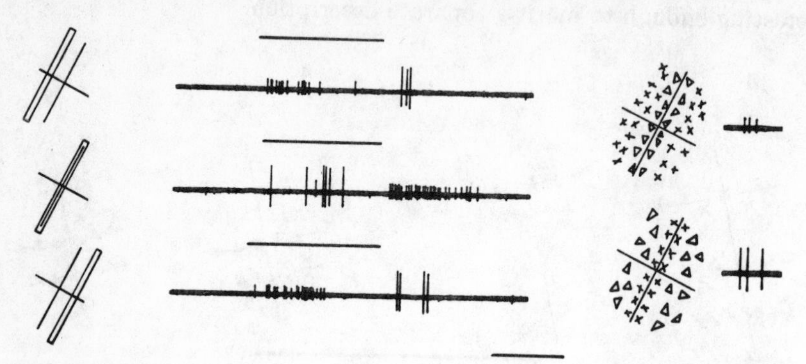

Text-fig. 17. Records of two simultaneously observed cells which responded reciprocally to stationary stimuli. The two receptive fields are shown to the right, and are superimposed, though they are drawn separately. The cell corresponding to each field is indicated by the spikes to the right of the diagram. To the left of each record is shown the position of a slit, $\frac{1}{4} \times 2\frac{1}{2}°$, with respect to these fields.

Both cells binocularly driven (dominance group 3); fields mapped in the left (contralateral) eye; position of fields 2° below and to the left of the area centralis. Time, 1 sec.

Relationship between receptive field organization and cortical layering. In a typical penetration through the cortex many different field types were found, some simple and others complex. Even within a single column both simple and complex fields were seen. (In Text-fig. 13 and 14 complex fields are indicated by the symbol 'Cx'; in Text-fig. 15, fields 1–3 were complex and 4–11 simple, all within a single column.) An attempt was made to learn whether there was any relationship between the different field types and the layers of the cortex. This was difficult for several reasons. In Nissl-stained sections the boundaries between layers of the cat's striate cortex are not nearly as clear as they are in the primate brain; frequently even the fourth layer, so characteristic of the striate cortex, is poorly demarcated. Consequently, a layer could not always be identified with certainty even for a cell whose position was directly marked by a lesion. For most cells the positions were arrived at indirectly, from depth readings and lesions made elsewhere in the penetrations: these determinations were subject to more errors than the direct ones. Moreover, few of the penetrations were made in a direction parallel to the layering, so that the distance an electrode travelled in passing through a layer was short, and the error in electrode position correspondingly more important.

The distribution of 179 cells among the different layers is given in the histograms of Text-fig. 18. All cells were recorded in penetrations in which at least one lesion was made; the shaded portions refer to cells which were individually marked with lesions. As shown in the separate histograms, simple-field cells as well as those with complex fields were widely distributed throughout the cortex. Cells with simple fields were most numerous in layers 3, 4 and 6. Especially interesting is the apparent rarity of complex fields in layer 4, where simple fields were so abundant. This is also illustrated in Text-fig. 15, which shows a sequence of eight cells

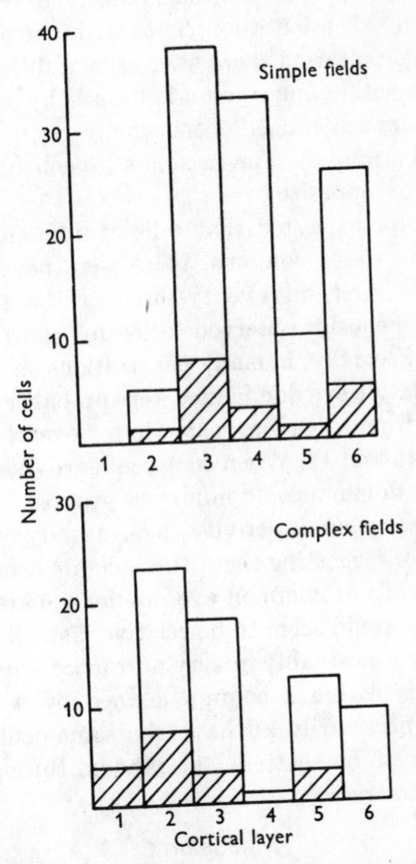

Text-fig. 18. Distribution of 179 cells, 113 with simple fields, 66 with complex, among the different cortical layers. All cells were recorded in penetrations in which at least one electrolytic lesion was made and identified; the shaded areas refer to cells marked individually by lesions. Note especially the marked difference in the occurrence, in layer 4, between simple and complex fields.

recorded from layer 4, all of which had simple fields. These findings suggest that cells may to some extent be segregated according to field complexity, and the rarity with which simple and complex fields were mapped together is consistent with this possibility.

Ocular dominance

In thirty-four multiple recordings the eye-dominance group (see Part II) was determined for both or all three cells. In eleven of these recordings there was a clear difference in ocular dominance between cells. Similarly, in a single penetration two cells recorded in sequence frequently differed in eye dominance. Cells from several different eye-dominance categories appeared not only in single penetrations, but also in sequences in which all cells had a common axis orientation. Thus within a single column defined by a common axis orientation there were cells of different eye dominance. A sequence of cells within one column is formed by cells 1–11 of Text-fig. 15. Here eye dominance ranged from wholly contralateral (group 1) to strongly ipsilateral (group 6). The two simultaneously recorded cells 5 and 6 were dominated by opposite eyes.

While these results suggested that cells of different ocular dominance were present within single columns, there were nevertheless indications of some grouping. First, in twenty-three of the thirty-four multiple recordings, simultaneously observed cells fell into the same ocular-dominance group. Secondly, in many penetrations short sequences of cells having the same relative eye dominance were probably more common than would be expected from a random scattering. Several short sequences are shown in Text-fig. 13 and 14. When such sequences consisted of cells with extreme unilateral dominance (dominance groups 1, 2, 6, and 7) the undifferentiated background activity was usually also driven predominantly by one eye, suggesting that other neighbouring units had similar eye preference. If cells of common eye dominance are in fact regionally grouped, the groups would seem to be relatively small. The cells could be arranged in nests, or conceivably in very narrow columns or thin layers.

In summary, cells within a column defined by a common field-axis orientation do not necessarily all have the same ocular dominance; yet neither do cells seem to be scattered at random through the cortex with respect to this characteristic.

DISCUSSION

A scheme for the elaboration of simple and complex receptive fields

Comparison of responses of cells in the lateral geniculate body with responses from striate cortex brings out profound differences in the receptive-field organization of cells in the two structures. For cortical cells, specifically oriented lines and borders tend to replace circular spots as the optimum stimuli, movement becomes an important parameter of stimulation, diffuse light becomes virtually ineffective, and with adequate stimuli most cells can be driven from the two eyes. Since lateral geniculate cells supply the main, and possibly the only, visual input to the striate cortex, these differences must be the result of integrative mechanisms within the striate cortex itself.

At present we have no direct evidence on how the cortex transforms the incoming visual information. Ideally, one should determine the properties of a cortical cell, and then examine one by one the receptive fields of all the afferents projecting upon that cell. In the lateral geniculate, where one can, in effect, record simultaneously from a cell and one of its afferents, a beginning has already been made in this direction (Hubel & Wiesel, 1961). In a structure as complex as the cortex the techniques available would seem hopelessly inadequate for such an approach. Here we must rely on less direct evidence to suggest possible mechanisms for explaining the transformations that we find.

The relative lack of complexity of simple cortical receptive fields suggests that these represent the first or at least a very early stage in the modification of geniculate signals. At any rate we have found no cells with receptive fields intermediate in type between geniculate and simple cortical fields. To account for the spatial arrangements of excitatory and inhibitory regions of simple cortical fields we may imagine that upon each simple-type cell there converge fibres of geniculate origin having 'on' or 'off' centres situated in the appropriate retinal regions. For example, a cortical cell with a receptive field of the type shown in Text-fig. 2*C* might receive projections from a group of lateral geniculate cells having 'on' field centres distributed throughout the long narrow central region designated in the figure by crosses. Such a projection system is shown in the diagram of Text-fig. 19. A slit of light falling on this elongated central region would activate all the geniculate cells, since for each cell the centre effect would strongly outweigh the inhibition from the segments of field periphery falling within the elongated region. This is the same as saying that a geniculate cell will respond to a slit with a width equal to the diameter of its field centre, a fact that we have repeatedly verified. The inhibitory flanks of the cortical field would be formed by the remaining outlying parts of the geniculate-field peripheries. These flanks might be reinforced and enlarged by appropriately placed 'off'-centre geniculate cells. Such an increase in the potency of the flanks would appear necessary to explain the relative indifference of cortical cells to diffuse light.

The arrangement suggested by Text-fig. 19 would be consistent with our impression that widths of cortical receptive-field centres (or flanks, in a field such as that of Text-fig. 2*F*) are of the same order of magnitude as the diameters of geniculate receptive-field centres, at least for fields in or near the area centralis. Hence the fineness of discrimination implied by the small size of geniculate receptive-field centres is not necessarily lost at the cortical level, despite the relatively large total size of many cortical fields; rather, it is incorporated into the detailed substructure of the cortical fields.

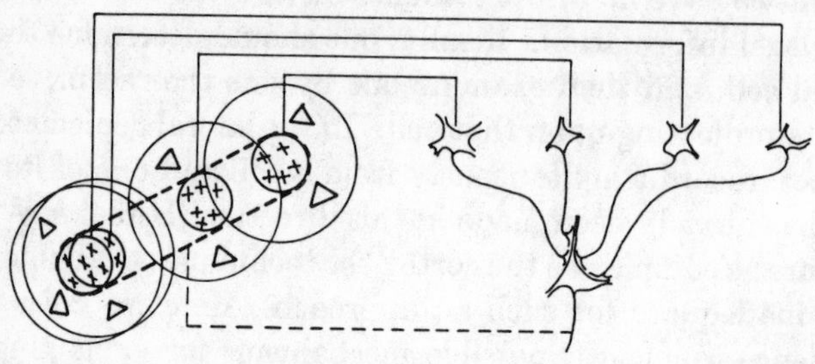

Text-fig. 19. Possible scheme for explaining the organization of simple receptive fields. A large number of lateral geniculate cells, of which four are illustrated in the upper right in the figure, have receptive fields with 'on' centres arranged along a straight line on the retina. All of these project upon a single cortical cell, and the synapses are supposed to be excitatory. The receptive field of the cortical cell will then have an elongated 'on' centre indicated by the interrupted lines in the receptive-field diagram to the left of the figure.

In a similar way, the simple fields of Text-fig. 2*D–G* may be constructed by supposing that the afferent 'on'- or 'off'-centre geniculate cells have their field centres appropriately placed. For example, field-type *G* could be formed by having geniculate afferents with 'off' centres situated in the region below and to the right of the boundary, and 'on' centres above and to the left. An asymmetry of flanking regions, as in field *E*, would be produced if the two flanks were unequally reinforced by 'on'-centre afferents.

The model of Text-fig. 19 is based on excitatory synapses. Here the suppression of firing on illuminating an inhibitory part of the receptive field is presumed to be the result of withdrawal of tonic excitation, i.e. the inhibition takes place at a lower level. That such mechanisms occur in the visual system is clear from studies of the lateral geniculate body, where an 'off'-centre cell is suppressed on illuminating its field centre because of suppression of firing in its main excitatory afferent (Hubel & Wiesel, 1961). In the proposed scheme one should, however, consider the possibility of direct inhibitory connexions. In Text-fig. 19 we may replace any of the excitatory endings by inhibitory ones, provided we replace the corresponding geniculate cells by ones of opposite type ('on'-centre instead of 'off'-centre, and conversely). Up to the present the two mechanisms have not been distinguished, but there is no reason to think that both do not occur.

The properties of complex fields are not easily accounted for by supposing that these cells receive afferents directly from the lateral geniculate body. Rather, the correspondence between simple and complex fields noted in Part I suggests that cells with complex fields are of higher order, having cells with simple fields as their afferents. These simple fields would all have identical axis orientation, but would differ from one another in their exact retinal positions. An example of such a scheme is given in Text-fig. 20. The hypothetical cell illustrated has a complex field like that

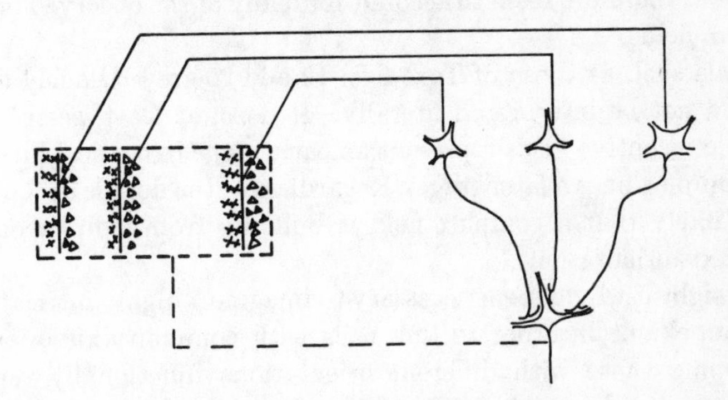

Text-fig. 20. Possible scheme for explaining the organization of complex receptive fields. A number of cells with simple fields, of which three are shown schematically, are imagined to project to a single cortical cell of higher order. Each projecting neurone has a receptive field arranged as shown to the left: an excitatory region to the left and an inhibitory region to the right of a vertical straight-line boundary. The boundaries of the fields are staggered within an area outlined by the interrupted lines. Any vertical-edge stimulus falling across this rectangle, regardless of its position, will excite some simple-field cells, leading to excitation of the higher-order cell.

of Text-figs. 5 and 6. One may imagine that it receives afferents from a set of simple cortical cells with fields of type *G*, Text-fig. 2, all with vertical axis orientation, and staggered along a horizontal line. An edge of light would activate one or more of these simple cells wherever it fell within the complex field, and this would tend to excite the higher-order cell.

Similar schemes may be proposed to explain the behaviour of other complex units. One need only use the corresponding simple fields as building blocks, staggering them over an appropriately wide region. A cell with the properties shown in Text-fig. 3 would require two types of horizontally oriented simple fields, having 'off' centres above the horizontal line, and 'on' centres below it. A slit of the same width as these centre regions would strongly activate only those cells whose long narrow centres it covered. It is true that at the same time a number of other cells would have small parts of their peripheral fields stimulated, but we may perhaps assume that these opposing effects would be relatively weak. For orientations other than horizontal a slit would have little or no effect on the simple cells, and would therefore not activate the complex one. Small spots should give only feeble 'on' responses regardless of where they were shone in the field. Enlarging the spots would not produce summation of the responses unless the enlargement were in a horizontal direction; anything else would result in invasion of opposing parts of the antecedent fields, and cancellation of the responses from the corresponding cells. The model would therefore seem to account for many of the observed properties of complex fields.

Proposals such as those of Text-figs. 19 and 20 are obviously tentative and should not be interpreted literally. It does, at least, seem probable that simple receptive fields represent an early stage in cortical integration, and the complex ones a later stage. Regardless of the details of the process, it is also likely that a complex field is built up from simpler ones with common axis orientations.

At first sight it would seem necessary to imagine a highly intricate tangle of interconnexions in order to link cells with common axis orientations while keeping those with different orientations functionally separated. But if we turn to the results of Part III on functional cytoarchitecture we see at once that gathered together in discrete columns are the very cells we require to be interconnected in our scheme. The cells of each aggregate have common axis orientations and the staggering in the positions of the simple fields is roughly what is required to account for the size of most of the complex fields (cf. Text-fig. 9). That these cells are interconnected is moreover very likely on histological grounds: indeed, the particular richness of radial connexions in the cortex fits well with the columnar shape of the regions.

The otherwise puzzling aggregation of cells with common axis orientation now takes on new meaning. We may tentatively look upon each column as a functional unit of cortex, within which simple fields are elaborated and then in turn synthesized into complex fields. The large variety of simple and complex fields to be found in a single column (Text-fig. 15) suggests that the connexions between cells in a column are highly specific.

We may now begin to appreciate the significance of the great increase in the number of cells in the striate cortex, compared with the lateral geniculate body. In the cortex there is an enormous digestion of information, with each small region of visual field represented over and over again in column after column, first for one receptive-field orientation and then for another. Each column contains thousands of cells, some cells having simple fields and others complex. In the part of the cortex receiving projections from the area centralis the receptive fields are smaller, and presumably more columns are required for unit area of retina; hence in central retinal regions the cortical projection is disproportionately large.

Complex receptive fields

The method of stimulating the retina with small circular spots of light and recording from single visual cells has been a useful one in studies of the cat's visual system. In the pathway from retina to cortex the excitatory and inhibitory areas mapped out by this means have been sufficient to account for responses to both stationary and moving patterns. Only when one reaches cortical cells with complex fields does the method fail, for these fields cannot generally be separated into excitatory and inhibitory regions. Instead of the direct small-spot method, one must resort to a trial-and-error system, and attempt to describe each cell in terms of the stimuli that most effectively influence firing. Here there is a risk of over- or under-estimating the complexity of the most effective stimuli, with corresponding lack of precision in the functional description of the cell. For this reason it is encouraging to find that the properties of complex fields can be interpreted by the simple supposition that they receive projections from simple-field cells, a supposition made more likely by the anatomical findings of Part III.

Compared with cells in the retina or lateral geniculate body, cortical cells show a marked increase in the number of stimulus parameters that must be specified in order to influence their firing. This apparently reflects a continuing process which has its beginning in the retina. To obtain an optimum response from a retinal ganglion cell it is generally sufficient to specify the position, size and intensity of a circular spot. Enlarging the spot beyond the size of the field centre raises the threshold, but even when diffuse light is used it is possible to evoke a brisk response by using an intense enough stimulus. For geniculate cells the penalty for exceeding optimum spot size is more severe than in the retina, as has been shown by comparing responses of a geniculate cell and an afferent fibre to the same cell (Hubel & Wiesel, 1961). In the retina and lateral geniculate body there is no evidence that any shapes are more effective than circular ones, or that, with moving stimuli, one direction of movement is better than another.

In contrast, in the cortex effective driving of simple-field cells can only be obtained with restricted stimuli whose position, shape and orientation are specific for the cell. Some cells fire best to a moving stimulus, and in these the direction and even the rate of movement are often critical.

Diffuse light is at best a poor stimulus, and for cells in the area of central representation it is usually ineffective at any intensity.

An interesting feature of cortical cells with complex fields may be seen in their departure from the process of progressively increasing specificity. At this stage, for the first time, what we suppose to be higher-order neurones are in a sense less selective in their responses than the cells which feed into them. Cells with simple fields tend to respond only when the stimulus is both oriented and positioned properly. In contrast, the neurones to which they supposedly project are concerned predominantly with stimulus orientation, and are far less critical in their requirements as regards stimulus placement. Their responsiveness to the abstraction which we call orientation is thus generalized over a considerable retinal area.

The significance of this step for perception can only be speculated upon, but it may be of some interest to examine several possibilities. First, neurophysiologists must ultimately try to explain how a form can be recognized regardless of its exact position in the visual field. As a step in form recognition the organism may devise a mechanism by which the inclinations of borders are more important than their exact visual-field location. It is clear that a given form in the visual field will, by virtue of its borders, excite a combination of cells with complex fields. If we displace the form it will activate many of the same cells, as long as the change in position is not enough to remove it completely from their receptive fields. Now we may imagine that these particular cells project to a single cell of still higher order: such a cell will then be very likely to respond to the form (provided the synapses are excitatory) and there will be considerable latitude in the position of the retinal image. Such a mechanism will also permit other transformations of the image, such as a change in size associated with displacement of the form toward or away from the eye. Assuming that there exist cells that are responsive to specific forms, it would clearly be economical to avoid having thousands for each form, one for every possible retinal position, and separate sets for each type of distortion of the image.

Next, the ability of some cells with complex fields to respond in a sustained manner to a stimulus as it moves over a wide expanse of retina suggests that these cells may play an important part in the perception of movement. They adapt rapidly to a stationary form, and continuous movement of the stimulus within the receptive field is the only way of obtaining a sustained discharge (Text-fig. 4*H*). Presumably the afferent simple-field cells also adapt rapidly to a stationary stimulus; because of their staggered fields the moving stimulus excites them in turn, and the higher-order cell is thus at all times bombarded. This seems an elegant means of overcoming a difficulty inherent in the problem of movement perception, that movement must excite receptors not continuously but in sequence.

Finally, the above remarks apply equally well to displacements of retinal images caused by small eye movements. The normal eye is not stationary, but is subject to several types of fine movements. There is psychophysical evidence that in man these may play an important part in vision, transforming a steady stimulus produced by a stationary object into an intermittent one, so overcoming adaptation in visual cells (Ditchburn & Ginsborg, 1952; Riggs, Ratliff, Cornsweet & Cornsweet, 1953). At an early stage in the visual pathway the effect of such movements would be to excite many cells repeatedly and in turn, rather than just a few continuously. A given line or border would move back and forth over a small retinal region; in the cortex this would sequentially activate many cells with simple fields. Since large rotatory movements are not involved, these fields would have the same axis orientations but would differ only in their exact retinal positions. They would converge on higher-order cells with complex fields, and these would tend to be activated continuously rather than intermittently.

Functional cytoarchitecture

There is an interesting parallel between the functional subdivisions of the cortex described in the present paper, and those found in somatosensory cortex by Mountcastle (1957) in the cat, and by Powell & Mountcastle (1959) in the monkey. Here, as in the visual area, one can subdivide the cortex on the basis of responses to natural stimuli into regions which are roughly columnar in shape, and extend from surface to white matter. This is especially noteworthy since the visual and somatic areas are the only cortical regions so far studied at the single-cell level from the standpoint of functional architecture. In both areas the columnar organization is superimposed upon the well known systems of topographic representation—of the body surface in the one case, and the visual fields in the other. In the somatosensory cortex the columns are determined by the sensory submodality to which the cells of a column respond: in one type of column the cells are affected either by light touch or by bending of hairs, whereas in the other the cells respond to stimulation of deep fascia or manipulation of joints.

Several differences between the two systems will at once be apparent. In the visual cortex the columns are determined by the criterion of receptive-field axis orientation. Presumably there are as many types of column as there are recognizable differences in orientation. At present one can be sure that there are at least ten or twelve, but the number may be very large, since it is possible that no two columns represent precisely the same axis orientation. (A subdivision of cells or of columns into twelve groups according to angle of orientation shows that there is no clear prevalence of one group over any of the others.) In the somatosensory cortex, on the other hand, there are only two recognized types of column.

A second major difference between the two systems lies in the very nature of the criteria used for the subdivisions. The somatosensory cortex is divided by submodality, a characteristic depending on the incoming sensory fibres, and not on any transformations made by the cortex on the afferent impulses. Indeed we have as yet little information on what integrative processes do take place in the somatosensory cortex. In the visual cortex there is no modality difference between the input to one column and that to the next, but it is in the connexions between afferents and cortical cells, or in the interconnexions between cortical cells, that the differences must exist.

Ultimately, however, the two regions of the cortex may not prove so dissimilar. Further information on the functional role of the somatic cortex may conceivably bring to light a second system of columns, superimposed on the present one. Similarly, in the visual system future work may disclose other subdivisions cutting across those described in this paper, and based on other criteria. For the present it would seem unwise to look upon the columns in the visual cortex as entirely autonomous functional units. While the variation in field size from cell to cell within a column is generally of the sort suggested in Text-figs. 9 and 15, the presence of an occasional cell with a very large complex field (up to about 20°) makes one wonder whether columns with similar receptive-field orientations may not possess some interconnexions.

Binocular interaction

The presence in the striate cortex of cells influenced from both eyes has already been observed by several authors (Hubel & Wiesel, 1959; Cornehls & Grüsser, 1959; Burns, Heron & Grafstein, 1960), and is confirmed in Part II of this paper. Our results suggest that the convergence of influences from the two eyes is extensive, since binocular effects could be demonstrated in 84 % of our cells, and since the two eyes were equally, or almost equally, effective in 70 % (groups 3–5). This represents a much greater degree of interaction than was suggested by our original work, or by Grüsser and Grüsser-Cornehls (see Jung, 1960), who found that only 30 % of their cells were binocularly influenced.

For each of our cells comparison of receptive fields mapped in the two eyes showed that, except for a difference in strength of responses related to eye dominance, the fields were in every way similar. They were similarly organized, had the same axis orientation, and occupied corresponding regions in the two retinas. The responses to stimuli applied to corresponding parts of the two receptive fields showed summation. This should be important in binocular vision, for it means that when the two images produced by an object fall on corresponding parts of the two retinas, their separate effects on a cortical cell should sum. Failure of the images to fall on corresponding regions, which might happen if an object were closer than the point of fixation or further away, would tend to reduce the summation; it could even lead to mutual antagonism if excitatory parts of one field were stimulated at the same time as inhibitory parts of the other. It should be emphasized that for all simple fields and for many complex ones the two eyes may work either synergistically or in opposition, depending on how the receptive fields are stimulated; when identical stimuli are shone on corresponding parts of the two retinas their effects should always sum.

Although in the cortex the proportion of binocularly influenced cells is high, the mixing of influences from the two eyes is far from complete. Not only are many single cells unequally influenced by the two eyes, but the relative eye dominance differs greatly from one cell to another. This could simply reflect an intermediate stage in the process of mixing of influences from the two eyes; in that case we might expect an increasing uniformity in the eye preference of higher-order cells. But cells with complex fields do not appear to differ, in their distribution among the different eye-dominance groups, from the general population of cortical cells (Text-fig. 12). At present we have no clear notion of the physiological significance of this incomplete mixing of influences from the two eyes. One possible hint lies in the fact that by binocular parallax alone (even with a stimulus too brief to allow changes in the convergence of the eyes) one can tell which of two objects is the closer (Dove, 1841; von Recklinghausen, 1861). This would clearly be impossible if the two retinas were connected to the brain in identical fashion, for then the eyes (or the two pictures of a stereo-pair) could be interchanged without substituting near points for far ones and vice versa.

Comparison of receptive fields in the frog and the cat

Units in many respects similar to striate cortical cells with complex fields have recently been isolated from the intact optic nerve and the optic tectum of the frog (Lettvin, Maturana, McCulloch & Pitts, 1959; Maturana, Lettvin, McCulloch & Pitts, 1960). There is indirect evidence to suggest that the units are the non-myelinated axons or axon terminals of retinal ganglion cells, rather than tectal cells or efferent optic nerve fibres. In common with complex cortical cells, these units respond to objects and shadows in the visual field in ways that could not have been predicted from responses to small spots of light. They thus have 'complex' properties, in the sense that we have used this term. Yet in their detailed behaviour they differ greatly from any cells yet studied in the cat, at any level from retina to cortex. We have not, for example, seen 'erasible' responses or found 'convex edge detectors'. On the other hand, it seems that some cells in the frog have asymmetrical responses to movement and some have what we have termed a 'receptive-field axis'.

Assuming that the units described in the frog are fibres from retinal ganglion cells, one may ask whether similar fibres exist in the cat, but have been missed because of their small size. We lack exact information on the fibre spectrum of the cat's optic nerve; the composite action potential suggests that non-myelinated fibres are present, though in smaller numbers than in the frog (Bishop, 1933; Bishop & O'Leary, 1940). If their fields are different from the well known concentric type, they must have little part to play in the geniculo-cortical pathway, since geniculate cells all appear to have concentric-type fields (Hubel & Wiesel, 1961). The principal cells of the lateral geniculate body (those that send their axons to the striate cortex) are of fairly uniform size, and it seems unlikely that a large group would have gone undetected. The smallest fibres in the cat's optic nerve probably project to the tectum or the pretectal region; in view of the work in the frog, it will be interesting to examine their receptive fields.

At first glance it may seem astonishing that the complexity of third-order neurones in the frog's visual system should be equalled only by that of sixth-order neurones in the geniculo-cortical pathway of the cat. Yet this is less surprising if one notes the great anatomical differences in the two animals, especially the lack, in the frog, of any cortex or dorsal lateral geniculate body. There is undoubtedly a parallel difference in the use each animal makes of its visual system: the frog's visual apparatus is presumably specialized to recognize a limited number of stereotyped patterns or situations, compared with the high acuity and versatility found in the cat. Probably it is not so unreasonable to find that in the cat the specialization of cells for complex operations is postponed to a higher level, and that when it does occur, it is carried out by a vast number of cells, and in great detail. Perhaps even more surprising, in view of what seem to be profound physiological differences, is the superficial anatomical similarity of retinas in the cat and the frog. It is possible that with Golgi methods a comparison of the connexions between cells in the two animals may help us in understanding the physiology of both structures.

Receptive fields of cells in the primate cortex

We have been anxious to learn whether receptive fields of cells in the monkey's visual cortex have properties similar to those we have described in the cat. A few preliminary experiments on the spider monkey have shown striking similarities. For example, both simple and complex fields have been observed in the striate area. Future work will very likely show differences, since the striate cortex of the monkey is in several ways different morphologically from that of the cat. But the similarities already seen suggest that the mechanisms we have described may be relevant to many mammals, and in particular to man.

SUMMARY

1. The visual cortex was studied in anaesthetized cats by recording extracellularly from single cells. Light-adapted eyes were stimulated with spots of white light of various shapes, stationary or moving.

2. Receptive fields of cells in the visual cortex varied widely in their organization. They tended to fall into two categories, termed 'simple' and 'complex'.

3. There were several types of simple receptive fields, differing in the spatial distribution of excitatory and inhibitory ('on' and 'off') regions. Summation occurred within either type of region; when the two opposing regions were illuminated together their effects tended to cancel. There was generally little or no response to stimulation of the entire receptive field with diffuse light. The most effective stimulus configurations, dictated by the spatial arrangements of excitatory and inhibitory regions, were long narrow rectangles of light (slits), straight-line borders between areas of different brightness (edges), and dark rectangular bars against a light background. For maximum response the shape, position and orientation of these stimuli were critical. The orientation of the receptive-field axis (i.e. that of the optimum stimulus) varied from cell to cell; it could be vertical, horizontal or oblique. No particular orientation seemed to predominate.

4. Receptive fields were termed complex when the response to light could not be predicted from the arrangements of excitatory and inhibitory regions. Such regions could generally not be demonstrated; when they could the laws of summation and mutual antagonism did not apply. The stimuli that were most effective in activating cells with simple fields—slits, edges, and dark bars—were also the most effective for cells with complex fields. The orientation of a stimulus for optimum response was critical, just as with simple fields. Complex fields, however, differed from simple fields in that a stimulus was effective wherever it was placed in the field, provided that the orientation was appropriate.

5. Receptive fields in or near the area centralis varied in diameter from $\frac{1}{2}$–1° up to about 5–6°. On the average, complex fields were larger than simple ones. In more peripheral parts of the retina the fields tended to be larger. Widths of the long narrow excitatory or inhibitory portions of simple receptive fields were often roughly equal to the diameter of the smallest geniculate receptive-field centres in the area centralis. For cells

with complex fields responding to slits or dark bars the optimum stimulus width was also usually of this order of magnitude.

6. Four fifths of all cells were influenced independently by the two eyes. In a binocularly influenced cell the two receptive fields had the same organization and axis orientation, and were situated in corresponding parts of the two retinas. Summation was seen when corresponding parts of the two retinas were stimulated in identical fashion. The relative influence of the two eyes differed from cell to cell: for some cells the two eyes were about equal; in others one eye, the ipsilateral or contralateral, dominated.

7. Functional architecture was studied by (*a*) comparing the responses of cells recorded in sequence during micro-electrode penetrations through the cortex, (*b*) observing the unresolved background activity, and (*c*) comparing cells recorded simultaneously with a single electrode (multiple recordings). The retinas were found to project upon the cortex in an orderly fashion, as described by previous authors. Most recordings were made from the cortical region receiving projections from the area of central vision. The cortex was found to be divisible into discrete columns; within each column the cells all had the same receptive-field axis orientation. The columns appeared to extend from surface to white matter; cross-sectional diameters at the surface were of the order of 0·5 mm. Within a given column one found various types of simple and complex fields; these were situated in the same general retinal region, and usually overlapped, although they differed slightly in their exact retinal position. The relative influence of the two eyes was not necessarily the same for all cells in a column.

8. It is suggested that columns containing cells with common receptive-field axis orientations are functional units, in which cells with simple fields represent an early stage in organization, possibly receiving their afferents directly from lateral geniculate cells, and cells with complex fields are of higher order, receiving projections from a number of cells with simple fields within the same column. Some possible implications of these findings for form perception are discussed.

We wish to thank Miss Jaye Robinson and Mrs Jane Chen for their technical assistance. We are also indebted to Miss Sally Fox and to Dr S. W. Kuffler for their helpful criticism of this manuscript. The work was supported in part by Research Grants B-2251 and B-2260 from United States Public Health Service, and in part by the United States Air Force through the Air Force Office of Scientific Research of the Air Research and Development Command under contract No. AF 49 (638)–713. The work was done during the tenure of a U.S. Public Health Service Senior Research Fellowship No. SF 304-R by D.H.H.

REFERENCES

Bishop, G. H. (1933). Fiber groups in the optic nerve. *Amer. J. Physiol.* **106**, 460–474.

Bishop, G. H. & O'Leary, J. S. (1940). Electrical activity of the lateral geniculate of cats following optic nerve stimuli. *J. Neurophysiol.* **3**, 308–322.

Bishop, P. O., Burke, W. & Davis, R. (1959). Activation of single lateral geniculate cells by stimulation of either optic nerve. *Science*, **130**, 506–507.

Burns, B. D., Heron, W. & Grafstein, B. (1960). Response of cerebral cortex to diffuse monocular and binocular stimulation. *Amer. J. Physiol.* **198**, 200–204.

Cornehls, U. & Grüsser, O.-J. (1959). Ein elektronisch gesteuertes Doppellichtreizgerät. *Pflüg. Arch. ges. Physiol.* **270**, 78–79.

Daniel, P. M. & Whitteridge, D. (1959). The representation of the visual field on the calcarine cortex in baboons and monkeys. *J. Physiol.* **148**, 33*P*.

Ditchburn, R. W. & Ginsborg, B. L. (1952). Vision with stabilized retinal image. *Nature, Lond.*, **170**, 36–37.

Dove, H. W. (1841). Die Combination der Eindrücke beider Ohren und beider Augen zu einem Eindruck. *Mber. preuss. Akad.* **1841**, 251–252.

Erulkar, S. D. & Fillenz, M. (1958). Patterns of discharge of single units of the lateral geniculate body of the cat in response to binocular stimulation. *J. Physiol.* **140**, 6–7*P*.

Erulkar, S. D. & Fillenz, M. (1960). Single-unit activity in the lateral geniculate body of the cat. *J. Physiol.* **154**, 206–218.

Grüsser, O.-J. & Sauer, G. (1960). Monoculare und binoculare Lichtreizung einzelner Neurone im Geniculatum laterale der Katze. *Pflüg. Arch. ges. Physiol.* **271**, 595–612.

Hayhow, W. R. (1958). The cytoarchitecture of the lateral geniculate body in the cat in relation to the distribution of crossed and uncrossed optic fibers. *J. comp. Neurol.* **110**, 1–64.

Hubel, D. H. (1957). Tungsten microelectrode for recording from single units. *Science*, **125**, 549–550.

Hubel, D. H. (1958). Cortical unit responses to visual stimuli in nonanesthetized cats. *Amer. J. Ophthal.* **46**, 110–121.

Hubel, D. H. (1959). Single unit activity in striate cortex of unrestrained cats. *J. Physiol.* **147**, 226–238.

Hubel, D. H. (1960). Single unit activity in lateral geniculate body and optic tract of unrestrained cats. *J. Physiol.* **150**, 91–104.

Hubel, D. H. & Wiesel, T. N. (1959). Receptive fields of single neurones in the cat's striate cortex. *J. Physiol.* **148**, 574–591.

Hubel, D. H. & Wiesel, T. N. (1960). Receptive fields of optic nerve fibres in the spider monkey. *J. Physiol.* **154**, 572–580.

Hubel, D. H. & Wiesel, T. N. (1961). Integrative action in the cat's lateral geniculate body. *J. Physiol.* **155**, 385–398.

Jung, R. (1960). Microphysiologie corticaler Neurone: Ein Beitrag zur Koordination der Hirnrinde und des visuellen Systems. *Structure and Function of the Cerebral Cortex*, ed. Tower, D. B. and Schadé, J. P. Amsterdam: Elsevier Publishing Company.

Kuffler, S. W. (1953). Discharge patterns and functional organization of mammalian retina. *J. Neurophysiol.* **16**, 37–68.

Lettvin, J. Y., Maturana, H. R., McCulloch, W. S. & Pitts, W. H. (1959). What the frog's eye tells the frog's brain. *Proc. Inst. Radio Engrs, N.Y.*, **47**, 1940–1951.

Maturana, H. R., Lettvin, J. Y., McCulloch, W. S. & Pitts, W. H. (1960). Anatomy and physiology of vision in the frog (*Rana pipiens*). *J. gen. Physiol.* **43**, part 2, 129–176.

Minkowski, M. (1913). Experimentelle Untersuchungen über die Beziehungen der Grosshirnrinde und der Netzhaut zu den primären optischen Zentren, besonders zum Corpus geniculatum externum. *Arb. hirnanat. Inst. Zürich*, **7**, 259–362.

Mountcastle, V. B. (1957). Modality and topographic properties of single neurons of cat's somatic sensory cortex. *J. Neurophysiol.* **20**, 408–434.

O'Leary, J. L. (1941). Structure of the area striata of the cat. *J. comp. Neurol.* **75**, 131–164.

Polyak, S. (1957). *The Vertebrate Visual System*, ed. Klüver, H. The University of Chicago Press.

Powell, T. P. S. & Mountcastle, V. B. (1959). Some aspects of the functional organization of the cortex of the postcentral gyrus of the monkey: a correlation of findings obtained in a single unit analysis with cytoarchitecture. *Johns Hopk. Hosp. Bull.* **105**, 133–162.

Riggs, L. A., Ratliff, F., Cornsweet, J. C. & Cornsweet, T. N. (1953). The disappearance of steadily fixated visual test objects. *J. opt. Soc. Amer.* **43**, 495–501.

Silva, P. S. (1956). Some anatomical and physiological aspects of the lateral geniculate body. *J. comp. Neurol.* **106**, 463–486.

Talbot, S. A. (1942). A lateral localization in the cat's visual cortex. *Fed. Proc.* **1**, 84.

Talbot, S. A. & Marshall, W. H. (1941). Physiological studies on neural mechanisms of visual localization and discrimination. *Amer. J. Ophthal.* **24**, 1255–1263.

Thompson, J. M., Woolsey, C. N. & Talbot, S. A. (1950). Visual areas I and II of cerebral cortex of rabbit. *J. Neurophysiol.* **13**, 277–288.

von Recklinghausen, F. (1861). Zum körperlichen Sehen. *Ann. Phys. Lpz.* **114**, 170–173.

EXPLANATION OF PLATES

Plate 1

Coronal section through post-lateral gyrus. Composite photomicrograph of two of the sections used to reconstruct the micro-electrode track of Text-fig. 13. The first part of the electrode track may be seen in the upper right; the electrolytic lesion at the end of the track appears in the lower left. Scale 1 mm.

Plate 2

A, coronal section through the anterior extremity of post-lateral gyrus. Composite photomicrograph made from four of the sections used to reconstruct the two electrode tracks shown in Text-fig. 14. The first part of the two electrode tracks may be seen crossing layer 1. The lesion at the end of the lateral track (to the right in the figure) is easily seen; that of the medial track is smaller, and is shown at higher power in *B*. Scales: *A*, 1 mm, *B*, 0·25 mm.

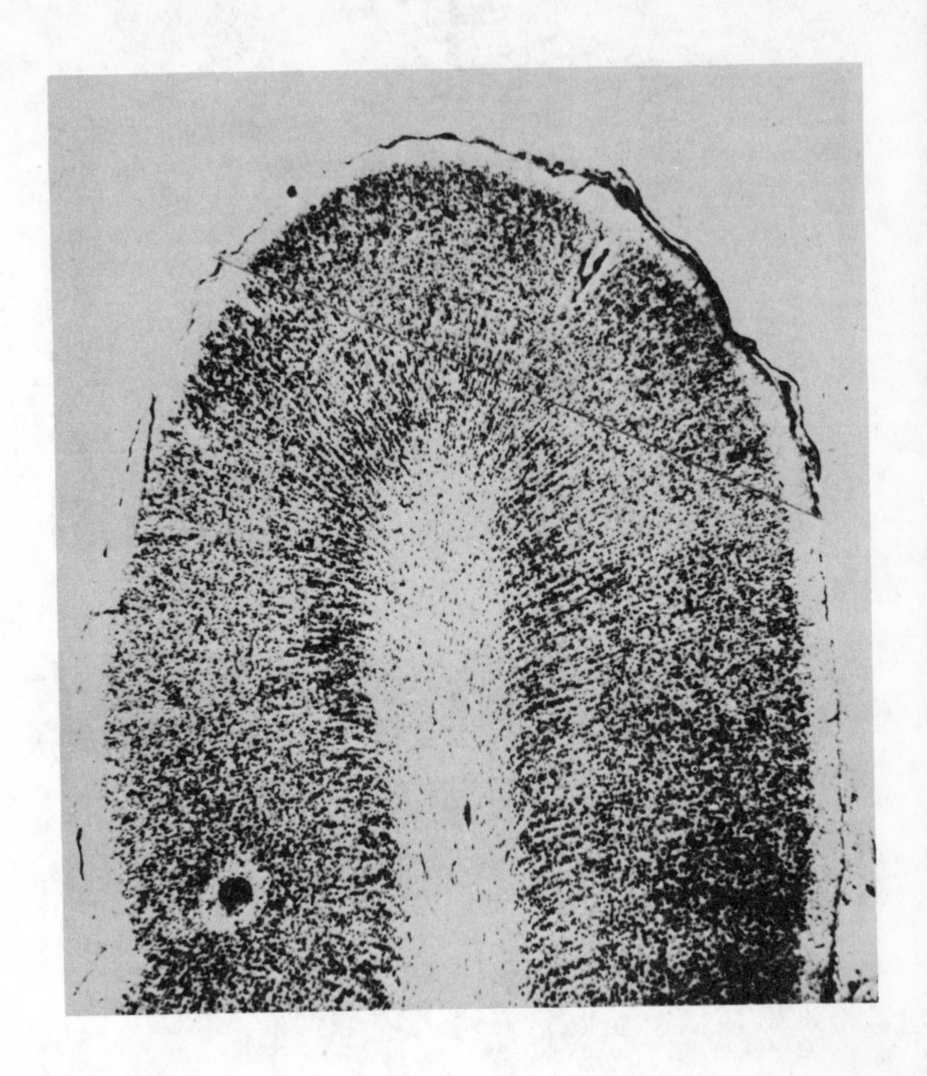

D. H. HUBEL AND T. N. WIESEL

(*Facing p.* 154)

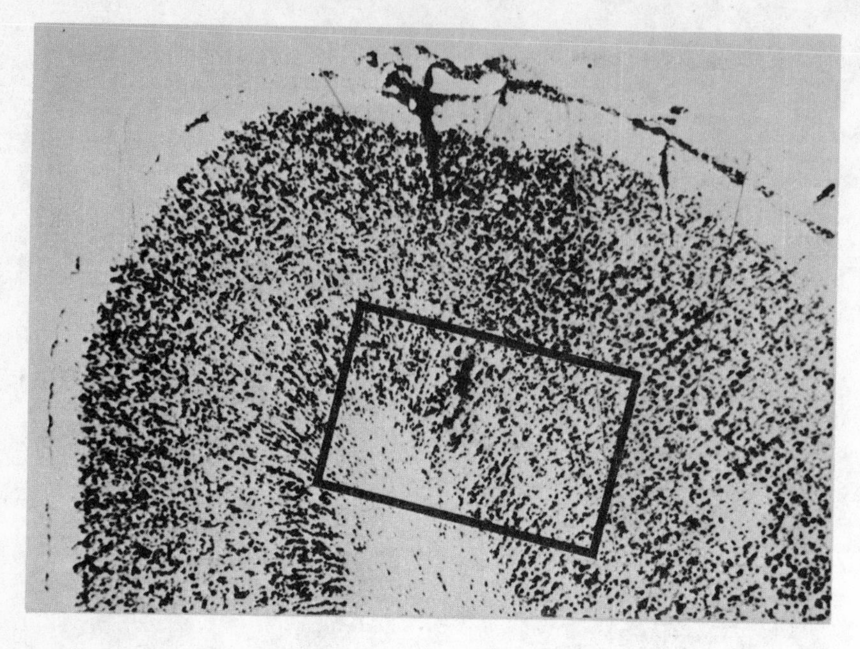

A

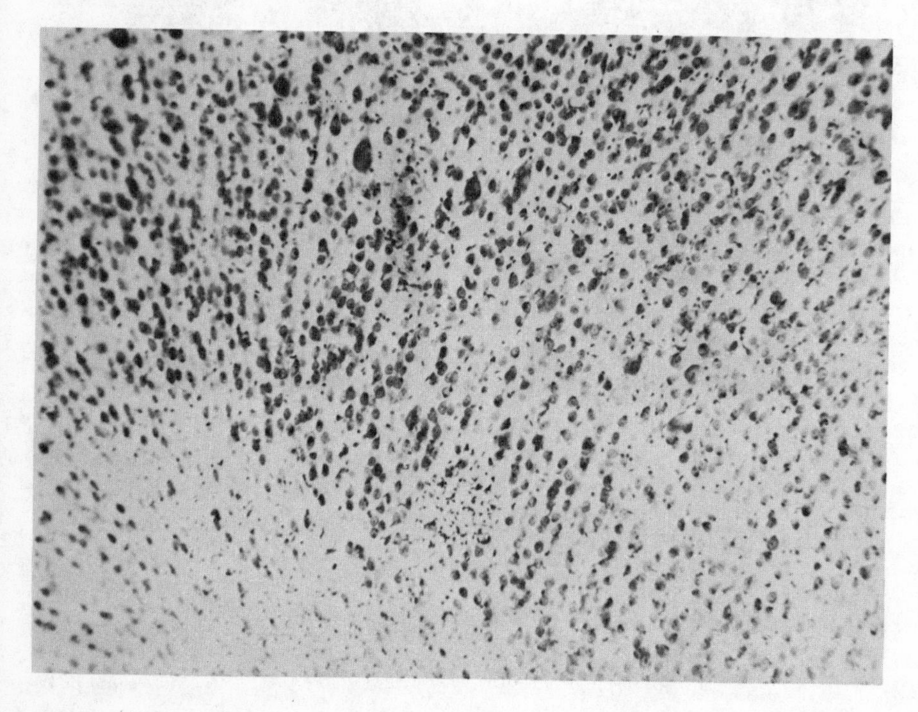

B

D. H. HUBEL AND T. N. WIESEL

SCIENCE, December 18, 1959, Vol. 130, No. 3390, pages 1709-1710

Evidence That Cut Optic Nerve Fibers in a Frog Regenerate to Their Proper Places in the Tectum

Abstract. The frog's retina projects into the superficial neuropil of the opposite tectum in four functionally different layers of terminals. Each layer displays a continuous map of the retina in terms of its particular function. The four maps are in register. The fourth-dimensional order is reconstituted after section and regeneration of the optic fibers.

Sperry (*1*) pointed out that the results of his experiments on optic nerve regeneration in adult frogs were consistent with specific reconnection of the optic fibers. He proposed that each individual neuron grew back to its original terminus in the tectum, for the behavior after visual recovery was as if the nerve had not been cut. In addition to the behavioral evidence, he produced scotomata in predicted quadrants by fairly large tectal lesions in frogs that had regrown their optic connections. The implications of his proposal are so odd that, while his elegant experiments were accepted, the interpretation was much disputed. Furthermore, the experiments with tectal lesions cannot be considered conclusive, since, by destroying part of the tectum, the ability of the animal to respond is also impaired. The purpose of this communication is to give electrophysiological evidence in support of Sperry's hypothesis.

We have developed a technique for recording single fibers in the frog's optic nerve and single terminal bushes in the tectum (*2*). In this work we have found that normally the frog's tectum has the following organization. The fibers of each optic nerve cross completely in the optic chiasma and enter the opposite colliculus after dividing into two bundles. One is rostromedial; the other, caudolateral. They sweep over the surface and are distributed in several layers in the outer neuropil that forms the superficial half (250 μ) of the tectal cortex (Fig. 1). Most tectal cell bodies lie below this neuropil and send their main dendrites through it up to the pial surface. The axons of the majority of these cells form a narrow stratum that lies immediately above the compact layers of cell bodies. The optic fibers end in a systematic way both along the surface and in the depths of the superficial neuropil, mapping the retina in a pattern that is constant from animal to animal. There are four layers of these optic fiber terminals, which we have thus far identified only physiologically. Each displays a continuous map of the retina with respect to each of the four following operations on the image at the receptors. The four maps are in register with each other and show position on the retina according to the cartography of Gaze (*3*).

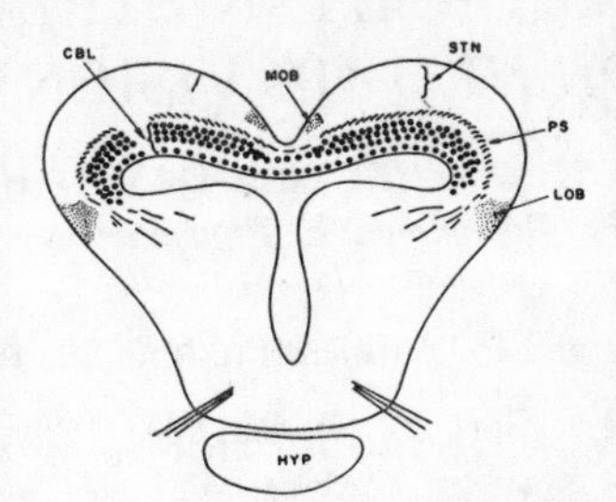

Fig. 1. Transverse section of the tectum of the frog at the level of the oculomotor nerves. CBL, cell-body layers; MOB, medial optic bundle; STN, superficial tectal neuropil; PS, palisade stratum; LOB, lateral optic bundle; HYP, hypophysis.

The first layer of terminals is formed by those elements each of which is sensitive to moving or maintained contrast within its receptive field. The sharper the contrast, the better the response. These are equivalent to Hartline's (*4*) and Barlow's (*5*) "on" fibers. The second layer is made up of terminals of units each of which detects a moving or recently stopped boundary within its receptive field, provided there is a net positive curvature of the edge of the darker phase. Such a fiber will not respond, for example, to a straight-edge boundary moving across its receptive field or to a preestablished edge within that field. Both of these strata represent the endings of the unmyelinated fibers of the optic nerve.

The third layer is made up of terminal bushes from "on-off" fibers.

The fourth layer is composed of endings from "off" fibers.

The layers of endings are distinct in depth, and with the exception of the first and second layers they rarely merge at the transition zones. In this conspicuous order, both along the surface and in the depths, the area of the retina "seen" from any point in the superficial neuropil is, at most, 10° in radius. Most of the ganglion cells whose terminals appear at that point are crowded toward the middle of that area.

For the purpose of testing Sperry's hypothesis of the specific regrowth of the optic fibers after section of the optic nerve, we cut one optic nerve in several adult frogs (*Rana pipiens*), ensuring the complete separation of the two stumps. At the end of 2 months the first signs of visual recovery were apparent, but full use of the eye did not occur for another month. When the visual recovery seemed complete, we exposed the colliculi and tested the initially deafferented colliculus for mapping of the retina. We found that the map had been regenerated along the surface, although the ganglion cells from whose terminals we were recording at any point were now spread over an area about two times as large as normal. The separation of operations in depth was also restored, and there was no sign of confusion between the operational layers.

The specific regrowth of the terminals to their proper stations cannot be explained by saying that an initial orderly array of fibers in the optic nerve crudely orders the fibers again at the time of regeneration. The fibers in the nerve simply are not in order *ab initio*. Any two contiguous fibers can come from the most widely separated points on the retina (*2, 6*).

This finding strongly supports Sperry's hypothesis that optic-nerve fibers grow back to their original destinations. They do so in an even more highly specific way than he proposed; the regrowth of the termini is also proper in depth (*7*).

Note added in proof. After this manuscript was prepared we noted that R. M. Gaze, of the University of Edinburgh, has presented to the Physiological Society similar findings in *Xenopus laevis* (*8*). He, however, has not studied the reconstitution of the distribution in depth of the optic fibers.

H. R. MATURANA

Research Laboratory of Electronics, Massachusetts Institute of Technology, Cambridge, and *Department of Biology, Medical School, University of Chile, Santiago*

J. Y. LETTVIN

Department of Biology and *Research Laboratory of Electronics, Massachusetts Institute of Technology*

W. S. MCCULLOCH

W. H. PITTS

Research Laboratory of Electronics, Massachusetts Institute of Technology

References and Notes

1. R. W. Sperry, "Mechanisms of neural maturation," in S. S. Stevens, Ed., *Handbook of Experimental Psychology* (Wiley, New York, 1951), pp. 236-280.
2. J. Y. Lettvin and H. R. Maturana, "Frog vision," *Mass. Inst. Technol. Research Lab. Electronics Quart. Progr. Rept. No. 53* (1959), pp. 191-197.
3. R. M. Gaze, *Quart. J. Exptl. Physiol.* **43**, 209 (1958).
4. H. K. Hartline, *J. Gen. Physiol.* **130**, 690 (1940).
5. H. B: Barlow, *J. Physiol. (London)* **119**, 69 (1953).
6. H. R. Maturana, thesis, Harvard University (1958).
7. The work reported here was supported in part by the U.S. Army (Signal Corps), the U.S. Air Force (Office of Scientific Research, Air Research and Development Command), and the U.S. Navy (Office of Naval Research), and in part by the U.S. National Institutes of Health.
8. R. M. Gaze, *J. Physiol. (London)* **146**, 40P (1959).

5 August 1959

43

J. Neurophysiol. (1963) **26**, 1003-1017

SINGLE-CELL RESPONSES IN STRIATE CORTEX OF KITTENS DEPRIVED OF VISION IN ONE EYE[1]

TORSTEN N. WIESEL AND DAVID H. HUBEL

Neurophysiology Laboratory, Department of Pharmacology, Harvard Medical School, Boston, Massachusetts

(Received for publication June 26, 1963)

INTRODUCTION

IN THE FIRST PAPER OF THIS SERIES (11) we showed that in a kitten 2–3 months of monocular light and form deprivation can produce a marked atrophy of cells in the lateral geniculate body. The changes were confined to layers receiving projections from the deprived eye. Despite the atrophy, most of the cells recorded had normal receptive fields. The present paper extends this physiological study of monocularly deprived kittens to the next level in the visual pathway, the striate cortex. We wished to learn whether one could influence cortical cells from the deprived eye, and whether the receptive fields were normal.

At the cortical level any long-term effects of tampering with one eye might be expected to show up as a change in normal patterns of binocular interaction. It may therefore be useful to begin by summarizing some previous findings on binocular interaction in the normal cat. Approximately four-fifths of cells in the cat striate cortex are binocularly influenced. For any given cell the receptive fields mapped in the two eyes are similar in arrangement and occupy corresponding retinal positions. Although stimuli to corresponding retinal points thus produce qualitatively similar responses, the strengths of the responses from the two eyes are not necessarily equal: some cells respond best to the contralateral eye; others prefer the ipsilateral. Figure 1, reproduced from a previous study (7), shows the distribution of 223 cells according to eye-dominance. It will be seen from this histogram that on the whole the contralateral eye is decidedly the more influential. In the second paper of this series (9) we showed that very young kittens resembled adults in all of these respects. This was true even of animals with no previous patterned visual experience.

We have been particularly interested in learning whether the distribution of cells among the different ocular-dominance categories would be appreciably altered in recordings from monocularly deprived kittens. The histogram of Fig. 1 represents the lumped results of 45 penetrations, and cannot necessarily be used to predict the distribution of cells in a single penetration. To

[1] This work was supported in part by Research Grants GM-K3-15,304 (C2), B-2260 (C2), and B-2253-C2S1 from the Public Health Service, and in part by Research Grant AF-AFOSR-410-62 from the U. S. Air Force.

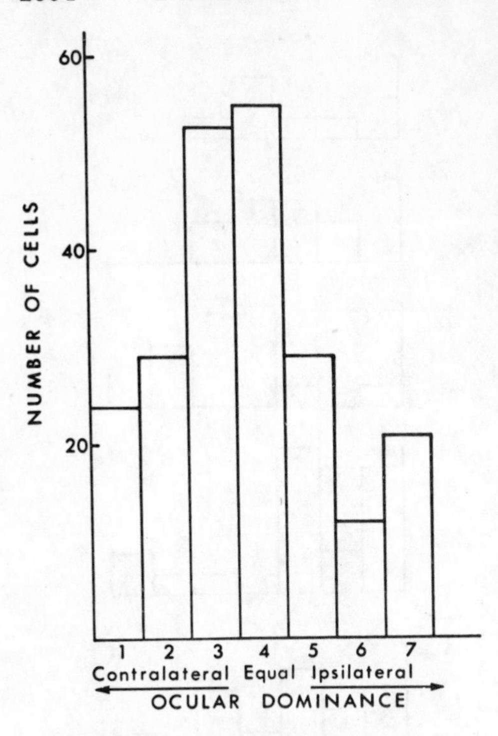

FIG. 1. Ocular-dominance distribution of 223 cells recorded from striate cortex of adult cats in a series of 45 penetrations (7). Cells of group 1 were driven only by the contralateral eye; for cells of group 2 there was marked dominance of the contralateral eye, for group 3, slight dominance. For cells in group 4 there was no obvious difference between the two eyes. In group 5 the ipsilateral eye dominated slightly, in group 6, markedly; and in group 7 the cells were driven only by the ipsilateral eye.

interpret the results of the present paper we need to know whether this distribution varies markedly from one penetration to the next. It might be much narrower and less constant from penetration to penetration if, for example, there were a strong tendency for cells with the same eye-preference to be grouped together in the cortex. To have a better idea of the variation in distribution from penetration to penetration we have therefore taken 12 separate consecutive penetrations from the series used to construct Fig. 1, and plotted each of their ocular-dominance distributions separately (Fig. 2). It appears from these histograms that there is no very marked tendency for cells to be anatomically grouped by eye-preference: penetrations *2* and *4* represent the extremes in the two directions, of dominance by the ipsilateral eye (*2*) or the contralateral eye (*4*), and even in these two penetrations both eyes make a substantial contribution. In judging an individual penetration it is therefore probably reasonably safe to take the distribution of Fig. 1 as the normal, regarding as probably significant only departures much greater than those of penetrations *2* and *4*.

METHODS

Seven kittens and one adult cat were used. The animals were monocularly deprived either by lid closure or by a translucent eye cover (11). Closure of lids prevented form vision and reduced the level of diffuse retinal illumination by about 4–5 log units. The translucent occluders also prevented pattern stimulation, but reduced the diffuse illumination by only about 1–2 log units. Four of the kittens were deprived from the time of normal eye-opening; the others had some prior visual experience. Deprivation was for periods of 1–4 months. Before some experiments the lids of the closed eye were separated or the translucent eye cover was removed. The normal eye was then covered with an opaque contact occluder and the animal's visual behavior was observed.

Experimental procedures for preparing the animals and for stimulating and recording are given in previous papers (5, 6, 7, 11). Extracellular recordings were made with tungsten

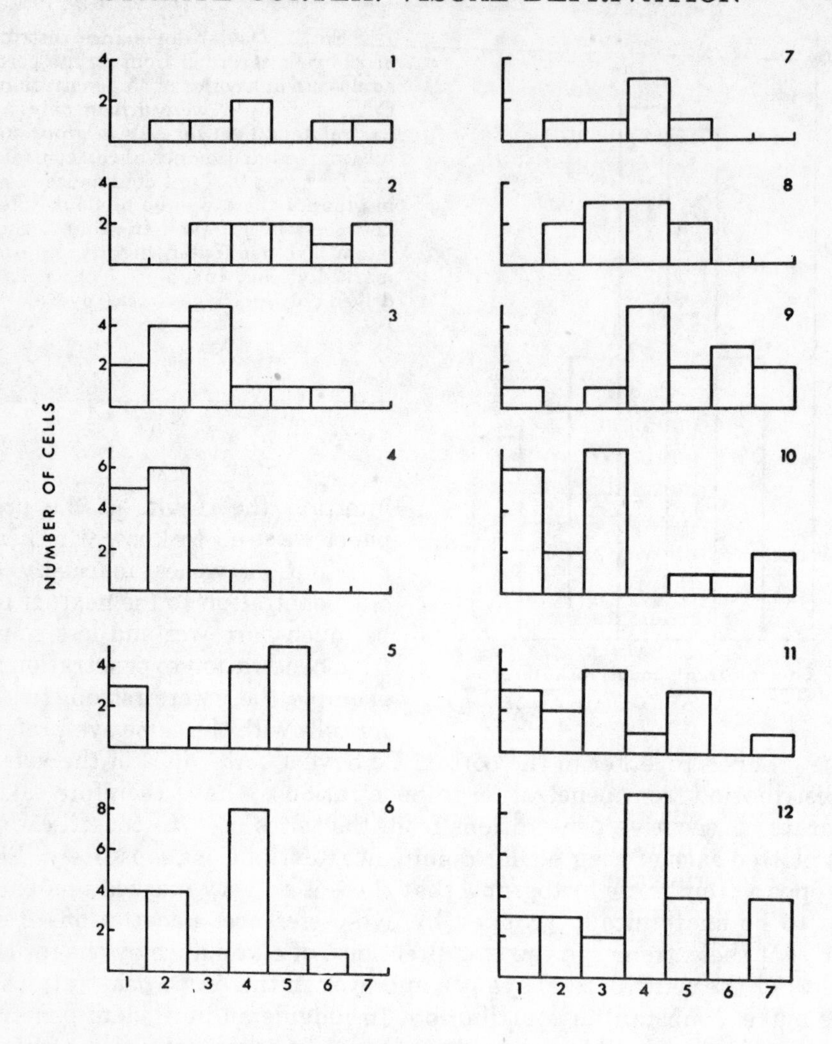

FIG. 2. Ocular-dominance distribution of 12 separate consecutive penetrations from the same series as Fig. 1 (7).

microelectrodes in the part of the lateral gyrus receiving projections from the retinal area centralis. In one animal electroretinograms and evoked cortical potentials were recorded, the electroretinograms with chlorided silver electrodes placed on the limbus of the upper outer quadrant of the cornea, the evoked potentials with similar electrodes placed on the dura over homologous points on the two lateral gyri (Horsley-Clarke frontal plane zero, 1 mm. from the midline); the indifferent electrode was the frame of the Horsley-Clarke head holder. Responses were evoked with a Grass photostimulator (model PS 2) set at maximum intensity and held 6 in. from the animal's eyes. The eyes were stimulated separately by covering each in turn with a patch of thick black rubber.

All brains were subsequently examined histologically in order to reconstruct electrode tracks marked by electrolytic lesions.

Results

Behavioral effects of monocular deprivation

Prior to some experiments in animals deprived from birth, the obstruction was removed from the right eye (by separating the lids or removing the translucent contact occluder), and an opaque occluder was placed over the normal left eye. Pupillary light reflexes were normal, and there was no nystagmus. No visual placing reactions could be obtained, though tactile placing was normal.

As an animal walked about investigating its surroundings the gait was broad-based and hesitant, and the head moved up and down in a peculiar nodding manner. The kittens bumped into large obstacles such as table legs, and even collided with walls, which they tended to follow using their whiskers as a guide. When put onto a table the animals walked off into the air, several times falling awkwardly onto the floor. When an object was moved before the eye there was no hint that it was perceived, and no attempt was made to follow it. As soon as the cover was taken off of the left eye the kitten would behave normally, jump gracefully from the table, skillfully avoiding objects in its way. We concluded that there was a profound, perhaps complete, impairment of vision in the deprived eye of these animals.

Physiological findings in kittens deprived of vision in one eye from birth

Of the 84 cortical cells recorded in kittens deprived from birth, 83 were completely uninfluenced by the deprived eye. The dominance of the normal eye in these cells was all the more striking since all but one of the five penetrations were made in the hemisphere contralateral to, and hence normally strongly favoring, the deprived eye (see Fig. 1).

The ocular-dominance histogram of a kitten whose right eye was sutured at 8 days for a $2\frac{1}{2}$-month period is shown in Fig. 3. Of the 25 cells examined in a single penetration of the left striate cortex, 20 were driven exclusively by the normal (ipsilateral) eye, and none could be influenced from the deprived (contralateral) eye. The 5 remaining cells could not be driven from either eye, and would have gone unnoticed had it not been for their spontaneous activity. The presence in most of these penetrations of a small number of unresponsive cells is worth stressing, since in normal adult cats it has been possible to drive all cells with appropriate visual stimuli (7). The electrode track of this penetration, reconstructed from the histological slides, is shown in Fig. 4. The 20 cells whose receptive fields were mapped in the normal eye all responded to line stimuli, and each strongly favored one orientation and failed to respond to a slit, edge, or dark bar placed at right angles to the optimum. The receptive fields were arranged in the usual "simple" or "complex" manner (in the sense that we have previously used these terms (7)) and varied in their orientation in a way consistent with a columnar arrangement. Unresolved background activity was present throughout the penetration, and, like the isolated units, it was strongly influenced by the normal eye but not at all by the deprived one.

In another 3-month-old kitten whose right eye had been closed from birth, penetrations were made in homologous regions in both hemispheres. The ocular-dominance histograms of these penetrations are shown in Fig. 5, and again illustrate a failure to drive any cell from the abnormal eye. Again there were several cells that could not be driven from either eye.

The one cell in the series that could be driven from the deprived eye was recorded from a third kitten, lid-sutured from birth to an age of $2\frac{1}{2}$

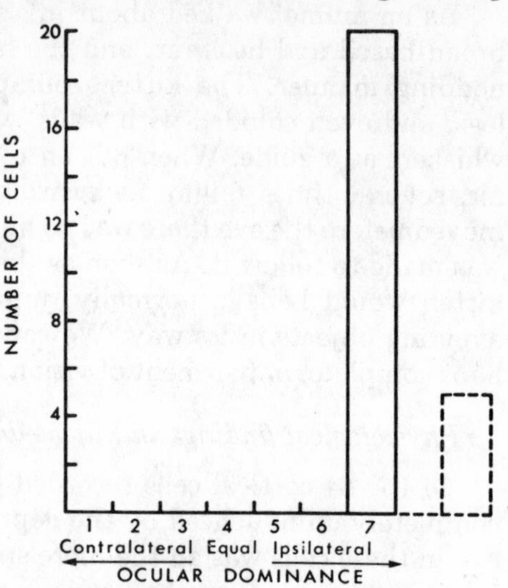

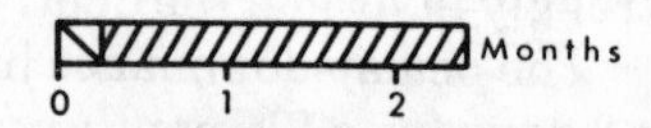

FIG. 3. Ocular-dominance distribution of 25 cells recorded in the visual cortex of a $2\frac{1}{2}$-month-old kitten. Experimental procedures are indicated beneath; during the first week the eyes were not yet open; on the eighth day the lids of the right eye were sutured, and they remained closed until the time of the experiment (cross-hatched region). The left eye opened normally on the ninth day. Recordings were made from the left visual cortex, contralateral to the eye that had been closed. Five of the cells, represented by the interrupted column on the right, could not be driven from either eye. The remaining 20 were driven only from the normally exposed (left, or ipsilateral) eye, and were therefore classed as group 7.

months. This cell was unusual in having abnormal fields in both eyes. Unlike other cells in the same penetration, which were normal except for their unresponsiveness to stimulation of the deprived eye, this cell had no particular orientation-preference, and the responses were more sluggish than those of the other cells. The receptive fields were in roughly corresponding parts of the two retinas.

To try to evaluate the relative importance of form deprivation as against light deprivation we raised one kitten from birth to an age of 2 months with a translucent contact occluder over the right eye. This prevented patterned retinal stimulation and reduced the general retinal illumination by about 2 log units, as opposed to 4–5 for the sutured lids. The ocular-dominance distribution of 26 cells recorded from the left hemisphere (contralateral to the occluded eye) is shown in Fig. 6. Just as in the lid-suture experiments, all cells were driven exclusively from the normal (left) eye, except for three which could not be driven at all. The cells that could be driven had normal receptive fields. From the electrode track reconstruction shown in Fig. 7, the electrode is seen to have traversed three columns, both shifts in orientation being small, discrete, and anticlockwise. Such orderly sequences have been seen in penetrations in adult cats (8), and support our impression that in parts of the striate cortex the columns are arranged in an orderly manner. As in the first two kittens there was continuous unresolved activity throughout the penetration, briskly responsive to stimuli of the left eye, but with no hint of a response from the right.

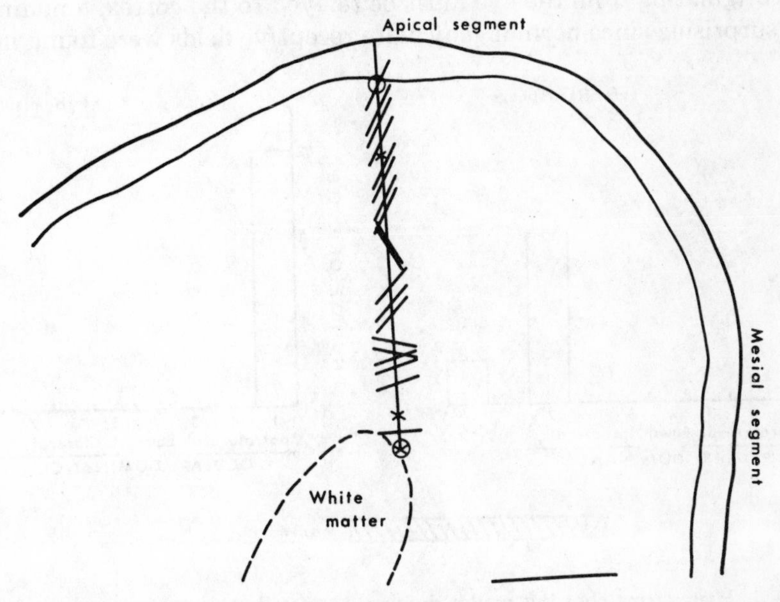

FIG. 4. Reconstruction of a microelectrode penetration through the postlateral gyrus of the left hemisphere. This $2\frac{1}{2}$-month-old kitten had its right eye covered from birth by lid suture. Lines intersecting the electrode track represent cortical cells; directions of these lines indicate the receptive-field orientations. Crosses indicate cortical cells uninfluenced by light stimulation. Simultaneous recordings from two units, which occurred three times in this penetration, are each indicated by only one line or cross. A lesion was made while recording from the first unit, and another at the end of the penetration: these are marked by small circles. The ocular-dominance distribution of units recorded in this penetration is shown in Fig. 3. All fields positioned 5–6° to the left of the area centralis, slightly below the horizontal meridian. Scale, 0.5 mm.

In order to have a gross over-all impression of the retinal and cortical activity in this kitten, bilateral electroretinograms and evoked potential recordings were made (Fig. 8). The corneal electroretinograms evoked from either eye by a brief flash showed the normal a- and b-waves (Fig. 8, *A* and *B*). Successive responses were almost identical, and with stimulation rates of 1/sec. there was no sign of fatigue. (The amplitude was, if anything, greater from the deprived eye, but this could easily be due to minor differences in stimulating conditions.) This is consistent with the findings of Zetterström (12), that the electroretinogram develops normally except for a somewhat delayed time course in kittens raised in darkness. On the other hand, cortical potentials evoked from the two eyes were far from equal. This was true for either hemisphere (Fig. 8, *C–F*). Responses from the previously occluded eye showed an initial positive component, but the later negative wave, so prominent in responses from the normal eye, was almost completely lacking. Moreover, the latency to stimulation of the deprived eye was 35–40 msec., as opposed to 25–30 msec. from the normal eye. That any cortical wave was evoked from the deprived eye indicates that some impulses originating from the eye must be relayed to the cortex, a finding that is not surprising since normal geniculate receptive fields were found in these

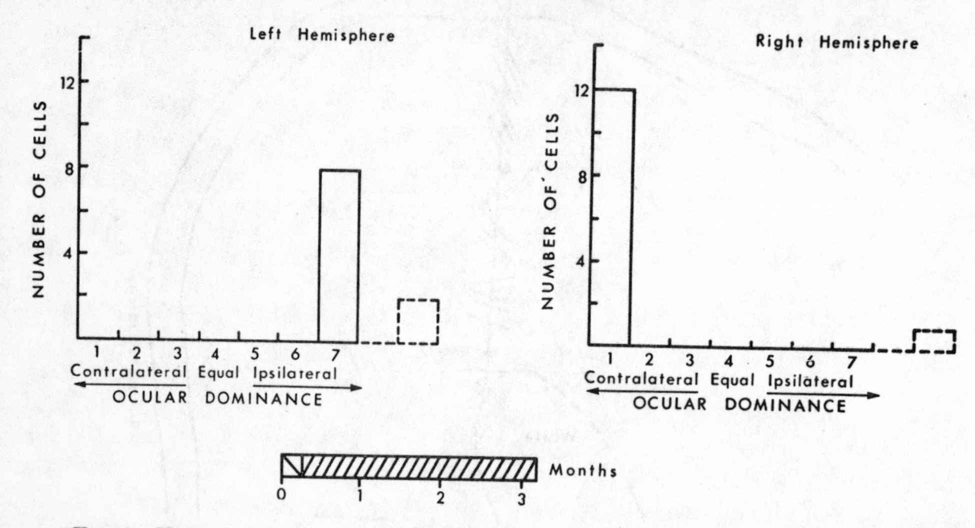

FIG. 5. Histograms showing ocular-dominance distribution of cells recorded in two penetrations, one in the left visual cortex and one in the right. A 3-month-old kitten with right eye closed by lid suture at 8 days (i.e., prior to normal eye-opening). Of a total of 23 cells, 3 were not influenced from either eye (interrupted lines). The remaining 20 could be driven only from the left (normally exposed) eye: 8 were recorded in the left hemisphere, and are therefore classed as group 7; 12 were recorded in the right hemisphere, and are classed as group 1.

animals. The striking differences in the negative phase support the single-unit findings in indicating that form deprivation and perhaps also moderate light deprivation during the first 3 months after birth can cause marked changes in the normal physiology of the striate cortex.

Deprivation of kittens with previous visual experience

In the first paper of this series (11) we showed that the geniculate atrophy resulting from 2–3 months of visual deprivation is much less if the animal has had 1 or 2 months of normal visual exposure, and that in the adult cat no detectable atrophy results from 3 months of monocular lid closure. The effects of delayed deprivation on the responses of cortical cells closely paralleled these anatomical findings. Figure 9 shows the ocular-dominance histo-

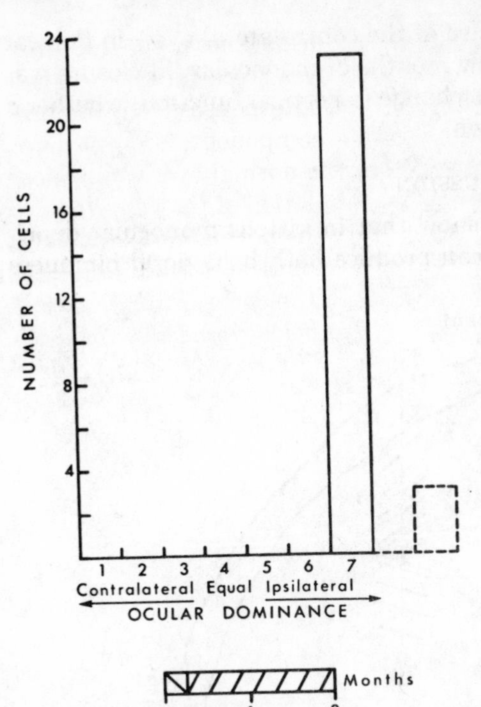

FIG. 6. Ocular-dominance distribution of 26 cells recorded in the visual cortex contralateral to the deprived eye. This 2-month-old kitten had had its right eye covered from the time of normal eye-opening by a translucent contact lens; the left eye was normally exposed. Twenty-three cells were driven by the normally exposed (left, or ipsilateral) eye, and were therefore assigned to ocular-dominance group 7. Three cells could not be activated by either eye (interrupted lines).

grams from a kitten deprived by lid closure at the age of 9 weeks, for a period of 4 months. For both hemispheres the eye that had not been occluded was again abnormally dominant. Particularly abnormal was the large number of cells driven exclusively by the normal eye. Now, however, some cells (11 of the 34) could be driven also by the deprived eye, and while the deprived eye was dominant in only 3 of these, it clearly exerted far more influence than the deprived eyes of kittens operated on at birth. The responses of all cells seemed normal, and, in contrast to experiments done in kittens deprived from birth, there were no cells that could not be driven.

A second kitten was light-deprived by lid closure at the age of 2 months for a period of only 1 month. The ocular-dominance distribution of a penetration contralateral to the deprived eye was clearly abnormal (Fig. 10), though it was less so than that of the previous kitten. Again, all cells were responsive to patterned-light stimulation and had normal receptive fields.

In one kitten the nictitating membrane was sewn across the right eye at 5 weeks, for a 3-month period. It will be recalled that this animal showed no geniculate atrophy (11). Nevertheless the ocular-dominance distribution was clearly abnormal (Fig. 11), suggesting that a decrease in effectiveness of the deprived eye in driving cortical cells is not necessarily of geniculate origin. Once again, the ocular-dominance distribution of cortical cells was less distorted than that of a kitten deprived by a translucent occluder from birth, for an even shorter time (Fig. 6).

Finally, a single penetration made in the left hemisphere of an adult cat whose right eyelids had been sewn for 3 months was completely normal. Here again it will be recalled that the lateral geniculate bodies were histologically normal (11). The ocular-dominance distribution of 26 cells (Fig.

12) shows the usual over-all dominance of the contralateral eye—in this cat the eye that had been deprived. A few months of monocular lid closure was thus not enough to cause any obvious change in cortical function; whether a longer period would have is not known.

Discussion

The results of the present paper show that in kittens monocular deprivation of light and form from birth can produce both behavioral blindness

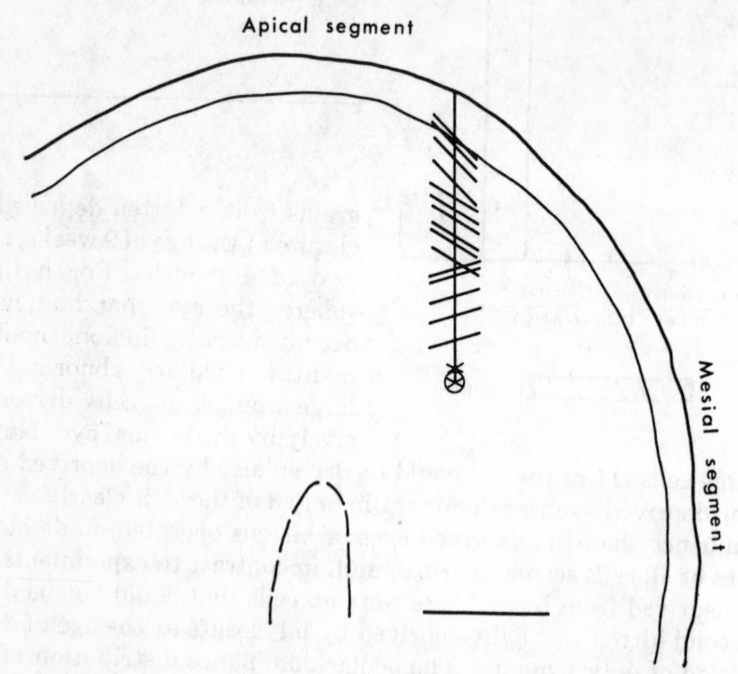

FIG. 7. Reconstruction of a microelectrode penetration through the left postlateral gyrus; ocular-dominance distribution of this penetration is given in Fig. 6. Conventions as in Fig. 4. Of the 17 recordings indicated, 8 were single-unit and 9 were 2-unit. Scale, 0.5 mm.

in the deprived eye and a failure of cortical cells to respond to that eye. These findings must be viewed in the light of those reported in the preceding paper (9), that the specific responses seen in cortical cells of normal adult cats are present also in newborn and very young visually inexperienced kittens. We conclude that monocular deprivation produces physiological defects in a system that was once capable of functioning.

In view of the anatomical findings described in the first paper of this series (11), the results of the present study may at first seem paradoxical: that deprivation by lid closure from birth should produce in the geniculate marked atrophy of cells with only mild physiological effects, and in the cortex just the reverse, no obvious anatomical changes but profound physiological deficits. The reasons for the differences in morphological effects in the two structures are easy to imagine. Each geniculate cell receives visual input almost exclusively from one eye, and the cells receiving afferents from a given eye are aggregated in separate layers, so that any anatomical change is easy to see, the more so since it tends to contrast with the normality of

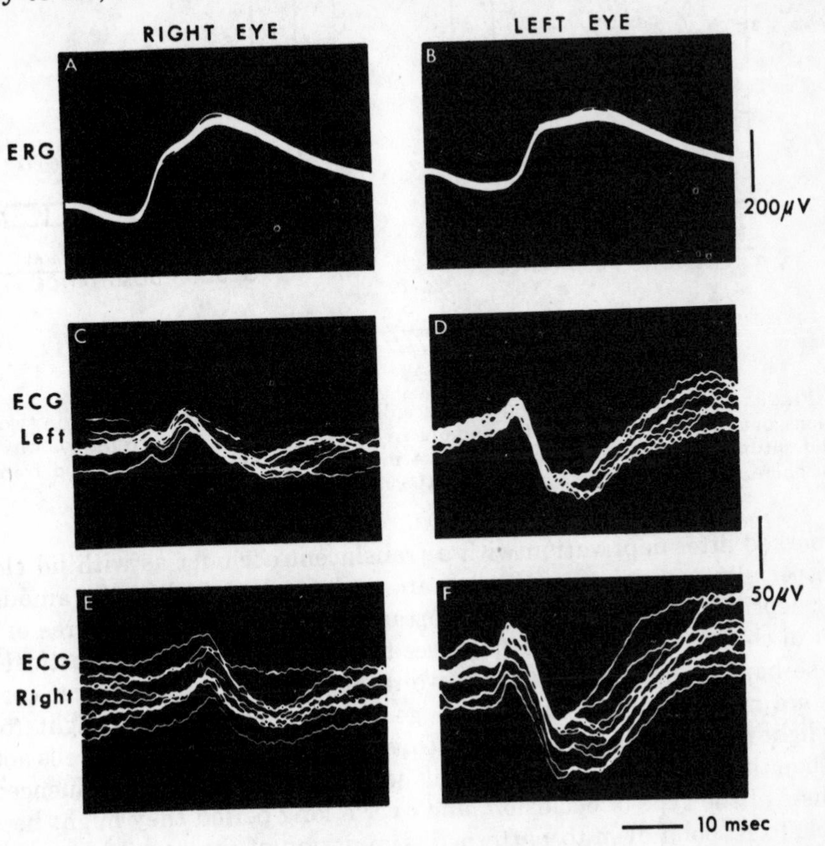

FIG. 8. Corneal electroretinograms and cortical evoked potentials (ECGs) in a 2-month-old kitten whose right eye had been covered by a translucent contact lens from the time of normal eye-opening (same kitten as in Figs. 6 and 7). *A, C, E:* stimulation of right (previously occluded) eye. *B, D, F:* stimulation of left (normal) eye. *A, B:* electroretinograms. *C, D:* cortical evoked potentials, left hemisphere. *E, F:* cortical evoked potentials, right hemisphere. Positive deflections upward.

the adjacent layers. The majority of cortical cells, on the other hand, have a binocular input (7), and the relatively few cells that are fed exclusively from one eye are intermixed with the others. One would therefore not expect a selective atrophy of the monocularly driven cells to stand out histologically.

It remains to account for the striking unresponsiveness of cortical cells to stimulation of the deprived eye, compared with the relative normality of geniculate responses and receptive fields. The cortical impairment was just

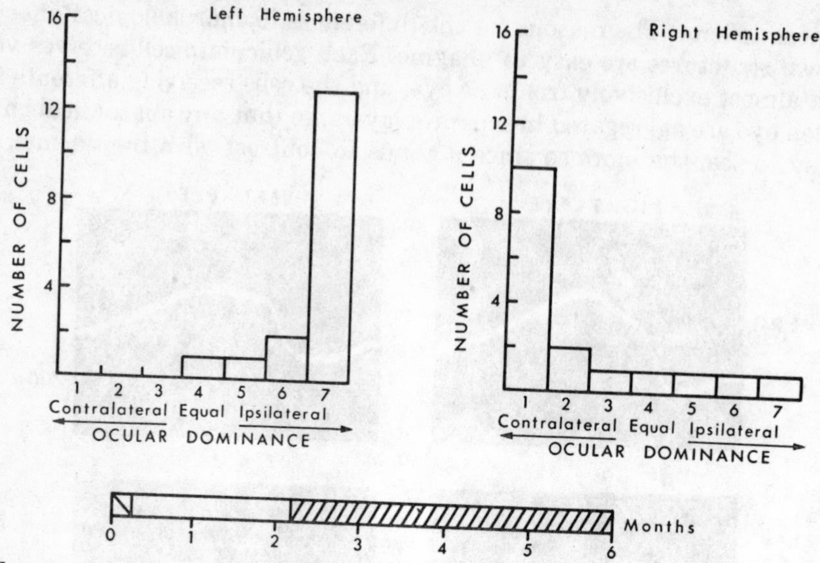

FIG. 9. Histograms of ocular-dominance distribution of 34 cells recorded in two penetrations, one in the left visual cortex and one in the right. Kitten whose right eye was closed by lid suture at 9 weeks, for a period of 4 months. Seventeen cells recorded from each hemisphere. All cells were influenced by patterned-light stimulation.

as marked after deprivation with a translucent occluder as with lid closure an especially surprising result since in the lateral geniculate the amount of light deprivation seemed to be important in determining the degree of anatomical change (11). These differences between geniculate and cortical cells can perhaps be best understood if we recall that in the normal cat cortical cells are much less responsive than geniculate cells to diffuse light (5, 6). Any light reaching the retina would thus help to keep geniculate cells active. On the other hand, most cortical cells would be practically uninfluenced regardless of the type of occlusion, and over a long period they might become unable to respond even to patterned stimulation of the deprived eye.

In discussing morphological changes in the lateral geniculate following visual deprivation (11), we pointed out that the maintained activity persisting in the retina under these conditions of lid suture or translucent occlusion is insufficient to prevent the atrophy. The same obviously applies to the present cortical recordings: the functional pathway up to the cortex was evidently not maintained by whatever activity persisted in the lateral geniculate body.

Though we have no direct way of knowing the exact site of the abnormality responsible for cortical unresponsiveness, the main defect must be central to the lateral geniculate body, since geniculate cells respond well to stimulation of the deprived eye. Nevertheless these geniculate impulses have no apparent effect on cortical cells, even though the cortical cells fire per-

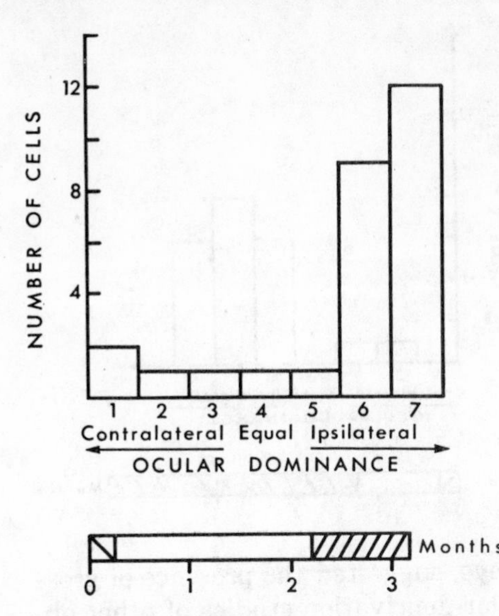

FIG. 10. Ocular-dominance histogram from a kitten whose right eye was closed by lid suture at the age of 9 weeks, for 1 month. Twenty-one cells were recorded from the left visual cortex. All cells were influenced by patterned-light stimulation.

fectly well to stimulation of the normal eye. This suggests that the abnormality is in the region of the synapse between the axon terminals of geniculate cells (those receiving input from the deprived eye) and the cortical cells on which these terminals end.

Though the most abnormal feature in the physiological studies was the ocular-dominance distribution, there was another consistent difference from normal penetrations: in every experiment in kittens deprived from birth there were a few cells that could not be driven from either eye. This probably cannot be explained by the immaturity of the kittens, since in a previous study of even younger animals all of the cells could be driven by appropriate stimuli (9). A more likely explanation is that the unresponsive cells were connected exclusively with the covered eye—at least the proportion of nondriven cells was about what one would expect on that assumption. Had it not been for their maintained activity these cells would probably have gone undetected, and one wonders if the maintained firing does not reflect some other, nonvisual input.

While the site of the physiological defect may be within the cortex, as suggested above, it would probably be wrong to assume that abnormalities at the geniculate level did not also contribute to the unresponsiveness of the cortical cells. Many of the geniculate cells were atrophic; in recordings from the geniculate there was an over-all decrease in activity in the layers connected with the deprived eye, and a few fields seemed abnormal (11). The cortical evoked responses to a flash made it clear that impulses in at least some geniculate fibers associated with the deprived eye reached the cortex of both hemispheres, but the marked increase in latency suggests that this input may have been abnormal. Thus it is probable that defects at several levels, from retina to primary visual cortex, contributed to the observed physiological abnormalities. At present we have no direct way of assessing their relative importance.

The behavior of our monocularly deprived kittens, under conditions in

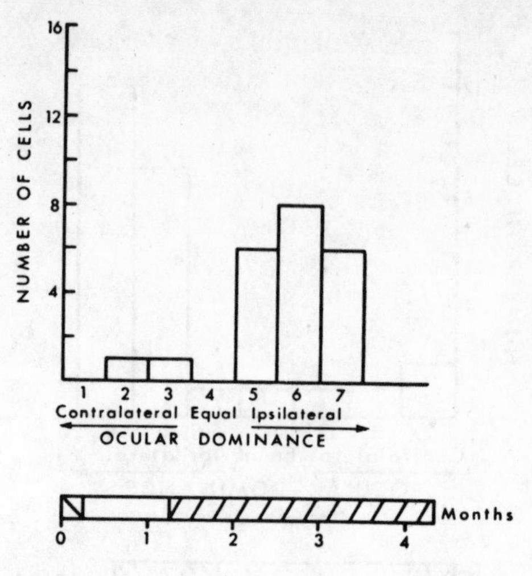

FIG. 11. Ocular-dominance histogram of 22 cells recorded from the left visual cortex; kitten with nictitating membrane sutured across the right eye at an age of 5 weeks, for 3 months. The left (ipsilateral, normal) eye dominated markedly, and all cells were influenced by patterned-light stimulation.

which they could use only the deprived eye, suggested the presence of gross visual deficits. This confirms the binocular-deprivation studies of other observers in several different mammalian species (2, 4, 10, 3). In monocularly deprived animals the visual pathway beyond the point of convergence of impulses from the two eyes was presumably intact, since cortical cells were actively driven from the normal eye, and since the animals were able to see with that eye. This suggests that also in bilaterally deprived animals the defect may not necessarily be at a far central level: e.g., the defect need not be in visually guided motor function, or the result of some emotional disturbance. There may indeed be such defects, but our results make it likely that abnormalities exist at a more peripheral level as well.

In interpreting the blindness resulting from raising animals in darkness it has generally been assumed that some of the neural connections necessary for vision are not present at birth, and that their development depends on visual experience early in life. Our results suggest the alternative possibility, that certain connections are intact at birth and become defective through disuse. One must, however, make one reservation: deprivation of one eye may be quite different in its morphological and physiological effects from binocular dark-raising. Conceivably if one eye is not stimulated, the fate of its projections in the central visual pathway may partly depend on whether or not the other eye is stimulated. This question of a possible competition between the eyes can only be settled experimentally, by studying animals that have been binocularly deprived of vision. For example, Baxter (1), comparing records from the visual cortex of normal and dark-raised kittens, could find no striking difference in the size or shape of the evoked potentials in the two groups. This finding contrasts sharply with the difference we saw in cortical potentials evoked from the normal, as opposed to the deprived eye (Fig. 8), and it remains to be learned whether or not the discrepancy reflects a difference in the two types of preparation.

The susceptibility of very young kittens to a few months of visual deprivation apparently does not extend to older animals, since there is a detectable lessening of effects when deprivation is begun at 2 months, and an absence of behavioral or physiological effects in adults. This is a clear demonstration of a pronounced difference between kittens and adults in susceptibility to deprivation, a difference one might have expected from the profound visual defects observed after removal of congenital cataracts in man, as opposed to the absence of blindness on removal of cataracts acquired later in life.

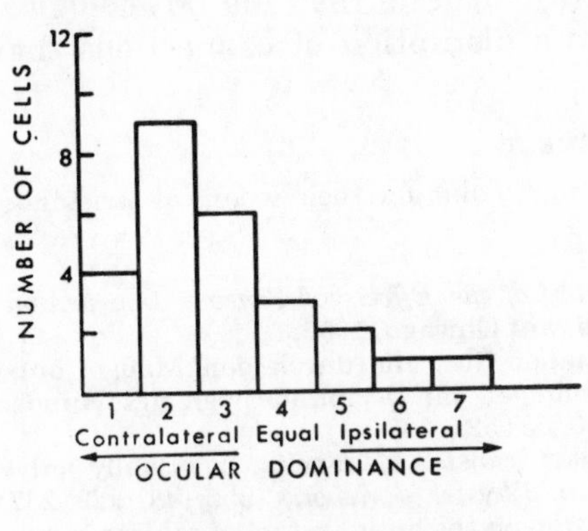

FIG. 12. Ocular-dominance distribution of 26 cells recorded in the left visual cortex of an adult cat whose right eye had been covered by lid suture for a period of 3 months. As in the normal cat, the contralateral eye—in this case the deprived eye—dominated.

SUMMARY

1. Single-unit recordings were made from striate cortex of kittens in which one eye had been deprived of vision either from birth or subsequently, and for various periods of time.

2. Kittens deprived from birth for 2–3 months showed profoundly defective vision in the deprived eye. Visual placing and following reactions were absent, and there was no hint of any ability to perceive form. Pupillary light reflexes were nevertheless normal.

3. In kittens deprived from birth, either by suturing the lids of one eye or by covering the cornea with a translucent contact occluder, the great majority of cortical cells were actively driven from the normal eye, with normal receptive fields. On the other hand, only 1 cell out of 84 was at all influenced by the deprived eye, and in that cell the receptive fields in the two eyes were abnormal. A few cells could not be driven from either eye.

4. In one 2-month-old kitten monocularly deprived with a translucent contact occluder, the corneal electroretinograms were normal in the two eyes. On flashing a light in the previously occluded eye the slow-wave potentials evoked in the visual cortex of the two hemispheres were highly abnormal, compared with responses from the normal eye.

5. One to two months of normal visual experience prior to monocular deprivation by lid suture or with a translucent occluder reduced the severity of the physiological defect, even though the ability of the deprived eye to influence cortical cells was still well below normal. On the other hand, 3 months of deprivation by lid closure in an adult cat produced no detectable physiological abnormality.

6. We conclude that monocular deprivation in kittens can lead to unresponsiveness of cortical cells to stimulation of the deprived eye, and that the defect is most severe in animals deprived from bith. The relative normality of responses in newborn kittens (9) suggests that the physiological defect in the deprived kittens represents a disruption of connections that were present at birth.

ACKNOWLEDGMENT

We express our thanks to Jane Chen and Janet Tobie for their technical assistance.

REFERENCES

1. BAXTER, B. L. *An Electrophysiological Study of the Effects of Sensory Deprivation* Doctoral dissertation (unpublished). University of Chicago, 1959.
2. BERGER, H. Experimentell-anatomische Studien über die durch den Mangel optischer Reize veranlassten Entwicklungshemmungen im Occipitallappen des Hundes und der Katze. *Arch. Psychiat. Nervenkr.*, 1900, *33:* 521–567.
3. CHOW, K. L. AND NISSEN, H. W. Interocular transfer of learning in visually naïve and experienced infant chimpanzees. *J. comp. Physiol. Psychol.*, 1955, *48:* 229–237.
4. GOODMAN, L. Effect of total absence of function on the optic system of rabbits. *Amer. J. Physiol.*, 1932, *100:* 46–63.
5. HUBEL, D. H. Single unit activity in striate cortex of unrestrained cats. *J. Physiol.*, 1959, *147:* 226–238.
6. HUBEL, D. H. AND WIESEL, T. N. Receptive fields of single neurones in the cat's striate cortex. *J. Physiol.*, 1959, *148:* 574–591.
7. HUBEL, D. H. AND WIESEL, T. N. Receptive fields, binocular interaction and functional architecture in the cat's visual cortex. *J. Physiol*, 1962, *160:* 106–154.
8. HUBEL, D. H. AND WIESEL, T. N. Shape and arrangement of columns in cat's striate cortex. *J. Physiol.*, 1963, *165:* 559–568.
9. HUBEL, D. H. AND WIESEL, T. N. Receptive fields of cells in striate cortex of very young, visually inexperienced kittens. *J. Neurophysiol.*, 1963, *26:* 994–1002.
10. RIESEN, A. H., KURKE, M. I., AND MELLINGER, J. C. Interocular transfer of habits learned monocularly in visually naïve and visually experienced cats. *J. comp. Physiol. Psychol.*, 1953, *46:* 166–172.
11. WIESEL, T. N. AND HUBEL, D. H. Effects of visual deprivation on morphology and physiology of cells in the cat's lateral geniculate body. *J. Neurophysiol.*, 1963, *26:* 978–993.
12. ZETTERSTRÖM, B. The effect of light on the appearance and development of the electroretinogram in newborn kittens. *Acta physiol. scand.*, 1955, *35:* 272–279.

BIBLIOGRAPHY

References to papers reproduced are in bold-faced type and those important for further reading are in capital letters.

Adolph, A. R., 1968. Thermal and spectral sensitivities of discrete slow potentials in *Limulus* eye. *J. Gen. Physiol.* **52**: 584–599.

Adrian, E. D., and Y. Zotterman, 1926. The impulses produced by sensory nerve-endings. Part II. The response of a single end-organ. *J. Physiol.* **61**: 151–171.

Armstrong, C., and L. Binstock, 1965. Anomalous rectification in the squid-giant axon injected with tetraethylammonium chloride. *J. Gen. Physiol.* **48**: 859–872.

Attardi, D., and R. W. Sperry, 1963. Preferential selection of central pathways by regenerating optic fibers. *Exptl. Neurol.* **7**: 46–64.

Atwood, H. L., and A. Jones, 1967. Presynaptic inhibition in crustacean muscle: axo-axonal synapse. *Experientia* **23**: 1–6.

Auerbach, A. A., and M. V. L. Bennett, 1969. A rectifying electrotonic synapse in the central nervous system of a vertebrate. *J. Gen. Physiol.* **53**: 211–237.

Axelsson, J., and S. Thesleff, 1959. A study of supersensitivity in denervated mammalian skeletal muscle. *J. Physiol.* **147**: 178–193.

Baker, P. F., A. L. Hodgkin, and T. I. Shaw, 1962a. Replacement of the axoplasm of giant nerve fibres with artificial solutions. *J. Physiol.* **164**: 330-354. **(Methods reproduced.)**

Baker, P. F., A. L. Hodgkin, and T. I. Shaw, 1962b. The effects of changes in internal ionic concentrations on the electrical properties of perfused giant axons. *J. Physiol.* **164**: 355-374.

Barlow, H. B., and W. R. Levick, 1965. The mechanism of directionally selective units in rabbit's retina. *J. Physiol.* **178**: 477–504.

Baylor, D. A., and J. G. Nicholls, 1969. Chemical and electrical synaptic connexions between cutaneous mechanoreceptor neurones in the central nervous system of the leech. *J. Physiol.* **203**: 591–609.

Bazemore, A., K. A. C. Elliott, and E. Florey, 1957. Isolation of Factor I. *J. Neurochem.* **1**: 334–339.

Behrens, M. E., and V. J. Wulff, 1965. Light-initiated responses of retinula and eccentric cells in the *Limulus* lateral eye. *J. Gen. Physiol.* **48**: 1081–1093.

Bennett, M. V. L., 1966. Physiology of electrotonic junctions. *Ann. N. Y. Acad. Sci.* **137**: 509–539.

Bernstein, J., 1902. Untersuchungen zur Thermodynamik der bioelektrischen Ströme. *Pflügers Arch. ges. Physiol.* **92**: 521–562.

Binstock, L., and L. Goldman, 1969. Current- and voltage-clamped studies on *Myxicola* giant axons. Effects of tetrodotoxin. *J. Gen. Physiol.* **54**: 730–740.

BITTNER, G. D., AND J. HARRISON, 1970. A reconsideration of the Poisson Hypothesis for transmitter release at the crayfish neuromuscular junction. *J. Physiol.* **206**: 1–23.

BOISTEL, J., AND P. FATT, 1958. Membrane permeability change during transmitter action in crustacean muscle. *J. Physiol.* **144**: 176–191.

Boyd, I. A., and A. R. Martin, 1956a. Spontaneous subthreshold activity at mammalian neuromuscular junctions. *J. Physiol.* **132**: 61–73.

Boyd, I. A., and A. R. Martin, 1956b. The end-plate potential in mammalian muscle. *J. Physiol.* **132**: 74–91.

Brazier, M. A. B., 1959. The historical development of neurophysiology. *In* H. W. Magoun, Ed., *Handbook of Physiology,* Sec. 1, "Neurophysiology," Vol. I, Ch. 1, pp. 1–58. Washington: Amer. Physiol. Soc.

Brightman, M. W., and T. S. Reese, 1969. Junctions between intimately apposed cell membranes in the vertebrate brain. *J. Cell. Biol.* **40**: 648–677.

Bullock, T. H., and G. A. Horridge, 1965. *Structure and Function in the Nervous Systems of Invertebrates,* 2 vols., 1719 pp. San Francisco: Freeman.

Carpenter, D. O., and B. O. Alving, 1968. A contribution of an electrogenic Na^+ pump to membrane potential in *Aplysia* neurons. *J. Gen. Physiol.* **52**: 1–21.

Chandler, W. K., A. L. Hodgkin, and H. Meves, 1965. The effect of changing the internal solution on sodium inactivation and related phenomena in giant axons. *J. Physiol.* **180**: 821–836.

Cole, K. S., 1968. *Membranes, Ions and Impulses.* Berkeley and Los Angeles: University of California Press, 569 pp.

COLE, K. S., AND H. J. CURTIS, 1939. Electric impedance of the squid giant axon during activity. *J. Gen. Physiol.* **22**: 649–670.

Cone, R. A., 1968. The early receptor potential. *Proceedings of the International School of Physics Enrico Fermi* (in press).

Coombs, J. S., D. R. Curtis, and J. C. Eccles, 1957. The generation of impulses in motoneurones. *J. Physiol.* **139**: 232–249.

Coombs, J. S., J. C. Eccles, and P. Fatt, 1955a. Electrical properties of the motoneurone membrane. *J. Physiol.* **130**: 291–325. **(Methods reproduced.)**

Coombs, J. S., J. C. Eccles, and P. Fatt, 1955b. The specific ionic conductances and the ionic movements across the motoneuronal membrane that produce the inhibitory post-synaptic potential. *J. Physiol.* **130**: 326–373.

Curtis, H. J., and K. S. Cole, 1942. Membrane resting and action potentials from the squid giant axon. *J. Cell. Comp. Physiol.* **19**: 135–144.

Dale, H. H., W. Feldberg, and M. Vogt, 1936. Release of acetylcholine at voluntary motor nerve endings. *J. Physiol.* **86**: 353–380.

Davis, W. J., 1970. Motoneuron morphology and synaptic contacts: determination by intracellular dye injection. *Science* **168**: 1358–1360.

Del Castillo, J., and B. Katz, 1954a. Quantal components of the end-plate potential. *J. Physiol.* **124**: 560–573.

Del Castillo, J., and B. Katz, 1954b. Statistical factors involved in neuromuscular facilitation and depression. *J. Physiol.* **124**: 574–585.

DEL CASTILLO, J., AND B. KATZ, 1955. On the localization of acetylcholine receptors. *J. Physiol.* **128**: 157–181.

Del Castillo, J., and B. Katz, 1956. Biophysical aspects of neuromuscular transmission. *Progr. Biophys.* **6**: 121–170.

Dowling, J. E., 1963. Neural and photochemical mechanisms of visual adaptation in the rat. *J. Gen. Physiol.* **46**: 1287–1301.

DOWLING, J. E., AND F. S. WERBLIN, 1969. Organization of retina of the mudpuppy, *Necturus maculosus.* I. Synaptic structure. *J. Neurophysiol.* **32**: 315–338.

Dudel, J., 1965. The mechanism of presynaptic inhibition at the crayfish neuromuscular junction. *Pflügers Arch. ges. Physiol.* **284**: 66–80.

DUDEL, J., AND S. W. KUFFLER, 1961a. The quantal nature of transmission and spontaneous miniature potentials at the crayfish neuromuscular junction. *J. Physiol.* **155**: 514-529. **(Methods reproduced.)**

DUDEL, J., AND S. W. KUFFLER, 1961b. Mechanism of facilitation at the crayfish neuromuscular junction. *J. Physiol.* **155**: 530-542.

Dudel, J., and S. W. Kuffler, 1961c. Presynaptic inhibition at the crayfish neuromuscular junction. *J. Physiol.* **155**: 543-562.

Eccles, J. C., 1964. *The Physiology of Synapses.* Berlin: Springer-Verlag and New York: Academic Press, 316 pp.

Eccles, J. C., R. Eccles, and M. Ito, 1964. Effects produced on inhibitory postsynaptic potentials by the coupled injection of cations and anions into motoneurones. *Proc. Roy. Soc.* *B***160**: 197-210.

Eccles, J. C., R. Eccles, and F. Magni, 1961. Central inhibitory action attributable to presynaptic depolarization produced by muscle afferent volleys. *J. Physiol.* **159**: 147-166.

Eccles, J. C., M. Ito, and J. Szentágothai, 1967. *The Cerebellum as a Neuronal Machine.* New York: Springer-Verlag, 335 pp.

Eccles, J. C., F. Magni, and W. D. Willis, 1962. Depolarization of central terminals of Group I afferent fibres from muscle. *J. Physiol.* **160**: 62-93.

Eckert, R., 1963. Electrical interaction of paired ganglion cells in the leech. *J. Gen. Physiol.* **46**: 573-587.

Edwards, C., and D. Ottoson, 1958. The site of impulse initiation in a nerve cell of a crustacean stretch receptor. *J. Physiol.* **143**: 138-148.

EYZAGUIRRE, C., AND S. W. KUFFLER, 1955a. Processes of excitation in the dendrites and in the soma of single isolated sensory nerve cells of the lobster and crayfish. *J. Gen. Physiol.* **39**: 87-119.

Eyzaguirre, C., and S. W. Kuffler, 1955b. Further study of soma, dendrite, and axon excitation in single neurons. *J. Gen. Physiol.* **39**: 121-153.

Fatt, P., and B. L. Ginsborg, 1958. The ionic requirements for the production of action potentials in crustacean muscle fibres. *J. Physiol.* **142**: 516-543.

Fatt, P., and B. Katz, 1951. An analysis of the end-plate potential recorded with an intra-cellular electrode. *J. Physiol.* **115**: 320-370.

Fatt, P., and B. Katz, 1952. Spontaneous subthreshold activity at motor nerve endings. *J. Physiol.* **117**: 109-128.

Fatt, P., and B. Katz, 1953. The effect of inhibitory nerve impulses on a crustacean muscle fibre. *J. Physiol.* **121**: 374-389.

Frankenhaeuser, B., and A. L. Hodgkin, 1957. The action of calcium on the electrical properties of squid axons. *J. Physiol.* **137**: 218-244.

Frisch, L., Ed., 1965. Sensory Receptors. *Cold Spring Harbor Symposia* **30**, 649 pp.

Fuortes, M. G. F., and G. F. Poggio, 1963. Transient responses to sudden illumination in cells of the eye of *Limulus. J. Gen. Physiol.* **46**: 435-452.

Fuortes, M. G. F., and S. Yeandle, 1964. Probability of occurrence of discrete potential waves in the eye of *Limulus. J. Gen. Physiol.* **47**: 443-463.

Furshpan, E. J., and D. D. Potter, 1959. Transmission at the giant motor synapses of the crayfish. *J. Physiol.* **145**: 289-325.

Furukawa, T., and E. J. Furshpan, 1963. Two inhibitory mechanisms in the Mauthner neurons of goldfish. *J. Neurophysiol.* **26**: 140-176.

Gaze, R. M., M. J. Keating, G. Szekely, and L. Beazley, 1970. Binocular interaction in the formation of specific intertectal neuronal connexions. *Proc. Roy. Soc. (B)* **175**: 107-147.

Geduldig, D., and D. Junge, 1968. Sodium and calcium components of action potentials in the *Aplysia* giant neurone. *J. Physiol.* **199**: 347-365.

Gouras, P., 1968. Identification of cone mechanisms in monkey ganglion cells. *J. Physiol.* **199**: 533-547.

Granit, R., 1955. *Receptors and Sensory Perception.* New Haven, Conn.: Yale University Press, 366 pp.

Gray, E. G., 1962. A morphological basis for pre-synaptic inhibition? *Nature* **193**: 82-83.

Grinnell, A. D., 1966. A study of the interaction between motoneurones in the frog spinal cord. *J. Physiol.* **182**: 612-648.

Grundfest, H., 1961. Excitation by hyperpolarizing potentials. A general theory of receptor activities. *In* E. Florey, Ed., *Nervous Inhibition,* pp. 326-341. New York: Pergamon.

Hagins, W. A., H. V. Zonana, and R. G. Adams, 1962. Local membrane current in the outer segments of squid photoreceptors. *Nature* **194**: 844-847.

Hagiwara, S., H. Hayashi, and K. Takahashi, 1969. Calcium and potassium currents of the membrane of a barnacle muscle fibre in relation to the calcium spike. *J. Physiol.* **205**: 115-129.

Hagiwara, S., K. Kusano, and S. Saito, 1960. Membrane changes in crayfish stretch receptor neuron during synaptic inhibition and under action of gamma-aminobutyric acid. *J. Neurophysiol.* **23**: 505-515.

Hagiwara, S., and H. Morita, 1962. Electrotonic transmission between two nerve cells in leech ganglion. *J. Neurophysiol.* **25**: 721-731.

Hagiwara, S., and K. Naka, 1964. The initiation of spike potential in barnacle muscle fibers under low intracellular Ca^{++}. *J. Gen. Physiol.* **48**: 141-146.

Hagiwara, S., A. Watanabe, and N. Saito, 1959. Potential changes in syncytial neurons of lobster cardiac ganglion. *J. Neurophysiol.* **22**: 554-572.

Hartline, H. K., 1938. The response of single optic nerve fibers of the vertebrate eye to illumination of the retina. *Amer. J. Physiol.* **121**: 400-415.

Hartline, H. K., 1940. The nerve messages in the fibers of the visual pathway. *J. Opt. Soc. Amer.* **30**: 239-247.

HARTLINE, H. K., AND C. H. GRAHAM, 1932. Nerve impulses from single receptors in the eye. *J. Cell. Comp. Physiol.* **1**: 277-295.

Hartline, H. K., and F. Ratliff, 1957. Inhibitory interaction of receptor units in the eye of *Limulus. J. Gen. Physiol.* **40**: 357-376.

Hartline, H. K., H. G. Wagner, and F. Ratliff, 1956. Inhibition in the eyes of *Limulus. J. Gen. Physiol.* **39**: 651-673.

Hecht, S., C. Shlaer, and M. H. Pirenne, 1942. Energy, quanta, and vision. *J. Gen. Physiol.* **25**: 819-840.

Hille, B., 1967. The selective inhibition of delayed potassium currents in nerve by tetraethylammonium ion. *J. Gen. Physiol.* **50**: 1287-1302.

Hodgkin, A. L., 1937. Evidence for electrical transmission in nerve, Parts I and II. *J. Physiol.* **90**: 183-232.

Hodgkin, A. L.,1938. The subthreshold potentials in a crustacean nerve fibre. *Proc. Roy. Soc. (B)* **126**: 87-121.

Hodgkin, A. L., 1939. The relation between conduction velocity and the electrical resistance outside a nerve fibre. *J. Physiol.* **94**: 560-570.

HODGKIN, A. L., 1958. The Croonian lecture: Ionic movements and electrical activity in giant nerve fibres. *Proc. Roy. Soc. (B)* **148**: 1-37.

HODGKIN, A. L., 1964. *The Conduction of the Nervous Impulse.* Ryerson, Toronto: Liverpool University and Springfield, Ill.: Charles C Thomas, 108 pp.

Hodgkin, A. L., and A. F. Huxley, 1939. Action potentials recorded from inside a nerve fibre. *Nature* (London) **144**: 710.

Hodgkin, A. L., and A. F. Huxley, 1952a. Currents carried by sodium and potassium ions through the membrane of the giant axon of *Loligo. J. Physiol.* **116**: 449–472.

Hodgkin, A. L., and A. F. Huxley, 1952b. The components of membrane conductance in the giant axon of *Loligo. J. Physiol.* **116**: 473–496.

Hodgkin, A. L., and A. F. Huxley, 1952c. The dual effect of membrane potential on sodium conductance in the giant axon of *Loligo. J. Physiol.* **116**: 497–506.

Hodgkin, A. L., and A. F. Huxley, 1952d. A quantitative description of membrane current and its application to conduction and excitation in nerve. *J. Physiol.* **117**: 500–544.

Hodgkin, A. L., and A. F. Huxley, 1953. Movement of radioactive potassium and membrane current in a giant axon. *J. Physiol.* **121**: 403–414.

Hodgkin, A. L., A. F. Huxley, and B. Katz, 1952. Measurement of current-voltage relations in the membrane of the giant axon of *Loligo. J. Physiol.* **116**: 424–448.

Hodgkin, A. L., and B. Katz, 1949. The effect of sodium ions on the electrical activity of the giant axon of the squid. *J. Physiol.* **108**: 37–77.

Hodgkin, A. L., and R. D. Keynes, 1955. Active transport of cations in giant axons from *Sepia* and *Loligo. J. Physiol.* **128**: 28–60.

Hodgkin, A. L., and W. A. H. Rushton, 1946. The electrical constants of a crustacean nerve fibre. *Proc. Roy. Soc. (B)* **133**: 444–479.

Hubel, D. H., and T. N. Wiesel, 1962. Receptive fields, binocular interaction and functional architecture in the cat's visual cortex. *J. Physiol.* **160**: 106–154.

HUBEL, D. H., AND T. N. WIESEL, 1968. Receptive fields and functional architecture of monkey striate cortex. *J. Physiol.* **195**: 215–243.

Hubel, D. H., and T. N. Wiesel, 1970. The period of susceptibility to the physiological effects of unilateral eye closure in kittens. *J. Physiol.* **206**: 419–436.

Hutter, O. F., and W. Trautwein, 1956. Vagal and sympathetic effects on the pacemaker fibers in the sinus venosus of the heart. *J. Gen. Physiol.* **39**: 715–733.

Huxley, A. F., and R. Stämpfli, 1949. Evidence for saltatory conduction in peripheral myelinated nerve fibres. *J. Physiol.* **108**: 315–339.

Kandel, E., W. T. Frazier, and R. E. Coggeshall, 1967. Opposite synaptic actions mediated by different branches of an identifiable interneuron in *Aplysia. Science* **155**: 346–349.

Kandel, E. R., W. T. Frazier, R. Washizu, and R. E. Coggeshall, 1967. Direct and common connections among identified neurons in *Aplysia. J. Neurophysiol.* **30**: 1352–1376.

Kaneko, A., 1970. Physiological and morphological identification of horizontal, bipolar and amacrine cells in goldfish retina. *J. Physiol.* **207**: 623–633.

Katz, B., 1937. Experimental evidence for a non-conducted response of nerve to subthreshold stimulation. *Proc. Roy. Soc. (B)* **124**: 244–276.

Katz, B., 1950a. Action potentials from a sensory nerve ending. *J. Physiol.* **111**: 248–260.

KATZ, B., 1950b. Depolarization of sensory terminals and the initiation of impulses in the muscle spindle. *J. Physiol.* **111**: 261–282.

Katz, B., 1962. The Croonian lecture: The transmission of impulses from nerve to muscle, and the subcellular unit of synaptic action. *Proc. Roy. Soc. (B)* **155**: 455–477.

Katz, B., 1966. *Nerve, Muscle, and Synapse.* New York: McGraw-Hill, 193 pp.

Katz, B., 1969. *The Release of Neural Transmitter Substances.* Ryerson, Toronto: Liverpool University, 60 pp.

Katz, B., and R. Miledi, 1965. The measurement of synaptic delay, and the time course of acetylcholine release at the neuromuscular junction. *Proc. Roy. Soc. (B)* **161**: 483–495.

Katz, B., and R. Miledi, 1967a. Release of acetylcholine from nerve endings by graded electric pulses. *Proc. Roy. Soc. (B)* **167**: 23–28.

Katz, B., and R. Miledi, 1967b. The timing of calcium action during neuromuscular transmission. *J. Physiol.* **189**: 535–544.

KATZ, B., AND R. MILEDI, 1967c. A study of synaptic transmission in the absence of nerve impulses. *J. Physiol.* **192**: 407–436.

Katz, B., and R. Miledi, 1969. Tetrodotoxin-resistant electric activity in presynaptic terminals. *J. Physiol.* **203**: 459–487.

Katz, B., and R. Miledi, 1970. Further study of the role of calcium in synaptic transmission. *J. Physiol.* **207**: 789–801.

Kehoe, J., and P. Ascher, 1970. Re-evaluation of the synaptic activation of an electrogenic sodium pump. *Nature* **225**: 820–823.

Kennedy, D., W. H. Evoy, and J. T. Hanawalt, 1966. Release of coordinated behavior in crayfish by single central neurons. *Science* **154**: 917–919.

Kennedy, D., A. I. Selverston, and M. P. Remler, 1969. Analysis of restricted neural networks. *Science* **164**: 1488–1496.

Keynes, R. D., and P. R. Lewis, 1951. The sodium and potassium content of cephalopod nerve fibres. *J. Physiol.* **114**: 151–182.

Kravitz, E. A., S. W. Kuffler, and D. D. Potter, 1963. Gamma-aminobutyric acid and other blocking compounds in Crustacea. III. Their relative concentrations in separated motor and inhibitory axons. *J. Neurophysiol.* **26**: 739–751.

Kravitz, E. A., P. B. Molinoff, and Z. W. Hall, 1965. A comparison of the enzymes and substrates of gamma-aminobutyric acid metabolism in lobster excitatory and inhibitory axons. *Proc. Nat. Acad. Aci.* **54**: 778–782.

Krnjević, K., and S. Schwartz, 1967. The action of γ-aminobutyric acid on cortical neurones. *Exptl. Brain Res.* **3**: 320–336.

Kuffler, S. W., 1953. Discharge patterns and functional organization of mammalian retina. *J. Neurophysiol.* **16**: 37–68.

KUFFLER, S. W., AND C. EDWARDS, 1958. Mechanism of gamma-aminobutyric acid (GABA) action and its relation to synaptic inhibition. *J. Neurophysiol.* **21**: 589–610.

Kuno, M., 1964a. Quantal components of excitatory synaptic potentials in spinal motoneurones. *J. Physiol.* **175**: 81–99.

KUNO, M., 1964b. Mechanism of facilitation and depression of the excitatory synaptic potential in spinal motoneurones. *J. Physiol.* **175**: 100–112.

Lettvin, J. Y., H. R. Maturana, W. S. McCulloch, and W. H. Pitts, 1959. What the frog's eye tells the frog's brain. *Proc. Inst. Radio Engrs.* **47**: 1940–1951.

Loewenstein, W. R., C. A. Terzuolo, and Y. Washizu, 1963. Separation of transducer and impulse-generating processes in sensory receptors. *Science* **142**: 1180–1181.

Loewi, O., 1921. Über humorale Übertragbarkeit der Herznervenwirkung. I. Mitteilung. *Pflüg. Arch. ges. Physiol.* **189**: 239–242.

Loewi, O., and E. Navratil, 1926. Über humorale Übertragbarkeit der Herznervenwirkung. X. Mitteilung. Über das Schicksal des Vagusstoffs. *Pflügers Arch. ges. Physiol.* **214**: 678–688.

Lorente de No, R., 1947. *A Study of Nerve Physiology,* Vols. 1 and 2 of Studies from the Rockefeller Institute for Medical Research, Vols. 131 and 132.

Martin, A. R., 1966. Quantal nature of synaptic transmission. *Physiol. Rev.* **46**: 51-66.

MARTIN, A. R., AND G. PILAR, 1963a. Dual mode of synaptic transmission in the avian ciliary ganglion. *J. Physiol.* **168**: 443–463.

MARTIN, A. R., AND G. PILAR, 1963b. Transmission through the ciliary ganglion of the chick. *J. Physiol.* **168**: 464–475.

Matthews, B. H. C., 1931. The response of a single end organ. *J. Physiol.* **71**: 64–110.

Maturana, H. R., J. Y. Lettvin, W. S. McCulloch, and W. H. Pitts, 1959. Evidence that cut optic nerve fibers in a frog regenerate to their proper places in the tectum. *Science* **130**: 1709-1710.

Mellon, F., Jr., 1968. *The Physiology of Sense Organs.* London: Oliver and Boyd; San Francisco; Freeman, 107 pp.

Michael, C. R., 1968. Receptive fields of single optic nerve fibers in a mammal with an all cone retina. *J. Neurophysiol.* **31**: 249-282.

Miledi, R., 1960. The acetylcholine sensitivity of frog muscle fibres after complete or partial denervation. *J. Physiol.* **151**: 1-23.

Miledi, R., 1962. Induction of receptors. *In* J. L. Mongar and A. V. S. de Reuck, Eds., *Ciba Foundation Symposium on Enzymes and Drug Action,* pp. 220-235. London: Churchill.

Miller, W. H., 1957. Morphology of the ommatidia of the compound eye of *Limulus. J. Biophys. biochem. Cytol.* **3**: 421–428.

Moore, J. W., M. P. Blaustein, N. C. Anderson, and T. Narahashi, 1967. Basis of tetrodotoxin's selectivity in blockage of squid axons. *J. Gen. Physiol.* **50**: 1401-1411.

MOUNTCASTLE, V. P., 1957. Modality and topographic properties of single neurons of cat's somatic sensory cortex. *J. Neurophysiol.* **16**: 37–68.

Mueller, P., and D. O. Rudin, 1968. Action potentials induced in biomolecular [*sic*] lipid membranes. *Nature* **217**: 713–719.

Nakajima, S., and K. Onodera, 1969a. Membrane properties of the stretch receptor neurones of crayfish with particular reference to mechanisms of sensory adaptation. *J. Physiol.* **200**: 161–185.

Nakajima, S., and K. Onodera, 1969b. Adaptation of the generator potential in the crayfish stretch receptors under constant length and constant tension. *J. Physiol.* **200**: 187-204,

Nakajima, S., and K. Takahashi, 1966. Post-tetanic hyperpolarization and electrogenic Na pump in stretch receptor neurone of crayfish. *J. Physiol.* **187**: 105–127.

Narahashi, T., N. Anderson, and J. W. Moore, 1967. Comparison of tetrodotoxin and procaine in internally perfused squid giant axons. *J. Gen. Physiol.* **50**: 1413–1428.

Noble, D., 1966. Applications of Hodgkin-Huxley equations to excitable tissues. *Physiol. Rev.* **46**: 1-50.

Obara, S., 1968. Effects of some organic cations on generator potential of crayfish stretch receptor. *J. Gen. Physiol.* **52**: 363-386.

Obata, K., M. Ito, R. Ochi, and N. Sato, 1967. Pharmacological properties of the postsynaptic inhibition by Purkinje cell axons and the action of γ-aminobutyric acid on Deiters neurones. *Exptl. Brain Res.* **4**: 43–57.

Otsuka, M., M. Endo, and Y. Nonomura, 1962. Presynaptic nature of neuromuscular depression. *Japan. J. Physiol.* **12**: 573–584.

Otsuka, M., L. L. Iversen, Z. W. Hall, and E. A. Kravitz, 1966. Release of gamma-aminobutyric acid from inhibitory nerves of lobster. *Proc. Nat. Acad. Sci.* **56**: 1110–1115.

Penn, R. D., and W. A. Hagins, 1969. Signal transmission along retinal rods and the origin of the electroretinographic *a*-wave. *Nature* **223**: 201–205.

PINSKER, H., AND E. R. KANDEL, 1969. Synaptic activation of an electrogenic sodium pump. *Science* **163**: 931–935.

RALL, W., 1967. Distinguishing theoretical synaptic potentials computed for different soma-dendritic distributions of synaptic input. *J. Neurophysiol.* **30**: 1138–1168.

Ritchie, J. M., and R. W. Straub, 1957. The hyperpolarization which follows activity in mammalian non-medullated fibres. *J. Physiol.* **136**: 80–97.

Sherrington, C., 1906. *The Integrative Action of the Nervous System.* New Haven, Conn: Yale University, 433 pp.

Sherrington, C., 1940. *Selected Writings of Sir Charles Sherrington,* D. Denny-Brown, Ed. New York: Hoeber, 532 pp.

Smith, T. G., W. K. Stell, and J. E. Brown, 1969. Conductance changes associated with receptor potentials in *Limulus* photoreceptor. *Science* **162**: 455–456.

Smith, T. G., W. K. Stell, J. E. Brown, J. A. Freeman, and G. C. Murray, 1969. A role for the sodium pump in photoreception in *Limulus. Science* **162**: 456–458.

Sperry, R. W., and E. Hibbard, 1968. Regulative factors in the orderly growth of retino-tectal connexions. *In* G. E. W. Wolstenholme and M. O'Conner, Eds., *Growth of the Nervous System.* Boston, Mass.: Little, Brown.

Stretton, A. O. W., and E. A. Kravitz, 1968. Neuronal Geometry: determination with a technique of intracellular dye injection. *Science* **162**: 132–134.

Takeuchi, A., and N. Takeuchi, 1960. On the permeability of the end-plate membrane during the action of transmitter. *J. Physiol.* **154**: 52–67.

TAKEUCHI, A., AND N. TAKEUCHI, 1965. Localized action of gamma-aminobutyric acid on the crayfish muscle. *J. Physiol.* **177**: 225–238.

Tasaki, I., 1959. Conduction of the nerve impulse. *In* H. W. Magoun, Ed., *Handbook of Physiology,* Sec. 1, "Neurophysiology," Vol. I, Ch. III, pp. 75–121. Washington, D. C.: Amer. Physiol. Soc.

Tasaki, I., 1968. *Nerve Excitation.* Springfield, Ill.: Charles C. Thomas, 201 pp.

Tasaki, I., I. Singer, and T. Takenaka, 1965. Effects of internal and external ionic environment on excitability of squid giant axons. *J. Gen. Physiol.* **48**: 1095–1123.

Tasaki, I., and T. Takeuchi, 1941. Der am Ranvierschen Knoten entstehende Aktionsstrom und seine Bedeutung für die Erregnungsleitung. *Pflügers Arch. ges. Physiol.* **244**: 696–711.

Tasaki, I., and T. Takeuchi, 1942. Weitere Studien über den Aktionsstrom der markhaltigen Nervenfaser und über die elektrosaltatorische Übertragung des Nervenimpulses. *Pflügers Arch. ges. Physiol.* **245**: 764–782.

Tomita, T., A. Kaneko, M. Murakami, and E. L. Pautler, 1967. Spectral response curves of single cones in the carp. *Vision Res.* **7**: 519–531.

Wald, G., P. K. Brown, and I. R. Gibbons, 1963. The problem of visual excitation. *J. Opt. Soc. Amer.* **53**: 20–35.

Watanabe, A., and H. Grundfest, 1961. Impulse propagation at the septal and commissural junctions of crayfish. *J. Gen. Physiol.* **45**: 267–308.

Watanabe, A., S. Obara, T. Akiyama, and K. Yumoto, 1967. Electrical properties of the pacemaker neurons in the heart ganglion of a stomatopod, *Squilla oratoria. J. Gen. Physiol.* **50**: 813–838.

Werblin, F. S., and J. E. Dowling, 1969. Organization of the retina of the mudpuppy, *Necturus maculosus:* II. Intracellular recording. *J. Neurophysiol.* **32**: 339–355.

Wiesel, T. N., and D. H. Hubel, 1963a. Effects of visual deprivation on morphology and physiology of cells in the lateral geniculate body. *J. Neurophysiol.* **26**: 978–993.

Wiesel, T. N., and D. H. Hubel, 1963b. Single-cell responses in striate cortex of kittens deprived of vision in one eye. *J. Neurophysiol.* **26**: 1001–1017.

Wiesel, T. N., and D. H. Hubel, 1965. Comparison of the effects of unilateral and bilateral eye closure on cortical unit responses in kittens. *J. Neurophysiol.* **28**: 1029–1040.

Willows, A. O. D., 1968. Behavioral acts elicited by stimulation of single identifiable nerve cells. *In* F. D. Carlson, Ed., *Physiological and Biochemical Aspects of Nervous Integration,* pp. 217–243. Englewood Cliffs, N. J.: Prentice-Hall.